THE
HISTORY OF THE
BRITISH FLORA

Biogeographical conclusions only find definite confirmation when it is possible to base them on a geological foundation, which historical geology can alone provide.

E. V. WULFF, 1943

THE HISTORY OF THE BRITISH FLORA

A Factual Basis for Phytogeography

Sir HARRY GODWIN, F.R.S., F.G.S., F.L.S.

EMERITUS PROFESSOR OF BOTANY, UNIVERSITY OF CAMBRIDGE
FELLOW OF CLARE COLLEGE

SECOND EDITION

CAMBRIDGE UNIVERSITY PRESS
CAMBRIDGE
LONDON · NEW YORK · MELBOURNE

Published by the Syndics of the Cambridge University Press
The Pitt Building, Trumpington Street, Cambridge CB2 1RP
Bentley House, 200 Euston Road, London NW1 2DB
32 East 57th Street, New York, NY10022, USA
296 Beaconsfield Parade, Middle Park, Melbourne 3206, Australia

This edition © Cambridge University Press 1975

Library of Congress catalogue card number: 73-84324

ISBN: 0 521 20254 X

First published 1956
Second edition 1975

Printed in Great Britain
at the University Printing House, Cambridge
(Euan Phillips, University Printer)

CONTENTS

CONTENTS

vi

PREFACE TO THE SECOND EDITION

After a lapse of seventeen years it is natural that the original preface should require addition and fresh explanation. In this period there have been remarkable developments in many areas of the wide field of Quaternary study: radiocarbon dating and oxygen-isotope analysis are now extensively employed, high quality optical microscopy and electron microscopy assist both pollen analysis and seed identification, whilst knowledge of Pleistocene geology has increased enormously, permitting far more insight into the character and sequence of the interglacial periods and of conditions during the glaciations. These many advances have accelerated the accumulation of historical plant records which now amount to ten times the 1955 total, in terms at least of numbers of index cards. The greater complexity and improved standards are reflected in the 'Festschrift' volume *Studies in the Vegetational History of the British Isles* (Cambridge University Press, 1970), presented to me by my former students and colleagues: a tribute that I am happy to acknowledge with gratitude and appreciation. Their volume includes such up-to-date accounts of methods and techniques now available in the study of fossil pollen and macroscopic plant remains that I am relieved of the need to alter greatly the introductory remarks on these matters as originally set out in the first edition.

It was explained in the original preface that this book arose as a by-product of general investigation into the whole field of Quaternary History, but what began in that manner is now established as a study worth pursuing in its own right. The thesis that biogeographic theory must be subject to the factual evidence of historically attested plant and animal records is now generally accepted and, as it turns out, the broad phytogeographic conclusions drawn in 1956 are substantially confirmed, though also expanded, by the great volume of new records made since then.

I have had to avoid the temptation to discuss the exciting new developments fostered by research work in many directions during the last few years: great as their promise is, their present contribution to our recording of the history of the British flora is still small and we have been forced essentially to collect and survey the factual evidence secured by older, not necessarily out-dated, methods. It is gratifying to consider how great advantages a future review will be able to command.

This extension of the 1956 preface allows me to renew my thanks to all those whose assistance was acknowledged then, with recognition of their often greatly advanced title and style. To them must be added further names in various categories. Among former students and colleagues in research who offered particular help I must now add Professors J. J. Donner and F. Oldfield, Drs F. G. Bell, H. J. B. and H. H. Birks, D. M. Churchill, M. Dąbrowski, J. M. Dickson, F. A. Hibbert, N. T. Moar, R. G. Pearson, L. A. Stevens, R. P. Suggate, C. Turner, J. Turner and Vishnu Mittre, Miss E. A. Pringle, Miss L. Phillips, Mr A. Brown, Mr P. D. Sell and Mr R. Sims. In particular Dr Colin Burrows has allowed me to incorporate many unpublished Late Weichselian records from North Wales, and Mrs D. G. Wilson much similarly unpublished work done under the supervision of Dr R. G. West, on material from several deposits of the Cromer Forest Bed series. Many of the skilled laboratory assistants in the Sub-department of Quaternary Research have been outstandingly helpful, most especially Mrs Dickson (Miss C. A. Lambert) and Miss M. Ransom. Miss S. Hallam has helped invaluably in the preparation of line drawings and figures, as also has Mr F. T. N. Elborn again with photography.

In a slightly different category are many concerned with perfection and use of new research tools in relation to Quaternary studies in Cambridge. Thus in electron microscopy Professor C. T. Chambers, Drs P. Echlin and R. Angold and Mr B. Chapman have assisted both pollen and seed identification. Dr E. H. Willis built our first radiocarbon dating apparatus, thanks to a generous grant from the Nuffield Foundation, and I happily acknowledge his fruitful co-operation in numerous pioneer projects of applying radiocarbon dating to problems of Quaternary study and vegetational history. Dr V. R. Switsur, his successor, has been equally obliging and effective in this decisively significant field, and both he and Dr Willis have received extremely strong support from devoted and able technical assistants. More recently Dr N. J. Shackleton's development of oxygen-isotope analysis to problems of palaeotemperature and geophysical process has also greatly strengthened our understanding of the Pleistocene.

Outside the immediate circle of the Cambridge Sub-department of Quaternary Research I owe thanks to many research workers not previously acknowledged, from elsewhere, especially Professors C. D. Pigott, F. W. Shotton, and W. A. Watts, Drs D. M. Anderson, W. W. Bishop, A. P. Carr, R. H. Clarke, J. Coles, G. R. Coope, D. G. Gaunt, M. P. Kelly, M. P. Kerney, W. Kirk, P. D. Moore, M. E. Morrison and Mr J. Fox.

For permission to reproduce, additionally to those of

the first edition, figures and photographs from their published work, I am much indebted to Professors C. D. Pigott and W. A. Watts, Drs H. J. Tallis, H. Tauber, H. Tralau, C. Turner, J. Turner, and R. G. West, Mrs Davies (Dr F. G. Bell) and Mrs D. G. Wilson. I am grateful for permission to reproduce these illustrations, as well as some of my own published figures, to the publishers of the following journals: *Arkiv för Botanik, Danmarks Geologiske Undersøgelse, Geological Magazine, Journal of Ecology, Nature, New Phytologist, Philosophical Transactions* and *Proceedings of the Royal Society, Svensk Botanisk Tidskrift,* and *Ulster Journal of Archaeology.* Likewise I am obliged to Messrs Longmans, Green and Co., and to Methuen and Co., in respect of illustrations from recently published books.

In the intricate task of designing our new 80-channel punch-cards we were able to draw upon the generous help and great experience of Dr F. H. Perring, Head of the Biological Records Centre of the Nature Conservancy, and he has at all times been exceedingly kind in proffering help with problems of processing our data. The presentation of our results has demanded also an extensive use of maps, and we have had the great advantage of being able to employ for this purpose the standard map of the British Isles used in the *Atlas of the British Flora,* and to draw upon the Biological Records Centre for its most recent records of modern distribution of British plants. For these substantial advantages I am grateful to all concerned. Likewise, since retirement from the Cambridge Chair of Botany I have had the task of compilation and analysis greatly helped by accommodation in the Botany School provided by Professor P. W. Brian and Dr R. G. West, and in the University Botanic Garden by its Director, Mr J. S. L. Gilmour.

I owe a very special debt of thanks to Mrs Joy Deacon who, thanks to financial support from the Natural Environment Research Council, has been constantly employed as my research assistant over the last five years upon the exacting and laborious task of collecting, entering and analysing the plant records. The character of this operation has been indicated in chapter I and in two separate scientific publications in which she deals respectively with the creation and employment of the data bank (Deacon, 1972) and with the exercise of applying ordination technique to analysis of the distribution of particular categories of present and of our recorded fossil flora (Birks & Deacon, 1973), the substantial basis in fact of chapter VI: 9. Mrs Deacon also prepared all the maps that represent the former ranges of a large number of components of the British flora. Her competent assistance has been invaluable. I am also very much indebted to Mrs Audrey Robinson for the skill and assiduity with which she has prepared virtually all the typescript of the book.

To these many collaborators, mentioned in the first or second preface, and others perhaps omitted inadvertently, I remain permanently grateful.

H. GODWIN

Cambridge, 1972

PREFACE TO THE FIRST EDITION

The writing of this book is to a large extent a by-product of the study of Quaternary History on which the author has been engaged since about 1930. It will increasingly appear throughout the book why an accumulation of knowledge of the origin and history of the British flora was an inevitable accompaniment of such study, and how therefore the writing of an account such as this was its natural consequence.

First contact with the problems of Quaternary History came in the joint attempt which my wife and I made to apply to the British Isles the technique of pollen analysis then newly worked out in Scandinavia. It very quickly became apparent that the new technique was capable of providing a very consistent picture of the vegetational history of these islands through the period since the last recession and final disappearance of the ice-sheets of the latest glaciation. This regular pattern of vegetational change has, by the extended research of the last twenty years, become securely knit into the general fabric of Quaternary History, through numerous correlations with geological, archaeological and climatic events of the past few thousand years, and, as is now well known, it has come to serve to some extent the purpose of a background time scale for those events.

Considerable stimulus to such co-ordinated research was given to Cambridge workers by the establishment, under the chairmanship of the late Sir Albert Seward, of the Fenland Research Committee, which ultimately achieved an effective, though obviously still incomplete, outline of the history of the East Anglian Fenlands from Late-glacial time to the Romano-British period. The constant appeal to geological evidence in this work was repeated in the later Cambridge investigations of the Mesolithic settlement sites in the ancient deposits of Lake Pickering, in East Yorkshire.

During the course of studies of this kind, as well as the stratigraphic investigations of lakes and bogs sampled in the course of pollen-analytic survey of the country, there are necessarily encountered, quite apart from the great numbers of pollen grains counted, all kinds of coarser plant material, the identification of which forms part of the routine of establishing a sound knowledge of local stratigraphy. Very often fruits and seeds have been recovered from layers closely dated by pollen analysis, by geological events or by archaeological material. It was at once apparent that such dated records could afford direct evidence of the former presence at specified places of many individual species within the British flora. Not only have such incidental records been conserved, but there

has been a sustained and deliberate attempt always to collect macroscopic plant material for identification wherever it occurred suitably in deposits datable by one means or another. This was of course precisely what Clement Reid had already done at the end of the nineteenth century, although with much less precise means of dating, and what Knud Jessen, following Scandinavian tradition, was still so ably doing in his Quaternary research upon the floras of Denmark and of Ireland.

An early investigated site of great significance for British floristic history was that of the rich Late-glacial deposits of Bodmin Moor, where, as in the later investigations at Nazeing in the Lea Valley, strong evidence emerged that many of the species recognized must have had geographical ranges very different from those of today. The evident importance of such evidence led to the systematic collection of all available well-dated records of plant remains from any part of the Quaternary period.

The following pages give an account of the sources of records, the methods used in collecting and assessing material, and the nature of the background of geological, climatic and biotic events against which such records have to be considered. These records are set out in a suitable condensed form, together with such brief comments as arise from their consideration. Finally we have set down the general conclusions which appear thus far to emerge from examination of the results as a whole: an attempt in fact to set down what we can now confirm as the pattern of the processes by which the present flora of the British Isles came to its present composition, and how its components came to have their existing ranges.

I owe thanks to very many willing helpers who have contributed their labour and skill to the identification of macroscopic plant material, most notably Dr M. H. Clifford, Miss A. P. Conolly and Miss Jean Allison, who formerly acted as research assistants to me. Many further records come from work, mostly unpublished, which past and present research workers kindly allow me to include: in this category are Mrs W. Tutin, Mr S. Seagrief, Mr P. A. Tallantire, Miss E. M. Megaw, Miss K. Pike, Mr J. N. Jennings, Mr D. Walker, Mr R. G. West, Miss S. L. Duigan and Mr A. G. Smith. I am also indebted to those who have given assistance in the routine of pollen-counting, most especially to Miss R. Andrew, whose experience and discernment has led to the recognition of many pollen types of the greatest interest and importance.

From outside Cambridge also, research workers have kindly given me records for inclusion or pollen diagrams for consideration, and here my debt is especially great to Mr G. F. Mitchell of Trinity College, Dublin, and Professor K. Jessen of Copenhagen whose published records for the Quaternary flora of Ireland I have employed as at once fully reliable and the only large available source of information on this subject. I am likewise much indebted to Dr K. B. Blackburn, Mr H. A. Hyde, Miss M. Beatson, Dr S. M. Walters and many others who have brought records or plant material to my notice. I owe particular thanks to the generosity of Dr H. Helbaek of the Danish National Museum in Copenhagen, whose elegant work is providing such a sound basis for our knowledge of the past history of the old world's cultivated plants: my accounts of the cereals and other crops are based almost entirely on his accounts, published alone or jointly with Professor Jessen.

For permission to reproduce figures from published papers or books I am much indebted to Dr H. Helbaek, Professor K. Jessen, Dr A. L. Backman of Helsinki, Dr J. Iversen of the Danish Geological Survey, Professor G. Manley, Professor J. G. D. Clark, Dr K. B. Blackburn, Mr R. G. West, Sir Arthur Tansley and many of my own collaborators in publication. I am grateful also to the Royal Society of London, the British Ecological Society, the Kongelige Danske Videnskabernes Selskab, the Societas Scientiarum Fennica, the Societas pro Fauna et Flora Fennica and the Royal Geographical Society, for permission to reproduce figures from their scientific periodicals. To the Royal Society of London, the Royal Irish Academy, and my co-editors of the *New Phytologist* I am further indebted for the loan of process blocks already made.

Many colleagues have kindly provided figures or photographs not hitherto published: these are Miss S. L. Duigan, Dr F. J. North, Mr J. Challinor, Mr H. A. Hyde, Dr A. L. Backman. In the preparation of my own photographs I am happy to acknowledge the unsparing and skilled assistance of the laboratory's photographer, Mr F. T. N. Elborn.

The task of compiling records from the literature has itself been no light one: it was given a good beginning by Mr C. W. Phillips, Director of the Archaeological Section of the Ordnance Survey during a long convalescence, but a very large part I owe to my students and research assistants. Mrs Carson, Mrs M. E. Robinson and Miss A. Ward have not only done much of this work but have given untiring secretarial assistance in the preparation of the book; without them it could indeed hardly have been achieved. Finally, I wish to express my gratitude for the encouragement given by my wife to this project and her cheerful tolerance of the inconveniences incidental to its performance.

It is perfectly apparent that in so far as the main thesis of this book is established, so far must it follow that it represents only a first approximation to the truth and a demonstration of the practicability of a particular line of approach. As this approach is further exploited and as further additions continue to be made to the existing list of British records, we shall endeavour to keep adequate note of them. All possible help to this end will be welcomed, and the author will likewise be happy to offer what assistance he can to investigators engaged in the problems associated with the origin and history of British flora.

I

INTRODUCTION

There was something of a crusade or at least a challenge in the first edition of this book, that in the course of writing revealed itself as an essay on the proposition that biogeography must, and indeed could, be based upon the identification of plant fossils and upon knowledge of the whole environmental circumstances of the immediate geological past. This view, subsumed by the quotation from E. V. Wulff on the title page, suggested in effect that an ounce of historical geological fact was worth a ton of the historical speculation, up to that time prevalent, based on nothing more than comparison of present-day distribution ranges. Compilation of the book proved, in the event, to provide such an unexpected quantity of evidence that the feasibility of the historical–geological approach was established, and it is now taken for granted that this is the proper and profitable approach to historical biogeography, and Quaternary geological research now has this consciously in mind as one of its major objectives. This said, we may continue our introduction in its original words.

Aside from a natural and affectionate interest with which everyone must regard the wild plants which have formed the background of life in our countryside, a particular scientific interest has always been attached to the history of the British flora. It has been in effect a test piece, a *cause célèbre*, in the science of biogeography for at least a half-century. The flora and fauna of the British Islands have probably been more exhaustively studied than those of any comparable area of country, and our knowledge of the distribution of species within them is detailed and extensive, although admittedly still incomplete. Moreover the knowledge of the distinctive distribution of plant species within these islands has been very profitably supplemented by knowledge of their distribution areas in other countries, especially in the neighbouring part of the European mainland.

There are those phytogeographers who pretend to an interest in these distributional data for their own sake, but scientific interest has of course centred upon the mechanisms by which such distributions have been produced and are maintained. Although there is a wide difference of opinion on the rates of spread of plants it is generally conceded that the distributional problem must be to a large extent an historical one, and it emerges fully from such quotations as the following from Deevey (1949): 'One of the most popular topics in all biogeography is the question whether the whole of the fauna and flora of the British Isles immigrated in Post-glacial times, or whether some fraction survived from an earlier time.' It has indeed been the purpose of biogeography to fit the facts of present distribution to an explanation in terms of migrational and evolutionary shifts induced by past climatic changes and geological processes. We may recognize this in the explanation so brilliantly expounded by Edward Forbes almost a century ago for the discontinuous distribution of Arctic–alpine species on the mountain ranges of Europe. Outliers of the arctic flora stranded on our mountain peaks were explained by him as the residue from a period of prevalent low temperatures during the Great Ice Age, during which arctic floras migrated across the great lowland plains of Europe. With the rise in temperature and the retreat of the ice sheets, the arctic flora was forced to ascend the mountain heights and to retreat northwards, leaving behind an indication of former events in the striking discontinuity of the relics. Darwin in the *Origin of Species* followed this line of explanation, and it will be seen at once what interest such a hypothesis gives to the biogeography of the British Isles.

There has been no very serious lack of distributional data and indeed since the days of Hooker and Darwin the essential character of this evidence has not altered: what has altered very greatly has been our knowledge of the whole background of climatic and geological changes through the Pleistocene period, the existence of repeated great glaciations and of intervening mild periods of character and length comparable with that period since the last glacial, the 'Post-glacial period' in which we live. Biogeographic hypotheses have indeed multiplied to keep pace with our knowledge of this climatic–geological history of the last one million years: 'warmth-demanding' species are explained as having immigrated in the Post-glacial Climatic Optimum, 'steppe' species as being the result of westward migration in the supposedly continental climate of the Neolithic and Bronze Ages of Britain, and a substantial complement of species familiar as weeds and ruderals is explained as due to introduction by Neolithic and later peoples, now known to have been pastoralists and agriculturists. There is nothing inherently improbable about this kind of explanation and indeed it may very well be along such lines that our final theory will shape itself: what is lacking is any positive evidence, as opposed to this purely circumstantial reasoning, that a given hypothesis actually does apply to a given species and that a given species was indeed present in these islands, or in given parts of these islands at the periods demanded by the hypothetical explanation.

It is not in the least surprising that geologists or botanists with geological interest or training should have been the persons to realize and emphasize this deficiency. Quite outstanding in this was the late Clement Reid whose vision stimulated a career of original and distinguished research on the history of the British flora, a labour culminating in his book *The Origin of the British Flora* published in 1899, immediately after the appearance of almost identical material in the *Annals of Botany*. He wrote in his introduction that 'This problem of the origin of our flora is one which can be solved, I think, by the historical method, and that seems to be the proper mode of attacking it', and he proceeds in the body of his book to record the stratigraphic position and the presumed geological age of all the known plant-bearing deposits of Post-miocene Age in the British Isles, to list the species found in them, and finally to analyse for each plant thus represented the extent of its known record. The vast bulk of such records was the result of his own determinations, as indeed was much of the concomitant geological investigation.

The late F. F. Blackman, one of the founders of modern plant physiology, once explained that 'a great man is a man who says something no one has ever said before, but something which, once said, everyone immediately accepts as true'. This sense of greatness certainly applies to Clement Reid's recognition of the way to certainty in problems of biogeography, and it is all the more surprising therefore that his lead was not more strongly followed. It is true that his wife, Mrs E. M. Reid, and Miss M. E. J. Chandler continued to add very appreciably to our records, but they became increasingly concerned with Tertiary material, and indeed justifiably so in view of the character of their great publication on the flora of the London Clay.

The decline in the interest of Clement Reid's approach was really due to the unexpressed realization that the background scale of geological, archaeological and climatic events was still too ill defined and coarse to serve adequately the purpose of correlation and explanation required of it, but this is a situation which has very greatly altered in the last thirty years or so. Our knowledge of Pleistocene geology has increased enormously, not only in the British Isles themselves but in the regions of the Alpine and Scandinavian glaciation areas, and indeed throughout Europe. Under the leadership of Abbé Breuil much progress has been made in correlating Palaeolithic cultures of prehistoric man with the glacial stages, and in the Post-glacial period with the definition of the Neolithic in a precise and restricted way has come the recognition of Mesolithic cultures bridging the gap between Palaeolithic and Neolithic. Faunistic and climatic evidence has accumulated and has to some degree been brought into line with the new geological evidence. In many ways, however, the most important advance has been the development of the technique of pollen analysis or 'palynology' in application to the field of Quaternary investigations. It was made an effective instrument of

research by Lennart von Post, who reconstructed by its means the successive vegetational stages of the Post-glacial period in Sweden, and correlated them with geological, archaeological and climatic events. The remarkable uniformity of the pattern so disclosed could only be explained by the assumption of a general climatic control of vegetational change, and it was this fact, coupled with the widespread occurrence of pollen-bearing deposits that gave to pollen analysts the power to furnish a time scale against which other events of the Quaternary period could be viewed and measured. For a large part of the period concerned, the forest trees (whose pollen alone was at this stage seriously considered) were the main vegetational dominants and thus stood as the most direct indices of regional climate. At the same time they disclosed the character of the environment in which contemporary animals and plants had to exist, among them the societies of prehistoric man.

In this manner the numbered zones based on pollen analysis in Sweden came to have an outstanding chronological significance, and this was heightened when a correlation was made between them and the chronology based on the varved lake deposits of de Geer. The systematic exploitation of pollen analysis spread rapidly across Europe and was introduced to the British Isles and to English-speaking countries by Erdtman in the mid-1920s. The far-sightedness of the Irish Quaternary Research Committee led to Dr Knud Jessen of Copenhagen being invited in the mid-1930s to investigate Irish deposits by this means, and fruitful indeed have proved his researches there and those of his pupil G. F. Mitchell. Meanwhile from about 1930 my wife and I had undertaken similar studies in England and Wales, as a result of which by 1940 I was able to publish a tentative pollen zonation for the Post-glacial deposits of this country, one which corresponded largely to those of Scandinavian countries and which Jessen was able to adopt without serious alteration for Ireland.

It was apparent from the outset that any objects, vegetable, animal or archaeological, found in deposits capable of being ranged within the palynological system could thereby be dated. As so many of the deposits primarily investigated for results by pollen analysis yielded also macro-fossils it was soon apparent that it was desirable to co-ordinate micro- and macro-fossil determinations wherever possible. Jessen himself had steadily pursued and advocated this aim, as is witnessed by the impressive collection of his dated plant records both for Denmark and for Ireland. It was he who also shewed that interglacial deposits could yield a similar pollen zonation to that of the Post-glacial (Flandrian) period, and his classic studies with V. Milthers of the interglacial deposits of Denmark and north-west Germany illustrate the technique of dating macroscopic plant and animal remains by the application of pollen analysis. Not only were the beds investigated by pollen analysts often rich in determinable fruits and seeds, but investigators were led to develop boring tools suitable for sampling peats and muds, and they found a fruitful source of study in the stratigraphy of peat mires, estuaries and lake deposits which they now examined with a precision altogether new to geologists. Thus, although they seldom penetrated more than 10 to 12 metres, they encountered in their borings great numbers of determinable macro-fossils and were also enabled to recognize and to collect from layers of particular interest which otherwise would have remained hidden indefinitely. Indeed, after two or three seasons' work in Ireland, Jessen could tell with considerable accuracy, from the lie of the country, localities where late-glacial (Late Weichselian) deposits were likely to lie beneath the surface, and he had, in fact, similarly detected many of the Danish and Holstein interglacial deposits referred to above.

In recent years the province of pollen analysis has been greatly widened by improvement in the technique of pollen identification by the aid of high-power oil-immersion microscopy. These improvements have stemmed from the work of Wodehouse in the United States of America, Erdtman in Sweden and Iversen in Denmark. The improved standard of identification has permitted the recognition of a considerable number of additional types, some generic but some at species level, among herbaceous plants and among woody plants hitherto unrecognizable by pollen-grain characters. Thus palynologists can now recognize pollen of *Juniperus*, *Populus*, *Artemisia*, *Thalictrum*, etc., at the genus level, and species such as *Centaurea cyanus*, *Plantago lanceolata* and *Pastinaca sativa*. We are still at the stage where such knowledge is spreading through the ranks of working palynologists but it is already apparent that a vast new source of datable Quaternary plant records has been made available, and indeed the characterization which Iversen was enabled to give to the Late Weichselian period by its aid has been amply confirmed and extended in many parts of Europe, among them the British Isles.

This work is essentially restricted to the evidence of fossil remains. It excludes direct historical reference, partly because its object is to establish the natural condition of the British flora and vegetation before human interference had become all-important, and more particularly because the examination of historical records is a wide and specialized task outside the capacity of the author. This deliberate restriction of purpose naturally excludes all reference to known introductions of species in the historic period and equally to known extinctions within this period, such, for instance, as the loss of *Sonchus palustris* by improved drainage. Despite this exclusion of written historical evidence, there are included records of fossil material from Roman, Norman and mediaeval times.

Appropriate as this original introduction remains, there are several considerable differences in the second edition that need brief exposition. The first edition was written during 1951 and 1952, the data for it having been indexed throughout the previous nine or ten years. With the prospect of rewriting it was decided to use a different system of registration, permitting more mechanical sort-ing of information, the incorporation of more categories of information, and one suited to take in new material for a long time ahead as part of the continuing activity of the Sub-department of Quaternary Research in Cambridge. Dr Franklyn Perring, of the Biological Records Centre of the Nature Conservancy, gave assistance to Dr R. G. West and Mrs J. Deacon in designing an 80-channel punch-card for this purpose, to serve as the heart of a data bank for British Quaternary plant fossil records and to be handled by an ICL group–select sorter. The operation of the system has been fully described (Deacon, 1972) so that we need only mention the salient features of it as they affect our present exercise.

By September 1970, when input of information was arbitrarily ended, Mrs Deacon had completed (entered and punched) about 60 000 cards together with 550 site-cards each containing a description of all relevant details of stratigraphy and provenance for each site that has yielded plant records. The volume of information now available by mechanical analysis of the punch-cards is such as necessitates a different form of published presentation from that in the first edition.

The particular advantage of the data-bank storage and mechanical sorting system is that it allows a quick extraction of the total fossil record of each taxon, of the fossil flora of a given stage, pollen zone or cultural phase (either for the British Isles as a whole, or region by region) and of broad categories of pollen frequency of particular genera or species. This last facility has been of particular value in the production of maps indicative of variations in pollen frequency of the major tree genera through the Late Weichselian and Flandrian periods.

Because of the great increase in recent publications we have fortunately been able to incorporate all Irish records in the data bank and to treat them conformably with those of the rest of the British Isles.

The entry for each species recorded in the first edition provided, in abbreviated form, the site, name, county, vice-county number, type of fossil material, age attribution (palynological, geological, archaeological) and author's name and date for the primary recording. Thus each entry gave a direct reference to all the primary sources of its Quaternary record. Helpful as such publication has proved to be, the space needed to print all the data now available makes this procedure impossible. Accordingly in this volume we have restricted ourselves merely to the description of the information extracted from the data bank of punch-cards. It will naturally be possible by enquiry of the Sub-department of Quaternary Research in Cambridge for individual research workers to seek the primary data for particular species. Our verbal, cartographic and diagrammatic representations of the Quaternary history of British plants remain equally based on the facts of former known occurrences, and are equally susceptible to checking.

The loss of some information from the individual species records will, we believe, be compensated largely by the provision of a comprehensive list of all the sites which have yielded fossil plant material (chapter IV: 4). By turning to this the reader seeking information on the source of a given plant record will find given the locality of the site (county and National Grid reference), a concise comment on the main features of Quaternary significance at the site, and the author and date of the most recent comprehensive publication concerning it. This index also indicates, where appropriate, any reinterpretation of the original dating attributed to the deposits.

At the time of preparation of the first edition the best estimate of present distribution of the British flora lay in the vice-county records, and maps corresponding with this were used for comparison of past and present distribution whilst each fossil record was given its vice-county number. Since that time there has appeared, under the editorship of F. H. Perring and S. M. Walters, two volumes of the monumental *Atlas of the British Flora* which illustrates to a remarkably complete extent the distribution of vascular plants on a 10 km grid over the whole of the British Isles. By courtesy of the editors and the Biological Records Centre we have been permitted to make use of the same maps (or the most recent versions of them) to illustrate this second edition. We have abandoned the references to vice-counties and each plant record has been indexed on its National Grid reference, so that fossil and present distribution patterns have a common basis, and the same mechanical printing-out processes will be available for both. To the great advantages thus presented to us we must add that of the biogeographic–ecological information about each species conveyed by the *Atlas* maps and the overlays that accompany them. A similar advantage, though involving fewer species, has come from the continued publication of additional accounts in the 'Biological Flora of the British Isles' as part of the *Journal of Ecology*.

The last two or three decades have seen very considerable changes in the significance of pollen analysis in relation to any study of the British flora, and the process of change is still active (Godwin, 1968a). Pollen analyses have been made at far more sites, analyses have been extended to interglacial and interstadial deposits, and the standards of pollen identification have been greatly improved so that many more species can be recognized than was formerly possible. Alongside these changes there has been a notable shift in the objectives of palynological study. A great deal of the impetus behind the early development of the subject stemmed from the fact that it offered a quasi-chronological reference system for the period between the last glaciation and the present, and thus met a serious need in the many subjects, biological, geological and archaeological, concerned with events of this time. Accordingly reference to the classic pollen zones for England and Wales played an important and necessary role in the analysis of evidence in the first edition. This continues but against a strong continuing shift of purpose and significance. The role of palynology as supplying a chronological base is being progressively taken over by radiocarbon dating which has however broadly confirmed the absolute chronologies attributed earlier to the pollen zones for England and Wales. Where radiocarbon dating cannot extend, as in past interglacial periods, palynology still affords not only the means of identifying the consecutive interglacials, but of dividing them into chronozones (West, 1970a) (see chapter III: 1). A very substantial proportion of recent plant records, both for macroscopic and microscopic remains, is now directly dated by radiocarbon age-determinations of which several hundred now relate to the British Quaternary record, and a great many more are supported indirectly by reference to the radiocarbon dates, so that we are able now to refer directly to the absolute ages (in radiocarbon years) before the present in which biogeographic events occurred. It is largely through radiocarbon dating that we are able to co-ordinate the increasing number of sites of recovered plant remains from the vastly important period of the latest, Weichselian, glaciation.

As radiocarbon dating progressively takes over, especially for the Flandrian (Post-glacial) period, the role of supplying a chronological reference system, so palynology reverts increasingly to the role of supplying identifications of plant taxa and plant communities at past known periods of time. It is here that evaluation of data for the second edition faces its most considerable problems, for the pollen record necessarily covers both early determinations of very modest specificity and a few recent determinations, such as those of H. J. B. Birks, that not only provide new recognition at specific level but cast doubt upon some past identifications. Furthermore, in a high degree pollen is subject to transport by wind from great distances, and affords, in comparison with macroscopic remains, only equivocal evidence of the *local* presence of a taxon at the site of recovery. Thus in many instances a strong element of subjectivity accompanies records based wholly upon pollen and these feature more than they did in our original historical record.

Pollen analysis now increasingly proceeds by the recognition of pollen-assemblage zones that are independently dated at each site and are subsequently correlated over regions, and these fossil assemblages are progressively being identified with those of existing plant communities. So far as this proves practicable, and it is yet in a very early stage, it may allow recognition of the former occurrence of communities outside their present ranges with far-reaching implications for the history of the component plant taxa. The compilation of this second edition comes therefore, in respect of pollen evidence, awkwardly and at a time of transition, although admittedly one of great promise. It increasingly offers the means of reconstructing *vegetational* history, but it must be stressed that whilst this is clearly of greatest importance to our purpose, that purpose remains the evaluation of evidence for *floristic* history, and in this sense palaeo-autecological rather than palaeo-synecological.

INTRODUCTION

The last twenty years have witnessed the most remarkable development of British Quaternary geology, a development made particularly evident by the recent creation, expansion and current activity of the Association for Quaternary Geology, a body already with several hundred members. A measure of the advance in our knowledge is given by R. G. West's *Pleistocene Geology and Biology* (1968), a book to which repeated reference must be made, and on which we have strongly depended, particularly in the writing of chapter III: 1. When the first edition was being written several British interglacial plant beds were known, but very few had been investigated thoroughly. The application to them of palynological techniques had barely started and the resolution of the sequence of British glacial deposits and their correlation had been hardly begun. Accordingly we had to be content to register a great part of the relevant records as merely 'Interglacial' and to draw conclusions simply that given species had been in the British Isles in an interglacial. An exception was the reference to the Cromerian Interglacial for the Cromer Forest Bed deposits, regarded by Clement Reid as upper Pliocene, which were now treated as 'the earliest interglacial', and we referred a few records respectively to the succeeding Hoxnian and Ipswichian Interglacials. Such has been the progress of our knowledge, we now confidently regard the Cromer Forest beds as belonging to the Middle Pleistocene, preceded by the Pastonian and Ludhamian Interglacials and followed by the Hoxnian, Ipswichian and the Flandrian, then called 'Post-glacial' or 'Holocene'. Thanks to a great volume of detailed field and laboratory study we can now refer plant remains with confidence at least to these last four interglacials, and a chronozone system established by West enables us effectively to sub-divide each of them. Our punch-card index takes full account of these possibilities. As a consequence our analyses may come within sight of a comparison of the vegetational and floristic history throughout the several interglacials. At the time when the punch-card was being designed the latest glacial stage in this country was called 'Weichselian'. Subsequently, to achieve consistency in nomenclature, this has given place to 'Devensian' (*Proc. Geol. Soc.*, 1969): rewriting had gone too far to allow us to change but we need expect no confusion from retention of 'Weichselian' throughout the book.

In correspondence with this new gain in quality of recording the summarizing list-head diagrams deal with the interglacial records on a wider scale with more resolution than before (chapter IV: 1). They also treat the period of the last glaciation somewhat differently and on a more extended scale. This is a consequence partly of the greatly increased number of plant beds now referable with certainty to this time, partly of the ability given by radiocarbon dating to place these organic deposits into absolute age ranges and partly by the decision to treat deposits of the Allerød interstadial not as belonging to a 'Late-glacial' period, but as the concluding part of the Weichselian glacial stage.

Already in the first edition only brief reference was made to the methods of field-sampling and laboratory treatment of plant materials either for pollen analysis or the extraction and identification of macroscopic plant remains. The many refinements and advances in these directions have been fortunately accompanied by authoritative publications of which we may mention West's *Pleistocene Geology and Biology* (1968), Faegri and Iversen's *Textbook of Pollen Analysis* (1964), Dimbleby's *Plants and Archaeology* (1967), particularly for wood identification, the papers respectively by Andrew and Dickson in *Studies in the Vegetational History of the British Isles* (1970), and Körber-Grohne's *Geobotanische Untersuchungen auf der Feddersen Wierde* (1967), for macroscopic plant remains. These are now so generally accessible that instructions on techniques have been removed from the present volume, although critical comments upon the interpretation of effects of sampling and preparation remain.

II

COLLECTION AND IDENTIFICATION OF PLANT REMAINS

1. GENERAL CONSIDERATIONS

An advantage, not without its own dangers, in the investigation of the Quaternary fossil flora found in this country, is that we may fairly expect to find most of the fossils represented in our living flora, which is so much more restricted than that of the continental mainland that the task of recognition is greatly simplified. Thus, for instance, in the Flandrian period only one species of alder, *Alnus glutinosa*, is known to have occurred in the British Isles, so that *Alnus* pollen may, with high probability, be referred to this species, although on the nearby continent palynologists would have to reckon with the possible presence of *A. incana*, *A. viridis*, etc. Similarly, in the Flandrian period so far as is known, we need reckon only with the one species of pine, *Pinus sylvestris*, and may almost entirely disregard the possibility that *P. montana*, *P. cembra*, etc. enter the palaeoecological picture. The corresponding danger lies in the tendency to overlook fossils representing extinct or foreign species: although several of these are recorded, it may well be that others have been overlooked.

Taxonomic evidence in the study of the flowering plants often presents its own problems. It may be possible to recognize the recently defined components of a former aggregate of species; thus *Empetrum hermaphroditum* and *E. nigrum* are recognizable by the different pollen sizes, and *Nasturtium officinale* and *N. microphyllum* have recognizably different seeds. On the other hand fossil records made before the recognition of the new species must generally remain referable only to the old aggregate. In many instances, naturally, the newly established forms are not recognizable from the plant organs preserved as fossils.

It almost goes without saying that the conditions of preservation of plant remains must play a very large part in controlling the identification of recent fossil material, by whatever plant organs or parts it may be represented. It is seldom that direct steps can be taken to meet this difficulty, but in the case of fruits and seeds, especially those with soft and fleshy outer layers liable to decay, it may be possible to macerate fresh material to varying degrees and so match the fossil material. It is naturally commoner to encounter the stones (pyrenes) of drupes than the whole fruit, the cypselas of Compositae seldom occur as fossils with a pappus, and nutlets of Cyperaceae most commonly occur without the utricle, a circumstance occasionally helpful in identification (as in *Cladium mariscus*), but more often, as in species of *Carex*, not.

A danger easier to point out than to avoid is that due to the wide and often overlapping range within related species of the characters upon which identification of a fossil may be attempted. Where a type collection contains only a few achenes from each species of *Ranunculus*, and those samples each from one or two individual plants, it may appear deceptively easy to refer a fossil achene to one particular species of the genus. Were, however, the type collection to be extended to include samples from more individuals from a more diverse ecological and geographical range, the identity of the fossil might appear less certainly established. Where one has the good fortune to be able to call upon taxonomic specialists capable of supplying such a range of type material the position may be safeguarded, if not simplified.

It is inevitable that amongst the records included in the index there is a very great range of reliability. This follows naturally from the varying ease or difficulty of recognition of different species in the material available, from the varying extent to which critical characters of the specimens have been preserved or destroyed, and naturally upon the skill and experience of the investigator. Only in a modest proportion of cases have the actual fossil specimens been preserved for future reference: these include much material investigated by the Reids and M. E. J. Chandler (British Museum) and material in the hands of investigators still active. It is naturally a matter of great difficulty or impossibility to assess the degree of reliability of the identifications recorded.

In assessing the results of our survey it must always be borne in mind that the sampling has been far from random and far from comparable in deposits of different ages. Thus the earliest workers tended to concentrate on deposits rich in fruits and seeds with the result, pointed out in the next section, that they tended to exclude the deep acid peat of raised mosses and blanket bogs. More recently there has been a very pronounced interest in the nature of the climate, fauna and flora of the Late Weichselian period and a very large number of sites have been investigated for information on this period alone. A tradition begun early and very properly sustained has been that of recording collections of seeds, fruit, wood and other plant material associated with human settlements and submitted by archaeologists to botanical colleagues. This has led to a substantial total of records for the periods extending from the Neolithic (*sensu stricto*) to Romano-British, but it will be appreciated that these contain a large proportion of crop plants, weeds and plants of disturbed ground: they do not represent,

6

generally speaking, the same types of habitat as do those beds investigated for purely geological purposes. Moreover, as we approach the Iron Age and Roman periods we discover that when we seek to correct this emphasis by examination of natural organic deposits, this proves remarkably difficult, for drainage and peat-cutting have everywhere destroyed the uppermost layers of our peat bogs and fens and have removed the evidence thus sought. It must further be noted that we must expect some heavy over-weighting of the marsh and aquatic plants in most of the natural organic deposits, an over-emphasis less frequent in the material from archaeological sites, and that other communities, for example dense woodland on heavy clay or beechwood on chalk, will naturally tend to escape representation.

These warnings, however, chiefly concern the temptation to make quantitative comparisons between one period and another, or to argue from the *absence* of a given species. Both are dangerous with the material so far available, and the strength of our conclusion must inevitably rest on the records of *presence* of particular species at particular times and places. The only direction in which we may escape this restriction is in conclusions based upon regional pollen analyses designed to disclose a general pattern of vegetational alteration. Here, however, specific identification plays a somewhat subordinate role and there enter fresh limitations in interpretation. These techniques and interpretations of pollen analysis, however, we shall consider separately.

2. SEEDS, FRUITS AND LEAVES, ETC.

The evidence of historical phytogeography is based upon the recognition of plant remains of two main categories of size: those visible and to some extent recognizable by the naked eye – the so-called macroscopic remains; and those requiring high-power microscopy for their study – the microscopic remains. The study of each of these groups has had a very separate history and a distinctive role. The macroscopic remains have long been the object of interest and record, including as they do such evident vegetable structures as wood, bark, leaves, bud-scales, fruits and seeds: upon occasion are found such identifiable organs as the prickles of *Rosa*, cones or bract scales of conifers and Amentiferae, and less frequently occur stamens or whole flowers as with the male flowers of *Myriophyllum alterniflorum* found in the late-glacial muds of Hawks Tor, Cornwall, or the intact flower of *Trifolium campestre* found by A. P. Conolly in the Bronze Age deposit at Minnis Bay, Kent. Although referred to the category of macroscopic structures, the identification of all these remains is greatly facilitated in fact by low-power binocular microscopy, and by the examination of suitable sections under low- or high-power magnification.

The recording of macroscopic plant remains has played an important part in the investigation of recent geological formations. On the Continent Nathorst, Andersson and Steenstrup showed the potentialities of such investigations and in this country Clement Reid, E. M. Reid and M. E. J. Chandler achieved an enviable distinction in thus establishing the floras of Tertiary and Quaternary times in Britain and the nearby continent. Many of these identifications can be carried to the specific level, and they are mostly significant in establishing the presence of a particular species in a given locality and given deposit. The relatively large size of these remains militates against their transport for more than small distances. It may in general be taken that it is to the macroscopic remains, seeds and fruits in particular, that we look for records of specific identity, and we may take it that in the majority of instances these remains come from a vegetation growing quite close to the site of eventual discovery. Where transport has taken place, as by river action, some clue to this will generally appear in the geological situation, and where the remains are derived from older beds this will often be apparent from their abraded or more carbonized condition, as for example with the fruits of *Carpinus* encountered in Late Weichselian deposits at Barnwell (Cambridgeshire), and in the Lea Valley (north of London).

The collection of fruits and seeds, before the introduction of pollen analysis, naturally turned upon the primary need to get together as numerous and representative a collection of plant material as possible from each given deposit. There was a natural tendency to consider as the unit of supply any more or less organic bed lying between mineral layers, notwithstanding that different types of organic mud and peat might be represented in the organic layer and that it might very well embrace a long period of time and a big climatic range during its formation. There was also a tendency to collect fruits and seeds from those deposits, only when they were abundant and easily separable from the matrix: C. Reid advises for this purpose collection from fluviatile sandy loams, deposits which, as he points out, are likely to contain representatives of several vegetation types upstream along the river banks. The advantages of rapid appreciation of general floristic character given by collecting from this type of deposit are offset by some disadvantages. Thus it is made harder to recognize the plant communities of former times (so getting guidance to microstratigraphy of the deposit), as well as to obtain more general botanical information. Similarly there is an unnecessarily strong tendency for certain types of community and flora to be unrepresented: thus we find in C. Reid's book the astonishing comments that *Eriophorum vaginatum* and *Cladium mariscus* are both unknown as fossils in the British Isles, whereas in fact remains of the one occur in great abundance, to the annoyance of peat-cutters, throughout raised mosses and blanket bogs everywhere, and fruits and rhizomes of the other are abundant wherever fen deposits occur in Ireland, England and Wales, from Boreal times onwards.

3. WOOD AND CHARCOAL

Wood and charcoal contribute the most abundant of macroscopic plant fossils, and the recognition of the trees and shrubs from which they came is of particular interest because these plants have been the vegetational dominants through a large part of our Flandrian and earlier interglacial history.

The conspicuous nature of tree stools, so frequently encountered in peat-cutting, peat-drainage and stream-eroded sections in low land as well as mountain peat mires, has produced a wealth of record in past centuries. In far too frequent instances, however, these records are of dubious value for our present purpose. In the first place they are very seldom related to a datable horizon or deposit. In the second place they are often very uncertainly identified by a casual or field inspection, and whilst field inspection will suffice when made by a careful and cautious investigator, confusion may easily arise between the similar bark of *Betula*, *Alnus* and *Corylus*, or between wood (devoid of bark) of *Salix*, *Populus* and *Betula*. It has very often happened that 'fen-oaks', reported to me from the Cambridgeshire fens, have proved upon inspection to be yews (*Taxus*); they were called 'oaks' by the fenmen merely on account of their remarkable hardness. Where tree stools occur *in situ* they often afford, beside the wood characters, evidence of habit, size and bark, which assist recognition: thus the piles of a late Bronze Age trackway in Somerset could be seen from their wood structures to belong to *Acer*, but the furrowed bark showed them clearly to be hedge maple (*A. campestre*) and not sycamore (*A. pseudoplatanus*). It is, moreover, common to find associated with a shrub layer, remains of leaves, fruits, bud scales or other parts of the prevalent trees. Thus at Wood Fen near Ely, the buried pine stools are accompanied by pine needles and cones, and the buried yews by fruit stones and microsporophylls (H. & M. E. Godwin & Clifford, 1935). One need hardly remark that pollen series nearby almost always reveal striking maxima of the corresponding pollen type, as is startlingly shown in the numerous instances of the fossil record of the vegetational succession from fen carr with *Alnus* and *Betula*, through fen wood with added *Quercus*, through the transitional pine-wood stage, to raised bog. This is a sequence repeatedly encountered in British as well as other west European mire sections (see Godwin, 1943, and chapter III: 6). Where shrinkage of the deposits has been caused through drainage, it often happens that the layer of cones, pollen, etc. which represents the woodland floor now lies a decimetre or more below the base of the tree stool; for the latter, propped up on its deeper roots, will have been unable to keep pace with the general subsidence (Woodwalton 'A'; Godwin & Clifford, 1938).

Wood identification has been made relatively simple by the increased interest in and knowledge of wood anatomy; for the purposes of British Quaternary history, of course, a very large part of the specimens to be identi-fied come from the small range of woody species within the living flora of the British Isles.

The utilization of wood by man from very early times has naturally meant that excavation of human settlement very commonly yields fragments of timber or bark, either cut, fashioned or built into structures, or perhaps surviving merely as fragments of firewood. In the latter case the wood is commonly converted to charcoal, in which form of course it has much improved qualities of persistence under conditions likely to destroy organic materials. Conversion to charcoal by no means destroys the structure of wood, and even fine microscopic structures can be recognized by suitable techniques of fracture, illumination and microscopy. Archaeologists have been quick to realize the value of learning the identity of charcoals associated with the industries and cultures disclosed by excavation, and many dated records of arborescent plants have thus been made. They have nevertheless suffered from some difficulty in finding experts willing to undertake the difficult and somewhat unrewarding task of these identifications. Some quite remarkably valuable results have nevertheless been obtained from this source, as for instance the identification of charcoal of beech (*Fagus sylvatica*) in Bronze Age charcoals from South Wales, a discovery going far to establish the native status of this tree in Britain (Grimes and Hyde, 1953).

At the same time the inherent difficulties of charcoal identification must be always borne in mind and much less weight must be attached to charcoal records than to corresponding records based upon well-preserved wood. Thus one is not disposed to allow charcoal identifications of the wood of *Aesculus hippocastanum* (horse chestnut) from the Neolithic site at Nympsfield in Gloucestershire and other ancient prehistoric settlements to overthrow the reasonable established belief that this tree has only been introduced to Britain at a much later period.

The large quantities in which wood is to be found in many prehistoric sites, especially such as crannogs, trackways, hut foundations, hearths and so on, have led to attempts to assess, from the relative frequency of the various species represented, something of the specific composition of the arborescent communities of the time and region concerned. Such attempts must always be hampered if not stultified by the fact that man, from the moment that he first began to use wood for any purpose at all, exercised selection of the material of different species. Not only do different species grow in situations making them more or less accessible, but they offer different resistance to cutting or to collection. Their fallen branches and trunks decay at variable rates, and above all they have very different suitabilities for the numerous purposes of prehistoric man. There is excellent evidence of selection of oak for piles, hazel and willow for wattle, yew for spear shafts, and it is not unlikely that even for firewood some selection operated. It is worth remembering also that in some instances wood may have been brought to prehistoric sites for the sake of the bark rather than the timber itself: the great sheets of birch-bark flooring, the innumerable birch-bark rolls at the Star Carr Mesolithic site give point to this suggestion, as also the common use of pine-bark floats by prehistoric fishing communities.

4. POLLEN AS AN INDEX OF PRESENCE

In the study of Quaternary history, and especially in relation to historical biogeography, identification of fossil pollen grains has come to occupy two distinct roles, each of great importance. The primary impulse to the development of the study of pollen, *palynology* as it is now generally called, came from the remarkable success of pollen analysis as a means of elucidating the sequence of former vegetation types and climates, and from its proven value in thereby providing a chronological scale against which to measure events of the Quaternary period. This aspect of palynology is dealt with separately in chapter III: 9.

Although pollen analysis, devoted to these broad aspects of Quaternary research, concerned itself for the first decade or two very largely with the pollen of the forest trees which naturally preponderated in the Post-glacial deposits of temperate north-western Europe, the analysts also encountered in greater or less amounts, pollen and spores of other than arboreal plants. These extra pollen types they identified as they could, but without particular concentration or aim. Meanwhile, however, the interest in pollen as a cause of hay-fever had led to increased concern with pollen morphology, and the work of Wodehouse in the United States of America and of Erdtman in Sweden set altogether new standards of microscopic examination and illustration of pollen grains. Both produced books having great influence on the direction of the study (Wodehouse, 1935; Erdtman, 1943). As the movement thus begun gathered way, pollen analysts found that the analysis of non-arboreal pollen types was important, indeed essential, to the study of two fields of Quaternary investigation; first the study of the Late-glacial period in which prevailed vegetation types not wholly or not at all dominated by trees, and secondly the investigation of the destruction of forest vegetation by Neolithic and later pastoralists and agriculturists, who thus created grasslands, heaths and similar vegetation types where woodland had prevailed before. The creative insight and technical skill needed to recognize and develop these themes was provided by J. Iversen of the Danish Geological Survey, who gave so many productive conceptions to the study of palynology.

The intensive study of pollen-grain morphology initiated by Iversen and Troels-Smith in Denmark, Faegri in Norway and Erdtman in Sweden, involved the use of the highest grade optical equipment and the constant employment of oil-immersion objectives for examination of the pollen-grain membranes. How great has been our progress in these techniques can be seen by consulting the two major scientific journals *Pollen et Spores* and *Grana Palynologica*. The use of high-power phase-contrast microscopy has also proved valuable, as in the identification of pollen of the chief genera of cereals and wild grasses by Körber-Grohne (1964). Thanks to the introduction of techniques for cutting ultra-thin sections, of transmission electron microscopy of pollen-grain sections, of the techniques of making carbon replicas of pollen surfaces, and more recently of scanning electron microscopy, we have now attained considerable knowledge of the structure of the pollen-grain exine at both microscopic and sub-microscopic levels. Whilst this knowledge gives more confidence in the interpretation of optical microscopy it has not yet greatly contributed to improvement in the standards of pollen identification nor has stereoscan electron microscopy yet been applied consistently to the study of fossil pollen. None the less through the last decade or so British palynologists have regularly identified and recorded a considerable range of pollen of herbaceous plants and shrubs often to the level of species or an aggregate of species. Whilst such identifications are facilitated by the keys and descriptions provided by the literature they all rest essentially upon comparison with type material made up as a permanent collection of pollen-slides, in which fresh pollen has been so treated as to render it comparable with fossil material. A great deal hinges upon the completeness and authenticity of any such reference collection of type material, which ought to include specimens from several sources, with appropriate taxonomic authority behind them. One such collection is in charge of Miss R. Andrew in the Cambridge Botany School, and another has been developed by Dr A. G. Smith in the Queen's University, Belfast. The latter, which has given replicate series outside Belfast, has the special advantage that all the preparations are mounted in silicone oil, a medium that makes them specially suitable for size measurement as an aid to diagnosis.

In considering pollen grains as indices of the former presence of particular genera or species in given localities, it must be constantly borne in mind that, in contrast with the larger and more fragile macroscopic remains, pollen grains may be transported vast distances by air and by water and will withstand conditions of transfer from old deposits to those newly forming. We need recall only that Erdtman was able to collect pollen of American origin from air over the middle Atlantic, that the ice of glacier fields yields vast amounts of pollen to the glacier streams and that the water of the North Sea has quantities of pine pollen suspended in it, for the possibilities of distant transport to be realized. It is only when the given pollen type forms a substantial or generally consistent component of a pollen rain known from general evidence to be of local origin that we can use it as proof of the former presence of the species identified. Churchill (in Godwin & Switsur, 1966) has shewn that a large proportion of the pollen collected by moss cushions growing on Signy Island in the Antarctic consists of the southern beech, *Nothofagus*, that can only have been wind-carried from

South America, more than 800 km away. Likewise Srodon (1968) shewed that the pollen of surface samples in Spitzbergen consisted very largely of pollen of pine and birch from Scandinavia, some 750 km away, whilst pollen of more thermophilous plants, though in smaller amount, originated in still more distant places. At high altitudes this distant component equalled or exceeded that produced locally. Whilst transport by water into lakes and fluviatile deposits has to be taken into serious account in the reconstruction of local vegetational history the distances involved are not great. In the case of marine oceanic sediments such as those forming in the Argentine Basin off the coast of South America (Groot *et al.*, 1967), the pollen appears not to have suffered redeposition or very long distance transport, but such deposits figure hardly at all in the British Quaternary record. Ever since Iversen (1936) pointed out that boulder clays contained pollen derived from the destruction of older interglacial beds, and that this pollen was incorporated in late-glacial deposits alongside contemporary pollen, palynologists have been aware of the serious dangers of secondary incorporation of pollen, more especially in mineral sediments. Although Iversen was able, in the special case he investigated, to show how the indigenous and derived pollen components could be separated in the late-glacial layers, this is by no means always possible. We may thus find, as in the Stort and Lea Valleys north of London, that the Late Weichselian organic silts contain a substantial proportion of pollen or spore types evidently of Tertiary origin, such as *Taxodium*, *Pinus haploxylon*, *Gleichenia* and *Mohria*, along with types generally characteristic of the Late Weichselian, such as *Armeria*, *Artemisia*, *Helianthemum* and *Polemonium*; and

one may be left in uncertainty as to the category in which to place the many other forms encountered, and whether indeed some may not have been derived from deposits of intermediate age. Intercalated layers of peat in such situations are of great value for they can be counted on to yield essentially contemporary pollen with at most a very small derivative component, brought in by wind.

When, during deposition, the pollen of the local vegetation was sparse, we may naturally count upon distant wind transport to have provided a relatively larger part of the total pollen: thus above the mountain tree-line one finds the warmth-demanding temperate trees of the valleys relatively better represented in the pollen catch than in the coniferous forest belt next lower in altitude, but the increased representation is no index to local presence.

These qualifications disclose the limitations in the value of identification of fossil pollen as an index of former presence of species as compared with macrofossils, but they have the qualities associated with their defects, namely ubiquity and frequency. We can count upon recovering recognizable pollen from far more situations than those yielding seeds: not only will peats and organic muds of all types contain them, but also most fine-grained mineral sediments; and a mere fraction of a cubic centimetre of material will often yield scores or hundreds of grains.

Let it be said in concluding this section that the methods used in pollen analysis reveal and concentrate also such material as skeletons of various algae, fungal spores and fruit bodies, and the spores of the Bryophyta and Pteridophyta. The last category has been included in our floristic record.

III

THE BACKGROUND SCALE OF PLEISTOCENE EVENTS

1. GLACIAL AND INTERGLACIAL PERIODS

The background against which we have to view vegetational and floristic changes of the Pleistocene is one dominated by the cycles of severe climatic change that led to repeated periods of glaciation, probably affecting the entire world directly or indirectly, and the alternating periods of milder conditions, the interglacial periods. The Pleistocene is indeed strongly characterized and set off from the preceding Pliocene period by these recurrent glacial events, as also by the general relatively lower temperatures that succeeded the decreasing warmth of the middle and later stages of the Tertiary era. The key to the British Pleistocene sequence lies in the deposits of East Anglia, a region close enough to the margin of successive ice-advances to register their effects adequately, but where the erosive processes weakened marginally so that the evidence of earlier glacial and interglacial events frequently escaped destruction. It has the advantage also of contiguity to the Netherlands where similar deposits have been assiduously investigated, and of a low altitude that brought its coasts within reach of the interglacial marine transgressions. Indeed the Plio-Pleistocene boundary is recognizable there in between the Coralline and Red Crag deposits, characterized largely by their marine molluscan faunas, whilst the succeeding early Pleistocene consists of alternating deposits of terrestrial and coastal marine facies in which study of pollen analyses on the one hand and of marine faunas on the other can usefully be combined. Table I shews Sparks and West's subdivision of the Pleistocene based primarily on the East Anglian sequence and dependent, not only upon the characterization of stone-content, fabric and stratigraphic sequence of the tills and associated cold period deposits, but upon study of the evidence of plant and animal remains (and especially pollen analyses) of the more or less organic deposits of the intervening mild periods, in each of which conditions were such as to permit optimally the general growth of mixed deciduous forest for a considerable length of time.

The lower Pleistocene sequence lacks any evidence for actual glaciation but the pollen analyses, particularly those of a deep borehole at Ludham, allow the recognition of the Ludhamian and Antian Interglacials and the cold Thurnian and Baventian stages succeeding them. The same is true for the mild Pastonian and Cromerian of the Middle Pleistocene, but in both the Baventian and Beestonian there is also evidence of the local prevalence of permafrost. The massive freshwater, estuarine and peat deposits of the Cromer Forest Bed series have supplied, since the time of Clement Reid, abundant identifiable macrofossil remains: they include the Cromerian Interglacial and to an uncertain degree preceding stages also. Above the Cromerian Interglacial deposits are the tills of the earliest of the three glaciations to invade East Anglia, the massive boulder clays of the Lowestoft and Cromer tills. Organic beds were formed in lake basins on the surface of the Lowestoft till, as was first convincingly shewn at the brickpits at Hoxne, where the careful stratigraphic and pollen-analytic evidence (with an interstratified Late Acheulian industry) combined to warrant this becoming the type site for the Hoxnian Interglacial. Subsequent work has shewn an even more extensive interglacial sequence at Marks Tey, near Colchester (Turner, 1970) and the fine laminations of the lake muds there, most probably annual, have given a most important indication that the duration of the Hoxnian Interglacial must have been of the order of 40000 years (Shackleton & Turner, 1967). The Gipping glaciation succeeded and gave place to the Ipswichian Interglacial, of which the type site is Bobbitshole, near Ipswich, also fully characterized by pollen analyses and macroscopic identifications of biological material.

The Ipswichian Interglacial gave place about 70000 years ago to the latest Glacial Period, the Weichselian, in which the New Drift represents the actual extension of British ice-sheets: only marginally and in a late phase did the ice actually reach East Anglia to deposit the Hunstanton Boulder-clay, but evidence of periglacial cold conditions after the Ipswichian are widespread. This last glaciation has been succeeded by the deposits of the period in which we live, commonly spoken of as the 'post-glacial' but better referred to in conformity with geological usage as the Flandrian, thus accepting the fact that we cannot disprove the possibility, indeed the strong probability, that this period is no more than another interglacial stage in the still uncompleted Pleistocene era. It follows equally that we abandon the former designation 'Holocene'. The term 'post-glacial' is however so generally understood that it is convenient and not misleading to retain its use in general description.

Till-fabric analyses allowed the Gipping glacial stage of East Anglia to be linked with that of the English Midlands and the biogenic deposits investigated by the Birmingham school of Quaternary geologists and others have yielded pollen-analytic and faunal evidence that amply confirm the Hoxnian correlations suggested by the

TABLE I

Stage	Climate	Type site or area
UPPER PLEISTOCENE		
FLANDRIAN (Postglacial)	Temperate	Postglacial mires and lake deposits
WEICHSELIAN (DEVENSIAN)	Cold, glacial	Glacial deposits of Cheshire Plain
IPSWICHIAN	Temperate	Interglacial lake deposits, Bobbitshole, Ipswich
WOLSTONIAN (Gipping)	Cold, glacial	Glacial deposits at Wolston, Warwicks.
HOXNIAN	Temperate	Interglacial lake deposits, Hoxne, Suffolk
MIDDLE PLEISTOCENE		
ANGLIAN (Lowestoft and Cromer tills)	Cold, glacial	Glacial deposits at Corton, Suffolk
CROMERIAN	Temperate	Lake deposits at West Runton, Norfolk
BEESTONIAN	Cold	Silts and fluviatile gravels at Beeston, Norfolk
PASTONIAN	Temperate	Tidal deposits at Paston, Norfolk
BAVENTIAN	Cold	Marine sands and silts at Easton Bavents, Suffolk
LOWER PLEISTOCENE		
ANTIAN	Temperate	Marine deposits of the
THURNIAN	Cold	Ludham borehole,
LUDHAMIAN	Temperate	Norfolk
WALTONIAN		Crag at Walton-on-the-Naze, Essex

stratigraphic sequence. Correlation with Irish Pleistocene deposits is necessarily less direct since the Irish glaciers stemmed from centres within the country itself and contact with the ice-advances in Great Britain was limited to the extreme north and east. None the less, the extremely significant interglacial deposits may in several instances be related pollen-analytically to the English interglacials, particularly those of Gort, Kilbeg and Baggotstown to the Hoxnian, so that they afford a key to the equivalence of the succeeding glacial stages, the Munster General Glaciation to the Gipping, with the Midland General Glaciation corresponding to the Weichselian, the Ardcavan Interglacial beds being in all probability, Ipswichian.

The whole of England south of the Thames was substantially untouched by any of the ice-sheets even at their maximum extension, but periglacial conditions during the glacial stages are witnessed by widespread solifluxion deposits and patterned ground, these naturally being most evident for the last glaciation. At each glaciation the amount of water locked up in the ice-sheets of the world was sufficient to cause a general (*eustatic*) lowering of the world's ocean level by a few hundred feet, and with melting, at the return of warm conditions, a return to the original height. Thus each interglacial was marked by the creation of coastal features, particularly beaches, and as there seems to have been an overall fall in ocean level through the Pleistocene the successive interglacial beaches lie at different heights, as do the terraces of the Thames and Somme so far as they were under control of sea-level.

Sea-level changes have profound effects upon river courses; at times of low relative sea-level the valleys cut down to deeper base levels, and as sea-levels rise the valleys aggrade, being filled at their mouths with marine or estuarine deposits and thence forming aggradation terraces sloping gradually up stream. In reaches beyond the influence of the changes of sea-level, the climatic cycles themselves greatly affect the character of the valleys: under glacial conditions frost-shattering, solifluxion and bare ground supply such immense quantities of material that aggradation preponderates; with melting of the snow and ice for a while there is tremendously powerful erosion, and then with the establishment of general vegetational cover the rivers settle down to steady down-cutting through the material accumulated in the glacial phase. It will be noted that thus the effects of the climatic cycle are opposed in the upper and lower reaches of the main valley systems and there is a region of great complexity where the two influences interact. Nevertheless, now that the principles have been recognized the sequence of river terraces has been made comprehensible, and where the terrace gravels contain the tools of prehistoric cultures, these cultures have been fairly consistently related to the terrace sequence. Outstanding studies of this nature have been made by Abbé Breuil in the Seine valley, and by King and Oakley in the Thames. It will be apparent that of the numerous plant-bearing deposits which occur in the great valleys beyond reach of the glacial ice, we may refer many to their appropriate glacial or interglacial stage.

In addition to the eustatic cycle of change in relative level of land and sea, there have been two types of regional effect. In those regions heavily loaded with ice, many thousands of feet thick, the earth's crust was locally depressed: its recovery is witnessed by the raising of sea-beaches hundreds of feet above sea-level after removal, in the next interglacial, of the ice-load responsible for this *isostatic* warping. In certain regions, of which the southern North Sea appears to be one, there has been a continuing tendency to tectonic subsidence during the Tertiary era and this may still be continuing. It may have been involved in the creation of the great interglacial seas of the Netherlands and North Germany, the Holstein Sea, the Eem Sea and, last of all, the Zuider Zee. It may also be that all of these, whose deposits lie at very similar altitudes, owe their origin to compensatory downwarping in a region marginal to the isostatic uplift of Scandinavia. No English marine deposits have been certainly correlated with these transgressive episodes of the eastern flank of the North Sea where they fit decisively into the Pleistocene sequence.

The last fifty years have seen one considerable step forward in our approach to the problems of interglacial floras, and this has been chiefly brought about by the application of pollen analysis methods. The advance began by the demonstration that an overall climatic shift had affected the whole of Europe in the Post-glacial period and that great vegetational alterations reflected the change from the cold of the end of the last glaciation, through progressive amelioration to the warmth of the Post-glacial Climatic Optimum and the following climatic deterioration. The pioneer researches of Milthers and Jessen upon the interglacials of Denmark clearly showed that a remarkably similar climatic sequence had characterized each of these, but they were able to show that the continuation of climatic worsening led in the interglacials to a return of arctic tundra conditions before the renewal of glaciation. Extension of palynology to other European interglacials has served to confirm that the picture is generally valid, and the acceptance of it greatly clarifies the interpretation of older accumulated data based upon macroscopic faunal and floral remains. The concept has been elegantly extended by Iversen (1958) and others: the main features of the full glacial–interglacial cycle are reproduced in fig. 147. It reflects the ultimate climatic control of all the components of the ecosystem of temperate latitudes which moves from (1) a 'Cryocratic' glacial stage with frost-shattered, unweathered soils moved by solifluxion and melt water and colonized by open communities of arctic–alpine plants in conditions of minimal competition and absence of shade or shelter, to (2) a 'Protocratic' stage of colonization of the unleached calcareous soils by herbaceous and then by arboreal vegetation with competition and shade increasing and the soils becoming progressively deeper and tending to neutrality, to (3) a 'Mesocratic' stage of climax woodland upon slightly acid brown earth mull soils, and general shading and root competition dominating all undergrowth, to (4) a 'Telocratic' or terminal stage, envisaged as 'retrogressive' in a vegetational sense, with soils now heavily podsolized and acidic bearing moor and heathland or acidic woodland such as the pine–birch woods with *Vaccinium* communities on leached forest soil. This last stage gives place in turn to the cryocratic

stage of the next glacial episode. Intensive study of Danish Interglacial deposits by Andersen (1966) led to a proposal that the latest part of Iversen's 'Mesocratic' stage should be cut off as an 'Oligocratic' stage in which, without any fall in mean temperature, natural processes of soil development have induced soil degeneration with podsolization and corresponding vegetation change. Turner and West (1968) have carefully discussed the implications of this and other schemes as a basis for subdivision of the interglacial periods and, now that there are numerous pollen diagrams from four different British interglacials to go upon, propose that in each there may be recognized four main sub-periods of vegetational development that are recognizable from the pollen diagrams and that may be regarded as natural biostratigraphic zones. They serve as means of correlation between the several sites of each interglacial and its type site, thus acting as chronozones. The zones are as follows:

I *Pre-temperate zone*: forest of boreal trees, especially *Betula* and *Pinus*, develops and closes after the late stages of a glacial period, but light-demanding herbs and shrubs persist.

II *Early-temperate (mesocratic) zone*: shews the expansion and dominance of deciduous mixed oak woodland, typically with *Quercus, Ulmus, Fraxinus* and *Corylus*, on rich forest soils.

III *Late-temperate (oligocratic) zone*: its opening marked by the expansion of *Carpinus* and *Abies*, sometimes *Picea* and, in the Flandrian, *Fagus*. This late establishment and relative decline of mixed oak forest is associated with soil degeneration rather than specific climatic worsening.

IV *Post-temperate zone*: shews a return to dominance of boreal trees, especially *Pinus, Betula* and *Picea*, with extinction of the mixed oak forest, but with opening of the woodlands and development of damp ericaceous heath communities on acid soils.

The boundary separating these four interglacial sub-stages from the glacial stages at either end is the level of equality of total arboreal and non-arboreal dry land pollen.

For the purposes of our analyses, plant records have been referred, wherever possible, to this interglacial sub-stage system. Many older records cannot safely be zoned within the known interglacial stage from which they come, and although we retain our system of reference to the established pollen-zone system for the Flandrian, it can be readily related to this interglacial zone system on the basis that the opening of IG sub-stage I corresponds to the England and Wales pollen zone III/IV boundary, that of IG sub-stage II to the England and Wales IV/V boundary, that of IG sub-stage III to the VI/VIIa boundary, and that of IG sub-stage IV to the VIIb/VIII boundary, although this last is made uncertain by the scarcity of suitable recorded sites and by the complication caused by man's progressive alteration of the environment.

In order to make comparison between the vegetational and floristic history of this country and that of the adjacent continental mainland it is essential to correlate the respective Pleistocene stages. The generally accepted view is as follows. The Dutch Reuverian in late Pliocene is equivalent to the Coralline Crag, and the Italian early Pleistocene Villafranchian–Calabrian corresponds to the Red Crag. Still within the lower Pleistocene are the Dutch Tegelian, with its highly productive organic beds, corresponding to the Ludhamian, and the Waalian, also a temperate stage, corresponding to the Antian. Thereafter the Dutch Needian and German Holsteinian correspond to the Hoxnian, and the Eemian of those countries to the Ipswichian as the penultimate interglacial. The Flandrian is of course named from the Netherlands marine transgressional series of the present interglacial. Of the three major glaciations of the British Isles, the earliest, the Lowestoftian (Anglian), corresponds to the Elster of north-western Europe and the alpine Gunzian; the Gippingian (Wolstonian) with the Saalian (including the Warthe stadium) and Mindelian; and the latest glaciation, the Weichselian, has the same stage name here and in north-western Europe: it is the equivalent of the Alpine Würm. The extra-alpine equivalents of the pre-Gunzian glaciations are still uncertain. We may note that this correlation makes what used to be called 'the Great Interglacial', the Mindel-Riss, to be represented in Britain by the Hoxnian.

Finally we may note that the fossil fauna of interglacial deposits quite commonly afford evidence of the climate and age of the time of deposition. Foraminifera, Crustacea, Bryozoa and marine Mollusca have contributed greatly to knowledge of Pleistocene conditions, especially during formation of the Crag deposits. Freshwater and land molluscs, insects and mammals have yielded information, especially for the terrestrial deposits of the later interglacials, although a great deal remains to be done, particularly with regard to the taxonomy of some groups in the older stages. Increasingly in the last decade or two fossil remains of animals have been collected and studied as part of detailed Pleistocene stratigraphy. This is especially so for the Foraminifera, Mollusca and Insecta: a useful and succinct account of the fauna of the British Pleistocene is given by West (1968).

2. RADIOCARBON DATING

When we come to consider deposits of Weichselian and Flandrian age we find ourselves for the first time to be in possession of a purely physical method for directly establishing the age of deposits and establishing correlations. This is the technique of radiocarbon dating whose conception and early development we owe to Dr W. F. Libby. The method rests essentially upon the fact that the effect of bombardment of the nitrogen of the upper atmosphere by cosmic ray neutrons is to produce carbon of atomic weight 14, a radioactive isotope of carbon. The radioactive carbon atoms slowly disintegrate, reverting to normal carbon of atomic weight 12. Experimental determinations give various similar values for the half-life, and a mean value of 5730 ± 40 years is that now generally adopted. The rate of this disintegration, like that of all radioactive bodies, is immutable and independent of changes in the physical or chemical environment. The radiocarbon formed in the upper atmosphere is quickly oxidized to carbon dioxide and as such enters the great reservoirs of carbon dioxide, the oceans, and from water and air alike is built up by photosynthesis into the organic compounds which form the structure of all plants and animals. Once thus incorporated, free exchange with the carbon of the atmosphere is largely precluded, so that the proportion of radioactive carbon in a substance such as the cellulose of a cell wall steadily diminishes in relation to normal carbon, reaching 50 per cent of its initial value at formation after 5730 years. Since the rate of cosmic radiation is broadly constant radiocarbon is being both created and destroyed at a constant rate so that the total amount in existence and the mean concentration in the atmosphere also remains constant. It was quickly realized that given methods suitable for the assay of radiocarbon in the proportions present in the natural products of photosynthesis, these circumstances might provide an absolute time scale of the greatest value for the later part of Pleistocene time, and Libby and his colleagues built and used electronic counting systems suitable to this purpose. So effective has the method proved that now in 1970, some eighty laboratories in all parts of the world are specifically devoted to carbon dating and publish annually some thousands of datings, most of which appear in the journal *Radiocarbon*.

The natural concentration of radiocarbon in the atmosphere is actually extremely small, amounting to no more than 10^{-12} that of normal carbon. In the counters now in use this yields only a few counts per minute per gram of carbon sampled, and as this is small in relation to the rate of adventitious radiation from outside, chiefly the penetrating cosmic radiations, the most immediate technical difficulty has been screening off or compensating for such extraneous radioactivity. The more effective this can be made, the more sensitive the age determinations become. Despite considerable sophistication and variety in the apparatus employed it is rarely practicable for age determinations to exceed 50000 years. Various factors that might at first glance seem likely to impair the method actually prove unimportant. The extra 'dead' carbon injected into the atmosphere by the burning of fossil fuel, a considerable total, can be measured and allowed for or a standard can be employed that is unaffected by it. Likewise the vast amounts of radioactive carbon liberated by atomic bomb explosions affect only modern samples, and although large short-term variations of cosmic radiation constantly occur the buffering effect of the great bicarbonate reservoirs in the world's oceans preserves radiocarbon concentrations of high constancy in the atmosphere. Fractionation effects between ^{12}C and ^{14}C resulting from variation in the physical and chemical systems involved in synthesis of different organic materials are seldom large and can be discounted by mass-spectrometric analyses. There also seems to be very effective mixing in the atmosphere and samples of woods grown in different parts of the world give very similar radiocarbon assays.

When direct tests were first made by dating wood of known age from the tree-rings of giant redwoods (*Sequoia*) it seemed that correspondence was close, but with improved exactitude of measurement it became clear that during the last 1300 years the mean atmospheric radiocarbon activity had varied by almost 2 per cent above and below present values. Subsequently Suess and others have taken advantage of dendrochronologically dated wood of *Sequoia* and the still older *Pinus aristata* to test further the basic assumption of constancy of atmospheric radiocarbon activity (Suess, 1967). In fact a large discrepancy is proved, such that conventional radiocarbon dates are too young by about 200 years round 1500 B.C., a difference increasing to about 800 years by 4000 B.C. Since however no less than 80 determinations have been made for the period between 4100 and 1500 B.C., it is practicable to apply for this period a calibration for conversion from 'radiocarbon years' to 'solar years'. Such correction applied to carefully chosen Egyptian material has given good correspondence with independent chronologies accepted by archaeologists. None the less, it is doubtful whether the high atmospheric activities occurred beyond about 5000 B.C. since radiocarbon assay and lake-varve countings yield similar results for about 8300 B.C. Accordingly we follow here the common practice of citing all ages determined by radiocarbon dating in 'radiocarbon years'. Thus far the method has been substantially vindicated but it has to be remembered that there are many potentialities for error in the use of unsuitable material, such for instance as calcareous lake muds subject to the so-called 'hard-water error', marine molluscs subject to exchange of the original carbonate of their shells with the marine environment and for which the original radiocarbon activity must be uncertain, and woods and charcoals liable to contamination by descending rootlets or organic substance in solution. In the last case chemical purification may remove adsorbed material before radiocarbon assay, and such pre-treatment is obligatory in dealing with very old samples where a very small intrusion of recent material would invalidate the result. No such laboratory cure is possible for the instances of misassociation in which the organic material does not in fact represent the event it is proposed to date. Indeed the most careful study of the provenance of each radiocarbon sample is always needed if the great cost and labour of laboratory dating are not to be misused (Godwin, 1959; Polach & Golson, 1966).

3. THE WEICHSELIAN

Geological evidence of the events of the last glacial period in Britain, though abundant, is extremely complex and confusing, so that it offers only a disjointed and often disputed scale of reference for the sparse organogenic deposits of the Early and Middle Weichselian that are to be found in the Midlands and East Anglia. The line of maximal advance of the ice across Britain is a composite one, the resultant of successive advances of ice-lobes at different times, now dominated by ice from the Irish Sea fed from Scotland and the Lake District, now by the Welsh glaciers, and again by ice from the north-east. The line of maximum extension is generally conceded to be as shewn in fig. 1 and within this lie several well-marked stages of minor standstill during the general retreat of Late Weichselian time. The melt waters from the ice-fields and run-off from a landscape sealed with permafrost cut wide channels in the pre-existing Pleistocene infill of the river valleys and built into them terraces of coarse gravels and sands, generally lower in level than those of the earlier glaciations. Such terraces in the Severn, Avon, Cam and related streams have yielded interstadial organic deposits reflecting temporarily milder conditions, not however such as prevailed in the interglacials. Commonly these plant beds are found in sequences where cryoturbations, ice-wedge casts and other phenomena indicative of freeze–thaw lie above and below, thus emphasizing the apparent climatic fluctuation. Over the whole of southern Britain, indeed outside the ice-field limits, similar phenomena are widespread and testify to the periglacial climatic conditions that prevailed even down to the extreme south-west of Cornwall (West, 1967). No doubt similar conditions held for that part of Ireland outside the extent of the Midland General Glaciation, and such upland ice-free areas as projected above the ice, like the Mountain Limestone uplands of Derbyshire and till surfaces once ice-covered but now exposed by glacial retreat. The sea-levels of the early and middle Weichselian are generally supposed to have been about 100 metres lower than those of today, a situation in which all the southern North Sea basin would have been dry land and the Irish Sea and English Channel reduced to narrows of only a few miles width.

The difficulties of correlating the scattered biogenic beds of the main Weichselian have been greatly simplified by the advent of radiocarbon dating which not only provides internal co-ordination but allows comparison with similar deposits in Holland and Denmark where detailed stratigraphic and pollen-analytic studies have allowed a tentative carbon-dated temperature scale to be attempted from the end of the Eemian through the Weichselian period (van der Hammen *et al.*, 1967). Radiocarbon dates around 60000 to 50000 B.P. have been obtained for the Dutch Amersfoort interstadial and the later and warmer Danish Brørup interstadial in which cool temperate woodland established itself; thereafter the generally cold tundra conditions were apparently

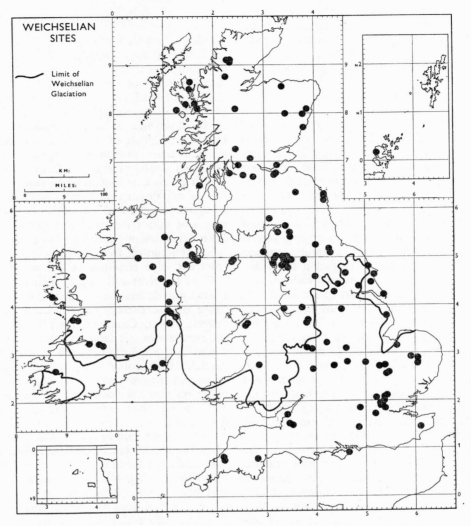

WEICHSELIAN
SITES

Limit of
Weichselian
Glaciation

K M:

MILES:
0   9   100

Fig. I. Map of the British Isles shewing sites of plant deposits of Early, Middle and Late Weichselian age. The presumed maximum of the latest (Weichselian) glaciation is shewn. Stages of halt or readvance in the recession of ice from this position are shewn in fig. 152.

unbroken until the Late Weichselian, save for two short and cool interstadials around 30 000 and 38 000 years B.P. at Hengelo and Denekamp possibly equated with the Paudorf Interstadial of Austria (fig. 12).

The last twenty years have witnessed not only the discovery and description of a substantial number of British sites of Early and Middle Weichselian age, but much better understanding of those previously known. The plant-bearing deposits at Sidgwick Avenue, Cambridge, are not carbon dated and are attributed to the earliest Weichselian, partly upon the biological evidence and partly because they occur in the braided channel system that meandered over the surface of the intermediate terrace of the River Cam (Lambert, Pearson & Sparks, 1963). Equivalent terrace deposits of the River Wissey at Wretton in Norfolk have yielded a remarkable wealth of evidence extending at the base from organic

beds covering a large part of the Ipswichian Interglacial through a large part of the Weichselian (Sparks & West, 1970). Of major importance also has been the recognition of a well-developed interstadial at Chelford on the Cheshire plain inside the area of maximum glaciation, overlain by till of the main Irish Sea glaciation and with a radiocarbon age, secured by de Vries only after isotopic enrichment of the sample, of 61 000 years (Simpson & West, 1958). The biological evidence reinforces the absolute age determination in suggesting equivalence to the Brørup of Denmark. A series of sites from the English Midlands (Upton Warren, Fladbury, Tame Valley, Brandon) have produced a remarkable wealth of plant and animal records, especially of insects: they have radiocarbon ages between 42 000 and 28 000 years and are referred by the Birmingham Research school to 'Upton Warren Interstadial Complex' that is clearly Middle

PLATE I

Nazeing, Essex. A deep channel in the gravels of the valley contains calcareous muds (Late Weichselian in age) at the base, and above them fen-peats (Flandrian) up to the sealing layer of river-borne clay. The staff shewn is 3 m in length. Such localities afford the best possible sites for collecting macroscopic plant material and samples for pollen analysis.

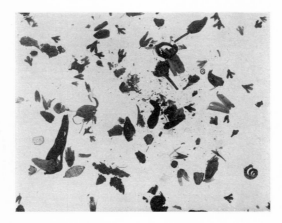

Debris sieved off for identification after maceration of calcareous lake mud by dilute nitric acid. This material is of Late Weichselian age from Nazeing, Essex and includes inflorescence and calyces of *Armeria maritima*, female cone scales of *Betula nana*, capsules of *Salix* and fruit valves of *Draba incana*.

PLATE II

Stools and timber of pine (*Pinus sylvestris* L.) extracted during agricultural operations from the peat at Wood Fen, near Ely. These pines were growing on mesotrophic peats along the fenland margin during the Sub-boreal period.

View of the submerged forest at Ynyslas, near Borth, on the shore of Cardigan Bay. This peat deposit is a seaward continuation of the lower layers of the great raised bog behind the coastal ridge of shingle and dune. The stumps include *Alnus*, *Betula*, *Quercus* and *Pinus*. (Phot. Mr Challinor.)

PLATE III

Face of Hawks Tor kaolin pit, Bodmin Moor, Cornwall, shewing large angular granite boulders moved by solifluxion down the hill slope under periglacial conditions.

View of the face of the Hawks Tor kaolin pit, Bodmin Moor, Cornwall, shewing muds and peats overlying the weathered granite surface. During the Late Weichselian period the bands of gravelly soil (pale in the photograph) were brought down the valley slopes into the basin and caught up the lower muds and peats in cryoturbatic movements.

PLATE IV

Wicken Sedge Fen, Cambridgeshire. View from a windmill across the boundary ditch into fen dominated by *Cladium mariscus* and invaded by bushes of sallow (*Salix atrocinerea* and *Frangula alnus* chiefly). In this neighbourhood great thicknesses of *Cladium* peat are encountered with occasional wood of *Alnus*, *Salix*, etc.

Alder fen-carr growing over lake mud in Heron's Carr, Barton Broad, Norfolk, with sparse undergrowth of *Carex acutiformis* and the marsh fern, *Thelypteris palustris*. A community of this type has given rise to great thicknesses of wood-peat in valley deposits in many parts of the British Isles.

[facing p. 17

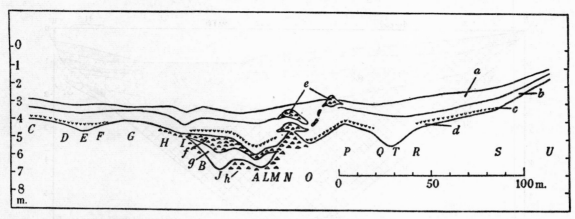

Fig. 2. Section through the deposits above the weathered granite surface in the Hawks Tor kaolin pit, Bodmin Moor, Cornwall; *g* is the layer of organic muds and peats formed in the Late Weichselian warmth oscillation (Allerød period = zone II); *f* is the layer of soli- fluxion earth brought down from surrounding slopes in the ensuing period (zone III) of returning cold. Above *f* are peats of the Flandrian period into which lenses of gravel (*e*) have been interjected subsequently by tin-streaming.

Weichselian and embraces the period of the Dutch Hengelo and Denekamp interstadia. Within the same radiocarbon age range are other sites described by Bell (1969) from somewhat further south (Earith, Marlow, Great Billing, Syston). The Lea Valley Arctic Plant Bed (E. M. Reid, 1949) has a radiocarbon date of 28 000 years, and the Barnwell Station bed may well be of similar age (see fig. 12).

It seems probable that the severest conditions of the British Weichselian now succeeded, for we have from Dimlington radiocarbon dates of 18 240 and 18 500 for plant material in silts overlaid by the three superimposed Drab, Purple and Hessel Tills of East Yorkshire. The earliest stages of the Late Weichselian appear to be represented by the plant-bearing peat erratics of Colney Heath with a radiocarbon age of 13 560 ± 210 B.P. and pollen spectra indicating open vegetation with only scattered northern trees.

There is no doubt that around 15 000 years B.P. there set in a rapid and widespread increase of temperature to or beyond the values sustained at the present day. This increase is dramatically registered in sediment cores from the deep oceans, as for instance in the Caribbean cores recently reported upon by Rona and Emiliani (1969) when after registering extremely low values the oxygen isotope 'palaeotemperatures' display a dramatic rise reaching to modern values within a few thousand years. It is true that the values for the $^{16}O/^{18}O$ ratios depend not only upon fractionation during deposition of the calcareous material tested, but also upon the isotopic composition of the sea-water as it is affected by the release into the oceans from melting ice-sheets. Both effects are however a registration of the same increase of temperatures, and they constitute an unequivocal means of correlating the marine and terrestrial sequences of glacial and interglacial periods (Shackleton, 1967). The Late Weichselian in north-western Europe covers the early part of this temperature rise and is generally set between the radiocarbon ages of 15 000 and 10 250 years B.P. It is strikingly characterized and sub-divided by the results of a pronounced climatic oscillation that affected glacier retreat, periglacial stratigraphy and vegetational cover in such a decisive manner that these phenomena were successfully used to sub-divide the Late Weichselian and to identify what was formerly described as the 'Lateglacial Period' in distinction from the preceding 'Fullglacial' and the succeeding 'Post-glacial'. The oscillation is known as the 'Allerød' from the Danish site at which it was first recognized by Hartz and Milthers. Here a deep basin contains laminated clays with sandy layers and pebbles brought down from the flanks of the basin by solifluxion in a cold climate. The clay is however divided into an upper and a lower portion by a layer of organic mud formed when the surrounding slopes were stabilized by vegetation and when there was sufficient warmth to permit the abundant growth of aquatic plants. Macroscopic remains established that the birches grew in this temperate phase, whilst sub-arctic plants characterized the periods before and after it. A similar threefold stratigraphy was widely recognized in Europe with the older and younger 'Dryas clays' separated by an organic layer of mud or peat formed under more genial conditions (figs. 2 and 3). The cold freeze–thaw conditions are often marked by cryoturbation, which may contort and involve the upper mineral layers with the organic deposits, or in deep water by the deposition of clays that may be varved as a result of seasonal melting and freezing processes (fig. 155).

The biological evidence was greatly strengthened by the application of pollen analysis to deposits of this age, especially since the extension of the recognition of pollen of many herbaceous plants. It has become evident that the threefold lithological division corresponds to pronounced vegetational shifts that have been made the

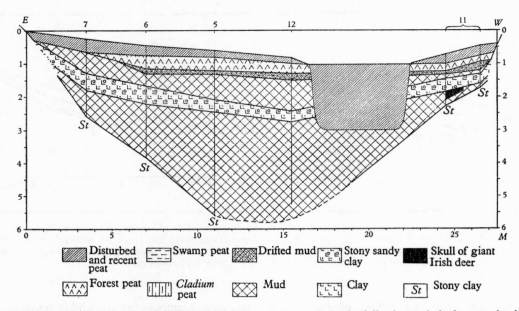

Fig. 3. Cross-section of the north-western bog at Ballybetagh, a Late Weichselian site in the Wicklow Mountains, showing a stratigraphy typical of the Allerød oscillation. During the mild Allerød oscillation open-water organic muds were laid down in the basin and in these were incorporated remains of the giant Irish deer (*Cervus megaceros*). In the following period of returned cold a strong solifluxion earth was laid over these muds, and this was itself succeeded by deposits of muds and peats in the temperate conditions of the Flandrian period. A very characteristic flora was recovered from the early deposits of the basin. (After Jessen & Farrington, 1938.)

basis of recognition of three zones (I to III) through most of western Europe. In Denmark the early and late cold periods had an open tundra-type vegetation dominated by grasses and sedges, but with abundant dwarf shrubs and herbs characteristic of fresh open soils, whilst the intervening temperate phase had 'park-tundra' vegetation with copses of tree-birches. Further south the temperate period was characterized by birch woods or even pine forest. The end of zone III is indicated generally by the re-expansion of woodland, so that arboreal pollen exceeds that of herbs, and by the closure of vegetational cover so that solifluxion ceases and organic deposits form in the valleys and lake basins above the mineral layers.

There is strong evidence from Scandinavia that zone I corresponds to slow retreat of glaciers from the Pomeranian belt of terminal moraines, followed by accelerated retreat during zone II setting free from ice the Danish islands and southern Sweden, the so-called 'Gotiglacial' retreat stage. Zone III corresponds with the halt or minor readvance of the ice during which were built the Norwegian 'Raa', the Middle Swedish moraines and the Finnish Salpausselka. To the north of this moraine line lie the pro-glacial lake basins containing varved clays that, thanks to de Geer and his successors, have provided a chronology independent of but coinciding with the radiocarbon scale. Both systems indicate that the zone III/IV boundary, the close of the Weichselian stage, was close to 10250 years B.P. We have very little knowledge of the corresponding glacial retreat stages in the British Isles, but Donner's research suggests that zone III may correspond to the Scottish Loch Lomond readvance, whilst elsewhere glaciation was limited to small corrie glaciers in the higher mountains. The result of radiocarbon dating samples from many sites at the zone boundaries indicated by the strongly coinciding biological and geological evidence has been to confirm decisively the synchroneity of the zones in north-western Europe: the I/II boundary is close to 11950, II/III to 10750, and III/IV to 10250 years B.P. (Godwin & Willis, 1959a).

Useful as the Allerød oscillation may be as a Late Weichselian marker in western Europe, it has to be admitted that despite earnest search, there is very little evidence for it in the rest of the world and the explanation of it must to some degree be a local one (Mercer, 1969). Likewise in the extreme north-west of Britain evidence for its effects is sometimes lacking (H. J. B. Birks, 1969).

In a limited area of Europe there appears to be evidence of a minor climatic oscillation prior to the Allerød and named, after the Danish site where it was first identified, as the Bølling oscillation. There is no clear evidence however that it can be recognized in Britain, although the period of its occurrence in zone I is covered by several detailed palynological series that yield various less clearly defined changes of different character or radiocarbon age.

The rapidly rising temperatures of the Late Weichselian began extensive melting of the world's ice-sheets that was immediately reflected in a eustatic rise of ocean-level, a rise that in recent years has been charted by an increasing number of radiocarbon age determinations of samples that stood during formation at sea-level (or a known height from it) and that occur where subsequent tectonic movements are unimportant. The broad conclusion

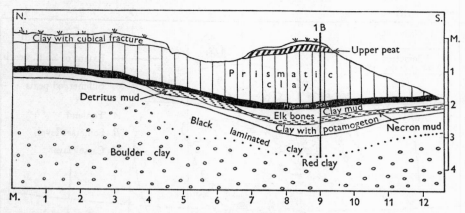

Fig. 4. Section through deposits at Neasham, near Darlington, shewing Late Weichselian deposits resting in a hollow in boulder clay. The organic muds with bones of elk (*Alces alces*) were deposited in the Allerød mild phase of zone II. (After Blackburn, 1952.)

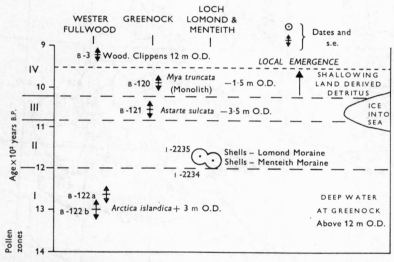

Fig. 5. Radiocarbon dates for the Scottish Late-glacial sea in the Firth of Clyde, mostly derived from marine shells in the position of growth. They indicate a marine transgression to at least 25 m above present O.D. some 13000 years ago and subsequent regres-sion (after Bishop & Dickson, 1970). Numerous plant beds are known from these deposits and corresponding early pollen zones are shewn.

already reached in 1958 (Godwin, Suggate & Willis) still stands, that from 14000 B.P. sea-level was rising at a rate of approximately 1.0 m per century. No deposits from the earlier part of this time range are accessible to us in Britain, but there is now proof that by 13000 B.P. the encroaching sea had invaded the west coast of Scotland still depressed under its local burden of ice. Thus were formed deposits of estuarine clays with marine Mollusca, and in the later stages the rich and long familiar plant beds from Paisley and Greenock (fig. 5). These deposits of the Scottish Late-glacial sea, subsequently elevated by isostatic recovery to at least 25 m above sea-level, have been correlated with the zone III glacier margin only a few miles distant by carbon-dated shells from the Late-glacial sea incorporated in the moraine of the Loch Lomond readvance at Drymen and in the Menteith moraine

(Bishop & Dickson, 1970). These results accord satis-factorily with the conclusions based upon palynological study of deposits close to the maximum of transgression at Garscadden Mains (Mitchell, 1952).

4. FLANDRIAN GEOLOGICAL EVENTS

The Flandrian period is not long in geological time and began no more than ten thousand years back. Nevertheless within this span deposits have been laid down in many situations adequate for the application of the principle of superposition to be applied to the dating of their plant-bearing beds, and if the scale and magnitude of the processes concerned in their formation seem unfamiliar or trivial to the geologist, there is a compensation in that

2-2

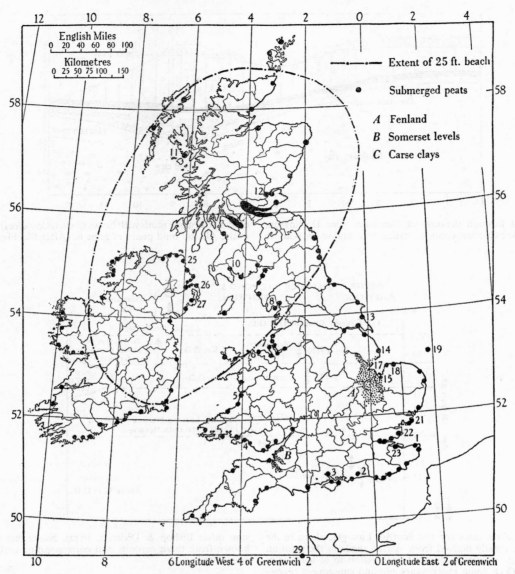

Fig. 6. The map shews the distribution of submerged peats along the coasts of the British Isles (the numbers refer to sites specially mentioned in 'Coastal peat beds of the British Isles and North Sea', Godwin, 1943), the region of isostatic elevation as indicated by the extent of the '25 ft raised beach', and three regions shewing inter-digitation of Flandrian freshwater and marine deposits. It is to be noted that there are abundant sites of submerged peats even within the area of isostatic elevation. In part the submerged peats result from eustatic rise in sea-level, and in part (in East Anglia) from local downwarping.

they begin to enter the range of knowledge of the ecologist, soil scientist and historian. Furthermore of course all Flandrian events lie easily within the span of radiocarbon dating: this means of checking and extending strati-graphically-based conclusions has been extensively exploited in the last decade or so.

With the cessation of glaciation and the disappearance of periglacial conditions many types of physiographic modification were no longer operative; glacial erosion and the formation of moraine, esker, drumlin, varved clays and loess ceased and frost-splitting, polygon formation, soil-striping and solifluxion phenomena generally disappeared or were restricted to a few high mountain localities. The open soil so characteristic of the Weich-selian and so conducive to erosion and soil movement, was rapidly covered in the warm climate of the Flandrian by a close mantle of vegetation that sealed the ground surface and reduced surface erosion by wind and water to negligible amount. Glacial lakes and kettle holes no longer received vast volumes of suspended mineral material and the layering of laminated clays was replaced by the deposition of organic muds (gyttjas). Glacial over-flow-channels and the great shallow erosion valleys of the south of Britain were now supplied by tiny streams or none

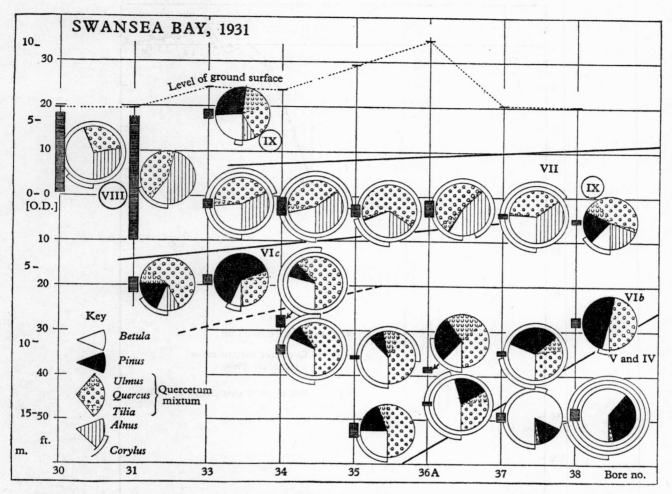

Fig. 7. Diagram to shew the results of pollen analyses of samples taken at different depths from borings made in Swansea Bay, South Wales. The vertical lines shew the separate bore-holes and the shaded rectangles the analysed samples at their proper depth in relation to mean sea-level. The circles represent the percentage content of the different tree-pollen genera, reckoned as a percentage of the total tree pollen (see inset key). The sloping lines indicate the forest-history zones into which the respective samples fall.

These results clearly shew the rapidity of marine transgression during the second half of zone VI, and the fact that by the end of that zone the sea-level had been restored to within 20 ft (6 m) of its present height in relation to the land. If this coast is conceded to be outside the region of recent isostatic movement this result implies that eustatic restoration of ocean level was almost complete by the end of zone VI.

and either became quite dry or more commonly were occupied by shallow pools or fens which caused them to fill up with muds and peats until they were flush with the countryside around; this we see in the New Forest valley bogs and such sites as the fen at Cothill (A. R. & B. N. Clapham, 1939) and the Ouse–Waveney channel at Lopham (Tallantire, 1953). The immature soils of the Glacial and Late-glacial periods, under the combined influence of changed climate and the new vegetation, began their more or less rapid progress towards mature climax types, and over much of the countryside brown-earth soils became prevalent with gleyed formations in locations where ground water was seasonally present.

Only slight modifications of the topography of Weichselian time have been caused by continued weathering and erosion. Nevertheless certain important physiographic processes have continued and certain new factors have become operative. We have already described the influence of the glaciation in abstracting water from the oceans and producing eustatic fall of sea-level, probably of the order of 100 m. With the Flandrian climatic amelioration this water was restored to the sea and a general rise in ocean level took place. This event re-created the North Sea by flooding what had been an extensive lowland plain traversed by the lower reaches of the Rhine, Thames and the other rivers of western

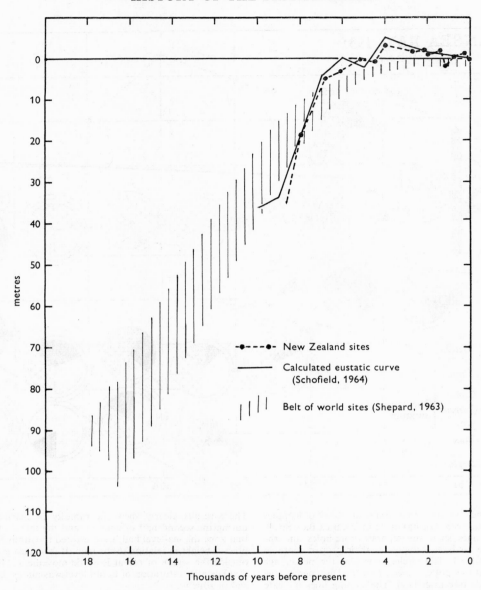

Fig. 8. Radiocarbon dates for material that grew at or close to sea-level shew a rapid and extensive eustatic rise in ocean level between 18000 and 6000 years B.P. (Shepard, 1963). There is some evidence from analysis of the Scandinavian isostatic recovery and from Antipodean data that present-day ocean levels had possibly been attained or exceeded after about 6000 B.P. (Schofield, 1964).

Germany and eastern England: it also created, or enlarged from small dimensions, the sea between Great Britain and Ireland. A phenomenon of such consequence as this to migration and repopulation of the British Isles by plants and animals merits close study, which is however complicated by the interplay of the eustatic rise of ocean level and tilting and warping of the land due mainly to isostasy. It is apparent that the whole of Scotland and north eastern Ireland come within an area of isostatic uplift recognizable primarily by the presence of the so-called 'Neolithic' or 'Twenty-five foot' raised beach that in fact occurs at all levels up to about 12 m above present sea-level. On the other hand there is reason to regard the southern North Sea basin, with our adjacent East Anglian coast, as an area of subsidence. Between the two is a broad belt running from south-west to north-east across Great Britain (and probably southern Ireland) in which isostatic effects are apparently absent: it is here that we may expect registration of the eustatic rise to be most clearly exhibited. Indeed we find that deposits off-shore from south-east Devon (Clarke, 1970) indicate between pollen zones IV/V and VIC a rise from −140 to −55 ft (−42 to −17 m), values that correspond well to those calculated independently by Schofield (1964) and the early radiocarbon datings of the eustatic rise based on evidence from many parts of the world (fig. 8). Like the

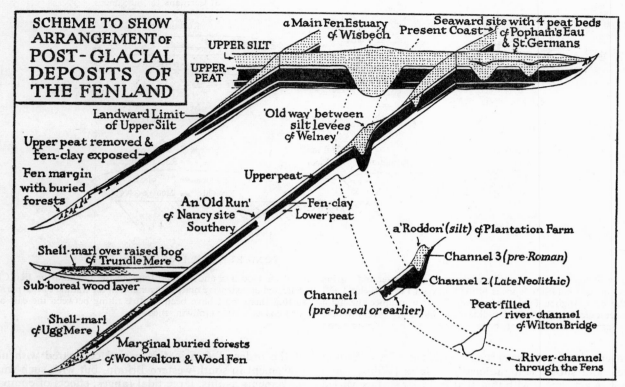

Fig. 9. Purely diagrammatic scheme to shew the general lines on which the Flandrian deposits of the English Fenland are arranged.

earlier results of pollen-analytic age determination of submerged peats in Swansea Bay (Godwin, 1940b) they shew a very rapid rise in sea-level (3 to 5 ft; 0.9 to 1.5 m) per century during the Boreal period (fig. 7). Such a rate of rise, if maintained, must have attained present sea-levels within the Atlantic period, and indeed all down the west coast of England and Wales we find that deep estuaries have been filled with marine deposits up to modern sea-level or thereabouts, and thereafter large raised bogs have grown upon the flat clay surfaces more or less uninterruptedly to the present day, as in the Somerset Levels and the Dovey estuary and the Lanca-shire coast. In these areas radiocarbon dates for the marine–freshwater contact are, broadly speaking, about 6000 years B.P. There is some evidence for subsequent marine transgression in the Bristol Channel region and on Cardigan Bay, but it was slight in extent.

When we enter the region of isostatic uplift the later stages of the eustatic rise are registered by two different geological phenomena. Like the estuaries of western England and Wales, the valleys of the Forth and Clyde were filled with marine deposits, the so-called 'Carse-clays', upon the level surface of which systems of raised bogs then grew, notably the great complex of Flanders Moss. Careful stratigraphic investigation and radio-carbon dating at key horizons indicate that the Carse-clay formed between about 8500 and 5500 B.P. (Godwin, 1960a; Newey, 1966). It now seems clear that the other

phenomenon, the formation of the post-glacial raised beach, took place at very much the same time, and, as W. B. Wright suggested, was the consequence of the balance between isostatic land rise and the last stages of eustatic rise of ocean level. Peat below the raised beach of Ballyhalbert, Northern Ireland has been dated approxi-mately to 8000 years B.P., a date according with the pollen-analytic zone reference, and material from the upper beach surface at Dalkey Island yielded a date of about 5300 B.P., in conformity with the archaeological evidence of occupation by transitional Mesolithic/Neolithic people. In the last 5000 years continued isostatic elevation has brought the raised beach to its present height. In the outer parts of the region of isostatic uplift, the Solway Firth and the north-west Lancashire coast, the surfaces of estuarine clays laid down at the end of the eustatic rise now lie 10 or 15 ft (3.0 or 4.5 m) above sea-level, possibly somewhat above their height at formation.

If we turn from the region of isostatic uplift to East Anglia we find evidence for downwarping. The Fenland basin shews, resting upon Jurassic clays, a freshwater peat that is continuous to the surface at the landward margin, but seawards is divided into an upper and lower by intervention of an estuarine deposit, the Fen Clay, and the upper peat is succeeded by a coastal belt of marine silts occupied and in part deposited in Romano-British time (fig. 9). The stratigraphic studies of the Fenland Research Committee established that the Fen Clay

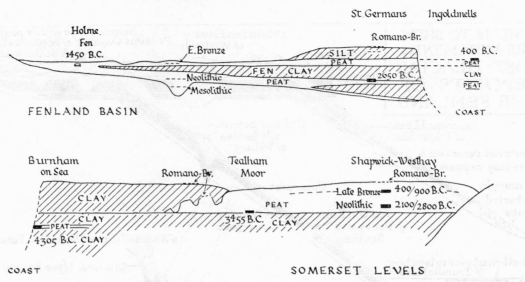

Fig. 10. Schematic representation of the main freshwater and marine layers in the Flandrian deposits of the Somerset Levels (below) and of the East Anglian Fenlands (above). In each instance the main archaeological horizons and radiocarbon dates are given. Whilst the Somerset coastal peat bogs escaped marine transgression between 3500 B.C. and Roman time, in the Fenland the Fen Clay indicated an extensive marine incursion in the middle of this period, so that there must have been relative tilting between the east and west coasts. (After Godwin, 1960a.)

transgression lay between the Neolithic and Early Bronze Ages, and subsequent radiocarbon dates (Willis, 1961) shew, in agreement with this, that its lower surface varied from 4700 at the seaward side to 4200 B.P. at its landward limit whilst its upper surface was about 4200 to 4000 B.P. Carbon dates not much in excess of 4700 suggest that peat formation throughout the basin was induced by the banking up of freshwater during the early stages of the Fen Clay transgression, but in deep river valleys it has been registered from the Late Weichselian onwards and both Mesolithic and Neolithic artifacts have been found in peat beneath the Fen Clay in one such site, Shippea Hill, radiocarbon dated respectively to 7600 and c. 5300 B.P., and tied in to the respective pollen-zone boundaries VIC/VIIa transition and VIIa/VIIb transition (Clark & Godwin, 1962). The history of marine transgression contrasts sharply with that in the Somerset Levels as is diagrammatically shewn in fig. 10. In the western coastal area the eustatic rise was completed by about 5500 B.P.; fen and raised bog succeeded without further marine invasion until Roman time, as is witnessed by a rich series of archaeological horizons from early Neolithic to Late Bronze Age and Iron Age within the raised-bog sequence, and a large number of corresponding radiocarbon dates. Thus where marine transgression had ceased in the west at the top of the main eustatic rise it continued in the east for more than 1000 years, a strong indication of relative downwarping of the east. This tendency has been documented by consideration of the present level of a number of peat deposits known to have formed close to sea-level at approximately 6500 B.P. on the British and Dutch coasts (Churchill, 1965). Churchill shews a depression in the southern North Sea

of the order of 6 m in this period compared with null movement in south-western Britain, but with such small movements as this, large tidal ranges, effects of compaction of sediments and the deduction of height of formation in relation to contemporary sea-level, there is room for argument and reappraisal.

It is not surprising with this geological history, to encounter all round the coasts of Britain evidence of marine transgressions in the shape of 'submerged forests' or coastal peat beds (fig. 6), nor to find that they represent almost every part of the Flandrian. Their presence must have had much to do with the persistence, and perhaps also the origin, of much folk-lore that concerns land lost beneath the sea, such as Atlantis, or Cantref-y-Gwaelod, the lost land of Wales. They certainly testify to the 'dry-land' stage of the North Sea, for freshwater peat ('moorlog') recovered abundantly by the trawlers of past years, has yielded pollen spectra which shew them to have formed just before or at the earliest expansion of the warmth-demanding forest trees, that is to say zones IV and V, a result conformable with the radiocarbon age of 8415 ± 170 B.P. for peat recovered in 37 m of water from the Leman and Ower bank, off the east coast of Norfolk, from which was dredged in 1932 a barbed point of the Mesolithic hunter culture since made familiar in the excavations at Star Carr, east Yorkshire. It can be understood how attractive to palaeobotanists and especially to budding palynologists these coastal peats have been, but their promise is misleading. They generally embrace only a short time span, they are often truncated and the contained pollen is apt to be extremely corroded: further the pollen spectra mostly reflect the local vegetational succession from saltmarsh to brackish and freshwater fen, thence to fen woods, so that the regional pollen rain is obscure, and only in favourable circumstances and after much research can one securely relate the peat bed to mean sea-level at the time of formation.

In regions removed from the sea, minor physiographic changes have also continued through the Flandrian period. The natural processes of vegetational development have caused the infilling of lakes in turn with limnic, telmatic and terrestrial organic matter and shallow lakes, held up by moraine or other blocks, have become filled with muds and peats and have often finally gone over to the domed structure of a raised bog. The influence of climatic shift in accelerating or setting back this process can be traced in the stratigraphy of lake, fen and bog deposits, so procuring means of correlation from site to site as well as providing valuable indices of the nature of past climatic change. At the same time such processes as delta formation have gone forward and have no doubt also shown response to climatic variation. Phases of dryness are indicated by the cutting of lower shore-lines, the erosion of previously deposited layers (as described in the margin of Windermere by Pennington (1947)), and by the laying down of layers of reworked necron-mud 'breccia', or even of mineral material from the strand. Finally, when clearance of forests by prehistoric man loosened the soil surface locally or more widely, the event was recorded by wind- and water-borne mineral layers in lake and mire deposits, as witness the sandy layers at the Mesolithic occupation level at 'Old Decoy' in the Fenland and the layers of gravelly hill-wash in the blanket-bog peats of Bodmin and Dartmoor. No doubt we may expect with care also to recover the layers of ash like those that characterize the occupation levels in Danish lakes with adjacent Neolithic settlements. It is notable that in many parts of England the uppermost deposit in our river valleys is a layer of freshwater clay, the result of flooding during the period of agriculture, and in the Somerset Levels within historic time structures were in fact operated to trap this flood clay for improving the peat areas. Such clay layers very usefully seal in and preserve underlying peats and muds.

We have not, so far, been able to recognize any substantial sequence of annual lamination of our organic lake sediments such as Welten (1944) so elegantly demonstrated in Switzerland, nor have we the advantage of incorporated layers of volcanic dust or ash such as occur in the peat bogs of New Zealand and Tierra del Fuego and for that matter as near as the Vulkanische Eifel of Germany where they have been shown to be of Late Weichselian age. Here and there deposits of calcareous tufa have been described, as at Blashenwell in Dorset and in Flintshire: unfortunately these beds which so often disclose remarkable leaf impressions are extremely difficult to date, since they are lacking in pollen content and difficult to ascribe with certainty to any given climatic cause or period. The usefulness of radiocarbon dating in such conditions is shewn by the fact that a date of 6540 ± 150 has been obtained from bone in the levels bearing mesolithic flint at Blashenwell.

5. CLIMATIC CHANGES

There is no difficulty in agreeing that the Pleistocene period is essentially characterized by the widespread development of ice-sheets that resulted from a general and progressive lowering of temperature in later Tertiary time (Hollingworth, 1962) and that the major stages into which it is divided in this country and elsewhere correspond with cold stages, with major glaciation alternating with temperate interglacials. It is equally accepted that the climatic changes of the Pleistocene are remarkable for the combination of a large range of temperature change and brief time span, and we have already outlined their consequences, in the major depositional and erosional processes by which the movements of the ice-sheets and the associated changes in ocean, river and soil regimes were recorded.

In recent years the global character of these cyclic changes has been convincingly shewn by geophysical measurements of deep ocean sediments from many parts of the world. An instance of this is given in fig. 11, reconstructed from the data of Rona and Emiliani (1969) for two Caribbean cores, samples from which were dated by radiocarbon assay or thorium-230/protactinium-231 analysis. Oxygen isotope ratios measured on foraminiferan tests throughout the cores indicate a striking cyclic change of ocean temperature that almost certainly represents the three latest glacial–interglacial cycles. As we have explained earlier, direct thermal fractionation in shell carbonate formation is not solely involved in the control of the oxygen isotope ratios, but nevertheless they certainly reflect world temperature changes, a conclusion strongly supported by consistent parallel changes in faunistic composition through measured cores.

It is very striking to see how consistently the oxygen isotope values from deep ocean cores reveal a steep rise in temperature from about 15000 B.P., whilst suitable undisturbed cores also shew a thermal maximum about 7000 to 3000 B.P. with a subsequent decline of 1 to 2 °C, all features corresponding to the terrestrial biological evidence. Likewise many cores shew an oscillation of the isotope ratios corresponding with the temperature changes known to be associated with the Allerød oscillation just prior to 10000 years B.P. With more effective radioisotopic dating, the earlier Pleistocene stages will be similarly tied to the record of ocean temperatures and correlated with associated changes in marine life, particularly the foraminiferan. Of less easily definable causation, but remarkable sensitivity, are the changes in rates of sedimentation of calcium carbonate in equatorial ocean sediments: those for instance given by Wiseman (1956) shew remarkable correspondence to the postglacial thermal rise, the thermal maximum and subsequent decline. More recently still the possibility has come into prospect of employing oxygen isotope analysis of cores from glacier ice to yield temperature records of great sensitiveness and long duration (Johnsen, Dansgaard & Clausen, 1970).

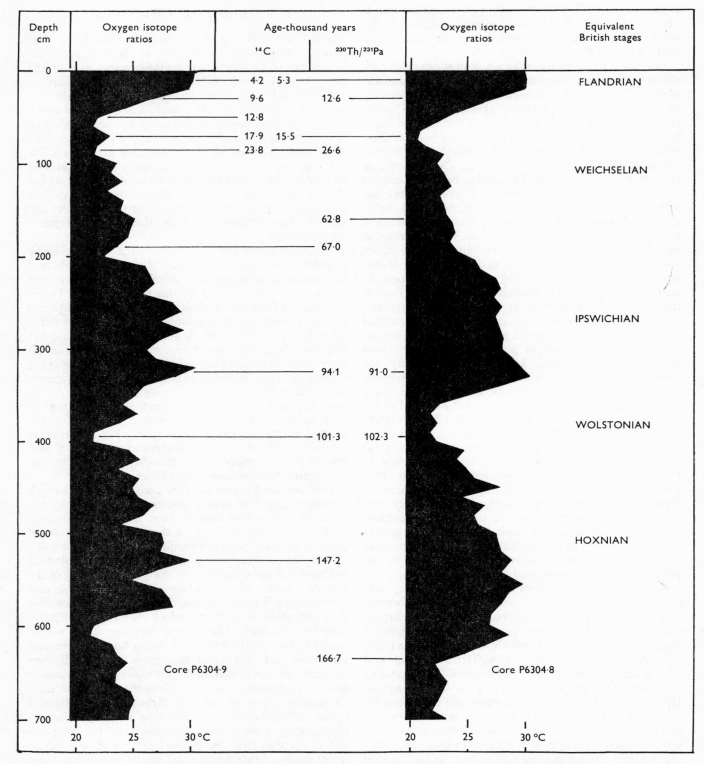

Fig. 11. Palaeotemperature determination by oxygen isotope measurement at intervals through ocean sediments in two Caribbean cores. The isotope ratios exhibit an amplitude corresponding to a range of about 22 to 32 °C for the surface ocean water. The profiles were dated in the upper samples by radiocarbon assay, and throughout by thorium-230/protactinium-231 analysis. The curves closely correspond and in all likelihood represent the sequence of the three latest glacial-interglacial cycles. (After Rona & Emiliani, 1969.)

This evidence of purely physical changes in the earth's crust is powerfully supplemented by and interwoven with the evidence of biological change, as witnessed by the system already described for sub-dividing the major inter-glacial stages, mainly upon pollen-analytic evidence, an outcome we could imagine as possibly foreseen by von Post who in 1946 entitled his Vega medal lecture 'The prospect for pollen analysis in the study of the earth's climatic history'. It was here that he defined the principles of revertence and regional parallelism in post-glacial vegetational history and proposed a division of the Post-glacial into three periods, the first of increasing warmth, the second of culminating warmth and the third of decreasing warmth. By characterizing the dominant ele-ments of the middle period as mediocratic and of the earlier and later periods as terminocratic, von Post went some distance towards showing that the same pattern of changing warmth had affected not only many parts of Europe, but many widely spaced regions of the globe. He shewed that a similar sequence also characterized the vegetational history of the interglacial periods. We have seen how this view has been developed by Iversen and Andersen into the concept of a cyclical evolution of climate, soil and vegetation through each glacial–inter-glacial sequence; climatic change however remaining the primary cause of the cycle. Only for the latest Pleistocene stage, the uncompleted Flandrian Interglacial, have more intricate systems of climatic sub-division been put for-ward, a consequence of the far more abundant evidence available for Flandrian time. Early in this century the painstaking analysis of the stratigraphy of Scandinavian peat bogs and lakes had permitted the very distinguished botanists Blytt and Sernander to elaborate a system of Post-glacial climatic periods that quickly came into wide-spread use, and which was afterwards related to the results of pollen analysis. These periods are briefly characterized as follows:

Sub-atlantic	Cold and wet	Oceanic
Sub-boreal	Warm and dry	Continental
Atlantic	Warm and wet	Oceanic
Boreal	Warm and dry	
Pre-boreal	Sub-arctic	

Von Post's early pollen analyses in southern Sweden showed that the Pre-boreal was a period of open tundra or birch forest, and the Boreal period was marked by early dominance of pine, soon giving way to the expan-sion of elm and oak, and by high values for hazel. The Atlantic was the zone of dominant mixed-oak-forest with alder and the linden (*Tilia*), and the Sub-boreal was similar save for sporadic appearance of spruce, beech and horn-beam. In the Sub-atlantic the three last-mentioned trees increased greatly, especially the beech in Scona: in general also pine and birch tended to recover some of their earlier importance. With greater or less success the Blytt and Sernander scheme was now extended outwards from Sweden to other parts of Europe (see especially Gams & Nordhagen, 1923); but, although often yield-ing a consistent picture of climatic and vegetational history, it was not finally found applicable to the great mass of new data for pollen analysis now available. None the less it had proved so convenient that it has persisted in use as a purely chronological system largely divorced from its original climatic significance: the radiocarbon ages attributed to it are shewn in the correlation table of fig. 31.

The pollen-zone system begun by von Post was intended by him 'mainly to serve as a means of determining geo-logical time'. It reflected the movements of the north-west European forest belts which, under the compulsion of changing climate, moved to their northern limits at the time of the warmth-maximum and then retreated in the following period of diminishing warmth. At the same time the altitudinal belts ascended and then suffered a descent on the major mountain systems. As conceived by von Post the pollen zones were intended to be synchronous units of post-glacial time, and indeed they went a long way to meet the urgent need for a pervasive chronological scale. All over Europe there were developed regional pollen-zone systems with this quasi-chronological charac-ter, though naturally there were great differences in the expression of the forest succession between the north and south of Europe, and some, though smaller ones, be-tween east and west. The remarkable consistency ex-hibited by most of these pollen-zone systems must derive from the fact that they reflected response of the major climax forest communities to the regional climatic changes, although, as we explain in a later section, differ-ential rates of migration and establishment and the destructive effects of prehistoric man have complicated the issue in particular times and places, and have modified the intended synchroneity of the zones.

Although the pollen-zonation systems are climatically based it remains difficult to employ them to describe past climates except in general terms, but we can call upon a great deal of other biological evidence to give us closer resolution of the problem; and increasingly direct physical measurement is becoming available, as is demonstrated for instance in the proceedings of the recent symposium on 'World climate from 8000 to 0 B.C.' organized by the Royal Meteorological Society (1966). The biological evidence is considered in section 7, but we may none the less make certain generalizations about it here, particularly to emphasize the very great difficulties of interpreting biological data in climatic terms. Much of it is based upon a very inadequate knowledge of cli-matic ranges and tolerances of plant and animal organ-isms. Nor can we be certain that the organisms of past periods were of similar constitution to those of today, and indeed within one and the same species different ecotypes have undoubtedly been present at different times, just as they now exist in different geographical localities or different habitats. Moreover, we are forced, in ninety-nine cases out of a hundred, to argue from the known geographical range of a species at the present day, whilst remaining uncertain whether the species in question is

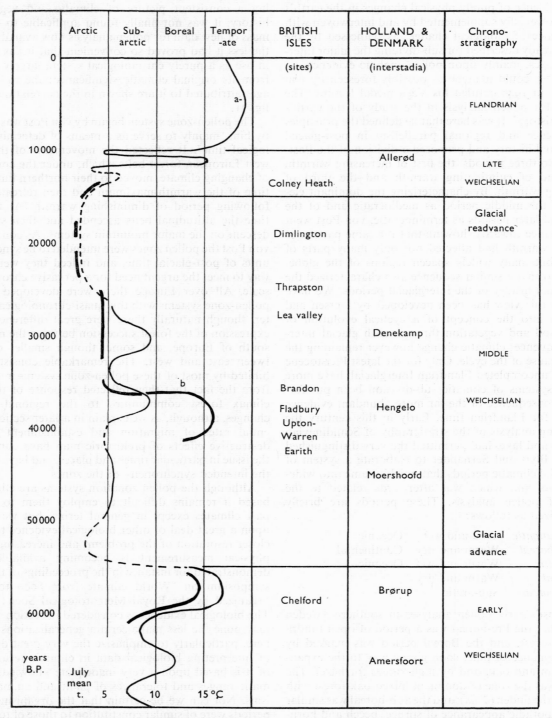

Fig. 12. Conjectural curves for mean summer temperature through the Weichselian and Flandrian, curve *a* based mainly on West European floristic evidence (Zagwijn & Paepe, 1968) and curve *b* based upon fossil insect faunas from lowland England (Coope *et al.*, 1971). Radiocarbon dating allows the positioning of key Weichselian sites in England, and of interstadia in Denmark and the Netherlands. The Early and Late Weichselian correspondence is clear; that of the Middle Weichselian less so.

truly in equilibrium with present climate, or indeed that its range is truly limited by climate. Except in very few instances we are ignorant of how climatic factors operate to determine climatic boundaries and are apt to apply to past circumstances the control that is now apparently operative in our own region. Thus to English botanists the beech is a continental species only forming self-regenerating forest in the south-eastern part of this country, but in central Europe and Switzerland it is regarded as an oceanic species. A comprehensive survey of the ecology of the tree shows that in its western range the beech is restricted by soil degeneration, northwards by lack of high summer temperatures necessary for flower initiation, southwards and eastwards by summer drought, whilst locally in sub-maritime areas it suffers heavily from spring frosts destroying the exposed stigmas of the flowers. It is not easy to say which of these reasons should be invoked in any given case of a former change of the frequency of beech as disclosed, for example, by pollen analysis.

It is probable that the northern range of many species is restricted by summer demands for high temperature, as seems to be the case for the mistletoe (*Viscum album*), the beech and many more. On the other hand Iversen's admirable survey of *Viscum*, *Hedera* and *Ilex* as climatic indicators shows that the two latter species, in common with other woody plants of generally oceanic range such as *Fagus* and *Taxus*, are also severely restricted by susceptibility to winter cold. This is a direct effect upon the living cells and is readily exhibited in the extensive damage or outright killing of such plants in the severe winters affecting oceanic countries from time to time. Such susceptibility not only restricts northern extension of a given species but excludes it from regions and periods of continental climate. In chapter VI: 9 and throughout the species records there are many instances of these and other processes of climatic control of the ranges of organisms.

Despite the difficulties that the records of individual species present, collectively the results may be extremely convincing as for instance in demonstrating that during the post-glacial thermal maximum, mean summer temperatures were some 2 °C higher than the present in western Europe, and in shewing the sudden rise of mean summer temperature between the closing Weichselian and the thermal maximum.

It is upon biological evidence that reconstructions are attempted, such as that of fig. 12, in which radiocarbon age determinations have allowed the placing of important British biogenic deposits successively through the Weichselian and in firm correlation with deposits of neighbouring western Europe. One of the temperature curves shewn is based primarily upon the evidence of floristic and vegetational analysis, the other on faunistic (chiefly Coleopteran) evidence, but they are in general agreement, and jointly convey a tentative outline of the British climate over the last 70000 years.

The conjunction of evidence from biological and physical sources is becoming increasingly effective. At the same time there has been increasingly thorough analysis of the climatic evidence supplied by prehistoric archaeology and by historic records. This development owes its present impetus to the persistence and insight of H. H. Lamb, who has now been able to advance further the study of past climates by reconstructing their patterns in terms of modern meteorological concepts. These concern the main air flow of the upper westerly winds in a circumpolar vortex with troughs and ridges deviated round regions of relatively high and low temperature. Disequilibrium in this flow of upper air is the cause of generation of the cyclonic and anticyclonic systems responsible for the main weather patterns of the temperate northern hemisphere. The circumpolar vortex may change in strength, in latitude, in centricity, and in amplitude and wavelength of the trough and ridge system. By reference to the meteorological records of recent years it is easy to demonstrate how such alterations affect the pattern of surface weather and correspondingly Lamb has been able to analyse many of the known major features of past climate in terms of the upper atmospheric circulation that must have produced them (Lamb, 1968; Lamb, Lewis & Woodroffe, 1966). It is not unreasonable to see far-improved possibilities of reconstructing former world climatic patterns by effectively co-ordinating the multifarious primary data produced from so many scientific and historical sources. This is a far cry from the simple conception of a progression of temperature changes uniformly affecting the whole globe through a glacial–interglacial cycle, but one corresponding far more closely with reality.

6. MIRE STRATIGRAPHY

It is in our lakes and peat mires that particularly long, continuous and sensitive records of former climates are to be found, not only in the stratification of remains of organisms that are thermal indicators, but in the sequences of mineral or organic layers with which they are filled or composed.

It will simplify our task if at the outset some of the basic terms and concepts are briefly considered. For the category of peat formations as a whole we propose to employ the term 'mire', using it in the same sense as that of the cognate Swedish *myr*, to denote such units as bogs or fens in their entirety with characteristic morphology, hydrography, stratigraphy and vegetation.

It is an accepted tenet of ecology that, even though climate conditions remain constant, bare areas of ground, rock on the one hand or open water on the other, become colonized by vegetation types which so react upon the site and so modify it that new vegetation types succeed the old in an orderly and regular progression. Thus, in what is termed the 'hydrosere' of shallow lakes of small extent, gradual filling-up of the lake by the accumulation of lake muds produced by free-floating plants eventually so raises the bottom level that the first rooted but floating plants can colonize it. These, whilst excluding their predecessors, also contribute to the natural succession by raising the bottom level further, so that they in turn are displaced by the outer reed swamp, perhaps of the lake sedge (*Scirpus lacustris*). By like process each reed-swamp type tends to be displaced by that of a stage suited to shallower water, so that on the Norfolk Broads for instance we find the lake sedge succeeded by reed mace (*Typha angustifolia*), reed (*Phragmites communis*) and sedge (*Cladium mariscus*). This natural succession leads to 'sedge fen' communities just at or above water-level, and these in turn suffer invasion, when the ground has become sufficiently high and dry, by such bushes as the sallow (*Salix atrocinerea*), then by trees such as birch (*Betula pubescens*) or alder (*Alnus glutinosa*). Finally, we have good reason to suppose that these are supplanted in the climate of East Anglia today by fen woods dominated by oak (*Quercus robur*) or perhaps pine (*Pinus sylvestris*). This temporal sequence is to be seen demonstrated in the form of belts or zones round the margin of open water, but it is also finally recorded in the successive layers of deposits beneath the most advanced plant communities. It has to be recognized that the infilling of a lake inevitably tends to proceed in such a manner as this, although the communities of one climate or of one type of geological region will differ from those of others. It is of great interest to recognize, however, that within a given climatic region, as the various hydroseres progress they tend to converge to one or other of a smaller number of alternative and stable vegetation types which are called the climatic-climax types. In the East Anglian hydrosere, which we have sketched above, the climax type is probably mixed-oak woodland.

From this consideration it must be evident that under stable climatic conditions the country will eventually come to be covered by these stable climax communities or by transient communities moving in regular development towards them. When, as in the British Isles, there is a substantial climatic gradient across the country as well as altitudinally, there will thus be a substantial regional modification of natural climax vegetation types. This is a fact of great importance in relation to the distribution of mire types in these islands. The accumulation of peat is the consequence of more or less maintained waterlogging and this is principally affected by precipitation factors, evaporation factors and drainage factors. In the extreme west of Ireland, roughly perhaps beyond the isohyet of 40 inches, the rainfall is so high and uniformly distributed and the humidity so constantly high that peat formation is independent of the accumulation of drainage water. Not only upon the flat surfaces of plateaux, but even upon slopes of as much as 15°, a mantle of peat, often several feet in thickness, covers the ground. This is the so-called 'blanket bog', which forms great stretches of desolate bog over Connemara, Mayo and Sligo.

Only more steeply sloping surfaces are exempt from the blanket bog, and for these the natural vegetation is pre-sumably a woodland type, probably dominated by *Quercus petraea*, and even then the high rainfall causes the soil to be much leached, acidic in character and poor in electrolytes. The valleys receive much water and small lakes and waterlogged basins abound, but these are always in process of natural colonization and succession towards the climax blanket bog. Though often base-rich when they were first formed, they tend by the accumulation of plant material in them, or by drainage from the acid soils round them, to become poor in electrolytes and to have a restricted and typical range of plants growing in them. The blanket-bog surface, isolated by the peat beneath from the mineral soil, constantly subject to precipitation in excess of evaporation and in receipt of no drainage water, suffers extreme poverty of mineral substances and plant nutrients. It is on this account characterized by 'oligotrophic' communities in which *Calluna vulgaris* (ling), *Scirpus cespitosus* (deer grass), *Eriophorum vaginatum* (cotton grass), *Molinia caerulea* (purple moor grass) and to a smaller extent *Sphagnum* moss are dominants. Like the next mire type to be discussed, the raised bog, the blanket bog is termed an 'ombrogenous' mire, to indicate the fact that its existence is directly determined by the rainfall and evaporation to which it is subject. It is evident that the natural conditions for blanket-bog formation will be found in the extreme west and especially the north-west of Britain, but they also occur further south and east where increasing altitude induces suitable climatic factors for their growth: thus blanket bogs are found over the upper parts of the Wicklow Mountains, the Welsh mountains, the Lake District, the Pennines and Dartmoor.

Under conditions of lesser oceanicity such as are encountered for instance in the great central plain of Ireland east of the 40 inch isohyet, the climax mire type is that known as the raised bog (*Hochmoor*). The restricted oceanicity permits the development of these ombrogenous mires only upon almost flat surfaces or in valley bottoms: even gentle slopes provide drainage sufficient to permit other climax communities, chiefly oak woodland, to exist over large areas alongside the raised bogs. Raised bogs take their name from their convex, cushion-like shape, which is due to the sloping margins where lateral extension of the *Sphagnum* moss, the chief peat-forming agent, is prevented by the collection of relatively base-rich drainage water from surrounding mineral soil. These sloping margins (German, *Rand*) are better drained than the bog centre and carry different communities, often pronouncedly zoned and dominated by such species as *Calluna*, *Molinia caerulea*, *Scirpus cespitosus* and *Betula pubescens*. The drainage water from the *Rand* and from the adjacent mineral soil falls into' a wet marginal area known as the 'lagg' where the higher base-status of the water is reflected in the occurrence of the more eutrophic species characteristic of fens. The central portion of the raised bog has a very gentle curvature indeed and carries highly characteristic communities dependent in their nature upon the conditions of the bog and its

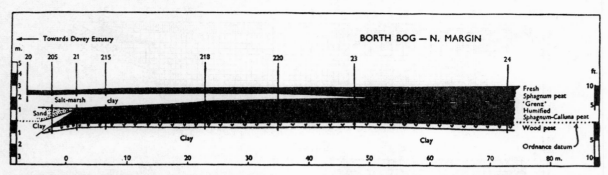

Fig. 13. Section through the raised bog adjoining the River Dovey estuary, Cardiganshire. A marine clay is overlaid by a layer of wood peat containing abundant tree stools (including those of pine); this is overlaid by *Sphagnum–Calluna* peat. Intruding into the raised-bog peat at a level above the *Grenzhorizont* is a wedge of salt-marsh clay with remains of *Juncus maritimus*, presumably representing the latest substantial eustatic rise of ocean level. The photograph (plate II) shews how the basal wood peat is exposed on the neighbouring shore as a 'submerged forest'.

climatic and drainage relationships. In the more oceanic climates such as those of central Ireland the undrained bog will still be actively growing and will present a mosaic of open pools and of hummocks of growing sphagna. This system is spoken of as the 'regeneration complex', and it produces a characteristic banded peat in which the vegetation of the pools and of the hummocks can be traced in constant alternation with one another. Highly constant species of the regeneration complex fall within such oligotrophic genera as *Calluna*, *Erica*, *Eriophorum*, *Andromeda*, *Oxycoccus*, *Drosera*, *Narthecium* and *Rhynchospora*, as well as equally typical mosses and liverworts in profusion, and such species as *Scirpus cespitosus* and *Molinia caerulea*. The species are those of the blanket bog, but with different frequencies, especially so in the case of the preponderant sphagna.

Raised bogs in continental climates tend to have their surfaces bush- or tree-covered, and in this country bogs artificially or naturally drained pass into a 'Standstill complex' in which dominance is taken over by *Calluna* or *Eriophorum vaginatum*, sphagna are less important and peat growth is reduced in rate or ceases, and the peat type produced is more humified than under regeneration complex. Such *Calluna–Eriophorum* bogs are drier altogether than the active *Sphagnum*-dominated raised bogs.

Raised bogs very commonly occur in former periglacial regions, often upon the sites of lakes created by the laying down of terminal moraines across glaciated valleys. The stratigraphy of such bogs shows all the early stages of colonization of more or less eutrophic lakes, the openwater lake muds being progressively replaced, marginally at first, by reed swamp, fen, and perhaps by fen wood. Only at these later stages of vegetational development, when peat formation has carried the ground surface above reach of the lake waters, was ombrogenous peat formation able to begin. From this stage onwards *Sphagnum* growth can be seen to have dominated the bog's development. In other instances, as in the Somerset levels, raised bogs originated upon the flat, poorly drained surface of sheets of brackish-water clay, the

ombrogenous peat formation then being preceded by long stages of *Phragmites* fen, *Cladium* fen and fen-woods. In many other instances again stratigraphy reveals that, although a raised bog may have originated above a lake, it has subsequently spread laterally over adjacent land, and then the stratigraphy shows the natural woodland on mineral soil replaced by the fen-woods of the encroaching lagg before the transition to the ombrogenous *Sphagnum*-peat formation. In the excellent vegetational studies made by Steffen on the shores of the Baltic near Königsberg we may find displayed together the zonation of living lake, fen and raised-bog communities and the stratification in the raised-bog deposits entirely corresponding with the zonation: such correspondence has indeed long been recognized.

In climatic conditions of reduced rainfall or diminished humidity the growth of ombrogenous peat no longer takes place. Periods of summer drought are too long and too severe for bog-building *Sphagnum* to flourish and peat formation is therefore *topogenous*, that is to say, restricted to drainage basins in which the precipitation of a large catchment area is concentrated. To such mires we give the comprehensive term 'Fen'. The concentration of drainage water in fens is associated in varying degrees with enrichment in dissolved electrolytes, so that the peat accumulating in these basins or valleys is relatively base-rich, possibly maintaining neutrality or alkalinity. This is the case in the East Anglian fens and those of the Norfolk Broads where areas of chalk or calcareous boulder clays occupy much of the catchment area. In areas of this kind the vegetational sequence from open water is of the kind already described, through lake communities, reed swamp, fen, or fen woodland. The base-rich waters prevent the development of *Sphagnum* moss and permit the characteristic dominance by grasses, sedges and rushes in the middle stages of the succession and of woody plants in the later stages, up to an apparent climax of mixed-oak woodland. In other instances, as in the valley bogs of the New Forest or of the Greensand in Surrey and in Norfolk, the drainage water is acid and not very rich in

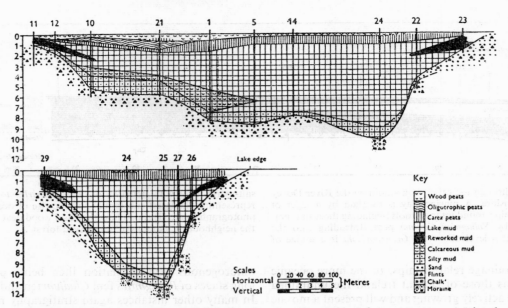

Fig. 14. Sections secured by boring through the lake deposits at Hockham Mere, Norfolk. Note in both the wedging out into the lake of the reworked material of zone VIC and, in the upper section, the oligotrophic *Sphagnum* peat representing the latest stage of the stratigraphic evolution of the mere.

dissolved bases so that acidic peat accumulates in them and supports local *Sphagnum* communities resembling those of raised bogs, but restricted in extent and associated more or less with indicators of mildly eutrophic conditions: such mires fall within the category of acid fens. Naturally, in the original unexploited condition of this country the fenlands would lie as channels or islands within the general carpet of the climax mixed-oak forest, which covered most of the higher ground.

We may thus visualize in very simplified outline the condition of the mires in the British Isles (or any region of equivalent climatic range) in the following way: (*a*) an extreme oceanic region with widespread blanket bog as climax type and woodland restricted to rather strongly sloping ground, calcareous and especially acid fens abundant as seral stages; (*b*) a moderately oceanic region with raised bogs in flat valley bottoms, woodland elsewhere except for local calcareous and acid fens as seral stages: (*c*) a least oceanic region with calcareous and acid fens in drainage basins, tending to develop to a climax type of woodland not very different from that generally covering the upland. Local factors of climate, topography and geology will naturally tend to cause complex interpenetrations of these three areas.

It is against this background of normal vegetational succession and of regional climatic control that we must consider the interpretation of mire stratigraphy as an index to the conditions of past times, and we may naturally expect to find consequences of two kinds, both changes advancing or retarding the succession, and changes moving the boundary of one mire type into a region hitherto not able to support it. Both types of effect have often been noted.

In lakes and fens, since the process of vegetational succession is bound up with *Verlandung*, climatic changes involving increased wetness will be associated with the setting-back of the normal sequence: thus for instance, after the reed-swamp stage has been attained open-water muds may recur; after sedge fen, reed swamp may return; fen woods may be set back to fen or to some earlier stage of the hydrosere. Moreover, provided that the local conditions are not such as to cause flooding by calcareous water, raised bog may replace the woodland as an ultimate mire condition.

Conversely, climate changes making for dryness of the mires accelerate the natural successions and even telescope them. Drying of a lake may be represented by such an extreme phenomenon as the exposure and erosion of lake muds, or by the contraction or omission of stages normally part of the hydrosere. Thus in the second half of zone VI, the spring-fed Hockham Mere in Norfolk was reduced in area to no more than one-fifth of its former size, the exposed lake bottom dried out and its layered muds were reworked into mud breccia. When the water-level was restored in early Atlantic time the reworked layer was overlaid by fresh necron mud. Pollen analyses from the margins and the centre (where accumulation had been continuous) serve to date the dry phase with accuracy.

Very little is yet known of the ecology of the blanket bog, and the peat type it produces is often homogeneous and featureless. All the same we may note that the bottom of the Irish blanket bog almost everywhere overlies wood peat resting on the mineral soil, and that this blanket-bog formation undoubtedly indicates a climatic change of considerable size and towards increasing

PLATE V

Peat cuttings on north Dartmoor shewing the characteristic aspect of blanket bog.

Section through blanket bog at Mynydd Beili-glas, South Wales, shewing numerous remains of birch trees (*Betula*). Such tree layers beneath blanket bog are abundant throughout the mountain and far western regions of the British Isles. (Phot. F. J. North.)

PLATE VI

View of the southern end of the raised bog complex at Tregaron, Cardiganshire, from the hillside to the east. The bog is on the site of a former glacial lake dammed by the terminal moraine seen as a ridge of cultivated land to the left of the photograph. The bog surface shews radial lines of peat diggings.

View of the regeneration complex of vegetation on the surface of an actively growing raised bog. Open pools with floating *Sphagnum* (*S. cuspidatum*) alternate with actively growing hummocks made of tussock-forming *Sphagnum* and these, invaded by *Calluna*, eventually cease growth and in turn become converted into pools. The west bog at Tregaron, Cardiganshire.

PLATE VII

Face of the Hawks Tor kaolin pit, Bodmin Moor, Cornwall. The rails are laid on the surface of the kaolinized gravel above which are layers of organic mud and peat. The lowest dark layer (seen behind the slope of spoil) is the organic mud and peat formed in the warm stage (zone II) of the Late Weichselian; above it is the band of solifluxion gravel brought down into the valley in the succeeding cold phase (zone III). Thence to the surface are Flandrian peats.

Peat-cutting on the raised bog at Tregaron, Cardiganshire. Just above the water-line a paler line shows a band of greasy humified peat, the 'retardation layer'. This was probably formed in a dry climatic period and may correspond to one of Granlund's later recurrence surfaces. The turves lying on the surface have been cut across the retardation layer; in drying, the peat of this layer has contracted more than the fresh *Sphagnum* peat above and below, giving a strongly waisted shape to them. (Phot. F. J. North.)

PLATE VIII

An unusual instance of an opportunity for correlating faunal and floristic history. A cluster of elk droppings (*Alces alces*) in the aquatic *Sphagnum* peat of Bronze Age date at Ugg Mere, near Woodwalton, Huntingdonshire. The droppings are conspicuous because they carry a weft of fungal mycelium.

Example of circumstances allowing correlation between archaeology, bog stratigraphy and pollen analysis. Face of a peat-cutting on Shapwick Heath, Somerset, shewing (*a*) behind the label, a Middle Bronze Age spear in position of discovery, (*b*) labels shewing main peat horizons determined in the field, and (*c*) the cleaned peat face from which samples can be directly taken for pollen analysis. The shaggy appearance of the peat face above the spear level is due to roots of *Cladium mariscus* which constitute a thick layer above the flooding surface marked by the second white label from the base.

PLATE IX

Hiller peat-borer being extracted from lake deposits.
Meare Pool, Somerset.

Hiller peat-borer extracted, chamber open and contents being
examined. Meare Pool, Somerset.

PLATE X

Female cone scales of the dwarf birch (*Betula nana*) recovered from Late Weichselian deposits at Nazeing, Lea Valley, Essex.

Peat face exposed near the Old Toll Gate House, Westhay, Somerset shewing a Late Bronze Age trackway made of coppice shoots of hazel (*Corylus avellana*) laid lengthwise. Metric scale and positions of pollen samples are shewn.

PLATE XI

Leaves of willow (*Salix* cf. *phylicifolia*) lying on the bedding planes of calcareous
mud of Late Weichselian age from Nazeing, Essex.

Somerset moors near Westhay. Peat digging exposing a section where the old humified *Sphagnum*
peat is overlaid by *Scheuchzeria* peat (pale band) and this in turn by unhumified regeneration-complex
peat. *Scheuchzeria* is unknown in this part of England, in the living state.

PLATE XII

Scheuchzeria palustris peat from the upper flooding horizon of the raised bogs at Westhay in the Somerset Levels. The papery rhizomes are very conspicuous and are associated with aquatic *Sphagnum* spp.

Cladium–Hypnum peat from the lower flooding horizon at Ashcott Heath, Somerset The large red rhizomes of *Cladium* are here seen cut transversely.

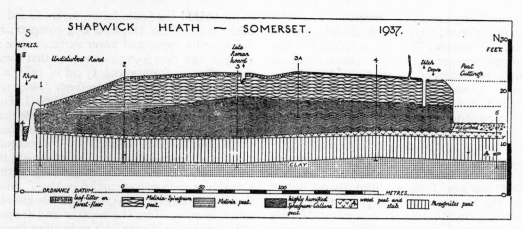

Fig. 15. Section constructed from boring records through the derelict raised bog at Shapwick Heath, Somerset. The valleys were filled with estuarine clay at the close of the Boreal period and this carried in turn *Phragmites–Cladium* fen, fen-woods with much *Betula*, and ombrogenous raised bog. The latter is shewn sharply divided into the upper unhumified *Sphagnum* peat and the lower highly humified peat, the boundary between the two being marked by many Late Bronze Age trackways. The bog apparently ceased growth in late Roman time and just below the present surface several late Romano-British hoards have been discovered.

oceanicity. The remains of late Bronze Age type in the wood layers indicate the period of this change, which must greatly have affected all the British Isles.

The raised bogs appear to be especially sensitive to climatic change (as might be expected from their ombrogenous character and intermediate status) and even in the upper layers, which formed when climax equilibrium with climate had been reached, they show stratigraphic phenomena of great interest. The most striking of these is the division of the ombrogenous peat into an upper young *Sphagnum* peat, which is fresh, generally unhumified and pale in colour, and a lower old *Sphagnum* peat which is dark chocolate-brown, highly humified and dense. This twofold division is evident throughout the raised bogs of north-western Europe and is naturally very familiar to peat-cutters all over this region, for the two grades of peat have very different qualities and economic values. The lower is good burning peat, the upper is suitable only for litter, fertilizer, etc. This twofold character is emphasized by the fact that the boundary between the two peat types is often one of great sharpness, and to this surface or 'boundary horizon' the term *Grenz-horizont* was given by C. A. Weber, the German pioneer in bog stratigraphy. The top of the old *Sphagnum* peat, especially in Germany, is marked by layers of pine stubs, of birch, of *Calluna* or cotton grass, all indicative of conditions of dryness at the bog surface, but these phenomena are less pronounced in British raised bogs.

The original interpretation of the difference between the upper and lower *Sphagnum* peat was that the warm and dry conditions of the Sub-boreal period, as defined by Blytt and Sernander, had caused both the drying of the bog surfaces and also secondary alteration by progressive humification of the older *Sphagnum* peat laid down throughout the earlier warm and wet period of the Atlantic. In this view the Sub-boreal peat was either missing or very thin, and the old *Sphagnum* peat had once had the appearance of the young *Sphagnum* peat above the *Grenze*. It became clear, however, that considerable peat thickness had in fact been added to the bogs throughout the period of time covered by the Sub-boreal period, and that there was no evidence of a gradient of humification steeply increasing downwards as would be expected on the hypothesis of secondary alteration of the old *Sphagnum* peat. It is much more probable that the old *Sphagnum* peat formed in much the condition which it exhibits today, and that its essential differences from the young *Sphagnum* peat are due to an origin under different climatic conditions and from different bog communities. As to what those communities may have been we are still uncertain, for the high degree of humification makes recognition of species difficult. Nevertheless it seems clear that very decayed species of *Sphagnum* are abundant, whilst leaves, twigs and flowers of *Calluna* and tussocks of *Eriophorum vaginatum* are probably far more abundant than in the upper *Sphagnum* peat in general. We recall the fact that *Calluna* and *Eriophorum* communities dominate the standstill or erosion phases of the dry raised bogs today and give rise to similar humified peat, and feel impelled to conclude that the bogs in the phase of the old *Sphagnum* peat must have been generally drier than after the *Grenze*. This condition moreover prevailed throughout, not only in the Sub-boreal period, but also through most or all of the preceding Atlantic period, which is commonly thought to have been both warm and wet.

In north-west Germany, Denmark and Sweden it is evident from discoveries of archaeological objects that the *Grenze* corresponds closely with the time of transition from the Bronze to the Iron Age for which a date of about 500 B.C. is generally allowed. From the widespread distribution of the twofold stratification of the *Sphagnum* peat it is evident that a considerable climatic 'deterioration' affected western Europe at about that time. We may deduce 'deterioration' (that is cool and wet conditions)

33

because the upper *Sphagnum* peat is generally such as now actively forms in the raised bogs in cool Atlantic climates like that of central Ireland. The young *Sphagnum* peat shows great preponderance of coarse unhumified mats of the tussock-building species of *Sphagnum*, such as *Sphagnum papillosum, S. magellanicum* and *S. imbricatum*, although darker bands with *Calluna, Scirpus* and *Eriophorum* continue to show the banded arrangement which records the process of vegetational change in the regeneration complex of an actively growing raised bog.

This simple and attractive picture of a bog horizon due to climatic change and available over big distances as a correlation level has now been modified almost out of recognition by the labours of several generations of investigators. It was shown that the *Grenzhorizont* was only one, though a very pronounced one, of a series of similar boundary horizons, at each of which regrowth of the bog surface began afresh after a period of arrest in presumably drier conditions. These are the Swedish 'recurrensytor' or 'recurrence surfaces', of which Granlund (1932) identified no less than five within the raised bogs of southern Sweden, attaching to them the dates 2300, 1200 and 500 B.C., and A.D. 400 and 1200. Nilsson later increased the number to nine. Although it was conceded that renewed bog growth was indeed much concerned with climatic changes the nature and mechanism of the operation of such changes has remained very obscure in ecological terms (Godwin, 1954). Detailed pollen-analytic studies revealed that it was no simple matter to correlate the recurrence surfaces from one region to another and van Zeist produced pollen-analytic evidence that a major recurrence surface extending across the raised bog near Emmen was of substantially different ages along its extent, a result repeated by several other investigators (see Frenzel, 1966). The difficulty of relying upon pollen zonation for a chronological comparison in the later stages of the post-glacial made it apparent that the resources of radiocarbon dating should be brought to the problem. Dates from recurrence surfaces in the north-western German bogs investigated by Overbeck gave a range of dates not very close to those of Granlund, not strongly self-consistent and with the most pronounced recurrence surface not of the age 800 B.C. as Weber had supposed, but about 100 B.C. (Overbeck *et al.*, 1957). Remarkably decisive evidence has recently been published by Lundquist (1962) reporting the results of dates obtained in Stockholm for Högmossen (Gästrikland) and Lidamossen (near Eskilstuna). The most prominent recurrence surface in each was surprisingly recent and not only was its age different in the two bogs, but in each the age varied by several hundred years between one site and another. The ages of samples taken just above the recurrence surface in all instances were one to four hundred years younger than those taken immediately below it, a result also found at the main recurrence surface in some British raised bogs, and indicative of temporary cessation of growth at the end of 'old *Sphagnum*-peat' formation. Extreme slowness of peat-growth also

characterized the 'retardation layer' found in all three raised bogs of the Tregaron Bog complex: radiocarbon dates for its upper and lower surfaces were respectively 768 ± 90 and 1477 ± 90 years B.P. so that some 700 years were taken to form about 13 cm of peat, a rate barely one twentieth of that of the fresh *Sphagnum* peat that lies above the main recurrence surface (J. Turner, 1964).

It is quite apparent that it is less simple than had been supposed to use recurrence surfaces as indices to climatic shift: none the less, so striking a phenomenon warrants still further investigation, and this might well be associated with detailed analysis of all stages of growth of an individual bog, such as was carried out by T. Nilsson and his Swedish colleagues (1964) on Ageröds mosse in Schona, south Sweden. Here the number and frequency of the radiocarbon dates were such as to allow the construction of a record of changing growth rate of peat through the whole post-glacial time and (taken with analysis of dated phases of bog stratigraphy) a hypothetical curve for changing wetness of the bog throughout the same period. This study reveals a very pronounced phase of dryness in the later half of the Boreal (*c.* 6500 B.C.) followed by increased wetness in the early Atlantic, a return to moderate dryness between about 4600 and 3500 B.C., followed by progressively increasing wetness to a maximum about A.D. 650. It is noteworthy that despite the upward growth and increased convexity of the bog between 3500 B.C. and A.D. 650 it becomes not drier but steadily wetter, a result all the more positively attributable to climate. The dated peat accumulation rates shew no clear relationship to the observed recurrence surfaces, except for the period of greatly reduced peat growth over a period between about 980 and 250 B.C., i.e. at the end of the Sub-boreal period and including the time of Granlund's recurrence surface III.

As table 2 shews, the main recurrence surface in the old raised bogs of the Somerset Levels falls consistently in the same period. It is represented there by the construction of numerous Late Bronze Age wooden trackways, built as a response to the major and progressive flooding of the dried bog surfaces (Dewar & Godwin, 1963). The table shews that at other British sites a major recurrence surface is also of this date.

The impact of lake and mire stratigraphy upon current concepts of climatic history can be very well seen in regard to present views on post-glacial periods of dryness. European biologists and archaeologists have tended in the past to lay much stress upon the existence of a 'xerothermic' period, corresponding with the Bronze Age and the Sub-boreal of Blytt and Sernander. It formed the basis of Gradmann's 'Steppenheidetheorie', according to which the climate of this time was so dry as to prevent forest growth on the lighter soils of central Europe, so that there existed a corridor of open steppe along which the fauna, flora and prehistoric peoples of south-eastern Europe migrated into the oceanic west. The detailed results of pollen analysis have reduced the possibility of such an episode. The stratigraphy of the raised bogs of the

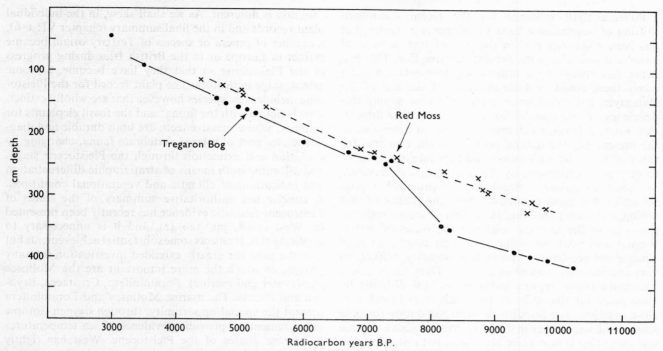

Fig. 16. The rate of peat growth in two raised bogs, Tregaron Bog, Cardiganshire and Red Moss, Lancashire, has been determined by radiocarbon assay of samples. (Hibbert, Switsur & West, 1971 and unpublished data.) At each site there is a strong break in stratigraphy about 7300 years B.P., close to the Boreal/Atlantic transition. The rate of growth of ombrotrophic peat after this date is remarkably similar in the two bogs: approximately 4.5 cm per century.

TABLE 2. *Radiocarbon dates of a prominent recurrence surface in England and Wales (years B.P)*

	Below	(Trackway)	Above
SOMERSET			
Meare Heath track	3061	2840	
		2852	
Shapwick Heath track	3310	2470	2220
			2197
Westhay track		2800	
Blakeway track	2790	2600	
Viper's track		2520	
		2630	
Viper's platform		2410	
		2410	
		2460	
Nidons track	2642	2585	2628
	2482	2590	
OUTSIDE SOMERSET			
Chat Moss	3061		2661
Tregaron	3029		2669
	2879		2624
Whixall	3238	2307	
Llan Llywth	3178		3230
Flanders Moss	2712		
AVERAGE	2959	2604	2575

British Isles has also given strong evidence in the same sense, for the old, highly humified, *Sphagnum–Calluna–Eriophorum* peat which can be shewn to have formed

between the opening of the Atlantic period and the close of the Sub-boreal shews no sign at all of a sub-division into an earlier wetter half and a later drier half. The high degree of humification throughout suggests that the bogs were substantially drier than they are now (when left undrained) in the cooler Sub-atlantic period, and if the closing stages of the Sub-boreal shew tree layers or concentrations of *Calluna* or cotton grass, it suggests rather a short closing stage of dryness than one covering a long period of time. Indeed a similar conclusion is likely also for the other dry period of the Blytt and Sernander scheme, the Boreal period, for it is only in the concluding sub-zone of this period that the low lake levels, layers of tree stumps and so forth can be recognized, and not in the period as a whole.

On the other hand, mire stratigraphy, supported by radiocarbon dating, very strongly supports the reality of the sudden climatic deterioration at the opening of the Sub-atlantic period. Not only does the widespread phenomenon of the main recurrence surface shew it, but the Late Bronze Age forest layers beneath the western Irish blanket bogs shew that what had been deciduous forest climate at this time changed into the highly oceanic blanket-bog climate. Thus two key levels at least of the somewhat discredited Blytt and Sernander scheme find an increasing degree of confirmation: they are the dry terminal stages of the Boreal and Sub-boreal periods contrasting sharply with the ensuing wetness respectively of the Atlantic and Sub-atlantic periods.

3-2

Pollen-analytic evidence for the recent downward shifting of vegetational belts supplements evidence that has been much reported in the past of tree stubs from mountain sites above the present-day tree line. This evidence has always been difficult of interpretation. Formerly there could be little evidence of the age of the stub layer, but radiocarbon dating will now supply this deficiency. There remain, however, two further difficulties. First, it is often extremely difficult to determine at the present day the natural position of the tree line, for this differs greatly with aspect and exposure and is liable to very great depression by the grazing that is prevalent in mountain pastures. Secondly, the progressive podsolization has most commonly been the means of first killing and then preserving the stools of trees originally growing on the mineral soil, and the onset of waterlogging and peat formation, although bound up with climate and accelerated by a shift to oceanity, is likely to vary considerably from place to place. Thus, for instance, Lundquist (1962) reports radiocarbon ages differing by 1800 years for the stubs of two such trees found very close together. The Swedish radiocarbon dating for pine stumps collected either in the birch zone or above the tree line altogether is now quite abundant, but instead of disclosing an origin within a restricted period representing the conclusion only of the hypsithermal, the dates are scattered through from more than 6000 to less than 1000 B.C.; those in the south, somewhat surprisingly, being of generally greater age than those in the north. We have to remember, in trying to understand such results, that climatic causes will have affected not only the growth of the trees, but their death, preservation and recent reappearance.

7. BIOLOGICAL EVIDENCE

The casts and corpses of plants and animals supply evidence of the background of Pleistocene events in two main and not altogether unrelated ways. They may, in a relatively small percentage of cases, act as zone or stage fossils and, far more often, they afford evidence of the past climatic and edaphic conditions. The use of the remains of an organism as a zone fossil in the geological sense is only possible if certain conditions are satisfied. The periods of time to be distinguished must be large enough to allow the given species to be evolved, widely spread and established, and again extinguished. That is to say that the biological changes concerned must be rapid in relation to the scale of geological events which are to be characterized. Within the Flandrian period, from which come a large part of the plant records considered here, this condition is not satisfied; and for animals as for plants there is abundant evidence of uncompleted distribution movements within this period and little or none of extinction or of creation and wide dissemination of new forms. When, however, the time scale is greatly lengthened by taking in the whole post-Tertiary span, the situation is different. As we shall shew, in the individual plant records and in the final summary (chapter VI: 1–6), a number of genera or species of Tertiary origin became extinct in Europe or in the British Isles during progress of the Pleistocene so that they have become, to some extent, stage indicators. The plant record for the Pleistocene includes few species however that are wholly extinct. This is not so with the fauna; and the fossil elephants for instance, whose massive teeth are both durable and diagnostic, are part of a large vertebrate fauna, changing by evolution and extinction through the Pleistocene stages and affording both means of stratigraphic differentiation and indications of climatic and vegetational conditions. A concise but authoritative summary of the status of Pleistocene faunistic evidence has recently been presented by West (1968, pp. 326–42), and it is unnecessary to duplicate this. It reflects some substantial achievements but also the need for greatly extended investigation in many groups, of which the more important are the Mollusca (freshwater and marine), Foraminifera, Crustacea, Bryozoa and insects. The marine Mollusca and Foraminifera present the special opportunity, through oxygen-isotopic measurement, of providing evidence of sea temperatures in marine phases of the Pleistocene. West has rightly drawn special attention to the recent development of research upon fossil insect faunas from Pleistocene deposits, more particularly through Coope, Shotton and Osborne at Birmingham and Pearson at Cambridge. We now have substantial records from Hoxnian, Ipswichian, Middle and Late Weichselian as well as Flandrian sites. The present-day distribution of Coleoptera, especially in the successive altitudinal zones of Fennoscandia, is well known and affords a means of inferring the vegetational and climatic conditions in which particular British faunas were deposited. This has proved of special interest in the study of mid-Weichselian interstadial deposits where there is corresponding evidence from extensive plant material (cf. fig. 12). In the case of the Fladbury interstadial and the Late Weichselian deposits at Colney Heath, the climatic inferences from the two sources are broadly similar, but the beetle evidence suggests reference to a lower, more wooded zone than the plant evidence appears to do. We have, however, in the beetles a similar diverse range of biogeographical types as that presented by the flora. A field for further research lies in the evaluation of insect faunas as indices of local and regional climate, a subject to which Coope (1967, 1968a, b, 1969, 1970) has directed a great deal of attention.

As with evidence from plant material, one feels most secure in drawing conclusions as to former climate when one knows the mechanisms by which the climatic effects actually operate upon the given species in the field. A most illuminating instance of this is to hand in the publication by Degerbøl & Krog (1951) of the history of the European pond tortoise (*Emys orbicularis*) in Europe (fig. 17). Breeding populations of this reptile are now found in Spain, Portugal, southern France, Italy, the Balkans, south Russia, east Czechoslovakia and most of Poland

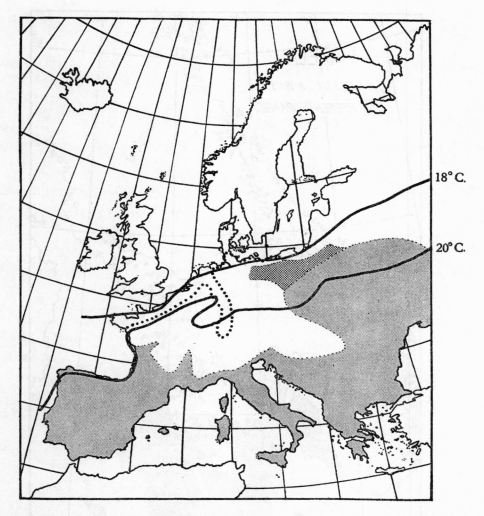

Fig. 17. The light stippling shows the extent of breeding populations in Europe of the pond tortoise, *Emys orbicularis*; the more heavily stippled region in Prussia and Poland is where the species is nearly extinct at the present day; and the dotted line shews the northern limit of occurrence of non-breeding animals. The isotherms shewn are mean July temperatures reduced to sea-level: the unreduced 20 °C July isotherm for France largely follows the northern limit for *Emys*. Abundant Danish records of breeding animals in the middle of the Flandrian period shew that that age had higher summer temperatures; there is a Flandrian record of the pond tortoise also from East Anglia. (After Degerbøl & Krog, 1951.)

except for the north-west. In north-east Germany sparse and declining populations persist.

A very large number of fossil records of *Emys* has been made from Danish peat bogs, especially during the Second World War, and these have mostly been dated by pollen analysis. They are strikingly abundant in the eastern part of the country and almost absent from the west. The analyses show that the pond tortoise first began to spread in Denmark in the Boreal period, was abundant in the Atlantic and Sub-boreal periods but disappeared immediately. The cause of the disappearance of the tortoise is confidently attributed to the climatic deterioration of the Sub-atlantic period in which the cool and damp summers were highly unsuitable to the animal, especially

for the development of the eggs. The present distribution suggests that a summer temperature of at least 18 °C, and more probably 19 or 20 °C, must have prevailed in Denmark during the period of its abundance. Very probably also the climate was continental rather than oceanic in character. This evidence of the extent and intensity of the climatic optimum, is, as the authors point out, substantially in line with that furnished by plant remains such as fruits of *Najas marina* and particularly the pollen of *Hedera*, *Viscum* and *Ilex*, which has yielded in the hands of Iversen (1944a) convincing evidence that in Atlantic and Sub-boreal time the mean summer temperature in Denmark was at least 2 °C higher than at present. A similar figure had been suggested by Andersson

37

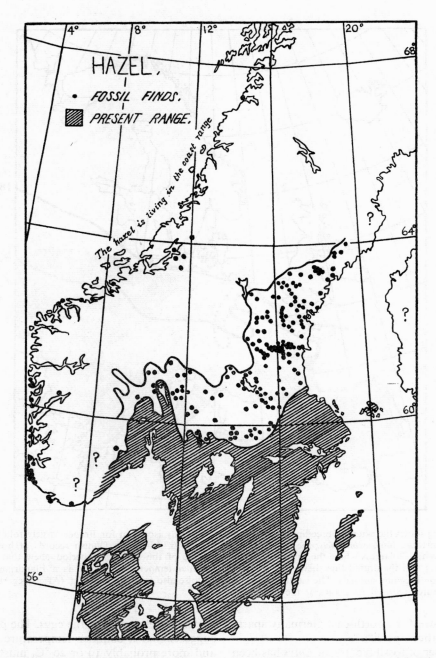

Fig. 18. Map shewing the distribution of the hazel (*Corylus avellana*) in Scandinavia at the time of the Flandrian thermal maximum and at the present day. (After G. Andersson.)

for Sweden on the basis of discoveries of *Corylus* (hazel) nuts widely beyond the present boundary of that species (see fig. 18).

The evidence has now become sufficient to establish the pond tortoise as a fossil indicator of the Flandrian warm period in countries near its northern limit, i.e. in Scania, Denmark and possibly eastern England. In a comparable manner the aggregation of remains of

lemming, arctic hare, arctic fox and reindeer have long been recognized as indices of arctic or sub-arctic conditions. More recently it has become apparent that the association of large herbivorous mammals, such as bison (*Bison bonasus arbustotundrarum*), horse (*Equus caballus*), reindeer (*Cervus tarandus*) and giant deer (*Cervus megaceros*), is highly characteristic of the open steppe-tundra of the Weichselian in north-west Europe. The giant Irish

38

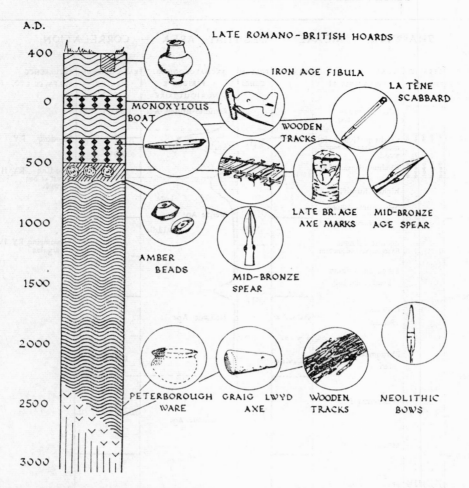

Fig. 19. Schematic representation of the relation of the various archaeological discoveries of the Shapwick–Meare–Westhay region of the Somerset Levels to the stratigraphic sequence normally found throughout the raised bogs of that area. (After Dewar & Godwin, 1963.)

deer has indeed not been found in younger deposits than this, and Mitchell & Parkes' review (1949) of its fossil distribution shews how very consistently it is associated with the Allerød warm period in Ireland and probably the British Isles as a whole. We remain however very uncertain as to the process of its extinction, an uncertainty that extends to a great deal of the megafauna of the Pleistocene. Among the great range of research and conjecture upon this topic particular interest attaches to the thesis developed by P. S. Martin that a sudden wave of extinction of large animals occurred in the late Pleistocene as prehistoric man multiplied his numbers and his skill as a big-game hunter, so introducing into the background pattern of Pleistocene events the first global overkill (Martin, 1967).

Although, because it constitutes part of the background of Pleistocene events, we have written of this great variety of biological evidence, for all the groups concerned (as for the higher plants that are our main concern) the Pleistocene evidence promises light upon such

issues as the response of species to climatic change, the nature of speciation, extinction and determination of geographic range. Light upon each will illuminate the whole.

8. ARCHAEOLOGY

The association of plant fossils with well-characterized remains of prehistoric man has two main sources of interest, first the possibility of utilizing the archaeological correlation as a dating mechanism, and secondly the possibility of recognizing from the association the former use made by man of natural plant products and the distributions of plants that he may wittingly or inadvertently have fostered during his occupations and migrations. In this country such studies were admirably established by the work of Jessen and Helbaek (1944) upon prehistoric weeds and cereals in the British Isles and in north-western Europe. This work has not been superseded but now forms part of a great volume of

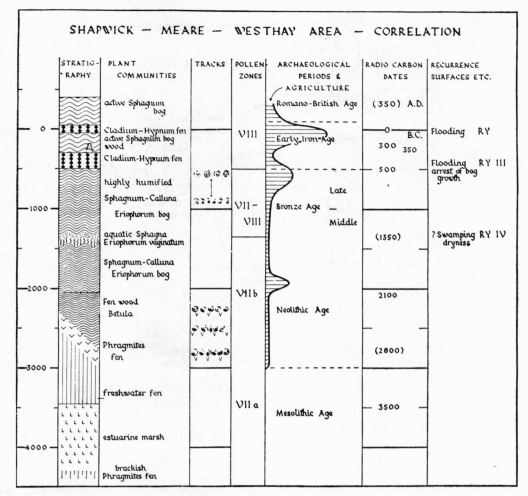

Fig. 20. Correlation table for events in the raised bogs of the Shapwick–Meare–Westhay region of the Somerset Levels (Dewar & Godwin, 1963). The curve for agricultural activity is based upon the frequency of indicator pollen in various diagrams (e.g., see fig. 116): the greatest activity corresponds with the occupation of the Iron Age lake villages. Two flooding horizons appear to correspond with Granlund's recurrence surfaces, RY IV and RY III: a third falls about A.D. 0.

research into the evolution and dissemination of domesticated plants and animals in all parts of the world, but most particularly in the Middle East where so many of them originated. The literature of this is approachable through publications such as Hutchinson (1965), Ucko and Dimbleby (1969), Körber-Grohne (1967), K. & H. Knörzer (1971) and Willerding (1969*a*, *b*).

It is now recognized that highly significant conclusions both for plant history and for archaeology follow the detailed and systematic investigation of archaeological material by trained botanists. A striking instance of the value of such work is found in the analyses made of the stomach contents of Iron Age corpses recovered from burials in the Danish raised bogs: from one such it has been possible for example to reconstruct 'Tollund man's last meal' as consisting of a large proportion of barley, a fair amount of linseed, *Polygonum lapathifolium* and *Camelina linicola* together with smaller quantities of *Chenopodium album*, *Polygonum convolvulus*, *Spergula arvensis*, *Brassica campestris*, *Viola arvensis*, *Galeopsis* sp., etc. It seems difficult to resist the conclusion that the meal was a gruel made of roughly ground-up seeds and there is ancillary evidence to suggest that many species now known as weeds were generally collected, if not deliberately grown, by these people. Instances of this comprehensiveness are not common, but we may note from the many British records of plant utilization those of a barrow-load of sloe-stones and of a form of *Vicia sativa* (broad bean) from the Glastonbury lake village, of fragments of seeds of *Sambucus nigra* from a piece of Iron Age pottery, of the lining of Roman coffins with shoots of *Buxus sempervirens* (box) and of the use of *Taxus baccata* (yew) wood in the oldest known wooden artifact in the world, the spear found by Hazzledine Warren in the Hoxnian Interglacial at Clacton.

The use of archaeological correlations to furnish a time

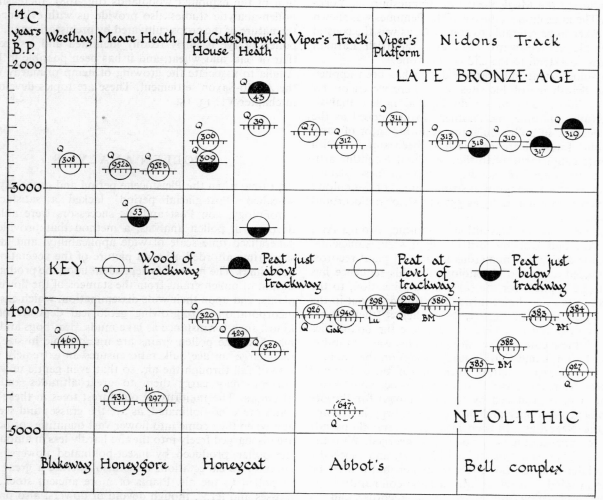

Fig. 21. Radiocarbon dates for prehistoric wooden trackways excavated from the raised bogs of the Shapwick–Westhay region of the Somerset Levels. The dates fall into two clearly separate age groups in conformity with the archaeological evidence and with their stratigraphic situation (see figs. 19 and 20).

scale for plant records has retained its usefulness much more for the later phases of human cultures than for the earlier, especially because Neolithic and post-Neolithic man often accumulated stores of food and provided settlement debris, both of which afford a rich source of plant material. It is otherwise with the earlier cultures. Not only are plant records very sparingly associated with Palaeolithic and Mesolithic cultures in the British Isles, but whereas at one time the artifacts of these cultures were of great value in differentiating the sequence of geological events, their role in this respect has been largely taken over by the extending power of stratigraphic and biological evidence, particularly that of pollen analysis in the differentiation of succeeding interglacial stages. The association of the Clactonian flake industry and the Acheulian hand-axe industry with the Hoxnian Inter-glacial is certain, the latter tantalizingly associated with an extensive forest-clearance episode. There is representa-

tion of the Levallois cultures in the succeeding glacial stage and very sparsely in the Ipswichian. The Weich-selian glacial stage is associated with cave occupations of the upper Palaeolithic that have been referred to Mousterian, 'Proto-Solutrian/Aurignacian' and Late-Magdalenian, but they afford no opportunity for direct correlation with the plant record. With the opening of the Flandrian this situation alters and at various sites there has been highly profitable application of pollen analysis to different Mesolithic occupation sites. The earliest and most remarkable was that of the Proto-maglemosian hunter-fisher culture at Seamer in East Yorkshire referred to pollen zones IV and V (Clark, 1954, 1970). The long-known Mesolithic at Thatcham, Berk-shire was shewn by Wymer (1962) and Churchill (1962) to have extended from the beginning of zone V to zone VI. The Creswellian of the Derbyshire–Nottinghamshire border in part at least extends into zone VI, to which is also referred the Maglemosian at Brandesburton, Yorkshire. The microlithic industry of the Pennines was shewn by Walker to be referable to zone VIIa, as was the coastal mesolithic at Westward Ho!, and the Irish Larnian appears to extend to the end of this zone.

To some extent radiocarbon dating has been applied to the British mesolithic sites, but where we encounter the Neolithic, radiocarbon dating and pollen analysis powerfully reinforce one another. So far indeed as the 'elm decline' in pollen diagrams is an indicator of early Neolithic forest clearance, we have been supplied with evidence supplementary to that of actual Neolithic artifacts of the spread of Neolithic culture in these islands, so that we now have proof of strong spread of this culture in western Britain as early as 5500 to 5000 (radiocarbon) years B.P.

The sequence of Mesolithic, Neolithic, Bronze Age, Iron Age and Romano-British might seem sufficiently exact for dating our numerous associated plant records, and indeed a general chronology for this sequence has had to be applied, for the sake of simplification, to the British Isles as a whole: this will be found set out in the correlation table of fig. 31. Nevertheless, there is no shadow of doubt that here as elsewhere the boundaries between these cultures are 'sloping'. This was admirably shewn by Grahame Clark's study (1965) on the application of radiocarbon dating to the expansion of farming cultures from the Near East over Europe, and by the isochron map produced by Renfrew (1970) for the development of copper and bronze metallurgy in Europe and the Near East. In each instance several thousands of years separate the extreme areas concerned. With the smaller distances across the British Isles, and especially in the later cultures, this effect is smaller, but it must be borne in mind as a factor affecting the comparative age of the archaeologically dated plant records, and as qualifying the broad correlations given in fig. 31. It is acknowledged in the diagram (fig. 170) that summarizes the results of radiocarbon dating of forest-clearance and agricultural episodes in various parts of the British Isles. As we shall shew, pollen analysis and identification of macroscopic remains now complement one another effectively in revealing the patterns of prehistoric plant and animal husbandry, and in some situations such as that found in the derelict raised bogs of the Somerset Levels. The mire stratigraphy, pollen analysis and radiocarbon dating can be combined to give a background into which artifacts of, successively, the Neolithic, Bronze, Iron and Roman periods can be consistently fitted. This is illustrated in figs. 19, 20 and 21. The trackways are not only indicative of successional and climatic shift towards flooding, but they afford evidence of the practice of coppicing hazel as early as Neolithic time, and of the local presence of trees such as the beech, less certainly inferred from the pollen evidence. The pollen analyses yield evidence of the agricultural activity of Neolithic, Late Bronze Age and Early Iron Age occupants of the neighbourhood and demonstrate corresponding modifica-

tion of the exploited woodlands (fig. 168). Increasingly pollen-analytic studies also provide us with evidence of the nature of the crops cultivated by prehistoric man. The pollen of rye is fairly readily identified and less easily that of oats and wheat, and it has been possible in East Anglia to associate the growing of hemp primarily with the Anglo-Saxon settlement. These are topics developed in chapter VI: 13–15.

9. POLLEN ANALYSIS

At a time when the Pleistocene period and especially the so-called 'Post-glacial period' lacked a satisfactory chronology, von Post and his successors were able to develop, in pollen analysis, a method that provided a generalized time scale of wide applicability, and at the same time afforded a broad picture of the vegetation of past times. The method is dependent upon the prodigious output of pollen grains from the stamens of the flowering plants, and upon their wide dissemination, which ensures incorporation into growing geological deposits of all kinds, such for instance as lake muds, peat bogs and salt marshes. The pollen grains are microscopic in size and their large surface/bulk ratio ensures an extremely slow rate of fall through the air, so that even gentle upwards currents may carry them to great altitudes and far distances. The majority of our forest trees in these latitudes are wind-pollinated, as are the grasses and sedges, and when they come into flower vast quantities of pollen are discharged freely into the air; hardly less in amount is the pollen produced by insect-pollinated flowers, and, contrary to expectation, these also liberate a great deal of pollen to the air. Plants of more ancient stock, the mosses and ferns, though devoid of flowers, also propagate themselves by microscopic wind-borne spores, which appear alongside the pollen grains in the records of the pollen analyst. It is difficult to realize the abundance of pollen production, but the yellow bloom of conifer pollen on lake surfaces, the sufferings of people susceptible to hay-fever, and counts of deposited pollen of the order of 5000 grains cm^{-2} $year^{-1}$ (Hyde, 1952) give some notion of its magnitude, and of the rate of sedimentation into contemporary geological strata. At average rates of accumulation of mud in lakes or peat in bogs. Hyde's figures would provide something of the order of 50–100 grains mm^{-3} of deposit.

To the advantages of abundance and ubiquity, there are added those of ready characterization and of durability. Many features of grain size, grain shape, pore frequency, pore distribution, pore size, pore shape, and of wall structure and wall pattern will often allow identification of pollen grains as belonging to a given family, genus or even species. When pollen grains are incorporated in material which is permanently waterlogged and devoid of oxygen, the outer membrane of the wall, the exine, in which these manifold characteristics are

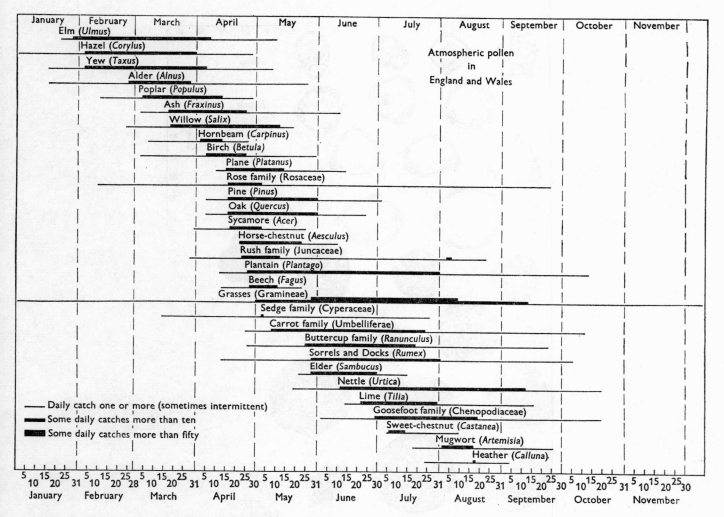

Fig. 22. Pollen calendar for England and Wales drawn up by H. A. Hyde on the basis of daily analysis of air-borne pollen at a range of stations through 1943. This diagram gives a clear idea of the pollen types most common and abundantly air-borne at the present time, and of their seasonal frequencies.

present, is apparently preserved indefinitely, although the protein contents of the grain and the cellulose inner wall, the intine, rapidly disappear.

If one adopts a technique to segregate the buried pollen grains from their embedding matrix, or to make them conspicuous under microscopic examination, it will be possible from even a few cubic millimetres of deposit to establish the nature of pollen sedimentation during its formation. From this it is possible to determine the relative frequencies in which the different pollen types are present, and thence, given certain assumptions, to reconstruct the vegetation type of the region at the time of deposition.

Thence, in von Post's succinct phrasing, 'By establishment of these frequency figures layer by layer through the pollen-bearing strata it becomes possible to follow former plant geographic changes from place to place and from one time-period to another.' In the vertical sequences in deposits at different places throughout a district, it now becomes possible to establish fundamental points of similarity, and by these in turn to establish horizons of equal age.

The particularly intensive investigations of Flandrian deposits over the last fifty years have increasingly confirmed von Post's threefold division of this period on a climatic basis into (a) the period of increasing warmth, (b) the period of maximum warmth, and (c) the period of decreasing warmth. It is apparent that, in the period of increasing warmth following the retreat and decay of the last ice-sheets, birch and then pine forests advanced northwards across the European plains to take over territory which had been 'tundra' or 'park tundra' or naked

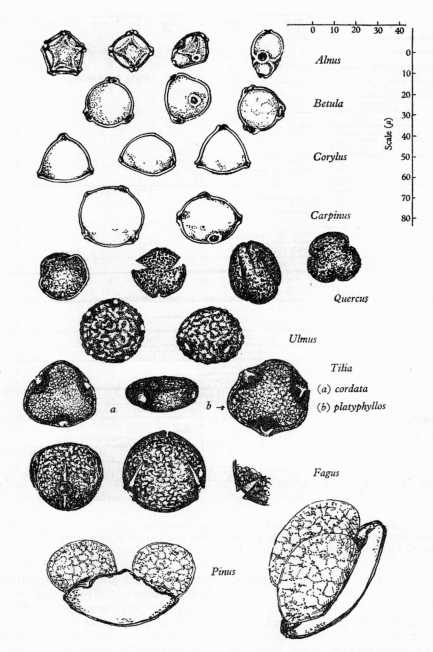

Fig. 23. Drawings of the chief types of tree pollen found in British Flandrian deposits drawn to a common scale of size.

glacial drift. After them followed the hazel and the warmth-loving deciduous trees, such as oak, elm and lime (and, somewhat later, the alder); during the climatic optimum these trees reached their greatest extent and importance, extending not only further north than they now occur, but also into higher belts upon the mountains. Towards the end of the period of maximal warmth the beech and hornbeam, and in central and southern Europe the spruce and silver fir, spread extensively at the expense of the mixed-oak forest. Finally, in the period of diminishing warmth, the pine and birch shew a tendency to return to their former preponderance.

Each climatic region exhibits its own distinctive course of vegetational evolution, but since all regions are affected by the broad drift of climatic alteration they also show a parallelism with one another; and, it is the recognition

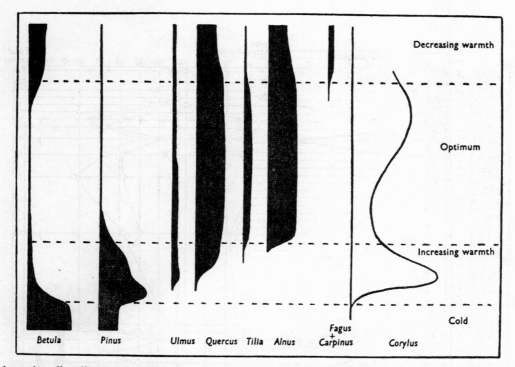

Fig. 24. Schematic pollen diagram characteristic of East Anglia showing the general drift of forest composition and the manner in which this reflects the threefold Flandrian division into periods of increasing warmth, maximum warmth and deterioration.

of this which permits pollen zonation schemes to be extended from one region to the next without apparent distortion of the underlying time scale. When once a reliable zonation has been thus established for a given area, quite a small peat or mud sample may suffice to refer a given layer to its period in it by pollen analysis. A lump of mud no longer than a wheat grain, adherent to a flint tool or a potsherd, may easily contain the few hundred pollen grains necessary to identify one of the major phases of forest history; and the task naturally becomes simpler where fossil material is encountered stratified into organic deposits which will yield a vertical series of samples for pollen analysis.

It has become the practice to express the results of analyses of vertical series of samples in the form of more or less conventionalized 'pollen diagrams' in which the chief pollen types are represented by histograms or curves, and these are either indicated by an internationally agreed set of symbols, or by a spaced out (dissected) arrangement (see figs. 25, 26, etc.).

Because of the compelling need of Quaternary geologists and archaeologists for a true time stratigraphy, and because von Post was himself a geologist, the pollen-zone system that he established and that has subsequently been initiated in Europe is *in intention* a time-stratigraphic system, as emerges from his masterly analyses of the progress and prospect of pollen analysis. He set out to establish more or less synchronous pollen zones, and in consequence of this his zones when compared across

large territories do not shew *identity* in different latitudes, but on the contrary disclose and necessarily accept *regional parallelism* in vegetational history. Pollen zonation in Britain began in 1940 with the definition of a similar system limited to England and Wales and numbered upwards like the Danish diagrams from zones I, II and III of the Late Weichselian (Godwin, 1940a). Even then the results shewed a strong indication of regional parallelism, the north and west clearly distinct throughout the Flandrian from the south and east. Jessen (1949) had already introduced for Ireland a similar zone system that he subsequently modified to bring it closer to the English one, the absence of *Tilia*, *Carpinus* and *Fagus* from the native Irish flora naturally making for increased difficulty in zone definition. The English scheme has likewise been extended to Scotland despite the diminished frequency of thermophilous elements in the forests. With the vast increase in number and complexity of available data throughout England and Wales the major zones as originally defined continue to be recognized and used, and they necessarily form the binding chronology for our historical plant record of the British Isles over the last fifteen thousand years. Radiocarbon dating has strongly confirmed the synchroneity of the Late Weichselian pollen-zone boundaries in Britain and the European mainland (Godwin & Willis, 1959; Godwin, 1960a), and we are in process of witnessing radiocarbon age determinations of the Flandrian zone boundaries in different regions. Although the earliest exercise of this kind, at

45

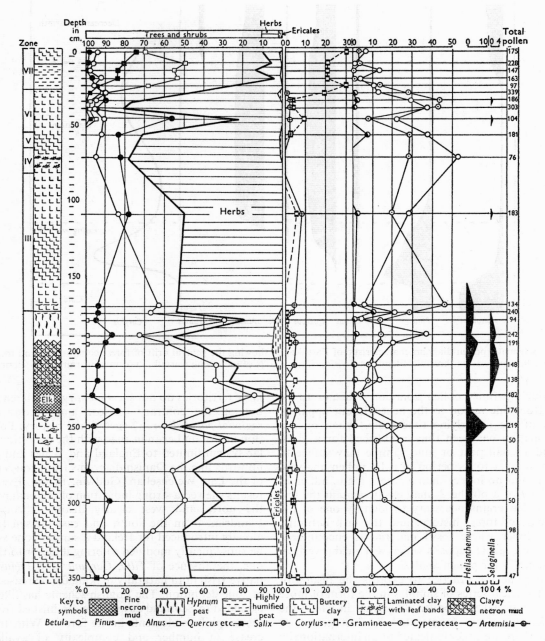

Fig. 25. Pollen diagram through deposits at Neasham, Co. Durham. The various pollen types are calculated as a percentage of the total pollen of all kinds. The total tree and shrub component is higher in the mild zone II deposits (in relation to herbs) than in the deposits of zones I and III. Note the Late Weichselian abundance of *Artemisia*, *Helianthemum* and *Selaginella*. (Blackburn, 1952.)

Scaleby Moss in Cumberland, shewed a broad synchroneity of the British, German and Dutch zone boundaries, and the most recent, Red Moss, yielded generally concordant results (fig. 27), there have now appeared fairly large deviations from synchroneity, though not sufficiently to impair the value of the standard zone system as it is here employed in ordering and analysing the already accumulated data of the last few decades. We are now moving into a stage of pollen-analytic research where methods and criteria are modified in relation to new objectives. Increasingly identification of past plant communities is sought in the pollen spectra, pollen diagrams are divided into pollen-assemblage zones with a strictly local significance, and in sites suitable for

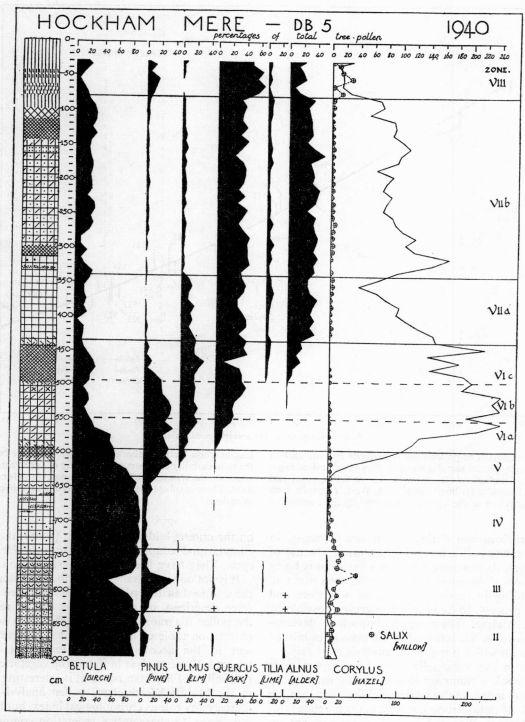

Fig. 26. Tree-pollen diagram at Hockham Mere in the north of the Breckland of East Anglia. For comparison with the non-tree pollen diagram (fig. 114) note especially the 340 cm level where the strong decline of the *Ulmus* curve indicates the boundary between zones VIIa and VIIb. The pollen zones are fairly characteristic for East Anglia, but zone VIII is difficult to recognize, especially since *Fagus* is unusually absent.

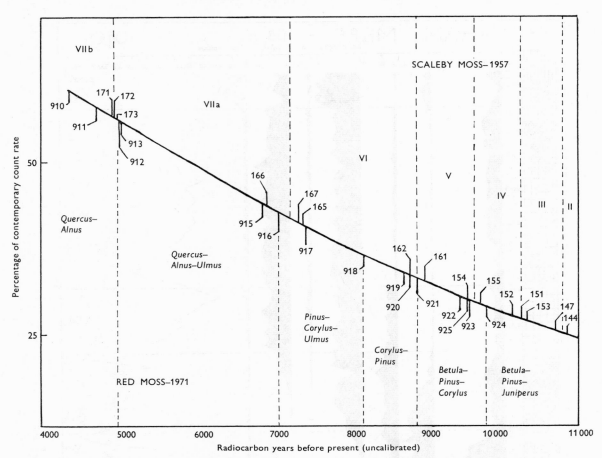

Fig. 27. Determination of pollen-zone boundaries by radiocarbon dating of critically placed samples along profiles in two raised bogs, Scaleby Moss, Cumberland (Godwin, Walker & Willis, 1957) and Red Moss, Lancashire (Hibbert, Switsur & West, 1971). In each instance the activities of the samples (indicated by the Cambridge Radiocarbon Dating Laboratory Q-numbers) have been placed on the radiocarbon decay curve of a half-life of 5568 years. The Scaleby zones are the standard British pollen zones: those for Red Moss are local. There is substantial agreement between the two sets of determinations.

sequential radiocarbon dating results are expressed in terms of absolute pollen incorporation rates. The use of pollen analysis to reconstruct former vegetation is being enhanced, and problems of correlation between sites and regions is being increasingly treated as a distinct and subsequent process, to be checked wherever possible by radiocarbon dating. These vastly important developments are as yet barely reflected in the mass of published information on which our present analysis must rely.

Although in the Weichselian and Flandrian periods pollen analysis is yielding up to some extent its role as a chronological index, in the interglacial deposits that are out of reach of radiocarbon dating it actively retains this function. Though the zone systems employed are increasingly given local prefixes, they are accepted as a basis for wide correlation as well as for identification of individual glacial stages. It is largely by pollen-analytic means that the early Pleistocene interglacial stages have been recognized and both the Hoxnian and Ipswichian of Britain have been divided into four sub-stages (I to IV)

on the criteria laid down by West (1968), that reflect the climatic and edaphic progress through the interglacial cycle. They have been already described on p. 13.

It is not now necessary to describe the criteria on which the standard British pollen zonation is based: it has been often described, its main features can be realized from the pollen diagrams shewn throughout the book and the correlation table of fig. 31, whilst in chapter VI: 11–15 and in the accounts of the individual tree genera (chapter V) we analyse in detail the vegetational changes through the Flandrian zones. It is interesting to note that in the Middle Weichselian, pollen analysis has played scarcely any role as chronological index, but has throughout served to characterize vegetation conditions during episodes correlated by other means, chiefly stratigraphic and physical.

There are now available excellent accounts of the methods and apparatus used in the field investigations that precede and accompany the taking of samples for pollen analysis (e.g., West, 1968) and likewise publications

PLATE XIII

a   *b*

a, Carbonized tubers of the onion couch, *Arrhenatherum bulbosum* (Gilib.) Schultz, found along with six-rowed barley at a Late Bronze Age site, Rockley Down, Wiltshire. The abscission layers, ribbed surface and scars of roots are recognizable on several specimens.

b, Fresh tubers of the onion couch, *Arrhenatherum bulbosum* (Gilib.) Schultz, for comparison with the carbonized tubers shewn in *a*.

PLATE XIV

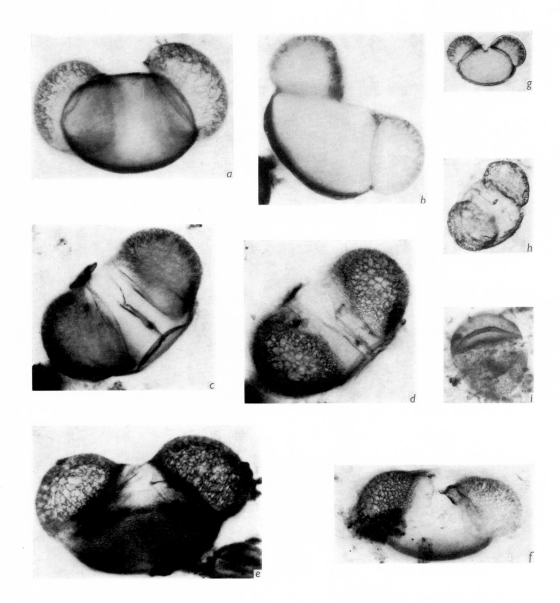

Pollen grains of Pinaceae: *a, b, Abies* (Hoxne, IG); *c–f, Picea* (Hoxne, IG); *g, h, Pinus sylvestris* (Cwm Idwal, zone VI); *i, Pinus haploxylon* (Eastwick, derived Tertiary type). Magnification ×400.

PLATE XV

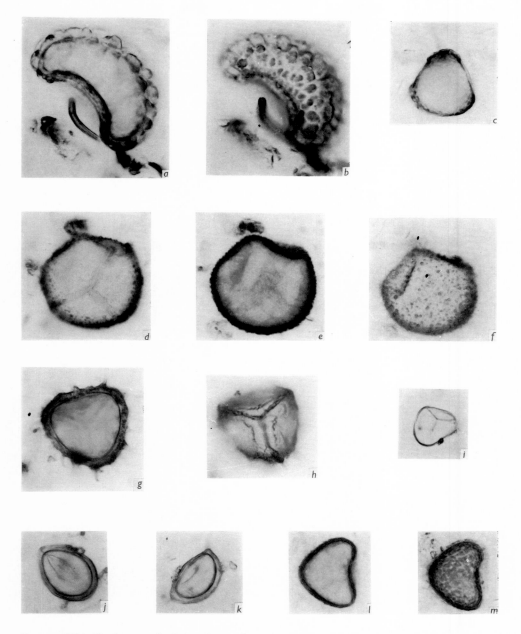

Spores of Pteridophyta: *a*, *b*, *Polypodium vulgare* (Cwm Idwal, zone VIIa); *c*, *Lycopodium selago* (Cwm Idwal, zone VI); *d, e, f*, *Osmunda regalis* (Drake's Drove, zone VIIb); *g, h*, *Selaginella selaginoides* (Whitrig Bog, zone II); *i*, *Pteridium aquilinum* (Cwm Idwal, zone VI); *j, k*, *Isoetes lacustris* (Cwm Idwal, zone VI); *l, m*, *Botrychium lunaria* (Whitrig Bog, late zone III). Magnification ×600.

PLATE XVI

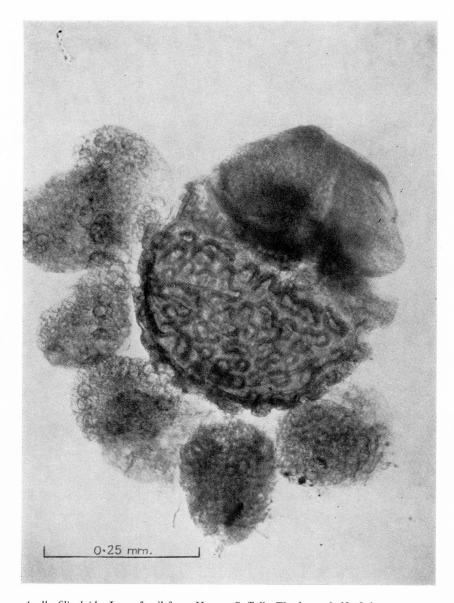

Azolla filiculoides Lam. fossil from Hoxne, Suffolk. The lower half of the megaspore is surrounded by eight massulae. Beneath the sculptured perisporium can be seen the collapsed inner wall of the megaspore, and glochidia can be seen projecting from the two lower massulae. Two of the swim-floats of the megaspore are visible.

[facing p. 49

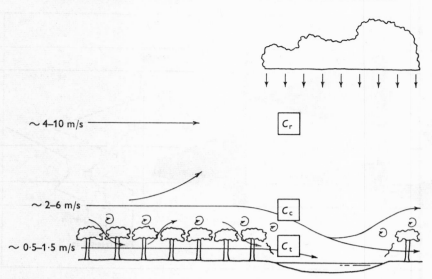

Fig. 28. Model of a tentative scheme (Tauber, 1965) of the processes of pollen transfer from source into a small lake in a forested area. Three main components are indicated: C_t, through the trunk space at low velocity; C_c, above the forest canopy, moving faster; C_r, pollen from a distance brought down by rain. Tauber shews that the evaluation of these components, which differ greatly in different situations, is a necessary condition for the interpretation of pollen-analytic data.

descriptive of the preparation of samples for microscopic examination and counting (Faegri & Iversen, 1964; West, 1968; Andrew, 1970). Whilst there is no point in replicating this information it seems useful to present a few special observations upon presentation and interpretation of pollen-analytic data. The rates of accumulation and decay of the pollen-bearing deposits are so various that it is not practicable to express the frequency of counted pollen on any absolute standard of sample taken, and it is therefore customary to express the amount of each pollen type as a percentage either of the total pollen count, or of the total tree-pollen count, depending upon whether the vegetation was substantially forest-dominated or not. In either instance the consequence is that the changes in frequency from level to level require careful interpretation, since absolute change in frequency of any one type automatically affects the recorded frequency of all others. Of course the *relative* frequencies of any two types of contrasting significance will be unaffected by this method of calculation. Thus if oak actually increased and pine actually decreased over a given period the relative movement of the oak and pine pollen curves would reveal this change, no matter what shifts there might be in the other tree pollen curves, or what local effects might intervene such as the development or decay of alder or birch thickets affecting the representation of these tree genera.

There still remains the interpretation of the pollen diagram in terms of changing vegetational composition, and in this task experience and ecological insight are essential. Pollen productivity varies much from one genus of tree to the next, some, like the pine and hazel, being much over-represented, others like oak, beech and lime under-represented (Andersen, 1970). We must have re-

gard to the effects of local conditions of vegetation upon the bog surface or lake shore, the effects of varying soil type, exposure and altitude all operating in the past as now to diversify the vegetational scene. Where mineral matter is present in the samples, one must always be prepared to recognize that pollen in them may have been derived from older deposits, and in certain deposits differential destruction of the pollen grains may have strongly distorted the original picture. This is particularly so in analyses made in mineral soil (Godwin, 1958).

Differential pollen dispersion by the wind can greatly affect the composition of the pollen deposition in any given site and it is important to consider to what extent in each instance we are concerned with local, regional or distant sources of production. A useful quantification of this approach is that of Tauber (1965) (fig. 28), and the growing volume of research on aeropalynology casts welcome light upon such problems as the possible trans-Atlantic importation of pollen of genera like *Ambrosia* and *Ephedra* to this country (see Hyde, 1969; Hirst & Hurst, 1967). Similar considerations of transport and origin are naturally involved with water-borne pollen that is sedimented in lakes, though studies of this are still few.

In our preoccupation with the use of pollen analysis to establish a general system of vegetational history, we must not overlook its employment as a subsidiary means of testing our conclusions about local stratigraphy and local vegetation deduced from geological evidence and the macroscopic remains of plants and animals. The ecological succession from salt marsh to freshwater fen, and fen-woods of alder, willow and oak is very clearly reflected in the pollen diagram of a thin buried peat bed containing 'buried forest' between brackish-water clays

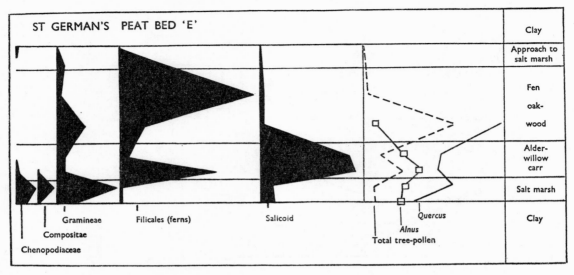

Fig. 29. Pollen diagram of buried peat bed at St German's near King's Lynn, shewing how from tree pollen and non-tree pollen together it is possible to reconstruct the local vegetational succession from salt marsh to fen oak-wood and subsequent retrogression. This sequence is typical of many 'submerged forests' on British coasts.

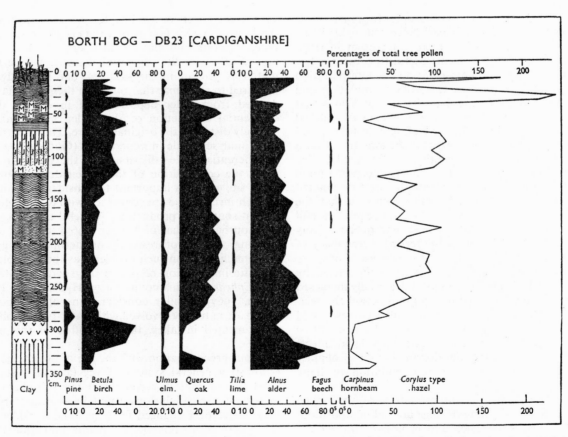

Fig. 30. Pollen diagram through Borth Bog, Cardiganshire. At the base of the diagram the pollen curves reflect the local changes in tree dominance in the sequence of peaks – alder, birch, pine. Stumps of all of these occur in the wood peat with pine at the transition to ombrogenous peat (see also fig. 13). Note the cessation of the *Tilia* curve and the rise of *Betula* at the boundary horizon (160 cm); also the low but recurring values for *Fagus* and *Carpinus*. The silty *Juncus maritimus* layer interrupts the upper, less humified *Sphagnum–Calluna* peat.

at St German's in the East Anglian Fenland (fig. 29). Likewise the pollen diagram through the base of Borth Bog, Cardiganshire (fig. 30) shews the sequence from salt marsh, through freshwater fen, and fen-woods with maxima of alder, birch and pine in turn, to ombrogenous *Sphagnum* bog. This sequence is one which has been commonly recorded in the marshes of the north-west German and Dutch coasts, and it corresponds closely with the vegetational succession and zonation demonstrated for the southern shores of the Baltic by Steffen.

Within the category of local changes of vegetation there also lie the manifold alterations, of varying scale, brought about by prehistoric man from early Neolithic time onwards. Over the last five millennia these activities have been so significant that they substantially hinder or prevent the application to this period of the Flandrian pollen zonation. The compensating advantage is however that pollen analyses allow us considerable insight into the effects of forest-clearance and the nature of early plant and animal husbandry, issues that are fully dealt with in chapter VI: 13–15.

The particular contribution of British research to Quaternary palynology has been recently summarized in Godwin (1968*a*).

10. PRIMARY DIVISIONS OF THE LAST FIFTEEN THOUSAND YEARS

The preceding sections of this chapter have been devoted to demonstrating the manner in which many lines of approach may be taken to the problem of setting in their true time relation the varied events of the Pleistocene period. It remains to present a synthesis for the last fifteen thousand years, the period covering the Late Weichselian and the Flandrian and the stretch of time for which our evidence is most abundant and which bears the most immediate significance for the history of the present-day British flora.

Of the many types of evidence available it is not to be assumed that any one has yet an absolute validity: apart from the radiocarbon assay method, all of them give largely relative chronologies. When considered together they may nevertheless be fitted into a single scheme of correlation in which an internal self-consistency guarantees the validity of the whole. To this ultimate scheme the table which we have set out in fig. 31 is merely a preliminary step to be succeeded, as the evidence grows, by progressively closer approximations.

Any compilation of this kind affords scope for prolonged discussion, but it will suffice for our present purpose to draw attention to some major features only. To keep conformity we represent all dates as (uncalibrated) radiocarbon years before the present.

The chronology itself was originally derived largely from varve counts made in Scandinavia, but it has to a surprising degree been confirmed by recent radiocarbon dating. The methods coincide in providing a date close

to 10 000 years B.P. for the end of the Weichselian and beginning of the Flandrian. Radiocarbon evidence suggests a duration of about 1200 years for the mild Allerød stage (zone II), but the placing of the opening of zone I, the boundary between the Middle and Late Pleistocene at 15 000 B.P., has been a matter of arbitrary decision.

We have already indicated in section 8 some of the factors concerned in the representation of successive archaeological cultures in the correlation scheme.

Under the heading 'Vegetation' we have given for each zone the designation used by Jessen for Ireland (Jessen, 1949) and a tentative sequence for northern Scotland, and between them terms appropriate for the British Isles as a whole are put down as a means of convenient reference.

The idea of the diagram to represent the proportional cover by forest throughout the last 13 000 years has been borrowed from Clark (1952), and is meant to have only broad qualitative significance.

In the columns containing geological evidence we have set the Scottish Late-glacial sea into the Late Weichselian on the evidence of radiocarbon dating (Bishop & Dickson, 1970) reinforcing evidence obtained at Garscadden Mains by Mitchell (1952) (see fig. 5).

Under 'Climate' the shaded arc represents that part of Flandrian time where in my view the evidence suggests most strongly a sustained thermal optimum, but since the 'Climatic Optimum' can perfectly well be conceived in a wider sense as embracing not only the Atlantic period but also the Sub-boreal after it and most of the antecedent Boreal, a dotted arc has been inserted to show this also. It must of course be borne in mind that the course of climatic change must have been far from regular in intensity or kind.

I have drawn attention in the table to two or three further points which need emphasis, the phase of dryness at the end of zone VI, the rapid and extensive eustatic rise in sea-level which caused the North Sea to expand and isolate these islands from the rest of the Continent during the Boreal period, and the minor eustatic rise of sea-level in Romano-British and Iron Age time.

The climatic improvement that was already in progress by 10 000 B.P. caused the rapid retreat and decay of the European ice-sheets, characterized in Scandinavia as the Finiglacial stage, during which Atlantic water entered the Baltic across central Sweden to form the cold salt sea named after its most typical mollusc *Yoldia hyperborea*. In the British Isles the vegetation, so rich in herbs, of the last stage of the Late Weichselian now rapidly gave place to birch woods formed by the extension of formerly dispersed copses, and in southern and eastern England pine freely accompanied birch. This is the basis for definition of zone IV, which is taken roughly to correspond to the Pre-boreal climatic period of Blytt and Sernander. It is fairly clear from pollen analysis of 'moorlog' that the present floor of the southern part of the North Sea basin was above ocean level at this time, a supposition borne out by the discovery of a Mesolithic barbed point of the Star Carr type in peat dredged from the Leman and Ower

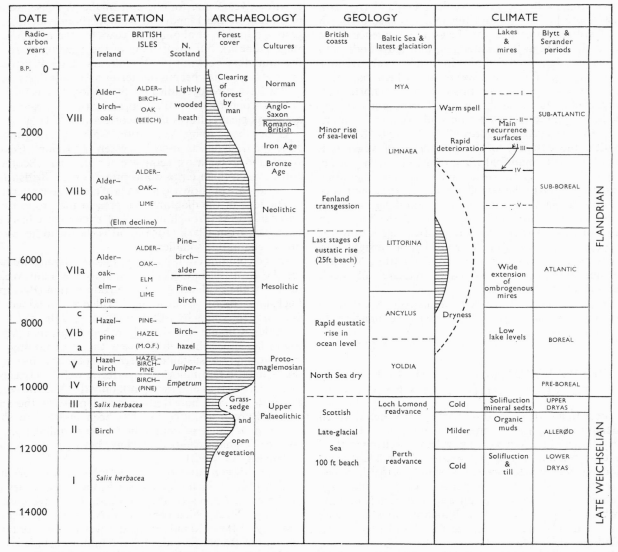

DATE	VEGETATION			ARCHAEOLOGY		GEOLOGY		CLIMATE		
Radiocarbon years B.P.	BRITISH ISLES — Ireland	BRITISH ISLES	N. Scotland	Forest cover	Cultures	British coasts	Baltic Sea & latest glaciation	Lakes & mires	Blytt & Serander periods	
0 —				Clearing of forest by man	Norman		MYA	Warm spell		
VIII	Alder-birch-oak	ALDER-BIRCH-OAK (BEECH)	Lightly wooded heath		Anglo-Saxon / Romano-British / Iron Age	Minor rise of sea-level	LIMNAEA	Main recurrence surfaces I, II, III, IV	SUB-ATLANTIC	FLANDRIAN
2000 —								Rapid deterioration		
4000 — VIIb	Alder-oak	ALDER-OAK-LIME (Elm decline)			Bronze Age / Neolithic	Fenland transgression			SUB-BOREAL V	
6000 — VIIa	Alder-oak-elm-pine	ALDER-OAK-ELM-LIME	Pine-birch-alder / Pine-birch		Mesolithic	Last stages of eustatic rise (25ft beach)	LITTORINA	Wide extension of ombrogenous mires	ATLANTIC	
8000 — VIb c / a	Hazel-pine	PINE-HAZEL (M.O.F.)	Birch-hazel			Rapid eustatic rise in ocean level	ANCYLUS	Dryness / Low lake levels	BOREAL	
V	Hazel-birch	HAZEL-BIRCH-PINE	Juniper-		Proto-maglemosian	North Sea dry	YOLDIA		PRE-BOREAL	
10000 — IV	Birch	BIRCH-(PINE)	Empetrum							
III	Salix herbacea			Grass-sedge and open vegetation	Upper Palaeolithic	Scottish Late-glacial Sea	Loch Lomond readvance	Cold / Solifluction mineral sedts.	UPPER DRYAS	LATE WEICHSELIAN
12000 — II	Birch							Milder / Organic muds	ALLERØD	
I	Salix herbacea					100 ft beach	Perth readvance	Cold / Solifluction & till	LOWER DRYAS	
14000 —										

Fig. 31. Correlation table shewing the main events of the Late Weichselian and Flandrian periods in the British Isles. Such schemata are subject to progressive amendment but this version is not substantially altered from that of the first edition.

banks in 120 feet (36.5 m) of water, and by a small number of consistent radiocarbon dates.

Continued climatic improvement is shown by the succeeding Boreal period, which includes the British pollen zones v and vi based upon the replacement of birch by pine, the phenomenal expansion of hazel, and the first establishment of elm and oak as important forest elements. During a large part of this period the Baltic was a freshwater lake of greater extent than at present, discharging to the Atlantic through the Svea River in central Sweden: this is the Ancylus Lake Stage. There is very clear evidence that it was during the Boreal period that there took place the most rapid and sustained restoration of water from the world's melting ice-sheets to the oceans. It was this restoration of ocean level which isolated Great Britain from the continental mainland of Europe, and either separated Ireland from Great Britain or at least greatly widened any existing sea barrier. The dating of this event for biogeographical purposes has an importance difficult to exaggerate, but radiocarbon dating has already allowed much progress towards this end (Godwin, 1960a; Schofield, 1960, 1964; Jelgersma, 1966).

The one major climatic change, that of increasing warmth, induced the several apparently unconnected sequences of Flandrian events. It caused rapid melting of the world's ice-fields and the water so restored to the oceans raised sea-level everywhere by something of the order of 100 metres; the improved climate caused the northward migration of vegetation long confined to more southerly regions, at first the biota of open tundra or park tundra, and later that of closed forests. It is of paramount importance to know what range of migration had already been achieved by the time that the general rise of sea-level deprived our western islands of easy overland access from the rest of Europe. So far as the present plant population of the British Isles is a matter not of perglacial survival or of recent human introduction, much depended upon the speed of dissemination and establishment in the time available whilst the North Sea, English Channel and Irish Sea were still dry land. The interest of this situation is heightened by the fact that we can now recognize that whilst its earlier stages favoured species of open habitats, the later stages were within a period of rapidly intensifying forest dominance. Thus, where thermal requirements or slowness of dispersal delayed extension of particular species at this time, these delayed species must have been substantially restricted to types adapted either to forest conditions or to the isolated and diminishing 'islands' of fen, lake, cliff and so forth within the thickening sea of woodland. We shall evidently require to know as much as possible of the exact climatic attributes of the period of open land connection, and to acquire as much direct evidence as possible of the actual invasion of plants within this period, most notably the known thermophiles, which by their presence signify the suitability of conditions for the spread of others of their kind.

There is evidence from many parts of the British Isles that the last stage of the Boreal period was one of dryness in which lakes tended to dry and bogs and fens to become tree-clad. This phase was followed, through a sudden transition, by the Atlantic period which corresponds with the pollen zone VIIa. In most of Great Britain the transition was marked by the onset of widespread development of ombrogenous bogs, both raised bogs and blanket bogs, by the sudden expansion of the alder (perhaps as a result of the provision of abundant locally wet areas in the general cover of deciduous forest), and by the expansion of the lime. In Ireland and north-western England the alder's expansion was much more gradual, and the lime of course did not reach Ireland. About this time the eustatic rise of ocean level finally reached its maximum, giving rise to the Carse Clays of the Scottish Lowlands, the clay and silt infilling of many estuaries around our coasts and the cutting of the 25-foot raised beach of northern Great Britain and Ireland. No doubt the beach formation was associated with a considerable period of approximate balance between the slow final eustatic rise of sea-level and slow isostatic uplift of the land. In Scandinavia the Atlantic period corresponds to the Littorina stage of the Baltic when this inland sea was warmer and salter than at any stage in its history. The beaches of this Baltic stage have been shown by recent Danish research to be multiple in character and to extend into the Neolithic period, but there is no evidence from the British Isles to suggest that the main eustatic rise of sea-level continued to this late time.

Half-way through zone VII is placed the boundary between pollen zones VIIa and VIIb, an horizon marked by widespread recession of the elm, and one generally taken to correspond with the Atlantic/Sub-boreal transition. This horizon marks the establishment in the British Isles of Neolithic agriculturists where hitherto only Mesolithic hunter and fisher peoples had been present, and numerous radiocarbon determinations place it about 5500 to 5000 B.P.

Although there are few geological indices characterizing the Sub-boreal period, the close of it is often marked by horizons or layers indicative of dryness in lakes and bogs, and correlations with both Neolithic and Bronze Age settlements now occur in some abundance. At a time estimated as about 3200 B.P. a phase of increasing wetness has led in some parts of the country to flooding horizons in the raised bogs, and it is possible that the most pronounced of the Irish recurrence surfaces is of this date.

At about 2500 B.P., which broadly corresponds to the establishment of Iron Age culture in Britain, there is much evidence of sudden climatic worsening and in raised bogs everywhere there is a sharp contact between the old highly humified, Calluna-rich peat of Atlantic and Sub-boreal age and the pale, fibrous, quickly grown and undecayed Sphagnum peat of the cool and moist Sub-atlantic period. In many places the boundary is an extremely sharp one, marked by flooding layers or precursor peat above layers of tree stools, Calluna heath or dense tussocks of cotton grass.

It is at this level that the boundary has been drawn between the pollen zones VII and VIII, making use of the decrease of linden, the expansion of birch (together with that of beech and hornbeam in southern England and Wales). It proved in practice extremely hard to define, and the vegetational changes on which it is based have turned out to be much more strongly determined by human activity than directly by climatic shift. Fortunately by this time radiocarbon dating is available to calibrate the pollen diagrams and after the opening of the Sub-atlantic period most datings of plant material rely upon archaeological correlations, but some further assistance is available in a last and minor eustatic rise in sea-level during Iron Age and Roman time, a rise which is especially evident in the uppermost clays and silts in the regions of the Wash and Somerset Levels, but which can also be found recorded more widely. Within such raised bogs as remain uncut there are registered further climatic changes within the Sub-atlantic period, as for instance by the 'retardation layer' in Tregaron Bog and by minor recurrence surfaces, but these are still very little investigated.

It will be appreciated that, within the framework of the period here outlined, much correlation of a finer grade is possible where intensive local studies have been made. Marine transgression and retrogression induced by the interplay of tilting and eustatic changes may produce a local basis for correlation, as may equally a detailed knowledge of bog stratigraphy or pollen zonation. Details of such local correlation schemes will be found in the papers and books cited.

IV

RECORDED SITES

1. GENERAL COMMENTS

It will have become apparent from the preceding discussion that the factors to be assessed in placing a deposit or a discovery at its proper place in the Pleistocene period may be numerous and complex, and that the final conclusion may be based upon more or less elaborate investigation and argument. For this reason Clement Reid, in writing his *Origin of the British Flora*, included a section in which he briefly sketched the stratigraphy and the evidence for dating provided by each site which had yielded plant material. The vast multiplication of sites since that time forces us to adopt an extremely curtailed entry for our own list of recorded sites but these, given in alphabetic sequence, contain the county and National Grid reference together with a reference to the most recent definitive publication or publications dealing with the site. The particular value of this lies in the fact that the individual plant records are now given only to sites and age (see chapter v), so that by looking at the site record the reader can rapidly discover the published sources for most records. A very short note for each recorded site summarizes the main character of the deposits and the nature of the correlations on which the plant record depends. In most instances the conclusion will be that set out in the primary record; but in certain instances where re-investigation has been made, or our knowledge of the general or local circumstances has widened, the author has been able to correct or to sharpen the original dating. Thus in certain cases cited by Clement Reid his records may now appear under a new heading. An excellent example of improved dating is the case of the Barry Docks in south Wales where C. Reid was able to provide a rich list of plant remains from the submerged peat beds exposed in the dock excavations: subsequent pollen analysis of submerged peats in Swansea Bay have now allowed the various layers at Barry to be placed in their appropriate pollen zones with considerable probability. The records from the peat beds at Crossness and Tilbury in the Thames estuary have similarly become more closely datable after the analysis of the stratigraphic evolution of the Fenland basin and the coastal deposits along both sides of the southern North Sea. In each of these instances radiocarbon dating in closely related sequences has supported the new age reference, and in the individual site records we have added any relevant radiocarbon ages; these are given in uncalibrated radiocarbon years B.P., without the probable error, but with the laboratory code number as published in *Radiocarbon*.

In yet other cases consideration of the total original floral or faunistic lists may more sharply define the age of the deposit than when it was first described. Thus the *Chara* marls with remains of *Cervus megaceros* in the Isle of Man, and the lacustrine deposits at Corstorphine near Edinburgh are now referable in all probability to the Late Weichselian period. We have omitted records referred merely to the 'Post-glacial' and the old records described by Reid as 'Neolithic' since the original description does not allow us to narrow the record to the modern use of the term.

The range of pollen zones cited is that which we ourselves estimate in terms of the standard zonation. The contractions *M* and *P* indicate that the site yields respectively macroscopic and pollen records.

It has been convenient to employ some now outdated portmanteaux terms, such as 'Cromer Forest Bed series' and 'Arctic Plant Bed': such usages are explained in the following chapter, 'The Plant Record'. In a number of instances sites have been included from which plant remains have been identified but which, because of uncertainty of age attributed, have been excluded from the book though not from the data bank.

2. SCOTTISH SITES

In considering the peat deposits of Scotland and their contained plant fossils we meet with a situation historically different from that in the English, Welsh and Irish deposits. In Scotland alone the peat bogs had attracted systematic investigation in the period prior to the development of pollen analysis. Painstaking observation and recording had been made by Geikie (1881), by Lewis (1905–7, 1911), and then by Samuelsson (1910). These investigators dealt with fossil remains that might be identified directly in the field or with a simple laboratory preparation, and they naturally produced schemes of correlation based upon macrostratigraphy. Samuelsson, following his teacher Sernander, was able to work from the basis of the competent and rapid elucidation of the Post-glacial history of Scandinavia, where also the earliest systematic results of pollen analysis were being compiled (see von Post, 1916). Considerably later than this, another Swede, G. Erdtman, this time a pupil of von Post, visited Scotland to apply to its Quaternary deposits the new pollen-analysis technique. His results appeared in 1924, 1925 and 1928.

The extremely energetic researches of Lewis have given

TABLE 3

Lewis' strata	Lewis' names		Sernander and Samuelsson's correlations
Recent peat	Upper Turbarian		Sub-atlantic
Forest			
Peat bog and 'Arctic' plants	Upper Forestian		Sub-boreal
Forest			
Peat bog plants	Upper peat bog		
Arctic plant bed	Second Arctic bed	Lower Turbarian	Atlantic
Peat bog plants	Lower peat bog		
Forest	Lower Forestian		Boreal or Atlantic
Arctic plant bed	First Arctic plant bed		Pre-boreal

us a very comprehensive picture of the main pattern of macrostratigraphy in the peat beds in different parts of Scotland and the Isles and the greater part of our records of sub-fossil species. It is all the more to be regretted, therefore, that we are still uncertain of the age of Lewis' main horizons. Samuelsson and Erdtman not only have proposed a correlation of Lewis' horizons with the Blytt and Sernander climatic periods, but they consider that the 'Arctic' species which he mentions have sometimes been wrongly identified; they also think that species put by Lewis into this category do not always have the validity he claims for them as indicators of arctic tundra conditions.

The above schema will show the correlation and terminology proposed for Lewis' sequence in the Southern Uplands of Scotland, a sequence which can be applied to the rest of Scotland, with certain provisos, namely that Samuelsson considers the Upper Forestian to be only a single layer in the Grampians, and that the bed called by Lewis 'First Arctic bed' in Shetland and Skye is in fact his Second Arctic bed, overlaid not by the Lower but by the Upper Forestian.

Lewis had followed Geikie in regarding the 'Arctic plant beds' as due to distinct glacial periods separated by temperate interglacials or interstadials in which peat and forest beds were formed. There is now, however, little doubt that Sernander's interpretation of the whole sequence as a Post-glacial one is correct and that therefore the nature of the 'Arctic plant beds' needs some reconsideration.

Whilst the lower Arctic bed is more or less accepted by the Swedish investigators, they discard the upper. Authentic Late Weichselian deposits have subsequently been described in many parts of Scotland, many reinforced by radiocarbon dating.

Until detailed investigations by pollen analysis or radiocarbon datings have been made at sites representative of Lewis' main stratigraphic types, it will be impossible to give proper weight to the dating of these Scottish records, and it is apparent that checking of the macroscopic identifications is also advisable. On this account it has been thought best to omit altogether results based upon Lewis' investigations for Scotland and the Pennines.

It is surprising that the application of radiocarbon dating to buried pine stumps in Scotland should not so far have disclosed any aggregation of dates that might correspond with Lewis' forest horizons (see account for *Pinus sylvestris*, p. 110).

Early pollen analyses in Scotland had sampling only at wide vertical intervals and it has proved difficult to utilize them in the plant record, but subsequently there has been a wealth of detailed and informative research closely tied into the geological background and to an absolute chronology. The cover now extends into the western and northern isles, and the Shetlands have even yielded evidence of interglacial conditions.

3. IRISH SITES

With regard to Ireland the situation is far different from that when the first edition was written. There has been such extensive publication that it has been possible to incorporate Irish data alongside those for the rest of the British Isles. The difficulties of extending to Ireland the standard pollen zonation were circumvented, in his later publications, by Mitchell who set up a local system taking account of succeeding phases, prehistoric and historic, of human modification of the Irish vegetational pattern. For purposes of co-ordination within the present volume it has accordingly been necessary to make our own estimate of the locus of the zone VII/VIII boundary in a number of Irish pollen diagrams. Since the first edition there has also been a wealth of extremely valuable new information upon plants of the Hoxnian interglacial in Ireland from sites such as Gort, Kilbeg and Baggotstown, and it has become evident that then, as now, Irish vegetation had a special stamp of its own.

4. SITE RECORD

Abbot Moss, Cumberland NY 5143 (Walker, 1966) MP
Stratigraphical succession through freshwater muds, fen peat, *Sphagnum* peat. Late Weichselian to Flandrian VIIb.

Abbots Way, Somerset   ST 4242 (Coles & Hibbert, 1968) *MP*
Raised bog stratigraphy. Neolithic track in Flandrian VIIb: 4800 (Q-647), 4750 (Q-431).

Abergwesyn, Brecon   SN 8552 (Jerman, 1935) *M*
Oak piles at Bronze Age site.

Abernethy Forest, Inverness   NH 9617 (H. H. Birks, 1970) *MP*
Raised bog stratigraphy and regional assemblage zones. Late Weichselian III to Flandrian VIIb.

Abersoch, Caernarvons. SH 3128 (Ridgeway & Leach, 1946–7) *M*
Mesolithic flint workshop.

Abingdon, Berks.   SU 4997 (Jessen & Helbaek, 1944; Helbaek, 1953b) *M*
Neolithic, Bronze Age and Anglo-Saxon.

Abington Pigotts, Cambs.   TL 3144 (Fox, 1922–3) *M*
Early Iron Age settlement.

Aby Grange, Lincs.   TF 4379 (Suggate & West, 1959) *MP*
Shallow lake deposits of clay and peat. Late Weichselian I to III: zone II 11 205 (Q-279).

Acklam Wold, Yorks.   SE 7861 (Jessen & Helbaek, 1944) *M*
Early Bronze Age.

Addington, Kent   TQ 6659 (Burchell & Erdtman, 1950) *P*
Flandrian VIIb.

Admiralty Offices, London   TQ 3080 (Reid, quoting Abbott, 1899) *M*
Interglacial deposits of uncertain age: not entered.

Agher, Meath   N 9752 (Mitchell, 1951, 1956) *P*
Raised bog stratigraphy. Flandrian VI to VIII.

Aghfarell, Dublin   O 0521 (Jessen & Helbaek, 1944) *M*
Late Bronze Age.

Airdrie, Lanarks.   NS 7665 (Reid, 1899) *M*
Probably Weichselian interstadial (Coope, 1962).

Albert Dock, Woolwich, Middx.   TQ 4280 (Reid, 1899) *M*
Submerged forest bed. Flandrian VIIb to VIII (Godwin, 1943).

Albury, Surrey   TQ 0548 (Godwin & Willis, 1964) *M*
Carbonized grains from Late Bronze Age storage pit: 2460 (Q-760).

Aldro, Yorks.   SE 8062 (Jessen & Helbaek, 1944) *M*
Early and middle Bronze Age.

Aldwick Barley, Cambs.   TL 4038 (Cra'ster, 1965) *M*
Iron Age pits.

Allachy Moss, Aberdeens.   NO 4891 (Durno, 1959) *P*
Post-boreal bog stratigraphy.

All Cannings, Wilts.   SU 0661 (Gose & Sandell, 1964) *M*
Iron Age.

Allenton, Derbys.   SK 3732 (Reid quoting Bemrose, 1899) *M*
Interglacial deposits of uncertain age: not entered.

Allt na Feithe Sheilich, Inverness.   NH 8526 (H. H. Birks, 1969) *MP*
Fen peat with stubs and blanket bog with regional assemblage zones. Flandrian VIIa to VIII. Pine 6960 (K 1419).

Altartate, Monaghan   H 52 (Mahr, 1934) *P*
Iron Age cauldron. Flandrian VIII.

Amberley Wild Brooks, Sussex   TQ 0213 (Godwin, 1943) *P*
Marine transgression into valley deposits. Flandrian VIIb, 2620 (Q-690), and VIII.

Amble, Northumberland   WU 2604 (Jessen & Helbaek, 1944) *M*
Late Bronze Age.

American Square, Middx.   TQ 27 (Lyell, 1912) *M*
Roman ditch.

Amesbury, Wilts.   SU 1541 (Helbaek, 1953b) *M*
Middle Bronze Age.

Amoy Bog, Balleymoney, Antrim   C 9525 (Jessen & Helbaek, 1944) *M*
Late Bronze Age.

Angel Road, Essex   TQ 3692 (Warren, 1912) *M*
Lea Valley Arctic Plant Bed. Middle Weichselian.

Annaholty, Tipperary   R 6769 (Mitchell, 1951) *P*
Raised bog stratigraphy. Flandrian VI to VIIb.

Anston Cave, Derbys.   SK 5293 (J. B. Campbell, unpub.) *P*
Late Weichselian: 9750 to 9940 (BM-439, 440 a, b).

Apethorpe, Northants.   TL 0295 (Sparks & Lambert, 1961) *MP*
Valley fill with mollusca. Late Weichselian III to Flandrian VIII.

Arbury Road, Cambridge   TL 4658 (D. Walker, unpub.) *M*
Roman.

Archerfield, East Lothian   NT 5084 (Jessen & Helbaek, 1944) *M*
Early Bronze Age.

Ardcavan, Wexford   T 0724 (Mitchell, 1948b) *MP*
Deposit of uncertain age, not entered.

Ardlow Inn, Cavan   N 5889 (Mitchell, 1956) *MP*
Raised bog stratigraphy. Flandrian VIIa to VIII.

Ardsbeg, Donegal   B 8738 (Jessen, 1949) *MP*
Raised bog stratigraphy. Flandrian VI to VIIb.

Arminghall, Norfolk   TG 2504 (Clark, 1936b) *M*
Bronze Age oak monument.

Arniston, Edinburgh   NT 3259 (Jessen & Helbaek, 1944) *M*
Late Bronze Age.

Aros Moss, Kintyre   NP 6722 (Nichols, 1967) *P*
Bog over 50-foot raised beach. Flandrian IV to VIII.

Ashcott Heath, Somerset   ST 4437 (Dewar & Godwin, 1963) *MP*
Raised bog stratigraphy. Flandrian VIIb. Neolithic bow, 4625 (Q-598).

Ashford, Kent   TR 0142 (Jessen & Helbaek, 1944; Helbaek, 1953b) *M*
Late Bronze Age.

Ashgrove, Fife   NT 3599 (Henshall, 1963–4) *MP*
Bronze Age cist burial.

Ashwell, Herts.   TL 2639 (Jessen & Helbaek, 1944; Helbaek, 1953b) *M*
Early Iron Age.

Aston, Derbys. SK 4129 (Reaney, 1966) *M*
Beaker burial and Neolithic hearth.

Athlone, Roscommon N 0341 (Erdtman, 1928) *MP*
Sequence from marl to *Sphagnum* peat. Flandrian VI to VIIb.

Aughrim Td., Kerry R 0941 (Jessen, 1949) *P*
Raised bog stratigraphy. Flandrian VIIb and VIII.

Austerfield, Yorks. SK 6696 (Gaunt, Coope, Osborne & Franks, 1970) *MP*
Silt layer in terrace gravels. Ipswichian III.

Avebury, Wilts. SU 0969 (Gray, 1934) *M*
Neolithic and Bronze Age.

Aveley, Essex TQ 5580 (West, 1969a) *MP*
Fluviatile deposit with elephants. Ipswichian IIb and III.

Avonmouth, Devon ST 5178 (Seddon, 1965) *MP*
Submerged peat beds. Flandrian VI and VIIa.

Bacton, Norfolk TG 3433 (Reid, 1899; Duigan, 1963) *MP*
Cromer Forest Bed series.

Badger Hole, Somerset ST 5347 (Campbell *et al.*, 1970) *P*
Upper Palaeolithic with mammals. Weichselian, >18000 (BM-497).

Baggotstown, Limerick R 6736 (Watts, 1964) *MP*
Kettle hole deposit. Hoxnian I to III and early Wolstonian.

Bagmere, Cheshire SJ 7964 (H. J. B. Birks, 1965a) *MP*
Sequence from open water to raised bog. Late Weichselian I to Flandrian VIII.

Bagshot, Surrey SU 9163 (Oakley *et al.*, 1939) *M*
Age uncertain: not entered.

Ballaugh, I.O.M. SC 3393 (C. A. & J. H. Dickson & Mitchell, 1970) *MP*
Kettle hole deposits. Late Weichselian II to Flandrian VIII.

Ballinderry, Down J 1767 (Jessen & Mitchell, 1942) *P*
Late Bronze Age.

Ballingarry Downs, Limerick R 7828 (Mitchell, 1953) *M*
Fourth to ninth century.

Ballintoy, Antrim D 0344 (Childe, 1936) *M*
Promontory fort.

Ballinvariscal Td., Kerry Q 8805 (Mitchell, 1951) *P*
Raised bog stratigraphy. Flandrian VIIb and VIII.

Ballon, Carlow S 8367 (Jessen & Helbaek, 1944) *M*
Late Bronze Age.

Ballowell Cairn, Cornwall SW 3531 (Helbaek, 1953b) *M*
Middle Bronze Age.

Ballybetagh, Dublin O 2122 (Jessen & Farrington, 1938) *MP*
Sequence from freshwater muds to raised bog. Late Weichselian II to Flandrian VII. *Megaceros* in II.

Ballycroghan, Down J 4982 (Smith, 1956) *P*
Early Christian. Flandrian VIII.

Ballydugan, Down J 4742 (Singh, 1970) *MP*
Lake muds. Late Weichselian I to Flandrian V.

Ballyhalbert, Down J 3636 (Morrison & Stephens, 1960) *P*
Peat below raised beach. Flandrian VIC, 8120.

Ballykeerogemore Td., Wexford S 7425 (G. F. Mitchell & W. A. Watts, unpub.) *M*
Hoxnian interglacial.

Ballymacombs More, Londonderry H 9991 (Jessen, 1949) *P*
Sequence from open water muds to raised bog. Flandrian VI to VIII.

Ballymakegogue Td., Kerry Q 7715 (G. F. Mitchell, unpub.) *M*
Late Hoxnian to early Wolstonian.

Ballymena, Antrim D 1003 (Jessen & Helbaek, 1944) *M*
Middle Bronze Age.

Ballynagilly, Down H 7383 (Pilcher, 1970) *P*
Sequence from freshwater muds to raised bog. Series of radiocarbon dates (UB-242–260).

Ballynakil Td., Westmeath N 1836 (Mitchell, 1956) *P*
Raised bog stratigraphy. Flandrian VI to VIII.

Ballyscullion, Antrim H 2949 (Mitchell, 1956) *P*
Raised bog stratigraphy. Flandrian VI to VIII.

Ballyvaloo Lower Td., Wexford T 3122 (Mitchell, 1951) *MP*
Kettle hole deposit. Flandrian IV and V.

Bann Estuary, Londonderry C 8235 (Jessen, 1949) *P*
Brackish water muds and *Phragmites* peat. Flandrian VIIa to VIIb.

Barhapple Loch, Wigtown NX 3174 (Helbaek, 1953b) *M*
Early Christian.

Barnwell Station, Cambridge TL 4859 (Bell & Dickson, 1961) *M*
Middle Weichselian deposit with Reindeer and Mammoth. Earlier plant records by Chandler (1921) revised. 19500 (Q-590).

Barons Court, Tyrone H 3782 (Davies, 1950) *MP*
Bronze Age. Flandrian VI to VIIb.

Barra, Outer Hebrides NF 7001 (Blackburn, 1946) *P*
Blanket peat. Flandrian IV to VIII.

Barrowell Green, Middx. TQ 3094 (Reid & Chandler, 1923b) *M*
Peat in terrace gravels. Middle Weichselian.

Barrow Nook, Yorks. TA 0721 (Jessen & Helbaek, 1944) *M*
Bronze Age.

Barry, Glam. ST 1167 (Reid, 1899; Godwin, 1943) *M*
Submerged peat beds, upper Flandrian VIIb, Neolithic; lower VIb and VIC (cf. *Swansea Bay*).

Barton Broad, Norfolk TG 3120 (Jennings, 1952) *MP*
Basal wood peat below recent lake muds, following peat cutting. Flandrian VII and VIII.

Barway Causeway (Fordy), Cambs. TL 5375 (Godwin, 1941) *MP*
Fen peat, brushwood causeway and Bronze Age artifacts. Flandrian VIIb, 2560 (Q-310).

Baskfield, Glenluce, Wigtown NX 1957 (Jessen & Helbaek, 1944) *M*
Middle Bronze Age.

Battlegore, Somerset ST 0741 (Helbaek, 1953*b*) *M*
Late Bronze Age.

Bawdrip, Somerset ST 3339 (H. S. L. Dewar, unpub.) *M*
Roman.

Beaghmore Td., Tyrone H 6983 (Pilcher, 1969) *P*
Raised bog stratigraphy. Flandrian VI to VIII. Clearance phases 6000 to 775 (UB-87 to UB-96).

Beaulieu, Hants. SU 3540 (Piggott, 1943*a*) *M*
Early Bronze Age.

Beckhampton, Wilts. SU 0969 (Helbaek, 1953*b*) *M*
Neolithic.

Bedham Hill, Sussex TQ 0218 (Keef, 1940) *M*
Neolithic.

Beeston, Norfolk TG 1843 (Reid, 1899) *M*
Cromer Forest Bed series.

Belfast, Antrim J 3576 (Mitchell, 1951) *MP*
Brick pit section. Late Weichselian I to III.

Belfast Lough, Antrim J 3576 (Morrison, 1961) *MP*
Submerged peat bed. Flandrian VIa.

Bell Tracks, Somerset ST 4242 (Coles & Hibbert, 1968; Coles, Hibbert & Clements, 1970) *MP*
Fen wood overlain by raised bog. Flandrian VIIb, Neolithic wooden tracks. 4840 (GaK-1600) to 3975 (BM-382-4, Q-927).

Bembridge, I.O.W. SZ 6489 (Reid & Chandler, 1925) *M*
Age uncertain: not entered.

Bere Regis Down, Dorset SY 8494 (Helbaek, 1953*b*) *M*
Middle Bronze Age.

Bermondsey, Middx. TQ 3680 (Kennard & Warren, 1903) *M*
Silt containing Roman pottery.

Bettisfield Moss, Salop SJ 4734 (Hardy, 1939) *MP*
Raised bog stratigraphy. Flandrian V to VIII.

Bielsbeck, Yorks. SE 8638 (Stather, 1906–9) *M*
Age uncertain: not entered.

Bigberry Camp, Harbledown, Kent TR 1157 (Jessup & Cook, 1936) *M*
Iron Age.

Bigholm Burn, Dumfries. NY 3181 (Moar, 1969*a*) *MP*
Stream section with peat and gravel. Late Weichselian II to Flandrian VIII, 11820 to 7520 (Q-694, 5, 7, 701).

Bincombe, Dorset SY 6884 (Jessen & Helbaek, 1944; Helbaek, 1953*b*) *M*
Late Bronze Age.

Birrens, Dumfries. NY 1979 (Jessen & Helbaek, 1944) *M*
Roman.

Bishops Canning, Wilts. SU 0666 (Proudfoot, 1965) *M*
Neolithic and Bronze Age.

Bishopston, Glam. SS 5989 (Williams, 1940*b*) *M*
Roman.

Blacklane, Devon SX 6268 (Simmons, 1964) *P*
Bog stratigraphy. Flandrian IV to VIIb.

Blackpill, Glam. SS 6893 (von Post, 1933, in Godwin, 1940*b*) *P*
Forest bed exposed on foreshore. Flandrian VIII.

Blakeway Farm Track, Somerset ST 4545 (Clapham & Godwin, 1948) *MP*
Neolithic, 4280 (Q-460).

Blanch, Yorks. SD 5760 (Jessen & Helbaek, 1944) *M*
Middle Bronze Age.

Blanchland, Northumberland NY 9048 (Raistrick & Blackburn, 1931) *MP*
Sequence from wood peat to blanket bog. Flandrian VIIa and VIIb.

Blashenwell, Dorset SY 9580 (Reid, 1899) *M*
Neolithic.

Bleaklow, Cheshire SK 0996 (Conway, 1954) *P*
Blanket bog stratigraphy. Flandrian VIIa to VIII.

Bleasdale, Lancs. SD 5745 (Varley, 1938) *M*
Middle Bronze Age.

Blea Tarn, Westmorland NY 2904 (Pennington, 1964–5) *P*
Freshwater muds. Flandrian IV to VIIb.

Blelham Bog and Tarn, Lancs. NY 3600 (Pennington & Bonny, 1970; Evans, 1970) *MP*
Freshwater muds and raised bog. Late Weichselian I to Flandrian VIIb. Series of dates (I-3589 to 3598, Q-758).

Blind Tarn, Lancs. SD 2696 (Pennington, 1964–5) *P*
Freshwater muds. Flandrian VIIb and VIII.

Bloak Moss, Ayrs. NS 3646 (Turner, 1964) *P*
Raised bog stratigraphy. Flandrian VIIb and VIII. Series clearances. 3320, 3170, 3050 (Q-724 to 726).

Bloxworth Down, Dorset SY 8894 (Jessen & Helbaek, 1944; Helbaek, 1953*b*) *M*
Middle Bronze Age.

Bobbitshole, Suffolk TM 1441 (West, 1958) *MP*
Freshwater deposit. Late Wolstonian to Ipswichian IIb. Mollusca.

Borth Bog, Cardigans. SN 6089 (Godwin, 1943; Moore, 1968) *MP*
Clay of eustatic maximum overlain by tree layer and raised bog with estuarine clay transgression. Flandrian VIIa to VIII, 2900 (Q-712).

Boscombe Down, Wilts. SU 2038 (Gose & Sandell, 1964) *M*
Iron Age.

Boscowen-un, Cornwall SX 0990 (Helbaek, 1953*b*) *M*
Middle Bronze Age.

Boskill Td., Limerick R 6949 (Mitchell, 1951) *P*
Freshwater deposit with skull of *Megaceros giganteus*. Late Weichselian and Flandrian IV.

Boston, Lincs. TF 3243 (F. T. Baker, unpub.) *P*
Late Bronze Age.

Bourton-on-the-Water, Glos SP 1821 (Dunning, 1932) *M*
Saxon.

Bovey Tracey, Devon. SX 8178 (Reid, 1899) *M*
Late Weichselian.

Bowness Common, Cumberland NY 2260 (Walker, 1966) *MP*
Sequence from freshwater muds to *Sphagnum* bog behind
raised beach. Flandrian VIIa to VIII.

Boyndlie, Aberdeens. NJ 9162 (Jessen & Helbaek, 1944) *M*
Early Bronze Age.

Bradford Kaim, Northumberland NU 1631 (Bartley, 1966) *MP*
Sequence from lake muds to raised bog. Late Weichselian I
to Flandrian VIII.

Brancaster, Norfolk TF 7945 (H. & M. E. Godwin, 1934) *P*
Submerged peat bed. Flandrian VI and VIIa.

Brandesburton, Yorks. TA 1147 (Clark & Godwin, 1956) *MP*
Organic deposits in gravel. Late Weichselian, Flandrian IV,
VIIa, VIIb. Mesolithic.

Brandon, Warwicks. SP 3976 (Kelly, 1968) *MP*
Two terrace deposits. Early Wolstonian and Middle
Weichselian: 32270, 30766 (Birm-27, 10).

Branston, Staffs. SK 2221 (P. A. Tallantire, unpub.) *M*
Late Weichselian and Flandrian.

Breagthwy, Mayo G 2518 (Mitchell, O'Leary & Raftery,
1941) *P*
Raised bog stratigraphy. Flandrian VIIa and VIIb.

Bredon Hill, Glos. SO 8219 (Hencken, 1939) *M*
Iron Age.

Breiddin Hill, Montgomery SJ 2915 (St J. O'Neil, 1937) *M*
Roman.

Bridgend, Glam. SS 9079 (Fox, 1937) *M*
Bronze Age.

Bridlington, Yorks. TA 1766 (Reid, 1899) *M*
Peaty marl. Late Weichselian.

Brigg, Lincs. SE 9906 (Smith, 1958a) *MP*
Marine clay and peats. Flandrian VIIb. Dug-out canoe and
trackway. 2796 to 2552 (Q-77 to 79).

Broad Gate Fell, Northumberland NY 9084 (Blackburn, 1953) *P*
Bog stratigraphy. Flandrian IV to VIII.

Broadway, Worc. SP 1038 (Smith, 1946) *M*
Roman.

Brook, I.O.W. SZ 3983 (Clifford, 1936) *MP*
Plant bed with Mesolithic flints. Flandrian VIIa.

Brook, Kent TR 0844 (Kerney, 1964) *MP*
Detritus mud and silt with mollusca. Late Weichselian I
and II, 11500 (Q-618).

Broombarns, Perthshire NO 0818 (Godwin & Willis, 1961) *M*
Peat below Carse Clay, 8354 (Q-422).

Broughton, Peeblesshire NT 1136 (Reid, 1899) *M*
Age uncertain: not entered.

Brownhead Cairn, Bute NR 8925 (Jessen & Helbaek, 1944) *M*
Middle Bronze Age.

Brownley Hill, Lancs. SD 5559 (Moseley & Walker, 1952) *P*
Blanket peat. Flandrian VIIa to VIII.

Broxbourne, Herts. TL 3707 (Warren *et al.*, 1934) *MP*
Lea Valley Arctic Plant Bed and later Mesolithic overlain
by Flandrian, zone VI.

Bryn Celli Ddu, Anglesey SH 5267 (Hyde, 1930) *M*
Megalithic cairn.

Bryngwyn, Flints. SJ 1073 (Stapleton, 1908) *M*
Bronze Age.

Buckie, Banffshire NJ 4265 (Jessen & Helbaek, 1944) *M*
Early Bronze Age.

Bunny, Notts. SK 5785 (Wilson, 1968b) *M*
Roman.

Bure Valley, Norfolk TG 3415 (Jennings, 1955) *MP*
Fen and brush-wood peats with clays of two marine transgressions (cf. Fenland sequence). Flandrian V to VII/VIII.

Burnham-on-Sea, Somerset ST 3149 (Godwin & Willis, 1959b)
M
Dredged *Phragmites* peat formed under brackish water
conditions, 6262 (Q-134). Flandrian VIIa.

Burnmoor Tarn, Cumberland NY 1804 (Pennington, 1965) *P*
Freshwater muds. Flandrian VIIa and VIIb.

Burn of Benholme, Kincardineshire NO 7969 (Donner, 1960) *P*
Late Weichselian II silt and zone III solifluxion.

Burren Td., Mayo M 1497 (G. F. Mitchell, unpub.) *M*
Hoxnian interglacial.

Burtree Lane, Durham NZ 2618 (Bellamy, Bradshaw, Milling-
ton & Simmons, 1956) *MP*
Freshwater deposit with *Paludella squarrosa* peat. Late
Weichselian III and Flandrian IV.

Bury Wood Camp, Wilts. SU 3443 (Grose & Sandell, 1964) *M*
Iron Age.

Cadder, Lanarks. NS 6172 (Jessen & Helbaek, 1944) *M*
Middle Bronze Age.

Caerau, Caernarvons. SH 4249 (St J. O'Neil, 1936) *M*
Roman.

Caerhun, Caernarvons. SH 5569 (Baillie-Reynolds, 1936) *M*
Roman.

Caerleon, Mon. ST 3391 (Helbaek, 1964) *M*
Roman.

Caerwent, Mon. ST 4890 (Lyell, 1911) *M*
Roman.

Caerwys, Flints. SJ 1272 (Reid, 1899) *M*
Flandrian tufa, age uncertain: not entered.

Cahercorney, Limerick R 55 (Mitchell, 1951) *M*
Freshwater silts and muds with *Megaceros giganteus*. Late
Weichselian I to III.

Cairnpapple Hill, West Lothian NS 9875 (S. Piggott, 1947-8)
M
Middle Bronze Age.

Caistor-by-Norwich, Norfolk TG 2303 (P. D. Broad, unpub.)
M
Roman.

Calvay Island, Outer Hebrides NF 7728 (Harrison & Blackburn, 1946) *MP*
Peat deposits. Flandrian VI to VIII.

Cambridge, Cambs. TL 4559 (Godwin & Willis, 1964) *M*
Prehistoric yew bow, 3680 (Q-684).

Canbo Td., Roscommon M 8996 (Mitchell, 1956) *MP*
Sequence from freshwater muds to raised bog. Flandrian IV to VIII.

Canna, Inner Hebrides NG 2406 (Flenley & Pearson, 1967) *P*
Shallow basin deposits, Late Weichselian to Flandrian VIII, *fide* H.G.

Cannons Lough, Londonderry C 9310 (Smith, 1961*a*) *P*
Freshwater muds and marginal fen peat. Late Weichselian II to Flandrian VIIb.

Canterbury, Kent TR 1557 (Williams & Frere, 1948) *M*
Roman.

Capel Cynon, Cardigans. SN 3849 (Davies, 1905) *M*
Bronze Age.

Car Dyke, Cottenham, Cambs. TL 4567 (J. Allison, unpub.) *M*
Roman.

Carmyllie, Forfar NO 5542 (Jessen & Helbaek, 1944) *M*
Middle Bronze Age.

Carnwarth Moss, Lanarks. NS 9647 (Fraser & Godwin, 1955) *P*
No stratigraphy, Flandrian VI to VIII.

Carrantuohill, Kerry U 8085 (Welten, 1952) *P*
Shallow blanket peat, Flandrian VIII.

Carrigalla, Limerick R 6541 (Jessen & Helbaek, 1944) *M*
Viking.

Carrowreagh, Roscommon M 7592 (Jessen, 1949) *MP*
Sequence from freshwater muds to *Sphagnum peat.* Late Weichselian to Flandrian VIII.

Casewick, Lincs. TF 0709 (Reid, 1899) *M*
Age uncertain: not entered.

Cassington, Oxon. SP 4809 (Atkinson, 1946–7) *M*
Bronze Age.

Casterly Camp, Somerset SU 1153 (Helbaek, 1953*b*) *M*
Early Iron Age.

Castle Cary, Stirlings. NS 7878 (Jessen & Helbaek, 1944) *M*
Roman.

Castle Eden, Durham NZ 4338 (Reid, 1920) M
Freshwater clay thought to correspond to Cromer Forest Bed series.

Castlelackan Demesne, Mayo G 1837 (Jessen, 1949) *MP*
Sequence from freshwater muds to raised bog. Flandrian VI to VII.

Catcott Burtle, Somerset ST 3944 (Godwin & Willis, 1964) *M*
Prehistoric yew bow, 3270 (Q-669).

Chagford, Devon SX 7087 (Fox, 1954) *M*
Iron Age.

Chalbury, Dorset SU 0206 (Whitley, 1943) *M*
Early Iron Age.

Chalgrave, Beds. TL 0127 (D. Walker, unpub.) *M*
Twelfth century A.D.

Chapel of Garioch, Aberdeens. NJ 7274 (Jessen & Helbaek, 1944) *M*
Early Bronze Age.

Chapel Point, Lincs. TF 5671 (Godwin & Clifford, 1938; Smith, 1958*a*; Godwin & Switsur, 1966) *MP*
Coastal sequence of clays and peats. Flandrian VIIb to VIII, Iron Age. 3943 to 2455 (Q-686–688, 844).

Charlesworth, Derbys. SK 0092 (Franks & Johnson, 1964) *P*
Landslip. Flandrian VI to VIII.

Chastleton Camp, Oxon. SP 2429 (Jessen & Helbaek, 1944) *M*
Early Iron Age.

Chat Moss, Lancs. SJ 7096 (H. J. B. Birks, 1963–4, 1965*a*) *MP*
Sequence from freshwater muds and silts to raised bog. Late Weichselian I to Flandrian VIII. Flooding horizon: 3070 (Q-682), 2645 (Q-683).

Chelford, Cheshire SJ 8172 (Simpson & West, 1958) *MP*
Organic muds in drift deposits. Weichselian interstadial, 60 800 (Gro. 1475).

Chester, Cheshire SJ 4066 (Newstead & Droop, 1932) *M*
Roman.

Chesterford, Cambs. TL 4944 (J. S. Henslow, unpub.) *M*
Roman.

Chewton Mendip, Somerset ST 5953 (Williams, 1947) *M*
Bronze Age.

Chippenham, Cambs. TL 6669 (Jessen & Helbaek, 1944) *M*
Early Bronze Age.

Chippenham, Wilts. ST 8774 (King, 1966) *M*
Neolithic.

Chrichie, Aberdeens. NJ 5943 (Jessen & Helbaek, 1944) *M*
Late Bronze Age.

Christchurch, Hants. SZ 1792 (Piggott, 1938) *M*
Middle Bronze Age.

Christ's Hospital, London TQ 27 (Lyell, 1912) *M*
Roman.

Chycarne, Cornwall SW 4831 (Helbaek, 1953*b*) *M*
Middle Bronze Age.

Cissbury Camp, Sussex TQ 1402 (Curwen & Ross-Williamson, 1931) *M*
Iron Age.

Clacton, Essex TM 1718 (Pike & Godwin, 1953; Turner & Kerney, 1961) *MP*
Interglacial, with marine transgression and Palaeolithic industry. Hoxnian IIB to III.

Clapham, Sussex TQ 0905 (Curwen, 1934*b*) *M*
Neolithic and Late Bronze Age.

Claremorris, Mayo M 3475 (Erdtman, 1928) *MP*
Sequence from marl to moss peat. Flandrian V to VIIb.

Clatteringshaws Loch, Kirkcudbright NX 5576 (H. H. Birks, 1969) *MP*
Pine stumps below raised bog. Flandrian VIIa to VIII, 5080 (Q-878).

Clea Lakes, Down J 5055 (Smith, 1959*b*) *M*
Early Christian.

Clee Hills, Salop SO 5975 (St J. O'Neil, 1934) *M*
Early Iron Age.

Clettraval, Outer Hebrides NF 7775 (Scott, 1935) *M*
Neolithic and Iron Age.

Clonsast, Offaly N 5819 (Mitchell, 1956; McAulay & Watts, 1961) *P*
Sequence from wood peat to raised bog. Flandrian VIC, 8264 (Q-19) to VIII; 1910, 1620 (D 26–29).

Cloonacool, Sligo G 4917 (Jessen, 1949) *MP*
Sequence from fen to raised bog. Flandrian IV and VIIa to VIII.

Cloonlara Td., Mayo G 4101 (Mitchell, 1956) *P*
Raised bog stratigraphy. Flandrian VIII.

Close y Garey, I.O.M. SC 2675 (Reid, 1899) *M*
Marl and peat deposits in a kettle hole. Early Flandrian, not entered.

Clough Mills, Antrim D 0618 (Jessen, 1949) *MP*
Sequence from freshwater muds and silts to *Sphagnum* peat. Flandrian V to VIII.

Cocker Beck, Notts. SK 54 (Lamplugh *et al.*, 1908–9) *M*
Alluvial silt. Flandrian VIIb.

Cockerham Moss, Lancs. SD 4548 (Oldfield & Statham, 1964–5)
Sequence of marine clay, wood peat and *Sphagnum* peat. Flandrian VIIb to VIII.

Coelbren, Glam. SN 8411 (Morgan, 1907) *M*
Roman Camp.

Coire Bog, Ross. NH 5985 (H. H. Birks, 1969) *MP*
Pine and birch stumps exposed in blanket bog peat dated 6980 (Q-887), 6731 (Q-888), 5005 (Q-889). Regional assemblage zones. Flandrian VIIa to VIII.

Cold Fell, Cumberland NY 6458 (Precht, 1953) *MP*
Blanket bog stratigraphy. Flandrian VIII.

Collier Gill, Yorks. NZ 7800 (Simmons, 1969) *MP*
Blanket bog stratigraphy. Flandrian VIIa to VIII.

Collingbourne Ducis, Wilts. SU 2453 (Helbaek, 1953*b*) *M*
Middle and Late Bronze Age.

Colney Heath, Herts. TL 2005 (Godwin, 1964) *MP*
Peat erratics in terrace gravels. Late Weichselian I: 13 560 (Q-385).

Combe Beacon, Somerset ST 3011 (St G. Gray, 1935) *M*
Middle Bronze Age.

Combwich, Somerset ST 2642 (Godwin, 1941) *MP*
Wood peat below marine clay: 6460 (Q-35).

Compton Downs, Berks. SU 5381 (Allison, 1947) *M*
Roman.

Cool Bog, Limerick R 7441 (Jessen, 1949) *MP*
Freshwater muds and clays with *Megaceros giganteus* skull. Late Weichselian I to Flandrian v.

Coolteen, Wexford T 02 (G. F. Mitchell, unpub.) *M*
Freshwater muds. Late Weichselian I to III.

Coolwest Td., Limerick R 24 (Mitchell, 1951) *P*
Raised bog stratigraphy. Flandrian VIII.

Coom Rigg Moss, Northumberland NY 6979 (Chapman, 1964) *P*
Sequence from fen to raised bog. Flandrian VI to VIII.

Cooran Lane, Wigtowns. NX 4784 (H. H. Birks, 1969) *MP*
Pine stumps exposed in blanket peat. Late Weichselian III to Flandrian VIII. 7541, 7471, 6805 (Q-871, 873, 874).

Coppice Gate, Somerset ST 4339 (Dewar & Godwin, 1963) *P*
Transition from *Sphagnum* bog to telmatic fen. Flandrian VII/VIII transition. Middle Bronze Age spearhead.

Copthall Avenue, London TQ 27 (Lyell, 1912) *M*
Roman.

Cordal, Kerry RO 0 (Mitchell, 1951) *P*
Blanket bog stratigraphy. Flandrian VIIa to VIII.

Corlona Bog, Leitrim H 2812 (Mitchell, 1956) *P*
Raised bog stratigraphy. Flandrian VIIa to VIII. Trackway: 3395 (Gro-272).

Corstorphine, Midlothian NT 1972 (Reid, 1899; Newey, 1970) *M*
Freshwater muds. Late Weichselian.

Corton, Suffolk TM 5497 (Reid, 1899) *M*
Bed of lignite and clay. Cromer Forest Bed series.

Corton Beds, Lowestoft, Suffolk TM 5391 (West & Wilson, 1968) *M*
Plant bearing clays in Corton Sands. Anglian interstadial, mollusca.

Cothill, Berks. SP 4600 (A. R. and B. N. Clapham, 1939) *MP*
Sequence from freshwater muds and silts to fen. Flandrian IV to VIIa.

Coumenare Td., Kerry Q 4705 (Mitchell, 1951) *P*
Blanket peat, Flandrian VIIb and VIII.

Cowden Glen, Renfrewshire NS 9899 (Reid, 1899) *M*
Peat deposit. Flandrian, not recorded.

Cow Down, Wilts. SX 8840 (Callow & Hassall, 1968) *M*
Iron Age. Series of radiocarbon dates.

Cowlan, Yorks. SE 9665 (Jessen & Helbaek, 1944) *M*
Early Bronze Age.

Craigentinny, Edinburgh NT 0956 (Jessen & Helbaek, 1954) *M*
Late Bronze Age.

Craig-y-Llyn, Glam. SN 9202 (Hyde, 1940) *MP*
Raised bog stratigraphy. Flandrian IV to VIII.

Cranes Moor, Hants. SU 1802 (Seagrief, 1960) *MP*
Valley deposits of peat and mud. Flandrian IV to VI.

Crayford, Kent TQ 57 (Kennard, 1944) *M*
Brick earth with Palaeolithic flints and rhinoceros bones. Age uncertain: not entered.

Crehelp, Wicklow N 9102 (Jessen & Helbaek, 1944) *M*
Middle Bronze Age.

Crianlarich, Perthshire NN 3825 (Reid, 1899) *M*
Peaty loam. Late Weichselian.

Crick, Mon. ST 4990 (Savory, 1940) *M*
Middle Bronze Age.

Crinnagh River, Kerry   V 9482 (Jessen, 1949) *P*
Blanket bog stratigraphy. Flandrian VIIa and VIIb.

Cromer, Norfolk   TG 2142 (Reid, 1899) *M*
Cromer Forest Bed series.

Cromer Forest Bed series (see p. 82).

Cross Fell, Cumberland   NY 7035 (Godwin & Clapham, 1951) *MP*
Blanket peat and clay. Flandrian VI.

Crossness, Essex   TQ 4781 (Reid, 1899) *M*
Submerged peat beds. Flandrian. Roman.

Crowhurst Park, Sussex.   TQ 7714 (Straker & Lucas, 1938) *M*
Roman.

Culbin Sands, Moray   NJ 0063 (Jessen & Helbaek, 1944) *M*
Bronze Age.

Culbour Hill, Somerset   SS 8448 (Helbaek, 1953*b*) *M*
Early Bronze Age.

Curleywee, Kirkcudbright   NX 4577 (Moar, 1969*a*) *MP*
Freshwater muds and silts. Flandrian IV to VIIa.

Curragh, I.O.M.   SC 3794 (Erdtman, 1925) *P*
Peat. Flandrian IV, VI and VIIa.

Cushenden, Co. Antrim   D 2332 (Jessen, 1949; Godwin & Willis, 1961) *P*
Freshwater silts in relation to raised beach. Flandrian VI. 4740 (Q-373).

Cwm Idwal, Caernarvons.   SH 6459 (Godwin, 1955) *MP*
Sequence from freshwater muds and silts to *Sphagnum* peat. Flandrian IV to VIIb.

Dalnagar, Perthshire   NO 0860 (Durno, 1965) *P*
Raised bog stratigraphy. Flandrian VI to VIII.

Danes Moss, Cheshire   SJ 9070 (Tallis & Birks, 1965) *MP*
Sequence from freshwater muds and silts to raised bog. Flandrian VI to VIIb.

Daviot, Aberdeens.   NJ 7528 (Kilbride-Jones, 1935) *M*
Late Bronze Age and Iron Age.

Deans Bridge, Edinburgh   NT 27 (Jessen & Helbaek, 1944) *M*
Late Bronze Age.

Deansgate and Hunts Bank, Manchester, Cheshire   SJ 8397 (Roeder, 1899) *M*
Roman.

Decoy Pool, Norfolk   TL 923909 (Jennings, 1952) *P*
Fen peats and marine clay. Flandrian IV to VIII.

Decoy Pool Wood, Somerset   ST 4240 (Godwin, 1948) *MP*
Raised bog stratigraphy with flooding horizons. Flandrian VIIb and VIII.

Dernaskeagh, Sligo   G 7408 (Jessen, 1949) *M*
Raised bog stratigraphy. Flandrian VIIb.

Derriaghy, Antrim   J 2969 (G. F. Mitchell, unpub.) *MP*
Late Weichselian I to III.

Derryadd, Armagh   D 0541 (Jessen, 1949) *MP*
Wood peat below terrace sand. Flandrian V to VII.

Derrybrien, Galway   M 5902 (Prendergast & Mitchell, 1960) *P*
Raised bog stratigraphy. Flandrian VIIb and VIII.

Derrycassan, Cavan   H 2511 (Jessen, 1949) *MP*
Drift mud and *Phragmites* peat. Flandrian IV, VI and VIIa.

Derrytagh North, Armagh   D 0541 (Jessen, 1949) *MP*
Raised bog stratigraphy. Flandrian VIIa to VIII.

Derryvree, Fermanagh   H 3639 (Colhoun *et al.*, 1972) *M*
Plant debris in laminated sand lens in till. Middle Weichselian: 30 500 (Birm-166).

Deverel Barrow, Dorset   SY 6895 (Helbaek, 1953*b*) *M*
Late Bronze Age.

Devil's Bit, Kerry   V 9384 (Jessen, 1949) *MP*
Phragmites peat. Flandrian VI.

Devoke Water, Cumberland   SD 1596 (Pennington, 1964) *P*
Freshwater muds and silts. Late Weichselian II to Flandrian VIIb.

Dimlington, Yorks.   TA 3921 (Bell, 1969) *M*
Organic deposit in till: 18 240 (Birm-18).

Dinas Emrys, Caernarvons.   SH 6149 (Seddon, 1961) *MP*
Dark Ages occupation.

Din Lligwy, Anglesey   SH 4786 (Baynes, 1908) *M*
Roman.

Dogger Bank (Godwin, 1943; Behre & Menke, 1969) *MP*
Submerged peat beds. Flandrian V and VI.

Dog in Doublet Sluice, Hunts.   TL 2798 (Godwin & Clifford, 1938) *P*
Sequence of lower peat, fen clay, upper peat. Flandrian VIIb and VIII.

Dolygaer, Glam.   ST 1575 (Ward, 1902) *M*
Early Bronze Age.

Dorchester, Oxon.   SU 5595 (Duigan, 1953) *M*
Plant remains, Palaeolithic implements and mammoth bones at base of terrace gravels. Wolstonian.

Doune Castle, Forfar   NO 4750 (Jessen & Helbaek, 1944) *M*
Middle Bronze Age.

Downham Market, Norfolk   TF 6003 (Jessen & Helbaek, 1944) *M*
Anglo-Saxon.

Downton, Wilts.   SU 1721 (Gose & Sandell, 1964) *M*
Late Neolithic.

Dozemare Pool, Cornwall   SX 1974 (Conolly, Godwin & Megaw, 1950) *MP*
Sequence from freshwater muds and silts to *Sphagnum* peat. Flandrian V to VIII.

Drake's Drove, Somerset   ST 4844 (Godwin, 1956) *MP*
Roman clay transgression into peats. Flandrian VIIa to VIII.

Dromsallagh, Limerick   R 1715 (Mitchell, 1956) *MP*
Sequence from freshwater *Phragmites* swamp to *Sphagnum* bog. Flandrian VIIa to VIII.

Dronachy, Fife   NT 2291 (Reid, 1899) *M*
Freshwater muds. Late Weichselian.

Drope, Glam.   ST 1075 (Reid, 1899) *M*
Age uncertain: not entered.

Drummcaladerry, Donegal   C 2142 (Mitchell, 1951) *MP*
Sequence from wood peat to blanket bog. Flandrian VIIb and VIII.

Drumskellan Td., Donegal   C 4927 (E. C. Colhoun, unpub.) *M*
Laminated leaf debris beneath raised beach. Flandrian VI and VII: 6950 (UB-206).

Drumurcher, Monaghan   H 5519 (Mitchell, 1951) *MP*
Mud, peat, gravel sequence. Late Weichselian III, Flandrian IV.

Drymen, Stirlings.   NS 4992 (Donner, 1956–7; Y. & A. Vasari, 1968) *MP*
Freshwater muds. Late Weichselian I to Flandrian VIII.

Drymmiewood, Balbirnie, Fife.   NO 2903 (Jessen & Helbaek, 1944) *M*
Late Bronze Age.

Duartbeg, Sutherland   NC 1638 (Moar, 1969b) *P*
Freshwater muds and silts overlain by *Sphagnum* peat. Flandrian V to VIII. Series of radiocarbon dates (erroneous).

Dublin   O 1535 (G. F. Mitchell, unpub.) *M*
Mediaeval.

Dunloy, Antrim   D 0219 (Jessen & Helbaek, 1944) *M*
Neolithic.

Dunruodeagh, Tyrone   H 6284 (Jessen & Helbaek, 1944) *M*
Middle Bronze Age.

Dunshaughlin, Meath   N 9852 (Mitchell, 1953) *MP*
Freshwater muds. Late Weichselian I to Flandrian VIII. Neolithic, Bronze Age, Iron Age.

Durrington, Wilts.   SU 1544 (Helbaek, 1953b) *M*
Middle Bronze Age.

Dyserth, Flints.   SJ 0579 (Glenn, 1915) *M*
Neolithic and Bronze Age.

Earith, Hunts.   TL 3875 (Bell, 1970a) *M*
Lenses of fine grained sediment in gravel. Middle Weichselian: 42 140 (Birm-88) and >45 000 (Birm-86).

Easterton, Moray   NJ 1168 (Jessen & Helbaek, 1944) *M*
Neolithic.

East Moors, Glam.   ST 17 (Hyde, 1936; Godwin, 1943) *MP*
Coastal peats above and below sea-level. Flandrian VIIb and VIII.

Easton Bavents, Suffolk   TM 5076 (Funnell & West, 1962) *P*
Interglacial sands and clay. Antian and Baventian with marine fauna.

East Tarbet, Kintyre   NR 6552 (Reid, 1899) *M*
Clyde Beds. Late Weichselian.

East Walton, Norfolk   TF 7416 (Bell, 1969) *MP*
Lens of organic matter in chalk rubble. Late Weichselian.

Eastwicke, Herts.   TL 4311 (unpublished) *MP*
Late Weichselian.

Eday, Orkney   HY 53 (Jessen and Helbaek, 1944) *M*
Neolithic.

Edentober Td., Louth   J 3131 (Mitchell, 1951) *P*
Raised bog stratigraphy. Flandrian VIIb and VIII.

Eglwys-Fach, Denbighs.   SH 8070 (Gardner, 1913) *M*
Bronze Age.

Ehenside Tarn, Cumberland   NY 0007 (Walker, 1966) *MP*
Freshwater muds. Flandrian V to VIII. Inconclusive archaeology.

Elan Valley, Cardigans.   SN 8575 (Moore & Chater, 1969a; Moore, 1970) *MP*
Sequence from freshwater muds and silts to raised bog. Late Weichselian I to Flandrian VIII.

Elder Bush Cave, Derbys.   SK 1055 (Bramwell, 1964) *M*
Ipswichian travertine with *Hippopotamus*.

Elie, Fife   NO 4900 (Reid, 1899) *M*
Age uncertain: not entered.

Elland, Yorks.   SE 1222 (Bartley, 1964) *MP*
Lenses of organic matter in gravel and sand. Flandrian VI to VIIb.

Ellerside Moss, Lancs.   SD 3480 (Oldfield & Statham, 1963) *P*
Raised bog stratigraphy over marine clay. Flandrian VIIa to VIII.

Elstead, Surrey   SU 8942 (Seagrief & Godwin, 1960) *MP*
Sequence from freshwater muds and silts to *Sphagnum* peat. Flandrian IV to VIIa.

Elworthy, Somerset   ST 0835 (Helbaek, 1953b) *M*
Late Bronze Age.

Ely, Cambs.   TL 5480 (Godwin & Willis, 1961) *M*
Oak tree from base of Fen Clay: 4495 (Q-589).

Embleton's Bog, Northumberland   NU 1629 (Bartley, 1966) *P*
Raised bog associated with Bradford Kaim. Flandrian IV to VI.

Emlaghlea, Kerry   V 4565 (Jessen, 1949) *P*
Sequence from wood peat to raised bog. Flandrian IV to VIII.

Enbourne Valley, Bucks.   SU 4265 (unpublished) *P*
Age uncertain: not entered.

Endsleigh Street, London   TQ 3182 (Reid, 1899) *M*
Interglacial site of uncertain age: not entered.

Esthwaite Basin, Lancs.   SD 3694 (Franks & Pennington, 1961) *MP*
Lake muds. Late Weichselian II to Flandrian VII.

Ewe Crag Slack, Yorks.   NZ 6989 (Erdtman, 1928) *MP*
Valley peat. Post-boreal.

Exe River Channel, Devon   SX 97 (Clarke, 1970) *P*
Intertidal muds, submerged extension of river channel. Flandrian IV to VI.

Fairwood, Glam.   SS 5989 (Williams, 1945) *M*
Bronze Age.

Falcarragh, Donegal   B 1943 (Mitchell, 1951) *P*
Raised bog stratigraphy. Flandrian VIIb.

Fallahogy Td., Derry   C 90 (Smith & Willis, 1961–2) *P*
Sequence from freshwater muds and silts to *Sphagnum* bog. Flandrian V to VIII. Dated Neolithic clearance phases (Q-555–558, Q-653, 4).

Fargo Plantation, Wilts. SU 1141 (Stone, 1938) *M*
Early Bronze Age.

Farnham, Surrey SU 8446 (Oakley, Rankine & Lowther, 1939) *M*
Raft of peat in terrace gravels. Wolstonian.

Faskine, Lanarks. NS 7665 (Reid, 1899) *M*
Age uncertain: not entered.

Fenns Moss, Salop SJ 4835 (Godwin & Willis, 1960; Tallis & Birks, 1965) *M*
Raised bog. Flandrian VIIb and VIII. Carbon dated recurrence surfaces (Q-392, 393, 434, 435).

Ffridd Faldwyn, Montgomery SO 2296 (St J. O'Neil, 1943) *M*
Early Iron Age.

Fifield Bavent, Wilts. SU 0125 (Jessen & Helbaek, 1944; Helbaek, 1953b) *M*
Early Iron Age.

Fillyside, Edinburgh NT 27 (Reid, 1899) *M*
Age uncertain: not entered.

Findon, Sussex TQ 1209 (Pull, 1953) *M*
Neolithic and Iron Age.

Finsbury Circus, London TQ 3282 (Reid, 1921) *M*
Roman.

Fir Bog, Angus NO 4189 (Durno, 1961) *P*
Raised bog stratigraphy. Flandrian VI to VIII.

Flanders Moss, Stirlings. NS 5595 (Turner, 1964) *P*
Sphagnum bog over Carse Clay. Flandrian VIIa to VIII. Clearance phase 1858 (Q-570) to 1755 (Q-571).

Flaxmere, Cheshire SJ 5572 (Tallis & Birks, 1965) *MP*
Sphagnum bog. Flandrian VIII.

Flixton, Yorks. TA 0381 (Clark, 1954) *MP*
Sequence from freshwater muds to fen peat. Late Weichselian I, II, 10413 (Q-66), to Flandrian VII. Protomaglemosian.

Flows of Dergoals, Wigtown NX 2459 (Erdtman, 1928) *P*
Age uncertain: not entered.

Ford, Northumberland NT 9437 (Jessen & Helbaek, 1944) *M*
Early and Middle Bronze Age.

Forth/Clyde Canal, Stirlings. (Jessen & Helbaek, 1944) *M*
Roman.

Foulshaw Moss, Westmorland SD 4682 (Smith, 1959a), *MP*
Estuarine clay overlain by fen and *Sphagnum* peat. Flandrian VIIa to VIII.

Four Ashes, Cheshire SJ 9108 (Bell, 1968) *M*
Organic deposits in terrace gravels. Middle Weichselian.

Freshwater West, Pembrokes. SR 8999 (Wainwright, 1963) *MP*
Submerged forest bed. Flandrian VIIa or VIIb: 5960 (Q-530). Mesolithic.

Frogholt, Kent TR 1737 (Godwin, 1962) *MP*
Fluviatile peat. Flandrian VIIb and VIII: 2490 (Q-348), 2640 (Q-349), 2980 (Q-354).

Fugla Ness, Shetland HU 3191 (H. J. B. Birks & M. E. Ransom, 1969) *MP*
Peat bed exposed in cliff face. Late Hoxnian.

Galway's Bridge, Kerry V 9179 (Jessen, 1949) *MP*
Fen and wood peat. Flandrian VII and VIII.

Gamrie, Banffshire NJ 7962 (Jessen & Helbaek, 1944) *M*
Early Bronze Age.

Garral Hill, Banffshire NJ 4445 (Donner, 1956–7; Godwin & Willis, 1959a) *MP*
Late Weichselian stratigraphy I, II and III. 11888 to 10808 (Q-100 to 104).

Garrowby Wold, Yorks. SE 7957 (Jessen & Helbaek, 1944) *M*
Early Bronze Age.

Garscadden Mains, Renfrew. NS 5371 (Mitchell, 1952; Donner, 1956–7) *P*
Varved clays and muds of Scottish Late-glacial sea. Late Weichselian I to III (see *Garvel Park*).

Gartmore, Stirlings. NS 5097 (Donner, 1956–7) *MP*
Freshwater muds, Flandrian IV to VIIa.

Garton Slack, Yorks. SE 9859 (Jessen & Helbaek, 1944) *M*
Early and Middle Bronze Age.

Garvel Park, Greenock, Renfrewshire NS 2975 (Bishop & Dickson, 1970) *M*
Marine clays of the Clyde Beds. Late Weichselian. 13020 to 9231 (Birm-3, 120–122).

Gayfield, Edinburgh NT 27 (Reid, 1899) *M*
Peat deposit. Late Weichselian.

Gilling, Yorks. NZ 1805 (Jessen & Helbaek, 1944) *M*
Early Bronze Age.

Girton College, Cambridge TL 4658 (Jessen & Helbaek, 1944) *M*
Anglo-Saxon.

Girvan, Ayrs. NX 1998 (Godwin & Willis, 1962) *M*
Silty sand beneath 25 foot raised beach: 9020 (Q-640).

Glannalappa East Td., Kerry R 0940 (Jessen, 1949) *MP*
Raised bog stratigraphy. Flandrian VIIb.

Glasson Moss, Cumberland NY 2360 (Walker, 1966) *MP*
Fen and *Sphagnum* peat on marine clay. Flandrian VIIa.

Glasson Shore, Cumberland NY 2660 (Walker, 1966) *MP*
Freshwater muds beneath warp deposits. Late Weichselian III to Flandrian V.

Glastonbury Lake Village, Somerset ST 4940 (Godwin, 1956) *MP*
Sequence from freshwater muds to fen carr. Flandrian VIIa to VIII. Iron Age Village.

Glenballoch, Perthshire NN 3119 (Jessen & Helbaek, 1944) *M*
Late Bronze Age.

Glenballyemon, Antrim D 1928 (Jessen, 1949) *P*
Sequence of wood, fen and blanket peat. Flandrian VI to VIII.

Glenluce Sands, Wigtowns. NX 1957 (Jessen & Helbaek, 1944) *M*
Bronze Age.

Gloucester, Glos. SO 8318 (Medland, 1894–5) *M*
Roman.

Goathland, Yorks. NZ 6090 (Erdtman, 1928) *P*
Valley peat. Flandrian VI and VIIA.

Goats Water, Lancs. SD 2697 (Pennington, 1964) *MP*
Freshwater muds. Late Weichselian III to Flandrian VIIb.

Godmanchester, Hunts. TL 2470 (C. A. Dickson, unpub.) *MP*
Roman.

Goodland, Td., Antrim D 3044 (Case, Dimbleby, Mitchell, Morrison & Proudfoot, 1969) *MP*
Raised bog stratigraphy. Flandrian VIIb and VIII. Clearance phases.

Goodmanham, Yorks. SE 8842 (Jessen & Helbaek, 1944) *M*
Early and Late Bronze Age.

Goole, Yorks. SE 7423 (A. G. Smith, 1958a) *P*
Peat. Flandrian VIII.

Gordano Valley, Somerset ST 4372 (Jefferies, Willis & Yemm, 1968) *MP*
Sequence from freshwater mud to fen peat. Late Weichselian III to Flandrian VIIb. Marine transgression in VI.

Gorey, Wexford T 1659 (Jessen & Helbaek, 1944) *M*
Late Bronze Age.

Gorseald, Flints. SJ 1476 (Williams, 1921) *M*
Bronze Age.

Gorsey Bigbury, Somerset SX 6847 (Helbaek, 1953b) *M*
Early Bronze Age.

Gort, Galway M 4502 (Jessen, Anderson & Farrington, 1959) *MP*
Organic deposit beneath gravel and clay of subsequent glaciation. Late Anglian to Hoxnian IV.

Gortagenerick, Cork V 6547 (Mitchell, 1951) *MP*
Raised bog stratigraphy. Flandrian VIIb and VIII.

Gortalecka, Clare R 2595 (Watts, 1963) *P*
Freshwater muds and clays. Late Weichselian I to III.

Gosport, Hants. SU 6101 (Godwin, 1945b) *P*
Submerged peat. Flandrian IV.

Gough's Cave, Somerset ST 4653 (Campbell *et al.*, 1970) *P*
Upper Palaeolithic.

Goyle Hill, Aberdeens. NO 6882 (Durno, 1959) *P*
Raised bog stratigraphy. Flandrian VI to VIII.

Goyt Moss, Staffs. SK 0071 (Tallis, 1964) *P*
Raised bog stratigraphy. Flandrian VIII.

Graig Llywd, Caernarvons. SH 7278 (Warren, 1921) *M*
Neolithic axe factory.

Graveney, Kent TR 0562 (Greenhill, 1971; Evans & Fenwick, 1971) *P*
Mediaeval boat, 1080 (BM-660).

Grays, Essex TQ 6278 (West, 1969a) *MP*
Silty deposits in brickearth. Ipswichian.

Great Bedwyn, Wilts. SU 2764 (Allison & Godwin, 1949) *M*
Middle Bronze Age.

Great Billing, Northants. SP 6182 (Bell, 1969) *M*
Organic deposits in terrace gravels. Middle Weichselian 28 225 (Birm-75).

Great Wymondley, Herts. T 2120 (Westell, 1937) *M*
Roman.

Greenhills, Dublin O 1030 (Jessen & Helbaek, 1944) *M*
Late Bronze Age.

Greenside Tarn, Westmorland (D. Walker, unpub.) *MP*
Late Weichselian.

Griffinrath Td., Kildare N 9735 (Mitchell, 1951) *P*
Chalk mud with *Megaceros giganteus*. Late Weichselian I to III.

Grillagh, Longford N 0870 (Mtchell, 1951) *P*
Raised bog with trackway. Flandrian VIIA to VIII.

Grindle Top, Pickering, Yorks. SE 7983 (Jessen & Helbaek, 1944) *M*
Middle Bronze Age.

Hackness, Yorks. SE 9494 (Dimbleby, 1952) *P*
Bronze Age.

Hackney Wick, Essex TQ 39 (Warren, 1916) *M*
Lea Valley Arctic Plant Bed. Middle Weichselian.

Hailes, Edinburgh NT 27 (Reid, 1899) *M*
Late Weichselian and Flandrian deposits. Flandrian not entered.

Hailes, Glos. SP 0530 (Clifford, 1944) *M*
Early Iron Age.

Haldon Hill, Devon SX 9885 (Clark, 1938a) *M*
Neolithic.

Halstead, Essex TL 8130 (Helbaek, 1953b) *M*
Roman.

Ham Hill, Somerset ST 2825 (St George Gray, 1926) *MP*
Early Iron Age and Roman.

Handley Down, Wilts. ST 9620 (Helbaek, 1953b) *M*
Early and Middle Bronze Age.

Handley Wood, Dorset ST 9917 (Helbaek, 1953b) *M*
Middle Bronze Age.

Happisburgh, Norfolk TG 3832 (Reid, 1899) *M*
Cromer Forest Bed series.

Harrow Hill, Sussex TQ 0810 (Holleyman, 1937) *M*
Neolithic.

Hartford, Hunts. TL 2572 (Godwin, 1959) *MP*
Seams of peat in terrace gravel. Late Weichselian.

Hartlepool, Durham NZ 5032 (Trechmann, 1947–9) *M*
Submerged peat, age uncertain: not entered.

Haslingfield, Cambs. TL 4052 (Jessen & Helbaek, 1944) *M*
Anglo-Saxon.

Hassocks, Sussex TQ 3015 (Jessen & Helbaek, 1944; Helbaek, 1953b) *M*
Late Bronze Age and Roman.

Hatfield, Herts. TL 2107 (Sparks, West, Williams & Ransom, 1969) *MP*
Organic deposit in terrace gravels. Late Anglian to Hoxnian IIIb.

Hatfield Moors, Yorks. SE 7105 (Smith, 1958a) *P*
Raised bog stratigraphy. Flandrian VIIA to VIII. 2215 to 1381 (Q-483 to 487).

Haugh Head, Northumberland NT 9928 (Collingwood & Cowen, 1948) *M*
Middle Bronze Age.

Hawes Water, Lancs. SD 4776 (Oldfield, 1960) *MP*
Freshwater muds. Basal sequence clay, organic mud, clay. Late Weichselian I to VIIb.

Hawks Hill, Surrey TQ 1655 (Smith, 1907) *M*
Anglo-Saxon.

Hawks Tor, Cornwall SX 1474 (Conolly, Godwin & Megaw, 1950) *MP*
Late Weichselian stratigraphy, I to III, 11071 (Q-2). Hiatus and Flandrian peat VI to VIII, 8450 to 7770 (Q-1, C-340).

Hayes Common, Kent TQ 3866 (Hogg, St J. O'Neil & Stevens, 1941) *M*
Neolithic.

Hazard Hill, Devon SX 7559 (Fox, 1951) *M*
Late Bronze Age.

Hedge Lane, Essex TQ 3392 (Warren, 1916) *M*
Lea Valley Arctic Plant Bed. Middle Weichselian.

Hedney's Bottom, Ranworth Marsh, Norfolk TG 3514 (Jennings, 1952) *MP*
Fen peat over upper marine clay. Flandrian VIII.

Hellsevean, Cornwall SW 5140 (Jessen & Helbaek, 1944) *M*
Early Historic.

Helsington Moss, Westmorland SD 4689 (Smith, 1959a) *MP*
Sequence of estuarine clay, fen, raised bog. Flandrian VIIa to VIII.

Helton Tarn, Westmorland SD 4184 (A. G. Smith, 1959a) *MP*
Freshwater muds. Basal sequence, clay, clay and mud, clay. Late Weichselian I, II, 10760 (Q-92), to Flandrian VIII. Marine transgression to VIIa.

Hembury Fort, Devon ST 1104 (Helbaek, 1953b) *M*
Neolithic and Early Iron Age.

Heysham, Lancs. SD 4061 (Moseley & Walker, 1952) *MP*
Submerged forest bed. Flandrian VIIa.

Highfield, Wilts. SU 1429 (Helbaek, 1953b) *M*
Early Iron Age.

High Penard, Glam. SS 5689 (Williams, 1941) *M*
Roman.

Hightown, Lancs. SD 2903 (Travis, 1926) *MP*
Shore peats. Flandrian VIIb or VIII.

High White Stones, Cumberland NY 2809 (Raistrick & Blackburn, 1932) *P*
Blanket peat. Flandrian VIIa.

Histon Road, Cambridge TL 4461 (Sparks & West, 1959) *MP*
Organic shelly muds and silts in terrace gravels. Ipswichian IIb to IV.

Hitcham, Bucks. SU 9282 (Jessen & Helbaek, 1944) *M*
Early Bronze Age.

Hitchin, Herts. TL 2029 (Reid, 1897, 1901) *MP*
Loam and shell marl in brickearth. Mammals and Palaeolithic implements. Hoxnian. Pollen analyses by R. G. West. Hoxnian IIc.

Hockham Mere, Norfolk TL 9393 (Godwin & Tallantire, 1951; Sims, 1973) *MP*
Drained lake basin. Late Weichselian II to Flandrian VIII.

Holcroft Moss, Lancs. SJ 6793 (H. J. B. Birks, 1965b) *MP*
Sequence from freshwater muds and silts to *Sphagnum* peat. Flandrian VIIa to VIII. Clearance phases.

Holdenhurst, Hants. SZ 1295 (Piggott, 1937) *M*
Neolithic and Early Bronze Age.

Hollingbury Camp, Sussex TQ 3108 (Curwen, 1931) *M*
Early Iron Age.

Holme Fen, Hunts. TL 2289 (Mittre, 1959; Turner, 1962) *P*
Fen peat to *Sphagnum* bog. Flandrian VIIb and VIII. 4190 to 3400 (Q-403 to 406). Clearance phases.

Holmpton, Yorks. TA 3623 (Reid, 1899) *M*
Loam-filled hollow in boulder clay. Late Weichselian, *fide* H.G.

Holt, Denbighs. SJ 4053 (Jessen & Helbaek, 1944) *M*
Middle Bronze Age.

Holywell Row, Mildenhall, Cambs. TL 6678 (Jessen & Helbaek, 1944) *M*
Anglo-Saxon.

Honiton, Devon ST 1601 (Shaw, 1933b) *M*
Middle Bronze Age.

Hornsea, Yorks. TA 2047 (Reid, 1899) *M*
Peats probably Late Weichselian and Flandrian. Flandrian not entered.

Hownam Rings, Roxburgh NT 7921 (Piggott, 1947–8) *M*
Third century A.D., hut.

Hoxne, Suffolk TM 1776 (West, 1956) *MP*
Lake muds between Anglian and Wolstonian till, Hoxnian I to IV. Silts, drift-muds and brecciated clays, Late Anglian and Early Wolstonian.

Hullbridge, Essex TQ 8194 (Reader, 1910–11) *M*
Neolithic.

Hungate, Yorks. SE 5951 (Godwin & Bachem, 1962) *M*
Roman, Anglo-Saxon, Norman and Mediaeval.

Hungry Bentley, Derbys. SK 1850 (Jessen & Helbaek, 1944) *M*
Middle Bronze Age.

Hunsbury, Northants. SP 7458 (Percival, 1934) *M*
Early Iron Age.

Huntspill, Somerset ST 3742 (Godwin, 1943) *MP*
Fen- and raised-bog peat between lower and upper estuarine clay. Flandrian VIIb to VIII. Roman at peat surface.

Hurdle Drove, Suffolk TL 6776 (Jessen & Helbaek, 1944) *M*
Bronze Age.

Hurn, Hants. SZ 1296 (Piggott, 1943b) *M*
Bronze Age.

Hurst Fen, Suffolk TL 7276 (Clark, 1960) *M*
Neolithic.

Hutton Bushel, Yorks. SE 9784 (Jessen & Helbaek, 1944) *M*
Bronze Age.

Hutton Henry, Durham NZ 4236 (Beaumont, Turner & Ward, 1969) *MP*
Peat beneath glacial till. Ipswichian IIb to Early Weichselian.

Ibsley, Hants. SU 1509 (Helbaek, 1953*b*) *M*
Middle Bronze Age.

Ickornshaw Moor, Yorks. SD 9540 (Godwin & Willis, 1964) *M*
Mesolithic: 8100 (Q-707).

Ilford, Essex TQ 4586 (West, Lambert & Sparks, 1964) *MP*
Organic deposits in brickearth. Late Wolstonian to Ipswichian IIb. Mollusca.

Immingham, Lincs. TA 0217 (Godwin & Willis, 1961) *M*
Submerged peat in estuarine clay. 6681 (Q-401).

Ingoldmells, Lincs. TF 5769 (Godwin & Clifford, 1938; Smith, 1958*a*) *P*
Two coastal peats separated by salt-marsh. Flandrian VIIb to VIII. Top of upper peat 2630 (Q-687, 8), part of Chapel Point series.

Ipswich, Suffolk TM 1744 (Reid Moir, 1917, 1920) *M*
Neolithic.

Irstead, Norfolk TG 3620 (J. Allison, unpub.) *P*
Flandrian VII.

Island Magee, Antrim D 4404 (Jessen, 1949) *P*
Estuarine clay, Flandrian VIIa.

Isleham Fen, Cambs. TL 6475 (Godwin, 1941) *P*
Fen peat, Flandrian VIIb.

Itford Beddingham, Sussex TQ 4408 (Jessen & Helbaek, 1955) *M*
Middle Bronze Age.

Itford Hill, Sussex TQ 4405 (Helbaek, 1957) *M*
Late Bronze Age.

Jamestown, Dublin O 2121 (Jessen & Helbaek, 1944) *M*
Late Bronze Age.

Jenkinstown, Louth J 3130 (Mitchell, 1951) *MP*
Freshwater and estuarine deposits; end of main eustatic rise. Flandrian VI to VIIa.

Johnstown Td., Wicklow N 9808 (Mitchell, 1951) *MP*
Organic drift in terrace deposits. Late Weichselian.

Kelsey Hill, Yorks. TA 2326 (Reid, 1899) *M*
Age uncertain: not entered.

Kentmere, Westmorland NY 4502 (Walker, 1955*b*) *MP*
Lake deposits over clay, detritus gyttja, clay and pebbles. Late Weichselian I to Flandrian VIII.

Kentra Bay, Outer Hebrides NM 6468 (Poore & Robertson, 1949) *P*
Uncertain Late Flandrian age, not recorded.

Kerry Hill, Montgomeryshire SO 1386 (Daniel, Evans & Lewis, 1927) *M*
Bronze Age.

Kilbeg, Waterford S 4602 (Watts, 1959*b*) *MP*
Freshwater muds deeply buried. Hoxnian I to IV.

Kildromin, Limerick R 7242 (Watts, 1967) *M*
Interglacial deposit. Hoxnian I to III.

Killarney, Kerry V 9790 (Mitchell & Watts, 1970) *P*
Arbutus pollen in Flandrian VIIa.

Killerby Carr, Yorks. TA 0681 (Clark, 1954) *MP*
Basal deposits of calcareous clay mud and solifluxion sands and gravels. Late Weichselian I to III.

Killy Willy Lough, Cavan H 2917 (Jessen, 1949) *MP*
Freshwater muds. Flandrian VI and VIIa.

Kilmacoe Td., Wexford T 3122 (Mitchell, 1951) *MP*
Kettle hole deposit. Flandrian IV to VII.

Kilmacoe II, Wexford T 3122 (Mitchell, 1951) *MP*
Freshwater muds. Late Weichselian I to III.

Kilmaurs, Ayrs. NS 4141 (Reid, 1899) *M*
Interglacial deposit. Age uncertain: not entered.

Kilmoyle, Ballycastle, Antrim D 1141 (Jessen & Helbaek, 1944) *M*
Early historic.

Kilmoyly North Td., Kerry Q 7926 (Mitchell, 1951) *MP*
Raised bog stratigraphy. Flandrian VIIa to VIII.

Kilpaison Burrows, Pembrokes. SM 8900 (Jessen & Helbaek, 1944) *M*
Middle Bronze Age.

Kinder Scout, Cheshire SK 0987 (Conway, 1954) *P*
Blanket bog stratigraphy. Flandrian VIIa to VIII.

Kingsdown Camp, Mells, Somerset ST 7249 (Maby, 1930) *M*
Roman. Plant identifications by J. C. Maby.

Kinlochewe, Ross. NH 0262 (Durno & McVean, 1959) *MP*
Blanket bog stratigraphy. Flandrian VIII.

Kippen, Stirlings. NS 6494 (Newey, 1966) *MP*
Peat below Carse Clays of Forth. Flandrian V and VI.

Kirkby, Lancs. SD 2389 (Greswell, 1958) *P*
Peat over marine clay. Flandrian VIIa.

Kirkby Thore, Westmorland NY 6425 (Walker & Lambert, 1955) *MP*
Sequence detritus mud, *Sphagnum–Polytrichum* peat, detritus mud. Flandrian V to VIIa.

Kirkmichael, I.O.M. SC 3190 (Reid, 1899; C. A. & J. H. Dickson; Mitchell, 1970) *MP*
Organic muds and peats in Late Weichselian stratigraphy of cliffs. I, II, III, 12210, 11310, 10270 (GRO 1616, 1639, Q-673) respectively.

Kirmington, Lincs. TA 1011 (Reid, 1899; Watts, 1959*a*) *MP*
Estuarine peat and silt. Hoxnian III (West).

Knackyboy Cairn, Isles of Scilly SV 9215 (Helbaek, 1953*b*) *M*
Late Bronze Age.

Knockasarnet, Kerry V 9594 (Jessen, 1949) *MP*
Wood peat. Flandrian VIIb.

Knockiveagh, Down J 1938 (Smith, 1957) *P*
Pits beside Neolithic cairn. Flandrian VIIb.

Knocknacran, Monaghan H 2830 (Mitchell, 1951) *M*
Chalk muds and peats. Late Weichselian I to Flandrian V. Late Weichselian II, 11310 (C-355) with *Megaceros giganteus*.

Lackford, Suffolk TL 7970 (Jessen & Helbaek, 1944) *M*
Anglo-Saxon.

Lady Bridge Slack, Yorks. NZ 8001 (Simmons, 1969) *MP*
Fen and wood peat. Flandrian v to VIII.

Laidwinley, Angus NO 4768 (Durno, 1959) *P*
Blanket peat. Flandrian VIIa to VIII.

Lake Derravaragh, West Meath N 3969 (G. F. Mitchell, unpub.) *M*
Fen peat and Larnian artifacts. 5360 (I-4234).

Lambourn Downs, Berks. SU 3381 (Jessen & Helbaek, 1944; Helbaek, 1953b) *M*
Early Bronze Age.

Lancing, Sussex TQ 1706 (Jessen & Helbaek, 1944) *M*
Late Bronze Age.

Langdale Combe, Westmorland NY 2608 (Fell, 1954; Walker, 1965) *MP*
Kettle hole succession from open water to *Sphagnum* bog. Flandrian IV to VIIb. Neolithic axe factory, VIIb.

Largs, Ayrs. NS 2058 (Jessen & Helbaek, 1944) *M*
Late Bronze Age.

Larkfield, Kent TQ 7058 (Grove, 1963) *M*
Mediaeval.

Leasowe, Cheshire SJ 2390 (Travis, 1929) *M*
Upper and lower coastal peat beds. Flandrian VII or VIII.

Leigh Td., Tipperary S 2215 (Mitchell, 1956) *MP*
Sequence from freshwater muds and clays to *Sphagnum* peat. Basal Late Weichselian stratigraphy with *Megaceros*, and Flandrian VI to VIII.

Leman and Ower Bank, North Sea (Godwin, 1943) *P*
Peat dredged from sea floor. Flandrian v. 8422 (Q-105) Mesolithic.

Letterston, Pembrokes. SM 9429 (Savory, 1948) *M*
Middle Bronze Age.

Leuchars, Fifes. NO 4622 (Percival, 1934) *M*
Cinerary urn 'c. 1800 B.C.'

Lexden, Essex TL 9725 (Shotton, Sutcliffe & West, 1962) *MP*
Channel filling with megafauna. Age uncertain, not recorded.

Lidbury Camp, Wilts. SU 1653 (Cunnington, 1917) *M*
Early Iron Age.

Lindow Moss, Cheshire SJ 8281 (H. J. B. Birks, 1965b) *MP*
Stratigraphic sequence from open water muds to *Sphagnum* bog. Flandrian VIIa to VIII. Clearance phases.

Lintlaw, Berwicks. NT 8258 (Jessen & Helbaek, 1944) *M*
Late Bronze Age.

Linton Mires, Yorks. SD 9962 (Raistrick & Blackburn, 1938) *MP*
Succession from freshwater muds to *Phragmites* fen peat. Late Weichselian I to Flandrian VIIa.

Lion Point, Essex TM 1915 (H. & M. E. Godwin, 1936) *P*
Foreshore peat above Beaker occupation. Flandrian VIIb.

Lisnolan Td., Mayo M 2484 (Jessen, 1949) *MP*
Succession from freshwater muds and fen peat to *Sphagnum* peat. Late Weichselian III to Flandrian VIIb.

Lissue Td., Antrim J 3236 (Mitchell, 1951) *MP*
Ditch deposits of 9–10th century raths.

Little Downham, Cambs. TL 5284 (Jessen & Helbaek, 1944) *M*
Middle Bronze Age.

Little Lochans, Wigtowns. NX 0757 (Moar, 1969a) *MP*
Stratigraphic sequence from freshwater muds to raised bog peat. Late Weichselian I to Flandrian VIIa.

Littleport, Norfolk TL 5996 (Shawcross & Higgs, 1961) *MP*
Fen peat with *Bos primigenius*. Flandrian VIIb.

Little Solisbury, Wilts. ST 7464 (Jessen & Helbaek, 1944; Helbaek, 1953b) *M*
Early Iron Age.

Little Thetford (Barway), Cambs. TL 5376 (Godwin & Willis, 1959b) *M*
Piled Bronze Age track. 2560 (Q-310).

Littleton Bog, Tipperary S 1854 (Mitchell, 1965) *P*
Stratigraphic sequence from freshwater muds and silts to raised bog peat. Late Weichselian to Flandrian VIII. Clearances.

Little Welnetham, Suffolk TL 8860 (West, 1953) *M*
Hoxnian with *Azolla*.

Little Wilbraham, Cambs. TL 5458 (Jessen & Helbaek, 1944) *M*
Anglo-Saxon.

Llanarth, Mon. SO 3811 (O'Neil & Foster-Smith, 1936) *M*
Charcoal with 13th century pottery.

Llanbleddian, Glam. SS 9974 (Grimes, 1938) *M*
Bronze Age.

Llandow, Glam. SS 9473 (Fox, 1943b) *M*
Bronze Age.

Llangeitho, Cardigans. SN 6260 (Hyde, 1939) *MP*
Raised bog stratigraphy. Flandrian VIIb. Archaeology.

Llanllwch, Carmarthen. SN 3818 (Thomas, 1965) *P*
Stratigraphic sequence from open water to raised bog. Flandrian VI to VIII. Clearances. Recurrence surface 3230, 3178 (Q-458, 9).

Llantwit Major, Glam. SS 9668 (Fox, 1941) *M*
Middle Bronze Age.

Llwyn Dwythwch, Caernarvons. SH 5757 (Seddon, 1962) *MP*
Basal sequence, clay, clay-mud, clay. Late Weichselian I to III. Freshwater muds to *Sphagnum* peat. Flandrian IV to VIII.

Llyn Fawr, Glam. SM 9104 (C. Fox, 1937) *P*
Iron Age hoard, Flandrian VIIb/VIII.

Llyn Gynon, Cardigans. SN 8064 (Moore & Chater, 1969a, b) *MP*
Raised bog stratigraphy. Flandrian VI to VIII.

Llyn Teifi, Cardigans. SN 7667 (Godwin & Willis, 1960) *M*
Pine stump layer on mineral soil, overlain by blanket peat (445 m) 5261 (Q-394).

Loch Borralan, W. Ross NC 2610 (Pennington *et al.*, 1972) *P*
Freshwater muds and silts. Late Weichselian I to III.

Loch Cill Chriosd (Skye), Inner Hebrides NG 6020 (H. J. B. Birks, 1969) *MP*
Freshwater muds, clays and silts. Late Weichselian I to Flandrian IV; 9655 (Q-959).

Loch Clair, W. Ross NG 7771 (Pennington *et al.*, 1972) *P*
Freshwater muds and silts. Flandrian IV to VIII.

Loch Coir' a' Ghobhainn (Skye), Inner Hebrides NG 4118 (H. J. B. Birks, 1969) *MP*
Late Weichselian sequence, clay, clay-mud, clay and freshwater muds. Late Weichselian I to Flandrian IV; 10254 to 8650 (Q-955-8).

Loch Craggie, W. Ross NC 3205 (Pennington *et al.*, 1972) *P*
Late Weichselian pollen sequence, I to III.

Loch Creag, Perthshire NN 9044 (Donner, 1962) *P*
Freshwater muds. Flandrian IV to VIII.

Loch Cuithir, Inverness. NG 4759 (Y. & A. Vasari, 1968) *MP*
Freshwater muds and silts. Late Weichselian III to Flandrian VIII.

Loch Droma, Ross and Cromarty NH 2755 (Kirk & Godwin, 1962-3) *MP*
Late Weichselian silts and peat, I, II, III; 12810 (Q-457).

Loch Dungeon, Kirkcudbright NX 5284 (H. H. Birks, 1969) *MP*
(*a*) Freshwater muds and clays. Flandrian IV to VIII.
(*b*) Blanket bog stratigraphy. Flandrian IV to VIIb. Pine stump 7165 (Q-876).

Loch Einich, Inverness. NH 9100 (H. H. Birks, 1969) *MP*
Blanket bog stratigraphy. Flandrian V to VIIb. Pine and birch stumps, 5880, 4150 (Q-881, 883).

Loch Fada, Inverness NG 4948 (Y. & A. Vasari, 1968; H. J. B. Birks, 1969) *MP*
Freshwater muds and silts. Late Weichselian I to Flandrian VIII; 7500 (Q-961).

Loch Kinord, Aberdeens. NO 4399 (Y. & A. Vasari, 1968) *MP*
Freshwater muds and clays. Late Weichselian I to Flandrian VIII.

Loch Mahaick, Perthshire NN 7006 (Donner, 1958) *P*
Freshwater muds and clays. Late Weichselian I to Flandrian VIII.

Loch Maree, Ross and Cromarty NH 0064 (Durno & McVean, 1959; H. H. Birks, 1972) *MP*
(*a*) Regional peat deposits. Flandrian VI to VIII.
(*b*) Lake muds and clays. Flandrian IV to VIII. 9085 to 4206 (Q-1005 to 1010).

Loch Mealt, Skye, Inner Hebrides NG 5065 (H. J. B. Birks, 1969) *MP*
Freshwater muds and clays. Weichselian I to Flandrian IV. 10820 (Q-929).

Loch Moedal, Skye, Inner Hebrides NG 6511 (H. J. B. Birks, 1969) *MP*
Freshwater muds and clays. Weichselian I to Flandrian IV. 9482 (Q-960).

Loch Nan Cat, Perthshire NH 6442 (Donner, 1962) *P*
Freshwater muds and clays. Flandrian IV to VIII.

Loch Park, Aberdeens. NO 7798 (Y. & A. Vasari, 1968) *MP*
Freshwater muds and clays, typical Late-glacial sequence. Late Weichselian I to Flandrian VIIb.

Loch Scionascaig, W. Ross NC 1213 (Pennington *et al.*, 1972) *P*
Freshwater muds and silts. Flandrian IV to VIII. 7880 to 4020 (Y 2362 to 2364).

Loch Tarff, Inverness NH 4210 (Pennington *et al.*, 1972) *P*
Freshwater muds and clays. Late Weichselian to Flandrian IV.

Lockerbie, Dumfries. NY 1179 (Bishop, 1963) *P*
Detritus band in clays and silts. Late Weichselian I. 12940 (Q-643).

Londesborough, Yorks. SE 8645 (Jessen & Helbaek, 1944) *M*
Middle Bronze Age.

Longford Pass, Tipperary S 2561 (Mitchell, 1953) *M*
Typical Late Weichselian stratigraphy with *Megaceros*.

Longlee Moor, Northumberland NU 1519 (Bartley, 1966) *MP*
Stratigraphic sequence of freshwater muds, fen peat, *Sphagnum* peat. Late Weichselian to Flandrian VI.

Long Range, Kerry V 9383 (Jessen, 1949; Watts, 1963) *MP*
Stratigraphic sequence of freshwater muds, fen peat, raised bog. Late Weichselian I to III, Flandrian IV to VIII.

Longstone Edge, Derbys. SE 2171 (Jessen & Helbaek, 1944) *M*
Early Bronze Age.

Long Wittenham, Berks. SU 5594 (Jessen & Helbaek, 1944; Helbaek, 1953*b*) *M*
Middle Bronze Age.

Lopham Little Fen, Norfolk TM 0479 (Tallantire, 1953) *MP*
Stratigraphic sequence from freshwater muds and silts to fen peat. Late Weichselian I to Flandrian IV.

Lough Bannus, Donegal H 1067 (Erdtman, 1928) *MP*
Submerged wood peat with pine stumps resting on swamp peat and gyttja. Flandrian VI and VIIa.

Lough Cranstal, I.O.M. SD 4502 (G. F. Mitchell, unpub.) *M*
Estuarine clay. Flandrian VI to early VIIa.

Lough Faughan, Down J 4541 (Morrison, 1955*b*) *M*
Crannog, early Christian.

Lough Firrib, Wicklow T 0599 (Jessen, 1949) *MP*
720 and 495 m. Blanket bog over wood peat. Flandrian VIIa to VIII.

Lough Goller, Clare R 1396 (Watts, 1963) *P*
Freshwater muds and clays. Late Weichselian I to III.

Lough Gur, Limerick R 6541 (Mitchell, 1954) *MP*
Freshwater muds. Flandrian VI to VIII. Clearances.

Loughlougan, Antrim D 1808 (Jessen & Helbaek, 1944) *M*
Middle Bronze Age.

Low Wray Bay, Windermere, Westmorland NY 3701 (Pennington, 1963; Godwin, 1960*a*) *MP*
Freshwater muds and silts with triple Late Weichselian sequence I to III. Zone II 11878 (Q-284).

Ludham, Norfolk TG 3819 (West, 1961*b*) *P*
Pleistocene Crag deposits shew alternation of temperate and glacial vegetation, Ludhamian, Thurnian, Antian, Baventian, Pastonian.

Lunds, Yorks. SD 7895 (Walker, 1955c) *P*
Typical Late Weichselian sequence on drumlin slope. Late Weichselian II, III, 11950, 10600 (Q-57, 61). Hiatus to Flandrian.

Luxford Lake, Wareham, Dorset SY 9287 (Jessen & Helbaek, 1944; Helbaek, 1953b) *M*
Late Bronze Age.

Lyles Hill, Antrim J 3576 (Jessen & Helbaek, 1944) *M*
Late Bronze Age.

Lymington, Hants. SZ 3295 (Hawkes, 1936) *M*
Late Iron Age.

Lytheat's Drove, Somerset ST 4241 (Dewar & Godwin, 1963) *P*
Raised bog stratigraphy. Flandrian VIII. Bronze Age amber beads.

Mablethorpe, Lincs. TF 5085 (Godwin & Clifford, 1938) *MP*
Peat of coastal series with pine stools and oak trunk cut by metal axe.

Magdalen Bridge, Edinburgh NT 27 (Jessen & Helbaek, 1944) *M*
Late Bronze Age.

Magheralaghan, Down J 4443 (Singh, 1970) *MP*
Freshwater muds. Basal sequence clay, clay mud and marl, clay. Late Weichselian I to Flandrian V.

Maiden Castle, Dorset SY 6686 (Jessen & Helbaek, 1944; Helbaek, 1953b) *M*
Neolithic, Bronze Age and Iron Age.

Malham, Yorks. SD 8866 (M. E. & C. D. Pigott, 1959) *MP*
Sequence from clays and silts to raised bog. Late Weichselian I to Flandrian VIII.

Malton, Yorks. SE 7871 (Jessen & Helbaek, 1944) *M*
Roman.

Mapastown, Louth N 9891 (Mitchell, 1953) *MP*
Basin deposits in outwash gravels of last glaciation. Typical Late Weichselian stratigraphy, I to III.

March Roddon, Cambs. TL 4395 (Godwin & Clifford, 1938) *MP*
Fen clay and upper fen peat beneath Roman roddon. Flandrian VIIb and VIII.

Margam, Glam. SS 8087 (Godwin & Willis, 1961) *M*
Submerged peat bed. Dated 6184 to 3402 (Q-274, 275, 265).

Marks Tey, Essex TL 9124 (Turner, 1970) *MP*
Freshwater muds and clays exposed in brick pit. Late Anglian to Hoxnian IV. Long sequence of annual laminations.

Marlow, Bucks. SU 8686 (Bell, 1969) *MP*
Terrace deposit. Middle Weichselian.

Marnhull, Dorset ST 7718 (Williams, 1950) *M*
Iron Age and Roman.

Marros, Pembrokes. SN 2107 (Mitchell *et al.*, 1970) *MP*
Sandy silty muds beneath head deposits on raised beach. Ipswichian.

Martins Town, Dorset SY 6589 (Helbaek, 1953b) *M*
Early Bronze Age.

Maumbury Rings, Dorset (St G. Gray, 1913) *M*
Roman.

Maze Court, Down J 2262 (Jessen & Helbaek, 1944) *M*
Middle Bronze Age.

Meare Heath, Somerset ST 4541 (Godwin, 1941; Dewar & Godwin, 1963) *MP*
Sequence from fen to *Sphagnum* peat. Trackway at recurrence surface dated 2850, 2840 (Q-52). Yew bow (Neolithic) 4640 (Q-646).

Meare Lake Village ST 4442 (Godwin, 1941) *MP*
Sequence from freshwater muds to raised bog with Iron age hearths overlain by estuarine clay. Flandrian VIIb to VIII.

Meare Pool, Somerset ST 4442 (Godwin, 1956) *MP*
Lake deposits associated with raised bog complex. Flandrian VIIa to VIII.

Meltham Moor, Yorks. SE 0608 (Bartley, 1964) *P*
Blanket bog stratigraphy. Flandrian VIIa to VIII. Clearance phases.

Meon Hill, Hants. SU 5202 (Liddell, 1935) *M*
Iron Age.

Mersea Island, Essex TM 0314 (Warren, 1914) *M*
Roman.

Merseyside, Lancs. SJ 39 (Travis, 1913) *M*
Age uncertain: not entered.

Methwold Fen, Norfolk TL 6696 (Godwin & Clifford, 1938) *P*
Fen margin peat associated with marine transgression. Flandrian VI to VIIb. Middle Bronze Age.

Mickleden, Westmorland NY 2606 (Walker, 1965) *P*
Sequence from freshwater muds to *Sphagnum* peat in kettle hole. Flandrian IV to VIIb.

Mildenhall Fen, Cambs. TL 6678 (Godwin & Clifford, 1938; Godwin, 1941; Jessen & Helbaek, 1944) *MP*
Late Bronze Age occupation level in fen peat. Flandrian VIIb and VIII.

Mileens Td., Kerry V 97 (Mitchell, 1951) *P*
Peat with Bronze Age trough. Flandrian VIIb.

Milmorane Td., Cork W 2772 (Mitchell, 1951) *P*
Blanket bog. Flandrian VIIb and VIII.

Minnis Bay, Kent TR 2769 (Conolly, 1941) *M*
Late Bronze Age.

Mockerkin Tarn, Cumberland NY 0823 (Pennington, 1965) *P*
Freshwater muds and silts. Flandrian VIIa to VIII. Extensive clearance.

Moel Hebog, Caernarvons. SH 5747 (Jessen & Helbaek, 1944) *M*
Early Bronze Age.

Monelpie Moss, Aberdeens. NO 2783 (Durno, 1959) *P*
Raised bog stratigraphy. Post-boreal.

Moneystown, Wicklow T 1999 (Mitchell, 1951) *MP*
Sequence from fen to *Sphagnum* peat. Flandrian VI to VIII.

Monzie, Perthshire NN 8725 (A. Young & M. C. Mitchell, 1939) *M*
Iron Age.

Moorthwaite Moss, Cumberland NY 5151 (Walker, 1966) *MP*
Sequence from freshwater muds to *Sphagnum* peat. Late Weichselian I to Flandrian VIII.

Morar, Inverness. NM 6892 (Lacaille, 1951) *P*
Post-boreal peat with stone tools.

Moreton Morrell, Warwicks. SP 3056 (Shotton, 1967) *P*
Peat, Flandrian VI to VIIa.

Moss Lake, Lancs. SJ 3591 (Godwin, 1960*b*) *MP*
Raised bog over lake deposits with characteristic Late Weichselian stratigraphy. Weichselian I to Flandrian VIII. Erroneous carbon dates (Q-217, 218, 220, 221).

Moss Maud, Aberdeens. NO 6299 (Durno, 1961) *P*
Flandrian VI to VIII.

Moss of Cree, Wigtowns. NX 4361 (Moar, 1969*a*) *MP*
Raised bog stratigraphy. Flandrian VIIb and VIII.

Moss Swang, Yorks. NZ 8003 (Simmons, 1969) *MP*
Sequence from wood to sedge fen peat. Flandrian V to VIIb. Clearance phases.

Mother Grundy's Parlour, Derbys. SK 5274 (Godwin & Willis, 1962) *M*
Charcoals associated with Creswellian artifacts dated 8800 to 6705 (Q-551 to 554).

Mounts Bay, Cornwall SW 5128 (Reid & Flett, 1907) *M*
Coastal peat, possibly Neolithic.

Moville, Donegal C 6137 (Mitchell, 1951) *M*
Flandrian VI to VIIb.

Muckle Cairn, Angus NO 3776 (Durno, 1959) *P*
Blanket bog stratigraphy. Flandrian VIIa to VIII.

Muckle Moss, Northumberland NY 8169 (Pearson, 1960) *P*
Sequence from freshwater muds to raised bog. Flandrian V to VIII. Clearance phases.

Mullaghmore Td., Down J 24 (Mitchell, 1953) *M*
Late Bronze Age to Early Iron Age.

Mullaghnashmount, Londonderry C 6308 (Jessen & Helbaek, 1944) *M*
Middle Bronze Age.

Mundesley, Norfolk TG 3136 (Reid, 1899; Duigan, 1963) *MP*
(*a*) Cromer Forest Bed series.
(*b*) Cromerian, *fide* West.

Mundesley River Bed, Norfolk TG 3136 (Reid, 1899) *MP*
Ispwichian, *fide* Miss Linda Phillips.

Munhin Bridge, Mayo F 8323 (Jessen, 1949) *P*
Sequence from freshwater muds to raised bog. Flandrian IV, VIIb and VIII.

Musselburgh, Edinburgh NT 3472 (Jessen & Helbaek, 1944) *M*
Middle and Late Bronze Age.

Myndd Epynt, Brecons. SN 94 (Dunning, 1943) *M*
Middle Bronze Age.

Myndd Llysworney, Glam. SS 9675 (Hyde, 1950–2) *M*
Iron Age.

Myndd Rhiw, Caernarvons. SH 2329 (Houlder, 1961) *M*
Neolithic axe factory. Erroneous date (Q-387).

Nairn, Inverness. NH 8656 (Knox, 1954) *P*
Band of peat with Mesolithic implements. Flandrian VI and VIIa.

Nant Ffrancon, Caernarvons. SH 6363 (Seddon, 1962; C. Burrows, unpub.) *MP*
Lake deposits. Late Weichselian I to Flandrian VIII.

Nar Valley, Norfolk TF 6914 (Stevens, 1960) *P*
Lacustrine silts and estuarine clays betweeen tills. Late Anglian to Hoxnian III.

Nazeing, Herts. TL 3807 (Allison, Godwin & Warren, 1952) *MP*
Organic muds, marl and peat above terrace gravels of Lea Valley. Late Weichselian to Flandrian VIIa. Molluscs and vertebrates.

Nechells, Warwicks. SP 0687 (Kelly, 1964) *MP*
Interglacial silts and muds. Late Anglian to Hoxnian IIIb.

Needham Market, Cambs. TM 0855 (Jessen & Helbaek, 1944) *M*
Early Bronze Age.

Netherley Moss, Kincardineshire NO 8694 (Durno, 1961) *P*
Flandrian VI to VIII.

New Barn Down, Sussex TQ 1302 (Curwen, 1934*b*) *M*
Neolithic.

Newbold, Yorks. SK 3773 (Jessen & Helbaek, 1944) *M*
Middle Bronze Age.

Newferry, Antrim H 9997 (Smith, 1961*b*) *MP*
Sequence of fen peat, muds and diatomite with Neolithic (Bann flake) industry and hearths. 5290 (B-36). Flandrian V–VIII.

Newlands Longside, Renfrews. NF 5865 (Jessen & Helbaek, 1944) *M*
Late Bronze Age.

Newmark, Yorks. TA 0929 (Jessen & Helbaek, 1944) *M*
Early historic.

Newstead, Roxburghs. NT 5634 (Tagg, 1911) *M*
Roman.

Newtown, Waterford S 2110 (Watts, 1959*b*) *MP*
Peat beneath boulder clay. Late Hoxnian.

Newtownbabe Td., Louth J 3030 (Mitchell, 1951) *MP*
Fen peat resting on freshwater muds with *Megaceros* and solifluxion deposits. Late Weichselian II and Flandrian V to VIIa.

Nicholaston, Glam. SS 5189 (Williams, 1940*a*) *M*
Megalithic tomb.

Nichols Moss, Westmorland SD 4282 (Smith, 1959*a*) *P*
Estuarine clay below fen and raised bog peat. Flandrian VIIa to VIII. Clearance phases.

Nordelph, Cambs. TF 5500 (Godwin & Clifford, 1938) *MP*
Fen clay and upper fen peat beneath Roman roddon. Flandrian VIIb and VIIb/VIII.

North Berwick, East Lothian NT 5485 (Jessen & Helbaek, 1944) *M*
Early Bronze Age.

North Ferriby, Yorks. SE 9927 (E. V. & C. W. Wright, 1947; Wright & Churchill, 1965) *MP*
Three sewn boats embedded in estuarine clay overlying fen peat. 3500 (Q-837), 3120 (Q-715), 2700 (BM-58).

Northfleet, Kent TQ 6275 (Burchell & Piggott, 1939) *P*
Mesolithic or Neolithic. Flandrian VIIa or VIIb.

North Gill, Yorks. NZ 7200 (Simmons, 1969) *MP*
Blanket bog stratigraphy. Flandrian VIIa to VIII. Clearance phases.

North Leigh, Oxon. SP 3813 (Morrison, 1955a) *M*
Roman.

North Sea Moorlog (Godwin, 1943) *MP*
Submerged peat. Flandrian V or VI (see also *Dogger Bank*).

Notgrove, Glos. SP 1121 (Clifford, 1937) *M*
Neolithic.

Nursling, Hants. SU 3515 (Seagrief, 1959) *MP*
Sequence from freshwater muds to wood peat. Late Weichselian III to Flandrian VI.

Nuthamstead, Herts. TL 4134 (Williams, 1946) *M*
A.D. 1300.

Nympsfield, Glos. SO 8001 (Clifford, 1938) *M*
Neolithic.

Oakhanger, Hants. SU 7735 (W. F. & W. M. Rankine & Dimbleby, 1960) *MP*
Mesolithic chipping floor dated 6300 (Zeuner, 1955). Flandrian VIIa.

Oakmere, Cheshire SJ 5677 (J. Tallis & H. J. B. Birks, 1965), *M*
Sphagnum peat. Flandrian VIIb.

Oban 1, Argyll. NM 8727 (Donner, 1956–7) *MP*
Freshwater muds. Flandrian IV to VIIb.

Oban 2, Argyll. NM 8529 (Donner, 1956–7) *MP*
Sequence from freshwater muds to *Carex* peat. Flandrian IV to VIIa.

Oban 3, Argyll. NM 9029 (Donner, 1956–7) *MP*
Freshwater muds. Flandrian IV to VIIb.

Oban 4, Argyll. NM 9127 (Donner, 1956–7) *MP*
Freshwater muds resting on gravel. Flandrian V to VIIb.

Oddington, Oxon. SP 5514 (Jessen & Helbaek, 1944) *M*
Middle Bronze Age.

Okement Hill, Devon SX 6087 (Simmons, 1964) *P*
Stratigraphy not reported. Flandrian VIIa to VIII.

Old Buckenham Mere, Norfolk TM 0492 (Godwin, 1968b) *P*
Freshwater clays and muds. Late Weichselian to Flandrian VIII. Clearance phases.

Old Croft River, Cambs. TL 5294 (Godwin & Clifford, 1938) *P*
Roman roddon. Flandrian VIII.

Old Keig, Aberdeens. NJ 6119 (Childe, 1934) *M*
Late Bronze Age.

Old Kilpatrick, Lanarks. NS 4572 (Jessen & Helbaek, 1944) *M*
Roman.

Oldtown, Kilcashel, Roscommon M 39 (Mitchell, 1956) *MP*
Raised bog stratigraphy. Flandrian VIIb and VIII.

Orwell Estuary, Suffolk TM 2238 (Reid Moir, 1930) *P*
Age uncertain: not entered.

Ostend, Norfolk TG 3632 (Reid, 1899) *M*
Cromer Forest Bed series.

Oulton Moss, Cumberland NY 2551 (Walker, 1966) *MP*
Sequence from freshwater muds to *Sphagnum* peat. Late Weichselian II to Flandrian VIIb. Clearance phases.

Outerston Hill, Edinburgh NT 27 (Jessen & Helbaek, 1944) *M*
Late Bronze Age.

Overstrand, Norfolk TG 2441 (Reid, 1899) *M*
Cromer Forest Bed series.

Overton Down, Wilts. SU 1368 (Gose & Sandell, 1964) *M*
Bronze Age.

Overtown, nr Beith, Ayrs. NS 3454 (Reid, 1899) *M*
Age uncertain: not entered.

Oxbow, Yorks. SE 33 (Gaunt, Coope & Franks, 1970) *MP*
Sands, silts and gravels below flood plain of River Aire. Middle Weichselian, 38600 (NPL 163B) and Flandrian VI and VIIb/VIII, 4280 (St-3057).

Painsthorpe Wold, Yorks. SE 8257 (Jessen & Helbaek, 1944) *M*
Early Bronze Age.

Pakefield, Suffolk TM 4390 (C. & E. M. Reid, 1914) *M*
Cromer Forest Bed series.

Papcastle, Cumberland NY 1131 (Jessen & Helbaek, 1944) *M*
Middle Bronze Age.

Parkmore, Antrim D 2021 (Morrison, 1959) *P*
Stratigraphical succession from fen to *Sphagnum* peat. Flandrian VI to VIII. Clearance phases.

Parkstone, Dorset SZ 0491 (Reid, 1899) *M*
Age uncertain: not entered.

Parsons Park, Cornwall SX 1971 (Conolly, Godwin & Megaw, 1950) *P*
Sequence from freshwater muds to *Sphagnum* peat. Flandrian IV to VII.

Peacock Farm, Somerset ST 4442 (Godwin, 1956) *MP*
Stratigraphy at margin of Meare Pool. Flandrian VIIa to VIII.

Peacock's Farm, Cambs. TL 6484 (Godwin & Clifford, 1938) *MP*
Sequence of lower fen peat with Mesolithic and Neolithic, Fen Clay, upper peat (Early Bronze Age) and estuarine roddon silt (Roman). Flandrian VI to VIIb. (See *Shippea Hill*.)

Pear Tree Hill, Norfolk TF 4302 (Godwin, 1941) *MP*
Sequence of Fen Clay, upper fen peat and silts with traces of Roman occupation. Flandrian VIIb and VIII.

Pedngwinian Point, Cornwall SX 6520 (Helbaek, 1953b) *M*
Middle Bronze Age.

Peel Hill, Lanarks. NS 7044 (Durno, 1965) *P*
Flandrian VIII.

Pennine Peat, Yorks. (Burrell, 1924) *M*
Blanket peat. Flandrian vIIb or vIII.

Penrhos Lligwy, Anglesey SH 4886 (Baynes, 1920) *M*
Roman.

Pentre-ifan, Pembrokes. SM 7525 (Grimes, 1948) *M*
Bronze Age.

Peppard, Oxon. SP 5106 (Peake, 1913) *M*
Neolithic.

Peterborough, Hunts. TL 1998 (Lethbridge, Fell & Bachem, 1951) *P*
Canoe embedded in nekron mud. Flandrian vIII.

Pevensey, Sussex TQ 6505 (Salzmann, 1908) *M*
Roman.

Pewsey, Wilts. SU 1659 (Grose & Sandell, 1964) *M*
Early Bronze Age.

Pickering, Yorks. SE 7983 (Jessen & Helbaek, 1944) *M*
Middle Bronze Age.

Pilling Moss, Lancs. SD 4046 (Oldfield & Statham, 1964–5) *P*
Raised bog over marine clay. Flandrian vIIb.

Plantation Farm, Cambs. TL 6484 (Godwin & Clifford, 1938) *P*
Contiguous with Peacock's Farm, Shippea Hill. Early Bronze Age horizon in upper peat. Flandrian vIIb.

Plas Newydd, Anglesey SH 5269 (Hemp, 1935) *M*
Neolithic.

Playden, Sussex TQ 9222 (Cheney, 1935) *M*
Middle and Late Bronze Age.

Ploopluck, Kildare N 8719 (Jessen & Helbaek, 1944) *M*
Middle Bronze Age.

Plumpton Plain, Sussex TQ 3612 (Jessen & Helbaek, 1944; Helbaek, 1953*b*) *M*
Late Bronze Age.

Plush, Dorset ST 7102 (Jessen & Helbaek, 1944; Helbaek, 1953) *M*
Late Bronze Age.

Plymouth, Devon SX 4555 (Dennell, 1970) *M*
Mediaeval.

Plynlimmon, Cardigans. SN 7987 (Moore, 1968) *P*
Raised bog. Flandrian vIIb and vIII. Clearance phases.

Ponders End, Essex TQ 3595 (Warren, 1912) *M*
Lea Valley Arctic Bed.

Poole, Dorset SZ 0190 (Erdtman, 1928) *MP*
Peat deposit with high values of pine pollen and pine needles. Boreal.

Portland 'Beehives', Dorset SY 6577 (Helbaek, 1953*b*) *M*
Early Iron Age.

Portrush, Antrim C 8521 (Jessen, 1949) *MP*
Peat resting on dune sands and overlain by raised beach deposit. Flandrian vIb and vIc.

Portstewart, Antrim C 8138 (Jessen & Helbaek, 1944) *M*
Middle Bronze Age.

Port Talbot, Glam. SS 7690 (Godwin & Willis, 1964) *M*
Submerged peat bed dated 10350 to 8970 (Q-660 to 663). Flandrian IV to VI.

Postbridge, Devon SX 6377 (Simmons, 1964) *P*
Blanket peat. Flandrian v to vIIb. Clearance phases.

Postwick, Norfolk TG 2907 (Clarke, 1938) *M*
Iron Age.

Prae Wood, Herts. TL 1507 (Jessen & Helbaek, 1944; Helbaek, 1953*b*) *M*
Early Iron Age.

Prestatyn, Flints. SJ 0682 (Newstead & Neaverson, 1938) *M*
Neolithic and Roman.

Prickwillow, Cambs. TL 5982 (Godwin, 1941) *P*
Lower fen peat, Fen Clay, upper fen peat. Flandrian vIIb.

Puddle Hill, Dunstable, Beds. TL 0023 (Fields, Matthews & Smith, 1964) *M*
Neolithic.

Puriton, Somerset ST 3241 (J. Allison, unpub.) *M*
Roman.

Pyle, Glam. SS 8382 (Hyde, 1953) *M*
Neolithic and Bronze Age.

Pyotdykes, Angus NO 3434 (Coles, Coults & Ryder, 1964) *M*
Late Bronze Age.

Quarley Hill, Hants. SU 2543 (Hawkes, 1939) *M*
Early Iron Age.

Quinton, Warwicks. SO 9884 (Bell, 1970*a*) *MP*
Silt deposit, Middle Hoxnian.

Racks Moss, Dumfriess. NY 0372 (Nichols, 1967) *P*
Stratigraphic sequence from fen peat (Flandrian VI), marine clay (VIIa) to fen peat (VIIb). Clearance phases.

Radley, Berks. SU 5398 (Jessen & Helbaek, 1944; Helbaek, 1953*b*) *M*
Late Bronze Age and Early Iron Age.

Radyr, Glam. ST 1280 (Grimes & Hyde, 1935) *M*
Iron Age.

Raheelin, Leitrim G 9645 (Jessen, 1949) *MP*
Sequence from wood peat to raised bog. Flandrian vIIb and vIII.

Ralaghan, Cavan N 6897 (Jessen & Farrington, 1938) *MP*
Fen and wood peat resting on freshwater muds and solifluxion deposits. Late Weichselian II to Flandrian vIIa.

Rasharkin, Antrim C 9812 (M. E. S. Morrison, unpub.) *P*
Stratigraphy not reported. Flandrian v to vIII. Clearance phases.

Rathcoffey South Td., Kildare N 8931 (Mitchell, 1951) *M*
Chalk mud with *Megaceros giganteus*. Late Weichselian II.

Rathjordan, Limerick R 64 (Mitchell, 1951) *MP*
Peats associated with crannog. Flandrian vIIb. Clearance.

Ratoath, Meath O 0252 (Mitchell, 1941*a*) *MP*
Freshwater muds over basal sequence of clay mud, chalk mud with *Megaceros giganteus*, clay with stones. Late Weichselian I to Flandrian v.

Ravensdale Park Td., Louth J 3131 (Mitchell, 1951) *P*
Peat beneath Megalithic cairn. Flandrian VIIa to VIII. Clearance.

Reach Lode, Cambs. TL 5568 (Godwin, 1941) *P*
Wood peat overlain by sedge fen peat thought to correspond to fen clay deposits further north. Flandrian VIIb.

Red Bog, Kildare N 9816 (Mitchell, 1951) *MP*
Kettle hole deposit. Flandrian VIIb.

Red Bog, Louth H 2940 (Mitchell, 1956) *P*
Sequence from freshwater muds to fen peat. Flandrian VIIa to VIII. Dated 2625–1450 (D-6, 7, 8, 10).

Redbourne Hayes, Lincs. SE 9900 (Smith, 1958a) *P*
Upper and lower peat separated by estuarine clay. Flandrian VIIb and VIII.

Redcote, Yorks. SE 3034 (Raistrick & Woodhead, 1930) *M*
Raft of plant remains in terrace gravels. Age uncertain: not entered.

Redhall, Midlothian NT 27 (Reid, 1899) *M*
Age uncertain: not entered.

Redhills, Goldhanger, Essex TL 9008 (Reader, 1907–9) *M*
Roman or Celtic.

Red Moss, Lancs. SD 6310 (Hibbert, Switsur & West, 1971) *MP*
Sequence from freshwater muds to *Sphagnum* peat. Dated 9798 to 4370 (Q-910 to 925). Chronozones. Late Weichselian to Flandrian VIIb.

Red Tarn, Westmorland NY 2603 (Pennington, 1964) *P*
Freshwater muds and clays. Flandrian IV to VIIb. Neolithic clearance.

Reffley Wood, Norfolk TF 6521 (Jessen & Helbaek, 1944) *M*
Early Bronze Age.

Rhyl, Flints. SJ 0081 (Neaverson, 1936) *M*
Peat exposed on foreshore. Flandrian VIIb/VIII.

Riggs, Yorks. SE 0373 (Jessen & Helbaek, 1944) *M*
Early Bronze Age.

Ringinglow Bog, Derbys. SK 2783 (Conway, 1954) *P*
Blanket peat. Flandrian VIIa to VIII.

Ringniell Quay, Down J 5367 (Morrison, 1961) *P*
Lagoon clay beneath Mesolithic/Neolithic occupation. Dated 7500 and 7345 (Q-632). Flandrian VIc.

Rishworth Moor, Cheshire SD 9817 (Bartley, 1964) *P*
Blanket peat. Flandrian VIIb. Clearance phases.

Risley Moss, Lancs. SJ 6691 (J. Tallis & H. J. B. Birks, 1965) *M*
Sphagnum cuspidatum peat. Flandrian VIII.

Rivenhall, Essex TL 8217 (Helbaek, 1953b) *M*
Roman.

Rockbarton, Limerick R 6239 (Mitchell & O'Riordain, 1942) *P*
Freshwater muds and fen peat beneath Bronze Age occupation. Flandrian VI and VII.

Rockbourne Down, Hants. SU 1118 (Helbaek, 1953) *M*
Middle Bronze Age.

Rockley Down, Wilts. SU 1573 (Allison & Godwin, 1949) *M*
Late Bronze Age.

Rockstown Upper Td., Wicklow T 01 (Mitchell, 1953) *M*
Blanket Bog. Flandrian VIIb or VIII.

Rodbaston, Staffs. SJ 9211 (Shotton & Strachan, 1956) *MP*
Kettle hole deposit of gyttja and peat. Late Weichselian III, 10670 (Y-464) to Flandrian VIIa.

Roddans Port, Down J 6373 (Morrison & Stephens, 1965) *MP*
Typical Late Weichselian stratigraphy overlain by fen peat and raised beach deposits. Late Weichselian I to Flandrian VI. 12110 to 9090 (Q-358 to 371).

Roden Downs, Berks. SU 5381 (Allison, 1947) *M*
Roman.

Rodmarton, Glos. ST 9497 (Clifford & Daniel, 1940) *M*
Neolithic.

Roke Down, Dorset SZ 19 (Jessen & Helbaek, 1944; Helbaek, 1953b) *M*
Late Bronze Age.

Romaldkirk, Durham NZ 9923 (Bellamy, Bradshaw, Millington & Simmons, 1966) *MP*
Lake muds, marls and clays. Late Weichselian I to Flandrian VI. *Bos primigenius.*

Romney Marsh, Kent TR 0430 (S. C. Seagrief, unpub.) *MP*
Flandrian VIIa.

Rosgoch Common, Radnorshire SO 1948 (Bartley, 1960) *P*
Phragmites detritus in Late Weichselian. Fen peat overlaid by *Sphagnum* peat. Late Weichselian I to Flandrian VIII.

Rotherly, Wilts. ST 9419 (Helbaek, 1953b) *M*
Roman.

Rothwell, Northants. SP 8181 (Jessen & Helbaek, 1944) *M*
Anglo-Saxon.

Roundstone I, Galway L 7241 (Jessen, 1949) *MP*
Submerged peat. Flandrian VIIa to VIII.

Roundstone II, Galway L 7241 (Jessen, 1949) *MP*
Sequence from freshwater muds to raised bog. Basal sequence clay, mud, clay and mud. Late Weichselian I to Flandrian VIII.

Roxburgh Street, Greenock NS 2776 (Reid, 1899) *M*
Mud from Clyde Beds. Late Weichselian.

Rudh'an Dunain Kave, Ross. NG 3917 (Scott, 1934) *M*
Iron Age.

Rudston, Yorks. TA 0967 (Jessen & Helbaek, 1944) *M*
Early Bronze Age.

Rusland, Lancs. SD 3385 (Greswell, 1958) *P*
Submerged peat. Flandrian VI/VII transition.

Saddlebow, Cambs. TF 5913 (Churchill, 1970) *P*
Upper fen peat dated 2495 to 1875 (Q-805 to 807, 549, 550).

St Albans, Herts. TL 1507 (Helbaek, 1953b) *M*
Iron Age and Roman.

St Bees, Cumberland NX 9611 (Walker, 1966) *P*
Organic detritus. Late Weichselian I, II, 10350 and 10510 (Q-304), and III.

St Erth, Cornwall sw 5435 (G. F. Mitchell, unpub.) *P*
Sands and clays resting on Palaeozoic slates and overlain by head deposits. Late Pliocene.

St Germans, Norfolk TF 5914 (Godwin, 1943) *P*
Four peat beds separated by estuarine clays, and two with tree stubs. La Tène to Anglo-Saxon glass bead. Foraminifera. Flandrian VIIa to VIII.

St John's College, Cambridge TL 4658 (Jessen & Helbaek, 1944) *M*
Anglo-Saxon.

St Just, Cornwall sw 8437 (Helbaek, 1953*b*) *M*
Middle Bronze Age.

St Kilda, Outer Hebrides NF 0999 (McVean, 1961) *P*
Blanket bog. Flandrian VI to VIIa.

Salhouse Broad, Norfolk TG 2714 (Jennings, 1952) *P*
Peat, Flandrian VII.

Salthouse, Norfolk TG 0844 (Henderson, 1914–15) *M*
Bronze Age.

Sand Le Meer, Yorks. TA 22 (Reid, 1899) *M*
Submerged peat bed. Age uncertain: not entered.

Scaleby Moss, Cumberland NY 4363 (Godwin, Walker & Willis, 1957; Walker, 1966) *MP*
Sequence from freshwater muds to *Sphagnum* peat. Late Weichselian II to Flandrian VIIb, 10835 to 4935 (Q-144 to 172).

Scotland (Erdtman, 1924) *P*
The earliest pollen analytical investigations in the British Isles at sites in Ross-shire, Skye, Lewis, Sutherland, Orkneys and Shetlands. Historically important but generally impracticable to zone, not entered.

Scrubitty Coppice, Dorset ST 9717 (Helbaek, 1953b) *M*
Middle Bronze Age.

Seamer, Yorks. TA 0281 (Clark, 1954) *MP*
Sequence from calcareous freshwater muds to fen. Late Weichselian I to Flandrian VII. (See *Flixton* and *Star Carr*.)

Seathwaite Tarn, Lancs. SD 2598 (Pennington, 1964–5) *P*
Freshwater muds and clays. Late Weichselian II to Flandrian VIIb.

Segontium, Caernarvons. SH 4862 (Hayter, 1921) *M*
Roman.

Selmeston, Sussex TQ 5106 (Clark, 1934) *M*
Mesolithic and Neolithic.

Selsey, Sussex SZ 8592 (West, 1961*a*) *MP*
Freshwater muds and estuarine clays beneath raised beach. Late Wolstonian to Hoxnian IIb. Transgression in IIb.

Selsey (Medmerry Farm), Sussex SZ 8693 (White, 1934) *M*
Anglo-Saxon.

Shacklewell, Middx. TQ 3584 (Reid, 1899) *M*
Interglacial deposit of uncertain age: not entered.

Shapwick Heath, Somerset ST 4339 (Godwin, 1943; Clapham & Godwin, 1948) *MP*
Sequence of estuarine clay, fen and wood peat, raised bog. Flandrian VIIa to VIII with two recurrence surfaces and

Roman hoard in VIII. Late Bronze Age wooden track at main flooding surface, 2010 (Q-39). Neolithic axe, 3550 (Q-423).

Shapwick Station, Somerset ST 4241 (Godwin, 1967*a*) *MP*
Stratigraphy at site of monoxylous boat (2295, Q-357) associated with flooding horizon over humified ombrogenous bog peat. Flandrian VIIb to VIII. Elk droppings.

Shippea Hill, Cambs. TL 6183 (Clark & Godwin, 1962) *MP*
Mesolithic and Neolithic levels in lower fen peat dated 7100 (Q-586, 587) and 5380 (Q-583, 584) respectively. Early Bronze Age level in upper fen peat. Flandrian VI to VIIb.

Shortalstown, Wexford T 1130 (Colhoun & Mitchell, 1971) *MP*
Muds and silts with *Megaceros giganteus*. Late Weichselian I and II 12160 (I-4963). Estuarine sand beneath underlying till. Ipswichian IIa.

Short Ferry, Lincs. TF 0877 (Smith, 1958*a*) *P*
Peat with dugout canoe, 2796 (Q-79). Flandrian VIIb.

Shower, Tipperary R 1616 (Mitchell, 1956) *MP*
Fen succeeded by raised bog. Flandrian VI to VIII.

Shrewsbury, Salop. SJ 4912 (Barker, 1961) *M*
Roman.

Shustoke, Warwicks. SP 2491 (Kelly & Osborne, 1965) *MP*
Peat in river alluvium, 4830 (NPL-39). Flandrian VIIa to VIIb.

Sidestrand, Norfolk TG 2639 (Reid, 1899) *M*
Cromer Forest Bed series.

Sidgwick Avenue, Cambridge TL 4457 (Lambert, Pearson & Sparks, 1963) *MP*
Peat in terrace gravels. Early Weichselian. Bison and molluscs.

Silchester, Hants. SU 6462 (Reid, 1901–9) *M*
Roman.

Silverdale Moss, Lancs. SD 4777 (Oldfield, 1960) *P*
Band of marine clay in raised bog dated 5734 to 6590 (Q-256, 260, 261). Flandrian IV to VIIa.

Sinfin Moor, Derbys. SK 3530 (Chapman, 1969) *P*
Lacustrine clay in shallow depression. Late Weichselian or early Flandrian. Marginal peat VIIb or VIII.

Skelsmergh Tarn, Westmorland SD 5396 (Walker, 1955*b*) *MP*
Freshwater muds. Basal sequence clay, chalk mud, clay and pebbles. Late Weichselian I to Flandrian VIIb.

Skendleby, Lincs. TF 4369 (Phillips, 1935) *M*
Neolithic, Bronze Age and Roman.

Skipsea, Yorks. TA 1855 (H. & M. E. Godwin, 1933) *P*
Peat exposed on sea cliff. Flandrian VI to VIIa. Mesolithic implements.

Slack, Yorks. SE 1417 (Dodd & Woodward, 1922) *M*
Roman.

Slapton Ley, Devon SX 8244 (Crabtree & Round, 1967) *P*
Freshwater muds and clays. Flandrian VIII.

Slieve Gallion, Londonderry H 7987 (Pilcher, 1970) *P*
Sequence from fen peat to blanket peat. Late Weichselian III to Flandrian VIIb. 9405 to 2670 (UB-271 to 282).

Sloggan Bay, Antrim J 0991 (Erdtman, 1928) *P*
Sequence from gyttja to moss peat. Flandrian v to viia.

Small Down Camp, Somerset ST 6640 (Helbaek, 1953*b*) *M*
Early Iron Age.

Snailwell, Cambs. TL 6467 (Jessen & Helbaek, 1944) *M*
Late Bronze Age.

Snibe Bog, Kirkcudbright NX 4781 (H. H. Birks, 1969) *MP*
Sequence from fen to *Sphagnum* peat. Regional assemblage zones. Flandrian IV to VIII.

Somerset, Londonderry C 8531 (Jessen, 1949) *MP*
Fen peat and brackish water muds in terrace sands. Flandrian viia to viib. Diatoms.

Somersham, Hunts. TL 3677 (Jessen & Helbaek, 1944) *M*
Early Bronze Age.

Somerton, Oxon. SP 4928 (Jessen & Helbaek, 1944) *M*
Early Bronze Age.

Southampton, Hants. SU 3912 (H. & M. E. Godwin, 1940) *MP*
Submerged forest bed. Flandrian IV to VI.

Southbourne, Dorset SZ 1590 (Calkin, 1947) *M*
Neolithic.

Southchurch, Essex TQ 9186 (Francis, 1931) *M*
Wooden track. Flandrian viib.

Southelmham, Suffolk TM 3483 (Chandler, 1889) *M*
Band of loam and clay exposed in brick pit with *Elephas*, *Equus* and mollusca. Hoxnian.

Southery, Norfolk TL 6294 (Godwin & Clifford, 1938; Godwin, 1941) *P*
Upper fen peat. Skeleton with Bronze Age necklace and pin. Flandrian viib.

South Ferriby, Lincs. SE 9821 (Smith, 1958*a*) *P*
Peaty clay with Roman artifacts overlying estuarine clay. Flandrian VIII.

South Walsham Marshes, Norfolk TG 3613 (J. N. Jennings, unpub.) *MP*
Flandrian vii.

Southwick, Sussex TQ 2704 (Curwen, 1932) *M*
Roman.

Speeton, Yorks. TA 1475 (West, 1969*b*) *P*
The Speeton Shell Bed. Ipswichian iib.

Spottiswood, Berwicks. NT 6049 (Jessen & Helbaek, 1944) *M*
Late Bronze Age.

Spring Bridge, Antrim C 8927 (Jessen, 1949) *MP*
Clayey diatomite on banks of river Bann. Flandrian viia. Corresponds to brackish water muds at Somerset.

Springhead, Kent TQ 5458 (Wilson, 1968*a*) *M*
Roman.

Stanborough, Hants. TL 2211 (Sparks, West, Williams & Ransom, 1969) *P*
Clays and muds. Hoxnian II and III.

Stanford, Norfolk TL 8594 (Jessen & Helbaek, 1944) *M*
Anglo-Saxon.

Stannon, Cornwall SX 1381 (Conolly, Godwin & Megaw, 1950) *P*
Organic detritus. Basal sequence gravel, mud, sandy clay. Late Weichselian and Flandrian VI to VIII.

Starr Carr, Yorks. TA 0280 (Clark, 1954) *MP*
At Flixton, and Seamer *Cladium* peat and detritus mud with Mesolithic occupation platform, 9557 (Q-14). Flandrian IV to VI.

Star Villa, Somerset ST 4358 (R. L. Jefferies, unpub.) *M*
Roman.

Stockton, Dorset ST 9838 (P. D. Broad, unpub.) *M*
Roman.

Stoke Newington, Middx. TQ 3286 (Reid, 1899) *M*
Age uncertain: not entered.

Stone, Hants. SZ 4598 (Reid, 1899; West, 1961*a*) *MP*
  (*a*) Ipswichian.
  (*b*) Brackish water deposits on foreshore. Ipswichian iib.

Stoney Bridge, Outer Hebrides—NF 7728 (Harrison & Blackburn, 1946) *P*
Wood and blanket peat. Flandrian viib and viii.

Stonham Aspall, Suffolk TM 1359 (Smedley & Owles, 1966) *M*
Roman.

Storrs Moss, Lancs. SD 4875 (Oldfield, 1966) *P*
Raised bog with 'Lake Dwelling' and Neolithic flints. Late Flandrian viia.

Stow Bedon Mere, Norfolk TL 9596 (P. A. Tallantire, unpub.) *MP*
Late Weichselian.

Strath Blane, Stirlings. NS 5381 (Jessen & Helbaek, 1944) *M*
Middle Bronze Age.

Streatley, Beds. TL 0921 (Dyer, unpub.) *M*
Middle Bronze Age.

Strichen Moss, Aberdeens. NJ 8954 (Fraser & Godwin, 1955) *P*
Stratigraphy not reported. Flandrian VI to VIII.

Stump Cross, Yorks. SE 0864 (Walker, 1956) *M*
Mud with charcoals and Mesolithic flints 6500 (Q-141), Flandrian v to viib.

Stuntney, Cambs. TL 5578 (Clark & Godwin, 1940) *P*
Fen peat with Late Bronze Age hoard. Flandrian viib/viii.

Stutton, Suffolk TM 1533 (Sparks & West, 1963) *P*
Brickearth in terrace gravels. Ipswichian III. Molluscs.

Sudbrook Camp, Mon. ST 5087 (Nash-Williams, 1939) *M*
Early Iron Age.

Sutton Courtenay, Berks. SU 5194 (Jessen & Helbaek, 1944) *M*
Anglo-Saxon.

Swaffham Bulbeck, Cambs. TL 5562 (Jessen & Helbaek, 1944) *M*
Middle Bronze Age.

Swaffham Drain, Cambs. TL 5369 (Godwin, 1941) *P*
Basal wood and fen peat overlain by calcareous marl; sedge peat corresponding to Fen Clay. Flandrian viib. Neolithic axe site.

Swallow Cliff Down, Wilts. ST 9725 (Clay, 1925) *M*
Early Iron Age.

Swansea Bay, Glam. SS 6893 (Godwin, 1940*b*) *MP*
Submerged peats. Flandrian V and VI.

Swanwick, Hants. SU 5109 (Fox, 1928) *M*
Late Bronze Age.

Syston, Leicestershire SK 6111 (Bell, 1969) *MP*
Organic detritus in terrace gravels. Middle Weichselian.
37420 (Birm-78).

Tadcaster, Yorks. SE 4543 (Bartley, 1962) *MP*
Freshwater muds. Basal sequence clay, calcareous muds,
clay. Late Weichselian I to VIIa.

Talbenny, Pembrokes. SM 8412 (Fox, 1943*a*) *M*
Early and Middle Bronze Age.

Tan Hill, Wilts. SU 0864 (Jessen & Helbaek, 1944; Helbaek,
1953*b*) *M*
Late Bronze Age.

Tattenamona Td., Fermanagh H 2334 (Mitchell, 1956) *P*
Raised bog. Flandrian VI to VIII.

Taw Head, Devon SX 6085 (Simmons, 1964) *P*
Blanket bog. Flandrian VI to VIII.

Tealham Moor, Somerset ST 44 (Godwin & Willis, 1959*b*) *M*
Fen peat overlying estuarine clay, 5412 (Q-120).

Teesdale, Yorks. NY 92 (J. Turner *et al.*, unpub.) *MP*
Cow Green Reservoir and Widdybank Fell area: valley
and blanket bog peats. Late Weichselian III, Flandrian IV to
VIII. 10070 to 3150 (GaK-2913 to 2920).

Telscombe, Sussex TQ 4103 (Jessen & Helbaek, 1944; Helbaek, 1953*b*) *M*
Middle Bronze Age.

Temple Mills, Essex TQ 3785 (Warren, 1916) *M*
Lea Valley Arctic Plant Bed.

Termonfeckin, Louth O 1680 (Jessen, 1949) *P*
Peat below raised beach. Flandrian VIIa. Upper surface
eroded.

Thatcham, Berks. SU 5266 (Churchill, 1962, 1963) *MP*
Calcareous marls in Thames valley deposits, Mesolithic,
Flandrian IV to VI, 10365–9480 (Q-650–652, 658, 659,
677).

The Loons, Orkney HY 2524 (Moar, 1969*c*) *P*
Freshwater muds. Flandrian IV to VIII.

Thetford, Suffolk TL 8779 (Clarke, 1914–15) *M*
Early Iron Age.

The Trundle, Sussex SU 8811 (Curwen, 1931) *M*
Neolithic and Early Iron Age.

Thickthorn Down, Dorset ST 9612 (Helbaek, 1953*b*) *M*
Early Bronze Age and Neolithic.

Thorne Waste, Yorks. SE 7214 (Turner, 1962) *P*
Raised bog. Flandrian VIIb and VIII. Clearance phases
3170–1855 (Q-479, 477, 481, 482). (See *Hatfield Moors*.)

Thorny Down, Dorset ST 9915 (Helbaek, 1953*b*) *M*
Middle Bronze Age.

Thrang Moss, Lancs. SD 4976 (Oldfield, 1960) *P*
Fen succeeded by raised bog. Flandrian VI to VIIb. Clearance
phases.

Thrapston, Northants. SP 9980 (Bell, 1969) *MP*
Organic detritus in terrace. *Carex* peat erratic 8920 (Birm-
87) not entered. Organic silts lens 25780 (Birm-113).
Middle Weichselian.

Three Holes, Cambs. TF 5100 (Godwin, 1941) *P*
Upper fen peat beneath Roman roddon. Flandrian VIIb/VIII.

Thriplow, Cambs. TL 4446 (D. H. Trump, unpub.) *M*
Iron Age.

Tievebullaigh, Antrim D 1927 (Jessen, 1949) *MP*
Wood peat succeeded by blanket bog. Flandrian VIIb and
VIII.

Tilbury, Kent TQ 6376 (Reid, 1899; Godwin, Willis &
Switsur, 1965) *MP*
Peats alternating with marine and freshwater clays of
Thames estuary. Flandrian VIIa, b; 7120 to 2390 (Q-790 to
793, 810, 811).

Togherbane, Kerry Q 7928 (Jessen, 1949) *MP*
Wood and fen peat. Flandrian VI to VIII.

Toll Gate House Trackway, Somerset. ST 4443 (Godwin,
1960*c*) *P*
Track resting on humified peat and overlain by aquatic
Sphagnum peat; 2600 (Q-306). Flandrian VIIb and VIII.

Toome Bay, Londonderry H 9891 (Mitchell, 1956) *P*
Peat with charcoal layers and Mesolithic occupation, 7680
(Y-957). Flandrian VI to VIIb.

Toreen More West, Kerry V 6548 (Jessen, 1949) *MP*
Raised bog overlying wood peat. Flandrian VIIa and VIIb.

Torrs, Luce, Wigtowns. NX 15 (Jessen & Helbaek, 1944) *M*
Late Bronze Age.

Totternhoe, Beds. SP 9922 (Hawkes, 1940) *M*
Early Iron Age.

Trafalgar Square, London TQ 3281 (Franks, 1960) *MP*
Organic deposits in fossiliferous gravels. Ipswichian IIb.

Traprain Law, East Lothian NT 5874 (Cruden, 1939–40) *M*
Roman.

Treanscrabbagh Td., Sligo G 84 (Mitchell, 1951) *P*
Sequence from freshwater muds to raised bog. Flandrian
IV to VIII.

Treanybrogaun, Mayo M 1898 (Mitchell, 1951) *P*
Raised bog. Flandrian VIIa to VIII.

Trebarveth, Cornwall SX 2677 (Helbaek, 1953*b*) *M*
Early Iron Age and Anglo-Saxon.

Tregaron Bog, Cardigans. SN 6861 (Godwin & Mitchell,
1938; Turner, 1964) *MP*
Sequence from freshwater clays and muds to raised bog.
Flandrian IV to VIII 10240 to 2942 (Q-930 to 947). Main
recurrence surface 2669 and 2624 (Q-388). Clearance phases.

Treneglos, Cornwall SX 2286 (Ashbee, 1958) *P*
Bronze Age.

Tresawsen, Merther, Cornwall SW 8644 (Helbaek, 1953*b*) *M*
Middle Bronze Age.

Treworrick, St Ives, Cornwall   SX 9877 (Helbaek, 1953*b*) *M*
Middle Bronze Age.

Trimingham, Norfolk   TG 2738 (Reid, 1899) *M*
Cromer Forest Bed series.

Trundle Mere, Hunts.   TL 2090 (Mittre, 1959; Turner, 1962) *MP*
Raised bog succeeded by marl. Flandrian VIIa to VIII. Clearance.

Tuddenham, Suffolk   TL 7371 (Jessen & Helbaek, 1944) *M*
Anglo-Saxon.

Turner's Puddle Heath, Dorset   SY 9293 (Piggott & Dimbleby, 1955) *P*
Bronze Age.

Turriff, Aberdeens.   MJ 7249 (Jessen & Helbaek, 1944) *M*
Early Bronze Age.

Tusculum, Berwicks.   NT 5485 (Cree, 1908) *M*
Iron Age.

Twickenham, Middx.   TQ 1473 (Reid, 1899) *M*
Age uncertain: not entered.

Twyford Down, Hants.   SU 4825 (Stuart & Birkbeck, 1936) *M*
Early Iron Age.

Ty Newydd, Anglesey   SH 3473 (Phillips, 1936) *M*
Early Bronze Age.

Ugg Mere, Hunts.   TL 2486 (Godwin & Clifford, 1938) *MP*
Fen and brushwood peat succeeded by Fen Clay, *Sphagnum* peat and calcareous marl. Flandrian VIIb to VIII. Elk droppings in *Sphagnum* peat, 3260 (Q-546).

Upavon, Wilts.   SU 1354 (Helbaek, 1953*b*) *M*
Early Bronze Age.

Upper Boyndlie, Aberdeens.   NJ 6463 (Jessen & Helbaek, 1944) *M*
Early Bronze Age.

Upper Hare Park, Cambs.   TL 5562 (Jessen & Helbaek, 1944) *M*
Middle Bronze Age.

Upton Warren, Worcs.   SO 9367 (Coope, Shotton & Strachan, 1961) *MP*
Bands of silt in terrace gravel, 41 500 (GRO 595) and 41 900 (GRO 1245). Vertebrates, molluscs, ostracods, insects.

Urswick Tarn, Lancs.   SD 2674 (Oldfield & Statham, 1963) *P*
Peat, Flandrian VI to VIII. Clearance.

Verulamium, Herts.   TL 1507 (Helbaek, 1953*b*) *M*
Roman.

Viper's Track, Somerset   ST 4240 (Godwin, 1960*c*) *MP*
Track in raised bog peat at main recurrence surface dated 2520 (Q-7) and 2630 (Q-312). Flandrian VIIa to VIII.

Wall, Staffs.   SK 1006 (Gould, 1964–5) *MP*
Organic deposits beneath Roman road. Age uncertain: not entered.

Wallasey, Lancs.   SJ 2893 (Travis, 1926) *M*
Coastal peat, age uncertain: not entered.

Wallington, Surrey   TQ 3163 (Robarts, 1905) *M*
Neolithic.

Waltham Cross, Herts.   TL 3700 (Reid, 1949) *M*
Lea Valley Arctic Plant Bed.

Walton-on-Naze, Essex   TM 2421 (Warren, 1912; H. & M. E. Godwin, 1936) *MP*
Mesolithic, Neolithic and Beaker on submerged land surface. (See *Lion Point*.)

Warborough Hill, Stiffkey, Norfolk   TF 9745 (Clarke & Apling, 1935) *M*
Iron Age.

Warcock Hill, Marsden, Yorks.   SE 0411 (Woodhead & Erdtman, 1926) *P*
Blanket bog overlying wood peat. Flandrian VIIb. Neolithic and Bronze Age.

Wareham Bog, Dorset   SY 9190 (Seagrief, 1959) *MP*
Freshwater muds and rootlet peat. Flandrian IV to VII.

Washerley, Durham   NZ 0545 (Raistrick & Blackburn, 1932) *MP*
Blanket peat with stumps of birch. Flandrian V to VIIa.

Water Wold, Yorks.   SE 8750 (Jessen & Helbaek, 1944) *M*
Middle Bronze Age.

Wellhaugh Flow, Northumberland   NY 6385 (Blackburn, 1947) *P*
Sequence from fen peat to moss peat. Flandrian VI to VII.

Welney Wash, Cambs.   TL 5294 (Churchill, 1970) *P*
Roddon silts (Roman) overlying upper fen peat. Flandrian VIII. Carbon dates dubious (Q-819, 820, 823, 829).

Wessenden, I, II, III, Staffs.   SE 0406 (Tallis, 1964) *P*
Blanket peats. Flandrian VIIb and VIII.

West Clandon, Surrey   TQ 0551 (Frere, 1944) *M*
Early Iron Age.

West Flanders Moss, Stirlings.   NS 5795 (Newey, 1966) *P*
Peat from Flandrian IV to VI, overlaid by Carse Clay in Flandrian VIb (see *Kippen*). Wood peat over clay, 5492 (Q-533), succeeded by raised bog. (See *Flanders Moss*.)

West Harling, Norfolk   TL 9785 (Clark & Fell, 1953) *M*
Early Iron Age.

West House, Yorks.   NZ 6009 (Erdtman, 1928) *P*
Blanket peat overlying wood peat and muddy detritus. Flandrian VI and VIIIa.

West Hyde, Middx.   TQ 0390 (Davis, 1967) *MP*
Peat lenses in gravel. Flandrian VI to VIIa.

West Kennet, Wilts.   SU 1168 (Helbaek, 1953*b*) *M*
Neolithic.

West Kilbride, Ayrs.   NS 2048 (Jessen & Helbaek, 1944) *M*
Middle Bronze Age.

West Overton, Wilts.   SU 1367 (Grose & Sandell, 1964) *M*
Early Bronze Age.

West Parley, Dorset   SZ 0997 (Drew, 1929) *M*
Early Iron Age.

West Rudham, Norfolk   TF 8127 (Hogg, 1940) *M*
Neolithic.

West Runton, Norfolk TG 1842 (Reid, 1882; Duigan, 1963) *MP*
    (*a*) Cromer Forest Bed series.
    (*b*) Cromerian.

West Stow Heath, Suffolk TL 8170 (Jessen & Helbaek, 1944) *M*
    Anglo-Saxon.

Westward Ho!, Devon SS 4229 (Churchill & Wymer, 1965) *MP*
    Kitchen midden with Mesolithic flints overlain by fen wood peat, 6585 (Q-672). Flandrian VIIa.

West Wittering, Hants. SZ 7897 (Reid, 1899) *M*
    Peaty clay exposed on foreshore. Ipswichian.

Westwood, Newport, Fifes. NO 4228 (Jessen & Helbaek, 1944) *M*
    Middle Bronze Age.

Wetherhill, Muirkirk, Ayrs. NS 6927 (Jessen & Helbaek, 1944) *M*
    Middle Bronze Age.

Weymouth, Dorset SY 6778 (Reid, 1899) *M*
    Age uncertain: not entered.

Whattall Moss, Salop. SJ 4331 (Hardy, 1939) *MP*
    Organic mud overlain by *Sphagnum* peat. Flandrian IV to VIII. Dugout canoe.

Wherram Percy, Yorks. SE 8564 (Jessen & Helbaek, 1944) *M*
    Middle Bronze Age.

Wherry Rocks, Cornwall SW 3628 (Reid & Flett, 1907) *M*
    Age uncertain: not entered.

White Bog, Killough, Down J 5336 (Stelfox *et al.*, 1972) *M*
    Chara marl and silts. Late Weichselian II.

Whitehawk Camp, Brighton, Sussex TQ 3105 (Curwen, 1934*a*) *M*
    Neolithic.

White Park, Antrim J 0143 (Jessen & Helbaek, 1944) *M*
    Neolithic.

Whitrig Bog, Berwicks. NT 6234 (Mitchell, 1948*a*; Conolly & Dickson, 1969) *MP*
    Marl and peat. Late Weichselian II to Flandrian IV and VIIb and VIII.

Whixall Moss, Salop. SJ 4936 (Hardy, 1939; Turner, 1964) *P*
    Fen peat overlain by *Sphagnum* peat. Pine stump at recurrence surface dated 2307 (Q-467). Flandrian VIIa to VIII. Clearance phases.

Wicken Fen, Cambs. TL 5570 (Godwin, 1941) *MP*
    Alternation of fen and wood peat reflecting marine transgression. Oak from wood peat dated 4380 (Q-129).

Widdybank Fell, Durham NY 8230 (Hutchinson, 1966) *M*
    Subfossil leaves of *Betula nana* found in peat near to living colony. Flandrian VIIa to VIII. (See *Teeesdale*.)

Wigton, Cumberland NY 2549 (Walker, 1966) *MP*
    Fluviatile deposit. Flandrian VIIb.

Willsborough, Londonderry C 2542 (Mitchell, 1951) *P*
    Peaty clay beneath estuarine clay. Flandrian VIa.

Wilton Bridge, Cambs. TL 6785 (Godwin, 1941) *P*
    Fen peat in original channel of River Ouse. Flandrian VI to VIII.

Winchester, Hants. SU 4826 (Hawkes, Myers & Stevens, 1929) *M*
    Early Iron Age.

Windermere, Cumberland NY 3701 (Pennington, 1947; Godwin, 1960*a*) *MP*
    Laminated clays in Late Weichselian I and III, freshwater muds in Late Weichselian II and Flandrian IV to VII. Zone II 11870 (Q-294).

Windmill Hill, Wilts. SU 0971 (Helbaek, 1953*b*) *M*
    Neolithic and Roman.

Windyhall, Rothesay, Bute NS 0864 (Jessen & Helbaek, 1944) *M*
    Middle Bronze Age.

Wingham, Kent TR 2457 (Godwin, 1962) *MP*
    Peat and organic mud. Flandrian VIIb and VIII. 3105 (Q-106) to 2340 (Q-110). Clearance phases.

Winklebury Camp, Wilts. ST 9421 (Helbaek, 1953*b*) *M*
    Iron Age.

Winterbourne Stoke, Wilts. TQ 2405 (Helbaek, 1953*b*) *M*
    Early Bronze Age.

Witchey Bridge, Somerset ST 3743 (D. Walker, unpub.) *MP*
    Flandrian VI.

Witherslack Hall, Westmorland SD 4386 (Smith, 1958*b*) *P*
    Lake muds. Basal sequence clay, clay-mud, clay. Late Weichselian I to Flandrian VIII.

Withington, Glos. SP 0315 (O'Neil, 1966) *M*
    Middle Neolithic.

Wolvercote Channel, Oxon. SP 4809 (Duigan, 1956) *M*
    Channel deposits. Late Hoxnian.

Woodcuts, Dorset ST 9618 (Pitt-Rivers, 1884) *M*
    Roman.

Wood Fen, Cambs. TL 5585 (H. & M. E. Godwin & Clifford, 1935) *MP*
    Marginal fen site with tree layers at Fen Clay margin, progression to ombrogenous peat. Flandrian VIIb to VIII.

Woodhead, Cheshire SK 1199 (V. M. Conway, 1954) *P*
    Eriophorum vaginatum peat over wood peat. Flandrian VIIb and VIII.

Woodgrange, Down J 4445 (Singh, 1970) *MP*
    Freshwater muds and clays. Basal sequence clay, marl and clay mud, clay. Late Weichselian I to Flandrian V.

Woodhenge, Wilts. SU 1564 (Cunnington, 1929) *M*
    Early Bronze Age and Iron Age.

Woodwalton Fen, Hunts. TL 2384 (Godwin & Clifford, 1938) *MP*
    Sequence of wood peat, Fen Clay, *Sphagnum* peat. Flandrian VIIb and VIII. 3415 (Q-545).

Wookey Hole, Somerset ST 5347 (Reid, 1911) *M*
    Age uncertain: not entered.

Worlebury Camp, Somerset ST 3162 (Helbaek, 1953*b*) *M*
Early Iron Age.

Wortwell, Norfolk TM 2784 (Sparks & West, 1968) *MP*
Organic deposits. Ipswichian Ia and IIb. *Elephas antiquus* and mollusca.

Wreay, Cumberland NY 4350 (Walker, 1966) *MP*
Aggradational river deposits. Flandrian VIIa.

Wretton, Norfolk TL 6899 (Sparks & West, 1970) *MP*
Fluviatile deposits in terrace gravels. Ipswichian II and III. Mollusca.

Wybunbury Moss, Cheshire SJ 6949 (J. H. Tallis & H. J. B. Birks, 1965) *MP*
Freshwater muds and *Sphagnum* peat. Flandrian VIIa to VIII.

Wylye, Wilts. SU 0037 (Gose & Sandell, 1964) *M*
Bronze Age.

Yare Valley, Norfolk TG 3306 (Jennings, 1955) *P*
Detritus mud overlain by estuarine clay. Flandrian IV to VIIa.

Yesnaby, Orkney HY 2315 (Moar, 1969*c*) *MP*
Freshwater marl and clay. Late Weichselian I to Flandrian IV.

Ynyslas, Cardigans. SN 6089 (Godwin, 1943) *MP*
Peat and brackish water clays exposed on foreshore. Peat, seaward extension of Borth Bog, 6026 (Q-380).

York Hill, Glasgow, Dunbartons. NS 5865 (Jessen & Helbaek, 1944) *M*
Roman.

Ysceifog, Flints. SJ 1976 (Fox, 1926) *M*
Bronze Age.

V

THE PLANT RECORD

1. INTRODUCTION: EXPLANATION OF CONVENTIONS

The plant record that follows is restricted to the vascular plants and its arrangement is that of the *List of British Vascular Plants* prepared by J. E. Dandy (1958) for the British Museum (Natural History) and the Botanical Society of the British Isles. The list numbers are given with each citation of a genus or species included in the list. Species which do not appear in the list are included after the listed species of the same genus and with extra sequential numbers. Genera and species that cannot be so treated are included in an appropriate taxonomic position but have no number, thus *Dulichium* follows the *Carex* citations and *Trapa natans* is included in the Onagraceae.

The greatly increased volume of information has imposed much more severe contraction than we used in the first edition. Authorities are nowhere cited for individual fossil records, but for each there is a site reference, and from the Site Record (chapter IV: 4) and Bibliography, author and publication can be traced. In the event of difficulty, individual records can be recovered by a request to the data bank. All records are given in sequence of glacial and interglacial stages through the Pleistocene, using the following contractions:

GLACIAL	INTERGLACIAL
	F Flandrian
W Weichselian (Devensian)	
	I Ipswichian
Wo Wolstonian	
	H Hoxnian
A Anglian	
	C Cromerian
Be Beestonian	
	P Pastonian
Ba Baventian	
	An Antian
T Thurnian	
	L Ludhamian

Cs = Cromer Forest Bed series – *vide infra*

The early and late deposits of each glacial stage are shewn by a prefixed *e* (early) or *l* (late), but for the Weichselian a middle period is indicated by *m*. The head-record does not give sub-divisions of the glacial stages or of the full Weichselian, but in the text there may be references to the sub-stages I to IV of both the Hoxnian and Ipswichian. Zones I to III of the Late Weichselian

and the Flandrian zones IV to VIII are however consistently given in the head-record. A proportion of records is referred, usually following the original author, to transitional zones or to combined zones, the citation appearing in the form 'IV/V' etc. Where local or assemblage zones have been employed by the author we have had to decide, upon the evidence available, the appropriate position in the standard zone system that here serves as a basic time scale. An interrogation mark (?) before a site name means that the fossil identification is uncertain. Where there is a clear need, the record-head lists are divided into two sections for pollen and macroscopic remains: generally they are combined.

Multiplicity of records has forced us to represent the evidence for all those taxa with more than twenty or thirty records simply as histograms (fig. 32). Although this means the omission of individual site names, these are less necessary when the number of records is large, and details can, in any event, be found by reference to the data bank. The histograms represent successive stages from either the Pastonian or Cromerian to the end of the Flandrian: they generally represent the evidence for macroscopic remains and pollen separately, and sometimes they indicate distinctively the merely tentative identifications. All precise chronological attributions are set out on the base line, but this evidence is supplemented at the top of the histogram by longer horizontal lines, each of which represents a single record spanning a longer period, such as the Cromer Forest Bed series, an interglacial as a whole, or the undivided Late Weichselian. With these we also shew the records for transitional or combined pollen zones. Such horizontal lines of course imply only that the record might derive from any part of the indicated time span.

The records, listed or histogram, give no indication of the numbers of individuals at a site or published illustration. The criterion is simply an authenticated presence at a given site and time. To some extent the inevitable deficiencies of such contraction are made good in the accompanying text, but the written accounts have the much wider purposes of general comparison between stages and zones, reference to biogeographic and ecological attributes of the species or genus, as well as appraisal of the history of the taxon in this country.

Brief explanation is necessary on the treatment and nomenclature of records in a few special categories. Thus all records derived from the extensive unpublished work of West and Wilson on the Cromer Forest Bed series are referred to under the heading 'Norfolk coast' and are not

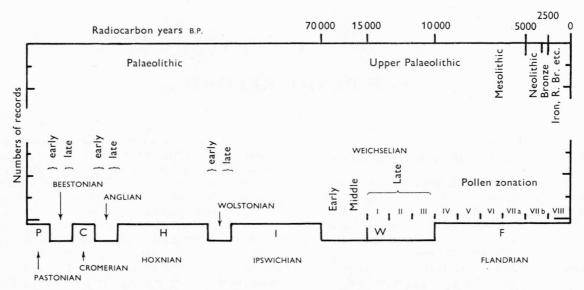

Fig. 32. Key to the chronological system of the record-head diagrams given in chapter V. For fuller explanation on usage see text.

here separately referred to individual sites on that coast. The information of recorded age is here limited to the identified stages.

The earlier published work by Reid in the same region is referred to under his site names, but the age is recorded as 'Cs', namely 'Cromer Forest Bed series', since it is generally impracticable to be sure what part of the full sequence, embracing possibly the Cromerian, Beestonian and Pastonian stages, each site fossil record represents. Dr S. L. Duigan's published records are generally attributable to a stage and her site names are quoted.

This system of citation will allow the reader to recognize in the record heads the three main sources of records from the Cromer Forest Bed deposits as a whole; Cs implies Reid, *Norfolk Co.* West and Wilson, C with specific site, Duigan.

The Lea Valley 'Arctic Plant Bed' described by S. Hazzledine Warren is an important source of plant records: it has been found at several sites and a recent radiocarbon date reinforces the view that it is referable to the Middle Weichselian: the original term is however sometimes conveniently retained.

An exception to the general convention is also made for records of the cereals: since virtually the whole of the records for these come from an archaeological context and these all are within Flandrian zones VIIb and VIII,

it has seemed helpful to arrange all the records under the headings of successive archaeological cultures.

The presentation of information cartographically also requires a few words of explanation. The 'clock-face' diagrams for the Flandrian pollen zones used in the first edition have been abandoned in face of the present multiplicity of sites. They have been replaced by the 'spot' diagrams for individual tree genera, in which the size of spot at each site indicates the relative frequency of pollen as a percentage of total tree pollen (excluding *Corylus*). The gradation of spot size has been determined by the data bank categories which are +, 0–2, 2–5, 5–10, 10–20, 20–30, 30–40, etc. For the Late Weichselian zones however it has been possible still to use clock-face diagrams to represent major components of vegetation as percentages of the *total* pollen count, but this is employed only in chapter VI, outside the plant records, as also are the exercises in biogeographic ordination, separately explained there.

Finally may we reiterate the warning already given, that these records all signify *presence*, and that they provide no effective basis for arguments based upon *absence*, nor can they be employed, save with great circumspection, in drawing conclusions as to relative abundance in different periods.

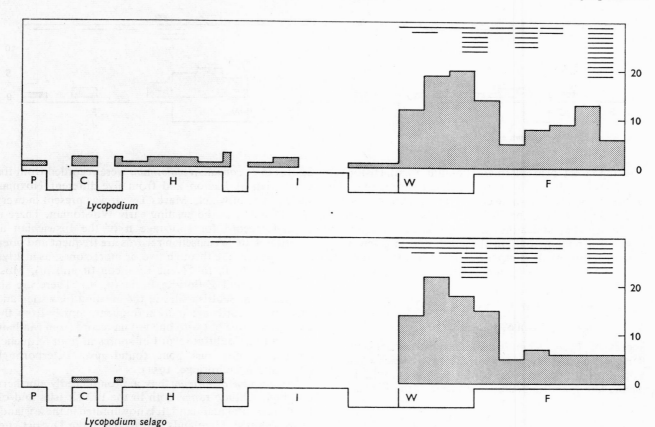

Lycopodium

Lycopodium selago

2. COLLECTED RECORDS

PTERIDOPHYTA

LYCOPODIACEAE

Lycopodium L.   I

The spores of *Lycopodium* are trilete and tetrahedral, with heavy walls strongly and coarsely reticulate or rugose.

Of the species of *Lycopodium* now growing in the British Isles *L. selago* and *L. inundatum* have much the most characteristic spores; those of *L. alpinum* and *L. clavatum* are more difficult to separate, although the reticulum in *L. alpinum* is coarser and the muri are deeper than in *L. clavatum*. When a spore record is given above for the genus only it is probable that it refers to one or other of these two species. *L. annotinum* is recognizable by the exceptional coarseness of its spore reticulation.

The generic record given in the histogram has to be considered alongside the now substantial evidence for the separate species. The genus is also represented in the Ludhamian, Thurnian, Antian and Baventian. It is thinly represented through the subsequent glacial and interglacial stages, is present in Early and Middle Weichselian, abundantly in the Late Weichselian, and in smaller frequency throughout the Flandrian. It is strongly represented in the latest zones by Scottish records, and in her research on the upland tarns of the Lake District Mrs Tutin draws attention to the spread of *Lycopodium* spp. and *Selaginella* in the grassland that followed clearance of the forests at high altitude (Pennington, 1964).

In the accounts that follow for the individual species of *Lycopodium* and for *Selaginella*, an extraordinary parallelism is evident, noticeable most easily in the record diagrams. All exhibit high site frequency in the Late Weichselian and a diminished frequency in the Flandrian and they all tend to have better Hoxnian representation (notably in western Ireland) than Ipswichian. They share of course the qualities of being northern–montane plants of dwarf stature intolerant of overshading and capable of growth on peaty substrata.

Lycopodium selago L. 'Fir clubmoss' (Plate XV*c*) I : I

There have been isolated recordings of *Lycopodium selago* spores from the Cromerian and Hoxnian interglacials and from the late Anglian. As elsewhere in north-west Europe the spores are frequently recorded from the British Late Weichselian and records continue, especially in the north, through all the Flandrian zones.

L. selago is general and common throughout the Arctic,

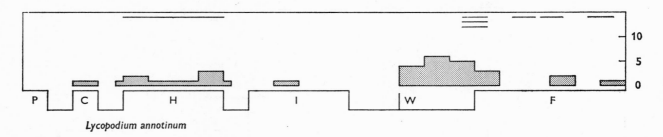

Lycopodium annotinum

reaching 81° 43′ N in Ellesmere (Polunin, 1940). Historical records shew that it formerly occurred throughout the British Isles save for some counties in southern, eastern and central England, but it has now lost a great deal of its former area and although common in Scotland, in England it remains only in the north and in a few sites in the south-west. It is more abundant in the mountains and almost always in open habitats.

Lycopodium inundatum L.   'Marsh clubmoss'   1 : 2

*l*W *Tadcaster*; F vIIb/vIII *S.W. Lancs. Co.*

Spores of *Lycopodium inundatum* were recognized by Bartley (1962) from zone I of the Late Weichselian in Yorkshire and there is an earlier record (Travis, 1926) for the late Flandrian submerged peat of the Lancashire coast.

Lycopodium annotinum L.   'Interrupted clubmoss' 1 : 3

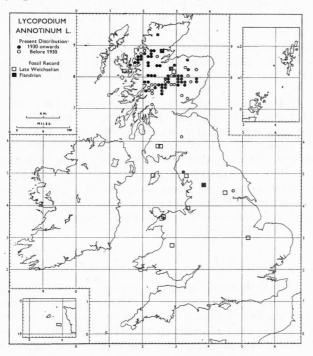

Fig. 33. Present and past distribution of *Lycopodium annotinum* L. in the British Isles.

Spores of *Lycopodium annotinum* were recorded from the Cromerian at Bacton and from five different Hoxnian sites, at one of which, Marks Tey, it was present in every sub-stage and in the ensuing early Wolstonian. There is a single record for sub-stage II of the Ipswichian at Wretton. Late Weichselian records are frequent and often extend at one site through two or more zones, as at Llyn Dwythwch (I, II, III), Nant Ffrancon (II and III), Moss Lake (I, II) and Esthwaite Basin (II, III). There are six records from Scottish sites in the opening Flandrian and thereafter records are more infrequent, mostly from the Cairngorm area but with one vI/vIIa record from Malham Tarn, and a doubtful record of spores in high frequency from a monoxylous boat found near Peterborough (Bachem, in Lethbridge, 1951).

Lycopodium annotinum has a pronouncedly northern and high altitude range both in the British Isles and on the Continental mainland. It is now limited in these islands to the Scottish Highlands and to one Lake District site. Thus the fossil records represent considerable loss of area since the Late Weichselian, though it persisted until recent historic time in Snowdonia and Yorkshire. It has not been found (apart from the doubtful Peterborough record) in the southern half of England since interglacial time, although it occurs throughout the Late Weichselian in Denmark.

Lycopodium clavatum L.   Stag's horn moss, 'Common clubmoss'   1 : 4

Spores of *Lycopodium clavatum* have been recorded from three different sites in sub-stage IV of the Hoxnian, although one of these was only to '*clavatum* type'. Records from the Late Weichselian are numerous and widespread. All zones are represented and there are repeated occurrences at several sites, some extending into the Flandrian, for example Tadcaster (zones I, II, III), Witherslack and Esthwaite Basin (II, III), Elan Valley (II, III, III/IV) and Loch Dungeon (III, IV, IV/V, V/VI, vIIa, vIIb/vIII). The same local persistence is found in the Flandrian alone, as at Allt na Feithe Sheilich (IV, vIIb/vIII) and Langdale Coombe (IV/V, VI, vIIb). This feature is no doubt indicative of the overriding significance for survival of local factors in the environment, especially high altitudes, that overrule climatic change.

Lycopodium clavatum is a species that has present or recent historic records throughout the British Isles,

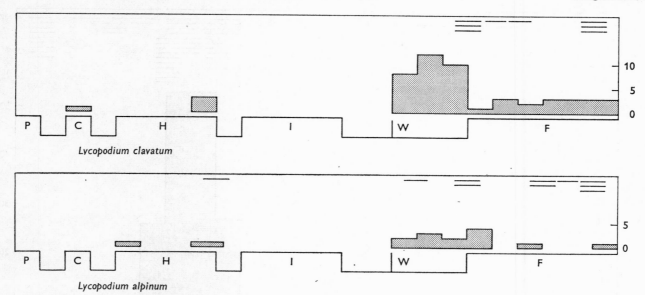

Lycopodium clavatum

Lycopodium alpinum

although it has always been rare in the lowlands. It is a boreal circumpolar plant of heaths, moors and mountain grassland. The fossil record indicates prevalence in the Late Weichselian and persistence from that time to the present day in regions where it still grows.

Lycopodium alpinum L.   (*L. complanatum* auct.)   1 : 5

Although site records for spores of *Lycopodium alpinum* are less abundant than those of the other species, they present a strikingly similar pattern, with small numbers of records in the early and late Hoxnian and early Wolstonian, a strong representation in Late Weichselian zones I, II, III and Flandrian IV, and a degree of persistence in the later Flandrian. There is again continuity or reappearance in several sites as at Loch Fada (zones I, II), Loch Meodal (I, IV,) Lochan Coir' a' Ghobhainn (I/II, III, III/IV), Loch Cill Chriosd (II, III) and Abernethy Forest (III/IV, IV).

L. alpinum is a circumpolar arctic–montane plant common throughout the Scandinavian mountains with a range in the British Isles very like that of *L. selago*; it is common on moors and mountain grasslands ascending to 4000 ft (1200 m) but it is very rare in the lowlands. The interglacial records from East Anglia apart, the fossil sites disclose no evidence for altered range.

SELAGINELLACEAE

Selaginella selaginoides (L.) Link (Fig. 34; plates XVg, h, XXi)
2 : 1

Both megaspores and microspores of *Selaginella selaginoides* have been identified in the fossil state, the former from amongst the smaller fruits and seeds, the latter in slides made up for pollen analysis. Both are easily recog-

nizable: the trilete rounded microspores have long curved spines on the distal surface quite unlike any other fossil spore of Quaternary origin. It has long been known that high frequencies of *Selaginella* spores characteriz late-glacial deposits and Firbas (1934) has recorded values for them eleven times greater than the total tree pollen.

S. selaginoides is placed by Hultén in the category of circumpolar arctic–montane plants and his Scandinavian map of it shews it as common throughout western Scandinavia from the North Cape to Stavanger in the south: it is, however, rare in southern Sweden and Denmark. Similarly in the British Isles *Selaginella* occurs only in the north, in Great Britain northwards of Merioneth, Cheshire, Derbyshire and Lincolnshire and in Ireland northwards of Co. Clare and Co. Wexford. It is particularly common on British mountains (ascending to 3500 ft (1060 m)) and in North America appears to be almost confined to them (Clapham, Tutin & Warburg, 1962). In the Alps it is described by Braun-Blanquet and Rübel as occupying a wide range of communities, soils and altitudes although chiefly sub-alpine to alpine. They record it from pastures, open woods, dwarf scrub, wind exposed ridges in the *Elynetum*, in fens and so forth.

It is not surprising that the majority of British records should come from the glacial stages. It is recorded from the Beestonian, Anglian and Wolstonian stages and especially from the well investigated Weichselian.

Although there are numerous records of *Selaginella* from the Hoxnian interglacial, all save one (sub-stage IV at Fugla Ness) are from western Ireland. It seems remarkable that this dwarf arctic–montane plant should persist through the wooded temperate part of the interglacial, but it may be recalled that at the present day the coasts of western Ireland harbour a remarkably large component of arctic–alpine species, possibly responding to the cool moist summer conditions. Watts (1959b) draws attention

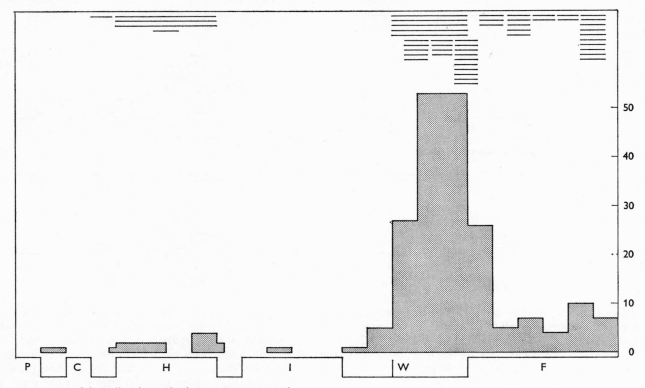

Selaginella selaginoides (micro- & megaspores)

to the fact that in the Hoxnian at Kilbeg there is a remarkably high proportion of *Selaginella* megaspores that appear to have been shed and fossilized whilst still immature: they are smaller and less round than the ripe spores. Just as the English Hoxnian yields no record of *Selaginella*, so its East Anglian Ipswichian deposits yield only one.

It is quite otherwise in the Weichselian, as is indicated in the distribution map (fig. 34). One Early Weichselian and five Middle Weichselian records are followed by records in great abundance throughout zones I, II and III supported by composite and transitional zone records widespread across the country far south of the present range of *Selaginella*. The case is paralleled by an abundance of Weichselian records from the lowland plains of Europe from which the species has since retreated northwards to Scandinavia or southwards and upwards into the Carpathian and Alpine mountains. This is precisely similar behaviour to that of *Betula nana*, as is pointed out by Tralau (1963) whose European map we reproduce.

There can be little doubt that *Selaginella* was a perglacial survivor in this country, that it had a wide range in Late Weichselian time and that the restriction to its present distribution area took place in the Flandrian. As with *Betula nana* again it is to be noted that Flandrian records from zone VI and later come from the Lake District or further north. In view of the pre-Weichselian evidence it seems not unreasonable to suppose that *Selaginella* has persisted through both interglacial and glacial stages in this country at least from the mid-Pleistocene. As with *Betula nana* it is not easy to attribute

the cause of Flandrian retraction, either to progressive restriction of open habitats or more directly to the influence of climatic change.

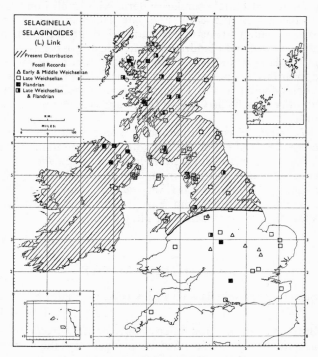

Fig. 34. Present and past distribution of *Selaginella selaginoides* (L.) Link in the British Isles.

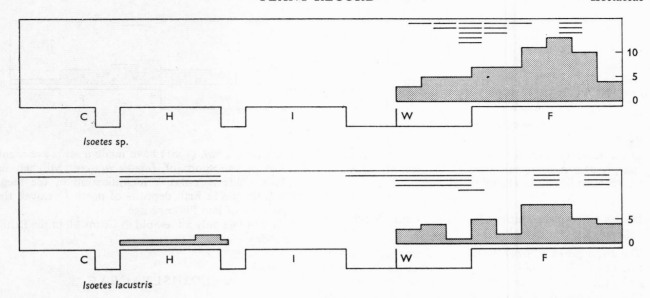

Isoetes sp.

Isoetes lacustris

ISOETACEAE

Isoetes sp. 3

Although the megaspores of individual species of *Isoetes* are very distinctive, the microspores of the commonest British species, *I. lacustris* and *I. echinospora* are separable only by size and this in practice is often difficult so that fossil records frequently do not go beyond generic level. They shew prevalence of *Isoetes* throughout all zones of the Late Weichselian or Flandrian in accordance with the record for the species separately. From time to time very high frequencies may be recorded as for instance in zone VIIa at Kentmere, where they rise to three times the total tree pollen (Walker, 1955*b*).

M. E. J. Chandler (1946) has pointed out that many of the identifications of *I. lacustris* made by C. Reid, and by herself and E. M. Reid following his lead, are for spores much too large to be attributed to *Isoetes*. This removes the records previously made for the Cromer interglacial and for the Weichselian deposits at Barnwell and the Lea Valley.

Isoetes lacustris L.   Quill-wort (Fig. 35; plate XV*j, k*) 3 : 1

The great majority, if not all, of the identifications of *Isoetes lacustris* made until recent years rest upon recognition of the megaspores which are covered with short, blunt tubercles, though this surface sculpturing may vary from slight to pronounced (fig. 35). The microspores are substantially larger than those of *I. echinospora*. It is common for mega- and microspores to be present together in a given deposit and for the microspores to be abundant enough to warrant separate representation in the pollen diagram where they commonly demonstrate a passing phase of abundance in the limnological evolution of the lake investigated.

I. lacustris has been found in all four sub-stages of the Hoxnian at Gort in western Ireland and in the later Hoxnian and early Wolstonian also at Kilbeg. There is one record from the Cromer Forest Bed series. Records are numerous in the Late Weichselian and through the succeeding Flandrian often with strong persistence in individual sites. In some Scottish lake deposits the Vasaris (1968) demonstrate the odd fact that microspores alone constitute the early record for *I. lacustris* and that the curve for megaspores appears later and after the microspore maximum.

I. lacustris is described by Hultén as a species with an incompletely boreal circumpolar distribution and he shews it as occurring scattered throughout Scandinavia from far within the Arctic Circle. In the British Isles it is found in oligotrophic lakes and tarns with unsilted bottoms; it occurs in Devonshire, in the mountain districts of Wales, Shropshire, south-east Yorkshire, the Lake District and more widely in Scotland and Ireland. Iversen (1929) has made the point that the plant can grow well in eutrophic waters but that it suffers severely in such conditions from the rank growth of other water plants: in the Late Weichselian period it need not therefore have been necessarily confined to the oligotrophic situations which it now favours. The adverse response of the plant to silting is shown in Pennington's study of upland tarns

Fig. 35. Megaspores of *a, Isoetes lacustris* L. and *b, I. echinospora* Durieu from the Irish Hoxnian. (After Watts, 1959*b*.)

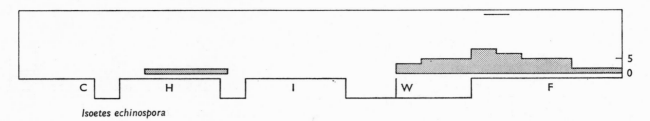

Isoetes echinospora

in the Lake District, where its spore frequency falls off sharply at a minerogenic wash-in horizon attributed to anthropogenic causes (Pennington, 1964).

Isoetes echinospora Durieu (*I. setacea* Lam.)   Quill-wort
  (Fig. 35)   3: 2

The megaspores of *Isoetes echinospora* are readily recognized by the long and slender spines that cover the surface or by the broken bases that remain in sub-fossil material (fig. 35). These have been recorded, along with the microspores (smaller than those of *I. lacustris*) from sub-stages II, III and IV of the Hoxnian at Gort, and the early Wolstonian at Kilbeg, both western Irish sites. This record parallels that for *I. lacustris* and indeed the two have very similar ecological requirements and may grow together (Watts, 1959b). The similarity in the fossil record extends to the Late Weichselian and Flandrian in which again *I. echinospora* is continuously represented from zone I to the present day. The phenomenon of local persistence in individual lakes is extremely evident as we may see at Loch Meodal (zones I, II, III, IV), Llyn Dwythwch (I, II, III, VIIa), Nant Ffrancon (II, III, IV, V), Cwm Idwal (IV/V, VI, VIIa) and Esthwaite Basin (IV, V, VI, VIIa, VIIb). We have here (and for *I. lacustris*) a situation analogous to that pointed out for species of *Lycopodium* and for *Selaginella* (see pp. 83–4), where local edaphic conditions are such that they override the influence of even substantial climatic changes.

I. echinospora has a Scandinavian range closely similar to that of *I. lacustris* (Samuelsson, 1934). In the British Isles it is chiefly a Scottish species, but isolated occurrences have been recorded from Wales and from the south of England. On the other hand we may recall that in the Nant Ffrancon, megaspore identifications allowed Seddon to conclude that whereas *I. echinospora* was present in the Late Weichselian, it was joined by *I. lacustris* only in zones V, VI and VIIa, i.e., the time of the hypsithermal interval: he points out that Lang had made a similar observation for lakes in the Vosges and Black Forest (Seddon, 1962). Welten (1967) has given a very thorough account of the past and present distribution of *I. echinospora* on the European mainland.

Isoetes hystrix Bory   3: 3
mW ?Upton Warren

In deposits of Middle Weichselian age in the English midlands, that have a radiocarbon age of about 42000

years, Coope *et al.* (1961) have made a tentative identification of the spore of *Isoetes hystrix*. Mitchell has similarly made a tentative identification of the megaspore from the St Erth deposits of north Cornwall that are possible of late Pliocene age.

I. hystrix has only a toe-hold in Cornwall in the British Isles today.

EQUISETACEAE

Equisetum sp.   4

The spherical non-aperturate spores of *Equisetum* often have an outer crumpled envelope. They have been frequently encountered in lake, marsh and fluviatile deposits and their records make up the bulk of those given in the accompanying histogram, although occasionally rootlets, rhizomes or stem diaphragms have also been identified.

Equisetum has been found through all sub-stages of the Hoxnian and Ipswichian interglacials and in the late Anglian and late Wolstonian. Site records are however particularly abundant in the Late Weichselian and early Flandrian, in response no doubt to the prevalence of sites suitable to early hydroseral invasion. Records continue through the Flandrian to the present. Examination of spore frequencies zone by zone reveals generally low levels with odd localities exhibiting very high values as at Longlee Moor and Elan Valley in zone IV, Holcroft Moss in zone VIIa, and Apethorpe, Frogholt and Wingham in zones VIIb and VIII, the three last valley sites in the East Midlands and south-eastern England.

Equisetum hyemale L.   Dutch rush   4: 1
F VIIb/VIII *?Kentmere*

Walker (1955b) tentatively identified rhizomes of *Equisetum hyemale* in late Flandrian lake deposits in the Lake District.

Equisetum fluviatile L. (*E. limosum* L.)   'Water horsetail'
                                                                                      4: 5

lW *Lock Park, ?Neasham*; F III/IV, IV, V, VI, VIIa, VIIb *Loch Park, Lock Kinord*; V, VI, VIIa, VIIb, VIII *Loch Fada*; VI *Cloughmills*; VIIa *Ardlow Inn, New Ferry*; VIII *Bann Estuary*

In Irish fen and marsh peats Jessen has recorded the stems or rhizomes of *Equisetum fluviatile* from zone VI onwards and K. B. Blackburn has made a tentative identification

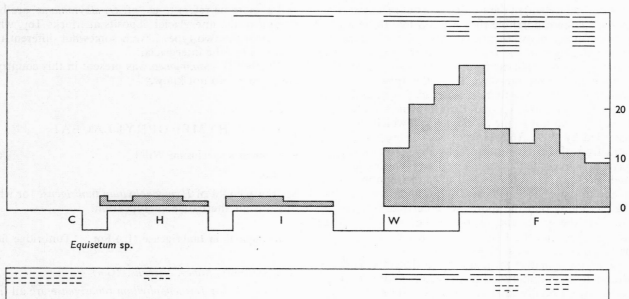

Equisetum sp.

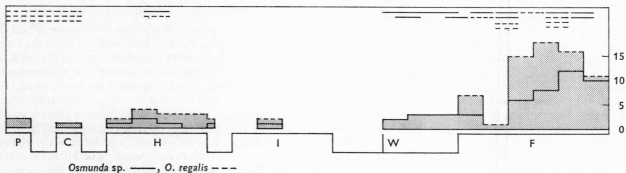

Osmunda sp. ———, O. regalis – – –

of it from the Late Weichselian in the north of England. Subsequently the Vasaris recorded it throughout very long periods in two lakes in Aberdeenshire and one in Skye: at one of the former, Loch Park, it was recorded from zone I through every zone to VIIa. *E. fluviatile* is common throughout the British Isles today and has a wide extra-British range extending into the Arctic.

Equisetum palustre L. 'Marsh horsetail'   4: 6

F ?IV *Nazeing*

The calcareous lake-marl of site X at Nazeing was richly penetrated by the rhizomes of *Equisetum palustre*, often with clusters of the characteristic tubers at the nodes, now largely in the form of casts. These stems had grown in from higher layers at a time which must have pre-dated zone VI, probably the phase of dry marsh surface in zone IV.

Equisetum telmateia Ehrh. (*E. maxima* Lam.) 'Great horsetail'   4: 10

F ?*Finsbury Circus*

Reid (1921) made a tentative identification of the stems of *Equisetum telmateia* at a Roman site in London.

OSMUNDACEAE

Osmunda sp.   5

Osmunda regalis L.   Royal fern (Plate XV*d–f*)   5: 1

The leaves and rhizomes of *Osmunda regalis* have long been recognized as macro-fossils, and by their means the plant's presence has been established in the Cromer Forest Bed, the Stoke Newington interglacial deposits, and in the rather recent Post-glacial peat layers exposed on the Lancashire and Cheshire coasts. The lateral patch of thickened cells that forms the 'annulus' in the *Osmunda* sporangium is also sometimes preserved fossil, as in a Pastonian record by Wilson (unpublished).

It is now usual to recognize and record the spores of *Osmunda regalis*: the record is often given only as the genus, the author having no other species in mind, but whilst this is probably valid for the Weichselian records there is evidence that other species may have been present in the Hoxnian (see account of *O. claytoniana* that follows).

Records of *Osmunda* spores have been made from the Ludhamian, Antian, Pastonian and Cromerian interglacial stages and from the intervening Thurnian and Baventian. The frequencies are fairly high (5 to 10 per cent of total pollen) in the Ludhamian but in later sections of the Ludham borehole they are too low to be a good index of indigeneity. *Osmunda* is recorded from the first three sub-stages of the Hoxnian and *O. regalis* from all four: both are reported from the following early Wolstonian and from sub-stage II of the Ipswichian. The point has to be made that Jessen *et al.* (1959) and Watts (1959*b*) specifically referred their Hoxnian material to *O. regalis* and not to the *O. claytoniana* which alone was recorded by West from Hoxne; Turner (1970) recorded both from the Hoxnian at Marks Tey. *Osmunda* spores have been recorded from a few Late Weichselian sites such as Stranraer (zones I and III) and Cannons Lough (II and III) but the frequencies are low and do not exclude the possibility that the spores were derived. Site records both for the genus and *O. regalis* become increasingly frequent through the Flandrian, so that the species may be taken as certainly native at least from zone IV.

O. regalis is native throughout the British Isles today, although much diminished by collectors and by drainage activities, especially in southern and eastern Britain. In Europe it does not occur north of the southern parts of Scandinavia, nor in North America north of Newfoundland and Saskatchewan. It grows on the better drained localities in fens, bogs, heaths and in woods on peaty soil. It is at home in the birch thickets which mark areas of locally better drainage on the surface of raised bogs, and it is notable that the abundant Irish and Somerset records come from regions of *Sphagnum* bogs (and often from such bogs themselves), where the fern still flourishes.

Osmunda claytoniana L.   5 : 2

H *Hoxne*, ?*Marks Tey*

In his investigation of the Hoxnian type site West (1956) reported that the *Osmunda* spores he had found in the later part of sub-stage II were not comparable with spores of *O. regalis* but closely resembled those of *O. claytoniana*, a species now found in eastern North America and, as a different variety, in east and south-east Asia. The interest of this attribution lay in the fact that *O. claytoniana* had already been found or was found about this time by Szafer and other workers in the Polish Masovian I (=Holsteinian) deposits and that it was subsequently found in the earlier Dutch interglacial, the Tiglian (=Lower Pleistocene) by Zagwijn (1960). The possibility seemed to present itself that *O. claytoniana* was one of a category of species widespread in the Tertiary period but now very disjunct and indeed extinct in Europe. Since both *O. claytoniana* and *O. cinnamonea* (another north-eastern American species) had been identified in Russian interglacial deposits it became a matter of importance to consider the possibilities of confusion in recognition of spores of the three species.

Although Andersen (1961) took the view that the ranges overlapped too much to make separate identification possible, this is not the conclusion of Turner (1970) who is satisfied on the basis of further work that both *O. regalis*

and a spore he refers to as 'cf. *O. claytoniana*' are both present in the interglacial deposits at Marks Tey, where moreover the two types have a somewhat different time range within the interglacial.

Whether *O. cinnamonea* was present in this country at any time we do not know.

HYMENOPHYLLACEAE

Trichomanes speciosum Willd.   6 : 1

III ?*Gort*

See the account of *Hymenophyllum tunbrigense* for which this was an alternative identification.

Hymenophyllum tunbrigense (L.) Sm.   'Tunbridge filmy fern'   7 : 1

H *Baggotstown, Kilbeg, Nechells, ?Gort*

Our records for *Hymenophyllum tunbrigense* are all from the middle Hoxnian, three Irish and one from the English Midlands. All are spore identifications but that at Kilbeg is supported by recognition also of the sporangium. In considering their material from Gort, Jessen *et al.* (1959) took the view that whilst it was easy to separate spores of *H. tunbrigense* from those of *H. wilsonii* (which was also present), they could not certainly be separated from those of *Trichomanes speciosum*, also a member of the Hymenophyllaceae. Watts however, working on Hoxnian material from Baggotstown and Kilbeg, found that spores of *Trichomanes speciosum* were rounder in outline and had less heterogeneous surface sculpture, so that a positive identification of *H. tunbrigense* was possible. Sporangia as well as spores were found.

H. tunbrigense grows luxuriantly in wet bryophyte communities and indeed has the habit of a moss. It requires a perpetually moist environment and is most abundant in oceanic regions such as the Killarney Woods of south-western Ireland.

According to Christ (1910) the two *Hymenophyllum* species have west European areas widely disjunct from their main area in the southern hemisphere. *H. tunbrigense* is not only strongly Atlantic in its European range but also in the British Isles, a fact giving special interest to the Hoxnian record from north Birmingham where it was apparently 'a common fern in the damp largely coniferous forests of the later part of the Interglacial' (Kelly, 1964).

Hymenophyllum wilsonii Hook. (*H. peltatum* auct.) 'Wilson's filmy fern'   7 : 2

H *Gort*

From sub-stage III of the Hoxnian interglacial at Gort, Jessen *et al.* (1959) were able to identify spores of *Hymenophyllum wilsonii* that are recognizable by their larger size and coarser sculpture from the otherwise similar spores

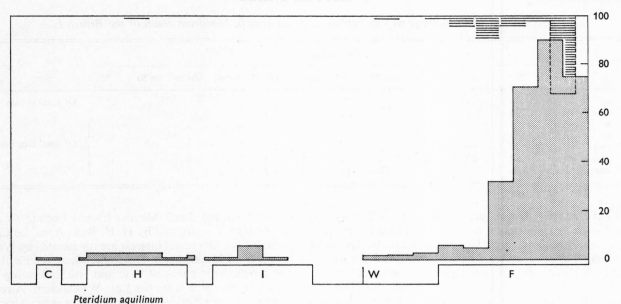

Pteridium aquilinum

of *Trichomanes speciosum* and of *H. tunbrigense*. *H. wilsonii* has, like *H. tunbrigense*, a very strongly oceanic range, but is slightly less susceptible to dryness than that species and occurs somewhat more commonly in the British Isles.

DENNSTAEDTIACEAE

Pteridium aquilinum (L.) Kuhn   Bracken (Plate XV*i*)
8: 1

Pteridium aquilinum does not possess the bean-shaped spores typical of most ferns but has a trilete, tetrahedral spore between 30 and 40 μm in diameter, with a rather finely granulate surface. Since Dr Conway's convincing account of the record of bracken spores in the bogs of the southern Pennines (1947) it has been widely recorded by British palynologists. Sometimes as in the late Anglian and middle Hoxnian of Western Ireland leaf remains were also recorded (Jessen *et al.*, 1959), and macroscopic remains, including rhizomes, have been more or less certainly recorded from Bronze Age and Roman sites.

There are single site records for *Pteridium* spores from the Ludhamian and Cromerian interglacials and more substantial records from all sub-stages of the Hoxnian, accompanied at Gort in sub-stage III by leaves, as also in the subjacent layers here referred to the late Anglian. The leaves are figured and described as having dichotomously branched veins, an incurved membranaceous margin and a lower surface covered with long multicellular hairs. Bracken spores have again been recorded from three sub-stages of the Ipswichian and the local presence of the plant need not be doubted. The spore records however for the early and late Wolstonian may well be due to derived material, and derivation or long-

distance transport are likely also to be responsible for the sparse spore records in all three Late Weichselian zones. It is apparent from the record-head diagram that the great Flandrian expansion of site records did not occur until zone VI, at which time the records are widespread throughout the British Isles. In zone IV however *Pteridium* in low frequency has been recorded from six sites, in Skye, the Cairngorms, Tyrone and Hampshire: at the last site, Crane's Moor, spore frequencies in the following zone reached 5 to 10 per cent of total tree pollen and it seems reasonable to regard *Pteridium* as locally established there along with *Pinus* (see p. 107).

The spore frequency table (table 4) usefully supplements the site frequency diagram for the Flandrian. It is evident that spore frequency remains low throughout zones VI and VIIa, bracken evidently confined substantially to the forest ground layer and natural clearances and margins. In VIIb however, and still more in VIII, both the modal frequency and the spread of site frequencies have greatly increased, a consequence without doubt of forest clearance. Dr Conway expressed the nature of the change in her account of the Ringinglow bog site:

First, in VII, a tall dense oak forest with associated elm and lime – both casting heavy shade – and possibly a hazel shrub layer as well, with a resultant light intensity too low for bracken; then, in VIII, after a drastic change in climate, a light oak–birch woodland of the well-known *Quercus sessiliflora* type of north hillsides, on leached soils, with a high internal light-intensity and good development of bracken. Finally, the forest clearance of 'VIII mod.' has encouraged the spread of bracken over many parts of the derelict hillsides, and its spores appear in quite large numbers in the pollen-samples.

All over the British Isles it is of course familiar that bracken has dominated sites of felled woodland where no alternative exploitation has followed and even against considerable pressure of grazing. There *Pteridieta* appear

TABLE 4. *Mean spore frequency of* Pteridium aquilinum *in Flandrian zones of the British Isles*

						Percentages							
	+	0–2	2–5	5–10	10–20	20–30	30–40	40–50	50–60	60–70	70–80	> 80	
IV	I	2	.	.	.	.	.	.	.	.	.	.	Of total pollen
IV	.	3	.	.	.	.	.	.	.	.	.	.	
V	.	3	I	I	.	.	.	.	.	.	.	.	
VI	5	18	4	I	.	.	.	.	.	.	.	.	Of total tree pollen
VIIa	3	33	10	8	.	.	.	.	.	.	.	.	
VIIb	.	28	20	21	6	2	I	I	.	.	.	I	
VIII	.	6	9	15	18	10	2	2	.	.	.	I	

to have had no natural predecessors: their history can be traced in such pollen diagrams as those of the East Anglian Breckland, and they are reflected in figures of *Pteridium* spore frequency in zones VIIb and VIII on the Kentish Downs exceeding the total tree pollen count.

It is not surprising that the detailed analysis of vegetational composition through prehistoric forest clearances should regularly shew *Pteridium* making a quick response to the forest opening. This is well shewn in the studies by J. Turner (1964, 1965) and those of many other workers subsequently.

P. aquilinum is a cosmopolitan species occurring all over the world except temperate South America and the Arctic: in Scandinavia it just enters the Arctic Circle: in the British Isles it is widespread and common, ascending to 2000 ft (610 m). The susceptibility of the young fronds and sporeling plants to frost is a recognized ecological feature, and it is clear that in suitable conditions the plant spreads freely by spores (Conway, 1953).

ADIANTACEAE

Cryptogramma crispa (L.) R. Br. ex Hook. Parsley fern
9: I

H *Clacton*; lW *Garth Farm, Flaxmere, Loch Cill Chriosd, Loch Meodal, Llyn Dwythwch, Nant Ffrancon, Lochan Coir' a' Ghobhainn*; F III/IV *Loch Dungeon, Bagmere*; IV *Nant Ffrancon*, IV, VIII *Llyn Dwythwch*, IV/V, VI, VIIb *Mickleden*

In 1962 B. Seddon published his identifications of fossil spores of *Cryptogramma crispa* from Late Weichselian deposits in Snowdonia. They are bluntly triangular in outline with a boldly verrucate surface with a pattern diminishing in size towards the triradiate scar, and they are approximately 70 μm in diameter. At Llyn Dwythwch it was present in zones I, II, III, IV and VIII, and at Nant Ffrancon in zones I, II and III. It amounted to 80 per cent of all other pollen and spores in zone III at Llyn Dwythwch and 60 per cent at Nant Ffrancon, so that it must have been at least locally abundant. The record for zone VIII tied in with communities of the parsley fern persisting at Dwythwch until the present day. Subsequently it has been recorded from Skye by H. J. B. Birks at Loch Cill Chriosd

(zones I and III), Loch Meodal (I) and Lochan Coir' a' Ghobhainn (I/II, III), and by H. H. Birks from the Cairngorm (III/IV). Of special interest are the records by Walker (1965) from Mickleden in Langdale, where the plant also still grows, from zones IV/V, VI and VIIb, indicating continuity of presence from the Late Weichselian. A record from Bagmere, Cheshire in zone III/IV is of interest because of the low altitude of the site: Birks (1965a) cites alongside this an unpublished record by Tallis from the Late Weichselian at Flaxmere similarly situated.

Cryptogramma crispa is a plant of late snow patch vegetation both in Norway and in Scotland, and it is common as a colonist of screes and boulder-clay surfaces. It grows on acid mountain soils and is locally common in North Wales and the mountains of the Lake District and Scotland. It has a few stations in eastern Ireland and Praeger conjectured that these might have been local colonists by spores airborne from Great Britain. The fossil record suggests survival since the Late Weichselian in mountain regions where the plant still grows but is inadequate to indicate whether it had formerly a lowland range also. It is described by Hultén as circumpolar arctic–montane with some occurrences in central Europe.

Adiantum capillus-veneris L. Maidenhair-fern   II: I

F VIIb ?*Glastonbury Lake Village*

BLECHNACEAE

Blechnum spicant (L.) Roth. (*Lomaria spicant* (L.) Desv.)
Hard-fern   13: I

lW *Loch Meodal, Lochan Coir' a' Ghobhainn*; F III/IV *Loch Dungeon*; IV *Loch Meodal, Lochan Coir' a' Ghobhainn*, VIIb *Moville*, VIII *Manchester*

From all three late Weichselian zones and from the earliest Flandrian H. J. B. Birks has recorded spores of *Blechnum spicant* at each of two lakes in Skye. Mrs Birks has also a spore record of it from the composite zone III/IV in Galloway. N. Macmillan (unpublished)) has a tentative record of its leaf from Donegal in zone VIIb and

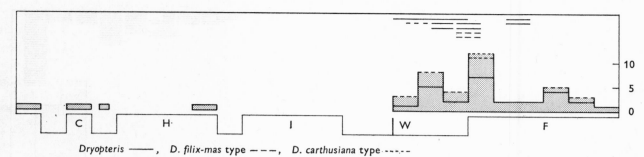

Dryopteris ———— , D. filix-mas type – – – , D. carthusiana type ----.--

there is an older record of its stems from a Roman site at Manchester.

At the present day *Blechnum spicant* occurs widely throughout the British Isles, particularly in the hilly and mountainous regions. It is a calcifuge, and of strongly oceanic range, indeed it is one of the species whose presence characterizes the Scottish dwarf shrub heaths in contrast with those on the mainland of western Europe (Gimingham, in Burnett, 1964). Nevertheless it occurs freely in western Norway to about 69° N latitude.

ASPLENIACEAE

Phyllitis scolopendrium (L.) Newm. (*Scolopendrium vulgare* Sm.) Hart's-tongue fern   14: 1

F *Dursley*

Leaves of the hart's-tongue fern, *Phyllitis scolopendrium*, were identified by C. Reid from a calcareous tufa of uncertain age in Gloucestershire.

ATHYRIACEAE

Athyrium filix-femina (L.) Roth   18: 1

H *Kilbeg*; F III/IV, IV *Loch Cill Chriosd*; VIIb/VIII ?*Hightown*, ?*Leasowe*

Spores of *Athyrium filix-femina* have been identified from undifferentiated Hoxnian material at Kilbeg, western Ireland, and from the Late Weichselian to early Flandrian in Skye. There are tentative identifications from the shore peats of south-west Lancashire.

Athyrium alpestre (Hoppe) Rylands *sensu lato* 'Alpine lady-fern'   18: 2

IW ?*Loch Cill Chriosd*, ?*Lochan Coir' a' Ghobhainn*

From two Late Weichselian sites in Skye H. J. B. Birks has recorded spores of a type that includes *Athyrium alpestre*, *Woodsia ilvensis* and *W. alpina*, all three mountain plants of very restricted present range in Britain. *A. alpestre* today centres on the Scottish highlands generally above 2000 ft (610 m) in block scree with prolonged snow lie, and often with *Cryptogramma crispa*

(Birks, 1969). In Scandinavia it is strongly restricted to the mountain ranges of the west where however it is found to the extreme north.

Cystopteris fragilis (L.) Bernh.   'Brittle bladder-fern'
                                                     19: 1

IW *Loch Cill Chriosd, Loch Fada,* ?*Tadcaster*; F IV ?*Tadcaster*

H. J. B. Birks has recorded spores of *Cystopteris fragilis* from two Late Weichselian sites in Skye, Loch Cill Chriosd (zones I, II and III) and Loch Fada (zone III). Determinations by Bartley from a Yorkshire site (zones I, II and IV) were only made tentatively to the genus *Cystopteris*. *C. fragilis* is the commonest of the British species today. It is a cosmopolitan species of very wide range from Greenland to Kerguelen, that is 'rather common in Scotland, northern England, Wales and northern and western Ireland' (C.T.W.). It occurs commonly throughout Scandinavia to the far north.

ASPIDIACEAE

Dryopteris Adans.   21

There are eight British species of *Dryopteris* and numerous hybrids. The bulk of spore identifications are only carried to generic level, although H. J. B. Birks has distinguished two groups referred to as the *D. filix-mas* and *D. carthusiana* types, the former including *D. filix-mas*, *D. borreri, D. abbreviata, D. aemula, Thelypteris limbosperma* and *T. robertiana*, and the latter *D. carthusiana, D. dilatata, D. cristata* and *Cystopteris dickieana*.

Beyond the fact that ferns of the *Dryopteris* type were present in the Late Weichselian, biogeographic conclusions are not practicable.

THELYPTERIDACEAE

Thelypteris palustris Schott (*Dryopteris thelypteris* (L.) A. Gray) 'Marsh fern'   24: 2

Spores: H *Gort, Kilbeg,* ?*Nechells, Marks Tey*; F VI, VIIa, VIIb/VIII *New Ferry*, VIIa *Clough Mills*, VIIb/VIII *Hightown*, VIII *Godmanchester*

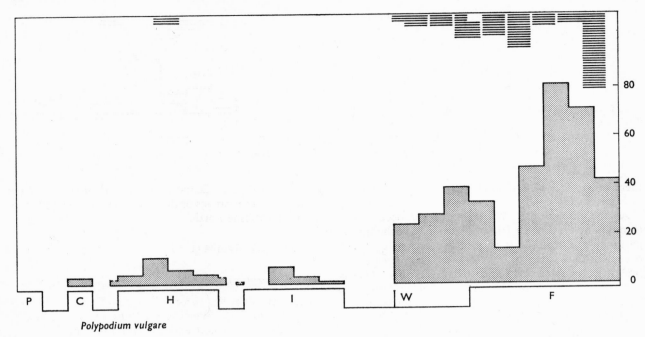

Polypodium vulgare

Macros: F v *Cool Bog*, vi *Castlelachan, Linton Mires*, viia *Togherbane*, viib *Woodwalton*, viib, viii *Canbo*, viii ? *Ardlow Inn*

When the exine is preserved the spores of *Thelypteris palustris* are readily recognizable for the exine is covered with short spines. Records of such spores are, however, rare, since the exine is so readily lost from fossil grains. There is very little doubt, however, that the marsh fern has been locally abundant in former times in the British Isles, for thick mats of its rootlets are often found and indeed may occupy characteristically a layer representing the vegetational transition from fen to raised bog (as at Woodwalton Fen). It seems likely that the extravagantly high percentages of naked fern-spores commonly found in fen-wood peats are attributable, in part at least, to this fern.

Spores referred by Jessen *et al.* to *T. palustris* type were recorded from all four sub-stages of the Hoxnian at Gort. From sub-stage ii there are unqualified records from Kilbeg and Marks Tey, and the tentative spore identifications at Nechells are for both sub-stage ii and sub-stage iii. In the Flandrian, macroscopic identifications preponderate and are enough to establish presence of the species from zone v through to viii.

T. palustris has a distinctly scattered and southern distribution in Scandinavia, and in Great Britain it shows a similar tendency, being local in the north of England and rare in Scotland: it is local throughout Ireland.

Thelypteris phegopteris (L.) Slosson (*Dryopteris phegopteris* (L.) C. Chr.)   Beech fern   24: 3

lW *Loch Meodal*; F iii/iv, iv *Loch Cill Chriosd*, iv *Loch Meodal*

From Skye H. J. B. Birks has recorded spores of *Thelypteris phegopteris* from all the Late Weichselian zones and the opening Flandrian. The beech fern is much commoner in the north and west of Britain than in the south and east: it is less common also in Ireland. It extends throughout Scandinavia to the far north.

Thelypteris dryopteris (L.) Slosson (*Dryopteris linnaeana* C. Chr.)   Oak fern   24: 4

I *Selsey Bill*, lW *East Walton, Loch Meodal, Esthwaite Basin, Lochan Coir' a' Ghobhainn, Loch Fada*: F iii/iv *Loch Dungeon*, iii/iv, iv *Loch Cill Chriosd, Lochan Coir' a' Ghobhainn*, iv *Loch Fada, Loch Meodal, Esthwaite Basin*

Apart from the sub-stage i/ii transition of the Ipswichian on the English south coast, all records of spores of *Thelypteris dryopteris* are from northern sites of Late Weichselian or early Flandrian age. There is strong local persistence as in Skye at Loch Meodal (zones i, ii, iii, iv), Loch Fada (ii, iii, iv), Lochan Coir' a' Ghobhainn (i/ii, iii/iv, iv), Loch Cill Chriosd (iii/iv, iv) and in the Lake District at Esthwaite (i, ii, iii, iv).

The oak fern has a strongly northern range in the British Isles today and the fossil record suggests probable persistence in Scottish and Lake District localities at least since the Late Weichselian. Its spores have been recorded from Late Weichselian deposits also on the Continent. It is a very wide ranging species found throughout Scandinavia to the North Cape and as far north as Iceland.

TABLE 5. *Mean spore frequencies of* Polypodium vulgare *in Late Weichselian and Flandrian zones of the British Isles*

| | | | | | Percentages | | | | | | |
	+	0–2	2–5	5–10	10–20	20–30	30–40	40–50	50–60	60–70	70–80	> 80	
I	5	**10**	3	1	.	.	.	.	.	.	.	.	⎫
II	11	**13**	1	.	.	.	.	.	.	.	.	.	
III	11	**15**	3	.	.	.	.	.	.	.	.	.	⎬ Of total pollen
IV	6	**12**	2	1	1	.	.	.	.	.	.	.	⎭
IV	2	**3**	1	1	1	.	.	.	.	.	.	.	⎫
V	2	**5**	1	.	.	.	.	1	.	.	.	.	
VI	8	**15**	5	3	1	1	.	.	.	.	.	1	⎬ Of total tree pollen
VIIa	7	**21**	20	9	2	.	1	.	.	.	.	.	
VIIb	5	16	20	13	3	.	.	.	1	.	.	.	
VIII	2	**12**	10	9	3	.	.	.	.	.	.	.	⎭

POLYPODIACEAE

Polypodium vulgare L.   'Polypody' (Fig. 36; plate XV a,b)
25 : 1

The very heavily thickened spores of *Polypodium vulgare* preserve remarkably well in muds and are easily distinguishable from the spores of other ferns. As pollen and spore analysis have come to be applied to interglacial deposits spores of polypody have been consistently found in them. They have been found in the Ludhamian and Cromerian, in all sub-stages of the Hoxnian and sub-stages II, III and IV of the Ipswichian. Site records occur in small number in the glacial stages, the Thurnian, late Anglian, and early and late Wolstonian but for small durable and recognizable objects such as these spores a secondary origin is possible, and absence from Middle and Early Weichselian is perhaps due only to the extreme infrequency of pollen analyses of this age. In the Late Weichselian, site records are however frequent, equally so in all zones in relation to site numbers analysed, and local presence can be inferred although as is shewn in table 5, the spores seldom exceed 5 per cent of the total of all pollen and spores. Site distribution for the Late Weichselian as a whole covers the whole of the British Isles although the biggest numbers are in north-western Scotland and the English Lake District (fig. 36). It is possible that this reflects some climatic differentiation of the country at this time, or the invasion of birch–pine woodland in the south and east.

Polypodium site records are numerous throughout the Flandrian, but the proportion of total analysed sites is never particularly high: it may well be that the fern is generally restricted in its habitats and that its spores are too heavy for distant transport. Spore frequencies, as the table shews, tend to remain low, and sites of high frequency are generally at high altitude as at Loch Clair (zone IV), Stump Cross (zones V, VI and VIIa) and Devoke Water (zone VIIb). In the basal sediments of the ditch round the great Neolithic mound at Maes Howe in the

Shetlands there were high frequencies of spores of polypody conjectured to have grown in niches upon the newly erected structure (Godwin, in Childe, 1954–6). The mean spore frequencies rise in zones VIIa and VIIb in the latter, possibly the effect of forest opening.

The polypody is now found throughout the British Isles and ascends to 2800 ft (860 m) in Co. Kerry. In western Scandinavia it is frequent to latitudes above 70° N and according to Hultén has a boreal circumpolar distribution. It is interesting that the sub-species found in this country appear to have distributions corresponding to those they have in Europe as a whole. It seems more likely that the Late Weichselian polypody belonged to

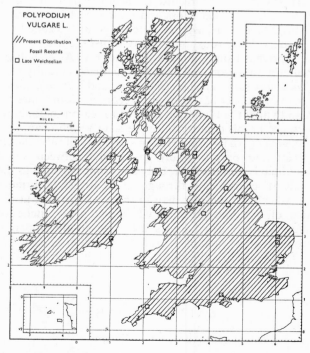

Fig. 36. Present and past distribution of *Polypodium vulgare* L. in the British Isles.

sspp. *vulgare* or *prionodes* rather than to *serrulatum* which is now southern and western (Perring & Sell, 1968). Identification can be partly based on sporangial characters, and indeed Watts has used sporangial annulus numbers to identify the tetraploid subspecies *vulgare* from the Hoxnian at Kilbeg (Watts, 1959*b*).

Although the polypody is a common woodland epiphyte it has to be recalled that it is also frequent in mature coastal dunes and rock ledges and in the Scottish dwarf shrub heath held to be ancient and often above the forest limit (Gimingham, in Burnett, 1964).

MARSILEACEAE

Pilularia globulifera L.   Pillwort   26: 1

H *Ballykeerogemore, Kilbeg*; H/eWo *Kilbeg*; eWo *Kilbeg*; I *Shortalstown*; IW *Chat Moss, Moss Lake*; F III/IV, IV, V *Moss Lake*, IV *Scaleby Moss*, VI *Nant Ffrancon*, VIII *Ehenside Tarn*

In 1953 Mitchell recorded one megaspore of *Pilularia globulifera* from Hoxnian material at Kilbeg and in his later investigations at the same site Watts (1959*b*) recognized a large number from the transitional end stage of the interglacial and the succeeding early glacial. He describes the megaspores as having the appearance of minute acorns with a constriction equally separating the spore into a smooth 'fruit' and a rough 'cupular' portion. Watts found microspores along with the megaspores. Megaspores were later recorded from sub-stage II of the Ipswichian at Shortalstown.

There is one site record from zone III of the Late Weichselian at Chat Moss and close by, at Moss Lake, Liverpool. The megaspores were found at the zone III/IV boundary and throughout zone IV whilst the microspores were present from late zone III to zone V. As at Kilbeg, *Pilularia* was associated with an aquatic community with *Elatine hexandra, Isoetes echinospora* and *Littorella lacustris*. As *Pilularia* has been also recorded from zones IV, VI and VIII of the Flandrian there is no reason to doubt its presence in this country since the Late Weichselian.

P. globulifera is within Hultén's category of sub-atlantic species and occurs locally in Europe between south Scandinavia and northern Italy and from Portugal to the Urals. It occurs locally throughout the British Isles, more sporadically in the east than in the west, growing at the edges of, or submerged shallowly in, acid pools, ditches or lakes. Watts (1959*b*) draws attention to the apparent anomaly of its present southerly restriction of range in south Sweden and the climate in which it grew at Kilbeg in the early Wolstonian.

SALVINIACEAE

Salvinia natans (L.) All. (Plate XXVI*f*)

C *Norfolk Co.*; IWo *Bobbitshole*; I *Bobbitshole, Wortwell, Wretton*

From sub-stages I and II of the Ipswichian at Bobbitshole, West (1958) recorded and illustrated megasporangia and megaspores of *Salvinia natans* in some frequency (plate XXVI). These discoveries were followed by others from sub-stage II of the same interglacial at Wortwell (Sparks & West, 1968) and at Wretton (Sparks & West, 1970). These identifications were accompanied by others of such aquatic plants as *Najas minor, Trapa natans* and *Stratiotes aloides*, all now extinct in the British Isles and indicative of considerable continentality of climate. Subsequently Wilson and West (unpublished) have identified *Salvinia* from the Cromerian interglacial, a discovery that perhaps increases the likelihood that the record for the late Wolstonian at Bobbitshole was secondarily derived.

At the present day *S. natans* is native of central and south-eastern Europe, with a range extending westwards to Holland and Spain. In Holland the reality of its native status is confirmed by recent investigations by Zandstra (1966), who found the entire massulae, macro- and microsporangia in large numbers in deposits of Atlantic age at three sites within the Rhine delta. It is noteworthy that all our sites for fossil *Salvinia* are from the most continental part of Britain, where however the plant has apparently not re-established itself in the Flandrian and is not naturalized.

AZOLLACEAE

Azolla filiculoides Lam. (Fig. 37; plate XVI)   27: 1

Pa *Ludham, Norfolk Co.*; C *West Runton, Norfolk Co.*; A *Corton*; H *Little Welnetham, Nechells, Hatfield, Kildromin, Baggotstown, Hoxne, Clacton, Marks Tey*; eWo *Hoxne*

Whilst washing out on a very fine sieve the interglacial lake mud from Hoxne, Suffolk, R. G. West recovered a quantity of the remains of *Azolla* fructifications. Not only the megaspores were recognizable, but also the massulae with their glochidia, which rather surprisingly survived the drastic treatment preparatory to pollen analysis (plate XVI). He was able to shew that in the tuberculate pattern of the megaspore, and the septation of the glochidia, this material very closely resembles present-day *A. filiculoides* (fig. 37). These discoveries were followed by many others, the earliest from two sites each in the Pastonian and Cromerian interglacials. Most striking however has been the recognition of *A. filiculoides* from a majority of the Hoxnian sites in the British Isles, almost always in the warmer sub-stages II and III and often, as at Nechells, Hatfield, Baggotstown and Hoxne, in both of them. At Nechells it was recorded both by Duigan and subsequently by Kelly. At Marks Tey it was recorded from sub-stage IV. Although reported from the early Wolstonian at Hoxne and from the Corton Beds in the Anglian glacial stage, these may be regarded as due to derived material.

Azolla filiculoides is an aquatic plant of placid waters over which it may extend as a continuous floating carpet.

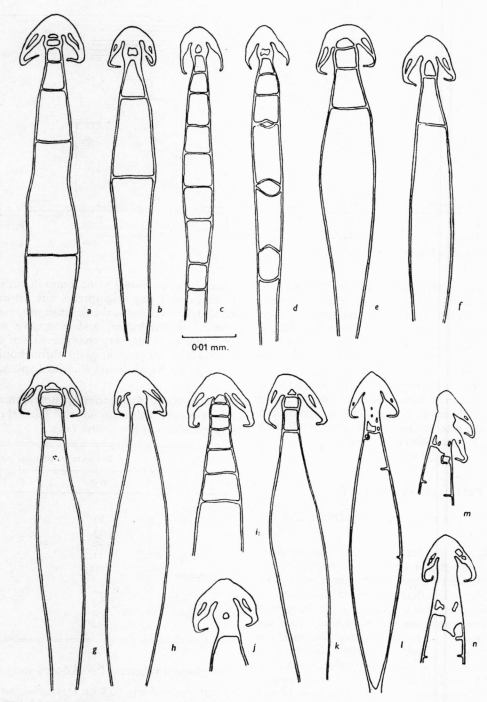

0·01 mm.

Fig. 37. Types of glochidia from the massulae of living and fossil species of *Azolla*: *a, b, A. filiculoides* Lam. var. *rubra* (R. Br.) Strasburger from Swan River, Western Australia; *c, d, A. mexicana* Presl. from Illinois, U.S.A.; *e–h, A. filiculoides* Lam., *f*, from Chile, others from Britain; *i–k*, fossil *A. filiculoides* Lam., *i* and *j* from Hoxne, Suffolk, *k* from Neede, Netherlands; *l–n, A. caroliniana* Willd. from Florida, USA. (After West, 1953.)

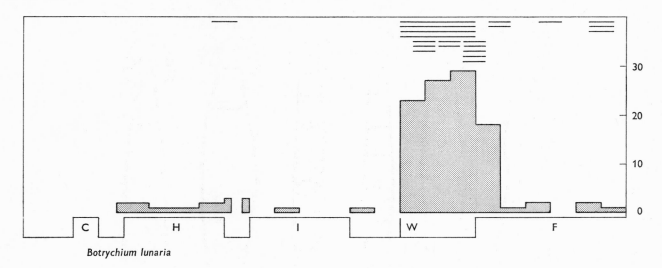

Botrychium lunaria

At the present day *A. filiculoides* occurs only in the New World, from Guatemala to Alaska, in the Andes and southern South America. It has been reintroduced into Europe and is naturalized in Britain, where, however, it has a rather sporadic southern occurrence and is somewhat susceptible to hard winter weather. Its fossil record in the British Isles is paralleled by that on the European mainland where it has been found in deposits ranging from the Tiglian to Holsteinian. Since it has not been found in the Eemian (Ipswichian) it is presumed to have been extinguished in the intervening glacial stage, in Britain the Wolstonian, from the whole of Europe.

OPHIOGLOSSACEAE

Botrychium lunaria (L.) Sw.   Moonwort (Plate XV*l, m*)
28: 1

The spores of *Botrychium* were first recognized by palynologists in the late-glacial deposits of north-western Europe and in the Allerød deposits at Bromma (plate XV). Iversen (1946) concluded that they most resembled *B. lunaria*. The bulk of records constituting the histogram may be taken to represent this species although they are often presented merely as the genus, and although the attribution must be more doubtful in the older deposits.

The abundance of site records through the Late Weichselian zones I to III and Flandrian zone IV is very evident. It matches the evidence from the late Anglian, early and late Wolstonian and Early Weichselian glacial stages and it contrasts with the poor representation in the interglacials though it is present in all four of the Hoxnian sub-stages and most of the Flandrian zones. On this evidence there is no reason to doubt its continuous presence in this country from the mid-Pleistocene. Spore frequencies are generally low, though possibly higher in zone I; occasional sites such as Moss Lake yielded values of 5 to 10 or 10 to 20 per cent of total pollen (see table 6).

B. lunaria is a boreal plant found in both hemispheres, at the present day widespread but rather rare in the British Isles: it ascends the Scottish mountains to 3350 ft (1020 m) in Perthshire, and it is, like so many Late Weichselian species, very characteristic of open habitats. *B. lunaria* is a common plant throughout Scandinavia save only for Swedish and Finnish Lapland.

TABLE 6. *Spore frequencies of* Botrychium lunaria *and* Ophioglossum vulgatum *in the Late Weichselian and early Flandrian of the British Isles*

	Percentages of total pollen				
	+	0–2	2–5	5–10	10–20
Botrychium					
I	4	12	2	.	1
II	9	13	.	1	.
III	7	12	3	.	.
IV	3	11	.	.	.
Ophioglossum					
I	2	2	.	.	.
II	5	3	.	1	.
III	7	4	.	.	.
IV	2	4	1	.	.

Ophioglossum vulgatum L.   Adder's tongue fern   29: 1

As with *Botrychium*, so with *Ophioglossum* it was investigations of late-glacial deposits in western Europe that led to the general recognition of spores of this genus, and here again it may be presumed that most of the records refer to *O. vulgatum* although often the identification has been only at generic level.

Ophioglossum is recorded from all sub-stages of the Hoxnian and Ipswichian interglacials and from all zones of the Flandrian. It is however relatively more frequent in sites of the Late Weichselian and earliest Flandrian, just

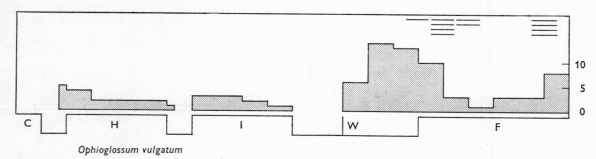

Ophioglossum vulgatum

as with *Botrychium lunaria*, although in a smaller total number of sites. As with that species, spore frequencies are, however, low (see table 6). Again the fossil record indicates long persistence as a native in this country.

O. vulgatum L. has a very wide distribution in the world, and occurs throughout the British Isles, more abundantly in the south-east and less common in Scotland. Its tolerance of cold climatic conditions is shewn by its presence in Iceland and Alaska, but in Scandinavia it does not extend beyond the Arctic Circle. It is a species of at least moderately open habitats and on the whole of basic soils. A rise in site frequency in zone VIII of the Flandrian seems to reflect (as the *Botrychium* records did not) a response to woodland clearance and spread of grassland.

SPERMATOPHYTA

GYMNOSPERMAE

PINACEAE

Tsuga spp. Hemlock

L *Ludham*; T *Ludham*; A *Ludham, Easton Bavents*; Ba *Easton Bavents, Ludham*; Pa *Ludham*; Cs *Happisburgh*; I *Marros*

For our knowledge of the presence of *Tsuga* in British Pleistocene deposits we are almost wholly dependent on two papers published by West (1961*b*) and Funnell & West (1962), and the evidence is entirely confined to that of pollen identification with its attendant uncertainties where pollen is infrequent.

Tsuga pollen is characterized by a ring-shaped rudimentary bladder on the distal side of the grain and by a coarsely verrucate surface extending over the body of the grain and the air sac. West follows Rudolph (1936) in distinguishing two pollen types, that of *T. canadensis* with a narrow equatorial fringe (the air sac) and that of *T. diversifolia* where the fringe is broad. The former type corresponds closely with pollen of *T. canadensis* (L.) Carr. and of the closely related *T. caroliniana* Engelm., both of which are species at present restricted to the eastern United States of America, the latter confined indeed to the Blue Ridge of the Alleghany Mountains at high altitude. The other pollen type is that exhibited by

T. diversifolia Maxim., an east Asian species, and by *T. mertensiana* (Bong.) Carr., a tree now limited to the coastal belt of western North America.

The early Pleistocene deposits traversed by the Ludham borehole were separable on the evidence of pollen, foraminifera and sea-level changes into five zones, later interpreted as the Ludhamian, Antian and Pastonian temperate interglacial stages separated by the cold glacial stages of the Thurnian and Baventian. Although *Tsuga* pollen was recorded from the glacial stages it was in low frequency and in the context of a sub-arctic park vegetation it can only be regarded as derived, or perhaps windblown from a distance. It is otherwise in the Ludhamian where *Tsuga* constitutes up to about 7 per cent of the total pollen, and in the Antian where its values are between 10 and 20 per cent at Ludham, and between 40 and 50 per cent at Easton Bavents. In these interglacials *Tsuga* is part of a pollen assemblage with high arboreal pollen frequencies and it accompanies *Pinus, Picea, Betula, Quercus, Ulmus* and *Alnus* with other trees in smaller amount. It evidently indicates a temperate mixed coniferous–hardwood forest such as the hardwood–hemlock–spruce forest of the Great Lakes region of the USA, where *T. canadensis* still grows.

At Ludham both interglacials shew a strong preponderance of pollen of *canadensis* over *diversifolia* type, but the ratio in the earliest Ludhamian is only about 3:1, whilst at Easton Bavents in the Antian it is approximately 2:1. A relatively small proportion of grains was undeterminable so that native status must be conceded to whatever taxon is represented by the pollen of *diversifolia* type.

The low frequencies in which *Tsuga* pollen occur in the Pastonian make its presence in that interglacial uncertain and the same holds for records from the Cromer Forest Bed series at Happisburgh, where however Miss Duigan (1963) points out factors suggestive of local derivation of the hemlock pollen. Pollen of *Tsuga* found by Mitchell (1970) in sub-stage IV of the Ipswichian at Marros in Wales may well be secondary.

Szafer described macroscopic remains of *Tsuga* from the Pliocene deposits of Króskienko in Poland. Very detailed examination of needles and cones caused him to refer them to two species that he recognized as very close to or identical with *T. canadensis* and *T. caroliniana* respectively. We should note that Mitchell (unpublished)

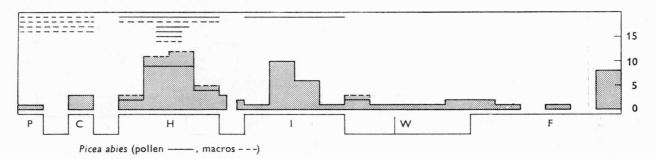

Picea abies (pollen ——— , macros – – –)

has also found *Tsuga* pollen at St Erth in beds probably of late Pliocene age. The fossil history of *Tsuga* in the Netherlands closely parallels that in East Anglia. It is present with *Carya* and *Pterocarya* in the Tiglian (Ludhamian) and with *Pterocarya* in the Waalian (Antian) in frequencies lower than those of eastern England but indicative of presence in a similar woodland type, and in the next succeeding interglacial its pollen is present only in low frequency. It remains an open question whether the rigour of the Baventian glacial stage may not indeed have finally extinguished all *Tsuga* species in Europe: in any event this was not long delayed thereafter.

Picea abies (L.) Karst.   Spruce (Plate XIV*c–f*)   31:1

Macroscopic remains of the spruce are locally frequent as fossils and readily recognizable. The leaves have pointed apices and are rhombic in section with stomata on all four faces, though leaves grown in shade are also flattened. The twigs bear the characteristic projecting peg-like rhombic leaf bases, the wood is recognizable microscopically and the cones have the typical linear-lanceolate trifid bract scales. The pollen also is highly recognizable, its very large grains having hemispherical air sacs that engage on the body of the grain with no constriction of the kind seen in *Abies* and *Pinus*. The plausible assumption is made that we are concerned throughout the British Pleistocene only with the one species, *P. abies*.

The case of the spruce emphasizes strongly the need to base deductions of presence of a species mainly upon macroscopic remains, for where vast European (and American) forests of spruce supply pollen to long distance air mass dispersal, low frequencies of *Picea* pollen can occur wherever local sources of tree pollen are sparse or water-flotation induces pollen concentration. Thus although *Picea* pollen in low frequency is recorded through the Middle and Late Weichselian and through the Flandrian, this provides no proof of presence of the tree in the British Isles, and similar records through the Ipswichian interglacial are, in the absence of macroscopic evidence, at least suspect.

In the early Pleistocene, for which we rely upon West's analyses of the Ludham borehole (1961*b*), these difficulties are somewhat offset by larger relative frequencies, at least in the temperate stages, so that 10 to 30 per cent of spruce pollen in the Ludhamian is part of a pollen assemblage that suggests affinity of the vegetation with northern temperate coniferous–deciduous forest, an attribution we may extend also for the Antian and perhaps the Pastonian, especially if we have regard to the low pollen productivity of the spruce. The intervening colder 'glacial' stages, the Thurnian and Baventian, boast only low spruce pollen frequencies, but *Picea* has decreased in them less than the thermophilic trees and it may possibly still have been growing in this country.

The situation in the middle Pleistocene is different, for Clement Reid commonly found spruce cones at various sites in the Cromer Forest Bed series, so that when Thomson (in Woldstedt, 1950) and later Duigan (1963) began pollen analysis in these deposits it was unsurprising that they recorded a *Picea* maximum in what now seems to be the post-temperate stage of the Cromerian *sensu stricto*. The presence of *Picea* in the Hoxnian is unequivocal, macroscopic remains having been recorded from every sub-stage: those at Gort were especially varied and abundant but they also occurred in Birmingham and East Anglia. This record has been paralleled by the pollen evidence which West (1956) presented for Hoxne as shewing low frequencies of spruce in sub-stage II, higher frequencies in III indicative of replacement of mixed oak forest by conifers, both *Picea* and *Abies*, in the late-temperate sub-stage, and values of spruce up to 20 per cent of total tree pollen in sub-stage IV where conifers were dominant. It appears from the pollen-frequency analysis of Hoxnian sub-stages in table 7 that a similar vegetational course was found elsewhere in the British Isles, with the highest frequencies however in sub-stage II at Nechells and East Winch. The pollen values may well under-represent the actual frequency of the tree. At Nechells the spruce maximum preceded that of *Abies* and followed the decline of such thermophiles as *Tilia*, *Fraxinus* and *Ilex*. The spruce is a tree of acidic mor soils and Kelly suggests that it was the leaching and increasing acidity of the soils that allowed *Picea* to take over dominance from the mixed-oak-forest trees in the later part of the interglacial.

Table 7 shews that spruce pollen has been recorded for all sub-stages of the Ipswichian interglacial, but always in low frequencies except in East Anglia where, at Beetley in Central Norfolk, Miss L. Phillips has recently found as much as 7 per cent A.P. in the *Carpinus* zone of sub-stage III. Moreover at an adjacent exposure where *Picea* pollen

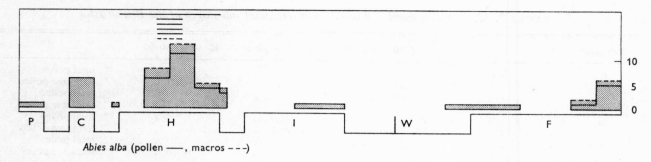

Abies alba (pollen ——, macros – – –)

reaches 9 per cent A.P., very large numbers of needles of spruce, positively identified as *Picea abies*, have been recovered. These records have been tentatively referred to sub-stage IV of the Ipswichian, although they might possibly belong to a Weichselian interstadial, such as that at Chelford in Cheshire which yields incontrovertible evidence of growth of spruce 60 800 years ago. Here, in organic muds between two deposits of till, Simpson and West (1958) have recorded needles, wood, cones, bark and a tree stump of *Picea* and pollen frequencies of spruce in an assemblage that strongly suggests forest of the type now growing in the conifer–birch region of Fennoscandia. This reference is strongly backed by the beetle fauna investigated by Coope (1959). Pine, birch and spruce pollen occur in similar frequencies to those of Chelford at various Danish and North German last interstadial sites including the Brörup Hotel Bog, and indeed also in the concluding phases of the preceding (Eemian) interglacial.

TABLE 7. *Mean pollen frequency of* Picea *in the Hoxnian and Ipswichian interglacials of the British Isles*

Sub-stage	+	0–2	2–5	5–10	10–20	
Hoxnian						
I	1	.	.	.	.	% total pollen
	1	.	.	.	.	% total tree pollen
II	2	.	1	.	.	% total pollen
	.	5	1	.	.	% total tree pollen
III	.	2	2	.	.	% total pollen
	.	2	1	2	2	% total tree pollen
IV	.	1	1	1	.	% total pollen
	.	.	.	1	.	% total tree pollen
Ipswichian						
I	1	.	.	.	.	% total tree pollen
II	2	6	2	.	.	% total tree pollen
III	.	3	3	.	.	% total tree pollen
IV	.	1	1	.	.	% total tree pollen

The severest phase of the Weichselian glaciation now appeared to kill out spruce from the British Isles and it has not reappeared as a native tree. The vicissitudes of its Pleistocene history here still involve considerable conjecture, more especially with regard to inferences of former climate. Thus the spruce is generally regarded as a strongly continental tree and Firbas describes its European range as follows: 'Along with its prevalence in the continental east with cold winters and at high altitudes in the mountains, and with its striking avoidance of mild-wintered oceanic regions there must also be noted considerable demands for soil moisture and an especial intolerance of valleys or depressions liable to summer drought.' This sorts so ill with the Hoxnian prevalence of spruce in western Ireland that Jessen *et al.* conjecture 'we may have to attribute to this species a greater richness of atlantic biotypes than is now found within it'. Likewise the presumed absence of spruce from the whole Ipswichian has provoked conjecture that early post-glacial rise in sea-level or other cause may have totally prevented immigration to Britain during an interglacial when it was common on the nearby Continent. If barriers to migration are indeed invoked then heed must be given to the recent documentation of the late Flandrian extension of spruce in Fennoscandia, where radiocarbon dates have precisely shewn the lateness and slowness of its extension towards the west and north (Aario, 1965; Aarolahti, 1966; Hafsten, 1970; Moe, 1970). This tardiness may be due to the gradient of oceanicity westwards and certainly at the present day the tree appears to regenerate very poorly in such western climates as that of the British Isles.

Abies alba Mill. (*A. pectinata* DC.) Silver fir (Fig. 38; plate XIVa, b)

Macroscopic remains of *Abies* have been found at many sites in the Hoxnian interglacial and they include wood, shoots, cones, cone-scales and seeds. The careful investigations by Jessen *et al.* (1959) of their Irish material leave little room for doubt that it is all referable to the central European fir, *Abies alba* Mill. Although it had previously been usual to refer European fossil *Abies* material of this age to *A. fraseri* Poret, a species now native in the high mountains of the eastern USA, these authors shewed that all the characters relied on for this identification (especially occurrence of resin-canals in the leaves and stomata on their upper surfaces) fall within the natural range of *A. alba*. We have accordingly treated all Pleistocene identification of *Abies* in this country as this species, including those of the more widely occurring pollen grains. These are large and bisaccate with the bladders larger than hemispheres which, like ...ose of *Pinus*, are contracted to less than their maximum diameter at the level of insertion on the body of the grain. The exine of the dorsal

TABLE 8. *Mean pollen frequency of* Abies *in Hoxnian sub-stages of the British Isles*

Sub-stage	+	0–2	2–5	5–10	10–20	20–30	30–40	40–50	50–60	
I	.	.	.	.	.	.	.	.	.	% total pollen
	.	.	.	.	.	.	.	.	.	% total tree pollen
II	.	3	.	.	.	.	.	.	.	% total pollen
	.	2	.	.	.	.	.	.	.	% total tree pollen
III	.	.	4	.	.	.	.	.	1	% total pollen
	.	1	.	4	.	.	.	2	.	% total tree pollen
IV	.	1	1	2	.	.	.	.	.	% total pollen
	.	.	.	.	.	.	.	.	.	% total tree pollen

Some sites have values expressed as percentage of total pollen, others of total tree pollen.

wall of the grain is also extremely thick and rough, and the body of the grain measures between 80 and 100 μm across.

Abies pollen has been found in low frequency in the Ludhamian, Thurnian and Pastonian stages (West, 1961*b*), but in the Cromerian we have the first strong evidence for its native presence: though only in small amount, it was found by Duigan (1963) in no less than nine of the twenty-six deposits that she investigated, and several of these, representing at least six sites, seem likely to represent the Cromerian itself and not merely the Cromer Forest Bed series. Macroscopic remains have however not been recorded.

It is otherwise in the Hoxnian where abundant evidence of *Abies* is found in sub-stages II, III and IV, not only in the Irish sites, but also in the English Midlands and East Anglia. We repeatedly find substantial pollen frequencies associated with the macroscopic remains, as especially at Gort but also at Kilbeg and Nechells. It has become clear that high *Abies* pollen values are typical of the late-temperate phase of the Hoxnian, following a maximum of *Carpinus*, and indeed they characterize this interglacial. The attached table (table 8) shews that the mean pollen frequencies in the Hoxnian sub-stages are very low in II, highest in III and diminished in IV. Particularly high frequencies (70 per cent) are recorded from Clacton (Pike & Godwin, 1953) where the *Abies* phase was first recognized, from Rivenhall (Turner, 1966) and from the North Sea site, Silver Pit, where however the conditions of marine deposition may have led to abnormal concentration of the saccate grains as happens also with *Pinus* (Fisher, Funnell & West, 1969). The large pollen grains fall rapidly in air and the tree is thought to be under-represented by its pollen production, so that we must accept *Abies* as a substantial component of the later Hoxnian woodlands. Kelly describes sub-stage III of the Hoxnian of north Birmingham in these terms: '*Abies* became the dominant conifer, soon after its appearance, in a forest largely of *Abies*, *Picea* and *Pinus* with more occasional *Betula*, *Quercus* and *Taxus*, *Alnus* still occupying the wetter sites.' The map given by Straka and Walter (after Selle) based on European interglacial records, shews a belt of *Abies* stretching at this time from western Ireland well into eastern Russia between the latitudes of 60° and 50° N. This is altogether north of its present central European range where it only attains about 50° N, and is far more constricted both westwards (about 0° in the Pyrenees) and eastwards (about 28° E in the Carpathians) as against the Hoxnian extension beyond 50° E.

It seems likely that the early Wolstonian and the late Anglian records for *Abies* represent derived Hoxnian material, and the few pollen records for the Ipswichian, Weichselian and Flandrian must all be due to secondary derivation or long-distance wind transport. *Abies* charcoal from the late Bronze Age site of Hurst Fen is presumably imported. The wood identifications from Roman Silchester were confidently ascribed by Reid to staves of wine casks imported from Spain, an attribution confirmed by similar discoveries of staves of wine barrels brought from Bavaria or the Danube basin into northern Germany in Roman time (Hopf, 1956). The absence of resin canals from the wood favoured this use.

The absence of the silver fir from Britain in the Ipswichian and Flandrian accords with evidence from the European mainland. In the Flandrian it only expanded in the terminocratic stage, and its northwards progress has been carefully demonstrated and mapped (Walter & Straka, 1970).

At the present day the silver fir is a purely montane tree occupying a vegetational belt that rises in altitude from north to south, but even in its northernmost localities it grows far above the low altitudes it occupied in the Hoxnian. Firbas takes the view that its present range is climatically determined and ecological observation suggests a demand for a cool, moist environment and absence of spring frost. It was the great discrepancy between the Hoxnian and modern ranges of *Abies* that encouraged the belief that the interglacial species might have been *A. fraseri*, a species subsequently lost from Europe, but now that this seems excluded we must face either the assumption that 'this formerly widespread fir had other ecological demands than that of today' (Turner, 1970) or accept that the later Hoxnian of western Europe had a climate much more oceanic and probably warmer than that of today. This alternative conclusion has strong support from the evidence of the whole floristic and

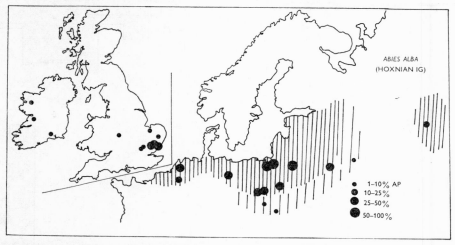

Fig. 38. Pollen frequencies of *Abies alba* Mill. in Europe during the Hoxnian interglacial, and in the British Isles during the Hoxnian sub-stage III (modified from Walter & Straka, 1970).

vegetational assemblage, not only in the far west, but in what are now the most continental areas of the British Isles (see p. 448). Climatic shifts of this magnitude may have substantially supplemented those of the glacial cycles in causing the evolution and isolation of related species of *Abies* that occur in the southern and eastern fringes of the present area of *A. alba*.

Larix decidua Mill. (*L. europaea* L.) Larch   32 : 1

J. C. Maby notes that this species is represented by a single cone, little charred and that it may be the result of nineteenth-century disturbance of the site: this is highly probable.

Thomson, in analysing samples of the Cromer Forest Bed series collected by Woldstedt, gives low frequencies of a grain 'cf. *Larix*' in a pollen assemblage of high arboreal/non-arboreal pollen ratio: Miss Duigan did not identify it in her many analyses of deposits of the same age.

Pinus sylvestris L. Scots pine, Scots fir (Figs. 39, 40; plates II, XIV*g*, *h*)   33 : 1

Pinus is well represented in the historical records of the British Isles, for not only is its pollen easily recognizable but it is abundantly produced and carried far afield. This fossil pollen, through its fluctuating frequency in numerous pollen diagrams, permits us to establish the changing status of the pine throughout our vegetational history; but because it is liable to long-distance transport, is not a safe guide to the local presence of pine in any given place. However, abundant macroscopic remains of wood, bark, needles and cones discovered *in situ* provide trustworthy records of past occurrence at or near the site of their recovery.

On the European mainland, Quaternary research workers must face the possibility that fossil remains of

Pinus may belong to one of several species, particularly *P. sylvestris*, *P. montana*, *P. mugo*, *P. cembra*, *P. nigricans* if we exclude the Mediterranean region, and only exceptionally are remains encountered which can be referred with certainty to a given species. In the British Isles by contrast we have no reason to suppose that during the Flandrian period any species other than *P. sylvestris* has been present. For the interglacials, though *P. sylvestris* was probably the species mostly represented, we cannot be certain that others did not also occur.

P. sylvestris occupied in the past, as it does today, two types of habitat which make it particularly liable to preservation in the deposits of growing peat bogs. It is, in the first place, highly characteristic of the transitional stages of vegetational succession which correspond with the initiation of raised bog above fen. The late stages of fen consolidation lead to the establishment of fen-carr and then fen-wood, at first rich in *Alnus* and *Betula pubescens*, but afterwards containing *Quercus* and *Fraxinus*. In regions favouring development to raised bog, however, at this stage (or before oak and ash establish themselves), base-tolerant *Sphagna* colonize the fen-wood floor, the soil of which becomes increasingly acidic. At this stage *Pinus sylvestris* may largely replace the other trees, only to succumb itself to the progressive waterlogging caused by continued accumulation of *Sphagnum*. This sequence has been most ably described by Steffen (1931) in his reports of the vegetation of the Baltic coast near the Kurischer Haf and the pine-dominated stage is referred to as 'Kiefernzwischenmoor'. No clear example of living pine in this plant community is known to me in the British Islands today, but our peat deposits, like those of the north-west German plain and of the Netherlands, give repeated evidence that it had this vegetational role (among others) in the past. We could hardly find a clearer example than that at Ynyslas on the coast of Cardigan Bay, where basal layers of Borth Bog are

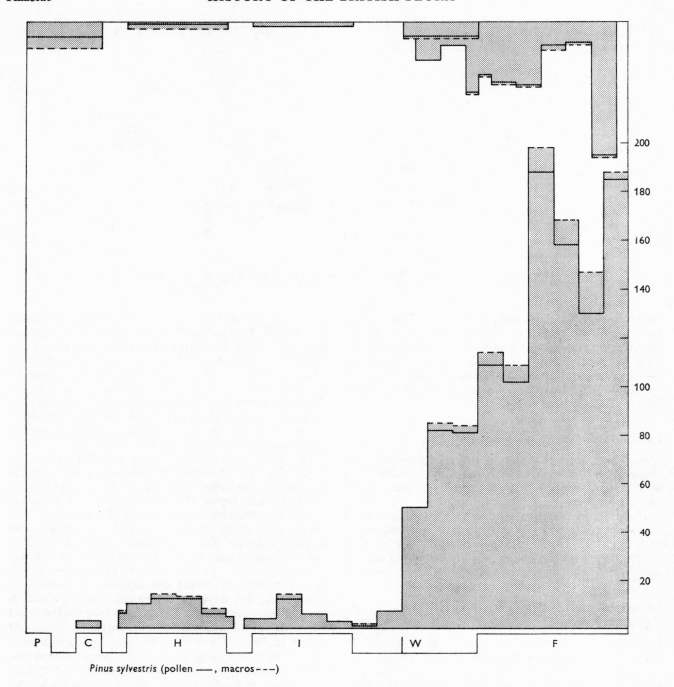

Pinus sylvestris (pollen ——, macros ---)

exposed at its seaward edge in the form of a 'submerged forest', where tree stools rooted in peat are constantly washed and eroded by the tides (plate II). Pollen analyses through this peat shew in sequence from below upwards maxima of pollen of alder, birch and pine. Trees of these genera and also of *Quercus* have been identified in the shore peats, and their root systems are spread in the shallow horizontal systems so characteristic of trees grown in fen woods where the high water table restricts the tree roots to the thin aerated surface soil (Godwin &

Newton, 1938). The same sequence is to be found inland in the basal layers of the Borth Bog itself, and in the sections exposed, by canalizing the River Leri, a magnificent series of pine stubs is seen to lie in this layer immediately below the old humified *Sphagnum* peat (see fig. 30). It will be noted that the pine pollen maximum at this level of the stub layer is not very large and this can only mean that most of the locally produced pollen was carried away by the wind and that *Pinus*, apart from these 'Zwischenmoorwälder', played a small part at this time in the general forest cover of the region. A very clear example of the local sequence through pine wood of this character is shown in the analyses by Brinkmann of sections in the Sehestedter Aussendeichsgebiet near the River Jade in Lower Saxony. The sequences of peat types and of non-tree pollen types here, as in our British example, fully substantiate the ecological interpretation outlined above. The raised bogs of the Somerset Levels, like those at Borth and on the north-west German coast, lie upon estuarine clay with intervening calcareous fen, and shew the same layers of pine, and the same local pine pollen maxima at the transition to oligotrophic bog. In the fenlands of East Anglia an approximation to the same conditions prevailed at that stage in the history of the the region, roughly the Sub-boreal period, when marine retrogression and climatic dryness continued to set the marginal peat lands free from flooding by calcareous water, and enabled them to move towards the development of raised bogs. Thus at Wood Fen, north of Ely, the classic site investigated by Skertchly, during the Bronze Age the fen woods of *Quercus* and *Taxus* became invaded by *Sphagna* and were replaced by woods in which *Pinus* preponderated (H. & M. E. Godwin & Clifford, 1935). Plate II shews the piles of pine stumps extracted from this fen during cultivation operations and after lowering of the fen water table by drainage. In the neighbourhood of Woodwalton Fen also, on the fen margin south of Peterborough, it has been shewn (Godwin & Clifford, 1938) that extensive growth of pine took place at the same stage and immediately before raised bog communities established themselves in the area. An absence of exact correlation in level between the layer of pine stubs and the maximum of pine pollen in a given section is explained by the fact that the strut action of the pine roots does not allow the stump to descend quite freely as the surrounding peat contracts following drainage. Both at Wood Fen and in the Woodwalton region, bark, pine needles and cones accompany the preserved trunks, as well as the hard resin-impregnated knots (often called 'cones' by the fenmen) formed by decay of the soft sap-wood from the central core of the trunk and the bases of its whorls of branches.

At Terneuzen, a coastal site in south-west Holland, Munaut employed a combination of dendrochronology and radiocarbon dating to demonstrate that the buried pine forest, overgrown by *Sphagnum* bog, was a single generation stand lasting about three centuries only, a conclusion conformable to the ecological explanations already suggested by the British coastal sites (Godwin, 1968c).

A second habitat type for the pine favouring preservation of its remains is that of the drier portions of raised bogs and blanket bogs. Although the central regeneration complex of British raised bogs is too wet for any but stunted and unhappy specimens of *Pinus*, such do in fact still occur and, upon the better-drained slopes of the marginal 'rand', examples may still be found of fairly dense growth of pines. Similarly they may occur where ancient peat-cuttings have given drainage conditions comparable with those of the uncut bog margin.

In these situations the pine will regenerate, and although in some instances the source of the colonies is undoubtedly sub-spontaneous spread from introduced trees, there is no reason against the view that some indigenous *P. sylvestris* has persisted upon the surface of our raised bogs from the earliest stages of their formation. Pines in this type of habitat were not necessarily always so sparse as they now are, and it is wise to recall the present existence of the densely tree-clad 'muskegs' of Canada, and the rich colonization by spruce and pine of the margins of the drier, more continental type of raised bog found in eastern Sweden. In this country, as elsewhere in Europe, periods of climatic dryness have often been marked by invasion of the bog surfaces by birch and pine, the latter tolerating remarkably well the high acidities developed by the weathering *Sphagnum* peat. A substantial part of the evidence for the dryness of the Sub-boreal period has always been the layers of pine stubs of this period exposed in the sections of the European bogs, and these are plentiful in all parts of the British Isles where acidic peat bogs form. Jessen (1949) reports that 'near the end of the period growths of pine and birch spread at many places out on to the bogs' and he cites as instances Cloughmills, Ballyscullion, Carrowreagh, Cloonacool, Castlelackan and Aughrim. Hardy (1939) reported a pine-stump layer in Whixall Moss, Shropshire, associated with a Middle Bronze Age palstave, and shewed that in the nearby moss at Bettisfield in addition to this pine layer a similar but less pronounced development had taken place in the middle of zone VII. Sub-boreal pine layers were similarly recognized by Whitaker (1921) and by Erdtman (1928) in the mosses of the Cleveland Hills, but all these records would benefit by radiocarbon dating where this is still possible (see p. 36). It has been remarkable that wherever pollen analyses have been made through such pine layers in the raised bogs, the pollen curves have shewn only small maxima of pine pollen, and this is undoubtedly due to the fact, as both Erdtman and Jessen explain, that the increase of pine was local, confined to the bog surfaces, and did not affect the general forest composition of the countryside. It may also be noted that *Pinus* does not appear to have been used in this country by prehistoric peoples for the construction of corduroy roads, although trackways of pine wood have been recorded from Germany. Here birch, alder and oak have been the timbers most employed.

Tree remains occurred with such consistency in Scottish peat deposits that they were ascribed by Geikie to two periods, the 'Upper Forestian' and 'Lower Forestian' respectively. Later workers such as Erdtman (1928) and Samuelsson (1910) concluded that the results afterwards obtained by Lewis conform reasonably with Geikie's schema and they regarded the Upper Forestian as corresponding with the Sub-boreal and the Lower Forestian with the Boreal in the climatic terminology of Blytt and

Sernander (fig. 40). The application of radiocarbon dating to buried tree layers has however not produced the expected clustering of dates at these times, either in Scandinavia or in the British Isles. For Scotland 'no climatic conclusions may be drawn from the occurrence of these pine stumps to support the Blytt–Sernander schema of climatic periods' (Switsur et al., 1970). It seems likely that local factors of aspect, elevation, drainage, etc., have played such a dominant role in causing death and preservation of the trees that only large numbers of samples would disclose any underlying climatic pattern. Since however the transition from the late Boreal to the early Atlantic is associated in this country with a great extension of ombrogenous peat formation, in part due to the conversion of fen to raised bog by the 'Zwischenmoorwälder' referred to earlier, and in part by the general waterlogging ('paludification' of von Post) of hitherto dry and forest-clad country, it would be surprising if pine stumps were not frequently found at this time.

The winged pollen grains of *Pinus* are readily recognizable, although many difficulties intervene between determination of relative frequency of the grains in a given sample, and stating the significance of this in terms of the frequency of the tree in the vegetation of that time. In the first place *Pinus* has a pollen productivity which is high in relation to that of most other trees. In the second place, although the large size of the grain must tend to increase the rate of sedimentation, this is offset by the reduction of density caused by the lateral air sacs, and the observed rate of fall in still air remains of the same order as that of unwinged smaller grains of angiosperm trees. On this evidence Erdtman rejects the notion that *Pinus* pollen is particularly liable to long-distance transport and explains the many instances of this phenomenon concerning *Pinus* grains as due to the fact that extensive pine woods frequently neighbour unwooded or sparsely wooded regions, where the pine pollen will inevitably appear to be heavily over-represented. A factor of very great importance must also be the vast area of pine dominated forests in northern and north temperate latitudes: these must ensure a large representation of pine in all situations where long-distance transport of pollen can express itself. Some notion of the distances over which pine pollen may be carried can be gathered from the fact that conifer pollen (including *Pinus*) reaches south-west Greenland in substantial amounts from the American mainland and that pollen of *Pinus* was picked up by Erdtman from air samples taken at all stages in his voyage from south Sweden to New York across the Atlantic. Uncertainty as to the weight to be given to such long-distance transport greatly hinders our appreciation of the exact localization of pine during the Weichselian period. Further, the pine pollen grains are resistant to decay and are often preserved in mineral sediments from which other pollen types have vanished. The air sacs give the grains such buoyancy that they float easily on water surfaces and at suitable seasons pine pollen forms a yellow bloom on the surface of lakes near to pine forests. There seems

some evidence from pollen diagrams that pine pollen frequency may vary very greatly in deposits of identical age taken from different parts of a lake basin. This has been seen in the cores of the bed of Windermere (Pennington, 1947) and in sampling of the former Lake Pickering in Yorkshire (Godwin & Walker, in Clark, 1954), and it seems attributable to local accumulation by drifting.

It has been noted for a long time now that where estuarine clays occur between peat layers, the clays invariably yield very high values for the percentage of contained *Pinus* pollen, values greatly in excess of those in the adjacent peat layers. This is strikingly evident in the 'Fen Clay' deposits of the England Fenland, in the basal brackish-water clay which underlies the Somerset Levels and in the clays of the Dutch and north-west German coasts. Various explanations have been proposed to account for the phenomenon, among them long distance wind transport, surface drifting, differential destruction and the incorporation of secondary pollen by erosion of pine wood peat. There is agreement only with the fact that the recorded figures grossly over-emphasise the frequency of pine in the contemporary forest cover and it is possible that different causes or combinations of causes have operated in different places.

When we turn to the assessment of the fossil record we find that although there are no pre-Cromerian records of macroscopic pine, its pollen occurred in high frequency in the Ludhamian where it dominated the preponderantly arboreal pollen spectrum, and again in the succeeding cooler Thurnian although arboreal pollen played a smaller total role (West, 1961b). In the Antian interglacial pine was more or less equal with other tree genera, both coniferous and deciduous. In the Baventian glacial stage, again with a low arboreal/non-arboreal pollen ratio, pine was dominant or co-dominant in the tree-pollen assemblage (Ludham and Easton Bavents). In these four stages we cannot be certain that the indigenous pine was *P. sylvestris*, nor in the Cromerian where there are large and sometimes dominant frequencies of pine pollen (West Runton, Duigan, 1963). From the Cromer Forest Bed series as a whole however Reid recovered from five separate sites macroscopic remains, including cones, which he attributed to *P. sylvestris*, and the probability is that this is the species largely or solely present in the early and middle Pleistocene.

This conclusion certainly holds for the Hoxnian interglacial where not only are there pollen records from all four sub-stages and the preceding late-glacial, but there are abundant correlations of these with macroscopic remains of cones, seeds, leaves, shoots, bark and wood (Gort, late Anglian sub-stages II, III, IV; Kilbeg, sub-stage II; Fugla Ness, sub-stage IV). Although in East Anglia pine pollen was present in low frequencies through the middle sub-stages of the Hoxnian it was more abundant in sub-stage I, and especially sub-stage IV; in western Ireland it retained high frequencies throughout the interglacial, though still strongly terminocratic. There are pollen records of pine from both early and late Wolstonian deposits. In the Ipswichian the pollen diagrams again shew pine present throughout but more frequent in sub-stages I and IV where broad-leaved trees except *Betula* are absent. At two sites (Shortalstown and Wretton) macroscopic remains also occur.

In the interstadial deposits at Chelford (Early Weichselian) not only was pine the dominant pollen, but cones and trunks were abundant along with leaves and seeds. With birch and spruce the dominant pine constituted a forest type like that now growing in Finland (Simpson & West, 1958). For the Middle Weichselian, as for the Wolstonian, records are restricted to pollen that is reasonably regarded as derived or transported from big distances. This situation is somewhat modified in the Late Weichselian where however the bulk of the pollen records are to be similarly explained. On the European mainland it seems possible, according to Firbas, that *P. sylvestris* survived the last glaciation north of the Alps only in France and a few favoured places in southern central Europe: thereafter, however, it spread rapidly so that we find it represented in zone II of the late-glacial by finds of cones and wood in east Prussia, Holstein and Denmark whilst pollen analyses indicate that in Holland and north-west Germany the latter part of the Allerød warm phase was characterized by some replacement of birch forest by pine. It is, however, uncertain as yet whether native pine was growing in the British Isles at this time. When Jessen and Farrington first published the evidence for sites of Allerød age in Ireland, they took the discovery of skeleton stomata of *Pinus* type, along with consistent but low pine pollen frequencies, as establishing the presence of *P. sylvestris* at this time. Subsequently Faegri pointed out that fossil stomata of *Juniperus* were not distinguishable from those of *Pinus*, and Jessen, in view of the proved presence of *Juniperus* in Ireland, abandoned the claim for native presence of *Pinus* in Ireland at this time. A zone II record for Windermere is based merely upon a tracheid fragment and the record of pine needles in this zone at Loch Kinord (Vasari & Vasari, 1968) is accompanied only by very low pollen percentages. This amounts to considerable uncertainty as to the presence of pine in the British Isles as early as the Allerød interstadial, but at Brook in Kent pine pollen frequencies rise to 40 to 50 per cent of the arboreal pollen so that one might guess that here, with no marine barrier intervening, we perhaps had a western extension of the pine woodlands of the Netherlands demonstrated by van der Hammen (1951), Polak (1963) and others. We are helped to accept such a conclusion by zone III records from the south of England where at Nursling, Hants. pine leaves and wood were identified with over 50 per cent of pine pollen (Seagrief, 1959), and at Elstead, Surrey where there were similar macroscopic finds and pine constituted about 30 per cent of the total arboreal pollen (Seagrief & Godwin, 1960).

Pollen frequencies throughout the Flandrian zones are displayed in the maps of fig. 39. In zone IV pine pollen frequencies of 10 to 20 per cent A.P. are common throughout Great Britain and this must surely indicate widespread local presence. Values in the south-east are much higher and are accompanied at Elstead, Nursling and Wareham by macroscopic remains. Pollen percentages are uniformly low in Ireland but remarkably high in north-western Scotland. (A high frequency at Glasson, on the Solway Firth, is for the combined zone IV/V.) By zone V pine had evidently spread northwards and was present abundantly over England and Wales on soils above all kinds of parent rock. There is shewn a tendency for greater abundance in the south and east, but now there is an expansion in the southern half of Ireland parallel with that in the rest of the British Isles, and locally as at Derryfadda and Cool Bog pollen frequencies are above 30 per cent A.P.

By the end of zone VI pine pollen frequencies are now high throughout the British Isles save for the southern half of Scotland and extreme south-western England. There are notably high frequencies throughout Ireland and at many sites the expansion was early in the zone. It was now that the area of the present-day Caledonian pine forests was first heavily colonized, and that pine woodlands extended to North Wales as well as the south. It is to be noted that macroscopic identifications widely accompany the high pollen frequencies.

In zone VIIa there is a sharp drop in pine pollen frequency throughout England, Wales and possibly central Ireland. High values however persist in north-western Scotland and in the Cairngorm area where H. H. Birks' pollen assemblage zone C4, with extremely high pine pollen frequencies, indicates dominance of that tree, and the establishment of the native Scottish pine woods of today. Pine stumps on the Cairngorm have been dated 5970 ± 120 and 6980 ± 100 B.P. and on the Spey–Findhorn watershed at 6960 ± 130 B.P. There is a north-western area of pine dominance embracing Loch Maree, Loch Clair and Loch Scionascaig but values are low in the Western Isles and to the north-east where *Betula* is the more important tree. In Ireland pine pollen values remain high in zone VIIa save for the central plain; there are abundant macroscopic remains, and Mitchell writes of this time as the *Alnus–Quercus–Pinus* period.

In zone VIIb pine pollen frequencies have become low throughout the British Isles save in the two Scottish areas just mentioned and in south-western Ireland. As macroscopic remains are found even where pollen frequencies are low we must assume persistence in locally favourable sites, as for instance in the marginal woods of the East Anglian Fenland. High pollen frequencies at Rodbaston in the Midlands and Wingham in Kent must have a similar explanation. Zone VIII exhibits further recession of pine everywhere except in the Cairngorm region. Scattered sites with high pollen frequencies reflect either locally favoured pine stands (and macroscopic remains still occur fairly widely) or situations such as high altitude or open grassland where long distance pollen transport is important.

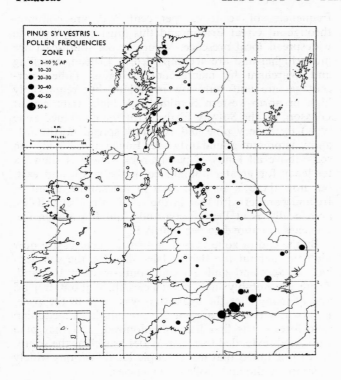

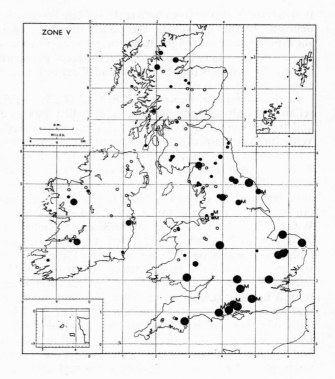

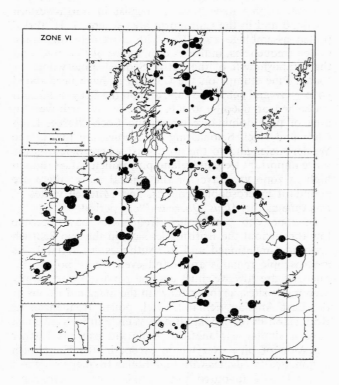

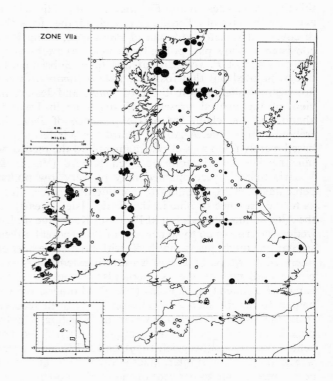

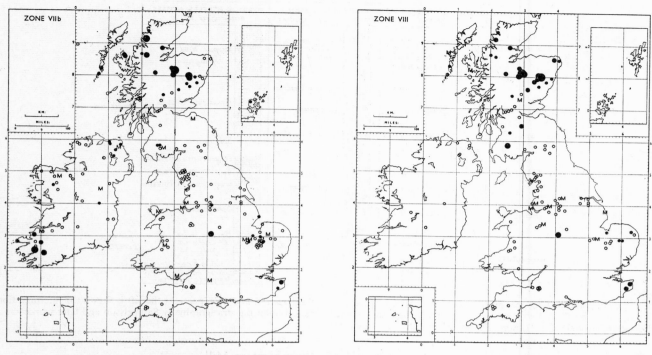

Fig. 39. Pollen frequencies of *Pinus sylvestris* L. in the British Isles through the six Flandrian pollen zones.

There can be no doubt that pine was strongly susceptible to human clearance. It seems likely that even were there no selective felling of pine, it was affected by general forest clearance from the earliest Neolithic onwards (Pilcher, 1970). Not only is pine very liable to fire, as seems to be shewn by the charcoal layers at Usselo and elsewhere in Holland in the Allerød period, but it was valuable as timber for many purposes. The macroscopic identifications indeed include three Neolithic, five Bronze Age, three Iron Age and one Mediaeval site.

Finally we have to note that in many pollen diagrams that extend to the present day increased pine pollen frequencies in the upper layers reflect the modern phase of replanting of woodlands: this is commented on for instance by Tallis (1965) for the Pennines and by Pilcher (1970) for northern Ireland.

In summary then it would appear that *Pinus sylvestris* has been present in the British Isles throughout all the Pleistocene except perhaps the severest parts of the glacial stages, and to judge from the Weichselian records it seems that it very quickly reappeared with amelioration of climate in the interstadial and terminal periods. So far as the evidence goes it suggests a generally similar role for pine in the four latest interglacials. In England and Wales at least it has been strongly terminocratic with an early phase of dominance before the expansion of the mixed oak forest and a later phase during decline of these woodlands with cooling of climate and progressive podsolization. Although *Pinus* tolerated both climatic and soil conditions of these terminal periods it must not be assumed

that it could not have flourished in the intervening hypsithermal: the tree regenerates well on thin chalk soil of southern England today but only in the absence of competition from the broad-leaved forest trees. Wherever through the middle stages of the Flandrian local conditions of altitude, exposure or poor acidic soils have been present, pine has grown within the main deciduous forest and only in the Flandrian interglacial has this picture of local survival of pine been obscured by the extensive felling and habitat modification by man, so that it is hard to say which, if any, populations of pine growing now in England and Wales are direct descendants of native trees and which owe their origin to extensive reintroduction of planted pine from the Continent.

The picture in Ireland differs in that pine was generally more abundant than in England and Wales both in the Hoxnian and in the middle Flandrian but subsequent disforestation probably affected it even more severely.

Our knowledge of the pine in Scotland is almost confined to the Flandrian. It is evident that in the Cairngorm region there was a great expansion of pine forest about 7000 B.P. and that the tree has not subsequently lost its dominance there except as a result of clearance. For this part of Scotland therefore pine forest must be conceived as the mediocratic element, a status in line with the view that the native Scottish pine forests are indeed the natural western extension of the great pine-forest belt of northern Europe.

Although the expansion of *Pinus* through the British Isles has certainly been metachronous, effective discussion

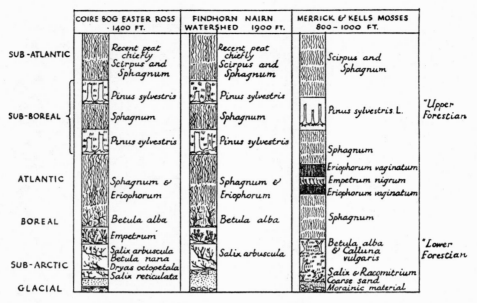

Fig. 40. Diagram shewing peat profiles from three Scottish localities as described by F. J. Lewis (1905, 1906). They shew the 'Upper Forestian' (pine) and 'Lower Forestian' (birch) separated by *Sphagnum–Eriophorum* peat. At the left is shewn the interpretation placed upon the stratigraphy by Samuelsson and later workers.

of this must await more numerous and apposite radio-carbon datings. These may shed light upon the behaviour of pine in that part of Scotland between about latitude 56° 45′ N and 55° 15′ N, broadly speaking the Scottish central plains and lowlands. By the end of zone VI, though pine was abundant north and south of this zone it has not apparently been dominant within it, and thereafter the zone remained in the oak–alder–birch forest that still seems its natural climax vegetation. If this gap in the range of former Scottish pine forest can be shewn to be real it will strengthen the case for considering that the tree reinvaded from the north-west as well as from the south, a fact possibly to be associated with the separate taxonomic status of the Scottish native pine as *P. sylvestris* var. *scotica*.

Pinus sylvestris has a wide European–Siberian range, extending from eastern Asia to Norway, and from the arctic forest limit in the north, southwards to the Russian steppe, the Caucasus, the Balkans, the Alps and the Sierra Nevada. The Pleistocene history of pine on the European mainland is very much in accord with that now emerging for it in the British Isles.

Pinus pinea L.   Stone pine   33 : 3

F VIII *Carrowburgh, Chew Park*

The very characteristic cones of the stone pine have been found in the debris of Roman wells and it has been conjectured that these were imported into Roman Britain as altar fuel. It is native of the Mediterranean and Portugal.

Pinus montana Mill.   33 : 4

Cs *Norfolk Co.*

Clement Reid (1899) reports that although *Pinus montana* was recorded by Heer and figured by Saporta from the Cromer Forest Bed, he himself has been unable to find evidence of it there. He indicates that possibly small cones of *P. sylvestris* may have been mistaken for those of *P. montana.*

Pinus haploxylon (Plate XIV*i*)   33 : 5

Cs *Bacton, West Runton, Mundesley*; *l*W *Nazeing*

Pinus haploxylon is a pollen type described by Rudolph from the late Tertiary of Bohemia. It has hemispherical air-sacs set upon the body of the grain, which is of the same diameter, so that there is no constriction at the junction between them (plate XIV). It has been conjectured that *P. haploxylon* may have been derived from an extinct species in the section *Strobus*, for this mainly north American group now has only one European species, *P. peuce*, whose pollen grains are however larger than those of *P. haploxylon*, though of similar morphology otherwise.

The Late Weichselian grains from the Lea Valley disclose by their condition that they are secondarily derived. At one site, Ostend, however Duigan (1963) found that no less than 4 per cent of the total tree pollen was *P. haploxylon* in a pollen spectrum dominated heavily by arboreal pollen, and it occurred in two other sites in the Cromer Forest Bed series that she investigated. It would therefore be wise to keep an open mind on the possibility that we may have here to reckon with an extinct species of pine still to be identified as native in the Pleistocene.

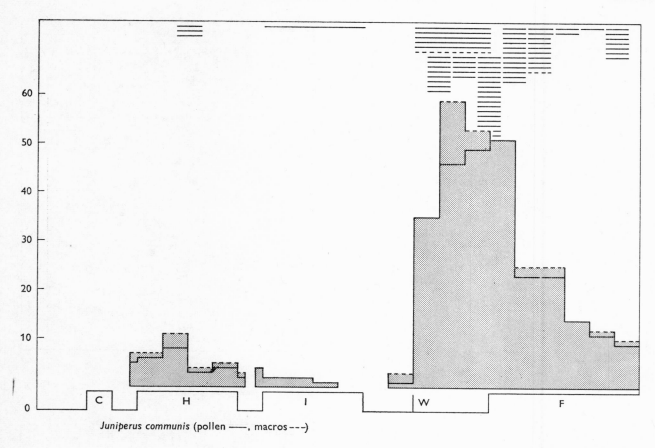

Juniperus communis (pollen ——, macros ---)

CUPRESSACEAE

Juniperus communis L.   Juniper (Fig. 41)   34: 1

The stems, 'fruits' and seeds of *Juniperus* are readily recognizable in a sub-fossil condition and from the Hoxnian interglacial there have been even identified the galls caused on juniper by the gnat *Oligotrophus juniperinus* (Jessen *et al.*, 1959). The heavily thickened stomata may also be recoverable when the remaining leaf-tissue has disappeared and though they resemble those of *Pinus sylvestris* they may occur at levels where juniper is the only likely source. *Juniperus* belongs to the group of Gymnosperms that have spherical pollen grains which open by the swelling of a middle exine layer splitting the exine almost completely into hemispherical halves. In juniper there is a small circular pore through which the slit passes and which is readily seen in intact grains. The two halves of the burst grain are often found more or less rolled into spindle shapes. The stalked spherical gemmae on the surface of the grain are very apt to be detached in fossilization or preparation and in any event they are sparser than in the pollen *Taxus* to which there is considerable resemblance (Bertsch, 1961; Jessen *et al.*, 1959). In the last twenty years improved standards of microscopy have led to the widespread inclusion of a juniper curve

in British pollen diagrams over most of the middle and late Pleistocene.

The record diagram shews abundant occurrences of both pollen and macroscopic remains in all the sub-stages of the Hoxnian interglacial with numerous instances of the presence of both together, especially in the Irish sites of Gort, Kilbeg and Baggotstown. Ipswichian records are less frequent, represent only three of the four sub-stages and are based on pollen entirely, but of course sites of this interglacial are fewer and largely in the most continental part of the British Isles. There are three Middle Weichselian records, the pollen-based one from a cave-deposit with a radiocarbon age of 18 000 years. The three Late Weichselian zones and the earliest Flandrian zone supply vast numbers of records reinforced by many more from transitional or combined zones, and in some instances such as Mapastown, zones II and III, pollen and macro-remains occur together. This extraordinary Late Weichselian prevalence of site records is matched by records from the late Anglian and early and late Wolstonian. After zone IV site records diminish in numbers through the Flandrian but the evidence of continuous presence is substantial. Aside from a possible doubt for the severest parts of the glacial stages, the evidence is strong for indigeneity in this country from the middle Pleistocene.

The mean pollen frequencies of *Juniperus* zone by zone

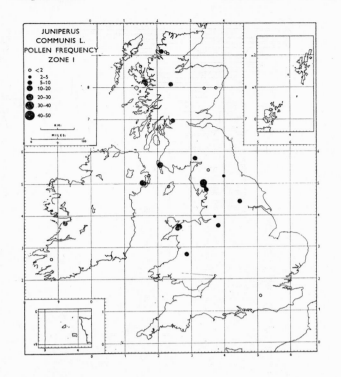

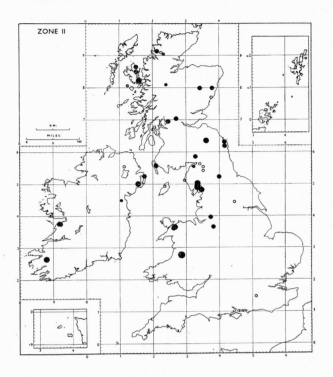

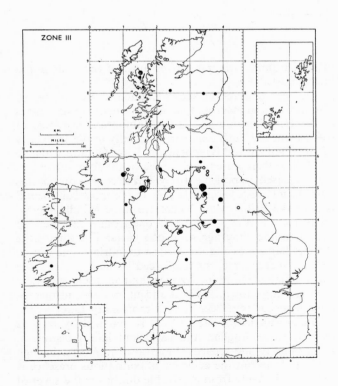

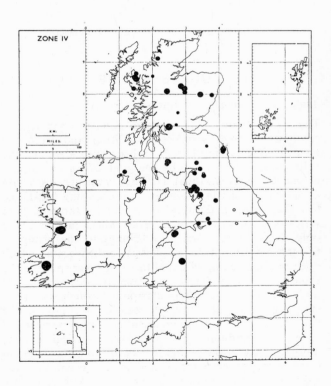

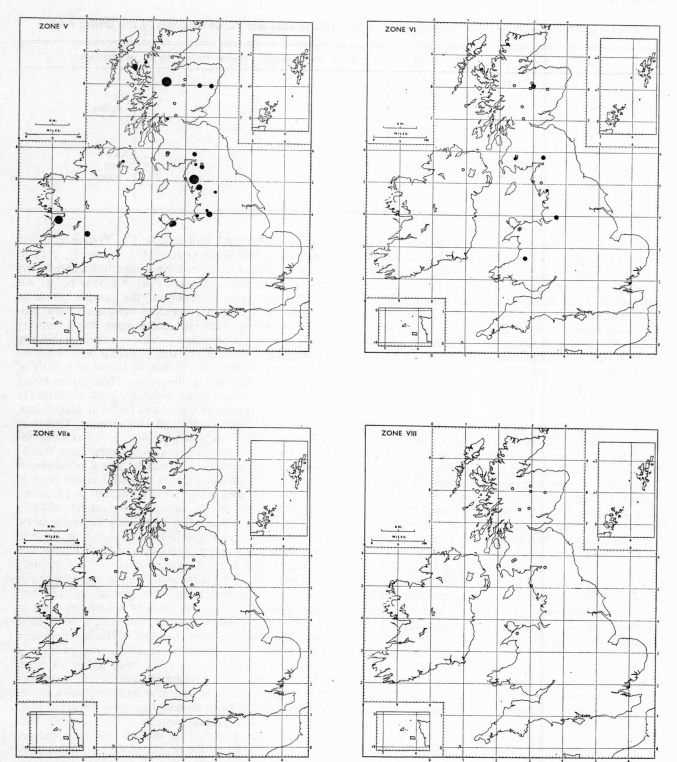

Fig. 41. Pollen frequencies of *Juniperus communis* L. in the British Isles through the eight pollen zones of the Late Weichselian and Flandrian.

TABLE 9. *Mean pollen frequency of* Juniperus *in Late Weichselian and Flandrian zones of the British Isles*

Zone	+	0–2	2–5	5–10	10–20	20–30	30–40	40–50	50–60	60–70	70–80	Percentage
I	1	11	7	10	5	.	.	.	.	.	.	total pollen
II	2	8	6	15	9	1	.	.	.	.	.	
III	2	15	14	7	2	.	.	.	.	.	.	
III/IV	0	1	1	2	7	4	2	.	.	.	.	
IV	2	5	5	11	11	1	.	.	1	.	.	
IV	.	.	2	3	5	2	.	1	.	.	.	total tree pollen
V	.	2	6	.	.	.	.	.	.	1	1	
VI	4	2	2	3	.	.	.	.	.	.	.	
VIIa	3	8	.	.	.	.	.	.	.	.	.	
VIIb	4	3	.	.	.	.	.	.	.	.	.	
VIII	4	1	.	1	.	.	.	.	.	.	.	

through the Late Weichselian and Flandrian are displayed in table 9. It is evident that juniper was present in substantial frequency in zone I but that frequencies increased in the milder zone II with a modal 5 to 10 per cent of total pollen. Zone III saw modal frequencies decrease greatly, but in zone IV and the transition zone III/IV they rose to the highest level, with especially high values at western Irish and high altitude sites. Zone V witnessed a sharp fall save at the high altitude sites of Loch Tarf and Bigholm Burn where the bulk of tree pollen was due to distant transport and local juniper stands were correspondingly well represented. In zones VI to VIII only low frequencies have been recorded. The distribution maps (fig. 41) supplement the tabular data by shewing that throughout the zones juniper has particularly characterized the north and west of the British Isles: where it has been recorded from the south and east values have always been low, primarily no doubt a climatic effect. The recession in frequency of zone III is especially noticeable in Scotland, and only two sites, one in the Lake District and one in Antrim, shew any increase. The zone IV increase is widespread, but nowhere so great as in western Ireland. Zone V shews general decrease, though individual widely separate sites have high values indicating that local circumstances are there outweighing the overall diminution. The values for all subsequent zones are low. This collected information has naturally to be seen against the behaviour of juniper pollen in individual pollen diagrams, where, since Iversen first drew attention to its importance, it has been given a strongly diagnostic role in the Late Weichselian and early Flandrian. The large, though temporary peak in juniper frequency at the transition between these stages he attributed to the fact that although present in zone III it was kept stunted and non-flowering by the sparse snow cover and severe climate. With the climatic amelioration it responded by proliferation, profuse flowering and regeneration until the slightly slower expansion of birch woodland suppressed the shade-intolerant shrub. A pronounced maximum at this time has been repeatedly found in the British Isles and radio-carbon dating has fixed its age at Scaleby Moss, Cumberland (*c.* 10 200 to 9600 B.P.). Red Moss, Lancashire

(*c.* 9500 to 9800 B.P.), Ballynagilly, Co. Tyrone (*c.* 10 000 to 9400 B.P.) and Slieve Gallion, Co. Tyrone (*c.* 9600 to 9000 B.P.). British authors have generally accepted Iversen's interpretation, although Pilcher (1970) points out that the 600 years duration of the juniper phase in northern Ireland and Cumberland indicates a slow progressive warming of climate, and suggests the possibility that juniper might have had a decisive role in stabilizing and developing soils at this time. There is also considerable local variation in the pattern of bursts of activity of juniper during the early pollen zones. Thus, just as Polak and van der Hammen found the major peak of juniper in the Netherlands in zone I, so it was found at Ballydugan, Co. Down by Singh (1970) and at Tadcaster, Yorkshire by Bartley (1962). The variability was particularly demonstrated by Watts (1963) in his study of three Late Weichselian sites in western Ireland. In the Carboniferous Limestone area of Co. Clare it shewed a minor peak in I, was low in II and had a large maximum in IV; in the boulder-clay area of the same county minor peaks in both I and II were succeeded by a fairly large one in IV, whilst in a sheltered valley in the Old Red Sandstone of Kerry there was a major peak in the first half of II as well as a larger peak in IV. Another aspect of this variability is the apparent increase in magnitude of the juniper rise of zone III/IV across Ireland from east to west (Pilcher, 1970). The suppression of juniper sometimes seen in zone II after an initial expansion in zone I, must have an explanation quite other than that of shade suppression by birch woodland given for its fall in zone IV since at many sites displaying this effect zone II has so much pollen of forbs as to warrant the attribution to 'grass–sedge tundra'. Possibly such communities, however determined, would deny juniper the open disturbed soils needed for seedling establishment.

The importance of the Late Weichselian role of juniper is strongly recalled by the details of vegetational history in the late Anglian and early Hoxnian at Gort where Jessen *et al.* (1959) report pollen values as high as 68 per cent, accompanied here as in zone I of the Late Weichselian in southern Germany, by a corresponding maximum in the pollen of *Hippophaë* (Bertsch, 1961). They suggest

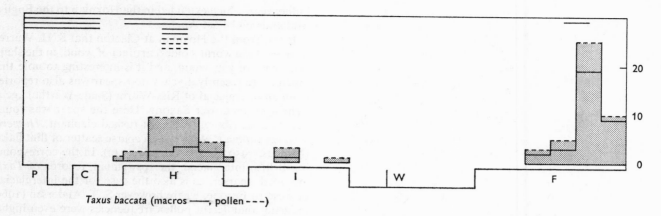

Taxus baccata (macros ——, pollen - - -)

that the local vegetational sequence has been *Betula nana* heath → *Juniperus–Hippophaë* scrub → forest, very much as in the Late Weichselian, and they point to some increase in juniper frequencies in the terminal stages of the Hoxnian as does Watts (1959*b*) for Kilbeg.

Within the Flandrian juniper was generally suppressed by the expanding birch forest, but locally, as in the Cairngorm, *Corylus* was its main competitor (H. H. Birks, 1970). It tended to survive where woodlands were open, as on the thin soils of hard limestone surfaces, but its major refuge must have been in montane areas above the tree-line where it exists today in extensive communities thanks to considerable resistance to sheep-grazing, wind and suitability to the cool, wet conditions. It may indeed have been favoured to some extent by late upland tree felling, for Florin (1963) has noted that in south Sweden heavy grazing in pine areas on sandy soil caused substantial increases of juniper during the Late Bronze Age and Early Iron Age. On the other hand we must accept the view of Tallis (1964) that the growth of peat mires from the Atlantic period onward has led to wide extinction of juniper in such montane areas as the Pennines, where its stems, along with those of birch and willow occur frequently in the wood layers beneath the blanket bog. Juniper is very susceptible to fire and its frequency in Scottish heaths has been reduced by the practice of repeated burning (Burnett, 1964).

Juniper occurs throughout the British Isles though most commonly in the northern half of Scotland, and the English Lake District. It has considerable ecological amplitude outside the area of dominance of lowland deciduous forest, occurring in pine and birch woods of the Scottish Highlands, and on heaths and moors, whilst it may be the dominant in communities of the limestones of southern England and above the tree-line in northern montane areas like the Cairngorm and the English Lake District. In Scotland it is an important component of the high altitude and euoceanic dwarf-shrub heath, the ssp. *nana* characterizing the *Juniperetum nanae* of McVean and Ratcliffe (1962). A typical Scottish occurrence of juniper also is in the acidic coastal dunes northwards of the Dornoch Firth. *Juniperus* has a wide extra-British range

extending far into the Arctic Circle. Although the pollen is referable only to the genus and *J. communis* is taxonomically variable, comprising what are often distinguished as ssp. *communis* and *nana*, for the purposes of this account we have treated all records as of the one Linnean species.

TAXACEAE

Taxus baccata L.   Yew (Figs. 42, 43)   35: 1

Taxus baccata, the yew, has very recognizable wood, leaves, seeds and male cones, and all of these have been found fossil. The stools of the tree are by no means infrequent in stub layers, for, although the species is not now listed as a natural component of fen woods, it grows quite well planted upon fen peat, and its stratigraphic position along with *Quercus*, *Pinus*, *Alnus*, etc. in wood peats in the East Anglian fens and in many 'submerged forest' beds shows that it must formerly have held this status. In the drainage operations of the Cambridgeshire fenlands many of the 'fen oaks' uncovered during cultivation turn out in fact to be stools of *Taxus*: they may be as much as 2·0 m in diameter and of several hundred years growth. Their wood is so hard that the fenman's axe rebounds from them and they are often clearly rooted in peat layers above older layers of stumps as at Wood Fen, near Ely, where male cones also indicate the contemporary ground surface. Despite its frequency in these wood peats the susceptibility of the tree to bad drainage is shown by the strongly one-sided development of roots and tilted crown shewn by many of the stools (fig. 43). The submerged pine forests of the Dutch–Belgian border described by Munaut include an upper horizon with numerous yew stubs as well as substantial pollen frequencies of *Taxus* (see Godwin, 1968) and there is clear inference in the pollen analytic–stratigraphic studies of the Neolithic trackways of the Somerset Levels that *Taxus* grew in local fen-carr dominated by *Alnus* and *Betula* (Coles & Hibbert, 1968; Coles, Hibbert & Clements, 1970).

The identification of *Taxus* pollen is thoroughly dis-

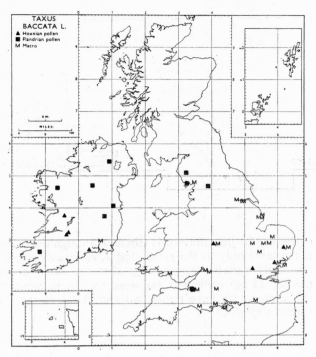

Fig. 42. Evidence of pollen and macroscopic remains of *Taxus baccata* L., the yew, in the British Isles during the Hoxnian and Flandrian interglacial periods.

cussed by Jessen, Andersen and Farrington (1959). The spherical grains split open in the same manner as those of *Juniperus* but less widely, and the minute heterogeneous gemmae are irregularly and often densely crowded on the surface, often seeming to touch one another. Occasionally, no doubt near a local source, *Taxus* pollen may be present in frequencies of 60 or 70 per cent of the total tree pollen, as in the Hoxnian at Gort.

There are three records of macroscopic remains of *Taxus* from the Cromer Forest Bed series, but within the interglacials the Hoxnian represents the great floraison of the yew. Pollen records come from all the Hoxnian sub-stages and there are macroscopic remains from three of them, with pollen and macroscopic remains together at Nechells, Hoxne, Marks Tey, Kilbeg and Gort. At Gort high pollen frequencies occur first in the local stage 3 in correlation with high values of *Fraxinus* and *Alnus* which are attributed to the prevalence of fen woods on the extensive limestone outcrop. The subsequent decline in *Taxus* is attributed to the extension of acidic terrestrial peat that favoured such species as *Picea*, *Abies* and *Rhododendron ponticum*. A substantial recovery in stage 5 is associated with some restoration of the conditions of stage 3. This emphasis on an edaphic control is confirmed by the relatively low frequencies of *Taxus* pollen in the same interglacial sub-stages at Kilbeg which is situated in the middle of an area of very acidic rock (Watts, 1959*b*). No doubt, as these authors recognize, the yew was particularly well suited by the mild oceanic climate, the

influence of which extended in the Hoxnian to the English Midlands and East Anglia.

It was from the Hoxnian at Clacton that S. H. Warren recovered the world's oldest artifact of wood, in the shape of a spear of yew wood, and it is interesting to note that much more recently a yew wood spear was also reported from an interglacial of Riss-Würm (Saale-Warthe) age at Lehringen in Lower Saxony. Here the spear was found between the ribs of a straight-tusked elephant, *Hesperoloxodon antiquus*, along with a sparse scatter of flint flakes (Movius, 1950; Jacob-Friesen, 1949). In the corresponding British interglacial, the Ipswichian, records of *Taxus* are much sparser, as is also the case for the interglacials of western Europe, where however S. T. Andersen (1969) indicates that *Taxus* pollen frequencies were even higher in the Herreskovian than in the succeeding Holsteinian (=Hoxnian).

It is unsurprising that a plant that avoids all regions of strong winter frosts should have no records from any part of the Weichselian, and the isolated records from the late Anglian and early Wolstonian are very likely due to secondary incorporation.

The earliest Flandrian records are for macroscopic remains in zone VI at Clonsast and Derrycassan in Ireland and for pollen in the beginning of that zone at Urswick Tarn at the southern end of the English Lake District. Records of both kinds increase greatly in frequency thereafter to a pronounced maximum in zone VIIb. Numerous archaeological sites contribute to the latest zones: Neolithic 7, Bronze Age 4, Iron Age 3, Roman 3 and early historical 1. This partly reflects the use made by prehistoric man of yew wood, not only for spears, as already indicated, or for axe hafts as in the Dutch Neolithic, but *par excellence* for bows, of which the two oldest examples, from Meare Heath and Ashcott Heath in the Somerset Levels, were respectively dated 2690 ± 120 B.C. and 2665 ± 120 B.C. (Dewar & Godwin, 1963). Yew pegs were

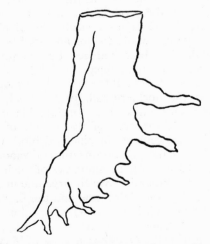

Fig. 43. Diagram shewing root system of a yew (*Taxus*) excavated at Green Dyke, near Woodwalton Fen, Hunts. The tilted crown, a common feature of yew stools in the Fenlands, is due to growth of the tree on the margin of drainage channels.

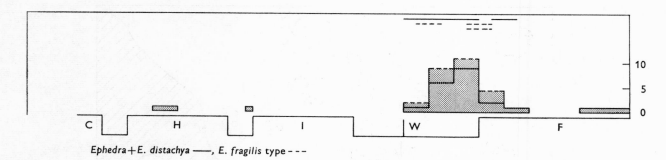

Ephedra+E. distachya ——, E. fragilis type - - -

employed for fastening the Neolithic 'Bell' trackway in the same area, a use that implies easy derivation from the local fen woods mainly of alder and birch in which the pollen evidence indicates that *Taxus* also grew. Withies of *Taxus* wood were also employed as stitches fastening the main timbers of the 'sewn' boats at North Ferriby, referred by radiocarbon dating to a period between about 1600 and 750 B.C. (Wright & Churchill, 1965). It must be recalled further that the tree has a strong tendency to be associated with religious and cultural practices, and it was certainly planted in this country in mediaeval times.

T. baccata has a Mediterranean–montane and west and central European distribution range. In the British Isles it extends north to Perth and Argyle and in Scandinavia to 63° N on the Norwegian coast and to 61° N in Sweden. There is little doubt that this limit is set by the tree's intolerance of winter cold, a fact probably to be related to the late appearance and extension of the tree in the British Isles as on the north-west German plain, where it only appeared 'at the transition from the middle to the later part of the warm period' (Firbas, 1949). The yew is also without doubt strongly hygrophilous and at the present day regenerates freely on various soils in this country (especially upon chalk and other limestones) and besides constituting a subordinate tree layer in beech woods, forms at places such as Kingley Vale pure yew woods, or yew–ash woods of considerable age and extent. There is no effective evidence to support Salisbury and Jane's contention that the yew has diminished in the south of England since Neolithic time (Godwin & Tansley, 1941).

EPHEDRACEAE

Ephedra (Fig. 44; plate XXIII*v*)

Not until the early 1950s was much attention directed towards the recognition of *Ephedra* in Pleistocene deposits, but in 1951 Iversen published a brief note indicating that pollen of *Ephedra* cf. *distachya* had been found by Brorson Christensen in Scania, by Krog in Copenhagen and by himself in Bornholm, in every instance within Late-glacial deposits. Upon the strength of this evidence Iversen suggested that *Ephedra* should be added to the list of steppe elements in the Late-glacial flora along with

plants such as *Centaurea cyanus*, *Hippophaë rhamnoides* and *Artemisia* species. It had also been identified by Lang (1951) in similar association from southern Germany. At about the same time Miss Andrew had independently recognized an *Ephedra* grain in the Late Weichselian deposits at Nazeing in the Lea Valley though this was then thought of as secondarily derived. Later she recognized three grains from different layers of the Late Weichselian at Whitrig Bog, Berwickshire and records from this period have subsequently multiplied all over Europe.

The pollen grains are highly recognizable: they are elongate ovoid with a number of thickened ribs of exine running meridionally down the grain from end to end. They project as strong ridges and in each valley between runs a more or less wavy fossa (colpus) from which short lateral branches extend at right angles up the flanks of the ribs; their alternate disposition is associated with the undulations of the main channel. Authors such as Welten (1957) and Steeves and Barghoorn (1959) have given attention to the recognition of the species of *Ephedra* by pollen-grain characters but British palynologists have been unable to go beyond the recognition of two types, one the *E. distachya* type with 3 to 8 ribs (but mostly 4 to 6), the colpi moderately zig-zag with long unbranched side branches arising normally, and the other, the *E. fragilis* type with many more ribs (9 to 12 or 15), with nearly straight colpi almost devoid of lateral branches.

The record-head diagram emphasizes the clear association of *Ephedra* with the Late Weichselian (especially zones II and III) and early Flandrian (zone IV). The greater numbers of records are for *E. distachya* type although early records often do not specify this: *E. fragilis* came to be recognized later. Despite a single late Wolstonian record there is yet no real parallel to the Late Weichselian occurrences in any earlier glacial stage. Miss Andrew encountered one grain of *E. distachya* type in the upper (zone VIII) layers of the raised bog at Shapwick Heath to which it must have been wind carried from a considerable distance. The *E. fragilis* type has been recorded by H. J. B. Birks from the Late Weichselian and early Flandrian in Skye, thus Loch Cill Chriosd (zones I, II, III, III/IV), Loch Fada (II, III, IV), Loch Meodal (II) and Loch Coir' a' Ghobhainn (I/II and IV). H. H. Birks records it also from zone III/IV in the Cairngorm.

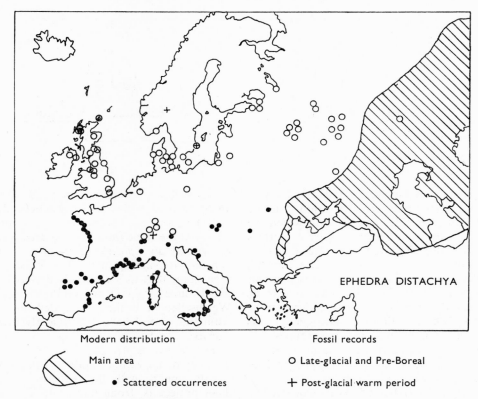

Fig. 44. Modern European distribution of *Ephedra distachya* and fossil pollen records of *Ephedra*
cf. *distachya* (adapted from Iversen, 1964).

The regrettable absence of macrofossils from the British and indeed the whole European scene throws us back upon an evaluation of the pollen record as evidence of local presence of *Ephedra*. The situation has been very comprehensively considered by H. J. B. Birks (1969) in the light of the abundant evidence for long-distance wind transport of *Ephedra* at the present day in southern Europe and in North Africa, of the present ranges of *Ephedra* species with pollen in the two respective types and of the pollen assemblages in which the fossil *Ephedra* has been found.

As early as 1951 Iversen committed himself to the view that *E. distachya* was part of a steppe-element, along with *Centaurea cyanus*, *Artemisia*, *Sanguisorba minor* var. *muricata*, in the European Late-glacial. Iversen in his note envisages the possibility that the Scandinavia records might be due to wind transport from some distance, but the present consensus of opinion is that on the Continental mainland it must have occurred in local stands in favourable places, and this may perhaps hold also for the British Isles. *E. distachya* L., the species most likely to be represented by the pollen type, is a steppe plant with its major area of continuous occurrence in central Asia, southern Russia, the Caspian and northern Black Sea regions, but as the map given by Walter & Straka (1970) makes apparent, it has many disjunct areas in southern

and Mediterranean Europe, many along coastal dune systems, including the Biscay coast of western France as far north as Brittany (fig. 44). These many separate present occurrences are most readily explicable as disjunct from a continuous Late Weichselian range which the contemporary flora and fauna indicate may well have gone far north across Europe. As the North Sea and English Channel were dry land it is hard to exempt England from this presumed past range. Where of course the local pollen production is low, as it was in the Late Weichselian, the long-distance transport component gains in relative importance and it will be hard to identify the marginal area fed by air-borne pollen from the central region where the plant is abundant.

The case with *E. fragilis* is different. Whereas *E. distachya* is an Old World species with a substantial central and south European range, *E. fragilis* is a more southerly Mediterranean plant, and pollen of this type is produced by a majority of North American species. Birks has calculated that in Skye and in Norway and Sweden the ratio of *E. fragilis* to *E. distachya* type grains is far higher than in Denmark, and deduces from this and the presence of *Ambrosia* pollen, that the Scottish Late Weichselian and early Flandrian pollen of *E. fragilis* type originated in North America.

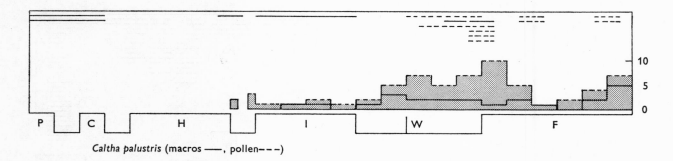

Caltha palustris (macros ——, pollen – – –)

ANGIOSPERMAE

RANUNCULACEAE

Caltha palustris L. Kingcup, Marsh marigold, May blobs (Plate XVII*i*)   36: 1

The seeds of *Caltha palustris* are easily recognizable by their somewhat irregular cylindrical shape, usually constricted in the middle, and by the surface which is black and smooth, showing a pattern of irregular, small, rectangular areas. There is a *Caltha* type of pollen grain (Faegri & Iversen, 1964) and although this includes *Aquilegia vulgaris* (H. J. B. Birks, 1969), there is little doubt that the majority of pollen records derive from *Caltha* itself nor again that most represent *C. palustris*. No one has distinguished the fossil seeds of the subspecific *C. radicans* T. F. Forst from those of the aggregate species.

The seeds are twice recorded from the Cromer Forest Bed series at the Hoxnian/Wolstonian transition, three times in the Ipswichian and in small frequency through the Flandrian. They occur in the early Wolstonian glacial stage, in the Early, Middle and Late Weichselian. The pollen site records, as may be seen from the diagrams above, correspond remarkably closely with those for seeds, with one Hoxnian sub-stage IV record, sites in the Ipswichian I, III and IV sub-stages and in all zones of the Flandrian, with also those late Wolstonian and Early, Middle and Late Weichselian records. For both seeds and pollen, site records are specially abundant in the Late Weichselian: although this reflects the special attention given to this period and the frequency of aquatic sites investigated, it strengthens the probability that *Caltha palustris* survived the glaciations in this country. This view is consonant with its biogeographical grouping as a 'Boreal–circumpolar plant with largely continuous distribution' (Hultén) and with its range throughout Scandinavia to the North Cape.

Trollius europaeus L. 'Globe flower'   37: 1

F IV *Loch Cill Chriosd*, V–VIII *Loch Dungeon*, VIIA *Altt na Feithe Sheilich*, ?*Treanscrabbagh*

Apart from a tentative seed identification by Mitchell from zone VIIA of the Flandrian, all the records given above are pollen identifications due to Dr and Mrs Birks and come from Scottish sites. They indicate continuous presence of the plant in Scotland through the Flandrian, a conclusion corresponding with the biogeographic status of the plant as 'North European with a boreal montane tendency' (Hultén), and its strongly northern montane range in the British Isles.

Clematis vitalba L. Traveller's joy, Old man's beard
                                                        45: 1

H *Clacton*; H/I *Stoke Newington*; F VIIA *Oakhanger*

By reference to K. Pike's pollen analyses in the Clacton channel deposits, it seems probable that the beds from which *Clematis vitalba* came fall within the warm middle part of that interglacial. Stoke Newington may be also Hoxnian or Ipswichian. These are fruit identifications but there is a single Flandrian record from Oakhanger of wood, carbon dated to 6300 B.P., associated with Mesolithic artifacts and pollen spectra of early VIIA. It is a species of decidedly southern present-day distribution in England and Wales.

Ranunculus sp. Buttercup (Fig. 153)   46

It very often happens that the accidents of partial preservation or the idiosyncrasy of particular specimens make it impossible to refer the fossil achenes of *Ranunculus* decisively to particular species.

The pollen grains of *Ranunculus* show some variability in structure, and the genus appears in three different places in the identification key of Iversen and Faegri. Although the tricolpate is commonest, as often happens in other genera, this type has given rise to pericolpate grains also. *Ranunculus* pollen has been identified routinely in recent British pollen diagrams.

Fruits have been found in all sub-stages of the Ipswichian, and in all zones of the Flandrian, with more site records in zones VIIB and VIII than previously, a result possibly associated with increased clearance and agriculture. There is also one Cromerian record. In the glacial stages they are frequent in all zones of the Late Weichselian.

There is a substantial pollen record also to be taken into account; many authors identify '*Ranunculus*' and

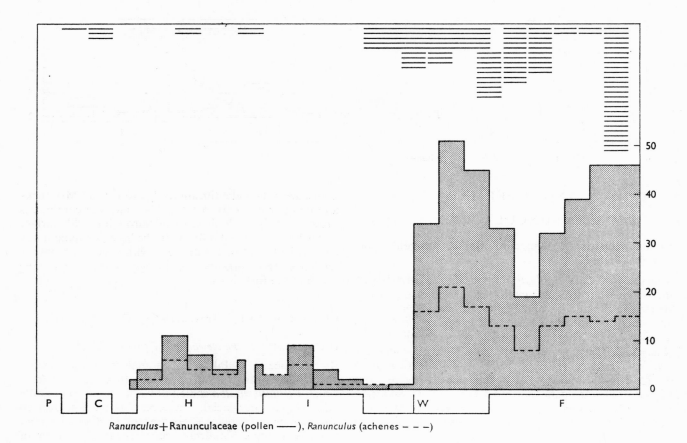

Ranunculus+Ranunculaceae (pollen ——), Ranunculus (achenes – – –)

rather more pollen of 'Ranunculaceae' but in effect there is no serious difference in the choice of name, and, as the pollen records under the two heads are so similar and follow so closely the distribution of macrofossils of the genus, we have reported them all under the generic heading. There are isolated records from the early glacial and interglacial stages and then from all sub-stages of the Hoxnian and Ipswichian and from all zones of the Flandrian. There are three Anglian and twelve Wolstonian records, and after single records in the Early and Middle Weichselian, very large numbers of sites in all zones of the Late Weichselian.

These Weichselian records are for pollen in low frequency, seldom over 5 per cent of total pollen and the same applies to zones IV to VIIa of the Flandrian as percentages of total tree pollen. In VIIb and especially zone VIII individual sites shew frequencies up to 20 per cent of total tree pollen or more. At Bigholm Burn similarly atypical very high frequencies are recorded in zones III and IV, no doubt reflecting purely local conditions (Moar, 1969a).

The Quaternary fossil record without doubt reflects the predilection of the genus for open and moist situations and prevalence in cool temperate climates of the northern Hemisphere, as indeed appears from the records of individual species that follow.

Ranunculus acris L.   'Meadow buttercup' (Plate XVIIg)
46: 1

Ranunculus acris is a west European–north Siberian species found throughout Scandinavia to the extreme north and throughout all the British Isles. Fossil achenes of *R. acris* have been found in small numbers in the four latest interglacial stages, Cromerian (1), Hoxnian (2), Ipswichian (8) and Flandrian (20). They were also present in the two latest glacial stages, Wolstonian (2) and Weichselian (15). It is to be noted that there are five records from the Middle Weichselian and five from the Late Weichselian zone III when climatic conditions had become severe after the Allerød amelioration. It is interesting (though not recorded in the heading diagram) that the Late Weichselian fruit records are paralleled by records by H. J. B. Birks of pollen of '*R. acris*' type, which however includes pollen of many other species.

The number of site records in the Flandrian increases sharply in the zones characterized by forest clearance and eleven of the seventeen records from these zones are actually from settlement sites, two Neolithic, six Roman and three Mediaeval. This is not surprising in view of its frequency in meadows and pastures where it is strongly resistant to grazing, and of the pollen-analytic evidence given by Iversen (1969) that in Denmark Neolithic forest

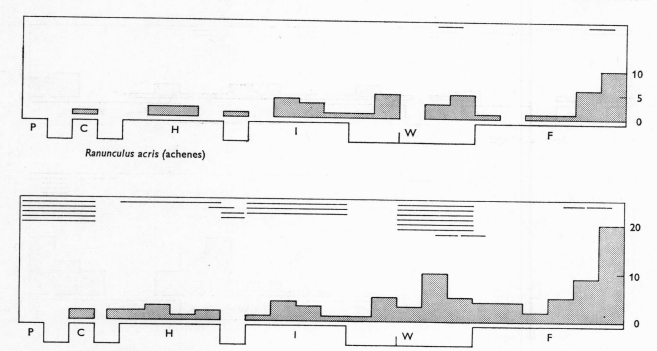

Ranunculus acris (achenes)

Ranunculus repens (achenes)

browsing after burning was habitually indicated by low frequencies of *Ranunculus* pollen of *acris–repens* type preserved in the mor forest soil.

Ranunculus repens L. 'Creeping buttercup'
(Plate XVII*h*)   46: 2

In addition to five records from the 'Cromer Forest Bed series' there are two specifically from the Cromerian Interglacial. Thereafter it is represented in every sub-stage of the Hoxnian and Ipswichian and in every zone of the Flandrian. It is present in the late Anglian glacial stage, in early and middle Wolstonian, and throughout the Weichselian, with ten records in the milder Late Weichselian zone II. There thus seems a very strong presumption of long persistence in this country through the Pleistocene.

As with *R. acris*, there is a great increase in the numbers of site records in the two or three latest Flandrian zones and again there is a very strong association with human occupation, viz., Mesolithic (2), Bronze Age (2), Iron Age (2), Roman (4), Norman (1) and Mediaeval (5). A number of Weichselian and Flandrian records are attributed to '*R. acris* or *repens*' but they all accord with the similar distribution already shewn for these two species.

The species belongs to Hultén's category of 'circumpolar plants strongly spread by cultivation'. It has a very wide range in Europe, extending to the North Cape and occurs throughout the British Isles in disturbed soils and in both arable and pasture.

Ranunculus bulbosus L. 'Bulbous buttercup'   46: 3

Ba/P *Norfolk Co.*; H *Clacton*; I ?*Allenton, West Wittering,* ?*Histon Rd*; *e*W *Sidgwick Av.*; *m*W *Barnwell St.*; *l*W *Nant Ffrancon*; F*vii*b *Elland,* *viii* ?*Cottenham, Mersea Is., Newstead*

R. bulbosus is a plant of southern distribution in Scandinavia, and although it is found throughout the British Isles it is more frequent in England. Though the fossil record is slighter it resembles that of other species of the genus in being represented both in interglacials and during the Weichselian glacial. The last three Flandrian records are all from Roman occupation sites.

Ranunculus sardous Cranz. 'Hairy buttercup'   46: 7

I *Allenton, West Wittering, Ilford*; *e*W *Sidgwick Av.*; *m*W *Earith*; F *viii Godmanchester, Silchester, Bermondsey, Caerwent*

The identifications at Allenton and West Wittering and the four Flandrian sites, all made by Reid, have been supplemented by recent determinations of West, Lambert and Bell. They establish *R. sardous* as an interglacial plant, and the four Flandrian records indicate that it was a weed on Roman occupation sites. There is no indication of its presence in salt-marsh habitats such as it very commonly occupies today, except that Bell (1969) notes it as a component of the halophytic group often conspicuous in full-glacial deposits such as those at Earith.

The two Weichselian records are interesting in view of the very southerly distribution of this 'Sub-atlantic' species in Scandinavia where it seldom grows naturally north of 58° N.

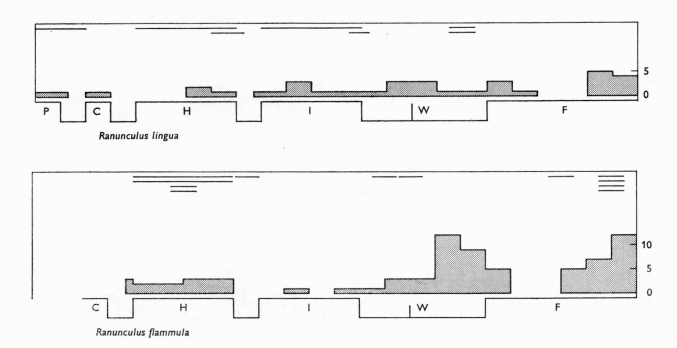

Ranunculus lingua

Ranunculus flammula

Ranunculus parviflorus L. 'Small-flowered buttercup'
46: 9

H *Clacton*; I *West Wittering, Bobbitshole, Trafalgar Sq.*; F VIII *Cottenham, Silchester, Godmanchester, Caerwent*

These records establish the presence of *R. parviflorus* in the middle part of two interglacials. The four Flandrian records are all from Romano-British occupation sites and the identifications are by three different authorities. All the sites are from Southern England within the present southern distribution of the plant in Britain: it has a southern and south-western European distribution.

Ranunculus lingua L.   Great spearwort (Plate XVII*f*)
46: 11

Although it has become rather uncommon in the British Isles today, perhaps in consequence of drainage, this species extends over most of the country. It grows in all parts of Europe except north Sweden, north Finland, Norway, Spain, Portugal, Greece and Turkey. In Scandinavia it is placed by Samuelsson (1934) in the category of 'south Scandinavian' water plants. The above records establish the presence of *R. lingua* from the end of the Baventian onwards. It was present in the Pastonian and Cromerian stages and in the middle, warm, sub-stages of the Hoxnian, all stages of the Ipswichian and most of the Flandrian pollen-zones. It was however also present in the opening Beestonian, the late Wolstonian and Early, Middle and Late Weichselian glacial stages. Despite the fact that about one quarter of the identifications are only tentative, the persistence of the plant in this country through a large part of the Pleistocene seems certain.

Ranunculus flammula L.   Lesser spearwort (Fig. 160*t, u*; plate XVII*e*)
46: 12

This plant, placed by Hultén in his west European–middle Siberian group is widespread through the British Isles, and in Scandinavia has a northern limit above 68° N, reaching up into the birch region in the mountains (Jessen, 1949). Both in England and Ireland it has an extraordinarily long and complete fossil record. The thirteen Hoxnian records cover all sub-stages, and in the Ipswichian two stages.

There is one record as early as the Baventian and three site records of fruits in the late Anglian are matched by a late Anglian record from Baggotstown of its pollen at a frequency of 10 per cent (Watts, 1964). There are four records in the Wolstonian glacial stage and no less than thirty-one in the Weichselian, where it occurs in the early and middle sub-stages, though more abundantly in the late. It occurs in zone IV of the Flandrian and from the end of zone VI onwards, at which time it shews a rather strong association with archaeological sites ranging in age from Mesolithic to Mediaeval.

Jessen has pointed out that the achenes of *Ranunculus flammula* are very like those of *R. reptans* L., and Conolly *et al.* suggest that one of the Cornish specimens from the Late Weichselian might perhaps represent the Scottish variety of *R. reptans*, *R. scoticus* E. S. Marshall. Despite this there are so many unqualified identifications that it cannot be doubted that this species has been continuously present in the British Isles, at least from the Anglian glacial stage.

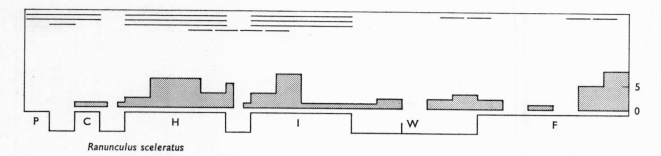

Ranunculus sceleratus

Ranunculus reptans L.   46: 13

*e*W *?Brandon*; *e*I *Marros*; *m*W *?Great Billing*, *?Brandon*;
*l*W *?Dunshaughlin*, *?Knocknacran*, *Ratoath*

From three Late Weichselian localities in eastern Ireland
Mitchell has recorded achenes of *Ranunculus* 'too small
to belong to any species growing in Ireland today'. They
are about 1.2 × 0.9 mm and 'as a label...to indicate that
achenes of this size do occur' he refers them to *R. reptans*
L. Many northern species of *Ranunculus* have small
achenes and are difficult to identify by fruit characters
alone and there is also a fairly wide range within each
species. The identification remains tentative but seems
likely, none the less, to indicate the Late-glacial presence
in Ireland of a northern *Ranunculus* no longer living
there. The three English records are similarly qualified.

R. reptans is a boreal–circumpolar plant now rarely
found in Great Britain and restricted to a few sites on
gravelly lake margins in the Lake District and Scotland.

Ranunculus sceleratus L. 'Celery-leaved crowfoot'
(Plate XVII*d*)   46: 15

The habitats of this species undoubtedly favour incorporation and preservation of the achenes and the record of its
presence in this country is very full. It occurs in low
frequency in the Pastonian and Cromerian, but in the
Hoxnian and Ipswichian interglacials it is present in all
sub-stages. Single records come from the Beestonian and
Anglian glacial stages, it occurs in the early and late
Wolstonian and in every stage of the Weichselian.
Present in zones IV and VI of the Flandrian, it is present
more frequently in zones VIIb and VIII. Most of these later
Flandrian records are associated with archaeological sites,
thus Bronze Age 2, Bronze and Iron Age 2, Romano-
British 4, Norman 1, Mediaeval 2. It seems likely that the
plant flourished by the margins of water-holes and drains
in settlement sites. In five sites, one Hoxnian III, two
Ipswichian III and two Flandrian (zone VIIb), there is
evidence of brackish water conditions, and salinity can be
attributed also in more general terms of some of the
Weichselian records. On this evidence it seems likely that
the plant has been continuously present in the British
Isles from the Pastonian stage on.

This is a boreal–circumpolar montane plant which

although widespread through the British Isles, is much
commoner in the south and east as also in Scandinavia,
possibly reflecting the greater frequency in this lowland
region of 'slow streams and ditches and shallow ponds of
mineral-rich water with a muddy bottom' (Clapham,
Tutin & Warburg, 1962).

Ranunculus (Batrachian) spp. (Fig. 153*b*; plate XVII*c*)
(See p. 124)   46: 16–23

It has seldom been possible to identify separately by
means of their fossil achenes the several species of the
batrachian ranunculi, and the category of remains identified under the general heading is wide: it appears in older
accounts as *Ranunculus aquatilis* L.

Remains of batrachian ranunculi are a useful index to
a shallow-water habitat, but the difficulty of specific
identification makes the fossil record of limited value.

There are 5 site records from the Cromer Forest Bed
series, 24 from the Hoxnian and 19 from the Ipswichian.
All sub-stages of the Hoxnian and Ipswichian are represented, as are all zones in the 46 records from the
Flandrian.

There are 5 site records from the Anglian glacial stage
(4 late), 6 from the Wolstonian (3 early and 2 late) whilst
the 82 Weichselian records include 1 Early and 2 Middle
Weichselian with 64 in the Late Weichselian, distributed
zone I, 8; zone II, 25; zone III, 20 and the rest in transitional
zones or not attributed to a zone.

This extensive representation through glacial and interglacial stages is not surprising in view of the several
species of water buttercup conflated in this category, but
the point is clear that the group as a whole has been
consistently represented over the full range of alternating
climatic conditions.

Ranunculus hyperboreus Rottb. (Fig. 150*d*)   46: 25

Ba *Norfolk Co.*; *e*C *Norfolk Co.*; A *Lowestoft*; *l*A *Hoxne*;
Wo *Dorchester*; W *Thrapston*; *e*W *Wretton*; *m*W *Marlow*,
Waltham Cross, *Earith*, *Barnwell Station*; *l*W *Ballaugh*, *Kirk-
michael*

This is a circumpolar arctic–montane species of extreme
northern range not found south of 61° N in Scandinavia
and not at present living in the British Isles, but this series

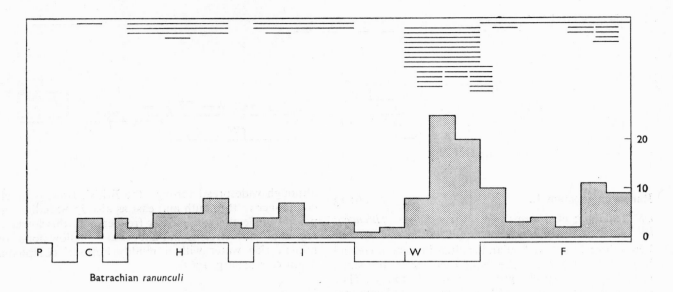

Batrachian *ranunculi*

of records conclusively shews that it was present in the Baventian and in each of the three latest glacial stages, and it has been found by Wilson and West in the Corton interstadial of the Lowestoft glaciation along with other cold-indicating plants. These sites are very strongly associated with the glacial conditions by climatic indices based on geological structures and faunistic evidence, and to the respective stages by firm stratigraphic, palynological and radiocarbon evidence. There is no clear indication that the species survived through an interglacial in this country, though West and Wilson have recorded it from sub-stage I of the Cromerian, and an additional record from Bembridge (Reid and Chandler, 1925) might, from the presence of *Picea* wood, be thought possibly to be of interglacial age; the uncertainty of the age of this example of *Ranunculus hyperboreus* is commented on by Jessen (1949). Dickson, Dickson and Mitchell (1970) cite Danish and Polish Weichselian records. Traulau (1963) illustrates the European distribution of the species.

It is a plant that grows 'on moist clay and occasionally enriched soils or in small pools, also in moss-rich habitats around springs' (Böcher, Holmen & Jakobsen, 1968) and in Scandinavia is said to occur in mountains where it favours pools fouled by livestock at the summer grazings (Lid, 1952).

Ranunculus nemorosus DC.   46: 26

Cs *Pakefield*; *mW Barrowell Green*

The Lea Valley record was made by E. M. Reid and M. E. J. Chandler in revision of C. Reid's earlier determination, and subsequently checked by E. M. Reid on original specimens. Both records however remain uncertain, especially as re-examination of the fossil material in the British Museum (Wilson, unpublished) suggests that they might be *R. repens*.

This European boreal–montane species is not known living or from the Post-glacial period in the British Isles.

Ranunculus aconitifolius L.   46: 27

mW Angel Rd, Barnwell Station, Earith

The positive identification by M. E. J. Chandler from the Barnwell Station beds has been confirmed by F. Bell (1968) and is backed by E. M. Reid's tentative identification of the species from the Lea Valley. They indicate that this arctic–alpine species grew in this country in full-glacial time and has since become extinct. There is no indication whether this may or may not refer to *Ranunculus platanifolius*.

Ranunculus   46

It is noticeable that certain species of *Ranunculus, R. scleratus, flammula, acris, bulbosus, repens* and possibly *bulbosus, parviflorus, sardous*, share two striking characteristics in their fossil record: they tend to have a long and persistent record through both glacial and interglacial stages, and they show a very strong response to the Flandrian clearance stages together with high frequency in historic and prehistoric settlement sites.

None of these species is shade tolerant and this accords with their prevalence in periglacial conditions and occupation in clearance areas. The persistence through warm forested interglacials however may be associated with a strong preference for river- and stream-side situations and damp habitats generally where forest dominance would be incomplete. Such habitats must also have favoured generous representation in the fossil record (see p. 7).

Thalictrum sp. (Plate XXI*r, s*)   50

Pollen grains have been readily identified as belonging to the genus *Thalictrum*, but not to individual species. The fresh pollen of even a well-defined species such as *T. flavum* is extremely variable in size and suggests a hybrid constitution. *Thalictrum* pollen has been found at various sites throughout all pollen zones of the Flandrian, in the Cromerian and all four sub-stages of the Hoxnian and Ipswichian interglacials: the earliest interglacial records however are from the Thurnian and Baventian. There is one identification in the Baventian, and *Thalictrum* occurs at several sites in the late Anglian glacial stage as well as early and late in the Wolstonian. It is found in the Early and Middle Weichselian, but is particularly common in the Late Weichselian where it is found at 44 sites in zone I, 60 in zone II and 59 in zone III at frequencies up to 20 per cent of the total pollen, although most often between 2 and 5 per cent. Similar frequencies obtain at the 45 sites for zone IV at the opening of the Flandrian, but thereafter frequencies as well as numbers of site records remain low. The records yield no evidence for geographical variation in frequency of site records within the zones, or of pollen frequency.

It is evident that the bulk of records must have come from plant communities growing in open periglacial conditions, a conclusion supported by the associated pollen-flora at three sites, and it seems likely therefore that the pollen came in large part from *Thalictrum alpinum* and *T. minus*, both of which species have yielded abundant records of fruits from the corresponding periods and conditions. The *Thalictrum* pollen records for the closely vegetated central parts of the interglacials may well derive from the robust fen species *T. flavum*, which yields corresponding fruit records.

TABLE 10. Thalictrum *sp.* (*pollen*). *Site records for four quarters of Great Britain and for Ireland*

Zone		SW	SE	NW	NE	I
*l*W	I	14	7	11	7	7
	II	18	8	14	12	10
	III	19	6	13	10	9
*e*F	IV	11	7	16	6	6

Thalictrum flavum L. 'Common meadow rue' (Plate XVII*b*)   50: 1

Cs *Beeston, Corton, Pakefield, Ostend, Sidestrand, Mundesley;* H *Wolverton, Saint Cross, Clacton, Nechells;* Wo *Dorchester;* I *Wortwell, Bobbitshole, West Wittering;* eW *Sidgwick Av.;* mW *Hedge Lane,* ?*Barrowell Green, Waltham Cross, Ponders End, Angel Rd, Broxbourne, Temple Mills;* lW *Dronachy;* *Colney, Nazeing, Nant Ffrancon, Hartford, Oulton Moss;* F IV *Star Carr,* VI *Nursling,* VII *Hightown,* VIIa *Ehenside Tarn,* VIII *Silchester*

Thalictrum flavum has characteristic and abundant achenes which preserve their features in fossil material, and the fen and waterside situations where it grows favour incorporation of the fruits. These have been found at six sites in the Cromer Forest Bed series (Reid) and four Hoxnian and three Ipswichian sites, including sub-stage III of the Hoxnian and sub-stages I and II of the Ipswichian. It has been found, though at single sites only, throughout the Flandrian, so there is reasonable presumption that it has persisted through the four latest interglacials. It was present in the Wolstonian glacial, and occurred at one site in the Early Weichselian, five in Middle Weichselian and five in the Late Weichselian (two in *l*W I and three in *l*W II). There is accordingly some ground for supposing that *T. flavum* persisted through the glacial stages, at least in southern Britain, though we may note the Nant Ffrancon record from high altitude in Allerød time (*l*W II). This is a conclusion of special interest because its present-day distribution is strongly centred in south-eastern England, whilst it grows in only scattered sites elsewhere in England, and in Wales, Scotland and Ireland. From Ireland there are as yet no fossil records.

Thalictrum alpinum L. 'Alpine meadow rue' (Fig. 45; plate XVII*a*)   50: 2

Ba *Norfolk Co.;* P/B *Norfolk Co.;* B *Norfolk Co.;* lA *Gort;* H *Kilbeg;* eW *Sidgwick Av.;* mW *Ponders End, Angel Rd, Syston, Temple Mill, Earith, Barnwell, Great Billing, Brandon, Upton Warren;* lW *Leigh, Colney, Nazeing, Derriaghy, Shortalstown, Newtownbabe, Littleton,* ?*Johnstown, Ballybetagh, White Bog, Mapastown, Nant Ffrancon, Hawks Tor, Neasham, Drumurcher, Knocknacran, Coolteen, Whitrig Bog*

Thalictrum alpinum is included in Matthews' category of arctic–alpine plants and only just fails to qualify for Polunin's definition of a truly arctic plant (Polunin, 1949). It is generally distributed in arctic and sub-arctic Europe, extending north to 72°, and 74° in Nova Zemlya. In the British Isles it is now restricted to mountain localities over 2000 ft; it is common only in the Highlands of Scotland, whilst in Wales it is restricted to Snowdonia, and in England to the Lake District and the region of Upper Teesdale. In Ireland it is again local upon high mountains, never descending to sea-level as do so many other highland species (Praeger, 1934). Its present restricted and highland range contrasts very strikingly with the wide range and lowland occupation indicated by the records given above. West and Wilson have recently identified *T. alpinum* fruits from the Baventian and Beestonian glacial stages, but with the exception of these, a mixed 'Hoxnian' sample from Kilbeg and a single Irish record from the end of the Anglian glacial, all come from the Weichselian glacial stage, one from its onset, nine from the middle and eighteen from the Late Weichselian (fig. 45). They are generally well supported by pollen-analytic, geological and faunistic evidence and many are directly dated by radiocarbon ages that range from > 45 000 to 11 000 years B.P. The preponderance of Irish sites is probably due to active investigation of the Late Weichselian there, and the abundance of zone II records

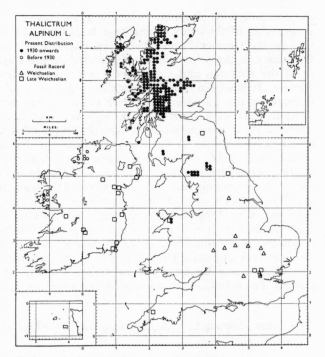

Fig. 45. Present distribution of *Thalictrum alpinum* L. in the British Isles, and Weichselian identifications of its fruits.

(11 as against 2 in zone I and 4 in zone III) may indicate only that the organic muds of that zone offer the best chances of incorporation and survival of the fossil fruits. The restriction of fossil records to the glacial stages and the fact that the species was present in at least four such stages is of particular interest. There can be little doubt that open soil conditions and freedom from closed cover of ranker growing species are requisite for the growth of this species; both conditions were satisfied in the Middle and Late Weichselian lowlands and the prevalent soli-fluxion must have given the plant very suitable soil conditions.

The striking absence of records in the Flandrian can hardly be due to chance and there seems every reason to regard *T. alpinum* as a plant of widespread British distri-bution in the Middle and Late Weichselian period which suffered progressive and severe restriction to northern and highland areas in the post-glacial period of increased warmth, dense forest covering and of widespread peat bog formation in elevated and atlantic situations. The history of this species in the British Isles accords fully with the European record of widespread occurrence in the Euro-pean lowland during the Full- and Late-glacial with extinction during the reforestation of the Pre-boreal (Tralau, 1963), and indeed with the highly disrupted cir-cumpolar arctic–montane northern hemisphere range illustrated by Hultén (1958).

Thalictrum minus L. 'Lesser meadow rue'   50:3

Be *Norfolk Co.*; Cs *Sidestrand, Mundesley*; C *Norfolk Co.*; ?*Beeston*; eWo *Brandon*; W *Thrapston*; eW *Sidgwick Av.*; mW *Marlow, Upton Warren, Barnwell St., Earith*; lW ?*Hailes, Kirkmichael, Windermere, Newtownbabe, Nant Ffrancon, Rod-baston*; F VI/VII *Drumskellan*, VIIb *Wingham*

Although there is a wide range of form and size in achenes of different species of *Thalictrum*, the identification is queried in few of these twenty records. The Norfolk coastal sites associated with the Cromer Forest Bed have yielded the five oldest records. There is an early Wol-stonian record from Brandon with abundant faunistic and geological evidence of periglacial conditions, and two well-dated middle Flandrian records from Drum-skellan and Wingham respectively. The remaining glacial stage records are all Weichselian, the two Middle Weich-selian from Earith and Upton Warren have radiocarbon ages about 42000 years and that from Barnwell a more doubtful age of 19500 B.P. Four of the Late Weichselian sites come from zone II and two also have suitable radio-carbon ages (Kirkmichael and Windermere); Rodbaston, with associated mollusca and insects and radiocarbon age of 10670, is a zone III record.

Although '*Thalictrum minus*' is a highly polymorphic aggregate it is currently split into three primary habitat groups characterizing respectively dry limestone situa-tions, coastal sand dunes and stream- or lake-side sands and gravels. It is impossible to refer the fossil material to any of these individual groups, all of which however favour conditions widely encountered in full- and late-glacial times.

NYMPHAEACEAE

C. & E. M. Reid (1908) have recorded from the Cromer Forest Bed series seeds which have very pronounced morphological features and which apparently belong to two unknown species of this family: illustrations are given of both.

Nymphaea alba L.   White water-lily (Plate XXIV*a*, *b*)
                                                        55: 1

Within the genus *Nymphaea* identifications of both seeds and pollen have been numerous, but authors have differed in the confidence with which they have referred fossil material to species or sub-species. The greater part of the fossil seed identifications have been made to species level but five Late Weichselian and three Flandrian records have been only to the genus: the probability is so great that these too belong to *N. alba* that these are all conflated in the record-head diagram for seed identifications. With pollen the convention has been in the opposite sense, the great majority of records being to the genus, with a few site records in the Late Weichselian and Flandrian given as *N. alba*. It appears in fact that investigators have been readier to identify seeds as '*alba*' than they have pollen.

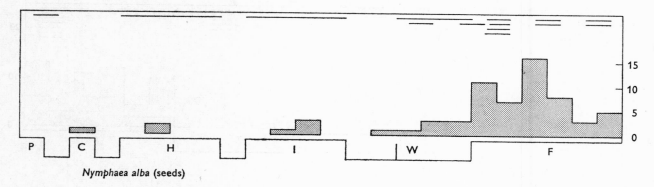

Nymphaea alba (seeds)

There is considerable variation in the white water-lilies in Europe (see Y. Heslop-Harrison, 1955) and this is partly represented in the British populations. However, the identification of pollen of *N. candida* from the Late Weichselian at Elan Valley (Moore, 1970) is only provisional, as is that for pollen of *N. occidentalis* (Ostenf.) Moss from the mid-Flandrian at Kentra Bay (Lacaille, 1951). We have accordingly treated all pollen site records as *N. alba* in the construction of the record-head diagram and this has to be borne in mind in drawing conclusions from it.

The seed record shews *Nymphaea alba* present in the Cromer Forest Bed series, but also on the Norfolk coast more precisely from the Pastonian–Beestonian transition and the Cromerian interglacial. Four records from the Hoxnian and five from the Ipswichian are paralleled by substantial numbers of site records from all zones of the Flandrian. These Flandrian records follow less frequent records from zones I to III of the Late Weichselian with one Middle Weichselian occurrence. These apart, there are no records from glacial stages.

The pollen record is remarkably similar to that for fossil seeds with one record each for the Ludhamian and Pastonian interglacials, records from all sub-stages of the Hoxnian and two of the Ipswichian and again very frequent site records in every zone of the Flandrian. Pollen site records are in all zones of the Late Weichselian but much more frequently than for seeds. There are single site records from the late Anglian and early Wolstonian glacial stages with none from Early and Middle Weichselian, but too much must not be made of this apparent near absence from the glacial stages since few pollen analyses are available from deposits attributable to them. All the same this pattern is in accord with what one would expect from the present-day distribution of the plant which Samuelsson places within the category of south Scandinavian Atlantic distribution, its northern Swedish boundary roughly corresponding to that of the oak. It is interesting that Samuelsson explains the presence of *N. alba* × *N. candida* hybrids outside the northern limit of *N. alba* as the results of a former more extensive distribution of the latter species in the Postglacial Warm period. As the species extends from Abo in south Finland north-easterly up to 65° N in an island in

the White Sea, it evidently endures winter cold, and British Late-glacial records need not be questioned.

At the present day *Nymphaea alba* is widespread throughout the British Isles and the table given below shews the absence of any signal variation in frequency of the fossil sites (seeds plus pollen) in the five major regions of the country through the last 15 000 years.

TABLE 11. *Sites of fossil* Nymphaea alba *in the British Isles*

	SE	SW	NE	NW	Ir.
Late Weichselian	4	4	6	5	6
Pre-Boreal + Boreal	4	11	9	14	16
Post-Boreal	5	13	7	14	8

The most that might be claimed is a trend to higher numbers in the west during the Flandrian.

Nuphar lutea (L.) Sm.   Yellow water-lily, Brandy-bottle
(Plate XXIV*c–f*)   56: 1

The pollen grains of *Nuphar* are not only highly recognizable as such but those of *N. lutea* differ from those of *N. pumilum* in having longer and more numerous spines upon the exine. It is probable that the great majority of pollen records so far observed belong to *N. lutea*. In the record-head diagram we have therefore treated all pollen records together as *N. lutea*, although the commonest attribution is to '*Nuphar*'. Likewise all fossil seed records have been treated as *N. lutea*, although it is to be noted that five early records, from the Baventian to the Anglian stages, are only attributed to the genus.

It is at once evident that the pollen and seed records shew a high degree of correspondence. There are records for both in the Pastonian and Cromerian interglacials (which supplement and give more precision to the seven records from the Cromer Forest Bed series) and for the middle sub-stages of the Hoxnian and Ipswichian, including the Shetland record from Fugla Ness. Likewise both pollen and seeds are recorded frequently in all zones of the Flandrian.

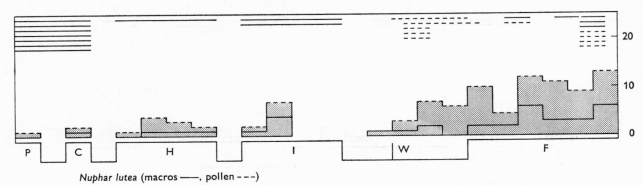

Nuphar lutea (macros ——, pollen - - -)

By contrast glacial stage records are infrequent with one each in the Baventian, Anglian and Middle Weichselian: they are more frequent, especially pollen in low frequency, in all zones of the Late Weichselian. There seems a strong case for regarding *N. lutea* as having persisted in the British Isles right through the Pleistocene, although restricted in the glacial stages. This is not surprising in view of its status as a 'West European–north Siberian' plant (Hultén) that is abundant in Scandinavia up to the Arctic Circle and occurs scattered as far as 68° N.

Nuphar pumila (Timm) DC.   Least yellow water-lily
56: 2

H ?*Kilbeg, Nechells*

In a repetition of Mitchell's investigations at Kilbeg, Co. Waterford, Watts (1959*b*) gives a tentative identification of pollen of *N. pumila* from sub-stage I of the Hoxnian interglacial, and Kelly (1964) positive pollen identifications from sub-stages II and III of the Hoxnian at Nechells, Birmingham. *N. pumila* has a far more restricted northern and montane range, both in Britain and Europe than has *N. lutea* (Heslop-Harrison, 1955).

Brasenia purpurea (Michx.) Casp.

From the Hoxnian (sub-stage III) at Gort in western Ireland, Jessen (1959) has recorded in low frequency pollen identical with that of *Brasenia purpurea*. He comments that seeds of this species are common in interglacial deposits from the European continent as far east as Central Russia and reach back into the Pliocene. This is the first record from the British Isles of a plant now no longer found in Europe, presumably extinguished by the severity of the last glacial stage (see Leopold, 1967) along with *Dulichium spathaceum* (p. 399).

CERATOPHYLLACEAE

Ceratophyllum sp.   Hornwort   57

Macros: Pa *Norfolk Co.*; H *Clacton*; *m*W *Waltham Cross*; *l*W I, II, III *Loch Park*, III, III/IV *Drymen*, III ?*Loch Fada*; F IV, V, VI, VIIa, VIIb *Loch Park*, IV *Drymen, Whitrig Bog, Ballaugh*, V *Randay Mere, Loch Fada*, VI, VIIa, VIIb *Loch Cuithir*

Pollen: *l*A *Nechells*; H *Nechells*; *l*W *Kilmacoe, Mapastown*; F IV *Whitrig Bog*

Occasionally a fossil fruit has been referred to the genus *Ceratophyllum* but the macroscopic records of it are mostly for the characteristic and persistent leaf-spines. Each consists of a single thickened cell, narrowly elongate and sharply pointed, bearing at its base the impression of attachment to adjacent cells. They may occur in pollen slide preparations so abundantly as to permit representation of a 'spine-curve' and indeed in the Nechells deposits Kelly (1964) presents such a curve and demonstrates that its peak coincides with recovery of abundant seeds of *C. demersum*. Not surprisingly in a submerged aquatic there is evidence of much local persistence as from the late Anglian through sub-stages I and II of the Hoxnian at Nechells, from every zone at Loch Park from I to VIIb, and from VI to VIIb at Loch Cuithir in Skye. The delicate pollen grains are infrequently recorded.

Ceratophyllum demersum L.   57: I

The long-spined fruits are very easily recognizable fossils. The earliest records are three from sites in the Cromer Forest Bed series and one from the Cromerian itself. The Hoxnian records are all English: they cover sub-stages I, II and III as well as the transitional stages to the preceding and succeeding glacial stages, and at Hoxne Reid and West have respectively recorded it from sub-stage III. Again in the Ipswichian there are records from sub-stages I, II and III as well as the preceding late Wolstonian. Two late Anglian records parallel the records for the early and late Wolstonian, but the species is only sparsely represented in the Weichselian with no evidence of presence in the coldest part. There are records throughout the Flandrian, and at Hockham Mere they embrace Late Weichselian, early Flandrian and the Boreal zones: in zone VIIa there was also recorded a fruit of *C. demersum* var. *apiculatum*.

It seems apparent that *C. demersum* has persisted through all the succeeding interglacials, including the present Flandrian, and was present in the less severe terminal parts of the glacial stages.

C. demersum is a eutrophic plant scattered throughout the midlands and south-eastern half of England, but rare

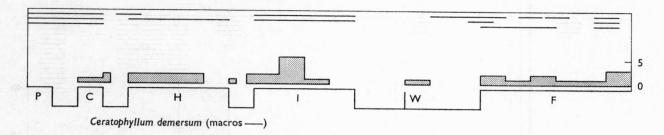

Ceratophyllum demersum (macros ——)

in Wales, Scotland and Ireland. It is fairly common in southern Fennoscandia, but its fossil range shews that 'it immigrated into northern Finland in early Post-glacial time and acquired a considerable distribution there, but retired before the post-glacial deterioration of the climate had asserted itself'. Samuelsson (1934), whose summary this is, considers that the restriction was due to the overgrowing and impoverishment of the lakes, but he may be underestimating the control by climatic change. Evidence by H. J. B. Birks (1969) based on leaf-hair identification from lake deposits in Skye, suggests restriction of range southwards since the Late Weichselian, a conclusion also tentatively indicated by *Sagittaria sagittifolia*.

Ceratophyllum submersum L.   57 : 2

C *Norfolk Co.*; I ?*Selsey Bill*; F VI *Kirkby Thore*, VII, VIIb/VIII *Hightown*, VIII *Dunshaughlin*, VIII *Finsbury Circus*

Ceratophyllum submersum is a plant of much more southern range than *C. demersum*, extending only as far north as Denmark on the European mainland, and apparently chiefly southern and eastern in England, whilst absent from Ireland. This has to be viewed against the fact that all the Flandrian records are from the milder part of the interglacial as are the tentative records from sub-stage IIa and IIb of the Ipswichian at Selsey Bill. As with *C. demersum* so here also there are fossil records from Fennoscandia far north of the present range of the plant, and it is possible that the Dunshaughlin record reflects a similar restriction in the British Isles. Jessen subscribes to the view that both edaphic and climatic factors are concerned in this post-glacial limitation of range.

PAPAVERACEAE

Papaver sp. Poppy   58

*l*W *Tadcaster, Esthwaite, Longlee Moor*

It is only recently that British palynologists have identified fossil the small pollen grains of *Papaver* and Bartley's identifications are to '*Papaver* type' (Tadcaster) and 'cf. *Papaver*' (Longlee Moor). The records embrace the Late Weichselian zone I (Tadcaster), zone II (Esthwaite and Longlee Moor) and zone III (Esthwaite). The identifications of seeds of the alpine poppy, *Papaver* sect.

Scapigerum, suggest that this may have been the plant from which this pollen came, as also perhaps the pollen from the Late Weichselian (zones II and III) at Hockham Mere referred to 'cf. *Meconopsis*'.

Papaver rhoeas L.   'Field poppy'   58 : 1

F VIIb *Minnis Bay*, VIII ?*Aldwick Barley, Silchester*, ?*Bunny, Godmanchester, Hungate, Dublin*

All seven fossil seed records for *P. rhoeas* are associated with human settlement, the first two late Bronze Age and Iron Age respectively, the next three Roman and the last two mediaeval. The plant has a strong trend to increased abundance in the south-east, both of Great Britain and Ireland.

Papaver argemone L.   'Long prickly-headed poppy'   58 : 5

F VIII *Silchester*

The only record of *P. argemone* is a Roman record by Reid (1901).

Papaver somniferum L.   'Opium poppy'   58 : 6

F VII *Fifield Bavant, Silchester, Caerwent, Dublin*

These records all come from settlement sites, in order Iron Age, Roman (2) and Mediaeval. They do not contradict the usual view that *P. somniferum* is an introduced species. Its seeds were found in large quantity in the Neolithic settlement site at Niederwil along with those of other species taken to have been drug plants (Waterbolk and van Zeist, 1966).

Chelidonium majus L.   Greater celandine   62 : 1

I *West Wittering, Wretton*; F VIII *Caerwent, Hungate, Shrewsbury*

Although the two mediaeval sites at Hungate and Shrewsbury and the Roman site at Caerwent reflect the common association of this plant with human settlement, the two records from the Ipswichian interglacial must be entirely natural and suggest that it may well have been native in the Flandrian also. It is a species fairly widespread in Denmark and south Sweden, with scattered occurrences northwards to the top of the Gulf of Bothnia.

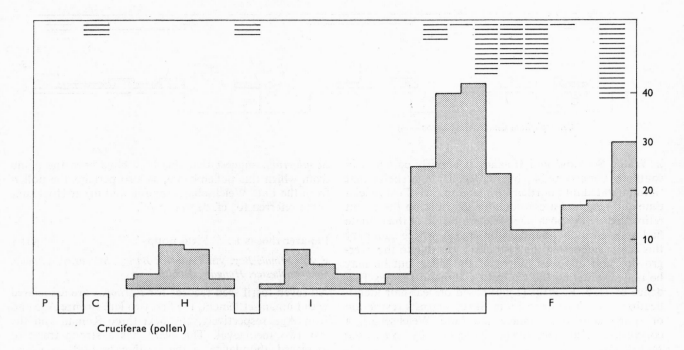

Cruciferae (pollen)

Hypecoum procumbens L.

Cs *Corton*

Seeds of *Hypecoum procumbens*, which now is found as an annual of sandy soil in the Mediterranean region and southern France, were recovered commonly from the Cromer Forest Bed series (C. & E. M. Reid, 1908).

FUMARIACEAE

Fumaria sp.   66

H *Clacton*; mW ?*Barnwell Station*

Though both records are only at generic level, the second remains tentative only.

Fumaria officinalis L. 'Common fumitory' 66:8

F vⅡb *Cocker Beck*, vⅢ ?*Aldwick Barley, Silchester, Mersea Island*

The Cocker Beck record is to the Neolithic in its broadest sense, that from Aldwick Barley, Iron Age and the remaining two, Roman, except that they all are associated with human settlement or cultivation.

CRUCIFERAE

It is relatively easy to identify pollen to the family Cruciferae and difficult to recognize within it taxa of lower rank. Accordingly the category 'Cruciferae' figures in most modern pollen diagrams. It is scarcely surprising that there should be considerable numbers of site records in all sub-stages of the Hoxnian and Ipswichian and in all zones of the Flandrian. What are particularly striking are the very high values in the Late Weichselian and the opening Flandrian. There can be little doubt, especially in view of the macroscopic records that follow, that these values reflect the preferences of so many Cruciferean plants for the disturbed and open habitats prevalent during late-glacial time. We may note in this connection not only late Anglian and early Wolstonian sites but the great increase of sites in zone vⅢ where clearance and arable cultivation affected so much of the country. In general the Cruciferean pollen is less than 2 per cent of the total pollen at all times before zone vⅡb, but thereafter in the south of England occasional sites (such as Frogholt, Wingham, Old Buckenham, Apethorpe and Glastonbury) yield very high frequencies in correspondence with local agricultural activity.

 It is less common for seeds to be referred merely to 'Cruciferae', but of the 23 such records, no less than 15 are Late Weichselian, in correspondence with the pollen record.

Brassica sp.   67

F vⅢ *Silchester, Car Dyke, Caerwent, ?Hungate*

The three first-mentioned sites are all Roman, and that at Hungate late Anglo-Saxon. From Hungate the identifications include not only *Brassica* sp., but also *Brassica* cf. *napus* (rape, cole and swede) and *Brassica* cf. *oleracea* (wild cabbage). The records for the following three species, *B. rapa* L., *B. nigra* (L.) Koch, *Sinapsis arvensis* L., are likewise all closely associated with human activity.

Brassica rapa L. (*Brassica campestris* L.)   Turnip, Naven
                                                                67: 3

F viib *Itford Hill*, viii *Pevensey, Dublin*

These sites are settlements respectively of Late Bronze
Age, and the Roman and Mediaeval periods.

Brassica nigra (L.) Koch   Black mustard   67: 4

I *Clacton*; F viii ?*Glastonbury Lake Village, Silchester*

The interglacial record suggests the possibility of a native
status for *Brassica nigra*, and this accords with Hultén's
phytogeographic assessment of it.

Sinapis arvensis L. (*Brassica arvensis* (L.) Kuntze)   Char-
    lock, 'Wild mustard'   70: 1

F viib *Portsteward*, viii *Newstead, Dublin*

These records are respectively from Neolithic, Roman
and Mediaeval sites, the earliest from a pottery impression
(Jessen & Helbaek, 1944).

Sinapis alba L. (*Brassica alba* (L.) Boiss.)   White
    mustard   70: 2

F viii *Silchester*

An age at least as great as Roman is established by
Reid's record for this species so commonly regarded as
an introduced alien.

Diplotaxis tenuifolia (L.) DC.   Perennial wall rocket
                                                  72: 2

W *Thrapston*, mW *Marlow, Earith, Barnwell Station*

The three first-mentioned records have been recently published by Bell (1969) from extremely well-authenticated
Weichselian sites, and the fourth is a record from hitherto
unidentified material of Miss Chandler's from Barnwell
and from a correction of a seed originally attributed by
Miss Chandler to 'Geranium x' (Bell & Dickson, 1971).
This evidence contradicts doubts of the plant's native
status and sorts oddly with the strongly south-eastern
emphasis on its English range and its casual occurrence
outside this region.

Raphanus raphanistrum L.   Wild radish, White charlock,
    Runch   74: 1

F viib *Avonmouth*, viii *Mersea Is., Pevensey, Caerwent, Silchester, Star Villa, Godmanchester, Isca*

The first-named site is a submerged coastal peat. All the
rest are Roman settlement sites, the last a carbonized
grain store found at the Roman fortress of Caerleon. This
last identification was based upon 'three whole siliqua
joints and twelve fragments'. Helbaek regards the
R. raphanistrum as a weed contaminant of the cereal
crop.

Lepidium sativum L.   Garden cress   79: 1

F viii *Larkfield*

This is a seed record from the mud of a mediaeval house:
the wild forms are native in Egypt and Western Asia.

Coronopus squamatus (Forsk.) Aschers (*Senebiera corono-
    pus* (L.) Poir.: *Coronopus procumbens* Gilib.)   Swine-
    cress, Wart-cress   80: 1

F viii *Silchester, Pevensey, Caerwent, Bunny*

All four records are from Roman occupation sites.

Isatis tinctoria L.   Woad   82: 1

F viii *Somersham*

In this Anglo-Saxon pottery impression of the seed of
the woad plant the record confirms historical account of
the use of this species. Jessen & Helbaek (1944) point
out that although the species evidently originates from
southern and eastern Europe, it is only from archaeological finds in western Europe that it has been shewn to
be a plant made use of in prehistoric economy. It continued in use as a cultivated dye-plant until recent time.

Thlaspi arvense L.   'Field penny-cress'   84: 1

eWo *Brandon*; F viib *Frogholt, Minnis Bay*, viii *Silchester,
Dublin*

The records at Frogholt and Minnis Bay are associated
with late Bronze Age occupation, though the second of
these sites has been suspected of contamination. Silchester is a Roman, and Dublin a mediaeval settlement
site. These strong associations with human settlement
accord with the present status of *T. arvense* as a ruderal
and weed of arable ground, but the record from the early
Wolstonian glacial stage suggests that it may also have
had a native status in periglacial conditions. It has been
commonly found as a weed in the Danish Iron Age.

Thlaspi alpestre L.   'Alpine penny-cress'   84: 4

H ?*Gort*

A tentative identification was made in the interglacial
deposits at Gort in western Ireland by Jessen *et al.* (1959),
from an horizon that corresponds with sub-stage iv.
T. alpestre is a local sub-alpine or alpine plant on basic
soils with a very disjunct range in Great Britain but not
now living in Ireland.

Capsella bursa-pastoris (L.) Medic.   Shepherd's purse
                                                  86: 1

lWo *Ilford*; mW *Upton Warren*

These records from the last and penultimate glacial stages
at once confirm the native status of this cosmopolitan
weed and suggest that some forms of it at least survived
the glaciations in Britain.

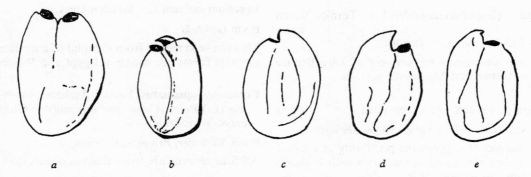

a   *b*   *c*   *d*   *e*

Fig. 46. Seeds of *Subularia aquatica* L.: *a–d*, from Late Weichselian (zone II) muds at Hawks Tor kaolin pit, Bodmin Moor, Cornwall; *e*, herbarium material (all × 30).

Cochlearia L.   Scurvy-grass   88

F VIIb ?*Brigg*

Dr A. G. Smith, from Late Bronze Age levels at Brigg, Lincs., identified pollen as 'cf. *Cochlearia*'.

Cochlearia officinalis L.   Scurvy-grass   88: 1

*l*W *Dromsallagh, Ballybetagh, Mapastown, Drumurcher, Knocknacran*

The aggregate species *Cochlearia officinalis* is a circumpolar species with such a far northern range that its occurrence in the Late- and Full-glacial is quite understandable. It is evident from the next entry that some of the Full-glacial records refer to the mountain plant *C. alpina* Wats., but of those above the Drumurcher record is to the aggregate species.

Cochlearia alpina Wats.   'Mountain scurvy-grass'   88: 2

*W*o *Farnham*; *m*W ?*Temple Mills*, ?*Waltham Cross, Angel Rd, Ponders End*; *l*W ?*Colney Heath*; F VI ?*Elland*

All these identifications, except those from Farnham, Angel Rd and Ponders End, are tentative only.

British botanists have separated *Cochlearia alpina* from *C. officinalis* L. and in Britain *C. alpina* is a mountain plant of a type one would expect to encounter in the conditions of the Full-glacial.

Cochlearia danica L.   'Danish scurvy-grass'   88: 5

*m*W *Waltham Cross*, ?*Barrowell Green*

Of these two records from the Lea Valley Arctic Plant Bed the second is strongly qualified.

This European shore plant has a southerly distribution in Scandinavia and tends to be less common in Scotland and Ireland than in Britain, so that these Full-glacial records clearly require confirmation.

Subularia aquatica L.   Awlwort (Fig. 46)   89: 1

*l*W *Hawks Tor, Elan Valley*

To the identifications of seeds from zones I and II of the Late Weichselian at Hawks Tor, Cornwall (Conolly *et al.*, 1950) there have now been added records from zone III and the II/III, III/IV transitions in the Elan Valley, Cardiganshire (Moore, 1970).

This species belongs to Hultén's group of 'incompletely boreal–circumpolar plants with a gap in east Siberia': it has a present distribution in the British Isles of a northern and western character, often in highland lakes. A postglacial restriction of range seems to have taken place; though the species is still present in Cardiganshire, it grows no nearer to Cornwall.

Alyssum sp.   91

I ?*Clacton*; *m*W *Earith, Barnwell Station*

The first-mentioned record was tentatively assigned to *A. saxatile* (Reid and Chandler, 1923*a*). The two Weichselian records are modern and from well-documented sites: that from Barnwell Station is a redetermination of material recorded by Chandler as *Cochlaeria officinalis* (Bell & Dickson, 1971).

Draba sp.   94

*l*A *Hoxne*; *e*Wo *Hoxne*; *m*W *Great Billing*; *l*W *Kirkmichael*

These identifications, all from deposits of glacial stages, have to be considered alongside the abundant records for *Draba incana*. They reflect uncertainty in some instances of a seed identification from a deposit where fruit valve identifications have allowed recognition of *D. incana*, for example, Late Weichselian zone I at Kirkmichael.

In addition to the above records, Mitchell has a tentative identification of a *Draba* seed from the deposits at St Erth, Cornwall, that are of unknown, possibly late Pliocene, age.

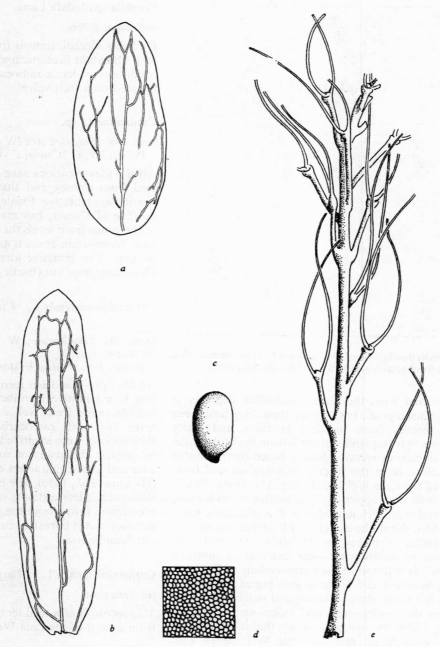

Fig. 47. Fossil remains of *Draba incana* recovered and drawn by S. Duigan from interglacial deposits in the Summertown terrace of the Thames Valley at Dorchester near Oxford. Samples supplied by Dr K. Sandford. *a*, *b*, valves of the fruit (×12); *c*, seed (×12); *d*, cell pattern of seed surface (×75); *e*, infructescence (×6).

Draba incana L. 'Hoary Whitlow grass' (Figs. 47, 48; plate XX*o*) 94: 3

*l*H ?*Wolvercote*; Wo *Dorchester, Farnham*; eW ?*Sidgwick Av.*; mW *Ponders End, Earith, Angel Rd, Waltham Cross, Barnwell Station, Upton Warren, Brandon, Great Billing, Thrapston*; *l*W *Nazeing, Kirkmichael, Newtownbabe, Mapastown, Ballybetagh, Whitrig Bog*

C. Reid (1916) describes as very abundant at three localities of the 'Arctic bed' in the Lea Valley the 'thin, tough, delicately veined oval valves of the pod of this plant' and comments that they were probably the 'small petal-like objects' mentioned by Lewis in his earlier identifications from the same deposits (fig. 47). M. E. J. Chandler found similar structures in the Barnwell Station beds which are held to be of the same age. Several more of these valves

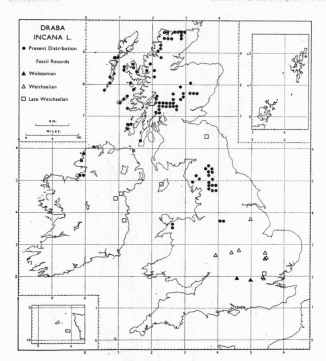

Fig. 48. Present distribution of *Draba incana* L. in the British Isles, with fossil occurrences in the last two glacial periods.

were recovered from the Late Weichselian deposits at Nazeing (Allison *et al.*, 1952). Since then there have been tentative records from the late Hoxnian and Early Weichselian, two unqualified records from the Wolstonian and very numerous records, mostly based on fruit valve identifications, from the Middle Weichselian and from all zones of the Late Weichselian (fig. 48). These Weichselian records are supported by a wealth of geological, faunistic and radiocarbon evidence. It is extremely striking that there should be as yet no Flandrian records.

This species is placed in Hultén's 'Amphiatlantic arctic–montane plants with some localities in southern mountains'. It is found in the northern half of England, in Wales, Scotland and Ireland, growing at sea-level in Scotland, but more often on moors and mountains where it occupies dry rocky screes and ledges up to altitudes of 3550 ft. There can now be no doubt that it has suffered extinction over large areas since the Weichselian with great retraction and disjunction of range, and the evidence of Conolly and Dahl (1970) suggests that control may have operated through susceptibility to high summer temperatures.

Erophila verna (L.) Chevall.   Whitlow grass   95: 1

*e*W *Sidgwick Av.*

This Early Weichselian record from a Cambridge site is based on recognition of a fruit valve.

Erophila spathulata Lang   95: 2

*m*W *Great Billing*

Dr Bell's identification is from a Northamptonshire site with abundant faunistic and geological evidence of cold conditions: it has a radiocarbon age of 28 220 B.P., and is based on a fruit valve.

Cardamine L. sp.   97

Seed: *e*W ?*Sidgwick Av.*, *l*W *Colney Heath*; F *North Ferriby*
    Pollen: *l*W, F ?*Canna*; F VIII ?*Rhosgoch*

All the identifications save the Late Weichselian zone I seed from Colney and that from zone VIIb at North Ferriby are tentative. Flenley, in his pollen analyses from the Isle of Canna, has made provisional identification of the genus from levels that may be considered as in the Late Weichselian zones II and III and Flandrian zones IV to VIIb. The tentative identification from Rhosgoch is Flandrian, zone VIII (Bartley, 1960).

Cardamine pratensis L.   Cuckoo flower, Lady's smock   97: 1

Seeds: *l*H ?*Wolvercote*; W *Thrapston*; *m*W *Great Billing*; *l*W *Kirkmichael*
    Pollen: F V, VI *Muckle Moss*, VIIa *Rhosgoch*

All four pre-Flandrian records are for seeds, though the first is a tentative identification only: the Weichselian records come from well authenticated geological contexts, two with radiocarbon dates. The two seeds at Kirkmichael were identified by their flattened shape and by cellular characters of surface and sub-surface layers over and above characters of gross morphology and size (Dickson *et al.*, 1970). The pollen from Muckle Moss and Rhosgoch is less certainly identified. *Cardamine pratensis* is common throughout the British Isles in suitably damp situations: it is boreal–circumpolar and extends throughout Scandinavia.

Cardamine amara L.   'Large bitter-cress'   97: 2

*l*W ?*Hartford*

This record is of a seed tentatively identified as *C. amara* from zone II of the Late Weichselian (Godwin, 1959).

Cardamine impatiens L.   'Narrow-leaved bitter-cress'   97: 3

I ?*Clacton*

A seed was tentatively identified by Reid and Chandler (1923*a*).

This species, placed by Hultén in the west European–middle Siberian group, has a local distribution in England and Wales as in south Scandinavia.

Cardamine flexuosa 'Wood bitter-cress'   97: 4

F vɪɪb ?*Frogholt*

At a site in Frogholt, Kent and from a level dated 2860 B.P. there is a tentative seed identification.

Barbarea sp.   98

mW *Hedge Lane, Earith*; lW *Nazeing*

Throughout several Late Weichselian layers at Nazeing, belonging to zone ɪɪ or earlier, were frequent seeds of *Barbarea* which could be referred to either *B. vulgaris*, R. Br. or *B. stricta* Andrz. These species cannot be distinguished by seed characters and the Nazeing specimens are therefore placed in the genus.

Barbarea vulgaris, R. Br. 'Winter-cress', 'Yellow rocket' (Plate XVIIj)   98: 1

H *Clacton*; I *Wretton, Histon Rd*; F vɪɪɪ *Godmanchester*

Although common on ground cultivated or disturbed by man, this species has natural habitats on river banks and the edges of fen. Its occurrence in the Hoxnian and Ipswichian Interglacials emphasizes its natural status. At Godmanchester it came from a Roman occupation site.

   B. vulgaris R. Br. is given by Hultén as fairly common to the top of the Baltic and both this species and *B. stricta* occur locally far into the Arctic Circle.

Cardaminopsis petraea (L.) Hiit. (*Cardamine petraea* L.; *Arabis hispida* L.f.; *Arabis petraea* (L.) Lam.) 'Northern rock-cress'   99: 1

lW ?*Abbot Moss, Ballybetagh*, ?*Knocknacran, Neasham*

After careful consideration Jessen concluded that several seeds of the Cruciferae from Ballybetagh may tentatively be referred to *Cardaminopsis petraea* (L.) Lam. Identifications from the zone ɪ/ɪɪ transition at Abbot Moss by Walker, and from zone ɪɪɪ at Knocknacran by Mitchell were provisional; Blackburn's record from zone ɪɪɪ at Neasham was unqualified. These Late Weichselian occurrences are consonant with the present range of the plant which is a species frequent on the higher mountains of northern and western Scotland. It has a pronouncedly disjunct range in Scandinavia and is placed by Hultén in his group of European boreal–montane plants with detached northern and southern areas. It now grows in two stations only in Ireland and the records jointly demonstrate a retraction in range since the Late Weichselian.

Arabis sp.   100

eWo *Marks Tey*; mW *Earith, Thrapston*; lW *Kirkmichael*

The two British species of *Arabis* have seeds that are hard to distinguish from one another, and the record from zone ɪ of the Late Weichselian at Kirkmichael is given as *A. hirsuta* or *A. stricta* on the basis of four seeds with typical translucent wing broadest at the apex, size, shape

and characteristic cell pattern. The authors (Dickson, Dickson & Mitchell, 1970) do not finally exclude other related North European species of the genus.

Arabis hirsuta (L.) Scop. 'Hairy rock-cress'   100: 4

lW *Mapastown*

Mitchell has identified as *Arabis hirsuta* a small (1.1 × 0.75 mm) dark brown seed, narrowly winged all round and with a well-marked reticulate cell pattern; this he found in Allerød age material in Co. Louth. A second similar but smaller seed from the same deposit might also belong to *A. hirsuta*.

   A. hirsuta occurs on chalk, limestone and dry banks throughout Great Britain and Ireland. In western Scandinavia it extends the full length of the peninsula and rises to 1230 m in the mountains; in the Alps also it extends to 1950 m.

Arabis stricta Huds. 'Bristol rock-cress'   100: 6

lW ?*Newtownbabe*

From Allerød layers at Newtownbabe, Mitchell has described a 'dark brown, small (1.1 × 0.8 mm) reticulate crucifer seed, narrowly winged only at the apex'. Though somewhat smaller than recent seeds of this species, it is referred tentatively to *Arabis stricta* which it otherwise closely resembles.

   At the present day, *A. stricta* has a very restricted and disjunct range, being found in the British Isles only on Carboniferous limestone near Bristol, and on the Continent in the mountains of Spain, south France and the Jura. Pring (1961) describes the plant as very intolerant of competition.

Rorippa sp.   102

A seed from the Middle Weichselian moss-layer at Dimlington with a radiocarbon age of 18 240 B.P. is only carried to generic level (Bell, unpublished).

Rorippa nasturtium-aquaticum (L.) Hayek Watercress   102: 1

lW ?*Hawks Tor*; F vɪɪb *Wingham*, vɪɪɪ *Apethorpe*

The Late Weichselian record from Hawks Tor is a tentative identification but there are unqualified records from the late Flandrian at Apethorpe and at two levels at Wingham with radiocarbon ages of 3100 and 2340 years respectively.

Rorippa microphylla (Boenn.) Hyland. 'One rowed watercress'   102: 2

H *Kilbeg*; I *Bobbitshole, Selsey, Ilford, Wretton, Histon Rd*; lW *Mapastown*; F vɪɪɪ *Apethorpe*

There are several records of *Rorippa microphylla* from different parts of sub-stage ɪɪ of the Ipswichian at Bobbitshole, and the other Ipswichian records are also from this sub-stage except Histon Rd, which is sub-stage IV. They are all from fluviatile deposits.

From zone II of the Late-glacial in Co. Louth, Mitchell has recorded two seeds which very closely resemble those of *Rorippa microphylla* (although somewhat larger). They show about 100 polygonal depressions on each face, as against the 25 or so characteristic of *Rorippa nasturtium-aquaticum*.

R. microphylla occurs throughout the British Isles, and Howard and Lyon (1952) indicate that it extends only as far north on the Continent as south Sweden and Gotland. It has been so recently recognized, however, that its range is very imperfectly known.

Rorippa microphylla × nasturtium-aquaticum Airy Shaw

F VIIb ?*Wingham*, VIII *Apethorpe*

Mrs C. A. Dickson has identified as this hybrid a seed from zone VIII at Apethorpe, and more tentatively one from zone VIIb at Wingham, Kent.

Rorippa islandica (Oeder) Borbas (*Nasturtium palustre* (*Leyss*) DC.) 'Marsh yellow-cress' (Plate XXe)

102: 4

lA *Hoxne*; H *Kildromin*; eWo *Brandon*; lWo *Selsey*; I *Wortwell, Histon Rd*; mW *Thrapston, Marlow, Waltham Cross*; lW *Nazeing, Colney, ?Hartford, Shortalstown, Ballaugh, Mapastown, Oulton Moss, Ballaugh, Moss Lake*; F IV *Moss Lake, Ballaugh*, VIII *Hungate*

Seeds of *Rorippa islandica* have been found in both the Hoxnian and Ipswichian interglacials but the great concentration of finds is in the Middle and Late Weichselian (all three zones) and the early Flandrian, where at Ballaugh Mrs Dickson even identified a flower. There is one record also from mediaeval time at Hungate.

The record gives a strong impression of widespread occurrence in periglacial conditions and persistence through interglacials in fluviatile situations. This accords with the biogeographic status of the species as 'boreal–circumpolar' (Hultén) and with the fact that it occurs scattered to the north of Scandinavia, though more frequent, as in Britain, to the south.

Rorippa amphibia (L.) Bess. Great yellow-cress 102: 5

H *Baggotstown*

Watts (1964) has identified two seeds of *R. amphibia* from sub-stage I of the Hoxnian interglacial deposits at Baggotstown, Co. Limerick as part of an extensive assemblage of basic fen species. It is a plant with a thinly scattered distribution in Ireland and in Great Britain it is rather limited to the south and east.

Erysimum cheiranthoides Treacle mustard 105: 1

W *Thrapston*; F VIII *Bunny*

Although the Roman record of this plant from Bunny, Notts., accords with the common belief that it may be an introduced plant in Britain, the Weichselian record by Bell from Thrapston suggests a native status.

Alliaria petiolata (Bieb.) Cavara & Grande (*Sisymbrium alliaria* (L.) Scop., *Alliaria officinalis*, Bieb.) Garlic mustard, Jack-by-the hedge, Hedge mustard 107: 1

F VIII *London*

This single Roman record (Lyell *et al.*, 1906) carries back the ascertained history of this characteristic hedgerow species as far as Roman time; further records would be of great interest.

Sisymbrium officinale (L.) Scop. Hedge mustard

108: 1

F VIII *Bunny*

This seed record from a Roman well (Wilson, 1968) is our only evidence for this widespread ruderal and weed plant.

Descurania sophia (L.) Webb ex Prantl. (*Sisymbrium sophia* L.) Flixweed 111: 1

lW ?*Hartford*; F VIII ?*Hungate, Ehenside Tarn*

The Hartford and Hungate (mediaeval) records are both qualified, but the Roman record from Ehenside is not.

RESEDACEAE

Reseda luteola L. Dyer's rocket, Weld 112: 1

F VIII *Hungate*

Professor H. C. D. de Wit questioned the likelihood that the four seeds recovered from 12–13th century layers at Hungate, York were those of *R. lutea* (see first edition). They were sent to him and pronounced to be *R. luteola*, an attribution which re-examination in Cambridge fully confirms. De Wit thinks it probable that the family Resedaceae is not naturally part of the British flora and that *R. luteola* was probably introduced intentionally as a dye-plant, *R. lutea* adventitiously.

Resedea lutea L. Wild mignonette 112: 2

F VIII ?*Newstead*

A tentative record of *R. lutea* was made by Reid (Tagg, 1911) from the Roman site at Newstead: it has to be considered in the light of the comments given above on *R. luteola*.

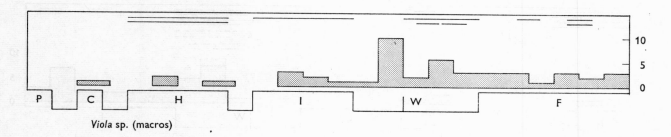

Viola sp. (macros)

VIOLACEAE

Viola sp. (Plate XVII) 113

There are very few records of pollen of *Viola*, no doubt partly because so little is shed into the air: there are two from the Flandrian, one from zone VIII at Apethorpe (Sparks & Lambert, 1961) and a queried identification from zone VIIb in the Isle of Canna (Flenley *et al.*, 1967). Although it is easy to refer fossil seeds to the genus *Viola*, and they are indeed common, the present British flora contains some fourteen species, together with hybrids, and investigators have found great difficulty in identification at the species level. Thus the great majority of records are referred only to the genus or to *Viola palustris* L., the species at once most recognizable by size, shape and the well-marked, alveolate surface pattern, and the most likely to occur in the wet situations most favourable for seed incorporation. The record-head diagrams are remarkable in that they shew such similar fossil records for the two taxa, which possibly indicate no more than different degrees of confidence among the responsible authors. Watts (1959*b*) gave criteria for distinguishing *Viola* seeds to taxa within the genus and the succeeding accounts show that others have followed this lead. It is evident from the diagrams above and from these accounts that the genus has been present in past interglacial stages and throughout the Flandrian. It evidently had some prevalence in the Weichselian with several records in the severe climatic conditions of the Middle Weichselian and indications of presence in earlier glacial stages. To judge from the present-day Scandinavian ranges this need not rule out *V. palustris*, *V. riviniana*, *V. rupestris*, *V. tricolor* or *V. canina*, but might bring into consideration arctic–montane species such as *V. biflora* not now in our flora.

Viola sub-genus Viola

I *Wretton*

Viola sub-genus Viola, section Uncinatae

lA *Hoxne*; H *Nechells*

These have been recorded by Turner (1968*a*) and Kelly (1964) as '*V. hirta* or *odorata*'; they have the largest seeds of British species.

Viola sub-genus Melanium

lA *Hoxne*; *e*W *Sidgwick Av.*; *m*W *Waltham Cross*; F VIIb *Streatly*

The seeds of this group are generally more slender than others in the genus: it is interesting in view of the glacial stage records that *V. lutea*, *V. tricolor* and *V. arvense* are included in this sub-genus, species that favour open conditions and bare ground and have wide present ranges in the British Isles.

Viola odorata L.   Sweet violet   113:1

I ?*Clacton*

A tentative identification only.

Viola hirta L.   'Hairy violet'   113:2

Cs ?*Pakefield*; H *Clacton*

V. hirta has a notably southern range in Scandinavia and is markedly south-eastern in the British Isles.

Viola riviniana Rchb.   'Common violet'   113:4

H and *e*Wo ?*Baggotstown*; H *Kilbeg*

Watts (1964) comments that this species probably cannot be separated, as seed, from *V. reichenbachiana* (see below).

Viola reichenbachiana Bor. or **V. riviniana** Rchb.
                                                      113:4/5

lW *Kirkmichael*; F VIIb *Frogholt*

Viola reichenbachiana Bor. (*V. silvestris* Reich.)   'Pale wood violet'   113:5

Cs ?*Pakefield*; H *Clacton*

Viola canina L.   'Heath violet'   113:6

*m*W ?*Temple Mills*, ?*Broxbourne*, ?*Hedge Lane*; lW *Kirkmichael*; F VIIb *Frogholt*, VII *Fifield Bavant*

The three first records are tentative only and are from the Lea Valley Arctic Plant Bed. The Kirkmichael records are from the zone I/II transition and zone II. The Frogholt (Kent) record is associated with cultivation indicators in the late Bronze Age, and Fifield Bavant is a Wiltshire Iron Age site.

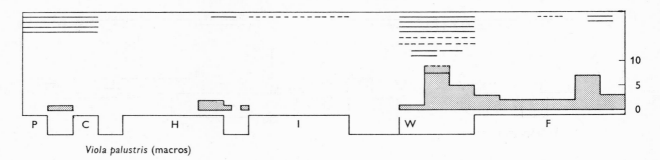

Viola palustris (macros)

Viola palustris L.   113: 9

The large number of records attributed to *V. palustris* is indicated in the record-head diagram above. It is a plant of bogs, fens, marshes and wet heaths, wide ranging both in the British Isles and Scandinavia and, as Watts says, has little value as a climatic or ecological indicator.

Viola lutea Huds.   'Mountain pansy'   113: 11

mW Barrowell Green, Lea Valley

There is nothing remarkable in finding *Viola lutea* in the Middle Weichselian of the Lea Valley, but the sites concerned lie well south of the present limit of the species in Britain.

Viola tricolor agg.   Wild pansy   113: 12

H *Clacton*; *mW Temple Mills, Broxbourne, Waltham Cross, Hedge Lane, Angel Rd*

It is of considerable interest to find this species now so typically a cornfield weed or ruderal, present in so many Middle Weichselian sites as well as in one interglacial. The identifications from the Lea Valley Arctic Bed were made in the first place by C. Reid but were checked later by E. M. Reid. (See also *Viola* sub-genus *Melanium* and *Viola lutea*.)

POLYGALACEAE

Polygala sp.   Milkwort   114

I *Austerfield*; *lW Loch Mealt, Esthwaite*; F vi, viiA *Crane's Moor*, viiA *Loch Dungeon*, viiB *Esthwaite*, viiA/b *Clatteringshaws Loch*, viii *Snibe Bog*

All the records for *Polygala* are at the generic level and based on recognition of the characteristic pollen. The earliest is from sub-stage iii of the Ipswichian. Franks and Pennington (1961) report it from all zones of the Late Weichselian at Esthwaite and again in zone viiB of the Flandrian in association with a woodland clearance horizon. The records for Crane's Moor in the New Forest are from Flandrian zones vic and viiA (Seagrief, 1960). The remaining records are all Scottish and made by Mrs Birks: they comprise zone i of the Late Weichselian at Loch Mealt, the Flandrian transition between zones

v and vi, zone viiA and viiB/viii transition at Loch Dungeon, zones viiB/viii and viii at Snibe Bog, and the viiA/b transition at Clatteringshaws Loch. Thus both in Scotland and the English Lake District we have presence in the Late Weichselian and the succeeding Flandrian with mid-Flandrian presence in the New Forest.

The British species include both calcicoles and calcifuges but are all dwarf herbs of grassland or heaths. *P. vulgaris* L. and *P. serpyllifolia* Hose are widely distributed throughout the British Isles, but both are rather strongly restricted in Scandinavia, *P. serpyllifolia* to the southwest coast of Norway and to Denmark, *P. vulgaris* extending through southern Sweden and Norway with scattered sites northwards on the west coast to the Arctic Circle.

GUTTIFERAE

Hypericum sp.   St John's wort   115

C *Norfolk Co.*; H *Ballykeerogemore*; I *Wortwell*; F iv *Seamer, Beaghmore*, viiB/viii ? *Blelham*

Hypericum seeds have been recovered from the Cromerian in Norfolk and from the Hoxnian at Ballykeerogemore and from sub-stage ii of the Ipswichian at Wortwell. All the British species of *Hypericum* except *H. pulchrum* tend to be infrequent or absent in the north of Scotland and to shew increased abundance in southern counties. Seed identification within the genus is fully considered by Watts (1959b) who writes 'Seeds of *Hypericum* show a strong generic similarity to one another, while, at the same time, they are distinct enough for exact specific determinations to be possible'.

Hypericum androsaemum L.   Tutsan   115: 1

H *Gort*

Jessen *et al.* (1959) have recorded from sub-stage ii of the Hoxnian interglacial at Gort a single seed with a strongly prominent raphe and surface features typical of *Hypericum androsaemum* L. The species is present throughout the British Isles except the far north but is far commoner in the west and south, as would be expected of a plant of the 'Atlantic–Mediterranean element extending to the Caucasus'.

Hypericum perforatum L.   'Common St John's wort'
115: 5

I *Clacton*; F viib ?*Minnis Bay*, viii *Silchester*

The tentative Minnis Bay seed identification is from a late Bronze Age deposit and that from Silchester is Roman. The Clacton record is from the middle Ipswichian.

Hypericum maculatum Cranz (*H. dubium* Leers)   'Imperforate St John's wort'
115: 6

I *West Wittering*; F vii/viii *Prestatyn*

The Prestatyn record is from the submerged coastal peat of the North Wales coast and is late Flandrian in age.

Hypericum tetrapterum Fr. (*H. quadrangulum* L.)   'Square-stemmed St John's wort'
115: 8

Cs *Pakefield*; A *Corton*; H *Kilbeg*; I *Shortalstown, Bobbitshole, Ilford, Wretton*; F viia/b *Shustoke*, viib *Frogholt*, viii *Wingham, Godmanchester*

The Corton record is from the interstadial in the Anglian glaciation and that from Kilbeg is from sub-stage I of the Hoxnian. These apart, the records indicate very mild climatic conditions, for the Ipswichian records are all from sub-stage II and the Flandrian records from zone viia/b transition (*c.* 5000 B.P.) onwards. The three latest are all concerned with human activity, at Frogholt in the late Bronze Age and at Wingham and Godmanchester in Roman time. Hultén shews the species as having a very restricted southern distribution in Scandinavia and he puts it in his sub-atlantic geographical group.

Hypericum humifusum L.   'Trailing St John's wort'
115: 9

F viib *Minnis Bay*

Miss Conolly (1941) identified a seed of *H. humifusum* from the late Bronze Age coastal peat bed at Minnis Bay, though she now regards the deposit as possibly subject to recent contamination.

Hypericum pulchrum L.   'Slender St John's wort'
115: 11

H *Kilbeg*; I *Wretton*; *l*W *Loch Cill Chriosd*; F iv/v *Loch Cill Chriosd*, v/vi, viib/viii, viii *Loch Dungeon*

The Wretton identification is of seed from sub-stage III. The Flandrian records are all of pollen of '*H. pulchrum* type' that embraces two species, *H. montanum* and *H. tetrapterum*, as well as *H. pulchrum*.

Hypericum hirsutum L.   'Hairy St John's wort' 115: 12

Cs *Pakefield*; H *Nechells*

The Pakefield record cannot be located in time more closely than to the Cromer Forest Bed series, but the Nechells record is from the warm sub-stage III of the Hoxnian.

Hypericum elodes L.   'Marsh St John's wort'   115: 14

*l*W *Ballaugh*

The record from Ballaugh, Isle of Man is for a seed from the solifluxion deposits of Late Weichselian zone III. Dickson *et al.* (1970) point out that size, cell-pattern and wall thickness distinguish it from seeds of other British *Hypericums* and recall that Watts (1959*b*) has already stressed its distinctiveness. It is a plant with a strongly southern and western distribution in the British Isles today and it is absent from Sweden: its presence in the cold post-Allerød phase is striking but as the authors of the record point out, many of the species they record from the Late Weichselian in the Isle of Man have a markedly southern range in Sweden or do not occur there at all.

CISTACEAE

Helianthemum sp. (Plate XXI*l, m*)   118

Since the publication by Iversen of his paper (1944*b*) on *Helianthemum* as a fossil plant in Denmark, pollen of this genus has been widely recognized in late-glacial deposits. Iversen gave reasons for referring the Danish fossil material to the species *H. oelandicum* (L.) Willd. At much the same time Welten (1944) and Lüdi (1944) showed that the treeless Late-glacial period of Switzerland was characterized by frequent pollen of the related *H. alpestre*.

Whilst identification of pollen to the genus is certain, the difficulties of recognizing the species have been such that few records at this level exist, viz., 'cf. *canum*' 2, '*chamaecistus*' 2, 'cf. *chamaecistus*' 1 and '*chamaecistus* type' 1. In view of this, for the purposes of the record-head diagram we have aggregated these with all pollen site records given merely as '*Helianthemum*'. This histogram startlingly emphasizes the prevalence of *Helianthemum* pollen sites throughout all zones of the Late Weichselian and the succeeding zone IV of the Flandrian. It was clearly present also in Early and Middle Weichselian time and in the two glacial stages that preceded it. The representation of the genus through most stages of the Hoxnian and Ipswichian, its presence in the Cromerian and throughout the Flandrian is notable, although not necessarily due to the same species as yielded the abundant records within the glacial stages.

TABLE 12. *Sites of fossil* Helianthemum
in the British Isles

	SW	SE	NW	NE	Ir.
Weichselian	3	1	0	0	0
Late Weichselian I	10	7	2	6	4
II	16	8	8	13	9
III	14	8	6	8	6

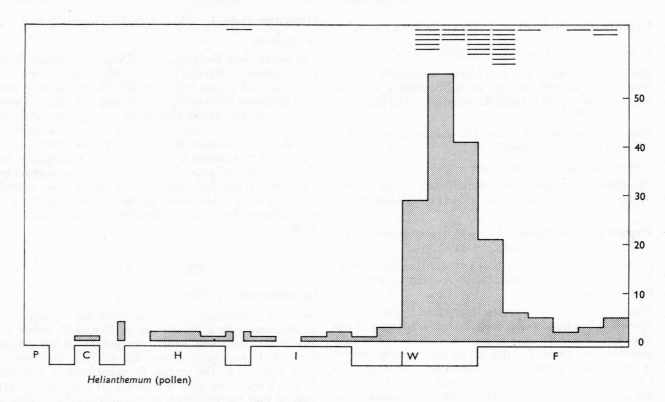

Helianthemum (pollen)

Table 12 shews the distribution of Weichselian site records in the five main geographical areas of the British Isles.

It is evident that *Helianthemum* was prevalent in all parts of these islands in the Late Weichselian; any apparent trends are probably due to varying numbers of sites investigated in the different regions.

In all sites the pollen frequencies are low, though this has to be considered with the fact established by Proctor and Lambert (1961) that contemporary pollen spectra from *Helianthemum* communities substantially under-represent the frequency of *Helianthemum* itself.

Helianthemum is an example of a plant genus widespread in Late Weichselian times, but with a remarkably restricted present northern range. Whether one considers the endemic *H. oelandicum*, *H. canum*, or the commoner *H. chamaecistus*, we find the Scandinavian distribution scarcely extending further north than 62°. We might wonder whether some factor other than temperature has prevented this plant, capable of flourishing in the Late Weichselian, from extending its Scandinavian range northward in the Flandrian period. This is a matter which might be related to intolerance of competition, to the calcicolous habits of the three species likely to be concerned, or to the operation of some aspect of climate (such as day-length and quality of light) not so far considered.

A history of *Helianthemum* in Ireland probably bears on this problem, for its apparent widespread occurrence in the Late Weichselian contrasts strikingly with its present rarity in that country, where *H. chamaecistus* occurs in one vice-county only and *H. canum* in only two. It is interesting in this regard to note that *Helianthemum* pollen, along with that of *Rumex* and *Empetrum*, has been cited as evidence for the persistence of late-glacial plants into the Flandrian, for at Slieve Gallion they occur in peat with a radiocarbon age of 7000 B.P. (Pilcher, 1970).

Helianthemum chamaecistus Mill.   Common rockrose

118: 1

H/Wo ?*Kilbeg*; *m*W *Badger's Hole, Upton Warren*; *l*W ?*Loch Cill Chriosd*

The two Middle Weichselian records are given as unqualified records of the pollen: the Kilbeg pollen record (Watts, 1959*b*) though cited as 'cf. *H. chamaecistus*' is backed by the approval of Dr Iversen; the Scottish record is given by H. H. Birks as '*H. chamaecistus* type' which embraces pollen both of *H. chamaecistus* and *H. appeninum*. The Kilbeg pollen comes from the unforested phase at the onset of the Wolstonian glacial, and occurs as part of a pollen assemblage dominated with grasses and sedges, but with *Artemisia*, *Thalictrum*, *Plantago* spp., *Lycopodium* spp., *Empetrum*, *Calluna*, *Jasione*, *Succisa* and *Juniperus*; in fact very much the character of the Late Weichselian pollen assemblages which contain *Helianthemum*.

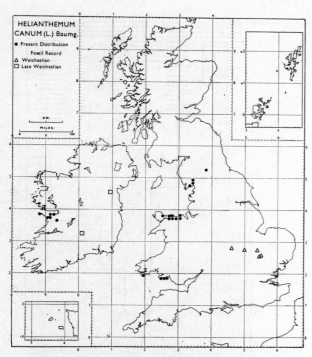

Fig. 49. Present and Weichselian distribution in the British Isles of *Helianthemum canum* (L.) Baumg.

Helianthemum canum (L.) Baumg. 'Hoary rockrose'
(Fig. 49; plate XXVIII*f*)   118: 3

W *Thrapston*; e*W Sidgwick Av.*; m*W Barnwell Station, Earith, Great Billing*; l*W Longfordpass, Teesdale, Mapastown, Neasham*; F IV *Neasham*

The two records from the Late Weichselian and zone IV of the Flandrian at Neasham are based on pollen cited as 'cf. *canum*' and pollen similarly recorded occurred at Mapastown and in Teesdale. All the remainder, including also Mapastown, rest on the much more secure basis of identification of leaf or capsule valve. The Barnwell Station record is by Mrs Dickson from unsorted material collected by Miss Chandler (fig. 49).

These records are of particular importance in view of the prevalence of *Helianthemum* pollen in Weichselian deposits, of the difficulty of referring such pollen to *H. canum* or *H. chamaecistus* respectively, and of the present restricted range of *H. canum* in the British Isles. The fossil evidence now appears to be strong that *H. canum* at least was widespread in late-glacial time and that the present distribution is due to post-glacial disjunction. In Upper Teesdale, one of the few areas where *H. canum* persists, Turner and Hewitson have shewn the persistence of *Helianthemum* pollen into the Flandrian phase of maximum local forest development and the commencement of widespread peat accumulation, an observation parallel to that of Pilcher at Slieve Gallion (see account of *Helianthemum* sp.).

Helianthemum pollen likewise occurs at various Flandrian levels in the peats of the raised bogs lying at the foot of Whitbarrow Scar (N. Lancs.), another present-day locality for *H. canum* (Smith, 1958*b*).

H. canum occurs very locally in the British Isles on rocky limestone pastures in Glamorgan, North Wales, Yorkshire, the English Lake District, and vice-counties Clare and west Galway in Ireland. It has been suggested that some degree of differentiation has occurred in these different populations. Its continental range is given by Hultén as 'boreal–montane with separated northern and southern areas': in Scandinavia it only reaches Öland (see account of the genus).

FRANKENIACEAE

Frankenia laevis 'Sea heath'   121: 1

P *Norfolk Co.*

An unpublished record by West and Wilson from the early Pastonian interglacial stage is the only evidence yet available for this inhabitant of the shingly margins of coastal salt marshes. The site is within the plant's present restricted south-eastern range.

ELATINACEAE

Elatine hexandra (Lapierre) DC.   122: 1

P *Norfolk Co.*; A *Corton*; H *Newtown*; H/Wo *Kilbeg*; I *Shortalstown*; l*W/F Moss Lake*; F IV, V, VI *Moss Lake*

The seeds of *Elatine hexandra* are very easily recognized, being about 0.7 mm long and gently curved, polygonal in section with each face occupied by a single row of reticulations, one end of the seed tapered and the other closed by a smooth conical lid (Watts, 1959*b*).

Through from the end of the last glacial stage, seeds of *E. hexandra* were found at Moss Lake, Liverpool in each Flandrian zone up to VI, and it occurred also in the mild sub-stage II of the Ipswichian and in the Pastonian interglacial. This seems to accord with the present European–Atlantic range of the plant with only a limited presence in southern Scandinavia. It is interesting on this account to find it recorded from the Corton interstadial and the late Hoxnian–early Wolstonian at Kilbeg as well as the Weichselian–Flandrian transition mentioned above.

Elatine hydropiper L.   122: 2

Cs *Pakefield*; A *Corton*; H *Clacton*; I *Wretton*

The seeds of *E. hydropiper* are distinguishable from those of *E. hexandra* by their sharp terminal curvature into the shape of a letter J. Apart from the record by Reid from the Cromer Forest Bed series the records are recent determinations made in Cambridge. The Corton record is interstadial, but the Clacton one is from sub-stage III and

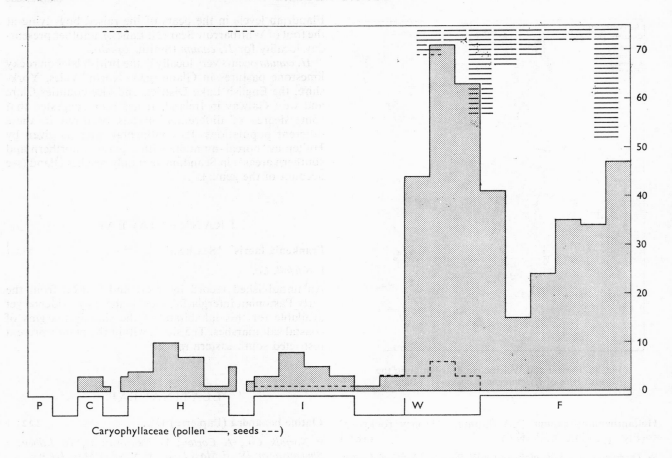

Caryophyllaceae (pollen ——, seeds ---)

the Wretton one from sub-stage II of their respective interglacials.

Although, like *E. hexandra*, having a rare and local occurrence in Britain, *E. hydropiper* has a more northerly range in Scandinavia and is listed by Hultén as 'west European–middle Siberian'. Both species favour ponds and small lakes, habitats essentially transient (see Watts, 1959*b*).

CARYOPHYLLACEAE

For a good many years British palynologists have recorded pollen as Caryophyllaceae for lack of more precise identification. The site records aggregated in the diagram above shew, as would be expected, records through all the interglacial sub-stages and all the Flandrian zones. What is particularly striking is however the great number of Late Weichselian and earliest Flandrian zone sites so recorded, both for zones themselves and transitions between zones. This has to be viewed against the accounts that follow for individual genera and species, where a similar effect is often evident, shewing, as with the Cruciferae, that the open conditions, fresh soils, and lack of competition more than compensated for any disadvantage there might be to these plants in periglacial surroundings. The Anglian

and Wolstonian glacial stage records are to be noted. Also requiring comment is the substantial rise in the numbers of site records in the later Flandrian zones, an effect, as again with the Cruciferae, presumably due to forest clearance and the expansion of agriculture. This explanation is borne out by consideration of the pollen frequencies. In the Late Weichselian these are mostly below 2 per cent of the total pollen until zone III, where they are mostly between 2 and 5 per cent. Values in zones V, VI and VIIa of the Flandrian never rise over the range of 5 to 10 per cent of total tree pollen, but in zones VIIb and VIII three sites fall in the 10 to 20 per cent range and one in 20 to 30 per cent. These high frequencies accompany other clear signs of agricultural activity.

A small number of seeds that have not been identified below family level are shewn in the diagram: they are not out of accord with the records based on pollen.

Silene sp.   123

W *Thrapston*; *e*W *Sidgwick Av.*; *m*W *Lea Valley*; *l*W ?*Hawks Tor*

The first three sites given above are for seeds. At Hawks Tor, equally in zones II and III, there are tentative identifications of both seeds and pollen to the genus *Silene*. H. H. and H. J. B. Birks currently recognize pollen of

'*Silene* type' and have recorded it from the Late Weichselian (zone I/II – Lochan Coir' a' Ghobhainn), early Flandrian (zone IV – Allt na Feithe Sheilich) and mid-Flandrian (zone VIIa – Coire Bog). The type includes *S. maritima, acaulis, otites, nutans, gallica*.

Silene vulgaris (Moench) Garcke (Agg.)   123: 1, 2

eWo Marks Tey, Brandon, Baggotstown; I *?Wretton*; *mW Brandon, Hedge Lane, Angel Rd, Ponders End, Waltham Cross, Temple Mills, Broxbourne, Barrowell Green, Barnwell Station, Marlow, Syston, Earith, Great Billing*; *lW Nant Ffrancon, Ballaugh, Kirkmichael*; F VIII *Shrewsbury* [See also the following accounts for the two sub-species *vulgaris* and *maritima*]

In the course of her investigations into Middle Weichselian deposits Miss F. Bell undertook extensive and critical research into the morphology of the seeds of a number of Caryophyllaceae, invoking in some instances low-power scanning electron microscopy for the resolution of the characteristic surface cell-patterns. *Silene* was among the genera reviewed and she employed the taxonomic usage of *Flora Europaea*, which puts within the *S. vulgaris* (Moench) Garcke aggregate the two sub-species *maritima* (With.) A. & D. Love and *vulgaris*. She points out that though these taxa are separable in Britain and may there be regarded as two species, in Scandinavia they intergrade. The two British ecospecies are separable at the present day on the seed character that *maritima* has a very low proportion of tuberculate seeds, and *vulgaris* a very high one. The constancy of these ratios breaks down outside Britain and Miss Bell rightly emphasizes that one cannot assume that present-day differentiation into two ecospecies had occurred in Weichselian time: with this in mind and the fact that at her own Middle Weichselian sites tuberculate and non-tuberculate seeds were present in similar frequency she is persuaded not to attempt sub-division of the aggregate species. Accordingly the sites Marlow, Syston, Earith and Great Billing are included in the list above, together with all sites where the identifications have been '*S. vulgaris* or *S. maritima*', and all the Lea Valley sites (Angel Rd, Hedge Lane, Ponders End, Temple Mills, Barrowell Green) plus Barnwell Station, from which '*Silene coelata* Reid' was recorded, or seeds of either *S. vulgaris* or *maritima*.

What is outstandingly apparent from the fossil record is the extraordinary abundance of the aggregate species *S. vulgaris* in full Weichselian time, where it is repeatedly found with geological conditions, and faunistic evidence indicative of fresh soils, open habitats, salinity and severe periglacial climate. The records of the present ranges of the two sub-species (see separate accounts) suggest that this is within their ecological tolerance. Seeds have been found from the Wolstonian glacial stage and of the sub-species *maritima* from the Anglian. It is recorded from one Early and five Late Weichselian sites. By contrast only the Wretton and West Wittering records represent an interglacial stage and the latter is for '*maritima*' from a coastal situation. There are no Flandrian records except for the Roman sites at Bermondsey and Finsbury Circus and the mediaeval site at Shrewsbury. It is evident that neither sub-species could have withstood the competition of forest conditions in the interglacials except in special situations or following clearances.

Silene vulgaris (Moench) Garcke subsp. vulgaris   Bladder campion   123: 1

eWo Baggotstown; I *?Wretton*; *mW* (see aggregate species account); *lW Drumurcher, Mapastown*; F IV *?Devil's Bit*, VIII *Bermondsey*

We have removed from this section all the Middle Weichselian records given in the first edition and those treated by the authors as belonging to the species '*vulgaris*'. This is not to say this taxon was not then present. It is to be noted that Mitchell's record for the Late Weichselian zone III at Drumurcher is based on the stronger evidence of seed plus capsule.

This sub-species extends as a common plant up to the Arctic Circle and falls within Hultén's category of boreal–circumpolar plants.

Silene vulgaris (Moench) Garcke subsp. maritima (Wild) A. & D. Love   Sea campion   123: 2

A *Lowestoft*; I *West Wittering*; *eW Sidgwick Av.*; *mW Ponders End*; F VIII *Finsbury Circus*

If we accept the likelihood that the sub-species *maritima* had already been differentiated in Britain by the Weichselian glacial stage, there is no need to retract the view that 'there appears to be a consensus of experienced opinion that *S. maritima* is represented in the Lea Valley "Arctic bed" deposits' (first edition). However, all these sites have now been recorded under the *S. vulgaris* aggregate species save one for Ponders End where a fruit capsule (originally recorded as *S. coelata*) was found.

S. maritima not only extends round Scandinavian shores to the region of the North Cape, but in Britain it occurs in high altitude mountain populations. It flourishes in open, unshaded habitats and tolerates salinity so there is considerable likelihood that it was a periglacial survivor in Britain.

Silene coelata Reid

Clement Reid, in the *Q.J.G.S.* for 1915 (published 1916), described and figured capsules and calyces of a *Silene* so different from those of any known species of *Silene* that they had to be placed in a separate species, *S. coelata* Reid. The species was, however, based upon a capsule without seeds from Ponders End and two calyces from Temple Mills.

When E. M. Reid (1949) took up the problem of distinguishing seeds of *S. coelata*, *S. cucubalus* and *S. maritima* from one another she submitted the fossil material to W. B. Turrill, the outstanding British authority on this group of species: he reported, 'I should be prepared to accept all the seeds received from you...as seeds of *Silene maritima*...the seeds named *Silene coelata* do not appear to me to differ from those of *S. maritima*.'

Dr S. M. Walters and Miss C. M. Lambert succeeded in shewing that the species has in fact no validity, for the capsule, like the seeds, could be safely referred to *Silene maritima*, whilst the 'calyces' proved on close examination to be the very decayed cupules of *Fagus*, which must be regarded as derived from deposits older than the Arctic bed.

Subsequently Dr F. Bell has examined all Reid's material labelled as *S. coelata* and referred it to the aggregate species *S. vulgaris*, which includes the taxon *S. maritima* mentioned above.

Silene gallica L.　　　　123: 6

F vIIb ?*Graig-Lwyd*, vIII *Larkfield*

The Graig-Lwyd record is from the Neolithic axe-factory site and that at Larkfield is mediaeval. This species is regarded as widely introduced and had a strongly southern and eastern distribution in the British Isles.

Silene acaulis L.　Moss campion　　123: 7

mW ?*Ponders End*, ?*Waltham Cross*, ?*Derryvree*; lW *Drumurcher*; *Loch Fada, Loch Cill Chriosd, Lochan Coir' a' Ghobhainn*

The Middle Weichselian records are seed determinations: the Late Weichselian Scottish records are by H. J. B. Birks based on pollen identification and cover zones I, II and III.

These records accord with the circumpolar arctic–montane range of the species and its present disjunct montane occurrence in Britain.

Silene nutans L.　Nottingham catchfly　123: 10

I *Wretton*

The seed identification is from Ipswichian zone IIb river terrace deposits (Sparks & West, 1970).

Silene dioica (L.) Clairv. (*Melandrium rubrum* (Weig.) Garcke)　Red campion　　　123: 13

H *Clacton*; Wo *Brandon*; I *West Wittering, Wretton*; mW *Hedge Lane*; lW *Hailes, Flixton, Neasham, Ballybetagh*; F vI ?*Drymen*, vII *Irstead*, vIIa/b *Shustoke*, vIIb *Wigton, Frogholt*

The Clactonian and Ipswichian records demonstrate presence in the warm middle sub-stages, Wretton having both pollen and seed records. All other of these records are based on seed identifications and they shew *S. dioica* present in all zones of the Late Weichselian and in zones vI, vII, vIIa/b and vIIb of the Flandrian.

H. H. and H. J. B. Birks record pollen of '*S. dioica* type' from Scotland, including the Hoxnian interglacial from Fugla Ness, Late Weichselian from Lochs Fada,

Cill Chriosd, Dungeon, Lochan Coir' a' Ghobhainn, and from Cooran Lane, Flandrian zone IV from Lochan Coir' a' Ghobhainn and zones IV, v/vI and vIIa at Alt Na Feithe Sheilich. The pollen types includes *S. dioica, S. alba, S. conica* and *S. noctiflora*, but it seems very likely, in view of the habitats and ranges today, that the records really represent *S. dioica*.

In strong contrast with *Silene alba* this species is well represented in two interglacials, two glacial stages, especially the latest, and throughout the Flandrian up to zone vIII. These records make a good case for regarding the species as a native persistent through the glaciations. It is a European boreal–montane species which extends as a common plant along western Scandinavia as far as the North Cape and it is not difficult to envisage it as a periglacial survivor. Although now commonly a British woodland species, it is also a 'characteristic plant of "bird-cliffs" in north-western Scotland and an occasional member of the flora of sheltered ledges with deep soil in mountainous regions' (Baker, 1947), whilst its Scandinavian habitats are damp meadows rather than woodlands (Hultén, 1950).

Silene alba (Mill.) E. H. L. Krause (*Melandrium album* (Mill.) Garcke)　White campion　123: 14

F vIIb *Fargo Plantation*, vIII *Silchester, Pevensey*, ?*Car Dyke*

All these records are associated with settlement sites, the first early Bronze Age, the rest Roman.

Baker (1947) writes of this species that it 'probably arose from an ancestor more closely resembling *Melandrium dioicum* by eco-geographical divergence, and has spread from a centre in the Middle East, the Mediterranean, and central Europe (where it grows in natural habitats) as a follower of agriculture'. He traces the recent expansion of this species from southern and eastern Britain by its progressive hybridization with *M. dioicum*, which by contrast is a native species primarily of woodland habitats.

The records so far available accord with Baker's view that the species was introduced into Great Britain during the Neolithic period and has become distributed through the country with the spread of agriculture.

Silene furcata Rafin.　　　　123: 15

W *Thrapston*; mW *Great Billing*

Dr Bell has identified from two Weichselian sites seeds that do not match those of any British species. She separates them from seeds of the closely related *S. wahlbergella* Chowdhuri by the narrow wing. It is a species that occurs widely in arctic and sub-arctic Europe, favouring dry to slightly damp, rich soil.

Silene wahlbergella Chowdhuri (*Melandrium apetalum* (L.) Fenzl)　　　　123: 16

eWo ?*Marks Tey, Broome*

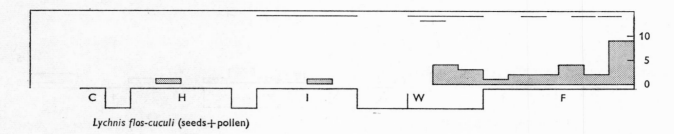

Lychnis flos-cuculi (seeds+pollen)

From the early Wolstonian glacial stage at Marks Tey, Turner and Mrs Dickson have made a tentative identification of *S. wahlbergella*, the seeds of which have prominent swollen wings and cells with strongly sinuous margins. Turner (1970) mentions that Mrs Dickson has also identified this species from glacial deposits at Broome.

Like *S. furcata*, *S. wahlbergella* is a circumpolar plant occurring in northern Scandinavia, the USSR and arctic America, but no longer in the British Isles.

Silene paradoxa L.   123: 17

mW Ponders End

The identification (Reid, 1949) of this south European plant from deposits of the Lea Valley Arctic Plant Bed seems questionable.

Lychnis sp.   124

mW Lea Valley, ?Temple Mills, Hedge Lane; lW ?Hawks Tor; F VIIb Prestatyn, VIII Caerwent, Godmanchester

The Caerwent and Godmanchester records are Roman. It is possible that the authors envisaged this genus as then embracing *Silene alba* and *S. dioica*. Material formerly recorded as '*Lychnis*' from Barnwell Station is deleted after revision by Bell and Dickson (1971).

Lychnis alpina L. (*Viscaria alpina* (L.) Don) 'Red alpine catchfly'   124: 1

W Thrapston; mW Upton Warren, Earith, Four Ashes, Syston, Angel Rd; lW Kirkmichael

In describing the fossil seeds from Thrapston, Earith and Syston, Miss Bell emphasizes the prominent hilum surrounded by a collar (a distinction from *Silene*), size and surface cell-pattern. In the latter character her material resembles Scandinavian rather than British material, possibly an indication of subsequent depletion of the Weichselian populations which, to judge from the fossil record, were widespread far beyond the present range of the species. It is an arctic and sub-alpine species, extending north to 73° in east Greenland (Polunin, 1940) and in Britain it is now restricted to a small number of far-separated high mountain sites in Scotland and the Lake District. The most recent geological horizon in which the plant has been found is the Late Weichselian zone I at

Kirkmichael, Isle of Man. The data very strongly indicate wholesale extinction and disjunction by the conditions of the Flandrian interglacial. In Britain *L. alpina* is exclusive to heavy metal soils and it is given as serpenticolous in Sweden: not only was the Weichselian range evidently wider, but soil requirement must have been less restrictive. It seems possible that 'the special edaphic preference allowed populations to survive in a few areas where heavy metal soils and permanently open ground coincide' (Dickson *et al.*, 1970).

Lychnis viscaria L. 'Red German catchfly'   124: 2

lW Kirkmichael

A single seed from the late Weichselian zone III at Kirkmichael, Isle of Man was found alongside seeds of *L. alpina* from which it was distinguishable by the acute tubercles borne on radially and concentrically orientated cells (Dickson *et al.*, 1970).

This species is now rare and local in Britain, limited to cliffs, dry rocks and rock debris at a few sites in north Wales and the eastern side of Scotland. It is common in southern Sweden and Finland with scattered occurrences into the Arctic Circle.

Lychnis flos-cuculi L. Ragged robin   124:3

The records given diagrammatically are mostly for seeds identified positively. They include a record from the mild sub-stage of the Hoxnian and Ipswichian interglacials and records from the end of Late Weichselian zone I through the Flandrian. Thus the historical pattern resembles that of *Silene dioica* and it may equally be regarded as a long persistent native, but it differs in shewing response to human settlement by records from the Iron Age (Wingham and Glastonbury), the Roman (Wingham and Silchester) and the Mediaeval periods (Hungate). Two of the four pollen records are Scottish (IV and V/VI zones) and two from Neolithic trackway sites in Somerset, that from the Bell Track based both on seed and pollen (Coles & Hibbert, 1968). *Lychnis flos-cuculi* occurs locally over much of northern Scandinavia, and is fairly common in central and southern Sweden and in southern Finland; it is common throughout the British Isles.

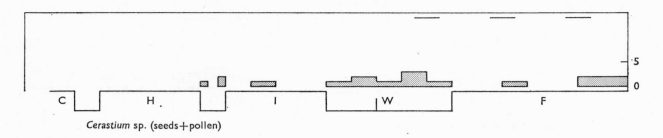

Cerastium sp. (seeds+pollen)

Lychnis triflora R. Br. 124: 4

*m*W ?*Hedge Lane*

This identification was based on comparison with seeds supplied by Porsild from Greenland where the species *Lychnis triflora* is endemic. E. M. Reid writes of the identification as rather doubtful, and Sørensen points out 'that seed-characters are quite unreliable', presumably in relation to the taxonomy of *L. triflora vis-à-vis L. furcata* (Raf.) Fernald (Polunin, 1940).

Agrostemma githago L. (*Lychnis githago* (L.) Scop.) Corn cockle 125: 1

F VIII *Isca, Silchester, Newstead, North Leigh, Bunny, Hungate, Dublin, Old Buckenham*

The first five of these sites are Roman settlements, Hungate and Dublin are mediaeval. The Old Buckenham site, the only record based on pollen identification, is from levels considered to be Iron Age, Anglo-Saxon and Norman, along with abundant evidence of arable cultivation (Godwin, 1968*b*).

The above records afford no evidence for a natural status of *Agrostemma githago*: it is evident that the abundant seeds in each instance are associated with a local collection of weeds of cultivation. There is some evidence that the corn cockle was formerly an exceedingly prevalent weed in cereal crops and that the high saponin content may have caused susceptibility to leprosy.

It is probable that the present-day infrequency of the plant in our cornfields has been produced by the efficiency of mechanical seed-cleaning during the past hundred years or so, but it is also possible that the plant has some close biological link with the rye, which now has become a rare crop in the British Isles.

Like other Mediterranean species, *Agrostemma githago* shews rapid germination at low temperatures, a fairly low maximal germination temperature and needs little incubation time, all properties appropriate to a weed of arable crops in this country (Thompson, 1970).

Dianthus sp. 127

I *Wretton*; *e*W *Sidgwick Av.*; F VIII ?*Hungate*

The Hungate record is mediaeval.

Dianthus carthusianorum L. 127: 3

*m*W ?*Upton Warren*

In the long list of plant identifications made in the Middle Weichselian deposits at Upton Warren some twenty seeds were tentatively referred to *Dianthus carthusianorum* by Coope *et al.* (1961). They write that 'of the three native species in Britain, only *D. gratianopolitanus* has seeds which are at all similar to the fossil ones'. *D. carthusianorum* is now native in southern, western and central Europe as far north as Holland and Denmark (55° N). In view of the present ranges of this species and *D. gratianopolitanus* and of the severity of Weichselian climatic conditions, further evidence of their fossil record would be of great interest.

Dianthus gratianopolitanus Vill. Cheddar pink 127: 7

*m*W ?*Earith*

From the abundantly documented Middle Weichselian terrace deposits at Earith, Cambs., Miss Bell has made a tentative identification of a seed of the Cheddar pink, a species now limited in Britain to the Carboniferous Limestone of the Cheddar Gorge. It occurs in France, Belgium and Germany but not further north.

Dianthus deltoides L. Maiden pink 127: 8

*l*W *Kirkmichael*

From zone I of the Late Weichselian in the Isle of Man Dickson *et al.* (1970) identified as *D. deltoides* two black seeds by reason of their strongly compressed shape, subacute apex and rounded base, low radially elongated tubercles radiating from the central hilum and the strongly sinuous margins of the tuberculate cells. The seeds are also smaller than those of any other British species of *Dianthus* except *D. armeria*.

D. deltoides has a scattered and mainly eastern range and is absent from Ireland and the Isle of Man.

Cerastium sp. 131

These records are to be considered as supplementing to some extent those for individually identified species. They indicate however that one or more species of the genus occurred freely during the Weichselian and in the preceding Wolstonian glacial stage.

Cerastium cerastoides (L.) Britton 'Starwort mouse-ear chickweed'   131: 1

lW Ballybetagh, Drumurcher

Careful revision by Mitchell of the fossil seeds recovered from Ballybetagh and described and illustrated by Jessen and Farrington (1938) has led to the conviction that some of them must be referred to *Cerastium cerastoides* and not to *C. vulgatum*. This separation is based upon the elongate shape (as against iso-diametric) of the superficial small decorative 'warts' with radiating ridges upon their flanks. Mitchell also identified as *C. cerastoides* two similar seeds from the Late-glacial at Drumurcher.

C. cerastoides is an alpine and sub-arctic species, chiefly European in range but extending to North America. It extends the full length of the Scandinavian peninsula, but is confined to the oceanic west. In the British Isles it is confined to high mountains in Cumberland and Scotland. It is unknown from Ireland and has evidently suffered much restriction of range in the Post-glacial period.

Cerastium arvense L. 'Field mouse-ear chickweed' (Figs. 50, 150c)   131: 2

lA Hoxne; Wo ?Dorchester; W Thrapston; eW Sidgwick Av.; mW Upton Warren, Great Billing, Earith, Oxbow, ?Derryvree, Marlow, Ponders End; lW Kirkmichael, White Bog, ?Knocknacran, Drumurcher, Ballybetagh

The flattened asymmetric seeds of *C. arvense*, often wider than long, are separable from those of the similar seeds of *C. alpinum* and *C. arcticum* by cell type and size (Bell, 1968). All sub-stages of the Weichselian are now recorded, as are all zones of the Late Weichselian, mostly supported by a wealth of independent evidence of geology, climate, fauna and age (fig. 50). The preceding Wolstonian and Anglian stages have single records, so that, despite the dangers of argument from absence, the total lack of Flandrian records becomes significant.

This species belongs to Hultén's category of 'circumpolar species strongly spread by culture' and he shews it common only in Denmark and southernmost Sweden, but with scattered occurrences extending into the Arctic Circle. It tends to be less frequent in Scotland than in England and is only local in Ireland. At the present day it is strongly associated with dry, sandy and disturbed soils.

Cerastium alpinum L. 'Alpine mouse-ear chickweed'   131: 4

Seeds: A *Corton*; mW *?Barrowell Green*; lW *?Branston*
 Pollen: lW *Lochs Cill Chriosd, Meodal, Fada, Lochan Coir' a' Ghobhainn*

The seed identifications, except that from the Corton interstadial, are all tentative: the Branston records (Tallentire, unpublished) are from all three zones of the Late Weichselian.

The pollen '*C. alpinum*' recognized by H. J. B. Birks

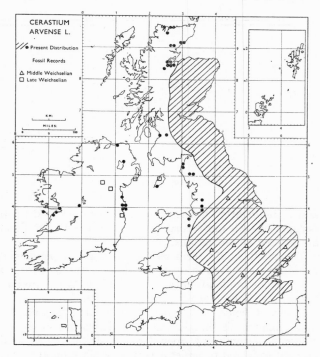

Fig. 50. Present and Weichselian distribution in the British Isles of *Cerastium arvense* L.

includes besides the type species, *C. cerastoides*, *C. arvense* and *C. arcticum* and the Scottish records again cover all three late Weichselian zones, often two or three at one site.

As with *C. arvense*, the records are restricted to glacial or late-glacial stages in accord with the fact that it is a most frequent circumpolar plant reaching the 'northernmost botanical locality on earth' at 83° 24' N (Polunin, 1940). It appears rather indifferent to substrate but in Britain is restricted to high altitudes, presumably a climatic control such as may be responsible for disappearance of the pollen type in the *Juniperus* sub-zone at Lochs Fada and Meodal (H. J. B. Birks, 1969).

Cerastium holosteoides Fr. (*C. vulgatum* L.) 'Common mouse-ear chickweed' (Plate XXg)   131: 7

H Kilbeg, Nechells; I Wretton; mW ?Barrowell Green; lW Newtownbabe, White Bog, Nazeing, Drumurcher, Knocknacran, Kirkmichael, Ballybetagh, ?Nant Ffrancon, ?Loch Fada; F VIIb, VIII Frogholt, VIII Ballingarry, Godmanchester

These records shew *C. holosteoides* to have been present in sub-stage II of the Hoxnian, and sub-stages II and III of the Ipswichian interglacials. A doubtful Middle Weichselian record is succeeded by several records each for zones II and III of the Late Weichselian, with the Nant Ffrancon record given as '*C. holosteoides* or *C. alpinum*'. All four Flandrian records are associated with agriculture or settlement, at Frogholt at levels radiocarbon dated 2490 and 2640 B.P., at Godmanchester at a Roman site

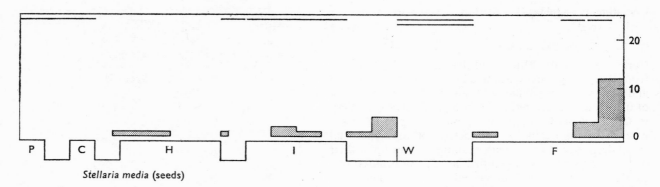

Stellaria media (seeds)

and at Ballingarry Downs a 4th- to 9th-century mediaeval occupation. All the records except that from Loch Fada are of seed identifications.

The plant would seem to have thrived in periglacial conditions, survived in open situations naturally through interglacials and to have responded to later human clearances. It is now very common in the British Isles, growing in fields, waste places, cultivated ground, by tidal streams, on shingle and sand dunes; all situations where the vegetation is open. It shews preference for well drained soils and occupies a very wide altitudinal range, reaching nearly 4000 ft in Scotland. It is found as a common plant throughout Scandinavia.

Cerastium fontanum L. 131: 13

W *Thrapston*

A seed of *C. fontanum* was identified by Miss Bell from Weichselian glacial deposits at Thrapston. The species is distributed throughout almost the whole of Europe but does not now grow in the British Isles.

Myosoton aquaticum (L.) Moench (*Stellaria aquatica* (L.) Scop.; *Malachium aquaticum* (L.) Fr.) Water chickweed 132: 1

Cs *Pakefield, Beeston*; I *West Wittering, Trafalgar Square, Selsey, Wretton, Bobbitshole, Histon Rd*; mW *Ponders End*; F IV/V *Moorlog*, VIII *Glastonbury, Apethorpe, Silchester, Finsbury Circus, Shrewsbury*

Myosoton aquaticum has two records from the Cromer Forest Bed series and one from sub-stage III of the Hoxnian interglacial. In the Ipswichian it has been recorded once from sub-stage I, four times from sub-stage II and twice from sub-stage III, shewing some persistence at the Wretton and Bobbitshole sites. In strong contrast with the records for *Cerastium* species, *M. aquaticum* has only a single record from a glacial sub-stage, the Weichselian. It occurred in North Sea moorlog (zones IV to V probably) and thereafter there are no records until zone VIII, where Silchester and Finsbury Circus are Roman and the Shrewsbury site mediaeval. This apparent response of a marsh plant to human settlement recalls that of numerous other species.

The plant has a very limited northern range in Britain, with a main distribution in the south-east, is absent from Ireland, and has only scattered occurrences in southern Scandinavia and southern Finland; this makes it especially desirable to have the Full-glacial record confirmed by further discoveries.

Stellaria sp. 133

Seed: Wo *Dorchester*; I *Shortalstown, Trafalgar Sq.*; mW *Syston*; lW *Drumurcher*; F VIIb *Prestatyn, Wingham*, VIII *Hungate, Bunny*

Pollen: H ? *Clacton*; F VIIb *Old Buckenham, Knockiveagh*, VIII *Moorthwaite*

Pollen: Stellaria type (H. H. & H. J. B. Birks) lW *Lochs Fada, Cill Chriosd, Cooran Lane, Lochan Coir' a' Ghobhainn*; F IV, V/VI, VIIa *Allt na Feithe Sheilich*, V/VI, VIII *Coire Bog*, V/VI *Loch Einich*, VIIb/VIII *Loch Dungeon*

The 'Stellaria type' pollen includes that of *S. nemorum, media, neglecta, graminea* and *alsine*, and the Scottish records under this head collectively cover, albeit thinly, all zones of the Late Weichselian and Flandrian. The other pollen records and seed records shew a concentration in the later Flandrian zones and at sites affected by clearance or settlement, a tendency evident also in the records for individual species.

Stellaria media (L.) Vill. Chickweed 133: 2

Stellaria media has been recorded from the Cromer Forest Bed series, and the Hoxnian and Ipswichian interglacials. There are records for both the Anglian and Wolstonian glacial stages and from the Early, Middle and Late Weichselian. In the early Flandrian it was present in the middle of what is now the North Sea but there are no records from the Boreal and Atlantic periods when we may suppose deciduous forest to have been widespread and almost uninterrupted, but with the inception of cultivation records are resumed and the species is represented right through from the Late Bronze Age to Mediaeval times. There are two records from the Bronze Age, one Iron Age, seven Roman, one Norman and three mediaeval records. It is apparent that *S. media* has been persistently native up to and through the Weichselian glaciation; it spread freely under conditions provided by

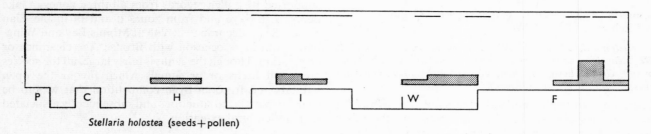

Stellaria holostea (seeds+pollen)

agriculture and human settlement, but whether from native or newly introduced stock the records cannot indicate. The plant occurs commonly as a weed well into the Arctic Circle and its Late-glacial occurrences and its presence in the Lea Valley Arctic bed are not surprising. The numerous interglacial records give striking proof of the plant's capacity to exist here independently of human influence.

Stellaria pallida (Dumort.) Piré (*Stellaria apetala* Ucria. 'Lesser chickweed'   133: 3

I *Clacton*

The interglacial record of *Stellaria pallida* must have come from the temperate forest period of the Clacton Interglacial.

Stellaria neglecta Weihe 'Greater chickweed' 133: 4

H *Nechells*; I *Selsey, Wretton*; F vɪɪb *Frogholt*, vɪɪɪ *Ehenside*

The Hoxnian record is from sub-stage ɪɪ, whilst both ɪɪ and ɪɪɪ are represented in the Ipswichian, indeed both at Wretton. *S. neglecta* is a plant of shaded habitats such as wood margins and stream banks. It has a sub-atlantic distribution and is infrequent and scattered in Scotland, southerly in England and Wales, and absent from Ireland.

Stellaria holostea L. 'Satin flower', Adder's meat, 'Greater stitchwort'   133: 5

Nine records for seeds of *Stellaria holostea* include one Hoxnian, four Ipswichian, one from the Cromer Forest Bed series, one Late Weichselian and two late Flandrian records. It is only recently that pollen has been systematically referred to this species and the impact is still very local. There is one Hoxnian pollen record, then a group of Scottish records covering all zones of the Late Weichselian, and finally eight records from zones vɪɪa, vɪɪb and vɪɪɪ of the Flandrian. Five of this latter group are from the Somerset Levels, three found at sites of excavation of Neolithic wooden trackways, an occurrence that is unsurprising since they were built over a fen-wood substratum.

S. holostea is a woodland plant widespread through the British Isles, though somewhat less frequent in north-west Scotland and western Ireland. It is common in Scandi-navia only in Denmark and southernmost Sweden, although sites are scattered to about 63° N: Hultén places it in his category of west European–middle Siberian plants.

Stellaria palustris Retz. (incl. *S. dilleniana* Moench and *S. glauca* With.) 'Marsh stitchwort'   133: 6

H *Clacton*; e*Wo Brandon*; l*Wo Selsey*; W *Thrapston*; m*W Upton Warren, Broxbourne, Hedge Lane*; F ɪv/v *Moorlog*, vɪɪ/vɪɪɪ *Hightown*

Stellaria palustris was found in the Hoxnian interglacial at Hoxne and in the early and late Wolstonian glacial. Thereafter there are four Weichselian records, the older identifications supported by more recent ones; one discovery from the early Flandrian in the North Sea moorlog and one from late Flandrian in the submerged forest of the Lancashire coast. These records point to the possibility that the species has long been native here, more especially because of its Middle Weichselian presence. Its present distribution is in line with this, for it occurs in south Finland commonly and has scattered localities up to and within the Arctic Circle. It is rare in Ireland and absent, or almost so, from Norway, and is one of the infrequent British species in Hultén's category of central Asiatic–Continental plants.

Stellaria graminea L. 'Lesser stitchwort' 133: 7

lA ?*Hoxne*; H *Nechells*; e*W Sidgwick Av.*; m*W Earith, Syston*; l*W Colney Heath*, ?*Drumurcher*; F vɪɪb *Frogholt*, vɪɪɪ *Apethorpe, Ehenside, Godmanchester, Silchester, Bermondsey, Pevensey, Caerwent, Hungate, Bunny, Dublin*

The Hoxnian identification is from the mild sub-stage ɪɪ, but the other pre-Flandrian records are from Glacial stages, including Early, Middle and Late Weichselian. All the Flandrian records are late, the earliest from Frogholt at c. 2490 B.P. The nine remaining include no less than seven from Roman settlements and the Dublin record is mediaeval. The species is common throughout the British Isles and extends abundantly north of the Arctic Circle in Scandinavia, so there is no need to question the likelihood that it was a perglacial survivor in the British Isles. It can be observed to favour artificial habitats to some extent.

Stellaria alsine Murr. (*S. uliginosa*, Grimm) 'Bog stitchwort' (Plate XX*b*)   133: 8

I *?Histon Rd*, *?Marros*; *m*W *Hackney Wick*, *Ponders End*; *l*W *?Nazeing*, *Nant Ffrancon*, *Aby*, *?Hartford*; F viib *Avonmouth*, *Graig Llwyd*, *Frogholt*, viib/viii *Rockstown*, *Godmanchester*, viii *Silchester*, *?Hungate*

There are two tentative identifications of *Stellaria alsine* from the Ipswichian, one from sub-stage III, the other possibly later. Two records from the Lea Valley Arctic Plant Bed are followed by five from the Late Weichselian, one, from zone II at Aby, a leaf identification followed by a zone III seed identification at the same site. The Flandrian records are all zone viib or later, and apart from Avonmouth and Rockstown, they are all associated with cultivation or settlement; thus Graig Llwyd (Neolithic), Frogholt (Late Bronze and Early Iron Age), Godmanchester, Silchester (Roman) and Hungate (Mediaeval).

There is clearly good reason for regarding the plant as native since at least Middle Weichselian time. It is a species belonging to Hultén's category of 'west European–middle Siberian' plants and he shews it as common in southern Scandinavia as far north as 61°, and with scattered Norwegian occurrences extending to the Arctic Circle. It is recorded from all vice-counties in Britain and it reaches high altitudes.

Stellaria crassifolia Ehrh.   133: 9

W *Thrapston*; *m*W *?Barnwell Station*, *Marlow*; *l*W *Penkridge*, *?Kirkmichael*, *?Dromsallagh*

The unqualified identifications above are all by Bell who describes the seeds as having elliptical elongated shape, slightly curved outline and finely rugose surface. The Barnwell Station record is a revision by C. A. Dickson of a seed previously recorded by Chandler as '*Arenaria sedoides*'. There now seems firm ground for assuming that *S. crassifolia*, no longer in the British flora, was present in the Weichselian and perhaps even in all three Late Weichselian zones. It is a plant that grows commonly in Scandinavia, extending on the western coast right to the North Cape, and it extends southwards into Germany. It is one of Hultén's 'boreal–circumpolar plants with substantially continuous distribution' and in the Arctic grows 'in damp grassy areas about sea-shores and around habitations' (Polunin, 1940). The data strongly indicate that it must have suffered post-glacial extinction in the British Isles.

Sagina sp. Pearlwort   136

Seed: I *?Shortalstown*; *l*W *Nant Ffrancon*, F viib *Minnis Bay*, *Wingham*

    Pollen: *l*W *Lochs Mealt*, *Fada*, *Chriosd*, *Dungeon*, *Lochan Coir' a' Ghobhainn*, *Elan Valley*; F v/vi, viia, viib/viii *Loch Dungeon*

There is only a tentative interglacial seed identification, but the Late Weichselian seed record in North Wales is

supported by pollen records from all three zones in lake deposits in Skye and from zones II and III in the Elan Valley. Seeds occur in zone viib at Minnis Bay and Wingham, both in association with Bronze Age clearance or occupation. Though the genus is fairly large, all the species are small herbs or perennials which favour the open habitats and freedom from competition that was to be found in periglacial situations and subsequently associated with agricultural extension.

Sagina procumbens L. 'Procumbent pearlwort' 136: 4

*l*W *Ballaugh*; F iv *Ballaugh*, viib *Frogholt*

At Ballaugh (Isle of Man) Mitchell (1958) has recorded seeds of *Sagina procumbens* from zone II, the transition II/III and zone IV. In later investigations in the Isle of Man a large number of seeds from the Late Weichselian zone III at Kirkmichael, Late Weichselian II and III, and Flandrian IV at Ballaugh were identified as '*S. procumbens* or *maritima*' (Dickson *et al.*, 1970): they were white to light brown as if unripe and had a pronounced semicircular hilum. Hultén gives this plant as strongly spread by human activity and of European or west Siberian origin: it occurs to the far north of Scandinavia and abundantly throughout the British Isles.

Sagina saginoides (L.) Karst. or *S. subulata* (Sw.) C. Presl.   136: 6–9

*l*W *Hawks Tor*

A great number of seeds from the lake mud of zone II at Hawks Tor, Cornwall proved impossible to refer decisively to one or other of the two species *Sagina saginoides* or *S. subulata* (Conolly *et al.*, 1950). Both the species have a local and infrequent distribution in the British Isles, but *S. saginoides* has a strongly arctic–alpine range and in Britain is restricted to a few areas in highland Scotland. Of the two *S. subulata* alone is found in southern Europe and it does not extend far north.

Sagina nodosa (L.) Fenzl 'Knotted pearlwort' 136: 10

F iv *?Ballaugh*, viib *Frogholt*, viii *Bunny*

The Ballaugh record is for zone IV, that from Frogholt Iron Age (2490 B.P.) and the Bunny record is Roman. This record though sparse fits the pattern common in the Caryophyllaceae of abundance in Late Weichselian and early Flandrian and then in association with human settlement.

S. nodosa is generally distributed through the British Isles today but with a tendency to greater frequency in the west.

Minuartia sp.   137

*l*W *Mapastown*; F viii *Peterborough*, *?Wingham*

The Wingham record is a seed from the Roman level, the other records are based on pollen, one from Mapastown in Late Weichselian zone I and the other from zone VIII, Peterborough.

Minuartia verna (L.) Hiern   'Vernal sandwort' (Fig. 51)
137: 1

I *Histon Rd*, W *Thrapston*; mW *Great Billing, Syston, Marlow*

The Histon Rd, Cambridge site is from the warm substage III of the Ipswichian Interglacial, but the four other seed records, all by Bell (1968), are all Weichselian (fig. 51). The plant has a local and disjunct distribution in Britain, centred chiefly in the north Pennines, the Lake District and North Wales upon dry calcareous rocks, screes and pastures, often common on old lead workings. The circumpolar arctic–montane distribution is consonant with the full-glacial occurrences.

Minuartia rubella (Wahlenb.) Hiern   'Alpine sandwort'
   (Fig. 51)   137: 2

mW *Earith*

Some fossil *Minuartia* seeds from the Middle Weichselian deposits at Earith were found by Bell exactly to match those of *M. rubella*, although the separation from *M. verna* is evidently not always possible.

 M. rubella is a circumpolar plant of the Arctic regions of Europe now limited to rock ledges near the tops of a few Scottish mountains and in Shetland. Whether we are concerned with *M. rubella* or *M. verna* the English lowland records from the Middle Weichselian clearly indicate a big subsequent retraction of range (fig. 51).

Minuartia stricta (Sw.) Hiern   'Bog sandwort' (Fig. 51)
137: 3

W ?*Thrapston*; lW *Ballybetagh, Knocknacran, Dunshaughlin*

The three Irish records are all from zone III of the Late Weichselian. *Minuartia stricta* is a circumpolar–montane plant that occurs throughout the Scandinavian mountains and in Britain is limited to a few Scottish sites and the better known high-altitude Teesdale site on Widdybank Fell (fig. 51). It is not now known from Ireland and presumably suffered post-glacial extinction there.

Cherleria sedoides L. (*Arenaria sedoides* (L.) F. J. Hanb.)
   Mossy cyphal   138: 1

mW *Barnwell Station*

Mrs Dickson revised the identity of one seed recorded as *C. sedoides* from this site by Miss Chandler but not that of others.

 There is only a single Full-glacial record for *Cherleria sedoides*, a species of the Alps of central and southern Europe and the Pyrenees, which is now found in Britain only at high altitudes on Scottish mountains. The Barnwell Station site is below 50 ft (15 m) O.D.

Honkenya peploides (L.) Ehrh. (*Arenaria peploides* L.)
   'Sea sandwort'   139: 1

I *Stone*

Possibly the absence of records of this species, now so common on British sand dunes and shingle beaches, from all but a single site and that an interglacial, is due to the lack of opportunities for incorporation in deposits where it might be preserved. It is a plant of far northern range.

Moehringia trinervia (L.) Clairv. (*Arenaria trinervia* L.)
   'Three-nerved sandwort'   140: 1

H *Clacton, Nechells*; I *Selsey, Bobbitshole, Ilford, Wretton*; lW ?*Flixton*; F IV/V *Moorlog*, VIIa/b *Shustoke*, VIIb *Cockerbeck, Frogholt, North Ferriby*

Both the Clacton and Birmingham records are from the middle sub-stages of the Hoxnian, but for the Ipswichian, records are more numerous and persistent; thus Selsey has sub-stages I and II, Wretton II and III as well as Bobbitshole II and Ilford II. The North Sea moorlog record is early Flandrian and then, as so often in Caryophyllaceae, there is a run of records reflecting anthropogenic influence, Cockerbeck (Neolithic), Frogholt and North Ferriby (Late Bronze Age).

 It now seems evident that *Moehringia trinervia*, a woodland herb of well drained rich soils, is a persistent native species although its behaviour in the glacial stages is uncertain. It is one of Hultén's 'west European–south Siberian plants' which has scattered occurrences in Scandinavia as far north as 70° but is only common south of about 61° N, just as it is more frequent in England and Wales than in Scotland and Ireland though present throughout the British Isles save the Outer Hebrides, Orkney and Shetland.

Arenaria sp.   141

W *Thrapston*; mW *Lea Valley, Barnwell Station*; lW *Drumurcher*

To the original record for seed at Barnwell Station by Miss Chandler, Mrs Dickson and Miss Bell now add revised records that originally were given as *Arenaria biflora* L., a non-British species, and *Arenaria gothica* Fr.

Arenaria serpyllifolia L.   'Thyme-leaved sandwort'
141: 1

Cs *Pakefield*; lA *Nechells*; H *Nechells*; lWo *Ilford*; I *Ilford*; lW *Kirkmichael*; F VIII *Newstead, Larkfield*

Seeds of *Arenaria serpyllifolia* were recorded by the Reids from the Cromer Forest Bed series. At Nechells Kelly has shewn the species to persist through the end of the Anglian glacial stage through zones I, II and III of the Hoxnian interglacial, i.e. to its warmest phase. At Ilford the plant is also persistent from the end of the Wolstonian glacial stage into zone I of the Ipswichian (West, 1964) and it appears again in the Late Weichselian zone I in

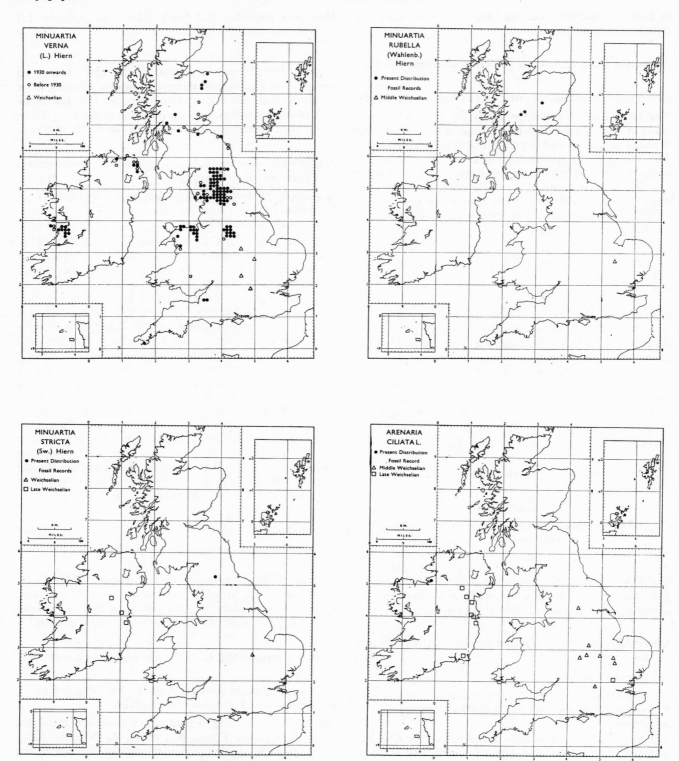

Fig. 51. Present and Weichselian distribution in the British Isles of four species now of restricted and disjunct montane range, *Minuartia verna, M. rubella, M. stricta* and *Arenaria ciliata*.

the Isle of Man. We have no Flandrian records until it appears in human settlements, Late Bronze Age at Minnis Bay (see following account 141: 1a), Roman at Newstead and Mediaeval at Larkfield where identification was based upon a calyx. Given bare ground and open situations such as occur naturally on river banks, dunes and cliffs *A. serpyllifolia* can evidently persist through the mild sub-stages of our interglacials and is equally at home in periglacial conditions, finally responding, like other Caryophyllaceae, to the numerous bare areas and diminished competition produced by forest clearance, settlement and arable cultivation.

Arenaria serpyllifolia occurs throughout the British Isles, though commonest in the south and east, a situation parallel to that in Scandinavia where it is common only south of 62° N, though occurring locally north beyond the Arctic Circle. It is one of Hultén's west European–south Siberian plants.

Arenaria serpyllifolia ssp. **macrocarpa** (Lloyd) Perring and Sell   141: 1a

*l*W *Kirkmichael*; F viib *Minnis Bay*

Dickson, Dickson and Mitchell (1970) record from the Late Weichselian zone I, in two sites at Kirkmichael, Isle of Man, macroscopic remains referable to the sub-species *macrocarpa* of *A. serpyllifolia*, a little-known taxon scattered round the British coasts (Perring & Walters, 1969; Perring & Sell, 1968). The fossil material included thirteen seeds and incomplete capsules, one still with three seeds inside. The seeds are distinguishable from those of other sub-species by their size. Seeds from the Late Bronze Age site at Minnis Bay recognized by Miss Conolly as *A. serpyllifolia* are also referable to this sub-species. Both sites are coastal.

Arenaria leptoclados (Reichb.) Guss. 'Lesser thyme-leaved sandwort'   141: 2

*l*W *?Ballaugh*

Mitchell (1958) made an identification from zone II at Ballaugh as cf. *A. leptoclados*.

Arenaria ciliata L. 'Irish sandwort' (Fig. 51)   141: 3

W *Thrapston*; *m*W *Earith, Great Billing, Barnwell Station*

Arenaria ciliata agg. (*A. ciliata* L., *A. norvegica* Gunn., *A. gothica* Fr.)

Wo *Farnham*; *e*W *Sidgwick Av.*; *m*W *Syston, Marlow, Brandon, Derryvree, Oxbow*; *l*W *Nazeing, ?Shortalstown, Mapastown, Ratoath, Knocknacran, Drumurcher, Dunshaughlin, Ballybetagh*; F IV *Nazeing*

Exceptionally thorough consideration has been given by Miss Bell to the morphology of seeds of *Arenaria* employing both optical and scanning electron microscopy upon extensive and well documented collections of seed from six countries for *Arenaria ciliata*, four for *A. nor-*

vegica (including three Scottish and one English site), and three for *A. gothica*.

It appears that *A. ciliata* seeds can be distinguished from the other two species by its larger size and by the more expanded tubercles more constantly present on each spur of the epidermal cells. Miss Bell refers to *A. ciliata* seeds from her own Weichselian sites at Thrapston, Earith and Great Billing. From Barnwell Station she finds some seeds labelled by Miss Chandler as *Arenaria* sp. are certainly *A. ciliata* and that one labelled *A. gothica* is probably this species also. As Miss Bell's work casts doubt upon previous records for *A. gothica* and *A. norvegica* and has set new standards for identification it has seemed proper to place under '*A. ciliata* agg.' not only records made in this form by Miss Bell and Mrs Dickson, but the ten last-cited records above, most of which were made some time ago. Jessen and Farrington moreover treated *Arenaria ciliata* as including what was then sub-sp. *norvegica*.

The present British populations are all extremely local and from the *Atlas of the British Flora* appear to be: *A. ciliata*, the mountains of Co. Sligo; *A. norvegica* ssp. *norvegica*, a very few sites in western and northern Scotland, including the Shetlands; *A. norvegica* ssp. *anglica* Halliday, the Craven district of Yorkshire. *A. gothica* Fr. appears to be absent from the British Isles.

Arenaria norvegica is given by Hultén as an 'Amphiatlantic–arctic plant': it occurs in Greenland and Iceland and scattered along the Scandinavian mountains to the north.

To whatever taxon within *A. ciliata* agg. the fossil seeds belong, it is apparent that an extremely wide lowland range in Middle and Late Weichselian times has now become one of extremely disjunct montane kind, presumably through the adverse character of the ensuing Flandrian with its higher temperatures and widespread close vegetational cover (fig. 51).

Spergula arvensis L. (*S. vulgaris* Boenn.; *S. sativa* Boenn.) Corn spurrey   142: 1

Seed: H *?Kilbeg*; F viib *Cockerbeck, Minnis Bay*, viii *Mersea Is., Queen's Park, Silchester, Carrigalla, Lissue*
    Pollen: *l*W *Loch Mealt*; F viib *Ladybridge Slack*, viib/viii *Coire Bog*, viib *Tilbury*, viii *Old Buckenham*

All the seed identifications save the much-queried Hoxnian one are from archaeological sites, Cockerbeck (Neolithic), Minnis Bay (Late Bronze Age), Mersea Is., Queen's Park and Silchester (Roman), Carrigalla (Anglo-Saxon) and Lissue (9th–10th c. A.D.). Again apart from H. J. B. Birks' Late Weichselian zone II record from Skye, all the pollen records are from zone viib or later, that at Old Buckenham mere occurring at a level thought to be Anglo-Saxon where *Secale* (rye) and *Linum* (flax) were present along with other typical weeds of arable cultivation. At the present day in Britain *Spergula arvensis* is almost restricted to cultivated ground. Jessen and Helbaek

(1944) have drawn attention to the very frequent occurrence together of seeds of *Linum* and *Spergula arvensis*, suggesting that the latter from Iron Age times onwards was a common weed in flax crops. Dutch investigators have reported *Spergula* pollen from the time of the Bronze Age forest clearance along with that of *Spergularia*.

Spergularia sp.   143

*l*W *?Mapastown*; F v *?Dogger Bank*, VIII *?Hungate*

All records are tentative. That from Hungate is a seed from a mediaeval context: the others are pollen records, the older from Late Weichselian zone II (Mitchell, 1953) and the North Sea record (*Spergularia* type pollen) from Flandrian zone v with indication of a brackish environment, recalling that there are salt-marsh species of *Spergularia* (Behre & Mencke, 1969).

ILLECEBRACEAE

Herniaria sp.   'Rupture-wort'   146

*l*Wo *Ilford*; *m*W *Earith*; *l*W *Colney Heath*

Two seeds from the late Wolstonian at Ilford and one from zone I of the Late Weichselian at Colney Heath were referred by Miss Lambert to the genus *Herniaria*, as was a single seed from the Middle Weichselian at Earith by Miss Bell. They found it impossible to specify whether *H. glabra*, *H. ciliolata* or *H. hirsuta* might be concerned as their seeds were indistinguishable on surface patterning and otherwise also closely similar.

Herniaria glabra L.   'Glabrous rupture-wort'   146: 1

*e*Wo *Brandon*

Despite the reservations mentioned in the previous account Kelly (1968) has recorded seeds of *Herniaria glabra* from the early Wolstonian deposits at Brandon, Warwickshire.

Herniaria glabra is a rare plant of scattered and southern distribution in Britain and characteristic of dry sandy places. It is frequent in Denmark, southern Sweden, and the Baltic region as far as the Gulf of Finland with scattered occurrences up to the Arctic Circle. This Scandinavian range makes it probable that the Weichselian records for the genus relate to *H. glabra*, for the other British species have distinctly more southern ranges as is well shewn in their British distribution today.

Scleranthus sp.   148

H *?Nechells*; I *Histon Rd*; *l*W *Stannon Marsh*; F VIIa *?Oakhanger*, VIIb *?Hackness*

A few pollen records for the genus *Scleranthus* have been made, from sub-stage II of the Hoxnian and Ipswichian interglacials, from zone II of the Late Weichselian and

from Mesolithic and Bronze Age sites in the Flandrian. These are no more than '*Scleranthus* type' identifications however.

Scleranthus annuus L. *sensu lato*.   'Annual knawel'

                                        148: 1

H *Clacton*; *e*Wo *Brandon*; *l*Wo *Ilford*; *m*W *Upton Warren*; *l*W *?Kirkmichael*; F IV *?Moss Lake*

The fruits, which constitute all but one record of *Scleranthus annuus*, are indehiscent nutlets enclosed by the persistent sepals and receptacle.

C. Reid recorded a fruit from what is now taken to be sub-stage III of the Hoxnian interglacial. All the other records are from glacial stages save the tentative pollen record for the Flandrian zone IV at Liverpool.

S. annuus is a dwarf plant characteristic of sandy and gravelly soils in cultivated and waste land, which evidently found favourable conditions in periglacial situations. It belongs to Hultén's category of west European–middle Siberian plants and in Scandinavia is common to 62° N, though there are scattered localities for it into the Arctic Circle. It occurs throughout the British Isles.

Scleranthus perennis L.   'Perennial knawel'   148: 2

*e*W *Chelford*; *l*W *Elan Valley, Colney Heath*; F IV *Elan Valley*, VI/VIIA *Addington*

In this country pollen of *Scleranthus perennis* was first recorded by Burchell and Erdtman (1950) from an archaeological site we can presume to be of Flandrian zone VI or VIIa. It has subsequently been recorded from Chelford and Colney Heath, Weichselian sites with radiocarbon dates respectively of 60 800 and 13 560 B.P. At the high altitude Elan Valley site in Cardiganshire, Moore (1970) has recorded the pollen from zones I, I/II, II/III, III and IV.

Scleranthus perennis is placed by Matthews in his category of Continental plants that seem to shew intolerance of oceanic western climate. In the British Isles it is a rare species found only in dry sandy fields in Norfolk and Suffolk, and on rock in Radnor. In Scandinavia it is common only below 60° N, though present locally at higher latitudes.

PORTULACACEAE

Montia fontana L.   Blinks   149: 1

In *Flora Europaea*, *M. fontana* L. is treated as 'a very variable species, in which the seed characters show correlation with geographical distribution and, less satisfactorily, with habit and habitat differences'. In accordance with this a key based largely on seed characters distinguishes four sub-species, and a very limited number of fossil seed records have been referred to these sub-species. The vast majority of records however are refer-

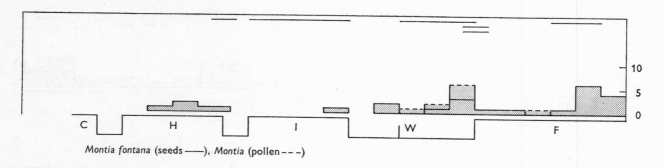

Montia fontana (seeds ——), *Montia* (pollen - - -)

able only to the aggregate *M. fontana* L. as are, with one exception, all the records based on the small dodeca-colpate pollen grains, that accordingly have been incorporated in the record-head diagram. This indicates presence in three sub-stages of the Hoxnian and one of the Ipswichian interglacial, and occurrences in all zones of the Flandrian. The early Wolstonian glacial record is matched by records from the Middle Weichselian and from all zones of the Late Weichselian, more particularly zone III and the transition to the Flandrian zone IV. There is thus a very strong suggestion of long persistence through glacials and interglacials: there are also five records suggesting some response to clearance and settlement at Cockerbeck and Graig Llwyd (Neolithic), Wingham (late Bronze Age), Mersea Island and Silchester (Roman). Hultén shews the sub-species *fontana* (as *M. lamprosperma* Cham.) as common in western Scandinavia to beyond 70° N, and he places it with *Chamaepericlymenum suecicum* and *Myrica gale* in a small category of circumpolar sub-oceanic species. Polunin writes of it as common around coasts in the Canadian Eastern Arctic and as extending almost as far north there as in northern Scandinavia. In Britain *M. fontana* L. occurs in all parts of the British Isles, more abundantly however in the north.

Montia fontana L. ssp. **fontana** (Fig. 52)   149: 1a

*e*Wo *Brandon*; *m*W *Brandon, Oxbow*; *l*W *Hawks Tor, Stannon Marsh, Scaleby Moss, Kirkmichael*; F IV *Ballaugh, Hawks Tor*

Numerous seeds were recovered from different levels and sites of the Late Weichselian and early Flandrian at Hawks Tor which agreed in their reticulation, absence of tubercles and size of seeds with seeds of *M. rivularis* Gmel., and the majority were black and shining so that they can be referred to *M. fontana* ssp. *fontana*. Along with the seeds there were many curious bract-like objects, nearly all occurring singly, but a few in pairs (fig. 52). These were found to be valves of *Montia* capsules (Conolly *et al.*, 1950).

The three Late Weichselian records embrace zones I to III. This record is consonant with the distribution range described in the account of the aggregate *M. fontana*.

Montia fontana ssp. **chondrosperma** (Fenzl) Walters (*M. verna* auct.)   149: 1b

I *Wretton*; F VIIA *Elland, Treanscrabbagh*, VIII *Godmanchester*

The records from Wretton are for both sub-stage II and III of the Ipswichian interglacial. All records represent mild climatic conditions, and this is likely to be significant

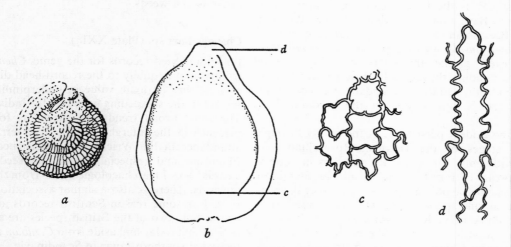

Fig. 52. Remains of *Montia rivularis* C.C. Gmel. from Late Weichselian layers (mostly zone II) at Hawks Tor kaolin pit, Bodmin Moor, Cornwall: *a*, seed (×25); *b*, valve of capsule (×25); *c, d*, cells from apex and from centre of capsule valve (×250).

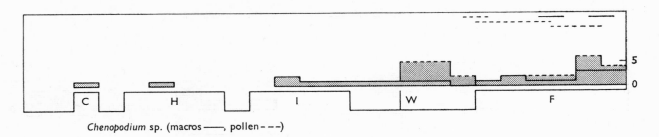

Chenopodium sp. (macros ——, pollen - - -)

in relation to the present range of the sub-species, which although extending in this country to Scotland has a very restricted and extremely southern range in Scandinavia and belongs to Hultén's category of Atlantic species.

Montia fontana ssp. variabilis Walters   149: 1c

F viib *Rockbarton*

One seed recovered by Mitchell from zone viib at the early Bronze Age deposit at Rockbarton, Co. Limerick was identified by S. M. Walters as ssp. *rivularis* Walters. This sub-species has a distribution in western and central Europe that is as yet not at all fully known.

CHENOPODIACEAE

The periporate pollen grains of the *Chenopodiaceae* are characterized by their very numerous small pores, usually more than fifty, and closer identification has not often been attempted save to the genera *Chenopodium* and *Atriplex*, of which there are separate accounts. None the less the family identification has substantial indicator value.

There are fossil records of this family from even such early stages as the Ludhamian (1), Thurnian (1), Antian (2) and Baventian (2), so that with the exception only of the Beestonian (for which there are seed records of *Chenopodium murale*, *Atriplex* sp. and *Atriplex hastata*), they cover every glacial and interglacial stage of the Pleistocene, extending through all sub-stages and zones, an impression confirmed by the abundant seed records for component genera and species in the accounts that follow.

We note the evident persistence of the family through the middle stages of the Hoxnian, Ipswichian and Flandrian with only a drop in site numbers in zone v reflecting spread of forest cover. In all the Flandrian zones *Chenopodiaceae* pollen records accompany the progress of the marine transgression at coastal sites from the Dogger Bank in v, Roddansport, Ringneill Quay, Kippen, Avonmouth *et al.* in vi, Westward Ho!, Helton Tarn and Bowness Common in viia, to Brigg and North Ferriby, Lion Point and Avonmouth in viib. Moreover,

examination of the relative pollen frequencies shews that these coastal and estuarine sites register remarkably high local percentages of *Chenopodiaceae* pollen (sometimes exceeding the total arboreal pollen), a clear reflection of the abundance with which the family grows in salt-marsh and other coastal habitats. Again and again in pollen diagrams from estuarine deposits the transition from freshwater to marine conditions is indicated by high, if transitory, values for this pollen type.

Equally striking is the very large number of site records from the Late Weichselian, following clear indices of presence in the Early and Middle Weichselian. It is apparent that members of the family were abundant at this time, a fact again borne out by generic and specific records, and this was certainly in part a consequence of their halophily in climatic situations that produced a wealth of saline habitats (Bell, 1969, 1970a). All the same the relative pollen frequencies remain low throughout, seldom exceeding 2 per cent of the dry land plant total.

Site numbers for *Chenopodiaceae* pollen do not shew the great increases in zones viib and viii that obtain with such plants as *Plantago lanceolata*, *P. major* or *media*, but nevertheless in those zones a large proportion of the records are associated with indications of agriculture or clearance: such records do not exhibit very high pollen frequencies but there is nevertheless a measure of the response one would expect in a family containing many ruderals and weeds.

Chenopodium sp. (Plate XXI*q*)   154

Pollen and seed records for the genus *Chenopodium* contribute about equally to the record-head diagram which, even so, has its main value in underpinning the results set out in the succeeding accounts for individual species. The same broad trends are shewn as for the family, presence in the interglacials (in river terrace deposits), abundance in the Weichselian, persistence through the Flandrian and (especially with the seed records) an association with archaeological sites from the Bronze Age onwards. There is also a similar association with coastal sites. For some reason Scottish records are lacking: of course only two of the British species are at all common in Scotland today and aside from *C. album* all have a very restricted southern range in Scandinavia.

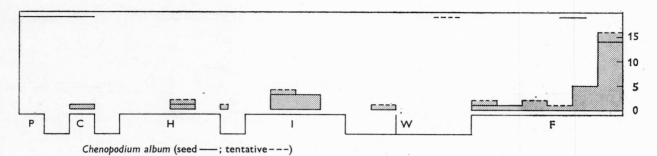

Chenopodium album (seed ——; tentative - - -)

Chenopodium bonus-henricus L. All-good, Mercury, Good King Henry   154: 1

I *Wretton*; F viib *Wallington*, viii *Silchester, Mersea Is.*

Seeds of *Chenopodium bonus-henricus* were recorded from both sub-stage II and sub-stage III of the Ipswichian interglacial at Wretton. Thereafter the next record is from the archaeological site at Wallington, near the Neolithic to Bronze Age transition. Two zone VIII records are from the Roman sites at Silchester and Mersea Island.

Good King Henry is often regarded as an introduced plant in Britain but the Wallington record is early for this and it could not well apply to the interglacial occurrences. It closely follows human settlements and may upon occasion have been used as a green vegetable.

Chenopodium polyspermum L. All-seed   154: 2

H *Clacton*; I *Bobbitshole*, ?*Histon Rd*; F viii *Godmanchester, Bunny*

Reid identified seed of *Chenopodium polyspermum* from sub-stage III of the Hoxnian interglacial, and it has been recorded from sub-stages II, III and IV of the Ipswichian: though the records for sub-stages III and IV are tentative only, in sub-stage III they were repeated in different excavations. The Flandrian records are both from Roman occupation sites. Within the British Isles it is a weed of strongly south-eastern distribution.

Chenopodium album L. Fat hen (Plate XVIII*q*) 154: 4

Pleistocene records of the seeds of *Chenopodium album* go back to the Cromer Forest Bed series and the Cromerian itself. They are in both the Hoxnian and Ipswichian interglacial and right through all the Flandrian, though with qualified identifications in zones VI and VIIa. In glacial stages the records are sparse and doubtful, the records for the late Wolstonian and the Weichselian zone II/III transition being tentative, whilst the Middle Weichselian record from Marlow is given by Miss Bell as 'C. album type'.

Not only has the plant been used as a green vegetable but until recently the seeds were widely used as food in times of famine and there is much evidence from Neolithic times onwards that the seeds were collected and stored, for they are often found preserved in earthenware pots. More to the point still, they occurred with seeds of other weeds in the stomach contents of the Danish bog-burials (Waterbolk, 1968; Willerding, 1970). There is indication in the high frequency of unripe fruit at some sites that the plant was brought into use as a vegetable, but it is certainly likely that seed was regularly gathered from weedy crop, abundant arable and fallow. The strong association with human settlement is brought out in the British records for zones VIIb and VIII which include Bronze Age (4), Iron Age (3), Roman (7), Anglo-Saxon (1), Norman (1) and Mediaeval (5).

The species is now cosmopolitan and widespread through the British Isles.

Chenopodium opulifolium Schrad. ex Koch & Ziz 154: 7

F viii ?*Hungate*

There is a tentative seed identification of this presumed non-British species from the Norman level in the Hungate, York.

Chenopodium murale L. 'Nettle-leaved goosefoot'
154: 11

Be *Norfolk Co.*; F viii *Silchester*

One seed record of *Chenopodium murale* comes from the Beestonian glacial stage on the Norfolk coast and another from the Roman site at Silchester.

Chenopodium urbicum L. 'Upright goosefoot' 154: 12

H *Clacton*; F iv ?*Ballaugh*, viii *Lissue*

Seed records for *Chenopodium urbicum*, a rather rare species regarded as doubtfully native in the British Isles, come from sub-stage III of the Hoxnian interglacial and from a mediaeval site at Lissue in Ireland where the plant is no longer found.

Chenopodium hybridum L. Sowbane   154: 13

H *Clacton*; I *Trafalgar Sq.*; F viii *Silchester*

Seed of *Chenopodium hybridum* was identified at Clacton by C. Reid and Chandler and again many years later by C. Turner, both from sub-stage III. It was found in sub-stage II of the Ipswichian in Trafalgar Square, London and lastly from the Roman site at Silchester. It is odd that, as with other *Chenopodium* species regarded as doubtfully

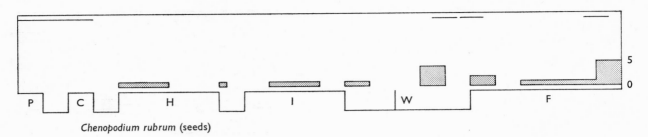

Chenopodium rubrum (seeds)

native, there should be interglacial records and then a gap until Roman or later time.

Chenopodium rubrum L. 'Red goosefoot'　154: 14

The seed record for *Chenopodium rubrum* is more continuous than that of most species in the genus. It is represented in the Cromer Forest Bed series and both the Hoxnian and Ipswichian interglacials, but it is unusual in having also Weichselian records including one from the Early Weichselian in Cambridge, and Late Weichselian ones from Kentmere, Ballybetagh, Shortalstown, Ballaugh and Drumurcher. At Ballaugh in the Isle of Man there was persistence too from zones II and II/III to IV, and a record for Jenkinstown (zone VI) bridges the gap to the numerous occurrences in zones VIIa and later.

C. rubrum often occurs near the sea and several of the recorded sites are coastal (Jenkinstown, Bowness Common and Leasowe). Its prevalence today on rubbish tips and in farm yards is reflected in the numerous archaeological site records, Neolithic (1), Iron Age (1), Roman (2), Anglo-Saxon (1), Norman (1) and Mediaeval (2).

In the British Isles, *C. rubrum* is almost limited to England and Wales (especially the south-east) and this throws into even stronger relief the Irish Late Weichselian occurrences. Though distinctly southern and continental in its Scandinavian range it does occur sporadically right round the Baltic.

Chenopodium botryodes Sm.　154: 15

H *Clacton*; I *Trafalgar Sq.*; F VIIb *Littleport*, VIII *Finsbury Circus*, ?*Hungate*

The seed records of *Chenopodium botryodes* are from sub-stage III of the Hoxnian and sub-stage II of the Ipswichian interglacial. Thereafter there is a Flandrian zone VIIb record from the Cambridgeshire fens, and two records from sites associated with human settlement, one Roman in north London, and a tentative one from mediaeval York.

C. botryodes is a very local plant of brackish habitats on the south-east coast of England. Of the recorded sites Clacton and possibly Littleport were coastal, whilst Trafalgar and Finsbury Circus were no great distance from the sea. It is a species that in Scandinavia is so southern that it occurs only in Denmark.

Chenopodium sub-genus pseudoblitum Benth. & Hook. (including C. rubrum, C. botryodes, C. glaucum)

C *Norfolk Co.*; I Wo *Selsey*; I *Selsey, Ilford, Histon Rd*; I W *Ballaugh*; F IV *Ballaugh*

In some instances Cambridge investigators have felt unable to carry identification of *Chenopodium* seeds below the level of the sub-genus *pseudoblitum*. This taxon has been recorded from the Cromerian and from sub-stages I, II and III of the Ipswichian interglacial. There is a single late Wolstonian record and at Ballaugh, where the seeds were very numerous, the records embrace Late Weichselian zones II and III and Flandrian zone IV.

Of the three British species in the sub-genus it seems, on grounds of present range and fossil record, most probable that the Ballaugh plants were *C. rubrum*, but this can only be conjectural.

Beta vulgaris L.　Sea beet　155: 1

I *Stone*

The seed of *Beta vulgaris* recorded from sub-stage IIB of the Ipswichian interglacial was, appropriately, recovered from estuarine marine clay. It occurs frequently on the coasts of England, Wales and Ireland, less commonly those of Scotland. It has a strongly southern restriction of range in Scandinavia.

Atriplex sp.　Orache　156

Rather more than half the records for *Atriplex* depend upon seed identifications: the rest are pollen identifications although the separation of this genus from other chenopods is not often attempted. The fossil occurrences shew presence in the Hoxnian (a tentative identification only) and in at least two sub-stages of the Ipswichian. The seed records from the Early Weichselian (Sidgwick Av.) and Middle Weichselian (Upton Warren and Earith) are followed by pollen records in the Late Weichselian and establish the presence of yet another halophilous taxon in the periglacial vegetation of the Weichselian glacial stage. The records indicate continued presence through the Flandrian associated, in the pollen records only, with two instances of strong local persistence, at Killerby Carr (zones I, II and V: Walker, in Clark, 1954) and at Hawks Tor (zones II, III, IV, V, VI: Conolly et al., 1950).

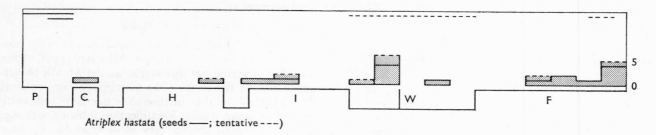

Atriplex hastata (seeds ——; tentative ---)

In the seed records it is unsurprising to find that there are at least four Flandrian and two interglacial records from coastal situations, as it is to find the role of *Atriplex* as a weed, and possibly a food plant, reflected in three Roman and two mediaeval records.

Atriplex littoralis L. 'Shore orache'   156: 1

F VI, VIIa ?*Jenkinstown*, VIII *Finsbury Circus*

Mitchell made two tentative seed identifications of *Atriplex littoralis* from estuarine deposits in County Louth from coastal deposits formed at the height of the postglacial marine transgression. Reid also identified it from the Roman site at Finsbury Circus.

A. littoralis is extremely local on the eastern coasts of Great Britain and Ireland today and absent elsewhere in Ireland.

Atriplex patula L. 'Common orache'   156: 2

Although a few of the seed identifications of *Atriplex patula* are tentative (and there is an entry for seeds referred to *A. patula* or *A. hastata*) there is enough evidence to suggest native status since the time of the Cromer Forest Bed series from which indeed there are three records. The Cromerian and Hoxnian interglacials each have a single record, but there are eight from the Ipswichian with five in sub-stages II and III where the plant must have been growing in riparian situations. Two Middle Weichselian and two Late Weichselian records indicate persistence through the glacial stages and there is an early Flandrian record from the North Sea moorlog. Zone VIIb and especially zone VIII shew the considerable increase in site frequency that is to be explained as due to the spread of clearance and arable cultivation, with numerous archaeological sites as follows, Bronze Age 2, Iron Age 3, Roman 8 and Mediaeval I. The seeds may indeed have been collected for food (see the following account, 156: 2, 3).

The common orache today is an inhabitant of cultivated ground and waste places, but the fossil record establishes a long history in this country before it assumed this role. It is a species found growing all over the British Isles, and its range on the west Norwegian coast beyond 70° N latitude is compatible with its apparent Weichselian survival.

Atriplex patula L. or A. hastata L.   156: 2, 3

I *Ilford*; F VI, VIII *Apethorpe*, VI *Ringneil Quay*, VIIb, VIIb/VIII *Frogholt*, VIII *Aldwick Barley, Ballycatteen*

Seed identifications of *Atriplex patula* or *A. hastata* come from sub-stage II of the Ipswichian and from zone VI of the Flandrian onwards. Only the Ringneil Quay site is coastal. At Frogholt it was found successively at levels carbon-dated to 2640 and 2490 B.P., i.e. near the late Bronze Age–Iron Age transition, associated with various weeds of arable cultivation. It presumably had such a role in the Iron Age site at Aldwick Barley too. However, in the excavation at Ballycatteen Fort, Co. Cork the fill of the souterrain (probably between the 11th and 16th century A.D.) included human faeces containing a ground-up mass of embryos and their envelopes, and seeds of the common orache that must have been eaten in some form of gruel (O'Riordain & Hartnett, 1943).

The Apethorpe records are of interest as indicating the persistence of *Atriplex* inland during the mid-Flandrian in fluviatile situations in the same way that this and other weeds and ruderals may have persisted in earlier interglacials.

Atriplex hastata L. 'Hastate orache'   156: 3

Although rather a large proportion of the seed identifications of *Atriplex hastata* are tentative, a long presence of the species through the Pleistocene is indicated, beginning with records in the Cromer Forest Bed series and the Cromerian interglacial. It was possibly present in the late Hoxnian, and certainly in sub-stages I and II of the Ipswichian. No less than six site records come from the Middle Weichselian although two, by Miss Bell, are only to 'A. hastata type': there is a tentative identification from the Early Weichselian and an unqualified record from zone II of the Late Weichselian. These Weichselian records echo the records from earlier glacial stages, the Beestonian and the late Wolstonian. Of the Flandrian records that begin in zone VI, no less than seven are from coastal sites and they include four Roman and one Iron Age settlement.

Atriplex hastata occurs right round the British coasts and widely inland, especially in the south-eastern half of England, mainly in association with arable cultivation. It is a plant that extends in western Norway, far into the

Arctic Circle, a range that conforms with its status as a Weichselian halophyte.

Atriplex glabriuscula Edmonst.　'Babington's orache'
　　　　　　　　　　　　　　　　　　　　　　156:4

*l*Wo *Selsey*; *l*W *Mapastown*; F VI *Jenkinstown*

Seeds of *Atriplex glabriuscula* have been recorded from three sites, two of them coastal. The Late Weichselian record is from zone II. It is a species that occurs right round the coasts of the British Isles, though southern in its Scandinavian distribution pattern.

Atriplex hortensis L.　Spinach　　　　　156:6

F VIII ?*Hungate*

Tentative identifications were made of seeds of *Atriplex hortensis* from both Norman and Mediaeval levels in the Hungate, York.

Suaeda maritima (L.) Dumort.　'Herbaceous seablite'
　　　　　　　　　　　　　　　　　　　　　　158

I *Selsey, Stone*; *e*W *Sidgwick Av.*; *m*W *Earith*; F IV/V *North Sea*, VI, VIIa *Jenkinstown*, VI/VIIa *Lough Cranstal*, VIIa *Westward Ho!*, VIIb *Littleport, North Ferriby*, VIII *Pevensey*, ?*Old Croft River*

As one would expect, the fossil record for seeds of *Suaeda maritima* is consonant throughout with its obligate halophily. The Early and Middle Weichselian records accompany those of other halophytes characteristic of the last Glacial stage. The three Ipswichian records from Selsey and Stone, all sub-stage II, are from coastal and estuarine deposits, and all the Flandrian deposits from zone IV/V on the Dogger bank to the Roman site at Pevensey and the Old Croft River in the Fenland are likewise coastal or estuarine. They confirm the constant presence of *Suaeda maritima* in this country at least from the Early Weichselian: no doubt it followed the eustatic and isostatic coastal shifts and moved inland during the glacial stages.

S. maritima occurs on all British coasts, but in Scandinavia is distinctly southern though it attains 66° N in scattered localities on the Norwegian coast.

Salicornia sp.　Glasswort, Marsh samphire　160

F *Littleport, Saddlebow*

The *Salicornias* are obligate halophytes and both of the Flandrian records from the East Anglian Fenland reflect this. That from Littleport (accompanied by *Suaeda maritima*) is from deposits close to the margin of the Fen Clay laid down in a marine transgression culminating about 2000 B.C. (zone VIIb), and that from Saddlebow is from deposits of the subsequent marine transgression in Roman time (zone VIII).

Corispermum sp.

Pa *Beeston*; *e*W *Wretton*; *m*W *Earith*

The only published accounts of fossil records of the non-British genus *Corispermum* are by Miss Bell (1968, 1970a) and they describe her discoveries at the Middle Weichselian site on the edge of the Cambridgeshire fenlands. The fossil fruits at this site though frequent are sufficiently eroded to make specific identification difficult, although of the European species they most resemble *C. intermedium* Schweigger, *C. marshalii* Steven and *C. hyssopifolium* L. These species grow on sandy and especially loessic soils from Germany to Hungary and add a strikingly southern and continental quality to the fossil assemblage.

There are earlier but unpublished discoveries of fruits of *Corispermum* from the Pastonian by Wilson and West, and from the Early Weichselian at Wretton by West.

Corispermum hyssopifolium has been identified in a Full Weichselian deposit in Belgium, where however the possibility of secondary derivation cannot be excluded (Paepe & Vanhoorne, 1967).

TILIACEAE

Tilia spp.　Lime, Linden (Figs. 23, 53)　　162

Macros: H *Nechells*; F VIIa ?*Brook*, VIIa/b *Shustoke*, VIIb *Whitehawk, North Ferriby, Llandow, Downton*, VIII ?*Meon Hill, The Trundle*, ?*Shapwick Station, Roden Downs*, ?*Goldhanger*

Although the Hoxnian and the late Flandrian both have records for *Tilia cordata* they are for wood and charcoal respectively and it seems, in view of the hazards of specific wood identification, reasonable to group these records as for the genus only. One record from zone VIIa/VIIb at Shustoke is based on recognition of leaf-scars, and one from North Ferriby on a fruit: the rest are of wood or charcoal and it is notable that with the exception of Shustoke and Nechells all are from archaeological sites: thus there are represented the Mesolithic (Brook), Neolithic (Downton and Whitehawk), Middle Bronze Age (Llandow), Late Bronze Age (North Ferriby), Early Iron Age (Meon Hill, The Trundle, Shapwick Station) and Roman periods (Roden Downs and Goldhanger). We may suppose that *Tilia* was brought to settlement sites for a variety of reasons, especially perhaps its value for leaf-fodder, for easy working timber and for bast fibre.

Pollen of *Tilia* has several properties important for interpretation of the historic record of the genus. It is highly recognizable, so that even in a very damaged or fragmentary state it can be recorded. This property is coupled with a remarkable resistance to decay, so that where conditions are in general unfavourable for pollen preservation, selective preservation acts in its favour. This tends to make *Tilia* pollen well represented in mineral soils and in pollen of secondary origin, as for instance in minerogene deposits of late-glacial age. Although the flowers of *Tilia* are insect-pollinated, the pollen production has been shown to be remarkably high: the evidence of the extent of its liberation into the air is conflicting but it is generally held to be so small proportionately that the pollen percentages recorded in peat and lake deposits substantially under-represent the actual frequency of the genus in the forest cover. As long ago as 1928, Tréla pointed out that the pollen of *T. cordata* and *T. platyphyllos*, the two species generally held to be native in Britain, could be easily distinguished. Using particularly two of his criteria, size and coarseness of reticulation, it was soon possible to shew that at least the bulk of fossil pollen from East Anglia was referable to *T. cordata* (Godwin, 1934). It was later pointed out that the large mesh of the *T. platyphyllos* grain had a distinctive optical appearance (Erdtman, 1943), a feature since explained by transmission and scanning electron microscopic studies of the *Tilia* exine (Godwin & Chambers, 1961, 1971). This knowledge, together with improved optical microscopy, now permits fairly easy recognition of the exine pattern typical of the two species. The position is however complicated by the independent discovery by Beug (1971) and Andrew (1971) that the *Tilia* pollen grain displays heteropolarity, the proximal and distal faces differing in coarseness of reticulation. By examining both faces of grains from suitably authenticated material, Miss Andrew has now shewn that heteropolarity is slight in pollen from pure *Tilia cordata* and *T. platyphyllos*, although grains from the presumed hybrid *T. × europaea* may shew *cordata* pattern on both faces, *platyphyllos* pollen on both faces, or very often *cordata* on one and *platyphyllos* on the other face. Moreover, judged on pollen criteria alone, some British material regarded as *T. cordata* may have some *T. platyphyllos* hybridity. This work has made it clear that fossil *Tilia* pollen can only be evaluated after examination of both polar faces. Miss Andrew's own examination of fossil *Tilia* pollen, and that of research students familiar with her work, has confirmed the presence in the British Flandrian of both *T. cordata* and *T. platyphyllos*, the former in far greater frequency (see separate accounts).

In view of this situation, for the record-head diagram and the distribution map (fig. 53) the only practicable course has been to aggregate all pollen records with those of macroscopic identifications under the genus *Tilia*, it being understood by the reader that none the less the results are mainly attributable to *T. cordata*.

These combined results then shew that the genus was present in the Cromerian, in sub-stages I and especially in II and III of the Hoxnian and sub-stages II and III of the Ipswichian. In the Flandrian again the early zones IV and V are represented by five records each, but after some sixty in zone VI, there are still larger frequencies in zones VIIa and b, with considerable diminution in zone VIII. Thus, despite the limitations of this index, *Tilia* can be seen as far more prevalent in the warm middle part of the Flandrian, just as it was in the two preceding interglacials. Consideration of data collected for *Tilia cordata* pollen alone gives essentially the same pattern though the numbers are much smaller throughout. This behaviour corresponds precisely with our expectation of a decidedly thermophilous tree, and one notes that the authors of the Weichselian (glacial stage) records (always for rare individual grains) regard them as due to derived pollen from older beds, a conjecture one may extend also to the early Wolstonian and late Anglian records, and possibly to that from the Thurnian.

The records for the Flandrian are so numerous that they allow us to carry further the analysis of site frequency and pollen frequency zone by zone in each of the 100 kilometre squares used in mapping for British biological records. Bearing the limitations of the method carefully in mind, we may derive from the Flandrian *Tilia* pollen map (fig. 53) a substantial outline of the history of *T. cordata* through the last ten thousand years in Britain.

The lime seems first to have established itself in the extreme south and east, for there are zone IV and V records for pollen in low frequency, chiefly in Hampshire, East Anglia and Yorkshire. During zone VI *Tilia* spread generally north and west, attaining its full Flandrian range and achieving substantial pollen frequencies (often 5 to 10 per cent) in the area south of the Humber–Bristol Channel line. Zones VIIa and VIIb saw not only many more sites with *Tilia*, but the highest recorded pollen frequencies with a notable concentration of very high frequencies in the extreme east, a feature underlined by the fact that there the values represented on the diagram are average values for several sites. As the entomophilous lime tends to be under-represented by its pollen we may suppose the East Anglian woods at least to have then contained a high proportion of *Tilia*. Over the country as a whole it will be seen that there is a progressive rise in pollen frequency through the zones to the maxima in VIIa and VIIb. There was a substantial decrease throughout the country in zone VIII both in respect of recorded sites and of pollen frequency at those sites, whatever the causes for this diminution may have been.

If we consider the zone VIIa/VIIb period, possibly 7500 to 2500 years ago, *Tilia* will be seen to be represented only by very low frequencies and in scattered sites throughout Scotland, Ireland, the Isle of Man, North Wales and the south-western peninsula of Devon and Cornwall. If it may be assumed that such small and infrequent records are due to long-distance pollen transport these will be the areas never colonized by the lime, and comparison with the *Atlas of the British Flora* will shew extremely close correspondence with the present distribution of native *Tilia cordata*. Even in East Anglia where standard publications suggest that native lime is absent, recent field work shews this not to be so despite extensive displacement of woodland since Neolithic times. It is further to be noted from fig. 53 that the zone VIIa/VIIb sites display the highest pollen frequencies in the

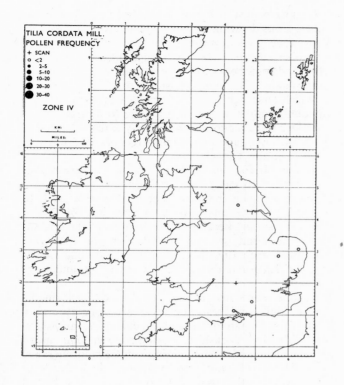

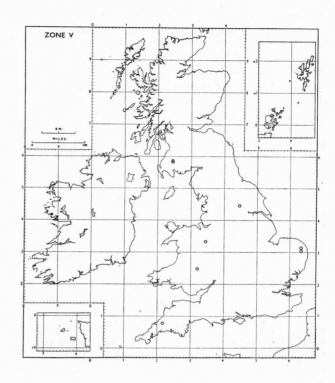

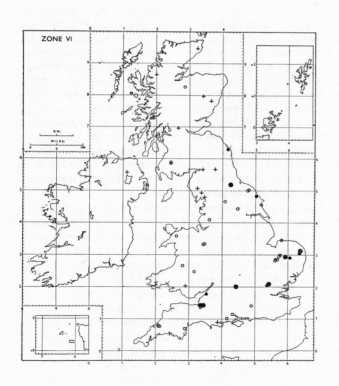

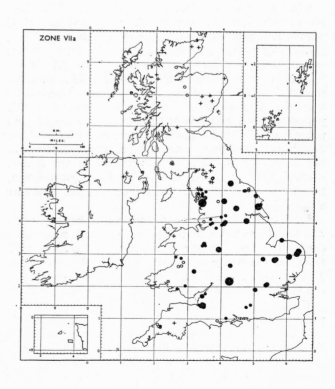

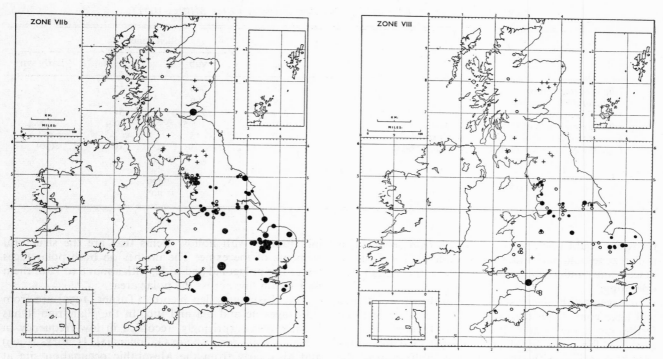

Fig. 53. Pollen frequencies of *Tilia cordata* Mill. in the British Isles through the six pollen zones of the Flandrian. To a small extent pollen of *T. platyphyllos* and hybrids may be included (see text).

major limestone areas of the Chalk, Carboniferous limestone, and the Jurassic and Liassic limestones. These still remain the centres of natural or semi-natural *Tilia* woodlands.

Throughout the deposits investigated by Kelly and Osborne (1965) at Shustoke, the *Tilia* pollen frequencies were of the order of 25 to 50 per cent of the total arboreal pollen, much exceeding the frequencies of *Quercus* and *Ulmus*. They considered this to indicate the local presence of woodlands dominated by the two species of lime upon well-drained fertile soils of the valley sides, and recalled that Iversen (1960) had concluded that similar forest occurred in the Danish climatic optimum in suitable situations. High *Tilia* pollen frequencies such as those at Shustoke have been recorded at a small number of British sites: ineffective pollen dispersal is given as the reason why they are not more widely recorded. Certainly the low pollen productivity of the limes must mean that untreated pollen diagrams undervalue the vegetational role of the genus, which may well have had at least local dominance in the middle Flandrian before it became subject to the anthropogenous factors. Iversen, writing again in 1969, cited Munaut *et al.* in northern France and Berglund in south-east Sweden for conclusions similar to his own for Denmark that 'Tilia not Quercus was the dominating tree in the climax forest of the Post-glacial warmth-period'. It should also be said that the French results mentioned above indicate local abundance of *Tilia* at the foot of slopes and near, but not in, the peaty alder

woods of the valley bottoms. There is some English evidence also for natural occurrence of *T. cordata* in fen margin woods.

Flandrian changes in the occurrence of *Tilia* pollen have been carefully examined by Mittre (1971) at sites on the western border of the East Anglian fenland, and the results given for Trundle Mere apply also to other neighbouring sites. The results given in table 13 shew very striking changes in *Tilia* pollen frequency from zone to zone, diminishing in early zone VIII to values only one-twentieth those of zone VIIa: a result equally recognizable with frequencies expressed either as percentage of total tree-pollen, or as of the total for *Ulmus, Quercus* plus *Tilia* pollen (a sum less influenced by pollen contribution from local fen-woods). Mittre's pollen diagram shews the decline in *Tilia* pollen frequency to have occurred abruptly at the zone boundaries, and for one of these Miss Turner, using primary data supplied by Mittre, has been able to shew that there is strong circumstantial evidence that it was associated with anthropogenic influences (Turner, 1962). At this and four other sites, including the contiguous site at Holme Fen, Dr Turner was able to shew that the decrease in *Tilia* pollen frequency was always associated with increases in the pollen frequency of grasses, *Plantago, Rumex*, other ruderal plants and of spores of the bracken. At the same time the pollen frequency of one or more tree genera diminished. These effects were interpreted as due to the clearance of woods dominated by *Tilia*. The results were naturally seen as

TABLE 13. *Pollen of* Tilia *at Trundle Mere, Hunts.* (*after Mittre,* 1971)

	Pollen zones			
	VIIa	VIIb	VII/VIII	Early VIII
Tilia cordata, grains	368	646	47	7
Tilia platyphyllos, grains	19	66	5	3
Total *Tilia*, grains	387	712	52	10
Tilia as percentage of total arboreal pollen	10.8	5.5	0.9	0.4
Tilia as percentage of total *Ulmus+Querus+Tilia*	> 20	c. 15	c. 2	c. 1
T. platyphyllos as percentage of total *Tilia* pollen	5	9	10	(30)
		5000 B.P.	3400 B.P.	

throwing doubt upon the value of the *Tilia* decline as a pollen zone boundary index, and as making it difficult to detect any underlying climatic effect that might be operating.

In large part the decreases in *Tilia* in British woodlands will have come from the preferential clearance of those fertile mor soils which it specially favours, as was suggested for *Ulmus* by Morrison (1959). One must however also reckon with strong selective utilization of the lime for leaf-fodder, timber and especially for bast fibre. Vuorela (1970) cites historical evidence of the importance of bast fibre in the mediaeval Finnish community, writing that halters, rope and fishing tackle were made of bast fibre and that taxes were paid in the form of bast. Certainly as late as 1297 in England the sale of lime tree bark (*corticis tilie*) was reckoned in forest returns, and this use must have extended back into remote prehistory. It may be conjectured that the coppice form of many surviving English *Tilia* woods reflects a former pattern of the economy of bast production.

Tilia platyphyllos Scop. 'Large-leaved lime' (Fig. 23)
162:1

H *Marks Tey*; F v/vi *Snibe Bog*, VIIa, b *Shippea Hill*, VIIa, b *Old Buckenham Mere, Holme Fen, Shustoke*, VIIa, b, VIII *Trundle Mere*, VIIb/VIII *Flagrass, Addington*

With two notable exceptions, that of a whole flower of *Tilia platyphyllos* recovered from the Neolithic level at the excavations at Shippea Hill, Cambridgeshire and that of over a hundred fruits and seeds from zones VIIa and VIIb at Shustoke, Warwickshire, all the records for this species are based upon pollen identifications, which, tentative at first, have now been given authority by the recent work of Beug and Miss Andrew (1971) taken in conjunction with electron microscopy.

In microscopic view of the pollen of *Tilia platyphyllos* it will be seen that in the centre of the wide meshes of the exine there is a central dot, brighter or darker than the field according to the depth of focus. This feature, absent in *T. cordata*, is apparently due to the wide bacu-

loid funnels with hollow stems that are the structural elements of the exine. The meshes on both polar faces tend to be polygonal and wide, with about 8 to 12 of them to a 10 μm line across the polar area.

The Hoxnian pollen records (Turner, 1970) are from sub-stages IIc and IIIa and IIIb. In the Flandrian it has been, rather surprisingly, recorded in low frequency in the composite zone v/vi in Galloway (H. H. Birks, 1969) and also early from the Mesolithic occupation site at Addington in Kent where unfortunately the pollen-yielding layer was very thin and superficial. The remaining Flandrian records, all from eastern England, are much more substantial. Mittre's results from Trundle Mere (table 13) make it clear that grains with *T. platyphyllos* characters were present throughout pollen zones VIIa, VIIb, VII/VIII and early VIII, though always far less abundantly than those of *T. cordata*. The same relative disproportion between the species was always found by Miss Andrew. In the longest sequence she examined, Old Buckenham Mere, *T. cordata* extended right through from about 5500 B.C. to about A.D. 1500, but *T. platyphyllos* only from 5500 B.C. to about 2000 B.C. At Shippea Hill, *T. platyphyllos* extended between levels shewn by radiocarbon assay to be 4700 and 3200 B.C., whereas *T. cordata* continued later. At Flagrass near March, *T. platyphyllos* pollen is present in low frequency through 100 cm of peat formed after *c.* 2000 B.C., and probably not after Roman time. Here again *Tilia cordata* was much more abundant.

At Shustoke, ten miles east of Birmingham, Kelly and Osborne (1965) made extensive floristic and faunistic study of an alluvial deposit, pollen analysis of which suggested a period including much of zones VIIa and VIIb, a conclusion supported by radiocarbon dating. From all parts of this were recovered immature fruits and separated seeds of *Tilia platyphyllos* whilst throughout the bed pollen of both *T. platyphyllos* and *T. cordata* type occurred. The '*platyphyllos*' type amounted to one third of the total in the older layers, but less in the upper. The authors suggested that on the well-drained fertile slopes there was a forest association of *Tilia* spp., *Quercus* and *Ulmus*, and in view of the low pollen productivity of the

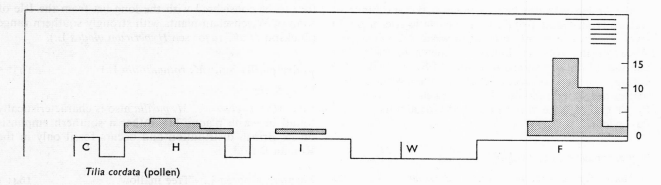

Tilia cordata (pollen)

limes they suggested the likelihood that *Tilia* was dominant in these woodlands.

These results and especially the coincidence of microscopic and macroscopic evidence at Shippea Hill and Shustoke leave no doubt of the native occurrence of *T. platyphyllos* in the middle Flandrian in eastern and central England.

Tilia platyphyllos is distributed throughout the greater part of central and southern Europe, but does not extend northwards beyond Denmark and south-western Sweden. In Britain the tree regenerates naturally and recent opinion holds it to be native in South Wales and the Welsh Marches, on the limestone of the southern Peak District, on the northern Magnesian limestone and in a few other scattered localities. Its European and British ranges indicate that it has higher thermal requirements than *T. cordata*.

Frenzel (1968) shews that in the last interglacial *T. platyphyllos* has been recorded substantially to the north and east of its present continental range. Our own middle Hoxnian records are from within the present native British range.

Tilia cordata Mill. 'Small-leaved lime' (Figs. 23, 53)
162: 2

As explained in the account of the genus, it may be safely taken that almost the whole of the *Tilia* pollen record for the British Isles refers to *Tilia cordata*, the winter linden. As is made clear in the accounts of *T. platyphyllos* and *T.* × *europaea* it is only recently that pollen of *Tilia cordata* has been definitively recognized. None the less there are enough of these specific records to underline the generic record. Thus there are specific site records in all sub-stages of the Hoxnian, and in sub-stages II and III of the Ipswichian, whilst the pattern of Flandrian behaviour for the genus is repeated in miniature.

Although the Flandrian recession of *Tilia cordata* and *T. platyphyllos* must now be largely attributed to anthropogenic causes it seems difficult to avoid the view that the lindens have behaved as the most warmth-demanding of our forest trees, an idea that is strengthened by consideration of the behaviour of *Tilia* on the European mainland. Firbas' maps for middle Europe shew a history

very similar to that in Britain: in the early warm period (Boreal) there are generally low values, but with sharply increasing extension at the end of this period, the lime along with the elms entered its period of heaviest representation, that is, the first half of the Atlantic period (Firbas zone VI). In the latter part of the warm period there was some recession of *Tilia* and finally the Subatlantic was a period of remarkable poverty in the genus.

That *T. cordata* has a very considerable northern range in Finland and Russia (exceeding those of the oaks and *Ulmus glabra*) need not exclude the probability that it has a much more restricted northern range in the oceanic west. This is indeed very appropriately stressed by Firbas who points out that it is the Atlantic Norwegian range sequence of northern limits, viz. *Ulmus–Quercus–Tilia*, which corresponds with the confirmed sequence of establishment of these genera in the early warm period of middle Europe. This sequence is indeed so strongly consistent in England and Wales that it has been made a substantial part of the criteria for separating the subzones 'a', 'b' and 'c' of zone VI.

Equally *Tilia cordata* plays the role of strongly thermophilous element in British interglacials and is so regarded by chief investigators of these deposits. It occurs mostly in sub-stages II and III, and where the pollen diagrams extend further as at the Hoxnian site at Marks Tey, it can be seen to disappear at the opening of sub-stage IV, a clear response to climate since anthropogenic activity on the Flandrian scale cannot have been present. The causes of the smaller and less consistent representation of *Tilia* in the Ipswichian as compared with the preceding Hoxnian have been considered by West (1969a): it is a point of special interest since on the continental mainland the Eemian (cf. Ipswichian) interglacial was characterized by a strong 'Linden phase', although indeed this was contributed to substantially by *Tilia platyphyllos* and *T. tomentosa* as well as *T. cordata* (Frenzel, 1968).

Native *Tilia cordata* appears to grow at the present day over most of England and Wales save the south-western peninsula, a pattern closely congruent with the fossil range through zones VI to VIII (fig. 53). It is not present in Scotland. In Ireland *Tilia cordata* has no claim to native status either in the Flandrian or in previous interglacials and one might conjecture that the expansion of the linden was too late to precede the severance of Ireland from Great Britain by the eustatic rise of ocean level. *Tilia* was certainly present in western England and Wales during zone VI, but, as is shewn in the pollen analyses of the submerged peat beds of Swansea Bay, the eustatic recovery of ocean level was then within a few metres of its final accomplishment and the Irish Sea must by the end of zone VI have attained virtually its present-day extent.

T. × europaea L. (*T.* × *vulgaris* Hayne)   162 : 1 × 2

H *Marks Tey*; F VIIa, VIIb, VIII *Old Buckenham Mere*, VIIa, VIIb, VIII *Shippea Hill*, VIIa, VIIb/VIII ? *Cooran Lane*, VIIb/VIII ? *Snibe Bog*

It was not until Miss Andrew had established the routine of separately examining both faces of each *Tilia* pollen grain that there emerged any effective means of recognizing pollen of the hybrid *T.* × *europaea* from that of the two parents. Accepting however that heteropolar grains with '*cordata*' pattern on one face and '*platyphyllos*' pattern on the other are an indication of hybridity, Miss Andrew was able to shew that at Old Buckenham Mere hybrid type pollen was present alongside that of both parents from about 5500 to 2000 B.C. and continued (in the apparent absence of *T. platyphyllos*) to A.D. 0. At Shippea Hill likewise the hybrid accompanied both parents. There are records also for a grain of intermediate character between *T. cordata* and *T. platyphyllos* at two sites in Galloway (H. H. Birks, 1969). In very low frequencies C. Turner has also found the hybrid pollen along with that of both parents in the two middle stages of the Hoxnian interglacial.

The fossil record reflects the comment upon the hybrid in *Flora Europaea* as 'Occasional as a natural hybrid in most regions of Europe where the parent species grow together', a view acceptable to British botanists such as Pigott (1969).

MALVACEAE

Malva sp. (Plate XXV*a*, *b*)   163

C *Barton*; F VIIb *Drake's Drove*

The large and highly characteristic pollen grains of *Malva* were recovered from the Cromerian by Duigan (1963) and from the peat bogs beside Meare Pool, Somerset.

Malva sylvestris   Common mallow   163 : 2

*l*W *Skelsmergh*; F VIII *American Square, Silchester*

The two Flandrian records for *Malva sylvestris* are both of Roman Age, and this accords with the plant's predilection for 'waste places', its designation by Hultén as 'strongly spread by man' and its southern distribution in the British Isles. The Late Weichselian zone II record from the English Lake District possibly deserves there-fore to be considered with the long list from the Isle of Man of Weichselian plants with strongly southern range (Dickson *et al.*, 1970: see *Hypericum elodes* L.).

Malva pusilla Sm. (*M. rotundifolia* L.)   163 : 5

F VIII *Silchester*

Like *Malva sylvestris*, *M. pusilla* also is characteristically found in waste places, has rather a southern emphasis on its British distribution and occurs fossil only in the Roman period.

Lavatera arborea L.   Tree mallow   164 : 1

F VIII ? *Caerwent*

Lyell (1911) gives only a qualified identification of this species from 'wood' found at the Roman site at Caerwent.

Althaea L. sp.   165

F VIII *Godmanchester*

Walker records pollen of *Althaea* from the Roman site at Godmanchester: possibly the medicinal value of *A. officinalis* may suggest that this was the species.

Althaea officinalis L.   Marsh mallow   165 : 1

F VIIa *Moss Lake, Burnham-on-Sea*, VIIb *Avonmouth*

The pollen grains of *A. officinalis* are distinguishable from those of other British Malvaceae by their smaller size and wider spacing of the spines of the exine.

A. officinalis is a coastal plant especially characteristic of the upper margins of salt and brackish marshes, a quality no doubt reflected in the high frequencies in which its pollen was recorded by Seddon (1965) from the peaty zone VIIb layers in the marine transgressive sequence at Avonmouth on the Bristol Channel. The Burnham-on-Sea record is also of pollen from a coastal peat bed, and at Moss Lake, where a single pollen grain was found in zone VIIa, coastal salt-marshes were then not far from the site.

The Marsh mallow has a strongly southern range in the British Isles, its most northerly occurrence lying only just north of the Moss Lake, Liverpool site.

LINACEAE

Linum sp.   Flax   166

Seed: *l*A ? *Hoxne*; *m*W *Hackney Wick*
    Pollen: *l*W *Old Buckenham, Esthwaite*; F v *Dogger Bank*, VIIb *Abbot Moss*, VIII *Holcroft Moss, Old Buckenham*

Although the bulk of records for flax are now referable to one or other of the three species *L. usitatissimum*, *L. catharticum* or *L. perenne*, whose accounts follow, there is only a handful of records to the genus. In the instance of Old Buckenham Mere (fig. 171) the association with arable cultivation of such crops as hemp, rye and other cereals in Anglo-Saxon and Norman time is so strong that it may be safely inferred that we are concerned with cultivated flax and the same is true for the Iron Age, Norman and Mediaeval phases of Holcroft Moss (H. J. B. Birks, 1965b) and of the VIIb record from Abbot Moss (Walker, 1966).

Linum usitatissimum L. Cultivated flax 166: 2

F VIIb *Windmill Hill, Handley Dawn, Westwood, Winterbourne Stoke, Newport, Agfarell,* VIII *Meare Lake Village, Ehenside Tarn,* ?*Silchester, Bermondsey, Pevensey, St John's, Girton College, Carrigalla* ?*Hungate, Lissue*

All the records are for seeds or seed impressions: it seems certain that they all refer to cultivated flax, as do the pollen records for three sites recorded under *Linum* sp. The former distribution of flax in this country has been fully considered by Jessen and Helbaek (1944) in their review of the evidence, mostly derived from plastilinia casts of impressions in dated pottery from different parts of the British Isles. They point out that the association of flax capsules with seeds of aquatic and marsh plants in lacustrine material, as found in three Scottish sites described by C. Reid (unfortunately not more closely dated than 'Post-glacial'), is very suggestive of flax-retting. The seeds from Handley Down, Winterbourne Stoke, Westwood and Agfarell are of particular importance in establishing that flax growing was already practised alike in England, Scotland and Ireland during the Bronze Age, and the Windmill Hill records match the earliest known European records. Willerding (1970) has recently assembled all middle European records for flax found in prehistoric archaeological contexts and has given a synoptic map of his results. These demonstrate a remarkable density of middle to late Neolithic records centred in the northern Alps and associated with pile dwellings: a few earlier Neolithic records come from the central European plain. Subsequently the Bronze Age has yielded fewer Alpine records and apparently none from the European lowland until the Bronze Age–Iron Age transition in Bornholm. Helbaek (1959) makes out a very strong case for regarding the progenitor of cultivated flax as a near East geographical race of the Pale flax, *Linum bienne*, that he has found as early as 5000 to 4500 years B.C. in settlement sites at Arpachiyah and Brak. It has the same chromosome number as *L. usitatissimum* and hybridizes with it. According to Helbaek's very experienced judgement, agriculture reached Europe from the middle East during the fifth millennium B.C., by way of the Aegean, Greece and Western Anatolia. From there it proceeded along the Danube giving the European Neolithic flax, the small-seeded winter-annual of the Swiss pile-dwellings and of the British Neolithic with its persistent derivatives in the Bronze Age in northern and western Britain. A second route of extension of agriculture from the Balkans was by way of the western Black Sea coast into Southern Russia where flax found highly suitable con-

ditions. It was, according to Helbaek, from here that the widespread Halstatt migrations of the first millennium B.C. brought into western Europe the second wave of *Linum* cultivation, this time of the large-seeded summer annual flax that reached Denmark in the Iron Age and may be represented by the *Linum* at the Meare Lake Village site in Somerset. Willerding's map shows clearly this secondary spread of Iron Age flax culture over the Central European plain.

It is impossible to say from the archaeological evidence what use was made of flax as a fibre plant in prehistoric time, but one must agree with Helbaek that a prior use of the oily seeds for food is likely, and that the summer annual flax was the type more suitable to fibre production.

The British records conclude with four Roman sites (Ehenside, Silchester, Bermondsey, Pevensey), three Anglo-Saxon (St John's, Girton College, Carrigalla) and two mediaeval (Hungate, Lissue). It is apparent that through five millennia there has been flax cultivation in these islands. As indicated in the account for the genus *Linum*, from Iron Age time onward there is a reflection of this in some pollen diagrams, and it will doubtless appear in others as time goes on.

Linum anglicum Mill. (*L. perenne*, ssp. *anglicum* Ockenden) 'Perennial flax' 166: 3

Seed: *l*Wo *Ilford*; I *Histon*; *e*W *Sidgwick Av.*; *l*W *Colney Heath*
Pollen: H *Marks Tey*; *l*W *Moss Lake*; F IV *Moss Lake*

The seed attributions given here are all by Mrs Dickson (formerly C. A. Lambert: see references under *L. praecursor*). Pollen identification was initiated in examination of deposits at Moss Lake, Liverpool (Godwin, 1959) where in zone IV three pollen grains clearly attributable to the genus *Linum* were found. They were large and had five or six pores of indefinite outline with the baculoid –gemmoid elements of the exine very strongly heteromorphic. In these characters the grains appeared identical with the pollen of *Linum anglicum* and unlike those of the other native British flaxes, *L. bienne* and *L. catharticum*. This discovery lent colour to the view that the flax seeds so prevalent in Weichselian and other glacial stage deposits were those of *L. anglicum*. Subsequently Turner (1970) has identified similar pollen from zone IVa of the Hoxnian at Marks Tey, a site from which, though at an unknown sub-stage, Miss Allison had earlier identified *Linum* seeds.

Linum anglicum is a perennial flax of calcareous grassland found very locally in eastern England from Durham to Essex and extending westwards into Westmorland. It occurs also in eastern and central France. It is now treated by Ockenden as a sub-species of *L. perenne* (*Flora Europaea*, vol. 2): it differs in its tetraploid chromosome number from the subsp. *perenne* and *alpinum*, which are diploids and have pollen grains with furrows, not pores. The following account (*L. perenne* agg.) may be taken as extending the account of *L. anglicum*.

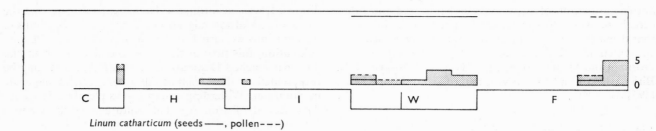

Linum catharticum (seeds ——, pollen ---)

Linum perenne agg.   Perennial flax   166: 3

Cs *Beeston*; H *Marks Tey, Hoxne*; eWo *Brandon, Marks Tey, Hoxne*; W *Thrapston*; mW *Brandon, Marlow, Earith, Syston*; lW *Nazeing* transferred from *L. praecursor* Reid; *Barnwell Station, Barrowell Green, Broxbourne, Hedge Lane, Waltham Cross, Angel Rd, Ponders End, Temple Mills, Upton Warren*

All the above identifications are based upon seeds except the Late Weichselian zone III pollen record from Nazeing (Allison *et al.*, 1952). As indicated in the account for *L. anglicum*, there is a rather strong presumption that a majority, if not all, of these records are attributable to *L. anglicum* and accordingly the records under both heads have been assembled in the above record. This gives a strong impression of the persistence of the perennial flax through the glacial stages and the early and late part of adjacent interglacials. The records for the Weichselian are remarkably abundant, but they do not include sites from Ireland, Scotland or Wales. Brandon, Upton Warren and Liverpool are sites west of the present strongly eastern range of *Linum anglicum* but it is surprising to what extent the fossil records are also concentrated in eastern England. The present British range of *L. anglicum* must surely be regarded as residual from a widespread presence in the Weichselian.

Linum catharticum L.   'Purging flax'   166: 4

There are now some eighteen seed records and seven pollen records for this small calcicolous therophyte in the British Pleistocene and it is very remarkable that with the exception of one occurrence in the late Hoxnian (Gort) they all come either from glacial stages or from late Flandrian sites associated with cultivation. The seed and pollen records agree closely with one another. Thus the late Anglian has yielded three seed and two pollen records, one of them (Baggotstown) with both together. There is one late Wolstonian pollen record at Ilford. The Weichselian records represent England, Scotland, Ireland and the Isle of Man and there are pollen records from the Chelford interstadial and from zone II of the Late Weichselian in Nant Ffrancon. Within the Flandrian there are seeds from the Late Bronze Age (Minnis Bay), the Roman period (Godmanchester, Ehenside) and mediaeval settlements (Dublin and Hungate), whilst there is pollen from Wingham, Kent with indications of Late Bronze Age–Early Iron Age cultivation.

There can be no doubt that the plant shews in these records its strong dependence upon fresh soils and open conditions, together with tolerance of climatic conditions of at least the early and late phases of the glacial periods. It is one of Hultén's 'west European–middle Siberian' plants that extends into the Arctic Circle in western Scandinavia, fruiting only in latitudes up to 61° N. It grows throughout the British Isles.

Linum praecursor Reid (invalid)

From practically all the sites in the Lea Valley Arctic Plant Bed C. Reid recovered seeds of *Linum* which he failed to match with those of existing species. These he placed in a new species *L. praecursor*. In his illustrated description of the seeds (Reid, 1916) C. Reid characterized the seeds as having a narrower and more oblong outline than those of cultivated flax, and mentioned that examples had been found also in interglacial deposits at Hoxne and Beeston, Norfolk.

When subsequently *Linum anglicum* Mill. was recognized as a native perennial flax Mrs Reid (1949) cast doubt upon the earlier diagnosis, pointing out that the fossil seeds resembled those of *L. anglicum* very closely save in point of size, a qualification that later workers have also removed.

Seeds of this type subsequently found in beds indicative of similar cold conditions, have accordingly often been referred to *Linum anglicum* (see above), but other authors have referred similar seeds to *Linum perenne* agg. and it is to this wider category that we have transferred all the early records given as *L. praecursor*. There is no reason to believe that two different taxa are involved (see Lambert, Pearson & Sparks, 1964 and West, Lambert & Sparks, 1963).

GERANIACEAE

Records for pollen referred either to Geraniaceae or to Geranium may be seen to have been recorded through the middle sub-stages of both Hoxnian and Ipswichian interglacials and the pollen zones of the Flandrian from VI onwards. There is so far one Cromerian record. The family is also represented, possibly by different species, in the late Wolstonian and Weichselian glacial stages.

We may view this record against the generally subtropical to temperate range of the family as a whole and of the considerable differences in ecological tolerances and present distribution of the British species of *Geranium* and *Erodium*.

Geranium sp. (Plate XXVe, *f*)   168

IG *Endsleigh*; W *Eastwick*; F VIII *Apethorpe*

Endsleigh is a seed record for an interglacial as yet of unknown age, Eastwick a pollen record from Middle or Late Weichselian, and Apethorpe a zone VIII seed record.

Geranium pratense L. or **G. sylvaticum** L.   Meadow cranesbill, Wood cranesbill   168: 1, 168: 2

*l*W *Colney Heath*

An erratic peat block with a radiocarbon date of 13560 B.P. yielded a pollen grain referred by Miss Andrew to either *G. sylvaticum* or *G. pratense*. Of these two species it is the former that has a decidedly more nothern range in Britain today and it might be thought the more likely source of this pollen.

Geranium sanguineum L. 'Bloody cranesbill' (Plates XVII*m*, XX*n*)   168: 7

*l*W *Nazeing*

A single seed of *Geranium sanguineum* and two of its empty carpels were washed out from the Late-glacial deposits at Nazeing. It is a calcicole species which has a somewhat local distribution in Scandinavia extending to about 61° N, and it is also found locally throughout the British Isles in pastures, on dry rocky cliffs and sand dunes. It is probable that the disjunction of its present distribution has been due to the Post-glacial forest period.

The finds of this species may be seen alongside the Weichselian records for pollen of Geraniaceae.

Geranium dissectum L. 'Cut-leaved cranesbill' 168: 11

F VIII ?*Newstead, Bunny*

The two seed records, one tentative, are both from Roman sites. *Geranium dissectum* occurs commonly throughout the British Isles, more particularly on waste and cultivated ground.

Geranium molle L. 'Dove's-foot cranesbill'   168: 13

F VIII ?*Newstead*

A tentative seed identification from a Roman site.

Geranium lucidum L. 'Shining cranesbill'   168: 15

F VIIb *Frogholt*

At a level corresponding to Late Bronze Age cultivation and authenticated by a radiocarbon date of 2860 B.P., a carpel of *Geranium lucidum* was recovered from the valley deposits.

Erodium sp.   169

P *Ludham*

There is a single pollen record from the Pastonian interglacial at Ludham (West, 1961*b*).

OXALIDACEAE

Oxalis acetosella L.   Wood-sorrel   170: 1

W *Hailes*; F VIIa/b *Shustoke*, VIIb ?*Ralaghan, Frogholt*, VIII *Loch Kinord, Crossness*

All the records are based on seed identifications. The fossil records establish the native status of *Oxalis acetosella*, but not much weight can be placed upon the Late-glacial record from Hailes for that bed contains, beside undoubted Arctic plants, *Alnus glutinosa* which must have come from the immediately overlying temperate bed, and the *Oxalis* seed may have done so too.

The Shustoke record from the zone VIIa/b transition has a radiocarbon age of 4830; the tentative Ralaghan record is also from VIIb as is that from Frogholt with a radiocarbon age of 2860. The Scottish record from Loch Kinord is zone VIII, as must be the Roman record from Crossness.

Wood-sorrel is common and often abundant throughout the British Isles, favouring woodland and other shady habitats and avoiding very heavy and wet soils.

BALSAMINACEAE

Impatiens sp.   Touch-me-not   171

H *Nechells*

Miss S. L. Duigan identified from the interglacial of Great Lister Street, Birmingham, three of the highly characteristic pollen grains of the genus *Impatiens*. It is not possible to say to what species they refer: of the species now present in the British Isles, *I. noli-tangere* is the only one generally regarded as native.

Impatiens parviflora DC. 'Small balsam'   171: 3

F ?*Coppice Gate Trackway*

This is a tentative pollen identification by Miss R. Andrew from zone VII/VIII at a site on Shapwick Heath, Somerset immediately above the level at which a middle Bronze Age spear head was recovered *in situ*. The find came from the very pronounced flooding horizon and in *Cladium mariscus* sedge-peat (Dewar & Godwin, 1963). This species is commonly regarded as introduced in Britain although freely naturalized.

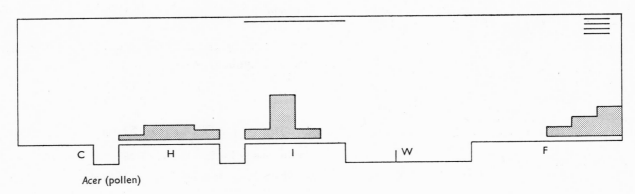

Acer (pollen)

ACERACEAE

Acer sp.   173

The tricolpate pollen of *Acer* is recognizable (at generic level) by its fine rugulate-striate sculpture: the record-head diagram includes pollen from Somerset diagrams recorded as *A. campestre* and interglacial pollen which wholly or in part represents *A. monspessulanum*. The diagram illustrates that the genus was present through all sub-stages of the Hoxnian and the first three of the Ipswichian, with no less than nine site records for zone II of the latter interglacial. The two records from the early Wolstonian are almost certainly of derived pollen: one is cited by the author as such. The absence, apart from these records, from glacial stages is not surprising. In the Flandrian there are no records before zone VIIb and they increase in later zones, perhaps the consequence of woodland clearances favouring *Acer campestre*.

An analysis of pollen frequencies in the sub-stages of the Hoxnian and Ipswichian interglacials (table 14) has some interest.

TABLE 14. *Interglacial pollen frequency of* Acer
(percentage total tree pollen)

	Scan	0–2	2–5	5–10	10–20	20–30
Ipswichian						
I	.	2	.	.	.	.
II	I	4	I	2	.	I
III	I	I	.	.	.	.
IV	.	.	.	.	.	.
Hoxnian						
I	I	.	.	.	.	.
II	I	2	.	.	.	.
III	.	2	I	.	.	.
IV	I	I	.	.	.	.

Whereas in the middle Hoxnian the pollen frequency is generally below 2 per cent of total tree pollen, once only exceeding this, in the sub-stage II of the Ipswichian two sites register 5 to 10 per cent and one 20 to 30 per cent. Having regard to the low pollen productivity of the

maples these values must indicate that trees of *Acer* were a substantial component of the natural woodlands, a conclusion borne out by the Eemian interglacial diagrams from the Netherlands and Denmark. West however makes the point that the irregularity with which the high frequencies occur between one site and another and at varying times implies, as for *Alnus*, that the maples must have had strongly local communities (West, 1964a). The Flandrian pollen frequency figures for *Acer* do not climb out of the 0–2 per cent class except for one site, where the soil below a Bronze Age barrow at Hackness yields 5–10 per cent, possibly the result of special factors of incorporation or preservation (Dimbleby, 1952).

Fruit or wood: H ?*Clacton*; I ?*Histon Rd, Trafalgar Sq.*; F VIIb *Llandow*, VIII *Stonham Aspel*

These macroscopic records include those of fruits from sub-stage III of the Hoxnian at Clacton, from sub-stage II of the Ipswichian at Histon Road (both tentative) and from a Roman site at Stonham Aspel. There is a charcoal record from the Middle Bronze Age at Llandow and one from unspecified macroscopic evidence in Ipswichian sub-stage II again at Trafalgar Square. Sparse as they are, these records fit well with the pollen record already considered.

Acer pseudo-platanus L.   Sycamore   173: I

F VIII *Dunshaughlin*

If this wood identification from an Irish site between A.D. 650 and 1000 is taken as secure then the sample is presumably from imported timber: Jones (1944) gives very comprehensive evidence of its rarity or absence in the British Isles in the mid-sixteenth century and the extension of its cultivation in the late sixteenth and early seventeenth centuries. Of course *Acer campestre* is not native to Ireland either.

Acer campestre L.   Common maple (Fig. 54)   173: 3

Cs *Pakefield*; F VIIb *Plas Newydd, Holdenhurst, Thriplow, Llandow, Downton, Pewsey, Coppice Gate Track*, VIII *Viper's Track, Meon Hill, Hembury Fort, Maiden Castle, Radyr, Glastonbury Lake Village, Sudbrook Camp, Hassocks, Pevensey, Caerwent, Crossness, All Cannings*

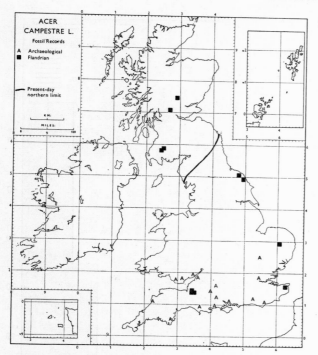

Fig. 54. Present northern limit in the British Isles of *Acer campestre* L. with its Flandrian records (mostly pollen) and macroscopic identifications from sites with an archaeological context in the same period.

The Pakefield record from the Cromer Forest Bed series is based upon leaf and fruit identification and has been confirmed by recent fruit record from new excavations at Pakefield by West and Wilson: all the rest are for wood or charcoal, and although the wood and charcoal of *Acer campestre* can only with difficulty be distinguished from that of other species of *Acer*, or not at all, there is every reason to think that all the Post-glacial records for this genus in Great Britain refer to *A. campestre*. The same applies to the pollen records. It should be noted that the stout vertical piles of the Late Bronze Age trackway at Coppice Gate, Shapwick Heath, Somerset, retained their bark which was undoubtedly that of *A. campestre*. All sixteen of the Flandrian records have been found in an archaeological context: perhaps it was used selectively or only in this way, having regard to its dry land habitats, were its remains thus preserved.

There is no British Flandrian record for the common maple earlier than the Neolithic period (zone VIIb). The Bronze Age, Iron Age and Roman records are all of similar frequencies. The diagram from Decoy Pool Wood, Somerset (fig. 168) is the only one in which there is anything approaching a continuous *Acer* pollen curve, and there it seems probable that the maple increased in the early part of zone VIII in consequence of forest clearance or thinning. Note that none of the macroscopic Flandrian records is further north than Anglesey and that there are five occurrences in Wales, thus con-

forming to what is regarded as the present native range (fig. 54).

A. campestre occurs in England and Wales from Westmorland and Durham southwards, attaining considerable frequency, especially on heavy limestone soils in southeast and central England. It is rare and probably introduced in Scotland and almost certainly not native in Ireland. On the European mainland *A. campestre* just enters southern Sweden (where it is rare) and extends from Denmark, Poland and south Russia to central Spain, Sicily and north Greece. It is placed by Matthews in the Continental–southern element of the British flora and its strongly southern restriction of range in this country agrees well with its recorded late appearance. It is apparently unaffected by frost in England, is notably resistant to shading and to coppicing and may make a very substantial woodland tree, although far more frequent today as undershrub or hedge plant.

Acer monspessulanum L.   173: 4

IG ?*Elderbush Cave*; I *Stone*, ?*Selsey*, ?*Bobbitshole*, *Trafalgar Square*

From the limestone tuff of Elderbush Cave, West has tentatively identified a leaf of *Acer monspessulanum*: the deposit is assumed to be interglacial. Clement Reid identified fruit from two Ipswichian sites at Stone and Selsey respectively, the latter tentative. The other records are all from sub-stage II of the Ipswichian and are based either upon fruit, or in the case of Trafalgar Square, fruit and leaf. These Ipswichian records for *Acer monspessulanum* inevitably raise the speculation that the high frequencies of *Acer* pollen already noted for this interglacial might well be due to this species. This is certainly the opinion of West (1964*a*) who writes 'Most sites with high frequencies of *Acer* pollen also contain fruits of *Acer monspessulanum*'. They are part of the evidence provided by many southern and continental plants that this part of the interglacial enjoyed considerable summer warmth. It is a plant of Mediterranean and central European distribution today (Walter & Straka, 1970) no longer native in the British Isles.

Acer platanoides L.   173: 5

H ?*Marks Tey*

Turner (1970) writes that '*Acer* pollen from the pre-temperate zone HO I appears to possess a coarser, more curly striate surface pattern. On ecological as well as morphological grounds it should be attributed to a different species (from *A. campestre*), perhaps *A. platanoides*.' The absence of *A. platanoides* from the present British flora is remarkable in view of its regeneration in this country and its presence in Scandinavia.

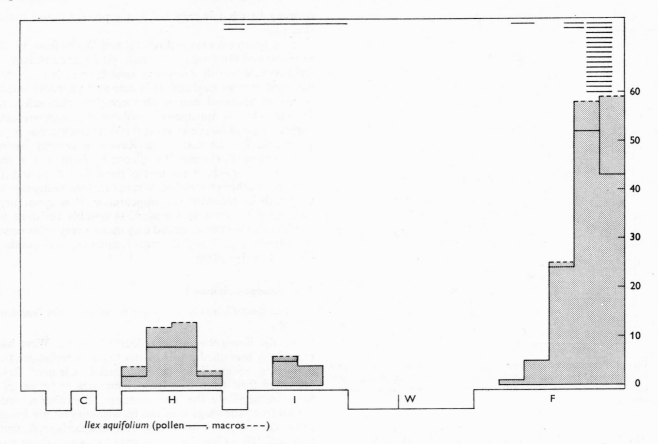

Ilex aquifolium (pollen ———, macros - - -)

HIPPOCASTANACEAE

Aesculus sp.   Horse chestnut   175

F viib   *?Avebury, ?Whitehawk Camp, ?Nympsfield Long Barrow, ?Bedham Hill, ?Gorsedd, ?Tusculum, ?Hembury Fort, ?Meon Hill, ?The Trundle, ?Cissbury, ?Red Hills, ?Cox Howman's Pit*

It is a remarkable fact that charcoal attributed to the genus *Aesculus* has been recorded many times from archaeological sites, chiefly in the south of England and of various ages from the Neolithic to Romano-British. Many of these are originally cited by the authors concerned as doubtful identifications and the rest should certainly also be so regarded. It is extremely easy to confuse charcoal of *Aesculus* with that of *Salix* and *Populus*, especially in young twigs such as appear frequently in residues of fire places. Professor F. W. Jane informs me that in fresh timber one can recognize *Aesculus* by the spiral thickening in some vessels and by the denser pitting on the walls between vessel and ray. These features are, however, very difficult to detect in many charcoal samples.

A. hippocastanum has its nearest acknowledged native habitats in the Balkans and the records cited must be discounted as evidence for its early presence in Britain.

AQUIFOLIACEAE

Ilex aquifolium L.   Holly (Figs. 55, 56, 57, 58; plate XXVo, p)   176: 1

Not only has the holly been very frequently identified macroscopically by its wood, charcoal, seeds and leaves, but its tricolpate pollen grains are easily recognizable by the distinctive sculpturing of the heavy exine, for the ectexine is made up of extremely stout separate clavate elements, distinctly variable in size and not joining by their swollen heads: the pilae are smaller round the furrows which thus shew a definite 'margo'.

It has been pointed out that because the holly is entomophilous and has a low pollen productivity, its pollen grains are unlikely to be subject to distant wind transport, a point of especial importance in view of the high climatic indicator value of the species. This also makes understandable the high degree of correspondence in the record-head diagram of the evidence based respectively on macroscopic and pollen evidence. It also explains the fact that in the fossil record high *Ilex* pollen frequencies are often found at levels containing macroscopic remains of the plant. Records from the glacial stages are absent except for scattered grains that may be regarded as derived (Thurnian (1), Baventian (1), late Anglian (1) and early Wolstonian (2)). There is also a single record

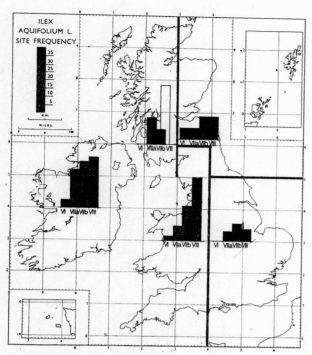

Fig. 55. Regional representation of *Ilex aquifolium* L. in the British Isles (macroscopic and pollen identifications) as a percentage of total site records zone by zone in the Flandrian. In north-eastern Scotland there are numerous records (open rectangle) for sites referred to the composite zone VII + VIII.

from the Ludhamian and another from the St Erth beds, but the Hoxnian shews site records for all four sub-stages, alike for pollen and macroscopic remains, and the records are more numerous by far in the milder middle sub-stages. For most of the Hoxnian sites the *Ilex* pollen frequencies average less than 2 per cent, but they are higher at Nechells, Kirmington, Baggotstown and especially at Gort, where they remain persistently large.

In the Ipswichian the site records fall entirely in the two middle sub-stages, again with values usually below 2 per cent, though reaching 5–10 per cent at Trafalgar Square where there were also macroscopic remains. There are no Weichselian records at all and within the Flandrian none before zone VI save a dubious North Wales pollen record in V and a similar Orkney record for the V/VI transition! The holly sites became very frequent in the Atlantic period (zone VIIa) and still more abundant in zones VIIb and VIII. The sixteen sites for macroscopic remains in zone VIII are mostly for wood and charcoal from archaeological sites (Iron Age 6, Roman 4, Anglo-Saxon 1 and Mediaeval 2), perhaps a reflection of the eminence of holly wood as fuel. There is little doubt either that the high frequency of site records for zone VIII is associated with clearances, the holly taking advantage of diminution of competition by forest dominants: this has been specifically shewn at Shapwick (Godwin, 1948), Holcroft and Chat Moss (Birks, 1963–4, 1965a, b), Plynlimmon (Moore, 1968), and Dinas Emrys (Seddon, 1961).

The Flandrian pollen records for *Ilex aquifolium* have been numerous enough to allow the analysis into five geographical regions presented in fig. 55. There is very strong contrast between the high values for site records in Ireland and the south-west quarter of Britain and the low values elsewhere, a contrast not due to the totals of sites investigated in the five regions. It is also evident that the natural expansion of the holly took place in all the regions at much the same time. Pollen frequencies for *Ilex* have been generally too low for regional or zonal trends to be apparent, but the highest frequencies have been found in Irish sites in Co. Kerry.

The interest of dated records of *I. aquifolium* is greatly enhanced by Iversen's penetrating analysis of *Viscum*, *Hedera* and *Ilex* as climatic indicators (Iversen, 1944a). By the painstaking and vigorous selection of sites, particularly meteorological, stations have been chosen and the performance near them of the three index species, ivy, holly and mistletoe, has been carefully established. As a convenient, reliable and informative climatic index Iversen has used the criteria suggested by Vahl, namely the mean temperature of the coldest month and the mean temperature of the warmest month. These two values give a good idea of the course of temperature through the year, the length of the vegetative period, the extreme temperatures (which in some instances have a very direct control of plant performance) and the relative oceanicity or continentality of the site under consideration. He has constructed for each species a temperature correlation graph with the temperatures of the warmest month as abscissa and the temperature of the coldest month as ordinate and each of the various stations recorded on the graph is given a symbol indicative of the plant's performance within reasonable proximity to it (usually within a circle of 20 km radius and inside a vertical height range of 40 m). Upon such a graph it is then easy to draw a 'thermal limit curve' such as that for *I. aquifolium* reproduced in fig. 56. This curve for the holly is remarkable in the vertical course followed at −0.5 °C, from which it appears probable that the plant is intolerant of winter frost; the upper part of the curve may indicate also some demand for summer warmth. During the very cold winters of 1939–42 direct observations made by Iversen in Denmark confirmed the holly's susceptibility to hard frost, a susceptibility primarily seated in the cambial tissue which is destroyed by low temperatures. In these years the hollies were very badly damaged and a large proportion of them killed (fig. 57). Iversen concluded, no doubt justifiably, that the eastern limit of *I. aquifolium* is chiefly determined not by length of the vegetative period but by the winter cold itself.

Iversen utilized this conclusion to deduce something of the climate of the last Interglacial period from which seven records of *Ilex* pollen were available, six Danish and one German. *Ilex* was found in the zones g, h and i of Jessen, during which conifers replaced the thermophilous deciduous trees, and this indicates continued winter mildness, indeed a climate milder than that of

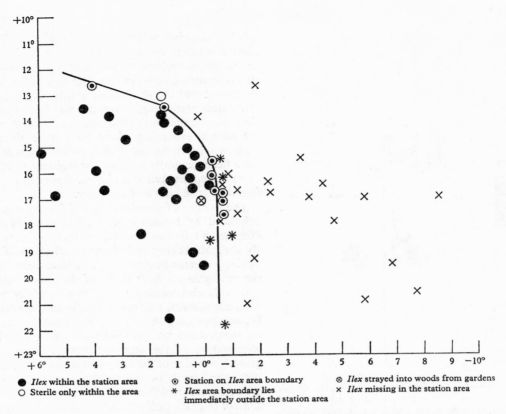

+10°
11°
12
13
14
15
16
17
18
19
20
21
22
+23°
+6°   5   4   3   2   1   +0°   −1   2   3   4   5   6   7   8   9   −10°

● *Ilex* within the station area   ⊙ Station on *Ilex* area boundary   ⊗ *Ilex* strayed into woods from gardens
○ Sterile only within the area   ✳ *Ilex* area boundary lies   ✕ *Ilex* missing in the station area
                      immediately outside the station area

Fig. 56. Thermal correlation curve for the holly (*Ilex aquifolium*) compiled by Iversen (1954). Ordinate, the mean temperature for the warmest month; abscissa, that for the coldest month. The holly is shewn intolerant of winter cold, but not demanding high summer temperatures (cf. *Viscum*, fig. 71).

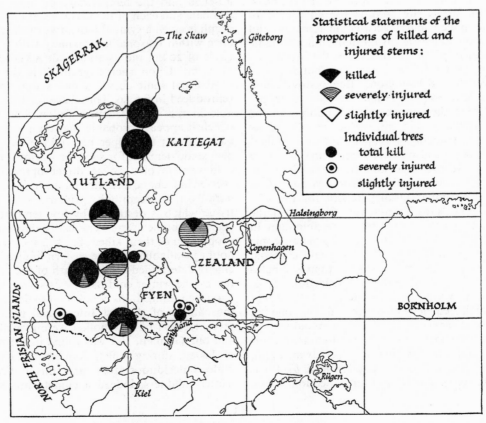

Fig. 57. Map prepared by Iversen (1944a) shewing the effect of the hard winters of 1939–42 upon the holly (*Ilex aquifolium*) in Denmark.

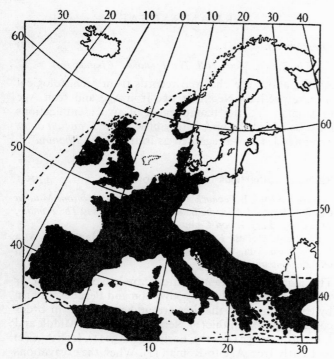

Fig. 58. Distribution of the holly (*Ilex aquifolium*) in Europe and North Africa (after E. Dahl), shewing the plant's absence from northern and continental regions.

today since the German station now lies outside the winter temperature limit of the holly. Since Iversen's conclusions were published the presence of *Ilex* has often been employed as a criterion of past climates, particularly for the Hoxnian and Ipswichian interglacials where the presence of holly, even in Yorkshire and East Anglia, certifies the winter mildness of the middle interglacial climate. This no doubt is bound up with the atlanticity of the British Isles, as consideration of the Flandrian *Ilex* pollen data makes apparent. It was possible for Iversen to write in 1944 that no *Ilex* pollen had been found in Danish post-glacial deposits, and although he has subsequently found it in the late Sub-atlantic in Draved forest (Iversen, 1969) it was evidently so infrequent as to keep Denmark at the extreme end of the range of diminishing frequency from Ireland (especially the south-west), western England, eastern England and south Scandinavia, a sequence that has evident close relation to the present European range of the tree (fig. 58).

In the first edition evidence was presented that seemed to shew that in Ireland the holly and ivy reached their maximum pollen frequencies in the Sub-boreal (zone VIIb), but in England in the Sub-atlantic (zone VIII), as if in response to the increased summer warmth of the Sub-boreal in Ireland and to the milder winters of the Sub-atlantic following the harsher Sub-boreal in England. It was however then noted that at Shapwick Heath (Somerset) the *Ilex* curve shewed a tendency to parallel the curves for weed and cereal pollen: as we have already

remarked, this effect has been apparent elsewhere so that responsiveness of holly to clearance is likely to confuse any explanation of changing abundance in climatic terms alone. It remains apparent none the less from fig. 55 that the site records diminish from South-East English zone VIIb to VIII. By contrast, in western England and Wales and in Ireland the site records increase from zone VIIb to VIII.

At the present day the holly is widespread in the British Isles except in the very open northern country of Caithness, Orkney and Shetland. It is described by Hultén as a European–Atlantic plant and his maps shew it in coastal sites in southern Norway to about 62° N latitude and almost or quite absent from Sweden and eastern Denmark.

CELASTRACEAE

Euonymus europaeus L.   Spindle tree   177: 1

F VII, VIII ?*Branston*, VIII ?*Cissbury*; VIIb *Old Buckenham*

The Branston record is for wood, that at Cissbury for charcoal of the Iron Age and Roman time, and that from Old Buckenham Mere is a positive pollen record.

BUXACEAE

Buxus sempervirens L.   Box (Plate XXV*q, r*)   178: 1

Macros: F VIIb ?*Whitehawk Camp*, VIII ?*Cissbury Camp*, *Roden Downs, Caerhun, Chesterford, Bartlow, Silchester*
    Pollen: H *Nechells, Kilbeg, Baggotstown*, ?*Histon Rd, Marks Tey, Gort*; F VIIA *Ellerside Moss*

A great deal of interest centres upon the status of the box, *Buxus sempervirens*, in Britain. It is classed by Matthews as a border-line species between the Continental–southern and Oceanic–southern elements, but by Walter and Straka (1970) as sub-Mediterranean. Small woods of it, generally taken to be natural, still occur at Boxhill, Surrey and at Boxwell in the south Cotswolds, and another mentioned by Ray in 1695, formerly occurred at Boxley, Kent. It has been several times recorded from Roman sites, and at Roden Downs in Berkshire J. Allison identified its leafy shoots as the material employed to line the coffin of a Roman burial: it was later discovered that Henslow had deposited in the Botany School herbarium, Cambridge, material of box which he had identified from two precisely similar situations in Roman burials at Chesterford and Bartlow, Cambridgeshire. It has sometimes been suggested that the plant was introduced by the Romans, but the identification of box charcoal from the Neolithic Whitehawk Camp at Brighton, although only tentative, makes a native status more probable. It should be noted that the Roman record from north Wales is from a boxwood comb possibly brought from a distance. There appear to be no records of the practice of lining coffins with box in the Roman burials of Italy or the Continent, so that the contention that the tree was introduced by the Romans to serve such a purpose has little support.

In contrast with the evidence of the macroscopic remains which is chiefly for the presence of box in the Flandrian, the pollen identifications, with one exception, all come from the Hoxnian interglacial where three English and three Irish sites are represented and a large part of the time span of the interglacial, with some sites having records in successive levels, as Baggotstown (sub-stage II/III and III) and Marks Tey (IIIb and IV), whilst at Nechells Kelly repeated the identification earlier made by Duigan.

Turner has pointed out that *Buxus* has been found in Holsteinian deposits of the Continental mainland in Denmark, Hanover and Berlin and that it may be regarded as a species characteristic of the late and post-temperate zones of the Hoxnian interglacial. He suggests also that it probably grew as a shrub within the *Abies* forest as it does in south-eastern Europe still. In sub-stage III its pollen is accompanied by that of *Abies, Vitis, Pterocarya* and *Ilex*, the latter in high frequency.

The one Flandrian pollen record in the English Lake District (Oldfield & Statham, 1963) is outside the present range of the box in England, but old enough for human introduction not to have been responsible for its presence. The adjacent limestone scars would have provided suitable habitats.

Pigott and Walters (1953) have drawn attention to the fact that the often quoted statement that box is rare in northern France and possibly not native is quite without foundation, and that there is no bar to regarding the southern English occurrences as the natural northern limit of the range of this species. Lang (1970) gives a map of the present distribution of the box and besides Mediterranean sites indicating late-glacial and early Flandrian presence he records fossil evidence indicating spread northwards from the Mediterranean refugia during the post-glacial warm period.

ANACARDIACEAE

Rhus sp.   Sumac

A *Ludham*

From the early interglacial, the Antian, West has recorded the pollen of *Rhus*, tricolporate grains with transverse costae. The temperate mixed deciduous–coniferous forest of this stage included also *Tsuga* and *Pterocarya*. Isolated grains of *Rhus* occur secondarily in various deposits and this Anglian record is not certain evidence of presence.

Rhus cotinus L. is native in the Balkans where it is regarded as a Tertiary relict, and Zagwijn records pollen of *R. cotinus* type in the Reuverian of the Netherlands. *R. coriaria* L., the sumac, is also a south European species.

RHAMNACEAE

Rhamnus sp.   179

F ?*Whitehawk, Meon Hill, The Trundle, ?Cissbury*

All the above are charcoal records from archaeological sites, respectively Neolithic, two Iron Age and Iron Age to Roman. To the tentativeness of two identifications must be added the uncertainty of whether *Frangula alnus* was intended to be regarded as in the genus *Rhamnus*.

Rhamnus catharticus L.   Buckthorn   179: 1

Macros: F VIIa, b *Peacock's Farm*, VIIb *Woodwalton, Maiden Castle, ?Newydd, Coombe Beacon, Mildenhall*, VIII *The Trundle, ?Cissbury, Bury Wood Camp*
    Pollen: C ?*West Runton*; F VI, VIIa, b *Shippea Hill*, VIIa *Helton Tarn*, VIIa, VIII *Urswick Tarn*, VIIb *Strichen Moss*, VIII *Old Buckenham*

The macroscopic identifications are all of charcoal or wood, accompanied at Woodwalton and Peacock's Farm by fruit stones. All the sites are archaeological, in order covering the Mesolithic (1), Neolithic (3), Early, Middle and Late Bronze Age (1 each), Early Iron Age (1), Iron Age (1) and Early Iron Age to Roman (1). When the excavations were reopened at Shippea Hill the two archaeological horizons found earlier at the Peacock's Farm excavation were seen again, and just as they had yielded fruit and wood at both levels, so now they gave *Rhamnus* pollen from both levels and covering the transition from zone VIc to VIIa, and that from VIIa to VIIb. The shrub presumably grew in the fen-woods of the wide valley beside which the settlements occurred. Aside from a doubtful Cromerian record, the remaining pollen records are from Helton Tarn, Urswick Tarn, Strichen Moss and Old Buckenham Mere. The pollen of *Frangula alnus* is not confusable with that of *Rhamnus* which in all these Flandrian records must mean *R. catharticus*. The concentration of records in the second half of the Flandrian and in the southern half of Britain corresponds to the present distribution of the species which does not extend into Scotland and is only abundant in southern and eastern England where it is strongly represented on the Chalk and Oolithic Limestone soils.

Frangula alnus Mill (*Rhamnus frangula* L.)   Alder buckthorn, Black dogwood (Plate IV)   180: 1

Cs *Pakefield*; H *Baggotstown, Gort, Hoxne*; eWo *Hoxne*; I *West Wittering, Wretton*; F VI *Romaldkirk*, VII *Betterhanger*, VIIa *Loch Kinord*, VIIb *Woodwalton, ?Moss Swang*, VIII, *Clea Lakes*

Pollen, fruit stones and wood of *Frangula alnus* are all readily recognizable and the records include all three categories. There is one record from the Cromer Forest Bed series, but pollen has been found in sub-stage I of the Hoxnian and both pyrenes and pollen in sub-stages II, III and IV, as in transitions between zones II and III. There can be no doubt that the alder buckthorn was a well established member of our flora in the Hoxnian from western Ireland to East Anglia. By contrast there are only two Ipswichian records, both seeds from West Wittering and Wretton respectively. Seed record from the early Wolstonian and pollen and seed record from the Hoxnian/ Wolstonian transition are probably of derived material: the same no doubt applies to two pollen records from the Weichselian, both from Esthwaite and one in soliflucted material.

The earliest Flandrian occurrence is in zone V in Dogger Bank peat. Thereafter pollen and macroscopic records keep step through the zones, with low figures in zones VI and VIII and high ones in VIIa and VIIb. The late Flandrian spread of *Frangula alnus* accords with its present distribution, almost absent from Scotland and Ireland and more abundant in the southern part of England and Wales. A zone VIIa macroscopic record from Loch Kinord (Vasari & Vasari, 1968) validates the native status of the species in Scotland, and its Hoxnian abundance in Ireland contrasts with its present dearth there. Andersen (1969) has cited *Frangula alnus* as a 'mor' plant in Denmark and has called attention to its remarkable abundance in the Herreskovian (Cromerian) and Holstein interglacials, especially early and late, with a later arrival and lower frequency in the Eemian.

VITACEAE

Vitis vinifera L.   Vine   181 : 1

Pollen: H *Marks Tey, Rivenhall*; F VIII *Bolton Fell Moss*
  Seed: eWo *Hoxne*; F VIII *London, Bermondsey, Silchester, Gloucester, Plymouth*

The pollen of *Vitis* has been recognized by C. Turner from sub-stages II and III of the Hoxnian interglacial at Marks Tey and from sub-stage III of the same interglacial at Rivenhall. The same investigator, re-examining deposits at Hoxne, identified a grape seed from the early Wolstonian level where it was presumably derived from material of the preceding interglacial. Turner (1969b) points out that seeds of *Vitis* have been recorded from the same interglacial in Holland, north Germany, Denmark and Poland and from the last interglacial in Berlin. He recalls that the wild sub-species *sylvestris* is a characteristic liane of damp woodland on alluvial soils in the large river valleys of central Europe, although not on the wettest soils: it could well have been in similar communities in East Anglia.

There are four records of seeds of *Vitis* from Roman sites where of course the fruit might have been imported. There is one record from mediaeval Plymouth, and K. Barker has reported a single pollen grain from a Yorkshire pollen sequence at a level he considers to be late Mediaeval, corresponding to the duration of the adjacent Lanercost Priory (A.D. 1150–1350). This site must lie near the northern limit of mediaeval viticulture in Britain.

Turner deduces from the presence of the wild vine in the Hoxnian that even the late-temperate sub-stage of it 'had a summer climate at least as warm, and probably warmer, and wetter than that predominant across north-west Europe today'. Van Zeist (1959) has given records of these individual pollen grains of *Vitis* from the Netherlands that suggest to him that the grape was cultivated there during the Bronze Age, but similar occasional grains identified from Giseux by N. Planchais (1967) are apparently late Boreal in age and convey an impression of natural occurrence separate altogether from the Sub-atlantic records accompanying those of cereals and indices of cultivation.

LEGUMINOSAE

In the Leguminosae identification has seldom been carried below family level. In terms of the British Flora alone it embraces a large number of genera and species with wide ecological tolerance. This may well explain why there are records through the Hoxnian, Ipswichian and Flandrian interglacials whilst also through the Weichselian, indeed with the highest numbers in the Late Weichselian and the opening zone of the Flandrian. Consideration of the records for individual genera and species may allow some conjecture as to components of this pollen record for the family.

Ulex sp.   187

Macros: lW ? *Hackney Wick*; F VIIb *The Trundle, Pentre-ifan, Simondston, Plumpton Plains, Itford Hill*, VIII *Hembury Fort, Bigberry Camp, Cissbury, Trebarveth*
  Pollen: F VIIb *Ehenside Tarn*, VIII *Oulton Moss*

Aside from the dubious record of a *Ulex* seed from the Middle Weichselian at Hackney Wick all the records for the genus (as indeed all those for *U. europaeus*) are from zone VIIb or later, and therefore fall inside the last five millennia. They are all derived from settlement areas, even the pollen records, and all apart from these are charcoals or wood which may be taken to represent firing. The sites include Neolithic (2), Neolithic to Bronze Age (1), Bronze Age (3), Iron Age (3), Iron Age to Roman (1) and Anglo-Saxon (1). They carry the implication that areas free from forest dominance existed and that pastoral or agricultural activities had probably already led to at least local forest clearance. It should be noted that apparently *U. europaeus* was deliberately planted on the Welsh hills in historic time to serve (cut and bruised) as a source of winter protein for livestock (*fide* R. Alun Roberts). Our records give no hint of the presence or status of the *Ulex* species before the impact of human settlement and although there are no records from Ireland, or from north of Chester, it is difficult to say that climatic controls have in any way limited its former occurrence. On the other hand van Zeist (1964) produces palynological evidence that in western Brittany *Ulex* was already present before the disforestation by prehistoric man led to its general increase, and he suggests that it was present upon wind-swept hill tops in treeless vegetation comparable with that now found in the so-called 'landes', and dominated by *Calluna* and other *Ericaceae*, *Ulex* and *Pteridium*. A possible analogue in Britain is the Lizard peninsula of Cornwall.

Although *Ulex europaeus* is frequent throughout the British Isles it barely extends into southern Norway and Sweden and the other British species, *U. gallii* and *U. minor*, have their northern limit of range here: they are all very typically western European in distribution.

Ulex europaeus L.   Furze, Gorse, Whin   187: 1

F vIIb *?Southbourne*, vIII *?Postwick, Warborough Hill, Pevensey, ?Goldhanger, Chester, Ehenside Tarn*

All these records, like those under the genus, are from archaeological contexts: Neolithic (1), Iron Age (2) and Roman (3, possibly 4).

Whilst it is perfectly possible to identify wood or charcoal as belonging to the genus *Ulex*, it is unlikely that, except in the case of very large stems, *U. europaeus* could be distinguished from the other British species, *U. gallii* and *U. minor*. Thus the above records do not have much better claim to represent *U. europaeus* than those listed under the genus only. Of course, the considerable probability remains that the bulk of the *Ulex* records *do* in fact belong to *U. europaeus* but there is no real certainty.

Sarothamnus scoparius (L.) Wimmer (*Cytisus scoparius* Link.)   Broom   188: 1

F vIIa *Oakhanger*, vIII *?Postwick, Red Hills, Essex*

From Oakhanger in a Mesolithic context dated 6300 B.P. is a record of wood of *Sarothamnus scoparius*; the doubtful Iron Age record at Postwick is for charcoal, as are records from the Red Hills of the Essex coast at Goldhanger and Carrewdon presumed to be Roman. As with *Ulex* these records presumably represent fuel.

Ononis sp.   Restharrow   189

F vIIb *?Viper's Track*, vIIb, vIII *Toll Gate Track, ?Godmanchester*

Although the pollen grains of the Leguminosae have not as yet been thoroughly examined there is ground for identifying as *Ononis* certain small tricolpate grains found in Late-glacial layers; some, however, resemble *O. spinosa* and others *O. repens* (see following accounts). The records above are at the sites of two Late Bronze Age tracks in Somerset and of a Roman site. It was conjectured that the Somerset pollen might have been *O. reclinata* (Dewar & Godwin, 1963).

Ononis repens L.   Restharrow   189: 1

lW II *Flixton*; F vIII *?Drake's Drove*

Both are pollen records and in view of the Weichselian one we may note that *Ononis repens* has a wider range in the British Isles than *O. spinosa* and has also a more extensive (though sparse) northern range in Scandinavia.

Ononis spinosa L.   Restharrow   189: 2

lW *?Stannon Marsh*

A Late Weichselian pollen record from Cornwall (Conolly *et al.*, 1950).

Medicago sp.   Medick   190

eW *Sidgwick Av.*

A seed referred to the genus *Medicago* was recovered from the Early Weichselian terrace gravels of Sidgwick Avenue, Cambridge (Lambert *et al.*, 1963).

Medicago lupulina L.   Black medick   190: 3

F vIIb *Itford*, vIII *Lidbury, Newstead*

These Flandrian identifications of seed of *Medicago lupulina* are all from archaeological sites, Late Bronze Age, Iron Age and Roman respectively.

Melilotus sp.   Melilot   191

H *?Clacton*

A tentative identification of a seed of *Melilotus* was made by Reid and Chandler from deposits at Clacton now known to be of the Hoxnian sub-stage III.

Trifolium sp.   192

Macros: Cs *?Beeston, ?Sidestrand, ?West Runton*; F vIII *Dublin*

Reid (1882) gave several records of seeds tentatively referred to the genus *Trifolium* from the Cromer Forest Bed series: a positive record is from mediaeval Dublin by Mitchell (unpublished).

Pollen: H *Marks Tey*; I *Marros*; eW *Chelford*; lW *Aby*; F vIIb *?Moss Swang, Shippea Hill*, vIIb/vIII *Loch Dungeon, Blelham Tarn*, vIII *Old Buckenham Mere, Godmanchester, Holcroft Moss, Lindow Moss*

The four oldest pollen site records are from horizons proximal to the glacial stages, Marks Tey in sub-stage IVb of the Hoxnian, Marros in late Ipswichian, Chelford in the interstadial and Aby in the post-Allerød zone III. By contrast all the Flandrian records are for zone vIIb or later. At least five are at archaeological sites or clearance phases, Neolithic (Shippea Hill), Iron Age (Old Buckenham Mere), Roman (Godmanchester) and Mediaeval (Holcroft and Lindow Mosses).

Trifolium pratense L.   Red clover   192: 2

F VIII *Larkfield*

Calyx and seed of *Trifolium pratense* were identified by Grove (1963) from a mediaeval site at Larkfield.

Trifolium striatum L. or **T. scabrum** L.   'Soft trefoil' or 'Rough trefoil'   192: 10, 11

H *Fugla Ness*

From sub-stage IV of the Hoxnian interglacial peat at Fugla Ness in the Orkneys, H. J. B. Birks (1969) recognized a pollen grain as either that of *Trifolium striatum* or *T. scabrum*. Neither species now occurs in the northern part of Scotland.

Trifolium scabrum L.   'Soft trefoil'   192: 11

F VII *Ehenside Tarn*

From what was presumed to be a Roman level in the deposits of Ehenside Tarn, Cumberland, Walker (1966) identified pollen of *Trifolium scabrum*.

Trifolium campestre Schreb.   'Hop trefoil'   192: 21

F VIIb *Minnis Bay*, VIII *?Godmanchester*

It was remarkable that the entire flower of this plant should have been preserved in the Minnis Bay deposits; it was associated with a large number of weeds and ruderal plants.

Lotus sp.   195

Lotus corniculatus L.   Birdsfoot-trefoil   195: 1

H *Fugla Ness; lW R. Annan, Loch Fada, Little Lochans, Elan Valley, Bradford Kaim, Moss Lake, Robert Hill, Culhorn Mains;* F IV *Bigholm Burn, Moss Lake, ?Loch Fada, ?Loch Coir' a' Ghobhainn,* V/VI *Allt na Feithe Sheilich,* VIIb *Frogholt,* VIIb/VIII *Wingham, Loch Dungeon,* VIII *Wingham, Frogholt*

It is only in the last few years that British palynologists have recognized pollen from the genus *Lotus* and it may be the choice of research sites by a few workers that has resulted in most of the records being from northern and montane sites. However, four of the five records after zone VIIb are from two sites in Kent, Wingham and Frogholt. Still more restricted are the sources for records of *Lotus corniculatus*: three are from south-western Scotland by N. Moar, and the others (three tentative) are from Skye by H. J. B. Birks. Aside from the interglacial record from Fugla Ness, Shetland, there is a conspicuous concentration of records, both generic and specific, to the three Late Weichselian zones and the succeeding zone IV of the Flandrian.

It is likely that *Lotus corniculatus*, now so widespread a constituent of British grasslands, was abundant in the Late Weichselian in Scotland and northern England at a time when these regions still held minor ice fields.

Iversen (1954) likewise recorded pollen of *Lotus* cf. *corniculatus* from the Late Weichselian zone III in Bornholm.

Lotus uliginosus Schkuhr or **L. hispidus** Desf.   'Large Birdsfoot-trefoil' or 'Hairy Birdsfoot-trefoil'

195: 3, 4

lW *?Moss Lake*; F IV *?Moss Lake*

Pollen referred by Miss R. Andrew to either *L. uliginosus* or *L. hispidus* came from Moss Lake, Liverpool both in Late Weichselian zone III and in Flandrian zone IV. They were substantially smaller than grains of *L. corniculatus*. Other floristic identifications from zone IV at Moss Lake suggested that shore conditions would then have been very suitable for *Lotus uliginosus*.

Astragalus sp.   200

F IV *?Little Lochans*

A pollen grain tentatively referred to the genus *Astragalus* was recovered by Moar from Flandrian zone IV at Little Lochans in Wigtownshire.

Astragalus alpinus L.   'Alpine milk-vetch'   200: 2

lW *Mapastown*

A pollen sample from zone I of the Late Weichselian at Mapastown when examined by Dr Iversen yielded a grain, which, after careful comparison with other species of *Astragalus* and *Oxytropis*, could be exactly matched only by *Astragalus alpinus*. This pollen is small and has a delicately and regularly reticulate polar area (Mitchell, 1953).

A. alpinus is a circumpolar arctic–montane plant found widely in the Arctic to the 65th parallel and extending to 76° N in Nova Zemlya. In the central Alps it ascends as high as 3000 m. In the British Isles it is now restricted to high altitudes (2350–2600 ft; 720–800 m) in three Scottish vice-counties and is absent from Ireland.

Oxytropis sp.   201

lW *?Loch Cill Chriosd*

H. J. B. Birks (1969) has made a tentative determination of a single pollen grain from his local zone LCC2 in Skye to the genus *Oxytropis*, probably referable to Late Weichselian zone I. Neither *O. halleri* nor *O. campestris* is now known to grow on Skye: both are rare in Britain and confined to high altitudes on Scottish mountains on basic-calcareous soils. The *Oxytropis* pollen is referred by Birks to a *Lycopodium–Cyperaceae* vegetation associated with long snow cover on calcareous soils.

Ornithopus perpusillus L.   Birdsfoot   202: 1

F VIII *Larkfield*

One of the distinctive fruits of *Ornithopus perpusillus* was recovered by Grove (1963) from the mediaeval site at Larkfield, Kent.

This small annual is generally distributed in Great Britain except northern Scotland and favours sands and gravels.

Onobrychis viciifolia Scop.   Sainfoin (Plate XXVIII*b*)

205: 1

*m*W *Earith*; *l*W *Hartford, Nazeing*

Miss Bell (1968) has reported that the pods or skeletons of the pods of *Onobrychis viciifolia* occurred in some quantity in the Middle Weichselian at Earith. They are very resistant to decay and leave a very recognizable vascular skeleton: this she also discovered in material from zone III at Nazeing that had been previously wrongly labelled. The fruit pod had earlier been found from Late Weichselian zone II deposits at Hartford. Pollen of this species has not been recorded in this country but Beug (1957) has found it in Late Weichselian and early post-glacial deposits at sites in the mountains of central Germany.

These records establish the early native status for this strongly calcicole plant that might otherwise be easily regarded as introduced.

Vicia sp.

206

Macros: H ?*Clacton*; *e*W *Sidgwick Av.*; F VIII *Worlebury Camp, Pevensey, Wookey Hole*

All of the macroscopic records for *Vicia* are of seeds, that at Clacton from sub-stage III of the Hoxnian interglacial and the from an Iron Age and others are from Roman sites respectively. It is possible that these archaeological contexts mean that a form of *Vicia faba* is inferred: of the Wookey Hole remains C. Reid writes 'charred bean seeds'.

Vicia hirsuta (L.) S. F. Gray   Hairy tare

206: 1

F VIII *Isca*

A single half-seed of *Vicia hirsuta* was identified by Helbaek from the charred grain store at the Roman site adjacent to Caerleon.

Vicia tetrasperma (L.) Schreb.   'Smooth tare'

206: 2

F VII*b* ?*Itford Hill*, VIII *Fifield Bavant, Isca, Caerwent*

Twenty-four whole or fragmentary seeds of *Vicia tetra-sperma* were identified with other weeds from Isca: it had earlier been recorded from Roman Caerwent. This plant is absent from Ireland, almost so from Scotland and present preponderantly in south-eastern England; but the Itford Hill record is Late Bronze Age, that from Fifield Bavant, Iron Age, so that if an introduction it is an ancient one. According to Helbaek the plant is met with abundantly in Danish and English grain deposits of the Middle Ages.

Vicia cracca L.   Tufted vetch

206: 4

F VIII *Isca*

A single seed of *Vicia cracca* was recovered from the burned Isca grain store.

Vicia sylvatica L.   Wood vetch

206: 10

*m*W *Temple Mills, Angel Rd*

C. Reid (1916), commenting on records from the Lea Valley Arctic Plant Bed, writes 'valves of the pod not uncommon at Angel Road and Temple Mills', and the identification can be regarded as assured. It has not been recorded elsewhere. It is a species placed by Hultén in his category of west Siberian–Continental plants and he shews it as occurring locally just inside the Arctic Circle. It is thinly scattered through the British Isles, including northern Scotland.

Vicia sepium L.   'Bush vetch'

206: 11

H ?*Fugla Ness*; *l*W *Moss Lake*; F IV *Moss Lake*

H. J. B. Birks has tentatively identified pollen from zone IV of the Hoxnian Interglacial in the Shetlands, and Miss Andrew has recognized it more positively at Moss Lake, Liverpool from the Late Weichselian zones II and III, and the succeeding Flandrian zone IV. *Vicia sepium* occurs throughout the British Isles and even in the Orkneys and Shetland Islands. According to Pigott (1958) it is commonly associated with *Polemonium coeruleum* in what appear to be natural British habitats.

Vicia sativa L.   'Common vetch'

206: 14

F VIII *Wickbourne, Isca*

From both Roman sites Wickbourne and Isca, Helbaek has identified seeds of *Vicia sativa*. This species is regarded as having been introduced into Europe from western Asia, but it is now naturalized and common in Britain.

Vicia angustifolia L.   'Narrow-leaved vetch'

206 :15

F VIII *Isca*

Fourteen seeds of *Vicia angustifolia* were identified by Helbaek from the burned plant material of the Roman grain store at Isca. *Vicia angustifolia* is distributed throughout the British Isles 'in hedges and grassy places' but can also be a cornfield weed.

Vicia faba L.   Broad bean, Horse bean

206: 19

F VIII *Glastonbury Lake Village, Meare Lake Village, Worle-bury Camp, Verulamium, Pevensey, Isca, Wookey Hole*

When C. Reid described the small beans found at the Glastonbury Lake Village he concluded that they 'agreed exactly in size and shape' with the variety from the Swiss lake-dwellings, described by Prof. Oswald Heer as *Faba vulgaris* var. *celtica nana*, and similar identification was later made by Percival for the beans from the Meare Lake Village, both Iron Age sites in Somerset. Worlebury is another Iron Age site yielding horse bean seeds. The four other sites are all Roman, and Helbaek comments that at Isca they seemed not to have been a crop but to have been residual strays in the cereal fields, to which fact he attributed their 'leaner' seeds.

Jessen and Helbaek (1944) recall that this plant has been found in a Neolithic content in the Channel Islands and Willerding also mentions a Bandkeramik (early Neolithic) find in Poland. Central European records were rare throughout the Bronze Age and evidently it became a popular crop only in the Iron Age, then reaching north-west Germany and, afterwards, Denmark. Jessen and Helbaek mention the possibility that *Vicia faba* reached Britain from the Iberian Peninsula by way of France and rightly point out the interest that further records might have.

There seems to have been no discovery in Britain comparable with the large quantities of bean haulms, found together with pods, seeds and pollen in the coastal 'Wurten' of the first to third century A.D. on the marshes of the north-west German coasts (Körber-Grohne, 1967).

Vicia or Lathyrus   206, 207

Macros: *mW Upton Warren*; F VIII *Godmanchester*

At the Upton Warren Middle Weichselian site Coope *et al.* (1961) based their identification on seed plus fruit valve: at the Roman site, Godmanchester, the record was a seed identification.

H. J. B. and H. H. Birks have identified a '*Vicia* type' pollen that includes several species of each genus: other authors including Moar, Simmons, Hibbert and Walker have referred to pollen as '*Vicia* or *Lathyrus*'. Because of the areas worked by these authors, in aggregate there is strong preponderance of Scottish records; indeed all up to zone VI are so save a single zone V record from the Dogger Bank. Not many of the British species of the two genera suggest themselves, on their present ranges, as the source of these Weichselian and early Flandrian records.

Lathyrus aphaca L. 'Yellow vetchling'   207: 1

F VIII *Isca*

See immediately following accounts.

Lathyrus nissolia L. 'Grass vetchling', 'Grass-leaved pea'   207: 2

F VIII *Isca*

Eleven seeds recovered by Helbaek (1964) from the burned Roman grain store at Isca, Caerleon were identified as *Lathyrus nissolia*. It is a Mediterranean plant with a local and strongly south-eastern distribution in England.

Many of the weed species identified by Helbaek were certainly introductions and it is possible that *Lathyrus nissolia* is also in this category.

Lathyrus sylvestris L. 'Narrowed-leaved everlasting pea'   207: 6

mW ?Angel Road

A seed from the Lea Valley Arctic Plant Bed was tentatively referred to *Lathyrus sylvestris* by Mrs Reid (1959). It is local and southern in Britain today.

Lathyrus cicera L. Grass pea   207: 13

F VIII *Isca*

From the same Roman grain store that yielded *Lathyrus nissolia* Helbaek recovered a single coatless seed of *L. cicera*, which he found impossible to refer decisively to the parent species *L. cicera* or to its domesticated descendant *Lathyrus sativus*. It is an abundant weed of pulse fields in southern Europe, and here seems an adventitious and transient immigrant.

Lens esculenta Moench. Lentil

F VIII *Isca*

From the burned Roman grain store at Isca, Helbaek (1964) identified about forty whole or fragmentary seeds, characterized by, among other features, size and the acutely lanceolate hilum. The lentil is widely cultivated in southern Europe and round the Mediterranean, and Helbaek conjectures that its presence here may represent 'the pathetic outcome of a serious attempt at introducing the plant as a crop in England' but adds that it may equally well have been accidentally brought in with seed corn. It is the only record known in England for a crop plant known in various German sites as early as the Neolithic but not at any time, according to Willerding, further north.

Pisum sativum L. Pea

F VIII *Wookey Hole, Silchester, Caerwent*

C. Reid, who identified seeds of *Pisum sativum*, the cultivated pea from Roman sites, commented that it was nevertheless 'rare on Roman sites'. *P. sativum* has had a long history of cultivation and may be presumed to be a Roman introduction here.

ROSACEAE

Pollen referred merely to the family Rosaceae figures in a great many British pollen diagrams. The aggregated results indicate, as might be expected, that the family is represented through the interglacials. There have been few pollen analyses in the glacial stages but Rosaceae

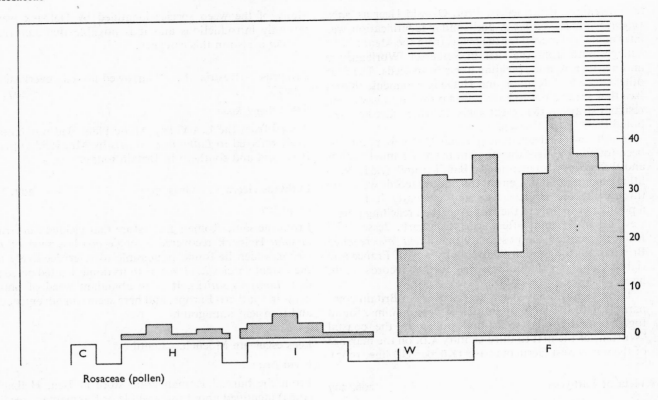

Rosaceae (pollen)

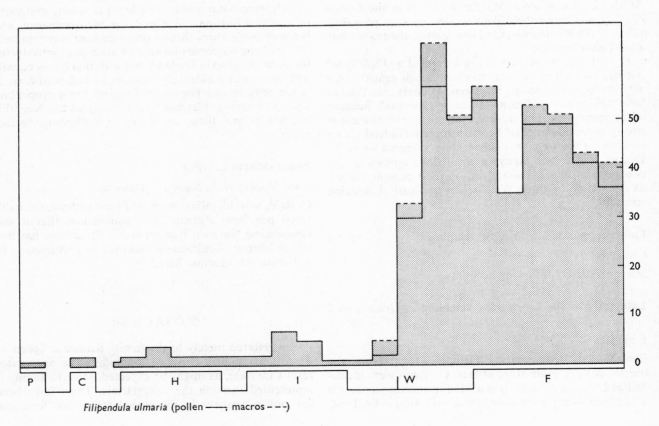

Filipendula ulmaria (pollen ——; macros - - -)

were evidently well represented in all zones of the Late Weichselian and early Flandrian, for the most part in frequencies below 2 per cent in the Weichselian itself, but including in Flandrian zone IV instances in the categories 5 to 10, and 10 to 20 per cent. Records for the individual species indicate that annuals and dwarf shrubs like those in the genus *Potentilla* may have contributed to this effect. In the Flandrian we may guess the high numbers of site records in zones V, VI and VIIa to have been differently derived, and since these are generally stages of forest dominance, high values for Rosaceae may presumably derive from trees of such genera as *Prunus*, *Sorbus* and *Malus*. The mean pollen frequencies are not high save in sites such as St Kilda where trees form only a small part of the total pollen catch and Rosaceae pollen in zone VIIa attains 10 to 20 per cent of the tree pollen.

Filipendula ulmaria (L.) Maxim. (*Spiraea ulmaria* L.) (Figs. 59, 60; plates XVII*n*, XXI*a*) Meadowsweet 210:2

The record-head diagram is for both pollen and macroscopic remains: these are so numerous that all records transitional between pollen zones have been omitted from the top of the diagram; although themselves numerous, they only confirm the trends of the primary zone numbers. The macroscopic records of *Filipendula ulmaria* are for carpels which have lost their outer coat and have an asymmetric twisted form, crescentic in side view, with a persistent style and a projection from the hollow of the concave side; they are quite recognizable in this condition, as also when they are entire (fig. 59). From the middle 1940s Late Weichselian deposits in many west European countries were found to yield a pollen grain identified as *Filipendula* cf. *ulmaria* and in England by 1952 the rather frequent occurrence of pollen of this type at Nazeing in layers yielding the fruits of *F. ulmaria* made it highly probable that the pollen indeed belonged to this species. As time has gone on the confidence with which pollen has been directly referred to *F. ulmaria* has increased alongside the belief that pollen of *F. vulgaris* is separately recognizable.

As the record-head diagram shews, *Filipendula ulmaria* has been present in the Cromerian and in all sub-stages of subsequent interglacials. It was moreover present in the late Anglian and both early and late Wolstonian glacial stages. More striking still is its presence in the Early and Middle Weichselian and its presence at very large numbers of sites in all zones of the Late Weichselian and early Flandrian, often in high frequency and not seldom accompanied by macroscopic remains. The average zonal pollen frequencies from I to VIII are exhibited in fig. 60 for Ireland and the four quarters of Great Britain. The fact that *average* values can be so high, even allowing for high pollen productivity (Faegri & Iversen, 1964), confirms the evidence of site frequency that the species was remarkably abundant in Late Weichselian time. This conflicts with the view sometimes expressed that it is 'a more or less thermophilous shade-tolerant herb' and

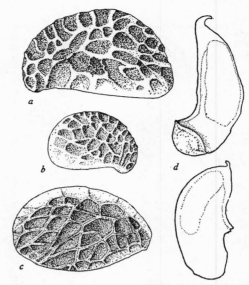

Fig. 59. Remains of rosaceous plants from Irish deposits (Jessen *et al.* 1959): *a*, *Rubus saxatilis*, pyrene, zone VI; *b*, *R. idaeus*, pyrene, zone VIIb; *c*, *R. fruticosus*, pyrene, zone VIIa; *d*, *Filipendula ulmaria*, two fruits, zone II.

justifies Miss Bell's reference (1969) to Keller's account of steppe vegetation in south-western Russia where *Filipendula vulgaris* (*F. hexapetala*) is a characteristic herb, and the conjecture that it may be this species which is so abundantly present in the Weichselian. It is however well also to consider the known prevalence of marshy habitats suitable to *F. ulmaria* in the Weichselian and the fact that it is a constant component of the tall-herb communities where other species, such as *Polemonium coeruleum*, have extensive Weichselian records.

It is notable that in western England and Ireland the pollen frequencies for *F. ulmaria* rise remarkably in zone VIII, possibly a consequence of generally improved lighting affecting the largely non-flowering woodland populations, but more likely to a generally increased wetness and wider prevalence of suitable marshy habitats.

F. ulmaria is placed by Hultén in the category of north European plants with a boreal–montane tendency, and he shews it as common throughout Scandinavia even in the extreme north. The records certainly establish continuous presence in the British Isles since the Cromerian, and it appears to be very strongly persistent through both glacial and interglacial stages.

Rubus sp. (p. 187)   211

With two exceptions only, wood identifications from a Mesolithic and an Iron Age site, all the site records for *Rubus* sp. are for fruit stones. It is unlikely that these records include any representing the montane or northern *Rubi* which have distinctive fruit stones and it seems probable that they derive chiefly from *Rubus idaeus* and *R. fruticosus*, a probability much enhanced by the great

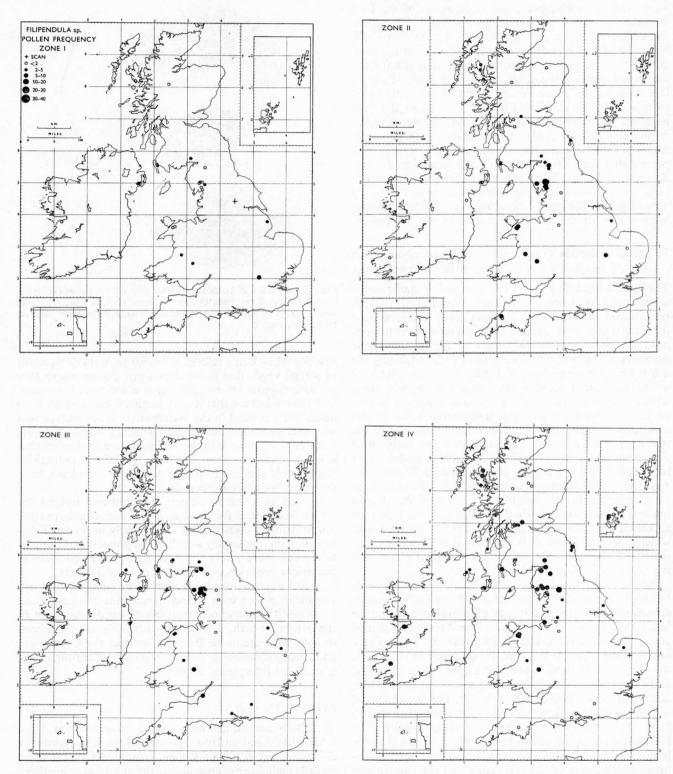

Fig. 60. Pollen frequencies of *Filipendula* sp. in the British Isles through the nine pollen zones of the Late Weichselian and Flandrian.

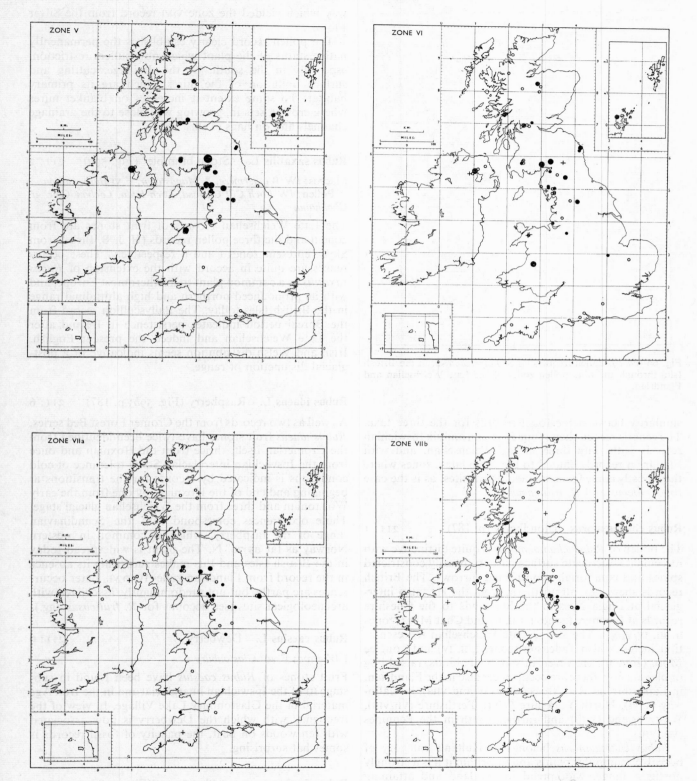

ZONE V

ZONE VI

ZONE VIIa

ZONE VIIb

Fig. 60 (cont.).

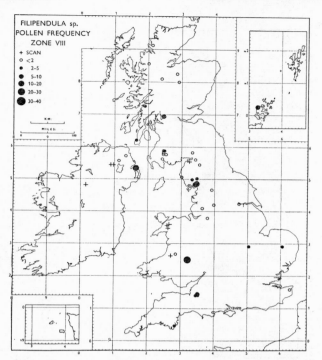

Fig. 60 (*cont.*). Pollen frequencies of *Filipendula* sp. in the British Isles through the nine pollen zones of the Late Weichselian and Flandrian.

similarity between the fossil records for the three taxa. The generic record shews four Hoxnian and six Ipswichian records with only three in the Weichselian, and with Flandrian records limited to the three latest zones where they mostly come from archaeological sites, as is the case for *R. idaeus* and *R. fruticosus*.

Rubus chamaemorus   Cloudberry (p. 187)   211 : 1

The pollen of *Rubus chamaemorus* is quite distinctive with its echinate–verrucate surface pattern, basally constricted spines and equatorially constricted furrows. The British records shewn are all northern, from the Hoxnian interglacial at Fugla Ness in the Shetlands to the Cheshire records at Bagmere (zones I, I/II, II) and Chat Moss (zones II, III, IV/V, VI). The strong Late Weichselian representation includes also Tadcaster (zones I, II, IV) and Longlee Moor (zone II). However, in suitable habitats, raised bog or blanket bog, there are occurrences also in the Flandrian, in Lancashire at Red Moss (zones IV and VIIa), in Galloway (V/VI), North Yorkshire (V/VI), Perthshire (VII/VIII), Westmorland (VIIb) and further south in the Pennines (VII, VIII).

Rubus chamaemorus belongs to Hultén's category of boreal–circumpolar plants and in Britain it has a strongly northern range, widespread in Scotland and attaining high altitudes, and extending south down the Pennines with only isolated sites in North Wales. It is absent from Ireland, the Isle of Man, and from south-western Gallo-

way which yielded the zone V/VI record from the Silver Flowe.

The pollen record clearly establishes the permanently native status of the plant which now suffers restriction, especially in the south, by the drainage, cutting and surface pollution of the mires which are its primary habitat. To some extent it increases on blanket mires where erosion sets in, growing well close to the drainage channels and on hags (Tallis, 1965).

Rubus saxatilis L.   'Stone bramble' (Fig. 59*a*)   211 : 2

Macros: *I*W *Windermere*, ? *Mapastown*; F VI *Ballyscullion*
   Pollen: *I*W *Loch Cill Chriosd, Loch Fada, Lochan Coir' a' Ghobhainn*

The Late Weichselian records of fruit stones are from zone II and the three pollen records (H. J. B. Birks) from Skye represent zones I and II respectively. These occurrences are quite in accord with the extension of *Rubus saxatilis* as a common plant to northernmost Europe, and with its pronounced northern and high altitudinal range in the British Isles today. The Ballyscullion record from the Boreal period indicates persistence in Ireland after the Late Weichselian and indeed the present English, Irish and Welsh distribution seems indicative of a post-glacial disjunction of range.

Rubus idaeus L.   Raspberry (Fig. 59*b*; p. 187)   211 : 6

As well as two records from the Cromer Forest Bed series, *Rubus idaeus* fruit stones have twice been identified from the Cromerian itself, thrice from the Hoxnian and once from the Ipswichian interglacial. Some tolerance of cold conditions is indicated by records for the transitions at beginning and end of the Hoxnian, by one from the early Wolstonian and three from the Weichselian glacial stage. These occurrences correspond with the Scandinavian range of the raspberry, which is common in western Norway as far as 70° N. The plant is widespread today in the British Isles and this makes surprising its absence in the record from Flandrian zones IV to VI. Later occurrences are partly, but not predominantly, associated with archaeological sites (see account for *R. fruticosus* agg.).

Rubus caesius L.   Dewberry   211 : 9

I *Wretton*; F VIII *Glastonbury*

Fruit stones of *Rubus caesius* have been found in sub-stage III of the Ipswichian interglacial and in the Iron Age material of the Glastonbury Lake Village. In view of the frequency with which the Dewberry is now associated with fen woods on peat, the paucity of fossil records is somewhat surprising.

Rubus fruticosus agg.   Blackberry (Fig. 59*c*)   211 : 11

The fruit stones of *Rubus* are well preserved and freely distributed by birds, but although the *R. fruticosus* aggre-

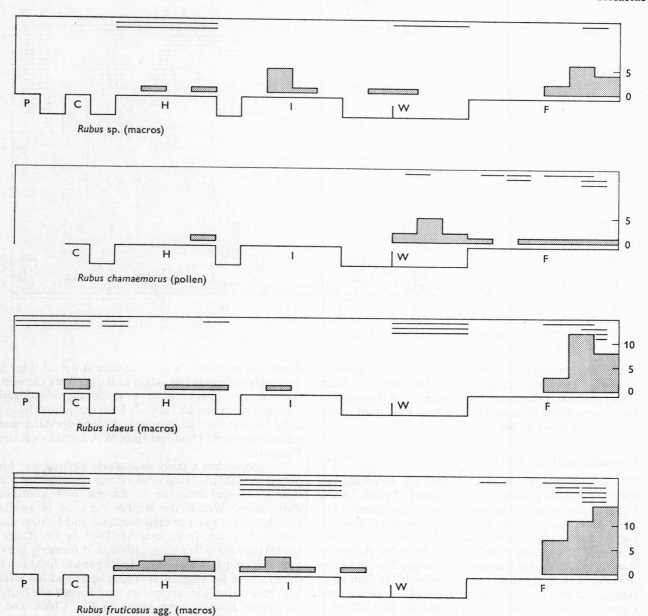

Rubus sp. (macros)

Rubus chamaemorus (pollen)

Rubus idaeus (macros)

Rubus fruticosus agg. (macros)

gate may be recognized, nothing of lower taxonomic rank within this has been identified. The record-head diagram (see above) shews not only three records from deposits of the Cromer Forest Bed series, but presence in all sub-stages of the Hoxnian and in at least three sub-stages of the Ipswichian interglacial. There is a single Early Weich-selian identification and one from the Flandrian zone IV/V transition from North Sea moorlog. Apart from this last, all Flandrian records, as for *Rubus* sp. and *Rubus idaeus*, are restricted to zones VIIa, VIIb and VIII. Table 15 shews the extent to which these late records are derived from archaeological sites, a clear indication, one would think, of the collection of the fruit for food.

TABLE 15. *British records of fruit stones of* Rubus *in an archaeological context*

	Rubus	Rubus idaeus	Rubus fruticosus agg.
Mesolithic	1	.	.
Neolithic	2	1	1
Bronze Age	1	3	4
Iron Age	1	.	2
Roman	1	4	8
Anglo-Saxon	.	.	2
Norman	.	.	1
Mediaeval	1	2	3

187

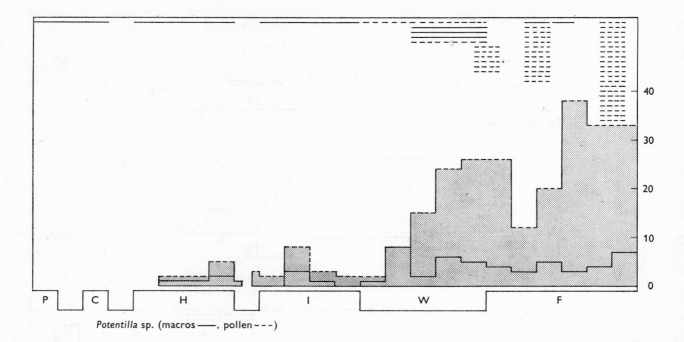

Potentilla sp. (macros ——, pollen - - -)

This preferential collection of fruit makes it impossible to deduce from the records that forest clearance from about 3000 B.C. so favoured these species that this explains the concentration of records in the later Flandrian zones, probable as this may be.

Potentilla sp. (Plate XVII*t, u*)   212

The ridged and corrugated achenes of *Potentilla* are among the most persistent and frequent of plant fossils, but in a proportion of instances specific identification has been impossible. Many British palynologists have recorded pollen of *Potentilla*, but H. H. and H. J. B. Birks give only '*Potentilla* type', which includes pollen of *Potentilla*, *Sibbaldia*, *Fragaria* and *Geum*. This indicates the reserve needed in interpreting not only the fruit component but also the pollen component of the record-head diagram, of which the Birks' records are part. There is good correspondence none the less between the records from macro- and microscopic evidence, indicating presence in the last three sub-stages of the Hoxnian, all four of the Ipswichian and throughout the Flandrian. Equally notable in view of the records for individual species of *Potentilla* are the frequent records in the Weichselian (especially the Late Weichselian) and in the early and late Wolstonian glacial stage.

Potentilla fruticosa L. 'Shrubby cinquefoil' (Plate XVII*r*)   212: 1

W Thrapston; *mW Barnwell Station*; *lW Nazeing*

A single achene of *Potentilla fruticosa* L. was found in zone III of the Late Weichselian deposits at Nazeing: its

shape and markings, with a prominent dorsal ridge and three pairs of transverse ridges radiating from the ventral scar of the gynobasic style made the identification certain. The species had already been recognized from the older 'Arctic bed' at Barnwell, Cambridgeshire, and it was subsequently identified from Weichselian deposits at Thrapston.

The species has a fairly wide world occurrence, being placed in Hultén's category of boreal–circumpolar plants that are boreal–montane in Europe with continuous distribution. Within the British Isles, as in southern Scandinavia, it has a severely localized and disjunct distribution. It grows from near sea-level in Co. Clare to 2000 ft (610 m) in Teesdale, although it formerly grew at 2300 ft (700 m) on Helvellyn. In its present localities it is found where the vegetation is kept open and has a fairly high base status, for example on scree in the Lake District, on grazed limestone pavement in Co. Clare, and on basalt rocks in the bed of the river in Teesdale.

The fossil records are far to the south of the present British range of the plant and suggest that the present disjunct occurrence is relict from a Weichselian period of more general and wider distribution (see Matthews, 1937, p. 66).

Potentilla palustris (L.) Scop. (*Comarum palustre* L.) *Potentilla comarum* Neshl. 'Marsh cinquefoil' (Plate XXVII*s*)   212: 2

To the numerous and widespread records of the achenes of *Potentilla palustris* there have recently been added a handful of pollen records by various research workers: at some sites such as Cross Fell (Cumberland) the fruits

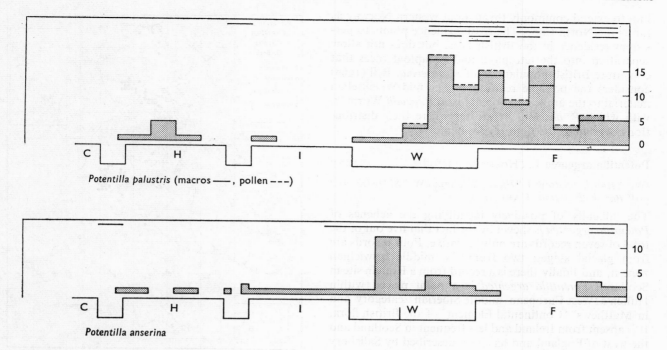

Potentilla palustris (macros ——, pollen ---)

Potentilla anserina

were accompanied by high pollen frequencies. The results demonstrate presence of the plant in three sub-stages of the Hoxnian and one sub-stage of the Ipswichian. Present in the late Anglian, and both early and late Wolstonian, it also occurred in the Early and Middle Weichselian, but it was evidently in the Late Weichselian that it became abundant, especially in the milder Allerød interstadial (zone II) with at least eighteen site records. Records remain frequent in the Flandrian, particularly the earlier zones: two of the later records are from archaeological sites, one Roman and one Norman. There seems a clear case for regarding *Potentilla palustris* as native right through the last Glaciation and ensuing post-glacial time, and some indication of persistence through earlier glacial and interglacial stages.

P. palustris belongs to Hultén's group of boreal-circumpolar plants with more or less continuous distribution and he shows it as a common plant throughout Scandinavia to the far north. Polunin describes it as circumboreal with a wide latitudinal amplitude, but chiefly temperate and sub-arctic, and cites Nova Zemlya as its northernmost record. It is a plant generally distributed throughout the British Isles though less frequent in England south of the line from the Humber to the Severn.

Potentilla sterilis (L.) Garcke (*P. fragariastrum* Ehrh.) 'Barren strawberry' (Plate XVII*q*)   212: 3

H *Baggotstown*; *m*W *Upton Warren*; *l*W *Nazeing*; F VII *Hightown*, VIIb *Frogholt*, VIII ?*Ehenside*

Achenes of *Potentilla sterilis* have been recorded from sub-stages I and II of the Irish Hoxnian, and then in zones VII, VIIb and VIII of the English Flandrian. By contrast it has been identified from the Middle Weichselian in the English Midlands and from zone I and the I/II transition at Nazeing in Essex.

Potentilla sterilis is a plant which is found in practically all parts of the British Isles, ascending to 2100 ft (640 m), but it thins out towards the extreme north: in Scandinavia it has an extremely restricted southern range. Hultén places it in the Sub-atlantic category, and Matthews (1937) puts it with the European species 'prevailingly west–central in their continental range'. In view of this limited northerly range the Weichselian records have special interest.

Potentilla anserina L.   Silverweed   212: 5

At individual sites the distinctive achenes of *Potentilla anserina* have been found in two sub-stages of the Hoxnian and all four of the Ipswichian Interglacial. Records are just as frequent in the Anglian and Wolstonian glacial stages whilst sites become very frequent in the Weichselian, especially so in the Middle Weichselian with twelve records. They occur in all zones of the Late Weichselian and in early Flandrian zone IV, but not thereafter until zones VIIb and VIII, where Roman and Bronze Age associations suggest that clearances may have multiplied sites available to this plant, now so characteristically ruderal in habitat though found also in many coastal and riverside localities, some of them brackish. It is easy to imagine how periglacial outwash and river terrace deposits favoured the pioneering qualities of the species which is also within Hultén's group of circumpolar plants strongly spread by culture, and is shewn by him to extend commonly throughout western Norway as far as the North Cape. The fossil evidence points to persistent residence in the British Isles, but does not allow separation into the tetraploid and hexaploid races that constitute British populations of *P. anserina*. Bell (1968) considers the possible relation of her mid-Weichselian material to the arctic and boreal taxon *P. egedii* Wormsk. with which *P. anserina* hybridizes where their distributions overlap in the Gulf of Bothnia.

Potentilla argentea L. 'Hoary cinquefoil'   212: 6

*l*Wo ?*Ilford*, ?*Selsey*; I ?*Ilford*, ?*Selsey*; e*W* ?*Sidgwick Av.*; m*W Barnwell Station*; F VIII *Newstead*

The difficulty of positively identifying the achenes of *Potentilla argentea* is shewn by the fact that five out of the total of seven records are only tentative. Four records are from glacial stages, two from the middle Ipswichian zone II, and finally there is a record from a Roman site in Scotland. *Potentilla argentea* is a plant placed within Hultén's 'west European–middle Siberian' category and in Matthews' 'Continental Element' of the British flora. It is absent from Ireland and less frequent in Scotland and the west of England and has been described by Salisbury (1932) as one of the East Anglian 'steppe' species. As such it would suitably accompany other 'steppe' species that have been shewn, for example by Bell (1969), to characterize the vegetation of the Weichselian glacial stage in Britain. It seems probable that at Ilford and Selsey it persisted in the interglacial because there were open habitats with light sandy soils.

Potentilla tabernaemontani Aschers. (*P. verna* auct.) 'Spring cinquefoil'   212: 11

m*W Hedge Lane*; l*W Hartford*

The two Weichselian records for *Potentilla tabernaemontani* are considered along with those of '*P. crantzii* or *P. tabernaemontani*' (below).

Potentilla crantzii or tabernaemontani (Fig. 61)   212: 12 OR 11

e*W*o *Brandon, Marks Tey*; l*W*o *Ilford*; I *Ilford*; m*W Brandon*; l*W Kirkmichael*

Several research workers have found it impossible to distinguish the achenes of *Potentilla crantzii* from those of *P. tabernaemontani*. Records under a joint head however include three from the Wolstonian and three from the Weichselian, including zones I and III at Kirkmichael. Both sub-stages I and II are represented by the Ilford Ipswichian records. The preponderance of records in glacial stages suggests that these may derive from *P. crantzii* with its arctic–sub-arctic range rather than *P. tabernaemontani* which only goes as far north as southern Scandinavia, and encourages doubts about the two unqualified attributions to *P. tabernaemontani* given above.

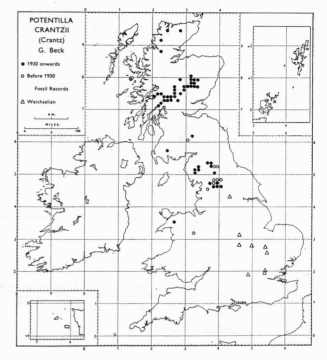

Fig. 61. Present and Weichselian distribution of *Potentilla crantzii* (Crantz) G. Beck in the British Isles.

Potentilla crantzii (Crantz) G. Beck. 'Alpine cinquefoil' (Fig. 61)   212: 12

W *Thrapston*; m*W Hedge Lane, Waltham Cross, Barnwell Station, Marlow, Great Billing, Syston, Oxbow*; F VIII ?*Ehenside*
     P. crantzii type: W *Thrapston*; m*W Earith*

Reid gave identifications of *Potentilla crantzii* from the Arctic Plant Bed of the Lea Valley, and Miss Chandler from corresponding deposits at Barnwell Station. These Middle Weichselian records have been since supplemented by Miss Bell who has made the most recent and much the most exhaustive study of achene morphology in British and west European potentillas. She adds the Weichselian records from Thrapston, Marlow, Great Billing and Syston. In some instances however she limits recognition to the '*P. crantzii* type' that includes *P. crantzii*, *P. tabernaemontani*, *P. arenaria* and *P. heptaphylla*, the two last non-British species. Her review of present-day populations of both *P. crantzii* and *P. tabernaemontani* shews that they have separate and distinctive achene morphology, and she points out that in the British Weichselian it is probable that distinctive populations also existed. Although Miss Bell can refer many fossils to *P. crantzii* she has not unequivocally identified *P. tabernaemontani*, a result that substantiates the doubt already expressed for Weichselian records of that species.

Potentilla crantzii is a plant in Hultén's category of 'Amphi-atlantic, arctic–montane plants also present in southern mountains'; he shews it as frequent in Scandi-

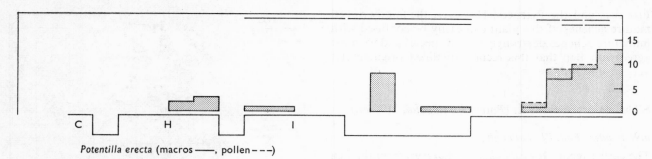

Potentilla erecta (macros ——, pollen – – –)

navia right round the North Cape and it is given by Polunin as fairly general in the Canadian Eastern Arctic. It is placed by Matthews in the arctic–alpine element of the British flora and it is much more abundant in Scotland than in England: it is absent from Ireland. It ascends to 2400 ft (730 m) in Snowdonia. The fossil records unequivocally place *P. crantzii* in the category of Weichselian species whose range has subsequently been restricted to the north (fig. 61).

Potentilla erecta (L.) Räusch. (*P. tormentilla* Neck.)
Common tormentil   212: 13

Potentilla erecta has been found in the two later substages of the Hoxnian and in the two earlier of the Ipswichian interglacial. There are eight Middle Weichselian records, although two of these are merely in the much wider *P. erecta* 'type' that includes *P. anglica* and *P. reptans*, since West's record for the Ipswichian substage ia is for '*P. erecta* or *reptans*'. After these records in the Late Weichselian it next is recorded in zone vi of the Flandrian, and far more frequently thereafter. It is notable that the few pollen records include high frequencies on Cross Fell (zone viia) and very high frequencies (up to 50 per cent of total tree pollen) in zones viia and viib on the Isle of Barra: both suggest local presence on blanket-bog where the species can grow commonly. The zone viii records include four from Roman sites and one from a mediaeval site.

Potentilla erecta is a plant common up to the Arctic Circle in Sweden, and within it in Norway, whilst in scattered localities it occurs throughout northernmost Scandinavia. The fossil record conforms to the implications of this range and both its Weichselian and late Flandrian occurrences accord with its preference for open habitats. It must also be borne in mind that taxonomic revision suggests that a complex of taxa is represented by '*Potentilla erecta*' in the British Isles. Iversen (1964) and Marmakowa (1968) have drawn attention to the incidence of pollen of *Potentilla erecta* in Danish pollen diagrams after forest fires.

Potentilla anglica Laicharding (*P. procumbens* Sibth.)
'Trailing tormentil'   212: 14

F viia ?*Branston, Boskill,* viib *Avonmouth,* viii *Ehenside, Hungate*

All the fossil records for fruits of *Potentilla anglica* come from rather late in the Flandrian, from zone viia onwards, with the two latest both in archaeological context: at Ehenside in a presumed Roman level and at Hungate in both Roman and Anglo-Saxon layers.

Although placed in Matthews' Continental–northern group, this species only extends into the extreme south of Scandinavia. The Hungate records reflect the tendency for the plant to occupy wayside or derelict ground although it is also sylvestral.

Potentilla reptans L.   'Creeping cinquefoil'   212: 15

H *Clacton; m*W *Upton Warren;* F viia *Clatteringshaws,* viii *Bunny, Caerwent*

Reid's record for *Potentilla reptans* can be referred to substage iii in the Hoxnian interglacial, and the Clatteringshaws Loch record for zone viia is similarly a middle phase of the Flandrian. Both Bunny and Caerwent are Roman sites and the occurrences suitable to a ruderal species.

Potentilla reptans, though widespread in the British Isles and belonging to Hultén's category of west European–south Siberian plants, is found in Scandinavia only in a few scattered localities north of 61° N, a feature to be remarked in considering the Upton Warren interstadial record.

Potentilla nivea agg.   212: 16

*m*W *Barrowell Green, Hedge Lane*

The identification of *Potentilla nivea* originally made by C. Reid from the 'Arctic bed' deposits in the Lea Valley was later confirmed by E. M. Reid, but Miss Bell has pointed out that the absence of ribbing that characterizes the achenes of this species is typical of all immature *Potentilla* fruits such as were frequent in these deposits. The records are thus very doubtful for this non-British aggregate species described by Polunin (1940) as having a circumpolar distribution, chiefly arctic and alpine and as extending as far north as 82° 28′ N in northern Greenland.

Potentilla nivalis Lapeyrouse   212: 17

*m*W ?*Barrowell Green*

Potentilla nivalis Lapeyrouse, no longer living in the British Isles, has a present distribution restricted to the Alps,

Provence and the Pyrenees. However, the faintly ribbed mature achenes of this plant can easily be confused with immature achenes of strongly ribbed species and we must agree with Bell that this record, qualified originally, is very insubstantial.

Sibbaldia procumbens L. (*Potentilla sibbaldi*, Hall fil.)

213: 1

mW Ponders End; *lW Ballaugh*

The earlier record for *Sibbaldia procumbens* from the Lea Valley 'Arctic bed' has now been supplemented by one from the Late Weichselian transition between zones II and III. Both records are based on identification of the achene, the apex of which is central and obtuse in *Sibbaldia*, not pointed as in *Fragaria* or ventrally inclined as in *Potentilla*.

Sibbaldia procumbens is a circumboreal plant chiefly sub-arctic and alpine extending to 73° 10′ N in east Greenland (Polunin, 1940). Matthews places it in his arctic–alpine element. Within the British Isles it is now found only on Scottish mountains. It has evidently suffered post-glacial restriction of range and extinction from the Isle of Man.

Fragaria vesca L.   Wild strawberry

215: 1

F VIIa *Cross Fell*, VII *Leasowe*, VIIb ?*Rockbarton*, VIII *Newstead, Silchester, Dublin*

The Cross Fell record for *Fragaria vesca* is from approximately 2350 ft, an altitude which is above the present limit for this species, but the period is within the Postglacial Climatic Optimum during which trees grew at this altitude on the Pennines at Cross Fell. The Leasowe record is from the coastal peat of the Lancashire coast. All the other records are from archaeological situations, Early Bronze Age, Roman (2) and Mediaeval.

Geum urbanum L. or Geum rivale L. Avens   216: 1, 3

Macros: H ?*Nechells*; *lW Nazeing, Flixton, Knocknacran*; F IV *Nazeing*, V/VI *Cross Fell*

Pollen: *lWo Bobbitshole*; I *Bobbitshole*; *lW Esthwaite*; F IV *Loch Cill Chriosd, Elan Valley*, IV, V, VI, VIIa, b *Esthwaite*

It is difficult to distinguish the fossil achenes of the two British species of *Geum*. There is included above a tentative identification of *G. rivale* from sub-stage II of the Hoxnian. The Nazeing fruits come from Late Weichselian zone I, the transition I/II and Flandrian zone IV, whilst the Flixton and Knocknacran records are from zone III of the Late Weichselian.

Pollen, which is difficult to distinguish from that of *Potentilla* and *Fragaria*, has been recorded from the late Wolstonian and Late Weichselian (zone II) interglacials, from sub-stages I and II of the Ipswichian interglacial and from the Flandrian zone IV in Skye and Wales, and zones IV to VIIb in the Esthwaite basin.

The two species have a similar world distribution

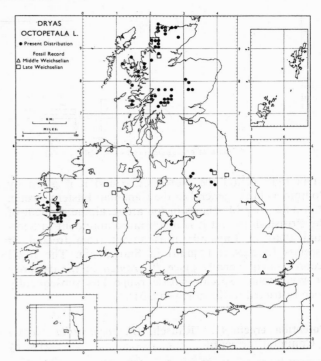

Fig. 62. Present and Weichselian (Middle and Late) distribution of *Dryas octopetala* L. in the British Isles.

but in Scandinavia *Geum rivale* has a considerably more northerly range and is common right up to the North Cape. Although both *G. rivale* and *G. urbanum* occur throughout the British Isles, *rivale* is commoner in the north and *urbanum* in the south. One might conjecture that the Late Weichselian records are more likely to derive from *Geum rivale*, especially since that species occurs commonly in the northern tall-herb communities that are otherwise well represented in Weichselian deposits.

Dryas octopetala L. 'Mountain avens' (Figs. 62, 63; plates XVIIo, XIXp)   217: 1

Macros: *mW Broxbourne, Barnwell Station*; *lW ?Crianlarich, ?Corstorphine, Loch Droma, Longfordpass, ?Yesnaby, Derriaghy, Neasham, Mapastown, Newtownbabe, Nazeing, Drumurcher, Kirkmichael*

Pollen: *lW Elan Valley, Kirkmichael, Mapastown, Loch Fada, Yesnaby, Widdybank Fell, Loch Cill Chriosd*; F IV *Loch Cill Chriosd*

The leaves of *Dryas octopetala* can be readily identified, even in fragmentary condition by their strongly recurved leaf-lobes, densely hairy lower surface and characteristic venation, and most of the records are based upon leaf recognition. However, upon occasion (as at Nazeing) fruits have also been recovered which are tentatively referred to this species. Of macroscopic remains there are Middle Weichselian records from the Lea Valley 'Arctic bed', and Late Weichselian records not only generally dated to this period, but specifically from

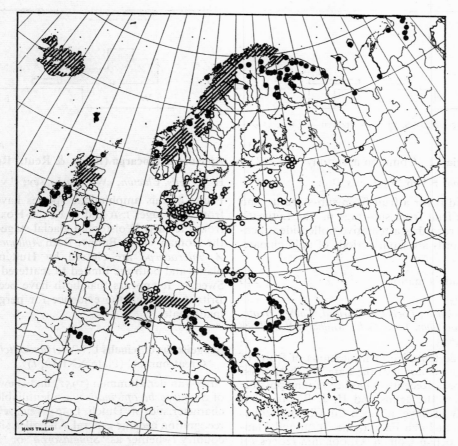

Fig. 63. Distribution of *Dryas octopetala* L. in Europe: recent range shewn by hatching and black dots, Weichselian plus Flandrian records by open circles. (After Tralau, 1963.)

zone II (4 records) and zone III (2 records) as well as from transition zones.

In recent years improved high-resolution microscopy has allowed several workers to identify *Dryas* pollen from the Late Weichselian. Often the same site has yielded records for more than one zone, thus Elan Valley (I, I/II, II/III), Yesnaby (II, III) and Loch Cill Chriosd (III and the ensuing Flandrian IV). The Yesnaby identifications were queried. At Kirkmichael the zone I deposits yielded pollen, fruit and leaf remains of *Dryas*, and indeed the fossil record for pollen corresponds remarkably closely with that for macroscopic remains. Upper Teesdale has yielded pollen from Late Weichselian zone II.

Dryas octopetala belongs to Hultén's group of circumpolar arctic–montane plants: it is found commonly throughout northern Scandinavia, Iceland and Spitzbergen, and northern Russia. It is found in northern England, Wales, Scotland and Ireland, usually on high ground, on mountain ledges on calcareous soils and open limestone rocky pastures. It ascends to 2800 ft (850 m) in Scotland and to 2300 ft (700 m) on Helvellyn, and descends nearly to sea-level on the limestone of Sutherland and Co. Clare.

The plant has long given its name to the older and younger 'Dryas clays' of Europe, deposits now recognized as corresponding respectively with zones I and III of the Late-glacial period. Tralau's map (fig. 63) shews dramatically the widespread Weichselian and post-Weichselian occurrence of *Dryas* over the central European lowlands and the present disjunction between the plant's arctic and alpine ranges.

D. octopetala was evidently widespread through the British Isles in Late Weichselian times, and then, as in the Middle Weichselian, occupied areas in the lowlands and far south of its present range (see fig. 62). It has evidently suffered great restriction of range during the Flandrian.

Dahl (1951) has suggested that the south-eastern geographical range of many Scandinavian arctic–alpine plants is correlated with the maximum July isotherm for that region, and subsequently Miss Conolly and he have shewn how the present Scottish range corresponds with the 23 °C summer maximum isotherm, and the Irish, Welsh and North British range with that for 25 °C.

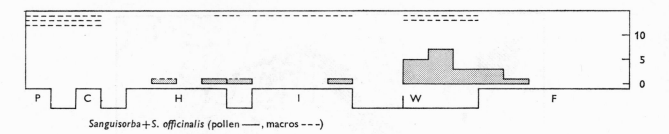

Sanguisorba+S. officinalis (pollen ——, macros ---)

Agrimonia eupatoria L.   Common agrimony   218: 1

H *Clacton*; I *Wretton*; F viib *?Lytheat's Drove*

The Clacton record is for a fruit from sub-stage III, and
that at Wretton a fruit from sub-stage II. The Lytheat's
Drove (Somerset) record is a tentative pollen identification. All these records are from periods of mild climate.

Alchemilla sp.   Lady's mantle   220

Macros: *l*W *St Bee's*
   Pollen: *l*W *Kirkmichael, Loch Cill Chriosd, Colney Heath,
Lochan Coir' a' Ghobhainn, Loch Fada, Malham*; F IV *Loch
Fada*, viib/viii *Wingham*

Walker (1966) made tentative identifications of fruits of
Alchemilla from zones I and II of the Late Weichselian on
the Cumberland coast. Recently it has been found possible
to identify pollen of the genus and H. J. B. Birks has
found it repeatedly in Skye, at Loch Cill Chriosd (zones
I, II, III and III/IV), at Loch Fada (zones II, III and Flandrian IV) and at Lochan Coir' a' Ghobhainn (zones I/II
and III/IV). Colney Heath is zone I, Malham zone II and
Wingham (Kent) yields the only recent record at a level
with a radiocarbon age *c.* 2720 B.P.

   In view of the northern and montane range of many
species of *Alchemilla* this record is unsurprising.

Alchemilla vulgaris L. agg.   220: 3

*m*W *Ponders End*; F viii *Newstead*

The Ponders End record is from the Lea Valley 'Arctic
bed' and the Newstead one from a Roman site.

Aphanes arvensis L. (*Alchemilla arvensis* (L.) Scop.)
Parsley piert   221: 1

Cs *Pakefield*; H *Clacton*; I *West Wittering, Ilford*; *e*W *Sidgwick Av.*; F viii *Wingham, Silchester*

It is of extreme interest to find that *Aphanes arvensis*, now
so typical of cultivated ground and waste places, was
present in three different interglacials as well as the Early
Weichselian, before its occurrence in the Iron Age at
Wingham and in Roman Silchester. The Weichselian
and Ipswichian records were specifically made to the
aggregate that includes the taxon *A. microcarpa* next
recorded: it is uncertain how far this applies to the other
records.

Aphanes microcarpa (Boiss. & Reut.) Rothm.   221: 2

H *Nechells, Clacton*; *e*Wo *Marks Tey*; F viii *Bunny*

Fruits of this amphimictic species have been identified
from sub-stages I, II and III of the Hoxnian interglacial,
from the early Wolstonian glacial stage and finally from
the Roman site at Bunny. Both *Aphanes arvensis* L. and
A. microcarpa are classed by Hultén as sub-atlantic
plants and both are limited to scattered southern sites in
Sweden. It appears that both have been long native in
Britain although the evidence for perglacial survival is
inadequate.

Sanguisorba officinalis L. (*Poterium officinale* (L.) A. Gray)
'Great burnet' (Plate XXIII*k, l*)   222: 1

After van der Hammen (1951) had shewn that the pollen
of *Sanguisorba officinalis* was identifiable and was indeed
characteristic of Dutch Late Weichselian deposits, its
recognition became general. All pollen site records for this
country (whether as '*Sanguisorba*' or '*S. officinale*') are
conflated in the record-head diagram, which shews substantial numbers of Late Weichselian and early Flandrian
records, but only single occurrences in the late Hoxnian,
late Ipswichian and the early and late Wolstonian glacial
stage. There is an oddly scattered occurrence of macroscopic identifications including three from the Cromer
Forest Bed series, two from the Hoxnian, one from the
early Wolstonian, one from the Ipswichian and two Late
Weichselian records respectively from Scotland and the
Isle of Man.

   S. officinalis is put by Hultén into the category of
Eurasiatic plants connecting with Scandinavia from both
the east and the south, a conclusion which evidently rests
upon the fact that this plant occurs naturally in widely
separate regions of Scandinavia, one in the Kola peninsula (on the Arctic Circle), one in south-western Norway and one in Gotland. It seems possible that this will
turn out to be a disjunction since Late-glacial or early
Post-glacial time. In the British Isles *S. officinalis* extends
as far north as Ayr and Berwick, and is very local in
Ireland.

   Whilst the fossil record indicates ancient persistence in
Britain it apparently indicates also some Flandrian retraction of range since Corstorphine and Ballaugh (I.O.M.)
are now outside the distribution limit of the species. The
record is somewhat comparable with that for *Poterium*

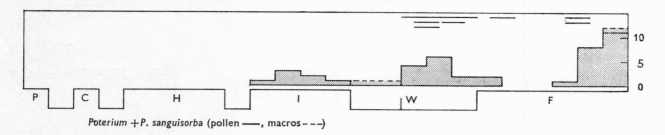

Poterium + P. sanguisorba (pollen —, macros – – –)

sanguisorba, save that *Sanguisorba officinalis* has so far no late Flandrian site records: no evident reason for this difference suggests itself.

Poterium sanguisorba L. (*Sanguisorba minor* Scop.)
Salad burnet (Plate XXIII*j*)   223: 1

The pollen of *Poterium sanguisorba* has a distinctive morphology and it has been independently recognized in the fossil state by different investigators. Van der Hammen first published this discovery for the Netherlands (van der Hammen, 1949), and has since shewn that pollen of this plant occurred, though sparsely, in the younger Dryas layers, and also in the preceding Late Weichselian layers of the Dutch Netherlands. It had meanwhile been recognized by Miss Andrew counting the material from the Histon Road, Cambridge interglacial deposit (Hollingworth *et al.*, 1950).

Subsequent pollen analytic records (sometimes as *Poterium sanguisorba* and sometimes as *Poterium*) have been aggregated in the record-head diagram and demonstrate clearly the plant's presence through all sub-stages of the Ipswichian, through all three zones of the Late Weichselian and Flandrian zone IV. Thereafter records are lacking until zone VIIa, but they are frequent in zones VIIb and VIII, no doubt in relation to forest clearance. The pollen records are supplemented by a tentative fruit identification from the Middle Weichselian (Waltham Cross), a receptacle from the Early Weichselian (Sidgwick Av.) and a fruit record from Roman Godmanchester.

P. sanguisorba is placed by Hultén in the category of sub-atlantic distribution and he shews it as having a local and scattered occurrence in southernmost Scandinavia, with very occasional outliers north of 60° N. Matthews writes of it as a widespread Eurasian plant. It occurs widely throughout England, Wales and south-east Ireland, but its scarcity in Scotland, with its very southern Scandinavian limits, make it difficult to understand the substantial Weichselian occurrences. None the less continuity of presence from the early Ipswichian seems evident. The gap in records for the middle Flandrian is associated with the plant's preference for grassland habitats (mainly calcareous) such as were infrequent before Neolithic disforestation started; the full Ipswichian record partly reflects locally open conditions at the sites investigated.

All present-day British material of *Poterium sanguisorba*

is referable to *Sanguisorba minor* ssp. *minor* (Nordborg, 1966) and not the ssp. *muricata*, with its tetraploid and octoploid races, which occurs in Scandinavia where the two cytotypes are distinguishable on pollen characters.

Rosa sp.   Wild rose   225

Macros: C *Norfolk Co.*; I *Stone, ?West Wittering, Grange, Selsey, ?Histon Rd, Wretton*; F vIIb *Peacock's Farm*, VIII *Glastonbury, Bunny, Hassocks, Crossness, Lissue, Dublin*
Pollen: *l*Wo *Hutton Henry*; *l*W *Elan Valley*

Macroscopic remains of *Rosa* include fruit stones prickles, stem and wood. Four of the Ipswichian records are for undefined parts of the interglacial, but two are from sub-stage II. All the Flandrian records save Peacock's Farm (zone VIIb) are from archaeological sites, Iron Age (1), Roman (4) and Mediaeval (2) from contexts that often suggest collection of the fruit for food.

The pollen records are as yet sparse and may relate to a pollen type of wide taxonomic derivation. This may explain the high-altitude Elan Valley records which are from Late Weichselian zones II and III (and the transition between them).

Rosa pimpinellifolia L. (*R. spinosissima* L.)   Burnet rose
                                                    225: 4
H *Clacton*

The record of a fruit stone of *Rosa pimpinellifolia* from the Hoxnian interglacial at Clacton (Reid and Chandler, 1923*a*) sorts with the coastal situation and the local marine transgression during the period covered by the deposits there.

Rosa canina L.   Dog rose   225: 8
H *Hoxne*; F vIIb *Walton on the Naze*, VIII *Silchester*

The Hoxnian records are from sub-stage III, repeated by successive workers on the site (Reid, West). Walton is a late Neolithic site and Silchester, Roman.

Prunus sp.   226

In recent years a few investigators, mostly concerned as it happens with Scottish or Welsh sites at high altitude, have identified pollen as *Prunus* or '*Prunus*-type': these yield a sparse but continuous record from the Late

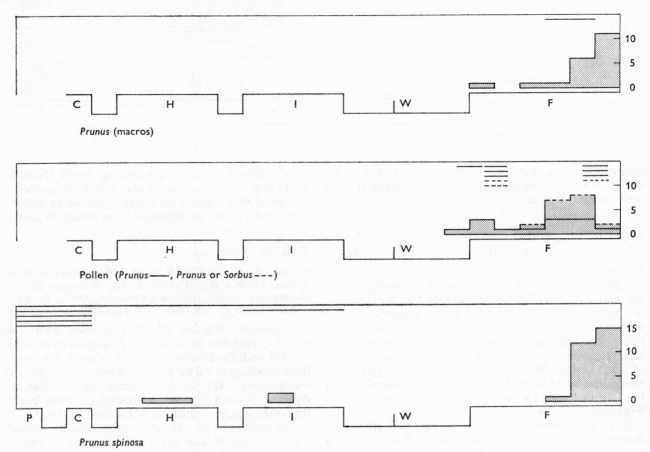

Weichselian through the Flandrian with a few records in Kent and Norfolk in zones VIIb and VIII at levels indicative of anthropogenous influences. One author (Simmons, 1969) has characterized pollen as 'Prunus or Sorbus' and these records, from Devon and North Yorkshire, are not unlike the records for Prunus pollen alone: they also include three zone VIIb sites associated with clearances.

The macroscopic remains attributed to Prunus have a range similar to that of pollen, but shew the effects of derivation from archaeological sites, seventeen of which make up the record for zones VIIb and VIII, three Neolithic, three Bronze Age, four Iron Age, one Iron Age to Roman, three Roman and three Mediaeval. All of these save a fruit stone of 9th–10th centuries A.D. from Lissue are for charcoal or wood or both, so that collection for firewood is indicated, probably selective and possibly Prunus spinosa: a majority are from South of England sites. The earliest records include tentatively identified fruit stones from Flixton (zone IV), Skipsea (zone VI) and St German's (zone VIIa).

Prunus spinosa L. (*P. communis* Huds.)   Blackthorn, Sloe
226: 1

The records for Prunus spinosa include four from the Cromer Forest Bed series, two from different sub-stages of the Hoxnian and two from sub-stage II of the Ipswichian as well as one (West Wittering) from an undefined part of that interglacial. All the remaining records, some for charcoal but mostly for fruit stones, are from archaeological sites that extend from the Mesolithic to 13th century Dublin: this closely resembles the record for Prunus sp. based mainly on charcoal or wood identifications and quite likely signifying Prunus spinosa.

The charcoal records indicate firewood usage, and carry the implication mentioned for Ulex that scrub had already followed forest clearance in the localities concerned. Bulleid and Gray have noted that in excavating Mound V of the Glastonbury Lake Village almost a wheelbarrow-full of sloe stones was recovered, and it seems evident that fruits of Prunus spinosa must have been generally collected by prehistoric man in Britain. Although it is generally assumed that the fruits were collected for food, the possibility of their use in dyeing should not be overlooked.

It is impossible to say from records of this kind, abundant as they are, whether absence of records before Mesolithic times corresponds with a real absence of the plant. In view of the interglacial records we may regard its absence from the early and middle Flandrian as more apparent than real.

P. spinosa is a plant of limited northern extension, going north of 60° only in a few scattered Swedish localities.

Prunus domestica L.   Cultivated plum   226: 2

F vIIb *Dyserth*, vIII *Maiden Castle, Crossness, Silchester, Bermondsey, Hungate, Dublin, Plymouth*

All records for *Prunus domestica* are based upon fruit stones from archaeological sites: Dyserth both Neolithic and Bronze Age (Glenn); Maiden Castle, Iron Age; Crossness, Silchester and Bermondsey, Roman; Hungate, late Anglo-Saxon and Norman; Plymouth, mediaeval; and Dublin, 13th century A.D.

It appears from these records that the cultivated plum has been present in Britain at least since Iron Age and Roman times. Reid wrote of the damson as being present in Roman Britain and some of the stones of the late Anglo-Saxon age from the fosse at the Hungate, York also corresponded with those of damson. Reid also recorded, beside those of the damson, stones of a large plum or prune from Roman Silchester, and larger plum stones were also present at the York Hungate in late Anglo-Saxon and Norman layers.

Prunus insititia L.   Bullace   226: 2b

F vIII *Maiden Castle, Silchester, Bermondsey, Hungate*

No great weight can be given to the reality of the separation of *Prunus insititia* from *P. spinosa* upon charcoal character, such as has sometimes been attempted, but identification of the fruit stones is more reliable and C. Reid writes (Bulleid & Gray, 1911) that both 'bullace and damson were eaten in Roman times'.

All the above records are based on fruit stones, one Iron Age, two Roman and one late Anglo-Saxon.

Prunus avium L.   Gean, Wild cherry   226: 4

C *Norfolk Co.*; I *West Wittering, Selsey*; F vIIb *Dyserth, Nympsfield*, vIII *Maiden Castle, Crossness, Silchester*

The records from Nympsfield and Maiden Castle are for charcoal, the rest for fruit stones, and at Selsey they were found earlier by Reid and afterwards (in sub-stage I of the interglacial) by West. All the Flandrian records are from archaeological sites Neolithic, Iron Age and Roman, with records by Reid at Crossness from both Bronze Age and Roman contexts.

Prunus cerasus L.   Sour cherry   226: 5

F vIIb *?Notgrove, Wooler*, vIII *Pevensey, Dublin*

All the records of *Prunus cerasus* are from archaeological sites, the first three representing Neolithic, Bronze Age and Roman cultures, but since these are based only on wood or charcoal that might well have come from other species of *Prunus*, they offer no reliable proof of the early occurrence of cultivated cherry. On the other hand the fruit stones reported by Mitchell from 13th century Dublin were common, along with those of *P. domestica* and *P. spinosa*, in a structure possibly a kind of fruit press, and have a clear identity.

Prunus padus L.   Bird cherry   226: 6

Macros: I *Selsey*; W *Airdrie*; F IV, vIIb *Drumurcher, v/vI Kilmacoe*, vI *Derrycassan*, vII *Dunshaughlin*, vIIa *Togherbane*, vIIb *Loch Kinord, Cocker Beck*, vIII *Aldwick Barley*
Pollen: *lW ?Lochan Coir' a' Ghobhainn, ?Loch Cill Chriosd*

The fruit stones of *Prunus padus* have been found right through the Flandrian: though most of the records are from Ireland, sites from Scotland, Nottinghamshire and Norfolk are included as well as a record for '*P. padus* or *P. spinosa*' from North Ferriby on the Humber. There is one Ipswichian record and one from Airdrie that is recently referred to an interstadial in the Weichselian. This last record accords with the tentative pollen identifications made by H. J. B. Birks in the Late Weichselian and early Flandrian in Skye. Such records are appropriate to a species that occurs freely to the extreme north of oceanic Scandinavia.

Prunus lusitanica Reid   226: 8

F vIII ?*Silchester*

Reid's record from Roman Silchester of a fruit stone of *Prunus lusitanica* is a dubious identification of an Iberian plant not otherwise reported fossil in this country.

Prunus amygdalus Batsch.   Almond   226: 9

F vIII *Plymouth*

From a mediaeval site at Plymouth Dennell (1970) has recorded a fruit stone of *Prunus amygdalus*.

Pyracantha coccinea M. J. Roem.   Firethorn (Fig. 64)
                                                            228: 1
I *Selsey, Wretton, West Wittering*

Both at Selsey and Wretton, West has identified the pyrenes of *Pyracantha coccinea*, in each instance from sub-stage IIb. In Clement Reid's own copy of *The Origin of the British Flora* he made an (unpublished) annotation that he had identified and figured a fruit stone of *Pyracantha coccinea* from West Wittering, also an Ipswichian site. This could well be the material afterwards referred by Mrs Reid to *Crataegus reidii*.

Pyracantha coccinea, now no longer native in Britain, grows in southern Europe and its range extends eastwards to the Caucasus and western Asia. In sub-stage II of the Ipswichian it is accompanied by many species indicative of considerable summer warmth including others like *Naias minor* no longer found in Britain.

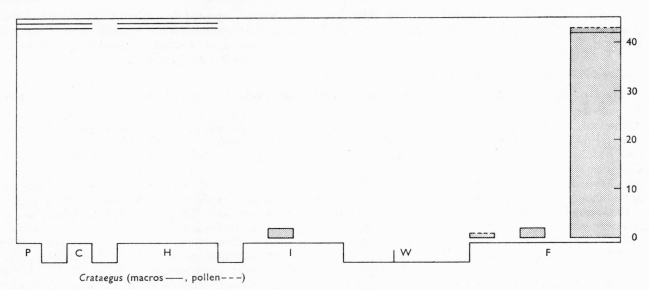

Crataegus (macros ——, pollen ---)

Crataegus sp.   Hawthorn (Fig. 64)   229

In the British Isles at the present day there are two distinct species of hawthorn, *Cratageus oxyacanthoides* Thuill. and *C. monogyna* Jacq. The former is an extremely shade-tolerant woodland species, the latter a plant much more characteristic of open habitats and scrub. It seems apparent from the investigations of Bradshaw (1953) that hybridization has been facilitated by woodland clearance as in the analogous situation with the North American *Crataegi*. Since differentiation between the species on wood alone is impossible all wood and charcoal records have been placed under the generic heading, and unless there is good evidence that the authors had deliberately identified fruit stones as those of *C. monogyna*, all fruit stone records are similarly treated. C. Reid (1899), in referring several fruit stones to *C. oxyacantha* L., makes it clear that this is a comprehensive species including the two above-named species; when he afterwards has referred fruit stones to *C. monogyna* it is clear that this was deliberate.

The macroscopic records for *Crataegus* thus aggregated in the record diagram include two each in the Cromer Forest Bed series, and the Hoxnian and Ipswichian interglacials.

The very numerous Flandrian records for the genus are very largely based upon charcoal identifications and are biased therefore by the factor of collection by prehistoric man. There is one record from zone vib to accompany Jessen's four records of *C. monogyna* in the Boreal period in Ireland, and one Mesolithic discovery. After this every cultural period from Neolithic to Norman time has records for hawthorn, often numerous. The prevalence of hawthorn charcoal is such that one cannot avoid supposing that hawthorn scrub must already have been present in prehistoric time and that it possibly first arose in its present-day form as a consequence of prehistoric forest clearance and pasturage.

The records are distributed as follows: Mesolithic 1, Neolithic 14, Bronze Age 16, Iron Age 13, Roman 14, Anglo-Saxon 2, Norman 1 and Mediaeval 2. The great majority of the records are for wood or charcoal, but there is an occasional pyrene, there is an Anglo-Saxon artifact of *Crataegus* wood, and the Roman site at Bunny yielded thorn, bud and dwarf shoot of hawthorn.

To the abundant macroscopic records we add three recent records of pollen from Flandrian zones IV, VIIb and VIII, the earliest, perhaps significantly, for peat from the North Sea Dogger Bank.

C. Reid (Bulleid & Gray, 1911) mentions that haws were collected for use as food and there is some Continental evidence for this.

Both *C. monogyna* and *C. oxyacanthoides* have a Scandinavian distribution area well within that of the deciduous forest trees; the latter is somewhat restricted in

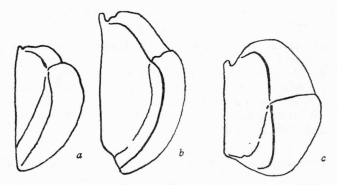

Fig. 64. Pyrenes of two extinct species of *Crataegus* from English interglacial deposits: *a*, *C. clactonensis* Reid & Chandler, from Clacton, Essex (×10); *b*, *reidii* Reid & Chandler, from West Wittering, Sussex (×11); *c*, for comparison, recent material of *Pyracanthus coccinea* (×10) (from Reid & Chandler, 1923a). It may possibly be that all of these types in fact fall within the range of *Pyracanthus coccinea* and that the erection of new (fossil) species has not been justified.

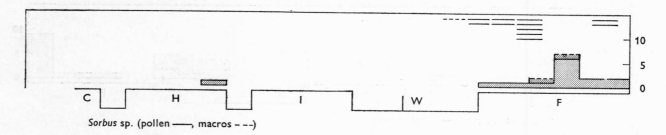

Sorbus sp. (pollen ——, macros – – –)

its northern extent and is placed by Hultén in the sub-atlantic distribution group whereas *C. monogyna* falls within the large 'west European–middle Siberian' group.

Crataegus monogyna Jacq.   Hawthorn   229: 2

H *Baggotstown, Clacton, Gort*; I *Wretton, Histon Rd*; F v *Ballybetagh*, v, vi ?*Kirkby Thore*, v/vi *Seamer*, ?*Kilmacoe*, vi *Cushenden, Portrush, Derrycassan*, viia *Westward Ho!*, viii *Glastonbury, Lissue*

The record from Wretton is a leaf identification; all the rest are for pyrenes (of characteristically round cross-section), woods and charcoals having been excluded. The Hoxnian records embrace sub-stages ii, iii and iv, the Ipswichian sub-stages i, ii and iii. The Flandrian records begin with Irish records from zones v and vi and more tentative records for v/vi and vi in Westmorland with a v/vi record from east Yorkshire. Thereafter follow a Mesolithic site (viia), the Iron Age lake village at Glastonbury and the 9th–10th century mediaeval site at Lissue.

The continuity of record in the middle (forested) sub-stages of the earlier interglacials is noteworthy, as in zones v to viia of the Flandrian: it suggests persistence in natural clearings and forest margins and indicates the sources for the development of hawthorn scrub that followed prehistoric and historic woodland clearances. Neither here nor in the record for the genus is there evidence of presence in Glacial stages, but evidently if excluded in these periods it returned readily to the British Isles.

Crataegus reidii Reid & Chandler (Fig. 64)   229: 6

I *West Wittering*

C. Reid discovered a fruit in the West Wittering inter-glacial which he identified as *Crataegus*. An examination made later by E. M. Reid and M. E. J. Chandler led to their establishment of it as a distinct species *C. reidii*. The plant has not been identified elsewhere. (See account for *C. clactonensis* below.)

Crataegus clactonensis Reid & Chandler (Fig. 64)
                                                        229: 7
H *Clacton*

E. M. Reid and M. E. J. Chandler established this new species *Crataegus clactonensis* for the fossil fruit of

Crataegus from the Hoxnian Interglacial. In describing it they point out its similarity to *C. pyracantha* Medic. and add that it is almost certainly extinct. It may well be that both *C. reidii* and *C. clactonensis* cannot be justified as species distinct from *Pyracantha coccinea* (fig. 64).

Mespilus germanica Hook.   Medlar   230: 1

F viii *Silchester*

The single find of medlar seed (*Mespilus germanica*) by C. Reid from Roman Silchester suggests a possible introduction of this plant by the Romans.

Sorbus sp. (Plate XX*j*)   232

Only one or two research workers have achieved the identification of *Sorbus* pollen and it happens that they were mostly concerned with Scottish sites. Accordingly particular care is needed not to argue from the absence of record sites. It is to be noted also that some of the records are for '*Sorbus* type' pollen. Seeds have been identified from sub-stage iii of the Ipswichian at Auster-field and from Flandrian zone vi at Malham; there is also a qualified identification from the Late Weichselian at Nazeing.

Sorbus aucuparia L.   Rowan, Mountain ash   232: 1

The record-head diagram shews macroscopic remains of *Sorbus aucuparia* from zones ii and iii of the Late Weichselian, although the zone ii identifications are tentative and there are also two records from zone iv of the Flandrian. These early records all depend on seed identifications but later records are mostly for wood or charcoal: if these suffice to distinguish *S. aucuparia* from other species of *Sorbus* they indicate presence of the plant from zone vi onwards. The nine site records from zone viii include eight from archaeological contexts, Iron Age (2), Roman (4) and Mediaeval (2).

H. J. B. Birks has recently recorded pollen of *Sorbus aucuparia* from Skye and there is one further northern record by Franks and Pennington. These are included in the diagram and confirm the Late Weichselian presence of the plant, a result not surprising in view of the abundance of this tree in the highest northern latitudes in western Europe, its high altitudinal range and its prevalence throughout Scotland at the present day. Similarly

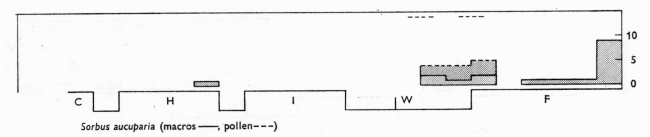

Sorbus aucuparia (macros ——, pollen---)

Mamakowa (1968) reports that in Denmark the majority of grains recorded from the pre-Boreal (early Flandrian) and later 'are undoubtedly *S. aucuparia*'. Pilcher (1969, 1970) reports that at Ballynagilly in Northern Ireland pollen referred to *Sorbus aria* or *S. aucuparia*, along with that of *Ilex*, is an indicator of forest regeneration after the Neolithic clearances between 3270 and 2350 B.C.

Sorbus aria (L.) Crantz   White beam   232: 5

Cs *Pakefield, Mundesley*; F viib *Ballynagilly*, viii *?Maiden Castle, Newstead*

The two records from the Cromer Forest Bed series are both based upon leaf identifications. That from the Neolithic level at Maiden Castle is a tentative charcoal identification and that from the Roman site at Newstead a wood identification tentatively to the *Sorbus aria* aggregate species. Although generally more abundant in southern England this Roman record from southern Scotland is consonant with scattered native occurrences of the tree in Scotland. The record from Ballynagilly is for pollen at the time of early Neolithic forest clearances and is supported by electron microscopy (Pilcher, 1969): the tree is now very local in Northern Ireland.

Sorbus rupicola (Syme) Hedl.   232: 5(7)

F iv *Loch Cill Chriosd*

H. J. B. Birks has identified an occasional pollen grain of *Sorbus rupicola* from Flandrian zone iv in Skye where the species still grows. Pollen of *S. rupicola* has also been recognized in the Danish Late Weichselian (Mamakowa, 1968; Jørgensen, 1963).

Sorbus torminalis (L.) Cranz   Wild service tree   232: 7

H ?*Hitchin, Hoxne*; F viii *Maiden Castle*

Three separate seed identifications of *Sorbus torminalis* have been made from the Hoxnian interglacial, two tentative by Reid and respectively from sub-stage II at Hitchin and sub-stage III at Hoxne, with a third unqualified and much more recent by West, also from sub-stage III at Hoxne. In addition, Salisbury and Jane cite the identification of charcoal of *Sorbus torminalis* from the Late Iron Age levels at Maiden Castle in the same paper as that in which *S. aria* charcoal is recorded from the Neolithic, and are apparently satisfied that the two species can be separated on charcoal characters.

It is interesting that *Sorbus torminalis*, not apparently native north of the Humber and generally southern in its European range, should have records for the warm sub-stages of the Hoxnian.

Pyrus sp.   233

F viib *Hurst Fen, ?Whitehawk, Trundle, Thickthorn, Long Wittenham, Coombe Beacon,* viii *Hollingbury Camp, Bigbury Camp, Harrow Hill, Hembury Fort, Meon Hill, Cissbury, Mells, Canterbury, Crossness, Dunshaughlin*

The numerous records for *Pyrus* are almost entirely based upon charcoal identifications and they represent collections made by prehistoric man for firewood. They represent culture from the Neolithic through to Romano-British and almost all come from the southern counties of England. It should be noted that many of the charcoal identifications are to the genus *Pyrus* in the wide sense, and may include *Sorbus*. We have placed in this generic category the wood from the Crossness submerged peat beds, which Marshall Ward had referred to *Pyrus communis* L.

No doubt there has been selective gathering of wood prized for fuel.

Malus sylvestris (L.) Mill. (*Pyrus malus* L.; *M. pumilla* Mill.)   Crab apple   234: 1

Cs *Pakefield*; F viib *Windmill Hill, Maiden Castle,* viii *Silchester, Caerwent, Bermondsey, Dublin*

Apart from that in Cromer Forest Bed series from Pakefield, all records for *Malus sylvestris* are from archaeological sites and Maiden Castle alone has records for charcoal of it in the Neolithic, Bronze Age and Iron Age, as well as seed impressions in Iron Age pottery. Windmill Hill has also yielded seed impressions in Neolithic potsherds. Silchester, Caerwent and Bermondsey are Roman sites that have yielded seeds and so has the mediaeval site in Dublin.

Jessen and Helbaek refer to the opinion of Heer, Neuweiler and others that apples were cultivated in Europe as early as the Neolithic period, but there is no British evidence on the matter. It is apparent from work by Helbaek (1952a) and others that apples were dried and stored for food by prehistoric peoples. The presence of apple pip impressions on no less than six separate shards at Windmill Hill warrants the assumption that the fruits were gathered for food in fair quantity.

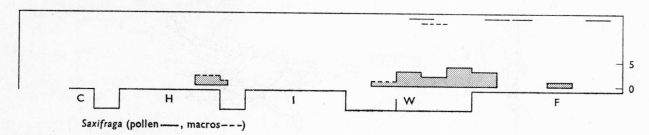

Saxifraga (pollen——, macros---)

CRASSULACEAE

lW Kirkmichael, Ballaugh; F IV *Ballaugh*

Pollen of the family Crassulaceae has been recorded from the Late Weichselian zone I at Kirkmichael in the Isle of Man (Dickson *et al.*, 1970), and from the succeeding zones II, III and IV at Ballaugh nearby (Mitchell, 1958). The narrower identification of pollen to *Sedum* sp. and to *S. rosea* yields results with similar age span (see below).

Sedum sp.   235

lW Kirkmichael, ?Lochan Coir' a' Ghobhainn

From Kirkmichael pollen of the genus *Sedum* has been recognized in zones I and III of the Late Weichselian, and there is a tentative identification from Skye in the transition zone I/II.

Sedum rosea (L.) Scop.   Rose-root, Midsummer-men
                                                                235: 1

Macros: *lW Nant Ffrancon*
   Pollen: *lW ?Loch Fada, ?Loch Craggie, ?Loch Mealt, ?Loch Cill Chriosd, ?Mapastown, ? Yesnaby, ?Cooran Lane*; F *? Yesnaby, ?Allt na Feithe Sheilich*

Burrows identified a seed of *Sedum rosea* from Late Weichselian zone II deposits in the Nant Ffrancon in Snowdonia where the plant still grows.

Pollen is less certainly identifiable and records are either tentative or referred to 'S. *rosea*' type as for Loch Craggie. Apart from that at Mapastown, all the records are from Scotland, embracing Skye, the Orkneys, Cairngorm, Sutherland and Galloway, and all within the present range of the species. The records represent all the Late Weichselian zones and early Flandrian with one high altitude site in the Cairngorm at the zone v/vI transition. There is often evidence of persistence from zone to zone, as at Loch Fada (zones I, II and III), Loch Craggie (I and III), Loch Mealt (II and III) and Yesnaby (III/IV and IV). As with the evidence for species of *Saxifraga*, that for *Sedum rosea* suggests long persistence from Late Weichselian times in situations still occupied by the plant.

Sedum rosea belongs to Hultén's group of circumpolar arctic–montane plants: it occurs in eastern and western North America and extends to 78° 30′ N in Spitzbergen. It has a strongly Atlantic distribution in Scandinavia where it extends as a frequent plant to the extreme north: it occurs also in the central European mountains. It

occurs in mountain habitats in the British Isles northwards from South and North Wales, Yorkshire, Lancashire and the Isle of Man: it is absent from the centre of Ireland.

Umbilicus rupestris (Salisb.) Dandy   Pennywort, Navel-wort   238: 1

I *Histon Rd*

Walker (1953) recorded a fruit of *Umbilicus rupestris* from sub-stage III of the Ipswichian interglacial in deposits exposed in the Histon Rd, Cambridge. It is a plant which has a strongly south-western range in the British Isles, and Cambridge lies on the edge of its sparse eastern localities. The Cambridge region also seems to lack congenial rocky habitats and further records of the plant would be welcome.

SAXIFRAGACEAE

Saxifraga sp. (Fig. 65)   239

Interglacial records for the genus *Saxifraga* are limited to two tentative macroscopic identifications, one a leaf. There is a tentative seed identification from the early Wolstonian at Ballymakegogue and unqualified identifications of seed from Earith (Middle Weichselian) and leaf from Nant Ffrancon (Late Weichselian II).

Pollen of *Saxifraga* has been recorded by a few workers at northern or upland sites, and it provides evidence of continuous presence in all zones of the Late Weichselian and zone IV of the Flandrian. There is a clear suggestion of continuous presence through the Flandrian. It is very noticeable also that individual sites supply strong evidence of continuity. Thus the pollen occurs at Yesnaby (Orkney) in zones I to IV, at Elan Valley (Cardigans.) in zones I, I/II, II, III and IV, at Blelham Tarn (North Lancs.) in zones I, III and possibly vIIb/vIII, and at Duartbeg (Sutherland) in zones IV, v/vI and vIIa.

It will be seen from the accounts for individual species that follow, that this situation is repeated for many of them.

Although the record is not for individual species the continuity of record indicates how important, for this genus at least, is the significance of the local ecological conditions in permitting continuous occupation of a site through changing overall climate.

Finally we may note that it is unlikely that the pollen

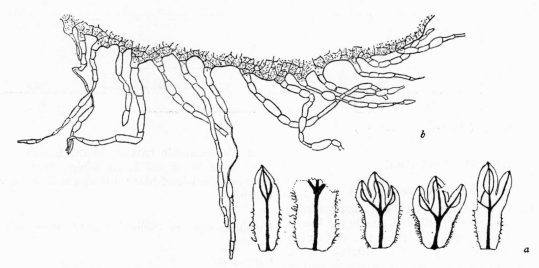

Fig. 65. *a*, Leaves of *Saxifraga rosacea* (originally referred to *Saxifraga hypnoides*) from Late Weichselian (zone III) deposits at Ballybetagh, near Dublin (× 6); *b*, the lower part of the edge of one of the leaves (× 75).

of such dwarf herbs is far distributed by wind: local movement by surface water is likely to be highly important and correspondingly the significance of the pollen record as evidence of local presence.

Saxifraga nivalis L. (Fig. 66)   239: 1

*l*W *Loch Fada, Mapastown, Loch Mealt, ?Kirkmichael, ?Ballaugh*; F IV *?Ballaugh*

All the records for *Saxifraga nivalis* are based on pollen identifications, those from the Isle of Man carried only to '*nivalis* type' (Mitchell, 1958). The records represent all zones of the Late Weichselian and early Flandrian zone IV.

S. nivalis is a local and rare species of high altitudes and wet mountain habitats, now restricted to a few Scottish sites, and very restricted occurrences in Caernarvon s. Sligo and the English Lake District. It still grows in Skye where there are two Late Weichselian records, but it has probably suffered much Flandrian extinction.

S. nivalis is placed by Hultén in the category of circumpolar arctic–alpine species.

Saxifraga stellaris L.   'Starry saxifrage' (Fig. 66) 239: 2

*l*W *Loch Fada, ?Kirkmichael, Stannon Marsh, Loch Mealt, Loch Cill Chriosd, Lochan Coir' a' Ghobhainn, Abernethy, Loch Dungeon, Widdybank Fell*; F V/VI *Loch Dungeon*, IV, VI, VIII *Widdybank Fell*

Except for the seed record from Stannon Marsh, the records for *Saxifraga stellaris* are all based on pollen identifications, and all except the Teesdale records and the tentative recognition at Kirkmichael are from Scottish sites worked by H. J. B. and H. H. Birks. They represent all zones of the Late Weichselian and transition to the early Flandrian. As with other records for *Saxifraga* spp. there is strong local continuity at individual sites: thus

Loch Fada in zones I, II and III, Loch Mealt in I and IV, Loch Cill Chriosd in zones I, II and III, Loch Dungeon in zones III/IV and V/VI, and Widdybank Fell in II, IV, VI and VIII.

Saxifraga stellaris occurs in northern Europe from Iceland to Russia and in the European mountain systems, as in those of the British Isles generally. It belongs to Hultén's group of amphi-atlantic arctic–montane plants with occurrences on mountains southwards.

Saxifraga hirculus L.   'Yellow marsh saxifrage' 239: 3

Macros: *l*W *Colney Heath, Ballaugh, Kirkmichael*
Pollen: *l*W *?Kirkmichael, ?Widdybank Fell*

There are seed records for *Saxifraga hirculus* from the Late Weichselian zone at Colney Heath (radiocarbon age 13 560 B.P.) and from zone III at Ballaugh and Kirkmichael (radiocarbon age 10 270 B.P.). The last-mentioned record is accompanied by that of pollen from zones I and II tentatively assigned to *S. hirculus*. Miss Turner's record from Upper Teesdale is recorded as *S. hirculus* or *S. hypnoides* and is from zone II.

Saxifraga hirculus is a boreal–circumpolar species with very rare and local occurrences in the British Isles, far from the sites of the fossil records. Polunin writes that in the Arctic it occurs in 'swampy areas and mossy tracts in marshes or drier patches of open clayey soil'. Many of the commonest associates of *S. hirculus* today in flushes supporting rich fen vegetation, have been found fossil with it in the Isle of Man (Dickson *et al.*, 1970).

Saxifraga sp. sect. Robertsonia   239: 4–7

H *?Baggotstown*

Watts (1964) made a tentative seed identification to *Saxifraga* sect. *Robertsonia* from sub-stage III of the

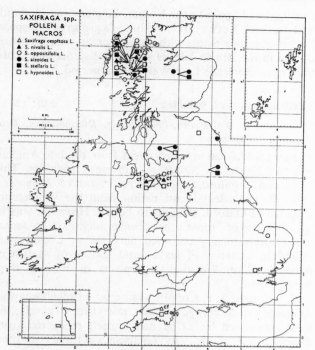

Fig. 66. Distribution in the British Isles of fossil identifications (some only tentative) of various species of *Saxifraga*. The records are variously based on pollen, seed and leaf identifications, and a large proportion of them are Late Weichselian (see text accounts for the individual species).

Hoxnian interglacial at Baggotstown, and conjectured that the fossils might belong to *S. spathularis* or alternatively to a dactyloid saxifrage.

Saxifraga hirsuta L. 'Kidney saxifrage'   239: 6

*m*W *?Great Billing*

From the Northamptonshire Weichselian site at Great Billing in deposits with a radiocarbon age of 28 220 B.P. Miss Bell has tentatively identified a seed of *Saxifraga hirsuta*.

This site is far outside the range of this Iberian species now restricted in the British Isles to shady rocks in Kerry and West Cork.

Saxifraga tridactylites L.   Rue-leaved saxifrage   239: 8

*m*W *Brandon*

Kelly has made a seed identification of *Saxifraga tridactylites* from Weichselian deposits near Birmingham that fall within the Upton Warren interstadial complex and have a radiocarbon age of 30 760 B.P.

S. tridactylites is classified by Hultén as a sub-atlantic plant and it has a southern distribution in Scandinavia, just as it is a lowland species more abundant in the southern half of the British Isles although occurring locally as far north as Caithness. It grows in xeric and

open grasslands as a rule, usually on basic soils, and these conditions were certainly present in the interstadial.

Saxifraga granulata L. 'Meadow saxifrage'   239: 9

*l*W *Nant Ffrancon, ?Nazeing*

There is an unqualified seed record of *Saxifraga granulata* from the Late Weichselian zone II in Snowdonia. Two seeds recovered from the Allerød layer (zone II) at Hawks Tor, Cornwall had already been tentatively referred to this species on the basis of size and tuberculate surface pattern.

S. granulata is fairly widely distributed in England and Wales but is limited to southern and eastern Scotland: it is absent from Ireland. It is regarded by Hultén as sub-atlantic in range and it occurs at scattered sites far north in Scandinavia though common only in the south.

Saxifraga cernua 'Drooping saxifrage'   239: 10

F IV *?Bradford Kaim*

Bartley (1966) identified pollen of 'Saxifraga cernua type' from Flandrian zone IV at Bradford Kaim, Northumberland. *S. cernua* itself is restricted to a few Scottish mountain sites above 3500 ft (1035 m) in Perth and Argyll.

Saxifraga cernua L. or Saxifraga rivularis L. 'Drooping saxifrage' or 'Brook saxifrage'   239: 10, 11

*l*W *?Whitrig Bog*

Miss Conolly has made tentative identification of a seed from the Late Weichselian zone III at Whitrig Bog, Berwickshire as either *Saxifraga cernua* or *S. rivularis*.

Saxifraga cespitosa L. 'Tufted saxifrage' (Fig. 66)   239: 12

*l*W *Nant Ffrancon, ?Hawks Tor, ?Cahercorney*

There are two seed identifications of *Saxifraga cespitosa*, both from the Late Weichselian zone II, but only that from the Nant Ffrancon is unqualified. The seed from the Allerød layer at Hawks Tor, Cornwall was only very tentatively referred to this species (Conolly *et al.*, 1950). From zone II at Cahercorney, Mitchell recorded a leafy shoot of a dactyloid saxifrage which was tentatively referred by D. A. Webb to either *S. cespitosa* or *S. rosacea*.

S. cespitosa is a plant of Arctic Europe that is limited to very few high mountain sites in the British Isles, one of which is, however, Snowdonia.

Saxifraga rosacea Moench. (Figs. 65, 66)   239: 14

*l*W *?Shortalstown, Ballybetagh, ?Whitrig Bog*

The Shortalstown record is a tentative seed identification from zone II of the Late Weichselian. The other two are from zone III.

The material from Ballybetagh (fig. 65) originally described (Jessen & Farrington, 1938) as leaves of

Saxifraga hypnoides sensu lato, was re-examined by D. A. Webb who reported: 'I am almost certain that these remains may all be ascribed to the rather polymorphic species, *S. rosacea* (Moench.).' (See also the account of *S. cespitosa*.)

S. rosacea is a mountain plant now found in the British Isles only in western Ireland, where, however, it may be abundant and where it ascends to 3500 ft (1060 m): it is extinct in Caernarvonshire. It occurs in Iceland, the Faeroes and the mountains of south-central Germany, west Czechoslovakia and in the Vosges.

Saxifraga hypnoides agg. Dovedale moss (Fig. 66; plate XX*f*)   239: 15

Pollen: *l*W *Loch Fada, Loch Mealt, Loch Cill Chriosd, Moss Lake, Lochan Coir' a' Ghobhainn, Loch Dungeon*; F IV *Loch Cill Chriosd, Loch Fada, Lochan Coir' a' Ghobhainn*

Macros: *l*H ?*Wolvercote*; W *Thrapston*; *m*W *Earith*; *l*W ?*Kirkmichael, Hawks Tor,* ?*Nazeing,* ?*Ballybetagh, Whitrig Bog*

Pollen of *Saxifraga hypnoides* has been recorded by H. J. B. Birks from all zones of the Late Weichselian at sites in Skye where the same pollen type persists from zone to zone. Thus at Loch Fada and Loch Cill Chriosd it is in zones I, II, III and IV, at Lochan Coir' a' Ghobhainn in I/II, III/IV and IV, and at Loch Mealt in zones I, II and IV. Miss Andrew had previously identified pollen at Moss Lake, Liverpool (zone II) and Mrs Birks recognized it from the zone III/IV transition at Loch Dungeon in Galloway.

Apart from the tentative record from the late Hoxnian, macroscopic remains shew an even greater restriction to the Weichselian. Miss Bell's seed identifications from Thrapston and Earith are to the '*S. hypnoides* type' that includes *S. hypnoides, S. rosacea, S. granulata* and *S. hirsuta*, and the Kirkmichael identifications (zones I, II and IV) and those from Nazeing and Ballybetagh (zone III) are all tentative.

Jessen and Farrington describe the discovery in zone III at Ballybetagh of several leaves which because of their form and nervation and the characteristic hairiness on the margins of their lower parts must be referred to *Saxifraga hypnoides* agg. They suggest that a small saxifrage fruit from the same sample may also belong to the same species. A single seed found in zone III deposits at Nazeing, Essex, was tentatively referred to the same species on the basis of 'the arrangement of the blunt spines into longitudinal rows'. Webb suggests that these Ballybetagh remains all probably refer to *S. rosacea* Moench (see preceding account).

S. hypnoides has a distinctly northern distribution in the British Isles. It is found widely in Wales and the border counties, Scotland and northern England (reaching 4400 ft (1340 m) on Ben Nevis); it has its southern limit on the Pennines of Staffordshire and Derbyshire, with a single isolated station in Somerset. Outside this country it is found in the Faeroes, and in western Europe from

Portugal and Spain to the Vosges, Belgium and Holland, and southern Norway. It is in Hultén's group of European boreal–montane plants with disjunct areas, and the distribution in the British Isles is similarly disjunct and probably relict from the Late Weichselian period.

Saxifraga sp. sect. **sedoides**   239: 12–15

*l*A *Hoxne*; H *Wolvercote*; *l*W *Whitrig, Dunshaughlin, Drumurcher,* ?*Mapastown*

Turner (1968*a*) has referred a leaf from the late Anglian level at Hoxne to *Saxifraga* sub-genus *sedoides*. This includes *S. cespitosa, S. hastii, S. rosacea* and *S. hypnoides*. Miss Duigan's leaf identification from the Wolvercote Channel at Oxford, thought to be late Hoxnian, was to *S. hypnoides* or 'some other species such as *S. rosacea*' and is conveniently recorded here. A. P. Conolly has abundant material of leaves and shoots of saxifrages from the Late Weichselian deposits of Whitrig Bog, Berwickshire.

Having seen A. P. Conolly's material, Mitchell was able to identify as *Saxifraga* sp., sect. *dactyloides* Tausch., the badly preserved leaves he had himself recovered from different zones of the Late Weichselian at three Irish sites, viz. Drumurcher and Dunshaughlin zone IV and Mapastown zones II and III. All the British species formerly placed in sect. *dactyloides* now are included in sect. *sedoides*.

Webb (1950) indicates that all species in the section are more or less drought susceptible chamaephytes that hybridize readily among themselves. Of the European geographical ranges one can only generalize by saying that the British species tend to be northern and disjunct.

Saxifraga aizoides L. 'Yellow mountain saxifrage' (Fig. 66)   239: 16

*l*W *Loch Fada, Loch Meodal, Loch Cill Chriosd,* ?*Longlee Moor,* ?*Mapastown, Cooran Lane, Abernethy, Loch Dungeon, Widdybank Fell*; F IV *Allt na Feithe Sheilich, Loch Cill Chriosd,* ?*Longlee Moor*

All the records for *Saxifraga aizoides* are based upon pollen identifications tentative for Longlee Moor and Mapastown (see *S. oppositifolia*). With the exception of these and that from Upper Teesdale, all are Scottish though they embrace Galloway, Skye and the Cairngorm: these are given by H. J. B. and H. H. Birks. As with other records for *Saxifraga* pollen there is much persistence from zone to zone; thus at Loch Cill Chriosd there are zones I, II, III and IV represented, at Loch Fada zones I and II, and at Longlee Moor zones II and IV. The transition zone III/IV is represented by the records from Loch Dungeon, Cooran Lane and Abernethy.

Saxifraga aizoides is placed by Hultén in his group of amphi-atlantic arctic–montane plants with occurrences on southern mountains. It is a mountain plant of wet situations and locally common, especially in Scotland north of the Highland line: it occurs in the Lake District, northern Pennines and locally in western Ireland and Antrim, but it seems no longer to grow in the Galloway mountains which yielded the Loch Dungeon and Cooran Lane records. Terasmae has recorded pollen of *S. aizoides* from the Late Weichselian of southern Sweden.

Saxifraga oppositifolia L. 'Purple saxifrage' (Fig. 66; plate XXVIII*a*)   239: 17

Macros: *m*W *Earith, Barnwell St.*; *l*W *Shortalstown, Mapas town*
   Pollen: *l*W *East Walton, Loch Craggie, Loch Borralan,* ?*Kirkmichael, Loch Fada, Loch Cill Chriosd, Loch Mealt,* ?*Ballaugh, Loch Sionascaig,* ?*Mapastown*; F IV ?*Nazeing, Red Moss*

A leaf, shoot and seed of *Saxifraga oppositifolia* were identified by Miss Chandler from the Middle Weichselian deposits at Barnwell Station and Miss Bell also recovered a leafy shoot and seed from deposits at Earith. Leaf identifications were made from the Late Weichselian zone I/II transition at Shortalstown, and Mitchell recovered from zone III at Mapastown a leaf and shoot together with a pollen grain referable to *S. oppositifolia* or *S. aizoides*.

   Several research workers have recently reported pollen of *Saxifraga oppositifolia* from all zones of the Late Weichselian, mostly, but not exclusively, from northern sites. As with pollen referred merely to the genus, so here also there is clear evidence for local persistence at individual sites; thus Loch Craggie (Sutherland) in zones I, II and III, Loch Fada (Skye) in zones I and III, Loch Mealt (Skye) in zones I and II, and Loch Cill Chriosd in zones I, II and III. The tentative identifications from Ballaugh (Isle of Man) have been made on different occasions, and are accompanied by similar determinations from Kirkmichael. The species no longer grows in the Isle of Man.

Chrysosplenium oppositifolium L. 'Opposite-leaved golden saxifrage'   242: 1

F VIIa/b *Shustoke*, VIIb *Frogholt*

There are two Flandrian records for the seeds of *Chrysosplenium oppositifolium*, the one from Shustoke, Warwickshire with a radiocarbon age of 4830 B.P., and the other from Frogholt, Kent with a radiocarbon age of 2860 B.P. In each case we are concerned with alluvial deposits that indicate ecological conditions of dampness and shade.

   C. oppositifolium occurs throughout the British Isles. It is a European-atlantic plant (Hultén), limited in the Scandinavian peninsula to the south-west coast of Norway.

Chrysosplenium alternifolium L. 'Alternate-leaved golden saxifrage'   242: 2

H *Nechells*

From sub-stage III of the Hoxnian interglacial at Nechells (Birmingham) Kelly recovered a seed of *Chrysosplenium*

alternifolium with characteristic lateral rib and shiny black surface with regular reticulate pattern. It is a plant of similar habitat to *C. oppositifolium* but much less frequent in the British Isles today and absent from Ireland.

PARNASSIACEAE

Parnassia palustris L.   Grass of Parnassus   243: 1

Macros: *e*Wo *Brandon*; *m*W *Upton Warren*; *l*W *Kirk michael, White Bog*
   Pollen: *l*W ?*Canna, Esthwaite, Elan Valley*; F IV *Elan Valley, Dogger*

Seeds of *Parnassia palustris* have been recognized from the early Wolstonian glacial stage, from the Middle Weichselian, and from zones I and II of the Late Weichselian. The pollen has been tentatively recognized from the Late Weichselian zone II in Canna, and from zone III in the Esthwaite Basin and the Elan Valley. In the Flandrian zone IV it recurs in the Elan Valley and it is found also in the North Sea submerged peat.

   Parnassia palustris is one of Hultén's boreal–circumpolar plants of almost continuous distribution; it is common throughout Scandinavia to the extreme north and it is widespread, if local, throughout the British Isles in wet moors and in marshes such as must have been abundant in periglacial Britain.

GROSSULARIACEAE

Ribes sp.   246

C ?*West Runton*

Miss Duigan made a provisional identification of pollen of *Ribes* from the Cromerian interglacial bed at West Runton, Norfolk.

DROSERACEAE

Drosera sp.   Sundew   247

Since it became practicable about twenty years ago to identify the very characteristic finely spinous pollen tetrads of *Drosera*, British records for the genus have accumulated despite the fact that the plants are dwarf, entomophilous and seldom more than locally abundant. The record-head diagram shews increasing abundance in the Flandrian from zone VI onwards, corresponding in large degree to the initiation and spread of ombrogenous peat bogs, especially after the Boreal–Atlantic transition, the zone VI/VII boundary. The earlier records are still sparse, representing however both Hoxnian and Ipswichian interglacials. There is also a record from zone II of the Late Weichselian, which is unsurprising in view of the high Scandinavian latitudes attained by the two

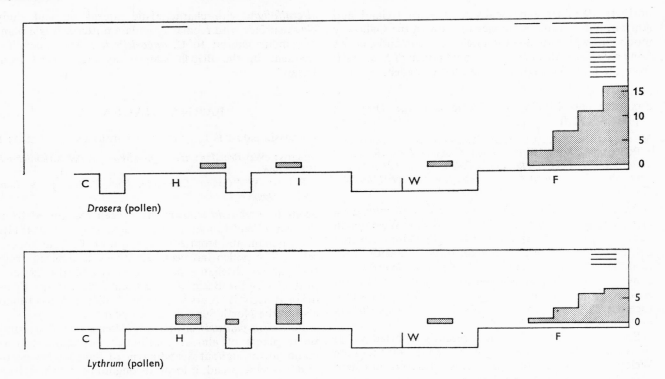

Drosera (pollen)

Lythrum (pollen)

British species *Drosera rotundifolia* and *D. anglica*, and the occurrence of all three British species throughout these islands.

Drosera intermedia Drev. & Heyne. 'Long-leaved sundew'   247: 3

F IV ?*Loch Cill Chriosd*, VII/VIII *Shapwick Station*

In a peat cutting near Shapwick Station, Somerset between two layers of *Cladium mariscus* peat, oligotrophic Sphagnum peat re-established itself during the Early Iron Age. With a highly typical raised bog flora, which included *Calluna vulgaris*, *Erica tetralix*, *Andromeda polifolia*, *Oxycoccus palustris*, *Rhyncospora alba*, *Eriophorum vaginatum* and various *Sphagna*, were found seeds of *Drosera intermedia*.

On Skye H. J. B. Birks has tentatively referred to *D. intermedia* pollen from Flandrian zone IV.

Drosera intermedia has a pronounced tendency to greater abundance in the west of the British Isles as it has also in southern Scandinavia.

LYTHRACEAE

Lythrum sp.   249

Three features suggest that pollen recorded as '*Lythrum* sp.' in British pollen analyses represents *Lythrum salicaria* rather than the other British species *L. hyssopifolia*: the pollen morphology itself; the present relative abund-

ance of the two species; and the habitats indicated by the pollen diagrams, most commonly *Magnocaricetum* or marginal reed-swamp environment. Accordingly we discuss these records in the account that follows for *L. salicaria*.

Lythrum salicaria L.   Purple loosestrife   249: 1

Pollen: H *Kilbeg*; F VI *Nant Ffrancon*, VIII *Bowness Common*
    Macros: H *Clacton*; F VIIa *Westward Ho!*

The pollen recorded above adds a Hoxnian sub-stage III record to two given for *Lythrum* sp. The seed identifications are from sub-stage III of the Hoxnian, and, for the Mesolithic, zone VIIa level at Westward Ho! where pollen of *Lythrum* was also recorded. If the record-head diagram for *Lythrum* sp. indeed represents *L. salicaria* this species was evidently present in both the Hoxnian and Ipswichian interglacials, but appeared only late in the Flandrian, increasing in its site records from zone VI to zone VIII. A rather large number of the Flandrian records are associated with an archaeological context, thus there are Mesolithic I, Late Bronze Age 4, Iron Age I, Roman I, perhaps a consequence of settlements being made beside water. The pollen records also shew much local persistence from zone to zone, thus Tollgate (VIIa, VIIb, VIII), Drake's Drove (VIIb, VIII) and Glastonbury Lake Village (VIIb, VIII). A different order of recurrence is shewn at Hoxne with records in Hoxnian sub-stage III, the early Wolstonian and Ipswichian sub-stage II. The records are certainly most abundant from southern England and the northernmost are in Kintyre and West-

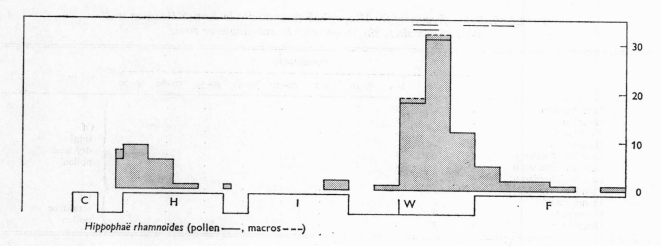

Hippophaë rhamnoides (pollen ——, macros – – –)

morland. This seems consonant with the present occurrence of the plant which is infrequent in Scotland and absent from much of the north, although common in England, Wales and Ireland. In Scandinavia too, it is common only in the south, although in scattered localities it attains the Arctic Circle.

Lythrum hyssopifolia L.   Grass poly   249: 2

F VIIb *Minnis Bay*

Miss Conolly (1941) recovered a macrofossil from the Late Bronze Age peat on the shore at Minnis Bay, Kent which unquestionably consisted of the calyx, style and capsule of *Lythrum hyssopifolia*.

It is interesting to find evidence of the presence of this very local and southerly British species from material as old as the Late Bronze Age and here associated with weeds of cultivation. Subsequently Miss Conolly has expressed the doubt that modern material may to some extent have contaminated the site.

Decodon sp. J. F. Gmel

I *Shortalstown*

Seeds referable to the genus *Decodon* are known in Europe from the Pliocene to the Lower Pleistocene, but these are assigned to an extinct species, *D. globosus*. A single seed recovered by Colhoun and Mitchell (1971), from what appears to be sub-stage II of the Irish Ipswichian interglacial at Shortalstown, closely resembled modern American specimens of *Decodon verticillatus*, the single living species of the genus. *Decodon* has been identified with certainty from the Waalian (late Lower Pleistocene) of the Netherlands and it seems therefore quite possible that it survived in Western Europe as late as the last interglacial. *D. verticillatus* is a plant of marshes and river margins.

ELEAGNACEAE

Hippophaë rhamnoides L. Sea buckthorn (Fig. 67; plates XXIII *e–i*, XXVI*a*)   252: 1

As early as 1924 von Post had identified fossil pollen of *Hippophaë rhamnoides*, and after a period in which it was held confusable with pollen of *Fagus*, it has been regularly and commonly recognized. The spheroidal pollen grains are moderately large and tricolpate with narrow tapering furrows and scabrate exine.

Throughout north-west Europe it has become apparent that *Hippophaë* was widespread in Late Weichselian time and particularly frequent, according to Firbas, at the transition to the Flandrian when the first great spread of woodlands occurred. Sandegren (1943) has presented a most instructive account of the occurrence of fossil *Hippophaë* in Scandinavia. From this it appears that in the Late Weichselian period, when the ice margin stood at the Central Swedish moraine line, *Hippophaë* was prevalent throughout the ice-free country to the south. When the final rapid retreat of the ice from this position took place in Pre-boreal and Boreal time *Hippophaë* spread, as forerunner of the forests, throughout Scandinavia and far into the north. With subsequent closure of the forests it became restricted to coastal and valley habitats where competition was less severe. In the south the coastal submergence diminished such habitats and the sea buckthorn, unable to penetrate the hinterland of closed forest, died out there. Further north along the Baltic, the constantly emerging coastline provided a wide belt of country for colonization exploited during the Sub-atlantic period, during which time also it immigrated into the newly emerged Åland islands. Thus the present distribution of the plant is explicable in terms of a former more extensive occupation area, not all of which was however occupied at any one time. Faegri has emphasized that even today *Hippophaë* has numerous localities in Norway in mountain habitats where tree competition is small, and it may well now characterize coastal dunes

TABLE 16. *Pollen frequency for* Hippophaë rhamnoides *in British Hoxnian and Weichselian sites: site frequencies by sub-stages or zones*

	Percentages										
	+	0–2	2–5	5–10	10–20	20–30	30–40	40–50	50–60	60–70	
Late Anglian	.	.	1	3	.	.	1	.	.	1	Of total dry land pollen
Hoxnian I	.	3	1	.	.	.	.	.	.	.	
Hoxnian II	.	3	.	.	.	.	.	.	.	.	
Hoxnian III	1	.	.	.	.	.	.	.	.	.	
Late Weichselian I	3	10	1	.	.	.	.	.	.	.	
Late Weichselian II	9	11	1	.	.	.	.	.	.	.	
Late Weichselian III	3	7	.	.	.	.	.	.	.	.	
Flandrian IV	1	2	.	.	.	.	.	.	.	.	
Flandrian IV	.	2	.	.	.	.	.	.	.	.	Of total tree pollen
Hoxnian I	1	4	.	.	.	.	.	.	.	.	
Hoxnian II	1	3	.	.	.	.	.	.	.	.	

simply because they supply such an exceptionally large part of our open native habitats on loose penetrable soils.

The Continental pollen records are supplemented by frequent identifications of the very characteristic stellate hairs of the leaves and stems, but these have so far only been recognized in the British Isles at the interglacial sites at Hoxne, Marks Tey and Nechells, and in zones I and II of the Late Weichselian at Roddansport on the coast of Antrim.

Not only do the British records confirm those for the Continental mainland, but, as the record-head diagram indicates, they shew that the sea buckthorn was very strongly represented in deposits associated with the Hoxnian interglacial and even more so with the late Anglian which immediately preceded it, during which time not only were both macroscopic and microscopic remains present together but the pollen occurred in high frequency. West has reported in two zones of the late Anglian at Hoxne pollen frequencies of over 30 and over 60 per cent of the total land plant pollen and at three other sites there were frequencies of 5 to 10 per cent. In view of the low pollen productivity of this entomophilous shrub one must deduce great densities of *Hippophaë* scrub in these situations. Sites of recorded *Hippophaë* are very numerous in sub-stage I of the Hoxnian itself and diminish towards the middle Hoxnian, no doubt with the closure of the woodlands.

There is a single early Wolstonian record but, in strong contrast to the Hoxnian, the Ipswichian provides only two very low frequency records and three for sub-stage IV at the approach of the Weichselian glacial stage. There is a single Middle Weichselian record from Badger Hole, with a radiocarbon age of 18 000 years; but with the Late Weichselian the site records become very abundant, more especially in zone II, and widespread over the country (fig. 67) though, as table 16 shews, pollen frequencies remain generally low. With the dwarf willows and dwarf birch the *Hippophaë* clearly was an important component of a shrub vegetation, no doubt locally dense in all inland periglacial sites. It had become less frequent in

zone III and diminished much more as forests closed in the early Flandrian; it failed to persist in mountain sites and remained solely along our shores.

H. rhamnoides occurs on eastern British coasts from Berwickshire to Sussex and on European coasts from Atlantic Norway at about 68° N, and the whole of the Baltic down to the English Channel; it also occurs in central Europe on river shingle, on the Black Sea coast, in temperate Asia and throughout the Himalayas. Its presence in quantity in periglacial situations is sometimes held to imply continentality of climate (West, 1968).

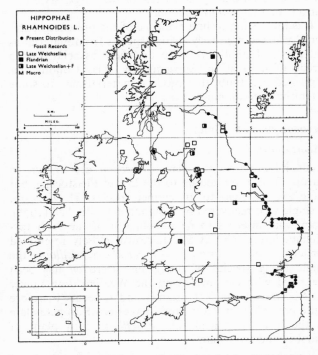

Fig. 67. Present and former distribution in the British Isles of *Hippophaë rhamnoides* L. The inland occurrences in the Late Weichselian contrast with the coastal restriction of today.

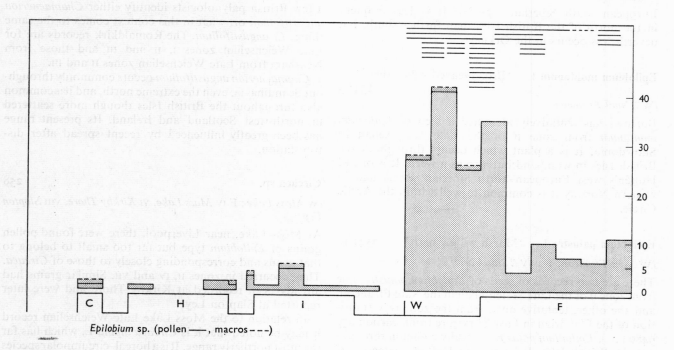

Epilobium sp. (pollen ——, macros - - -)

ONAGRACEAE

Epilobium sp. Willow-herb   254

Pollen grains of the genus *Epilobium* are very easily recognizable from their large pores and large size, but apart from a short list of records for *Chamaenerion angustifolium* (see following page) there are no pollen records for distinct species. There is a tendency among palynologists in north-west Europe to consider that the *Epilobium* pollen found so consistently in Late-glacial deposits is in part that of *Chamaenerion angustifolium*. Colour is given to this view by the prevalence of this species along with *C. latifolium* in arctic and sub-arctic regions and by the regularity with which Wenner (1947) found pollen of *Chamaenerion* in his analyses of surface samples amongst the present-day vegetation of north Labrador. We may certainly assume that the record-head diagram for *Epilobium* above includes *Chamaenerion*, both in cases where the author has deliberately indicated this and in cases where he has not. It is perhaps not remarkable that a genus with so many species and of such differing ecological requirements should be represented through the interglacials and all the Flandrian zones, but the great abundance of records in all zones of the Late Weichselian and zone IV of the Flandrian is extremely striking. Periglacial edaphic conditions must have been highly suitable to some species at least, and it will be seen that there are indeed specific Late Weichselian records for at least five of them.

Epilobium parviflorum Schreb. 'Small-flowered hairy willow-herb' (Fig. 68)   254: 2

*l*Wo *Ilford*; I *Wretton*; *l*W ?*Lopham Fen, Hockham Mere*; F VIIb *Frogholt*, VIII *Apethorpe*

All records for *Epilobium parviflorum* are based upon seed identifications. They include a late Wolstonian and an Ipswichian sub-stage II record. There are qualified identifications from both zone II and III of the Late Weichselian at Lopham, and not far away a single seed was found at Hockham Mere in the time span from Late Weichselian zone II to Flandrian zone IV (fig. 68). There are no further records until one at Frogholt from the late Bronze Age level (radiocarbon date 2860 B.P.) and another, tentative one at the same site but a higher level (radiocarbon date 2490 B.P.). The latest is the zone VIII record from Apethorpe. There is thus an indication of ancient indigeneity in the record and the Late Weichselian records are of special interest since *E. parviflorum* has a range in Scandinavia scarcely extending north of 60° and thinning out westwards: it is in Hultén's category of 'west

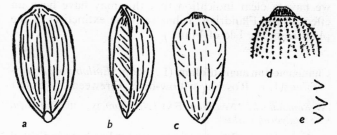

Fig. 68. Seed of *Epilobium* cf. *parviflorum* Schreb. from Late Weichselian or early Flandrian muds at Hockham Mere, Norfolk: *a*, front; *b*, side; *c*, back (all × 33); *d*, detail of back; *e*, spines.

European–south Siberian' species. It is less frequent in the north of the British Isles than further south, but nevertheless occurs in the Orkneys.

Epilobium montanum L.   'Broad-leaved willow-herb'
254: 3

*l*W ?*Nant Ffrancon*

Burrows has tentatively identified a seed of *Epilobium montanum* from zone II of the Late Weichselian in Snowdonia. It is a plant which occurs throughout the British Isles in woodland but also as a weed. It is one of Hultén's west European–north Siberian plants and in western Norway it is common to well within the Arctic Circle.

Epilobium palustre L.   'Marsh willow-herb'   254: 10

*l*W *Nant Ffrancon*; F IV/V ?*Star Carr*

There are two records of seeds of *Epilobium palustre*, one from zone II of the Late Weichselian in the Nant Ffrancon, and the other, tentative only, from the zone IV/V transition of the Flandrian in East Yorkshire (radiocarbon age 9480 B.P.). *Epilobium palustre* is locally common throughout the British Isles. It belongs to Hultén's category of boreal–circumpolar plants of largely continuous distribution and it is common throughout Scandinavia to the extreme north. Its calcifuge tendency apart, its range suggests the qualities of a Weichselian plant.

Epilobium alsinifolium Vill.   'Chickweed willow-herb'
254: 12

*l*W *Kirkmichael*, ?*Shortalstown*

There is an identification of *Epilobium alsinifolium* by Mrs Dickson from zone I of the Late Weichselian in the Isle of Man based upon careful study of the form and epidermal features of twelve seeds. There is also a tentative seed recognition from zone II in Ireland. *Epilobium alsinifolium* is in the category of European arctic–montane plants missing from the intervening lowlands. It occurs scattered throughout Scandinavia north of about 59° N and in the British Isles at scattered mountain sites chiefly in Scotland but also in northern England, Caernarvonshire and north-western Ireland. Like the European, the British range of *E. alsinifolium* appears very disjunct and we have a clear indication that this may have been an effect of the Flandrian: it has seen the extinction of the plant from the Isle of Man.

Chamaenerion angustifolium (L.) Scop. (Epilobium angustifolium L.)   Rosebay willow-herb, Fireweed   255: 1

*l*W *Romaldkirk*, ?*Neasham*; F VI ?*Neasham*, IV, VIIb *Red Tarn*, VIIb *Ladybridge Slack*

In the account of *Epilobium* (p. 209) it has been indicated that there is good evidence that the pollen recorded for that genus contains much from *Chamaenerion*. However,

a few British palynologists identify either *Chamaenerion angustifolium* or, what in this context comes to the same thing, *E. angustifolium*. The Romaldkirk records are for Late Weichselian zones I, II and III and those from Neasham from Late Weichselian zones II and III.

Chamaenerion angustifolium occurs commonly throughout Scandinavia, even the extreme north, and it is common also throughout the British Isles though more scattered in north-west Scotland and Ireland. Its present range has been greatly influenced by recent spread after disforestation.

Circaea sp.   258

*l*W *Moss Lake*; F IV *Moss Lake*, VI *Kirkby Thore*, VIII *Slapton Ley*

At Moss Lake, near Liverpool, there were found pollen grains of *Epilobium* type but far too small to belong to that genus and corresponding closely to those of *Circaea*. They occurred in zones II, IV and VII. Similar grains had already been recorded at Kirkby Thore and were later reported at Slapton Ley.

In relation to the Moss Lake Late Weichselian record it may be noted that it is *Circaea alpina* L. which has far the most northerly range. It is a boreal–circumpolar species which remains common in western Norway well within the Arctic Circle, whereas *C. lutetiana* and *C. intermedia* (the putative hybrid between the other two) have Scandinavian ranges barely attaining 62° N and then only in the west.

Circaea lutetiana L.   Enchanter's nightshade   258: 1

Cs *Pakefield*; *l*W *Loch Droma*; F VIIa ?*Westward Ho!*

The record of *Circaea lutetiana* from the Cromer Forest Bed series at Pakefield is based on macroscopic evidence. That from the Mesolithic site at Westward Ho! is based on pollen that was associated with a rich flora of dry fenwood character, such as the plant might accompany today. The Flandrian records given under '*Circaea* sp.' may well be referable to *C. lutetiana*, by far the commonest of the British species at the present day though less frequent in the northern half of Scotland than further south. *C. alpina* might be involved with the Weichselian record.

Trapa natans L.   Water chestnut

Cs *Mundesley, Pakefield, Ostend, Sidestrand, Kessingland, Bacton*; H *Quinton*; I *Wortwell, Trafalgar Sq.*

The highly characteristic fruits of *Trapa natans* L., the water chestnut, were described by C. Reid from several sites of the Cromer Forest Bed series, by the Rev. C. Green from Bacton, and fruits or spines were found later by Mrs Wilson from Pakefield and Ostend either in the Cromerian *s. stricto* or the Pastonian. It was found by Bell (1970) in the Hoxnian at Quinton along with other extinct species. Subsequently West identified the fruit

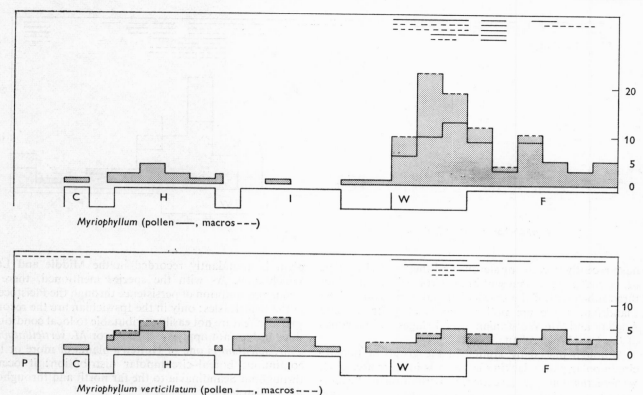

Myriophyllum (pollen ——, macros - - -)

Myriophyllum verticillatum (pollen ——, macros - - -)

spines of *Trapa* from sub-stage II of the Ipswichian inter-glacial at Wortwell, Norfolk together with *Naias minor* and *Salvinia natans*, all these species absent from the British Flandrian but present and fruiting on the Continental mainland. *Trapa* and *Naias minor* were also found by Franks (1960) in the same sub-stage of the Ipswichian at Trafalgar Square, London.

The fossil record of *Trapa natans* in Scandinavia is one of the lines of evidence establishing the record there of a Post-glacial Climatic Optimum and subsequent deterioration (Malmström, 1920; Samuelsson, 1934). The evidence here is of fruits found far north of the present range of a plant that seems to require high summer temperatures in order to bear fruit. Although the pollen grains are highly characteristic they are recorded only sparsely from the middle Flandrian of Scandinavia: they have been recently reported from the Dutch 'Uddeler Meer' deposits in the same period but they were absent from the following Sub-atlantic time (Polak, 1959).

The British interglacial records are paralleled by similar ones from Denmark and north-west Germany.

HALORAGACEAE

Myriophyllum sp.   Water-milfoil   259

There are numerous instances in which investigators have been unable to identify pollen, leaves, fruits or seeds

of *Myriophyllum* as belonging to particular species. The evidence set out in the record-head diagram shews the genus to have been present in the four last interglacial stages, including all zones of the Flandrian. There are two records each respectively in the late Anglian and early Wolstonian, but it is in the Late Weichselian that records abound and here macroscopic and pollen records contribute an equal weight of evidence. The record is best viewed in the light of that for the individual records for the three British species of *Myriophyllum* that follow.

It is of interest that these generic identifications include the comment by W. Tutin (Pennington, 1947) that in the Allerød layer of Windermere the *Myriophyllum* leaves were the most abundant of all plant remains, in places forming almost pure deposits: she favoured the view that they were most likely *M. alterniflorum*. Dr Blackburn's records for Neasham, Co. Durham, shew that all three British species of *Myriophyllum* were present together there in the Allerød period.

Myriophyllum verticillatum L.   'Whorled water-milfoil'
(Plates XVII*w*, XXI*t, u*)   259: 1

Macroscopic records for *Myriophyllum verticillatum* are few and scattered but nevertheless they establish presence of the plant in the Cromer Forest Bed series, the Hoxnian and Ipswichian interglacials, the Anglian and Wolstonian glacial stages as well as the Middle and Late Weichselian and the early and middle Flandrian. Pollen identifications

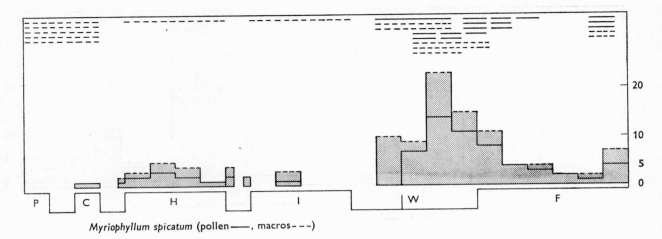

Myriophyllum spicatum (pollen——, macros---)

have recently become numerous and they now shew that *M. verticillatum* was present in the Cromerian, in the first three sub-stages of the Ipswichian and every zone of the Flandrian. There are pollen records also for the late Anglian and late Wolstonian glacial stages and all zones of the Late Weichselian.

Although placed by Hultén in the group of boreal–circumpolar plants lacking large gaps in their areas, *M. verticillatum* has only a scattered distribution in Scandinavia north of Denmark, and few of these local occurrences extend above the Arctic Circle. Samuelsson writes of its scattered outpost distribution as relict. There can be no doubt from the fossil record that *M. verticillatum* is anciently indigenous in the British Isles, and as with *Hippuris vulgaris*, it may well have been present all through the Pleistocene.

Myriophyllum spicatum L.　‘Spiked water-milfoil’
259: 2

It is very remarkable, from the record-head diagram, how consistently the records based upon macroscopic remains (mostly fruit) and pollen support one another, possibly a reflection of the occurrence of both in muds from the lakes or pools in which the plant grew. Only in the Middle Weichselian do the fruit records stand alone, but pollen analyses have been very few in such deposits. Records from broad or transition zones moreover fully support those with more exact age attribution and provide five records from the Cromer Forest Bed series. It is to be noticed that the seed and bract records given separately for forms ‘*squarrosum*’ and ‘*muricatum*’ have been included in the record-head diagram above: they are few and come from zones II, III and IV.

The results as a whole are strikingly similar to those of *M. verticillatum* and of *Hippuris vulgaris* in that they demonstrate presence in the four latest interglacial stages and in the glacial stages separating them: moreover although records are absent (as are plant-bearing recorded sites) from the middle sub-stages of the early glacials, the

plant is abundantly recorded in the Middle and Late Weichselian. As with the species mentioned, there is strong presumption of persistence through the Pleistocene in the British Isles; only in the Ipswichian are the records sparse, a feature not easily attributable to local conditions since it does not apply to *Hippuris* or *M. verticillatum*.

M. spicatum is one of Hultén’s plants of more or less continuous boreal–circumpolar distribution: it occurs throughout Scandinavia to the far north and throughout the British Isles.

Myriophyllum spicatum f. squarrosum Laestad (Fig. 69*a,b*)
259: 2*a*

*I*W *Kilteely, Ballybetagh, Ratoath*

In their Ballybetagh paper, Jessen and Farrington recorded bracts of *Myriophyllum verticillatum*, but Jessen later corrected this to *M. spicatum* f. *squarrosum*, pointing out that the broad, firm and close segments correspond to those of scale-leaves from the lower part of the stalk of the form mentioned (fig. 64). Similar scale-leaves were found at Kilteely, and Jessen writes: ‘As a consequence of this being cleared up *M. verticillatum* has been erased from the list of Ireland’s late and post-glacial flora.’ Since Mitchell (1951) records *M. spicatum* and not *M. verticillatum* from Ratoath, instead of the converse as origin-

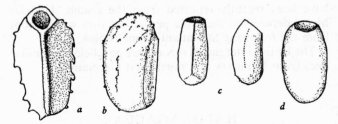

Fig. 69. Remains of aquatic plants from Irish deposits (Jessen, 1949): *a*, *Myriophyllum spicatum* f. *squarrosum*, nut from zone III (Dunshaughlin); *b*, *Myriophyllum spicatum* f. *squarrosum*, nut from zone II (Ballybetagh); *c*, *M. alterniflorum*, nuts in ventral and lateral aspect, zone II; *d*, *Hippuris vulgaris*, fruit stone, zone VI.

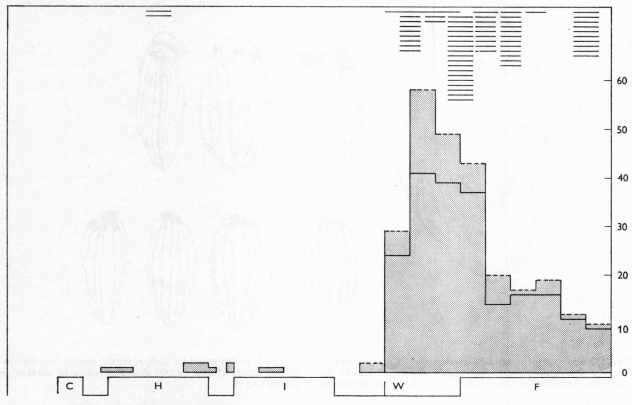

Myriophyllum alterniflorum (pollen ——, macros – – –)

ally recorded, we take it that a similar change has been made here. Jessen indicates that in Scandinavia and Finland the *squarrosum* form seems to occur particularly in the more northerly and higher regions.

The records all come from the Late Weichselian or early Flandrian (Kilteely zones II–IV, Ballybetagh II, IV and Ratoath II, III). The bracts were accompanied by fruits at Ballybetagh and by fruits and pollen at Ratoath.

Myriophyllum spicatum f. muricatum Ahlf.   259: 2b

*l*W *Neasham*

Miss Blackburn (1952) recovered from the organic muds of zone II at Neasham, near Darlington, fruits that she referred without comment to *Myriophyllum spicatum* f. *muricatum*.

Myriophyllum verticillatum or spicatum   259: 1, 2

*l*Wo *Ilford, Selsey*; I *Ilford, Selsey, Bobbitshole, Wretton Histon Rd, eW Sidgwick Av.*; mW *Syston, Broxbourne*; *l*W *Nazeing, Colney Heath, Roddansport*; F IV *Roddansport*

From the time of C. Reid there has been difficulty in separately identifying the fruits of *Myriophyllum verticillatum* and *M. spicatum*, and workers, particularly West (responsible for all the Wolstonian and Ipswichian records), have often preferred an alternative identifica-

tion. Especially in this period the records are repetitive through the sub-stages, thus Ilford (*l*Wo; I, I and II) Selsey (*l*Wo; I, and II). The Ipswichian is well represented in sub-stages I and II and it is present in IV, a feature perhaps to be associated with the infrequency of definitive records for *M. spicatum* in the same interglacial. The Late Weichselian records are either for zone I or the zone I/II transition or for both, as at Nazeing where shoot apices accompanied the fossil fruits. This accords with the heavy representation of *M. spicatum* in the late Weichselian. There is one Flandrian zone record from Roddansport following one from the zone I/II transition.

Myriophyllum alterniflorum DC.   Alternate-flowered water-milfoil (Figs. 69c, 70)   259: 4

The fruits of *Myriophyllum alterniflorum* DC. are recognizable, through their small size and lack of tubercles, from fruits of other species of *Myriophyllum*, and the pollen is very strikingly distinct in that it has two pairs of extremely protuberant, heavily ringed pores on each grain (see fig. 70b): the pollen grains of the other British species are three-pored and the pores are comparatively little emphasized. In the Hawks Tor Late-glacial deposits were also found the male flowers and the very characteristic anther cluster, the stamens full of easily recognizable pollen (fig. 70). Leaves alone are more difficult to refer to

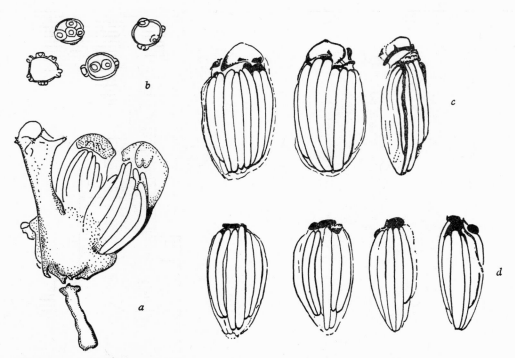

Fig. 70. Remains of *Myriophyllum alterniflorum* DC. recovered from the Late Weichselian muds and kaolin gravels at Hawks Tor kaolin pit, Bodmin Moor, Cornwall: *a*, flower with anther clusters ($\times 25$); *b*, pollen grains ($\times 500$); *c*, dorsal, ventral and lateral views of one anther cluster ($\times 25$); *d*, similar views of another anther cluster ($\times 25$).

separate species (but see comment under *M. spicatum* f. *squarrosum*).

As the record-head diagram shews, there are a few records from the Hoxnian interglacial, the preceding late Anglian and succeeding early Wolstonian: these are all Irish records (Gort, Kilbeg and Newtown). East Anglia supplies two pollen records from the late Wolstonian (Ilford and Bobbitshole), and there is an Ipswichian sub-stage II record from Trafalgar Square. The great bulk of records come however from the Late Weichselian and Flandrian with fruit records corresponding closely to those for pollen and transition zone records giving a largely congruent picture save that they shew high values at the III/IV and VII/VIII boundaries, at both of which renewed waterlogging provided abundant new habitats.

Like *M. spicatum*, *M. alterniflorum* is described by Samuelsson (1934) as one of the most ubiquitous fresh-water plants of Scandinavia: it grows commonly well within the Arctic Circle and is locally present in northern Lapland. It is generally regarded at the present day as exhibiting preference for oligotrophic waters, but Samuelsson, Iversen (1929) and others have pointed out that it may grow in very calcareous waters, a fact reconciling its frequent presence in the chalk muds of the Allerød and the early Flandrian periods.

The sparse representation of *M. alterniflorum* in the interglacial records contrasts with that for the other species of *Myriophyllum*, and neither of these shews the remarkably high frequency of site records in the Late Weichselian and early Flandrian, a feature matched in deposits of western Europe. Of the three species it is of course far the commoner throughout northern Scandinavia and even in the British Isles it becomes more frequent in the north. Hultén (1958) gives it as an amphi-atlantic species growing in north and western Europe, Iceland, South Greenland, eastern and arctic western North America: his map clearly shews a sub-atlantic or western distribution in Europe. Evidently *M. alterniflorum* has persisted in the British Isles from Middle Weichselian time and may well have been continuously present for far longer.

HIPPURIDACEAE

Hippuris vulgaris L.   Mare's tail (Fig. 69*d*; plate XVII*v*)
261: 1

The fruits (mericarps) of *Hippuris vulgaris* L. are among the most characteristic and (in older deposits) the commonest of Pleistocene plant remains (Fig. 69). Fossil pollen is rare but has been twice recorded in this country. The record-head diagram makes it evident that *Hippuris* was present in all the last five interglacial periods, represented in both the Hoxnian and Ipswichian at sub-stages I, II and III, whilst it has clearly been present throughout all zones of the Flandrian. Equally its fruits have been recognized from all the last four glacial stages, and in

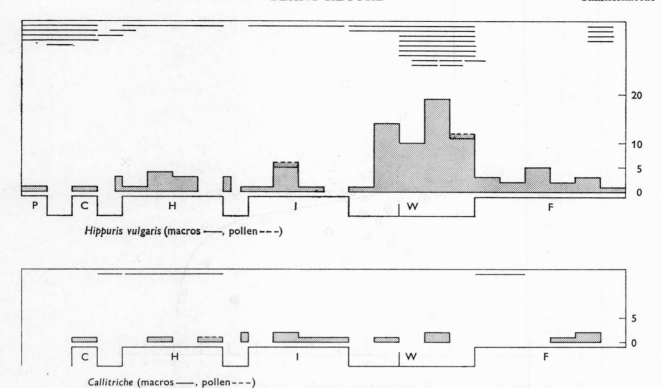

Hippuris vulgaris (macros ——, pollen ---)

Callitriche (macros ——, pollen ---)

the latest, the Weichselian, the records are extremely abundant, not only in all zones of the Late Weichselian but in Early and Middle Weichselian also. In this last-named period it occurs repeatedly in association with a rich periglacial flora and fauna. The evidence from the Weichselian, taken in conjunction with the fossil record as a whole, makes it seem highly likely that *Hippuris vulgaris* has persisted in this country throughout the Pleistocene.

Samuelsson (1934) classifies *H. vulgaris* as one of the ubiquitous Scandinavian species occurring without interruption throughout the whole extent of Norway, Sweden, Denmark and Finland, and Hultén shows it as common into the Arctic Circle and throughout the Kola peninsula. It is in Hultén's group of boreal–circumpolar plants without big gaps in the distribution area. Polunin (1940) describes it as a circumpolar and southern hemisphere plant present throughout the sub-arctic and lower arctic regions, reaching a northernmost locality at 76° 49′ N in east Greenland, but in the Canadian Eastern Arctic rare north of the Arctic Circle. It grows today in all parts of the British Isles.

CALLITRICHACEAE

Callitriche sp. Water starwort   262

Although the identification of species in the genus *Callitriche* depends substantially on fruit characters there are many instances in which specific determination of fossil fruits has not been possible. These are assembled in the record-head diagram together with a single pollen record by H. J. B. Birks from the Hoxnian interglacial at Fugla Ness.

The total record represents, though sparsely, every glacial and interglacial stage from the Cromerian and terminates with zone VIIb records from Skye and Kent respectively.

Callitriche stagnalis Scop.   262: 1

C Norfolk Co.; lW Coolteen, Mapastown

The fruits of *Callitriche stagnalis*, recognizable by their conspicuous narrow wings, have been recorded from the Cromerian interglacial in Norfolk; and in Ireland from zones II and III of the Late Weichselian. *C. stagnalis* is a species now common throughout the British Isles: it is a boreal–circumpolar plant of more or less continuous range.

Callitriche platycarpa Kütz. (*C. polymorpha* Lönnr.)

262: 2

eWo ?Brandon

Kelly (1968) has tentatively identified the fruit lobes of *Callitriche platycarpa* from the early Wolstonian glacial deposits at Brandon, Warwickshire.

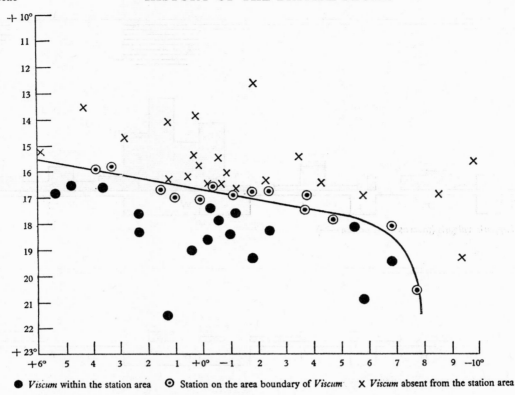

● *Viscum* within the station area   ⊙ Station on the area boundary of *Viscum*   ✗ *Viscum* absent from the station area

Fig. 71. Thermal correlation curve compiled by Iversen (1944*a*) for *Viscum album* L. The mean temperature for the warmest month is given along the ordinate, that for the coldest month along the abscissa. Each point represents a single selected meteorological station within, at, or just beyond, the boundary of the natural growth of mistletoe. It is apparent that *Viscum* will tolerate low winter temperatures, but requires fairly high summer temperatures.

Callitriche obtusangula Le Gall   262: 3

IW Kirkmichael

From both zone I (Ballyre) and zone III (Wyllin) of the Late Weichselian in the Isle of Man there have been recorded fruits of *C. obtusangula*. This is remarkable since this species is no longer native in the Isle of Man, nor does it seem, even at the present day, to grow in the British Isles north of this latitude.

Callitriche intermedia Hoffm.   262: 4

C *Norfolk Co.*; *IW Hawks Tor*

Fruits of *Callitriche intermedia* have been identified from the Cromerian interglacial and from zones I and II of the Late Weichselian in Cornwall. *C. intermedia* occurs throughout the British Isles, as thoughout Scandinavia, and Hultén in his provisional map of amphi-atlantic plants (1958) shews it as mainly European, extending however to Iceland and the southern coast of Greenland.

Callitriche intermedia or stagnalis   262: 1, 4

From sub-stage IV of the Hoxnian interglacial at Fugla Ness (Orkneys) Birks and Ransom (1969) have recovered four fruits that they refer to *Callitriche intermedia* or *C.* *stagnalis* after very full consideration of all available characters including surface cell pattern. From the site of recovery of three of the seeds there were found the pollen grains recorded under *Callitriche* sp. As these authors point out, these two species occupy very similar habitats and have very similar widespread ranges in the British Isles today.

Callitriche hermaphroditica L. (*C. autumnalis* L.) (Fig. 153*d*)   262: 5

C *Norfolk Co.*; *IW Mapastown, Coolteen, Hawks Tor*; F VIIa, b *Loch Park*, VIIb *Drumurcher*

The broad wing of the fruits and the short cell pattern make identification of *Callitriche hermaphroditica* simpler than that of other species of *Callitriche*. It has been found in the Cromerian; in the Late Weichselian it occurs in zones II and III at Mapastown, and elsewhere in zone II. The Flandrian records are from zones VIIa and b at Loch Park (Aberdeenshire) and zone VIIb at Drumurcher.

C. hermaphroditica falls within Hultén's category of boreal–circumpolar plants with fairly continuous distribution, and he shews it as having a scattered distribution throughout Scandinavia to the far north. It is rare and local in the British Isles, occurring in lakes and stagnant waters of northern England and Ireland, more commonly

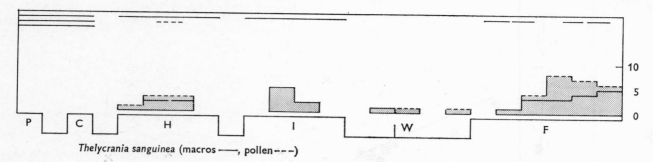

Thelycrania sanguinea (macros ——, pollen – – –)

in Scotland (including the Orkneys and Shetlands) and much more infrequently south of these areas. *C. hermaphroditica* has been recorded from the lower Dryas clays (zone I) at Allerød in Denmark in the original description of that site (Hartz & Milthers, 1901). It looks highly probable, especially in view of the Cornish record, that the species was widespread in the Late Weichselian and has since suffered much retraction of range.

LORANTHACEAE

Viscum album L.   Mistletoe (Fig. 71; plate XXV*l–n*)
263:1

C *West Runton*; H *Kilbeg, Nechells, Marks Tey*; F VI *Kirkby Thore*, VIIa *Old Buckenham Mere*, VIIa/b *Shustoke*, VIIb *Tilbury, Coppice Gate, Bell Track*, VII *Slapton Ley*

Iversen's determination of the thermal-limit curve for mistletoe at the present day (fig. 71) indicates that the plant is restricted northwards by a high temperature requirement, which is however rather less in the oceanic west than further east. It appears to tolerate much more winter cold than the ivy or holly. In Denmark discoveries have been made of the very recognizable pollen grains of *Viscum* in Boreal, Atlantic and Sub-boreal deposits outside the present range of the plant, so that former conditions of greater summer warmth can be inferred.

The time distribution of records of fossil pollen of *Viscum* is a striking one, for apart from a single Cromerian record they are equally divided between the Hoxnian interglacial and the second half of the Flandrian. All sub-stages of the Hoxnian are represented with strong local persistence, thus Kilbeg (sub-stage I), Nechells (IIb, IIc, IIIa) and Marks Tey (IIc, IIIa, IIIb, IVa). Watts (1959*b*) points out that the single grain at Kilbeg is probably not secondary and that it gains particular interest from the Flandrian absence of *Viscum* from Ireland. The summer warmth implications for the Hoxnian as a whole are evident. Too much however should not be made of the apparent absence of Ipswichian records.

The earliest Flandrian record is from zone VI in Westmorland close to the present northern limit of the species which is naturally absent from Scotland, Ireland and North Wales: all the zone VII and VIII records lie well within this area.

CORNACEAE

Thelycrania sanguinea (L.) Fourr. Dogwood (Plate XX*k*)
265:1

The very characteristic and robust fruit stones of *Thelycrania sanguinea* have been found at three sites in the Cromer Forest Bed series, and in the middle sub-stages of both Hoxnian and Ipswichian interglacials. The earliest Flandrian record is from the zone IV/V transition in North Sea moorlog, but thereafter the record is continuous. The record of the plant from the Full Weichselian at Earith is not merely tentative but is regarded as derived by Miss Bell (1969). We take a similar view of the records of pollen from Loch Park and Loch Kinord from the Scottish Late Weichselian (Vasari & Vasari, 1968), since these sites lie so far beyond the present-day range of the dogwood. With this exception the pollen records confirm those for the fruit stones.

Thelycrania sanguinea is placed in Hultén's group of sub-atlantic species, and he shews it as frequent in parts of Denmark but only scattered in the south Scandinavian peninsula: its southerly restriction is pronounced also in the British Isles, for it only reaches southern Scotland and is absent from most of northern Ireland. In view of this southern tendency it is of considerable interest to note the certain presence of the plant in Essex in zone V and in Ireland and South Wales in zone VI, it was thus able to participate in the Boreal extension of the deciduous forest trees. There seems no good evidence from the fossil record that the dogwood survived glaciation in this country.

Chamaepericlymenum suecicum (L.) Aschers. & Graebn.
  (*Cornus suecica* L.) 'Dwarf cornel'   267:1

H *Fugla Ness*; mW *?Ponders End*; lW *Esthwaite*

The fruit stone of *Chamaepericlymenum suecicum* has been identified from the cool terminal sub-stage IV of the Hoxnian interglacial in the Orkneys and tentatively from the Lea Valley Arctic Plant Bed. Its pollen has also been identified from Late Weichselian zones II and III of the Esthwaite Basin.

The dwarf cornel is one of a very limited group of 'circumpolar sub-oceanic' plants (Hultén). It is common throughout northern and western Scandinavia; local and rare in northern England, it is frequent in the western part

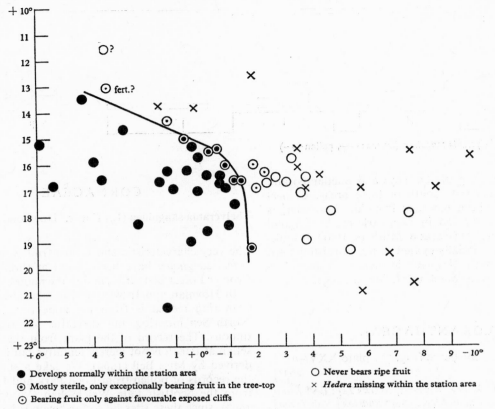

Fig. 72. Thermal correlation curve for the ivy (*Hedera helix*) compiled by Iversen (1944a). Ordinate, the mean temperature for the warmest month; abscissa, that for the coldest month. The ivy is shewn only slightly more tolerant of winter cold than the holly (cf. fig. 56).

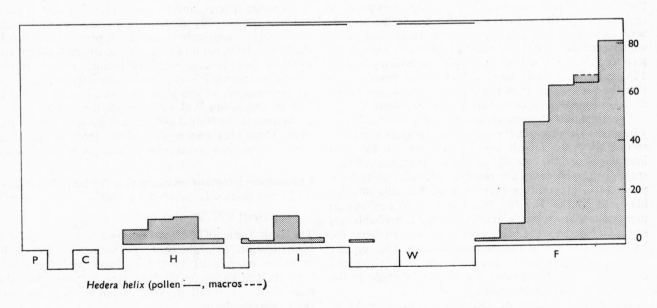

Hedera helix (pollen ———, macros ----)

of the Scottish Highlands. It seems likely that the plant has suffered restriction of range since the Weichselian, and we may note that, on the strength of Late Weichselian pollen identifications in north Germany, Holland and Denmark, Straka (1969) has already had no hesitation in claiming that the present distribution of the species in north-western Europe is relict from the last Glacial period.

ARALIACEAE

Hedera helix L.   Ivy (Figs. 72, 73, 74; plate XXV *h–k*)

268: 1

Macro-fossils of *Hedera helix*, the ivy, have been recorded as seeds and roots from five deposits of varying ages, but the recognition of the pollen grains has given us a very much more extensive knowledge of the plant's history.

H. helix has a tricolpate grain with rather short, acutely tapering furrows crossed equatorially by the elongated pore. The sculpture is a heavy reticulation; the exine is intectate.

Although the ivy is entomophilous and grows commonly on the woodland floor, it flowers on stems that reach the tree canopy and 'diffuses its pollen well' (Andersen, 1966). Its pollen record gains particular importance from the high climatic indicator value of the species.

We have already described in the account for *Ilex aquifolium* (pp. 173 ff.) Iversen's technique in constructing a thermal correlation graph to describe the climatic range of certain species in north-western Europe. That for the ivy, reproduced in fig. 72, appears to show a double control: moderately high summer temperatures and winter temperatures not falling below −1.5 °C as average for the coldest month. The very severe damage caused to the erect plants of ivy during hard winters has been independently determined and it evidently operates directly by killing the cambium. On the other hand the winter thermal control also operates by reducing fertility and by restricting growth to levels where snow-cover gives protection through the worst winter weather. In view of this susceptibility of the ivy, reinforcing the known facts of its geographical range, Iversen confirms Troll's transfer of it from the category of sub-oceanic to that of eu-oceanic plants. It is, of course, the only British member of the tropical family of Araliaceae, and we may regard it as a plant not altogether fully adjusted to our present climate, as witness its extraordinarily late flowering date which involves considerable destruction of inflorescences by the first autumnal frosts.

In the light of his conclusions as to the thermal range of *Hedera*, Iversen (1944*a*) examined the former frequency of the plant in Denmark during the various zones of the Post-glacial period, and the results are based upon a tree-pollen count of over 400 000. From this it is clear that the ivy was present only in very low frequency in the early Boreal (zone V), but increased greatly in the later Boreal (zone VI), and in the Atlantic period (zone VIIa) it must have been remarkably frequent since its pollen averages no less than 0.3 per cent of the total tree pollen; in the Sub-boreal the pollen values are suddenly severely reduced (to 0.06 per cent) and in the Sub-atlantic they are even lower (0.04 per cent). Iversen points out that this behaviour closely parallels that described by von Post for *Cladium mariscus* in middle Sweden. For direct compari-

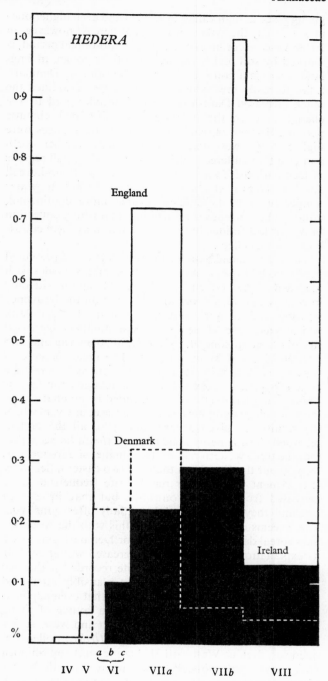

Fig. 73. Average frequency of pollen of ivy (*Hedera helix*) expressed as a percentage of total tree pollen, for succeeding pollen zones of the Flandrian period. Data for Denmark from Iversen (1944*a*), for Ireland from Jessen (1949), and for England (heavily weighted by results from Somerset) from unpublished analyses.

son with Iversen's results we abstracted (first edition) ivy pollen frequencies, zone by zone, from results then available from Ireland and England (fig. 73).

Like the Danish, the English and Irish records shew an occasional grain in zone IV, very low frequencies in zone V, giving place to high values in zone VI and even higher ones in zone VIIa, the Atlantic period. Whereas, however, in Ireland the succeeding zone VIIb, the Sub-boreal period, is marked by still higher frequencies of ivy pollen, in England there is a considerable recession and in Denmark a fall to really low values. In 1960 Iversen examined the much increased Danish data with special regard to the change between the Atlantic and Sub-boreal climatic periods. He shewed that in the Atlantic frequencies were higher in the west and in those regions further north and east that suffered marine transgression. In all regions of Denmark there was a severe decline in the Sub-boreal, an effect Iversen attributed mainly to the fall in winter temperatures as the climate became more continental, although he conceded the possibility that winter collection of ivy for leaf fodder by Neolithic man may have contributed to this effect.

The accumulated data for the British Isles are presented in the record histogram in terms of numbers of sites with recorded *Hedera* pollen. It is recorded from one site only in zone IV and only seven in zone V: at all the frequency was low and all were western except the Dogger Bank which was close to sea-level. Site numbers increased sharply to 48 in zone VI, 63 and 64 in zones VIIa and VIIb respectively, and 81 in zone VIII. The great increase in sites was accompanied by a substantial increase in relative pollen frequency in accord with the data given in fig. 73. This broad picture is usefully extended by comparison of the five main geographical regions of the country, in which the results are given as percentages of all the pollen-recorded sites, zone by zone (fig 74). It will be seen first that the three western regions have higher site frequencies throughout the Flandrian than the two eastern. Secondly it is evident that in four regions site frequencies have increased from zone VI onwards, but that in eastern England they have fallen substantially after zone VIIa. There seems every reason to link this with the Atlantic–Sub-boreal decline demonstrated for Denmark and to see its explanation in the effects of increased winter cold. It has to be noted that the eight site records for the two earliest Flandrian zones cannot be plausibly attributed to long-distance pollen transport and that even explanation in these terms has to predicate a source of living plants at no great remove to the south and west. A few sites have yielded outstandingly high *Hedera* pollen frequencies, such as Westward Ho! (zone VIIa) and Shippea Hill (zone VIIb), possibly reflecting its growth in dry birch–alder fen wood, one of its present-day British habitats and one indicated by van Zeist (1964) in his analyses at Spézet, western Brittany. A more remarkable case is that given by Dimbleby (in W. F. & W. M. Rankine & Dimbleby, 1960) where a buried soil surface of a Mesolithic site at Oakhanger, Hampshire shewed ivy pollen locally dominant to the exclusion of almost all other species. Dimbleby (1967) suggests that this may have been due to autumnal importation of ivy shoots for cattle fodder and cites the similar case of soil below a Bronze Age barrow at Portlesham, Dorset. It seems

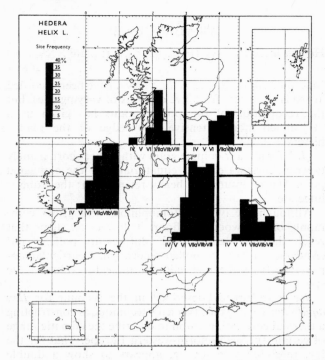

Fig. 74. Occurrence of the ivy, *Hedera helix* L., in the British Isles throughout the Flandrian period. The histograms represent, for each of five geographical regions, the percentage of all pollen-recorded sites in the region, zone by zone, in which *Hedera* was present. The open rectangles represent records for zone transitions or combined zones.

possible alternatively that ivy could have been brought in as part of burial ceremonial (see account of *Buxus*), and we recall that macroscopic remains of ivy were found at one Iron Age site (Bury Wood Camp) and two Roman (Tusculum and Crossness). If selective gathering of ivy is invoked to explain Sub-boreal decline in pollen frequency it has to be equally borne in mind that temporary and marginal forest clearances are likely to have been very beneficial to growth and to flowering of the ivy, an effect detected by R. Sims (unpublished) in the Mesolithic clearances at Hockham Mere, Norfolk by temporarily increased ivy pollen percentages.

It seems quite certain that the scanty site records shewn in the histogram for glacial stages are due to secondary pollen, but there is good evidence of the presence of *Hedera* in both the Hoxnian and Ipswichian interglacials. In the former it was especially common in sub-stages II and III, but in the latter high frequencies are limited to sub-stage II. Since numerous pollen series exist in the second half of the Ipswichian, we may consider the effect a real one, perhaps due to increased continentality in sub-stages III and IV in East Anglia, where the sites are mostly situated. The British results accord with those of S. T. Andersen (1966) for Danish interglacials, especially in that *Hedera* tends to have its maximum early in the interglacial cycle in contrast with *Ilex*; this he attributed chiefly to the development of acidic mor soils in the later part

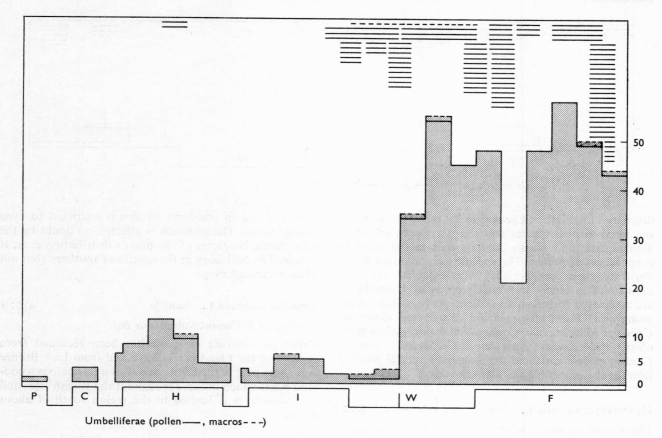

Umbelliferae (pollen——, macros– – –)

of the interglacial cycle, giving edaphic conditions better tolerated by *Ilex* than by *Hedera*. This however can scarcely be invoked to explain the *Hedera* decline in the highly calcareous environment of East Anglian sites, especially as the more oceanic Hoxnian shews far later response. A feature of some minor interest is the record from the Hoxnian deposits at Clacton, Essex where scrapings from a rhinoceros tooth yielded a pollen spectrum with no less than 37 per cent of ivy pollen. This seems to indicate grazing upon *Hedera* growing and flowering by river margins or perhaps beside clearings, frequented by these large mammalia and possibly caused by them (Pike & Godwin, 1953).

In summary we might conclude that the history of *Hedera* in the Flandrian appears to be simply explicable by expansion in response to higher winter temperatures as the hypsithermal conditions are established in zone VI: corresponding retreat with the return of cold winters in zones VIIb and VIII is felt only in the continental region of the British Isles, but is more strongly expressed on the Continental mainland. A broadly similar course seems to have been followed in the two preceding interglacials with the ivy shewing earlier decline in the Ipswichian than in the more generally oceanic Hoxnian. Neither rate of immigration nor response to soil development seems necessarily involved.

Hedera helix is common throughout the British Isles and on a wide range of soils, although avoiding water-logged, very dry and very acid situations. In Scandinavia it extends to about 60° N on the west coast of Norway and has a strongly coastal emphasis in its distribution. It does not extend into the Continental mainland east of the Baltic and north of the Black Sea. It is regarded by Hultén as sub-atlantic, but by Troll as eu-atlantic.

UMBELLIFERAE

Pollen of the Umbelliferae can seldom be referred to individual genera or species. The fruits, on the other hand, are very readily identifiable and only damaged or ill-preserved mericarps remain in the category of 'Umbelliferae'. Accordingly the bulk of the very large number of records in the family record diagram is for pollen. The earliest records of all are omitted: they shew the family present in the Ludhamian and Baventian glacials and the intervening Thurnian. Thereafter the family is represented in every interglacial, including all sub-stages of the Hoxnian and Ipswichian and all the Flandrian zones. It is also represented in all the glacial stages from the late Wolstonian, with very large numbers of records in the Late Weichselian. The fullness and continuity of the Pleistocene record are scarcely surprising in view of the

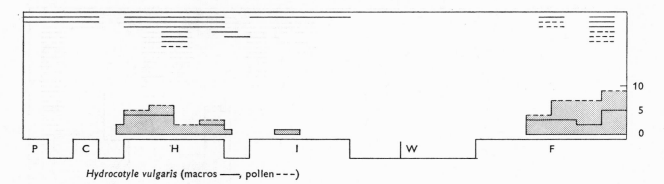

Hydrocotyle vulgaris (macros ——, pollen - - -)

diversity of habitat and geographical range embraced by the many British species and the prevalence of species with aquatic and marsh habitats that favour local incorporation of pollen. The specific accounts that follow illustrate these points: they also establish the large number of instances in which records of Umbelliferae are associated with Roman settlements (e.g. *Aegopodium podagraria, Foeniculum vulgare, Smyrnium olusatrum, Chaerophyllum aureum, Conopodium majus, Apium graveolens, Peucedanum graveolens, Coriandrum sativum, Conium maculatum*): had the Romans a special predilection for the umbelliferous plant flavours?

Hydrocotyle vulgaris L.   Pennywort, White-rot   269: 1

The conditions under which *Hydrocotyle vulgaris* grows favour preservation of its remains so that fruit and pollen commonly occur together. This happens for example in the Hoxnian at Kilbeg in sub-stages I, II, the transition II/III and the transition to early Wolstonian: the same is found in the later Flandrian shore peats at Hightown, and at the Bell Track in Somerset (zone VIIb). The fossil record is a striking one, shewing presence in every interglacial after the Cromer Forest Bed series, and with no records in glacial stages except for the late Anglian and Wolstonian which are contiguous with the Hoxnian in which the species is well represented in every sub-stage. By contrast, and for no obvious reason, *Hydrocotyle* is poorly represented in the Ipswichian, a case exactly paralleling that for *Myriophyllum* sp. and *M. spicatum* (pp. 211 ff.). In the Flandrian it has first been found in zone VI, thereafter increasing. This cannot be associated with the increasing growth of ombrogenous bogs as with some plants (e.g. *Drosera* sp.) for it is eutrophic, and in the Somerset Levels where it is often present in the pollen diagrams it is found in the deposits of Meare Pool or in horizons indicative of flooding with calcareous water. The record as a whole, as well as the delayed Flandrian expansion, prompts the query whether this may be a species that did not survive the glaciations, especially the Weichselian, in Britain.

H. vulgaris occurs however throughout the British Isles today, even in the Orkneys and Shetlands, despite the fact that it only just extends north of 60° N in coastal Norway, and in southern Sweden is restricted to even further south. The situation is affected no doubt by the sub-atlantic character of the plant's distribution area. It is placed by Salisbury in the species of southern (but not Mediterranean) range.

Sanicula europaea L.   Sanicle   270: 1

F IV *Loch Cill Chriosd*, VIIb *Minnis Bay*

Pollen of *Sanicula europaea* has been recorded from zone IV of the Flandrian in Skye and from Late Bronze Age shore peat of the Kent coast. It is a calcicolous woodland plant which occurs throughout the British Isles and in Scandinavia is limited to the region south of about 64° N.

Chaerophyllum temulum L.   'Rough chervil'   273: 1

I *West Wittering*

Although the status of *Chaerophyllum temulum* in the *natural* plant communities of the British Isles today is extremely obscure (see Clapham, Tutin & Warburg, 1962), it clearly had a natural status in the Ipswichian interglacial period.

Chaerophyllum aureum L.   273: 2

F VIII *Silchester*

It is of great interest to find that *Chaerophyllum aureum*, a plant of central and southern Europe encountered in Britain today as a casual, was already present (presumably of similar status) in Roman time at Silchester. Reid speculates that it may have been introduced in packing material for crockery, or have been employed in cookery.

Anthriscus caucalis Bieb. (*Anthriscus neglecta* Boiss. and Reut.; *Chaerophyllum anthriscus* (L.) Crantz) 'Bur chervil'   274: 1

F VIIb ?*Minnis Bay*

There is a tentative identification of a fruit of *Anthriscus caucalis* from the Late Bronze Age shore peat at Minnis Bay, Kent.

Anthriscus sylvestris (L.) Hoffm. Cow parsley, Keck
274: 2

Cs *Pakefield*; I *West Wittering, Histon Rd*

The fruit of *Anthriscus sylvestris* has been recorded from the Cromer Forest Bed series, from an undefined part of the Ipswichian, and, in Cambridge, from sub-stage IV of the Ipswichian. This species, of doubtful status in the British Isles today, was accordingly a certain native in past interglacial periods.

Torilis japonica (Houtt.) DC. (*T. anthriscus* (L.) C.C. Gmel., *Caucalis anthriscus* Huds.) 'Upright hedge-parsley'
277:1

Cs *Pakefield*; H *Clacton*; F VIIb *Frogholt*, VIII *Silchester, Finsbury, Caerwent, Dublin, Lissue*

Native interglacial status for *Torilis japonica* seems assured by one fruit record from the Cromer Forest Bed series and another from sub-stage III of the Hoxnian interglacial. In the Flandrian however all its records are associated with human activity, at Frogholt in the Late Bronze Age, at three Roman sites, and at two mediaeval sites in Ireland. This very common roadside umbellifer is historically a camp follower!

Torilis nodosa (L.) Gaertn. 'Knotted hedge-parsley'
277: 3

I *Stone*; F VIII *Aldwick Barley, Bermondsey, Plymouth*

After the Ipswichian interglacial, fruits of *Torilis nodosa* have been next identified in charred Iron Age material from Aldwick Barley and then in Roman and mediaeval time. It is a ruderal and weed of dry arable fields commoner in the south-east of the British Isles than elsewhere.

Coriandrum sativum L. Coriander
279: 1

F VIIb *Minnis Bay*, VIII *Silchester, Caerwent, Godmanchester*

It is interesting to find that fruits of the Mediterranean plant *Coriandrum sativum*, the coriander, had been introduced into this country not merely in Roman time, as shewn by the discoveries at Silchester, Caerwent and Godmanchester, but even in the Late Bronze Age. Presumably it was used for flavouring, then as now.

Smyrnium olusatrum L. Alexanders
280: 1

F VIII *Caerwent*

Smyrnium olusatrum was formerly cultivated as a pot-herb and the single record for its fruit is from a Roman site.

Conium maculatum L. Hemlock
282: 1

Cs *Pakefield*; C *Norfolk Co.*; F VIII *Bunny, Pevensey, Caerwent, Finsbury, Silchester*

Fruits of the hemlock, *Conium maculatum*, have been found in the Cromer Forest Bed series and, more recently, in the Cromerian itself.

The strong association of *Conium maculatum* with cultivated and disturbed ground is reflected in the fact that all five subsequent records of its fossil occurrence are from sites of Roman occupation. This dual interglacial–Roman appearance is typical of other Umbelliferae.

Bupleurum sp.
283

*e*Wo *Brandon*

From the early Wolstonian deposits at Brandon, Warwicks., Kelly (1968) has identified pollen of the genus *Bupleurum* of which the three British species are strongly southern or south-eastern when growing naturally.

Bupleurum tenuissimum L. 'Smallest hare's-ear'
283: 4

F VIII *Bunny*

Mrs Wilson has identified a fruit of *Bupleurum tenuissimum* from the Roman settlement at Bunny, near Nottingham, close to the probable north-eastern limit of the species today. It is a ruderal and salt marsh plant more abundant coastally than inland.

Apium sp.
285

Macros: Cs *Pakefield*; C *Norfolk Co.*; F VIII *Copthall Av., Plymouth*
Pollen: F VIIb/VIII *Loch Dungeon*, VIII *Wingham*

The records of identification to the genus *Apium* can be regarded mainly as supplementing the fuller records for three of the British species of the genus. Two early fruit records from the Cromerian and the longer span of the Cromer Forest Bed series are succeeded by one from a Roman site in London and a mediaeval one in Plymouth. There is a pollen record from the zone VIIb/VIII transition in Galloway, and at Wingham in Kent *Apium* pollen is present at the Iron Age level in very high frequency (exceeding the total tree pollen): at the succeeding Roman level these high values persist, no doubt indicative of the continued growth of one of the aquatic species nearby.

Apium graveolens L. Wild celery
285: 1

H *Nechells*; I *West Wittering, Histon Rd*; F VIIb *Rhyl, Cocker Beck, Prestatyn*, VIII *Silchester, Welney, Hungate, Caerwent*

Fruit records of *Apium graveolens* have been made for sub-stage I of the Hoxnian interglacial and sub-stage III of the Ipswichian: the sub-stage for West Wittering is unknown. Thereafter there are no records until the Flandrian zone VIIb at Cocker Beck and the VIIb or VIIb/VIII coastal peat beds at Rhyl and Prestatyn. These two last even at formation were near the coast and the Roman site at Welney was tidal: this accords with the frequent occurrence of *Apium graveolens* in brackish conditions. Not only at Welney is the plant in a Roman context:

this is true also for the Silchester, Hungate and Caerwent records.

Apium graveolens has a southern range in the British Isles, not growing further north than the latitude of the north coast of Ireland, and it is chiefly found near the coast. In Scandinavia it is likewise extremely southern in range, hardly going north of Denmark save sporadically (Hultén).

Apium nodiflorum (L.) Lag.   Fool's water-cress   285: 2

H *Nechells*; I *West Wittering*; W *Airdrie*; mW *Barrowell Green*; F v *Apethorpe*, vIIb *Frogholt, Cocker Beck, Minnis Bay*, vIIb, vIII *Wingham*, vIIb/vIII *Prestatyn, Leasowe*, vIII *Bermondsey, Finsbury Circus*

There have been fruit identifications of *Apium nodiflorum* from sub-stages I and II of the Hoxnian, and from un-defined parts of the Ipswichian and succeeding Weich-selian. There is also a record from the Lea Valley Arctic Plant Bed. Apart from one record in zone v all the Flandrian records are from zones vIIb or later, with a con-siderable number associated with human activity: thus Neolithic 1, Late Bronze Age 3, Iron Age 2 and Roman 3. There is evidence of repetition at different horizons at the same site at Nechells, Apethorpe, Frogholt and Wingham.

Apium nodiflorum is a plant of ditches and shallow ponds, rather commonly distributed through the lowland areas of England, Wales and Ireland, but uncommon in Scotland. It belongs to Matthews' Continental–southern element in the British flora and all Scandinavia is outside its present range so that the two Weichselian records are unexpected.

Apium inundatum (L.) Reichb. f.   285: 4

eWo *Brandon*; I *Wretton*; F vIIb *Helton Tarn*, vIII *Ehenside*

The fruit of *Apium inundatum* has been recorded from the early Wolstonian glacial and from sub-stage III of the Ipswichian interglacial; thereafter in the Flandrian the records are respectively from zones vIIb and vIII at sites in north-west England. *A. inundatum* is found locally but widely throughout the British Isles: common in Denmark, it does not extend far into southern Sweden. It is classified by Hultén as a European–Atlantic species.

Petroselinum segetum (L.) Koch   'Corn caraway'
   286: 2

H *Clacton*; I *Histon Rd.*; lW ?*Nazeing*; F vIII *Ehenside*

There is a fruit identification of *Petroselinum segetum* from one sub-stage III of the Hoxnian interglacial, but in the Ipswichian it was found at three different excavations by different workers, respectively in sub-stages II, III and IV. There is a qualified record from the Late Weichselian zone II/III boundary at Nazeing and the one Flandrian record is as late as zone vIII. It is a southern species occurring in Western Europe from England and the

Netherlands south to Portugal: in the British Isles it grows only south of a line from the Humber to Cardigan Bay. Whatever we think of the Nazeing record, it would seem that this very southern species successfully estab-lished itself here in at least three different interglacials.

Cicuta virosa L.   Cowbane (Plate XVIIx)   288: 1

Cs *Pakefield, Beeston*; Be *Norfolk Co.*; C *Norfolk Co.*; W *Thrapston*; mW *Hackney Wick*; lW *Aby, Nazeing*; W/F *Loch Dungeon*; F IV *Nazeing*, IV, v/VI, vIIa *Tadcaster*, IV *Star Carr*, VI *Hockham Mere, Killywilly Loch*

Identifications of fruits of *Cicuta virosa* have been made not only twice from the Cromer Forest Bed series, but also from the Beestonian glacial stage and the Cromerian interglacial stage that lie within its range. They have been recorded from the Weichselian (*sensu lato*), Middle and Late Weichselian. Thereafter fruits have been found in each Flandrian zone up to vIIa with repetitions at indi-vidual sites such as Nazeing and Tadcaster. Tadcaster (Yorks.) is the most northerly of these sites for fruit records, but there is a single pollen record (from rather high altitude also) at Loch Dungeon (Galloway) at the Late Weichselian/Flandrian transition.

Cicuta virosa is a plant which occurs commonly throughout Finland to 65° N and has numerous Scandi-navian localities throughout Sweden and beyond the Arctic Circle. This is agreeable with the fossil occurrences in Middle and Late Weichselian time with more or less continuity thereafter through the Flandrian. The present-day range of *C. virosa* in the British Isles is local and oddly discontinuous, with few localities or none in southern Ireland, south-western England and Wales and the northern half of Scotland. It is so poisonous to cattle that it has of course suffered deliberate eradication, but the fossil record still indicates that it is a northern (continental) species that has suffered disjunction through the Flandrian period.

Bunium bulbocastanum L.   292: 1

lW ?*Colney Heath*

From the peat erratics in the terrace of the River Colne, Hertfordshire Mrs Dickson tentatively identified a damaged but well-preserved fruit of *Bunium bulbo-castanum*. The age of the deposit is judged to be Late Weichselian on faunistic, floristic and geological grounds and this accords with a radiocarbon date of 13 560 B.P. Nevertheless *B. bulbocastanum* does not occur in Europe north of the Netherlands and it has only a few English stations, and these are in Hertfordshire, Buckingham-shire, Cambridgeshire and Bedfordshire, an area com-prising the Late Weichselian site.

Conopodium majus (Gouan) Lor. & Barr. (*C. denudatum* Koch)   Pignut, Earthnut   293: 1

F vIII *Silchester*

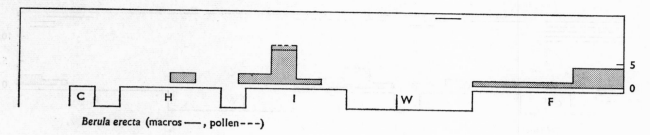

Berula erecta (macros ——, pollen ---)

There is a record from Roman Silchester of the fruit of *Conopodium majus*, a European–Atlantic plant occurring in the coastal region of southern Norway, sparsely in Denmark but widely throughout the British Isles.

Pimpinella saxifraga L. 'Burnet saxifrage' 294

H *Nechells*

From stage II of the Hoxnian interglacial in North Birmingham Kelly (1964) has recorded a mericarp of *Pimpinella saxifraga*, a calcicole with a wide distribution in Britain (though rare in northern Scotland and Ireland) and common in Scandinavia, even north of 65° N.

Aegopodium podagraria L. Goutweed, Bishop's weed, Ground elder, Herb gerard 295: 1

F VIII *Car Dyke*

The Roman canal at Cottenham, Cambridgeshire yielded fruit of *Aegopodium podagraria*, a species prevalent and noxious as a weed in Britain but seldom seen in natural habitats such as the woodland habitats it favours in southern Sweden. A species of wide European range, it is said to have been used as a pot-herb.

Sium sp. 296

IWo *Selsey*; I *Selsey*

Pollen referred to the 'Sium type' has been identified by West at Selsey not only from the late Wolstonian, but successively from sub-stages I (a and b) and II (a and b) at the same site, with frequencies amounting to 5 to 10 per cent of the total tree-pollen in the earlier occurrences. This seems one of the instances of local persistence of an aquatic plant for a long time in a favourable site, and it is to be noted that fruits of *Berula erecta* (*Sium erectum*) were also recorded at Selsey from the late Wolstonian and Ipswichian Ia, IIa and IIb (West & Sparks, 1960).

Sium latifolium L. Water parsnip 296: 1

Macros: C *Norfolk Co.*; eWo *Hoxne*; I *Wretton*; F VIIa *Bowness Common*
 Pollen: VIIb/VIII *S.W. Lancs. Co.*

Fruit records for *Sium latifolium* come from the early Wolstonian and from three interglacials, the Cromerian,

Ipswichian (sub-stage II) and Flandrian. The single pollen record (Travis, 1926) can hardly be relied on to specific level. *S. latifolium* is grouped by Hultén in the west European–middle Siberian plants: it only attains 61° N in Southern Scandinavia and is decidedly rare in the British Isles, nor does it extend much north of the Humber or a similar latitude in Ireland. The two Lancashire records, late though they are, seem to be outside the plant's present English range.

Berula erecta (Huds.) Coville (*Sium erectum* Huds.; *S. angustifolium* L.) 'Narrow-leaved water-parsnip' (Plate XVIII*a*) 297: 1

The fruits of *Berula erecta* have been twice found in sub-stage III of the Hoxnian interglacial and abundantly in the first three sub-stages of the Ipswichian and the preceding late Wolstonian. There is exhibited at this time remarkable persistence at individual sites: thus it occurs at Ilford in late Wolstonian and Ipswichian sub-stages Ia, IIa and IIb, at Selsey in late Wolstonian and Ipswichian Ia, IIa, and IIb, and at Wretton in Ipswichian IIb and III. Similarly later on it occurs at Nazeing in Late Weichselian zone II/III and Flandrian IV, at Apethorpe in Flandrian V, VI and VIII and at Wingham in VIIb and VIII. This phenomenon of local persistence in aquatics has been remarked elsewhere (e.g. *Apium nodiflorum, Cicuta virosa* and the genus *Sium* to which no doubt pollen of *Berula erecta* has contributed: West makes one tentative pollen identification to the species itself at Selsey in a sub-stage where fruits are also found).

The species is found in most parts of Britain today, decreasing in frequency northwards. Although frequent in Denmark and the adjacent tip of Schona, it is otherwise infrequent in Scandinavia and does not reach latitude 60° N. None the less the Late Weichselian and early Flandrian records from Nazeing are unqualified and are followed by continuous presence through the Flandrian. As with other aquatics the later records are associated with settlement sites, the two latest Roman.

Oenanthe sp. 300

Cs *Corton*; I *Histon Rd*; mW *Angel Rd*; IW *Dronachie*; F VII *Hightown*, VIII *Caerwent, Cottenham*

The fruit records for the genus *Oenanthe* have to be considered in conjunction with those separately given for all

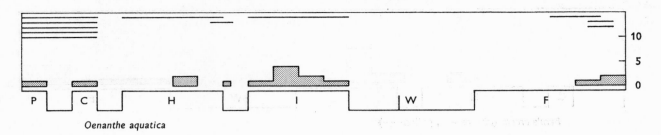

Oenanthe aquatica

seven of the British species. The Histon Rd record is from sub-stage III of the Hoxnian; those at Caerwent and the Car Dyke, Cottenham are Roman.

Oenanthe fistulosa L.   Water dropwort   300: 1

I *West Wittering, Histon Rd*; F VIII *Apethorpe, Silchester*

Oenanthe fistulosa fruit has been found in sub-stage III of the Ipswichian interglacial at Histon Rd, Cambridge and also in an undefined stage at West Wittering. Both Flandrian records are from zone VIII; Silchester is Roman.

O. *fistulosa* is one of Hultén's sub-atlantic plants: it is common in Denmark but has a very limited extension into southern Sweden. Likewise in the British Isles it is absent from northern Scotland and rather infrequent in Ireland, Wales and south-western England, a distribution pattern very similar to that of *Berula erecta* (p. 225).

Oenanthe pimpinelloides L.   300: 2

F *Ipswich*, VIII ? *Hungate*

Both records of fruits of *Oenanthe pimpinelloides* are late Flandrian, the one Neolithic and the other (tentative only) from a mediaeval site.

O. *pimpinelloides* is a plant of pronouncedly southern range in Britain: the record from Suffolk is approximately at its northern limit. It is put by Matthews in his 'oceanic–southern' element of the British flora.

Oenanthe silaifolia Bieb.   300: 3

F VIII ? *Hungate*

From the late Roman level at Hungate, York there is a tentative fruit identification of *Oenanthe silaifolia*, a very local English species that does not today extend as far north as Yorkshire.

Oenanthe lachenalii C.C. Gmel.   'Parsley water dropwort'   300: 4

IW *Greenock*; F VIII *Welney, Silchester*

Fruit of *Oenanthe lachenalii* has been recovered with a rich Late Weichselian flora, from Garvel Park, Greenock. Both Flandrian records are from zone VIII, that from Silchester is Roman.

O. *lachenalii* is a plant regarded by Hultén as sub-atlantic (with O. *fistulosa*) but placed by Matthews as

Continental–southern. Certainly it barely extends into southern Scandinavia and is absent from Scotland except for its western coasts. It tolerates brackish conditions and is almost limited to coastal sites in Ireland and much of western England and Wales: conditions at the Welney site were brackish. The Scottish record is possibly surprising but this site was coastal when the Late Weichselian deposits were being formed.

Oenanthe crocata L.   'Hemlock water dropwort'   300: 5

I ? *West Wittering*; F VIIb *Helton Tarn, Ipswich*, VIII *Glastonbury, Finsbury Circus*

Besides a qualified record from the Ipswichian there are four unqualified records of fruits of *Oenanthe crocata* from the Flandrian, two from zone VIIb (one Neolithic) and two from zone VIII (one Iron Age and one Roman).

O. *crocata* is an oceanic west European and Mediterranean species not found further north than Belgium and Scotland, but present throughout the British Isles.

Oenanthe aquatica (L.) Poir. (*Phellandrium aquaticum* L.)   'Fine-leaved water dropwort'   300: 6

The fossil record for *Oenanthe aquatica* includes records from the five latest interglacials including every sub-stage of the Ipswichian, and five records from the Cromer Forest Bed series. The glacial stages are represented by a single early Wolstonian record and a late Hoxnian/early Wolstonian record, both from Hoxne, one by West and the other by Reid. There are no Weichselian identifications and none in the Flandrian until zone VII. The Ipswichian and Flandrian records stand in sharp contrast and might be taken to indicate delayed re-establishment in the Flandrian after extinction in the Weichselian, a conjecture not at variance with the plant's present-day restriction in range in Scandinavia, where it does not grow north of 62° N, and in the British Isles, where it is absent from Scotland. O. *aquatica* belongs to the 'oceanic, west European' element of Matthews but to the 'west European–middle Siberian' group of Hultén.

Three of the Flandrian records are from archaeological sites, respectively Neolithic, Iron Age and Roman, and two records for about the VIIb/VIII transition (Bann Estuary and Newferry) lie somewhat north of the present extension of the plant in Ireland.

Oenanthe fluviatilis (Bab.) Coleman   300 : 7

F viib/viii *Mapastown*, viii *?Cottenham*

There is an Irish record of the fruit of *Oenanthe fluviatilis* from near the Flandrian zone viib/viii transition, and one (tentative only) from the Roman occupation level of the Car Dyke in Cambridgeshire. It is of interest to have an indication that *Oenanthe fluviatilis*, almost endemic to England, was present already in and possibly before Roman time.

Aethusa cynapium L.   Fool's parsley   301 : 1

Cs *Pakefield*; H *Clacton*; F vi *Apethorpe*, viia/b *Shustoke*, viib *Streatley*, viii *Aldwick Barley, Glastonbury, Pevensey, Caerwent, Silchester*

After a record from the Cromer Forest Bed series, fruit of *Aethusa cynapium* is next recorded from sub-stage iii of the Hoxnian interglacial. Thereafter it does not recur until zone vi of the Flandrian (Apethorpe) and the zone viia/b transition (Shustoke). Subsequent records are all associated with archaeological sites, one Bronze Age, two Iron Age and three Roman, a pattern fitting the characteristic habit as a weed of cultivated ground.

A. cynapium is one of Hultén's 'west European–middle Siberian' plants: it occasionally reaches the top of the Baltic but is common only to about 61° N. In the British Isles too, it becomes less frequent in Ireland, and much less so in Scotland, a pattern seen in other Umbelliferae with a late Flandrian expansion and response to anthropogenic influence.

Foeniculum vulgare Mill.   Fennel   302 : 1

F viii *Bermondsey*

The only fossil record for *Foeniculum vulgare* (fennel) is from a Roman occupation level in the Thames alluvium at Bermondsey. It is a culinary plant apparently naturalized on the coasts of England and Wales and casually in waste land: it is thought to be native in the Mediterranean region.

Silaum silaus (L.) Schinz. & Thell. (*Silaus flavescens*, Bernh., *S. pratensis*, Bess.) 'Pepper saxifrage'
                                                        303 : 1

H *Clacton*

There is a fruit identification of *Silaum silaus* from sub-stage iii of the Hoxnian interglacial. In the British Isles it is only found south-east of a line from Berwickshire to Devon.

Angelica sylvestris L.   Wild angelica   307 : 1

I *West Wittering*; IW *Cahercorney, Mapastown, Johnstown, Nant Ffrancon, Flixton*; F viii *Hungate*

There is a record of the fruit of *Angelica sylvestris* from an undefined part of the Ipswichian interglacial. More strik-

ing and more precisely dated are five Late Weichselian records, four from zone ii and one from iii. These recall Late Weichselian records for many other species of the Scandinavian and British montane 'tall-herb' communities that characterized the Allerød period.

A. sylvestris is a 'west European–middle Siberian' species (Hultén, 1958) common throughout the whole of Scandinavia, as of the British Isles. The fossil record indicates that it was widespread here already in the Late Weichselian. The Hungate record is Norman and probably reflects only the local marshy conditions.

Although a pollen '*Angelica* type' has sometimes been recorded, it embraces too many genera to warrant separate recording here.

Peucedanum sp.   309

IW *?Angel Rd*; F viii *?Hungate*

Two dubious fruit identifications of the genus *Peucedanum* have been made respectively from the Lea Valley Arctic Plant Bed and from late Roman deposits in the Hungate, York.

Peucedanum ostruthium (L.) Koch   Master-wort
                                                        309 : 3

F viii *Lissue*

Two slightly damaged fruits of *Peucedanum ostruthium* were recovered from the rath at Lissue, Co. Antrim.

P. ostruthium is a plant whose centre of distribution lies in southern Europe. It was formerly cultivated as a pot-herb in Great Britain and persists now as a naturalized plant in moist meadows and on river banks from Carmarthen, Stafford and Lancashire northwards to the Shetlands. In Ireland it occurs only in a few places in the north-east. The fossil record establishes its presence as early as the tenth century A.D.

Peucedanum graveolens Benth. & Hook. (*Anethum graveolens* L.) Dill   309 : 4

F viii *Silchester, Caerwent*

It is of considerable interest to find two records of the fruit of dill, *Peucedanum graveolens*, a Mediterranean plant used as a medicine and condiment, at two Roman sites in Britain.

Pastinaca sativa L. (*Peucedanum sativum* (L.) Benth.)   Wild parsnip (Fig. 75; plate XXIIIo, *p, q*)   310 : 1

Although the fossil record for *Pastinaca sativa* is sparse, it is convincing and informative, with pollen and fruit identifications fully supporting one another, and although the pollen records are often given as '*Pastinaca* type' it seems reasonable to consider them all as *P. sativa*. There was at one time some doubt as to possible confusion of the fruits with those of *Heracleum* sp., but careful study by Miss Duigan (1956) made it evident that they consistently differ in the shape of the vittae on the mericarps,

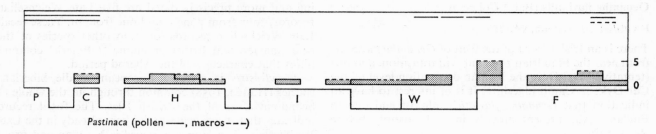

Pastinaca (pollen ——, macros ———)

those of *Heracleum* being always club-shaped (figs. 75, 76). The evidence is that *P. sativa* was present in the Cromer Forest Bed series, the Cromerian itself, all sub-stages of the Hoxnian and two of the Ipswichian interglacial. Its remains are reported through Early, Middle and Late Weichselian and then throughout the Flandrian, save (no doubt *pro tem.*) in zone v. It appears that the species is a long-persistent native at least in southern England whence most of the records come. Although now characteristically a weed or ruderal plant, like many such it found suitable habitats in the Weichselian and, no doubt more locally, through the interglacials.

P. sativa falls within Hultén's category of 'west European–middle Siberian' plants and he shews it as occurring up to 60° N in Sweden and infrequently in Scandinavia even into the Arctic Circle, although of course in artificial habitats. In the British Isles its natural range is concentrated in the south-east and it appears to be absent, save as an introduction, from Scotland and all of Ireland save the south-east.

Fig. 75. Fruits of parsnip, *Pastinaca sativa* L., from the Wolvercote Channel (after Duigan, 1956) × 3.3.

Heracleum sp.   311

H *Fugla Ness, Wolvercote*

Bell (1904) recorded a fruit of *Heracleum* from a Thames terrace at Wolvercote now regarded as late Hoxnian, and from the equivalent sub-stage iv of the same interglacial H. J. B. Birks has identified *Heracleum* pollen.

Heracleum sphondylium L. Cow parsnip, Hogweed, Keck (Fig. 76)   311:1

Macro: Cs *Pakefield*; H *Hitchin, Clacton, Wolvercote*; lW *Twickenham*; F viii ?*Newstead, Bermondsey, Silchester, Hungate*

Pollen: F iv *Loch Mealt, Loch Fada*, v/vi *Allt na Feithe Sheilich*

Fruits of *Heracleum sphondylium* have been recorded from the Cromer Forest Bed series, from sub-stages ii,

iii and iv of the Hoxnian interglacial, from the Lea Valley Arctic Plant Bed and from two Roman and one mediaeval site. The record from Roman Newstead was based upon stem and leaf-base identification (Tagg, 1911).

Pollen has been reported at two sites in Skye (H. J. B. Birks) and at a high altitude site in the Cairngorm (H. H. Birks). These pollen records go a little way to establish native continuity between the Late Weichselian and the Roman time.

Fig. 76. Fruits of hogweed, *Heracleum sphondylium* L. from the Wolvercote Channel (after Duigan, 1956) × 3.3.

H. sphondylium grows throughout the British Isles to-day but interpretation of the fossil record is complicated by the taxonomic variability of the species. The sub-species *australe* is the plant now growing in Britain and this hardly extends north of 61° N in Scandinavia save on dumps: the sub-species *sibiricum* has a far more extensive northerly range in Scandinavia. It is probable that the records have been essentially based upon comparison with ssp. *australe*.

Daucus carota L.   Wild carrot (Plate XVIII*b*)   314:1

F v *Nazeing*, viii *Godmanchester, Silchester, Dublin*

There have been identifications of fruit of *Daucus carota* from Flandrian zone v in the Lea Valley, from two Roman sites and from mediaeval Dublin; the first of these confirms the natural native status of the plant in Britain.

Identifications of pollen of '*Daucus* type' have been made at various Flandrian levels in Loch Dungeon, Galloway but besides '*D. carota*' this type embraces *Pimpinella saxifraga* and *Conium maculatum* (H. H. Birks).

Astrantia minor L.

H ?*Gort*

From sub-stage iii of the Hoxnian in western Ireland Jessen *et al.* (1959) made a tentative identification of the pollen of *Astrantia minor*. They write that 'the *Astrantia* species today are found in tall herb communities in mountain woods extending from the subalpine to the montane belt' and indicate that the calcicolous *A. minor*, not native in the British Isles, grows in the Pyrenees, Cevennes, western Alps and north Italian Apennines. *A. major* is naturalized in some British woods and meadows and fruits of this species have been recorded from the Masovian 2 interglacial in Poland.

CUCURBITACEAE

Bryonia dioica Jacq.   White or red bryony   315: 1

IG *Ipswich*; F VIII *Silchester*

Bryonia dioica, a plant of decidedly southern range in the British Isles and western Europe, has been recorded from the Roman period in Hampshire and from a deposit in Suffolk thought to be possibly interglacial.

EUPHORBIACEAE

Mercurialis perennis L.   Dog's mercury   318: 1

Macros: H *Nechells*; I *West Wittering, Wretton*; W *Thrapston*; F VIIa/b *Shustoke*, VIIb, VIII *Crossness*

Fruits of *Mercurialis perennis* have been found in sub-stage III of the Hoxnian and sub-stage II of the Ips-wichian, as well as in both the Ipswichian and Weich-selian undivided. At Crossness it was found at Bronze Age and Roman levels.

Pollen records for *Mercurialis* seem much more likely to represent *M. perennis* than the far less common and more restricted southerly *M. annua*. The records for pollen of *M. perennis* include one from sub-stage I of the Ipswichian at Selsey, and a Late Weichselian zone II record from Skye (H. J. B. Birks) that underlines Miss Bell's Weichselian fruit record. The early Flandrian records are also Scottish, zones IV and V from Bigholm Burn. From the Neolithic onwards rather a large propor-tion of the records have loose archaeological associations, probably indicative of the response of dog's mercury to opening and partial clearance of woodlands.

M. perennis is one of Hultén's Sub-atlantic–west Siberian plants: in Scandinavia it is strongly southern in range, barely reaching 61° N save in one west Norwegian coastal region at about 66° N, which perhaps has rele-vance for the British Weichselian records and supports the fossil record indication of possible perglacial survival in the oceanic west. Though generally common in Great Britain, *M. perennis* is rare and scattered in Ireland and absent from the Orkneys, Shetlands, Lewis and much of northern Scotland. We have seen no Irish fossil records.

Euphorbia sp.   319

Macros: Cs *Cromer*; eW *Sidgwick Av.*; mW *Earith*
    Pollen: Ba *Ludham*; lW *Brandesburton*; F VIII *Ellerside Moss*

Seeds referred to the genus *Euphorbia* have been recorded tentatively from the Cromerian. Apart from one late Flandrian, all the other records are from glacial stages, the earliest Baventian and the rest covering Early, Middle and Late Weichselian (see *E. cyparissias*, 319: 16).

Euphorbia hyberna L.   'Irish spurge'   319: 5

H *Clacton*

Euphorbia hyberna, one of the members of the restricted group of Hiberno-Lusitanian species, is now found within the British Isles only in Ireland and rarely in south-west England, but has a seed record from sub-stage III of the Hoxnian interglacial in south-east England.

Euphorbia stricta L.   'Upright spurge'   319: 8

H *Nechells, Clacton*

The seed of *Euphorbia stricta* has been identified from both sub-stage II and sub-stage III of the Hoxnian inter-glacial. It is a central and east European species restricted today to a very limited area of the British Isles in west Gloucester and Monmouth where it occurs in woods on limestone. Hegi mentions an occurrence however in alluvial alder woods that would fit the Nechells record very well (Kelly, 1964).

Euphorbia cyparissias L.   'Cypress spurge'   319: 16

I *Histon Rd*; mW *Earith*

The seed of *Euphorbia cyparissias* has been recorded from sub-stage IV of the Ipswichian interglacial in Cambridge, and both seeds and capsules from the Middle Weich-selian in Cambridgeshire.

The cypress spurge is a rhizomatous perennial regarded as sparse and only doubtfully native in the British Isles and often behaving as a casual or garden-escape. It is described by Hultén as a west European–Continental plant: though most frequent in Scandinavia in the south-east, it occurs in scattered localities considerably further north, a point in favour of accepting it as one of our present-day weeds that grew here in periglacial situations.

Euphorbia amygdaloides L.   Wood spurge   319: 17

Cs *Pakefield, Mundesley*; F VIIb *Frogholt*

Euphorbia amygdaloides is a Continental–southern species absent from Ireland and in Great Britain not found north of a line from the Ribble to the Wash. There are two seed records of it from different exposures of the Cromer Forest Bed series, and one from Frogholt in Kent in zone VIIb at an horizon radiocarbon dated to 2860 B.P.

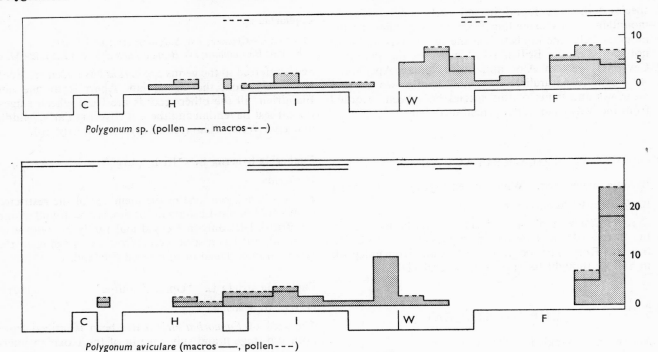

Polygonum sp. (pollen ——, macros - - -)

Polygonum aviculare (macros ——, pollen - - -)

POLYGONACEAE

W *Thrapston*; *l*W *Woodgrange, Loch Borralan*; F vIIb ?*Canna, Brigg*

Many pollen records in the Polygonaceae are taken to generic or specific level, but a handful refer only to the family and they represent all zones of the Late Weichselian and the late Flandrian.

Polygonum sp.   320

Both pollen and seed identifications contribute to the record of *Polygonum* but they give congruent results from which its seems that the genus was present in both the Hoxnian and Ipswichian interglacials as throughout the Flandrian. Records from the Wolstonian glacial stage are paralleled by those from the Early and Late Weichselian. Evidently the genus has long persisted naturally in this country, but these records have to be seen alongside the substantial specific records, accounts of which follow.

Polygonum aviculare L. *sensu lato* Knot grass (Plate XVIII*r*)   320: 1

The fruits of *Polygonum aviculare* constitute the bulk of the fossil record but there is a supplementation by conformable pollen identifications. There are records from the Cromer Forest Bed series and the immediately consequent early Anglian. From sub-stage III of the Hoxnian records are continuous through the Wolstonian, all stages of the Ipswichian and all parts of the Weichselian to zone II/III of the Late Weichselian. The numerous Middle

Weichselian records are mostly identifications by Miss Bell of the nutlets without perianth, and they are recognizable by their shape and the ornament of strings of tubercles that covers the three faces of the nutlet. There are no records in the Flandrian until the sudden expansion in zones vIIb and vIII where there is close association with archaeological situations, viz. Bronze Age 3, Iron Age 3, Roman 8 and Mediaeval 6. In these later occurrences, where often fruit and pollen records correspond, it is apparent that the knot grass has been playing the role of a weed or ruderal plant that has often been accidentally present in grain stores, probably in large part as *P. aviculare* L. (*P. heterophyllum* Lindm.). It is equally clear however that *P. aviculare* must have had a very long history of presence in natural habitats such as shores and river banks, surviving there through the forested interglacials and expanding more widely in the glacial stages.

P. aviculare sensu lato is a plant growing in cultivated ground today not only throughout the British Isles but throughout Scandinavia to its northernmost limits, being recorded as frequent even beyond 70° N latitude.

Polygonum aviculare L. *sensu stricto* (*P. heterophyllum* Lindm.)   320: 1*a*

*m*W *Earith, Marlow*

Among the large number of fruits of *Polygonum aviculare sensu lato* discovered in Middle Weichselian sites, Miss Bell at Earith and Marlow identified a number of nutlets as belonging to *P. aviculare sensu stricto* on the grounds of shape and the equality of the three faces of the trigonous fruits. She indicated an opinion that many nutlets

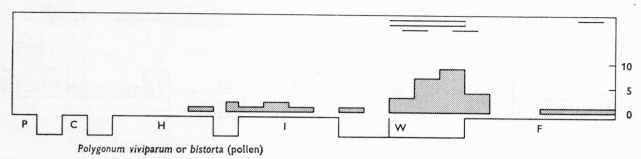

Polygonum viviparum or *bistorta* (pollen)

referred to the larger taxon might well also have belonged to the sub-species. She made a caveat that in size some of the *P. aviculare* fruits exceeded those of modern populations but ascribed this to possibly greater variability within the Weichselian genotypes.

Polygonum rurivagum Jord. ex Bor.   210: 1*c*

F VIII *Bunny*

This weed, mainly of western and south-central Europe, was recovered by Mrs Wilson from the Roman site at Bunny.

Polygonum aequale Lindm. (*P. arenastrum* Boreau)
                                                        320: 1*d*
F VIII *Bunny*

The Roman site at Bunny has yielded a record of *Polygonum aequale*, a taxon probably of wide distribution in Europe except the far North.

Polygonum maritimum L. 'Sea knotgrass'   320: 3

*l*Wo ?*Selsey*; *l*W ?*Moss Lake*; F IV ?*Moss Lake*

Tentative pollen identifications alone support records of *Polygonum maritimum* from the late Wolstonian at Selsey and from zones III and IV at Moss Lake, Liverpool.

Polygonum viviparum L.   320: 5

Wo *Farnham*; eWo *Hoxne*; W *Thrapston*; eW *Sidgwick Av.*; mW *Barnwell Station, Earith, Waltham Cross, Marlow, Syston, Brandon*; *l*W *Colney Heath, Whitrig Bog*

With the possible exception of the Brandon record cited as 'fruit' all the above records of *Polygonum viviparum* are based upon recognition of its freely shed bulbils, which are sometimes found in early stages of germination (Bell, 1968). They are all from glacial stages, from the early Wolstonian and from all parts of the Weichselian though most abundantly in the Middle Weichselian. In the Late Weichselian the records come from the colder zones I and III. The Weichselian sites in six instances have radiocarbon dates and these span the period between 45 000 and 13 000 years B.P.

 P. viviparum is a circumpolar arctic–montane plant found on the separate central European mountain sys-

tems. It is one of the most ubiquitous of arctic plants, being found throughout the Arctic to latitudes as high as 83° N (Polunin, 1940). In the British Isles it is most common in the Scottish Highlands and the mountains of northern England, but it also occurs in North Wales and occasionally in the isolated mountains of western Ireland. Although it commonly characterizes snow-patch vegetation along with *Salix herbacea* it has yet to be found in the Irish Late Weichselian where remains of this arctic willow abound.

It is apparent that during the last glacial stage *Polygonum viviparum* occurred widely in the English lowlands from which it has since withdrawn. There is no indication in the analyses of Conolly and Dahl (1970) that this withdrawal is directly due to the temperature change since the Late Weichselian.

The pollen of *P. viviparum* is not recognizable separately from that of *P. bistorta*: the combined record follows.

Polygonum bistorta L.   Snake-root, Easter-ledges, 'Bistort'   320: 6

F IV *Star Carr*

Polygonum bistorta fruit has been recorded from east Yorkshire in zone IV of the Flandrian. Within the British Isles today it is commonest in northern England and it is found in the Shetlands. It is very local in Ireland. Though native of northern and central Europe it is absent from most of Fennoscandia. Its pollen is not distinguishable from that of *P. viviparum* (*vide infra*).

Polygonum viviparum L. or **P. bistorta** L.   320: 5, 6

British palynologists generally hold that it is impossible to distinguish the pollen of *Polygonum viviparum* from that of *P. bistorta*, and refer such pollen to '*P. viviparum* type' or '*P. bistorta* type', failing the alternative attribution. We have aggregated all these records in our diagram. The records for this pollen type exist in sub-stage IV of the Hoxnian and the first three sub-stages of the Ipswichian. The main weight of the records is however in the glacial stages with two late Wolstonian occurrences, one in the Early Weichselian (Chelford) and many in the three Late Weichselian zones. Many of the latter are Scottish records and shew repetition in successive zones as in Skye: Loch Mealt (zones I, II, III), Loch Cill Chriosd

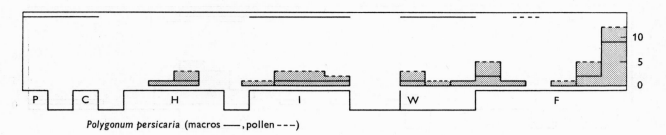

Polygonum persicaria (macros ——, pollen - - -)

(I, II, III, IV) and Lochan Coir' a' Ghobhainn (I/II, III); and in the Scottish mainland Loch Borralan (II, III) and Loch Scionascaig where the record continues into the late Flandrian (zones I, II, III, IV, VIIb, VIIb/VIII).

In view of the associated pollen spectra, at least in the sites on Skye, and the resemblances they shew to existing montane plant communities of which *Polygonum viviparum* is a constituent (H. J. B. Birks, 1969) it seems reasonable to regard the pollen record as mostly that of *P. viviparum*, but this of course remains conjectural.

Polygonum amphibium L.   'Amphibious bistort'
320: 8

Macros: Cs *Mundesley*; I *Osterfield*; *m*W *Upton Warren*; *l*W *Jenkinstown*; F VIII *Silchester, Bermondsey*

Pollen: *l*Wo *Selsey*; I ?*Histon Rd*; *l*W *Rhosgoch Common, Nazeing, Oulton Moss, Hockham Mere, Moss Lake*; F IV, V *Moss Lake*, VIIb, VIII ?*Ehenside Tarn*

Both fruit and pollen of *Polygonum amphibium* are recorded fossil although in two sites the pollen is referred to '*P. amphibium* type'.

The records jointly include one from the Cromer Forest Bed series, one from the late Wolstonian and one each from sub-stages III and IV of the Ipswichian interglacial. Weichselian records are more numerous, with an important fruit record from the Middle Weichselian and records from all zones of the Late Weichselian. Some of the pollen records shew the persistence often seen in aquatic plants, viz. Hockham Mere (zones II and III), Rhosgoch Common (I, III, IV) and Moss Lake (III, IV, V). Although these demonstrate persistence into the early Flandrian, records are absent thereafter until zones VIIb and VIII at Ehenside Tarn (pollen of *P. amphibium* type) and the two fruit records from Roman sites.

It seems likely from the fossil record that *P. amphibium* has persisted in this country through from at least the middle of the last interglacial. It is an aquatic or semi-aquatic plant generally distributed throughout the British Isles: in Scandinavia it is common only in the south, but it occurs in scattered localities to very high latitudes.

Polygonum persicaria L.   'Persicaria'
320: 9

The fruits of *Polygonum persicaria* are more positively identifiable than the pollen which appears in palyno-logical accounts either as 'cf.' or '*P. persicaria* type'. None the less the record-head diagram shews the two types of evidence as congruent. They jointly shew presence in the middle Hoxnian and throughout the Ipswichian interglacial, after a single fruit record from the Cromer Forest Bed series. There is a pollen record from the late Wolstonian, and the plant was present in Late Weichselian zones I, II and III and early Flandrian zones IV and V. The fruit was at Apethorpe (a riparian site) in zone V, and at Moss Lake in zones V/VI and VIIa. It seems reasonable to presume therefore persistence in the country from at least the opening of the Late Weichselian to the expansion in zones VIIb and VIII where the archaeological associations are numerous, there being fruit identifications of Iron Age (1), Roman (4), Norman (1) and Mediaeval age (4).

It is evident that *P. persicaria*, like *P. aviculare*, behaved as a weed once cultivation had been introduced to this country, but that in earlier times, in this instance interglacial and Late Weichselian, it certainly formed part of the natural vegetation. Simmonds (1945) writes that it is 'always in disturbed communities such as waste ground, arable land, ditches, roadsides, etc.' and of course one cannot say that it may not have been re-introduced.

P. persicaria occurs throughout Europe (including Iceland); in Scandinavia it is frequent only in the south, though scattered localities in the west extend to 70° N latitude.

Polygonum lapathifolium L. (*P. scabrum* Moench)   'Pale persicaria'
320: 10

C *Norfolk Co.*; *l*Wo *Selsey*, I *Selsey, Bobbitshole*; *m*W *Waltham Cross*; F VII *Dunshaughlin*, VIIb *Glenluce, Culbin Sands, Gorey*, VIII *Glastonbury Lake Village, Maiden Castle, Manchester, Birrens, Silchester*, ?*Ehenside Tarn*, ?*Lough Gur, Larkfield, Hungate, Dublin, Lissue*

All records for *Polygonum lapathifolium* are for nutlets. They include one from the Cromerian, and two from sub-stage II of the Ipswichian interglacial. Of glacial stage records there is only one from the late Wolstonian and one from the Middle Weichselian. These records, though sparse, suffice to establish the ancient native status of the pale persicaria and suggest its possible perglacial survival. The Flandrian records are numerous, late and associated in every case but two with an archaeological excavation.

These cultural associations are Bronze Age 3, Late Bronze Age to Iron Age 1, Iron Age 3, Roman 4, Anglo-Saxon 2, Norman 1, Mediaeval 4.

Although at present a corn-field weed, there is much evidence suggesting that it was formerly one of the numerous plants whose fruits were gathered for food by prehistoric man, especially when he lived on poor soil or under population pressure. In the stomach contents of both the Tollund and Grauballe bog-burials of the 3rd to 5th century A.D. fruits of *Polygonum lapathifolium* occurred in great numbers, many intact and very well preserved, and often having been ingested with the attached glandular perianths. It seems generally agreed that the weeds along with cultivated grain, chiefly barley, were made into gruel. It seems likely that the usage was similar in the British Isles.

P. lapathifolium has been very strongly spread by arable cultivation and occurs throughout the British Isles. In Scandinavia it is common only south of about 64° N but it occurs in scattered localities inside the Arctic Circle.

Polygonum nodosum Pers. (*P. maculatum* (Gray) Dyer ex Bab., *P. petecticale* (Stokes) Druce) (Fig. 77)   320: 11

Macros: C *Norfolk Co.*; H *Clacton*; lW *Ratoath*; F IV *Hockham Mere*, VIII *Lissue*
Pollen: F VIIb *Frogholt*, VIIb/VIII ? *Wingham*

Fruit records of *Polygonum nodosum* establish it as present in two interglacials, zone II of the Late Weichselian in Ireland, in a zone IV site in East Anglia, and at a mediaeval site also in Ireland. The two pollen records are both from the late Bronze Age or early Iron Age levels at sites with many other indices of arable cultivation. *P. nodosum* has a rather southern range in the British Isles, being absent from Scotland and northern Ireland.

Fig. 77. Fruit of *Polygonum nodosum* Pers. (*P. petecticale* Druce) from the Late Weichselian muds at Hockham Mere, Norfolk.

Polygonum lapathifolium L. or **P. nodosum** Pers.
320: 10, 11

lWo *Ilford*; I *Wretton*; lW *Moss Lake*; F IV *Moss Lake*, VIIb *Frogholt*, VIIb/VIII *Wingham*, VIII *Apethorpe*

The fruit records referred either to *Polygonum nodosum* or *P. lapathifolium* have a time distribution very like that of other weed species in the genus: again, occurrence in an interglacial stage and in two glacial stages including the Weichselian zone III with persistence into Flandrian zone IV at Moss Lake.

Polygonum hydropiper L.   Water-pepper   320: 12

Pa *Norfolk Co.*; lA *Hoxne*; mW ?*Ponders End*; F VIIb *Drumurcher, Frogholt, Crossness*, VIIb/VIII *Wingham*, VIII *Glastonbury Lake Village, Silchester, Bermondsey, Hungate, Shrewsbury*

Fruits of *Polygonum hydropiper* are recorded from the Pastonian interglacial and the late Flandrian. Between these are two records from glacial stages, the late Anglian and, tentatively, from the Middle Weichselian. In Flandrian zone VIIb the three records include two from the Bronze Age. At Wingham there is one record from the zone VIIb/VIII transition, and from zone VIII there are two Roman, one Iron Age, one Norman and two Mediaeval records.

P. hydropiper is primarily a plant of marshes and river banks, possibly a habit not out of character with its recovery from so many archaeological sites. It grows throughout the British Isles except the far northern isles, and in Scandinavia has a range very like that of *P. lapathifolium*.

Polygonum mite Schrank (*P. laxiflorum* Weihe)   320: 13

F VI *Treanscrabbagh*, VIIb *Helton Tarn*, VIII *Ehenside Tarn, Finsbury Circus*

The four fruit records of *Polygonum mite* are all from Flandrian zones VII and VIII: they include one Roman site.

Polygonum minus Huds.   320: 14

C *Norfolk Co.*; lW *Abbot Moss*; F VIIb *Helton Tarn, Drumurcher*, VIII *Ballingarry Downs*

Fruit records of *Polygonum minus* include one from the Cromerian interglacial, one from the zone I/II transition in north-west England, and three from zones VIIb and VIII of the Flandrian, the latest an Irish mediaeval site.

Polygonum convolvulus L.   Black bindweed   320: 15

Cs *Pakefield*; I ?*Histon Rd*; lWo *Ilford*; F VIIa/b *Shustoke*, VIIb *Helton Tarn, Dean Bridge, Culbin Sands, Itford Hill*, VIII *Ehenside Tarn, Aldwick Barley, Glastonbury Lake Village, Pevensey, Finsbury Circus, Silchester, Newstead, Isca, Hungate, Dublin, Lissue, Plymouth, Larkfield*

There are records of the nutlets of *Polygonum convolvulus* from the Cromer Forest Bed series, tentatively from sub-stages III and IV of the Ipswichian interglacial and from the later part of the succeeding glacial stage. Thereafter the many records are all Flandrian but none earlier than the zone VIIa/VIIb transition, and all save three are associated with an archaeological excavation. There are three from the late Bronze Age, two from the Iron Age, five are Roman and five mediaeval. The Flandrian record accords with the status of black bindweed as a plant of waste places and of arable ground, but the earlier records shew, as with other polygonums, an early native status. There is here however no evidence of presence in the Weichselian or early Flandrian and we cannot exclude the possibility of late reintroduction of the plant by Neolithic and later farmers.

P. convolvulus grows throughout the British Isles and in Scandinavia it is common as far north as the Arctic Circle and occurs locally well within it.

Fagopyrum esculentem Moench   Buckwheat   321: 1

F VIII *?Newstead*

According to Hegi *Fagopyrum esculentem* is an Asiatic plant brought westwards by man from the Pontic region, and largely cultivated in some countries, especially upon light sandy soils. The record by Tagg (1911) from the Roman site at Newstead is tentative only. For the European evidence see Munaut (1967).

Koenigia islandica L. (Fig. 78; plate XXIIIa–d)   322: 1

lW Loch Droma, Loch Mealt, Windermere, Abbot Moss, Woodgrange, Roddansport, Magheralaghan, Burn of Benholm, Loch Park, Loch Borralan, Bigholm Burn, Whitrig Bog, Loch Kinord, Drymen, Ballaugh, Oulton Moss, Loch Craggie, Kirkmichael, Ballyduggan, Cannons Lough; F IV Bigholm Burn

Koenigia islandica L. has pollen of a highly characteristic morphology, the ellipsoidal pores forming a pronounced pattern and the surface spines having an individual distribution pattern and size. None of the other species described by Hedberg (1946) as having *Koenigia*-type pollen occurs in or near to the British Isles. Miss Andrew first identified the pollen from Late Weichselian zone III deposits at Whitrig Bog in Berwickshire at a time when *Koenigia* was not known still to be a living member of the British flora, but since that time it has been confirmed that it grows freely and naturally at several localities in Skye and Mull in open sandy or silty habitats where the soil is liable to disturbance.

Likewise the fossil occurrences have greatly increased in number. Virtually limited to the Late Weichselian, there are two from zone I, seven from II, twelve from III, four from the transition zone III/IV and one from zone IV. They include numerous instances of local persistence as at Roddansport and Woodgrange (II, III, III/IV), Loch Borralan and Loch Park (II, III) and Bigholm Burn (III and IV). At the original site of pollen identification Miss Conolly has subsequently identified macroscopic remains of *Koenigia*. The distribution map of fossil occurrences (fig. 78) shews how widespread the species was in the Late Weichselian and that it must have suffered great retraction and disjunction of range subsequently. This is not in the least surprising since an annual plant of such dwarfness must be permanently tied to habitats with disturbed soils and absence of closed vegetational cover.

K. islandica is a plant of circumpolar distribution, extending north almost to 80° N in Spitzbergen: it occurs along the whole length of the high Scandinavian mountains and up to the North Cape, but is absent from central Europe. It grows also in the Southern Hemisphere. In the Arctic it is common on sandy or muddy sea or lagoon shores, on the margins of freshwater pools, and on 'open, muddy areas such as the surface of polygons' (Polunin, 1940). In the Storr in Skye it is described by Ratcliffe (1959) as confined to stony ground with some degree of flushing and to scree; he adds that *Koenigia* is

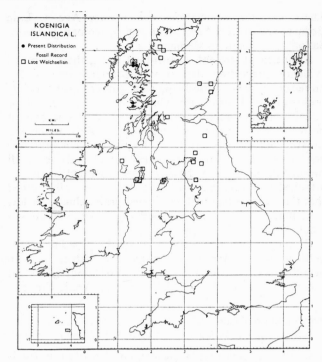

Fig. 78. Present and Late Weichselian distribution of *Koenigia islandica* L. in the British Isles.

not infrequently found in soils affected by solifluxion, a process prevalent everywhere in the Late Weichselian on suitable terrain. Birks (1969) has described how in modern *Koenigia* communities in Skye the pollen is absent from moss cushions but occurs in large quantity in the silt of flushes; he suggests that *Koenigia* pollen reaches lake sediments during phases of mineral inwash, an explanation agreeing with the fossil evidence. No doubt the continuous curve for *Koenigia* pollen with frequencies up to 5 per cent of total pollen, found by Moar (1969a) at Bigholm Burn, were also associated with concentration by water transport.

Conolly and Dahl (1970) point out that so dwarf a plant as *Koenigia* must be closely controlled by soil temperatures. They correlate its Scottish range with the 21 °C maximum summer temperature isotherm among the lowest values in the arctic–montane species they consider. This is at variance with the results quoted by Löve and Sarkar (1957) who report that the plant survives temperatures of 40 °C or more for several days.

Oxyria digyna (L.) Hill   'Mountain sorrel' (Figs. 79, 80)
                                 324: 1

Macros: *A Corton; Wo Farnham; eWo Marks Tey; W Thrapston; mW Angel Rd, ?Temple Mills; lW Corstorphine, Nant Ffrancon, Ballybetagh, Low Wray Bay, Drumurcher, Whitrig Bog*

Pollen: *lW Loch Fada, Loch Cill Chriosd, ?Mapastown, Lochan Coir' a' Ghobhainn, ?Kirkmichael, ?Ballaugh, ?Loch Dungeon; F VIIb/VIII Loch Dungeon*

The bicarpellary fruits of *Oxyria digyna* are easily recognizable when, as often happens, they retain all or part of their lateral wings (fig. 79). They have been recorded from the glacial stages exclusively, beginning with records from the Corton Beds interstadial and the Wolstonian, but with the remainder in the Weichselian. There is one undefined Weichselian record by Miss Bell, and two Middle Weichselian records, one tentative by C. Reid from the Lea Valley Arctic Plant Bed. Two records from Late Weichselian zone II follow and three from zone III: they are all by different investigators and represent England, Wales, Scotland and Ireland.

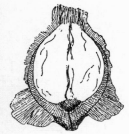

Fig. 79. Fruit of *Oxyria digyna* L. (×12) from Late Weichselian (zone III) deposits at Ballybetagh, near Dublin (Jessen & Farrington, 1938).

Although Faegri and Iversen's key permits resolution of pollen to the level of '*Oxyria* type', palynologists have recently made unqualified pollen records of *O. digyna*, especially H. J. B. Birks whose records for Skye include Loch Fada (zone I), Loch Cill Chriosd (zones I, II, III), and Lochan Coir' a' Ghobhainn (zones I/II, III, III/IV). Franks and Pennington also record it unqualified from the Esthwaite Basin in zone III, but all other pollen records are tentative, 'type' or '*Oxyria* or *Rumex crispus*'.

Oxyria digyna is an arctic–alpine plant of general arctic distribution reaching high altitudes (82° 48′ N in north Greenland) and favouring open soil or rock-crevices where competition is slight (Polunin, 1940). In the British Isles it occurs most commonly in the mountains of northern Scotland and northern England, but it has outposts also in north Wales, western Ireland and Tipperary (fig. 80). The research of Conolly and Dahl shews that its present range corresponds with the 23 °C maximum summer temperature summit isotherm: they map its present and fossil distribution and these make it evident that in the Weichselian, and probably in earlier glacial stages, *Oxyria* occupied the lowlands of the British Isles as well as the highlands to which it is now confined.

Rumex sp.   325

The record-head diagram includes records both of fruits and pollen referred to the genus *Rumex*: undivided stage records and those for transitions have been omitted, but the overall picture is quite unaffected, especially since the two types of identified material lead to similar conclusions. From the Pastonian and Cromer Forest Bed time onwards there is evidence of *Rumex* in all sub-stages of the interglacials and in early and late glacial stages. There is adequate evidence of presence in Early and Middle Weichselian and site records are thereafter abundant in all zones of the Late Weichselian and Flandrian, with a diminution in zones V and VI however that is evident in many families and genera of mainly light-demanding taxa. Not surprisingly in a genus rich in ruderals and weeds, the late Flandrian records include many from archaeological sites, viz. Bronze Age 4, Iron Age 2, Roman 5 and Mediaeval 4: in part this is a consequence of the fruit occurring in grain stores and in part to the way in which marsh and aquatic plants tend to be well represented in settlement areas.

Consideration of the relative frequency of *Rumex* pollen zone by zone through the Late Weichselian and Flandrian does something to enlarge our idea of the changing abundance of the genus (table 17).

It is evident that through the Late Weichselian, and outstandingly in zone I, sites with high or very high percentages of *Rumex* pollen are common, as if large communities of docks grew in many parts of the landscape. We can glimpse what the species may have been from the individual accounts that follow. Zone IV is transitional but through the middle Flandrian few sites have frequencies above 0 to 2 per cent of the total tree pollen, although the total number of sites registering dock pollen is not much reduced. It seems that extensive communities dominated by *Rumex* have become rare, though evidently there was persistence of some species, perhaps shade-tolerant in fen and riparian situations within the general forest cover. In zone VIIb and especially in VIII there is substantial re-expansion in the number of sites with high pollen percentages of *Rumex*, an undoubted consequence of woodland clearance and the spread of agriculture. Miss Turner's analyses of surface samples collected in regions dominated respectively by pastoral

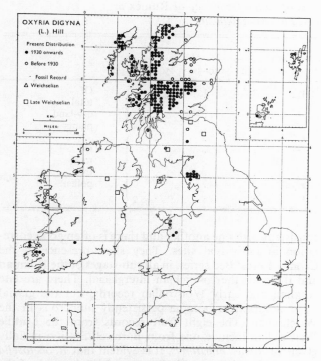

OXYRIA DIGYNA
(L.) Hill

Present Distribution
● 1930 onwards
○ Before 1930

· Fossil Record
△ Weichselian

□ Late Weichselian

Fig. 80. Present and fossil distribution of *Oxyria digyna* (L.) Hill in the British Isles.

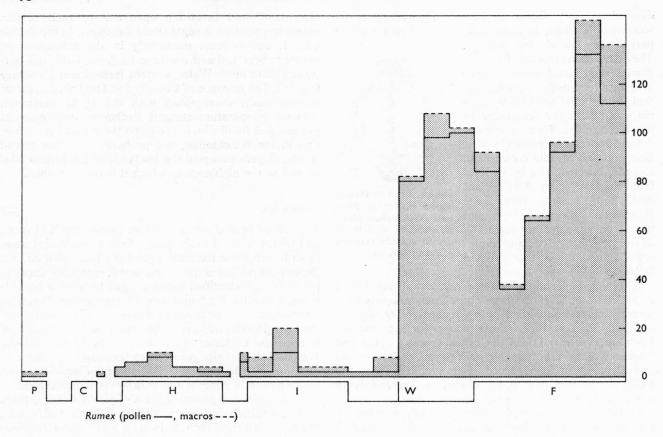

Rumex (pollen ——, macros - - -)

TABLE 17. *Site numbers with varying levels of pollen frequency of* Rumex *in British Late Weichselian and Flandrian deposits*

	Percentages									
	+	0–2	2–5	5–10	10–20	20–30	30–40	40–50	50–60	
I	I	4	6	4	7	4	2	I	.	Of
II	6	6	7	19	.	.	.	.	.	total
III	I	6	9	10	10	.	.	.	.	dry-land
IV	I	6	8	7	I	.	.	.	.	plants
IV	2	2	3	5	3	I	I	.	.	Of
V	.	8	4	.	.	.	.	.	.	total
VI	5	16	2	I	.	.	.	.	.	tree
VIIa	6	26	I	.	I	.	.	.	.	pollen
VIIb	7	30	5	2	I	I	.	.	.	
VIII	2	15	14	13	4	I	.	.	.	

or arable economy (Turner, 1964) indicate that docks are strongly associated with both, whilst present-day field experience suggests that there are many anthropogenic communities in which one species of dock or another is abundant.

Rumex acetosella L. *sensu lato*   Sheep's sorrel (Fig. 81*c*)
325: 1

As will be seen from the record-head diagram, fruit and pollen identifications of *Rumex acetosella sensu lato* are

consistent with one another and there is often good correlation at individual sites. They shew presence in the Cromer Forest Bed series, in all sub-stages of the Hoxnian and three of the Ipswichian interglacial. There are late Anglian and early Wolstonian records, but the glacial stage occurrences become persistently numerous in the Weichselian. The eight fruit records from the Middle Weichselian include one from Barnwell Station initially given for *Rumex maritimus*. It is evident that *R. acetosella* was prevalent (along with *R. acetosa*) in all zones of the Late Weichselian, but that it diminished greatly in zone IV

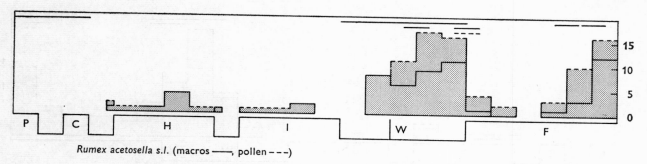

Rumex acetosella s.l. (macros ———, pollen - - -)

of the Flandrian, an effect not seen in most of the genera and families of the Late Weichselian herbs until zone v. Possibly *R. acetosella* responded to the closing of the herbaceous vegetational cover in zone IV in a way that taller species did not: the two records in zone v are both from high altitude sites in Snowdonia that still supported the sheep's sorrel in zone VIII. After zone v there are no records until the period of disforestation and most of the fruit records are from archaeological sites which include

one from the Bronze Age, seven Roman and four Mediaeval.

The fossil record indicates that *Rumex acetosella sensu lato* has persisted through both glacial and interglacial stages in this country from as far back at least as the Cromerian. It is a plant that occurs commonly throughout Scandinavia to its extreme northern limits and Jessen comments that at the present day in Ireland 'it is common in all districts...from sea-level up to the naked tops of

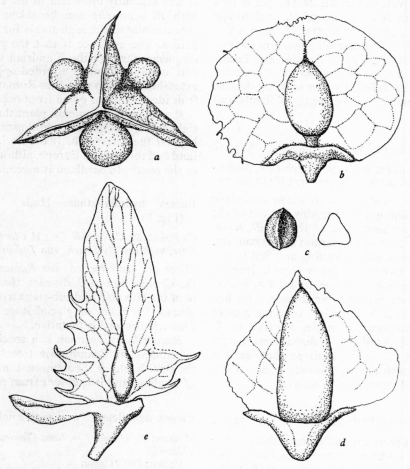

Fig. 81. Various remains of *Rumex* spp. from different Irish deposits (Jessen, 1949): *a*, *b*, two views of fruit and perianth of *Rumex crispus* (Ballybetagh, zone II); *c*, nut of *R. acetosella* (Ballybetagh, zone III); *d*, perianth fragment of *R. obtusifolius* (Dunshaughlin, early zone VIII); *e*, perianth of *R. hydrolapathum* (Killywilly, zone VI).

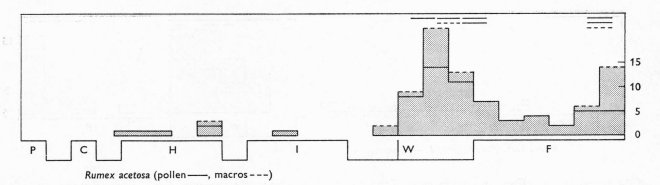

Rumex acetosa (pollen——, macros - - -)

the mountains, where it grows with other of the prevailing summit plants such as *Empetrum* and alpines such as *Salix herbacea* and *Carex rigida*'.

Rumex angiocarpus Murb.   325: 1*b*

W *Thrapston*; *m*W ?*Oxbow*

Miss Bell writes of her identification of fruits of *Rumex angiocarpus* from the Weichselian site at Thrapston that 'the persistent perianth is a feature of *R. angiocarpus* alone out of the species of the aggregate' (*R. acetosella*). She subsequently made a tentative identification of fruit from the Middle Weichselian site at Oxbow near Leeds. *R. angiocarpus* is one of a number of plants of southern range to which Miss Bell calls attention as being present in the Weichselian deposits of the south midlands and Fenland border.

Rumex tenuifolius (Wallr.) Löve   325: 1*c*

*l*W ?*Hawks Tor, Newtonbabe, Ballaugh, Ballybetagh, Mapastown, Drumurcher, Knocknacran, Cahercorney*; F VIII *Lissue*

Within *Rumex acetosella sensu lato* there are gymnocarpic species in which the shining nutlets readily fall out of the perianth segments at maturity: one of these is *R. tenuifolius*, the status of which is still uncertain in this country. On the strict criteria imposed by Perring and Sell (1968) it seems that *R. tenuifolius* may not now grow in Ireland, whereas all the fossil records save one are Irish or Manx, including the mediaeval record from Lissue. The last apart, all records are Late Weichselian, covering zones I to III. Between the gymnocarpic species the fruit of *R. tenuifolius* is distinguished only by its dimensions (length 0.9 to 1.3 mm, width 0.6 to 0.8 mm) so that, as was suggested for the Cornish record (Conolly *et al.*, 1950), the identifications need treating with reserve.

Rumex acetosa L.   Sorrel (Plate XX*l*)   325: 2

Identifications both of fruit and pollen contribute to the record-head diagram: we have omitted a smaller number of pollen records made either to '*R. acetosa* or *R. acetosella*', or to '*R. acetosa* type' which includes both these species. The omission does not modify the result pre-

sented. Of the records that Miss Bell has made from the Middle Weichselian at Earith she writes that 'the perianths...are well known fossils distinguished by a marginal area through which the lateral veins run horizontally' and that 'they may or not still bear a narrow tubercle'.

It appears that *R. acetosa* occurred both in the Hoxnian and the Ipswichian interglacial as well as the late Anglian. It was evidently abundant in the Late Weichselian, and with *R. acetosella* can be taken to have contributed substantially to the high totals for *Rumex* pollen in this period. The evidence is that the plant persisted in low frequency through the Flandrian until plant husbandry and human settlement afforded opportunity for renewed extension. In seven sites, four Roman and three mediaeval, fruit identifications come from occupation sites.

R. acetosa is a common plant throughout Scandinavia right to the North Cape and it ascends the mountains far beyond the forest belt. It occurs in Greenland and Iceland and throughout Europe, although normally montane in the south. In Scotland it ascends to 4050 ft.

Rumex hydrolapathum Huds. 'Great water dock' (Fig. 81*e*)   325: 4

Cs *Pakefield*; C *Norfolk Co.*; H *Clacton*; I *Ilford*; F IV ?*Star Carr*, VI *Killywilly Lough*, VIIa *Tadcaster*

There are fruit records for *Rumex hydrolapathum* from the Cromer Forest Bed series, the Cromerian, sub-stage III of the Hoxnian and sub-stage II of the Ipswichian interglacial. There are no glacial stage records and the early Flandrian record is tentative.

Rumex hydrolapathum is a species of rather southern distribution in Scandinavia (reaching about 61° N). In the British Isles it is commonest in the south-eastern half of England and it is absent from northern Scotland.

Rumex aquaticus L. non auct. angl. (Fig. 82)   325: 7

Macros: *l*W *Hawks Tor, Nant Ffrancon*, ?*Ballybetagh*; F IV, V ?*Nazeing*
   Pollen: *l*W ?*Canna*

A single complete fruiting perianth of *Rumex* found in the Late Weichselian Allerød muds at Hawks Tor,

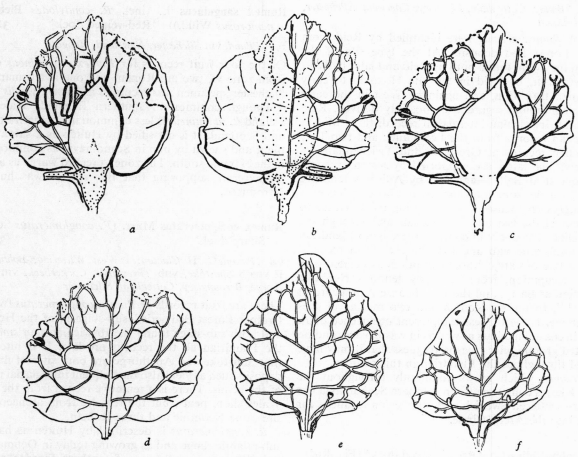

Fig. 82. *Rumex aquaticus* L. non auct. angl: *a, b, c*, three different views of an almost complete flower from the Late Weichselian (zone II) muds at Hawks Tor kaolin pit, Bodmin Moor, Cornwall; *d*, posterior inner perianth segment of the same flower with one sterile stamen; *e, f*, inner perianth segment from herbarium flowers of *R. aquaticus* L. non auct. angl., *e*, from the variety described as *R. hippolapathum* Fries, and *f*, from the variety *R. heleolapathum* Drej.

Cornwall, was sufficiently complete and well preserved to permit its clear distinction from *R. longifolius* DC. (= *R. domesticus* Hartm.) and *R. alpinus* L., and its recognition as *R. aquaticus* L. This identification was confirmed by J. E. Lousley. It was also found in zone II deposits in Snowdonia by Burrows (unpublished). The same species was more tentatively identified by Jessen from an undated (but possibly Late-glacial) layer at Ballybetagh, and again tentatively from poorly preserved material from the early Flandrian at Nazeing in Essex. Flenley and Pearson (1967) made a tentative identification of pollen of *R. aquaticus*, again from the Late Weichselian zone II, in the Scottish island of Canna.

In the British Isles the plant is known living only from the shores of Loch Lomond (Stirlingshire) where it was found by Mackechnie in 1935 (Lousley, 1939). All previous records were erroneous and referred to *R. longifolius* DC., to which the name 'aquaticus' has in the past been misapplied by British authors. *R. aquaticus* has a wide range in Europe but is only scattered in the west. It extends through northern Asia to Siberia, and in Europe extends to Finmark (70° 40′ N).

It seems probable that its British distribution has been restricted since the Late Weichselian.

Rumex longifolius DC. (*R. domesticus* Hartm.; *R. aquaticus* auct., non L.) 325: 8

F VIIb *Itford Hill*

Helbaek (1953*b*) records fruit and perianth of *Rumex longifolius* from the Late Bronze Age settlement at Itford Hill along with *R. crispus*. The record is of particular interest since *R. longifolius* does not now occur in Great Britain south of the Humber–Mersey latitude: it is absent from Ireland.

Rumex crispus L. 'Curled dock' (Fig. 81*a, b*) 325: 11

Cs *Sidestrand*; H *Hoxne*; eWo *Hoxne*; I *West Wittering*; mW *?Ponders End, Angel Rd, Hedge Lane*; lW *Garvel Park, Twickenham, Ballybetagh, ?Nazeing*; F VIIb *Barry, Itford Hill,*

VIII *Isca, ?Bunny, Cottenham, Pevensey, Caerwent, Silchester, Larkfield, Lissue*

Fruits of *Rumex crispus* were identified by Reid from Cromer Forest Bed deposits. At the type site of the Hoxnian they were found both by Reid and later by West in sub-stage III, by Reid in the late Hoxnian to early Wolstonian transition, and again by West in the early Wolstonian though regarded as possibly derived. Reid has also an Ipswichian interglacial record, three Middle Weichselian records from the Lea Valley and two Late Weichselian records at Greenock and Twickenham respectively. The latter are supported by Jessen's records from zone II at Ballybetagh and by Allison's tentative record from the zone II/III horizon at Nazeing. There is accordingly every reason for attributing native status to *R. crispus* in the last two interglacials and through the Weichselian, but oddly in contrast there is no Flandrian record until zone VIIb and the late Flandrian records, though numerous are, without exception, associated with human occupation, from the early tenuous Neolithic association at Barry, and the Late Bronze Age record at Itford Hill, to the six Roman and two mediaeval sites listed above. None the less the present occurrence of *R. crispus* in coastal habitats as well as in waste places and on cultivated ground, gives reason to guess that it may have survived throughout the Flandrian in this country.

R. crispus, which grows commonly throughout the British Isles, is common also in southern Scandinavia but there are scattered Swedish and Norwegian localities for it right into the Arctic Circle.

Rumex obtusifolius L.   'Broad-leaved dock' (Fig. 81*d*)
                                                           325: 12

Cs *Pakefield*; H *Hoxne, Clacton*; I *West Wittering*; IW *Close-y-Garey*; F VIII *Dunshaughlin, Manchester, Bermondsey, Newstead, American Sq., Silchester, Hungate, Plymouth, Lissue*

Fruits of *Rumex obtusifolius* have been recorded from the Cromer Forest Bed series, from sub-stage III of the Hoxnian at two sites and from an undefined time in the Ipswichian interglacial. There is one Late Weichselian (?zone II) record from the Isle of Man, and thereafter, as with *R. crispus*, the subsequent records are all late in the Flandrian and all save the record at Dunshaughlin, associated with archaeological sites, five of them Roman, one Norman and two mediaeval.

R. obtusifolius extends to about 64° N latitude in western Norway; scattered and temporary introductions are, however, found much further north than this in Sweden and Finland particularly. As with *R. crispus*, distribution is widespread through the British Isles today and despite the sparse evidence for Weichselian presence it seems likely that *R. obtusifolius* also has long been native in this country.

Rumex sanguineus L. (incl. *R. condylodes* Bieb. (*R. nemorosus* Willd.))   'Red-veined dock'   325: 14

F VI *Elland*, VIII *Silchester, Newstead, Hungate*

Of the four fruit records for *Rumex sanguineus* one is from zone VI, two are Roman and one is Norman. The Silchester specimen was recorded as *Rumex viridis* Sibth.

Though common in southern England, Wales and Ireland *R. sanguineus* is less common in northern England and Scotland. It is classified by Hultén as a sub-atlantic plant and shewn by him in Scandinavia as common only in the extreme south. The zone VI record warns us against prematurely supposing that this species was humanly introduced.

Rumex conglomeratus Murr. (*R. conglomeratus* Schreb.)   'Sharp dock'   325: 15

Cs *?Pakefield*; H *Clacton*; I *West Wittering, Bobbitshole*; F VIIa/b *Shustoke*, VIIb *?Prestatyn, Cockerbeck*, VIII *?Apethorpe, Bermondsey, Cottenham, Silchester*

There are fruit records of *Rumex conglomeratus* from the Cromer Forest Bed series, sub-stage III of the Hoxnian, and both sub-stage II and the Ipswichian *tout simple*. In the Flandrian all the records are late. It is interesting that Shustoke and Apethorpe are both sites of fluviatile deposits such as yield also the frequent interglacial records of the genus. There is a tentative record from the North Wales shore peats, but the rest are from archaeological sites, one Neolithic and three Roman.

R. conglomeratus is described by Hultén as having a sub-atlantic range and as growing today in Denmark but only just entering the tip of southern Sweden and the island of Gotland. Its Scandinavian, like its British range, greatly resembles that of *R. sanguineus*. There is no evidence of perglacial survival of either here.

Rumex maritimus L.   'Golden dock'   325: 18

The inner fruiting sepals of *Rumex maritimus* have long slender spines which make well-preserved fruits of this species readily recognizable. It has been recorded from the Cromer Forest Bed series and subsequently from the Cromerian itself. There are records from the first three sub-stages of the Hoxnian and all four sub-stages of the Ipswichian interglacial. These Middle Weichselian records accord with records for the transition from the Anglian to Hoxnian, Hoxnian to late Wolstonian and the early Wolstonian itself. Similarly at Moss Lake, Liverpool, records both for zone III and zone IV bridge the Weichselian–Flandrian boundary. Thereafter we have no Flandrian records until that at Mapastown (VIIb or VIII) and two of Roman age.

R. maritimus is rather rare in the British Isles and absent from the north of Britain: in Ireland it is now in two sites only. In southern Norway and Sweden it is found locally in coastal habitats and on waste ground as far north as 60° N, but it occasionally reaches somewhat

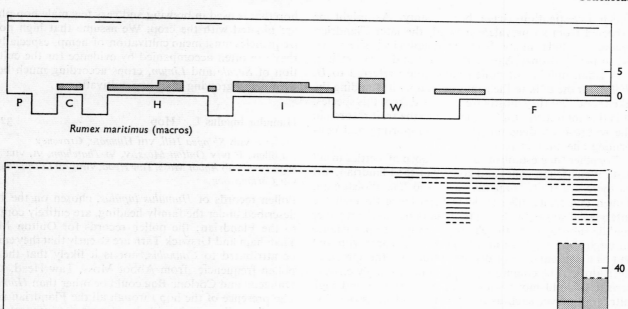

Rumex maritimus (macros)

Urtica dioica (macros ‑ ‑ ‑), Urtica (pollen —)

higher latitude in western Russia. Casual introductions occur further north, but in view of the easy identifiability of the fruits, the glacial stage records cannot be ignored. Thus the inference is strong for ancient persistence as a native plant.

Salisbury (1932) has drawn attention to the fact that although generally regarded as rare, this species, like *R. limosus*, may appear in very large amounts upon drying mud, the result of mass germination following drying of seeds long held dormant in the submerged lake floor. Such might well account for the abundance of the fruits recorded by C. Reid from the Roman Finsbury Park site.

URTICACEAE

Urtica urens L.   Small nettle   328: 1

Cs *Pakefield*; H *Clacton*; F vIIb *Frogholt*, vIII *Lidbury Camp, Godmanchester, Newstead, Caerwent, Pevensey, Bunny, Car Dyke, Silchester, Finsbury Circus, Dublin, Hungate, Ballingarry Downs*

The fruits of *Urtica urens* have been reported from the Cromer Forest Bed series and from the middle of the Hoxnian interglacial. The rest of the records are from Flandrian sites with strong archaeological association. Frogholt and Lidbury are Iron Age, eight are Roman and three mediaeval.

U. urens is a typical north temperate zone weed of light soils, somewhat local in the British Isles and strikingly more frequent in the east than the west, perhaps partly in relation to the distribution of suitable soils. The Flandrian record gives no indication against the view that it could have been accidentally re-introduced with crops in the Iron Age and later.

Urtica dioica L.   Stinging nettle (Fig. 160v; plate XXd)
                                                          328: 2

The shining nutlets of *Urtica dioica* have been found, as the record-head diagram shews, throughout the Pleistocene record from the Cromer Forest Bed time, excepting only the Anglian and Early Weichselian, for both of which investigations have been scanty. As might be expected from so prevalent a weed, the later Flandrian records include many from archaeological sites, viz. Neolithic 3, Bronze Age 3, Roman 8 and Mediaeval 4.

A small number of pollen records are referred to *U. dioica* but the bulk to the genus only, notwithstanding the likelihood that they mostly were derived from this species. It is this total figure that accompanies the fruit records in the record-head diagram, the two supporting and confirming one another remarkably.

Together they establish the abundance of nettles in all Late Weichselian zones and zone IV of the Flandrian, just as there are fruits from the early and late Wolstonian. Pollen and fruits alike indicate persistence through the interglacial stages, with greater abundance in the warmer middle sub-stages. Both again bear witness to persistence through the Flandrian and expansion in zones VIIb and VIII as a consequence of disforestation and the spread of agriculture. The common nettle is now strikingly characteristic of mild moist soils of high phosphate and high nitrifying power such as are often found in association with human settlements, but it has many quite natural habitats such as fen-woods and the 'tall-herb' communities that appear to have been well-represented in the Late Weichselian lowland landscape as in sub-montane situations today (see account for *Polemonium coeruleum*, 388: 1). Response to the modification of the habitat by human settlement seems to be indicated by the nettle as early as the time of Mesolithic culture, with fruit in the appropriate zone IV level at Star Carr, and pollen at the presumed early clearance level in zone VI at Aros Moss (Nichols, 1967).

The high latitudinal and altitudinal ranges of *U. dioica* are consistent with the evidence that it survived the last glaciation and possibly earlier ones in this country.

CANNABIACEAE

British records for the Cannabiaceae may be presumed to refer either to *Humulus lupulus*, a native plant extensively cultivated since the early sixteenth century, or to *Cannabis sativa*, an Asiatic species which we conclude was brought into general cultivation here in Anglo-Saxon time. It is only lately that it has become possible to make separate identification of the pollen of these two plants and very often pollen has been described indifferently as 'Humulus', 'Humulus-type' or 'Humulus–Cannabis'. Accordingly under *Humulus* we list fruit identifications and pollen records that predate introduction of hemp, i.e. records for Flandrian zones up to VIIb. Under *Cannabis* we record the few sites where pollen recognition of *Cannabis* has been positively made, together with those sites where pollen of the *Cannabis–Humulus* type, however described by the author, has been present in such high frequency as to suggest local growth of a crop. This is permissible since it is only the female

hop that is used in brewing and very few male hop plants are planted with the crop. We assume that high pollen frequencies must mean cultivation of hemp, especially as they are often accompanied by evidence for the cultivation of *Secale* and *Linum*, crops according much better with hemp growing than hop cultivation.

Humulus lupulus L.   Hop   329: 1

Fruit: F VIIb *Shippea Hill*, VIII *Hungate, Graveney*
     Pollen: F IV/V *Oulton Moss*, V, VI *Thatcham*, VI, VIIa *Urswick Tarn*, VIIb *Abbot Moss, Taw Head*, VIIb/VIII *Loch Dungeon*, VIII *Corlona Bog*

Pollen records of *Humulus lupulus*, chosen on the basis described under the family heading, are entirely confined to the Flandrian; the pollen records for Oulton Moss, Thatcham and Urswick Tarn are so early that they cannot be attributed to *Cannabis*, nor is it likely that the low pollen frequencies from Abbot Moss, Taw Head, Loch Dungeon and Corlona Bog could be other than *Humulus*. The presence of the hop through all the Flandrian zones accords well enough with its status as a natural component of moist woodlands, especially fen-woods, the provenance of the Shippea Hill record.

There is good historical evidence (Parker, 1934) that cultivation of hops only became common in England in the first half of the sixteenth century, but hops were evidently used in this country for brewing a good deal earlier, for there are records of substantial imports of them from the Netherlands. The timber boat recently excavated from Graveney Marsh, Kent contains considerable numbers of *Humulus* fruits and from the radiocarbon date of the boat, 1080 B.P., it might be conjectured to have carried an imported cargo of hops even as early as the eleventh century. The fruit identifications from late Anglo-Saxon deposits in the Hungate, York were only occasional and cannot be taken to indicate gathering or cultivation.

The hop is a natural liane in alder fen-woods (Tansley, 1939) and there is no reason to doubt its natural presence in other types of moist woodland.

Cannabis sativa L.   Hemp

Macros: F VIII *Bar Hill*
     Pollen: F VI, VIIa, VIIb, VIII ?*Urswick Tarn*, VIIb, VIII ?*Ellerside Moss*, VIIb/VIII ?*Skelsmergh Tarn*, VIII ?*Chat Moss*, ?*Ehenside Tarn, Lindow Moss*, ?*Holcroft Moss, Old Buckenham Mere*

The records given above without interrogation mark are for sites at which pollen of *Cannabis* was specifically determined, those with a question mark are for sites where pollen of the *Cannabis–Humulus* type or *Humulus* has been found in such high frequency and usually with such associated species as *Secale* and *Linum* as to warrant the assumption that we are concerned with crops of hemp grown nearby. D. M. Churchill was able to define a number of characters visible in oil-immersion microscopy that allow recognition of the pollen grains of *Cannabis* and separation from those of *Humulus* (Godwin, 1967*b, c*).

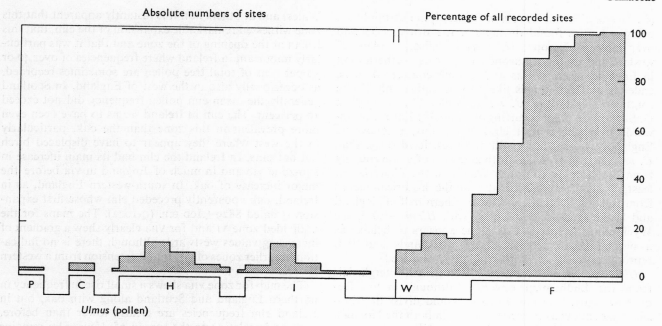

Absolute numbers of sites   Percentage of all recorded sites

Ulmus (pollen)

These were checked and applied by Miss Andrew to the pollen series at Old Buckenham Mere, Norfolk, where by this means it was shewn that there had been substantial local cultivation of hemp in late Saxon and Norman times along with flax and rye (fig. 171). The same criteria allowed the identification of *Cannabis* pollen in the later of the two clearance phases, probably Tudor, recognized by H. J. B. Birks at Lindow Moss, Cheshire, a result conformable with the fact that at that time cultivation of hemp was enforceable by law as a means of ensuring provision of cordage for the navy. All the pollen records given above are from zone VIII. Although that from Skelsmergh Tarn (Walker, 1955*b*) possibly indicates Romano-British cultivation, Chat Moss seems to be Anglo-Saxon and the rest are later.

These results are conformable with the rather scanty evidence from the rest of Europe and accord with the view that hemp was introduced to this country with the folk-migrations that brought the Anglo-Saxons westwards to Britain, and that their multiple ox-teams enabled cultivation of heavier and deeper soils suitable for the crop. Isolated finds, such as the hempen rope from the Roman well at Bar Hill, might well have been due to importation. A fuller account of the ancient history of hemp is given in Godwin (1967*b*, *c*).

ULMACEAE

Ulmus sp.   Elm (Figs. 23, 83)   330

Macros: Cs *Happisburgh*; I ?*Grays, Trafalgar Sq.*; F VI *Glenballyemon, Cushendun*; VI or VII *Blashenwell*, VI/VIIa, *Farnham*, VIIa *Island Magee, Cushendun*, VIIa + VIIb *Peacock's Farm*, VIIb *Notgrove, Nympsfield, Harrow Hill, Whitehawk Camp,*

Skendleby, VIII *Maiden Castle, Meon Hill, Trundle, Radyr, Sudbrook, Bunny, High Penard, Great Wymondley, Caerau, Rotherley, Dunshaughlin*

Identification of *Ulmus* in sub-fossil material in the vast majority of instances depends upon recognition on the one hand of wood or charcoal, or on the other, of pollen. Since in neither case can an effective recognition of species be made, we have aggregated all records under the genus, although some investigators have tentatively or without qualification referred their material to *U. glabra* Huds. or to *U. procera* Salisb.

If we confine our attention to records of charcoal, wood and leaves we find that it is recorded from the Cromer Forest Bed series and tentatively from the Ipswichian interglacial, although the latter has also yielded a fruit identification from sub-stage II in the Trafalgar Square excavations. The great majority of other macroscopic records are of wood or charcoal from archaeological sites. These embrace Mesolithic 2, Neolithic 4, Early Bronze Age 1, Iron Age 5, Roman 5, with one transitional Iron Age to Roman and one mediaeval. The Roman well record at Bunny, Notts. was of buds. None of these records is earlier than Flandrian zone VI.

Pollen of the genus *Ulmus* is readily recognizable with its thick rugose exine and five equatorial pores inconspicuously set at surface level: it is produced relatively freely and seems to preserve well. There have been many attempts to recognize the pollen of individual species but this has certainly not been achieved by any investigator of fossil material from the British Isles. The taxonomy and biogeography of British elms are much complicated by hybridization, planting and incomplete analysis of information, but the line suggested by *Flora Europaea* and the *Atlas of the British Flora* is clear. Three species are possibly native to the British Isles: *Ulmus glabra* Huds.

(*U. montana* With., *U. scabra* Miller) quite certainly so, being a species found throughout the British Isles and over most of Europe; *U. procera* Salisb., a plant of western and southern Europe and near its northern limit in England where it was at one time considered to be endemic and where it is likely to be native only in the south and east; and *U. minor* Miller (*U. carpinifolia* G. Suckow), a plant occurring over most of Europe southwards from the central Ural Mountains, Estonia and England, where it has a still more restricted range than *U. procera*, not growing north and west of a line roughly from the Humber to South Wales. In the Flandrian at least, it may be conjectured that the historical record from Ireland, Scotland and the northern half of England and Wales is concerned wholly with *Ulmus glabra*, the Wych elm, the species that alone appears to behave as a natural component in native woodlands and that reproduces freely and with regularity from seed.

Ulmus pollen is present in all the British interglacials from the Ludhamian onwards, although in low frequencies only. The numerous sites and divisibility into sub-stages allow us to recognize that in both the Hoxnian and Ipswichian, as in the Flandrian, *Ulmus* played the role of mediocratic component of the mixed oak-forest attributed to it in the early schemata of von Post (1930, 1946). It was present in all four sub-stages of the Hoxnian and Ipswichian but most consistently in the two warm middle ones, tending to appear first towards the end of sub-stage I, to have maximal frequencies in sub-stage II, reduced frequency in III and minimal amounts in IV. At most however, the frequencies are below 5 per cent in the Hoxnian and a little higher in the Ipswichian, and nothing like the values encountered in the Flandrian. Although *Ulmus* pollen has been recorded from the various glacial stages it has always, no doubt correctly, been regarded as derived. This we presume to be the case also for reports of *Ulmus* pollen from a moderate proportion of British Late Weichselian sites. In the succeeding Flandrian, a third of sites analysed in zone IV contain a trace of elm, and in zone V half of them. As the record diagram shews, from then elm is represented at 90 to 100 per cent of sites in every zone.

As may be seen from table 18 and from fig. 83, in zone IV elm pollen frequencies seldom rise out of the 0 to 2 per cent category, presumably the result of wind carriage from outside the British Isles; where the percentage is a trifle higher it is part of a very low total tree pollen count in exposed northerly sites. Zone V discloses not only more sites with elm, but a higher mean pollen frequency made more significant by the fact that the tree-pollen/non-tree pollen ratios were much higher. Records of 5–10 per cent of elm pollen, and occasionally more, scattered widely over Great Britain must surely indicate a general growth of the tree in favourable localities: there is a similar record from Newtownbabe in eastern Ireland. Although interpretation of zone VI is made more difficult by the fact that at some sites it has been divided into sub-zones (mostly in England and

Wales) and at others not, it is instantly apparent that this zone witnessed remarkable expansion of the elm, that this began at the opening of the zone and that it was particularly important in Ireland where frequencies of over 40 or 50 per cent of total tree pollen are sometimes recorded, as occasionally also in the west of England. In Scotland generally the mean elm pollen frequency did not exceed 10 per cent. The elm in Ireland seems to have been even more prevalent in this zone than the oak, particularly in the west where they appear to have displaced birch but not pine. In Ireland the elm had its main increase in sub-zone VIb and in much of England in VIa before the major increase of oak. In south-western England, as in Ireland, oak apparently preceded elm whose first expansion is dated 8829 ± 100 B.P. (Q-1022). The maps for the undivided zone VI and for VIIa clearly shew a gradient of increasing values westwards although there is no indication in earlier zones of any initial extension from a western source.

The map for zone VIIa shews a small rise in frequency in northern England and Scotland along with oak, but in Ireland elm frequencies are clearly lower than before, perhaps in relation to the spread of *Alnus*. Elm remains much more prevalent than oak over the central Irish plain and must have been at least codominant in the deciduous woodlands. The feature so apparent in individual diagrams, of the 'elm decline' to values half or less what they formerly were, is readily apparent from consideration of the map of zone VIIb. It evidently obtained throughout the British Isles, though specially evident in Ireland where zone VIIa frequencies were highest. Zone VIII displays little difference from VIIb although the decline apparently has continued save at the occasional site such as Lough Gur in western Ireland. Throughout the Flandrian *Ulmus* prospered particularly in the south and west, establishing itself sooner and rising earlier to maximum frequency in England, but attaining its greatest abundance in Ireland. It never reached more than modest frequency in Scotland, though clearly present widely.

On the Continental mainland the elm's history in comparison with that of other trees of the mixed oak forest was characterized, as in England, by early extension, an early maximum and early retrogression (see Firbas, 1949). In the course of, and towards the end of, the Pre-boreal period a continuous elm curve had established itself over middle Europe, and throughout Germany at least this was soon followed by a vigorous rise, shewing that by the opening of the Boreal period the elm must have been widespread. In correspondence to this it may be noted that numerous macroscopic finds of *U. glabra* have been recorded from Mesolithic settlements of Mullerup age (Boreal) in Denmark. At the change from the early warm period to the middle warm period (the opening of the Atlantic period) *Ulmus* appears to have reached its absolute maximum. Elm pollen frequencies remained high in the older part of the middle warm period (the Atlantic), but at the transition to the younger part (the Sub-boreal) there was a very sharp fall, and frequencies

TABLE 18. *Mean pollen frequency of* Ulmus *in the Flandrian of the British Isles*

	+	0–2	2–5	5–10	10–20	20–30	30–40	40–50	50–60	
					Percentages					
IV	3	9	1	.	.	.	.	.	.	Of total pollen
IV	1	21	2	.	.	.	.	.	.	
V	2	28	20	4	2	.	.	.	.	
VI (not sub-divided)	3	4	17	38	21	2	3	2	.	
VIa	1	2	8	17	13	11	1	.	.	Of total tree pollen
VIb	.	1	3	14	13	8	8	1	1	
VIc	.	2	6	17	20	10	1	1	.	
VIIa	3	9	18	66	63	10	1	1	.	
VIIb	2	28	65	78	9	.	.	.	.	
VIII	2	39	61	63	3	.	1	.	.	

remained low thereafter. Iversen and Faegri took the *Ulmus* fall as the boundary between the middle and late warm period, exactly as the present author was independently using it for Britain and the North Sea region, and as Jessen subsequently used it for Ireland.

A very great deal of careful study and consideration has subsequently been concentrated upon the 'elm decline' in the British Isles as in all western Europe, as may be seen, for instance, in the recent review by ten Hove (1968). The association with the earliest stages of Neolithic agriculture demonstrated initially by Iversen (1941), has been repeatedly confirmed. The phenomenon first demonstrated was a temporary forest clearing with burning of the felled timber and growing of cereals until soil exhaustion led to abandonment and recolonization of the site by returning forest in which however elm seldom recovered its original frequency. Troels-Smith drew attention to the fact that the main elm decline is preceded in many pollen diagrams by a fall which is not accompanied by the signs of weeds, grasses and cereals so typical of the major decline itself. He proposed that this was the product of a distinct form of animal husbandry in which tethered stock were fed upon leafy shoots of deciduous trees, principally elm, which were selectively gathered for this purpose. Subsequently a good deal of evidence has been found in favour of this view and the phenomenon has been very clearly demonstrated in north Lancashire by Oldfield and Statham (1963).

Iversen initially took the view that the elm decline probably had a climatic cause, particularly since it occurred in Denmark at a level where the thermophilous ivy shewed also a strong decline. This has proved a thesis difficult to sustain. Even were the Danish elm decline to have concerned *Ulmus minor*, a species near its northern limit, a similar decline occurs in many countries, Ireland for example, where the tree concerned was the much hardier *U. glabra*. A climatic deterioration is not made likely either by the associated increase of *Fraxinus*, for the ash is not less susceptible to low temperatures than elm, certainly than *U. glabra*.

However, radiocarbon dating has conclusively shewn that the elm decline was remarkably synchronous over wide regions of Europe about 5000 radiocarbon years B.P. It has been found difficult to imagine that so widespread and sudden an effect could have been caused by prehistoric man, and not only a climatic cause but epidemic disease has been advanced as alternative explanation although without direct evidence. What appears from the progress of intensive study is that the association of the elm decline with early Neolithic husbandry grows stronger and stronger, and one inclines to seek mechanisms by which small human populations might have induced such tremendous changes in forest composition. Iversen had suggested that perhaps old elms were killed by ring-barking and the young growth kept pruned as scrub for leaf-fodder. Such scrub was not capable of much pollen production and was very susceptible to browsing when cattle were later set free in the modified forest. In northern Ireland where the elm decline was associated with the appearance of grassland, there seemed no need to postulate the use of leaf-fodder, and the destruction of elm was thought by Morrison (1959) and Smith (Smith & Willis, 1961–2) to have followed forest clearance in areas of more fertile soil where elm was growing in pure stands or was strongly dominant. Whatever the means of primary clearance it is apparent that cattle grazing in cleared areas would substantially prevent elm regeneration. Again in northern Ireland, at Ballynagilly, there is most conclusive evidence from conjoint archaeology, palynology and radiocarbon dating that here it was indeed Neolithic farmers that caused the elm decline and that it was achieved within as little as thirty years (Pilcher, 1969). Pilcher also makes the notable observation that conversion of radiocarbon ages for the Irish elm decline (c. 3000 to 3350 B.C.) to the absolute scale permitted by bristlecone calibration gives a greatly increased time span (3700 to 4350 B.C.) more suitable for this major vegetational change.

The presence of two phases of the elm decline has been associated both in Denmark and the Netherlands with

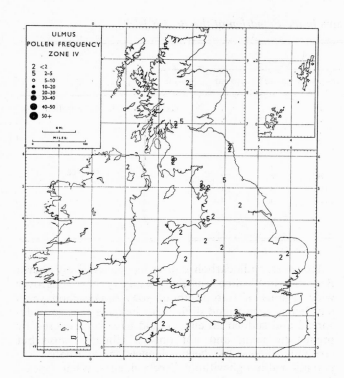

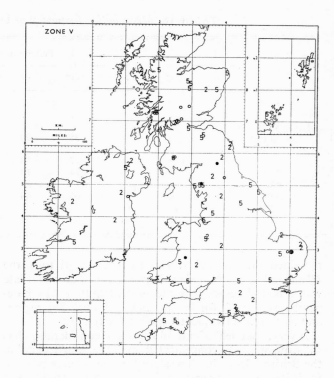

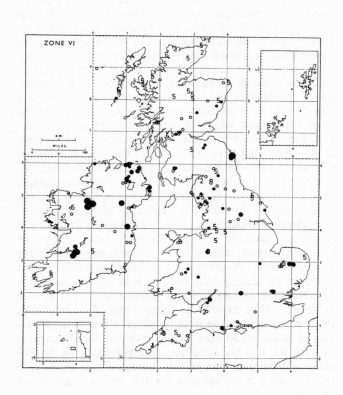

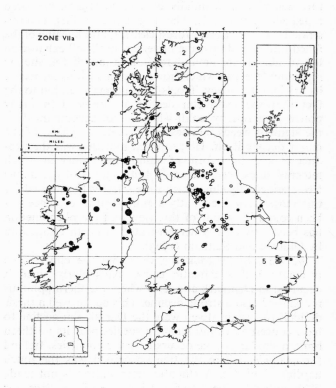

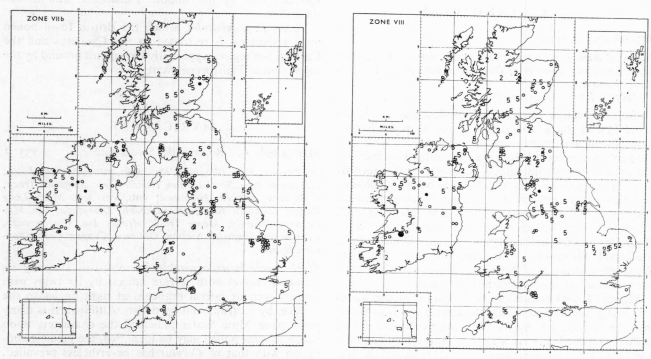

Fig. 83. Pollen frequencies of elm (*Ulmus*) in the British Isles through the six Flandrian pollen zones.
Note especially the prevalent 'elm decline' between VIIa and VIIb.

distinct cultures of prehistoric people (Troels-Smith, 1954; Waterbolk, 1954) but there is so far no evidence of this in the British Isles. Nor do we find here any indication of the alpine 'Piora' oscillation, a phase of lowered temperatures pointed out by Frenzel (1966) as probably occurring about the time of the elm decline. Despite this it would probably be unwise to dismiss altogether the idea of a climatic component in the changes of frequency of the elm at the Atlantic/Sub-boreal boundary, since climate can act so variously upon plant performance. Late spring frosts have very great influence upon the fruiting of elms, although *U. glabra* is less susceptible than *U. minor* and *U. procera* since it normally flowers later in the season. Pennington (1964) has contrasted the poor recovery of elm (after the elm decline) on poor soils as contrasted with calcareous, but the operation of climatic change through soil modification will be both slow and locally variable.

There have of course been indications of modification of the status of elm by man outside the main elm decline. Sims (1973) has been able to shew in his reconsideration of vegetational history at Hockham Mere that there was an earlier absolute non-selective decline of elm frequency perhaps attributable to mesolithic wood clearings. Again it has been shewn by Tallis in the Derbyshire Pennines and by Watts in Ireland that the last three centuries or so have witnessed much planting of elms as also of other trees. This accords with the comprehensive studies of Richens that take account of historical and place-name

evidence as well as careful numerical taxonomy to indicate the probable introduction and establishment of many distinct populations of elms, possibly from pre-Roman times onward, throughout southern Britain. It has been a tree of considerable economic importance, not least as a source of wood for coffins.

MORACEAE

Ficus carica L.   Fig   331 : 1

F VIII *Finsbury Circus, Silchester, Bermondsey, Verulamium, Plymouth, Dublin*

There are four separate identifications of the seeds of the Mediterranean fig, *Ficus carica* L., from Roman sites and two from a mediaeval context. For the latter it is historically known that the figs were imported (see Mitchell, 1972), and no doubt this was so equally for the Roman records, whether the plant was then established in this country or no.

Morus nigra L.   Black mulberry

F VIII *Silchester*

The mulberry, *Morus nigra*, found by Clement Reid in Roman Silchester, may be presumed to have been a Roman introduction.

247

JUGLANDACEAE

Juglans sp.   Walnut   332

Pollen: C *Bacton*; F vIIb, vIII *Taw Head*; vIIb/vIII *Clattering-shaws Loch*; vIII *Snibe Bog, Old Buckenham Mere, Peacock Farm Somerset*
    Macros: *Rotherley, Plymouth*

Juglans has rather characteristic pollen grains, oblate-spheroidal in shape with rather numerous pores irregularly placed and mostly not on the equator. Although Duigan's pollen record from the Cromerian may indicate native status in that interglacial none of the other records do so. The isolated grains from the peat bogs of Galloway and Devon certainly are due to long-distance transport and that at Old Buckenham Mere is at the Norman level. Pitt-Rivers identified charcoal from the Roman site at Rotherley and Dennels recognized remains (?shells) from mediaeval Plymouth. Pollen records from the Late Weichselian deposits of Nazeing, Flixton and Kent have been omitted as almost certainly derived.

Munaut (1967) discussing the Belgian evidence writes that it is generally agreed that the walnut was introduced into the Netherlands by the Romans, pointing out that pollen evidence is always after B.C., and referring to the presence of walnut shells and abundant pollen of *Juglans* together in the Roman pit at Destelbergen. The British history may well have been similar.

Pterocarya sp.

L.T.A. *Ludham*; H *Marks Tey*; eWo *Marks Tey*; I *Trafalgar Sq.*; lW *Esthwaite Basin*; F *Esthwaite Basin*, ?*Cross Fell*

The stephanoporate pollen grains of *Pterocarya* have been recognized from the older Pleistocene deposits in Britain. They constitute up to 2 per cent of the total tree pollen in the Ludhamian, Thurnian and Antian. They are likewise consistently present in sub-stages III and IV of the Hoxnian and in the early Wolstonian at Marks Tey, and sparsely in sub-stage II of the Ipswichian in London. Occasional grains were recorded from the Late Weichselian and early Flandrian of the Lake District. Whilst the latest records must be attributed to distant pollen transport or derivation, West is satisfied that the pollen spectra of the temperate early Pleistocene stages indicate the presence of *Pterocarya* in mixed deciduous and coniferous forest, and Turner links its persistent presence in sub-stages III and IV at Marks Tey (along with *Abies, Vitis* and *Buxus*) with similar occurrences in the Danish, German and Polish Holsteinian. The early Wolstonian record is presumably due to secondary incorporation. The Ipswichian interglacial record is too slight to be used as evidence of local presence, but it seems clear that in England, as in the European mainland, *Pterocarya* was native as late as the terminal part of the Hoxnian. Thus the British history would correspond to that of mainland Europe where, as summarized by Walter and Satrka, it was abundant in the late Pliocene and early Pleistocene,

less common in the middle Pleistocene and absent thereafter.

Today *P. fraxinifolia*, which the British fossil pollen most resembles, is restricted to the Caucasus and the Caspian Sea region, apparently having lost ground by the succession of Pleistocene events.

MYRICACEAE

Myrica gale L.   Bog myrtle, Sweet gale   333: 1

lW ?*Brook*; F IV *Wareham*; vIIa, vIIb, vIII *Shapwick Heath*; vIIb *Roundstone, Southampton, Trundle Mere, Shapwick Track, Emlaghlea*; vIIb/vIII *Hightown*; vIIb, vIII *Aros Moss, Long Range, Crinnagh River, Dernaskeagh, Derryfada*; vIII *Milmorane, Raheelin, Meare Track, Ardlow Inn, Carrowreagh, Cloonacool, Shapwick Station, Oldtown, Coppice Gate, Shapwick Heath*

Myrica gale is recognizable in sub-fossil state by its glandular leaves with serrate apices, by its twigs with stumpy spinous lateral branches at right angles to the surface, by its fruits and by pollen. Although it is many years since Mme Szafer indicated the characters upon which *Myrica* pollen may be identified, the greatest confusion with that of *Corylus* has nevertheless prevailed. Characteristically the *Myrica* grains are more nearly spherical, take a deeper brownish stain with safranin, have a serrate inner border to the pore thickening and possess a somewhat different wall structure. Under the electron microscope there are slight differences in the pattern of minute spinules on the grain surface and this may be detected under optimal conditions in oil-immersion visual microscopy. Nevertheless, in practice, where a deposit contains both *Corylus* and *Myrica* in fair quantity a rather large proportion of grains remain strictly indeterminable. In some pollen diagrams it has been observed that the joint *Corylus–Myrica* pollen curve repeatedly had pronounced maxima wherever macroscopic remains of *Myrica* were present in the peat of the sampling site, as for instance in the *Hypnum-Cladium–Myrica* layers of the flooding horizon above the Late Bronze Age wooden trackways of the Somerset Levels. British palynologists have followed their individual opinions in attempting or not attempting separation of the *Corylus* and *Myrica* pollen components, and in specifying whether or not the '*Corylus*' curve excludes *Myrica*: sometimes we are given the '*Corylus–Myrica*' curve. This being so, it has seemed best to limit our records to those based upon macroscopic remains.

Myrica gale today grows chiefly upon acidic fens and bogs, avoiding extreme conditions of acidity and oligotrophy on the one hand and eutrophy on the other. The records undoubtedly are determined to a large degree by the periods of formation of deposits of suitable type. It is striking that almost all of the record occurs after the end of zone vIIa, and that save for one record the Atlantic period, in which the old *Sphagnum-Calluna* peat was

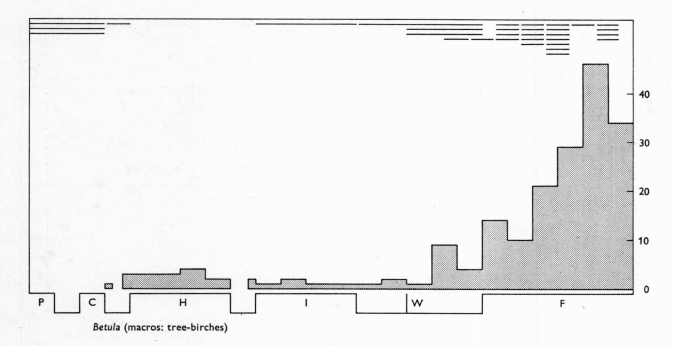

Betula (macros: tree-birches)

formed, has no indication of its presence. Perhaps the raised bogs were too much in the *Calluna–Eriophorum* dominated 'Standstill' phase for *Myrica* to have inhabited them, or possibly the warm climate operated directly to restrict the plant, for it belongs to Matthews' 'Oceanic–northern' element and Hultén's 'circumpolar sub-oceanic' group, whilst Matthews' map of its distribution (Matthews, 1937) shows that south of Britain it is closely restricted to northern and western coasts of France and the north coast of Portugal. It has a fairly high northern latitudinal range in Scandinavia, but its sub-fossil record differs strikingly from that of other plants of similar northern range in that it gives such sparse evidence of its presence in the Late Weichselian (one tentative zone II record from Kent) and one in the early Flandrian (zone IV in Dorset).

PLATANACEAE

Platanus sp.   Plane   334

F VIII *?Bigbury Camp*

The tentative identification of charcoal of *Platanus* from the Early Iron Age in Kent is highly suspect.

Our attention has been drawn by an observant student to the manner in which the expanding leaves of the plane trees commonly planted in this country shed their tomentum of branched hairs. We were able to confirm that these are very common in the air near parks and gardens, and that they even occur in laboratory air. They very closely resemble the 'candelabrum' hairs of the leaves and stems of *Verbascum* spp. There is no doubt in

the mind of Miss Andrew and myself that the Late Weichselian and Flandrian records made in Cambridge for such hairs and attributed to *Verbascum* were indeed a local contamination of *Platanus*.

BETULACEAE

Betula sp. (Figs. 23, 84; plate V)   335

At the time when C. Reid's *Origin of the British Flora* was published, and for many years afterwards, British botanists placed all our tree-birches within the species *Betula alba* L. Although the *B. alba* of some authorities corresponds with *B. pendula* Roth., the older British records under *B. alba* L. cannot be regarded as more closely identified than to the genus *Betula* and as such they are given: this is of course amply confirmed by the fact that most of them are based upon wood, bark and charcoal identifications which do not allow separation of the two species *B. pubescens* and *B. pendula*. All such records in the genus *Betula* may be taken to represent the tree-birches, for *B. nana* has distinctive fruits and is unlikely to have yielded determinable wood or charcoal without accompanying leaves, fruits or cone-scales.

It is apparent from the record diagram that tree-birches have been present in the British Isles from the time of the Cromer Forest Bed series: macroscopic remains of them have been found throughout the Hoxnian and Cromerian interglacials and they are even represented in the early and late parts of the intervening glacial stages, as in the Early and Middle Weichselian. They are more frequent in the middle Allerød stage of the Late Weichselian and

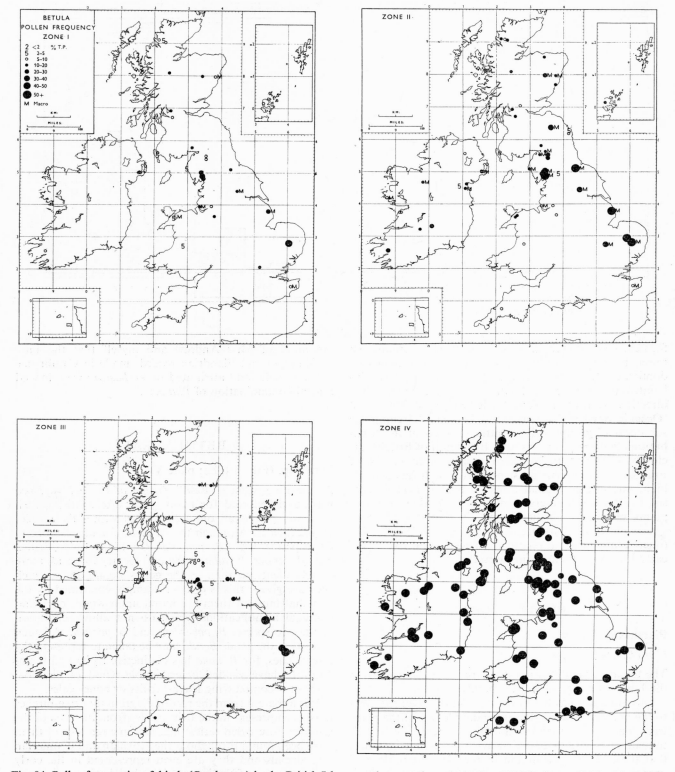

Fig. 84. Pollen frequencies of birch (*Betula* spp.) in the British Isles, zone by zone through the Late Weichselian and Flandrian. This overwhelmingly represents the tree-birches, especially through the Flandrian.

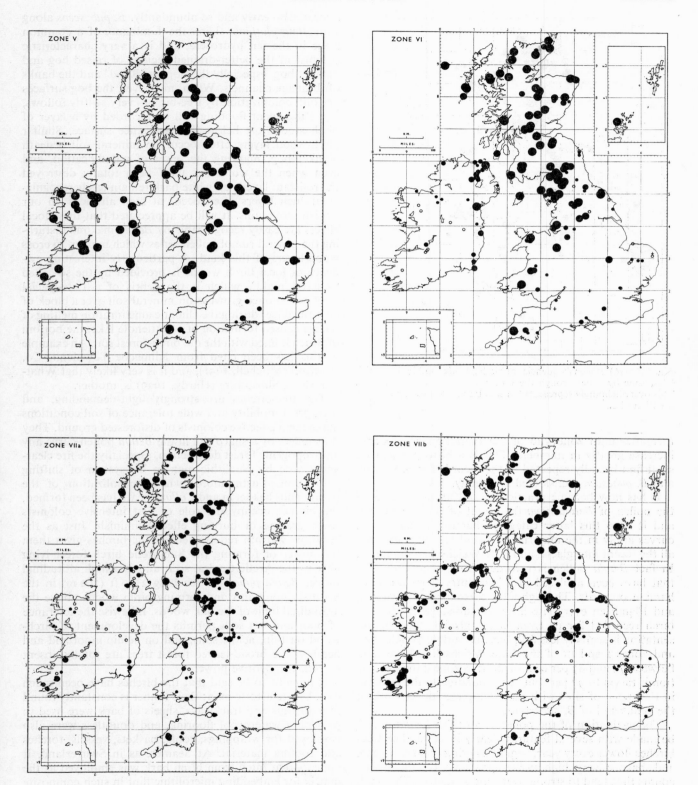

Fig. 84 *(cont.)*.

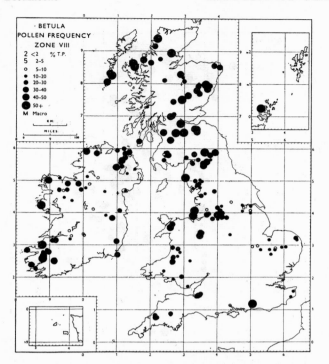

Fig. 84 (*cont.*). Pollen frequencies of birch (*Betula* spp.) in the British Isles, zone by zone through the Late Weichselian and Flandrian. This overwhelmingly represents the tree-birches, especially through the Flandrian.

thereafter until zone VIIb of the Flandrian site records increase greatly in numbers. This is a history which we shall find broadly repeated in the records for *B. pendula* and *B. pubescens* considered separately.

It has now become practicable and usual to separate the pollen of *Betula nana* from that of the tree-birches and it is to this latter category that the *Betula* pollen curves apply in the majority of Flandrian diagrams and all the recent interglacial and Late Weichselian diagrams. In fact *Betula* pollen is recorded from almost all sites that have been analysed in this country from the Ludhamian onwards. Throughout the Hoxnian, Ipswichian and Flandrian every site in every sub-stage or zone has birch recorded for it. Even in the glacial stages representation is scarcely less full, but the Middle Weichselian and zones I and III of the Late Weichselian have it in between 75 and 87 per cent of sites: in zone II the proportion is raised to 97 per cent corresponding to the increase in that zone of sites with macroscopic evidence both of the genus and of *B. pubescens*.

The evaluation of the pollen record for *Betula* has to be made with several features clearly in mind. The tree-birches flower every year, begin flowering at an early age, and have a very high pollen productivity: the pollen counts thus tend to strong over-representation. The small size of the grain gives a slow rate of fall and favours very wide dispersal, but this as it happens is not an effect of much significance in our records since the birches have

appeared so early and so abundantly. *B. pubescens* along with *Alnus glutinosa* shares the dominance of a common stage in the fen hydrosere, and is a very characteristic colonist of the better-drained regions of raised bog and blanket bog, especially the sloping 'Rand' and the banks of drainage channels. When, moreover, the bog surfaces dry out, colonization by tree-birches very swiftly follows, and subsequent flooding will be recorded by a layer of birch stools just below the recurrence surface. Similar birch-wood layers often occur on mineral soil beneath blanket bog, marking the onset of paludification, and even when the wood has itself been totally destroyed the resistant layers of white bark remain easily recognizable: such layers often occur at high altitudes on our British mountains. It will be appreciated that these local effects are easily recognizable by *Betula* maxima disturbing the general run of pollen series which happen to cross wood-layers of this kind. A particularly interesting evidence of local birch woods is provided by the so-called 'Allerød muld', which is the debris of a Late-glacial birch wood once growing on mineral soil over a block of entombed ice: during the climatic amelioration melting of the ice caused formation of a kettlehole lake, the bottom of which is lined with the old birch-forest soil. An example of this has been cited from Kilmacoe Townland, Co. Wexford (Mitchell, 1951), and it is very likely that Whattall Moss, Shropshire (Hardy, 1939) is another.

The tree-birches are strongly light-demanding, and their great mobility and wide tolerance of soil conditions make them effective colonists of disforested ground. They thus reflect by a sharp rise in the *Betula* pollen curve any anthropogenic forest destruction, especially the fire clearance made by Neolithic man in his system of shifting cultivation. Furthermore, when podsolization of the poorer soils has set in and *Calluna* heath has been formed, the birches occupy the role of first intensive colonists when grazing or burning effects diminish. Just as the very low warmth requirement of the birches gives them the power of forming the Sub-arctic birch forest lying next to the Arctic tundra, so it allows a high altitudinal range, *Betula pubescens* reaching 2500 ft (760 m) in the Scottish mountains and forming birch woods above the altitudinal limit of the oak woods. Likewise the tolerance of poor and wet soils permits the development of birchwood on thin, acidic soils, and on acidic moorland and heath where broad-leaved forest trees are totally absent or lack competitive power.

It remains to be said that the birches have been much used by prehistoric man: their timber is constantly found in platforms and trackways, sheets of bark were used to provide a waterproof flooring, and doubtless were also employed for wrappings, roofs, baskets, vessels, torches and roofing material. At Seamer, as in Switzerland, it appears that pitch from birch bark was employed as the matrix for embedding microlithic flint in such composite tools as spears and arrows. Curiously enough it seems probable that at Seamer, mesolithic man had also collected the fructifications of the polypore fungus, *Fomes fomen-*

tarius which is a typical parasite of birch in the north of Britain and has numerous continuing uses for tinder, wound-dressing, etc. These uses of the birch are cited to give a background to the numerous records of birch made in an archaeological context: these number Neolithic 8, Bronze Age 20, Iron Age 7, Roman 18 and Mediaeval 3.

In the early Pleistocene birch pollen, though present, is in low frequency and does not shew systematic variation in its abundance, but in the Cromerian and each of the three interglacial stages that have succeeded it birch has been the most decisively terminocratic of the forest trees, preceding pine at the onset and following it at the close of the two interglacials that extend to a sufficiently late phase. Macroscopic finds in the early and late glacial stages (both for the genus and *B. pubescens* and *B. pendula* separately) support this view, as equally they indicate persistence of the tree-birches through the milder interglacial stages. From the high pollen frequencies we judge that birch woods or birch copses in half open country must have characterized the early and late vegetational stages. The Chelford interstadial deposits of the Early Weichselian provide evidence of conifer–birch forests similar to those of Finland today, and indications that both *B. pubescens* and *B. pendula* were present. Outside the interglacials the Early and Middle Weichselian supply only slight evidence for local presence of tree-birches, as in a fruit recovered at Thrapston (Bell, 1968). The Late Weichselian presents a much fuller picture. Although in zone I (fig. 84) sites are infrequent, they are well scattered through the British Isles, values of 5–10 per cent of total pollen are common and there are several sites with much higher frequency, notably in East Anglia and the Lake District: at least local birch copses must be presumed even in central Scotland, Sutherland and western Ireland. In zone II birch pollen frequencies are generally increased and within the eastern side of Great Britain from the Thames to the Forth, and in the Lake District, values are so high as to make it certain that birch woodland was fairly continuous: as we have already indicated (p. 107) there were probably pine woods in Kent at this time. Macroscopic finds reinforce the view that birch was now widespread. In the colder zone III that followed birch did not lose in total area but its frequency was reduced everywhere except in eastern Lincolnshire and East Anglia: in these sites birch woodland may have remained but elsewhere it must have been much broken and reduced, as indeed the total pollen spectra everywhere indicate. The map for zone IV (together with the composite zone IV/V where these are unseparated) shews such high pollen frequencies for *Betula* that we must presume dominant birch forest throughout the British Isles: south-east from the Hampshire Basin to East Anglia and locally elsewhere perhaps, that dominance was shared by the pine. Zone V is marked by some solidification in the birch forests in the north and west, but now the birch has yielded dominance (mostly to the pine) in a larger area extending from Devonshire to the

Humber. Through zone VI this big south-eastern half of England and Wales and most of Ireland save the east saw birch reduced to minor proportions in the face of the developing mixed oak forest with hazel and the pine forest already there. In the Atlantic period (zone VIIa) that followed, the process of suppression of birch wood continued, generally by mixed oak forest with hazel and alder, but in the Cairngorm and Sutherland and much of Ireland by pine also. In the Western Isles of Scotland and the far north of Scotland, birch retained dominance. The picture altered little in zone VIIb in Great Britain, but birch now receded in all central and eastern Ireland as it had earlier in England and Wales.

It has to be noted that despite the recession in its frequency to very low values in the area south and east of the Pennines, north and west of this neither in zone VIIa, VIIb nor VIII did birch ever become negligible; the same situation held for western and north eastern Ireland as against the south-east and the central plain. Comparison of the maps shews that there was in zone VIII some general increase in birch pollen frequency, as well in the areas of former mixed oak forest dominance as elsewhere. There seems no way of deciding how far this is a response to clearance of the dominant forest, and how far to the sub-atlantic change of climate: the former is more likely. It is of interest, considering the sequence of zones, to note how the present important role of birch in all the forest regions of Scotland has been prefaced by a similar importance right through from the Pre-boreal and this is particularly true for the extreme north of Scotland where birch forest is now the natural climax (McVean, in Burnett, 1964): there is no fossil evidence of the possible history of the typically northern *B. pubescens* ssp. *odorata* (*B. pubescens* ssp. *tortuosa*) (Birks, 1969).

It is interesting to observe that multiplication of the pollen evidence has confirmed the fact, already apparent in 1940, that at every stage in the Flandrian history of England and Wales the north and west were set off from the south and east by their much higher values for *Betula* pollen. Results for Scotland are now seen to be conformable. It is a distributional feature strongly exhibited also on the Continental mainland where Firbas (1949) lays much stress upon it in the Late-glacial period and attributes it to the very small amount of warmth required by the birches: his isopol map for middle Europe in the Pre-boreal displays it quite dramatically, shewing average values below 10 per cent in central Europe, 60 per cent or more over the whole north German plain, and more than 80 per cent along the North Sea coast.

Both *B. pubescens* and *B. pendula* had already reached the north-west German plain by zone I of the Late Weichselian period, and throughout that period were abundant in middle Europe, although repressed by extension of pinewoods from the south during zone II, the Allerød period. Firbas concedes the possibility that the tree-birches may have survived the glaciation north of the Alps, but takes the view that speculation on the position of such refugia is totally nullified by the extremely rapid

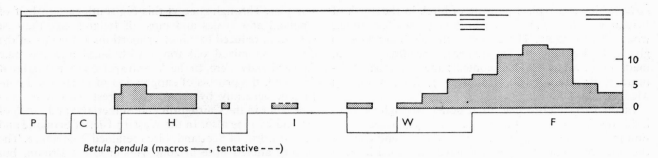

Betula pendula (macros ——, tentative - - -)

rate of dispersal of these trees. He points out that whereas in the Late Weichselian period and Pre-boreal the low warmth requirements of the birches have dominated their distribution pattern, in the ensuing warm period competition from other trees has become so severe that the occurrence of birch is now much more determined by response to soil factors than to climate. It is correspondingly the prevalence of heaths, raised bogs and fens in the north-west German plain that has maintained relatively high birch pollen values there throughout the climatic optimum.

The rather full and consistent picture of the history of the tree-birches in the British Isles is completed by the accounts given below under the headings B. pubescens, B. pendula and their hybrids with B. nana and with one another.

Betula pendula Roth. (*B. verrucosa* Ehrh., *B. alba* auct.) Silver birch (Figs. 85, 86, 160; plate XVIII*t*)   335: 1

The cone-scales of *Betula pendula* can be recognized by the broadly spreading and downwards curving lobes, and the fruits by shape and the broad wings which extend well beyond the apex of the ovary. Both types of organ are produced abundantly and are freely shed so that the fossil record is quite full. One record in the Cromer Forest Bed series is followed by several in the Hoxnian, with five in sub-stage I following three in the late Anglian. Several of these are Irish records, and Jessen *et al.* (1959) comment that *B. pendula* occurs in the earlier part of the interglacial when the soil was still base-rich. Somewhat surprisingly the Ipswichian has yielded only a single record, and that tentative, like the single records from the early Wolstonian and Early Weichselian, both of which might be of derived material. However, there is convincingly full record for the Late Weichselian, especially with records in all three zones (and transitions) from sites in Aberdeenshire, but also records from Sutherland, Westmorland, east Yorkshire, Essex, Surrey and possibly Norfolk (fig. 85). Clearly *B. pendula* was widespread throughout Great Britain and along with *B. pubescens* composed the birch woodlands and copses indicated by the high pollen frequencies (see account of the genus). Sites multiply throughout the Flandrian zones to a maximum in VIIa and VIIb: thereafter they sharply decrease, for no presently apparent cause. It is possibly significant

that we have no Irish records for *B. pendula* in the Late Weichselian or in the Flandrian until the single record at Roundstone in zone V.

B. pendula is a European–Siberian species of wide European and Asiatic range, extending to 67° or 68° N latitude in Scandinavia. Though present throughout the British Isles it is rare in north Scotland and commoner in southern England than *B. pubescens*.

Betula pubescens Ehrh.   Birch (Figs. 85, 86)   335: 2

The cone-scales of *Betula pubescens* are distinguishable from those of *B. pendula* by the more spreading (not descending or reflexed) shape of the lateral lobes, and the fruits by the narrower wings which hardly project above the apex of the ovary.

There are records of *B. pubescens* in every sub-stage of

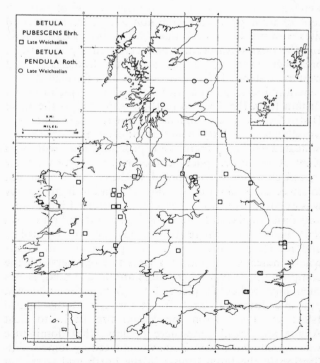

Fig. 85. Late Weichselian records in the British Isles of the two tree-birches *Betula pubescens* Ehrh., and *B. pendula* Roth. These records depend upon identifications of fruits and female cone scales.

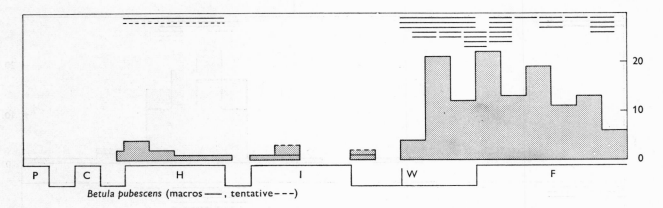

Betula pubescens (macros ——, tentative ---)

the Hoxnian interglacial, and as with *B. pendula* they are most frequent in the cool initial stage and the preceding Late-glacial. There is a smaller representation (as with *B. pendula*) in the Ipswichian, and again only doubtful presence in the Early Weichselian. In the Late Weichselian however, records of *B. pubescens* greatly outnumber those of *B. pendula* (fig. 85) and there can be little doubt that of the two British tree-birches it was *B. pubescens* which contributed most to the birch copses or birch woods of the Late Weichselian and early Flandrian periods, so strongly indicated by pollen diagrams. The rather frequent identifications of the hybrid between *B. pubescens* and *B. nana* in early Flandrian deposits lend support to this view.

It will be noted that although *B. pubescens*, like *B. pendula*, is a tree-birch of wide Eurasiatic range, it has distinctly the more northerly range of the two, has a higher altitudinal range, is more tolerant of cold and wet conditions, and is commoner in the north than in the south of the British Isles, especially as the sub-species *odorata* (Bechst.) E. F. Warb., already mentioned. To some extent the abundance of Flandrian records for macroscopic remains of *B. pubescens* must be due to the fact that it is this species that is so commonly a component of fen carr and of the drying surfaces of raised bogs. At the same time comparison of the record diagrams for the two tree-birches seems to disclose a significant difference between the early expansion of *B. pubescens* and the later expansion to a maximum in the middle Flandrian, of the more thermophile *B. pendula*.

Betula pendula × pubescens (= *B. × aurata* Borkh.)

335: 1 × 2

I *Bobbitshole*; eW *Chelford*; lW *Nant Ffrancon, Hartford, Nazeing, Flixton*; F v *Nazeing*, v, vi *Apethorpe*, viia *Llyn Dwythwch*, viib *Wigton*, viib/viii *Irstead*

Although the fruits and cone-scales of *Betula pendula* and *B. pubescens* have good diagnostic characters, there is within each a considerable range of variation and the underdeveloped terminal segments of the cones are specially variable. This makes it difficult to be certain of the hybrid between the two species. It is interesting that

in the Ipswichian interglacial, where records for tree-birch are so sparse and there is only one tentative identification of *B. pendula*, Dr West found fruits and cone-scales from sub-stages ia, ib and iia at the one site, Bobbitshole (West, 1958), that he could not precisely place in either species. He found himself similarly placed with some birch fruits found in the interstadial at Chelford where many others were definitely placed in *B. pendula* and *B. pubescens*.

In the Late Weichselian the presumed hybrid has been recorded several times, at Nant Ffrancon in all three zones (Seddon, 1962) and at Hartford in zone ii, as well as at Nazeing in zones ii, ii/iii and zone v of the ensuing Flandrian. So far as this evidence goes it supports the evidence for the presence of the two tree-birches in the Late-glacial of England and Wales. There is a scatter of records also in the Flandrian.

Betula nana × pubescens (= *B. × intermedia* Thomas ex Gaudin) (Fig. 160)   335: 3 × 2

lW *Old Buckenham, Lopham, Neasham, ?Nant Ffrancon, Elstead, Carrowreagh*; F iii/iv *Hockham Mere*, iv *Gosport*, iv, v *Knocknacran*, v ?*Apethorpe*

In deposits containing *Betula nana* there occur from time to time female cone-scales quite unlike those of *B. pendula*, *B. pubescens* or *B. nana*, and resembling in the upturned lateral lobes and broad base those of the hybrid between *B. nana* and *B. pubescens*. The fruits are somewhat less certainly recognizable. This hybrid is recorded from a limited region of Scotland at the present day and occurs in northern Scandinavia where the two parents flourish alongside. As would be expected from the fossil record of *B. nana* this hybrid has been found in Late Weichselian and early Flandrian deposits and has an equally widespread distribution.

Betula nana L. 'Dwarf birch' (Figs. 86, 87, 88, 160; plates I, X, XVIIIu, XIXq, XXq)   335: 3

The almost circular outline of the fruits of *Betula nana*, their small size and the narrow wing are characteristic, and equally so are the three diverging prongs of the

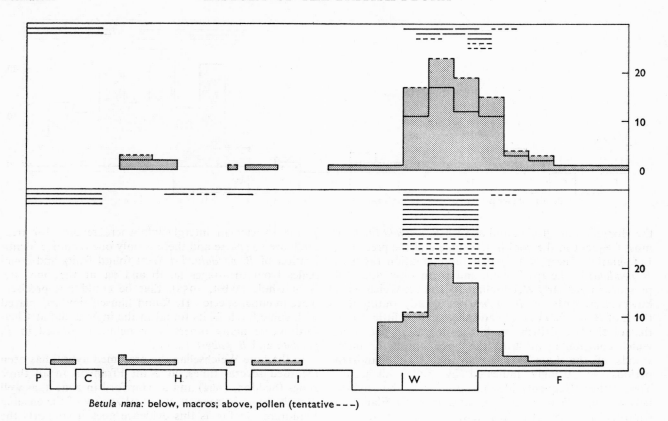

Betula nana: below, macros; above, pollen (tentative – – –)

female cone-scale. Upon occasion the male flowers have been recognized and the strongly crenate leaves are readily identifiable, along with the perithecia of the ascomycetous fungus which parasitizes them. British workers have not employed the methods used by Polish workers to characterize the morphology of fruits of the tree-birches (see Mamakowa, 1968). Increasingly however there have been attempts since the publication of his procedure by Terasmäe (1951) by palynologists to identify the pollen of *Betula nana* and to estimate its former frequencies. Initially the separation of *B. nana* pollen from that of the tree-birches has depended upon its circular outline in optical section and its less protuberant pores, together with an impression (over a large population) of smaller mean size. These criteria have indeed been made the basis for separate representation of the pollen frequencies of dwarf birch and tree-birch pollen (Bartley, 1962, 1966). Walker demonstrated the value of the ratio of grain diameter to pore depth as a numerical index to pore protuberance, and more recently H. J. B. Birks has given more precision to analysis by a combination of this parameter with statistics of mean diameter in populations of pollen prepared in a standard manner from type material of *B. nana, B. pendula, B. pubescens* ssp. *pubescens* and *B. pubescens* ssp. *odorata* (= *B. tortuosa* Ledeb.). The results establish that *B. nana* pollen may be distinguished from that of *B. pubescens* ssp. *pubescens* by the parameter grain diameter/pore depth, but not from that of *B.*

pubescens ssp. *odorata.* Grain size will however separate *B. nana* from *B. pubescens* ssp. *odorata*, but not from ssp. *pubescens* or *B. pendula*. Accordingly, reliable separation of *B. nana* pollen from that of the two tree-birches requires the use of both parameters: this he shews visually in the graphs reproduced in fig. 86. These conclusions have particular relevance to Scotland since two of Birks' samples of *B. pubescens* ssp. *odorata* were respectively from Assynt and Teesdale whilst the third was from Swedish Finland. This is the taxon which constitutes a large part of the natural northern birch forests and which might be presumed to have accompanied *B. nana* at least from the earliest Flandrian in Scotland. In view however of the certain early macroscopic records for *B. pendula* we also need more consideration of pore protuberance as a means of distinguishing pollen of that species from the pollen of *B. nana*. Birks remains extremely cautious in attributing pollen to *B. nana* and many of the tentative records in the record diagram are his and represent *B. nana* with high probability.

The Pleistocene record for *Betula nana* begins with macroscopic identifications made from the Beestonian by Dr West in deposits exhibiting frost polygons and ice-wedge phenomena and accompanied by such arctic-alpine species as *Salix polaris, S. herbacea* and *Oxyria digyna*. It was found in a similar assemblage of Wolstonian age at Farnham (Lambert, unpublished). There are also macroscopic records from the Cromer Forest Bed series

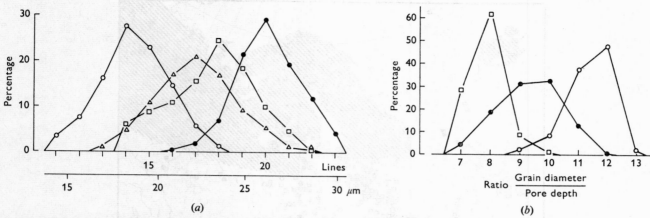

Fig. 86. (a) Size frequency curves for pollen of four species of Betula; (b) frequency curves for the ratio grain diameter:pore depth (i.e., relative projection of the pore), for these species of Betula: ○ B. nana, △ B. pendula, □ B. pubescens, ● B. tortuosa. (After H. J. B. Birks, 1968.)

that embrace the Beestonian and two site records for pollen from the Cromerian. From this time forward, as the record diagram shews, the presence of B. nana is repeatedly attested, often both by macro remains and by pollen. Thus at Nechells (Kelly, 1964) size frequency distribution diagrams for Betula pollen in sub-stages I and II of the Hoxnian gave a strongly bimodal curve suggesting the presence of B. nana alongside more abundant tree-birches, a conclusion borne out by pore protuberance measurements and identification of fossil birch fruits. There is as yet no evidence of B. nana in the early Anglian or in sub-stage III of either the Hoxnian or Ipswichian interglacials, but its persistence may be presumed. One can say this with confidence based on the abundant Weichselian records in which pollen and macroscopic identifications repeatedly reinforce one another, as in the Middle Weichselian at Oxbow and in the Late Weichselian and early Flandrian at Esthwaite Basin (zones II, III, IV), Nant Ffrancon (zones I, II, III) and at numerous sites in a single one of these zones. There is one pollen record for the Early Weichselian, and one from the Middle Weichselian where however no less than nine sites have macroscopic identifications. The Weichselian records come from all parts of the British Isles including the south of England and even Ireland where B. nana is no longer native (see fig. 87).

Thereafter in the Flandrian, records become progressively less frequent and increasingly restricted to northern sites. There are macroscopic identifications from zone V at Apethorpe (with a tentative pollen record), and from both V and VI at Nazeing, from zone VIIa at Widdybank Fell and from VIIb at Loch Cuithir in Skye. Of these we note that Apethorpe and Nazeing are sites where river terraces may have afforded habitats suitable for survival, and Loch Cuithir might be regarded as marginal to the present Scottish range of B. nana. Widdybank Fell presents its own special interest, for this site in Upper Teesdale has long been regarded as an area where

glacial or late-glacial species have survived. Research workers in the University of Durham have now recognized B. nana pollen from zones V, VI, VIIa, VIIb and VIII whilst one of them has not only found a perfectly preserved leaf in zone VIIa but has discovered a small colony still living on Widdybank Fell (Hutchinson, 1966). This dramatic proof of survival in situ parallels that of Overbeck and Schmitz (1931) and Overbeck and Schneider (1938–9), who shewed at the Lüneberger Heide in northwest Germany that B. nana persisted in successive layers

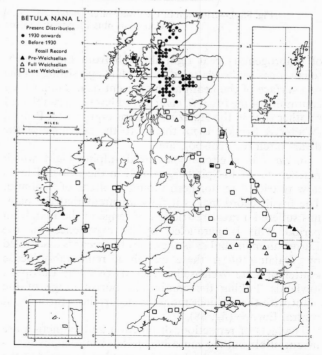

Fig. 87. Present, Weichselian (Middle and Late) and pre-Weichselian distribution in the British Isles of the dwarf birch, Betula nana L.

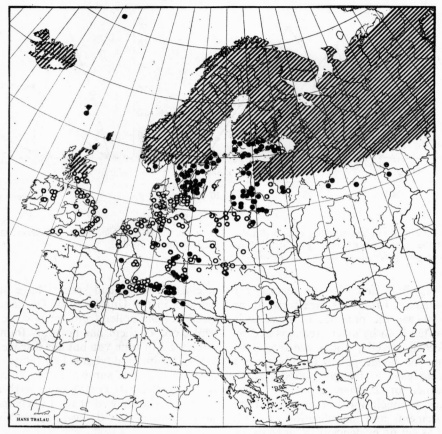

Fig. 88. Distribution of the dwarf birch, *Betula nana* L., in Europe: recent range shewn by hatching and black dots, Weichselian plus Flandrian records by open circles. (After Tralau, 1963.)

of ombrogenous peat right through from Late Weichselian times to the scattered living plants persisting on the bog surface of the heath at the present day. Similar long continuity of the fossil *B. nana* record has been established at other sites in the Hartz mountains and in west Prussia (Walter & Straka, 1970) thanks to the tolerance of the dwarf birch for peat-bog sites.

B. nana is a circumpolar arctic–alpine plant which reaches 78° N in Spitzbergen, and in the British Isles is now restricted to northern mountain areas of Scotland. The fossil record makes it fully apparent that the plant has suffered a great restriction of range in the Flandrian period. This is paralleled by its experience on the European mainland as is shewn by the records assembled by Tralau (1963): these exhibit a remarkably dense scatter of Late Weichselian records over the European lowlands, linking the continuous boreal range of the plant with the disjunct montane occurrences throughout central Europe (fig. 88).

The facts of retraction, disjunction and extinction are incontestible but the causative mechanism remains somewhat unclear. Conolly & Dahl (1970) have pointed to the correspondence between the present-day distribution of *B. nana* and the 22 °C maximum summer temperature

summit isotherm, with the inference that the plant may be unable to tolerate any higher value. It is possible that climatic control though present is more complex, and highly probable that growing competition from closed woodland during the Flandrian excluded the dwarf birch from great areas of the British Isles, although possibly not from the sparser woods of upland areas. Hutchinson (1966) indicated that whilst in southern Sweden the dwarf birch is typical of upland wooded bog or poor fen communities, over the rest of its range it is chiefly found in a sub-alpine dwarf shrub-heath with *Calluna*, *Vaccinium* spp., *Empetrum* spp. and northern willows, or in a tundra vegetation type on mineral soil with dwarf arctic–alpine plants. Whilst today the Scottish sites for *B. nana* are of the dwarf shrub-heath type, the Teesdale site has species indicative of the tundra type vegetation and it was presumably in this latter community that *B. nana* chiefly existed in the glacial stages in this country.

Betula humilis Schrank   335: 4

F iv *Dogger Bank*

Behre and Menke (1970) refer fruits and cone-scales of birch found in the submerged peat of the North Sea floor

Absolute numbers of sites   Percentage of all recorded sites

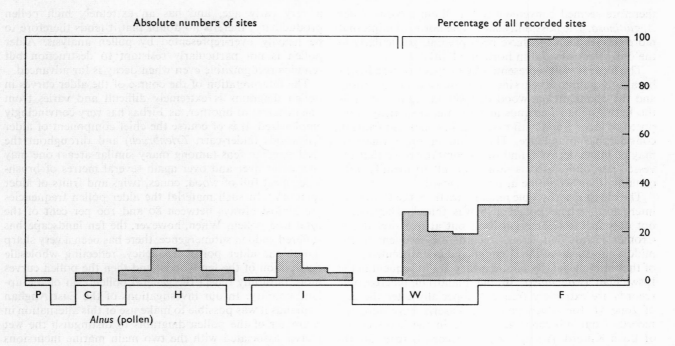

Alnus (pollen)

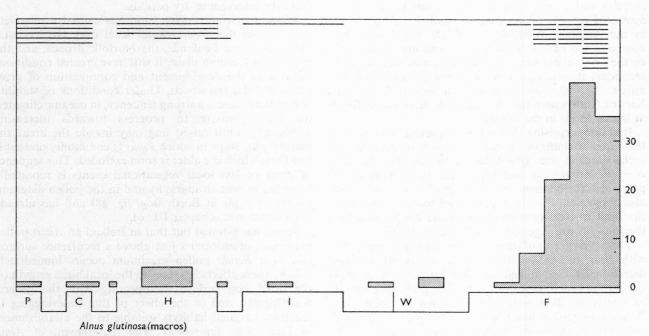

Alnus glutinosa (macros)

to *Betula humilis* Schrank, a shrubby birch found in the bogs and fens of central and eastern Europe but not in the British Isles or Scandinavia.

Betula tortuosa Ledeb. (= *B. pubescens* ssp. *odorata* (Bechst.) E. F. Warb.) (Fig. 86)   335: 26

See references given in the accounts of *Betula, B. pubescens* and *B. nana.*

Alnus glutinosa (L.) Gaertn. Alder (Figs. 23, 89; plate IV)   336: 1

Only one species of alder, *Alnus glutinosa* (L.) Gaertn., is now native in the British Isles, and it is probably safe to presume that all sub-fossil remains of *Alnus* in the Flandrian period belong to this species. We have indeed no positive evidence that any other species has been present here during any part of the Pleistocene, and it has therefore seemed convenient to list all our records under *A. glutinosa*. The possibility should be borne in mind that more species than this have been present, particularly in the older deposits and in northerly latitudes.

The alder is well represented by its macroscopic fossils because its habitat favours incorporation and preservation, and the abundant fen-wood deposits in regions such as the East Anglian Fenlands and Broads are largely composed of alder wood with twigs, bark, cones and fruits in considerable abundance. The wood in such situations may be much decayed, and when compressed so that the vessels are obliterated is more difficult to identify with certainty than would be at first supposed.

The oldest macroscopic remains are from the Pastonian interglacial, where pollen also was found. They occur not only in the Cromerian itself but at five sites in the Cromer Forest Bed series. Records are frequent in the middle sub-stages of the Hoxnian and in unspecified parts of that interglacial. By contrast they are infrequent in the Ipswichian (*vide infra*). In the Flandrian macroscopic records are extremely frequent, especially after the end of zone VI, for which zone the Vasaris have supplied records from sub-zones a, b and c in the Scottish site of Loch Kinord. A single zone V record is from Southampton and a zone V/VI record is from Loch Kinord. Scattered records of macroscopic finds have been made in the Anglian, Wolstonian, Middle Weichselian and zone II of the Late Weichselian. Doubts are raised, often by the original authors, whether these may not be due to secondary incorporation or, sometimes, uncertain recognition. Among these is a zone II record from Loch Kinord from where the Vasaris took other macrofossils at higher levels in the borings.

It is not surprising that a tree as prevalent as alder in the latter Flandrian should be much represented on archaeological sites (Neolithic 6, Bronze Age 14, Iron Age 10, Roman 12 and Mediaeval 4). In part it was preserved fortuitously in waterlogged sites, but it was also convenient for use in piling and trackway construction and was easily worked, as in the bucket that held the Late Bronze Age hoard at Stuntney, Cambs.

The pollen grains of *Alnus* are very readily recognizable, with four to six aspidate pores, somewhat elongate meridionally, and linked together by pairs of strongly curved thickened bands or 'arci'. The grains are smooth or nearly so, flattened, and polygonal in polar view. Praglowski (1962) has illustrated the means by which pollen of *A. glutinosa* can be distinguished from that of *A. incana*, and this ought to be recalled in pollen analyses of British interglacial deposits, and possibly those of our Full or Late Weichselian deposits also. A feature requiring further investigation is the tendency at certain localities and certain times for there to be a large and significant excess of four-pored as against five-pored grains, presumably a phenomenon associated with hybridity. Praglowski has also described the six- or seven-pored grains of a tetraploid, *A. glutinosa*.

A. glutinosa flowers every year, begins to flower at a very early age, and has an extremely high pollen productivity: there is no doubt that it tends therefore to be heavily over-represented by pollen analysis. Alder pollen is not particularly resistant to destruction but remains recognizable even when decay is far advanced.

The interpretation of the course of the alder curves in pollen diagrams is extremely difficult and varies from one instance to another, as Firbas has very convincingly emphasized. It is of course the chief component of alder fen-woods (alder-carr, *Erlenbruch*) and throughout the East Anglian fens (among many similar areas) one may encounter over and over again several metres of brushwood peat full of wood, cones, twigs and fruits of alder (plate IV). In such material the alder pollen frequencies are almost always between 80 and 100 per cent of the total tree pollen. When, however, the fen landscape has suffered sudden submergence, there has been a very sharp fall in the alder pollen frequency, reflecting wholesale destruction of the fen-woods, and then the pollen curves more faithfully reflect the general pollen rain of the upland country. In our investigations of the East Anglian Fenlands it was possible to make use of this alternation in character of the pollen diagrams to distinguish the wet phases associated with the two main marine incursions from the intervening dry periods.

When a marine transgression has caused a relatively slow rise in the ground-water level over large coastal areas, as in the Fenlands, the Norfolk Broads, and the north-west German plain, it will have created conditions suitable to the development and continuation of great areas of alder fen-woods. Under conditions of stability of coastline there is a strong tendency, in oceanic climates, for the hydrosere to progress towards increasing oligotrophy until raised bog may invade the area: the mesotrophic stage in which *Pinus* is commonly present is one from which the alder is soon excluded. This sequence of more or less local vegetational events is repeatedly recorded in peat stratigraphy and in the pollen diagrams (as for example at Borth Bog, fig. 28) and has already been mentioned (chapter III: 6).

Jessen has pointed out that in Ireland an *Alnus* pollen maximum often occurs just above a recurrence surface, just as a *Betula* pollen maximum occurs immediately below. These effects he traces to the local birch growth on the dried bog surface disappearing with the general waterlogging, and to the effect of that waterlogging in creating marginal habitats suitable to the establishment of local alder fen-woods. The requirement of *Alnus glutinosa* for high soil moisture may of course be met by a sufficiently high rainfall/evaporation ratio, but the operation of this factor through promoting leaching and podsolization can very quickly lead again to diminution of the alder. Furthermore we have the complication that whilst it is undoubted that in many regions the alder chiefly occupies fenlands, lake shores and river margins in a distribution pattern among the woods of the mineral soils, there is some evidence, as in the *Quercus petraea* woods of Wales, that it may under suitable conditions also exist as a normal component of the mixed oak forest itself. How far the one situation or the other prevailed at a given time and place in the past it is hard to say.

It is only after we have evaluated edaphic effects of the kind outlined above that we can proceed to consider the possible causes, migrational and climatic, that may have determined the broad pattern of the tree's behaviour in the British Isles as a whole.

In recent years pollen analysis has extended our knowledge of the alder in British interglacial deposits far beyond that based solely on macrofossils. Knowledge of the oldest stages rests largely upon the evidence of the Ludham borehole (West, 1961b) from which it appears that *Alnus* pollen was present throughout the Ludhamian, Antian, Baventian and Pastonian in frequencies of the same order as that of the main woodland dominants, that is about 5 to 20 per cent total pollen. The representation of *Alnus* was not notably less in the cool than in the more wooded temperate stages. Although not recorded in the Beestonian the alder was very abundant again in the temperate part of the Cromerian. In the Hoxnian it was remarkably abundant: represented in several sites in sub-stages I and IV, it is recorded in every site in the intervening zones, often in great abundance and with macrofossil remains. Of the sudden alder expansion in sub-stage IIc at Nechells Kelly suggests that it probably had a dual role, in wet oakwood with ash on the valley floor, and in alder fen carr. There were extremely abundant macrofossils in the middle sub-stages correlated with high and continuous alder pollen curves. In sub-stage III, where coniferous forest spread, the alder persisted in wetter sites, but presumably not in the general forest cover. Turner (1970) records a similar history at Marks Tey, but then a later and conspicuous decline in alder in sub-stage IV, despite the high rainfall; this is attributed to displacement by other plant communities, notably acid bog. The Irish Hoxnian follows a similar pattern of alder behaviour, as do the deposits of this interglacial elsewhere in west Europe, with high frequencies and local dominance in the temperate sub-stages.

In the Ipswichian interglacial, although *Alnus* is recorded from all four sub-stages and from almost every site in the two temperate middle ones, the situation is different from that of the Hoxnian in that repeatedly the pollen frequencies stay low (usually 1–5 per cent tree pollen). Only locally and where macrofossils occur is alder pollen found in high frequency, and West concludes that in this East Anglian region of lakes and wide slow rivers, alder was not generally abundant in the regional forest cover.

Alder pollen has been found in low frequency in deposits of the glacial stages (with an occasional macrofossil) but the possibilities of secondary derivation, of long distance transport and possible confusion with *Alnus incana* make inference of local presence highly uncertain. This remains true for even the Late Weichselian, where recorded sites are naturally, with total investigation, more numerous.

The most salient feature of the Flandrian history of *Alnus glutinosa* is reflected in the record diagram that shews the tree represented in virtually every examined site in the British Isles from zone VI to the present day. In fact the site records increase progressively through sub-zones a, b and c of zone VI in a manner reflecting the rise in alder pollen frequencies through these zones also. *Alnus* had been treated by von Post (1930, 1946) as a mediocratic element in his analyses of Danish Holsteinian, Eemian and post-glacial pollen diagrams, and this quality was admirably displayed in his tabulation analysis of post-glacial diagrams ranged south to north and east to west across Europe (von Post, 1930). *Alnus* emerges as a mediocratic element in the northern half of the mixed oak forest belt across the continent, shewing indeed a consistent northerly extension beyond that belt. The British Flandrian history of alder, which is consistent with this, can best be appreciated by consideration of the maps for successive zones and sub-zones (fig. 89). Sites with pollen frequencies of 0–2 and 2–5 total pollen occur at scattered zone IV sites over the British Isles: at two Snowdonian sites and one in Norfolk 5–10 is reached and at one Scottish and one Yorkshire site the composite zone IV/V yielded 10–20 per cent. In zone V sites are more numerous and in Galloway, Perthshire and the Lake District they yielded 5–10 per cent: Scottish sites with higher values of 10–20 per cent derived from the composite zone V/VI. This concentration of relative abundance recalls the similar phenomenon for *Corylus* (see pp. 271ff.) and might indeed suggest similar early presence in this area, the more plausibly since Firbas (1949) had already concluded that as early as the Pre-boreal period *Alnus* was present in small and local stands in the north German plain.

In only a modest proportion of sites has zone VI been subdivided and the maps for zones VIa and VIb are sparse: none the less whilst they reveal small difference from that of V, the map for zone VIc (combined V/VI and undivided VI) demonstrates substantial rises in abundance everywhere. In Scotland and the north of England there are many sites of percentages 5–10, 10–20 and 20–30, and north-eastern Ireland displays a similar trend with one site (Newtownbabe) of very high frequency. This northern and montane expansion of alder is presumably linked to the fact that the tree was widely available from an early date, and that it is a tree found today on west Scandinavian mountains as an altitudinal belt between the pine and birch belts (Praglowksi & Wenner, 1968).

The map for zone VIIa shews dramatic increases in alder pollen frequencies throughout the British Isles: even allowing for its high pollen productivity it must have become exceedingly common in the general forest cover of the country. It remained sparse however throughout central Ireland and in Scotland in that part of the highlands now dominated by pine, as in the far north and the Hebrides where birch preponderated. The map for zone VIIb shews maintenance of the high abundance seen in VIIa with some increase in the Irish frequencies, and that

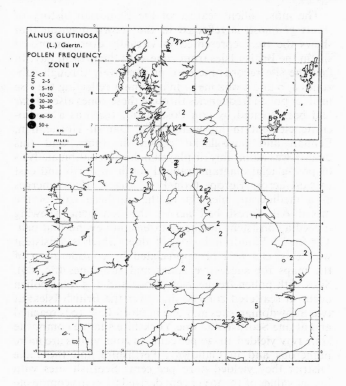

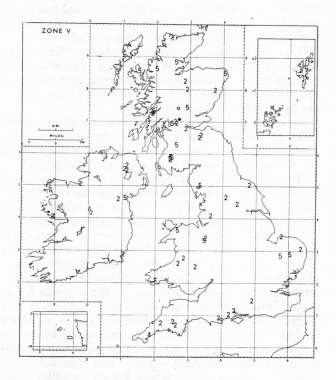

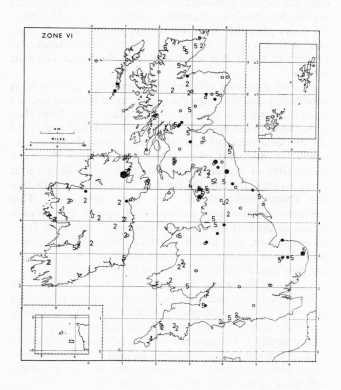

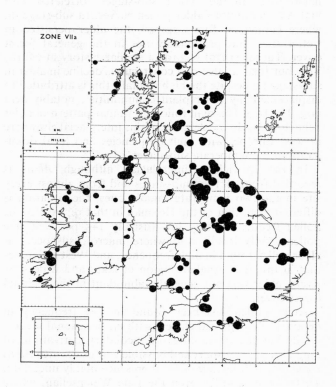

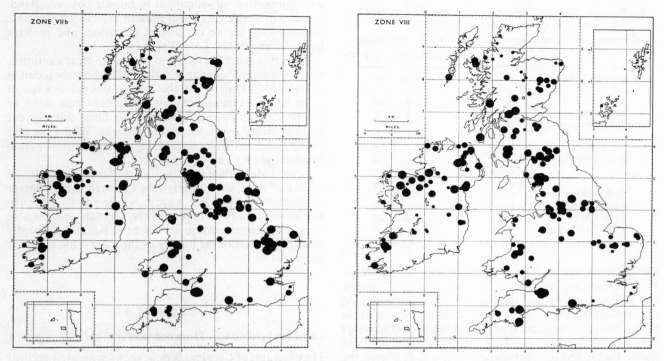

Fig. 89. Pollen frequencies of alder (*Alnus glutinosa* (L.) Gaertn.) through the six pollen zones of the Flandrian.
Note the early presence in some localities and the explosive expansion in zone VIIa.

for zone VIII illustrates a general fall in frequency that is very marked in East Anglia, presumably the consequence of very heavy forest clearance and drainage, and somewhat less pronounced but still strong in northern Scotland and south-eastern Ireland.

It is a very familiar feature of British pollen diagrams that at the opening of zone VIIa the alder pollen frequencies rise with startling suddenness to the high values they afterwards maintain: this is especially so in England and Wales but often, in Ireland and north-west England, there is a slow protracted rise throughout zone VIIa.

In explanation of this behaviour it must first be said that no such dramatic expansion as that at the zone VI/VII boundary could have taken place had not the alder already been widely present in favourable localities, a conclusion we may now take as independently established. Conditions then altered over the whole country, most probably by an increase in climatic wetness made more dramatic by its following the period of pronounced dryness that in the last part of zone VI lowered lake-levels in many parts of the country. We may visualize a general increase in waterlogging of all gently sloping ground, especially on clays, and the creation throughout the mixed oak forest of frequent wet glades full of willow and alder, like the swales in the mixed temperate forests of the eastern U.S.A. with arbor vitae (*Thuja occidentalis*), *Betula lenta* and *Acer rubrum*. A mosaic pattern of this kind still persists in the few remaining natural undrained European forests, such as Białowieza, and nothing else can

explain the abundance and constancy of alder in the middle Flandrian. It will be appreciated that in more rugged terrain such waterlogging would follow to a less degree and the rise of alder frequency is indeed far slower in a site such as that at Waen Ddû described by Mrs D. M. Anderson (unpublished), where the forest mainly grew upon the steep well-drained hill slopes of the limestone scarp adjacent to the sampling site. Given adequate base-status of the rocks and increased rainfall then alder would certainly have become established as a permanent component of hillside oak forests as in *Quercus petraea* woodland in Wales at the present day. In situations however where widespread formation of blanket mire on raised bog was the immediate consequence of increased atlanticity, as was the case rather widely in Ireland, pre-existing woodland was swamped and alder was relegated to marginal strips constituting but a small part of the total vegetational cover and of the woodlands in general where podsolization was also active. This is a consideration that has been earlier advanced by Jessen in respect of Ireland, and by Turner (1970) for the later stages of the Hoxnian interglacial. Van Zeist (1964) has also pointed out how in western Brittany some soils were too poor for alder and carried birch instead, whilst alder thrived on the fertile clays.

It is true, as Firbas pointed out, that the sudden expansion of *Alnus* at the zone VI/VII transition was associated with the concluding stages of the general eustatic rise in sea-level that was then affording the opportunity for

263

TABLE 19. *Radiocarbon ages for major rise in* Alnus *pollen frequencies in the British Isles*

	Station no.	Radiocarbon years B.P.
Scaleby Moss	Q-166	6998 ± 131
	Q-165	7425 ± c. 350
	Q-167	7354 ± 146
Red Moss	Q-916	7107 ± 120
Tregaron Bog	Q-936	7280 ± 180
	Q-937	7360 ± 240
	Q-938	7098 ± 140
Nant Ffrancon	Q-900	6884 ± 110
	Q-901	6790 ± 100
	Q-902	6726 ± 100
Ballyscullion	UB-120	6950 ± 85
Loch Maree	Q-1007	6513 ± 65
Loch Claire	I-4813	6520 ± 145
Loch Sionascaig	Y-2363	6250 ± 140

establishing fen-woods over very extensive coastal tracts. In calcareous regions such as the East Anglian Fenland, the Thames estuary and the Somerset Levels continuing or renewed transgressions kept extensive alder woods in being. The main expansion of alder inland could however only be connected causally with the eustatic rise, if it were shewn that it had brought about a major shift of climatic patterns of western Europe.

At the present day *Alnus glutinosa* is a wide ranging tree present throughout all Europe except Lapland, the Scandinavian mountains and eastern and central Spain: its range is very like that of *Ulmus glabra*. Autecological studies by McVean (1953) have indicated its considerable tolerance of cold and frost-hardiness, coupled with seedling sensitivity to late spring frosts, some sensitivity to wind exposure and great dependence on soil moisture. McVean pointed to the facility of dispersal by running water and wind drift over water surfaces, and he also demonstrated the existence in the British Isles of pronounced clinal variation. It has been pointed out by Mamakova (1968) that both *A. glutinosa* and *A. incana* are avoided by grazing animals, *A. glutinosa* accordingly is not used for leaf-fodder and may have enjoyed some advantage therefore in the early stages of forest exploitation. It does indeed often maintain high pollen frequencies through temporary Neolithic forest clearances, and plays a part in Sub-boreal upland secondary forest, along with oak, ash and birch where the elm has disappeared (Pennington, 1969).

It is apparent that the alder has responded in a complex and sensitive way to the range of opportunities presented to it during the progress of the Flandrian. Sensitive to thermal requirements it made no mass extension until the temperate stages of this interglacial; needing high soil moisture it responded both to local topographic conditions of high water supply and to general climatic wetness; tied to high base requirements it was excluded from poor and acidic soils and from oligotrophic mires; yet, given high atmospheric moisture, it was able to exist as

co-dominant or sub-dominant in hillside oak woodland. It seems probable that it was present in locally favourable stations long before its general extension, and perhaps from the end of the Late Weichselian.

With this great flexibility of response, great caution is needed in interpreting applications of radiocarbon dating to its history. Only where the sudden and massive rise of alder is well expressed can we expect consistent dates to follow from an overall climatic cause. Such dates appear to be provided by the records in table 19. Whilst the sites for England and Wales accord reasonably well, those from Scotland seem significantly later.

Other indices than this, such as the 'empiric' limit and 'rational' limit, shew as might be expected, a high degree of diachroneity (Pilcher, 1970) and the results from the Belfast laboratory have proved the very late and variable dates of alder expansion in northern Ireland. Evidently *detailed* analysis of the history of the alder remains a very laborious matter.

CORYLACEAE

Carpinus betulus L.   Hornbeam (Figs. 23, 90)   337: 1

The diagram for *Carpinus betulus* shews separately records based on macroscopic remains, charcoal, wood and the distinctive deeply ribbed fruits, and those based on the pollen which is quite distinctive with its slightly projecting *Corylus*-type pores, usually four in number, unequally spaced round the equator of the almost spherical grains.

There are pollen records for the hornbeam in the Ludhamian, Antian, Pastonian and Cromerian interglacial stages, backed by a fruit record from Pakefield in the Cromer Forest Bed series. In the Hoxnian, pollen site records are frequent, especially in sub-stage III, and at Hoxne itself, where the pollen frequency rises to its highest values, there is also a fruit identification. The situation is similar in the Ipswichian in that *Carpinus* is represented in sub-stages II, III and IV and most frequently in sub-stage III. As table 20 shews, in this sub-stage hornbeam pollen frequencies often are very high, as at Wretton and Histon Road, Cambridge and at Austerfield, three sites at all of which hornbeam fruits have also been found. It was very early apparent that this phase of the Ipswichian was the equivalent of zone g of the Danish Eemian investigated by Jessen and Milthers (1928) and that these high frequencies of hornbeam were particularly characteristic of the penultimate interglacial in western Europe. The British Flandrian records begin in zone V with pollen from North Sea moorlog where it might have been derived, and zone VI where the low frequency at a site in the Cairngorm suggests wind carriage. The zone VI site at Hockham Mere might be more indicative of local presence. In zones VIIb and VIII there is a dramatic multiplication of site records, backed by five unqualified and and two tentative macroscopic records in VIIb and six

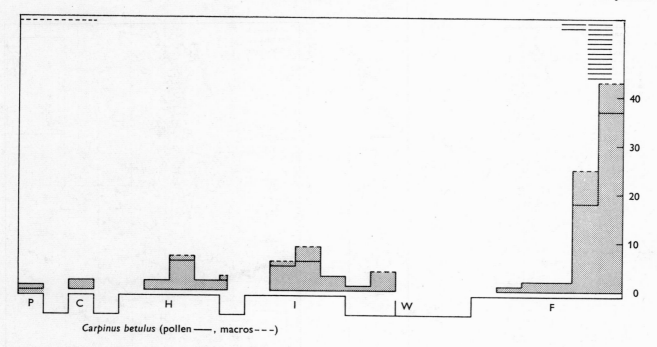

Carpinus betulus (pollen ——, macros – – –)

TABLE 20. *Mean pollen frequencies of* Carpinus *in the three latest interglacials in the British Isles*

	Percentages of total tree pollen								
	+	0–2	2–5	5–10	10–20	20–30	30–40	40–50	50–60
Hoxnian									
Sub-stage II	1	1	.	.	.	.	.	.	.
Sub-stage III	.	3	2	1	.	.	.	.	.
Sub-stage IV	1	1	.	.	.	.	.	.	.
Ipswichian									
Sub-stage II	.	1	2	2	.	.	.	.	.
Sub-stage III	.	.	.	.	3	.	.	1	2
Sub-stage IV	.	2	.	1	.	.	.	.	.
Flandrian									
Zone VIIb	7	12	.	.	.	.	.	.	.
Zone VIII	10	24	3	1	.	.	.	.	.

unqualified in VIII. The records for wood and charcoal are entirely from archaeological sites attributed respectively to the Neolithic (4), Bronze Age (2), Iron Age (5) and Roman (1) periods. As may be seen in table 20, though sites are numerous the pollen frequencies are always low, even the highest, at Wicken, Cambridgeshire, falling far short of the biggest Ipswichian values.

It is no doubt safe to dismiss as of secondary origin the rather numerous instances of records from the glacial stages although these include macroscopic records, of which no less than four are from the Middle Weichselian. The fruits of hornbeam are so durable and so long recognizable that we can be sure they were derived from deposits of the preceding interglacial, especially since they are often accompanied by fruit stones of the thermophilous

Thelycrania sanguinea and nuts of *Corylus*, as at Earith (Bell, 1969).

As the evidence accumulates, the behaviour of *Carpinus* in the three latest interglacials is seen to have particular interest, and its explanation has attracted a good deal of speculation, clarified to some extent by the fact that whereas in the Flandrian human activity has enormously affected the hornbeam forests, the Hoxnian and Ipswichian are presumably free from these effects. It is immediately apparent in each of the interglacials, *Carpinus*, although present in low frequency for a considerable time previously, did not become a notable forest component until a late stage in the cycle. It is especially apparent in the Hoxnian where it intervenes between the thermophilous mixed oak forest typical of most of

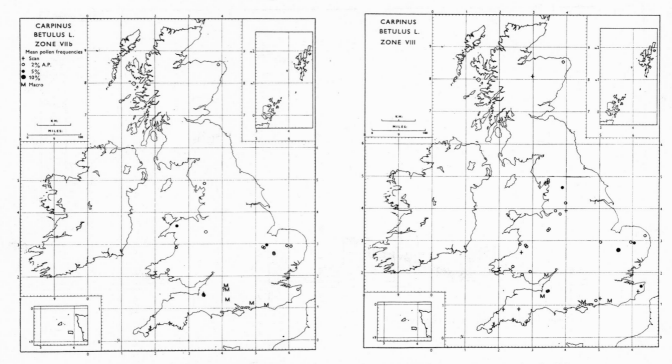

Fig. 90. Pollen frequencies and macroscopic identifications of hornbeam (*Carpinus betulus* L.) in the British Isles during the two latest zones of the Flandrian.

sub-stage III and the *Abies–Empetrum* or *Abies–Pinus* woodlands of sub-stage IV, and in the Ipswichian where however *Abies* was not present in the succeeding woodlands. *Carpinus* might be regarded as occupying a stage introductory to the telocratic proper, probably under the control of both climatic and edaphic factors.

The present range of the hornbeam in Europe is from southernmost Sweden to France, Italy and Greece. It is disjunct in its eastern extension from this area, possibly an effect of the last glaciation (Walter & Straka, 1970); it extends further east into the continent than *Fagus* but much less far than *Quercus robur*, and its interglacial occurrences extend that range north to the Gulf of Finland (60° N) and east beyond Moscow to 45° E. It appears to avoid the western fringe of Europe, and in Britain is regarded by Miller-Christy (1923) as native only in the south-eastern counties, and doubtfully in the lower Severn valley. The Flandrian records for zones VIIb and VIII (fig. 90) broadly confirm this picture, though the macroscopic remains suggest former presence in Sussex, Hampshire, Somerset and Monmouth and the pollen frequency records might be taken to indicate local hornbeam woods occasionally north of Miller-Christy's area. *Carpinus betulus* is classed by Matthews in his Continental element and Miller-Christy points to the infrequency of its seed ripening in trees planted in Scotland, as well as to its great irregularity in fruiting, even in Essex, where it grows most abundantly. It is also, like *Fagus*, susceptible to spring frosts, a phenomenon of atlantic and sub-

atlantic rather than continental climates. There is of course particular difficulty in deciding upon the present natural range of the hornbeam because of its tremendous exploitation by man and its endurance as coppice and as pollard trees under systems of 'lop and top' to provide its very estimable firewood.

Miller-Christy suggests that *Carpinus* will grow upon an extremely wide range of soils, but Tansley indicated that it is most commonly found in Britain on moist silty sands or silty clays. It avoids thin chalk soils, perhaps because of dryness. The British woodlands of today are however far too modified to serve as a total reference for considering the natural, and especially the interglacial, hornbeam forest. Oberdorfer (1967) indicates that on the European mainland the hornbeam is a lowland and a submontane tree of base-rich deep soils often influenced by ground water, and that it is constantly found in localities favouring some acidification. Of three major sociological groups of hornbeam–oak woodlands, the third characterized, 'Auenwälder', is found on damp loamy soils particularly, as Schoenichen (1933) points out, on glacial terraces, ground moraines and similar periglacial features. The East Anglian Ipswichian occurrences appear particularly likely to have been in situations of this kind.

In the past thirty years or so much effort has been put into documenting and explaining the so-called 'explosive expansion' of the hornbeam in central and western Europe in the latest stages of the Flandrian. Firbas (1949) concludes that it cannot have survived the last glaciation north of the Alps, and constructs distribution maps, enlarging the evidence of Szafer's earlier isopol maps, shewing invasion into central Europe from the east in the 'middle warm period', extension westwards in the 'late warm period', and finally in the 'post warm period' stronger extension yielding high pollen frequencies in the Baltic region of east Prussia (where pure hornbeam woods are still centred) but accompanied by much extension to the west and south-west. He makes it clear that the tree was actually present 1500 years or more in west and north Germany prior to its major expansion, a situation paralleling that already mentioned for British interglacials including the Flandrian. Emphasis has been laid by Firbas and many other investigators on the singular relationship of hornbeam to the activities of agricultural man. Ralska-Jasiewicz (1964) writes of its history in central Europe: 'This species was particularly devastated in periods of intensive settlement on account of the fertility of the occupied habitats. The fact that this tree was able to survive down to the present day as one of the main components forming oak–hornbeam forests, must be ascribed to its ability to regenerate in comparatively short time and to its great spread over certain territories.' She cites many authors to shew the tremendous tenacity of *Carpinus* under severe cutting and grazing regimes. Badly as the tree suffered there is a great deal of evidence from palynology that the forest clearances, by removing competition of the hornbeam's most formidable forest competitors, beech more especially, were actually advantageous to *Carpinus*. This is the view put forward by Müller (1947), Dąbrowski (1959), and Wiermann (1962, 1965) who shews three important gaps in the progress of settlement of north-western Germany: in the late Iron Age (*c*. B.C.), the 'Wüstungsperiode' (A.D. 400 to 550) and the Thirty Years' War (1618–48); in each the interruption of agriculture by war led to a mass expansion of both beech and hornbeam.

More especially in view of the interglacial evidence it is probably incorrect to imagine climatic and cultural control as mutually exclusive. It may be that we can most reasonably explain the late expansion of *Carpinus* in all the interglacials as the expression of a '*Clisere*' in the Clementsian sense, that is a temporal sequence of vegetation determined by a change of climate, a succession in which the interglacial climatic cycle at a given (late) stage provides the hornbeam with a combination both of the climatic conditions and the climatic soil type (transitional from mull to mor forest soils) that it needs to achieve dominance over, or co-dominance with, oak and beech. Whereas in the earlier interglacials this process was unhindered, in the Flandrian it was complicated by man, who may at times have set back and at times accelerated an expansion already basically in progress. The observed metachroneity of *Carpinus* expansion in western Europe (see Kubitzki & Munnich, 1960) would be specially understandable in these conditions.

Corylus avellana L.   Hazel, Cob-nut (Figs. 18, 23, 91; plate X)   338: 1

Pollen, wood and fruits of *Corylus avellana*, the hazel, have been very abundantly identified in Pleistocene deposits. The nuts are possibly the most commonly recognized plant remains of our peat and alluvial deposits: it is true that they preserve well in conditions of waterlogging, but they appear to have been incorporated in such great quantities that the former abundance of the plant cannot be doubted. It was the abundance of such evidence of fossil fruits that permitted G. Andersson to construct his remarkable map contrasting the present with the former distribution of the hazel in Scandinavia, and adding another substantial proof of the reality of the Post-glacial Climatic Optimum (fig. 18). The fossil nuts commonly shew that they have been opened by mice or by squirrels (which animals have their own recognizable techniques) and the aggregates of shells often suggest that they are rodent hoards; in other instances they suggest the shore-line detritus of stream or pond. In numerous mesolithic and later human settlements on the European mainland hazel nuts occur in such quantities that it is evident that they were collected for food, and indeed Schwantes has suggested that in the earlier cultures they had something of the role later taken by the cultivated cereals. It cannot be doubted that similar use was made of hazel nuts in prehistoric Britain. Danielsen (1969) indeed attributes to mesolithic man an important role in the early and rapid migration of the hazel into new territory.

In view of the range of shape of the fruits between central and northern Europe at the present day (*fide* Bertsch, 1951), it would be of much interest to analyse fossil material of known ages from the British Isles for this quality.

The microscopic anatomy of hazel wood is sufficiently characteristic to allow definite recognition in all tolerably well-preserved specimens of wood and in good charcoal. The radial lines of moderately sized vessels and the conspicuous aggregate rays are the most obvious features, and in many muds and peats the wood has a characteristic clear yellow colour when freshly exposed. The shiny smooth bark resembles that of *Alnus* and *Betula*.

Rather striking evidence for early utilization of the wood is provided by the Neolithic trackway recovered at Blakeway Farm in the Somerset Levels (Clapham & Godwin, 1948). It was constructed of faggots each laid as a single row of parallel and very straight rods of hazel closely side by side: they varied from 10 to 13 ft (3.0 to 4.0 m) in length, and examination of their annual rings shewed that they were not more than 8 to 17 years old. Such straight, unbranched and rapid growth could only have been produced by coppice growth, a view supported by closer analysis of the ring pattern of the butts. This wood has been given a radiocarbon age of 4460 ± 130 years B.P. A similar trackway later identified about 1000 yards away again consisted of hazel poles (plate X), but these

Absolute numbers of sites   Percentage of all recorded sites

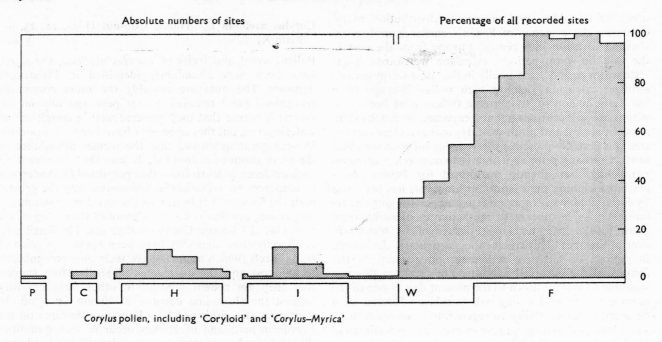

Corylus pollen, including 'Coryloid' and '*Corylus–Myrica*'

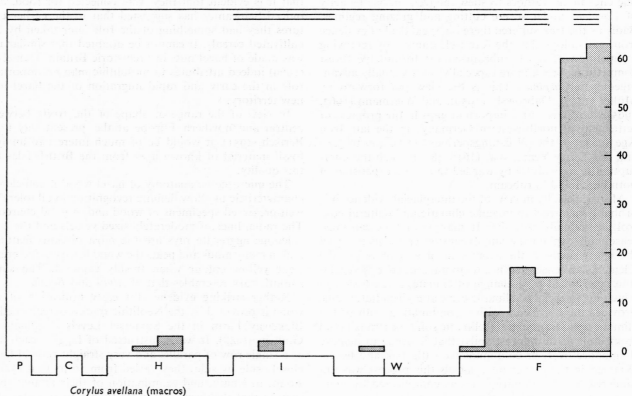

Corylus avellana (macros)

bore marks of Late Bronze Age axes and had a radio-carbon age of 2600±110 B.P. (Godwin, 1960a). Subsequently coppice-wood of hazel has been recognized in other Neolithic trackways at Chilton in the Somerset Levels (Annual Reports of the Sub-department of Quaternary Research, 1970). It seems likely from this evidence that hazel-coppicing was already practised upon the slopes of the Wedmore Ridge and the Mendip Hills during the Neolithic and Bronze Ages. There is, as would be expected, some evidence that hazel wood was also used by prehistoric man for hafting weapons and tools.

Macroscopic remains set out in the histogram record shew *Corylus* to have been present in the Cromer Forest Bed series and in both the Hoxnian and Ipswichian interglacials. Although there is a record of it from the Middle Weichselian at Earith, it is for a nut almost certainly derived from older material. The bulk of macroscopic records are Flandrian, the earliest in zone IV at Drumurcher in south-western Ireland, and the next in zone IV/V from North See moorlog. Very large numbers of macroscopic records have an archaeological context, thus there are Neolithic 24, Bronze Age 27, Iron Age 20, Roman 30, Anglo-Saxon 2, Norman 2 and Mediaeval 5. In particular the numerous prehistoric settlements on the English limestone soils have yielded *Corylus* charcoal and wood very abundantly.

The pollen grains of hazel are smooth and three-pored, sub-triangular in polar view and sub-oblate seen equatorially. Although the pores are aspidate they project only slightly from the general outline of the grain and seen in optical section lack the forked appearance of the wall at the pores of *Betula* grains. Although in the Balkans there is the presence of *Corylus colurna* to be reckoned with, no other species of *Corylus* than *C. avellana* can have contributed to British Post-glacial records, nor perhaps to those of our interglacials. There does, however, remain one important source of confusion, which is between the pollen of hazel and that of the sweet gale, *Myrica gale*. The grains of the latter are generally more circular in polar view, take a browner stain with safranin, and disclose, with good optical equipment, a more serrate inner margin to the pore than the grains of *Corylus*. Although these criteria enable one to refer a large percentage of fossil grains to one category or the other, there remains a troublesome residue of very uncertain affinity, and this fraction can easily become larger and of much significance. This is particularly so in many pollen series from more or less oligotrophic mires in oceanic regions where *Myrica* is prevalent. Very often the former abundance of sweet gale is apparent from the layers of twigs, leaves and fruits, and where the '*Corylus*' curve shows a pronounced maximum at such a level the curve is certainly suspect (see account for *Myrica gale*, p. 248).

As a result of this situation many British palynologists instead of recording '*Corylus*' pollen have recorded a category of '*coryloid*' pollen or of '*Corylus–Myrica*' pollen. Very often of course this represents wholly or largely pollen of *Corylus* itself, although occasionally the attempt is made to give separate categories in the one diagram. In the pollen record histogram the different categories have been summed, for unqualified '*Corylus*' amounts to 75 per cent of the records in all zones and the records are so numerous that they can safely be considered as essentially reflecting the history of the hazel. Their abundance is such that in the Late Weichselian and Flandrian area of the histogram, records from transitional or composite zones have been omitted and results are

expressed as percentages of all sites from which pollen records exist. It has to be noted that it has been a long-standing practice of British palynologists not to include *Corylus* in the sum of total tree pollen, although its frequency is expressed as a percentage of that total. This was a consequence of regarding the hazel entirely as an undershrub, not competing for dominance with the tall forest trees, but there have probably been conditions during which the hazel grew as a dominant in natural scrub or coppice. To simplify vegetational reconstruction in such periods the *Corylus* pollen has often been reckoned as part of the total arboreal pollen. Our own analyses of Flandrian *Corylus* frequencies (fig. 91) have been based on the earlier convention.

Site records for hazel pollen exist from the Ludhamian, Antian, Pastonian and Cromerian interglacials, and from all sub-stages of both the Hoxnian and Ipswichian, with greater frequency in the middle sub-stages where at individual sites there is a correlation of macroscopic or pollen identifications. Even in the earliest Flandrian zone hazel pollen is recorded from nearly 80 per cent of all sites, and this rises virtually to 100 per cent for zone VI and subsequently. This is to be considered with the evidence of macroscopic records from all Flandrian zones and with that given later for relative pollen frequencies. Hazel pollen has been recorded from Thurnian, Baventian, late Anglian, early and late Wolstonian deposits, and sparsely from the Early and Middle Weichselian, but there is no reason to regard this as being anything but of secondary origin. It may be otherwise in the Late Weichselian where hazel is recorded in successive zones I to III respectively at 33, 40 and 55 per cent of all sites. This makes so smooth a transition to the Flandrian that it warrants the conjecture, despite the absence of macrofossil records, that *Corylus* must have been present in or adjacent to the British Isles even in the Late Weichselian.

Hazel pollen is produced in great abundance and the bushes begin to flower at a very early age, so that in general the plant may be regarded as strongly over-represented in pollen spectra. Nevertheless it has been pointed out that under heavy tree canopy the hazel flowers sparsely if at all, and it has been suggested that the bulk of hazel pollen in the post-glacial pollen diagrams is derived from bushes growing marginally to woods, in clearings or even in pure hazel copses. With our increased awareness of the prevalence and extent of prehistoric forest clearance, the seral role of hazel in stages of regeneration has been widely recognized. Though hazel is eaten down by sheep, its foliage is unpalatable to cattle as may be seen on the Burren today and as was observed by Iversen in his Draved forest experiments: in view of this it is not surprising often to find, as in the Neolithic clearances at Fallahogy (Smith & Willis, 1961–2) and in those of the Late Bronze Age or Early Iron Age at Tregaron (Turner, 1964), that hazel pollen frequencies have actually increased during the occupation that followed forest clearing. At such sites undergrazing of partly cleared woodland is presumed.

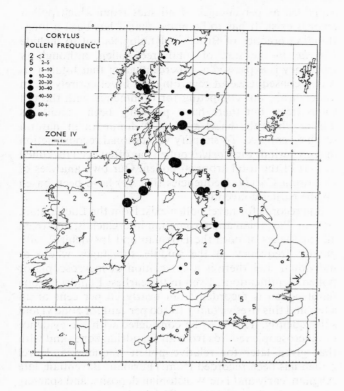

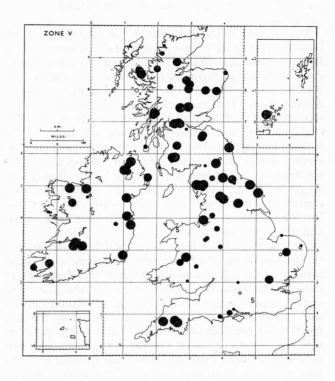

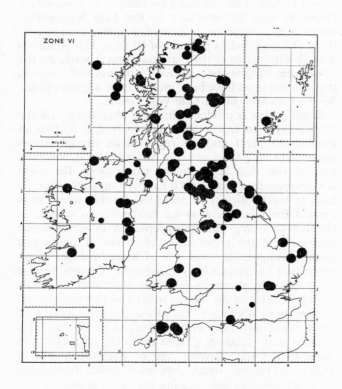

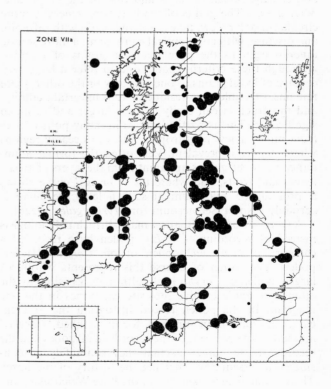

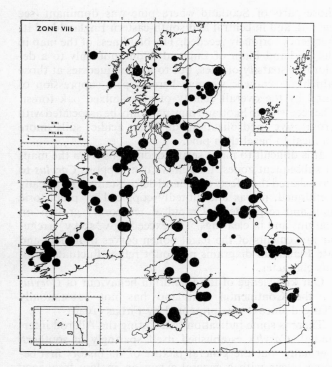 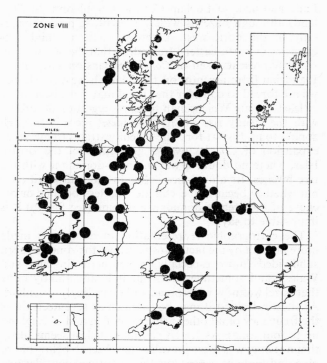

Fig. 91. Pollen frequencies of hazel (*Corylus avellana* L.) in the British Isles through the six pollen zones of the Flandrian. The values indicated are percentages of arboreal pollen excluding *Corylus* itself.

The pattern of behaviour of *Corylus* as indicated by its pollen curves is strikingly and diagnostically different in the succeeding interglacial stages of East Anglia. In the Cromerian it was in low frequency throughout. In the Hoxnian, though present earlier and later, it slowly rose to high frequencies in the later half of sub-stage II and the first half of sub-stage III: its values at Hoxne in H IIc exceeded those of the sum of all other types. The Ipswichian was characterized by a swift rise of the hazel pollen curve to high frequencies at the *beginning* of sub-stage II, after those of *Ulmus* and *Quercus*, but before that of *Alnus*, with a fall throughout sub-stage III to very low values in IV. In contrast with these three interglacials, the Flandrian shews, over the whole country indeed, extremely early and extremely high frequencies of hazel pollen, with an expansion preceding those of *Ulmus* and *Quercus*, and the attainment of values often greatly in excess of the totals of all other tree pollen. These differences of behaviour have been carefully considered by West (1961*b*, 1968), who thinks they might perhaps reflect the differing distances of the glacial refuges of *Corylus* during the successive glaciations, and possibly progressive evolution and sorting during the Middle and Late Pleistocene of ecotypes with differing ecological preferences. He adds that differences in forest structure and composition may also come into consideration. Similar issues arise with respect to many species, and a general consideration is given of them in Chapter VI: 16.

The pollen records for hazel in the Late Weichselian are of considerable interest although uncertain in meaning. At a great many sites, more especially northern ones, *Corylus* is present to the extent of 2–5 per cent of *total* pollen, and at one, Moorthwaite Moss, to the extent of 5–10 per cent. Site records of this kind become more frequent from zone I to zone III, with the implication of a spread of hazel at no great distance. Naturally where pollen frequencies have been expressed as a percentage of total *tree* pollen they yield higher values, so that in the Lake District and Pennines several (for all three zones) fall in the 10–20 per cent category and one in the 30–40 per cent.

The significance of these Late Weichselian records is better appreciated alongside those for zone IV. We may note especially that at Loch Fada, in Skye, both H. J. B. Birks and the Vasaris record *Corylus* at 2–5 per cent of total pollen before the dramatic increase in abundance in zone IV. Here the outlying situation and prevalence of westerly winds exclude long-distance transport as a likely explanation, nor does it seem reasonable to invoke secondary derivation or confusion in identification, where so many instances of high Weichselian frequency are concerned. The suggestion of local presence in favourable situations, especially in the north and west, must be strong.

Fig. 91 shews the sudden increase of the hazel pollen frequency in zone IV. In Skye Birks has defined a *Corylus–Betula* assemblage zone thought to be of this age, beginning about 10200 years B.P. at Loch Meodal and Loch

Fada and 9200 at Lochan Coir' a' Ghobhainn. In England, although hazel is widely present, it predominates only in the north-west. Likewise there are high frequencies in north-eastern Ireland, so that the picture consistently presents itself of *Corylus* expansion early in the Flandrian from a centre off the Scottish west coast, and we are forced to consider that hazel may have been there at least in Late Weichselian time, a suggestion according with the steady rise in sites for hazel records through zones I to III. Fig. 91 for zone V demonstrates tremendous expansion of *Corylus* over a great part of the British Isles, where no doubt it was often co-dominant with the tree birches. Its abundance is less in the south and east, but whether because the climate was less oceanic or because of the greater dominance of pine (see fig. 39) it is impossible to say.

The cartographic representation of zone VI is made more difficult by the fact that sometimes the zone has been subdivided (more especially in southern England and Wales), and sometimes not, but it is abundantly clear that hazel was now widespread and exceedingly abundant. At sites in Ireland it may be as much as seventeen times the frequency of all other arboreal pollen and in England up to four times. As Jessen had indicated, it seems likely that not only was there prevalent hazel scrub below birch, pine and perhaps aspen, but that it formed considerable stretches of pure hazel woodland. Within zone VI in England the maximum extension of hazel tended to occur in sub-zone VIa, whilst it receded conspicuously in VIc, as also in Ireland (Mitchell, 1951), perhaps in consequence of a phase of climatic dryness.

It is now so usual to encounter *Corylus* as an undershrub in oak or ash wood that we are apt to assume that it can have no other ecological niche, and indeed to forget that it can make a very fair-sized tree. There have repeatedly been suggestions that the boreal hazel grew beneath ash or aspen woodland now vanished without trace, but increased knowledge of the *Fraxinus* pollen record has shown that ash only became frequent after the *Corylus* maximum, and there is no real evidence for poplar woodlands either. On the other hand, when grazing, sparsity of parent trees and exposure to winds operate strongly, we may still get, in such oceanic regions as western Ireland, scrub substantially dominated by the hazel. In the absence of competition of the mixed oak forest trees it is easy to accept the likelihood that hazel copses flourished, especially upon fresh calcareous soils, and especially since the Boreal hazel maxima were highest in those regions where today hazel copse shews its best development. The favourable influence of oceanic conditions is no doubt associated with the need for mild winters in a shrub flowering so early as the hazel does, but it must be recalled that its stigmas are sensitive, like those of *Fagus* and the elms, to late spring frosts, and that such frosts are characteristic of sub-atlantic, although not of extreme atlantic climates.

In zone VIIa *Corylus* was still generally abundant though doubtless an undershrub in intact woodland: only in those parts of Scotland where pine was dominant (see fig. 39), and in central and south-eastern England had its frequency seriously receded. The blankness of the map in the Midland Plain of England is due not only to a deplorable sparsity of sites, but to low frequencies at those which are known: we may here suspect suppression of hazel by the overall density of alder–mixed oak forest. The recession in Scotland may naturally be associated with progressive podsolization on thin acidic soils more inimical to hazel than pine.

It is difficult to make secure deductions from the maps of subsequent zones, though we can suppose *Corylus* to have shared the effects of general deforestation. As already mentioned, it was undoubtedly coppiced by prehistoric man and responded by rapid if temporary growth on margins and in clearings. The exceedingly 'spiky' irregular movements of the hazel pollen curves in middle and late Flandrian diagrams no doubt reflect fluctuations of this character.

Our knowledge of the Flandrian behaviour of *Corylus* on the Continental mainland has not substantially changed since it was presented by Firbas (1949).

There is some indication that during the Allerød interstadial, *Corylus* established itself sparingly in southern Germany, but the general picture of its history first becomes clear with a general extension in low frequency during zone IV, the Pre-boreal period, which carried it right into the north-west German plain. This extremely rapid extension may have been due in part to rodents and to birds but was most probably assisted by flooding of the great northwards-running river system of middle Europe. In zone V there began the swift rise to the Boreal 'Corylus maximum', the values for which show a very pronounced rise towards the west. At this time hazel pollen values of 100 per cent or more are common in the young moraine region of the Baltic coast, on the strip of old moraine country along the German North Sea coast, and in the valleys of the western and south-western mountain regions. In the succeeding Atlantic period the hazel values are diminished and run more or less parallel with those for the mixed oak forest: in the Sub-boreal and early Sub-atlantic there are still more reduced hazel frequencies all over middle Europe although some ultimate recovery due to anthropogenic effects in the Mediaeval period may be recognizable.

Firbas' isopol maps strongly emphasize the preference of *Corylus* for an oceanic climate, exhibiting both much higher frequencies in the west during the Boreal, and less pronounced decrease in the west after the middle 'Wärmezeit', say after 3000 B.C. Subsequent work in eastern central Europe has confirmed the migration of hazel there from the west in the time of the hypsithermal, an effect possibly indicative of an increased oceanicity that was afterwards lost.

The abundance of evidence in its Pleistocene record gives *Corylus* a particular importance, especially in relation to far-reaching questions that affect many other plants. In particular the evidence for *Corylus* concerns the

TABLE 21. *Radiocarbon ages for early* Corylus *in the British Isles*

Site	Authority	First continuous curve		Beginning of big rise	
		Station	Years B.P.	Station	Years B.P.
Star Carr	Godwin & Willis (1959)	Q-14	9557 ± 210	.	.
Scaleby Moss	Godwin, Walker & Willis (1957)	Q-155	9740 ± 183	Q-161	9002 ± 194
		Q-154	9557 ± 209	Q-162	8809 ± 192
Port Talbot	Godwin & Willis (1964)	Q-661	9920 ± 170	.	.
Red Moss	Hibbert, Switsur & West (1971)	Q-924	9798 ± 200	Q-921	8880 ± 170
		.	.	Q-920	8790 ± 170

position and distance of glacial refugia, rates of migration and degree of synchroneity in appearance and sudden expansion, to which questions radiocarbon dating is now importing a welcome objectivity. The most recent survey (Hibbert, Switsur & West, 1971) indicates that the first consistent appearance of *Corylus* in north-western Europe was roughly synchronous around 9800 years B.P. There is however an apparent delay of some 800 years between the major *Corylus* expansion in south Sweden and coastal North Sea region respectively, with the English sites of an intermediate age round about 9000 B.P.

It would be premature however to see in this any final proof of consistent diachroneity since samples obtained even more recently by H. J. B. Birks from four sites within the one Scottish island of Skye have been shewn to range between 8650 and 10000 years B.P., i.e. may shew a range larger than that of the full extent of the cited west European sites. The Irish sites also shew a big time range but the datings are of very variable validity.

Corylus avellana is now native throughout the British Isles, save only in the Shetlands, and it reaches an altitude of more than 2000 ft (610 m). It has a wide European range and on the mainland of Europe shows a much greater northerly extension to the west, reaching almost 68° N latitude in Norway, 63° N in eastern Sweden, 62° N in Finland, and about 58° N in central Russia. It extends south to Spain, Sicily, Crete and Cyprus.

It favours basic soils, and will occupy both moist and dry situations upon them: it also occurs freely on damp neutral or moderately acid soils. It is particularly common as coppice, but shews nevertheless every sign of being a natural understorey shrub of British oak woodlands and of ash woods.

Ostrya sp. Mich. ex L.

L *Beccles, Stradbroke*; Th *Hoxne*; Ant *Hoxne*; P *Bacton*, ?*Yarmouth*

Pollen agreed by R. Andrew, H.-J. Beug and R. G. West as that of the genus *Ostrya* has been found in deposits from the Ludhamian to the Pastonian, and R. Beck has recognized it in deposits at Yarmouth tentatively referred to the Pastonian also. This is a record conformable with 'Ostrya-type' pollen on the Continental mainland

where it is found in the Miocene, Pliocene and older Pleistocene.

O. carpinifolia belongs to the sub-mediterranean geo-element native in southern Europe from France to Bulgaria and Austria.

FAGACEAE

Fagus sylvatica L.   Beech (Figs. 23, 92)   339: 1

The native status of the beech, *Fagus sylvatica*, has long been doubted, a phrase in Julius Caesar's account of south-eastern England suggesting that beech was absent from a territory now the heart of our natural beech forests. It seems likely, however, that the word *Fagus* had been employed by Caesar to mean sweet chestnut. Aside altogether from the pollen evidence, by now very strong, the records of macroscopic remains of beech are alone ample to establish its natural and pre-Roman status. There are four records of wood or charcoal from Neolithic sites, six from Bronze Age sites, five from the Iron Age and four from Roman sites.

Just as wood, bark, fruits and cupules of the beech are highly recognizable, so are the large pollen grains which are generally regarded as under-representing the frequency of beech in relation to other trees. This follows partly from the low pollen productivity of its flowers, partly from its habit of intermittent flowering, and partly from a known tendency for its pollen to sink directly to the forest floor and escape entry into ascending air currents. Possibly the chief obstacle in Britain to the elucidation of the history of *Fagus* by palynological evidence has been the natural rarity of suitable lake and bog deposits in the dry landscape of the chalk and oolite formations where our beech forests are centred.

The oldest Pleistocene records of *Fagus* in Britain are the sparse pollen grains recorded by Duigan (1963) from the Cromerian and the cupule she reports 'almost certainly from the Cromer Forest Bed series' at West Runton, following reports of *Fagus* leaves by Reid (1882) from the same series. Next in time are Hoxnian records for sparse pollen in sub-stages III and IV at Marks Tey in Essex, not thought likely to have been secondarily derived

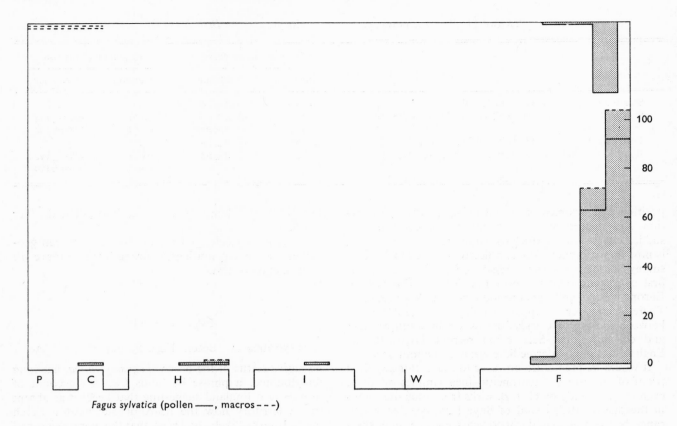

Fagus sylvatica (pollen ——, macros – – –)

(Turner, 1970), and five cupule fragments identified from sub-stage IV of the Hoxnian at Gort in western Ireland, and referred after careful consideration to *F. sylvatica* (Jessen *et al.*, 1959). Gort is of course far to the west of the present natural range of beech in Europe but it is clear from investigations summarized by Frenzel (1968: fig. 47) that it was present at sites in a large part of eastern and central Europe during the Holsteinian interglacial, more particularly in the *Abies–Carpinus* phase, although it represented a small component only of the total forest. Just as at Gort it occurred far west of its present range, so it extended at this time eastwards into Russia well beyond its present limits.

Evidence for the Ipswichian interglacial is limited to the record of occasional pollen from sub-stage III at the single site of Hutton Henry, Co. Durham (Beaumont, Turner & Ward, 1969). This is barely sufficient to suggest that *Fagus* existed in our forests even in low frequency during the phase of *Carpinus* expansion. It seems from Frenzel's summary (1968: fig. 54) that the European records come mainly from the *Carpinus* phase and are chiefly in east central and eastern Europe, indeed to the east of latitude 8° E; they extend eastwards into Russia but are generally lacking in central and western Europe. Frequencies are again low.

There are no records for *Fagus* in any of the glacial stages, and in the Flandrian, as in the preceding interglacials, records do not appear until late in the cycle. We have already seen that records of macroscopic remains are numerous but restricted to zones VIIb and VIII; nevertheless they have significance in interpreting the pollen records for earlier zones for they afford many instances of proof of local presence of the beech at sites where good continuous pollen records shew *Fagus* in only low frequencies. Thus beech was present in the Neolithic wooden trackway at Blakeway Farm, Somerset, radiocarbon dated at 4460 ± 130 B.P. and in the Late Bronze Age 'Bulleid's' track on Meare Heath, dated similarly at 2840, 2850 ± 110 B.P. (Godwin, 1960c), but at neither of these sites nor at the many other sites of pollen analyses in the raised bogs of this locality was beech pollen recorded other than intermittently and in low frequency. This makes it difficult to dismiss as due to distant transport by wind the sites in early or late zones where beech pollen is present though sparsely (fig. 92). Especially this is so for sites at Wareham and Crane's Moor in the Hampshire Basin where beech pollen was recorded sporadically in zone VIc, to be followed by substantial frequencies however of 2–5 per cent in VIIa at Wareham, and 0–2 per cent at Crane's Moor. Likewise we cannot easily disregard values of 0–2 per cent in zone VI in the Huntspill Cut near the Bristol Channel. Early as zone VI is for a first entry of beech into Britain it may be noted that its immigration would be expected first upon this coast, and that this period might suggest a possible entry at a time when the English Channel had less than its present width (fig. 162). By zone VIIa there are scattered

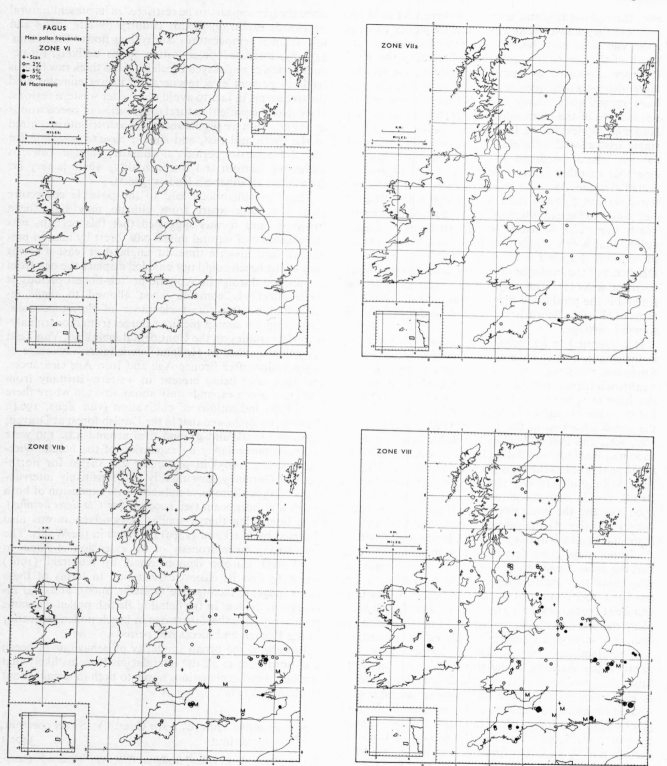

Fig. 92. Pollen frequencies of beech (*Fagus sylvatica* L.) in the British Isles through the four latest Flandrian pollen zones, with macroscopic identifications in zones VIIb and VIII.

sites throughout England though nowhere save at Wareham are they above the 0–2 per cent frequency class. In zone VIIb sites have become far more numerous, and south of the line from the Bristol Channel to the Wash they are supported by numerous macroscopic records. Here also pollen frequencies are higher and beech must have grown quite widely. It is harder to estimate the significance of the many sites now present through England and Wales: those in Scotland probably, and the single Irish record certainly, may signify wind transport to sites where woodland was sparse through altitude, clearance or other factors. Similar conclusions are suggested for the zone VIII map where both macroscopic remains (Iron Age and Roman) and high pollen frequencies testify to the expansion of beech south of the Severn–Wash line. The increased pollen frequencies suggest indeed that beech woods may have been locally present throughout England and Wales south of the Lake District, but the evidence of macroscopic remains would be most welcome. The single high frequency given for Ireland is the product of historic planting: the others, as most or all of the Scottish records, will be due presumably to wind carriage.

It is essential in considering the history of the beech in the British Isles to have in mind its history also on the European mainland, where fortunately there is a great wealth of information and where many competent authorities have analysed the situation comprehensively. When Firbas made his survey in 1949 he was able to shew that the beech appeared in the Böhmerwald about 5000 to 6000 B.C., and that it reached the Baltic coast by about 2000 B.C. Radiocarbon dating has subsequently provided additional proof of the early presence of beech in Europe: thus Willerding (1969a) cites beech charcoal from a Bandkeramik settlement at Kleinenhagen dated c. 4500 B.C. and the beginning of continuous *Fagus* pollen curves at the Fichtelgebirge (3900 B.C.), Oberhartz (3300 B.C.) and Gifhorn (3025 B.C.). It was apparent from the isopol maps given first by Szafer and then by Firbas that the great expansion of *Fagus* in Europe did not take place until the 'Nachwärmezeit', the Sub-atlantic. This has been later confirmed by many workers, for instance van Zeist (1959, 1964) for Holland and Brittany. It seems likely therefore that *Fagus* was present in isolated stands over very large areas before it achieved the remarkable general dominance it has had in historic time. This parallels the situation in the British Isles and of course raises the question of what were the factors that prevented its earlier spread, and those that allowed its subsequent sudden expansion. Firbas took the view that in the Postglacial Warm Period north of the Alps, the beech cannot have been restricted by too short or too cool a vegetative period, nor by excessive dryness, and he tended to lay emphasis upon winter cold, late spring frosts and possibly upon too much warmth in spring. The explanation in fact remains uncertain.

The Sub-atlantic expansion of the beech in England can scarcely have been due to diminishing temperature since the tree appears to be restricted in its present natural westwards extension in this country by late frosts and low summer temperatures controlling flower formation. Equally, increased precipitation seems unlikely, for accumulation of undecayed beech leaves as thick raw humus tends to prevent easy regeneration in wet climates, and the tree seems in fact to avoid our wetter western regions. If we take the present natural distribution of beech woods as our criterion we should recognize a distribution centred in the south and east, and should have to postulate for the Sub-atlantic period a climatic change towards increasing warmth and continentality, which is very far from what all other evidence suggests.

These considerations lead to the possible conjecture that something apart from climate may have been involved, and it was suggested that this perhaps was the clearance of natural mixed oak forest by man in prehistoric and historic time. In England at least it seems that whilst beech could not unaided displace the oak from dominance during the Atlantic or Sub-boreal periods, it *could* effectively compete when allowed to share recolonization after forest clearance of limestone soils. Increasingly a similar picture emerges from modern palynological studies on the Continental mainland. Van Zeist (1959) shews that in Holland beech takes advantage of regeneration after Bronze Age and Iron Age clearances, and that after being present in western Brittany from 2000 B.C. beech expands only about A.D. 300 where there have been indications of cultivation (van Zeist, 1964). Iversen (1969) shewed that in the Danish forest at Draved, removal of oak and ground fires around A.D. 130 were followed immediately by expansion of the beech. Wiermann (1962, 1965) in particular has shewn for northwestern Germany how rapidly and sensitively interruptions of agriculture have induced mass expansion of both beech and hornbeam (see account for *Carpinus betulus*). It is interesting that a similar responsiveness was also detected by Markgraf (1969) for *Picea* in the Wallis. The very strong metachroneity of the extension of *Fagus* in north-west Europe demonstrated by Kubitzki (1961) seems to find its easiest explanation in terms of these anthropogenic factors (see Godwin, 1966), and it is increasingly evident in the detail of British pollen diagrams how many sites shew increases in beech pollen frequency in the Roman and mediaeval periods.

In view of the broad similarity of behaviour of beech and hornbeam in all the interglacials it seems likely that similar causal explanations apply to both in terms set out in chapter VI: 6.

Castanea sativa Mill. (*C. vesca* Gaertn.)   Sweet chestnut, Spanish chestnut   340: 1

IG ?*Crayford*; F VIIb ?*Bedham Hill*, VIII *Cissbury Camp, Pevensey, Woodcuts, Rotherley, Christ's Hospital, ?Goldhanger, Nuthampstead*

The records for the sweet chestnut, *Castanea sativa*, depend upon identification of its wood or charcoal, which

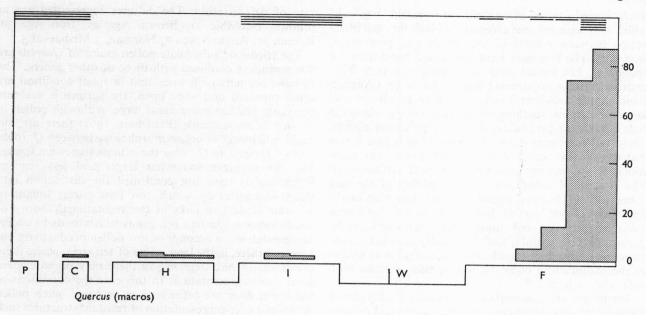

Quercus (macros)

Absolute numbers of sites   Percentage of all recorded sites

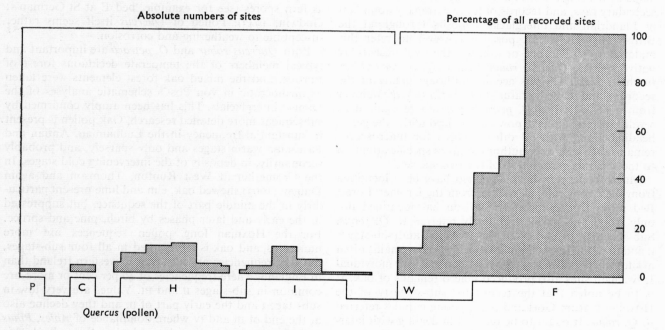

Quercus (pollen)

may under certain conditions be confused with samples of *Quercus* chancing to have very narrow rays. The identification of *Castanea* from the Crayford brick-earth has been challenged, but Kennard has stated that the Mollusca indicate a warmer climate than that of today, and there is no doubt of the interglacial age of the deposit.

The Flandrian records are all from archaeological sites, of which the oldest is Neolithic, a doubtful recognition at Bedham Hill. It was recorded at the Iron Age to Roman level at Cissbury Camp, and at five Roman sites and one mediaeval site about A.D. 1300.

Castanea sativa is a warmth demanding Tertiary species unknown since the Reuverian in north-western Europe save as an introduction, and although it is often regarded as a Roman introduction to Britain the archaeological evidence may merely indicate imported timber.

Quercus L. Oak (Figs. 23, 93; plate II)   341

Quercus may be recognized macroscopically by its timber, twigs, bud scales, leaves, fruit and cupules, and stools of oak are very abundant in mires of various kinds, especially in those representing late fen-wood stages in a hydrosere.

Such 'buried forests' often yield the so-called 'bog oak' which may have become exceedingly black through interaction between tannins in the woods and percolating iron salts. The bog oaks have a deserved reputation for toughness, but buried oaks in general may vary from extreme softness to extreme hardness. In the East Anglian Fenland the Sub-boreal oaks are often of remarkably large dimensions, reaching for instance a height along the bole of 90 ft (27.5 m) before the first large lateral branch. These tall trees grew upon the Gault Clay before it was waterlogged but other oaks, usually smaller, are commonly found as they grew upon the peat surface itself. The process of overthrow and embedding of the oak stems can still be seen occasionally in operation in the fen-woods of our Norfolk Broads. On account of prevalence or suitability oak timber was much employed by prehistoric people and it was abundantly used in the construction of dwellings, trackways and palisades as well as in the manufacture of boats, looms, tubs, shields, toolhafts and so forth.

The timber and charcoal of oak are very readily identifiable by the ring porous nature, huge vessels and massive secondary rays, and records of it are extremely abundant in Flandrian and interglacial deposits throughout the British Isles. It is not possible however to refer this material to either one or other of the two present-day native species, *Quercus robur* L. and *Q. petraea* (Mattuschka) Liebl. On this account although many of the records stand in the literature as *Q. robur* (a large number from a period when *Q. petraea* was not recognized as a valid species) all have now been placed under the genus heading along with all other records for macroscopic remains save where the author has given specific attention to the separate recognition of the two species.

Macroscopic remains of *Quercus* have been identified from the Cromerian, as well as from the Cromer Forest Bed series from which Mrs Wilson has identified not only wood but a fruit tentatively referred to *Q. robur*. Sub-stages II, III and IV of the Hoxnian and sub-stages II and III of the Ipswichian have yielded oak remains often at sites where pollen was also recorded and four individual Ipswichian sites also produced macro remains of oak. It is to be noted that the record for sub-stage IV of the Hoxnian is from Gort and is for young cupules referred to *Q. robur*. It ought to be recalled in dealing with interglacial records of *Quercus* that the non-British European oaks, such as *Q. ilex*, *Q. cerris*, and *Q. lanuginosa*, are not necessarily excluded, although it is practically certain that they are not concerned with records from the Flandrian.

Apart from the rather uncertainly dated record from North Sea moorlog assigned to zone IV or V, the earliest Flandrian records are from zone VI, and they remain relatively sparse in zone VIIa. Thereafter records are extremely abundant but it would be mistaken to attribute this changing frequency to alteration in the abundance of the tree; it is due rather to the increasing utilization of oak wood by prehistoric man, and possibly indicates the rate of deforestation. The cultures represented are as follows: Neolithic 21, Bronze Age 45, Iron Age 30, Roman 41, Anglo-Saxon 3, Norman, 3 Mediaeval 5.

The tricolpate subprolate pollen grains of *Quercus* are not nowadays confused with those of other genera. The furrows are narrow fissures that in fossil condition are often ruptured and wide open, the surface is scabrate verrucate and the polar areas large. Although pollen of *Q. ilex* is recognizable (Planchais, 1967) there are but slight differences in pollen morphology between *Q. robur* and *Q. petraea*. In *Q. robur* the exine is somewhat thicker and the verrucae somewhat larger and less regular. Palynologists have not confirmed the distinction into shape categories by which von Post (1924) sought to separate these two oaks in the vegetational history of south Sweden. *Quercus* is a genus which tends to underrepresentation on account of low pollen productivity but here and there, in the later stages of fen seres, where there are stools of oak trees *in situ* there may be an evident local over-representation. In this case the conditions of the forest floor are often such as to cause much pollen decay and over-representation of resistant structures such as fern spores (see for example, bed E at St German's; Godwin, 1943). Pollen of *Quercus* itself seems rather susceptible to weathering and corrosion.

Both *Quercus robur* and *Q. petraea* are important and typical members of the temperate deciduous forest of Europe, and the mixed oak forest elements were taken as mediocratic in von Post's schematic analyses of the Danish interglacials. This has been amply confirmed by subsequent more detailed research. Oak pollen is present in substantial frequency in the Ludhamian, Antian and Pastonian warm stages and only sparsely, and probably secondarily, in deposits of the intervening cold stages. In the Cromerian at West Runton, Thomson and again Duigan (1963) shewed oak, elm and lime present particularly in the middle part of the sequence, but suppressed in the early and later phases by birch, pine and spruce. For the Hoxnian long pollen sequences are more numerous, and oak is represented in all four sub-stages, but the frequencies are far lower in western Ireland than in England, where frequencies of 40 per cent or more are common in sub-stages II and III. Values are very low in sub-stage I and the early part of II, and they decline also at the end of III and IV where dominance of *Abies*, *Pinus* and *Betula* reflects the changed climatic and edaphic situation. In the special local conditions of Nechells, Birmingham, early dominance of oak in the temperate sub-stages was apparently reduced by widespread waterlogging and great increase of *Alnus glutinosa* (Kelly, 1964). The pollen record for the Ipswichian is similar although site records are more concentrated in the two middle sub-stages. Again very high *Quercus* pollen frequencies obtain, especially in the early part of sub-stage II, but in this interglacial the relative importance of oak is reduced by the great expansion of hornbeam in sub-stage III. As in the Hoxnian it is infrequent in sub-stage IV (West, 1958, 1968; Sparks & West, 1970). It seems unnecessary to regard as other than derivative the sparse records of oak pollen reported from the late Antian, early and late Wolstonian and Early Weichselian.

Within the Late Weichselian low percentages of oak pollen are recorded at 10 to 20 per cent of all investigated sites. Much of this is attributed to secondarily derived pollen, but it has to be recognized that these site percentages increase through zones IV and V of the Flandrian where this explanation is implausible. This then suggests the alternative explanation, viz. that these low *Quercus* pollen counts come from the general long-distance pollen rain and that the source supply is getting larger and nearer through the zone sequence I to V. With such an explanation in mind the pollen frequency maps for Flandrian zones IV and V (fig. 93) must be considered, with their countrywide low frequencies of oak pollen, to represent very largely distant pollen-carriage especially since the highest values are associated with open sites of low local pollen production. Local growth is more likely in zone IV at Gosport, Hants., where values attain 10 to 20 per cent of total tree pollen, and at Old Decoy, Norfolk, where the mean value is 5 to 10 per cent. In zone V, the *Quercus* pollen now is counted in spectra containing that of local birch, pine and hazel, but none the less displays a general rise in frequency mostly in the 2 to 5 per cent tree pollen category, but at several in that of 5 to 10 per cent. Two of the higher values are in the Lake District: similar ones for Scotland are of possibly later date as they represent the combined V–VI zone. There can be little doubt that these results point to some local establishment of oak in zone V over much of England and possibly also in eastern Ireland. Within zone VI, and subsequently, virtually all sites of pollen analyses contain *Quercus* pollen.

Zone VI was the zone of mixed oak forest establishment, and sub-division of the zone for England and Wales was based upon preponderance of elm in sub-zone a, and of oak in b, with substantial addition of lime and alder (with other criteria) in c. The maps of fig. 93 indicate a substantial rise in oak pollen frequency over Ireland, England and Wales, a movement progressing strongly through VIb and VIc. This rise though evident in Scotland, is far smaller than elsewhere in the British Isles: only in Galloway and a few other western sites do values exceed 10 per cent of total tree pollen. In eastern England the increase of oak appears limited by high frequencies of pine, whereas in the west oak faced the less severe competition of the birch and expanded much more. In the south-western peninsula of England the early establishment of oak is attested by two radiocarbon dates of 9053 ± 120 and 9061 ± 160 B.P. (Q-1021 and Q-1020) following one for *Corylus* of 9295 ± 180 (Q-1019).

The map for zone VIIa illustrates the continued consolidation of oak in all territories already occupied and expansion in modest frequency throughout Scotland. There are especially high frequencies in Galloway, the Lake District and north-eastern Ireland which here, as with other species and at other times, behave very simi-

larly. There are especially high values in south-western England, but in East Anglia and south-eastern England oak remains relatively less important possibly because of widespread expansion of *Alnus* and *Tilia*. In VIIa oak is less frequent in Ireland than in England and Wales. Jessen writes 'in many parts of Ireland the optimum conditions for the growth of oak occurred in Sub-boreal time, and at the end of this period it was almost as common in the west as in the north-east of the country', but the mapped evidence for zone VIIb shews that the effect was not large. The oak however now expanded considerably in eastern England, possibly favoured by diminished competition from *Ulmus* and *Tilia*, the latter a tree much under-represented by its pollen count, and prevalent in zone VIIa. Zone VIII shews merely a slight diminution in oak frequency in England and north-eastern Ireland. So far as *Quercus* was affected by forest clearances of zones VIIb and VIII this was by general felling of woodland, rather than selective clearance such as concerned *Ulmus* and *Tilia*. Unfortunately, few English pollen diagrams extend far into zone VIII and we can hardly expect to recognize in them any trace of the mediaeval encouragement of oak, as the staple of swine-pannage, leather-tanning, ship-building and miscellaneous building and farming industries.

On the European mainland *Quercus* displays a history conformable with that described for the British Isles. It is thought to have spread right across Germany during the later part of the Pre-boreal period. This rapid spread was no doubt made possible by bird dispersal and the range of soil conditions tolerated by the oaks. From the opening of the Boreal time onwards the oak has always been present in the mainland forests of Europe, expanding greatly in frequency and range in the Atlantic and Sub-boreal, and receding from its Continental border towards the west in the Sub-atlantic. In the south-west its recession was largely associated with extension of the beech. Throughout the warm period the oak was undoubtedly the principal component of the deciduous forests and its importance is certainly under-played by the pollen representation. It was always more important in the south and west than to the north and east where it made contact with the territory of the conifer forests.

The attested rapidity of spread of oak across the Continental mainland makes it easy to accept its presence in southern and eastern England in zone IV. The possibility of survival in western coastal refugia is generally discounted and it will be recalled that at the zone IV/V transition oak pollen was present to the extent of 10 per cent in moorlog on the Leman and Ower bank just off the Norfolk coast. Invasion from the south and east, with ocean levels still low, seems fairly certain.

Q. robur and *Q. petraea* occur throughout the British Isles, but whereas the former is the characteristic dominant of heavy and basic clay and loam soils, and thus preponderates in the Midlands, southern and eastern England, the latter is a characteristic dominant of siliceous soils in the north and west of the British Isles. Both

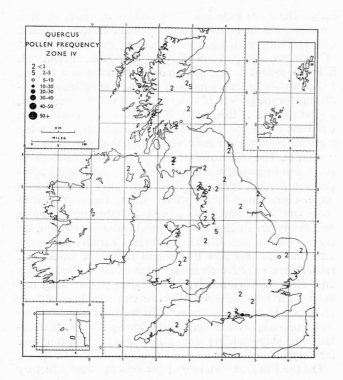

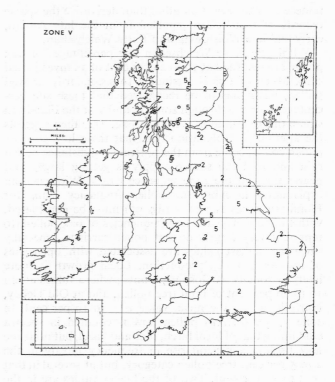

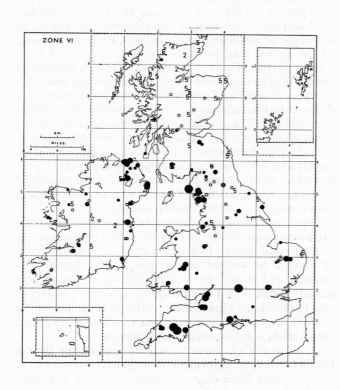

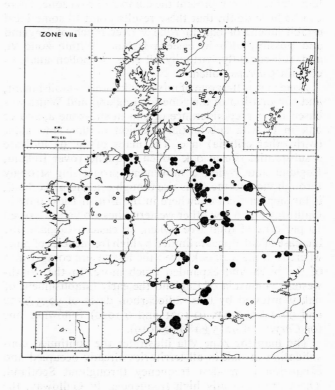

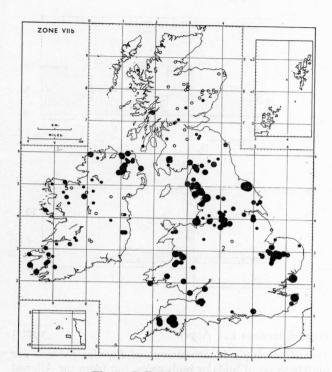 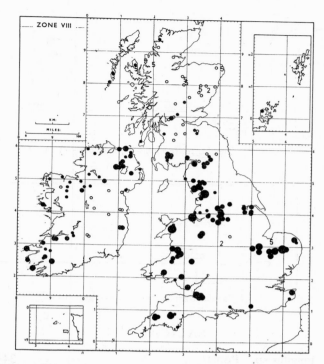

Fig. 93. Pollen frequencies of oak (*Quercus*) in the British Isles though the six Flandrian pollen zones.

species have been much planted and they freely hybridize. In some regions of contact between areas dominated respectively by the two species such as the Welsh marches, there is a belt of abundant hybrid forms. There is a considerable body of opinion that *Q. petraea* of the two is most favoured by an atlantic climate, and in a broad sense this idea is reinforced by the first thorough survey of the range of the two species now published in the *Atlas of the British Flora*.

Both species have a wide range on the Continental mainland, but *Q. robur* extends somewhat further north in Scandinavia and far east into the central Russian plains whereas *Q. petraea* follows an eastern boundary close to that of the beech, i.e., from east Prussia south-south-east to the Black Sea. Both extend into the Caucasus, Asia Minor, the Balkans, Italy and north Spain.

Quercus ilex L.   Evergreen oak, Holm oak, Ilex 341 : 2

F VIII ?*Hembury Fort, Pevensey*

The tentative identification of wood of *Quercus ilex*, the holm oak, from Hembury Fort (Early Iron Age), and the record of its wood from the Roman site at Pevensey (where it could easily have been introduced), do little to suggest a natural British status for this Mediterranean species.

It is to be noted that French research workers can recognize pollen of *Q. ilex* and have been able to demonstrate its substantial expansion in south-western France

during the Boreal period, apparently in response to increasing temperature (Planchais, 1967). It has apparently also been found in middle Flandrian deposits in Normandy (van Campo & Elhai, 1956). *Q. ilex* regenerates sub-spontaneously in certain warm localities on the chalk of southern England so that there seems every reason to look for its pollen in English interglacial deposits.

Quercus robur L.   Common oak, 'Pedunculate oak'

341 : 3

Cs ?*Norfolk Co.*; H *Gort*

Mrs Wilson's tentative identification from the Cromer Forest Bed series was for a fruit. That of Jessen *et al.* (1959) from sub-stage IV of the Hoxnian in western Ireland was based upon immature fruits with cupules (see the general account for the genus).

Quercus petraea (Mattuschka) Liebl. (*Q. sessiliflora* Salisb.)   Durmast oak, 'Sessile oak'   341 : 4

F VI/VIIA *Drumskellan*, VIIA *Tooreen More, Derrycassan, Ardlow Inn*, VIIb *Drumurcher, Elland*

The leaves of *Quercus petraea* have been recognized at the Boreal/Atlantic transition (locally dated 6950 B.P.) at Drumskellan, and from zone VIIb in northern England. Elsewhere the identifications were based upon recognition of the buds, and at Drumurcher of buds together with the gynaecium.

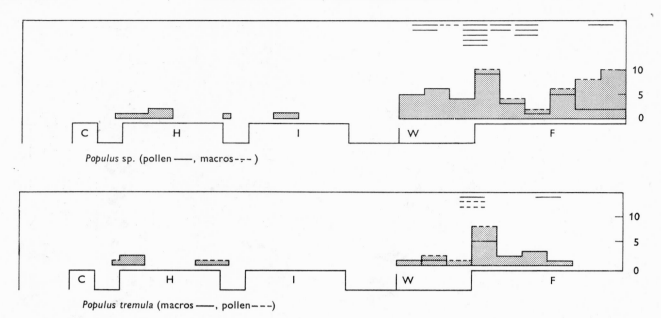

Populus sp. (pollen ——, macros ----)

Populus tremula (macros ——, pollen – – –)

SALICACEAE

Populus sp.   Poplar   342

Since oil-immersion microscopy has been applied to paly-nology it has become possible to recognize the delicate inaperturate pollen grains of *Populus* which have a non-tectate exine covered with very small granules. These techniques have provided Irish records from the late Anglian and sub-stages I and II of the Hoxnian, together with four records from Late Weichselian zone I and six from zone II. Pollen records elsewhere cover the early Wolstonian, sub-stage II of the Ipswichian, zone III of the Late Weichselian and all zones of the Flandrian, for which many are Scottish records by Mrs Birks.

Bud-scales, cone-scales and twigs provide similar evidence of presence in the Flandrian zones IV to VIIa, and less certainly in the Late Weichselian at Nazeing. At Westward Ho! the macroscopic evidence is reinforced by pollen identification in zone VI. Charcoal of *Populus* has often been recorded from archaeological sites of the Flandrian zones VIIb and VIII; these number Neolithic 1, Bronze Age 4, Iron Age 3, Roman 3 and Mediaeval 2. We have not separately recorded a number of charcoal identifications of '*Populus* or *Salix*' from other archaeological sites of this period.

It seems that the genus has been present in the British Isles at least from the beginning of the Late Weichselian, most probably represented by the aspen, *Populus tremula* (*vide infra*). Pollen of *Populus* is recorded for all three interglacials in Denmark, i.e. the three latest in the British sequence, with the pollen of light-demanding species in the early and late stages of each but suppressed in the middle forest-dominated part (Andersen, 1961).

Populus tremula L.   Aspen   342: 3

The macroscopic identifications of *Populus tremula* are mostly those of buds or bud-scales, but from the Allerød layer in the bed of Windermere the very recognizable catkin-scale was identified, a find that parallels similar discoveries in the Allerød of Denmark and north-west Germany. There are two macroscopic site records from sub-stage I of the Hoxnian and these are supported by pollen identifications at Gort of pollen of *P. tremula* type from sub-stage IV, the late Anglian and early Wolstonian. In his reconstruction of vegetation during the Irish Hoxnian, Watts (1967) suggests that in zone G3 *Betula pendula*, *B. pubescens* and *Populus tremula* probably were dominant: this is the stage in which *Pinus* 'begins to emigrate' and we take it to be within our sub-stage I.

Sparse pollen records of *P. tremula* in the Late Weichselian and earliest Flandrian reinforce the macroscopic evidence in suggesting that the aspen has been present in the British Isles since the beginning of the Late Weichselian, as was indeed indicated by the records for the genus.

The aspen is a tree of far northern range and such a history would be expected of it.

Populus nigra L.   Black poplar   342: 4

F VIIb *Findon*, VIII *Bourton-on-the-Water*

The two Flandrian records for *Populus nigra* are both of charcoal and are respectively of Neolithic and Saxon origin.

Salix sp.   Willow (Fig. 95; plate XI)   343

Among the willows recognized as native in the British Isles there are species so widely different in edaphic,

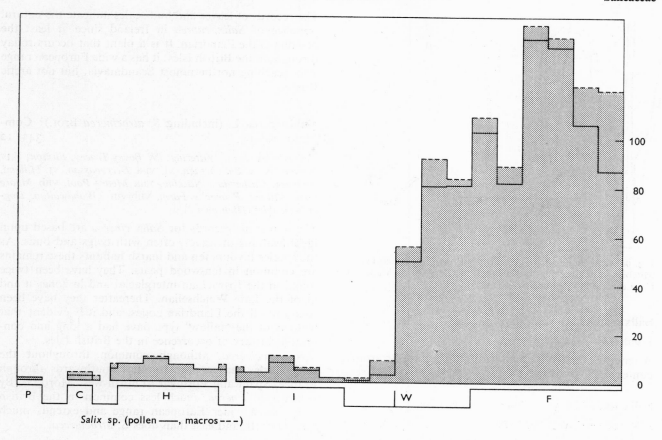

Salix sp. (pollen ——, macros ---)

climatic and geographical range that records which extend only to the genus level are not of much value: they will, however, be found to reflect the general picture given by the aggregate of the following records for identified species.

Charcoal and wood identifications cannot be successfully brought down to species level, nor, with one or two exceptions, such as *Salix herbacea*, is it possible for pollen either. Records of one or the other kind exist in every stage from the Ludhamian onwards, and where the interglacials are divided into sub-stages *Salix* has been found in all of them, quite commonly with correlations between the occurrence of macroscopic and microscopic remains. *Salix* occurs throughout the Weichselian and is present indeed in the majority of Late Weichselian sites with again frequent correlations of pollen and macroscopic records. In zones I to IV between 80 and 95 per cent of all pollen-analysed sites record *Salix*; thereafter the proportion falls to 45 per cent in zones VIIb and VIII, but zone VI has again the high value of 81 per cent, possibly the result of local fen and lake-side development of willow carr. Broadly the results of macroscopic identifications parallel those from pollen, but this category is strongly represented in wood and charcoal from archaeological sites in zones VIIb and VIII. Thus there are the following records: Neolithic 7, Bronze Age 5, Iron Age 14, Roman 14, Anglo-Saxon 1, Norman 1 and Mediaeval 1.

The large numbers of species in the genus and the fact that earlier analyses omitted *Salix* altogether make it difficult to draw useful conclusions from the analysis of pollen frequency zone by zone and region by region. In the late Weichselian, *Salix* pollen rarely displays frequencies above 10 per cent of the total pollen, nor is there much regional differentiation, though Watts (1963) uses a fall in *Salix* frequency as a criterion of the zone I/II boundary in western Ireland. In zones IV, V and VI *Salix* pollen frequencies are somewhat higher (10–20 per cent of total tree pollen) but, especially in VI, shew strong response to local conditions, as indeed remains the case in the later zones where frequencies in general are much lower. It is difficult to discern that the willows were favoured by forest clearance, and one should indeed recall that the osiers were always liable to exploitation for basketry, rope-making etc.

Salix alba L.   White willow   343 : 2

F VIII *Swallowcliffe Down, Goldhanger, Pevensey*

The three records for *Salix alba* from Iron Age or Roman sites are all based upon charcoal determinations (A. H. Lyell), and suffer from the natural untrustworthiness of such material in species diagnosis.

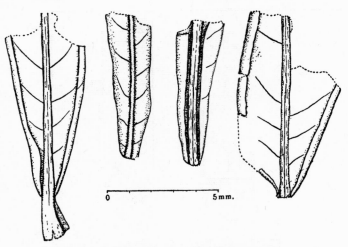

Fig. 94. Leaf fragments of *Salix viminalis* agg. recovered by G. F. Mitchell from Late Weichselian (zone II) deposits at Mapastown, Co. Louth, and identified by K. Jessen (× 6).

Salix fragilis L.   Crack willow   343: 4

F vIIb ?*Holywell*

A tentative charcoal identification in a Late Bronze Age context.

Salix viminalis agg.   Common osier (Fig. 94)   343: 9

mW ?*Barnwell St., Earith*; lW *Mapastown, Shortalstown*

In Mitchell's account (1953) of the Late Weichselian zone II material from Mapastown there is a careful statement by Jessen establishing the identity of leaf fragments found there as *Salix viminalis* agg. Miss Bell (1968) gives the linear shape, cuneate base and entire revolute margins of the leaves as reliable diagnostic characters allowing recognition of the species in the Middle Weichselian deposits at Earith that have a radiocarbon age of somewhat above 42 000 years. She specifically excludes the high northern species *S. rossica* and the central European *S. eleagnos* (*S. incana*, Schrank). A record by Miss Chandler of *S. lapponum* from Barnwell Station has been revised tentatively to *S. viminalis* by Bell and Dickson (1971).

Salix viminalis has a widespread distribution through central and southern Europe, but it is only doubtfully native in south Scandinavia although it now occurs there sub-spontaneously. In the British Isles it is widespread and common in the lowlands though rare in hilly areas.

These Middle and Late Weichselian records now strongly support Jessen's conjecture that this species, despite its present southerly range, may have survived the Weichselian glaciation in Britain and possibly elsewhere.

Salix caprea L.   Great sallow, Goat willow   343: 11

F IV, v *Ballybetagh*, VI, vIIa, *Derrycassan*, VI *Ballyscullion, Portrush, Cushendun, Barry Docks*, vIIb *Ardlow Inn*

The records of leaves and bud-scales indicate the natural presence of *Salix caprea* in Ireland since at least the opening of the Flandrian. It is a plant that occurs today throughout the British Isles: it has a wide European range also, reaching northernmost Scandinavia, but not arctic Russia.

Salix cinerea L. (including *S. atrocinerea* Brot.)   Common sallow   343: 12

I *Mundesley, West Wittering*; lW *Bovey Tracey, Flixton*; F IV *Flixton*, IV, v *Ballybetagh*, VI, vIIa *Derrycassan*, VI ?*Elland, Portrush, Cushendun, Nazeing*, vIIa *Meare Pool*, vIIb *Meare Lake Village, Peacocks Farm*, vIIb/vIII ?*Woodwalton, Dog-in-a-Doublet, Hightown*

The numerous records for *Salix cinerea* are based upon identifications of leaves, often with twigs and buds. As the species favours fen and marsh habitats these remains are common in fen-wood peats. They have been twice found in the Ipswichian interglacial and in zones II and III of the Late Weichselian. Thereafter they have been found in all the Flandrian zones, and it is evident that willows of the 'sallow' type have had a long and continuous history of occurrence in the British Isles.

S. atrocinerea, although common throughout the British Isles, has a present range southwards through France, Spain and Portugal to north-west Morocco. By contrast *S. cinerea*, much less common in the British Isles, has a wider European range and extends much further north and east than does *S. atrocinerea*.

Salix aurita L.   'Eared sallow'   343: 13

lW ?*Ballybetagh*, F IV, v ?*Ballybetagh*, IV/v ?*Dogger Bank*, v, VI *Moss Lake*, vIIb ?*Togherbane*, vIIb/vIII *Hightown*, vIII ?*Maiden Castle*

In view of the large range of leaf form in the willows and possibility of confusion with *Salix atrocinerea*, the identifications of *S. aurita*, based upon leaves or leaf impressions, can scarcely be more than tentative, as Jessen has pointed out in relation to his Irish records. The one charcoal identification from Maiden Castle is not more secure.

The records suggest that *S. aurita* may have been present in the Allerød period in Ireland, and during the early Flandrian both on the Dogger Bank and in Ireland. Later records extend throughout the Flandrian.

Salix aurita occurs throughout the British Isles, though commoner in the north and west. In western Norway it extends to the Arctic Circle but it is not found in Iceland or the Faroes.

Salix phylicifolia L.   'Tea-leaved willow' (Figs. 95, 96; plate XI)   343: 15

mW *Earith*, ?*Upton Warren*; lW *Nazeing, Neasham*, ?*Low Wray Bay*, ?*Derriaghy*, ?*St Bee's*, ?*Shortalstown*, ?*Ballybetagh*, ?*Kentmere*, ?*Drumurcher*; F IV *Nazeing, Neasham*

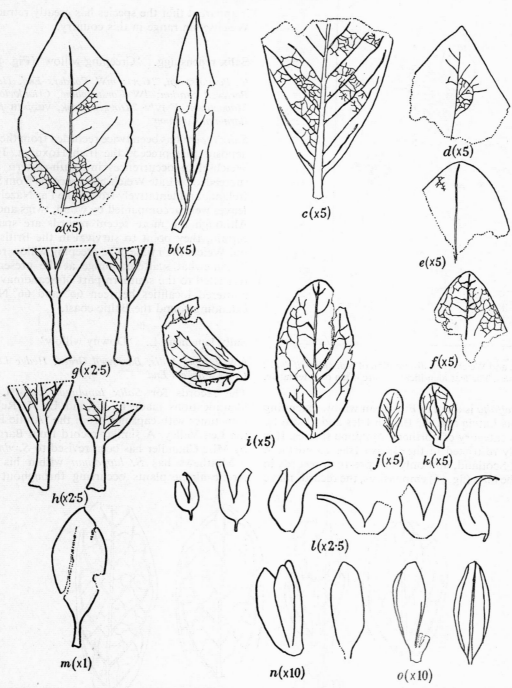

Fig. 95. Leaves and capsules of *Salix* from the Late Weichselian calcareous muds at Nazeing, Essex. Some of the leaves resemble *S. phylicifolia* (note the cuneate base and long petiole of *c, g, h*): many are mere prophylls not referable to species (*j, k*); *b* resembles *S. repens* and *m* resembles *S. myrsinites*, but neither *b* nor *m* is outside the variation range of *S. phylicifolia*, to which species the capsules (*l*) are also probably referable; *n* and *o* are shoots of *Erica tetralix* from the same deposit.

All the records of *Salix phylicifolia* come from the Middle or Late Weichselian or the immediately succeeding early Flandrian. All zones of the Late Weichselian are well represented and there is local persistence at sites such as Nazeing (zones I, II/III, IV) and Neasham (I, II, IV).

Although the leaf determinations (based upon the obovate shape, cuneate base and long petiole) are not easy, the recognition is helped by the frequency of the leaf remains and by the presence with them, in some instances, of the rather characteristic capsules, sometimes still in the infructescence.

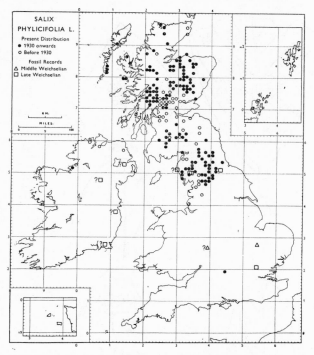

Fig. 96. Present and Weichselian distribution of *Salix phylicifolia* L. in the British Isles. The leaf identifications are often only tentative.

Salix phylicifolia is a north European willow extending south as far as Latvia and the British Isles and placed by Hultén in his category of northern–montane species. It is almost wholly restricted in the British Isles to northern England and Scotland, and only has pre-1930 records in Ireland. As the map (fig. 96) emphasizes, the records make

it apparent that the species has greatly retracted from its Weichselian range in this country.

Salix repens agg.   'Creeping willow' (Fig. 97)   343 : 16

lA *Baggotstown,* ?*Gort*; *m*W *Ponders End, Hackney Wick, Barnwell Station*; *l*W *Corstorphine, Crianlarich,* ?*Nazeing, Mapastown*; F IV/V ?*Dogger Bank,* VIIb/VIII *Hightown,* VIII *Barnham Common*

Salix repens has been twice recorded from the late Anglian deposits that precede the Irish Hoxnian. Three Middle Weichselian occurrences in south-eastern England are succeeded by Late Weichselian records from Scotland and Ireland, and tentatively from zone I at Nazeing where the leaves were accompanied by wood, twigs and bud scales. Although the more recent records are sparse and uncertain, they point to survival in the British Isles from full Weichselian time. The creeping willow retains a wide if somewhat scattered range at the present day. It is restricted to the southern part of Scandinavia except for scattered localities between 60° and 66° N and a belt extending round the Baltic coast.

Salix lapponum L.   'Downy willow'   343 : 17

*m*W *Temple Mills, Barrowell Green, Hedge Lane, Waltham Cross, Ponders End*

The records for *Salix lapponum* consist of repeated identifications made by C. and E. M. Reid of leaves, sometimes with capsules, from the 'Arctic Plant Bed' of the Lea Valley. A similar record from Barnwell Station by Miss Chandler has been revised to *S. viminalis*.

Matthews has *S. lapponum* within his category of Arctic–alpine plants occurring throughout central and

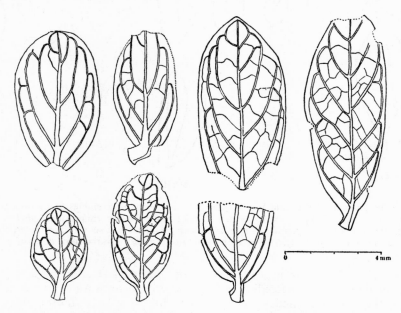

Fig. 97. Leaves and leaf fragments of *Salix repens* L. recovered by G. F. Mitchell from Late Weichselian (zone II) deposits at Mapastown, Co. Louth (× 6). Identification by K. Jessen.

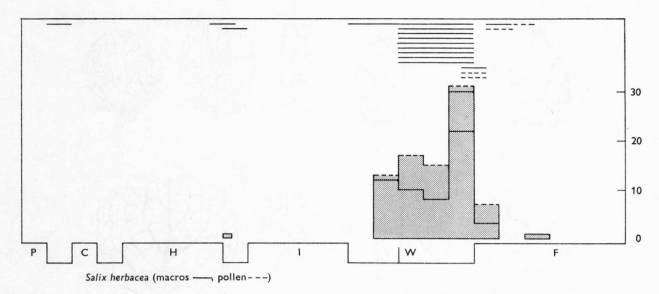

Salix herbacea (macros ——, pollen - - -)

northern Scandinavia and extending in mountain regions south to the Alps and Pyrenees, but absent from the lowlands of south Sweden and Denmark as from those of the British Isles. The Lea Valley of course lies far south of the restricted northern and montane range of the species in Britain.

Salix lanata L. 'Woolly willow'   343: 18
mW Barnwell Station

Leaf remains from the Middle Weichselian in Cambridge, identified by Miss Chandler as *Salix cinerea*, have been re-examined and are now referred to *S. lanata* (Bell & Dickson, 1971).

Salix lanata is now restricted to a very small number of montane sites on basic rock in the Scottish Highlands, so that the fossil record indicates a great retraction of range. It is placed by Matthews in his category of Arctic and Sub-arctic plants.

Salix arbuscula L.   343: 19
mW Barnwell Station

Salix arbuscula, which is now a rare plant of the Scottish mountains, has a single record from the full Weichselian deposits of the Cam valley in East Anglia.

Salix myrsinites L.   343: 20
lH/eWo, eWo Hoxne

The record made by Reid of *Salix myrsinites* from Hoxne in Suffolk has been referred by West to the transition between the Hoxnian interglacial and the succeeding glacial stage. He himself has separately recorded the species from early in the Wolstonian at Hoxne. The site is of course far outside the present British range of this species, one of Matthews' 'Arctic–alpine' element now found infrequently in the mountains of central and

northern Scotland. In the deposits at Hoxne it accompanies other Arctic–alpine plants.

Salix herbacea L. 'Least willow' (Figs. 98, 99)
343: 21

Salix herbacea has been chiefly recognized hitherto in fossil condition by its broad rounded leaves with their crenate–serrate margins and prominent venation. They

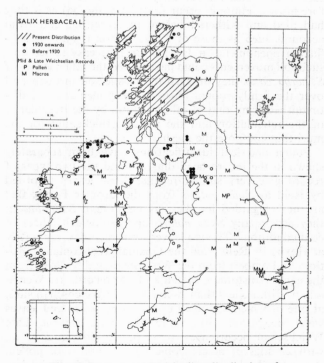

Fig. 98. Present and Weichselian distribution of *Salix herbacea* L. in the British Isles.

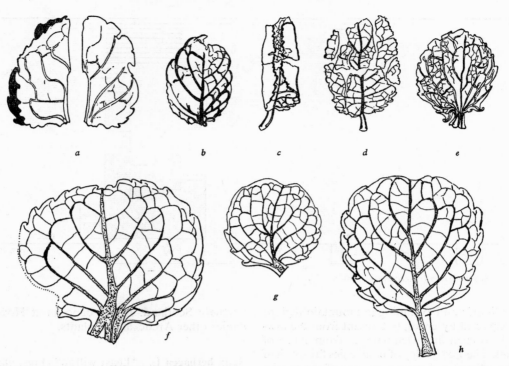

Fig. 99. Leaves of *Salix herbacea* L. from Late Weichselian (zone II) layers at Hawks Tor kaolin pit, Bodmin Moor, Cornwall (*a* and *e*, ×5; *b*, ×4; *c* and *d*, ×2.5). Leaves folded as in *c* are especially abundant and may well have dried in this form before falling to the ground and being incorporated in muds or solifluxion soils; *f*, *g*, *h*, leaves of *Salix herbacea* L. (×3.3) from Late Weichselian (zone III) deposits at Ballybetagh, near Dublin (Jessen & Farrington, 1938).

turn brown and fall off early in the season and therefore most commonly occur folded up (fig. 99) as if they had died and withered before preservation, and from accounts such as those of J. W. Wilson (1952) for solifluxion phenomena in Jan Mayen it is easy to see how such soil movements could have buried the fallen leaves of *S. herbacea* as they dragged the plastic soil through and over the firmly rooted bushes. The pollen may be distinguished from that of other willows by the irregularity of the reticulation and the nodular thickenings of the muri that compose the mesh.

Pollen and leaf identifications have been separately shewn in the record-head diagram and the two types of record give very similar results. Some leaves recorded by Reid as *Salix polaris* from the Cromer Forest Bed are now regarded by West as Beestonian *S. herbacea*: they accompany *Salix polaris* and geological indices of cold conditions. Though leaves were recorded separately at Hoxne by Reid from the Hoxnian/Wolstonian transition, by Lambert from the Wolstonian and by West from the early Wolstonian, the bulk of our records are Weichselian, with no less than twelve Middle Weichselian sites, at one of which, Oxbow (Yorks) both pollen and leaves were recognized. Still more abundant are the Late Weichselian occurrences, most particularly in zone III, with 22 sites of macroscopic records and 9 of pollen. Records of both types persist into zone IV of the Flandrian, and there is even a tentative record for zone VI from

a high altitude site in the mountains of Galloway. The prevalence of the species in the Irish Weichselian led Jessen to name zones I and III respectively the 'Older *S. herbacea* period' and the 'Younger *S. herbacea* period' but as the distribution map (fig. 98) indicates, it was widespread throughout the British Isles in the Weichselian when and where there was open ground.

S. herbacea ranges throughout the arctic and northern regions of Europe, Asia and North America, as well as the central European mountains, and is considered by Polunin (1939) as a truly arctic plant. In the British Isles it is a characteristic member of Watson's highland type, and is locally common on the summits of the higher mountains of the north and west. Its most southerly occurrence at the present day is at the Brecon Beacons, 2850 ft (870 m).

S. herbacea is a basiphilous species which favours fresh and open soils, and is often abundant in regions of solifluxion (J. W. Wilson) or scree formation. It is also strongly characteristic of late-snow patches, but, as pointed out by Polunin (1940), outside these habitats it is found in the Arctic in a wide range of habitats including dry sandy areas, herb slopes, grassy meadows and even marshes. In the Full-glacial and Late-glacial conditions of the British Isles stratigraphic evidence shows that habitats of this character, highly suitable to it, abounded. It is quite evident that *S. herbacea* belongs to the category of perglacial survivors with a wide British range in

Weichselian time, and which suffered an early and severe restriction to higher altitudes to the north and west by the conditions of the Flandrian period. This is an explanation that fully accords with the main body of European evidence summarized by Tralau (1961, 1963) though there remains the possibility that though *S. herbacea* was prevalent round the margin of the north European icefields, and on the foothills north of the Alpine glaciation, it was absent from the intervening central lowlands (Walter & Straka, 1970). If the distribution of *S. herbacea* is restricted by high summer temperatures as suggested by Dahl (1951) and later by Conolly and Dahl (1970), the Flandrian rise in these values may well have directly led to retraction and extinction of the species, although it is of course also very susceptible to competition of taller plants such as then became prevalent.

Salix reticulata L.   'Reticulate willow' (Fig. 100)

343 : 22

*m*W *Ponders End, Barnwell St.*; *l*W *Dronachy, Gayfield, Hailes, Corstorphine, Crianlarich, Whitrig Bog*

The rounded coriaceous and entire leaves of *Salix reticulata* with veins deeply impressed from the upper surface are easily recognizable in fossil form and have been recorded from British Middle and Late Weichselian deposits. Among the latter, however, there is only the single record from zone III at Whitrig Bog for which the actual zone could be ascertained.

S. reticulata has a very wide circumpolar distribution, extending to 79° N in Spitzbergen (but not including Greenland or Iceland). It is now restricted to five vice-counties in Scotland, where it occurs very sparsely between 2000 and 3600 ft (610–1100 m). It is absent from England and Wales.

There can be no doubt that since Middle Weichselian time, when the plant grew in south-eastern lowland England, *S. reticulata* has suffered a great reduction of range (fig. 100). The map given by Tralau (1963) shews that a similar retraction to that in Great Britain can be proved also for the European mainland, for Weichselian and post-Weichselian fossil records are spread right through the central and west European lowlands, though now it does not occur between the Scandinavian mountains and Lapland in the north and the Pyrenees, Alps and isolated localities in the Carpathians and Balkans in the south. Conolly and Dahl (1970) have correlated the present-day British range with the 21 °C maximum summer summit isotherm; in Scandinavia it is substantially higher.

Salix polaris Wahl. (Plate XXVIII*e*)   343 : 23

Be *Mundesley, Beeston, Ostend*; *l*H/*e*Wo, *e*Wo *Hoxne*; *m*W *Barnwell Station, Earith*; *l*W *Corstorphine, Dronachy, Hailes, Gayfield*

There is some difficulty in distinguishing the leaves of *Salix polaris* from those of *S. herbacea* (*vide infra*), and

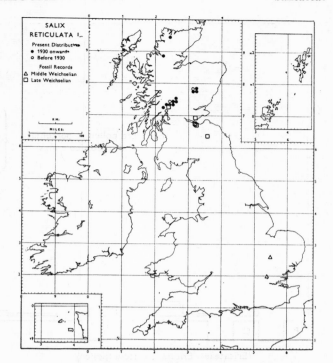

Fig. 100. Present and Weichselian distribution of *Salix reticulata* L. in the British Isles.

indeed the two species in some areas hybridize. However, whereas *S. herbacea* leaves have a toothed or notched margin with veins running out to the notches, and a short petiole, those of *S. polaris* have a simple margin with an entirely vein-free narrow outer strip and long petioles. *S. polaris* leaves, in distinction from those of *S. retusa*, are nearly always circular in outline.

Reid's records for the Cromer Forest Bed series are regarded by West as Beestonian. At Hoxne also Reid's records came from where the organic beds of the interglacial adjoin the overlying boulder-clay, and later West has identified *S. polaris* from the succeeding early Wolstonian. To Miss Chandler's Middle Weichselian record from Barnwell Station we now can add that of Miss Bell at Earith where there were also found leaves referred to '*S. polaris* or *S. herbacea*'. *S. polaris* leaves were also reported by Reid from four sites in the Scottish lowlands that can be broadly referred to the Late Weichselian. Thus all British fossil records refer to glacial stages, and it is evident that this plant, no longer native in the British Isles, occurred in low latitudes here in the middle of the last glacial stage and persisted in the Late Weichselian.

Hultén writes of *S. polaris* as a 'Eurasiatic–arctic montane' plant and maps its distribution along the length of the Scandinavian mountain chain to the highest latitudes. Tralau (1961, 1963) gives records and maps which make it clear that a similar post-Weichselian extinction has occurred over the length and breadth of the central European and east Russian lowlands and from south Scandinavia, though he is rightly critical of many of the

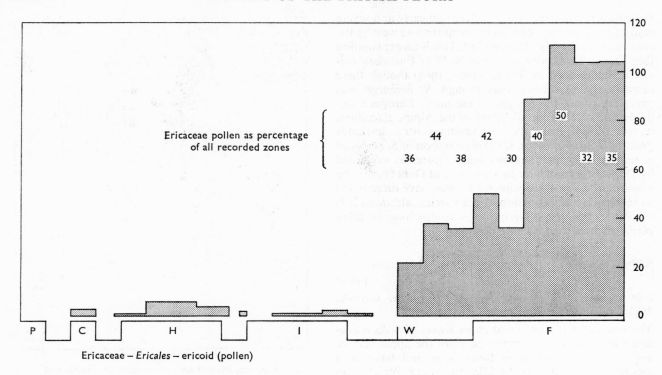

Ericaceae – *Ericales* – ericoid (pollen)

TABLE 22. *Frequency of Ericaceae pollen (including 'Ericales' and 'ericoid') in Late Weichselian and Flandrian of the British Isles*

						Percentages								
	Scan	0–2	2–5	5–10	10–20	20–30	30–40	40–50	50–60	60–70	70–80	80–100	>100	
I	3	**9**	2	2	.	.	.	.	.	.	.	.	.	Of
II	2	**10**	7	2	2	.	.	.	.	.	.	.	.	total
III	4	**10**	4	5	.	.	I	.	.	.	.	.	.	dry-land
IV	.	**11**	10	I	I	.	.	.	.	.	.	.	.	pollen
IV	.	**9**	I	5	.	I	2	.	.	.	.	.	.	
V	.	**8**	2	2	2	.	I	.	.	.	I	.	.	Of
VI	2	**15**	10	5	6	3	4	2	4	.	.	7	.	total
VIIa	2	**24**	13	5	7	4	6	2	.	3	.	4	4	arboreal
VIIb	3	**9**	4	4	8	2	4	7	3	2	2	6	3	pollen
VIII	2	7	5	5	7	6	2	6	I	3	5	**20**	14	

early identifications. Records of fossil *S. polaris* from the Netherlands present a history closely similar to that for the British Isles (Tralau & Zagwijn, 1962).

ERICACEAE

One must beware of reading too much into the pollen analyses of the gross category of the Ericaceae, abundant as such records are. The nomenclature of palynologists has always differed and it has altered as the years have allowed progress to a higher degree of identification of genera, and in some instances, of species. Accordingly we have aggregated all pollen records listed as 'Ericaceae', 'Ericales' and 'ericoid' for the purposes of the

record-head histogram and the accompanying frequency table.

Pollen of Ericaceae has been recorded from all glacial and interglacial stages except the Beestonian. It has been found in all sub-stages of the Hoxnian, sometimes in Ireland attaining high frequency in the middle stages thanks to large contributions from *Rhododendron*. It has been sparsely recorded from the three latest sub-stages of the Ipswichian and likewise the late Anglian and both early and late Wolstonian glacial stages. The volume of records increases greatly in the Late Weichselian where, in all three zones, 36 to 44 per cent of all British pollen zone records include an ericoid component: one may suspect that the figures would have been higher had not many authors separately recorded *Empetrum* pollen for

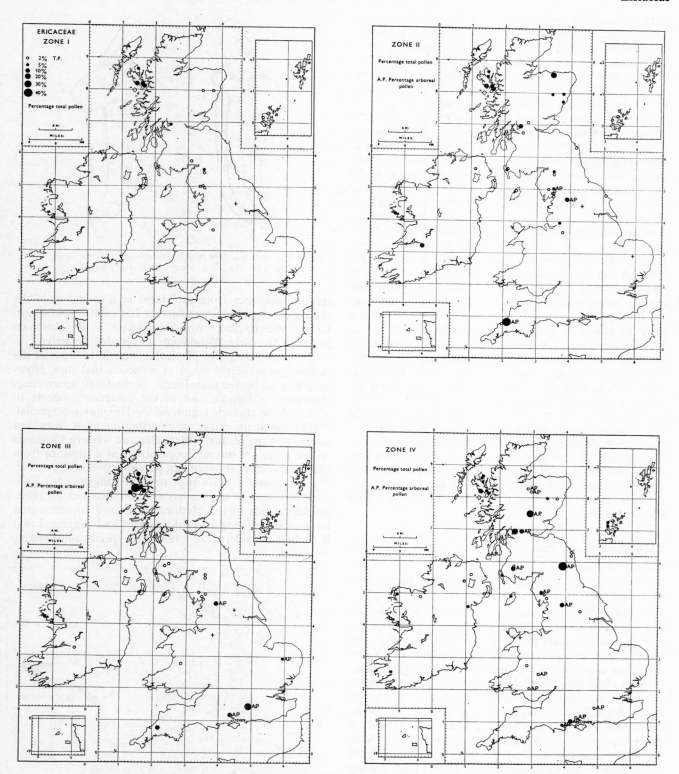

Fig. 101. Pollen frequency of the Ericaceae in the British Isles during the Late Weichselian and earliest Flandrian. The frequencies are based upon percentages of total pollen except when indicated as A.P. (arboreal pollen).

these zones. The succeeding Flandrian zone IV was no doubt affected similarly (fig. 101). It is to be noted that since there is little evidence for the development of ombrogenous mire or acidic peat in this period, the high ericoid zone records presumably reflect local developments of acidic communities on leached mineral soils or substrata deficient in lime from the outset, the crests of solifluxion ridges and blown sand for example. Table 22 shews that in these zones ericoid pollen only accounted for a modest proportion of the total dry land pollen as one would expect in such circumstances. Conditions evidently altered as the Flandrian progressed. Not only did the site records in zone VIIa amount to half of all total zone records, but the ericoid pollen was very often now present at high frequency. This was certainly associated with the widespread development of ombrogenous peat mires and in zone VI all seven of the sites with Ericaceae over 80 per cent of total arboreal pollen were northern Scottish sites where acidic wood peats formed over fen, whilst most of the other sites in that zone with high ericoid pollen frequencies were also in *Sphagnum* peat. No doubt the common choice of raised mire and blanket mire for pollen analytic sampling is reflected both in the high frequencies of site records and in the high pollen frequencies in so many of these sites. The local stratigraphic development of these mires is reflected in such features as the very high number of sites from the transition zone VII/VIII (not separately shewn in the histogram) which covers the pronounced recurrence-surface phenomena ending the dry terminal phases of the Sub-boreal when peat bogs were often clothed with *Eriophorum* and *Calluna*.

Although the topics concerned are dealt with elsewhere, it is appropriate to mention here the modern and comprehensive account by Mitchell and Watts of the history of the Ericaceae in Ireland during the Quaternary Epoch (Walker & West, 1970).

Rhododendron ponticum L. (Figs. 102, 103)   345

Macros: H *Gort*
    Pollen: lA ?*Gort*; H ?*Gort, Kildromin, Kilbeg, Baggotstown*; eWo ?*Baggotstown*

It was in 1948 that Jessen first reported the identification from Gort in western Ireland of seeds, capsules (fig. 102) and a probable bract of *Rhododendron ponticum* L. The unequivocal recognition was documented by Jessen, Andersen and Farrington in 1959 by which time it had become clear that the Gort deposits must be referred to the Hoxnian interglacial. At this time also it was recognized that *Rhododendron* formed the greater part of the ericaceous pollen found at Gort. The macroscopic remains are sparse in sub-stage II but chiefly from sub-stage IV of the interglacial, and the pollen maxima clearly concentrate in the second half of the interglacial. Though pollen is recorded from the late Anglian and the early Hoxnian at Gort the authors advise that it is probably contaminant from above. Subsequently Watts (1959b, 1964, 1967) has reported *Rhododendron* pollen in sub-

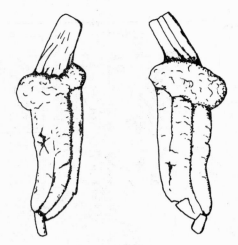

Fig. 102. Capsule of *Rhododendron ponticum* from the Irish Hoxnian (after Jessen *et al.*, 1959) × 5.

stantial frequency from sub-stages II, II/III and III of the same interglacial at three further Irish sites, Kildromin, Kilbeg and Baggotstown. Here again at Baggotstown pollen in the early Wolstonian is regarded as derived.

These records are of particular interest because of the present geographical range of a species that now grows naturally no nearer than Iberia, of the fossil occurrences elsewhere in Europe and of the evidence it seems to afford of the climatic regime in the Hoxnian interglacial. Whilst the main centre of distribution of *R. ponticum* today is on the southern Black Sea and western Caucasus region there are discrete populations at a distance from this area, such as those of var. *baeticum* in Spain, and var. *brachycarpum* in Syria (fig. 103). Writing in 1944 Cain drew attention to the numerous fossil localities for interglacial records in the Mediterranean and Alpine region between the existing areas of the plant and suggested that it 'is best considered as a relic of a species once widely

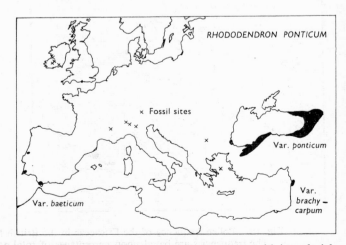

Fig. 103. Present range of *Rhododendron ponticum* with interglacial fossil localities, including four from the Irish Hoxnian. (After Cain, 1944.)

distributed in central European mountains'. The Irish records emphasize the breadth of that former distribution, though Watts (1959b) calls attention to the fact that the European fossil records are from various interglacials.

Although no longer native in Ireland, *Rhododendron ponticum* is so suited to Irish conditions that it regenerates freely both vegetatively and from seed, forming dense under-canopy in woodlands and extensive free standing thickets on peat bogs and mountain sides up to considerable altitudes, of at least 500 m. It evidently requires a mild wet climate, such as indeed it enjoys in its other Pontic and Iberian stations, and an acidic soil, itself a consequence in high degree of the oceanicity of climate. Watts has made the point that suitable soil conditions would accompany the telocratic phase of each interglacial cycle: the absence of *R. ponticum* from the British Isles in the Flandrian serves to emphasize that historical facts of survival and migration are involved to determine the present ranges of species no less than their ecological requirements.

Loisleuria procumbens (L.) Desv. (*Azalea procumbens* L.)
    'Loisleuria'   346: 1

IW Corstorphine, Nant Ffrancon

Reid identified macroscopic remains of *Loisleuria procumbens* from the Late Weichselian deposits at Corstorphine near Edinburgh and recently Burrows identified its leaves from the Allerød zone II level in North Wales. It is an arctic montane species common throughout the Scandinavian mountains and along the north of the Kola peninsula. In the British Isles it is virtually restricted to the highlands of central and western Scotland between 400 and 1200 m. Thus the two Weichselian sites are outside its present range both of latitude and altitude, a sufficient indication of Flandrian retraction. Conolly and Dahl correlate the present range of *Loisleuria* with the 22 °C summer temperature summit isotherm, thus suggesting, along with the evidence from *Betula nana* and a few other species, a summer temperature depression in zone II of perhaps as much as 5 degrees relative to the present.

Daboecia cantabrica (Huds.) C. Koch (*Menziesia polyfolia* Juss.; *Daboecia polifolia* D. Don) St Dabeoc's Heath (Fig. 173)   349: 1

H Burren, Baggotstown, Gort, Kilbeg, Fugla Ness

Daboecia cantabrica, one of the components of the Irish–Lusitanian flora, is a known fossil from Ireland and the Shetlands. In each instance the record comes from the Hoxnian interglacial. Jessen and Watts identified seeds from Gort in sub-stage IV and Watts (1959b) identified both seeds and pollen from sub-stage IV in Kilbeg, and macrofossils from sub-stage III at Baggotstown. Mitchell has an unpublished record from some part of the Hoxnian in the Burren and H. J. B. Birks has again found seeds of it in the Hoxnian interglacial at Fugla Ness.

The plant is now restricted in the British Isles to the counties of Mayo and West Galway, where it is locally common. There is a disjunction thence to its next occurrences in western France, north-west and central Spain and north Portugal. All the Irish sites, and more strikingly still the Shetland site, lie outside the present range of *Daboecia* (fig. 173).

Watts (1959b) emphasizes that at Kilbeg *Daboecia* grew 'in a herb-rich tree-less community reminiscent of Late-glacial conditions', with *Empetrum*, *Calluna*, *Sphagnum* and *Juniperus*, and in a climate sufficiently unfavourable to have eliminated trees. He draws attention to the fact that in its Spanish mountain stations it may be snow-covered for substantial periods so that it is more cold-tolerant than is generally supposed. All the same he thinks it unlikely to have survived the next ensuing glaciation in Ireland.

It is of interest that Oldfield, who has made a special study of the pollen of Ericaceae, has recorded pollen of *Daboecia* from Le Moura in south-west France very early in the Flandrian, not much after 7000 B.C. on radiocarbon evidence. He points out that had *Daboecia* survived the Weichselian glaciation in Spain and invaded the Pays Bas at the given date, immigration to Ireland by the western coasts would have been easily possible before the eustatic rise of sea-level severed the land connection (Oldfield, 1964a, b).

Andromeda polifolia L. 'Marsh andromeda'   350: 1

IW Corstorphine, Hailes, Windermere; F VIIb Woodwalton, Trundle Mere, Blakeway Farm Track, VIII Westhay Track, Shapwick Track, Viper's Track, Shapwick Station, Bowness Common

In *Andromeda polifolia*, as in *Vaccinium oxycoccus*, the leaves and shoots have such characteristic features that they are readily recognized in the field and they have been far more frequently found than the published results suggest. In the derelict raised bogs of the Somerset Levels *Andromeda* is frequently found in the unhumified upper *Sphagnum* peat and is particularly abundant in the 'Precursor peat' which marks the first stage of recovery of active *Sphagnum* bog growth after a phase of dryness and arrest. Consequently the stems and leaves of robust *Andromeda* plants will often be found lying in the aquatic *Sphagna* and mud peats which occupy the pools of the newly flooded recurrence surfaces, in somewhat shallower situations of course than the *Scheuchzeria* which also characterizes these levels. It may be taken that *Andromeda* grew at the pool margins. The Woodwalton, Trundle Mere and Bowness Common records also represent growth on raised bog. All these Flandrian records are from zone VIIb or VIII; thus there is a considerable time gap from the Late Weichselian zone II record from the English Lake District and the two Scottish Late Weichselian records. These earlier records must have been from sites with modest local development of acidic heath.

Although not of the extreme arctic category it is a

boreal–circumpolar species which occurs freely through-out northernmost Scandinavia. *Andromeda* is, sur-prisingly, not found in northern or central Scotland and it is remarkable that the two Scottish Late Weichselian sites lie close to its present northern limits, though temperatures must then have been much lower than now. The present British occurrence of the plant must have been much affected in the south by the destruction of raised bogs by cutting and drainage.

Arbutus unedo L.   Strawberry tree   353: 1

F vɪɪb *Carrantuohill, Killarney*

According to Oldfield (1959) *Arbutus unedo* produces the largest and smoothest tetrads of all the west-European Ericaceae, with exceptionally long furrows, striking 'endo-cracks' and hardly discernible ectexinal columellae. It appears however to have a very low pollen production because it is hard to collect pollen even from *Sphagnum* peat directly below mature trees. Welten (1952) gives a pollen diagram from the shallow blanket-bog surface of Carrantuohill, at 1010 m above sea-level, and in it at two levels 3 per cent of pollen of *Arbutus unedo* is re-corded. The deposit is only 15 cm in depth and may be extremely recent despite a tentative sub-zoning of the Sub-atlantic made by the author. More recently Mitchell and Watts (1970) report that Miss Vokes has 'shewn that *Arbutus* was present at Killarney perhaps as early as 4000 years ago'.

Sealy (1949) describes the ecological relationships of the strawberry tree very fully and discusses the implica-tions of its geographical range, which is Mediterranean and Iberian with extensions into south-western France and with outlying areas in Kerry, West Cork and Sligo in western Ireland.

Arctostaphylos sp.   354

lW Hockham Mere, F ɪv *Burren*

Pollen of *Arctostaphylos*, identified only to generic level, was found in zone ɪɪ deposits of Hockham Mere, Norfolk.

Arctostaphylos uva-ursi (L.) Spreng.   Bearberry (Fig. 104;   plate XVIII*i*)   354: 1

I *Histon Rd*; m*W Barrowell Green*, ?*Temple Mills*; *lW* ?*Bovey Tracey, Neasham, Brook, Nazeing*; W/F *Apethorpe*

Fossil records of fruit stones of *Arctostaphylos uva-ursi*, apart only from one from the late Ipswichian in Cam-bridge, are all from deposits of the last Glaciation. There are two from the Middle Weichselian of the Lea Valley (one tentative), and two from zone ɪɪ of the Late Weich-selian, one from the zone ɪɪ/ɪɪɪ transition and one from the ɪɪɪ/ɪv transition. The tentative record from south Devon is probably from a Late Weichselian deposit: though its age is not exactly known it contains *Betula nana, Pinus, B. pendula* and *Salix cinerea*.

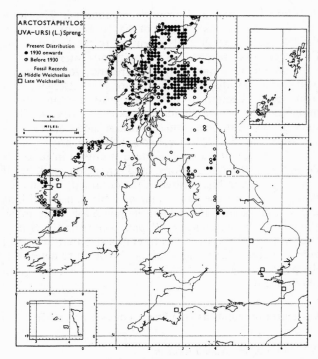

Fig. 104. Present and Weichselian distribution (Middle and Late) of *Arctostaphylos uva-ursi* (L.) Spreng. in the British Isles.

Most of the records for the bearberry lie well south of the present southernmost limit of the species in Britain. It is placed by Matthews in the category 'Arctic–alpine' and by Hultén in the group of boreal–circumpolar plants that are boreal–montane in Europe with continuous dis-tribution; nevertheless it shows some falling off in frequency from the south-east towards northern Scandi-navia. It may certainly be regarded as one of the British plants with a wide Late Weichselian distribution which has subsequently been restricted to higher latitudes. A possible clue to the process by which this has operated is offered by Conolly and Dahl's demonstration that its present range corresponds with the 24 °C maximum summer temperature summit isotherm in Ireland and Highland Scotland, and with that of 26 °C for Wales, northern England and adjacent Scotland.

Calluna vulgaris (L.) Hull   Ling, Heather (Fig. 105)
                                               356: 1

Calluna vulgaris might well claim to be the most easily recognized and most prevalent of Post-glacial sub-fossil plants. The crooked stems with wrinkled bark resist the decay which reduces the other components of humified *Sphagnum–Calluna* peat to a structureless mass, the tightly ranked leaves sit upon the terminal shoots in very characteristic manner and these, often bearing well-preserved flowers, occur in profusion in the less humified *Sphagnum* peats. In recent years palynologists have also regularly identified the pollen tetrads of *Calluna* and

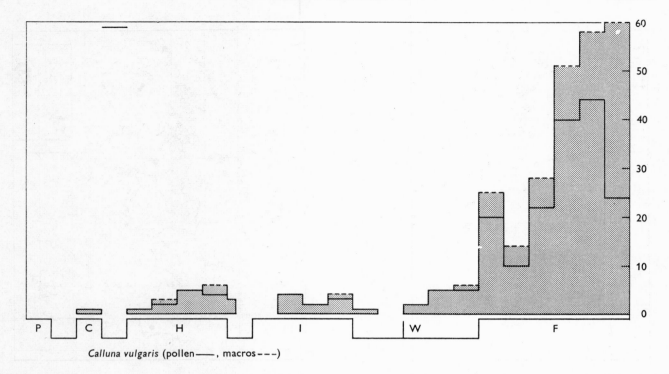

Calluna vulgaris (pollen——, macros - - -)

represent it regularly in the pollen diagrams. The record-head diagram abstracts these records together with those for macroscopic remains. The numbers are so considerable that we have omitted all records for transition zones and broad spans of time. The oldest interglacial record is Cromerian. In the Hoxnian it is represented in all sub-stages at sites from Ireland and south-eastern England: often, as at Gort, pollen identifications accompany those of macroscopic remains, and at Mark's Tey there is persistence through all sub-stages of the interglacial. It seems likely that *Calluna* heath had then developed in East Anglia as in the various sites that yielded records for sub-stages II, III and IV of the Ipswichian. *Calluna* is represented in the Corton Beds of the Anglian glaciation and in the Chelford interstadial of the Weichselian. There are pollen records for the early Wolstonian at Kilbeg, Hoxne and Mark's Tey but it is hard to be sure that these are not secondary. On the other hand pollen records occur in all zones of the Late Weichselian accompanied to some degree by macroscopic remains: it seems significant that most of the Late Weichselian sites are from Scotland or sites further south where also the preponderant rock formations are granite or sandstone and acidic peat accumulation could have begun easily. The ensuing Flandrian records may be taken largely to reflect the development of ombrogenous peat mires, more particularly after the late Boreal dry phase at the end of zone VI. Wherever raised or blanket bogs occur in the British Isles, remains of *Calluna* will be found and, as in general the growth of these ombrogenous mires began at the opening of the Atlantic period (zone VIIa), it may be taken that from this time onwards, at least, the plant has been a major part

of the vegetation of such habitats. Indeed, the highly humified conditions of the raised bog peat formed during Atlantic and Sub-boreal times suggests that they were in a relatively dry condition, with *Calluna* dominating their surface vegetation to an extent now seen only in conditions of artificial drainage. In southern and eastern England the spread of *Calluna* heath as a consequence of Neolithic and later disforestation of the lighter soils also contributed to high numbers of soil records in the later zones. In addition to this widespread and well known effect with its accompanying persistent podsolization of lighter soils, Iversen has shewn that in the Viking Age in Denmark heather-covered glades were induced in woodlands on poor soil by clear felling followed and maintained by burning to produce grazing. The practice was only recently abandoned. Pennington (1964) has also shewn that in the English Lake District, upland tarns often yield pollen diagrams where a sudden rise in the frequency of *Calluna* pollen accompanied a minerogenic wash-in horizon. This she attributes to the effects of disforestation enhancing podsolization and the spread of *Calluna* and afterwards leading to soil erosion and transport into the lakes of the soil of the 'A' horizon with its content of *Calluna* pollen. It would be dangerous to read much into the fluctuations in the site numbers of the record-head diagram where so many major artifacts of sampling are involved, but the persistence of *Calluna* in the British Isles from Late Weichselian and possibly Early Weichselian times must be assured.

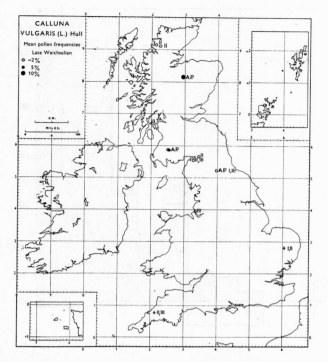

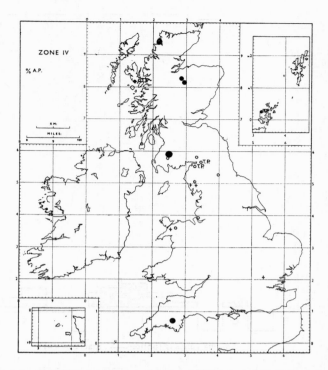

Fig. 105. Pollen frequency of ling, *Calluna vulgaris* (L.) Hull, in the British Isles during the Late Weichselian and the earliest Flandrian (zone IV). Frequencies are based upon percentages of total pollen (T.P. in fig.) except where indicated as A.P. (arboreal pollen) and upon a mean Late Weichselian occurrence unless the zones (I to III) are shewn.

Erica sp.   357

Macros: H *Fugla Ness*; *lW Elstead*; F VIII *Newstead, Dunshaughlin*

   Pollen: F VIIb *Esthwaite*

The record for leaf identification of *Erica* from sub-stage IV of the Hoxnian in the Shetlands needs to be taken with the more specific identifications (*E. tetralix*, *E. mackaiana*, *E. scoparia*). The Elstead record is also for a leaf in zone III of the Late Weichselian.

Erica tetralix L. Cross-leaved heath, Bog heather
(Figs. 95*n, o*, 106, 107*c*)   357: 1

A tentative identification of the carbonized leaves of *Erica tetralix* is given by Mitchell from the Cornish deposits at St Erth that are possibly of late Pliocene age. There are leaves, flowers and seeds of it in sub-stages II and IV of the Hoxnian interglacial at Gort, seeds from sub-stage IV at Fugla Ness in the Shetlands and macroscopic remains from an undefined part of the Hoxnian at the Burren in western Ireland. The next records, most or all of leaves, are from the Late Weichselian in which all zones are represented and the sites range from western Ireland and Cornwall to Essex and from Hampshire to Aberdeenshire. Throughout the Flandrian there are records of *E. tetralix* in every pollen zone and the same wide spread across the British Isles is maintained.

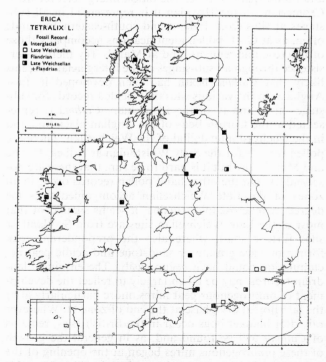

Fig. 106. Interglacial, Weichselian and Flandrian site records for cross-leaved heath, *Erica tetralix* L. in the British Isles.

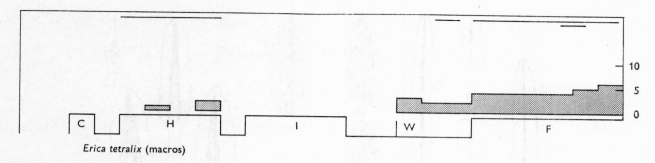

Erica tetralix (macros)

The records together speak unequivocally for persistence in all parts from the beginning of the Late Weichselian. Of particular interest, in view of the pronouncedly oceanic, west-European character of its present distribution, is the suggestion it conveys that even in Late Weichselian time the British climate was oceanic overall. There is of course a convergence of evidence that this was so in the Hoxnian interglacial. *E. tetralix* extends to 65° N in western Norway and also grows in Iceland.

Erica mackaiana Bab. (*E. mackaii* Hook.) (Fig. 107*b*)

357: 2

H *Ballykeerogemore, Burren, ?Gort, ?Kildromin, Kilbeg, Baggotstown, Fugla Ness*; F VI, VIIa *?Roundstone*

On two occasions Jessen found in Irish deposits leaves which 'with their elongated to ovoid-lanceolate shape and short, rolled-back margins agreed completely with *Erica Mackaii*'. In many instances the base of the marginal gland of the hairs was preserved. The first of these determinations was made at Roundstone, site I, Co. Galway within the area of the plant's present distribution, and the reference to pollen zones VI and VIIa makes it highly improbable that the plant can have been introduced into Ireland in the Flandrian period by human agency. The second determination was from the interglacial deposits at Gort now known to be Hoxnian. Subsequently however Jessen *et al.* (1959) have indicated that the leaves of *Erica ciliaris* are of the same type, and as seeds and flowers of that species occur in the Gort material whereas none of *E. mackaiana* are found, the fossil leaves are likely to be *E. ciliaris*. Whether the Flandrian leaves from Roundstone could be this species also is more doubtful for it is not generally believed native of Ireland, whereas *E. mackaiana* still grows locally at Roundstone itself. Gay (1957) has however suggested that the fossil leaves fall within the wide variation range of those of *E. tetralix* and suggests indeed that the Irish *E. mackaiana* populations are sterile hybrids between *E. ciliaris* and *E. tetralix*.

The seeds of *E. mackaiana* appear to offer a much more secure basis for identification (Watts, 1959*b*) and there are five Irish Hoxnian sites and one Shetland site whence the species has been identified, apparently covering the sub-stages II, II/III, III and IV. A damaged seed from Kilbeg formerly attributed to *Cicendia filiformis* (see first edition) is now seen to be *E. mackaiana*.

Erica mackaiana now grows near Roundstone in Co. Galway and Lough Nacung, Co. Donegal and then in north-west Spain, chiefly in the province of Asturias (Watts, 1959*b*). It belongs to the small group of Hiberno-Lusitanian species which have a distribution area curiously restricted to more or less scattered localities in western Ireland and to the Pyrenees or the western Mediterranean. Whilst the fossil record very securely establishes its wide presence in the Hoxnian interglacial, it does not make it clear when the present Irish populations arrived in that country.

Erica ciliaris L. (Fig. 173)

357: 3

H *Gort*

As indicated in the account of *Erica mackaiana*, leaves initially attributed to that species from the interglacial deposits at Gort were subsequently referred to *E. ciliaris* (Jessen *et al.*, 1959). Along with them were abundant seeds identical with modern seeds of *E. ciliaris*, and 'in several cases they were taken out of loculicidal capsules of decayed flowers'. The seed identifications were based upon the unique characters of the epidermis after Hansen (1950). The fossils came from stages 3 to 6 of the local sequence, which appear to represent the sub-stages II to IV of the Hoxnian interglacial. *Erica ciliaris* is a Lusitanian species occurring locally in western Ireland, Dorset, south Devon and west Cornwall, through western France and western Iberia just into adjacent north Africa (fig. 173). It appears to belong characteristically to moist sandy heaths and in France also to woods on poor soils. The Gortian record parallels that of several Irish Lusitanian species.

Erica cinerea L.   Bell-heather (Fig. 107*a*)

357: 4

H *Gort*; I *Histon Rd*; F V, VI, VIIa *Roundstone*, VIIb *Ipswich, Loch Park*, VIII *Loch Cuithir*

From washing samples in sub-stage IV of the Hoxnian interglacial at Gort numerous seeds of *Erica cinerea* were recovered, some of them from a decayed capsule with the remnants of its split corolla. There is a tentative identification from sub-stage III of the Ipswichian in Cambridge. In the Flandrian, Jessen (1959) reported leaves and flowers from zones V, VI and VIIa at Roundstone, Co. Galway; in Scotland the Vasaris (1968) have

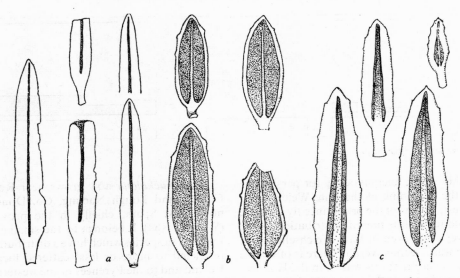

Fig. 107. Leaves of *Erica* spp. from various Irish deposits (Jessen, 1949): *a*, *Erica cinerea* (Roundstone, zones v and vi); *b*, *E. mackaiana* (Roundstone, zones vic and viia); *c*, *E. tetralix* (Roundstone, zone viia).

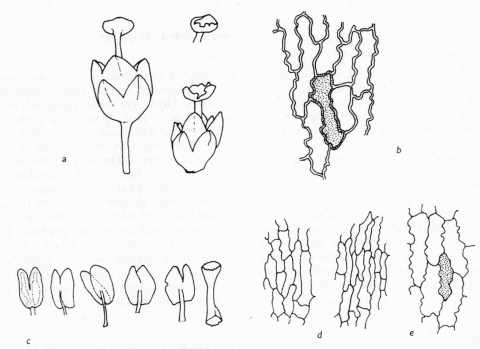

Fig. 108. *Erica scoparia* L. from Hoxnian interglacial at Gort, Co. Galway. *a*, Flowers (× 10), *b*, epidermal cells of seeds (× 100), *c*, five stamens and style (× 10), *d*, *e*, epidermal cells of modern seed (× 100). (After Jessen *et al.*, 1959.)

found leaves from zones viib and viii, and seeds also in viii. There is also a 'Neolithic' record, presumably zone viib by Reid from Ipswich. Thus the Flandrian record, though sparse, stretches widely across the British Isles, like the present-day distribution of this extremely oceanic plant, which Matthews places in his Oceanic–west European element and which in Scandinavia is limited to the extreme coastal fringe of southern Norway.

Erica terminalis Salisb. (*E. stricta* Don.)　　　357: 9

H ?*Marks Tey*; eWo ?*Marks Tey*

Turner (1970) has reported the discovery from Hoxnian deposits in Essex of a distinctive tricolporate pollen grain with the strongly thickened margins of the furrows that is characteristic of the Ericaceae. These grains however occur not in tetrads but singly, as is the case in only one

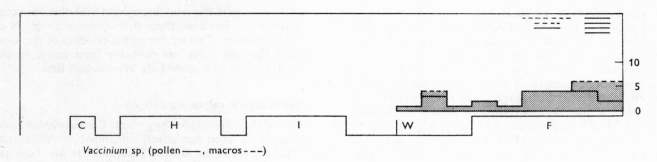

Vaccinium sp. (pollen ——, macros – – –)

European species of *Erica*, i.e. *E. terminalis*, a western Mediterranean plant not now growing nearer to England than Corsica, Sardinia and south-west Spain. The fossil pollen matches that of *E. terminalis* quite closely but the author cites the record as 'cf. *terminalis*', in the absence of confirmatory evidence from macroscopic remains. The pollen was sporadic in sub-stages II and III and occurred in 'some frequency' in IV: it was also sporadic in the succeeding early Wolstonian.

Erica scoparia L. (Figs. 108, 173)   357: 10

H *Burren, Baggotstown, Gort, Kilbeg, Fugla Ness*

From the upper layers of the Hoxnian interglacial deposits at Gort, Jessen, Andersen and Farrington recovered a number of complete and other incomplete flowers of an *Erica* that closely matched *Erica scoparia* in every respect except that the seeds were consistently larger and had thicker walls than living material (fig. 108). They accordingly referred the fossil material to var. *macrosperma* n. var. foss. Watts (1959*b*, 1964) also recorded the same *E. scoparia* from the later half of the same interglacial at Kilbeg and at Baggotstown, and Mitchell has reported it from the Hoxnian at the Burren (unpublished). A large block of peat, probably an erratic in the till at Fugla Ness, Shetland Islands, has been referred also to the Hoxnian by Moar (1964) and he reported from it seeds tentatively identified as *E. scoparia* var. *macrosperma*. These probably also came from sub-stage IV. The identification first tentatively made by Mrs Dickson has been confirmed by further work including comparison with the Gort material (Birks & Ransom, 1969).

Erica scoparia L. grows in south-western France, the western Mediterranean, Spain, Portugal, Madeira and the Canaries where it appears to favour dry acidic sites in xerophilous scrub, or lowland heath and open woodland on dry sandy soils (fig. 173). As in the Irish Hoxnian site, so here *E. scoparia* is accompanied by the other Mediterranean Lusitanian species *Daboecia cantabrica* and *Erica mackaiana*.

Erica arborea L.   357: 11

F VIII *Pevensey*

The identification of wood of *Erica arborea* was made by A. H. Lyell: the sample came from a Roman well blocked with vegetable matter, containing small pieces of Roman pottery. The record carries little or no weight as evidence for natural status of this species.

Vaccinium sp. (Fig. 109)   358

The record-head diagram is based both upon pollen and macroscopic remains, the latter including seeds, rootlets and in one doubtful case, a leaf. Out of twenty-seven pollen records, no less than nine are either tentative or are regarded as '*Vaccinium* type', but in cases such as the Cross Fell zone VIIa and VIIb records, macroscopic and pollen records support one another. Together the records cover every zone from the Early Weichselian to the present day and the recent Scottish records by Tutin and Bonney are supported by radiocarbon dates from nearly 9000 years B.P. There is a strong preponderance of records from Scotland and northern England and at several sites

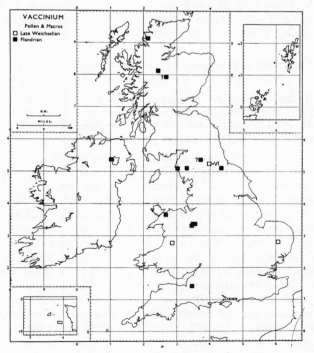

Fig. 109. Late Weichselian and Flandrian records for the genus *Vaccinium* in the British Isles.

clear indication of local persistence, thus Ronaldkirk (zones I, II, III and VI), Loch Clair (VI, VI/VIIa, VIIa, VIIb, VIII) and Ehenside Tarn (VIIa, VIIb and VIII). The records for *Vaccinium oxycoccus* which follow do not shew any preference for Scotland – rather the reverse – and it might be reasonable to regard these generic records as due to the other British species of *Vaccinium*, all of which unlike *V. oxycoccus* are strongly centred in the north.

A Middle Weichselian record for *V. uliginosum* formerly recorded (first edition) from Barnwell Station, Cambridge has been deleted upon revision by Bell & Dickson (1971).

Vaccinium vitis-idaea L.   Cowberry, Red whortleberry
358: 1

H *Gort*; F v, vi *Moss Lake*, viib *Loch Kinord*

A seed of *Vaccinium vitis-idaea* has been identified from sub-stage IV of the Irish Hoxnian. At Moss Lake, Liverpool leaves from Flandrian zones V and VI, at first referred to *Cotoneaster*, were later identified as *V. vitis-idaea*, and from Loch Kinord there is a macroscopic identification in zone VIIb. The Flandrian records are within the present range of the species.

Vaccinium myrtillus L.   Bilberry, Blaeberry, Whortleberry, Huckleberry
358: 2

lW *Loch Fada*; F VIII *Dublin*

The Vasaris have reported macroscopic remains of *Vaccinium myrtillus* from zone III of the Late Weichselian in Scotland and Mitchell has found its fruits along with those of other plants in association with a fruit press or similar structure in mediaeval Dublin.

Vaccinium oxycoccus L. (*Oxycoccus quadripetalus* Gilib. *O. palustris* Pers.)   Cranberry
358: 4

H *Baggotstown*; lW *Shortalstown, Drymen*; F IV *Drymen*, VIIa *Oulton Moss, Roundstone*, viib *Lonsdale Moss, Trundle Mere, Oldtown Kilcashel*, viib/viii *Meare Heath Track, Woodwalton*, viii *Shapwick Station, Ardlow Inn*

The fine stems and well-spaced, characteristic, ovate leaves of the cranberry (*Vaccinium oxycoccus*) are rather commonly observed by anyone taking the trouble to split open the fresh turves of unhumified Sub-atlantic peat on any of our raised bogs. Relatively few such identifications have been recorded, though all the records listed above are for remains of this kind. The Irish Hoxnian record is from the middle of the interglacial. The two Late Weichselian records represent zone II and the III/IV transition, the latter accompanied at Drymen by a Flandrian zone IV occurrence. Thereafter the records are for zones VIIa, viib and viii and mostly come from raised bog deposits in England or Ireland, reflecting in their range the wide occurrence of the plant despite its sparsity and recent diminution in the south and east with cutting and draining of its habitats.

V. oxycoccus is common throughout Scandinavia except the far north and even there it is common to 70° N in western Norway. This reinforces the evidence of the fossil record that the plant has probably been native in the British Isles at least since Late Weichselian time.

Chamaedaphne calyculata Mönch

Seeds of the oligotraphic bog plant, *Chamaedaphne calyculata*, have been tentatively identified by West (unpublished) from the British Pastonian. It has been recorded from the middle and late Pleistocene of Europe, and has a present-day circumpolar–boreal range that includes northern Europe.

EMPETRACEAE

Empetrum nigrum L.   Crowberry (Fig. 110; plate XIX*a*)
364: 1

Not only are the fruit stones of *Empetrum nigrum* readily recognizable, but so are the pollen tetrads, and the less frequently recorded leaves and twigs. Jessen has emphasized that the size of the fossil pollen from the Lateglacial in Ireland leaves no doubt that it belongs to *E. nigrum sensu stricto*, and not to *E. hermaphroditum* Hagerup, and the same applies to the vast bulk of British records, whether given for *E. nigrum* or for the genus. It will be seen at once from the record-head diagram how closely these two categories agree. They shew presence in the Cromerian, all sub-stages of the Hoxnian, sub-stages II to IV of the Ipswichian and every zone of the Flandrian. There is an equally substantial record for the glacial stages with records in the late Anglian, early and late Wolstonian, the Early Weichselian and, with high frequency, all zones of the Late Weichselian, with over 70 sites in zone II and over 60 in III as in the ensuing Flandrian zone IV. The records for undivided or combined zones confirm the Late Weichselian abundance and strengthen the evidence for frequency in the early Flandrian.

West (1968) also records that in the early Pleistocene 'the flora of the cold stages, the Thurnian and Baventian, is characterised by high frequencies of Ericales tetrads, many of which are of *Empetrum nigrum* type'.

Macroscopic remains, indicated on a separate histogram, despite the smaller frequency yield a comparable picture with records from all Hoxnian sub-stages, all Flandrian zones and, along with clear presence in Early and Middle Weichselian, abundant sites in zones II and III of the Late Weichselian and in Flandrian IV. There are also two early Wolstonian records. The Ipswichian is unrepresented, just as pollen records for that interglacial are sparse, a consequence no doubt of the concentration of Ipswichian sites in the continental south-east of England where soils also are preponderantly alkaline. By contrast there are numerous Hoxnian interglacial sites in western

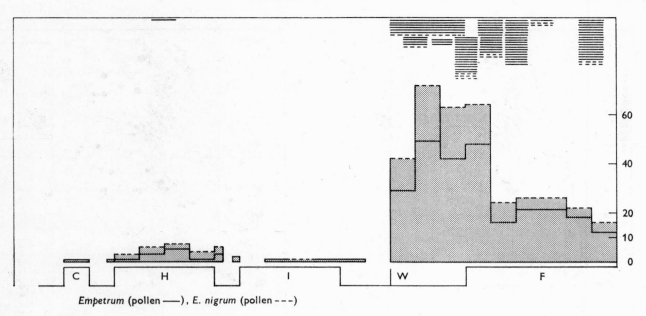

Empetrum (pollen ——), *E. nigrum* (pollen ---)

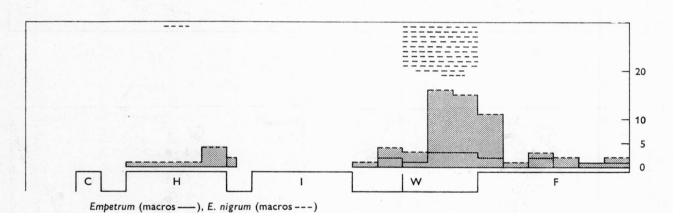

Empetrum (macros ——), *E. nigrum* (macros ---)

Ireland, that at Kilbeg yielding both pollen and macro remains in sub-stages II and III (as also in the succeeding early Wolstonian) and those at Gort and Newtownbabe shewing similar correlation in sub-stage IV. Correlations of this kind are extremely common in the records for the Late Weichselian and early Flandrian and often establish strong local persistence of the species, as they do at Kirkmichael (zones I, II, III), Roundstone (II, III, IV and V) and Moss Lake, Liverpool (also II, III, IV and V), sites respectively in the Isle of Man, western Ireland and northwestern England.

The records in total give an impression of remarkable continuity of presence through the Pleistocene glacial and non-glacial stages alike, though favoured by the terminal stages of both, expanding characteristically in the Weichselian Late-glacial before the advent of closed forest and in the telocratic section of the interglacials, notably the Hoxnian when acid heath entered and displaced the forest. Such a history might seem appropriate to a genus that has

a bipolar world distribution only to be explained by ancient distribution along trans-equatorial high mountain ranges (du Rietz, 1935). Palynologists of north-western Europe have long recognized *Empetrum* as a common and characteristic component of the dwarf-shrub vegetation so prevalent in the Late Weichselian, but in the Netherlands, the Hartz mountains and elsewhere high values of *Empetrum* pollen seemed to characterize zone III as against the rest of the Late Weichselian period (van der Hammen, 1949). This gives particular interest to the changes of mean pollen frequency set out for the British Isles in table 23.

It is evident that *Empetrum* was present in zone I and locally in fair frequency. This frequency increased in zones II, III and IV, attaining its largest values repeatedly at sites such as Loch Tarff, Loch Criagie, Garral Hill, Loch Clair, Scionascaig and Yesnaby in northern Scotland. The maps (fig. 110) make it evident that despite the higher pollen frequencies in the north and west the

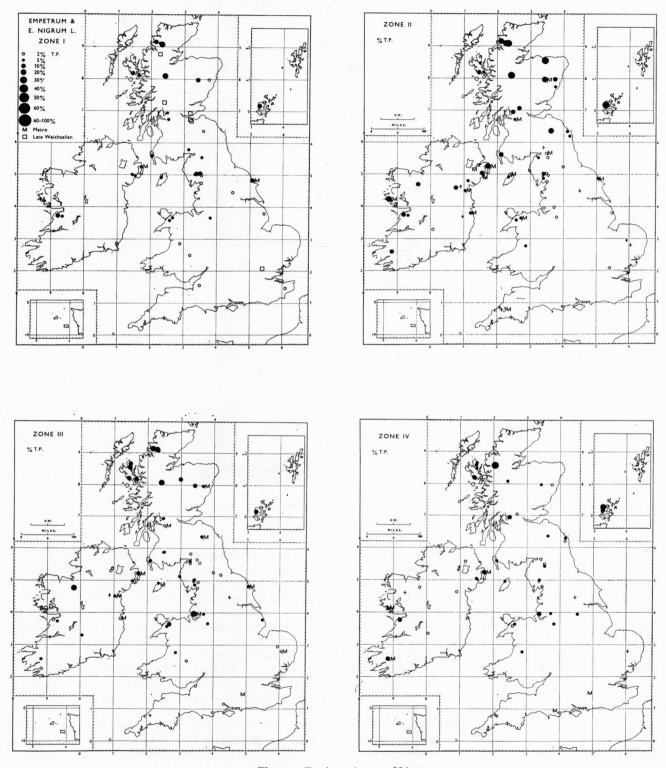

Fig. 110. For legend see p. 304.

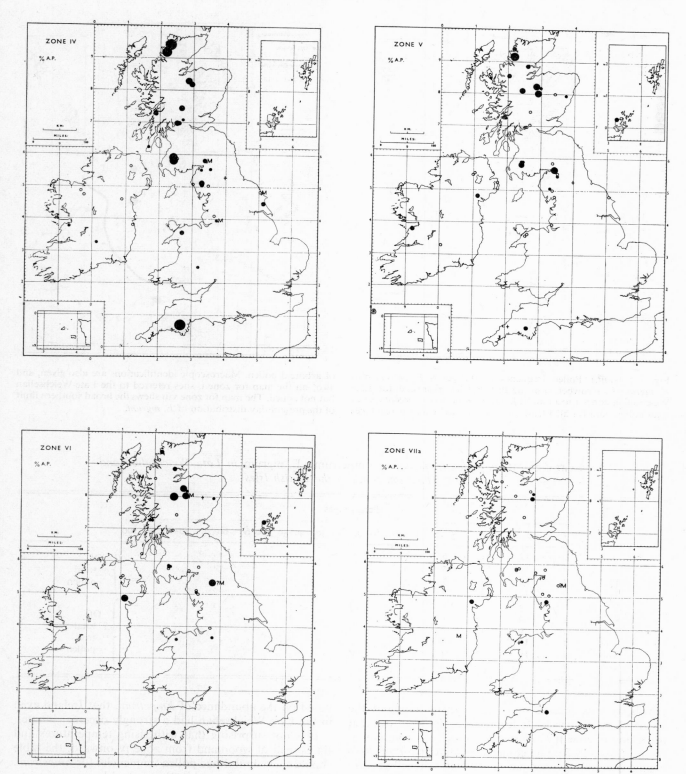

Fig. 110 (cont.).

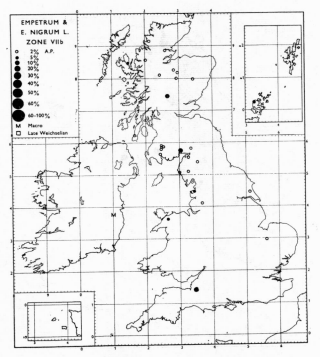

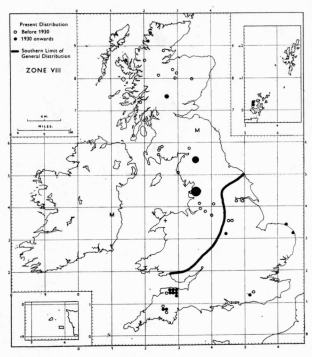

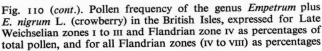

Fig. 110 (*cont.*). Pollen frequency of the genus *Empetrum* plus *E. nigrum* L. (crowberry) in the British Isles, expressed for Late Weichselian zones I to III and Flandrian zone IV as percentages of total pollen, and for all Flandrian zones (IV to VIII) as percentages of arboreal pollen. Macroscopic identifications are also given, and also, on the map for zone I, sites referred to the Late Weichselian but not zoned. The map for zone VIII shews the broad southern limit of the present-day distribution of *E. nigrum*.

TABLE 23. *Mean pollen frequency of* Empetrum + E. nigrum *in Late Weichselian and Flandrian zones of the British Isles*

							Percentages							
	+	0–2	2–5	5–10	10–20	20–30	30–40	40–50	50–60	60–70	70–80	80–100		
I	.	12	11	6	2	.	.	.	.	.	.	.	⎫	Of
II	6	14	16	5	5	4	.	.	.	.	.	.	⎬	total
III	4	13	17	8	3	.	.	.	.	.	.	.	⎪	pollen
IV	3	10	12	6	2	1	.	.	.	.	.	.	⎭	
IV	.	5	3	5	3	2	.	.	1	.	.	.	⎫	Of
V	2	4	0	1	1	.	.	1	.	.	.	.	⎬	total
VI	2	6	1	.	.	1	.	.	.	.	.	.	⎪	tree
VIIa	1	15	2	3	.	.	.	.	.	.	.	.	⎪	pollen
VIIb	.	10	1	3	.	1	.	.	.	.	.	.	⎪	
VIII	1	4	1	.	.	1	.	.	.	.	.	.	⎭	

crowberry was present in less amount throughout the whole of the British Isles. Jessen (1949) pointed out that in Ireland in zone II *E. nigrum* was especially common in the north-west where he conjectured that true *Empetrum* heaths probably existed, an effect that Mitchell (1941*b*) and Mitchell and Parkes (1949) correlated with the apparent absence of remains of giant Irish deer and reindeer from those parts. Later workers in Ireland have tended to discount Jessen's view but not the conclusion, partly based on the abundance of *Empetrum*, that Ireland even in the Late Weichselian had a strongly oceanic climate.

It is not surprising that with rising temperatures and the spread of woodland from zone IV onward, the table should shew diminished pollen frequencies for *Empetrum*. Numerous sites of very high local abundance persist and these can be identified as high moorland or bog where local tree cover was sparse, for example Loch Scionascaig and Duartbeg (Sutherland), Black Lane (Dartmoor) and

Nick of Curlywee (Galloway). In the later Flandrian zones high pollen frequencies are especially associated with the better drained areas of raised bogs, sites where *Empetrum* escaped tree-competition and enjoyed moist acid soil. Sites of this kind diminish the Flandrian retraction of range that *Empetrum* would otherwise shew, so that it is regarded more as a 'Glacial plant' than as a 'Glacial relict of disjunct distribution' (Walter & Straka, 1970). Very recent increases of *Empetrum* frequency (zone VIII) have been used by Tallis (1965) as indices of the development of drainage channels on the Pennine blanket bogs, during mediaeval times on the lower slopes and during the industrial revolution at higher altitudes.

Empetrum nigrum is classed by Hultén as a boreal–circumpolar plant of continuous range. It is abundant throughout Scandinavia. In the British Isles it is commonest in Scotland, northern England and Wales, but absent south and east of a line from the Humber to the Severn except for Devonshire and west Somerset. Brown (1971) has recently examined the Late Weichselian and early Flandrian pollen record of *Empetrum* as a basis for deducing the regionally varying oceanity of climate during those periods. Berglund and Malmer (1971) have, *per contra*, shewn that in southern Scandinavia the increase in *Empetrum* in the later part of the Allerød period is due more to progressive leaching and podsolization of the soils than to changing climate. Of course these effects are interacting.

Empetrum hermaphroditum Hagerup   364: 2

F VI, VIIa,b *Cwm Idwal*, VIII *Dinas Emrys*

Whilst the diploid *Empetrum nigrum sensu stricto* is dioecious, the monoecious *E. hermaphroditum* is tetraploid. The tetraploid is the dominant form in the north, in the Alps and eastern Pyrenees. In Britain, though its range is incompletely known, it mainly occurs at high altitudes in the Scottish Highlands. It is found also however on Snowdon, and Miss Andrew was able to recognize its pollen tetrads from Flandrian zones VI, VIIa and VIIb in the deposits of Llyn Idwal in the cwm above which *E. hermaphroditum* still grows (fig. 175) (Godwin, 1955). Seddon (1961) later identified it from zone VIII at the nearby site of Dinas Emrys, Caernarvonshire. Koperowa (1964) has also reported its pollen from the Late Weichselian in Poland.

Corema alba (L.) D. Don

Cs *Corton*; C *Pakefield*

Pyritized endocarps from the Cromer Forest Bed were referred by C. and E. M. Reid to an extinct species of the genus *Corema* (Empetraceae), of which the only living European species, *C. alba*, has a coastal range in Portugal, Spain and the Azores. More recently West and Wilson have recovered from the channel infilling at Pakefield, presumed to be of Cromerian age, a very rich plant

assemblage that includes material identical with *Corema alba* and it seems highly likely that the Reids' material belonged also to this species (21st Annual Report of the Sub-department Quaternary Research).

PLUMBAGINACEAE

C *Bacton, Mundesley*

From two sites on the Norfolk coast Duigan (1963) has reported pollen of the family Plumbaginaceae: at the same time she identified other pollen of the same age at generic level in the family.

Limonium sp.   365

C *Bacton, West Runton*; H *Histon Rd*; F VIIb *Chapel Point*

In addition to Duigan's identifications of pollen of *Limonium* at two sites in the Cromerian interglacial, there is a pollen record of it in sub-stage II of the Hoxnian in Cambridge along with certain other indicators of saline conditions. The Flandrian record from the Lincolnshire coast is for stem-bases recovered from the '*Triglochin* Clay' formed under salt marsh conditions. All the British species of *Limonium* have a southern or south-eastern range in these islands, a fact enhancing the interest of the three separate interglacial appearances of the genus.

Limonium vulgare Mill.   365: 1

F VIII ?*Slapton Ley*

There is a tentative pollen identification of *Limonium vulgare* from zone VIII deposits on the Devon coast (Crabtree & Round, 1967).

Armeria maritima (Mill.) Willd.   Thrift (Figs. 111, 112, 150; plates I, XIXr, XXe, f, XXVIIIc)   366: 1

The calyces of *Armeria* have often been recognized in subfossil state, and rather inconclusive attempts at specific identification have been based upon their characters. C. Reid, in identifying the Pleistocene *Armeria* remains, had referred them to *A. arctica* Wallroth, with the recommendation, however, that they ought to be compared with mountain forms of *Armeria* in Britain. E. M. Reid, after re-examination of material from the Lea Valley Arctic bed, concluded: 'It is certain that the affinity of the Lea Valley *Armeria* is with the arctic and alpine, and not with the lowland forms. It seems closest to *A. planifolia* (Syme), which is a Scottish alpine, as evidenced by the size, shape and ribbing of the calyx, and by the distribution of hairs.'

In 1940 Erdtman drew attention to the condition of floral dimorphism in *Statice armeria* L., in which some plants bore 'coblike' stigmas and coarsely reticulate, spinous pollen grain (type A), whilst others bore 'papillate' stigmas and had much more finely reticulate pollen

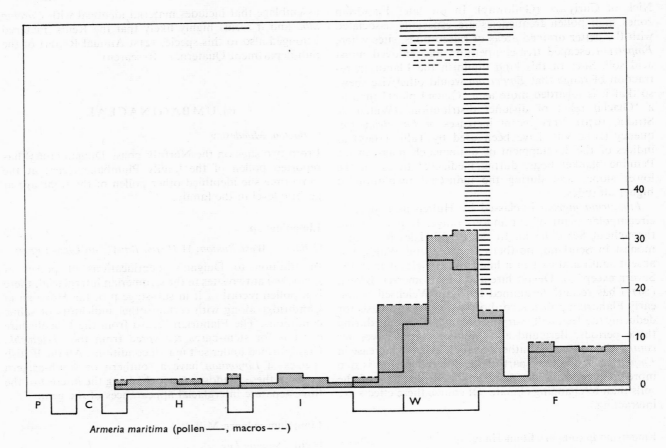

Armeria maritima (pollen——, macros----)

(type B). In the same year Iversen (1940) drew attention to the fact that the very homogenous genus *Armeria* can be separated into two distinct groups, the one (of European–Mediterranean range) is dimorphic with the two forms self-incompatible but cross-compatible, whereas the other (which includes the holarctic form of northern America and Asia), is monomorphic and self-compatible. *A. vulgaris* Willd.* is shewn to be dimorphic and to include the Atlantic and maritime variety, var. *maritima*, and the mountain variety, var. *alpina*. On the other hand, the nearest holarctic *Armeria*, *A. labradorica*, which now exists in northern Greenland, is monomorphic. In view of these conclusions it is of particular interest to find that where *Armeria* pollen has been found in the Late-glacial deposits in Cornwall and Nazeing, both types, A and B, of pollen have been present. The pollen may thus be referred in all probability to *Armeria maritima* Willd. which takes in *A. planifolia* Syme, and this is the course taken by Jessen, who, after consideration of Iversen's and Szafer's analyses, concludes that the Irish Late-glacial forms must be also referred to *A. vulgaris sensu lato* (Iversen). When Szafer (1945) examined fossil *Armeria* in the European Pleistocene, though dealing chiefly with Polish material, he examined also calyces from the Barn-

well Arctic bed in Cambridge and was able to find in them pollen of both A and B types. Baker has subsequently extended observation of this kind to the sub-fossil *Armeria* calyces from the Late-glacial at Whitrig Bog, Berwickshire (Baker, 1948a, b). Baker, as a result of his critical studies of the British *Armerias*, is strongly of the opinion that apart from *A. plantaginea*, all existing British *Armerias* fall within the single species *A. maritima* Willd. and that the present coastal and inland aggregates are the residue of a much more continuous population in Late Weichselian time.

As the record-head diagram indicates, the fossil records for *Armeria* are now very numerous and extensive in time, presenting strong evidence of continuous presence from the Cromer Forest Bed time onwards, with interglacial records in all sub-stages of the Hoxnian and Ipswichian and all zones of the Flandrian and with glacial stage records in the late Anglian, early and late Wolstonian, and (in great abundance) in the Early, Middle and Late Weichselian. The very high frequency of records in zone III of the Late Weichselian is emphasized by some fifty extra records at the transition from that zone to early Flandrian zone IV, chiefly by authors unwilling or unable to make a precise definition of the III/IV boundary.

* Lawrence (1940) has shewn that what Iversen names 'A. vulgaris Willd.' is properly called 'A. maritima Willd.', and this usage we have followed.

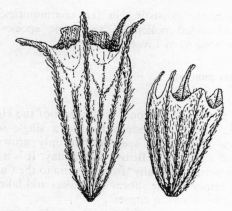

Fig. 111. Calyx tubes of *Armeria maritima* from Late Weichselian layers at Ballybetagh, Ireland: the left-hand specimen is hairy on the ribs only, the right-hand both on and between the ribs. (Jessen & Farrington, 1938.)

There are many instances where records of pollen and of macroscopic remains complement one another, as notably in the Middle Weichselian sites at Upton Warren, Earith and Marlow. There is also strong evidence of persistence at individual sites: thus it occurs at Nechells in the late Anglian and sub-stages I and II of the succeeding Hoxnian, as at Selsey in the late Wolstonian and sub-stages I and II of the Ipswichian.

It is apparent that four particular categories of habitat are strongly represented, two of them characteristic of the present-day British situation, two not. Coastal and estuarine sites occur repeatedly, as during the mid-interglacial marine transgression at Clacton (Hoxnian), at Selsey in the Ipswichian, at Jenkinstown in zone VI, and many sites in the Flandrian zones VIIb and VIII such as Chapel Point and Brigg (Lincolnshire), Slapton Ley (Devonshire), Canna, and mosses beside the Solway Firth. Likewise present-day high altitude populations have their counterparts in such high altitude sites as Cwm Idwal (zone VIIa), Loch Dungeon (III/IV and VIIb/VIII) and Loch Cuithir (VIII). These go some way to indicate that the mountain populations of today are indeed relict through the middle Flandrian forest period. There remain many lowland and inland sites, to which the Weichselian occurrences contribute almost entirely. In many instances, especially in the Middle Weichselian, *Armeria* is a consistent constituent of the distinctively halophilous vegetation that is now accepted as characteristic of some periglacial and sub-arctic landscapes and the evidence for which is fully discussed by Bell (1969). Finally however we have to accept that in the interglacials there appears to have been persistence of *Armeria* into the middle forested sub-stages on river terraces far inland, as at Nechells (Birmingham) in sub-stages I and II of the Hoxnian. We are too unsure perhaps of the contemporary coastline to cite Marks Tey (Hoxnian sub-stages I and II) and Histon Rd, Cambridge (Ipswichian sub-stages III and IV) similarly.

At the present day *Armeria maritima* extends right round the British coasts and occurs up to 1280 m on

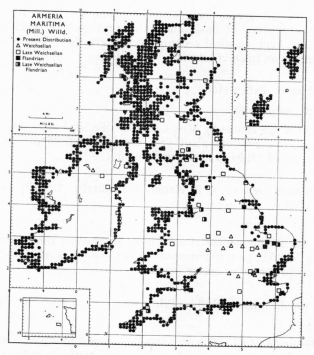

Fig. 112. Present, Weichselian and Flandrian distribution of the sea-thrift, *Armeria maritima* (Mill.) Willd., in the British Isles. The present-day restriction to coasts and mountains contrasts strikingly with the wide Weichselian range on inland and lowland sites.

mountains inland. It occurs in western and northern Europe from France to Iceland and to 70° N latitude in Norway. The map of its Weichselian distribution (fig. 112) makes it evident that the present coastal and inland aggregates are the residue of a much more continuous population at this time.

PRIMULACEAE

I *Trafalgar Square*; F *Canna*

Primula sp.   367

W *Thrapston*; mW ?*Barnwell Station, Upton Warren*; lW *Magheralaghan*

Seeds of *Primula* have been recorded from the Middle and undefined Weichselian pollen from the Late Weichselian zone III in Ireland. They are records that compare with the Weichselian records for *P. farinosa* and *P. scotica* in the accounts that follow.

Primula farinosa L.   'Bird's-eye primrose'   367: 1

W *Thrapston*; mW *Marlow*, ?*Barrowell Green, Upton Warren*

Since *Primula farinosa* is now limited in the British Isles to the north of England (with a few sites of recent extinction in southern Scotland), it is apparent that the species' range has greatly contracted in Flandrian time. It is characteristic of basic and calcareous soils, especially those kept seasonally damp and where the vegetation retains an open character.

Primula scotica Hook.   367: 2

mW Barnwell Station

The single record by Miss Chandler of a seed of *Primula scotica* shews the plant very far south of its present restricted range in northern Scotland.

Primula veris L. or P. elatior (L.) Schreb.   Cowslip or Oxlip   367: 3, 4

F VIII *Shrewsbury*

A seed from mediaeval material at Shrewsbury is referred by Mrs Dickson to either cowslip or oxlip, the only fossil record for either species so far.

Primula vulgaris Huds.   Primrose   367: 5

H *Gort*

Jessen *et al.* (1959) have identified from sub-stage IV of the Hoxnian interglacial at Gort two typically peltate, warty seeds with a long hilum: they appear to be of *Primula vulgaris*, a species generally distributed through the British Isles but one not tolerant of summer drought save in woodlands and on retentive soils.

Lysimachia sp. (Plate XXIg, h)   370

lW Elan Valley, Nazeing; F IV *Elan Valley, Dogger Bank*, V/VI *Postbridge*, VIIa *Bowness Common, Racks Moss*, VIIa, b, VIII *Ellerside Moss*, VIIb *Shippea Hill, Bell Track*, VIII *Helton Tarn, Holcroft Moss*

The pollen of *Lysimachia* has been recorded from various sites between Devonshire and Galloway in Great Britain and was found in Flandrian zone IV in Dogger Bank peat. The records are fairly continuous from zone II/III of the Late Weichselian to the present day and there is local persistence at the high altitude Elan Valley site and at Ellerside Moss.

Lysimachia nemorum L. or L. nummularia L.   'Yellow pimpernel' or Creeping Jenny   370: 1, 2

F VIIb *Frogholt*

Seeds referred to either *Lysimachia nemorum* or *L. nummularia* were found at Frogholt, Kent in a stream deposit at the foot of the Chalk Downs: the habitat would have suited either species.

Lysimachia vulgaris L.   'Yellow loosestrife'   370: 3

I *Selsey*; F VI *Moss Lake*

Seeds of *Lysimachia vulgaris* were identified from the Ipswichian interglacial at Selsey in sub-stages Ia, IIa and

IIb, presumably persisting in fen communities. Miss Andrew identified pollen of the same species from Flandrian zone VI at Liverpool.

Lysimachia punctata L.   370: 5

H *Gort*

Jessen *et al.* identified from sub-stage IV of the Hoxnian interglacial at Gort, western Ireland a single seed of *Lysimachia punctata*, a species that only grows as a naturalized alien in the British Isles today. It is a Pontic species occurring naturally 'from Austria to the Caucasus and Asia Minor on the shores of rivulets and lakes or in moist meadows and in copses'.

Lysimachia thyrsiflora L. (Naumburgia thyrsiflora (L.) Reichb.)   'Tufted loosestrife'   370: 7

H *Gort*

From the same sub-stage IV of the Hoxnian at Gort that yielded *L. punctata* Jessen *et al.* identified one seed of *Lysimachia thyrsiflora*, a species not now native in Ireland and only rarely found in northern England and southern Scotland.

Anagallis sp.   372

lW Elan Valley

From the Late Weichselian zone I/II transition at the high-altitude Elan Valley site in Wales Moore has recorded pollen of the genus *Anagallis*.

Anagallis tenella (L.) L.   Bog pimpernel   372: 1

H *Kilbeg*

Watts (1959b) has recorded abundant seeds of *Anagallis tenella*, the bog pimpernel, from sub-stage II to III of the Hoxnian interglacial at Kilbeg. He points out the special interest of the discovery in that the plant has a very narrow ecological and geographical range, being confined to wet acidic bog surfaces and having a strongly western European distribution area. Such species are indeed very characteristic of the Irish Hoxnian.

Anagallis arvensis L.   Scarlet pimpernel, Shepherd's weather-glass   372: 2

Macros: H ?*Nechells*; I *Wretton*; lW *White Bog*; F VIIb *North Ferriby, Frogholt*, VIII *Apethorpe, Silchester, Plymouth, Dublin*   Pollen: H *Fugla Ness*; F VIII *Drake's Drove*

The Hoxnian interglacial has yielded a tentative seed identification of *Anagallis arvensis* from sub-stage I, and from sub-stage IV in the Shetlands H. J. B. Birks has recorded its pollen. In the Ipswichian seeds were recovered from sub-stages II and III and in the Late Weichselian there is an unpublished seed record by Mitchell. Subsequent records save for one record at Apethorpe are all associated with archaeological settlement or clearances.

Two are Bronze Age, one Iron Age, one Roman and two Mediaeval. This fossil record for so characteristic a weed of arable land is interesting in indicating previous natural status in the Late Weichselian (along with many other weeds) and persistence into previous interglacials in fluviatile situations.

Anagallis arvensis occurs throughout the British Isles but becomes less common in Scotland and northern Ireland: in Scandinavia also its occurrence is concentrated in the south.

Anagallis minima (L.) E. H. L. Krause (*Centunculus minimus* L.)   Chaffweed   372: 4

F viib *Avonmouth*, viii *Wingham*

The seed of the tiny *Anagallis minima* has been recorded from Flandrian zone viib in Gloucestershire and in zone viii in Kent, at the latter site associated with Iron Age cultivation and with a radiocarbon date of 2340 B.P. The Avonmouth record accords with a certain preference for coastal localities.

Glaux maritima L.   'Sea milkwort', 'Black saltwort'
                                                                         373: 1

I ?*West Wittering, Stone*; W *Thrapston*; mW *Marlow, Earith, Upton Warren*; lW *Kirkmichael*; F viib *North Ferriby*

The seed records for *Glaux maritima* derive from two types of situation. Those from the two interglacials, Ipswichian (sub-stage II) and Flandrian, are from coastal sites where independent evidence shews halophytic conditions. By contrast the Weichselian sites are all inland, or were at the time of deposition, but equally provide evidence of local salinity. This was commented upon by Kelly (1964) for Upton Warren and Bell (1969) in respect of Thrapston, Marlow and Earith (see chapter VI: 7).

Although primarily a coastal salt-marsh plant found round all the British coastline, it also occurs in a few saline habitats inland.

Androsace septentrionalis L.

eWo *Brandon*; mW *Upton Warren*

Kelly (1968) identified a single crumpled seed of *Androsace septentrionalis* from the early Wolstonian deposits of the Brandon Channel, and this led him to re-examine material from the Middle Weichselian site at Upton Warren whence he recovered five well preserved seeds.

No longer native in the British Isles, it is placed by Hultén in the category of boreal–circumpolar plants that in Europe have a continuous boreal–montane distribution. A plant of dry, sandy and grassy places, intolerant of competition from taller species, it apparently became extinct in the Flandrian with loss of suitable habitats.

OLEACEAE

Fraxinus excelsior L.   Ash (Fig. 113)   376: 1

It was not until Erdtman's book had appeared in 1943, and Iversen's wartime study on 'Land occupation in Denmark's Stone Age' had reached this country, that *Fraxinus* pollen began to be regularly recorded, but, as the record-head diagram shews, records are now both numerous and extensive in time.

The wood and charcoal of the ash (*Fraxinus excelsior*) are readily recognizable and there is a very large number of records for the plant from settlement sites of Mesolithic and later times. The bulk of specimens so identified have been charred firewood, but some have formed part of structures such as trackways or foundations or the hafts of tools and weapons: others have occurred naturally in wood peat. It seems certain that all the Flandrian records, whether for pollen or macroscopic remains, refer to *Fraxinus excelsior* L., but for the earlier periods the possibility should be borne in mind that other species of *Fraxinus*, for example *F. ornus* L. or *F. oxycarpa* Willd., might be concerned.

The pollen record for *Fraxinus* is virtually restricted to earlier interglacial and Flandrian records: the pollen twice recorded from the Wolstonian is regarded by the author as likely to be derived, as the late Anglian occurrence may also be. Single records from the Ludhamian, Pastonian and Cromerian are followed by several records in each sub-stage of the Hoxnian (supported by a wood identification in sub-stage II at Hitchin). There are thirteen site records in the Ipswichian, of which all but one are in the middle, warmer sub-stages. Throughout the Weichselian, despite numerous and recent pollen analyses, there has been no single record of the tree, which indeed is recognized as frost-sensitive, particularly to late spring frosts. Its reappearance in the Flandrian is correspondingly tardy. There are only two records from zone v, one from Dogger Bank moorlog and one from Fergus Moss, Aberdeenshire; two records from the composite v/vi zone are both from Dartmoor. As the record-head diagram shews, a vast extension occurred in zone viia which was extended and sustained in zones viib and viii. Table 24 of the mean *Fraxinus* pollen frequencies zone by zone demonstrates effectively how the expansion in site numbers was accompanied by substantial increase in the proportion of total tree pollen contributed by the ash.

TABLE 24. Fraxinus *as percentage of total tree pollen in the Flandrian of the British Isles: numbers of records*

	Scan	0–2	2–5	5–10	10–20
VIII	5	15	25	29	3
VIIb	9	33	40	14	.
VIIa	4	39	8	3	.
VI	2	3	.	.	.
V	.	1	.	.	.

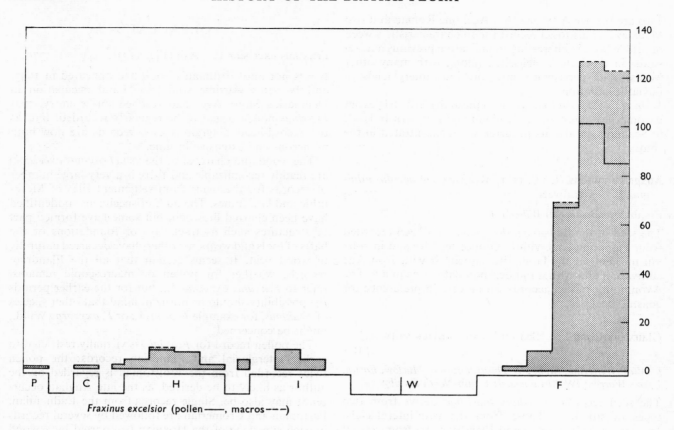

Fraxinus excelsior (pollen——, macros———)

The low pollen productivity means that in effect abundance of the tree rose even more.

The series of maps (fig. 113) shewing pollen frequency throughout the British Isles from zone VI onwards makes the same point. It illustrates also how widespread the tree had become already in zone VI, its wide and early occurrence (but low frequency) in Scotland, and the highest zone VIIb and VIII frequencies tending to be grouped in north-west England and Somerset where, on the mountain limestones occur most of our woodlands now dominated by ash.

The distribution of macroscopic records of ash shews similar concentration in the later Flandrian zones and out of sixty-seven no less than sixty-three are associated with archaeological sites, distributed as follows: Mesolithic 2, Neolithic 11, Bronze Age 14, Iron Age 13, Roman 18, Anglo-Saxon 1, Norman 2 and Mediaeval 2. Because the ash enters the later stages of woodland succession following calcareous fen (Kassas, 1951) there are several records of its wood from fenland deposits.

There is no doubt of the thermophilous nature of the ash, and Firbas took the view that it could not have persisted through the last glaciation north of the Alps. Its easy seed dispersal makes it improbable that its late Flandrian expansion was merely due to failure of dispersal, and one may doubt whether low temperatures alone delayed it. The clue was apparently offered by Iversen who in 1941 reported that in Denmark *Fraxinus*, although sparse in Post-glacial deposits, yielded nevertheless a curve of considerable value since it shewed a consistent rise at the boundary between the Atlantic and Sub-boreal periods when the *Ulmus* curve regularly diminishes. Iversen pointed out that charcoal analyses by Jessen confirmed the reality of the exchange of relative importance at this time between *Ulmus* and *Fraxinus*. It has since been generally acknowledged that the elm decline was induced by exploitation of the tree by Neolithic man, and that from this time onwards the woodlands were progressively thinned and then cleared. There is no doubt at all that in the British Isles *Fraxinus* is mostly seen in a seral role compatible with a pronounced intolerance of shade, so that on suitable soils it forms transient communities giving place naturally to climax woodland of beech and oak. There is much in the detailed pollen analyses as well as in the generalized pollen record to suggest that the expansion of the ash in the later Flandrian primarily reflects human clearance of the climax forest and the ash trees' strong invasive response. Thus in the South Pennines Miss Conway's pollen diagrams shew a big increase in *Fraxinus* pollen frequencies (to values of 10–20 per cent) during the historical period of forest clearance there after about A.D. 1100. Likewise, in Somerset there is indication in the Decoy Pool Wood diagram (fig. 168) that *Fraxinus* may have been favoured by prehistoric forest clearance along with *Acer* and *Ilex*. There then remains the fact that upon thin limestone soils

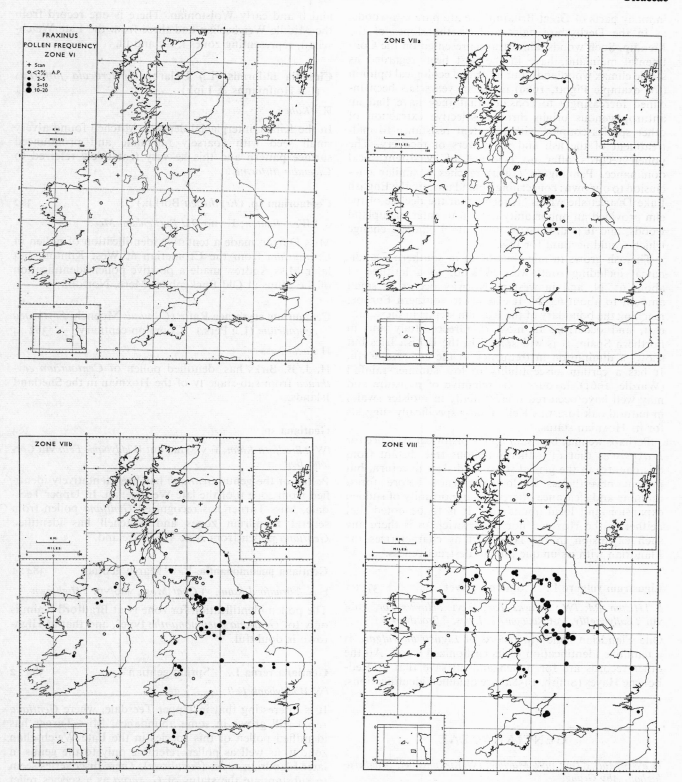

Fig. 113. Pollen frequency of the ash, *Fraxinus excelsior* L., in the British Isles through Flandrian zones VI to VIII.

in many parts of Great Britain there are pure ash-woods, as in the Derbyshire dales and on the slopes of the Mendip. Such woods, almost unrepresented on the Continental mainland, have in the past been regarded as stable climax woodlands, but informed ecological opinion (for example Pigott, 1969) in recent years has been inclined increasingly to consider that they have had an anthropogenous origin through selective extraction of other species coupled with the great resistance to maltreatment of the ash and its powers of recovery. The pollen records offer no evidence of persistent local dominance. Pennington (1964) reaches a similar conclusion to our own: concerning upland tarns in the English Lake District she writes 'It seems that the decline of the elm provided an opportunity for ash to enter the upland woods, and it is difficult to postulate a climatic change which would account for this.'

The ash grows in fair abundance in southern Scandinavia, including south-west Finland, to a latitude of about 61° N, and in western Norway occurs in small amount to about 64° N. It extends to southern Europe, reaching the boundary of the Russian steppe in the southeast, and occurring in most of Greece, Italy and in northern Spain. It is widespread in the British Isles but shows a diminution in frequency in the extreme north. It has a certain susceptibility to low summer rainfall (Wardle, 1961), favours soils retentive of moisture and may well have occurred preferentially in moister swales in natural oak forest as Kelly (1964) specifically suggests for its Hoxnian status.

Perhaps we may best sum up the historical evidence for *Fraxinus* as that of a thermophilous tree distant from this country in the glacial stages and slow to return, but present nevertheless in low frequency before forest thinning and clearance offered the opportunity of sudden expansion and local abundance. It is to be noted that neither in the Hoxnian nor the Ipswichian is there any such expansion in the later part as characterizes the Flandrian with its unique human destructiveness.

Ligustrum vulgare L.   Common privet   378: 1

I ?*Histon Rd*; lW ?*Hawks Tor*; F vi ?*Hawks Tor*, viib/viii ?*Ballynagilly*, viii *Redbourne Hayes*, ?*Broadway*

One Flandrian zone viii record of *Ligustrum vulgare* is a tentative identification of its carbonized wood. All the other records are for pollen and all save that at Redbourne Hayes (Smith, 1958a) are qualified identifications.

GENTIANACEAE

H eWo *Marks Tey*; mW *Marlow*; lW *Kirkmichael, Magheralaghan, Aby Grange*

Pollen, not identified beyond the level of the family Gentianaceae, was recorded at four levels in the Hoxnian and related deposits at Marks Tey, sub-stages iic, iiia

and b and early Wolstonian. There is one record from the Middle Weichselian and three from the Late Weichselian, representing zones i, i/ii and iii.

Cicendia filiformis (L.) Delarbre (*Microcala filiformis* (L.) Hoffmanns & Link)   380: 1

H ?*Kilbeg*

In the Kilbeg interglacial deposits Mitchell found a very small seed with coarsely reticulate and pronounced surface pattern and this he very tentatively referred to *Cicendia filiformis*.

Centaurium sp. (*Erythraea* Borkh.)   382

C ?*West Runton*; F viib *Old Buckenham Mere*

Miss Duigan made a tentative identification of pollen of *Centaurium* from the Cromerian at West Runton, but later Miss Andrew made a positive pollen identification of the genus at Old Buckenham Mere, Norfolk.

Centaurium erythraea Rafn. (*C. minus* Moench, *Erythraea centaurium* (L.) Pers.)   Common centaury   382: 4

H *Fugla Ness*

H. J. B. Birks has identified pollen of *Centaurium erythraea* from sub-stage iv of the Hoxnian in the Shetland Islands.

Gentiana sp.   384

lW ?*Bradford Kaim*; F vi, viia, viii *Widdybank Fell*, viii *Corlona Bog*

Pollen of the genus *Gentiana* has been tentatively identified from zone ii of the late Weichselian. In Upper Teesdale, Miss Turner has recognized *Gentiana* pollen from several Flandrian zones and Mitchell has identified *Gentiana* pollen from zone viii in Ireland.

Gentiana pneumonanthe L.   'Marsh gentian'   384: 1

F iv ?*Bradford Kaim*, vi *Chat Moss*, viib/viii ?*Hightown*

The pollen identification for zone iv at Bradford Kaim is only to '*Gentiana pneumonanthe* type', and that at Hightown is doubtful.

Gentiana verna L.   'Spring gentian'   384: 2

lW *Widdybank Fell*

It is interesting that in Upper Teesdale, where *Gentiana verna* still grows in some abundance, Miss Turner has identified pollen of this species in the Late Weichselian zone ii, as well as pollen referable only to the genus in later Flandrian zones (see above). These discoveries seem to substantiate the status of *G. verna* as a species relict from the late Weichselian and persistent in the locality through the Flandrian.

G. verna is limited today to two areas in the British

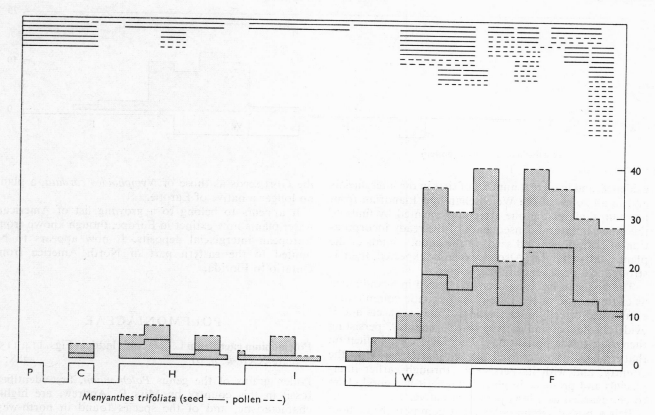

Menyanthes trifoliata (seed ——, pollen – – –)

Isles, northern England and western Ireland, where it occurs locally upon limestone. It belongs to Matthews' Alpine element of the British flora, occurring in the Pyrenees, Jura, Alps and Carpathians but not (apart from Britain) in mountains further north nor in the Arctic, so that the British populations are widely disjunct and their Pleistocene history the more significant therefore.

Gentiana cruciata L.   384: 4

mW Barnwell Station

The seed identified as *Gentiana cruciata* from the Middle Weichselian in Cambridge by Miss Chandler (see first edition) has been re-identified as that of *Campanula rotundifolia* L. (Bell & Dickson, 1971).

Gentianella campestris (L.) Börner   'Field gentian'
385: 1

lW ?*Tadcaster*, ?*Moss Lake*, ?*Esthwaite Basin*; F vIIb, vIII ?*Ehenside Tarn*

Although the pollen records for *Gentianella campestris* in the Late Weichselian cover all three zones (I and III at Tadcaster, II and III at Esthwaite) they are only tentative. So too are the seed identifications from Cumberland.

It is a species sparsely but generally distributed through the British Isles, more frequent however in the north. It extends commonly on the west coast of Norway to 70° N and grows in Iceland so that Late Weichselian records are not surprising.

MENYANTHACEAE

Menyanthes trifoliata L. Buckbean, Bogbean (Plate XVIIIf)   386: 1

The polished, kidney-shaped seeds of *Menyanthes trifoliata* are perhaps the most abundant and most easily recognizable of all Quaternary plant fossils. They occur not only in lake muds and in mesotrophic fens but in the pools of raised-bog surfaces, save perhaps under extremely oligotrophic conditions. They thus have every chance of incorporation in deposits of the type investigated in Quaternary research. The pollen is also recognizable and pollen records are as numerous as those of seeds. The rhizomes are encountered also but the identification is seldom recorded.

The extent and completeness of the fossil record of *Menyanthes* are remarkable. From the Cromer Forest Bed series onwards it is represented in every glacial and interglacial stage, it is in all four sub-stages of the Hoxnian and three of the Ipswichian. Thereafter, right through the Weichselian and Flandrian, there is no gap in the records. Throughout the seed and pollen records yield comparable

313

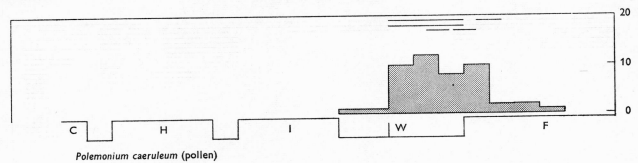

Polemonium caeruleum (pollen)

evidence, and in a great number of sites in the interglacials and in all zones of the Weichselian and Flandrian from I to VIII, finds of the one are complemented by finds of the other, a natural consequence of the ready incorporation of both pollen and seeds in the aquatic muds of the plant's native habitat. Also as would be expected, there is strong local persistence in many sites.

The bogbean has long been recognized in Scandinavia as characteristic of the Late-glacial aquatic communities established directly after melting of the ice-sheets and it evidently had a similar role in this country, persisting thereafter to the present day. There is a clear indication that it might have survived the last glacial period in the country, and with its persistence through earlier interglacials and presence in glacial stages, it has good claim to be regarded as a long persistent native.

It is a boreal–circumpolar plant common throughout Scandinavia to the extreme north and nothing in its present distribution conflicts with the history established by its fossil record.

Nymphoides peltata (S. G. Gmel.) Kuntze (*Limnanthemum nymphoides* (L.) Link) 'Fringed waterlily' 387: 1

H ?*Gort*

From sub-stage III of the Hoxnian interglacial at Gort Jessen *et al.* (1959) made a tentative identification of pollen of *Nymphoides peltata*. In the British Isles today *N. peltata* appears to be native only in central and eastern England. The Gort record, for which the generic identity is certain, may well relate to the north American species *N. cordata* (see following account), a seed of *N. peltata* was however found by Hartz from the last interglacial in Denmark.

Nymphoides cordata Ell. (Fern.) 387: 2

H *Gort*

Jessen and his associates in 1959 described from sub-stages II, III and IV of the Hoxnian beds at Gort in Co. Galway some thirty seeds that resembled, apart from their small size and clear epidermal characters, those of *Menyanthes trifoliata*. They were therefore placed within the newly erected species *M. microsperma*. More recently W. A. Watts (1971), with extensive experience of identifying sub-fossil seeds in North America, has recognized

the Gort seeds as those of *Nymphoides cordata*, a plant no longer a native of Europe.

It appears to belong to a growing list of American water plants now extinct in Europe, though known from European interglacial deposits. It now appears to be limited to the eastern part of North America from Ontario to Florida.

POLEMONIACEAE

Polemonium caeruleum L. Jacob's ladder (Figs. 114, 115; plate XXI*i–k, n, o*) 388: 1

Pollen grains of the genus *Polemonium*, first identified fossil in this country by Miss R. Andrew, are highly characteristic, and of the species found in north-west

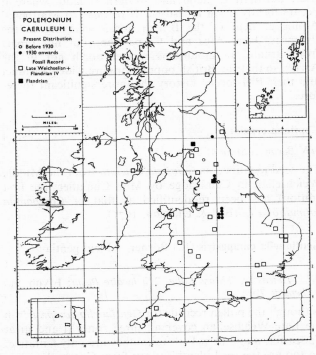

Fig. 114. Present, Weichselian and Flandrian distribution of Jacob's ladder, *Polemonium caeruleum* L., in the British Isles, shewing great Flandrian retraction and disjunction.

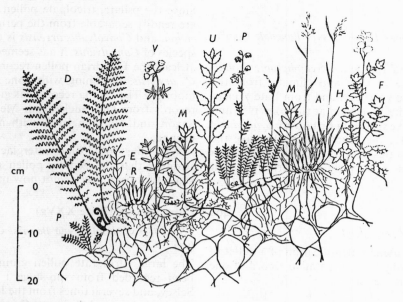

Fig. 115. Transect of tall-herb vegetation developed on a soil 'creep' terrace on limestone scree below a cliff in Bastow Wood, Wharfedale. This is a characteristic present-day community of *Polemonium caeruleum* (P) with A, *Arrhenatherum elatius*; D, *Dryo-* *pteris filix-mas*; E, *Epilobium montanum*; F, *Filipendula ulmaria*; H, *Heracleum sphondylium*; M, *Mercurialis perennis*; R, *Festuca rubra*; U, *Urtica dioica*; V, *Valeriana officinalis*. (From Pigott, 1958.)

Europe it is possible to distinguish *P. caeruleum*, the British species, by details of the grain sculpturing.

Pollen of *P. caeruleum* has been recorded in the British Isles from Early and Middle Weichselian and, in considerable frequency, in all zones of the Late Weichselian and zone IV of the Flandrian. The map (fig. 114) makes it apparent how widespread the plant was throughout England and Wales in this period: there are no Irish records but in Scotland there is one from Aberdeenshire and one in Flandrian zone V from Dumfriesshire. At Moss Lake, near Liverpool the pollen is recorded from zones V and VI, and at Malham Tarn, Yorkshire, where the plant still grows, it has been recorded as late as zones VI and VIIa, a clear indication of local persistence.

Although cultivated forms of *P. caeruleum* often become naturalized in the lowlands of Britain, they differ morphologically from the native plant which Pigott (1958) describes as 'undoubtedly native and locally abundant on upland screes and cliff ledges in northern England'. It is nevertheless a rare British plant and restricted almost totally to the Carboniferous Limestone. It seems evident that, like *Thalictrum alpinum*, *Betula nana* and others, it became reduced and disjunct in range after the Late Weichselian. Pigott emphasizes that in this instance it was more likely as a result of changes in vegetation and soil conditions than as a direct consequence of climatic amelioration. He points out that moist, fertile soils and freedom from competition are its prime requirements and that it is intolerant of sustained water loss. He conjectures that in the Late Weichselian it grew 'in natural meadows, particularly on lake shores, in fenland and open birch woodland, as it does at the present day in parts of Germany, the Baltic States and north Russia'. From consideration of the 'tall herb' communities in which *P. caeruleum* grows today it is apparent that many of the companion plants were present with it in the Late Weichselian vegetation (e.g. *Filipendula ulmaria*, *Urtica dioica*, *Valeriana officinalis*, *Melandrium dioicum*, *Heracleum sphondylium*; see fig. 115).

The collective species *P. caeruleum* has a circumboreal distribution, and is one of few Eurasian representatives in a predominantly American family. *P. caeruleum* occurs scattered throughout Scandinavia to the extreme north, and is placed by Matthews in the category of Northern montane species. It is a plant strongly under-represented in its pollen spectra and it would be unwise to base conclusions upon its apparent absence from any of the British glacial stages preceding the Weichselian. A similar Flandrian retraction of range to that in Britain is known also from central Europe.

BORAGINACEAE

IW Ballydugan, Woodgrange

Singh (1970) recorded pollen of Boraginaceae from two Late Weichselian sites in northern Ireland: at Ballydugan from zones I, II, III and III/IV and at Woodgrange from zone II. There is so far only slight indication of the smaller taxa these records might represent.

Symphytum sp.   392

C *West Runton*; *l*W *Colney Heath*; F IV/V *Langdale Combe*,
VIIb/VIII *Wingham*

Pollen identifications of the genus *Symphytum*, though
few, cover a big range of time and climate, from the
Cromerian to only 2700 years B.P. and from Late Weichselian zone I through the Flandrian at the zone IV/V and
VIIb/VIII levels. The native species in Britain are *S. officinale*
L. and *S. tuberosum* L., the latter tending to have the
more northerly range in this country: neither of these has
however anything but a strongly southern range in
Scandinavia.

Myosotis sp.   400

*e*W *Sidgwick Av.*; F VIII ?*Newstead*

There is an Early Weichselian identification of a nutlet of
Myosotis from Cambridge, and a tentative record of one
from the Roman site at Newstead.

Myosotis scorpioides L.   'Water forget-me-not'   400: 1

F VIII ?*Apethorpe*

From zone VIII of the Flandrian at Apethorpe, among
a long list of aquatic plant records, is a tentative identification of a nutlet of *Myosotis scorpioides*.

Myosotis caespitosa K. F. Schultz   'Water forget-me-not'
                                                          400: 4

F VIIa/b *Shustoke*

From deposits of site A at Shustoke that span the zone
VIIa/VIIb boundary, Kelly and Osborne report the fruit of
Myosotis caespitosa.

Myosotis arvensis (L.) Hill   'Common forget-me-not'
                                                          400: 8

F VIII *Shustoke*

Site B at Shustoke, probably of zone VIII age, yielded
a nutlet of *Myosotis arvensis*.

Lithospermum arvense (L.) Hill 'Corn gromwell',
'Bastard alkanet'   401: 3

F VIII *Aldwick Barley, Silchester,* ?*Newstead*

The nutlets of *Lithospermum arvense*, a typical weed of
arable ground, have been found fossil only in an archaeological context, the Iron Age at Aldwick Barley, and the
Roman at Silchester and, doubtfully, at Newstead.

CONVOLVULACEAE

Convolvulus arvensis L.   Bindweed, Cornbine   405: 1

I ?*Bobbitshole*; F V *Dogger Bank*, VIIa *Canna*, VIII *Old Buckenham Mere*

Since the psilate, tricolpate pollen grains of *Convolvulus*
are readily separable from the periporate grains of *Calystegia*, and *Convolvulus arvensis* is the only native British
species of *Convolvulus*, it has seemed reasonable to regard
at least the Flandrian pollen records for the genus as of
C. arvensis, beginning with zone V for the North Sea
moorlog. The Canna record was given as *C. arvensis*, the
records from Old Buckenham Mere come from Anglo-
Saxon and Norman levels with many weeds of arable
ground.

With the Ipswichian interglacial the possibility is
stronger that the generic pollen record might refer to
some European species not now native.

Calystegia sp. (Plate XXV*g*)   406

I *Selsey*; F VI *Nazeing*, VIIa *Bowness Common*, VIIa,b *Glastonbury*

The large periporate pollen grains of *Calystegia* have
been recorded from sub-stage I of the Ipswichian at
Selsey, and several times from the Flandrian. It is possible
that at Selsey and Bowness Common the coastal *Calystegia soldanella* might have been represented and not
C. sepium, the other and commoner British species.

Cuscuta sp.   Dodder   407

*l*W *Windermere*; F VIIa *Oakhanger*, VIIb/VIII *Blelham Tarn*,
VIII *Old Buckenham Mere*

There are Flandrian records for the pollen of *Cuscuta* from
soil samples of zone VIIa at the Mesolithic site of Oakhanger in Hampshire, and from the Iron Age level at
Old Buckenham Mere. The pollen from zone VIIb/VIII at
Blelham Tarn was tentatively referred to *C. epithymum*.

SOLANACEAE

Atropa bella-donna L.   Dwale, Deadly nightshade
                                                          410: 1

F VIII *Silchester, Caerwent*

Seeds of *Atropa bella-donna* have twice been recorded
from Roman sites. C. Reid who identified them writes that
it was 'one of the commonest plants in Roman Britain'
and draws attention to the use of the plant as a cosmetic,
thus explaining the constant occurrence of the seeds in
Roman house refuse. The capacity of the species to
spread from areas of cultivation was well demonstrated in
England immediately after the 1914–18 war, when local
crops had been grown as sources of atropine.

Matthews describes the plant as Continental–southern,
absent from Scotland and probably introduced in Ireland.

Hyoscyamus niger L.   Henbane   411: 1

F VIII *Aldwick Barley, Pevensey,* ?*Hungate*

Seeds of *Hyoscyamus niger* have been found fossil only in
archaeological sites, Iron Age at Aldwick Barley, Roman

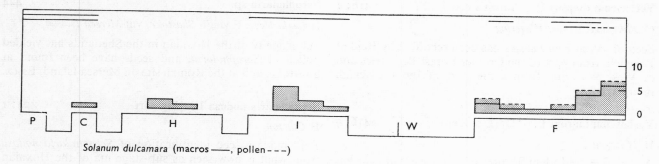

Solanum dulcamara (macros ——, pollen ---)

at Pevensey and, tentatively identified, mediaeval in the Hungate, York. Ødum (1965) has commented upon the association of this plant with archaeological sites in Denmark and on the remarkable longevity of its seeds in certain situations on these sites.

Solanum sp. 413

C *Norfolk Co.*; F vIIb *North Ferriby*, vIII *Bermondsey*

Seed identifications of *Solanum* have been made from the Cromerian interglacial, from zone vIIb on the Humber shore and from the Roman site at Bermondsey.

Solanum dulcamara L. Bittersweet, Woody nightshade
413: 1

There are records of the seeds of *Solanum dulcamara* in each of the last four interglacial stages beginning with two from the Cromer Forest Bed series and one from the Cromerian proper. Sub-stages II and III of the Hoxnian and II, III and IV of the Ipswichian are represented, and every zone of the Flandrian. The pollen records, though few, conform to the Flandrian seed record, notably in that Behre and Menke record the pollen from zone IV of the Dogger Bank moorlog which had much earlier provided a seed to C. Reid. The later Flandrian records are notably associated with archaeological sites, Neolithic 1, Bronze Age 2, Iron Age 1, Roman 3 and Mediaeval 1, possibly suggestive more of site preferences than of deliberate collection of the fruits.

Solanum dulcamara occurs frequently in Scandinavia, including south Finland, to about 61° N with scattered occurrences to the head of the Gulf of Bothnia: it has, however, a tendency to be much less frequent in the west, but in the British Isles this is not evident. It is absent from the north of Scotland. This range is conformable with the evidence that the plant immigrated early in the Flandrian, having been absent through the Weichselian as apparently from earlier glacial stages.

Solanum nigrum L. 'Black nightshade' 413: 3

Macros: C *Norfolk Co.*; F IV/V *Star Carr*, vII *Dunshaughlin*, vIIb *Frogholt*, vIII *Silchester, Caerwent, Hungate, Dublin, Ballingarry Downs*
Pollen: IV *Moss Lake*, vIIa,b *Old Buckenham Mere*

Apart from a seed record in the Cromerian, all records of *Solanum nigrum* are from the Flandrian. There is a seed record from zone IV/V at Star Carr and a pollen record from zone IV at Moss Lake: otherwise the records are all zone vII or later. Two records are from Roman sites and three from mediaeval ones, whilst the Frogholt record of 2860 B.P. accompanies much evidence of weeds of arable cultivation. Besides confirming its status as a weed and ruderal, the fossil record establishes an early native status in England for *Solanum nigrum*.

The plant today is concentrated in the south-eastern part of England, and is thought not to be native in Scotland or Ireland, a fact lending interest to the two mediaeval records of it from Ireland and to the fact that the earliest Flandrian records both lie near the presumed northern limit of its range. In Scandinavia *Solanum nigrum* is common only in Denmark and southern Sweden though there are scattered localities to higher latitudes along shores and in cultivated ground.

SCROPHULARIACEAE

Pollen of Scrophulariaceae occurs in low frequency in sub-stage II of the Ipswichian and in all the zones of the Late Weichselian and Flandrian. In reporting upon the high altitude Elan Valley site in Cardiganshire, Moore states that from zone I to zone IV there are both psilate and reticulate pollen grains in this category. There is one doubtful seed record from the Roman site at Shrewsbury.

Verbascum sp. Mullein 416

In the first edition it was reported that candelabrum-type branched hairs had been found in deposits of various ages: these were illustrated and identified, with some reservation, as leaf hairs of mullein. It was reported that these hairs had been found as contaminant in the laboratory air. Now, thanks to Dr J. H. Dickson, reporting the observation of a Glasgow student, it is apparent that these are in fact the leaf hairs of *Platanus* sp. that are shed from the maturing leaves in the spring. An avenue of plane trees grows near the Cambridge Botany School and has been the source of these rightly suspect records (plate XXVIc).

Verbascum thapsus L.   Aaron's rod   416: 1

Cs *Pakefield*; I *West Wittering*

Seed of *Verbascum thapsus* has been recorded by Reid at Pakefield from part of the Cromer Forest Bed series and at West Wittering from some part of the Ipswichian interglacial.

Verbascum nigrum L.   'Dark mullein'   416: 7

H *?Clacton*

A tentative seed identification of *Verbascum nigrum* has been made from sub-stage II of the Hoxnian interglacial at Clacton.

Linaria sp.   420

*l*W *Esthwaite Basin*

Franks and Pennington (1961) recorded pollen of the genus *Linaria* from the Late Weichselian zones II and III at Esthwaite Basin in the English Lake District. The following account shews the clear presence of *L. vulgaris* in the Weichselian: no other *Linaria* species have been recorded.

Linaria vulgaris Mill.   Toadflax (Plate XVIII*k*)
                                420: 4

H *?Clacton*; *e*W *Sidgwick Av.*; *m*W *Earith*; *l*W *Hartford, Nant Ffrancon, Nazeing, Elstead*

The readily identifiable, winged seeds of the common toadflax, *Linaria vulgaris*, were tentatively identified from sub-stage III of the Hoxnian interglacial at Clacton. Later, unqualified records have come from the Early, Middle and Late Weichselian. There is no zone I record, but zone II occurrences include one in North Wales, one in Huntingdonshire and one in Essex. There is a zone III record from Surrey.

  This evidence makes the native status of the plant quite evident despite the very strong association that it has today with waste places and man-made habitats.

  L. vulgaris occurs commonly in Sweden and Finland to about 63° N and to somewhat higher latitudes in western Norway: isolated records are given beyond this and well into the Arctic Circle.

Chaenorhinum minus (L.) Lange   'Small toadflax'
                                   421: 1

*l*W *Brook*

From Late Weichselian zones I and II at Brook in Kent Miss Lambert identified seeds of *Chaenorhinum minus*, thus adding to the total of plants now weeds of arable and waste land that were part of the native flora in southern Britain at the end of the Weichselian glacial stage. In Scandinavia this toadflax is common only in southern Sweden and parts of Denmark but there are scattered localities for it along the Gulf of Bothnia.

Scrophularia sp.   424

H *Fugla Ness*; F vIIa/b *Shustoke*, vIII *Mersea Is.*

Sub-stage IV of the Hoxnian in the Shetlands has yielded pollen of *Scrophularia*, and seeds have been found at Shustoke, and at the Roman site on Mersea Island, Essex.

Scrophularia nodosa L.   Figwort   424: 1

H *Clacton*

Reid made a seed identification of *Scrophularia nodosa* from what is now seen as sub-stage IIIa of the Hoxnian at Clacton.

Limosella aquatica L.   'Mudwort'   426: 1

Cs *Pakefield*

C. and E. M. Reid (1908) identified and illustrated seeds of *Limosella aquatica* from the Cromer Forest Bed series at Pakefield.

Digitalis purpurea L.   Foxglove   429: 1

F v/vI, vIIb/vIII *Loch Dungeon*

Mrs Birks has recently recorded pollen of *Digitalis purpurea* in the composite zones v/vI and vIIb/vIII at Loch Dungeon in Galloway.

Veronica sp.   430

*l*W III/IV *Lochan Coir' a' Ghobhainn*; F v/vI *?Allt na Faithe Sheilich*, vIIb/vIII *?Ballynagilly*

Mrs Birks has identified pollen of 'Veronica type' from deposits on the Cairngorm, her husband has found it at the Weichselian/Flandrian transition on Skye, and Pilcher has made a tentative identification also from the deposits at Ballynagilly, Northern Ireland.

Veronica anagallis-aquatica L.   'Water speedwell'
                                   430: 2

*l*W *Colney Heath, Hawks Tor*

Seeds of *Veronica anagallis-aquatica* come from zone I of the Late Weichselian at Colney Heath (13560 B.P.) and from zone II at Hawks Tor, Cornwall (11070 B.P.).

Veronica scutellata L.   'Marsh speedwell'   430: 4

*l*W *?Hawks Tor, Scaleby Moss*

There is a tentative seed identification of *Veronica scutellata* from zone II of the Late Weichselian at Hawks Tor (11070 B.P.) and an unqualified record from zone III at Scaleby Moss (10580 B.P.).

Veronica chamaedrys L.   'Common speedwell'   430: 7

Cs *Pakefield*

C. and E. M. Reid recorded seed of *Veronica chamaedrys* from the Cromer Forest Bed series at Pakefield.

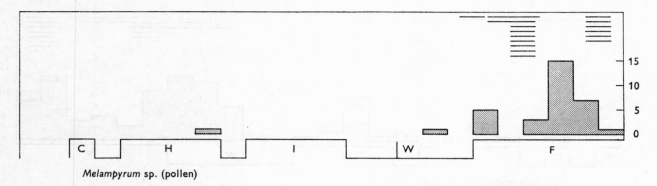

Melampyrum sp. (pollen)

Veronica hederifolia L. 'Ivy speedwell'   430 : 20

F viib *Great Bedwyn*, viii *Aldwick Barley, Silchester*

The three sites from which seeds of *Veronica hederifolia* have been recovered are respectively Middle Bronze Age, Iron Age and Roman. It is characteristically a weed of cultivated ground.

Pedicularis sp.   432

H *Fugla Ness*; lW *Loch Meodal, Esthwaite Basin*

Pollen of *Pedicularis* has been identified by H. J. B. Birks from sub-stage iv of the Hoxnian in the Shetlands, and from Late Weichselian zone ii in Skye. Franks and Pennington have also pollen records from zones ii and iii in the English Lake District.

Pedicularis palustris L.   Red rattle   432 : 1

mW *Upton Warren, Earith, Marlow*; lW *Brook, Kirkmichael*; F iv *Burtree Lane*, viii *Newstead*

Seeds of *Pedicularis palustris* have been recorded from the Middle Weichselian by Coope from Upton Warren, and by Miss Bell from Earith and Marlow. Further records come from zones i and iii of the Late Weichselian in Kent and the Isle of Man and from zone iv of the Flandrian in Northumberland. Finally there is a Roman record from Newstead.

Pedicularis palustris occurs throughout the British Isles today ascending to 850 m, and has a wide European range, in Scandinavia extending commonly to 69° N and in scattered localities still further north. The fossil record strongly indicates perglacial survival in Britain.

Pedicularis hirsuta L.   432 : 3

mW *Temple Mills*

A seed identification of *Pedicularis hirsuta*, a non-British species, was made from the Lea Valley Arctic Plant Bed at Temple Mills (Reid, 1949). It is one of the northern unicentric element in the Scandinavian mountains.

Pedicularis lanata Cham. & Schlect.   432 : 4

mW *Earith*

Miss Bell found a seed in the Middle Weichselian beds at Earith that differed in structure from all other European species she examined but corresponded closely with seeds of *Pedicularis lanata*, a plant of 'dry ground on heath, fell-field and sometimes marshes in Greenland'. This is one of a short list of Arctic plants no longer native in this country found by Miss Bell in the British Middle Weichselian.

Rhinanthus sp.   433

mW ? *Upton Warren, Earith, Marlow, Syston, Barnwell Station*; lW *Nant Ffrancon*; F viib *Loch Cuithir*, viii *Godmanchester, Shrewsbury*

Apart from a pollen record from zone viib of the Flandrian at Loch Cuithir, all records for *Rhinanthus* are based on seed identifications. The age distribution is remarkable with no less than five Middle Weichselian and one (zone ii) Late Weichselian occurrences and then none until the Roman record at Godmanchester and the mediaeval record at Shrewsbury. The Barnwell Station record is a recent correction of the record given by Miss Chandler as '*Geranium* Y'.

The several species of this genus are all annuals favouring grassy or open vegetation, some of them montane and northern in range. *Rhinanthus minor* L. is common throughout Scandinavia right up to the North Cape.

Rhinanthus minor L.   433 : 2

F viii *Dublin*

Mitchell reports a macroscopic identification of *Rhinanthus minor* from mediaeval Dublin.

Melampyrum sp.   Cow-wheat   434

The bulk of the pollen records for the genus *Melampyrum* are recent and depend on a few investigators of northern, chiefly Scottish sites. Even the Hoxnian sub-stage iv record is from the Shetlands, but there are records from the North York moors in zone v/vi and from Northern Ireland in zone iv. When the composite zones used by the Birks are taken into account there is a continuous record for the genus through the Flandrian. The Late

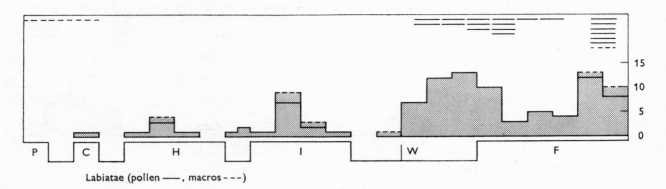

Labiatae (pollen ——, macros – – –)

Weichselian zone II record is from Skye and the III/IV zone record from Galloway. Presumably the records derive from either *M. pratense* or *M. sylvaticum*.

Mamakowa (1968) has pointed out that in Norwegian pollen diagrams *Melampyrum* pollen consistently occurs at levels where it seems fire has been used to clear ground. Pilcher at Slieve Gallion finds high pollen frequencies of it associated with clearance and increase in pollen of cereals and *Salix*. He finds its peak occurrences between 4600 and 3300 B.C. after an establishment in zone VI. He is inclined to think *Melampyrum* may have been abundant in fen communities.

Melampyrum pratense L.   'Common cow-wheat'

F VIII *Larkfield*   434: 3

From the clay daub of a mediaeval timber house in Kent, Grove has identified the capsule and seed of *Melampyrum pratense*.

Odontites verna (Bell.) Dum. (*Bartsia odontites* (L.) Huds., *Odontites rubra* Gilib.)   'Red bartsia'   436: 1

lW *Garvel Park*

C. Reid recorded a seed of *Odontites verna* from the Late Weichselian deposits of the Clyde valley at Greenock.

Bartsia alpina L.   438: 1

mW ?*Barnwell Station*

In reporting the identification of a seed of *Bartsia alpina* from the Middle Weichselian deposits at Barnwell Station Miss Chandler commented that the size was only half that of her type material though otherwise there was agreement. It is an amphi-atlantic arctic–montane plant that is common through western Scandinavia and the far north.

LENTIBULARIACEAE

Utricularia sp.   Bladderwort   442

F IV *Nazeing*, VIIa,b *Canna*, VIIa *Holcroft Moss, Peacock's Farm*, VIIb *Littleport*, VIIb/VIII *Snibe Bog*, VIII *Meare Pool*

A seed of *Utricularia* was identified by Miss Lambert from deposits at Littleport, Cambridgeshire. This apart, all the records are for pollen which is highly recognizable though fragile and produced only sparsely. The records, widely distributed through the country, are all Flandrian; the earliest from zone IV at Nazeing, the rest in zones VIIa to VIII. The three British species of bladderwort differ a good deal in ecological requirement and range and all might have contributed to the record.

VERBENACEAE

Verbena officinalis L.   Vervain   444: 1

I *Histon Rd, Trafalgar Square, Bobbitshole*; F VIII *Caerwent*

Vervain (*Verbena officinalis*) is a plant of southerly distribution type in the British Isles, apparently no longer found native in Scotland or the northern half of Ireland. It is interesting therefore that its fruits should have been found at three different south-eastern sites in Cambridge, London and Ipswich in the same temperate substage, IIb, of the Ipswichian interglacial.

The only Flandrian site is the Roman Caerwent. Waterbolk and van Zeist (1966), who have found it in the Swiss Neolithic settlement at Niederwil, suggest that it may have been brought in, along with other species, as a drug plant.

LABIATAE

The record-head diagram for the Labiatae depends largely upon pollen identifications and it does not greatly add to the plentiful records for individual genera and species that follow. None the less it shews persistence through the interglacial sub-stages and presence through all zones of the Late Weichselian and Flandrian. It shews the same reduction in the early Flandrian zones that is exhibited by other large families of shrubby or herbaceous plants susceptible to prevalent closed forest. There is slight and possibly dubious representation in the full-glacial stages, as might be expected of a family whose distribution is mainly centred upon the Mediterranean.

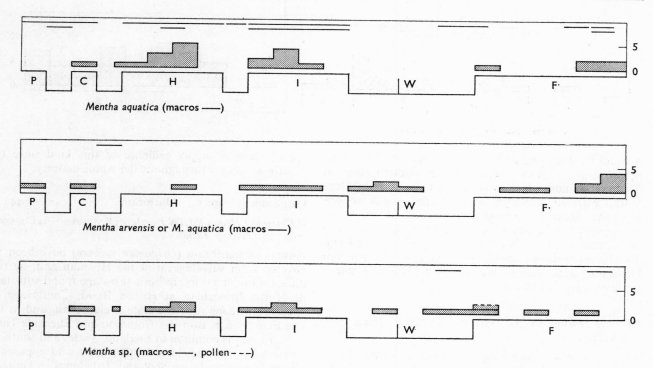

Mentha aquatica (macros ———)

Mentha arvensis or M. aquatica (macros ———)

Mentha sp. (macros ———, pollen – – –)

Mentha sp. Mint (Plate XVIII*l*)   445

Throughout the genus *Mentha*, and its species so far as they are recognizable, the fossil record is almost exclusively for the hard nutlets which have a good chance of surviving secondary deposition, as for instance in glacial and periglacial deposits. The record-head diagram for the genus is very similar to those for *M. aquatica* and for *M. arvensis* or *aquatica*. It shews presence in each interglacial from the Cromerian, a sparse but early Flandrian record and sparse occurrences in the Weichselian and earlier glacial stages. At Selsey the records shew persistence in five levels, later Wolstonian and sub-stages Ia, b and IIa, b, suggestive of the behaviour of a waterside plant. The one pollen record is for 'Mentha type' from Flandrian zone IV in North Sea moorlog.

Mentha arvensis L. 'Corn mint'   445: 3

eWo ?Brandon

Kelly (1968) has made a tentative fruit identification of *Mentha arvensis* from the silty channel deposits of the early Wolstonian at Brandon, Warwickshire.

This is a boreal–circumpolar species that occurs sporadically up to the Arctic Circle in Scandinavia, and is common to about 63° N.

Mentha arvensis L. or **M. aquatica** L.   445: 3, 4

A record-head diagram is given for the numerous fruit identifications made by various investigators that could not be more precisely placed than within *Mentha arvensis*

or *M. aquatica*. It will be seen that remains of this type have been recovered from every interglacial from the Pastonian onwards. It was found in the Corton Beds (Anglian), the late Wolstonian, and most notably once in the Early Weichselian, twice in the Middle Weichselian and once in zone I of the Late Weichselian. The Flandrian records are sparse but increase latterly and include two from Roman and one from a mediaeval site.

The records shew notable persistence at Ilford through six separate levels from the late Wolstonian to the end of sub-stage II, and in the Flandrian at Apethorpe in zones V, VI and VIII. At both sites we are concerned with fluviatile deposits.

Mentha aquatica L. 'Water mint'   445: 4

Fruit identifications of *Mentha aquatica* are numerous and, as the record-head diagram indicates, they occur in the Cromer Forest Bed series, the Cromerian itself and in each subsequent interglacial. At Baggotstown Watts has shewn remarkable persistence of this species from the late Anglian, through sub-stages I, II, II/III, III to the early Wolstonian. Likewise it was found at Selsey in the late Wolstonian and sub-stages Ia, Ib, IIa and IIb of the Ipswichian. These and less dramatic instances suggest, as was said for the genus record, the persistence of a riverside species with ample opportunity for incorporation of its fruits in the accumulating fluviatile sediments. Despite one record each from the Beestonian, the late Anglian and early and late Wolstonian there is no clear record from the Weichselian glacial stage. This might be taken to indicate that the Weichselian records in the previous

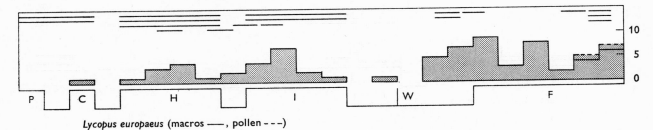

Lycopus europaeus (macros ——, pollen - - -)

account were attributable to *M. arvensis* rather than *M. aquatica*, a species of distinctly more southern range in Scandinavia and which is common there only in Denmark, extreme southern Sweden and south-western Norway. *Mentha aquatica* is however found today throughout the British Isles and it seems likely to have been present at least from the opening of the Flandrian. Two Roman records, one tentative, conclude the sequence illustrating again the tendency of marsh plants to occur in settlement sites of this age.

Lycopus europaeus L. Gipsy-wort (Fig. 160; plate XVIIIm)   446: 1

The nutlets of *Lycopus europaeus* are readily identifiable, particularly so when they retain, as is often the case, the wide inflated border. Of the numerous records that comprise the record-head diagram only two, for '*Lycopus-type*' pollen, are not for fruits.

From the Cromerian onwards every sub-stage or zone of the interglacials has one or several records. They are of course fewer in the glacial stages but occur in the early and late Wolstonian, the Middle Weichselian and commonly in the Late Weichselian zones II and III and early Flandrian IV. There is very pronounced local persistence at Selsey, Bobbitshole and Ilford, especially in the Ipswichian, and at Nechells and Baggotstown in the Hoxnian. Each instance must represent growth for a thousand years or more in the same area and through considerable climatic and vegetational changes, embracing two or three sub-stages. It is interesting that the pollen records from the late Flandrian were in deposits also yielding *Lycopus* fruits, and that the six zone VIII records include four Roman and one mediaeval record.

The persistence locally from late-glacial or into early-glacial deposits reinforces the direct evidence from the Weichselian of persistence in glacial stages, so that all in all there is a strong case for regarding *Lycopus europaeus* as having been present in the British Isles throughout at least the middle and late Pleistocene.

Lycopus europaeus belongs to Hultén's category of west European–south Siberian plants and he shews it as common in south Scandinavia to about 62° N. Scattered localities extend one or two degrees north of this, but fruits have been found fossil in Finland and round the Baltic to much higher latitudes, a distribution supposedly relict from the Flandrian thermal maximum. The British

records cannot supply evidence of this kind since the plant now occurs throughout the whole country.

Origanum vulgare L.   Marjoram   447: 1

H *Clacton*; I *Histon Rd*; lW *Hartford*; F VIIa *Bowness Common*, VIIb, VIII *Ehenside Tarn*

Nutlets of marjoram (*Origanum vulgare*) have been recovered from sub-stage III of the Hoxnian and, by two different investigators, in both sub-stage II and sub-stage III of the Ipswichian at Histon Road, Cambridge. It occurred in zone II of the Late Weichselian, and in the late Flandrian at Bowness Common and Ehenside Tarn.

Marjoram is common in England, Wales and southern Ireland, but becomes local further north and apparently absent in north-western Scotland. It belongs to Hultén's group of west European–south Siberian plants, and although commonest in south-eastern Scandinavia it extends sporadically in the west up to the Arctic Circle.

Thymus sp.   448

From the Isle of Skye right through the zones of the Late Weichselian and early Flandrian H. J. B. Birks has recorded pollen of '*Thymus* type', which however embraces pollen of several genera of the Labiatae. There is a tentative *Thymus* pollen record from zone II/III at Nazeing and one of the '*Thymus* type' from the composite VIIb/VIII zone at Loch Dungeon by H. H. Birks. The strong Weichselian representation gives particular interest to these records.

Thymus serpyllum L.   Wild thyme   448: 2

lW *Roxburgh St*

C. Reid recorded a nutlet of *Thymus serpyllum* from the Late Weichselian deposits at Roxburgh Street, Greenock, in the Clyde valley.

Calamintha ascendens Jord.   Common calamint   451: 2

lA *Gort*

From the beds at Gort just preceding development of the Hoxnian interglacial sequence, Jessen *et al.* (1959) recovered 12 small nutlets that exactly matched those of *Calamintha ascendens*. It is a southern central and south European species that has a strongly southern British range today, being absent from Scotland and the northern

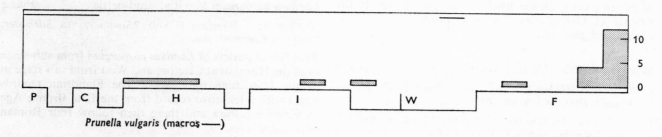

Prunella vulgaris (macros ——)

half of Ireland. Matthews places it in his Continental–southern element.

Acinos arvensis (Lam.) Dandy (*Calamintha arvensis* Lam., *C. acinos* (L.) Clairv., *Thymus acinos* L.)   Basil thyme
452: I

Cs *Pakefield*; H *Clacton*; *l*W ?*Hartford, Ballybetagh*; F VIII *Silchester*

Basil thyme (*Acinos arvensis*) has fruit records that cover a long period of time from the Cromer Forest Bed series to Roman Silchester. Between lie a record from sub-stage III of the Hoxnian interglacial at Clacton, a tentative Late Weichselian zone II record from Huntingdonshire and an Irish zone III record. *Acinos arvensis* is now a cornfield weed and ruderal whose present British range is strongly south-eastern: the fossil record suggests an ancient natural status for it.

Clinopodium vulgare L. (*Calamintha clinopodium* Benth.)
Wild basil
453: I

H *Clacton*; I *Wretton, Histon Rd*; *l*W *Ballybetagh*

Nutlets of *Clinopodium vulgare* are reported from sub-stage III of the Hoxnian interglacial at Clacton, from sub-stages II and III of the Ipswichian at Wretton and sub-stage III in Cambridge. At Ballybetagh Jessen has recorded it both from zone II and zone III of the Late Weichselian. It is a plant of dry calcareous soils common in England and Wales but limited in Ireland and Scotland to scattered south-eastern localities. It has a pattern of historical record shared by other Labiatae such as *Acinos arvensis* and *Origanum vulgare*.

Prunella vulgaris L.   Self-heal (Plate XVIII*n*)   457: I

The nutlets of self-heal, *Prunella vulgaris*, have been found in the Cromer Forest Bed series, at three different sites in sub-stages I, II and III of the Hoxnian interglacial, and again in sub-stage III of the Ipswichian. There is one Early Weichselian and one Late Weichselian record. Thereafter there is only a zone V record in the Flandrian before the numerous sites for zones VIIb and VIII, of which most have an archaeological origin. Thus there are six Roman and five mediaeval records. Some small tendency is shewn to local persistence as at Frogholt in successive layers of zone VIIb to VIII.

The historical evidence is suggestive of ancient native status for self-heal, and its present geographical range seems to make this not improbable. It occurs commonly throughout the British Isles, and on the oceanic west Norwegian coast it often occurs well into the Arctic Circle and throughout Finland extends frequently to 65° N; scattered localities are known throughout northern Lapland.

Betonica officinalis L. (*Stachys betonica* Benth.; *Stachys officinalis* (L.) Trev.)   Betony   458: I

F VIII *Mersea Is., Shrewsbury*

One Roman and one mediaeval site, both in the south of England, have yielded fruit records of *Betonica officinalis*. It is a plant that barely extends today into Scotland or Ireland.

Stachys sp.   459

H *Hoxne*; I *Bobbitshole*; *l*W *Hartford*; F VIIa *Loch Kinord*, VIIb/VIII *Branston*, VIII *Dublin, Shrewsbury*

Both Reid (tentatively) and West recorded nutlets of *Stachys* from sub-stage III of the Hoxnian at the type site and West similarly recorded it from sub-stage II of the Ipswichian. There is one Late Weichselian zone II record and several in the Flandrian, the two latest mediaeval.

Stachys arvensis (L.) L.   'Field woundwort'   459: 3

Cs *Pakefield*; H *Nechells*, I ?*Ilford*; F VIII *Silchester*

The earliest record of the nutlets of *Stachys arvensis* is from the Cromer Forest Bed series. Kelly found it in sub-stages IIb and c, IIIa and b of the Hoxnian at Birmingham and West made a tentative identification of it in sub-stage II of the Ipswichian. The only Flandrian record is from Roman Silchester.

It is interesting to have these records of natural interglacial status for this weed of arable non-calcareous soils.

Stachys germanica L.   'Downy woundwort'   459: 4

eWo ?*Hoxne*

C. Turner makes a tentative identification of the fruit of *Stachys germanica* from the late-glacial deposits preceding the interglacial itself at Hoxne. This is a species of extreme rarity in the British Isles now limited to the vicinity of Oxford: its European range is principally central and southern.

Stachys palustris   'Marsh woundwort'   459: 6

Cs *Beeston*; H *Hitchin, Clacton*; I *West Wittering*; *l*W *Hailes*; F vIIa *Lake Derravaragh*, vIIb *Itford Hill*

C. Reid is responsible for all the pre-Flandrian records of fruit of *Stachys palustris*, from the Cromer Forest Bed series, from sub-stages II and IIIa and b in the Hoxnian, from some unknown part of the Ipswichian and equally of the Late Weichselian in Scotland. The Irish Flandrian site is Mesolithic and the English one late Bronze Age. It is a historical pattern very like those of *Stachys arvensis, S. sylvatica* and the genus itself.

Stachys sylvatica L.   'Hedge woundwort'   459: 7

Cs *Pakefield*; C *Norfolk Co.*; *l*A *Baggotstown*; H *Baggotstown, ?Nechells, Hitchin, Clacton, Gort*; I *?Wretton*; F IV *Seamer*, vIIa/b *Shustoke*, vIIb *Cockerbeck, Frogholt*, vIII *Bunny, Caerwent, American Sq., Hungate*

The nutlets of *Stachys sylvatica* have been recovered from the Cromer Forest Bed series and the Cromerian interglacial itself. At Baggotstown it occurred in the late Anglian and also in the sub-stage II of the ensuing Hoxnian interglacial. There are two English records for this sub-stage, two for III and one for IV; albeit the Nechells records (IIc, IIIa and b) are tentative, as are West's records for sub-stages II and III of the Ipswichian. The Flandrian zone vIIb and vIII records all have an archaeological context: Neolithic I, Late Bronze Age I, Roman 3, Mediaeval I.

Stachys sylvatica is a west European–middle Siberian plant (Hultén) that reaches into the Arctic Circle on the Norwegian west coast. It occurs throughout Britain in semi-shaded habitats on moist and especially on rich soils so that it is often associated with settlements as in the historical record. There is good evidence of ancient native status but no direct indication of perglacial presence.

Ballota nigra L.   Black horehound   460: 1

Cs *Pakefield*; C *Norfolk Co.*; I *Histon Rd*, F vIII *Silchester, ?Caerwent*

Nutlets of *Ballota nigra* have been recorded in the Cromer Forest Bed series, the Cromerian, sub-stage III of the Ipswichian and twice (though once only tentatively) in Roman sites.

Lamium sp.   462

F vIIb *Minnis Bay*, vIII *Dalnagar*

Nutlets referred to *Lamium* were found in the Late Bronze Age deposits at Minnis Bay, and pollen from Flandrian zone vIII at Dalnagar.

Lamium purpureum L.   Red dead-nettle   462: 4

H *Clacton*; I *Wretton*; F vIIb *?Stuntney*, vIII *Silchester, Pevensey, Caerwent, Bunny*

Reid found nutlets of *Lamium purpureum* from sub-stage III of the Hoxnian at Clacton, and West from sub-stage III of the Ipswichian interglacial. The Flandrian records begin with a tentative record from the Late Bronze Age in Cambridgeshire and there then follow four Roman occupation sites.

Whilst the Flandrian records conform to the present habit of the plant favouring cultivated ground and waste places, the interglacial records shew a capacity for existence in a countryside far less, if at all affected by man.

Galeopsis sp.   Hemp-nettle   465

Macros: *m*W *Barrowell Green*; F vIIb *Wigton*, vIIb/vIII *Garscadden Mains*, vIII *Fifield Bavant, St Albans*
   Pollen: *Lindow Moss, Holcroft Moss*

The record of a nutlet of *Galeopsis* sp. from Barrowell Green should be seen alongside the Weichselian records for *G. tetrahit*. All the other *Galeopsis* records are from Flandrian zone vIIb or later. The Wigton vIIb record, like the Iron Age and Roman records by Helbaek, are given as *Galeopsis speciosa* or *tetrahit*. Pollen has been recorded by H. J. B. Birks from two Cheshire mosses at Norman and mediaeval levels respectively, in each instance with other indicators of cultivation.

Galeopsis angustifolia Ehrh. ex Hoffm.   'Narrow-leaved hemp-nettle'   465: 1

F vIII *Plymouth, Larkfield*

There are records from two mediaeval sites of fruits of *Galeopsis angustifolia*.

Galeopsis tetrahit L. *sensu lato*   Common hemp-nettle (Plate XVIIIo)   465: 4

Macros: *l*W *Twickenham, Flixton, Nazeing*; F IV *Star Carr, Nazeing*, vIIa *Elland, Cockerbeck*, vIII *Bermondsey, Silchester, Caerwent, Newstead, Lissue, ?Ballingarry Downs*
   Pollen: *l*W *?Esthwaite*; F vIIb *?Esthwaite*

Pollen tentatively ascribed to *Galeopsis tetrahit* has been recorded from the Esthwaite basin both in zone II and zone vIIb; all other records are for the nutlets. These have been found in Late Weichselian or early Flandrian at Flixton (zone II), Star Carr (zone IV) and Nazeing zones II/III and IV, and also at Twickenham, a site probably also of Late Weichselian age. The other records are later in the Flandrian, with Elland in zone vIIa, with one Neolithic, four Roman and two mediaeval sites.

It seems probable that the plant has been native since Late Weichselian time, having been favoured by the open soil conditions of that time as by those reintroduced by cultivation in a far later period. It is possible that the *Galeopsis* identified from the Full-glacial in the Lea Valley

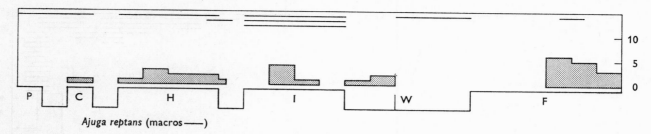

Ajuga reptans (macros ——)

might be this species also, and in this case it would have some claim to be reckoned as a perglacial survivor.

The tolerance of *G. tetrahit* for cold conditions is shewn by its common occurrence in cultivated ground in western Norway beyond 70° N, and its frequent occurrence right round the top of the Baltic Sea.

Nepeta cataria L.   Cat-mint   466: 1

*m*W ?*Hedge Lane*; F VIII *Bunny*

Nepeta cataria has been identified from the Roman site at Bunny, and there is a tentative record also from the Lea Valley Arctic Plant Bed by C. Reid who also made the record next given for *Glechoma hederacea*.

Glechoma hederacea L. (*Nepeta glechoma* Benth., *N. hederacea* Trev.)   Ground ivy   467: 1

*m*W *Hackney Wick*

The ground ivy, *Glechoma hederacea*, was identified from the Full-glacial Lea Valley Arctic Bed. Although now occurring in woodlands, hedges and copses in the British Isles, it also has a wide frequency on more open disturbed ground and upon such sites as grasslands. Hultén describes it as common in south Sweden and Denmark, locally throughout central Scandinavia and occasionally into the Arctic Circle.

Marrubium sp.   468

F VIII ?*Plymouth*

There is a tentative fruit record for *Marrubium* from mediaeval Plymouth.

Marrubium vulgare L.   White horehound   468: 1

I *Wretton*

Sparks and West identified nutlets of *Marrubium vulgare* from sub-stages II and III of the Ipswichian interglacial. It is a species of generally southern range and in the British Isles it is found mostly by roadsides and in waste places.

Teucrium scordium L.   'Water germander' (Plate XVIII*p*)
                                                   470: 2

I *Bobbitshole, Ilford*; F IV *Nazeing*

A single nutlet found in deposits at Nazeing, Essex was identified by characters of size, shape and surface-pitting

as *Teucrium scordium*. Subsequently its fruits were identified by West in sub-stage IIb of the Ipswichian interglacial both at Bobbitshole and at Ilford.

This is a plant very rare in the British Isles today, growing in a few wet calcareous localities scattered over England and Ireland. It is also found in Denmark, Germany, Belgium, Holland, France, Italy, Switzerland, eastern Europe and in Siberia. It is in Matthews' category of Continental–northern plants.

Teucrium scorodonia L.   Wood sage   470: 4

H *Gort*; F VII *Leasowe*

Nutlets of the wood sage, *Teucrium scorodonia*, have been recorded from sub-stage IV of the Hoxnian interglacial, and in the late Flandrian coastal peat beds of the Lancashire coast. This is a strongly sub-atlantic plant common throughout the British Isles but only present locally in the very south of Scandinavia.

Ajuga sp.   471

H *Wolvercote Channel*

Deposits of the Wolvercote Channel, possibly of late Hoxnian age, yielded a fruit record of the genus *Ajuga* (Bell, 1904; Duigan, 1956).

Ajuga reptans L.   Bugle   471: 2

Records for the nutlets of *Ajuga reptans* come from the Cromer Forest Bed series, the Cromerian itself, every sub-stage of the Hoxnian and the two middle sub-stages of the Ipswichian, as well as from less closely defined deposits. In the Flandrian the lateness of the records is remarkable, none being before zone VIIa, a fact the more remarkable in that the bugle is, to a considerable degree, a woodland plant unlikely to have been suppressed by the closed forest of the middle Flandrian. It seems probable that the *Ajuga* nutlet reported by Lewis from the basal wood peat in Flandrian zone VII on Cross Fell (600 m), along with *Fragaria vesca*, was *Ajuga reptans*. There is in these records an indication of response to milder climatic conditions that agrees with the geographical range of the species which within Scandinavia is frequent only in Denmark, though with scattered localities in Finland and round the Baltic to about 64° N, and found commonly throughout the British Isles.

By contrast we have one Early Weichselian record from

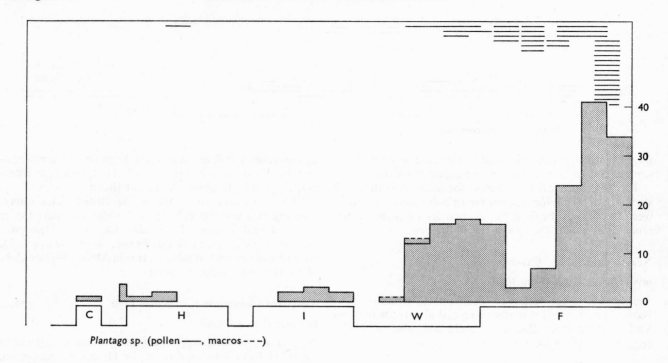

Plantago sp. (pollen ——, macros - - -)

TABLE 25. *Percentage frequencies of* Plantago *sp. pollen in British deposits*

					Percentages				
	+	0–2	2–5	5–10	10–20	20–30	30–40	40–50	
I	2	3	1	1	.	.	.	.	Of
II	1	4	3	1	.	.	.	.	total
III	.	9	1	.	.	.	.	.	dry-land
IV	2	4	.	.	.	.	.	.	pollen
IV	3	4	1	.	.	.	.	.	
V	1	1	.	.	.	.	.	.	Of
VI	2	5	1	.	.	.	.	.	total
VIIa	3	14	2	1	.	.	.	.	arboreal
VIIb	4	12	11	10	.	.	.	.	pollen
VIII	1	0	5	10	11	3	2	.	

Cambridge, two from the Middle Weichselian of the Lea Valley Arctic Plant Bed and one undefined Late Weichselian record from Hailes. These lead one to suspect perglacial survival of *Ajuga reptans*.

PLANTAGINACEAE

Plantago sp.   472

It was Iversen who in 1941 first drew attention to the feasibility of microscopic recognition of the pollen of Danish species of *Plantago*, and pointed out the very great indicator value which *P. major*, and more especially *P. lanceolata*, have for the progress of disforestation by prehistoric man from Neolithic times onwards. Pollen of the different north-west European species of *Plantago* can be recognized by the number of germ spores, the conspicuousness or faintness of the pore ring, and the vigour of the sculpture pattern on the grain surface.

The British fossil record of *Plantago* pollen referred only to the genus indicates that at the sites mentioned, besides pollen grains – referred definitely or tentatively to various species – others only referable to the genus have been noted.

The record-head diagram shews pollen of *Plantago* in the Cromerian, in sub-stages I, II and II/III of the Hoxnian, sub-stages II, III and IV of the Ipswichian and in every zone of the Flandrian. There are three pollen records from the late Anglian glacial stage, but none from Early and Middle Weichselian despite a seed record from the Middle Weichselian at Great Billing. This contrasts

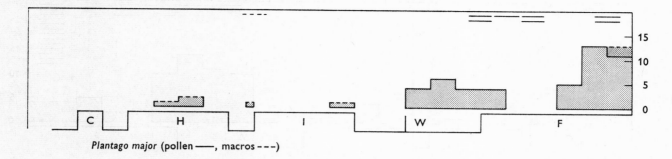

Plantago major (pollen ——, macros ---)

sharply with the abundance of records in the Late Weichselian, a reflection no doubt of the much greater numbers of pollen analyses of this period. The zone I records are backed by a seed-identification from Kirkmichael, Isle of Man. The records for zones II, III and Flandrian zone IV are still more frequent, and it is always assumed, no doubt rightly, that this reflects the response of *Plantago* spp. to the open vegetation and fresh soils that then prevailed. By the same token the sharply diminished numbers of records in Flandrian zones V and VI are taken to indicate the consequence of closure of forest cover, a suggestion gaining weight from the fact that sites in these zones are often of high altitude, such as Scionascaig, Red Tarn, the Nick of Curlywee or coastal sites such as St Kilda where the machair still supports *Plantago* swards. In zones VIIa, VIIb and VIII the site records expand to values bigger than those of the Late Weichselian in response to forest clearance and the spread both of pasture and arable cultivation. Table 25 sets out the mean *Plantago* pollen frequencies zone by zone. It will be seen that despite the numerous sites in zones II to IV the mean pollen frequency is low, from 0 to 2 per cent of the dry land plant total, and only an odd site registers as much as 5 to 10 per cent. This condition holds through to zone VIIa, though the frequency is given for the Flandrian as a percentage of total arboreal pollen, but in zone VIIb the mean frequency lies between 0 and 10 per cent whilst in zone VIII it is between 5 and 20 per cent with five sites shewing still higher frequencies. These figures reinforce the explanation already given for the increases in site record numbers, but emphasize the greater magnitude of the changes in zone VIII when the iron axe and ox-team ploughs had given weight to clearance activity.

One must of course remember that we are here dealing with a composite generic curve and that we might be concerned with a preponderance of one species at one period, and of another at a different time, as indeed the records of the individual species seem to shew.

Plantago major L. 'Great plantain', Waybread 472: 1

Pollen records for *Plantago major*, aside from a few queried identifications not included in the record-head diagram, are all Late Weichselian and Flandrian, follow-

ing in their frequency the pattern shewn by the genus, but we have to bear in mind the far more numerous pollen records given as '*Plantago major* or *media*' discussed in the next account and expressed in a separate record-head diagram. Seed identifications confirm the presence of waybread in the Hoxnian and Ipswichian, as in the late Wolstonian, and its transition to the Ipswichian. Likewise there are seed records from zone VIII at the Roman site of Bunny and the mediaeval site at Shrewsbury.

It is apparent that *Plantago major* is a species of ancient natural status in this country. It has so close a connection with man-made habitats that Iversen wrote of its Danish occurrences: '*P. major* is so closely associated with places with direct cultural influences that it would doubtless disappear from our flora if the culture ceased', but the historical record in Britain is now so continuous that it can be seen to have persisted here at least from the beginning of the Late Weichselian, and was able likewise to exist in past interglacial periods when man's influence was minimal. Pennington (1969) reminds us that *P. major* can endure much trampling, and, as its name implies, follows trackways, doubtless expanding its range as these were progressively extended by early man into the native vegetation.

P. major is naturalized over most of the world.

Plantago major L. or **media** L.   472: 1, 2

The pollen grains of *Plantago media* L. and of *Plantago major* L. have pores indistinctly ringed, and a very vigorous surface sculpture. They are not easily separated from one another and accordingly the identifications are often made as alternative between the two. The result, expressed in the separate record-head diagram, repeats and confirms the historical pattern shared by the two species individually, but it also strengthens the evidence for the prevalence of these plantains in the Late Weichselian and early Flandrian, whilst two Middle Weichselian records indicate the possibility of their full perglacial survival. The large number of composite zone VIIb/VIII records comes from the difficulty, found in Scottish sites particularly, of separating the two zones.

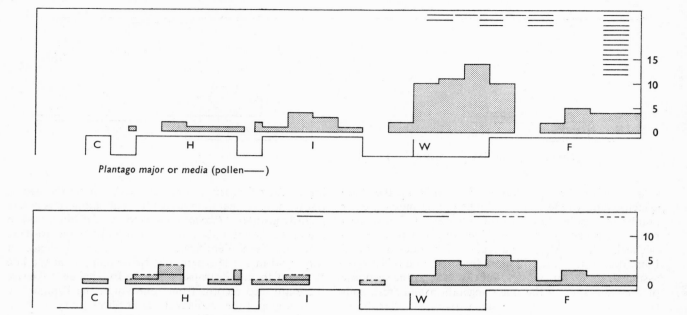

Plantago major or media (pollen———)

Plantago media (pollen ———, pollen cf.- - -)

Plantago media L. 'Hoary plantain' (Plate XXI*p*)

472 : 2

The record-head diagram for pollen of *Plantago media* includes a rather high proportion of grains only tentatively identified but it is perfectly clear from it that this plant has been present, like *P. major*, from the beginning of the Late Weichselian, as well as occurring in earlier interglacials including the Cromerian. The records for early and late glacial stages are mostly tentative except for the Weichselian.

Plantago media is a plant of neutral and basic soils occurring fairly generally throughout southern England and the Midlands, rarer in the north and regarded as introduced in Ireland. *P. media* grows in most regions of Europe except northern Scandinavia, in Siberia, Persia, and the Caucasus. The difference in its British range from that of *P. major* is not reflected in the historical pattern, but it will be noted that it shews none of the expansion of *P. major* in zones VIIb and VIII, no doubt because it is so much less a plant of arable ground. The persistence that it shews in Flandrian zone v is associated with records from several high altitude sites less affected by afforestation than the lowlands.

The earliest known association of plantain with forest clearance is in sub-stage II of the Hoxnian, where West at Hoxne and later Turner at Marks Tey found *Plantago* pollen along with that of other indicators in a vegetational phase extremely similar to that of a Neolithic clearance, though whether caused by Acheulian man who was locally present then or to natural causes remains unknown. This plantain pollen is identified by West as 'cf. *media*' and by Turner as '*media* type' and '*maritima*' (West, 1956; Turner, C., 1970).

Plantago lanceolata L.   Ribwort (Fig. 116)

472 : 3

The pollen grains of *Plantago lanceolata* L. are recognizable among those of other plantains by having a large number of pores (10 to 14), each surrounded by a thickened, projecting ring. Iversen (1941) has established beyond question the significance of the pollen curve of this species as indicator of the progress of disforestation by prehistoric man in Denmark, and in other countries of north-west Europe his results have been amply confirmed. Thus at Hockham Mere in the northern Breckland of East Anglia the creation of the Breckland heaths from former forest in Neolithic and later time was established by the pollen curves typical of these great heath communities, and the curve of *P. lanceolata* pollen was found to run closely parallel with theirs (Godwin, 1944) (see fig. 165). Similarly at Decoy Pool Wood on Shapwick Heath in the Somerset Levels (fig. 168) it was shewn that the pollen curves for weeds of cultivation and for cereal pollen could be used to reflect the variation in cultivation intensity from the Late Bronze Age to Late Roman time and the *P. lanceolata* pollen curve ran closely parallel with these (Godwin, 1948). Since then there have been a great many exercises directed towards tracing the character and progress of disforestation in the British Isles, and all of them have relied substantially upon the evidence of pollen of the ribwort plantain (see chapter VI: 14). The indicator value of *P. lanceolata* naturally rests upon the facts that it is so commonly associated today with pasturage and cultivation, is so intolerant of competition by woody plants and such a prolific producer of windborne pollen.

The record-head diagram bears witness to the dramatic increase in site numbers from below ten in early and

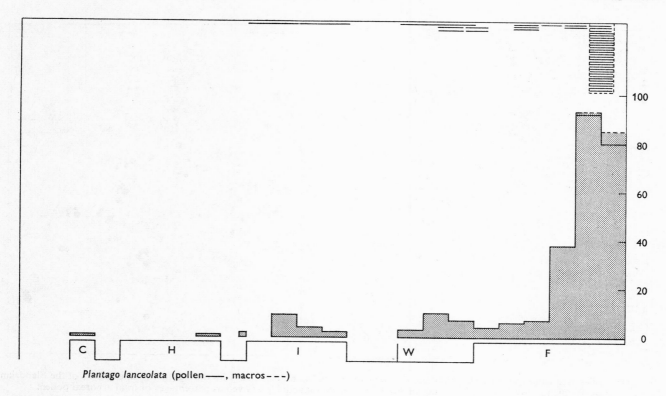

Plantago lanceolata (pollen——, macros - - -)

TABLE 26. *Pollen frequency of* Plantago lanceolata *in British deposits*

	+	0–2	2–5	5–10	10–20	20–30	30–40	40–50	50–60	60–70	
				Percentages							
I	4		1	.	.	.	.	.	.	.	Of
II	7	1	.	.	.	.	.	.	.	.	total
III	6	.	.	.	.	.	.	.	.	.	dry-land
IV	.	2	.	.	.	.	.	.	.	.	pollen
IV	.	1	.	.	.	.	.	.	.	.	
V	3	1	.	.	.	.	.	.	.	.	Of
VI	1	4	.	.	.	.	.	.	.	.	total
VIIa	6	18	.	.	.	.	.	.	.	.	arboreal
VIIb	7	29	23	9	4	1	.	.	.	.	pollen
VIII	7	4	19	33	10	3	2	1	1	.	

middle Flandrian to eighty, ninety or more in zones VIIb and VIII, a feature emphasized by the composite zone figures and by five sites for seed identifications, two late Bronze Age, one Roman and one mediaeval. It had at one time been thought, in Denmark at least, that *P. lanceolata* came into the country with agriculture, but we have a Late Weichselian capsule from Scotland and the consecutive pollen records now establish that it was present in every zone of the Late Weichselian in this country and remained through each zone of the Flandrian to the present day. Its fortunes throughout this period can be better realized by considering the table of its pollen frequency zone by zone, as set out in table 26.

These results are striking in two respects. First in that the pollen frequencies in the Late Weichselian are very low, indeed below those for the genus and for *P. maritima* (see p. 331). *P. lanceolata*, though present, was not abundant. In the second place, as with the generic record, as site frequencies increase in zones VIIb and VIII (see fig. 116) so do the pollen frequencies at those sites so that the mean pollen frequency was as high as 5 to 10 per cent of total tree pollen and at some sites it was far higher. *P. lanceolata* is widely distributed throughout the British Isles in open habitats and waste ground and is a highly constant member of some grassland communities: these figures indicate how it came to occupy this position.

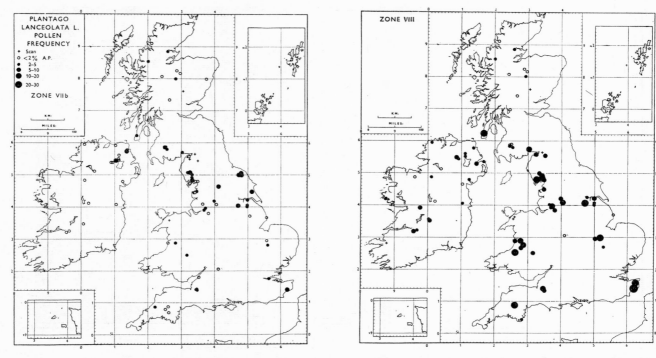

Fig. 116. Pollen frequency of the ribwort plantain, *Plantago lanceolata* L., in the British Isles during the two latest zones of the Flandrian when woodland were being opened and replaced by agricultural activities. Given as percentages of total arboreal pollen.

As would be expected, *P. lanceolata* figures in a very wide range of palynological publications and from which we may only mention a few leading issues. West (1969*a*; Sparks & West, 1970) has pointed to the occurrences of *P. lanceolata* pollen, sometimes in high frequency, in the forest-dominated stages of the Ipswichian interglacial: this, with the presence of other open ground indicators such as *P. media/major* and *Artemisia*, he relates to the aggradation of silt and sand in the river valleys, a clear indication of a mid-interglacial refugium for such species. This accords with the comment by Groenman van Waateringe that *P. lanceolata* is not necessarily indicative everywhere of man's activities since it still grows in communities of the *Agropyro–Rumicion crispi* alliance that occurs in naturally unstable contact zones such as coasts and river-sides. All the same *P. lanceolata* is a most consistent indicator of early forest clearance and as such accompanies the elm decline in most places. Following the initial work by Miss Turner (1964), Pilcher (1970) has sought with the aid of close radiocarbon dating to establish the duration of the phases of early woodland clearance at Beaghmore and Ballynagilly. After arrival about 2870 B.C., at both sites, *P. lanceolata* seems to have persisted about 200 years before restoration of forest cover came with abandonment of the early Neolithic settlement. Miss Turner (1964), by the collection of surface pollen samples in regions of preponderant arable and pastoral agriculture, established that high frequencies of *Plantago* and *Gramineae* pollen were highly typical of the latter, and

she used this as an index of the changing historical pattern of land usage in Cardiganshire, where she shewed a phase of pastoral farming between the opening of the Iron Age (*c.* 400 B.C.) and the establishment of the monasteries (*c.* A.D. 1200) which introduced widespread arable farming. Certainly the *Plantago* species most concerned was *P. lanceolata*.

In western Brittany van Zeist (1964) has found pollen of *P. lanceolata*, not only at the elm decline, dated to *c.* 3500 B.C., but considerably earlier, and although this could be due to the presence of Neolithic farmers at a very early date, he recognizes that 'it cannot be ruled out that in this region *P. lanceolata* could persist locally during the Post-glacial', a suggestion amply supported by the British record, and by van Zeist's own record of a single pollen grain from the Late Weichselian in west Brittany.

Plantago maritima L. 'Sea plantain'   472:4

The pollen grains of *Plantago maritima* have a faint surface sculpture and few pores, these with very indistinct rims, and their recognition in recent years had produced highly significant information. Thus it is evident from the record-head diagram that *Plantago maritima* has been found in every glacial and interglacial sub-stage or zone from the late Anglian to the present day. One could hardly wish for firmer evidence of persistence through all the vagaries of the Pleistocene climatic cycles, especially in view of the fact that in the Middle Weichselian four sites have

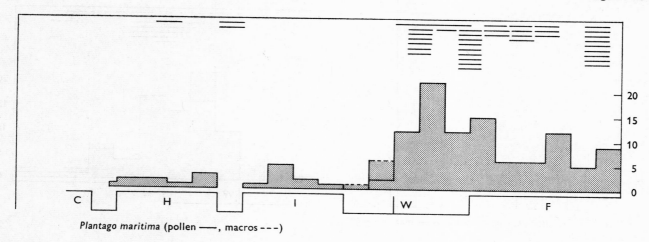

Plantago maritima (pollen ——, macros - - -)

yielded seeds of *P. maritima* and that at Earith and Marlow not only seeds but capsules were found. There is also a tentative seed identification in the Early Weichselian at Histon Road, Cambridge whilst at Marlow the macroscopic finds were accompanied by *P. maritima* pollen in high frequency. The straightforward pollen identifications have yielded numerous site records in all zones of the Late Weichselian in low but significant frequencies and it forms part of the long list of facultative and obligate halophytes that is so characteristic of the Weichselian period (Bell, 1968). No doubt persistence of *P. maritima* has been strongly associated with its preference for coastal habitats and we can envisage it as moving with the transgressing or regressing coastline of the great eustatic changes of ocean level. In the middle Ipswichian it occurred in marine situations at Selsey and Stone, and in the late Hoxnian at Shoeburyness. Iversen (1941) drew attention to its presence in marine muds right through the Atlantic period in Denmark, and we find it in similar situations in this country, as for instance in estuarine mud at the site of the late Bronze Age sewn boats on the Humber in zone VIIb.

Coastal sites were not however the only refugia of the species from the spread of forest in the Flandrian for it

has persisted to the present as inland mountain populations. Strong support for this view is given by the presence of *P. maritima* pollen throughout zones IV to VIIa at Cwm Idwal, where the species still grows above the high altitude lake whose sediments registered the past pollen rain. Examination of the pollen frequency of *P. maritima* zone by zone (table 27) shews that its contribution throughout the Flandrian, as in the Late Weichselian, remained low, generally below 2 per cent of the the total dry land or total tree pollen, but nevertheless, judged both by site numbers and by pollen frequency, it must have been the most important of the plantains during the Weichselian stage, having both the climatic and edaphic tolerances needed to take advantage of the absence of competition from taller closed vegetation. It exhibited neither of the great increases in site numbers and pollen frequency in the late Flandrian zones which characterized *P. lanceolata* and it was evidently, and not surprisingly, unaffected by the widespread forest clearances of that time.

P. maritima occurs round practically the whole coast of the British Isles and in mountain habitats in North Wales, the Pennines and Scotland. On the European mainland it extends in coastal habitats from the North Cape to Spain and occurs rarely in inland situations.

TABLE 27. *Pollen frequency of* Plantago maritima *in British deposits*

	Percentages					
	+	0–2	2–5	5–10	10–20	
I	3	7	1	.	.	Of
II	8	10	.	.	.	total
III	5	5	.	.	.	dry-land
IV	3	4	2	.	.	plants
IV	2	2	.	.	.	
V	2	2	.	.	.	Of
VI	1	3	.	.	.	total
VIIa	4	6	.	.	.	arboreal
VIIb	2	1	1	1	1	pollen
VIII	5	3	2	.	.	

Plantago coronopus L.   Buck's-horn plantain   472: 5

*l*W *Haweswater, Moorthwaite Moss*; F VI, VI/VIIa *Moorthwaite Moss*, VIIa *Scaleby Moss*, VIIa,b *Bowness Common*, VIIb *Beaghmore, Nichol's Moss, Brandesburton, Pilling Moss*, VIIb, VIII *Thorne Waste*, VIIb/VIII *Aros Moss, Slapton Ley*, VIII *Abbot Moss, Nant Ffrancon, Drake's Drove*

The pollen of *Plantago coronopus* is distinguishable from that of other species of *Plantago* by its infrequent pores (5–7) and the fact that they have a distinct projecting rim.

Aside from two records for zones I and II of the Late Weichselian all records of pollen of *Plantago coronopus* are Flandrian. They often shew much local persistence, as at Moorthwaite Moss, Bowness Common, Beaghmore and Thorne Waste. The majority of records from zones

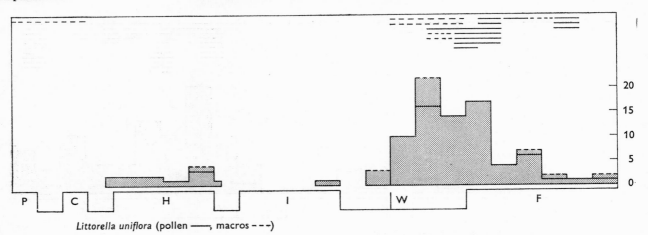

Littorella uniflora (pollen ——, macros - - -)

VIIb or VIII come from pollen spectra shewing indications of agricultural activity.

P. coronopus is essentially a maritime species and many of the record sites are close to the coast, but others such as Thorne Waste are in inland situations where the plant is favoured by the light sandy soils of the region. Other inland sites such as Nant Ffrancon or Beaghmore are at high altitude. The pollen record as a whole certainly suggests persistence from the Late Weichselian though more records for that period are desirable.

P. coronopus is regarded by Matthews as a component of the Continental–southern element of the British Flora and Dodds (1953) emphasizes its prevalence in the Mediterranean region. Hultén regards it as a sub-atlantic plant and shews it as having a natural distribution restricted to the southern coasts of Sweden and of Denmark. It occurs throughout most of the British Isles although absent from some inland counties.

Littorella uniflora (L.) Aschers. (*L. lacustris* L.) (Fig. 117; plate XXIV*i, j*)   473: I

The pollen of *Littorella uniflora* is as recognizable as the one-seeded 'bony' fruits, and both contribute substantially to the historical evidence. The record-head diagram shews fruit from the Cromer Forest Bed series, and there are records from every sub-stage of the Hoxnian interglacial. In the Ipswichian it is only recorded from sub-stage IV, possibly a consequence of the fluviatile character of most of the investigated deposits. In all Flandrian zones there are pollen records with a thin scatter of fruit records. In the glacial stages there are records in the late Anglian and early Wolstonian, but the Weichselian records are abundant with numerous pollen site records in all the Late Weichselian zones (and Flandrian zone IV), and thirteen fruit records, three in the Middle Weichselian. Many of these fruit records are

Fig. 117. Nut of *Littorella uniflora* (L.) Aschers. from zone VI, Ireland. (After Jessen, 1949.)

from Ireland. It seems evident that *Littorella* was prevalent in the shallow lakes of the Weichselian period and that it has persisted in suitable situations since then. No doubt the decrease in the Flandrian was associated with the natural infilling and conversion of such waters to fens or mosses.

Littorella uniflora occurs throughout the British Isles though more abundantly in the north. It has been placed by Matthews in the category of Continental–northern plants and by Hultén in the sub-atlantic group, although Jessen points out it has a northerly range for a sub-atlantic species. It is fairly frequent in Denmark and southern Sweden and is found locally and with diminishing frequency northwards to the Arctic Circle: its northern limit lies between 65° and 70° N.

CAMPANULACEAE

Pollen: C *West Runton*; H *Marks Tey*; eWo *Marks Tey;* F IV *Crane's Moor*, IV, V/VI *Black Lane*, VIII *Fallahoghy*

Campanula sp. (Plate XXII*a, b*)   475

The record-head diagram for *Campanula* is largely based on unqualified records of the genus but there are also a few records (H. J. B. & H. H. Birks) of '*Campanula type*' which comprises five species of *Campanula* and two of *Phyteuma*. It is evident that the genus occurred in both the Hoxnian and Ipswichian interglacials and in every zone of the Flandrian. There are two Middle Weichselian pollen records (backed by a seed record from Barrowell Green), but it is in the Late Weichselian, especially zone II, that site records are most frequent. There is also one seed record from the Weichselian *tout simple*. Site numbers remain high in zone IV but remain low thereafter through the Flandrian, a possible hint that the main contributors of zones I to IV were not the woodland species of the genus, but plants of open habitats.

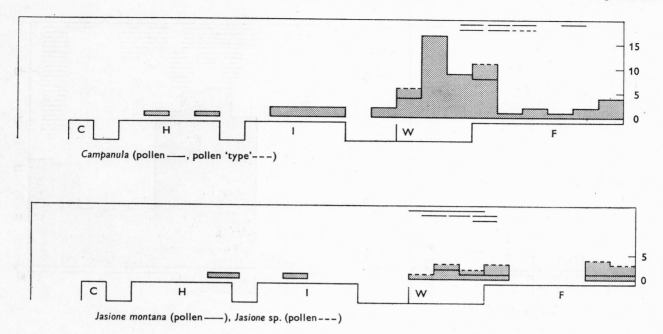

Campanula (pollen ——, pollen 'type' ---)

Jasione montana (pollen ——), Jasione sp. (pollen ---)

Campanula rotundifolia L. Harebell, Bluebell (in Scotland)
(Plate XX*h*)   475: 7

Macros: A *Corton Beds*; *e*Wo *Marks Tey*; W *Thrapston*; *m*W *Barnwell Station, Upton Warren, Earith, Syston*; *l*W ?*Kirkmichael*, ?*Nazeing*; F viib *Minnis Bay*

Pollen: *l*W ?*Llyn Dwythwch*, ?*Nant Ffrancon*, ?*Blelham Tarn*; F viii ?*Ehenside Tarn*

Miss Bell (1968) describes the criteria upon which seeds of *Campanula rotundifolia* may be identified and she has been responsible for all the Middle Weichselian records cited here, having revised Miss Chandler's record for Barnwell Station of *Gentiana cruciata* to *C. rotundifolia*. She reports these seeds as common in the Middle Weichselian deposits and in the undefined Weichselian at Thrapston, at which site she records also some seeds 'which approached the long narrow seeds of *Campanula uniflora*', a non-British species, a circumpolar arctic–montane plant of the Scandinavian mountains which would not be unexpected in the British periglacial scene. Seeds of *C. rotundifolia* were found also in the Corton interstadial of the Anglian glacial stage and in the early Wolstonian. Similarly, in the Late Weichselian seeds were tentatively recorded in zones i/ii and iii at Kirkmichael, whilst from zone ii/iii at Nazeing in the Lea Valley is reported a seed either of *C. rotundifolia* or *C. patula*. In the late Bronze Age shore peat at Minnis Bay there is the only Flandrian seed record and that a tentative one.

All pollen records for *Campanula rotundifolia* are qualified. They include Late Weichselian zone ii from Llyn Dwythwch, zones ii and iii on the other side of Snowdon in the Nant Ffrancon, and zone iii at Blelham Tarn.

The historical record gives a clear indication that *Campanula rotundifolia* was common in the British Isles during the last glacial stage, and probably was so in earlier ones. Evidence for persistence through the Flandrian is lacking, likely as this may be.

Campanula rotundifolia is a north temperate plant of almost continuous boreal–circumpolar distribution which occurs commonly throughout Scandinavia, even to the North Cape and Kola Peninsula. It is common also throughout the British Isles and no doubt owes some of its ubiquity to its indifference towards the lime content of the soil: it occurs equally on acid sands and limestones.

Legousia hybrida (L.) Delarb. (*Campanula hybrida* L.; *Specularia hybrida* (L.) A.DC.)   Venus's looking-glass
476: 1

*l*W ?*Elstead*

A tentative identification of a pollen grain of *Legousia hybrida*, a species of decidedly southern range, was made by Seagrief from zone iii of the Late Weichselian in Surrey.

Jasione montana L.   Sheep's-bit (Plate XXI*e,f*)   479: 1

Pollen of *Jasione* closely resembles that of the genus *Campanula* but is substantially smaller and fossil grains attributed to this genus are generally below 20 μm in diameter. There is only one British species and palynologists give their records as '*Jasione montana*', '*Jasione* sp.', or '*Jasione* type'. They are put together in the one diagram, which may reasonably be regarded as for *J. montana*. There is only one Hoxnian and one Ipswichian interglacial record, the former followed by a single late Wolstonian occurrence. The Late Weichselian records are more numerous and include various instances of local persistence, as at Chat Moss (zones ii and iii), Elan

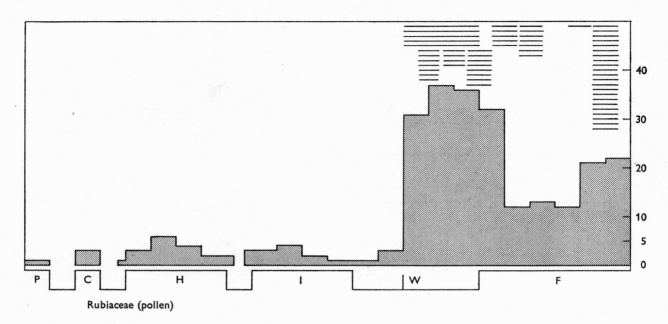

Rubiaceae (pollen)

Valley (zones I/II, II, II/III, III/IV, IV) and Old Buckenham Mere (*l*W and Flandrian VIIb and VIII). The last mentioned site is of interest as the *Jasione* present in the Late Weichselian reappears in zone VIIb at Roman and Anglo-Saxon levels along with weeds typical of arable cultivation on sandy soils.

The records are spread widely across the British Isles and indicate prevalence in the Late Weichselian and possible persistence subsequently.

Jasione montana is a European and Mediterranean species with a distinctly southern range in Scandinavia terminated northwards by scattered sites in Finland: it ascends to 1200 m in the Alps. It occurs locally throughout the British Isles on 'light sandy or stony lime-free soils, in rough pastures, on heaths, cliffs and banks' (Clapham, Tutin & Warburg, 1962).

RUBIACEAE

The stephanocolpate pollen grains of the Rubiaceae, with their six to ten long narrow furrows with unthickened margins, are highly distinctive and have been found in British Pleistocene deposits from the Antian onwards. They have been recorded in all sub-stages of the Hoxnian and Ipswichian interglacials and in all zones of the Flandrian. They occurred in the Baventian, late Antian, early and late Wolstonian, and similarly in Early, Middle and Late Weichselian. They are characteristically abundant however throughout all zones of the Late Weichselian and the opening zone IV of the Flandrian. Their frequency at this time is clearly related to the generally open character of the vegetation known to have been prevalent then, and the curve has now acquired a certain indicator value dependent upon this in interpreting the results of pollen analyses. It is likely that the bulk of the records come from the genus *Galium*, for which a separate record-head diagram has been given to correspond to the published records despite the very large overlap in significance stemming from the more or less alternative use of the two categories.

The considerable increase in site frequency in zones VIIb and VIII of the Flandrian must certainly be associated with disforestation and cultivation.

Sherardia arvensis L.   Field madder   481 : 1

F VIIb *Itford Hill*, VIII *Aldwick Barley*

In reporting the Late Bronze Age identification of fruit of madder from Itford Hill, Helbaek (1953*b*) comments that this is apparently the earliest European record of the plant. It was cultivated as a dye plant by the Greeks and Romans, and behaves in this country as a typical cornfield weed and ruderal. The Aldwick Barley record is from an Iron Age site. It is a plant now widespread throughout the British Isles.

Galium   485

The comments made for pollen recorded as Rubiaceae apply, as we have indicated, to the genus *Galium*, the separate record-head diagram for which displays a similar relative abundance in the Late Weichselian and an increase (relatively smaller) in zones VIIb and VIII. It is to be noted that of the six zone VIII records for fruits referred to '*Galium* sp.', five come from archaeological sites, one Iron Age, three Roman and one mediaeval. The identifications of fruits of *Galium verum* and of *G. aparine* that are recorded below shew the same concentration in archaeological sites.

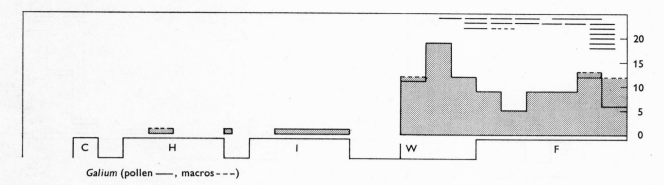

Galium (pollen ——, macros - - -)

Despite the relative infrequency of fruit records, the Late Weichselian has produced evidence for the presence in the British Isles of no less than four species, *Galium mollugo, G. verum, G. uliginosum* and possibly *G. aparine.*

Galium boreale L.   'Northern bedstraw'   485: 2

Cs ?*Beeston*

Galium mollugo L.   'Great hedge bedstraw'   485: 3

*l*W *Nant Ffrancon*

The identification of a fruit of *Galium mollugo* from the Allerød interstadial in North Wales is interesting in view of the fact that this species is far commoner in England, especially southern England than elsewhere in the British Isles.

Galium verum L.   Lady's bedstraw   485: 4

*l*W *Knocknacran*; F VIII *Silchester, Caerwent*

Fruits of *Galium verum* have been identified from the Allerød period in Ireland and from two Roman sites in England and Wales respectively.

Galium saxatile L. (*Galium hercynicum* Weigel) 'Heath bedstraw'   485: 5

F *Boskill*

This record of the Irish Atlantic period, based upon two mericarps covered with small acute tubercles, is the only evidence for the past history of this calcifuge plant now so widespread throughout the British Isles.

Galium palustre L.   'Marsh bedstraw'   485: 8

F *Elan Valley, Silchester*

Fruits of *Galium palustre* have been identified from the Flandrian zone IV/V transition at high altitude in Wales, and from the Roman site at Silchester.

Galium uliginosum L.   'Fen bedstraw'   485: 10

*l*W *Newtownbabe*

A small fruit of *Galium* from the Late Weichselian zone II at Newtownbabe has been assigned by Mitchell to *G. uliginosum,* a plant of wide European and British occurrence, somewhat rare however in Ireland. In Scandinavia it remains frequent well within the Arctic Circle.

Galium aparine L.   Goosegrass, Cleavers   485: 12

Cs *Pakefield*; *l*W ?*Nazeing*; F VIIb *Itford Hill*, VIII *Lidbury Camp, Winklebury Camp, Silchester, Larkfield, Dublin*

Galium aparine fruits have been recognized in the Cromer Forest Bed series, and tentatively from the Late Weichselian zone II/III in Essex. Thereafter every record is from an archaeological site, and these embrace the Late Bronze Age, the Iron Age and Roman and mediaeval periods. The plant which is abundant throughout the British Isles is now strongly associated with man-made habitats, but it also occurs in such natural situations as scree and coastal shingle. In Scandinavia it extends as far north as 69° N latitude.

Galium spurium L. (*G. vaillantii* DC.)   'False cleavers'
                                                   485: 13

F VIIb *Itford Hill*

In identifying a fruit of *Galium spurium* from the Late Bronze Age site at Itford Hill in Sussex, Helbaek (1953*b*) makes no qualification. It is a rare plant of arable fields in the southern half of England. It has been found from Neolithic and Bronze Age strata in Switzerland and from Iron Age deposits in Bornholm and Gotland in the Baltic.

CAPRIFOLIACEAE

Sambucus ebulus L.   Danewort   487: 1

F VIII *Godmanchester*

When Lambert (unpublished) identified the seed of *Sambucus ebulus* from the Roman site at Godmanchester it was the first and only record of a species of doubtfully native status found by road-sides and in waste places throughout most of the British Isles (save north-western

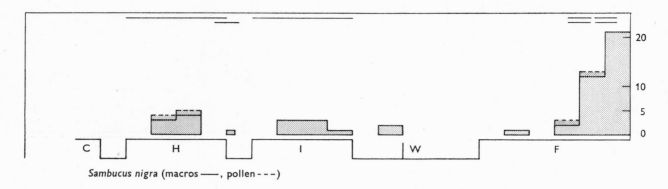

Sambucus nigra (macros ——, pollen - - -)

Scotland) but not encountered in natural plant communities. It is a plant of central and southern Europe, occurring only southwards of the Netherlands, and belongs to a considerable list of southern species only represented in this country (or first represented) by Roman records.

Sambucus nigra L.   Elder   487: 2

The macroscopic identifications of *Sambucus nigra* are mostly for seeds, but include also wood and charcoal referred merely to '*Sambucus*' but placed here in *S. nigra* as the only woody species presently native. It is apparent that the elder was present in the middle sub-stages of both Hoxnian and Ipswichian interglacials, and must have been locally abundant in the former, where at Birmingham its pollen attained 2 to 5 per cent of the total tree pollen. We may suspect that Reid's record for the end of the Hoxnian is due to derivation (with many other plant remains) in the brecciated muds of that time, and West specifically suggests that his 'early' Early Wolstonian record is due to derivation: we may so regard also the Ponders End and Angel Road records for the Lea Valley Weichselian. As with *Thelycrania* so here also the earliest Flandrian record is from the Dogger Bank, and although present in zone VIIa it is not frequent until zones VIIb and VIII. Throughout the Flandrian the records are strongly associated with archaeological situations, thus the records are Mesolithic 2, Neolithic 3, Bronze Age 6, Iron Age 6, Roman 8, Anglo-Saxon and later 4. It is of course possible that the fruits had been collected there for food and the species has furthermore a strong association with nitrophilous and broken soil habitats found with human settlements.

Sambucus nigra has a sub-atlantic distribution with a Scandinavian area rather severely restricted to the south although in western Norway it occurs infrequently as far north at 63° N latitude. It occurs throughout the British Isles except the extreme northernmost Scottish islands and is found in some natural woodland situations. As with *Thelycrania sanguinea* the fossil record seems to point towards late Flandrian expansion without good evidence of perglacial survival.

Sambucus nigra or **S. racemosa**   487: 2 or 3

H *Clacton*

The identification of two seeds from the middle Hoxnian as '*S. nigra* or *S. racemosa*' (Turner, 1970) reminds one of the possibility that *S. racemosa*, not now native in Britain, might well have been so in earlier interglacials: it is quite common in southern Scandinavia.

Viburnum sp.   488

Macros: F ?*Whitehawk Camp*
    Pollen: C ?*West Runton*; H *Kildromin*; F IV *Dogger Bank*, IV, V, VI *Bigholme Burn*, IV, V/VI *Duartbeg*, VIIa, b, VIII *Old Buckenham Mere*, VIIb *Bell Track*, ?*Redbourne Hayes*, ?*Short Ferry*, VIIb/VIII *Loch Creagh, The Loons*, VIII *Slapton Ley*

There are tentative records of charcoal of *Viburnum* from the Neolithic Whitehawk Camp in Sussex. This apart, all the records are for pollen and it has to be recognized that many investigators, especially the earlier, have hesitated to identify beyond the genus whereas others have recognized pollen of *V. opulus*, the commoner and more widely ranging of the two British species (see following account, 488: 3). The pollen records indicate presence in two older interglacials and a continuous Flandrian occurrence begins with three zone IV records, one from the Dogger Bank and two from Scotland. All three of these sites shew local persistence, the Dogger record to zone V, Duartbeg to zone V/VI and Bigholme Burn to VI; later on there is recurrence at Old Buckenham Mere in zones VIIa, VIIb and VIII and one may speculate whether this may not be due to persistence of *V. opulus* in fen woods near the lake. The records only become frequent in zones VIIb and VIII and are commonly associated with forest clearance. The Scottish records, not only from Galloway but from Perthshire and the Orkney, must surely represent *V. opulus*.

Viburnum lantana L.   Wayfaring tree   488: 1

Macros: H *Clacton*; I *West Wittering, Wretton*
    Pollen: H *Marks Tey*

All our records for *Viburnum lantana* are from interglacial deposits. The Hoxnian is represented by a seed

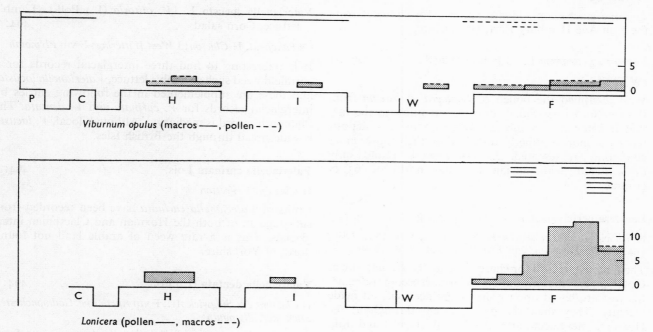

Viburnum opulus (macros ——, pollen - - -)

Lonicera (pollen ——, macros - - -)

identification from sub-stage IIIa at Clacton, and by pollen identifications made from sub-stage II (divisions a, b and c) and sub-stage III (divisions a and b) by Turner at Marks Tey.

V. lantana has a very strong south-easterly restriction of range in the British Isles and it is interesting that all the records come from the warm–temperate middle sub-stages of the interglacials.

Viburnum opulus L.   Guelder rose   488: 3

Seeds of *Viburnum opulus* have come from the Cromer Forest Bed series, sub-stages III and IV of the Hoxnian interglacial and sub-stages I and III of the Ipswichian. There are two Irish records from the Flandrian zones VI and VIIa, the later a Mesolithic site at Lake Derravaragh. All subsequent records are from archaeological sites, viz. charcoal from the Neolithic Whitehawk Camp, wood from a Late Bronze Age wooden trackway on Shapwick Heath, seed from the Iron Age Glastonbury Lake Village and seed from the Roman site at Crossness. It must not be thought that the fruits were collected for food for they are made most unpalatable by the valerianic acid they contain and are neglected even by birds.

The pollen records at Marks Tey and Clacton substantiate the Hoxnian seed records, and Mrs Birks has produced recurrent records for sites in Galloway through the Flandrian: thus at Loch Dungeon *V. opulus* is recorded in zone III/IV, IV/V, V/VI, VIIa and VIIb/VIII and at Cooran Lane in zones IV and V. In addition she has another high altitude record from the Cairngorm in zone V/VI. Her husband has produced a Late Weichselian zone II pollen record from Loch Meodal in Skye and there is also a tentative zone VIIb record from the Esthwaite Basin.

The fossil record as it stands, but especially taken with the evidence for '*Viburnum* sp.', rather strongly suggests long persistence of the guelder rose, especially in Scotland: we may expect this record to be much supplemented by future work.

Viburnum opulus is one of Hultén's 'boreal–circumpolar plants with incomplete range': it grows commonly in Scandinavia to 63° N and sparsely as far as the Arctic Circle. In the British Isles *V. opulus* is generally common except in Scotland where nevertheless it is not infrequent and is found up to Caithness.

Linnaea borealis L.   'Linnaea'   490: 1

lW Elan Valley, Loch Tarff

Pollen of *Linnaea borealis* has been recovered from zone III Late Weichselian deposits at high altitude in Cardiganshire and in Sutherland. It is a plant common to the extreme north of Scandinavia: in Britain it is rare and limited to a few areas of eastern Scotland ascending to 2400 ft (730 m). It is a boreal–circumpolar species of continuous boreal–montane range in Europe (Hultén): the Welsh record points to its Flandrian withdrawal from the south of Britain.

Lonicera sp.   Honeysuckle (Plate XXVc, d)   491

Lonicera has a distinctive tricolpate, more or less spheroidal pollen grain, with fairly thick echinate ectexine forming a tectum with small spines: the polar area is large and the bacula in the wall are coarse and distinct. There is a strong probability that all or most of the records given in the diagram refer to *L. periclymenum* L., pollen of which is often also specifically recorded (see below).

The one macroscopic identification is for charcoal from the Iron Age Hembury Fort, Devonshire.

Lonicera xylosteum L.   Fly honeysuckle   491: 1

eW *Chelford*

West identified the pollen of *Lonicera xylosteum* from the Chelford interstadial, drawing attention to the fact that it belongs to a substantial group in this deposit having a more southerly distribution than those more commonly represented. *L. xylosteum* is found only in scattered Scandinavian localities north of 63° N latitude.

Lonicera periclymenum L.   Honeysuckle   491: 3

H *Kilbeg, Baggotstown, Fugla Ness*; F VI, VIIA *Moss Lake*, VIIA, b *Nant Ffrancon*, VIIb *Bell Track*, VIII *?Goldhanger*

The tentative record for charcoal from the Roman site at Goldhanger is the only one based on macroscopic remains: the rest are based upon pollen identifications, all made recently. They shew the plant present throughout the Hoxnian interglacial, at Kilbeg in sub-stages I and II/III, at Baggotstown in II/III and at Fugla Ness in IV, results that support those for the genus in sub-stages II and III. The genus is once represented in the Ipswichian, but neither in the genus nor this species is there any record in any glacial stage. The species is recorded in the Flandrian from zones VI, VIIA and VIIb, results that fit perfectly into the fuller record for the genus where the late expansion of the plant is strongly apparent. This expansion occurs too early to be chiefly attributable to clearance, if indeed this would really favour the species, and makes it more likely to have been a response to increased climatic warmth. This may be why the earliest zone IV, V and VI presence is at Hawks Tor in Cornwall. *Lonicera periclymenum* occurs throughout the whole of the British Isles at the present day, but in Scandinavia, though common in Denmark, it is otherwise restricted to the west coastal fringes of Sweden and Norway south of 63° N. This accords with Hultén's classification of it as a European–Atlantic plant.

VALERIANACEAE

Valerianella sp.   494

F VIIb/VIII *Loch Dungeon, ?Skelsmergh Tarn*

Two pollen records, one tentative, have been made for the Flandrian zone VIIb/VIII transition of the genus *Valerianella*: there are records for the fruit of *V. dentata* from zone VIII. The British species all have distributions centred on the Mediterranean or central and southern Europe.

Valerianella locusta L. (*V. olitoria* (L.) Poll.)   Lamb's lettuce, Corn salad   494: 1

Cs *Pakefield*; H *Clacton*; I *West Wittering*; F VIII *Plymouth*

It is interesting to find three interglacial records for a cornfield weed such as lamb's lettuce, *Valerianella locusta*, and these are supplemented in the following entries by interglacial records for *V. carinata* and *V. dentata*. The Plymouth record is mediaeval. Although local, *V. locusta* is widespread through the British Isles.

Valerianella carinata Lois.   494: 2

H *Clacton*; I *Wretton*

Fruits of *Valerianella carinata* have been recorded from sub-stage III of both the Hoxnian and Clactonian interglacials. This is a rare weed of arable land not found north of Yorkshire.

Valerianella dentata (L.) Poll.   494: 5

H *Clacton*; eW *Sidgwick Av.*; F VIII *Apethorpe, Godmanchester, Silchester, Plymouth*

Records of the fruits of *Valerianella dentata* come from sub-stage III of the Hoxnian interglacial, from the Early Weichselian and from four sites in Flandrian zone VIII, of which two are Roman and one mediaeval. *V. dentata* shews a strongly south-eastern emphasis in its present-day British range and this is more pronounced now than before 1930: it barely persists in Scotland or Ireland and there only in the south-east. It is interesting to note that although the Flandrian zone VIII records are so late as to suggest the possibility of introduction with arable crops, there is an interglacial record, here as for other species of the genus, where such an explanation is impossible.

Valeriana sp.   495

It seems likely that only the two species now native in Britain, *Valeriana officinalis* and *V. dioica*, need come into consideration as constituting the pollen record for the genus and this accordingly must be interpreted in the light of substantial Pleistocene evidence for these species given in the separate accounts below. The parallelism with the record for *V. officinalis* is particularly striking.

Valeriana officinalis L. (incl. *V. sambucifolia* Mik.)   Valerian (Plate XXII*i–l*)   495: 1

Identifications of the fruits of *Valeriana officinalis* establish the presence of this plant in the Cromer Forest Bed series, the Hoxnian and Ipswichian interglacials and the early and late Flandrian. This record is greatly expanded and substantiated by the pollen record (both for genus and species), which indicates presence in the Ludhamian, Thurnian and Baventian and thereafter in every glacial and interglacial stage to the present day. The plant was remarkably abundant in the Late Weichselian, especially

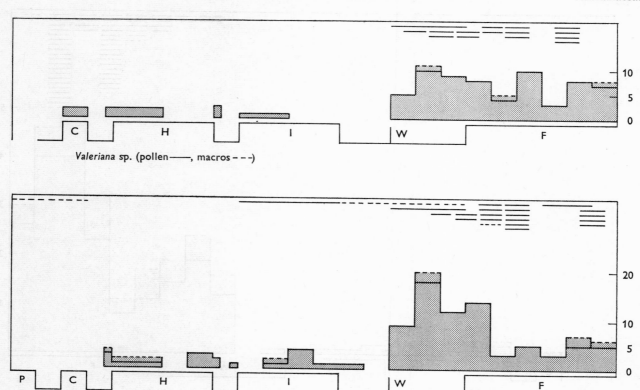

Valeriana sp. (pollen——, macros – – –)

Valeriana officinalis (pollen——, macros – – –)

in zone II, but was present in the earlier Weichselian. It continued abundant in the early Flandrian zone IV but continued thereafter in diminished frequency, save possibly for an increase in zone VI conjecturally attributable to the prevalent development of lake margin vegetation at this time. It would appear from this evidence that *V. officinalis* has been a persistent native plant throughout the Pleistocene and probably a perglacial survivor. It has an effective wind-dispersal mechanism and no doubt spread rapidly in the favourable conditions of the Late Weichselian and early Flandrian.

Valeriana officinalis occurs throughout the British Isles excepting only the Shetlands. *V. officinalis*, excluding *V. sambucifolia* Mik., is frequent in south Finland and at comparable latitude on the east coast of Sweden and there are scattered records in Scandinavia and Finland north as far as about 63° N. *V. sambucifolia* itself extends as a common plant in Norway as far as the North Cape, and this suggests the possibility that this is the plant to which some at least of the Weichselian records refer.

Valeriana dioica L.   'Marsh valerian' (Plate XVIII*d*)
495: 3

Macros: H *Nechells, Clacton*; I *Histon Rd*; *m*W *Barrowell Green*; *l*W *Flixton, Hartford, Nazeing*; F *Silchester*

Pollen: *l*Wo *Selsey*; *e*W *Chelford*; *l*W *Eastwick, Flixton*, ?*Colney Heath, Seamer, Skelsmergh, Nazeing, Rodbaston*;

F IV/V *Kentmere*, VIIb *Oulton Moss, Frogholt*, VIII *Old Buckenham Mere, Ehenside Tarn*

Fruit identifications establish the presence of *Valeriana dioica* in both the Hoxnian and Ipswichian interglacials, as in the Middle and Late Weichselian (zones I, II and II/III). Pollen records include the late Wolstonian and Early Weichselian as well as all zones of the Late Weichselian. Apart from the zone IV/V record at Kentmere, the remaining pollen records are in the latest two zones of the Flandrian. At Old Buckenham Mere it was present at the presumed Iron Age, Roman, Anglo-Saxon and Norman levels, no doubt a persistent denizen of the fen fringing the lake. Incidental points are that the Flixton (zone I) record is for a flower identification, that the fruit at Colney Heath (zone I; 13560 B.P.) is recorded as '*V. dioica* or *V. officinalis*', and that the Silchester fruit record is of Roman age.

Though less full than that for *V. officinalis* the record for *V. dioica* yields very similar evidence of long persistence in the British Isles, especially through the Weichselian glacial stage. This is notable in view of the limitation of present range of *V. dioica* to England, Wales and southern Scotland, and the present-day distribution in Scandinavia where the plant is frequent only in Denmark and adjacent southern Sweden and where its very scattered occurrences north of this do not reach 60° N.

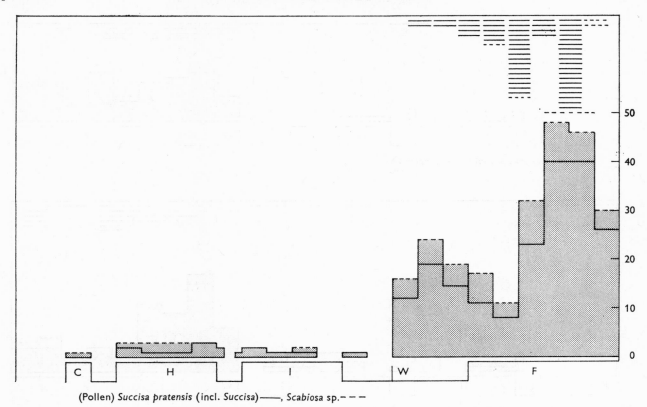

(Pollen) *Succisa pratensis* (incl. *Succisa*)———, *Scabiosa sp.*- - -

DIPSACACEAE

F VI, VIIa, b *Slieve Gallion*, **VIIa, b, VIII** *Beaghmore*, **VIIa, VIIb/ VIII** *Ballynagilly*

Pollen referred to the family Dipsacaceae has been identified by Pilcher at three adjacent sites in Northern Ireland, where the records from zone VI to zone VIII are remarkably co-ordinated by a series of seven radiocarbon dates running conformably between 7640 and 27000 B.P. It is thought likely that the bulk of the records is attributable to *Succisa pratensis* (see separate account).

Knautia arvensis L. 'Field scabious'   498 : 1

F VIII *Old Buckenham Mere*

Pollen of *Knautia* has been identified at the conjectural Anglo-Saxon level in the lake deposits of Old Buckenham Mere, Norfolk.

Scabiosa columbaria L. 'Small scabious'   499 : 1

Macros: Cs *Pakefield*; H *Clacton*; I *Wretton*; mW *Earith, Upton Warren*; lW *Hartford*
  Pollen: H *Marks Tey*; lWo *Selsey*; I *Selsey, Histon Rd*; lW ?*Blelham*; F v/vi *Loch Dungeon*

Taking the fossil fruit and pollen identifications together, they begin with a record from the Cromer Forest Bed series that is a reidentification of a record given originally as *Oenanthe lachenalii*. There follow records for sub-

stages III and IV of the Hoxnian, and for sub-stages I, III and IV of the Ipswichian. There is one late Wolstonian record, two from the Middle Weichselian, two from zone II of the Late Weichselian and a single Flandrian record from Galloway.

Scabiosa columbaria is absent from Ireland and from most of Scotland except the south-east, and it is very southern and scattered also in Scandinavia, not reaching 60° N latitude. These facts make more conspicuous the strong evidence for presence in the Weichselian and the strong suggestion of native persistence in Britain. No doubt the fresh base-rich soils and open vegetation of periglacial situations favoured the plant. It would be of interest to know more of its response to climatic control.

Succisa pratensis Moench (*Scabiosa succisa* L.)   Devil's bit scabious (Plates XXIIn, o; XXIIIs)   500 : 1

Macros: I *West Wittering*; mW *Barnwell Station*

As *Succisa pratensis* was once known as *Scabiosa succisa* the abundant pollen records for this plant appear not only under its full specific name and the equivalent contraction 'Succisa', but also as *Scabiosa pratensis* and as the contraction 'Scabiosa' which may usually (though not totally) be regarded as meaning this species. Accordingly the record-head diagram for pollen subsumes these categories: there is much similarity between those based on 'Succisa' and those based on 'Scabiosa' records, and the latter are in any event less numerous. There can be little doubt that

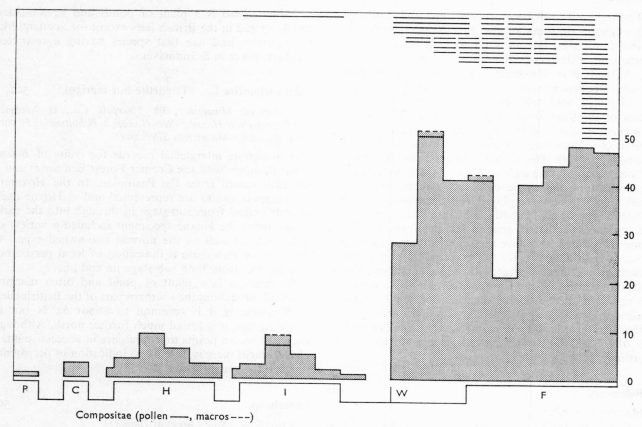

Compositae (pollen ——, macros ----)

in toto these records give a true picture of the history of *Succisa pratensis*. It needs supplementation however by a note that there are single pollen records also for the Baventian and Pastonian stages, and by regard to the fruit identifications given above from the Ipswichian interglacial and from the Middle Weichselian.

We find *Succisa pratensis* then in the Pastonian and Cromerian interglacials, in all four sub-stages of the Hoxnian and the first three sub-stages of the Ipswichian. It was present in the early and late Wolstonian and in Early and Middle Weichselian prior to a very strong representation in all zones of the Late Weichselian, particularly the milder Allerød period. There followed a recession through zones IV and V of the Flandrian, possibly a reflection of forest closure over formerly open ground. Thereafter, in zones VI, VIIa and VIIb, as confirmed also by records from the transition zones, there was a most remarkable expansion in recorded sites. Consideration of the primary records suggests that this probably reflects the widespread development of ombrogenous peat bogs, many of which have been the sites of systematic pollen analysis, and in whose surface vegetation the devil's bit scabious is consistently present. Such an explanation also recognizes the decrease in zone VIII when bog surfaces became wetter and less frequently yielded pollen records.

Succisa pratensis grows commonly today throughout the British Isles. It is frequent in southern Finland and Sweden and extends commonly along the Norwegian west coast to the Arctic Circle. This range is consonant with the strong implication of the fossil record that *Succisa pratensis* has been a long persistent native of the British Isles surviving successive glaciations in this country or quickly being wind spread in periglacial conditions.

COMPOSITAE

The fruits of the Compositae have remarkably good diagnostic value as is shewn by the numbers of specific records which follow. The pollen on the other hand, apart from a few notable exceptions such as *Centurea cyanus* and *Artemisia*, seldom allows finer distinction than that into three or four main categories. It is not surprising therefore that, as the record-head diagram indicates, there should be very few macroscopic records but large numbers of pollen records not carried below family level. Compositae pollen has been found in all stages of the Ludhamian series (Ludhamian, Thurnian, Antian and Baventian), in the Cromerian and subsequently in all sub-stages or zones of ensuing interglacials, the late Antian and both early and late Wolstonian. There is one Early Weichselian record, but in the Late

Weichselian records become very numerous and stay so throughout all zones of the Flandrian, though somewhat diminished in zone v possibly in response to the extension of dense woodland in that zone.

At the bulk of Weichselian and zone IV sites examined, pollen frequencies were below 5 per cent of the total dry land pollen, and subsequently most were below 2 per cent of the total arboreal pollen. None the less higher frequencies occurred in zones VIIb and VIII, no doubt in relation to disforestation and agriculture. At occasional sites very high frequencies obtained as at Frogholt (zones VIIb and VIII) and the Nick of Curlywee (zone IV).

Instead of the single category 'Compositae', some palynologists have employed the portmanteau groups 'Liguliflorae' and 'Tubuliflorae', but these are still so comprehensive that in general all sites from which the one is recorded also have the other. In consequence the record-head diagrams for the two categories are almost identical and indeed repeat, with reduced numbers, almost all the features of the Compositae curve. No doubt as a result of the particular sites investigated by authors using the dual category, there are fewer Ipswichian than Hoxnian records and fewer zone VIII than zone VII records in the Flandrian, but these effects are no more than artifacts of the method.

Bidens sp. 502

Macros: C *Norfolk Co.*
Pollen: H *Hoxne*; eWo *Hoxne*; F VI, VIIa, VII/VIII *Racks Moss*, VIIb/VIII *Aros Moss*, VIII *Slapton Ley*

The fact that there are only two species of *Bidens* now native in the British Isles and that they have very similar present ranges as well as fossil records gives particular value to the pollen records for the genus. The fruit record is from the Cromerian. At Hoxne, West records *Bidens* both from Hoxnian sub-stage III and the early Wolstonian, along with fruits of *B. tripartita* through the same periods. At Racks Moss in Galloway Nichols shews similar persistence through Flandrian zones VI, VIIa and VIIb/VIII, a record that fills in the middle Flandrian record for the two separate species.

Bidens cernua L. 'Nodding bur-marigold' 502:1

Macros: *l*Wo *Selsey*; I *Wortwell, Bobbitshole, Histon Rd*; *l*W *Stannon Marsh*; F VIIb, VIII *Wingham*, VIII *Car Dyke, Hungate*
Pollen: F IV ?*Moss Lake*, VIIb/VIII *Wingham*

Fruits of *Bidens cernua* have been identified from the late Wolstonian and from all sub-stages of the Ipswichian interglacial, being recorded at Histon Road from sub-stage III by Walker and from sub-stage IV afterwards by Sparks and Lambert. A similar persistence is shewn in the Flandrian at Wingham with fruits from zones VIIb and VIII and pollen from the intervening transition zone. It has been found profusely in the Late Weichselian in Cornwall. The latest records are Roman and mediaeval respectively.

Bidens cernua is a plant of ponds and stream-sides, widely spread in the British Isles except for Scotland like *B. tripartita*, and like that species having a restricted southern range in Scandinavia.

Bidens tripartita L. 'Tripartite bur-marigold' 502:2

Cs *Pakefield, Mundesley*; Pa ?*Norfolk Co.*; H *Nechells, Hoxne*; *l*H/eWo *Hoxne*; eWo *Hoxne*; I *Bobbitshole*; F VIIb/VIII *Garscadden Mains*, VIII *Silchester*

The numerous interglacial records for fruits of *Bidens tripartita* begin with the Cromer Forest Bed series and a tentative record from the Pastonian. In the Hoxnian, sub-stages II and III are represented and at Hoxne itself records extend from sub-stage III through into the early Wolstonian: the Hoxne specimens included a variety of four awns as well as the normal two-awned type. At Bobbitshole also there is indication of local persistence with records from both sub-stage IIa and IIb.

B. tripartita is a plant of pond and ditch margins plentiful throughout the southern part of the British Isles. In Scandinavia it is common to about 62° N but in scattered sites it is found much further north. Although the fossil record points to its presence in successive interglacial stages there is little or no indication of persistence through the glacial stages.

Senecio sp. 506

Macros: F IV ?*Star Carr*, VIII *Dublin*
Pollen: *l*W *Roddansport, Rosgoch, Romaldkirk, Hockham Mere*; F IV, v, VI, VIIa, VIIb/VIII *Red Moss*, VI, VIIa, b, VIII *Rosgoch*, VI, VIIb *Avonmouth*, VIII *Borth Bog, Parkmore*

A few investigators concerned with Late Weichselian and Flandrian sites only have recorded pollen as 'Senecio type'. These shew continuous presence in every zone from Late Weichselian I to the present day. They exhibit also a great degree of continuity at most of the sites investigated, especially Roddansport, Rosgoch and Red Moss. The frequency seldom rises above 2 per cent of total tree pollen in the Flandrian or of total dry land pollen in the Weichselian.

Senecio aquaticus Hill 'Marsh ragwort' 506:2

A *Norfolk Co.*; I *West Wittering*

Fruits of *Senecio aquaticus* have been recorded from the Anglian glacial stage and from the Ipswichian interglacial. It is a widespread British species with however a restricted southern range in Scandinavia.

Senecio sylvaticus L. 'Wood groundsel' 506:6

F VIII *Pevensey*

The fruit record from the Roman site at Pevensey is the only one for *Senecio sylvaticus*, a widespread British plant.

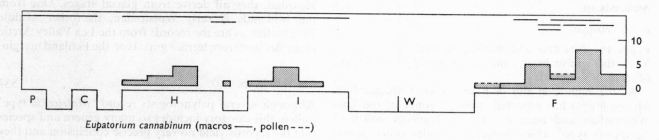

Eupatorium cannabinum (macros ——, pollen - - -)

Tussilago farfara L.   Coltsfoot   508 : 1

Macros: Cs *Pakefield*; I *Histon Rd*

Fruits of *Tussilago farfara* have been identified from the Cromer Forest Bed series by Reid and from sub-stage IV of the Ipswichian interglacial by Walker. The present range and ecological habit of this species allow one to anticipate records from Weichselian deposits in the future.

Gnaphalium sp.   515

F VIII *Dublin*

A fruit of *Gnaphalium* has been recovered by Mitchell from the mediaeval site at Dublin.

Antennaria dioica (L.) Gaertn.   Cat's-foot   517 : 1

F VIII *Dublin*

The only decisive record for *Antennaria dioica* at present is the fruit identification by Mitchell from mediaeval Dublin.

A pollen category '*Antennaria* type' used by H. J. B. Birks includes, besides *A. dioica*, five species of *Gnaphalium* and three of *Filago*. Under this heading he reports pollen from three sites in Skye, Loch Fada, Lochan Coir' a' Ghobhainn and Loch Cill Chriosd, which fully represent the Late Weichselian from zone I to the zone III/IV transition.

Aster tripolium L.   'Sea aster'   519 : 1

H *Kirmington*; I *West Wittering, Shortalstown, Stone*; F IV/V *North Sea*, VIIa *Westward Ho!*

The fruit of *Aster tripolium* has been identified in the Hoxnian and Ipswichian interglacials, twice in the sub-stage II and once at an uncertain period, but all of the sites are coastal ones where brackish conditions prevailed at a time of high sea-level. The Flandrian site on the Dogger Bank also provides evidence of coastal conditions as does the mesolithic site at Westward Ho! on the Devonshire coast. *Aster tripolium* grows right round the British coasts and in Scandinavia it sporadically occurs within the Arctic Circle, though common only south of this. The North Sea record encourages the thought that it may well have followed the shore line through the marine regressions and transgressions that accompany the glacial–interglacial cycle.

Although a pollen category '*Aster* type' provides records right through the Late Weichselian at high altitude in the Elan Valley in Cardiganshire (and sometimes in frequencies of 2 to 5 per cent) it must derive from some genus other than *Aster* (Moore, 1970).

Bellis perennis L.   Daisy   524 : 1

F VIII *Dublin*

The only definitive record of *Bellis perennis* is a fruit found by Mitchell in deposits of mediaeval Dublin.

One of the large categories of Compositae pollen recognized by R. Andrew, C. A. Lambert and others is a '*Bellis* type'. For this there is one Ipswichian record, two Late Weichselian zone III records and some nine Flandrian records distributed through zones IV, VI, VI/VII, VIIa, VIIb and VIII. At Apethorpe this pollen type recurs at frequencies up to 5 or 10 per cent of the arboreal pollen in zones III, VI and VIII. The lowly stature and consequently poor pollen dissemination of *Bellis perennis* make this an unlikely source at least for the Apethorpe records.

Eupatorium cannabinum L.   Hemp agrimony   525 : 1

The semi-aquatic habitat of *Eupatorium cannabinum* no doubt favours preservation of its characteristic cypselas so that its fossil fruits present a substantial record. Beginning in the Cromer Forest Bed series, it is represented in the three first sub-stages of both the Hoxnian and Ipswichian interglacials: in the Flandrian they occur from zone V (Apethorpe) to zone VIII. There are no Weichselian records and the two records for the late Hoxnian–early Wolstonian and the early Wolstonian both come from Hoxne where there is some possibility of derivation from earlier warm deposits in brecciated muds. Little weight can be attached to the two pollen records (Cothill) that were published as early as 1939.

Eupatorium cannabinum occurs throughout most of the British Isles, but is less common in Ireland and Scotland, especially the north: it is absent from Lewis, the Orkneys and Shetlands. It is one of Hultén's west European–south Siberian plants, common only in the southernmost part of Scandinavia where even local occurrences barely reach 62° N in Sweden or 63° N in Finland. This distribution pattern lends apparent significance to the absence of the plant from the Weichselian glacial stage and its late Flandrian extension.

Anthemis sp.   526

F VIII ?*Hungate*

From the Norman and mediaeval levels at Hungate, York there have been tentative identifications of fruits of *Anthemis*.

At the high-level site of Elan Valley, Cardiganshire, Moore (1970) has reported from all zones of the Late Weichselian and zone IV of the Flandrian pollen of '*Anthemis* type' which however includes other genera than *Anthemis*.

Anthemis cotula L.   Stinking mayweed   526: 2

eWo *Marks Tey*; F VIII *Bunny*, ?*Hungate, Dublin*

Cypselas of *Anthemis cotula* have been identified from the early Wolstonian, from the Roman site at Bunny, and twice in mediaeval occupation levels, though once tentatively. The Flandrian records conform with the status of the plant as a weed of arable land and waste places.

Achillea millefolium L.   Yarrow, Milfoil   528: 1

lA *Hoxne*; W *Thrapston*; eW ?*Sidgwick Av.*; mW *Earith, Marlow*; lW *Nant Ffrancon, Kirkmichael*

All the fruit identifications of *Achillea millefolium* come from glacial stage deposits, the earliest the late Anglian and the rest Weichselian. The Early Weichselian record is given as either *A. millefolium* or *A. ptarmica* and the record is repeated below: zones II and III of the Late Weichselian are represented respectively in Wales and the Isle of Man.

These records give convincing proof of the natural conditions in which one of our commonest weeds formerly flourished naturally. *A. millefolium* grows plentifully everywhere in the British Isles and throughout Scandinavia even to the far north, a range sorting well with the Weichselian records.

Achillea ptarmica L.   Sneezewort   528: 3

eWo *Brandon*; eW ?*Sidgwick Av.*

There is an early Wolstonian and a qualified Early Weichselian fruit identification of *Achillea ptarmica*, a species now widespread through the British Isles and one which though commoner in southern Scandinavia nevertheless occurs scattered to very high latitudes there.

Tripleurospermum maritimum (L.) Koch 'Scentless may-weed'   531: 1

H *Marks Tey*; eWo *Brandon*; mW *Earith, Upton Warren, Hedge Lane*

Tripleurospermum maritimum (L.) Koch is an aggregate that includes the coastal sub-species *maritimum* and the widespread weed sub-species *inodorum* (L.) Hyland. ex Vaarama: the fossil fruit records do not distinguish between them. Apart from a record in sub-stage IV of the

Hoxnian, they all derive from glacial stages. One from the Midlands is early Wolstonian, the other Middle Weichselian as are the records from the Lea Valley Arctic Plant Bed and from terrace gravels on the Fenland margin.

Matricaria sp.   532

Although several palynologists record '*Matricaria* type' pollen, this category includes so many genera and species that the records yield no very precise conclusion and they are not reported.

Matricaria recutita L. (*M. chamomilla* auct.) Wild chamomile   532: 1

H *Nechells*

Kelly has recorded the fruit of *Matricaria recutita* from sub-stage I of the Hoxnian in Birmingham.

Chrysanthemum segetum L.   Corn marigold   533: 1

F VIII *Ehenside Tarn, Bermondsey, Shrewsbury, Dublin, Hungate*

All the well-dated records for fruits of *Chrysanthemum segetum* are associated with archaeological occupation levels, two Roman and three mediaeval.

Chrysanthemum leucanthemum L.   Marguerite, Moon-daisy, Oxeye daisy   533: 2

W *Thrapston*; F VIII *Glastonbury, Silchester, Caerwent, Godmanchester*

Aside from the Weichselian record of *Chrysanthemum leucanthemum* all records of its fruits come from archaeological sites, one Early Iron Age and three Roman. It grows today throughout the British Isles and to the far north of Scandinavia, though commoner there in the southern half.

Chrysanthemum vulgare (L.) Bernh. (*Tanacetum vulgare* L.)   Tansy   533: 5

Cs *Beeston*

Artemisia sp.   535

The pollen grains of *Artemisia* are readily recognizable. They are tricolpate, spheroidal to sub-prolate and 17–29 μm in greatest width. Each furrow shews an equatorial os or pore. The conspicuously thick exine shews clearly the bacula which make it up and these fuse externally to form a surface layer which bears rather widely spaced small spinules about as broad as high. Many authorities have attempted a diagnosis of species upon pollen grain characters, but among British palynologists it appears that only the (recent) recognition of *Artemisia norvegica* can be considered successful. Accordingly the account which follows embraces all species in the genus and

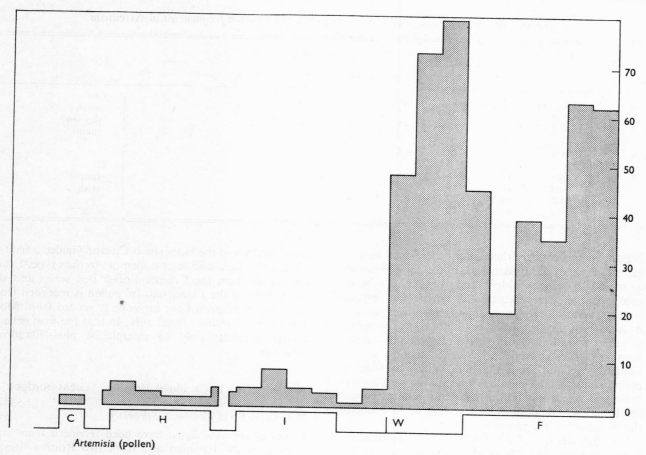

Artemisia (pollen)

includes many records from which *A. norvegica* has not been excluded.

The record-head diagram for pollen of *Artemisia* has such high numbers of sites that records for transition zones have been omitted. Also excluded are early occasional records of single sites from the Ludhamian, Thurnian, Antian and Pastonian, with two sites from the Baventian. It will be seen that the overall Pleistocene record is strikingly complete with occurrences through all sub-stages of the Hoxnian and Ipswichian, Early and Middle Weichselian and all zones of the Late Weichselian and Flandrian. The very high values for the Late Weichselian indicate that a large proportion of recorded sites registered *Artemisia*. Indeed the site numbers amount to no less than 79, 85 and 84 per cent respectively of all British zone I, II and III recorded pollen sequences or spectra. The fall shewn by the diagram in zones IV and V does not reflect smaller numbers of investigated sites, for the percentage of British sites with *Artemisia* falls to 33 per cent in zone IV and 17 per cent in zone V: it was presumably caused by the spread of closed woodland. Thereafter the percentage values do not rise above 22 per cent so that the higher values of total sites in zones VIIb and VIII merely reflect the increased representation of these zones in the British pollen records.

For the majority of the Late Weichselian and Flandrian sites we have abstracted the mean *Artemisia* pollen frequencies zone by zone (table 28). These are expressed in zones I to IV as percentages of the total dry-land pollen and shew *Artemesia* preponderantly in the 2–5 per cent range in zone I, 0–2 in zone II, 5–10 in zone III and back to 0–2 in zone IV. These figures express the remarkably high frequencies of *Artemisia* pollen familiar to pollen analysts in all four zones but especially that in zone III after a considerable recession in zone II, an effect commonly associated with the closure of vegetational cover and woodland growth in that zone.

The Flandrian zone pollen frequencies are expressed as percentages of total arboreal pollen (zone IV appearing in both scales, some sites having been presented on one basis, some on the other). It seems clear that throughout the Flandrian, sites with *Artemisia* reaching above 2 per cent of the arboreal pollen are rare until zone VIII with its great extension of arable cultivation. The zone VI record of 40 to 50 per cent of total tree pollen is from the mesolithic site at Thatcham and may well represent a local stand of *Artemisia vulgaris*. It will be noted how substantially the results now quoted substantiate those of fig. 117 in the first edition (p. 338).

To summarize, the genus has evidently been persistently

TABLE 28. *British Isles: numbers of sites with given pollen frequencies of* Artemisia

		Percentages								
	+	0–2	2–5	5–10	10–20	20–30	30–40	40–50		
I	1	11	**16**	7	3	·	·	·	}	Of total dry-land plants
II	6	**31**	18	1	·	·	·	·		
III	2	12	13	**14**	8	1	·	·		
IV	3	**14**	6	·	·	·	·	·		
IV	2	**11**	5	2	2	·	·	·	}	Of total trees
V	3	**8**	1	2	·	·	·	·		
VI	11	**16**	2	1	·	·	·	·		
VIIa	11	**14**	2	·	·	·	·	·		
VIIb	9	**37**	3	2	·	·	·	·		
VIII	12	**21**	17	5	·	·	·	·		

present in this country throughout the glacial and inter-glacial stages of the Pleistocene. In the Late Weichselian it spread extremely widely and formed a substantial component of the vegetation in zone I and especially zone III. Without reduction in its range it was reduced in its local frequency in zone II, no doubt by the competition of other species including trees. In zone IV both the number of sites and pollen frequency at those sites were suddenly reduced from the high values of zone III, a phenomenon which parallels that on the west European mainland where this fall in *Artemisia* pollen frequencies has been made a major index for what we now regard as the Weichselian/Flandrian boundary.

Through the rest of the Flandrian, *Artemisia* is repre-sented in about one fifth of all the pollen records, but in low frequencies except for the ultimate zone VIII where there was some evident advantage conferred by the spread of agriculture.

Danish workers have reported that *Artemisia vulgaris* remained a serious weed of cultivation in Denmark until comparatively recent time, when deep ploughing allowed eradication of the deep vertical branching root-stock. The rhizomatous habit of *A. campestris* might appear, however, to suggest that it could also have spread in a dangerous fashion into early arable land. We are unaware whether *A. maritima* may not, like *Armeria maritima* and *Plantago maritima*, at one time have had populations occupying inland habitats.

Without considering fully the geographical ranges of the species of *Artemisia* we may, nevertheless, note that at the present day *A. campestris* ssp. *borealis* and *A. vulgaris* ssp. *tilesii* play quite an important part in the vegetation of arctic tundra (Firbas, 1949). The aggregate *A. vulgaris* is indeed shewn by Hultén as fairly common on the west coast of Norway as far as 69° N and very common throughout southern Scandinavia.

Arctium sp.   Burdock   538

Macros: Cs *Pakefield*; H *Clacton*; F VIII *Dunshaughlin, Bunny*
  Pollen: H *Clacton*; F IV *Moss Lake*, VIII *Slapton Ley*

Sub-stage IIIa of the Hoxnian at Clacton yielded a fruit of *Arctium* to Reid and later pollen of *Arctium* type. Fruits also come from the Cromer Forest Bed series and the latest zone of the Flandrian. Its pollen is recorded from zone VIII at Slapton Ley: the zone IV record from Moss Lake is for 'Arctium type' only, so that the firm generic record amounts only to interglacial plus Flandrian zone VIII.

Arctium lappa L. (*A. majus* Bernh.)   'Great burdock'
                                              538: 1

H *Hitchin*; F VIII *Caerwent, Silchester*

Fruits of *Arctium lappa* have been recorded from sub-stage II of the Hoxnian and from two Roman sites, a distribution in time similar to that of the genus.

Carduus sp. (Plate XVIIIe)   539

Macros: Cs *Beeston*; lA *Hoxne*; eWo *Marks Tey*; I *Selsey, Wretton, Histon Rd*; W *Airdrie, Thrapston*; eW *Sidgwick Av.*; mW *Earith*; lW *St Bee's, Hartford, Nazeing*; F IV *Nazeing*, VIII *Glastonbury, Car Dyke*
  Pollen: lW *Loch Dungeon*; F VIIb/VIII *Loch Dungeon*

Fruit referred to the genus *Carduus* has been found in the Cromer Forest series, and from sub-stages II and IV of the Ipswichian interglacial. There are glacial stage records from the late Anglian and early Wolstonian and from Early, Middle and Late Weichselian, but Flandrian occurrences are limited to one zone IV site at Nazeing and two archaeological sites in zone VIII respectively of Iron Age and Roman provenance. Mrs Birks has pollen recorded as *Carduus* from Loch Dungeon. The three Ipswichian records seem to indicate persistence of the genus on the relatively open river terrace sites during the middle forest-dominated stages of the interglacial, whilst some species clearly lived through the Weichselian in similarly open situations. These conclusions should be regarded with those for the component *Carduus* species that follow and the records for *Cirsium*.

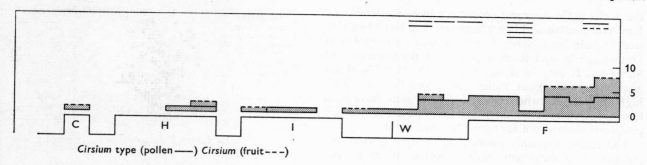

Cirsium type (pollen ——) Cirsium (fruit - - -)

Carduus pycnocephalus L. 539: 2

mW Hedge Lane

The record of a fruit of *Carduus pycnocephalus* is from what is regarded as one of the exposures of the Lea Valley Arctic Plant Bed. It is a plant of southerly distribution and one of the small group making the Hedge Lane flora so difficult to interpret. It is currently regarded as introduced in the British Isles.

Carduus nutans L. 'Musk thistle' 539: 3

Cs *Pakefield*; H *Hitchin, Clacton*; *mW Temple Mills, Hedge Lane*; *lW ?Hawks Tor*; F VIII *Silchester, Caerwent, Finsbury, American Sq., Bunny*

Fruits of *Carduus nutans*, the musk thistle, have been found in the Cromer Forest Bed series, in sub-stages II and III of the Hoxnian interglacial, and at two sites of the Lea Valley Arctic Plant Bed. There is a tentative late Weichselian zone II record from Cornwall. The only Flandrian records are five in Roman occupation sites. The all-over picture resembles that for the genus and for the species, *C. acanthoides* below.

Carduus acanthoides L. (*C. crispus* auct.) 'Welted thistle' 539: 4

C *?Norfolk Co.*; *lW ?Hawks Tor*; F VIII *Silchester*

The Cromerian and zone II Late Weichselian records for fruits of *Carduus acanthoides* are merely tentative, that from the Roman site at Silchester is unqualified.

Cirsium sp. 540

The record-head diagram includes fifteen records for fruits referred to *Cirsium* and a larger number of pollen records that have been given as 'Cirsium type'. In their own reports H. J. B. and H. H. Birks indicate that this pollen type includes eight species of *Cirsium* and three of *Carduus*, and no doubt an equally wide range is attributable generally to pollen so designated.

The total pollen record shews presence in four interglacials, including all zones of the latest, the Flandrian, and throughout the Weichselian. This record of persistence is confirmed by the age distribution of the fruit records, which are however remarkable (as for *Carduus*)

for the concentration in the later archaeological sites, Minnis Bay (late Bronze Age), Welney and Car Dyke (Roman), Lissue and Shrewsbury (mediaeval). These occurrences of familiar weeds do not surprise one.

Carduus or Cirsium

lWo Ilford; I *Ilford*; F VIIb *Littleport*

Fruits referable to either genus of thistle have been recorded at Ilford from the late Wolstonian and stages Ia and IIb of the Ipswichian. There is also a zone VIIb record from the East Anglian fenland.

Cirsium vulgare (Savi) Ten. (*C. lanceolatum* (L.) Scop., non Hill, *Cnicus lanceolatus* (L.) Willd.) 'Spear thistle' 540: 2

Cs *Pakefield*; lA *?Baggotstown*; H *Clacton*; I *West Wittering*; *?Histon Rd*; *lW Hartford*; F VIIb *Minnis Bay*, VIII *Bermondsey, Silchester, Bunny, Dublin*

The fruits of *Cirsium vulgare* occurred in the Cromer Forest Bed series, in sub-stage III of the Hoxnian and in sub-stages II and probably IV of the Ipswichian interglacial. There is a record of a 'Cirsium type' cypsela from the late Anglian in Ireland and from zone II of the Late Weichselian in East Anglia. All the Flandrian records are late and found in archaeological sites, one late Bronze Age, three Roman and one mediaeval. The historical pattern is one already recognized within both *Carduus* and *Cirsium*.

Cirsium palustre (L.) Scop. (*Cnicus palustris* (L.) Willd.) 'Marsh thistle' (Plate XVIII*f*) 540: 3

Cs *Pakefield*; H *Southelmham*; *lWo ?Ilford*; *lWo/I ?Ilford*; I *West Wittering*; *mW Hackney Wick*; *lW ?Nazeing, Flixton, Elstead*; F IV *?Nazeing*, VI *Elan Valley, Cross Fell*, VIIb *Prestatyn, Cocker Beck, Dunshaughlin*, VIIb/VIII *Leasowe*, VIII *Bermondsey, Silchester*

In addition to the numerous unqualified records of the cypselas of *Cirsium palustre* there are some in which an alternative identification is given: *C. arvense* in the case of the two Ilford records and *C. heterophyllum* in Nazeing records from Late Weichselian zones I, II/III and IV, the Flixton zone II record and the Cross Fell zone VI record. Despite these uncertainties the pattern already seen in the genus again emerges: recurrence in interglacials, the Late Weichselian and late Flandrian. The late Flandrian sites include both Neolithic and Bronze Age records at Dunshaughlin, and Roman records at Bermondsey and Silchester. It may be significant that the two Flandrian zone VI records, Elan Valley and Cross Fell, are both high altitude sites where woodland cover must have been less dense and both Prestatyn and Leasowe are submerged peat beds formed not far from the open coast.

The record certainly offers a basis for regarding *C. palustre* as a persistent native species, at least from Weichselian time.

It is a plant which occurs commonly in Scandinavia almost up to 65° N latitude, and with lower frequency well into the Arctic Circle. It is widespread throughout the British Isles today.

Cirsium arvense (L.) Scop. (*Cnicus arvensis* (L.) Hoffm.)
Creeping thistle   540: 4

IA *Nechells*; IWo *?Ilford*; IWo/I *?Ilford*; I *Wortwell*; IW *?Derriaghy, Apethorpe*; F VIIb *Helton Tarn, Minnis Bay*, VIII *Newstead, Silchester, ?Huntspill, Dublin, Larkfield*

The qualified fruit records given above include two from Ilford cited as *Cirsium arvense* or *palustre* and reported already under *C. palustre*. The Wortwell record is for sub-stage I of the Ipswichian and there are Late Weichselian records for the zone I/II transition (tentative) and zone III. Thus far the records are all associated with the presumably open conditions of the opening and closing phases of glacial stages. Thereafter the records are late in the Flandrian; apart from the zone VIIb record at Helton Tarn and the tentative zone VIII record from Somerset, all are associated with human settlement, Late Bronze Age, Roman (2) and Mediaeval (2).

C. arvense now grows throughout the British Isles and although commoner in southern Scandinavia it extends in scattered localities well into the Arctic Circle, ranges which offer no conflict with an historical pattern repeating that of the thistle genera and species.

Cirsium heterophyllum (L.) Hill (*C. helenioides* (L.) Hill, *Cnicus heterophyllus* (L.) Willd.)   'Melancholy thistle'
(Plate XVIII*f*)   540: 7

Cs *Pakefield*; mW *Barrowell Green*; IW *Abbot Moss*; *?Nazeing, ?Flixton*; F IV *?Nazeing*, VI *?Cross Fell*

There are three unqualified records for fruits of *Cirsium heterophyllum*, that at Abbot Moss representing the zone I/II transition in the Late Weichselian. All the remaining records were recorded as *C. heterophyllum* or *C. palustre* and also appear in the list for the latter species.

None of these records is out of character with the present geographical distribution of the plant which is 'Continental-northern' (Matthews) and is shown by Hultén as frequent throughout northernmost Scandinavia far inside the Arctic Circle. Conolly and Dahl (1970) have shown that the present northern range of *C. heterophyllum*

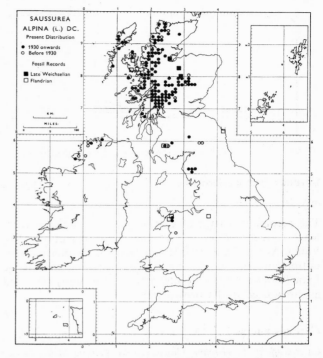

Fig. 118. Present, Late Weichselian and Flandrian distribution of *Saussurea alpina* (L.) DC. in the British Isles: the fossil records are all pollen identifications.

in Britain suggests a limitation by the 27 °C maximum summer temperature summit isotherm.

Onopordum acanthium L.   Scotch thistle, Cotton thistle
                                     542: 1

F VIII *Silchester*

It is interesting to find that *Onopordum acanthium*, generally taken to be an alien in Britain, has been identified by Reid from Roman Silchester.

Saussurea alpina (L.) DC.   'Alpine saussurea' (Fig. 118)
                                     543: 1

IW *Loch Droma, Loch Cill Chriosd, Loch Meodal, Loch Mealt, Loch Fada, Lochan Coir' a' Ghobhainn, Bradford Kaim, Llyn Dwythwch, ?Oulton Moss, Bagmere, Nant Ffrancon, Cooran Lane, ?Loch Dungeon*; F IV *Bradford Kaim, Lochan Coir' a' Ghobhainn, Allt na Feithe Sheilich*, V/VI *Coire Bog, Loch Einich*

It is only recently that British palynologists have been able to recognize the pollen of *Saussurea alpina*, following the lead of Seddon (1962) who identified it in Late Weichselian muds in Llyn Dwythwch on the flanks of Snowdon and in the Nant Ffrancon nearby. The records are all Late Weichselian or early Flandrian: zones I, II and III are all well represented, and although sites in Skye and the Scottish mainland preponderate, North Wales and northern England are also represented (fig. 118). There

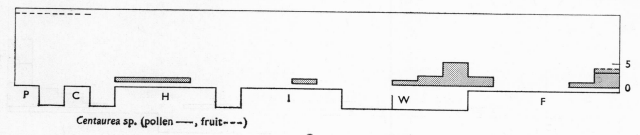

Centaurea sp. (pollen ——, fruit - - -)

is much continuity of representation at individual sites: thus Loch Cill Chriosd (zones I, II, III, III/IV), Loch Mealt (zones I, II, III), Lochan Coir' a' Ghobhainn (zones I/II, III, III/IV, IV), Bradford Kaim (zones II, IV) and Llyn Dwythwch (zones II, III). The latest of the records are from Coire Bog in Galloway and Loch Einich in the Cairngorm, both from transition zone V/VI. The British record is paralleled by Late Weichselian records from the Continental mainland.

Saussurea alpina is a Eurasiatic arctic–montane plant common throughout the whole of Scandinavia except Denmark, south Sweden and south Finland. In montane Scandinavia it occurs in the boreal conifer belt as well as above the tree-line, and in the Alps it occurs in grassland and north-facing scree between 1600 and 3000 m. In the British Isles it has a disjunct range in the north and west with a few widely separate occurrences in Ireland and the main distribution in north-western Scotland, rather closely resembling the distribution pattern of *Salix herbacea*. This species possibly has its range now limited by the mean summer summit isotherms of 23 °C for Ireland and most of Scotland, 24 °C for south Scotland and 25 °C for North Wales (Conolly & Dahl, 1970). The Late Weichselian range was evidently wider and at lower altitude, so that the species must have suffered considerable Flandrian retraction, although persistent close to some sites where it was found in the Late-glacial, as in Snowdonia and Skye.

Centaurea sp.   544

Pollen records for the genus *Centaurea* are fairly numerous and come from both Hoxnian and Ipswichian interglacials, and extend throughout the Late Weichselian and zone IV of the Flandrian in which stage they only reappear late as if responding to reopening of the vegetational cover. Indeed all the records are from pollen diagrams reflecting strong anthropogenic effects, and the fruit record in zone VIII is from the Roman level at Hungate, York. There is one fruit record from the Cromer Forest Bed series.

It should be noted that within the genus *Centaurea*, pollen of *C. scabiosa* and *C. cyanus* is readily recognizable so that pollen under the generic head must relate to other species of which far and away the most abundant today in Britain is *C. nigra*.

Centaurea scabiosa L. 'Greater knapweed' (Plates XXIIh, XXIIIt, u)   544: 1

Macros: Cs ?*Pakefield*; IW *Hartford*; F ?*Bermondsey*, ?*Hungate*

Pollen: C *West Runton*; H *Marks Tey*; I *Histon Rd*; IW ?*Lopham Fen, Windermere*, ?*Hockham Mere*, ?*Nazeing*; F IV *Elstead*, V *Decoy Pool*, VI ?*Nazeing*

Many of the records for *Centaurea scabiosa*, whether based on pollen or on the cypselas, are qualified. The fruit record from the Cromer Forest Bed series, originally cited as *C. calcitrapa*, is now regarded as *C. nigra* or *C. scabiosa*, and the Roman record from Bermondsey and the mediaeval one from York are both tentative. There are however unqualified pollen records from three interglacials prior to the Flandrian, those at Histon Road, Cambridge, shewing 2 to 5 per cent of arboreal pollen in sub-stage III of the Ipswichian and 5 to 10 per cent in sub-stage IV. The Late Weichselian pollen records for zones II and III are all tentative, except the zone II record for Windermere, but the fruit identification from zone II at Hartford is not. Some local persistence is indicated by the Nazeing records for zones II/III, IV and VI, the Hockham Mere and Lopham records for zones II and III and the Histon Road records already mentioned.

C. scabiosa is common in southern Scandinavia and has scattered localities (perhaps influenced by culture) further north as far as the Arctic Circle. In the British Isles it is common in the south but rare in Scotland. The fossil record, though sparse and qualified, does seem to indicate long persistence in this country.

Centaurea cyanus L. Cornflower, Bluebottle (Plate XXIIIm, n, r)   544: 3

Pollen: IW *Lopham Fen*; F VIIa *Moss Lake*, VIIb *Moss Swang, Ladybridge Slack*, ?*Glastonbury*, VIII *Old Buckenham Mere, Peterborough, Malham Tarn, Drake's Drove, Slapton Ley*

In 1947 Iversen published evidence, based upon pollen identification, that *Centaurea cyanus* had been a natural component of the Late glacial vegetation of Denmark and was indeed present in the three main zones of that time. Surprising as this seemed it was confirmed before long in other west European regions (e.g. Holland; see van der Hammen, 1951); and Tallantire has recorded a single grain of it from the Allerød period in East Anglia. Iversen has pointed out that the later Danish records from the Post-glacial are all from the Sub-atlantic period

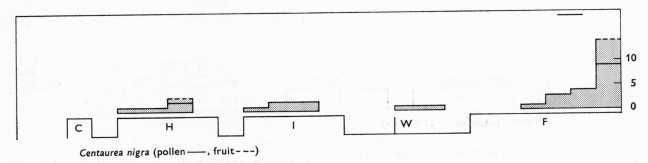

Centaurea nigra (pollen ——, fruit - - -)

where its association with pollen of rye indicates how its spread was favoured by cereal cultivation. In the British Isles our Post-glacial records are similarly late and associated with agriculture. At Old Buckenham Mere pollen of *Centaurea cyanus* occurs especially at the Anglo-Saxon and Norman levels that shew strong evidence for the cultivation of rye and the presence of many arable weeds. The late pollen records from Somerset likewise shew strong association with other weeds of agriculture. It remains uncertain how far the cornflower persisted in this country from the Late Weichselian or how far it was reintroduced with the crops, although the zone viia record from Moss Lake precedes Neolithic clearances and indicates persistence in the middle Flandrian. It is important to note, as Iversen suggests, that Szafer has identified an achene of *Centaurea* cf. *cyanus* from the penultimate glacial period of Poland.

C. cyanus is regarded as growing wild in the eastern Mediterranean, on arid mountain slopes in Italy, in Thessaly and Macedonia, and in western Asia in dry steppes (Iversen, 1947*a*). It follows cereal crops, however, to high latitudes, seed ripening as far north as Finmark. *C. cyanus* was formerly common throughout Great Britain but has much diminished in recent years, possibly as a result of improved seed-cleaning, possibly as a result of diminished culture of rye, a cereal with which it has a very close association.

Centaurea nigra L. (*C. obscura* Jord.) 'Lesser knapweed', Hardheads   544: 6

The fruit records of *Centaurea nigra* include one from sub-stage III of the Hoxnian interglacial, but the rest are all from human settlements, three Roman and three mediaeval: only one is tentative.

The pollen records are, with only two exceptions, unqualified but it may well be that 'C. *nigra* type' would be a more exact description: it is likely that the far greater abundance today of *C. nigra* than of other British species with similar pollen has influenced the finality of identification. With this caveat, we may note that the record-head diagram indicates presence through the first three sub-stages of both Hoxnian and Ipswichian interglacials, a pattern not repeated in the Flandrian where the records are increasingly concentrated in the later zones, especially in zones viib and viii where a large part of them is

associated with the effects of prehistoric clearances and agriculture. Thus at Wingham, Kent it occurs at the late Bronze Age to early Iron Age level and again, with a frequency of 5 to 10 per cent of the arboreal pollen, at the early Iron Age level carbon-dated to 2490 B.P.

These effects indicate responsiveness to open conditions (as with pollen of the genus) and are underlined by the fact that of the earlier Flandrian records two are from high altitudes, Nant Ffrancon (zone vi) and Cwm Idwal (zone viia), whilst a third is from the coastal Mesolithic site at Westward Ho! (zone viia). The two Late Weichselian records, like the more numerous records for the genus at this time, reflect the same tendency. Just as we may suppose *C. nigra* to have persisted in the middle Flandrian in lightly wooded upland refugia, so it seems probable that it persisted in earlier interglacials on the river terraces and gravel spreads represented by sites such as Marks Tey and Nechells in the Hoxnian and Ilford, Aveley and Histon Road, Cambridge in the Ipswichian.

Unlike the other British species of *Centaurea*, *C. nigra* is common throughout the whole of the British Isles, though possibly only introduced in the Shetlands. The fossil record gives grounds for regarding it as a long-persistent native plant.

Centaurea calcitrapa L.   Star thistle   544: 9

F viii *Caerwent*

Centaurea calcitrapa, which is so southern in its British distribution (northern limit in Norfolk), has a single Roman record from Monmouthshire.

Serratula tinctoria L.   Saw-wort   545: 1

F viii *Silchester*

The only record of *Serratula tinctoria* is a fruit record by Reid from Roman Silchester.

Lapsana communis L.   Nipplewort   547: 1

Cs ?*Pakefield*; H ?*Kirmington, Nechells, Hitchin, Clacton*; I *West Wittering, Selsey, Wretton*; mW *Earith*; F viii *Fifield Bavant, Finsbury Circus, Silchester, Lissue, Shrewsbury, Hungate, Dublin, Plymouth*

The cypselas of *Lapsana communis* are well represented in the interglacials, and only two records are qualified, including the oldest from the Cromer Forest Bed series. The four Hoxnian sites cover sub-stages II and III, as do the three Ipswichian. There is a clear record from the Middle Weichselian of age between 40000 and 50000 years B.P., but thereafter there are none until the many archaeological records from Flandrian zone VIII: one Iron Age, two Roman and five mediaeval.

Like *Centaurea nigra* just described, it is common throughout the British Isles and it has yielded the very similar historical pattern, common to so many Compositae, of presence in several or all interglacials, in the Weichselian and thereafter abundantly in a late Flandrian archaeological provenance.

Hypochaeris sp.   549

*l*W *Nant Ffrancon*; F VIII *Shrewsbury*

Fruit identified as of the genus *Hypochaeris* comes from zone II of the Late Weichselian in North Wales, and from a mediaeval site in Shrewsbury a fruit has been referred to *Hypochaeris radicata* or *H. glabra*.

Hypochaeris radicata L.   Cat's ear   549: 1

F VIII *Silchester, Hungate*

The Roman site at Silchester and the mediaeval level at Hungate, York have yielded fruits identified as *Hypochaeris radicata*.

Hypochaeris glabra L.   549: 2

F VIII *?Godmanchester*

A fruit from the Roman site at Godmanchester has been tentatively referred to *H. glabra*.

Leontodon sp. (*Thrincia* Roth)   550

*l*Wo *?Ilford*; W *Thrapston*; *l*W *Nant Ffrancon*

Identifications of fruits of *Leontodon* at the generic level, two of Weichselian age and a tentative Late-glacial record, are best considered in conjunction with identifications at the species level described in the following accounts.

Leontodon autumnalis L.   'Autumnal hawkbit'   550: 1

Cs *Pakefield*; *l*A *Hoxne*; H *Nechells*; *e*Wo *Brandon*; *e*W *?Sidgwick Av.*; *m*W *Upton Warren, Angel Rd, Hedge Lane, Marlow, Earith*; *l*W *Nant Ffrancon, Hawks Tor*; F VI *Roundstone*, VIIb *?Frogholt, Drumurcher*, VIII *Godmanchester, Car Dyke, Silchester, Dublin, ?Hungate*

Of the twenty fruit identifications recorded, only three are tentative. The records include one from the Cromer Forest Bed series and one from the early Hoxnian: these and the Flandrian records excepted, all relate to glacial stages, including the late Anglian and early Wolstonian as well as a remarkable number from the Weichselian. These last include not merely a tentative Early Weichselian record, and two from the Allerød (zone II) but five

records from unexceptional Middle Weichselian sites with rich accompanying fauna and flora. Thereafter follow Irish records in Flandrian zones VI and VIIb and a tentative record from VIIb in Kent. The remaining sites are zone VIII archaeological localities, three Roman and two mediaeval.

This record suggests long presence in the British Isles and possibly perglacial survival there. It is a plant which grows commonly throughout Scandinavia to latitudes within the Arctic Circle and on the western coasts even to the North Cape. It evidently took advantage of forest clearance and agriculture, but it is unusual in the strength of evidence for it in the glacial stages.

Leontodon hispidus L.   'Rough hawkbit'   550: 2

*l*W *?Kirkmichael*; F VIII *Silchester, ?Shrewsbury, ?Hungate*

Of the four records for fruits of *Leontodon hispidus* three are tentative: that from zone I in the Isle of Man and the two mediaeval records from Shrewsbury and York. The Roman record from Silchester is unqualified.

Leontodon taraxacoides (Vill.) Mérat (*Thrincia hirta* Roth, *T. leysseri* Wallr.)   'Hairy hawkbit'   550: 3

H *Hitchin*; F V *Apethorpe*, VIIb *?Frogholt*, VIII *?Shrewsbury*

Unqualified fruit identifications have been made of *Leontodon taraxacoides* from sub-stage II of the Hoxnian and zone V of the Flandrian. The zone VIIb record from Frogholt has an alternative identification of *L. autumnalis*, and the mediaeval record from Shrewsbury one of *L. hispidus*.

Picris echioides L. (*Helminthia echioides* (L.) Gaertn.)   'Bristly ox-tongue'   551: 1

F VIIb *Minnis Bay*, VIII *Apethorpe, Finsbury Circus, Bunny*

The oceanic–southern species, *Picris echioides* (ox-tongue) has been found in the Late Bronze Age and the Roman period, in each instance associated with settlement and possibly in the former case with threshing residues. Its fruits were also found in zone VIII at Apethorpe, not specifically in an archaeological context.

Picris hieracioides L.   'Hawkweed ox-tongue'   551: 2

Cs *Pakefield*; H *Clacton*; I *Wretton, Histon Rd*; F VIII *Newstead, Silchester, Shrewsbury*

Fruit records of *Picris hieracioides* come from the Cromer Forest Bed series, from sub-stage III of the Hoxnian and sub-stages II and IV of the Ipswichian interglacials, where we may regard the plant as having grown on open river terraces. In the Flandrian it reappears only in archaeological sites, two Roman and one mediaeval. The plant very closely resembles *P. echioides* in its present British range and shares its responsiveness to conditions associated with agriculture.

Tragopogon pratensis L.   Goat's beard   552: 1

F *Merseyside, Godmanchester*

The Merseyside record of *Tragopogon pratensis* fruit is from a coastal peat bed of late Flandrian age. The Roman record from Godmanchester was referred only to the genus, but most probably belongs to the one native British species, *T. pratensis*.

Lactuca sp.   Lettuce   554

F VIIa *Brook*

From the Mesolithic site in the Isle of Wight, Clifford has recorded six perfect fruits which are not fully comparable with those of any wild British species, although near to those of *Lactuca serriola*: they agree, however, with fruits of some varieties of cultivated lettuce, *L. sativa*.

Sonchus sp.   556

mW *Barrowell Green, ?Earith*

The only other Weichselian record for fruits of *Sonchus* is that from Nazeing for *S. arvensis* but there are records for other species from earlier glacial stages.

Sonchus palustris L.   'Marsh sow-thistle'   556: 1

F VIII *Silchester, Caerwent*

Sonchus palustris, which has an extremely sparse and southerly distribution in England and is all but extinct, has been recorded in Roman deposits, once north of the plant's present range.

Sonchus arvensis L.   'Field milk-thistle', 'Corn sow-thistle' (Plate XVIII*h*)   556: 2

lA *Nechells*; H *Nechells, Hitchin, Clacton*; I *Histon Rd*; lW *Nazeing*; F VIII *Silchester, Finsbury Circus*

Fruits of the corn sow-thistle' (*Sonchus arvensis*) have been found at the end of the Anglian glacial stage and in zone II/III of the Late Weichselian. They occurred in sub-stages I, II and III of the Hoxnian interglacial and sub-stage IV of the Ipswichian. In the Flandrian both records are from Roman sites, completing an outline of an historical pattern familiar in Compositae weeds. *S. arvensis* evidently belongs to the category of present-day weeds that had a natural British status in the Late Weichselian and may have persisted from that time. It occurs throughout Scandinavia in cultivated ground, even occasionally into the Arctic Circle, although it is common only in the southern part.

Sonchus oleraceus L.   Milk- or Sow-thistle   556: 3

Be *Norfolk Co.*; C *Norfolk Co.*; lA *Hatfield, Gort, Baggotstown*; eWo *Brandon*; F VI *Jenkinstown*, VIII *Bermondsey, Caerwent, Pevensey, Silchester*

There is only one interglacial record, Cromerian, for the fruits of *Sonchus oleraceus*, but it occurs in the Beestonian glacial, at three sites in the late Anglian and once in the early Wolstonian. There are no records from the Weichselian, but in the Flandrian it has reappeared in four Roman sites in England as well as in a zone VI site at Jenkinstown in Ireland. There is a case for associating every site with open vegetational conditions.

Sonchus asper (L.) Hill.   'Spiny milk-thistle', 'Spiny sow-thistle'   556: 4

H *Clacton*; F VIII *Frogholt, Dunshaughlin, Apethorpe, Godmanchester, Silchester, Finsbury Circus, Caerwent, Pevensey*

Sub-stage III of the Hoxnian has yielded a fruit record for *Sonchus asper* but, this apart, all records of it come from the late Flandrian. There is one record from the late Bronze Age level at Frogholt in Kent, and no less than five from Roman sites besides the two zone VIII records from Dunshaughlin and Apethorpe. It is of course natural that such consistent weeds of arable ground should accompany the cereal stores so often investigated from archaeological sites.

Hieracium sp.   558

lW *Hartford, Nant Ffrancon, Mapastown*; F VIII *?Hungate, Plymouth*

All three late Weichselian records for fruits of the genus *Hieracium* are from zone II, the milder Allerød period. The two Flandrian records are both mediaeval and that from York only tentative.

Hieracium pilosella   Mouse-ear hawkweed   558: 2

I *West Wittering*

Reid identified the fruit of *Hieracium pilosella* from the coastal site of West Wittering, at an undetermined phase of the Ipswichian interglacial.

Crepis sp.   559

C *Norfolk Co.*; eWo *Brandon*; lW *Derriaghy*

The spread in time of fruits referred to the genus *Crepis* is wide and in the pattern of most Compositae weeds.

Crepis mollis (Jacq.) Aschers. (*C. hieracioides* Waldst. and Kit.)   'Soft hawk's-beard'   559: 4

Cs *Pakefield*

Reid's identification of *Crepis mollis* fruit from the Cromer Forest Bed series is our only record of a plant now scattered and rare in northern Britain.

Crepis capillaris (L.) Wallr. (*C. virens* L.)   'Smooth hawk's-beard'   559: 6

mW *Angel Rd*; F VIIb *Minnis Bay*

Crepis capillaris, now such a widespread ruderal in the British Isles, was identified at the Late Bronze Age site of Minnis Bay, but it was previously present in the Lea Valley during the Weichselian period.

Crepis paludosa (L.) Moench.   'Marsh hawk's-beard'

*l*W *Garscadden Mains*, ?*Mapastown*   559: 8

Crepis paludosa, which is a northern plant in Britain, placed by Matthews in his Continental–northern group and extending beyond 70° N latitude in Norway, has been found in the Late Weichselian zone I in Scotland and has been tentatively recognized from zone II in Ireland.

Taraxacum sp.   Dandelion   560

H *Southelmham*; *l*H/*e*Wo *Hoxne*; *e*Wo *Brandon, Hoxne*; I *West Wittering*; *m*W *Angel Rd, Hedge Lane, Temple Mills, Ponders End, Upton Warren*; *l*W *Greenock, Corstorphine, Hailes, Garvel Park, Derriaghy, Hartford, Coolteen, Nazeing, Mapastown*; F IV *Nazeing*, VIII *Silchester*, ?*Hungate*

Although pollen of '*Taraxacum* type' has been recorded by a number of palynologists it embraces as many as fifteen genera and corresponds largely to the category 'Liguliflorae': it has not seemed profitable to detail the age records of it.

On the other hand the cypselas of dandelion are readily recognizable as referable to '*Taraxacum*', but of course taxonomy within the genus is complex. The records cited above are mostly given as for *T. officinale*, but since many were made before the sub-division into the four species that the *London Catalogue* now lists, and others are specifically referred to an aggregate, they are all kept for our purposes within the genus. There follow however accounts of identifications carried further, most of them made by Dr F. Bell and confirmed by A. J. Richards.

The interglacial records are from unspecified parts of the Hoxnian and Ipswichian respectively, and apart from three Flandrian records all the rest come from glacial stages. The Wolstonian records are all early. Weichselian records are abundant, with no less than five from the Middle and nine from the Late Weichselian. Four of the Middle Weichselian sites are in the Lea Valley, and at Nazeing the dandelion was found both at the zone II/III transition and in zone IV of the Flandrian. Scottish and Irish sites are well represented and the Late Weichselian records extend through from zone I/II to III. The late Flandrian records include one from Roman Silchester and doubtful identifications from the Norman and mediaeval levels at the Hungate, York.

This historical record indicates the very widespread occurrence through the last glaciation of a plant whose major present-day role is that of a weed or ruderal. Despite the absence of intervening records there is no reason to presume that *Taraxacum* wholly disappeared between then and the Roman period. It would seem also that this pattern of persistence has prevailed through earlier glacial and interglacial stages.

The more exact specific records that follow greatly strengthen the evidence for abundance of the genus in the British Isles during the last glacial period.

Taraxacum officinale Weber *sensu lato*   Common dandelion (Plate XVIII*g*)   560: 1

*m*W *Derryvree, Earith, Marlow, Syston*

Miss Bell has records for the fruits of *Taraxacum officinale*, section *vulgaria*, from the Middle Weichselian Irish site in Co. Fermanagh (radiocarbon age 30 500 B.P.) and from three English sites of similar age.

Taraxacum palustre (Lyons) Symons *sensu lato* 'Narrow-leaved marsh dandelion'   560: 2

*m*W *Earith*; *l*W ?*Hartford*

Fruits of *Taraxacum* of the section *palustria* have been identified from the Middle Weichselian at Earith, and tentatively from zone II of the Late Weichselian, not far away in Hartford, Hunts.

Taraxacum spectabile Dahlst. *sensu lato* 'Broad-leaved marsh dandelion'   560: 3

W *Thrapston*; *m*W *Derryvree, Earith*; *l*W/F *Loch Cuithir*

Miss Bell has identified fruits of *Taraxacum* of the section *spectabile* from three Weichselian sites and the Vasaris give a Scottish record from the Weichselian/Flandrian transition.

Taraxacum laevigatum (Willd.) DC. *sensu lato*   'Lesser dandelion'   560: 4

*m*W *Earith*

Miss Bell refers fruit of *Taraxacum* from Earith also to the section *Erythrosperma* that falls inside *T. laevigatum*.

Taraxacum sect. alpina G. Hagl.   560: 5

*m*W *Earith*

Along with the other species of *Taraxacum* now native in the British Isles that were identified from the terrace sands and gravels at Earith, Cambridgeshire, A. J. Richards was able to recognize fruit from the section *alpina*, which he finds to occur today in Glen Clova.

'Iva xanthifolia' type

IG *Trysull*; *l*W *Brandesburton*; F VIIb *Lytheat's Drove, Wingham, Frogholt*, VIII *Old Buckenham Mere*

A pollen grain similar to that of '*Ambrosia*' type but with a long furrow is recognized by North American palynologists and, like the '*Ambrosia*' type, it is occasionally found fossil in British deposits. Miss Andrew had recorded it from interglacial and Late Weichselian deposits, and from Flandrian zones VIIb and VIII, but the results are unpublished since it has been thought that these grains,

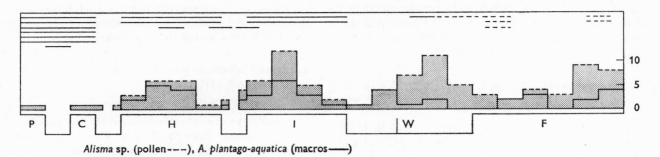

Alisma sp. (pollen – – –), A. plantago-aquatica (macros ——)

like those of the 'Ambrosia' type, are due to long-distance transport and are so uncertainly identified.

'Ambrosia' type

I *Bobbitshole*; *m/lW Skye*; *lW Little Lochans, Dumfriesshire*; F v *Nick of Curlywee*, viiA *Duartbeg*, viib *Copse Gate*, viii *Bigholm Burn*

Palynologists in Cambridge have recorded from time to time single grains which Miss Andrew describes as 'a medium-sized composite with sub-acute spikes and a very short furrow, to which belong *Xanthium strumaria*, together with North American species of *Ambrosia*, *Xanthium, Franceria* and some (short furrowed) *Iva* species'.

The fossil occurrences are so infrequent and in deposits of such various ages that one must conclude they have been brought in by long-distance wind transport, possibly across the Atlantic. The identifications were made by N. T. Moar, R. G. West, H. J. B. Birks and R. Andrew, but are mostly or wholly unpublished. N. T. Moar gives a useful discussion of the subject in his thesis (1964).

ALISMATACEAE

Alisma sp.   563

mW Hedge Lane

From the Arctic Plant Bed in the Lea Valley C. Reid reported fruits of the genus *Alisma* but the considerable bulk of fruit identifications are to *A. plantago-aquatica* (see below). On the other hand the finely reticulate pollen grains are not characterized below generic level. Records for these now have a similar abundance to those of macroscopic identifications for *A. plantago-aquatica* and the record-head diagram where both appear together shews so much correspondence in temporal distribution and frequency, and so many instances of correlation at particular sites that we may reasonably assume most of the pollen refers to this, far the commonest British species today.

Alisma plantago-aquatica L. Water-plantain (Plate XIX*b*)   563: 1

The fruits of *Alisma plantago-aquatica* have not only been recorded by Reid from six sites in the Cromer Forest Bed

series, but later by Wilson separately from the Pastonian, Beestonian and Cromerian on the Norfolk coast. They have been found in sub-stages I, II and III of the Hoxnian, all sub-stages of the Ipswichian, and most of the pollen zones of the Flandrian. They are recorded from the Anglian glacial stage, the terminal Hoxnian, early and late Wolstonian and Early, Middle and Late Weichselian, so that these macroscopic finds give very clear indication of persistence through both glacial and interglacial stages. This impression is very strongly reinforced by the generic pollen evidence, especially in view of the site correlations of fruit and pollen records in the Hoxnian (sub-stages I and II at Nechells and sub-stage III at Hoxne), in the late Wolstonian at both Selsey and Ilford, and in the Ipswichian (sub-stage I at Ilford and Selsey, sub-stage II at Histon Road, Wretton, Bobbitshole, Wortwell and Selsey, and sub-stage III at Wretton). Pollen records occur in each of the Late Weichselian zones, no fewer than nine in zone II and in every Flandrian zone. Pollen and fruits together present an impressive case for the long indigeneity of at least *Alisma plantago-aquatica* in the British Isles.

Although rather rare in the north of the British Isles and absent from the extreme north of Scotland, Orkney and Shetland (Clapham, Tutin & Warburg, 1962), the plant is placed in Hultén's group of boreal–circumpolar species, and although remaining scattered in Norway and northern Sweden, in the more continental region of Finland it is frequent up to and beyond 65° N as well as having a few localities in the Arctic Circle (Hultén, 1950).

Damasonium alisma Mill. (*D. stellatum* Pers.; *Actinocarpus Damasonium* R. Br.) Thrumwort   564: 1

I *Beetley*; *eW Wretton*; *mW Earith, Hedge Lane*

The identity of the fossil *Damasonium alisma* from Hedge Lane is unquestionable for the very characteristic attenuate carpels were present, and these are sufficiently fragile to rule out the possibility that the fossil had been derived from an older bed. Nevertheless the presence of this oceanic–southern species in an 'Arctic Bed' seemed very extraordinary. It was not until more than fifty years later that Miss Bell identified four seeds of highly characteristic surface sculpturing from the unimpeachably Middle Weichselian deposits at Earith. It has subsequently been recorded, but not yet published, from the

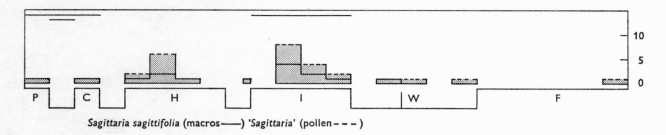

Sagittaria sagittifolia (macros———) *'Sagittaria'* (pollen– – –)

Ipswichian at Beetley and recorded from the Early Weichselian at Wretton (Sparks & West, 1970). Miss Bell points out that the small seed production of the plant argues greater abundance than the fossil seed numbers would indicate, and that it favours shallow water sites liable to seasonal drying out such as would occur in glacial outwash plains.

Not only does *D. alisma* have a sparse British range strongly limited to the south-east, but in Europe as a whole it is southern, extending south from France to Italy and north Africa. It is classified by Matthews in his 'oceanic–southern' element and one wonders why a species that withstood the rigour of the Middle Weichselian conditions was apparently unable to extend its northerly range in the ensuing Flandrian period.

Sagittaria sagittifolia L.   Arrowhead (Plate XXIV*g, h*)
565: 1

Besides the Cromer Forest Bed series record from Pakefield there are more recent records of fruits of *Sagittaria sagittifolia* from the Pastonian, Beestonian and Cromerian stages separately. There are records also from three sub-stages of both the Hoxnian and Ipswichian, together with a single Middle Weichselian record from Earith and a late Wolstonian from Ilford.

The periporate echinate pollen grains with indistinct pores have been recorded rather infrequently. Most of the pollen records come from the interglacials where they are often correlated with the fruit records, as at Nechells (Hoxnian sub-stages I, IIa and IIb) and at several sites in sub-stages II, III and IV of the Ipswichian. There are pollen records in Late Weichselian zones I and III. The only Flandrian record is for zone VIII, possibly a reflection of the fact that in this period few fluviatile deposits have been investigated, whereas they preponderate in the early interglacial investigations.

Sagittaria sagittifolia is distributed throughout England, save the far north and the south-west: it is infrequent in Wales and Ireland and absent from Scotland. It is placed by Hultén in his category of west European–south Siberian plants, but nevertheless in Scandinavia it barely enters Norway, and is more frequent round the Baltic: there are occasional sites for it within the Arctic Circle though in the most northern sites it is represented by sterile submerged phenotypes.

Despite the absence of Flandrian records there seems good reason to regard the arrowhead as long native in Britain. It is very remarkable that the two Late Weichselian pollen records should be from Loch Cill Chriosd in Skye (H. J. B. Birks, 1969) and one might ask if this perhaps points to a Flandrian extinction in Scotland, especially as there is also a Late Weichselian record from Oulton Moss, Cumberland (Walker, 1966).

BUTOMACEAE

Butomus umbellatus L.   Flowering rush   566: 1

Pollen: H *Gort*; I *Histon Rd*; *m*W *Marlow*; F VIIb *Plantation Farm*
   Macros: I *Ilford*; F VIIb *Plantation Farm*

The monocolpate pollen grains of *Butomus umbellatus*, with distinct columellae in the muri and a wide margo of smaller mesh round the pore, have been recognized in sub-stage III of the Hoxnian in western Ireland, in sub-stages III and IV of the Ipswichian in Cambridge, and in zone VIIb of the Flandrian at Shippea Hill in Cambridgeshire. From the same level at the latter site a calyx of *B. umbellatus* was recovered, and fruits were identified by West from sub-stages IIa and IIb of the Ipswichian at Ilford.

The present British range of *Butomus* is remarkably like that of *Sagittaria sagittifolia*, i.e. as a native chiefly in England, sparse in Wales and Ireland, but absent in Scotland and south-western England. Its Scandinavian range is also very similar to that of *Sagittaria* although it is described by Hultén as a Eurasiatic species with links to Scandinavia in the south and east. Here again the Middle Weichselian record and the range into the Arctic Circle north of the Baltic make it hard to understand why the plant has a southerly restriction of range in Britain today. Occurrence in three interglacials and a full glacial suggests long native status.

HYDROCHARITACEAE

Hydrocharis morsus-ranae L.   Frog-bit   567: 1

H *Nechells*; I *Ilford, Selsey, Bobbitshole, Wretton*

Although pollen of *Hydrocharis morsus-ranae* is recognizable it has not yet been recorded fossil here; the records

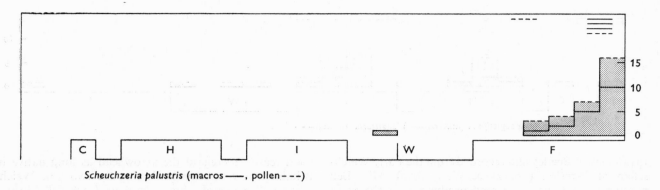

Scheuchzeria palustris (macros ——, pollen – – –)

are all for fruit described by Kelly as 'biconvex elliptical ...with characteristic coarse epidermal cells...the spiral thickenings of which become loosened to give a tangled hairy surface to the fruit'. All the records are of interglacial age, from sub-stage II of the Hoxnian and sub-stages I and II of the Ipswichian. Whereas at Ilford West *et al.* (1964) record it at the levels of Ia, I/II, IIa and IIb, at each of the other sites it is in sub-stage IIb, i.e. the warm part of the interglacial.

The frog-bit is placed by Hultén in his category of west European–middle Siberian plants and he shews it as rather southerly and continental in its Scandinavian range, absent from Norway and scattered in south Sweden and in Finland to the north of the Baltic, but commoner in the south. A similar range is exhibited in the British Isles where it is scattered through England save the north and extreme south-west, commoner in the south-east, absent from Scotland and limited in Ireland to the central region and in Wales to a few southern and eastern localities. The British range corresponds closely with those of other aquatics such as *Sagittaria sagittifolia* and *Butomus umbellatus*, both of which are recorded with *Hydrocharis* in the Selsey interglacial. It has to be recalled that the frog-bit most commonly occurs in calcareous waters.

Stratiotes aloides L.   Water soldier   568: I

Pollen: I *Bobbitshole, Selsey, Shortalstown*

Macros: Cs *Corton, Pakefield, Sidestrand, Beeston*; Pa *Norfolk Co.*; C *Norfolk Co.*; A *Corton*; H *Southelmham*; I *Wretton, Selsey*

Stratiotes aloides, the water soldier or water aloe, is the sole surviving species of a genus shewn by palaeontological research to have had a history extending far back into the Tertiary period. This history, based upon seed identifications, has been critically summarized by M. E. J. Chandler (1923). She describes eight extinct species alongside the present-day *S. aloides* and traces their presumed evolutionary relationships.

There seems no reason however to regard the British interglacial records as referring to any species but *S. aloides*. Seeds have been found in four sites of the Cromer Forest Bed series, in the Pastonian and Cromerian interglacial stages, in an undefined part of the Hoxnian and in

sub-stage II of the Ipswichian both at Wretton and Selsey. In the same sub-stage of the Ipswichian, pollen of *S. aloides* has been recorded from Bobbitshole, Selsey and Shortalstown. The records of seeds from the Cromerian and Pastonian of the Norfolk coast are accompanied by identifications of the leaf-margin spines, as pollen and seeds accompany one another at Selsey. West (1961*b*) comments that the water soldier was found along with *Cladium mariscus, Hydrocharis morsus-ranae* and *Lemna* cf. *minor* at an equivalent zone of the Ipswichian both at Bobbitshole and Selsey and he adds that their presence indicates a considerable summer warmth, greater than that of the present day in the area. This is a point of special interest in view of the record of *Stratiotes* in the interstadial deposits at Corton, Lowestoft.

Stratiotes aloides has a continental range extending from middle Sweden to northern Italy, with a distribution in Scandinavia resembling that of *Hydrocharis*. In the British Isles it is still more strongly restricted than that species to the continental south-eastern part of the British Isles, and it virtually never sets seed in this country. Its distribution is made more complex by the different present ranges of the male and female plants: in Britain today female plants preponderate though hermaphrodite flowers have been recorded in the south-east. Samuelsson (1934) take the view that in Scandinavia, where, as in the British Isles, only (or almost only) female plants are present, distribution has been by vegetative means, large and heavy though the young rosettes may be.

It is of interest that whilst the fossil record provides no evidence whatever of this thermophilous plant in the glacial stages, it occurs nevertheless in every interglacial from the Pastonian onwards.

SCHEUCHZERIACEAE

Scheuchzeria palustris L. (Fig. 119; plates XI, XII)

568: I

The pollen grains of *Scheuchzeria palustris* in their unusual 'dyad' formation are readily recognizable, but have not often been reported, and the fruits are also

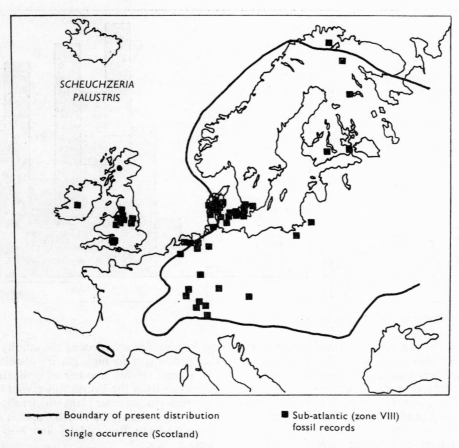

Fig. 119. Present distribution of *Scheuchzeria palustris* L. and its fossil records (pollen, rhizomes and seeds) in the Sub-atlantic period (Flandrian zone VIII for the British Isles). (After Tallis & Birks, 1965.)

recognizable but infrequent. On the other hand, where the plant has occurred, the elongate rhizomes clothed with pale papery leaf-bases, are generally of such abundance that they form a *Scheuchzeria* peat (plate XII). At many sites such as Danes Moss, Thorne Waste, Whybunbury Moss, Holcroft Moss and Flaxmere pollen records accompany those of the macroscopic remains. The oldest British record is for seed from the Middle Weichselian, and both fruits and seeds have been found at Westhay Level, Somerset in the zone VIII peat (Godwin, unpublished). The British records up to 1965 were effectively described by Tallis and Birks, who also consider the status and ecology of *Scheuchzeria* in Europe. When the plant was first described fossil from Britain it was in raised bog peats of the Somerset Levels where it particularly characterized the 'upper flooding horizon' of about A.D. 50 and where it formed part of the 'precursor peat' that developed by the widespread submergence, in shallow acid waters, of the previously dry bog surface (plate XI). In such situations it often accompanies plants like *Carex limosa* and *C. chordorrhiza* with aquatic *Sphagna* that constitute a floating mat round the margins of deeper pools on the regeneration complex of untouched

raised bogs like the Swedish Komosse. *Scheuchzeria* appears to respond to the flooding of dried out and uneven peat surfaces and a distinct layer of *Scheuchzeria* peat has been commonly found at the main bog recurrence surface in Danish peat bogs. As Birks and Tallis point out, it also commonly occurs at the flooding surface over wood peats.

In the British Flandrian *Scheuchzeria* has been found increasingly from zone VI onwards, with much the greatest frequency in zone VIII and the preceding transition from VIIb, this almost certainly reflecting the growth of raised bogs which in this country are seldom initiated, as ombrogenous structures, earlier than zone VIIa. The indication seems clear that the plant extended its range and abundance in the early part of zone VIII, very likely in response to the sub-atlantic climatic shift towards increased wetness, but since that time it has greatly diminished, even within the last century, in consequence of bog-cutting and drainage, so that it now persists only on Rannoch Moor. It has vanished from the only Irish site at Pollagh Bog, Co. Offaly, where it was only recently discovered living and where sub-fossil remains occurred in the peat to a depth of 175 cm.

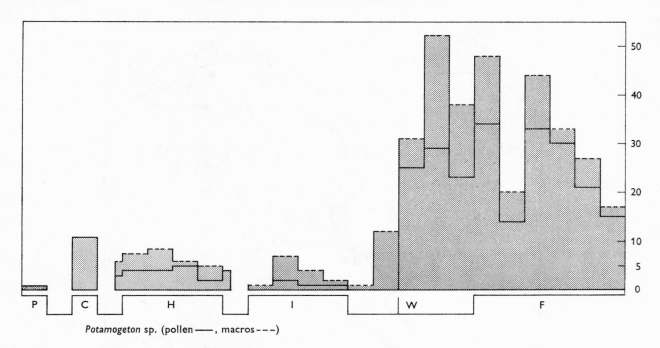

Potamogeton sp. (pollen ——, macros ---)

Scheuchzeria palustris is a plant of boreal–circumpolar distribution whose European range is indicated in fig. 119. Its virtual absence from Ireland seems to parallel its infrequency in western Scandinavia; the cause of this limitation is hard to guess although it seems more at home in the large bog pools typical of continental peat mires than in the smaller and shallower pools of those shewn to be characteristic of oceanic regions (Bulman, unpublished). The Middle Weichselian records are consonant with Late Weichselian and early Flandrian records from the Netherlands, and indicate the possibility that *Scheuchzeria* may have persisted through from the last glacial stage in Britain.

JUNCAGINACEAE

Triglochin maritima L. 'Sea arrow-grass'   574: 2

*m*W *Earith, Upton Warren*; F vɪɪb *Chapel Point*, vɪɪb *Bann Estuary*

Triglochin maritima is a highly characteristic species of the upper levels of British salt-marshes and its rhizomes have twice been found in Flandrian deposits of this kind. The so-called '*Triglochin* clays' of the Lincolnshire coast lie between the lower and the upper shore peats, in the latter of which occurs a Halstatt settlement level. By comparison with the Fenland basin (Godwin & Clifford, 1938) it seems probable that this record is to zone vɪɪb, as is the Irish record from the Bann estuary.

Two Middle Weichselian occurrences are evidently different since they are far inland. That at Upton Warren was tentatively explained by the local presence of brine springs, but these are absent at Earith and possibly both

may be regarded as reflecting the salinity found associated with permafrost in high glacial conditions. Bell (1969) lists a considerable number of obligate and facultative halophytes from the British full-glacial flora. *T. maritima* is a boreal–circumpolar plant which follows all the Scandinavian coasts up to the North Cape; it is found coastally right round the British Isles and also at a few inland brackish sites.

ZOSTERACEAE

Zostera marina L.   Eel-grass, Grass-wrack   576: 1

F vɪɪa *Island Magee*

Zostera noltii Hornem. (*Z. nana* auct.)   576: 3

*l*Plio *St Erth*

Mitchell has an unpublished record of the fruit of *Zostera noltii* from the estuarine beds at St Erth in Cornwall, possibly of late Pliocene age.

POTAMOGETONACEAE

Potamogeton spp.   Pondweed (Fig. 120)   577

The habitat of the potamogetons is such that their resistant organs are very well represented in our fossil records, and their hard 'pyrenes' or 'fruit stones' are among the most abundant plant identifications from all kinds of lake and channel deposits. In certain circumstances even the fragile stems and leaves may well be preserved. The pollen, although preserved in quantity, suffers

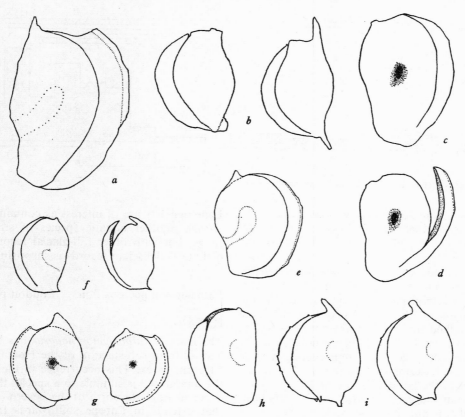

Fig. 120. Stones of various species of *Potamogeton* from various pollen-dated Irish deposits (Jessen, 1949): *a*, *P. praelongus* (zone II); *b*, *P. alpinus*; *c*, *P. natans* (zone VI); *d*, *P. natans*; *e*, cf. *P. perfoliatus* (zone VII); *f*, *P. pusillus* (zone VI); *g*, *P. polygonifolius* (zone V); *h*, *P. filiformis*; *i*, *P. obtusifolius* (zone II).

from the defect that it can be referred only to two main categories, those of the section Coleogeton (*Potamogeton pectinatus* and *P. filiformis*) and those of the section Eupotamogeton. Few British palynologists however make this distinction and all pollen of the genus is given unseparated in the record-head diagram. With the fruit stones the case is fortunately different, for their morphology and size allow with some trouble a fairly precise recognition of species to be made (fig. 120). Jessen (1955) published a most valuable key to these fruit stone identifications, enlarging and refining that given in his paper on Quaternary and flora-history of Ireland (1949), and Aalto (1970) has greatly extended and refined these observations. There remain of course difficulties due to the taxonomic complexities of the genus as well as to natural variation, immaturity and inadequate preservation (see Dickson, 1970).

The generic records are of limited value but the pollen and fruit records correspond satisfactorily: the absence of pollen records from the Middle Weichselian reflects the unsuitability of deposits of that age, in general, for pollen analysis. It is apparent that the genus was represented throughout all the interglacial stages and likely enough also through the glaciations: the records for individual species are far more illuminating.

Potamogeton natans L. *P. hibernicus* (Hagstr.) Druce 'Broad-leaved pondweed' (Fig. 120*c*, *d*; plate XIX*c*, *d*)
577: 1

The characteristic pyrenes of *Potamogeton natans* with their dimpled cheeks have been found so commonly from the Pastonian and Cromerian onwards, in both glacial and interglacial stages, that it must be considered a persistent native plant. The Early Weichselian record (Sidgwick Avenue) and the Middle Weichselian (Barrowell Green) merely anticipate an abundance of records in the Late Weichselian (especially in zones II and III) as in all succeeding zones of the Flandrian.

P. natans has a boreal–circumpolar distribution (Hultén) and extends in western Norway in some frequency to high latitudes within the Arctic Circle, a range that accords readily with its long uninterrupted presence in the British Isles, and its present occurrence over the whole extent of these islands.

Potamogeton polygonifolius Pourr. (*P. oblongus* Viv.; *P. spathularis* auct.) 'Bog pondweed' (Fig. 120*g*) 577: 2

C *Norfolk Co.*; H *Marks Tey*; *m*W *Barrowell Green*; *l*W *Ardcavan, Nant Ffrancon*; F IV, V *Wareham, Cranes Moor,*

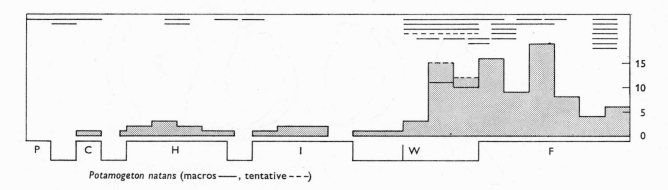

Potamogeton natans (macros ——, tentative - - -)

iv/v *Loch Dungeon*, v, viib *Roundstone*, v *Little Lochans*, v/vi *Ballyscullion*, vi *Castlelackan*, viia/viib *Freshwater West*, viib, viii *Loch Kinord*, viib *Wingham, Rhosgoch*, viib/viii *Dernaskeagh*, viii *Shapwick Track*

The small fruit stones of *Potamogeton polygonifolius* have been recorded once from the Cromerian and once from the Hoxnian interglacial: they next appear in the Middle Weichselian of the Lea Valley, in the Late Weichselian in Ireland and in all three zones in Snowdonia. Thereafter there is a fairly continuous array of records throughout the Flandrian, representing a wide geographical range. It seems certain that the plant has been native since, at the latest, the Middle Weichselian.

The Scandinavian range of *P. polygonifolius* has been given by Samuelsson and by Hultén, who shew it as common only in Denmark, south-west Sweden and south-west Norway, although there are scattered localities on the Norwegian coast as far north as 68° N. Although the plant tends to characterize oligotrophic waters it is not absolutely restricted thereto: nevertheless the fossil record may be limited in time and space on account of this, and this also explains why, in a distribution extending throughout the British Isles, it is sparser in the preponderantly limestone regions.

Potamogeton coloratus Hornem. (*P. plantagineus* Du Croz. ex Roem. & Schult.) 'Fen pondweed'   577: 3

H *Baggotstown, Hatfield, Nechells*; I *Wretton*; mW *Hedge Lane*; F v, vi *Apethorpe*, vi *Avonmouth*, ?*Nursling*, viib/viii *Branston*

Fruit stones of *Potamogeton coloratus* have been recorded from the Irish Hoxnian at Baggotstown (sub-stages i or i/ii, ii and ii/iii) and from the English Hoxnian at Nechells (sub-stage ii) and Hatfield (sub-stages ii and iii). In the Ipswichian at Wretton it occurred in sub-stages ii and iii, whilst in the Flandrian it has been found in zones v, vi and at the end of viib. All these records are from the temperate sections of their respective interglacials, and to set against this there is a single record from the Middle Weichselian of the Lea Valley, which is the more conspicuous since *P. coloratus* is a pondweed of sub-atlantic range that is so southern that in Scandinavia, apart from Bornholm, it barely enters southern Sweden. The Hedge

Lane deposits are of interest as containing *Damasonium alisma*, another aquatic species of strikingly southern range, but with other full-glacial records. The absence of Late Weichselian records is surprising.

Potamogeton nodosus Poir. 'Loddon pondweed'
    577: 4

H *Clacton*

The fruit of *Potamogeton nodosus* has been recorded by Turner from sub-stage iii of the Hoxnian interglacial at Clacton, Essex. This occurrence in the warmest part of the interglacial is suitable to a species that now does not occur in England north of the Severn–Thames line, and that extends in Europe southwards from Poland and Germany to the Mediterranean.

Potamogeton lucens L. 'Shining pondweed'   577: 5

H *Nechells*; I ?*West Wittering*; lW ?*Rodbaston*; F iv ?*Star Carr*, v/vi *Tadcaster*

Three out of the five records for fruit stones of *Potamogeton lucens* are qualified: the two unqualified identifications are from sub-stage ii of the Hoxnian and the zone v/vi transition in the Flandrian. It is a species of wide range both in the British Isles and in Scandinavia and the Late Weichselian and early Flandrian records with radiocarbon ages of 10670 and 9480 years B.P. respectively are not at all surprising, though the identifications are only tentative.

Potamogeton gramineus L. (*P. heterophyllus* Schreb.) 'Various-leaved pondweed'   577: 6

Cs *Beeston, Cromer, Trimingham, Overstrand, Mundesley*; Be *Norfolk Co.*; C *Norfolk Co.*; lA *Gort*; H *Southelmham, Gort*, ?*Fugla Ness*; I *West Wittering, Stone*; eW *Sidgwick Av.*; mW *Ponders End, Hedge Lane, Temple Mills, Angel Rd, Barnwell Station*; lW *Colney Heath, Hartford*, ?*Mapastown*; F vi ?*Elland*, viia, viib *Loch Park*, viib/viii *Wallasey*, viib/viii *Hightown*

The fossil record for *Potamogeton gramineus* begins with five identifications from the Cromer Forest Bed series, backed up by later ones from the Beestonian and Cro-

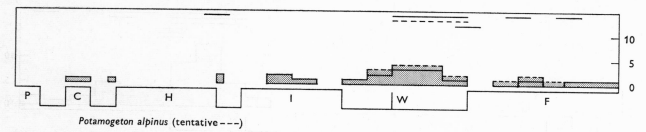

Potamogeton alpinus (tentative ---)

merian of the Norfolk coast. At Gort it has been found in the late Anglian and sub-stages III and IV of the Hoxnian, in which interglacial there is also one record from South-elmham and a tentative one from the Shetlands (sub-stage IV). From the Ipswichian there are two records not referable to sub-stages. One Early Weichselian record is followed by five from the Middle Weichselian and three that represent each zone of the Late Weichselian. Finally, five Flandrian occurrences span zones VI to VIII.

The record, though not prolific, exhibits such continuity as to leave little room for doubting the permanence of *P. gramineus* in the British flora since the middle Pleistocene, more especially since the Weichselian records are so numerous and representative.

Potamogeton gramineus is a boreal–circumpolar plant in Hultén's terminology, and it occurs frequently throughout Sweden into the Arctic Circle and locally throughout the north of the peninsula. It is much less frequent in the west of Scandinavia than in the east. It occurs locally throughout the British Isles though less commonly in Wales and not at all in Devon and Cornwall. Its preference for non-calcareous waters makes the abundance of East Anglian interglacial and glacial records somewhat surprising.

Potamogeton gramineus × lucens (*P. × zizii* Koch ex Roth)
577: 5 × 6

mW Hedge Lane, Barnwell Station; *lW* ?Mapastown

Potamogeton × zizii has been recorded from the Middle Weichselian deposits of both the Lea Valley and that of the Cam. There is a tentative identification also from Ireland in zone III of the Late Weichselian. Of the two parent species, *P. gramineus* has a strong Weichselian record and *P. lucens* a weak one.

Potamogeton alpinus Balb. (*P. rufescens* Schrad.) 'Reddish pondweed' (Figs. 120b, 121)
577: 7

Fruit stones of *Potamogeton alpinus* have been reported from the Cromerian, from the two middle sub-stages of the Ipswichian interglacial and from all Flandrian zones from V onwards. The glacial stage records are at least equally abundant, covering the late Anglian, early Wolstonian and Early, Middle and Late Weichselian. Although about a quarter of the identifications are qualified they do not impair the evidence of continuous presence, at least since the late Hoxnian, in this country.

There is also considerable evidence of local persistence, as at Hoxne (*lA*, *lH/eWo*, *eWo*), at Wretton (sub-stages II and III of the Ipswichian) and Dunshaughlin (Flandrian VIIa/VIIb, VIII). The record is given strength by the wide variety of authors (19 or 20) who have identified the plant. The records are mainly English and Irish, although the present range is general throughout the British Isles (fig. 121). *P. alpinus* is a boreal–circumpolar species shewn by Hultén as frequent in Scandinavia well into the Arctic Circle and occasional to the extreme north of the peninsula.

Potamogeton praelongus Wulf. 'Long-stalked pond-weed' (Figs. 120a, 122)
577: 8

The fruit stones of *Potamogeton praelongus* are readily recognizable and Miss Blackburn has also recorded leaves and stems of it from the Late Weichselian at Neasham. It is recorded from the Cromer Forest Bed series and

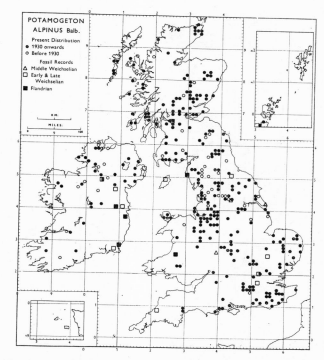

Fig. 121. Present, Weichselian and Flandrian distribution in the British Isles of *Potamogeton alpinus* Balb., a species apparently retaining a similar broad range throughout these periods.

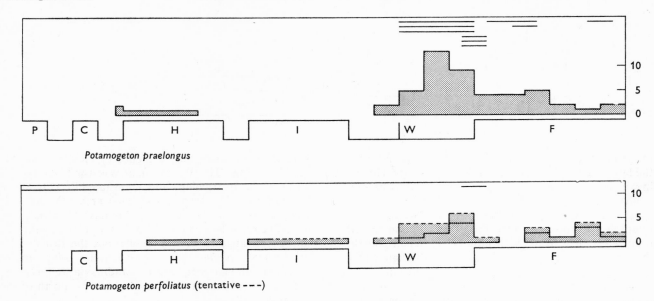

Potamogeton praelongus

Potamogeton perfoliatus (tentative - - -)

from the first three sub-stages of the Hoxnian interglacial, as well as twice from the late Anglian. It is however in the Weichselian that records are particularly frequent, with Waltham Cross and Hedge Lane from the Middle Weichselian and large numbers from the Late Weichselian, especially zone II. Thereafter records occur in all Flandrian zones and it is evident that *P. praelongus* has been in this country since the Middle Weichselian and

that its present range has to be viewed against its Late Weichselian prevalence (fig. 122).

P. praelongus occurs throughout the British Isles though with some emphasis on an easterly as against a western trend. It is placed by Matthews in his Continental–northern category, and by Hultén in the boreal–circumpolar group. *P. praelongus* occurs scattered throughout the whole of Scandinavia including the far north but, as in Britain, is more frequent on the continental side than on the oceanic.

Potamogeton perfoliatus L.   'Perfoliate pondweed'
(Fig. 120*e*)   577 : 9

There is often uncertainty in making an unqualified identification of the fruit stones of *Potamogeton perfoliatus* but, as the record diagram shews, tentative and positive records support one another in the overall picture. This slightly indicates presence in the Cromer Forest Bed series, and through both the Hoxnian and Ipswichian interglacials. A qualified Middle Weichselian record from Syston is followed by numerous records from all Late Weichselian zones and enough Flandrian zone records to indicate continuous presence. Local persistence is shewn at many sites: thus in the Ipswichian at Bobbitshole (sub-stages Ib, IIa, IIb) and Histon Road (sub-stages III, IV), in the Late Weichselian at Lopham Fen (zones I, II, III) and subsequently at Loch Cuithir (zones III, VI, VIIa, VIIb) and Bradford Kaim (III, VI).

Potamogeton perfoliatus is classed by Hultén as a boreal–circumpolar plant and he shews it as a fairly common plant through almost the entire north–south extent of lowland Scandinavia. On the other hand, although present locally in Norway, it shows a very striking increase in frequency to the east. A similar tendency is discernible in the British Isles where it is also of wide extent.

POTAMOGETON
PRAELONGUS
Wulf.

Present Distribution
● 1930 onwards
○ Before 1930
Fossil Records
□ Late Weichselian
■ Flandrian

Fig. 122. Present, Late Weichselian and Flandrian distribution in the British Isles of *Potamogeton praelongus* Wulf.: the plant was apparently as widely distributed in the Late Weichselian as now.

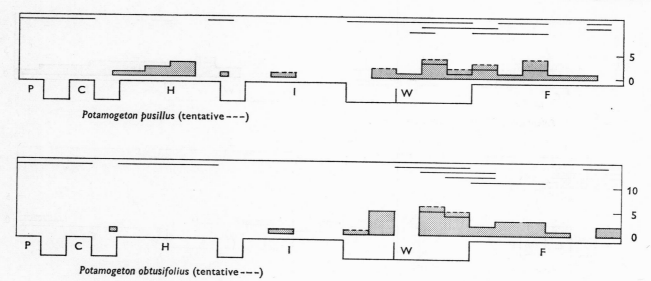

Potamogeton pusillus (tentative ---)

Potamogeton obtusifolius (tentative ---)

Potamogeton friesii Rupr. (*P. mucronatus* Schrad. ex Sond.) 'Flat-stalked pondweed'   577: 11

C *Norfolk Co.*; I *?Selsey*; *IW ?Hartford, Longlee Moor, ?Kentmere*; F III/IV *Drumurcher*

Fruit stones of *Potamogeton friesii* have recently been identified by Wilson from the Cromerian, *sensu stricto*, and there is a tentative record from sub-stage II of the Ipswichian. There are three records: two are tentative from the Late Weichselian zone II and at Kentmere both II and III are represented. An Irish record covers the Weichselian/Flandrian transition. *P. friesii* is an amphi-atlantic plant that, according to Hultén, occurs at the northern end of the Gulf of Bothnia and even at the Arctic Circle, but it is very local in Scandinavia. Widely but locally present in the British Isles it is, like many aquatics, much commoner in the south-eastern quarter of England. Thin as the fossil record is, it suggests that the present range derives from a Late Weichselian one.

Potamogeton pusillus L. (*P. panormitanus* Biv.)   577: 13

Although there are frequent qualifications in the identification of fruits of *Potamogeton pusillus* and numerous transitional or undivided zone records, there is a convincing overall picture of presence from the Cromer Forest Bed series onwards. It occurred in three sub-stages of the Hoxnian and at the type-site was present in sub-stage III, the early Wolstonian and the transition between Hoxnian and Wolstonian. It was tentatively seen in two zones of sub-stage II of the Ipswichian at Bobbitshole. Two tentative Middle Weichselian records are given by Bell and Reid respectively and thereafter through the Late Weichselian the plant has clearly been present up to the present day. At Neasham in zone IV Miss Blackburn identified its turions, the vegetative propagules.

P. pusillus has a British range very like that of *P. friesii*,

but is far commoner than that species in Scandinavia. It is classed by Hultén as boreal–circumpolar so that perglacial survival in Britain would not be surprising.

Potamogeton obtusifolius Mert. and Koch 'Grassy pondweed' (Figs. 120*i*, 160)   577: 14

In pre-Weichselian deposits *P. obtusifolius* is infrequently recorded although it was present in the Cromer Forest Bed series, the Hoxnian and Ipswichian interglacials and the late Anglian. There is however a tentative Early Weichselian record and four Middle Weichselian records, the latest by Bell from Earith. Zones II and III of the Late Weichselian supply numerous records and thereafter the species has been found in all the Flandrian zones except VIIb. It has evidently been present in this country at least since the middle of the last glacial stage. At many sites it shews strong local persistence or recurrence: thus Bagmere (zones II and IV), Ballaugh (II, III), Ballybetagh (II, V), Bradford Kaim (III, VI), Skelsmergh Tarn (II, VI) and Kirkby Thore (V, VI).

   P. obtusifolius is an amphi-atlantic plant that belongs to Matthews' Continental–northern group of plants and Hultén shews it as scattered throughout Scandinavia, chiefly in lowland sites, and extending sparsely within the Arctic Circle. Apart from Cornwall, Somerset and north-west Scotland it occurs locally throughout the British Isles.

Potamogeton berchtoldii Fieb. 'Small pondweed'

577: 15

All the British records for fruits of *Potamogeton berchtoldii* are quite recent identifications, although by several separate investigators and seldom qualified. They begin in the early Wolstonian and continue in the late Wolstonian and the two first sub-stages of the Ipswichian interglacial. After a single Early Weichselian occurrence

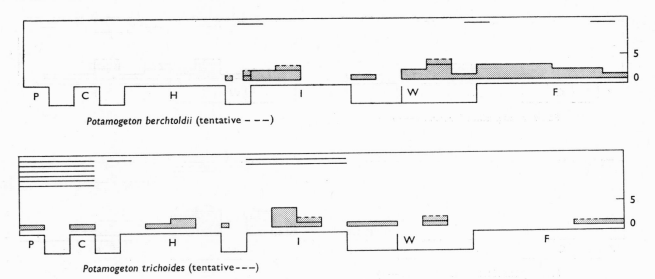

Potamogeton berchtoldii (tentative – – –)

Potamogeton trichoides (tentative – – –)

it has been recorded for all the Late Weichselian and Flandrian zones. As with so many aquatics there has been strong local persistence as at Ilford (*l*Wo, *l*Wo/I, I, Ia, IIb), Selsey (tentative – *l*Wo, I, Ia, Ib), Loch Park (I, II, III, IV, V, VI, VIIa, VIIb), Drymen (III/IV, IV, V), Loch Fada (V,VI), Loch Kinord (VI, VIIb + VIII) and Ehenside Tarn (VIIa, VIII). The records by the Vasaris (1968) from Loch Park in Aberdeenshire are particularly striking and make it easy to accept the general case of long persistence in this country.

P. berchtoldii is, according to Hultén, a boreal-circumpolar species, with a wide European range that extends right to the north of Scandinavia; it is also generally distributed throughout the British Isles.

Potamogeton trichoides Cham. & Schlecht. 'Hairlike pondweed'   577: 16

Five of the Cromer Forest Bed series records for *Potamogeton trichoides* were made by C. Reid as var. *tuberculata* in the 1882 paper, but were not afterwards mentioned in his comprehensive work. Subsequently Wilson and West have recognized *P. trichoides* fruits in the two separate Pastonian and Cromerian interglacials, and it has also been found in the middle sub-stages of both the Hoxnian and Ipswichian. Indeed at Hoxne, where it had been first reported by Reid, it was rediscovered by West at a similar horizon. In view of these interglacial records it is surprising that the Flandrian occurrences should be so sparse and so late. The Hoxne discoveries included not only sub-stage III of the interglacial but the transition from the preceding late Anglian and the succeeding early Wolstonian. There are records also from the Early, Middle and Late Weichselian and the fruits have also been found fossil in the Late Weichselian in Denmark. The overall record gives an indication of permanent occupation in the British Isles: reintroduction in the Flandrian of course cannot be ruled out.

P. trichoides is one of Hultén's west European–Continental species, which has a few scattered localities in Denmark and three or four only in artificial and eutrophic habitats in southernmost Sweden. In England it is chiefly southern and eastern: it is absent altogether from Ireland and almost so from Scotland and Wales. It is often found in brackish situations and several of the fossil localities suggest this kind of habitat.

Potamogeton acutifolius Link. 'Sharp-leaved pondweed'   577: 18

*l*A ?*Hoxne*; H *Clacton*; *e*Wo *Brandon*; I *Selsey, Bobbitshole, Austerfield, Histon Rd*; *e*W *Sidgwick Av.*; *m*W *Earith, Marlow*; *l*W ?*Scaleby Moss*; F v *Dogger Bank*, VI *Skelsmergh*

Fruit stones of *Potamogeton acutifolius* were tentatively identified from the late Anglian. They were found in sub-stage III of the Hoxnian and in all four sub-stages of the Ipswichian and at four separate sites. There is an early Wolstonian record for the pyrenes and leaf apex, and there are records also for Early and Middle Weichselian with a tentative Allerød phase record from Cumberland. Brackish water deposits on the North Sea floor have yielded a zone v record and there is also a Flandrian zone VI record from Westmorland.

P. acutifolius, placed by Hultén in his west European-middle Siberian plants, is restricted in Scandinavia to Sweden south of 60° N latitude and Denmark: in the British Isles it is uncommon and restricted to the south-eastern sector of England. None the less the Weichselian records are unqualified and support the possibility of perglacial and more ancient survival in this country.

Potamogeton trichoides Cham. & Schlecht. or **P. acutifolius** Link   577: 16 or 18

*l*Wo *Ilford*; I *Wretton*; *l*W *Ballaugh*; F IV *Ballaugh*

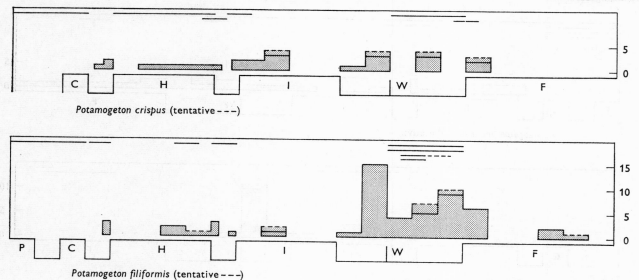

Potamogeton crispus (tentative – – –)

Potamogeton filiformis (tentative – – –)

A number of recent identifications of *Potamogeton* pyrenes have been made to '*P. trichoides* or *P. acutifolius*': these embrace the terminal part of the Wolstonian, substages II and III of the Ipswichian and, in the Isle of Man, zone II of the Late Weichselian followed by zone IV of the Flandrian. The two species are not generally confused in the main records, and these records of alternative identity sort equally well with the records for either separate species.

Potamogeton crispus L. 'Curled pondweed'   577: 19

As the record-head diagram shews, *Potamogeton crispus* has an almost complete record from the Cromer Forest Bed series to the middle Ipswichian, with middle and late Anglian records, three Hoxnian sub-stages, early and late Wolstonian, and sub-stages I and II of the Ipswichian all represented. The record is resumed for Early, Middle and Late Weichselian and ceases with three records for zone IV of the Flandrian. If it is accepted that the paucity of Flandrian records is due to the fact that fluviatile deposits of this age have seldom been examined, the record indicates long persistence in this country, probably through the glaciations. This despite the present range of the species, which belongs to Samuelsson's group of south Scandinavian water plants, and does not extend further north than Öland and is almost absent from Norway. Scattered sites have recently been reported from Finland. *P. crispus* occurs throughout the British Isles but is markedly commoner in the east and south than elsewhere.

Potamogeton filiformis Pers. (*P. marinus* auct.) 'Slender-leaved pondweed' (Figs. 120h, 123, 150; plate XIXf)   577: 20

Pyrenes of *Potamogeton filiformis* have been found in the Cromer Forest Bed series and sparsely in both the

Hoxnian and Ipswichian interglacials. By contrast records from the glacial stages are frequent, including the undivided Anglian and Wolstonian, late Anglian, early and late Wolstonian and Early Weichselian, but especially the Middle Weichselian with 15 sites, and frequent sites in all zones of the Late Weichselian and Flandrian zone IV.

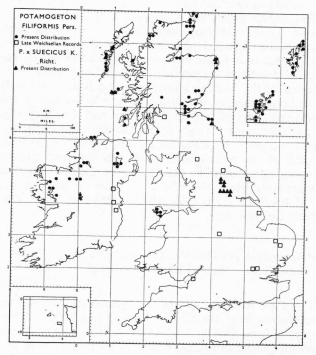

Fig. 123. Present, Late Weichselian and Flandrian distribution in the British Isles of *Potamogeton filiformis* Pers., a species retracted northwards since the Weichselian. Also shewn, the present range of *Potamogeton × suecicus* K. Richt., the hybrid between *P. filiformis* and *P. pectinatus*. The hypothesis that a former more southerly extension of *P. filiformis* is required to explain the present range of the hybrid is thus supported by the fossil evidence.

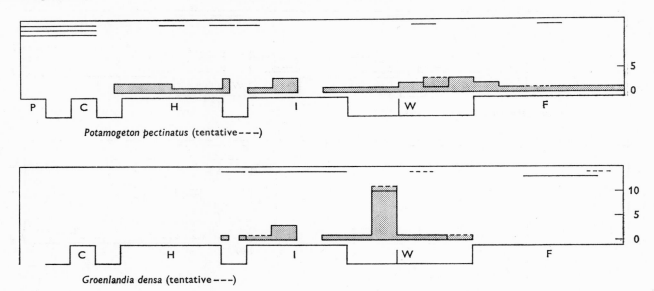

Potamogeton pectinatus (tentative - - -)

Groenlandia densa (tentative - - -)

By contrast there are only three later Flandrian records from zones VIIa and VIIb. Local persistence is notable at the Irish site of Baggotstown (lA, H II/III, III and eWo).

P. filiformis is an amphi-atlantic species, classified by Matthews as Continental–northern, that has been recorded from Late-glacial deposits in southern Sweden. It occurs in brackish water along the whole coastline of Scandinavia except the Kola peninsula, and has also a wide though scattered distribution in eutrophic, fresh waters inland. Its halophilic tendencies may partly explain why it has been found in almost every British full Weichselian plant-bearing deposit. Samuelsson regards it as typical of the Swedish alpine zone and it has a strongly northern range in the British Isles which takes in the northern part of Scotland, Ireland and Anglesey (fig. 123). The present range of P. × suecicus, the sterile hybrid between P. filiformis and P. pectinatus, extends south of the present area of P. filiformis and may be presumed to relate to the Middle or Late Weichselian when both parents shared a much more southerly range. The fossil record is evidently indicative of long occupation of Britain.

Potamogeton pectinatus L. (*P. interruptus* Kit.; *P. flabellatus* Bab.) 'Fennel-leaved pondweed' (Plate XXVIe)
577: 21

There are three records of the pyrenes of *Potamogeton pectinatus* from the Cromer Forest Bed series. Subsequently from the late Anglian onwards there is no glacial or interglacial sub-stage or zone to the present, save substage III of the Ipswichian, without record for it. A Pakefield record by West from the early Wolstonian includes both fruit stones and tubers for this pondweed. It shews also strong local persistence in the Hoxnian interglacial with records from sub-stages I and II at Nechells and from I, II, II/III as well as early Wolstonian from Baggotstown.

The remarkable continuity of the fossil record must indicate long native status and probably perglacial survival.

P. pectinatus is abundant in base-rich waters throughout the British Isles, and commonly occurs in brackish conditions. The latter fact accounts for its preponderantly coastal range in Scandinavia where it occurs round the whole Baltic coast, rather sparsely at the north of the Gulf of Bothnia, and may be associated with its Weichselian occurrences in Denmark as well as in this country. It is a circumpolar plant that is widespread throughout the world excepting parts of the Arctic. Distributed widely also through the British Isles, it is commoner nevertheless in the south-eastern sector of England, possibly in consequence of the base-rich river water of the lowland zone.

Groenlandia densa (L.) Fourr. (*Potamogeton densus* L.) 'Opposite-leaved pondweed' (Fig. 150f)   578: 1

The record for fossil fruit stones of *Groenlandia densa* is remarkable in demonstrating virtual continuity from the early Wolstonian, through the Ipswichian and the succeeding Weichselian, together with so many records from the Middle Weichselian that Bell writes of it as 'almost a constant' and hence very unlikely to be derivative. It has been found in broadly post-Boreal muds at Hockham Mere and at the VIIb/VIII transition at the neighbouring Old Buckenham Mere, but it is profitless to speculate on the absence from either the Flandrian or Hoxnian and earlier deposits.

The high Weichselian abundance is remarkable in view of the fact that *G. densa* does not grow in Scandinavia north of Denmark. It is however classed by Hultén as a sub-atlantic plant, and in Britain it extends sparsely into Scotland, although like so many Potamogetons and other aquatics, it is far commoner in the south-eastern part of the British Isles. Bell (1969 has discussed the implications of *G. densa* as part of a conspicuously

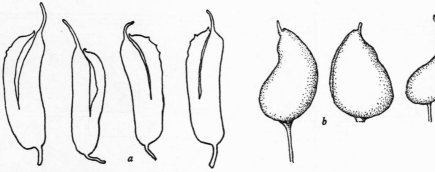

Fig. 124. *a*, Four fruits of *Zannichellia palustris* from sub-zone VIIa, Bann Estuary, Co. Londonderry; *b*, two fruit stones of *Ruppia maritima* (*R. spiralis* Dum) from sub-zone VIIa, Somerset, Co. Londonderry; *c*, fruit stone of *R. rostellata* Koch. (*R. maritima* L.) from sub-zone VIIa, Somerset, Co. Londonderry. (After Jessen, 1949.)

southern element in the British Weichselian flora and inclines to the view that, in this case, high summer temperatures may have been a major controlling factor.

RUPPIACEAE

Ruppia sp.   579

I *Shortalstown, Trafalgar Square, Selsey*; F v *Dogger Bank*, VI *Ringneill Quay, Jenkinstown*, VI/VIIa *Loch Cranstal*

Fruits reported either as 'Ruppia sp.' or 'R. maritima agg.' have been found three times in sub-stage II of the Ipswichian, twice in north-east Ireland and once in the Isle of Man late in Flandrian zone VI. All but one of these is cited as from a brackish water deposit, the two Irish Flandrian sites marking the top of the post-glacial eustatic rise of ocean level.

Behre and Menke (1969) have reported pollen of *Ruppia* from Flandrian zone v deposits in the North Sea where they have also reported fruits of *R. maritima*. Despite the Loch Cranstal record neither of the two British species of *Ruppia* appears to grow at present in the Isle of Man, and neither seems to occur in inland saline habitats. It is interesting in this connection that neither has been recorded from the extensive Weichselian halophyte floras (see Bell, 1969).

Ruppia spiralis Dum. (*R. maritima* auct. mult. non L.) (Fig. 124*b*)   579: 1

I *Stone, West Wittering*; F IV/v *North Sea*, v, VI *Moss Lake*, VI *Jenkinstown, Bann Estuary*, VI/VIIa *Cushendun, Westward Ho!*, VII *Hightown*, VIIa *Somerset (Ireland)*, VIIb *Littleport*, VIII *Puriton Drove*

Jessen (1949) has pointed out that the fruits of *Ruppia spiralis* may be recognized by the fact that style and pedicel lie in one line, whereas in *R. maritima* L. (*R. rostellata* Koch) they are out of line, on account of the extreme obliquity of the fruit.

There are two interglacial deposits of Ipswichian age and both in a coastal situation. The records from Moss Lake, Liverpool are from lake deposits shewing no direct

marine influence, but the remaining Flandrian records are directly related to marine incursions, the Irish ones specifically to the later stages of the post-glacial eustatic rise of sea-level.

R. spiralis occurs sparsely and intermittently in brackish ditches round the coasts of the British Isles: it has lost a good deal of its known historic range and is now known from the coast of north-east Ireland, Lancashire and south-east England. It occurs round the coasts of south Scandinavia discontinuously: it is very infrequent and scattered north of 61° N, but in western Norway actually extends to 68° N. It seems probable that *R. spiralis* followed the transgressing shore lines as they rose from their glacial low levels.

Ruppia maritima L. (*R. rostellata* Koch) (Fig. 124*c*) 579: 2

I *Stone, West Wittering*; F IV, IV/V, v *Dogger Bank*, VII *Hightown*, VIIa *Westward Ho!*, *Somerset (Ireland)*, VIIa/VIIb *Avonmouth*, VIIb *Littleport*

The fossil record for fruits of *Ruppia maritima* closely resembles that for *R. spiralis*, and indeed many of the sites are identical. Again there are two Ipswichian coastal sites and a range of Flandrian records from zone IV to VIIb, and again there is the close association with coastal marine situations and brackish conditions.

R. maritima occurs locally round British coasts, somewhat less commonly in the north. In Scandinavia its distribution resembles that of *R. spiralis* but has a somewhat higher northern range in Norway. Fossil records of *R. maritima* round the northern coasts of the Gulf of Bothnia shew that it formerly had a wider post-glacial range, and it is supposed (Samuelsson, 1934) that the retreat has been caused not by temperature change, but by the known diminution of salinity of the Baltic since the Littorina period.

It should be realized that over a large part of the British Isles coastal deposits of Late-glacial and early post-glacial age such as alone could contain fossil *Ruppia*, have been more or less deeply submerged by the post-glacial eustatic rise in sea-level and no conclusions must be drawn from the absence of early records.

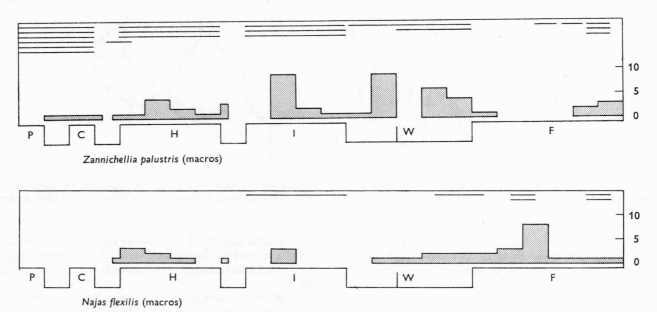

Zannichellia palustris (macros)

Najas flexilis (macros)

ZANNICHELLIACEAE

Zannichellia palustris L. (*Z. polycarpa* Nolte ex Reichb.; *Z. gibberosa* Reichb.; *Z. pedunculata* Reichb.; *Z. pedicellata* Fr.; *Z. maritima* Nolte ex G.F.W. Mey.; *Z. brachystemon* Gay ex Reut.) 'Horned pondweed' (Fig. 124*a*)　　　　　　　　580: 1

We have followed the most recent edition of the *London Catalogue* in treating *Zannichellia* as comprising in Britain the single species *Z. palustris* L., and accordingly conflate all the fossil records, many of which were attributed to *Z. pedicellata* Fr. (see the first edition).

The record-head diagram shews that *Z. palustris* has been more or less continuously present in this country since the middle Pleistocene. There are six records from the Cromer Forest Bed series now given more precision by records on the Norfolk coast from the Beestonian, the Cromerian and the early Anglian. Similarly records for all sub-stages of the Hoxnian are preceded by late Anglian and Anglian/Hoxnian transition records and are followed by three in the early Wolstonian. Sub-stage I of the Ipswichian interglacial so far has no records but there are no less than nine in sub-stage II, many but not all in coastal situations. Sub-stages III and IV are represented and are followed by one record in the Early Weichselian and six in the Middle Weichselian. The latter indicate both the likely perglacial persistence of the species, and the prevalent halophily of plant communities at inland sites in the Full-glacial time. Not surprisingly, there are also numerous Late-glacial records, especially in zone II. Flandrian records are relatively sparse, perhaps because the coastal deposits of the earlier zones are now submerged by the post-glacial rise of sea-level.

The Scandinavian distribution of the cosmopolitan aggregate *Z. palustris* is remarkable in being fairly con-

tinuous round the whole Baltic coast, but highly discontinuous on the Atlantic shore, with two or three isolated records on the coast of the Arctic Ocean and White Sea. Hultén's map includes a considerable number of fossil records round the northern Baltic and outside the present range of the species. The cause of the diminution here is uncertain: it may be partly associated with the retreat of the coastline since the Littorina period, partly with the growing up of lakes formerly suitable to the species, and partly with the curious phenomenon observed also in *Najas marina*, that the species formerly was less strongly associated with brackish water conditions than it now is. Although *Z. palustris* is found throughout the British Isles, it is far less common in Scotland, Ireland and Wales than in England and is most frequent south-east of the Humber–Severn line.

NAJADACEAE

Najas flexilis (Willd.) Rostk. & Schmidt (Figs. 125, 126, 128*a*; plate XIX*g*)　　　　　　　581: 1

The last few years have substantially enlarged the fossil record for *Najas flexilis* in this country, and in particular Irish sites at Gort, Kildromin and Baggotstown have jointly yielded records for the first three sub-stages of the Hoxnian interglacial as well as the preceding late Anglian and succeeding early Wolstonian. At Baggotstown it was found in four of these periods and at Gort in two. There are now three records from sub-stage II of the Ipswichian interglacial, including that at Histon Road, Cambridge where it was accompanied by the extinct *N. minor*. The record for the undivided Ipswichian is that from West Wittering, formerly recorded as *N. graminea* by Reid and now revised by West. A significant Middle Weichselian

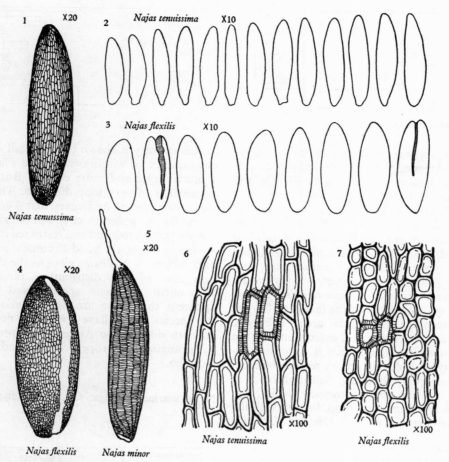

Fig. 125. Fruits of three species of *Najas* showing shape and surface cell-pattern of *N. minor*, *N. flexilis* and *N. tenuissima* (Backman, 1950). *N. tenuissima* and *N. minor* occur in the British Isles only in interglacial deposits; *N. flexilis* has a present disjunct western range in the British Isles.

record from Earith (Bell, 1969) is followed by Scottish records for all zones of the Late-glacial and by more widespread records for all the Flandrian zones. There is very clear evidence of persistence at some sites, thus at Loch Park (zones I, II, III, V, VI), Loch Kinord (II, III, III/IV, V, VI, VIIb) and Kentmere (VI, VIIa, VIIa/VIIb) and the fruits are locally abundant in the early Flandrian also in Loch Cill Chriosd (H. J. B. Birks). Here, as in Scandinavia *N. flexilis* was particularly abundant in zone VI, i.e. within the Boreal period. At this time it is recorded very frequently from sites in the north and west of Ireland, and indeed Mitchell reports that lake muds of Boreal age are invariably crowded with its slender shining fruits. The falling off of records after zone VI is undoubtedly to be taken as indicating a real diminution in frequency of the species, and this is attested also by the fact that the fossil records have a much wider geographical range than has the living plant. This appears to be true even though the habit of the plant is so inconspicuous that careful and deliberate search would possibly increase the number of its known localities in Britain where the *Atlas of the British Flora* shews it limited to

a few sites in western Ireland and Scotland with one in the English Lake District (fig. 126). Backman (1948) gives a present European distribution between the latitudes of about 57° and 62° N, but Luther (1945) reports it on the White Sea in European Russia, whilst in Asia it is found on the Mongolian highland where summer temperatures are high but winter temperatures are generally around −20° C. This evidence of resistance to low winter temperatures agrees with fossil evidence for presence in the Middle and Late Weichselian, but at the same time the Boreal frequency and range have been held to indicate the requirement for long or warm summers (Vasari, 1962). There is no doubt that in Scandinavia *N. flexilis* was formerly abundant far north of its present range for there are abundant fossil records in north Finland and Lapland. Whilst the subsequent restriction of range may have been affected by reduction in the numbers of calcareous or basic lake basins unaffected by drainage and other human interference, it seems very likely that reduction in summer temperatures since the hypsithermal has also been involved.

Najas flexilis is placed by Hultén in his category of

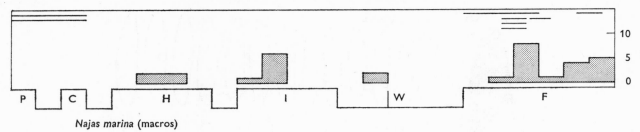

Najas marina (macros)

amphi-atlantic plants whose main bulk of present occurrences lie in North America: it there extends between British Columbia and Quebec in the north to Florida and California in the south. It is far less strongly represented in Europe and eastern Asia. This great disproportion led to a somewhat facile acceptance of the view that *N. flexilis* was a species of North American origin which had successfully crossed the Atlantic by a land bridge (since vanished), along with *Eriocaulon septangulare* and *Spiranthes romanzoffiana*, plants of very similar range. The notion that the western sites of *Najas flexilis* represent a relict Tertiary range is opposed by the fact that they lie, almost without exception, within the area of Europe covered by glaciation at its maximal extent, and by the fact that in its western interglacial sites it is not accompanied by species typical of the Tertiary record, whereas in more central European sites (e.g. Schwanheim near Mainz, Hessen) it *is* found along with such species, e.g. *Eucommia ulmoides* and *Hammemelis* sp. In a nutshell the present exiguous and extreme westerly range of *N. flexilis*

in this country is shewn by the fossil evidence to be the consequence of withdrawal in late Flandrian time from a much larger and more easterly British range, and an occupation of very long duration. This coincides with the view expressed by Deevey that *Najas flexilis* is clearly a species of general circumpolar distribution whose range has been reduced and restricted in Europe and Asia. As Deevey says, the good documentation of the Quaternary history of *N. flexilis* gives us the clue to the explanation also of the distribution ranges of the other Irish–north American species, and indeed makes it apparent that these are not essentially different from the species which have wide, but not unequal, ranges on both sides of the Atlantic, or species abundant and wide-ranging in Europe, but rare and easterly in North America.

Najas marina L. (Figs. 127, 128*b*, 129; plate XXVIII*d*)
581: 3

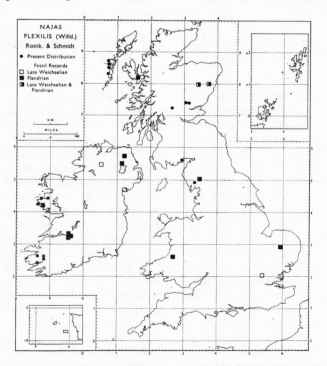

Fig. 126. Present, Weichselian and Flandrian distribution in the British Isles of *Najas flexilis* (Willd.) Rostk. & Schmidt: its present range is evidently contracted and disjunct.

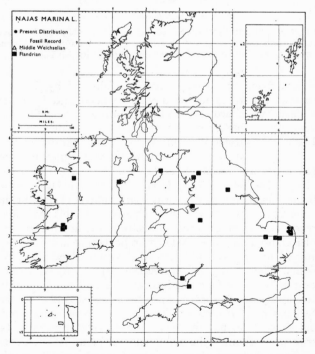

Fig. 127. Present, Middle Weichselian and Flandrian distribution in the British Isles of *Najas marina* L.; the very restricted modern range contrasts strongly with its wide mid-Flandrian distribution.

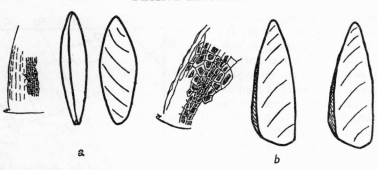

a   b

Fig. 128. Fossil fruits of *Najas* from Hockham Mere, Norfolk: *a*, *N. flexilis* Rostk. & Schmidt; *b*, *marina* L. The former species is represented by one fruit only (zone VIIa), the latter by many from zones VI and VII (see fig. 129).

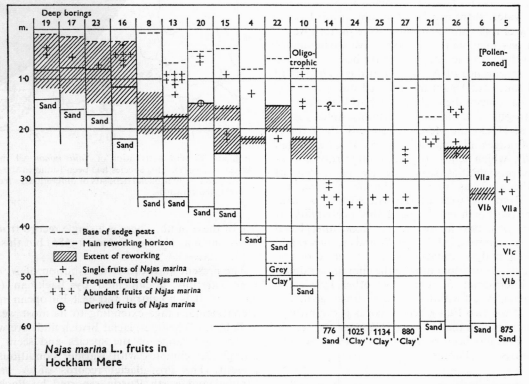

Fig. 129. Diagram shewing records of the fruits of *Najas marina* L. found in the lake muds at Hockham Mere, West Norfolk, in a series of borings to different depths. A period of low water levels at the end of zone VI produced a band of reworked lake mud shewn in the diagram by hatching. Fossil fruits are abundant in the reworked layer and in the muds of zone VIIa which overlie it, that is to say the period of the thermal maximum.

The characteristic broad 'bivalve' fruits of *Najas marina* have been recorded from two sites in the Cromer Forest Bed series, from sub-stages II and III of the Hoxnian interglacial and once from sub-stage I of the Ipswichian and six times from sub-stage II. There are no Late Weichselian records, but in the Middle Weichselian it was recovered both at Barnwell Station and Waltham Cross. In the Flandrian, a single record in zone V from Moss Lake precedes no fewer than eight in zone VI, a Boreal expansion like that for *N. flexilis*. A single record in zone VIIa is followed by more frequent records in zones VIIb and VIII; the last zone has three sites in East Anglia including Hockham Mere where the stratigraphy was investigated in detail and where the fruits were present in considerable abundance in several borings, with highest frequency in zones VI and VIIa (fig. 129). Jessen noted a similar abundance of fruits in zone V at Boskill, and at the zone V/VI transition at Red Bog in Lough Gur Townland.

The records as a whole make it apparent that *N. marina* has been native in the British Isles since early Boreal time, and it is of particular interest to note that in the Norfolk Broads where the plant still survives there are two fossil records of its fruits (fig. 127). Elsewhere the story has been one of severe Post-boreal restriction of range, and total extinction from Ireland and the Isle of Man. It is very apparent that the maximum British extension of the species corresponds with the Post-glacial Climatic Optimum, and it is therefore the more striking that in Scandinavia this period also witnessed an extension of *N. marina* far northwards of its present range. Whereas today it is limited to the extreme south of Norway, southern Sweden and southern Finland, fossil records extend along all the Baltic coast and even to the Arctic Circle. It has been shown that whilst Scania and Denmark were occupied in the Pre-boreal period, the northern Finnish localities were occupied at the time of the Ancylus Lake, which falls into the Boreal period.

Although the species is now restricted in Scandinavia to brackish water, there can be no doubt at all that the vast majority of its fossil occurrences were in fresh water. Samuelsson has suggested that progressive worsening of the nutritive status of the lakes may have been concerned with the disappearance of *N. marina* in the later part of Post-glacial time, but Jessen has pointed out that it disappeared from some of the Irish lakes whilst they were still strongly eutrophic and the same certainly holds for its diminution at Hockham Mere. Taken with the parallel restriction of range and diminution in frequency exhibited by *N. flexilis*, which tends to favour oligotrophic conditions, the probabilities seem heavily in favour of regarding the shift as due to climatic change, the withdrawal from the west at the same time as withdrawal from the north emphasizing this probability. It may also be noted that at the Histon Road interglacial site *N. marina* accompanied *N. flexilis* and *N. minor*, and that the pollen analyses indicated that the period of formation was *zone f*, corresponding to the climatic optimum of the last interglacial. Similarly five other Ipswichian interglacial records fall within the warm sub-stage II. The significance of the two Full-glacial records is doubtful.

The parallelism with the history of *Cladium mariscus* in the Baltic region has been remarked by various Scandinavian botanists (e.g. Samuelsson, 1934, p. 194).

Najas minor All. (Figs. 125, 130)   581: 4

Cs *Pakefield*; C *Norfolk Co.*; H *Clacton*; eWo *Hoxne*; I *West Wittering, Selsey, Bobbitshole, Trafalgar Sq., Wretton, Wortwell, Histon Rd*

Fruits of *Najas minor*, a species no longer native in the British flora, are readily identifiable by their distinctive surface cell pattern (fig. 125). Reid's record from the Cromer Forest Bed series has been followed by a more precise placement in the Cromerian itself (Wilson & West, unpublished). Reid's Hoxnian records from Clacton, attributable to sub-stage IIIa,b, have been followed by C. Turner's identifications from IIb and IIIa, whilst Reid's record from an undefined part of the Ipswichian at West Wittering has been subsequently supported by various authors who have found it at six different sites in southern and eastern England, all in sub-stage II and

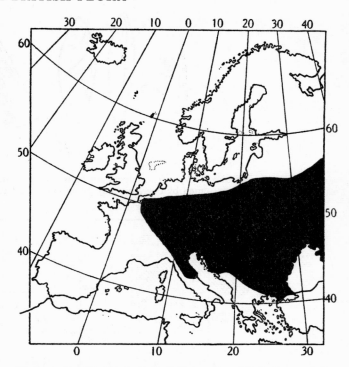

Fig. 130. Present distribution of *Najas minor* All. in Europe (after Backman, 1951). This species has been found in the south and east of England in interglacial deposits of different ages from the Cromer Forest Bed onwards.

five of these in IIb. There is a single record from the early Wolstonian at Hoxne (Turner, 1968b) but this could well be derivative.

The present and former distribution of *N. minor* have been carefully documented by Backman (1951) (see fig. 130). It is a central and east European species with a continental range extending to its most westerly limit in Belgium. The interglacial British finds are well outside the present range of the species and seem to suggest interglacial climates with warmer conditions than the present. The Post-glacial records of *N. minor* from Finland and north Russia reported by Backman shew that this species, like *N. flexilis* and *N. marina*, has suffered much restriction of range on the Continent in the later part of the Flandrian period.

It is to be noted that the assemblages of plants identified along with *N. minor* in the British interglacials are such as frequently accompany it today, and did so in Continental interglacials also (see Backman). The concentration of all the British finds, from all three interglacials, in the south-eastern corner of the British Isles is doubtless significant – as may be the greater frequency of Ipswichian records – for on the Continental mainland *N. minor* is regarded as one of the species particularly characteristic of the Eemian interglacial.

Najas tenuissima (A. Br.) Magnus   581: 5

Cs *Paston*

A fossil fruit of *Najas tenuissima* has been identified by
West and Wilson (unpublished) from the Cromer Forest
Bed series, though at a site where the exact chronostrati-
graphic position is still uncertain.

N. tenuissima at the present day is confined to eastern
Europe, but its European fossil record, in various interglacials including the Flandrian, shews that it had a wider
extension formerly (Tralau, 1962).

ERIOCAULACEAE

Eriocaulon septangulare With.   Pipe-wort (Fig. 131)
                                                        582: 1

H ?*Gort*; *e*Wo ?*Kilbeg*; F VIIa *Roundstone*

The pollen grains of *Eriocaulon septangulare* are readily
recognizable although the exine is thin and not very
resistant. The long spiral colpi encircling the grains and
the finely echinate surface are highly distinctive (fig. 131).
They were reported by Jessen from mud of zone VIIa at
Roundstone in Connemara, a discovery of particular
value and interest since this is long before the time of any
possible human introduction of the plant to Ireland.
Subsequently Jessen, Andersen and Farrington (1959)
reported a grain of 'Eriocaulon cf. septangulare' from
sub-stage III of the Hoxnian interglacial at Gort, and
Watts, also in 1959, reported that similar grains were
common at Kilbeg in stage D, which is taken to be early
Wolstonian.

E. septangulare is remarkable for its lop-sided amphi-
atlantic distribution with its major area of occurrence
in the eastern and middle-western states of the United
States and southern Canada, and a very restricted range
in Europe. In effect *E. septangulare* occurs along the
western coast of Ireland from Kerry to Donegal, often
abundantly but not in limestone areas, and also in Skye
and Colonsay off the Scottish coast.

Explanation of this striking disparity in range must
take account of the discovery by A. and D. Löve (1958)
that the diploid chromosome number for American
material is 32, whereas the corresponding number for
Irish plants is 64. In accord with this it seems that pollen
of *Eriocaulon* from Ireland, interglacial, Flandrian and
modern, correspond in size whilst pollen from American
type material is somewhat smaller. If this is confirmed it
would indicate, as Watts says, that the European stock of
E. septangulare is not only native in the Flandrian, but
survived the Wolstonian and Weichselian glaciations on
this side of the Atlantic. A case such as this must have
important implications for other species, such as *Najas
flexilis* and *Sisyrinchium bermudianum*, with strongly im-
balanced amphi-atlantic ranges (see p. 486).

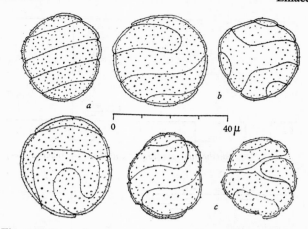

Fig. 131. *Eriocaulon septangulare*: *a*, *b*, recent pollen grains; *c*, two
fossil pollen grains from Roundstone, Co. Galway, sub-zone
VIIa.

LILIACEAE

The monocolpate reticulate pollen grains of the Liliaceae
have been reported from various sites, notably from the
Irish Late Weichselian and early Flandrian (G. Singh,
1970), the Hoxnian (Jessen *et al.*, 1959) and Middle
Weichselian (Bell, 1968).

Tofieldia pusilla (Michx.) Pers.   583: 1

*l*W *Elan Valley*

Moore (1970) identified the dicolpate pollen of *Tofieldia*
from the Late Weichselian zones I/II, II, II/III and III at
the high altitude site in central Wales, along with that of
other plants of restricted northern range. *T. pusilla* is not
found at all in Wales or Ireland and in England is limited
to Upper Teesdale.

Narthecium ossifragum (L.) Huds.   Bog asphodel 584: 1

It has recently become practicable to identify the pollen
of *Narthecium ossifragum* and, as the diagram overleaf
indicates, these records correspond well with those based
upon identification of the rhizomes and leaf-bases. There
is a seed record from sub-stage IV of the Hoxnian at the
type site where there was a tentative pollen identification
in this sub-stage and in the following early Wolstonian.
A single zone VI record in the Flandrian is followed by
numerous records in VIIa, VIIb and VIII, a reflection no
doubt of the development of the ombrogenous stages of
peat mires that became prevalent first in these zones. No
doubt the record is still very incomplete.

N. ossifragum is common over most of the British Isles
and is locally abundant on acidic bogs (including both
blanket bog and raised bog), wet heaths and moors, and
it occurs in wet sites on mountains as high as 3270 ft
(1000 m). It is described by Hultén as European–atlantic;
and by Matthews as oceanic–northern.

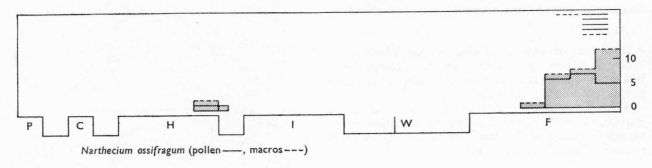

Narthecium ossifragum (pollen ———, macros – – –)

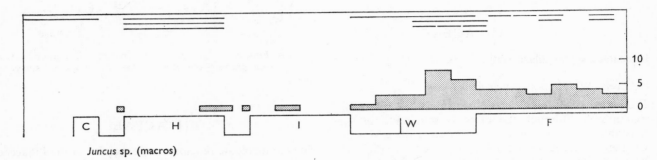

Juncus sp. (macros)

Endymion non-scriptus (L.) Garcke (*Scilla nutans* Sm.; *S. non-scriptus* (L.) Hoffmans, and Link) Bluebell, Wild hyacinth   600: 1

H ?*Gort*, ?*Kilbeg*; F vIIa, vIIb/vIII *Loch Dungeon*, vIIb ?*Frogholt*, vIIb, vIII ?*Old Buckenham Mere*

Jessen, Andersen and Farrington (1959) record from the Irish Hoxnian interglacial at Gort five liliaceous pollen grains that 'match perfectly the modern grains of *Endymion non-scriptus*' whilst differing from those of the two native British species of *Scilla*. Because of the similarity of the grains of *E. non-scriptus* and those of the non-British *E. hispanicus* and *S. bifolia* the identification is made only tentatively. The records fall in sub-stages II, III and IV of the interglacial, and Watts (1959b) has made a similar identification from sub-stage II or III at Kilbeg.

In the Flandrian there are unqualified records of *Endymion* pollen from zones vIIa and vIIb/vIII in Galloway. There are also records published as '*Scilla*' from Frogholt and Old Buckenham Mere, but Miss Andrew who determined them now prefers to regard them as '*Endymion* type', a category including *E. non-scriptus* (possibly the most likely) but overlapping into *Scilla* spp.

Endymion non-scriptus is a characteristic component of British woodlands, especially those of lighter and slightly acid soils, but it also occurs in non-wooded coastal habitats on deep, moist soils with high organic content, as at the Lizard (Malloch, unpublished). It is a species of strongly atlantic range in Europe (Roisin, 1969) prevalent throughout the British Isles, though it reaches its northernmost European stations here.

JUNCACEAE

Juncus sp. L.   605

In recent years it has been shewn, especially through the work of Körber-Grohne (1964), to be practicable to make specific identification of the seeds of *Juncus* (see Dickson, 1970) and the fossil records are now rapidly increasing. Older records and instances in which the diagnostic outer seed-coat is absent leave a substantial residue of generic records. The record diagram indicates that *Juncus* occurred in the Hoxnian and Ipswichian interglacials and throughout the Flandrian: similarly it was present in the Anglian and Wolstonian glacial stages and throughout the Weichselian. This is consonant with the fossil records for the many individual species of the genus.

The pollen tetrads of *Juncus* are held not usually to be preserved in a fossil state.

Juncus squarrosus L. 'Heath rush'   605: 1

lW *Kirkmichael*; F vIIb *Goodland Td.*, vIIb/vIII *Pennines*, vIII ?*Newstead*

Mrs Dickson (1970) refers a single seed with very thick testa from zone III of the Late Weichselian to *Juncus squarrosus*. Other records for it are late Flandrian.

Juncus gerardii Lois. 'Mud rush'   605: 5

mW *Earith*; F vIIa ?*Bowness Common*, vIIb *North Ferriby*, *Littleport*, vIII *Welney Wash*

The seeds of *Juncus gerardii* are readily recognizable (Dickson, 1970). All the Flandrian records are from estuarine-marine deposits, those in the Fenland including both the Fen Clay and the Roddon silts of Iron Age to

Roman date. The presence of this obligate halophytic species in the Middle Weichselian, dated about 40000 years before the present, is strongly indicative of saline soils in the Full-glacial (Bell, 1968, 1970a).

J. gerardii is an abundant species of the upper levels of coastal salt-marshes, with rare occurrences at inland saline sites. It is a boreal–circumpolar plant with a modern range agreeable to the Full-glacial presence in Britain.

Juncus bufonius L.   Toad rush   605 : 7

*l*A *Baggotstown*; H *Fugla Ness*; I *Marros*; *l*W *Kirkmichael, Ballaugh*; F IV *Ballaugh*, VIIb *Frogholt, Avonmouth, Prestatyn*

Seeds of *Juncus bufonius* have been found in sub-stage IV of both the Hoxnian and Ipswichian interglacial, and in the late Anglian and zones I and III of the Late Weichselian. At all these times fairly open vegetation was prevalent, as in the Pre-boreal stage of the Flandrian. The Frogholt site is one of strong early Iron Age clearance and the Avonmouth and Prestatyn sites are both coastal.

The toad rush is a cosmopolitan species now common on paths, roadsides and arable land, but it also occupies many natural habitats and the fossil record confirms its ancient native status.

Juncus inflexus L. (*J. glaucus* Sibth.)   Hard rush
                                                     605 : 8

H *Marks Tey*; F VIIb *Wingham, Glastonbury*, ?*Woodwalton*, ?*Mildenhall, Avonmouth*; VIIb/VIII ?*Meare Lake Village*

Seed of *Juncus inflexus* has been found in sub-stage IV of the Hoxnian interglacial but thereafter not until zone VIIb of the Flandrian. Though abundant in England, Wales and Ireland, the hard rush is local in Scotland: it is strongly southern in its Scandinavian range. The Flandrian records are at least partly associated with woodland clearance. The tentative Meare Lake Village record is given as either *J. inflexus* or *J. effusus*.

Juncus effusus L.   Soft rush   605 : 9

H *Kilbeg*, ?*Kildromin*, ?*Newtown*; H/eWo *Kilbeg*; I *Marros*; *l*W *Ballaugh*; F IV *Ballaugh*, IV, VI, VIIa, VIIb, VIII *Nant Ffrancon*, VI, VIIa, VIIb *Llyn Dwythwych*, VIIa *Llyn Gynon*, VIIb/VIII ?*Elland*, VIII ?*Newstead*

Irish Hoxnian records for the seeds of *Juncus effusus* cover sub-stages II, II/III and IV and at Kilbeg they reach the transition to the next glacial stage. There is an Irish record also for sub-stage IV of the Ipswichian, and zones II, II/III of the Late Weichselian, with early Flandrian IV, are represented in the Isle of Man. Sites in North Wales indicate strong local persistence from zone VI onward and, the tentative Newstead record apart, all the Flandrian sites are upland. Seeds tentatively referred to '*J. effusus* or *J. articulatus* type' have been recorded from sub-stages II and III of the Hoxnian at Baggotstown by Watts (1964).

J. effusus is a species of oceanic or sub-oceanic ten-

dencies widespread and common throughout the British Isles though southern in its Scandinavian range. This may have climatic significance for the repeated recognition of the plant in the Late Weichselian at Ballaugh. Long native presence in the British Isles is indicated.

Juncus conglomeratus L. (*J. communis* E. Mey.)
                                                     605 : 10

H *Kilbeg*; H/eWo *Kilbeg*; F VII/VIII *Pennines*

Juncus effusus L. or **J. conglomeratus** L.   605 : 9, 10

H *Fugla Ness*; *l*W *Ballaugh, Kirkmichael*; F IV *Loch Fada, Moss Lake*, V/VI *Loch Einich, Coire Bog*, VIIa *Clatteringshaws Loch*, VIIb *Goodland Td., Ashgrove, Avonmouth, Wingham*

All the seed records given as '*Juncus effusus* or *conglomeratus*' are quite recent: they reflect the close similarity of the two species and possibly suggest some similar qualification for the generally earlier records of *J. effusus*. Note that the Late Weichselian records from the Isle of Man have the alternative attribution by Dickson *et al.* (1970) whereas Mitchell's earlier records are to *J. effusus*. The ranges of the two species in the British Isles and in Scandinavia are similar, though *J. conglomeratus* favours the more acidic soils.

Juncus balticus Willd.   605 : 13

*l*W ?*Kirkmichael*, ?*Derriaghy*, ?*Ballaugh*

In zones I, II and III at Kirkmichael, and in zones II and III at Ballaugh, Dickson *et al.* (1970) tentatively identified *Juncus* seeds lacking a testa as *J. balticus*. Mitchell has made subsequent unpublished records of the same species from the Late Weichselian zones I/II and II at Derriaghy. Baker & Watts are cited by Dickson *et al.* (1970) as having both found seeds of *J. balticus* in North American Late-glacial deposits.

Juncus balticus is a creeping rhizomatous rush almost limited to dune-slacks and consequently abundant in coastal sites. It extends round the Scandinavian coast to the North Cape, but is rare in this country, occurring chiefly on the northern and eastern Scottish coasts and the Hebrides; there are outliers in Lancashire and Somerset. It is placed by Matthews (1937) in his category of oceanic–northern plants but is notably absent from Ireland and the Isle of Man.

Juncus maritimus Lam.   'Sea rush'   605 : 14

I ?*Stone*; *l*W ?*Derriaghy*; F VIIa *Bowness Common*, VIII *Borth Bog*

The seeds of *Juncus maritimus* have been tentatively identified from sub-stage II of the Ipswichian interglacial, from zones I/II and II of the Late Weichselian and positively from zone VIIa of the Flandrian. There is also an unequivocal record of the highly recognizable rhizomes from the River Dovey flank of Borth Bog, where they indicate a marine incursion into the raised bog at a radiocarbon date of 2900 years B.P. The plant is typical of salt-marsh vegetation near the high-water mark of spring tides, and at Stone and Bowness Common it was found, as at Borth, in coastal marine deposits.

J. maritimus is common round the British coasts except those of northern Scotland, and in Scandinavia it is restricted to Denmark and the south Swedish Baltic coast. In view of this, more positive evidence of this halophyte in the Late Weichselian is desirable.

Juncus subnodulosus Schrank   'Blunt-flowered rush'
605: 17

F vɪɪb ?*Wingham, Frogholt,* ?*Godmanchester*

Seed identifications have been made tentatively of *Juncus subnodulosus* at Wingham (radiocarbon age 3100 B.P.) and Godmanchester (Roman), whilst there is a positive identification from Frogholt (radiocarbon age 2860 B.P.). There was a substantial weed flora at all these sites and clear indications of local wetness.

Juncus acutiflorus Erhr. ex Hoffm.   'Sharp-flowered rush'
605: 18

*l*W *Rhosgoch Common;* F vɪɪb *Frogholt*

From a Late Weichselian deposit in Radnorshire Bartley (1960) recorded fruits, flowers and seeds of *Juncus acutiflorus*. Seeds were also recorded from Frogholt, Kent in layers dated 2490 years B.P. The paucity of records of a species common throughout the British Isles is partly explained by the difficulty in separating its seeds from those of *J. articulatus* (*vide infra*). It has a very pronouncedly southern range in Scandinavia, not extending north of Denmark, but Bartley's material is unequivocal evidence of its presence in Wales in the Late Weichselian.

Juncus acutiflorus Erhr. ex Hoffm. or J. articulatus L.
605: 18, 19

H *Kilbeg, Fugla Ness;* H/*e*Wo *Kilbeg;* I *Marros; l*W *Loch Fada, Loch Meodal;* F ɪv *Loch Fada, Loch Meodal, Lochan Coir' a' Ghobhainn,* vɪɪb/vɪɪɪ *Wall*

Seeds alternatively referred to *Juncus acutiflorus* or *J. articulatus* have been found in sub-stages ɪɪ/ɪɪɪ and ɪv at the two respective Hoxnian sites, and at Kilbeg they were found again in the terminal interglacial deposits. The Ipswichian record is for sub-stage ɪv. There are records of persistent occurrence at the Scottish sites of Loch Fada (zones ɪ, ɪɪ, ɪɪɪ, ɪv) and Loch Meodal (ɪɪ, ɪv) whilst there is another zone ɪv record from Skye at Lochan Coir' a' Ghobhainn. There is one late Flandrian record from the Midlands.

Juncus articulatus L.   'Jointed rush'
605: 19

F vɪ ?*Llyn Gynon,* vɪɪb *Avonmouth,* vɪɪb/vɪɪɪ *Hightown,* vɪɪɪ ?*Wingham*

The seeds of *Juncus articulatus* are hard to distinguish from those of *J. acutiflorus*, so that there are only two unqualified records for it, both as it happens late Flandrian coastal sites. Like *J. acutiflorus, J. articulatus* is common throughout the British Isles, but it has a substantially more northerly range than that species in Scandinavia.

Juncus bulbosus L. 'Bulbous rush'
605: 22

H *Kilbeg; l*W *Kirkmichael, Ballaugh;* F ɪv *Ballaugh,* vɪɪb *Goodland Td.,* vɪɪb, vɪɪɪ *Wingham;* vɪɪb/vɪɪɪ *Wall,* vɪɪɪ *Jenkinstown, Bunny*

Seeds of *Juncus bulbosus* were recovered by Watts from the Hoxnian sub-stage ɪɪ/ɪɪɪ at Kilbeg. In the Late Weichselian and Pre-boreal deposits of the Isle of Man seeds definitely referred to *J. bulbosus* come from zones ɪ, ɪɪ, ɪɪ/ɪɪɪ and ɪv as well as more eroded material referred to '*J. bulbosus* type' from zones ɪɪ, ɪɪɪ and ɪv. Both kinds were frequent. Flandrian records do not recommence until zone vɪɪb, often in contexts indicative of clearance and occupation as at Wingham in deposits dated 3100 B.P. (Bronze Age) and 2340 B.P. (Iron Age) and in the Roman well at Bunny, Notts.

This dwarf perennial rush is a boreal–circumpolar plant found throughout the British Isles though more common in the north and west, partly no doubt because of its preference for acidic soils. It reaches about 68° N latitude in Scandinavia. The fossil record clearly reflects the plant's need for open vegetation and suggests the likelihood of ancient native status.

Luzula sp.
606

H *Clacton;* W *Thrapston; m*W ?*Waltham Cross, Brandon, Great Billing; l*W *Kirkmichael, Colney Heath, Bigholm Burn, Ballaugh*

Seeds referred to the genus *Luzula* were found at Clacton by Reid at levels now referred to sub-stage ɪɪɪ. These apart, all the generic records are Weichselian. Separate authorities vouch for Middle Weichselian records and again there are three authorities for the Late Weichselian records that cover zones ɪ, ɪɪ, ɪɪ/ɪɪɪ and ɪɪɪ. The sites are certified by radiocarbon dates: Brandon 30760, Great Billing 28220, Bigholm Burn 10820 and Kirkmichael 10270 B.P., and parenthetically we may note an age of 41900 B.P. for *Luzula spicata* at Upton Warren. It seems likely that various species of *Luzula* were present in our Weichselian plant communities.

Luzula sylvatica (Huds.) Gaudin (L. maxima (Reichard) DC.)   'Greater woodrush'
606: 3

F vɪ ?*Treanscrabbagh,* vɪɪa/b *Shustoke,* vɪɪɪ *Dublin*

Seeds of *Luzula sylvatica* have so far only been found in the Flandrian.

Luzula spicata (L.) DC. 'Spiked woodrush' 606: 6

A *Corton*; mW *Upton Warren*; lW *Derriaghy, Drumurcher, Knocknacran, Whitrig Bog*

There are now six recorded sites for fossil seeds of *Luzula spicata*. All are from glacial deposits, the oldest the Corton Beds near Lowestoft. At Upton Warren (radiocarbon age 41 900 B.P.) they are part of a very extensive Middle Weichselian fauna and flora. The Late Weichselian records include one from zone II and three from zone III.

Luzula spicata is a circumpolar arctic–montane plant of screes and rocky situations on non-calcareous soils that appears now to be limited within the British Isles to the northern half of Scotland. In Scandinavia it is common throughout the far north but in the south it is restricted to the mountains. Polunin (1940) writes that '*L. spicata* in northernmost Canada occurs chiefly among the forbs and grasses that characterise the most favourable and sheltered, south-facing "flower slopes", and also in the outermost heathy zones of snowdrift areas that are bared relatively early in summer.' This may well help to characterize the British glacial sites. The fossil records in any event establish clear post-glacial restriction of range and extinction in Ireland.

Lazula campestris (L.) DC. Sweep's brush, 'Field woodrush' 606: 8

lW *?Nant Ffrancon, ?Ballaugh*

The two tentative seed identifications of *Luzula campestris* refer respectively to zones II and II/III of the Late Weichselian.

Luzula campestris (L.) DC. or **L. multiflora** (Retz.) Lejeune 606: 8, 9

F VIII *Godmanchester*

Luzula multiflora (Retz.) Lejeune 'Many-headed woodrush' 606: 9

lW *?Loch Droma, ?Loch Park, ?Nant Ffrancon, ?Knocknacran*; F VIII *? Ballingarry Downs*

There are only tentative identifications of the seeds of *Luzula multiflora*, but none the less they comprise records by four authors at four Late Weichselian sites embracing all three zones of that period.

Luzula multiflora occurs today commonly throughout Scandinavia and is found throughout the British Isles though more abundantly in the north and west.

Luzula multiflora (Retz.) Lejeune or **L. sylvatica** (Huds.) Gaudin 606: 9, 3

F VIII *Shrewsbury*

A seed from a mediaeval context.

AMARYLLIDACEAE

Allium schoenoprasum L. Chives 607: 8

mW *Oxbow, Marlow, Earith, Waltham Cross, Angel Rd*

The early identifications of seeds from the Lea Valley Arctic Plant Bed by Reid have now been followed by others by Miss Bell from the Middle Weichselian of East Anglia, the East Midlands and Yorkshire, in the last case by checking Gaunt's material. Miss Bell (1970a) comments that 'the seeds are distinguished from other *Allium* seeds of triangular shape and flattened faces by the cell pattern of polygonal, angular cells which have a raised surface and are separated by narrow grooves'.

Allium schoenoprasum is a plant of open rocky pastures on limestone, now very local in the British Isles. It is one of Hultén's incompletely boreal–circumpolar plants: its native range is hard to assess since it commonly escapes from cultivation, but it is frequent in south Finland and south-eastern Sweden, a range not out of character for a British Weichselian plant. This is a view supported by the fact that it forms part of the *Salicetum viminalis* of the Lena valley in Siberia. *Allium schoenoprasum* var. *sibiricum* occurs at far higher latitudes in Scandinavia, but this is probably not the plant found in Cornwall and recorded as *A. sibiricum* L. (fide C.T.W.).

IRIDACEAE

Iris sp. Iris 616

lW *Colney Heath*; F VIIa *Bowness Common*, VIIb/VIII *Wingham*

There are three pollen records referred to the genus *Iris*; that from Colney Heath is from zone I of the Late Weichselian.

Iris pseudacorus L. Yellow flag 616: 4

Macros: I *Wretton*; F VI *Killywilly Lough*, VIIa *Bridgewater Bay, Salthouse Broad*, VIIb *Shippea Hill, Helton Tarn, Frogholt, Prestatyn*, VIIb/VIII *Dog and Doublet Sluice*, VIII *Crossness, Glastonbury*

Pollen: F VIIb, VIII *Frogholt*, VIII *Old Buckenham Mere*

The large seeds of *Iris pseudacorus* have been found in sub-stage II of the Ipswichian interglacial, and in the Flandrian at many sites from Boreal time onwards. At Frogholt in Kent not only were seeds found at the level carbon dated at 2860 B.P., but at that level they were accompanied by pollen of *Iris* in high frequency, and similar pollen occurred higher in the sequence there in deposits dated 2490 B.P. No doubt this reflects a local stream-side population, and a local population is indicated also by the pollen frequencies between 2 and 5 per cent of total arboreal pollen in zone VIII at Old Buckenham Mere.

The yellow flag occurs throughout the British Isles: in Scandinavia, although common only in the southern

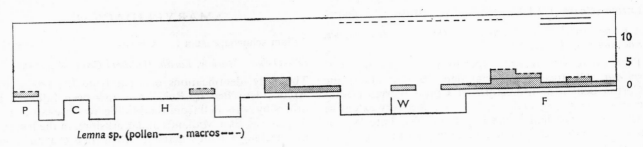

Lemna sp. (pollen——, macros— — —)

part, it extends in scattered localities to the head of the Gulf of Bothnia and even further north on the Norwegian coast.

ORCHIDACEAE

F VI *Thatcham*   624

In reporting on the Mesolithic site at Thatcham in the Thames valley Churchill (1962) indicated the presence in zone VI of pollen of Orchidaceae amounting to 3 per cent of total tree pollen.

Orchis sp.   642

F VIIb *Old Buckenham Mere*

Some of the orchids have separable pollen tetrads and it was pollen of this type that was recorded from zone VIIb at Old Buckenham Mere (Godwin, 1968b).

LEMNACEAE

Lemna sp.   Duckweed   650

Since the demonstration that the pollen grains of *Lemna* are recognizable in oil-immersion microscopy by the bottle-shaped spines about 1.5 mm long which decorate the monoporate grains (Beatson, 1955), there have been numerous records from British Pleistocene deposits. Different authors at different sites have recorded *Lemna* pollen from sub-stages II, III and IV of the Ipswichian interglacial, and it has been reported twice from the Late Weichselian, tentatively from Colney Heath and positively from Loch Tarff. Pollen records from English sites span all the Flandrian zones.

Fruits of *Lemna* have been found alongside fossil pollen not only in zones V and VI at Kirkby Thore (Walker & Lambert, 1955), but after on the Dogger Bank in zone V (Behre & Menke, 1969). The fruits have been recovered from the Pastonian interglacial and Bell (1968) records them from the Weichselian at Thrapston. There is a further Flandrian record from Littleport in zone VIIb.

Macroscopic and pollen identifications reinforce and complement one another, giving good reason to believe that the genus has been present in the British Isles through a large part of the Pleistocene. The Weichselian occurrences are of special interest in view of the infrequent and irregular flowering and fruiting of all four species of *Lemna* now native to this country. *Lemna minor*, the commonest of them, has the most northerly range in the British Isles, as in Scandinavia where it occurs sporadically even within the Arctic Circle. Janssen (1960) has shewn a continuous curve for *Lemna* pollen in zones I, II and III of the Late Weichselian of south-eastern Holland as well as for the succeeding early Flandrian.

Lemna trisulca L.   'Ivy duckweed'   650: 2

C ?*Norfolk Co.*; IW ?*Colney Heath*

Only tentative identifications have been made of the fruits of *Lemna trisulca*, one from the Cromerian and one from zone I of the Late Weichselian at Colney Heath, referred to this species rather than *L. minor* because of its greater length (0.9 mm). Two smaller fruits from the Colney Heath deposit were recorded as '*L. minor* or *L. trisulca*'.

Lemna minor L.   Duckweed   650: 3

C *Norfolk Co.*; I ?*Selsey*, ?*Bobbitshole*, ?*Ilford*, ?*Wretton*, *Histon Rd*

Fruits of *Lemna minor* have been recorded, albeit for the most part tentatively, at various sites and levels within the Ipswichian interglacial where they come from sub-stages I, II and III, at Bobbitshole in both IIa and IIb and at Wretton in both IIb and III. The wide floodplains where the East Anglian interglacial deposits were made evidently included still backwaters suitable to duckweed as to many other aquatics. Fruit of *L. minor* has also been recognized in the Cromerian interglacial and preceding accounts (650, 650: 2) have indicated the likelihood that *L. minor* was the species of the genus present in Britain during the Weichselian. It is a boreal–circumpolar plant of very wide occurrence save in the polar regions and the tropics.

SPARGANIACEAE

Sparganium   Bur-reed   652

Macros: C *Norfolk Co.*; H *Ballykeerogemore*, *Hitchin*; I *Stone, Trafalgar Sq.*; W *Jenkinstown*; F IV/V *Loch Dungeon*, VIIb *Barry, Loch Fada*, VIII *Ehenside Tarn, Loch Cuithir, Saddlebow*

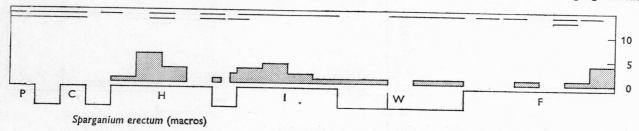

Sparganium erectum (macros)

Numerous identifications of the fruits of *Sparganium* have not been carried beyond generic level: they usefully supplement the records for the individual species that follow. Criteria for identifying the fruit stones of the individual species have been given by Cook (1961) and Dickson (1970).

The pollen closely resembles that of *Typha angustifolia*, but the separate identification has been asserted in some cases, and in some instances substantial pollen frequencies have been found along with recognizable fruit stones of *Sparganium*, as for instance in the Allerød period (zone II) muds at Hawks Tor, Cornwall (Conolly et al., 1950) and in many interglacial sites. Apart however from the special cases of *Sparganium erectum* ssp. *neglectum* and *S. minimum* all pollen records for *Sparganium* have been presented graphically along with those for *Typha angustifolia* (653: 2): see p. 382.

Sparganium erectum L. (incl. *S. ramosum* Huds.) Bur-
reed　　　　　　　　　　　　　　　　　652: 1

Fruit stones of *Sparganium erectum*, distinguished from those of other species of *Sparganium* by the presence of 6 to 10 longitudinal ribs, have been recorded from three sites of the Cromer Forest Bed series, from numerous sites in sub-stages I, II and III of the Hoxnian and from all sub-stages of the Ipswichian interglacial. There are records also for the end of the Anglian and both early and late Wolstonian, occurrences that match those at Sidgwick Avenue and Marlow for the Early and Middle Weichselian. There are positive Late Weichselian records from Hawks Tor, Scotland and Ireland (Dunshaughlin, zone III). There is a sparse Flandrian record but enough to bear out the suggestion, with the evidence already mentioned, that *S. erectum* has been a member of the British flora since at least Cromerian time.

The four British sub-species of *S. erectum*, readily distinguishable by their ripe fruits, apparently have few correlated morphological characters, few ecological differences and only slight differences in British or European geographical range (Cook, 1961).

Sparganium erectum L. ssp. **erectum** (*S. ramosum* ssp.
polyedrum Asch. and Graeb.)　　　　　652: 1a

F VIIb *Drumurcher*

Sparganium erectum L. ssp. **microcarpum** (Leuman)
Domin

F IV, V, *Ballybetagh*, VIIa *Portrush*, VIIb *Drumurcher*

In recording *Sparganium erectum* var. *microcarpum* from Irish deposits of zones IV, V and VII, Jessen notes that this variety has been recorded from Ireland by More. The later evidence of the Supplement to the *Atlas of the British Flora* shews it sparsely present throughout all counties of the British Isles. Ssp. *microcarpum* extends between the Arctic Circle and North Africa and extends eastwards to Siberia, a range consonant with the early and later Flandrian records.

Sparganium erectum L. ssp. **neglectum** (Beeby) Schinz and
Thell.　　　　　　　　　　　　　　　　652: 1b

Macros: H ?*Clacton*; lW ?*Hawks Tor, Dunshaughlin*
Pollen: F VIIa *Vipers Crannog*

Records for fruit stones of *Sparganium erectum* ssp. *neglectum* come from the middle Hoxnian and from zones II and III of the Late Weichselian; two of them are tentative only.

The pollen grains of *Sparganium erectum* ssp. *neglectum* are much less distinctly reticulate than those of other British species of *Sparganium*, or of the other sub-species of *S. erectum*. R. Andrew has recognized substantial concentrations in the early Atlantic reed-swamp layers below Shapwick Heath near Viper's Track.

The ssp. *neglectum* has a very restricted southerly range in Scandinavia: the British distribution seems also to have a southern emphasis.

Sparganium emersum Rehm. (*S. simplex* Huds.) 'Unbranched bur-reed'　　　　　　　　　　652: 2

lA *Gort*; H *Gort, Nechells, Clacton*; I *Bobbitshole*; eW *Sidg-
wick Av.*; mW *Barnwell Station*; F IV/V *Dogger Bank*, V
Ballybetagh, VI *Jenkinstown*, VIII *Dunshaughlin*, ?*Ehenside
Tarn*

Fruit stones of *Sparganium emersum* were described by Jessen from the late Anglian and sub-stages I and IV of the Hoxnian Interglacial at Gort, whilst Kelly records them from many successive levels (I, IIb and c, IIIa and b) from Birmingham, indicating long local persistence. They also were found in sub-stage III of the Hoxnian at Clacton, and in sub-stage II of the Ipswichian at Bobbitshole. There are records for the Early and Middle Weichselian and for

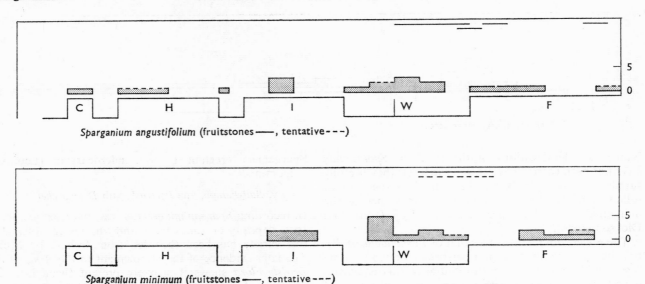

Sparganium angustifolium (fruitstones ——, tentative - - -)

Sparganium minimum (fruitstones ——, tentative - - -)

early, middle and late Flandrian, so that there is a good presumption of long native standing in this country.

S. emersum is a boreal–circumpolar species that is frequent round the Baltic to about 64° N and that has numerous scattered localities well within the Arctic Circle up to 69° N. It occurs throughout the British Isles.

Sparganium angustifolium Michx. (S. affine Schnitzl.; S. natans auct.) 'Floating bur-reed' (Figs. 132, 153f) 652: 3

Fruit stones of Sparganium angustifolium have been recorded from the Cromerian interglacial and tentatively at Baggotstown from both sub-stage I and sub-stage II of the Hoxnian. There are three records from the Ipswichian with recurrence at Bobbitshole in both IIa and IIb. The unqualified identification from the early Wolstonian at Hoxne is matched by one from the Early Weichselian at Sidgwick Avenue. These are followed by two tentative records by Reid from the Lea Valley Arctic Plant Bed that have to be considered with four others, not separately recorded here, that he referred to 'S. angustifolium or S. erectum'. There are several records for zones I and II of the Late Weichselian and evidence of continuity into the middle and late Flandrian. It seems fairly certain therefore that S. angustifolium was a survivor in this country through the Weichselian glacial period and possibly from much earlier time.

S. angustifolium is placed by Matthews in his group of Continental–northern species, and by Hultén in the incompletely boreal–circumpolar group. In Scandinavia it is common even within the Arctic Circle, and in the British Isles it is strongly northern and western in its present distribution so that the Weichselian records from England provide clear evidence of post-glacial restriction of range (fig. 132).

Sparganium minimum Wallr. 'Small bur-reed' 652: 4

Pollen: H ?Fugla Ness; IW ?Loch Fada, ?Lochan Coir' a' Ghobhainn, ?Loch Meodal; F IV ?Lochan Coir' a' Ghobhainn, ?Loch Meodal, ?Loch Cill Chriosd

H. J. B. Birks has made tentative identifications of the pollen of Sparganium minimum from the Late Weichselian and Pre-boreal of lakes in Skye, these embracing jointly zones I to IV. He has a similar tentative record for sub-stage IV of the Hoxnian in the Shetlands.

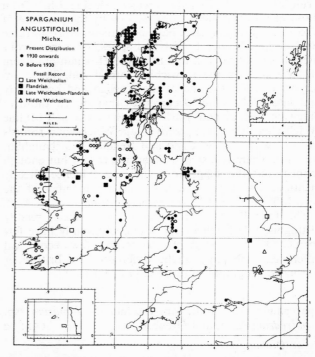

Fig. 132. Weichselian, Flandrian and present-day distribution of Sparganium angustifolium Michx., in the British Isles.

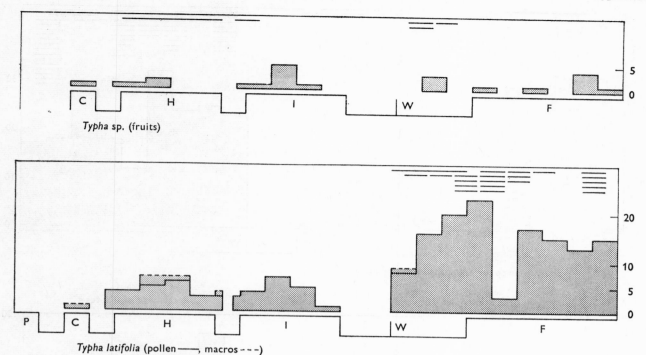

Typha sp. (fruits)

Typha latifolia (pollen ——, macros - - -)

By contrast the numerous records of sub-fossil fruit stones are for the most part unqualified. Three sites cover sub-stages II and III of the Ipswichian interglacial and five Middle Weichselian records are followed by records for each zone of the Late Weichselian. The Flandrian records, though sparse, suffice to indicate continuity of presence which thus seems likely from Ipswichian time at least.

S. minimum is placed in Hultén's category of boreal–circumpolar plants and he shews it as frequent in western Norway almost to the Arctic Circle (in Finland to about 64° N), whilst scattered localities inside the Arctic Circle extend its range to about 69° N. *S. minimum* occurs throughout the British Isles but is sparse in the southern half of Great Britain even if one takes account of older records less affected by drainage than the present ones.

TYPHACEAE

Typha sp.   Reedmace, Bulrush (Plate XXc)   653

The pollen of *Typha latifolia* occurs in flat tetrads which are readily recognizable, that of *T. angustifolia* as single grains which are extremely difficult to separate in fossil condition from grains of *Sparganium*. We have accordingly not presented individually the rather infrequent records given for '*Typha*' or for '*Typha angustifolia*'.

The small and fragile fruits of *Typha* yield however an adequate historical record which begins with the Cromerian, includes the first two sub-stages of the Hoxnian

interglacial and the first three of the Ipswichian. There are late Anglian and late Wolstonian records and six for the Late Weichselian, followed by a scatter through the Flandrian. This record has to be considered alongside the abundant evidence for *T. latifolia*.

Typha latifolia L.   'Great reedmace', Cat's-tail (Plate XXIIg)   653: 1

To the very full record of the pollen of *Typha latifolia* presented in the record diagram must be added records for the Thurnian and Ludhamian. Apart from the Early and Middle Weichselian it would seem that this plant has been present in the British Isles from at least Cromerian time. The absence of records from the Full-glacial may be due to the fact that deposits of this age have scarcely been examined by pollen analysis, to the small size and fragile character of the fruits, or to the real absence of the plant from England at this time. A few fruit records supplement the pollen evidence and at Shippea Hill, Cambridgeshire, rhizomes were recorded from early Bronze Age levels above the Fen Clay.

T. latifolia, though generally distributed throughout the British Isles, is less frequent in the north, a fact parallel with the situation in Scandinavia, where it occurs frequently only as far north as 61° N although extending in scattered sites in Finland and near the Baltic as far as the Arctic Circle. It is placed by Samuelsson in his Bothnian sub-group of south Scandinavian freshwater plants, and this southern tendency of range contrasts somewhat with the evidence for the species growing in the British Isles as early as zone I of the Late-glacial period.

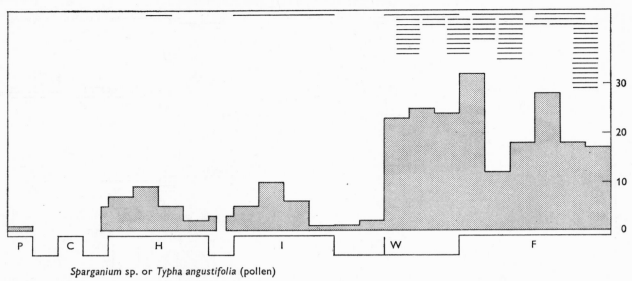

Sparganium sp. or Typha angustifolia (pollen)

Absolute numbers of sites   Percentage of all sites

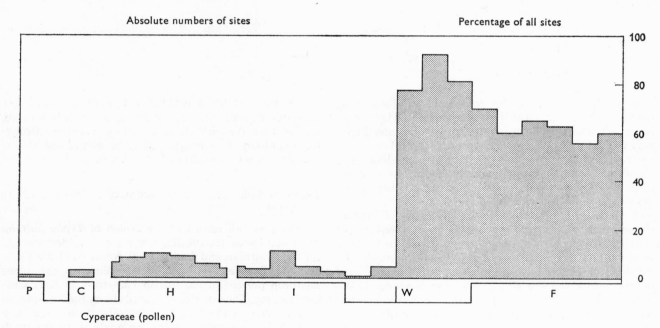

Cyperaceae (pollen)

Typha angustifolia L. 'Lesser reedmace'   653: 2

IA Marks Tey; IW Coolteen

Although in general fruits of *Typha* have only been identified at generic level, those of *T. angustifolia* have been
recorded from the late Anglian at Marks Tey and from
zones I and II of the Late Weichselian at Coolteen.
Although some pollen records are given for all the Late
Weichselian and Flandrian zones we have subsumed these
into the category '*Typha angustifolia* or *Sparganium*'.

Sparganium L. or **Typha angustifolia** L.   652 or 653: 2

Accepting the opinion of many experienced palynologists
that the pollen of *Sparganium* spp. and of *Typha angusti-*

folia cannot be separately identified, we have given a
record diagram for the joint pollen type. It necessarily
omits one record each from the Ludhamian and Antian
and two records from the Baventian.

The record is very similar to that of *Typha latifolia* and
is broadly congruent with those of the genus *Typha* and
the component species of *Sparganium*, and evidently this
group of aquatic plants has been very strongly persistent
here.

CYPERACEAE

The pollen grains of Cyperaceae are generally pear-shaped
with a smooth surface and tectate wall structure, and
rudimentary pores (lacunae) often hard to distinguish.

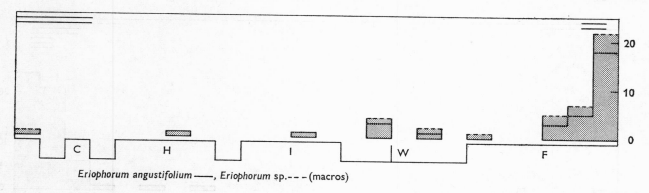

Eriophorum angustifolium ——, *Eriophorum* sp. – – – (macros)

TABLE 29. *Pollen frequency for Cyperaceae in British Late Weichselian and Flandrian sites: site frequencies by zones*

					Percentages							
	0–2	2–5	5–10	10–20	20–30	30–40	40–50	50–60	60–70	70–80	80–100 or more	
I	.	2	12	14	5	5	2	.	.	.	.	Of total pollen
II	.	2	8	21	11	5	2	1	2	2	1	
III	.	.	8	19	10	5	4	2	1	1	1	
IV	1	5	18	12	.	4	.	1	.	1	.	
IV	2	2	8	4	5	2	1	1	1	1	4	Of total tree pollen
V	4	2	6	6	1	1	2	1	.	1	5	
VI	7	9	20	12	2	2	1	2	2	1	8	
VIIa	5	24	21	19	6	8	4	4	1	.	3	
VIIb	6	10	29	19	8	7	5	1	1	3	3	
VIII	4	3	12	19	7	7	11	6	2	1	19	

Although Faegri and Iversen have gone some distance in the identification of pollen types within the Cyperaceae, sometimes to a genus or group of genera and sometimes to individual species, as *Scirpus maritimus, Cladium mariscus* and *Rhynchospora alba*, British pollen records seldom go so far. The pollen of this family, although not further separated into genera, is often abundant and furnishes a clue to general vegetational conditions, to sedge fen or *Eriophorum* moor in post-glacial conditions and to a grass-sedge tundra or park tundra in Late Weichselian time. On occasion, as in the raised bogs of the Somerset Levels, high maxima of Cyperaceae pollen correspond to layers of remains of *Cladium mariscus* in between oligotrophic *Sphagnum* peats.

The vast range of species represented makes it unremarkable that there should be numerous site records for Cyperaceae pollen right through the Pleistocene record. In the Late Weichselian 80 to 90 per cent of all sites have curves for this pollen type and even in the succeeding Flandrian the figure remains throughout the zones at about 60 per cent of all sites with pollen analyses. Somewhat more information is yielded by considering the frequencies in which Cyperacean pollen has been found through the Late Weichselian and Flandrian zones (table 29).

It is evident that the representation of Cyperaceae varies greatly from site to site, no doubt because of large stands of sedge vegetation near sites later to be pollen analysed. Sites such as Scaleby Moss and Bigholm Burn yielded very high frequencies in the Late Weichselian and the modal frequencies were higher in zone II than in the succeeding Flandrian when the landscape was largely wooded. The fall in modal frequency in zone VIIa might be conjecturally associated with the great expansion of alder woods at this time, the majority of which certainly invaded and suppressed the sedge-beds of fens and lake margins. Possibly the number of sites with high frequencies in zone VIII is correspondingly attributable to woodland clearances of wet sites.

Eriophorum sp.   654

The records of macroscopic remains referred merely to the genus *Eriophorum* are not numerous, and, as the record diagram indicates, correspond in age and frequency to those for *E. angustifolium* and *E. vaginatum*.

Eriophorum angustifolium Honck. Common cotton-grass   654: 1

Eriophorum angustifolium is usually distinguished in fossil condition from the densely tufted *E. vaginatum* by its

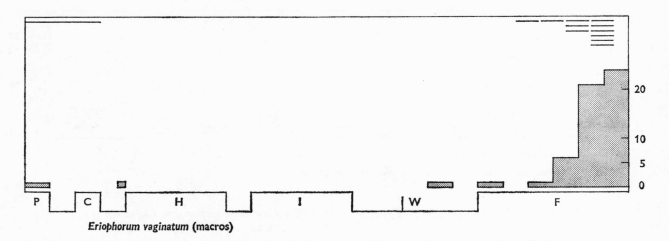

Eriophorum vaginatum (macros)

rhizomatous habit which produces separate aerial shoots, although these, like the shoots of *E. vaginatum*, also have very fibrous leaf bases. Furthermore, the flattened adventitious roots which occur abundantly in raised bog peat are of a characteristic fresh pink colour instead of black. Unfortunately neither stem bases nor roots indicate contemporary ground surface with exactitude.

E. angustifolium has been twice recorded from the Cromer Forest Bed series, and once from sub-stage III, in each of the Hoxnian and Ipswichian interglacials. There are three Middle Weichselian records and a qualified identification from zone II of the Late Weichselian at Hawks Tor. The only Flandrian records are from zones VIIa, VIIb and VIII, particularly the last. It may well be that its preponderance in zone VIII is due to the fact that its remains are better preserved and more recognizable in the relatively unhumified peats of this period, but it may have been favoured by the prevalence of pools on the bog surfaces of this time as contrasted with the drier conditions of the bog surfaces in zone VII. Although no records have been made for the period between zones II and VII it is most unlikely that the plant was absent from all parts of the British Isles during this time.

E. angustifolium is a circumpolar plant which occurs throughout the arctic and sub-arctic regions, and is described by Polunin as common throughout the Canadian eastern Arctic. It is found as far north as 83° N latitude in Greenland. It is placed by Matthews in his Arctic–alpine category, but it is a plant of very wide geographical and climatic range, as is indicated for instance by Hultén's map which shews it as frequent throughout the whole of Scandinavia. Although also recorded throughout the British Isles, it is now rare in some parts of the south and east, possibly as a result of the cutting of peat and drainage of suitable habitats.

Eriophorum latifolium Hoppe Broad-leaved cotton-grass   654: 3

I ?*Histon Rd*

There is a tentative identification of a fruit of *Eriophorum latifolium* from sub-stage III of the Ipswichian interglacial at Histon Road, Cambridge. The earlier Cambridge record, from the Middle Weichselian at Barnwell Station, has been deleted after re-examination of Miss Chandler's material.

Eriophorum latifolium, very local but widespread in the British Isles today, extends to 70° N latitude in Scandinavia.

Eriophorum vaginatum L.   Hare's-tail   654: 4

Wherever raised bogs and blanket bogs are cut for fuel the peat diggers are familiar with the tough fibrous clods which represent the tussock bases of *Eriophorum vaginatum*: they have many dialect names, such as 'mabs', and their resistance to the spade or loy is often considerable. The roots, in distinction from those of *E. angustifolium* which are pink, are black and rather easily recognizable, but naturally penetrate layers below the contemporary surface of growth. The resistance of cotton-grass fibre to decay is such that very humified peat originating from *Sphagnum–Calluna–Eriophorum*-dominated communities may lose all recognizable macrofossils save the cotton-grass, and may incorrectly be identified therefore as a 'cotton-grass' peat. Thus Woodhead and Erdtman (1926) write: 'The peat of the southern Pennines is formed largely of cotton grass, the chief peat former being *Eriophorum vaginatum*. *Sphagnum* peat occurs hardly at all...' This is a result that recent investigations have not borne out.

Remains are so common that Jessen records the species as present in the peat of 'all ombrogenous bogs in Ireland' and the same would doubtless apply to the rest of the British Isles. Thus our diagram is based only upon a small selection of the potential or actual records for the species. Two factors contribute strongly to influence the frequency with which the plant has been recorded from various stages of the Post-glacial period. Firstly the vast majority of bogs at present ombrogenous only reached

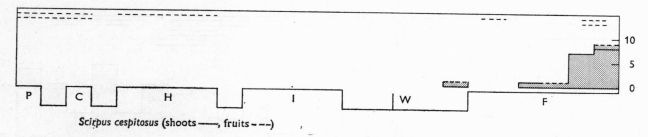

Scirpus cespitosus (shoots ——, fruits ----)

this condition in late Boreal or early Atlantic time, so that there were few opportunities for the plant to be recorded early in such deposits. Secondly, the oldest layers of ombrogenous bogs, especially those of zone VIIa, appear now in a condition of such advanced humification that even fibres of *Eriophorum vaginatum* are difficult to recognize in them. These two factors give a preponderance in our records to zones VIIb and VIII, but it is quite clear that the plant was prevalent also in VIIa, and present in zones VI, IV and zone II of the Late Weichselian. The latter record is for the so-called 'spindles', elongated aggregates of sclerenchyma that are the last stages, still recognizable however, of disintegration of the fibrous leaf-bases. Similar identifications have recently extended our records back to the late Anglian and even the Pastonian.

Although so ubiquitous a member of the vegetation of blanket bogs and raised bogs, *E. vaginatum* is by no means typical of the wettest habitats upon them, and indeed is an important tussock-builder of many different regeneration cycles, and shews by increased vitality and dominance its response to improved drainage conditions, such as those found on a sloping 'rand', or beside rivulets on the bog margin. This response is concerned in the status given to the species by the eminent pioneer of peat stratigraphy, C. A. Weber, who described the 'Wollgrastorf' as 'Landtorf' characterizing with *Calluna* peat the dry phase of the 'Grenzhorizont', or as 'Halblandtorf' characterizing the hydroseral stage following mesotrophic pine and birch woods and leading to the oligotrophic communities of the raised bogs. In the British Isles there are often layers rich in *Eriophorum vaginatum* in circumstances suggesting climatic or edaphic dryness of bog surfaces. Thus in the Somerset Levels the peat at the top of the old *Sphagnum–Calluna* peat is often rich in cotton grass, and the Late Bronze Age trackways which formed when that horizon was flooded afforded a purely local site on which *Eriophorum vaginatum* grew very densely. In the Hebrides the 'Upper Forestian' is often represented by layers of *Scirpus cespitosus* or *Eriophorum vaginatum* peat instead of the wood peat which characterizes this period on the mainland.

E. vaginatum is a Eurasiatic species found commonly throughout Scandinavia (including the far north) but falling off somewhat in abundance in south-east Sweden and Denmark. It is placed by Matthews in his category of arctic–alpine plants: it occurs in Nova Zembla but not in Iceland or Greenland. It occurs throughout most of the British Isles, but is decreasing in the south-east, no doubt in part because of drainage of suitable habitats.

Scirpus hudsonianus (Michx.) Fernald (*Eriophorum alpinum* L.; *Trichophorum alpinum* (L.) Pers.)   655: 1

I *?Histon Rd*

A tentative identification was made by Walker (1953) from sub-stage III of the Ipswichian interglacial in Cambridge of fruit of *Scirpus hudsonianus*. Formerly this species grew in Angus, but it is now extinct. It is a species widespread throughout central and northern Europe including the whole of Scandinavia.

Scirpus cespitosus L. (*Trichophorum cespitosum* (L.) Hartm.)   655: 2

As will be seen from the record-head diagram, most of the records for *Scirpus cespitosus* are based upon recognition of the shoots, i.e. the characteristic stem bases, clothed in dark brown glistening scale leaves: these are fairly common in *Sphagnum–Calluna–Eriophorum* peats of raised bogs and blanket bogs, and the British records, chiefly from Ireland and the Somerset Levels, by no means represent the full tally of instances where such have been found. It seems probable that these stem bases decay rather readily into an unrecognizable condition and on this account the plant may be over-represented in the records for the upper and less humified layers of zones VIII and VIIb as compared with those of the more humified peats of lower zones. Furthermore, of course, ombrogenous bog growth in the Flandrian was seldom well established before the Atlantic period, i.e. zone VIIa, and there is much similarity on this account with the records for the *Eriophorum* species. In a number of instances fruits of *S. cespitosus* have been also identified, records of this kind coming from the Cromer Forest Bed series, the Hoxnian, the Late Weichselian and early Flandrian.

There is no possibility of distinguishing in this fossil material between the two sub-species, *cespitosus* (*austriacus* (Palla) Brodd.) and *germanicus* (Palla) Hegi. Of these two the latter is stated to be the commoner in the British Isles and the former only occurs on Ingleborough and Ben Lawers, but the range of both is imperfectly known. The Scandinavian range of the two sub-species is vastly different, ssp. *cespitosus* being frequent from central Sweden and central Finland throughout the peninsula to the extreme north, and ssp. *germanicus* being frequent

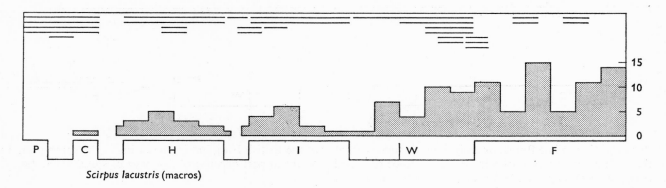

Scirpus lacustris (macros)

in Denmark and south-west Norway and Sweden, but not extending beyond 64° N, and being totally absent from Finland.

It would seem probable in view of the evidence of present British range that the numerous records of *Scirpus cespitosus* from the later part of the Flandrian period represent ssp. *germanicus*, but this will be difficult to establish.

Scirpus maritimus L. 'Sea club-rush' 655: 3

H *Kirmington*; I *Stone*; F vIIa *Barry Docks*, vIIa/b *Avonmouth*, vIII *Welney*

Fruits of *Scirpus maritimus* were found in the Kirmington and Stone interglacial deposits (the latter sub-stage II), in pre-Neolithic submerged peat at Barry Docks, in peat of similar age at Avonmouth and in the Roman excavations of the roddon site at Welney, Norfolk. All these sites shewed deposition in brackish waters.

Scirpus sylvaticus L. 'Wood club-rush' 655: 4

H *Nechells*; F vIII *Newstead*

At Nechells, Birmingham, Kelly (1964) has reported the fruits of *Scirpus sylvaticus* from sub-stages IIb, IIc, IIIa and IIIb, along with many other woodland herbs and plants of river banks and marsh. It was also recorded from the Roman site at Newstead near Melrose. *S. sylvaticus* is widely if thinly scattered through the British Isles excepting the north-west.

Scirpus americanus Pers. (*S. pungens* Vahl; *Schoenoplectus americanus* (Pers.) Volkart) 'Sharp scirpus'
655: 7

eW ?*Sidgwick Av.*; F IV ?*Ballaugh*

Mrs Dickson has twice recorded fruits of *Scirpus americanus*, once from an Early Weichselian site in Cambridge, and once from the Pre-boreal in the Isle of Man. She comments that the low asymmetric hump is characteristic of the fruits of this species, but that wide variation in related species makes the identification only tentative.

S. americanus occurs today in pool margins near the sea in Jersey but not elsewhere in the British Isles, though it is found in fresh, brackish and saline waters in north Germany and the Netherlands.

Scirpus lacustris L. (*Schoenoplectus lacustris* (L.) Palla) Bulrush (Fig. 160*p*, *q*, *r*; plate XIX*i*) 655: 8

The fruits of *Scirpus lacustris* are large, durable and easily recognizable. They have been recorded in abundance from practically every sub-stage or zone of the Pleistocene from the Baventian glacial stage to the present day. The habitat of the plant naturally favours incorporation and preservation in aquatic muds and littoral peat deposits. Presence in seven Middle Weichselian sites reinforces that in three earlier glacial stages and abundance in the Late Weichselian, suggesting unequivocally perglacial survival. One could not well have better evidence for ancient permanent residence in the British Isles.

S. lacustris is one of Hultén's Eurasiatic species and his maps shews that in Scandinavia it extends in scattered localities well into the Arctic Circle: in Finland it occurs fairly frequently to about 66° N, but in the Scandinavian peninsula it is restricted as a frequent or very frequent species to central and south Sweden and Denmark. It is widespread in this country.

Scirpus × carinatus Sm. (*S. lacustris × triquetrus*) 655: 8

mW ?*Waltham Cross*; lW *Lough Gur*

A tentative identification has been made by E. M. Reid of fruits of *Scirpus × carinatus* (the hybrid between *S. lacustris* and *S. triquetrus*) from the Late-glacial bed at Waltham Cross, and a positive one by Mitchell from zone II of the Irish Late Weichselian. The one parent, *S. lacustris*, has been several times recorded from deposits of this age; the other, unrecorded fossil and now very local in the British Isles, has a present southern and central European range, and is of course halophilous.

Scirpus tabernaemontani C.C. Gmel. (*Schoenoplectus tabernaemontani* (C.C. Gmel.) Palla) 'Glaucous bulrush' 655: 9

Cs *Mundesley, Castle Eden*; Pa *Norfolk Co.*; C *Norfolk Co.*; H *Clacton*; mW ?*Waltham Cross*; lW *Branston, Helton Tarn, Roddansport*; F IV/V *Dogger Bank*, V *Ballybetagh*, vIIa *Westward Ho!*, vIIb *Bann Estuary*, ?*Prestatyn*, vIII *Finsbury Circus, Glastonbury Lake Village*

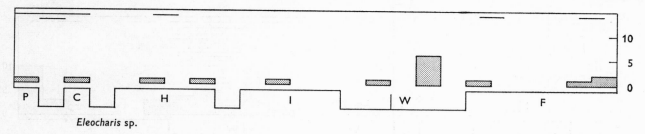

Eleocharis sp.

The fruits of *Scirpus tabernaemontani*, distinguishable by their biconvex shape, have been recorded from a number of sites in the British Isles, many of them coastal or estuarine. Reid recorded them from the Cromer Forest Bed series and the Durham coast, and later Wilson reported them from the Pastonian and Cromerian individually. Reid recorded fruits at Clacton (sub-stage III) and tentatively from the Middle Weichselian of the Lea Valley. These have now been supported by three Late Weichselian records, two specifically for zones II and III. Flandrian records occur from zone IV onwards. Though sparse, the fossil record suggests presence in this country from the Weichselian with possibly the long native status of *S. lacustris*.

S. tabernaemontani has a somewhat more restricted northern range in Scandinavia than has *S. lacustris*, but is found in scattered localities as far north as 65° N. Like *S. lacustris* it is generally distributed throughout the British Isles though in far smaller frequency.

Scirpus setaceus (*Isolepis setacea* (L.) R. Br.) 'Bristle scirpus'   655: 10

H *Kirmington, Hoxne*; *lH/eWo Hoxne*; *eWo Hoxne*; F VIIb *Frogholt, Bann Estuary*, VIIb/VIII *Leasowe*, VIII *Newstead*

Fruits of *Scirpus setaceus* have been found in the Hoxnian interglacial at Kirmington, and by Reid in what is now seen as sub-stage III at the type site where subsequently it was found by West in the last phase of the interglacial as well as in deposits of the next glacial stage. The four Flandrian records are late and two are from coastal sites. Although recorded throughout the British Isles *S. setaceus* is a plant of restricted southern range in Scandinavia where it is not found north of 60° N latitude.

Scirpus fluitans L. (*Eleogiton fluitans* (L.) Link) 'Floating scirpus' (Fig. 160w)   655: 12

Cs *Beeston*; H *Kirmington, Southelmham, Hitchin*; *lW Dronachy, Abbot Moss*; F IV *Gosport*, IV/V *Dogger Bank*, VIIa *Lonsdale, Westhay Track*

Scirpus fluitans fruits have been found in the Cromer Forest Bed series, and at three Hoxnian sites of which that at Hitchin can be referred to sub-stage II. The record from Dronachy falls within the Late Weichselian, and a record from Abbot Moss is referable to the zone I/II transition. There are two early Flandrian records and two from zone VIIa. This gives an indication, but scarcely

more, that the plant has a long record of continuous native presence.

S. fluitans belongs to Hultén's European–Atlantic group, and he shews it as restricted in Scandinavia to western Denmark and the western part of south Sweden. It has a wide but very local distribution in the British Isles.

Eleocharis sp. 'Spike-rush'   656

Fruits of the genus *Eleocharis* not specifically identified have been found in the Cromer Forest Bed series and separately on the Norfolk coast in Pastonian, Beestonian and Cromerian deposits. There are three Irish Hoxnian records and one East Anglian record for the Ipswichian. A single Middle Weichselian record from Marlow is followed by six from zone II of the Late Weichselian, four of them Irish. There are two early Flandrian records, one of them from the Dogger Bank, and four from the late Flandrian.

The generic record is made unremarkable by the fact that the records for four *Eleocharis* species are all quite similar (*vide infra*). It is particularly evident that the rather dwarf spike-rushes were well represented here in the open vegetation of the Weichselian glacial stage.

Eleocharis parvula (Roem. and Schult.) Link ex Bluff, Nees and Schau (Fig. 133)   656: 1

A *Corton*

West and Wilson (1968) have reported the identification of a nutlet of *Eleocharis parvula* from a clay lens in the Corton Beds that occur sandwiched in between the Cromer and Lowestoft Tills on the coast of Suffolk.

This is a dwarf halophyte that occurs on the Atlantic and Baltic coasts of western Europe, extending northwards to southern Scandinavia. It is extremely sparsely distributed on the coasts of the British Isles today.

Fig. 133. Fossil fruits of *Eleocharis*: *a*, *E. ovata*; *b*, *E.* cf. *carniolica* (from Irish Hoxnian, Watts, 1959b); *c*, *E. parvula* (from the Corton Beds of Suffolk, West & Wilson, 1968) (magnification × 23).

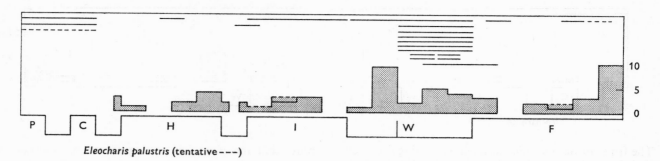

Eleocharis palustris (tentative – – –)

Eleocharis acicularis (L.) Roem. and Schult. (*Scirpus acicularis* L.) 'Slender spike-rush'   656: 2

Pa ?*Norfolk Co.*; H *Hoxne*; I *West Wittering, Stone*; F VIII *Silchester*

The records for *Eleocharis acicularis* are mostly of fruits found in interglacial deposits. These were tentatively identified from the Pastonian, and Reid and subsequently West recorded them from sub-stage III of the Hoxnian at the type site. Reid made two records also from the Ipswichian (not more closely specified) and from the Roman site at Silchester.

The plant often grows submerged in clear shallow water and flowers when littoral colonies are exposed. It is a species now generally scattered throughout the British Isles, and one which is fairly frequent in southern and central Sweden and Finland and scattered throughout the rest of Scandinavia up to 70° N latitude.

Eleocharis quinqueflora (F. X. Hartmann) Schwarz (*Scirpus pauciflorus* Lightf.; *Eleocharis pauciflora* (Lightf.) Link) 'Few-flowered spike-rush'   656: 3

Cs *Mundesley, Corton, ?Pakefield, ?Beeston*; H *Southelmham, Hoxne*; *l*H/*e*Wo *Hoxne*; *e*Wo *Hoxne*; I *West Wittering, ?Selsey*; *e*W *Sidgwick Av.*; *m*W *Upton Warren*; *l*W *Hailes, Dronachie, Corstorphine, Stair*; F VIIa *Lonsdale*

A large part of the records for fruits of *Eleocharis quinqueflora* were made before pollen analysis had been developed, so it is possible only to refer them to undivided glacial stages. This is so for Reid's four records from the Cromer Forest Bed series (two tentative), Miss Chandler's Hoxnian record from Southelmham, Reid's two records from the Ipswichian and his four records from the Scottish Late Weichselian. On the other hand, at Hoxne later work by West has not only given a new record for sub-stage III but has placed Reid's early records respectively in the same sub-stage and at the Hoxnian–Wolstonian transition. West has also recorded *E. quinqueflora* at Hoxne in the early Wolstonian. The fruits have been recorded both from the Early and Middle Weichselian but in the Flandrian there is only the one record, by Rankin, from zone VIIa in Lonsdale. Despite the range of the fossil record, many gaps require filling before continuity of presence can be really assured.

E. quinqueflora occurs scattered throughout the British Isles, rather commoner in the north than in the south. It is one of Matthews' Continental–northern species. It is placed by Hultén in his category of 'incompletely boreal–circumpolar' plants, and he shews it as scattered throughout the full length of Scandinavia to the extreme north, but somewhat more common in the south.

Eleocharis multicaulis (Sm.) Sm. 'Many-stemmed spike-rush'   656: 4

H *Kilbeg*; *l*W *Roddansport, Roundstone*; F *Roundstone*

The fossil record for *Eleocharis multicaulis* includes examples of local persistence. The first Hoxnian record at Kilbeg, by Mitchell, was substantiated by Watts for sub-stage II and for the composite sub-stage II/III. At Roddansport the fruits were found in the Late Weichselian in zone I/II and zone III, and at Roundstone I it was recorded not only from the Late Weichselian zone II but from Flandrian zones IV and V. In zone V it was also recorded from the Roundstone site II. There are so far no extra-Irish records for this species which, though scattered all through the British Isles, is noticeably more frequent in western Ireland and Scotland.

E. multicaulis is a European–Atlantic plant with a range in southern Sweden only just attaining 60° N.

Eleocharis palustris (L.) Roem. and Schult. 'Common spike-rush' (Plate XIX*h*)   656: 5

The fruits of *Eleocharis palustris* have been recorded from the Cromer Forest Bed series, and from three out of four sub-stages in both the Hoxnian and Ipswichian interglacials. In the glacial stages it has been recorded from the late Anglian and both early and late Wolstonian, whilst in the last glacial stage, after one record in the Early Weichselian, it has been frequently found in the Middle Weichselian, often in local abundance, and in all zones of the Late Weichselian. Apart from a partial hiatus in zone V it is recorded right through the Flandrian: of the ten records in zone VIII six are from Roman and three from mediaeval archaeological sites. The fossil record points strongly to a native status for *E. palustris* at least from the middle Pleistocene, and the discoveries of *E. palustris* ssp. *microcarpa*, not shewn in

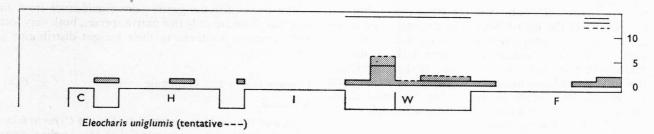

Eleocharis uniglumis (tentative - - -)

the record-head diagram, carry the evidence still further back in time.

This 'nearly cosmopolitan' species is widely distributed throughout the British Isles, and is described by Hultén as boreal–circumpolar. It is very frequent in Finland to 66° N and in scattered localities extends to 70° N.

Only one or two fossil fruits, those examined by Dr Walters, have been actually referred to ssp. *vulgaris* and these have been included in the diagram of the species *E. palustris*.

Eleocharis palustris ssp. microcarpa Walt.   656: 5*b*

Pa *Norfolk Co.*; Be *Norfolk Co.*; C *Norfolk Co.*; *m*W *Hedge Lane*

Walters has made it clear that distinction must be made between ssp. *microcarpa* Walt., which is the plant generally referred to as '*E. palustris*' in southern and eastern Europe, and the ssp. *vulgaris*, which is common over the greater part of northern and western Europe together with ssp. *microcarpa*. Mrs Wilson has recovered it from the Pastonian and Cromerian of the Norfolk coast and from the intervening glacial stage: these identifications were checked by Dr Walters, as was the Hedge Lane record from the Lea Valley Arctic Plant Bed.

The Critical Supplement to the *Atlas of the British Flora* shews ssp. *microcarpa* as apparently restricted in the British Isles to southern and eastern England and the west midlands, an area embracing that of the fossil sites.

Eleocharis uniglumis (Link) Schult.   656: 6

Bell (1968) has pointed to the small cells of the nutlet surface in *Eleocharis uniglumis* as responsible for the shiny appearance that distinguishes them from the punctate fruits of *Eleocharis palustris* in which the cells are larger and present concavities at the surface of the nutlet.

There is a single interglacial record (the Flandrian apart) for sub-stage III of the Hoxnian at Clacton, but it has been recorded from the preceding Beestonian glacial stage and the succeeding Wolstonian. In the Weichselian, records are numerous and well-attested, beginning with the Early Weichselian at Sidgwick Avenue and continuing with a series of Middle Weichselian records (Bell, 1968) with radiocarbon dates round about 40000 B.P. Checking of the Reids' material from the Lea Valley Arctic Plant Bed by Dr Walters has reduced the records there to two tentative identifications, at Hedge Lane and

Angel Road respectively. Scotland, Wales and Ireland are represented in records that include all three Late Weichselian zones, and there is also a record from the Mesolithic occupation level at Star Carr, Yorkshire. Thereafter there are no records until zones VIIb and VIII of the Flandrian where the sites are mostly either coastal or archaeological. It is very unlikely that *E. uniglumis* was absent from the British Isles through Flandrian zones V, VI and VIIa and the fossil data clearly suggest perglacial survival here.

E. uniglumis is a Eurasiatic species which is thinly distributed throughout the British Isles. In Scandinavia it extends along the Baltic coast to the head of the Gulf of Bothnia, is fairly frequent on the Norwegian coast to about 69° N, and occurs in scattered localities on the extreme northernmost part of the peninsula. Although frequent in coastal salt-marshes, *E. uniglumis* without doubt also occurs in freshwater fen. It has been suggested (Pigott & Walters, 1954) that the freshwater ecotypes may be of recent evolution, following the widespread occurrence in glacial time and local survival in relict habitats in the Flandrian. Both the open conditions and saline habitats of the Weichselian must have favoured the species.

Eleocharis carniolica Koch   656: 7

H ?*Kilbeg*, ?*Nechells*

Twice fruits of *Eleocharis* from the Hoxnian interglacial have, after consultation with Dr Walters, been referred tentatively to *E. carniolica*. At Nechells (Kelly, 1964) they were from sub-stage II of the interglacial, and at Kilbeg (Watts, 1959*b*) from a sample not referable to a sub-stage.

E. carniolica is a plant of south-eastern Europe not now found nearer to the British Isles than Austria, but it has a habit and favours habitats like those of British species of the genus.

Eleocharis ovata (Roth) R. Br. (Fig. 133)   656: 8

H *Kilbeg*

The small biconvex fruits of *Eleocharis ovata* with broad persistent stylar base were found in Hoxnian deposits at Kilbeg (Watts, 1959*b*) and the determination was confirmed by Dr Walters. *E. ovata* is an ephemeral plant of open wet-ground habitats, more or less circumboreal in its broad distribution and widespread in Europe, extending west to central Germany and the Paris region but not now found in the British Isles. The strongly continental emphasis of its European range today contrasts sharply with the strong oceanicity of western Ireland, no less marked in the Hoxnian than now. Watts conjectures that it is possibly an example of glacial extinction from the British Isles, since he notes that the areas of its absence in northern Europe 'correspond roughly with areas that have been glaciated'.

Blysmus compressus (L.) Panz. ex Link　'Broad blysmus' 657: 1

*l*W *Hartford*

From deposits at Hartford, Hunts., tentatively referred to zone II of the Late Weichselian, there were recovered fruits of *Blysmus compressus*, a rhizomatous plant of marshy open communities which occurs today scattered through England, Scotland and south-east Wales. It is widespread in Europe but does not extend north of about 61° N.

Blysmus rufus (Huds.) Link　'Narrow blysmus'　657: 2

H *Hoxne*; *e*Wo *Hoxne*; *m*W *Upton Warren*; F vIIa *Bowness Common*

Fruits of *Blysmus rufus* were recorded from Hoxne by Reid, and West later placed the level of Reid's recovery in sub-stage III, where he also found the plant, as he did in the early Wolstonian deposits there. Coope *et al.* (1961) recorded it from the Middle Weichselian of the Upton Warren deposits, radiocarbon dated 41 900 B.P. There is a single Flandrian record from brackish water deposits of zone vIIa on the Solway Firth, a discovery sorting with the present habitat of the plant, i.e. the short turf of the upper salt-marsh.

Blysmus rufus occurs round the shores of the British Isles excepting those of southern and eastern England; in Scandinavia it occurs in coastal situations right into the Arctic Circle on the west, but it is hardly present in the Gulf of Bothnia on the east. Scattered sites on the shores of the White Sea reflect tolerance of Arctic conditions. The rather common presence of *Blysmus rufus* in the Upton Warren deposits is matched by that of other halophytes including *Glaux maritima* and *Triglochin maritimum*, but there is no evidence of salinity in the mid-Hoxnian records.

Cyperus sp.　　　　　　　　　　　　　　　　658

Pa *Norfolk Co.*; H *Clacton*; I *?Grays*; *l*W *Abbot Moss*

Fruits referred to the genus *Cyperus* have been found in the Pastonian of Norfolk. At Clacton Reid made a tentative identification from deposits now seen to be substage IIa of the Hoxnian, and Turner and Kerney (1971) have later made a positive identification at the same site from IIb. There is a tentative Ipswichian record from Essex and a positive one from the composite zone I+II of the Late Weichselian in Cumberland.

The record for the genus gains significance from the fact that there are only two native species, both very local and extremely southern in their present distribution in this country.

Cyperus longus L.　Galingale　　　　　658: 1

A *Corton*

West and Wilson have reported the fruit of *Cyperus longus* from the Corton Beds interstadial of the Anglian glacial stage. The galingale is a plant whose distribution centres on the Mediterranean, becoming sparser northwards and now naturally present in this country mostly in scattered sites in southern England. The possibility that the fossil material of these sandy beds has been derived from older deposits has been seriously suggested.

Schoenus sp.　　　　　　　　　　　　　659

*l*W *Ballaugh*

Fruit of *Schoenus* sp. has been recorded from zone III of the Late Weichselian in the Isle of Man.

Schoenus nigricans L.　Bog rush　　　659: 1

F vII, vIII *Ards Beg*, vIIb *Roundstone*, vIIb/vIII *Munhin Bridge*, vIIb/vIII, vIII *Long Range*, vIII *Emlaghlea*

The black imbricate leaf-bases on the tussocks of *Schoenus nigricans* are rather readily recognizable, but all records of them so far have come from Ireland where they date from zones vII and vIII and are restricted to western regions of high rainfall, where the plant now commonly occurs in the oligotrophic blanket peat despite its calciphilous tendencies over the greater part of its range.

S. nigricans is a sub-atlantic (Hultén), or south-west European (Jessen), plant, which occurs in a few scattered localities only in Denmark and south Sweden, and until recently had its northernmost localities in south Norway. The concentration of records in the later part of the Post-glacial period is partly to be associated with the spread of the western Irish blanket bogs at this time, and partly perhaps with the plant's climatic requirements. Certainly although the plant now grows in all parts of the British Isles, its localities are densest in the far west of both Ireland and Scotland.

Rhyncospora sp.　　　　　　　　　　　660

*m*W *Barnwell Station*, *?Ponders End*

Fruits of the genus *Rhyncospora* have been twice recorded from the Middle Weichselian, although once only tentatively.

Rhyncospora alba (L.) Vahl　'White beak-sedge' 660: 1

*l*W *?Hawks Tor*; F v *Roundstone*, vIIb *Lonsdale*, vIIb, vIII *Meare Heath Track*, vIII *Bowness Common*, *Westhay Track*, *Decoy Pool Wood*, *Shapwick Station*, *Agher*

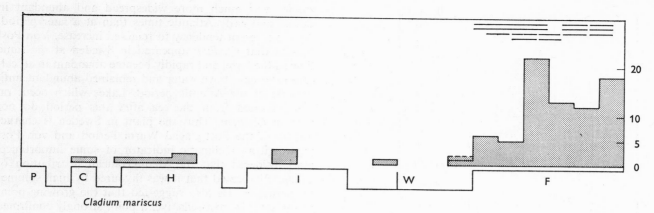

Cladium mariscus

The small scirpoid fruits of *Rhyncospora alba* have a characteristic appearance and in macerated peat samples often appear with their two valves sprung open at the base. The plant is a characteristic invader of the shallow water or mud surfaces of pools in ombrogenous bogs, and reproduces very freely from seed. Its fruits occur commonly in the aquatic *Sphagnum* peat layer of the bogs of the Somerset Levels, often characterizing the flooding horizons there.

The oldest of the British records is the qualified one from zone I of the Late Weichselian in Cornwall; there is also a positive zone V record from western Ireland, but otherwise the records are all late Flandrian, a reflection no doubt of the late development of ombrogenous bogs that similarly affected the records for *Eriophorum* spp., *Schoenus nigricans* and *Scirpus cespitosus*.

Rhyncospora alba is one of Hultén's incomplete boreal–circumpolar species: whilst fairly frequent in southern Scandinavia it extends in scattered localities as far north as the Arctic Circle.

Rhyncospora fusca (L.) Ait. f.   'Brown beak-sedge'
660: 2

F VIIb ?*Shapwick Track*

During excavation of the Late Bronze Age trackway (age 3310 B.P.) on Shapwick Heath a fruit was recovered which is tentatively recorded as *Rhyncospora fusca*, a very western and uncommon British plant which, however, grew until recently in the Somerset Levels where the find was made.

Cladium mariscus (L.) Pohl   Sedge (Fig. 134; plates IV, VIII, XII)   661: 1

Cladium mariscus is easily identifiable from its macroscopic remains and, given favourable circumstances, produces a readily recognizable type of peat, often reaching a thickness of some metres as in the Cambridgeshire Fens, the Somerset Levels and the lower part of Irish bogs, originating from colonization of calcareous lakes. The stout rhizomes, although decaying somewhat easily,

long remain recognizable by the bright salmon-pink colour of their half-decayed axes in which the parallel vascular strands remain distinct and recognizable (plate XII). The roots of *Cladium* are of two kinds (see Conway, 1936), the upper thin and fibrous, and those penetrating into deep anaerobic layers, three or four millimetres thick, spongy and unbranched: it is the latter which remain abundant and recognizable in *Cladium* peat, jet-black and shiny, usually flattened and often crumpled concertina-wise by contraction of the peat or mud after its initial penetration by the roots (plate VIII). The slender central stele remains visible inside the wide cortical cylinder.

The fruits are not easy to recognize so long as they remain within the utricle, but once broken out of this the nutlet is among the most easily recognizable of fossil 'seeds'. It has an urn shape with an obtusely pointed apex and three triangular cusps projecting at the base. It often happens (as for instance in the flooding horizons of the Somerset Levels) that a layer of *Cladium* peat closely corresponds with maxima of a large cyperaceous pollen, and there is every reason to think that this *is* the pollen of *Cladium*; in ordinary circumstances the pollen can barely be distinguished from that of other Cyperaceae.

C. mariscus fruit has been found in the Cromerian of the Norfolk coast. The Hoxnian records include a remarkable sequence from Baggotstown where it is represented in sub-stages I, II, II/III and III, and a record from sub-stage III at Hatfield. The Ipswichian provides three records from sub-stage II at Wretton, Selsey and Bobbitshole. Whereas these interglacial occurrences reflect temperate climatic conditions, a record from the late Anglian at Baggotstown, if not derivative, suggests survival there from much more inclement conditions. These are also implied by a record from the organic lenses in the gravels at Thrapston (Bell, 1969), one of which has a recent radiocarbon date of 25780 ± 870 B.P. and so is Middle Weichselian: of course here again secondary incorporation may be possible. This is made likely also for the zone III Late Weichselian record at Bradford Kaim, where Bartley reports a fractured *Cladium* nutlet. Another

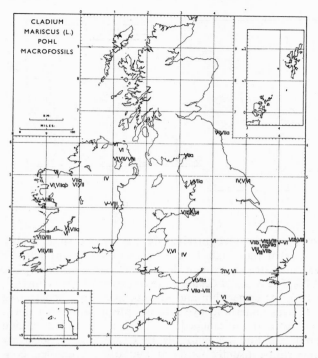

Fig. 134. Distribution in the British Isles of macrofossils of *Cladium mariscus* (L.) Pohl, in the various zones of the Flandrian.

zone III record, Skelsmergh Tarn, is only tentative, but there are no less than six records for Flandrian zone IV. Thereafter records are abundant through the Flandrian, most notably in zone VI (fig. 134).

Records for the later zones could certainly be multiplied if the numerous separate sites in the Cambridgeshire Fens and Somersetshire Levels were set down individually. In the Somerset Levels *Cladium* peat occurs abundantly at two different periods and correspondingly different situations: in zone VII it occurs along with *Phragmites* as a thick deposit of fen peat between the basal estuarine clays and the overlying oligotrophic *Sphagnum* peat, and it later occurs at many suitable sites as bands interrupting the general sequence of the upper part of the raised bog succession. These upper layers, which are particularly well marked on Shapwick Heath, represent episodes of flooding of the raised bogs by calcareous water brought down from the Mendip, the Quantocks and the Polden Hills.

Conway (1942) describes *Cladium mariscus* as 'a species of temperate to sub-tropical climates, heliophytic, basicolous, and intolerant of shade'. It occurs throughout Europe south of latitude 60° although scarce east of longitude 25° E. It is a plant of lowland habitats but occurs scattered throughout England, Wales and Scotland, generally scanty and becoming rarer although locally abundant still in East Anglia. It occurs throughout most of Ireland but is more abundant in the west than in the east. *C. mariscus* was made the subject of special study by von Post (1925) who shewed that in Scandinavia the species was much more widespread and abundant in Boreal and early Atlantic times than at a later period, despite a recent tendency to renewed increase. Von Post shewed that *Cladium* appeared in Sweden at the same time as the hazel and rapidly became abundant in all calcareous areas above water and remained abundant until the end of the Atlantic period. Lakes which occur on land elevated from the sea after this period do not contain *Cladium*. Thus the plant in Sweden is characteristic of the Post-glacial Warm Period and von Post regards it as a climatic indicator of some importance. He considered that the Mediterranean distribution of *Cladium* indicated that it was favoured by high summer temperatures. He also suggested that the growing-point of the stem is frost-sensitive, a point strongly confirmed by Conway's later experiments (Conway, 1938). Thus the plant appears to be intolerant of continental conditions and von Post suggested that it was the onset of continental conditions in the Sub-boreal period which chiefly led to the decline of *Cladium* in Scandinavia. In view of this suggestion it is of special interest to note that in the British Isles with their more oceanic climate there is no evidence for post-Atlantic climatic restriction of the species, and also that it may have become established here earlier than in Scandinavia.

There is no doubt that artificial drainage has been responsible for much of the most recent reduction in frequency of *Cladium* in the British Isles.

Carex laevigata Sm. (*C. helodes* Link)   'Smooth sedge' (Plate XIX*m*)   663: 1

Cs ?*Pakefield*; IW ?*Aby Grange*, ?*Nazeing*; F IV ?*Nazeing*, ?*Flixton*, VIIa *Bowness Common*

Carex laevigata fruits have been tentatively identified in the Cromer Forest Bed series.

From Nazeing in the Lea Valley, in zones III and IV, were recovered abundant *Carex* fruits with utricles which, after very thorough investigation, were concluded to correspond closely to those of *C. laevigata*, although the absence of any beak from the fossil fruits prevented certainty of recognition. E. Nelmes was disinclined to attribute this fossil *Carex* to *C. laevigata*, and after some hesitation refused to identify it with any living British species of sedge. Further tentative identifications of *C. laevigata* were made from zone II of the Late Weichselian at Aby Grange and from zone IV of the Flandrian at Flixton, Yorkshire. The only unqualified record is from zone VIIa at Bowness Common.

Although *C. laevigata* is generally distributed through most of the British Isles it is absent from the very north of Scotland and does not grow higher than 1050 ft (320 m). On the Continent it has a western distribution but only extends as far north as Holland and Belgium, a circumstance lending special importance to the possible presence of the plant in England in the Late Weichselian.

Carex distans L. 'Distant sedge'   663: 2

H *Hoxne*; I ?*Selsey, West Wittering*; mW *Hedge Lane*

Carex distans was tentatively identified by Reid from sub-stage III deposits of the Hoxnian interglacial, and West subsequently made a positive record at the same site and sub-stage. Reid made two records, one tentative, from the Ipswichian interglacial, and an unqualified record from the Middle Weichselian of the Lea Valley.

C. distans is a coastal sub-atlantic species which does not extend north of latitude 61° N in Scandinavia, although reaching the Shetland Isles of Britain.

Carex punctata Gaud. 'Dotted sedge'   663: 3

mW ?*Broxbourne, Angel Rd, Barrowell Green*

Carex punctata fruits were identified (once doubtfully) by Reid from the Middle Weichselian deposits of the Lea Valley.

C. punctata is a very local coastal sedge in Britain today, found as far north as Kirkcudbrightshire and Wigtown. It occurs round the Mediterranean and western European coasts, but barely extends into south Scandinavia.

Carex hostiana DC. (*C. hornschuchiana* Hoppe; *C. fulva* auct.) 'Tawny sedge'   663: 4

lW *Roddansport*

From a zone II Late Weichselian site in northern Ireland Morrison has recorded a fruit of *Carex hostiana*. *C. hostiana* is classified by Hultén as a sub-atlantic plant: it extends on the west Norwegian coast above the Arctic Circle, though much commoner in southern Scandinavia. It occurs throughout the British Isles.

Carex binervis Sm. 'Ribbed sedge'   663: 5

F VIIa ?*Nazeing*

Carex flava agg. (including *C. flava* L.; *C. lepidocarpa* Tausch; *C. demissa* Hornem.; *C. serotina* Mérat) 'Yellow sedge'

H *Nechells*; Wo *Farnham*; mW *Barnwell Station, Upton Warren, Temple Mills*, ?*Hedge Lane*; lW ?*Elstead*; F IV ?*Elstead*, IV/V *Dogger Bank*, VIIa *Jenkinstown*, VII *Hightown*

Although many of the records given above are cited by the authors as *Carex flava* L., the date at which they were made makes it most probable that they really refer to the *C. flava* aggregate which has only more recently been properly analysed: even E. M. Reid's 1949 paper is work completed for publication many years before it appeared. Other records like those from sub-stage IIIa and IIIb of the Hoxnian at Nechells and the Jenkinstown record have been explicitly referred to the aggregate species.

Carex lepidocarpa Tausch   663: 7

lW *Loch Kinord*, ?*Elstead*; F IV ?*Elstead*, IV/V/VI ?*Rhosgoch*, VIII *Bunny*

Carex lepidocarpa has been positively recorded from zone II of the Late Weichselian in Scotland and tentatively from zone III, as from zone IV of the Flandrian, at Elstead in Surrey. The zone VIII site from Bunny, Notts., is Roman.

Carex serotina Mérat (*C. oederi* auct.)   663: 10

F VIIb/VIII *Hightown*

Carex extensa Gooden. 'Long-bracted sedge' 663: 11

H *Clacton*; F VII/VIII *S.W. Lancs. Co.*

Like *Carex punctata* and *C. distans*, *C. extensa* is a plant of the European coastline. It does not extend north of about latitude 61° N: the fossil record is easily conformable with this range. Both sites are coastal: that at Clacton is from sub-stage IIIa of the Hoxnian interglacial, the other rather late Flandrian.

Carex sylvatica Huds. 'Wood sedge'   663: 12

H *Clacton*

Reid and Chandler's record of *Carex sylvatica* from Clacton is now referable to sub-stage IIIa of the Hoxnian interglacial.

Carex capillaris L. 'Hair sedge'   663: 13

mW *Barnwell Station*

Carex capillaris, which has been identified from the Middle Weichselian at Barnwell, is a species confined in the British Isles today to the mountains of northern England and Scotland. It is a plant of circumpolar, arctic and alpine range extending to 79° N in Greenland (Polunin, 1940); Hultén places it in the group of boreal–circumpolar plants boreal–montane in Europe, and it belongs to Matthews' Arctic–alpine element. It occurs commonly throughout western Scandinavia to the northernmost limits of the land.

Carex pseudocyperus L. 'Cyperus sedge'   663: 15

H *Nechells*; I *Wretton*; mW ?*Great Billing*; lW *Tadcaster*; F VI *Portrush*, VIIa *Tadcaster*, VIIb/VIII *Woodwalton*, VIII *Hockham Mere*

At Nechells, Birmingham the nutlets and utricles of *Carex pseudocyperus* occurred at such a sequence of levels (sub-stages I; IIa,b,c; IIIa,b) that one has to suppose the plant persisted locally there through a large part of the Hoxnian interglacial (Kelly, 1964). It was recorded also from sub-stage II of the Ipswichian at Wretton by Sparks and West. There is a tentative record for the Middle Weichselian and a positive one from the Late Weichselian. The Flandrian records are from zone VI and later.

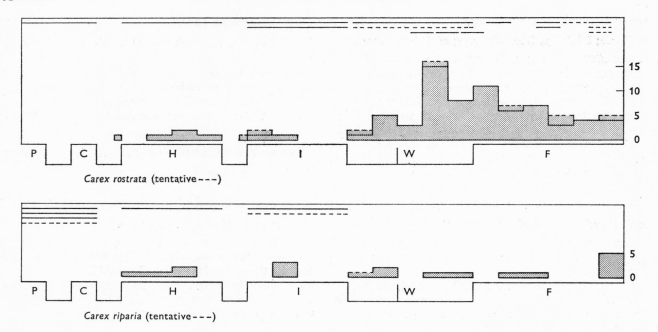

Carex rostrata (tentative ---)

Carex riparia (tentative ---)

C. pseudocyperus is a species now only reaching about 62° N latitude in Scandinavia, but with a very considerably more northern range in the fossil state, presumably from the Flandrian hypsithermal; it has been several times found as fossil at the head of the Gulf of Bothnia within the Arctic Circle. A record of this kind of course calls in question the significance of the two Weichselian records, especially as the species now is barely found in northern England and Scotland, though moderately common further south in the British Isles.

Carex rostrata Stokes (*C. ampullacea* Gooden.; *C. inflata* auct.) 'Beaked sedge', 'Bottle sedge' (Figs. 135, 160*k, l*; plate XIX*n*)   663: 16

The fruits of *Carex rostrata* are more commonly identified fossil than those of any other species of *Carex*: they have a small trigonous nutlet, often with a curved stylar base,

and a characteristic inflated ribbed utricle. They have been recorded from the Cromer Forest Bed series, from three sub-stages of the Hoxnian and two of the Ipswichian as well as from sites in each interglacial not referable to a particular sub-stage. There is a positive record for the late Anglian and a tentative one for the late Wolstonian: these are in line with numerous records for all parts of the Weichselian glacial stage with no less than fifteen unqualified records for Weichselian zone II. Since these are followed by a substantial number of records for every Flandrian zone it appears that perglacial survival of *C. rostrata* has been demonstrated whilst the interglacial records suggest still more ancient native status.

C. rostrata is one of Hultén's west European–middle Siberian plants. He shews it as frequent throughout the whole of Scandinavia to the furthest north: it grows throughout the British Isles though somewhat less common in the south and east.

Carex vesicaria L. 'Bladder sedge'   663: 17

Cs *Corton*; F IV *Dogger Bank*, V, VI *Kirkby Thore*, VI *Malham Tarn*, VI, VIIb/VIII *Elland*, VIIb *Wingham*

Apart from Reid's record of *Carex vesicaria* from the Cromer Forest Bed series, fruits of this species have been noted only from the Flandrian where the record however, if thin, is continuous and affords instances of local persistence.

Carex riparia Curt. 'Great pond-sedge'   663: 20

Fruits of *Carex riparia* have been recorded from the Cromer Forest Bed series and equivalent deposits at Castle Eden. Not only are they reported from Southelmham as Hoxnian *per se*, but at Nechells Kelly reports them

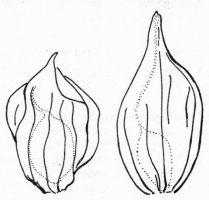

Fig. 135. Fruits with utricles of *Carex rostrata* Stokes, from the Late Weichselian (zone II) layers at Hawks Tor kaolin pit, Bodmin Moor, Cornwall (× 10).

from sub-stages I, IIa,b,c and IIIa,b, a remarkable local persistence. It was present also in sub-stage III at Hatfield. There are five Ipswichian records, these concentrated in sub-stage II. A tentative Early Weichselian identification is followed by two Middle Weichselian records from the Lea Valley, and records from both zone II and zone III of the Late Weichselian at Neasham. Flandrian records are sparse, two Irish records for zones V and VI preceding five zone VIII records, of which one comes from an Iron Age site and three from Roman settlement sites. Though the total record is sparse and lacks representation in glacial stages other than Weichselian, it suggests long native presence of *C. riparia* in the British Isles.

C. riparia is now generally distributed throughout the British Isles though rarer in the north: in Scandinavia it does not extend above c. 62° N.

Carex acutiformis Ehrh. (*C. paludosa* Gooden.) 'Lesser pond-sedge'   663: 21

Cs ?*Pakefield*; Be *Norfolk Co.*; C *Norfolk Co.*; IA *Hoxne*; *l*Wo *Selsey*; *l*Wo/I *Bobbitshole*; I *Selsey, Bobbitshole*; F VI, VIII *Apethorpe*, VIIa ?*Shapwick Heath*, VIII ?*Car Dyke*

Carex acutiformis was tentatively identified by Reid from the Cromer Forest Bed series, and later respectively from the Beestonian and Cromerian that form part of that series. At Hoxne it was found in the late Anglian and at Selsey in the late Wolstonian, but there are no Weichselian records. The Ipswichian is represented in repeated levels at two sites, at Bobbitshole where it is in the opening transition and in sub-stages Ib, IIa and IIb, and at Selsey where it is not only in late Wolstonian but in Ipswichian sub-stages Ia, Ib, IIa and IIb, no doubt in response to persistent riparian growth. The Flandrian records are sparse and not before zone VI, with the latest from the Roman canal at Cottenham, Cambridgeshire.

C. acutiformis is one of Hultén's west European–south Siberian plants. In Scandinavia it is frequent only in the south, having but scattered localities north of 60° N latitude. In the British Isles too, though present generally save for north-western Scotland and parts of northern and western Ireland, it is more abundant in the south and east, the area in fact from which all the fossil records have come.

Carex pendula Huds. 'Pendulous sedge'   663: 22

C *Norfolk Co.*; H *Clacton*; Wo ?*Dorchester*; F V *Apethorpe*

The Hoxnian record of *Carex pendula* by Reid is from sub-stage III. It is a sub-atlantic species which occurs in Scandinavia no further north than Denmark; in the British Isles it has a range similar to that of *C. acutiformis*.

Carex strigosa Huds.   663: 23

I *Ilford*; F IV/V *Star Carr*, VI *Kirkby Thore*, VIIb *Littleport*

The Ipswichian record of *Carex strigosa* fruit is from sub-stage II: the Flandrian records also correspond to

Fig. 136. Top of female spikelet and two fruits of *Carex panicea* from Late Weichselian (zone III) deposits at Ballybetagh, near Dublin (×4). (After Jessen & Farrington, 1938.)

temperate climatic conditions. Like *C. pendula, C. strigosa* is one of Hultén's sub-atlantic plants and in Scandinavia is restricted to Denmark. In the British Isles too it has a similar range, but one even more restricted, in its frequent occurrence to the south-east.

Carex pallescens L. 'Pale sedge'   663: 24

*m*W ?*Ponders End*

Carex panicea L. Carnation grass (Fig. 136)   663: 26

H *Clacton*; W *Airdrie*; *l*W *Twickenham, Johnstown,* ?*Elstead,* ?*Ballybetagh*; F *Burtree Lane,* VIIa, VIIb ?*Elland,* VIIb/VIII *Elland*; VIII ?*Silchester*

The only interglacial record for *Carex panicea* is from sub-stage III of the Hoxnian. The Weichselian records, two only tentative, embrace both zone II and zone III. At Ballybetagh Jessen recovered the top of a female spike which he says 'comes very close to *Carex panicea*' although in some characters 'the spike resembles rather that of the northern *C. sparsiflora* (Wg.) Steud'.

C. panicea is common and generally distributed in the British Isles, with a high altitudinal range, and in Scandinavia it occurs as far north as 70° N, although more frequent in the south.

Carex flacca Schreb. (*C. glauca* Scop.; *C. diversicolor* auct.) Carnation grass   663: 31

*m*W ?*Barnwell Station, Hedge Lane, Waltham Cross*; *l*W *Kirkmichael,* ?*Roddansport,* F IV ?*Roddansport,* VI *Neasham,* VIII *Bunny,* ?*Larkfield*

Carex flacca has been recorded from two Middle Weichselian sites in the Lea Valley and tentatively from

Barnwell. Though positively identified in the Late Weichselian at Kirkmichael, Morrison gives only tentative identification of it from north-eastern Ireland although the records cover zones I/II, II and III as well as Flandrian zone IV. There is one zone VI record, a positive Roman site record from Bunny and a mediaeval one from Larkfield. The fossil record gives fair indication of persistence through the last glacial stage to the present day of a species that is now common throughout Britain. *Carex flacca* is, like *C. strigosa* and *C. pendula*, within Hultén's category of sub-atlantic plants though less restricted than they are in Scandinavian range: it scarcely exceeds latitude 61° N, but on the Norwegian coast it occurs in scattered localities up to the Arctic Circle.

Carex hirta L.   Hammer sedge   663: 32

Cs *Pakefield*; H *Clacton*; F VIII *Larkfield*

Carex lasiocarpa Ehrh. (*C. filiformis* auct.)   663: 33

F IV, VI, VIIa, VIIb, VIII *Loch Kinord*, V, VI *Drymen*, VIIa *Ardlow Inn*

Records from two sites provide at once evidence of local persistence of *Carex lasiocarpa* and its presence in Scotland through the whole Flandrian (Vasari & Vasari, 1968). There is one Irish record from zone VIIa. It is a fairly common plant through the northern half of the British Isles.

Carex pilulifera L.   'Pill-headed sedge'   663: 34

lW ?*Mapastown*

The doubtful record of *Carex pilulifera* at Mapastown is from zone III.

Carex buxbaumii Wahlenb.   663: 42

F VIII ?*Silchester*

A single tentative identification of *Carex buxbaumii* from Roman Silchester is the only record of this sedge now restricted in this country to one or two Scottish localities.

Carex atrata L.   'Black sedge'   663: 43

mW *Waltham Cross*

Now that Miss Chandler's record from Barnwell Station has been revised (Bell & Dickson, 1971), the only record for *Carex atrata* is Reid's Middle Weichselian identification from the Lea Valley Arctic Plant Bed.

C. atrata is a circumpolar arctic–alpine plant restricted in Britain to high mountains in Snowdonia, the Lake District and Scotland.

Carex norvegica Retz. (*C. alpina* Liljeb., non Shrank; *C. halleri* auct.)   663: 44

lW *Kirkmichael*

Carex norvegica, recorded by Reid from the Late Weichselian of the Isle of Man, is a species now restricted within the British Isles to a very few Scottish sites. It is regarded by Hultén as belonging to the amphi-atlantic species of arctic–montane plants present also in southern mountains. It ranges through the mountains of Scandinavia from about 59° N to the North Cape, and is described by Polunin as circumpolar, chiefly alpine and low arctic.

Carex elata All. (*C. stricta* Gooden., non Lam.; *C. hudsonii* A. Benn) 'Tufted sedge'.   663: 46

F VIIb, VII/VIII *Woodwalton*

This species still grows commonly in the East Anglian fenland.

Carex acuta L. (*C. gracilis* Curt.)   'Tufted sedge'   663: 47

Wo ?*Dorchester*; F VIII ?*Ranworth*

Carex aquatilis Wahlenb. or **Carex bigelowii** Torr. ex Schwein 'Stiff sedge' (Plate XIX*l*)   663: 48, 663: 52

A *Corton*; I *Histon Rd*; eW *Chelford, Sidgwick Av.*; W *Thrapston*; mW *Earith, Great Billing, Barnwell Station*; lW *Old Buckenham Mere, Skelsmergh Tarn, Nazeing, Colney Heath, Newtonbabe, Ballybetagh, Johnstown, Derriaghy, Lopham*; F V *Cross Fell, Newtownbabe*, VI *Jenkinstown*, VIIa *Derravaragh*

Identifications of flat *Carex* nutlets within nerveless utricles have been almost always treated by authors as '*Carex aquatilis* or *C. bigelowii*': Bell (1968) refers three (Thrapston, Earith, Great Billing) to a type that includes these two species along with *C. nigra*. The few instances where the double appellation is not used are (with a single exception, the Late Weichselian record of *C. aquatilis* from Old Buckenham Mere) only tentatively named. Accordingly all the fossil records have been treated in the one category.

The single sub-stage III record from the Cambridge Ipswichian and the four Flandrian records are the only ones from the temperate interglacial stages. From glacial stages there is one record from the Corton Beds, and a great number from the Weichselian, Early, Middle and Late. The Late Weichselian zones are all represented and there is recurrence of the records at some sites, thus Nazeing (I and II/III), Newtownbabe (II and V), Ballybetagh and Lopham Little Fen (II and III). The Middle Weichselian record for Barnwell has been added by Bell and Dickson (1971) on revision from Miss Chandler's material referred only to the genus.

The alternative identification of *C. aquatilis* or *C. bigelowii* retains biogeographical significance from the fact that both species have a northerly distribution at the present day. The former is placed by Hultén in his category of northerly boreal–circumpolar plants and by Matthews into his arctic–sub-arctic element; *C. bigelowii* is in Hultén's circumpolar–arctic montane plants. Both have a strongly northern representation in Scandinavia. In the British Isles neither species is found south-east of a line from South Wales to North Yorkshire and both are plants characteristic of high mountains.

It follows therefore, that to whichever taxon the fossil fruits belong, the Weichselian presence was wide and subsequent retraction of range very considerable.

Carex nigra (L.) Reichard (*C. fusca* All.; *C. goodenowii* Gay) 'Common sedge' (Fig. 160*o*)   663: 50

*m*W *Barnwell Station, Hedge Lane, Waltham Cross*; *l*W ?*Hawks Tor,* ?*Hartford*; F IV *Gosport*, VIIb *Avonmouth*

Fruits of *Carex nigra* have been recorded from the Middle Weichselian of the Lea and Cam valleys, and it will be recalled that Bell placed *C. nigra*, along with '*C. aquatilis* or *C. bigelowii*', in a type she recorded from one undivided Weichselian site and two Middle Weichselian sites. The tentative Cornish records are for zones I and II of the Late Weichselian: the Huntingdonshire record is also for zone II.

According to Hultén *C. nigra* is an incompletely boreal–circumpolar species that is common in Scandinavia well into the Arctic Circle, and it is common throughout the British Isles. It seems probably a perglacial survivor.

Carex paniculata L. 'Panicled sedge' (Plate XIX*k*)   663: 54

H ?*Hatfield*; I ?*Ilford*; *l*W ?*Nazeing, Loch Cuithir*; F IV ?*Flixton*, V, VI, VI/VIIa, VIIa *Rhosgoch*, VIIa *Tadcaster, Ardlow Inn*, VIIa/b *Dernaskeagh*, VIIb *Avonmouth*, ?*Littleport*, VIIb, VIII *Loch Cuithir, Wingham*, VIIb/VIII *Hightown*, VIII *Canbo*

Records for fruits of *Carex paniculata* begin with tentative identifications in zones I and II/III in Essex, and an unqualified zone III determination from Skye, at a site subsequently yielding records also for zones VIIb and VIII. There is persistence also at Wingham and more strikingly at Rhosgoch Common. There is thus evidence of native status at least since the Late Weichselian in Britain.

C. paniculata is decidedly southern in its Scandinavian distribution, a fact according with the plant's greater frequency in the southern part of Great Britain.

Two interglacial records, from sub-stage III of the Hoxnian and sub-stage II of the Ipswichian, have been referred to '*C. paniculata* or *appropinquata*'.

Carex appropinquata Schumach (*C. paradoxa* Willd., non J. F. Gmel.)   663: 55

F VI, VIII *Apethorpe*, VIII *Godmanchester*

The Apethorpe identifications of *Carex appropinquata* were based upon both spikelets and leaves. The Godmanchester record is from the Roman occupation site. There are two interglacial records for '*C. paniculata* or *appropinquata*' (see preceding account, 663: 54).

C. appropinquata is a plant of calcareous peaty fens, very local in its British distribution which includes particularly the central Irish plain, east Yorkshire and East Anglia.

Carex diandra Schrank (*C. teretiuscula* Gooden.)   663: 56

*l*W ?*Kirkmichael*; F VIII *Canbo*

Dickson's tentative identification of *Carex diandra* at Kirkmichael covers Late Weichselian zones I and III: it does not now grow in the Isle of Man.

Carex otrubae Podp. (*C. vulpina* auct.) 'False fox-sedge'   663: 57

F VIII *Apethorpe*

Carex vulpina L. 'Fox-sedge'   663: 58

F VIII ?*Ehenside*

A record for *Carex vulpina* from the Middle Weichselian at Barnwell Station has been deleted after re-examination.

Carex disticha Huds. 'Brown sedge' (Fig. 160*m, n*)   663: 60

F IV *Gosport*

Carex arenaria L. 'Sand sedge'   663: 61

*m*W ?*Barnwell Station*

Carex divisa Huds. 'Divided sedge'   663: 62

*m*W *Barnwell Station*; F VII, VIIb/VIII *Hightown*

Carex divisa, a preponderantly coastal sedge in the British Isles today, has been recorded from shore peats in Lancashire and from the Middle Weichselian of the Cam valley. *C. divisa* belongs to the Continental–southern element of Matthews and has a distribution in Britain confined to the south-east.

Carex maritima Gunn. (*C. incurva* Lightf.) 'Curved sedge'   663: 64

H *Kirmington*; *e*Wo ?*Hoxne*; *m*W *Broxbourne, Hackney Wick, Barnwell Station*

Carex maritima was identified from the Kirmington interglacial, a decidedly coastal deposit, and tentatively from the early Wolstonian at Hoxne. There are three Middle Weichselian records from the Lea and Cam valleys.

C. maritima is an arctic–alpine species (Matthews) now found only on the coasts of extreme northern and north-eastern Scotland and north-east England. In Scandinavia it has a very strikingly western range which extends from north Jutland and Holland to the North Cape. Polunin writes of it as circumpolar and general in the Arctic, extending to 81° N. It is a plant of the damp hollows of fixed dunes, and in Scandinavia also of sandy stream margins. Its presence inland in the Weichselian recalls that of other coastal sedges such as *C. divisa* and *C. distans*.

Carex spicata Huds. (*C. contigua* Hoppe; *C. muricata* auct.) 'Spiked sedge'   663 : 67

IW Scaleby Moss

The identification of *Carex spicata* from the Late Weichselian at Scaleby Moss is from zone III with a radiocarbon age of *c.* 10 580 B.P. The site is near the present limit of the range of the plant, which, strongly south-eastern, barely enters Scotland or Ireland.

Carex muricata L. (*C. pairaei* F. W. Schultz) 'Prickly sedge'   663 : 68

Cs *Pakefield*; I *Stone, West Wittering*; F VIII *Silchester*

Carex muricata L. was identified by Reid from the Cromer Forest Bed series, twice from the Ipswichian interglacial and once from Roman material at Silchester. It is not known however if any of these records belongs to *C. spicata* Huds. or to *C. muricata* L.; both are south-eastern in the British Isles.

Carex echinata Murr. (*C. stellulata* Gooden.) 'Star sedge'   663 : 70

IW Hawks Tor, Loch Fada; F v/VIa *Kilmacoe*, VIIb/VIII *Rockstown*, VIII ?*Ehenside Tarn, Loch Fada*

Carex echinata has been recorded from zone II of the Late Weichselian in Cornwall, and at Loch Fada (Skye) in zone III, the composite zone III/IV and zone VIII. There are enough Flandrian records to suggest continuity of presence through from the Late Weichselian.

C. echinata, a plant typical of acidic peaty soils, occurs frequently throughout Scandinavia to 65° N and on the west coast almost to 70° N.

Carex remota L. 'Remote sedge'   663 : 71

Cs ?*Pakefield*; F VIII ?*Silchester*

Carex curta Gooden. 'White sedge'   663 : 72

eW ?*Chelford*; *IW* ?*Loch Kinord,* ?*Kirkmichael*; F IV ?*Loch Kinord*, IV, VI ?*Loch Cuithir*, VIII ?*Loch Fada*

There are only tentative identifications of the fruits of *Carex curta* which nevertheless comprise the early and Allerød phases of the Weichselian. *C. curta* is a sedge of acid soils occurring throughout Britain but commoner in the north whence in fact come all the Flandrian records.

Carex lachenalii Schkuhr. (*C. leporina* L., nom. ambig.; *C. lagopina* Wahlenb.)   663 : 73

Wo *Farnham*; mW ?*Barrowell Green, Temple Mills, Waltham Cross, Ponders End, Barnwell Station*

Carex lachenalii has been identified in the Wolstonian and at five Late Weichselian sites, one in Cambridge and four in the Lea Valley; of the latter, E. M. Reid has queried the identification at Barrowell Green.

As Clement Reid made separate identifications of *C. ovalis* Gooden. (see below) it seems highly likely that his identifications (*C. leporina*) correspond to *C. lachenalii* Schkuhr.

In contrast to *C. ovalis*, *C. lachenalii* has a distribution range extending into the far north and is indeed a widely distributed circumpolar species. Within the British Isles *C. lachenalii* is now found only in a very few Scottish localities.

Carex ovalis Gooden. (*C. leporina* auct.) 'Oval sedge'   663 : 74

F VIII ?*Silchester*

Carex pauciflora Lightf. 'Few-flowered sedge'   663 : 78

IW Hartford

The very narrow fruit shape tapered at both ends facilitates identification of *Carex pauciflora*. It was found in zone II of the Late Weichselian at Hartford, Huntingdonshire.

Carex capitata L.   663 : 79

mW *Temple Mills*, ?*Angel Rd*

Fruit of *Carex capitata* was recorded by Reid, once tentatively and once without qualification, from the Lea Valley Arctic Plant Bed. Miss Chandler's further Middle Weichselian record of it from Barnwell has been deleted after revision (Bell & Dickson, 1971).

These sites are widely outside the present range of the species which is not certainly known native in the British Isles, its only record depending on a single tuft on southern Uist in the Outer Hebrides.

It is placed by Hultén in his category of circumpolar arctic–montane plants found in the mountains of central Europe, and is written of by Polunin as 'chiefly sub-arctic and alpine, never high-arctic'. It occurs scattered throughout the mountain area of Scandinavia north of 60° N.

Carex pulicaris L. Flea-sedge   663 : 80

mW *Ponders End*; *IW Hawks Tor*; F IV/v *Dogger Bank*, VII *Ipswich*

Carex pulicaris has been identified in the Middle Weichselian, zone II of the Late Weichselian, and in Pre-boreal and late Atlantic stages of the Flandrian.

It is one of Matthews' Continental–northern species, but Hultén regards it as sub-atlantic in accordance with

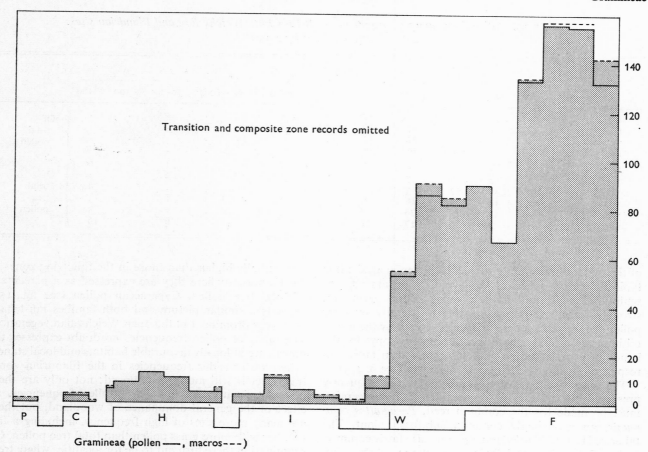

Transition and composite zone records omitted

Gramineae (pollen ——, macros ---)

its predominantly southern and western Scandinavian range. It is noteworthy that *C. pulicaris* extends well within the Arctic Circle in Norway. In Britain it has become rare in the south and east through drainage but is otherwise widespread. The fossil record, though sparse, indicates the likelihood that it was a perglacial survivor.

Carex dioica L. 'Dioecious sedge'   663 : 81

Cs *Corton*; C *Norfolk Co.*; W *Airdrie*; mW *Upton Warren, Waltham Cross, Barrowell Green*; lW *Greenock*; F VIII *Silchester*

Carex dioica, recorded by Reid from the Cromer Forest Bed series, has subsequently been identified by Wilson from the Cromerian. His general record for the Weichselian, his two Middle Weichselian records from the Lea Valley and his Late Weichselian record from the Clyde have been recently supplemented by identification from the Middle Weichselian at Upton Warren. Reid also recorded *C. dioica* from Roman Silchester.

C. dioica is one of Matthews' Continental–northern plants, and is placed by Hultén in his category of west European–north Siberian species. It extends through the full range of Scandinavia, as it does through the British Isles despite diminution associated with drainage in southern England. Though proof is lacking it seems likely

that present British populations have been derived from those of the Weichselian.

Dulichium arundinaceum (L.) (Plate XXVIII)

H *Quinton*

Fossil fruits of the North American sedge, *Dulichium arundinaceum*, have been identified by Dr F. G. Bell (1970a) from middle Hoxnian deposits exposed in a motorway cutting near Birmingham. In the same deposit were also seeds of *Picea abies*, *Najas flexilis* and *Trapa natans*, of which only the *Najas* survives in the British flora.

Dulichium is a circumpolar Tertiary genus that is represented in the Tertiary and Quaternary deposits of Europe though now extinct in this continent: its latest records are from the Eemian interglacial (cf. Ipswichian) in Denmark. Besides *D. arundinaceum* the fossil record includes the extinct *D. vespiforme* Reid, from the Tegelian (early Pleistocene) of Holland.

GRAMINEAE

The family Gramineae is represented in this country by many species occupying a very wide range of habitats. Pollen referable to the family occurs abundantly in much

TABLE 30. *Pollen frequency for Gramineae in British Late Weichselian and Flandrian sites: site frequencies by zones*

	Percentages													
	+	0–2	2–5	5–10	10–20	20–30	30–40	40–50	50–60	60–70	70–80	80–100	> 100	
I	.	.	1	3	9	12	15	4	.	.	.	.	.	Of total pollen
II	.	.	1	6	13	17	13	5	2	1	.	.	.	
III	.	.	.	5	22	13	12	1	1	.	.	.	.	
IV	.	.	.	6	16	15	.	3	.	.	.	.	.	
IV	.	.	.	5	8	7	4	1	2	.	1	.	6	Of total tree pollen
V	1	1	3	9	7	3	5	1	1	.	2	1	.	
VI	.	16	24	13	5	5	2	2	1	2	.	.	4	
VIIa	.	12	14	30	26	10	8	2	4	1	2	.	4	
VIIb	.	6	10	28	25	8	4	7	1	.	2	.	5	
VIII	.	1	3	7	17	21	11	6	2	1	6	.	16	

of the material commonly subject to pollen analysis: it is typically clear, spherical or ovoid, with almost unsculptured walls and with a single round germpore, strongly ringed. It is not possible to identify by their pollen the separate genera or species, except by the use of oil-immersion and phase-contrast microscopy in the hands of experienced palynologists, and then mostly in respect of the large pollen grains of cereals (see p. 405).

The vegetative parts of most grasses are not frequently preserved nor are they easily identifiable, but exception must be made for the common reed, *Phragmites communis*, an exceedingly common sub-fossil plant. The tuberized bases of *Molinia coerulea* are also fairly common and recognizable, especially in conjunction with the stout, severely twisted cord roots.

Fruits and spikelets of wild grasses figure occasionally, but increasingly, in the record of fossil plants. The bulk of grass fruit records are those of cereals found associated with human occupation sites. For discussion of these see the account under 'Cereals' and the accounts of the genera *Triticum*, *Hordeum*, *Avena* and *Secale* that follow it.

The site records for the family Gramineae shew, as would be expected, presence throughout the whole of the Pleistocene from the Ludhamian onwards. Although the bulk of such records are for grass pollen, there is a consistent supporting component of macroscopic remains, mostly for fruit. These are especially frequent in archaeological sites, so that in Flandrian zone VIII of the ten, there are Iron Age 1, Roman 4, Mediaeval 4. It is noticeable also that in the Early and Middle Weichselian macroscopic site records almost equal the pollen site records. In the Late Weichselian zones between 85 and 100 per cent of all recorded sites have yielded grass pollen, whereas in the Flandrian zones the proportion generally remains at about 70 per cent.

More informative is the analysis of pollen frequency zone by zone set out in table 30. It is noticeable at once that in the Late Weichselian, even though the frequencies are expressed as a percentage of *total* pollen, they are substantially higher than those in the following zones of the Flandrian where they are expressed as a percentage of total tree pollen. Cyperacean pollen (see fig. 156) presents a similar picture and both families must have been very prominent in the open Weichselian vegetation. The range of pollen frequencies no doubt expresses the prevalence of locally favourable habitats and local stands. When we consider frequencies in the Flandrian zones this effect is still more pronounced: not only are there more sites with very low grass pollen frequencies, an effect of the general dominance of woodland, but there are many more sites of high frequency, including a fair number with more grass pollen than total tree pollen. On examination these turn out to be for localities where trees were scarce, sites very often at high altitude though sometimes in exposed coastal situations such as South Uist and Calvay Island. These results indicate persistence of grassland in suitable sites through even the maximal afforestation of the hypsithermal interval. In zone VIII no less than sixteen sites have grass pollen in excess of total tree pollen: partly the sites are of the categories already mentioned but there are now added others from large peat bogs such as Moorthwaite Moss, Scaleby Moss, Darnig Moss, and areas where human disforestation had by this time been extreme, such as Frogholt on the Kentish Downs and Old Buckenham Mere in the East Anglian Breckland. In the wet sites suitable for collection of pollen series there was also always the chance of heavy local representation of the grasses typical of reed-swamp and fen. The higher modal values of zone VIII are no doubt due to the greatly accelerated disforestation by man from the Iron Age onwards.

Phragmites communis Trin. (*Arundo phragmites* L.)
Reed   665: 1

Phragmites communis Trin., the common reed, is so plentifully present in Quaternary deposits that it has certainly escaped exact recording in a vast number of sites which could have yielded evidence of its history, and

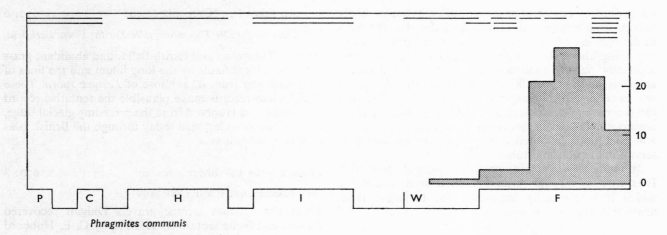

Phragmites communis

there is no doubt that the record diagram is very incomplete. It furnishes, none the less, a fair picture of the history of the plant in the British Isles.

An outstanding difficulty with this grass is the extraordinary power of vertical growth of the rhizomes through as much as metres of underlying mud or peat, so that only the presence of fallen leaves and foliar shoots will give certain proof of the plant growing at the same time as a given deposit. This, however, can be quite secure, as in the case of the Mesolithic occupation platform at Star Carr, Seamer, Yorkshire, where horizontal layers of *Phragmites* leaves and stems *within* the brushwood platform proved the site to have been only intermittently occupied by man, and shewed the reed to have been present at the end of zone IV and the opening of zone V.

Phragmites peat occupies a place of especial interest in the sequences of estuarine deposits, for in the development of a coastal halosere the uppermost salt-marsh communities (if not too swiftly buried in blown sand) are frequently displaced by *Phragmites* fen. The sequence observed upon halt or recovery after submergence is therefore very often: tidal clay or silt – *Phragmites* peat – alder, birch fen-wood – *Quercus* fen-wood. Such a sequence may by later coastal changes be brought into such a situation as to constitute a 'submerged forest', and the *Phragmites* layer indicates rather closely the former marine–freshwater contact. The *Phragmites* rhizomes freely penetrate the anaerobic estuarine clays and accompany their later stages of deposition, to yield a very characteristic brackish-water clay with *Phragmites* which is known all along the coasts of the southern North Sea. It is extremely foetid and in the English Fenland is called 'bear's muck' or 'devil's dung': the German equivalents are 'Hundeköd' or 'Darg' (see 'Coastal peat beds of the British Isles and North Sea', Godwin, 1943).

P. communis stems, leaves and nodes are recorded by Clement Reid as common at most localities of exposure of the Cromer Forest bed and there are specific records from Bacton and Mundesley, but it has to be said that the recent very thorough explorations of this series of deposits by West and Wilson have failed to find it: possibly Reid's

own caution about the dangers of intrusive modern material in these beds is applicable. There is one Hoxnian record of it in brackish-water conditions at Kirmington, and two Ipswichian records from Stone and West Wittering respectively. The only Middle Weichselian record rests on a fruit identification from Hackney Wick made by C. Reid and not mentioned by E. M. Reid in her later summary: it must be regarded doubtfully. There are however three Late Weichselian records, one not more precisely dated, one from zone III at Rhosgoch Common and one from Port Talbot where it occurred both in zone II and in the transitional zone III/IV opening the Flandrian. It has three records in Flandrian zone IV including one from the North Sea moorlog and one quite certain record from the Mesolithic occupation platform at Star Carr, Yorkshire. From this time on, as the diagram shews, the plant has been continually present in this country.

Phragmites communis is now common throughout the British Isles where suitable habitats occur. It is a widespread circumpolar species which occurs in frequency well within the Arctic Circle in Scandinavia and locally even reaches latitude 70° N.

Molinia caerulea (L.) Moench   'Purple moor-grass'
667: 1

F IV *Chat Moss*, VI ?*Cross Fell*, VIIa ?*Clonsast, Lonsdale*, VIIb *Cloonacool, Raheely, Roundstone, Oldtown Kilcashel*, VIIb/VIII *Moyar Wood, Ardsbeg, Clatteringshaws Loch*, VIIb, VIII *Meare Pool, Ardlow Inn*, VIII *Loch Firrib, Shapwick Track, Tregaron Bog, Aughrim Td., Carrowreagh, Meare Track*

Molinia caerulea generally grows in bog or fen sites where total waterlogging is absent, and this fact operates against its chances of survival as fossil material. Although at the present day it is abundant or co-dominant on vast stretches of blanket bog in the British Isles, the humification of plant material in peat of this mire type is so severe that *Molinia* remains can relatively seldom be found. Occasionally, however, its clusters of tuberized stem-bases and sharply twisted adventitious cord-roots are recognized in bog stratigraphic studies. It seems likely that *Molinia* occupied the surfaces of the high Pennines at the opening of zone VII as blanket bog formation was beginning there. It is common also on raised bogs, especially on the rand and the margins of drainage channels, and as the bulk of the fossil records come from one or other of these bog types it is not surprising that the records are concentrated in the late Flandrian when ombrogenous bogs in this country were actively growing and conditions favoured quick incorporation and preservation of roots and tubers.

Molinia caerulea is widespread throughout the British Isles as it is through Scandinavia: it is a species for which it is especially important not to argue from absence in the fossil record.

Sieglingia decumbens (L.) Bernh. (*Triodia decumbens* (L.) Beauv.) 'Heath grass'   668: 1

F VIII *Ehenside Tarn*

Glyceria sp.   669

lA *Hoxne*; F VIIb, VIIb/VIII *Elland*, VIIb/VIII *?Branston*

Glyceria plicata Fr.   669: 2

F VIIb *?Frogholt*

Fruits tentatively assigned to *Glyceria plicata* were found in the Kentish valley site at Frogholt both at the Late Bronze Age and Iron Age levels indicated by radiocarbon dates of 2860 and 2490 B.P.

Glyceria declinata Bréb.   669: 3

F VIIa/b *?Freshwater West*, VIII *?Apethorpe*, *?Dinas Emrys*

Glyceria maxima (Hartm.) Holmberg (*Poa aquatica* L.) Reed-grass   669: 4

F VIII *Apethorpe*

Festuca sp. Fescue   670

A *Corton*; H *Wolvercote Channel*; Wo *Dorchester*

Fruits of 'Festucoid type' have been recorded from the Corton Beds interstadial of the Suffolk coast. Duigan described and figured a flower of a fescue from what is now thought to be sub-stage IV of the Hoxnian, and fruit of *Festuca* also from the Wolstonian glacial stage at Dorchester in the Thames valley. These identifications, which have to be considered with the following attributions at specific level, have especial significance as a partial clue to the important grass component in British periglacial vegetation, suggested by the pollen data. It is to be noted that these records by Miss Duigan of grass fruits, with other of her records reported below, all depend on determinations of C. E. Hubbard of Kew.

Festuca rubra L. 'Creeping fescue'   670: 6

Wo *?Dorchester*; W *Thrapston*; mW *Earith*; F VIII *Larkfield*

At both Thrapston and Earith Bell found abundant grass caryopses, identifiable by the long hilum and the lines of cells radiating from it, as those of *Festuca rubra*. These Weichselian records make plausible the tentative record by Duigan and Hubbard from the preceding glacial stage, of a species as widespread today through the British Isles as it is in Scandinavia.

Festuca ovina L. Sheep's fescue   670: 8

WO *?Dorchester*; F VIII *?Newstead*

From the Thames terrace gravels Duigan recovered flowers and fruits tentatively referred by C. E. Hubbard to two sub-species of *Festuca ovina*, one to ssp. *ovina* and one recorded below as *F. tenuifolia* Sibth. The late Flandrian record is based only on leaf identification.

Festuca rubra L or **F. ovina** L.   670: 6, 8

mW *Barnwell Station*; lW *Kirkmichael*; F IV *Ballaugh*

The Middle Weichselian record for a caryopsis of '*Festuca rubra* or *F. ovina*' is a re-identification of material formerly ascribed to *Carex ustulata*.

Festuca tenuifolia Sibth.   670: 9

Wo *?Dorchester*

Festuca halleri Allioné   670: 13

mW *Temple Mills, Broxbourne, Waltham Cross*

At these Middle Weichselian sites in the Lea Valley Reid has recorded remains of *Festuca halleri*. This grass is no longer native in the British Isles. It occurs in the Alps between 2100 and 3400 m (Braun-Blanquet & Rübel, 1932), in the lower alpine belt above the tree-limit. Its strong association with highly acidic soils in the Alps makes the Lea Valley records seem somewhat remarkable.

Lolium perenne L. Rye-grass, Ray-grass   671: 1

F VIII *Verulamium, Isca*

Helbaek has identified the carbonized caryopses of *Lolium perenne* from the Roman sites at both Verulamium and Isca.

Poa annua L. 'Annual poa'   676: 1

W *?Thrapston*

Poa pratensis L. Meadow-grass   676: 10

lW *?Kirkmichael*

The tentative fruit identification from the Isle of Man is for Late Weichselian zone I, like those given in the two next following accounts for the same site, Kirkmichael.

Poa trivialis L.   676: 13

*m*W *?Earith*; *l*W *?Kirkmichael*

Poa pratensis L. or **P. trivialis** L.   676: 10, 13

*l*W *Kirkmichael*

Cynosurus cristatus L. 'Crested dog's-tail'   679: 1

F VIII *Larkfield*

From the mediaeval deposits at Larkfield, Grove has identified both the fruit and spikelet of *Cynosurus cristatus*.

Bromus sp.   683

F VIII *Lidbury, Wickbourne, Fifield Bavant, Glastonbury Lake Village, Winkelbury, Portland 'Beehives', Maiden Castle, Verulamium, Isca*

The first eight records given above are from Early Iron Age sites, the ninth Roman, whilst at Wickbourne fruits of *Bromus* were found at the level of both culture stages. All save the pottery fruit impression at Fifield Bavant are for carbonized grain, and most are records by Helbaek.

Bromus sterilis L. (*Anisantha sterilis* (L.) Nevski) Barren brome (Fig. 137*a*)   683: 15

F VIIb *Itford Hill*, VIII *Winkelbury, Fifield Bavant, Maiden Castle, Verulamium*

Jessen and Helbaek described the characteristic caryopses of *Bromus sterilis* as amongst the longest and slenderest of caryopses of all north European grasses. It has been recorded in the Late Bronze Age in Sussex, in Wiltshire and Dorset in Iron Age sites, and at Verulamium in the Roman period.

It is certainly indigenous in the Mediterranean region, the Balkans and adjacent parts of Europe, but it is so strongly associated with cultivation that its status over a wider range in Europe is quite uncertain.

Bromus secalinus L. 'Rye-brome'. Or **Bromus mollis** agg. Lop-grass (Fig. 137*b*)
683: 18 or 683: 10, 11, 12

F VIIb *Pond Cairn, Wickbourne*, VIII *Worlebury Camp, Glastonbury Lake Village, Fifield Bavant, Winkelbury, Maiden Castle, Portland 'Beehives', Wickbourne, Park St (St Albans), Verulamium, Larkfield*

Carbonized grains of 'chess' occur frequently in Iron Age and Roman deposits in this country and have sometimes been recorded as *Bromus secalinus*, sometimes as *B. mollis*. The most recent and authoritative assessment of the position, however, comes from Helbaek (1953*b*) who reports that these remains exhibit no decisive characters permitting separation of the two species from one another. We have therefore aggregated all former records under

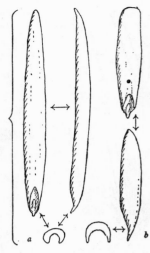

Fig. 137. *a*, Carbonized grain of barren brome (*Bromus sterilis*); *b*, carbonized grain of chess (*Bromus secalinus* or *B. mollis* agg.). Both of early Iron Age (× 6). (Jessen & Helbaek, 1944.)

the one heading. The thin, boat-shaped chess fruits are distinguished from those of *Avena* by being hairless and having a characteristic coarse, epidermal pattern.

Chess fruits have been recorded in Britain from one Middle Bronze Age, and many Iron Age and Roman deposits. Helbaek comments that chess appears to have been introduced (together with *A. fatua*) in imported spelt. Since both were apparently incorporated along with cereals in the gruel or porridge eaten by prehistoric man, it is somewhat difficult, as Helbaek points out, to decide whether they are to be regarded as mere weeds.

Bromus secalinus is a species hardly found outside areas of cultivation; it is less abundant in the north of Great Britain and in Ireland it is restricted to the east. The aggregate *B. mollis* is also far less common in the north than in the south of the British Isles.

Agropyron repens (L.) Beauv. Couch-grass, Scutch, Twitch   685: 3

F VIII *Larkfield*

Agropyron junceiforme (A. & D. Löve) A. & D. Löve
685: 5

*l*W *?Moss Lake*

The record from zone II of the Late Weichselian at Moss Lake, Liverpool is unusual in that it is based upon identification of fossil pollen. The germ-pore is large in relation to grain diameter. The pore margin is very projecting and the wall is unusually thick. *A. junceiforme* is a plant of coastal fore-dunes but might well have occupied Late Weichselian inland sites.

Hordeum marinum L.   687: 2

F VIII *Larkfield*

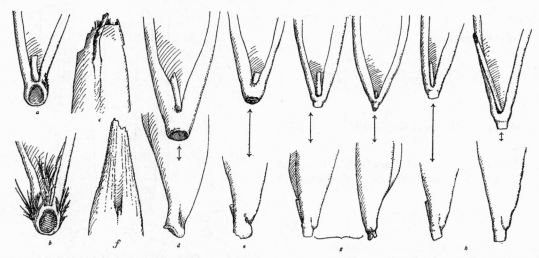

Fig. 138. Remains of various species of oat (*Avena* spp.) from British prehistoric sites; *a*, *b*, the wild oat (*Avena fatua*); *c*, *d*, *e*, *Avena sativa*; *f*, *g*, *h*, oats of the *A. strigosa* group. (After Jessen & Helbaek, 1944.)

Avena fatua L.   Wild oat (Fig. 138*a*, *b*)   692: 1

F VIII *Fifield Bavant, Maiden Castle, Forth–Clyde Canal, Camp Hill, Malton, Girton, Lough Faughan*

All the sites at which *Avena fatua* have been recorded are zone VIII archaeological sites, and in the order given they are Iron Age 1, Roman 3, Anglo-Saxon 1 and early Christian 1. The Girton record is for a pottery cast: the rest are for carbonized grain.

The carbonized grains of *Avena fatua*, the wild oat, are usually recognizable among other oats by the broadly V-shaped, deep ventral furrow. Commonly in these finds the wild oat accompanies other species of oat: it presumably behaved as a weed in some instances, but whilst not reaching the status of a true crop it may have been grown in mixed crops to a substantial extent.

A. fatua, which is regarded as the original form of *A. sativa*, has been discovered along with that species when it made its first appearance in central Europe in the Bronze Age: the evidence given above suggests that it did not reach Britain until somewhat later, but further discoveries may well modify this conclusion.

Arrhenatherum tuberosum (Gilib.) Schultz.   Onion couch (Plate XIII)   694: 1

Carbonized plant material from a Late Bronze Age ditch filling on Rockley Down, Wiltshire yielded, along with grains of barley, about fifty tubers of the onion couch, *Arrhenatherum tuberosum*, readily recognizable by their characteristic flagon and conical shapes, the apical abscission scar, the basal root-scars, ribbed surface and homogeneous, parenchymatous interior. There is no evidence that the association with barley grains implies that these tubers had been collected for food, but the absence of seeds of cornfield weeds is notable.

This strongly tuberized grass is a prevalent and serious weed of cultivated ground in the north and west of Britain; it is much scarcer in the east. *A. tuberosum* interbreeds freely with the non-tuberous *A. elatius*, producing intermediates, so that its claim to specific rank is doubtful (Clapham, Tutin & Warburg, 1962).

Holcus lanatus L.   Yorkshire fog   695: 1

F IV/V *Longfordpass*, VIII *Larkfield*

A floret of *Holcus lanatus* was recovered by Mitchell and determined by Jessen from the early Flandrian in Tipperary, and Grove recovered a spikelet of it from the mediaeval deposits at Larkfield.

Deschampsia flexuosa (L.) Trin.   'Wavy hair-grass'   696: 3

F VIIb *Graig-Lwyd*

When excavating the Neolithic axe factory site on Penmaenmawr, Caernarvonshire, Warren recovered a fruit of *Deschampsia flexuosa* from the culture level.

Calamagrostis canescens (Weber) Roth.   'Purple small-reed'   700: 2

F VIII *Bure Valley*

Agrostis sp.   701

mW *Marlow, Great Billing, ?Barnwell Station*

Bell (1968) identified fruits of *Agrostis* from two sites in the Middle Weichselian and with Dickson revised the record, given by Miss Chandler as *Carex atrofusca*, from a third Middle Weichselian site to 'either *Agrostis* or *Poa*'.

Phleum bertolonii DC. (*P. nodosum* auct.)   Cat's-tail   707: 1

F VIII *Fifield Bavant*

The caryopsis of *Phleum bertolonii* from Fifield Bavant was recognized by 'its short, truncate shape and coarsely reticulate epidermal sculpture'. Helbaek states that it occurs in Danish Iron Age grain.

Phleum arenarium L.   'Sand cat's-tail'   707: 5

F vııb ?*Graig Lwyd*

Phalaris sp.   713

F vııı ?*Meare Pool*

CEREALS

The history of the introduction and development of different crop plants to this country during the historic period is of such interest that we have not hesitated to include records of them in this account of the history of the British flora, different though their history may have been from that of the 'natural' flora. The cereals are of particular importance, but the task of recognizing the macroscopic remains, usually carbonized grains or impression cavities, is one calling for great skill and long experience. We have therefore restricted ourselves, in the accounts which follow, almost entirely to the results set out in the 1944 paper by K. Jessen and H. Helbaek 'Cereals in Great Britain and Ireland in prehistoric and early historic time' and in the 1953 paper by H. Helbaek 'Early crops in southern England'. The research described in these publications has established new standards for such work, and their conclusions represent by far our best modern estimate of the available evidence. A small number of records has subsequently been added by other workers and these have been included in our record lists: they scarcely alter at all the conclusions of Jessen and Helbaek and it has seemed pointless to modify the graphs (figs. 139 to 141) based upon their results that were used in the first edition.

The bulk of cereal records depend upon discoveries associated with human occupation sites, where the practice of parching grain to aid storage undoubtedly led to a great improvement in the chances of carbonization and preservation. In this way not only grains, but the far more diagnostic spikelets or ear fragments, have not infrequently been preserved as well, although like the grains they may have suffered some changes of shape and size during carbonization. A substantial part of the records for cereals is derived from the method perfected by Helbaek of taking plastilinia casts of the cavities in prehistoric pottery caused by the combustion of grains accidentally included in the clay during fabrication. This method has the advantage of representing a chance sampling of the material scattered about settlement floors in the various periods, and may be more safely employed as a guide to the real frequency of the various cereals than the stores of carbonized grains in which local circumstances of collection or preservation can easily lead to great over-weighting of one type against another. The impressions also have the advantage of durability and permanent association with the more or less certainly datable sherds in which they occur.

It was shewn by Firbas in 1937 that the pollen of cultivated cereal grasses is much larger than that of the other grasses (usually 35–50 μm, seldom more than 60 μm or less than 35 μm, as against a size range of 20–25 μm): the wall and pore dimensions of the cereal grains are correspondingly greater and the wall sculpturing is more pronounced. Only a very few wild grasses, such as *Elymus arenarius*, have pollen confusable with cereal pollen. It has thus become possible to detect by pollen analysis the time at which cereal cultivation began and spread in the neighbourhood of any investigated site, and numerous investigations, following those of Firbas in the Fichtelgebirge, have shewn that pollen of characteristic weeds of cultivation commonly accompanies the appearance and expansion of the cereal pollen curve. Thus one of the earlier British analyses of this kind, that from Decoy Pool Wood, Shapwick Heath, Somerset (fig. 168) shews cereal pollen first in Late Bronze Age layers and extending to middle Romano-British time, its fluctuations in quantity matched by similar variation in the pollen of *Plantago lanceolata*, *Artemisia* and other categories of weeds. This is now a common feature of pollen diagrams in this country from the Neolithic agricultural clearances onwards (see chapter VI: 14, 15).

The identification of separate genera of cereal grasses by pollen characters had to wait upon the development of high grade oil-immersion microscopy and the use of phase-contrast, a development very effectively demonstrated by Grohne in 1957 and Beug in 1961. This discrimination makes use of the grain size, its surface features, the size, projection and diameter of the germ-pore, the morphology of the pore-annulus, and the nature of the pollen exine layers, especially the dimensions, distribution and morphology of the bacula (columellae). Thus one can refer well-preserved pollen grains to the genera *Triticum*, *Hordeum*, *Avena*, *Secale* (as to some of the large-grained wild grasses) and from changes in their relative frequency something can be deduced of their cultivation history. It has however to be borne in mind that of the cereals in cultivation in north-western Europe only the rye has a large production of wind-borne pollen. It is for this genus only, the most recognizable, that the British pollen record has yielded substantial results. For the other genera so far published results merely supplement the conclusions based upon macroscopic remains such as are expressed in the graphs of figs. 139 and 140. These are based upon the identifications made or accepted by Jessen and Helbaek up to 1944, and shew in the continuous and broken curves respectively the proportion of sites in which each cereal type has been represented, and the proportion of the total cereal impressions of each period attributed to each cereal type considered. Although the curves are based upon small numbers and both are

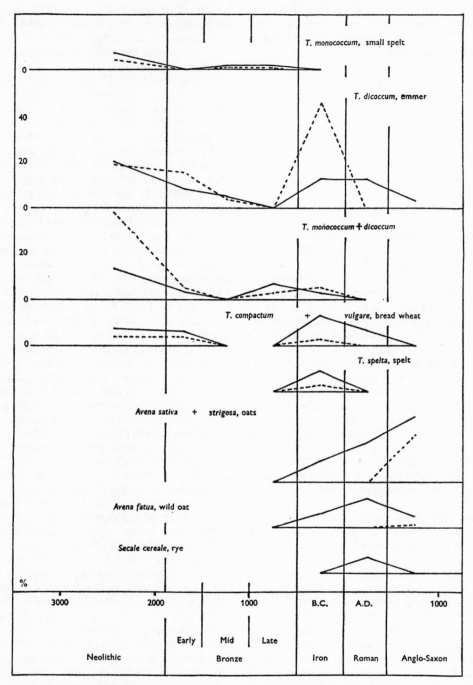

Fig. 139. Records of the various types of wheat (*Triticum*), oats (*Avena*), and rye (*Secale*) in the British Isles since Neolithic time: based on data given by Jessen and Helbaek (1944). The continuous curves shew for each period the proportion of all examined sites in which impressions or carbonized grains of a given cereal were found. The broken curves shew for each period the proportion of the total grain impressions attributable to the given cereal type.

subject to special difficulties in interpretation, certain broad features are at once apparent from them. Thus wheat and barley were of comparable importance in the Neolithic period, but at all stages of the ensuing Bronze Age, barley far exceeded wheat in frequency. Neverthe-

less, after the Bronze Age, wheat regained a good deal of its former importance at the expense of barley. Spelt first appeared in the Iron Age. In Anglo-Saxon time once more wheat apparently receded in importance. Oats appeared only in the Iron Age, but apparently increased

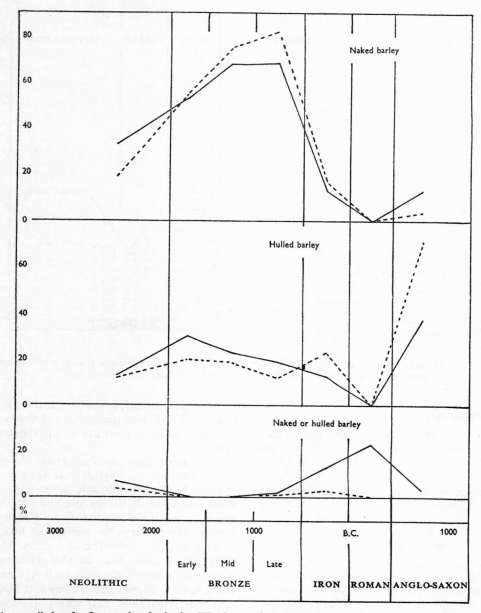

Fig. 140. Records compiled as for fig. 139, but for barley (*Hordeum polystichum* Doell.); also from data by Jessen & Helbaek (1944).

afterwards. Rye, shewn as having only been found in the Roman period, now also has three records in the Iron Age.

Of these effects it seems likely that the late entry of oats and rye and the displacement of wheat by barley in the Bronze Age are the most assured results. One is indeed tempted to seek a climatic explanation for the last-mentioned effect since both wheat and barley are present together right through from the Neolithic to the Iron Age and merely change in relative frequency. The Bronze Age is held to correspond in large part to the Sub-boreal climatic period, and it might be thought that on the dry soils where so many Bronze Age settlements occur barley

could better have tolerated hot, dry summers. Nevertheless, many other factors have also to be reckoned with, such as the invasion of new peoples with new traditions, new techniques of cultivation and possibly with the means and desire to exploit new soil types. Over and above this is the progressive modification of genetic types within the main cereal groups, something of which is discussed in the separate sections which now follow.

Finally, it may be remarked that it is surprising to find no record of grains of *Panicum miliaceum* L., millet, since this has been found in the Dordogne in the Bronze Age, in Holland abundantly in the Iron Age, and since the early Neolithic in central Europe.

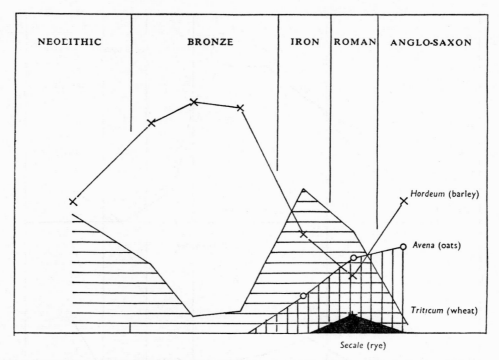

Fig. 141. The digram shews for each archaeological period the proportion of all examined sites in which impressions or cereal grains themselves have been recognized of the four main cereal types.

Avena spp. Various oats

F VIII I.A. *Meare Lake Village, Glastonbury Lake Village, Worlebury, Little Solsbury, Fifield Bavant, Maiden Castle, Portland 'Beehives'*; R. *Park Street (St Albans), Verulamium, Isca*; M. *Dublin*

The records include seven Iron Age, three Roman and one Mediaeval attribution.

Avena sativa   Cultivated oat (Fig. 138c, d, e)

F VIII I.A. *Fifield Bavant, Maiden Castle*; R. *Malton, Barhapple Loch, Forth and Clyde Canal, York Hill, Queen's Park, Camp Hill, Birrens, Castle Cary*; A.S. *Hawkshill, Sutton Courtenay, Abingdon, West Stow Heath, Downham Market, Rothwell, Barrington, Girton College, Haslingfield, Little Wilbraham, St John's College*; V. *Hellsevean*; M. *Larkfield, Lough Faughan*

Most of the high-yielding and extensively grown varieties of oats cultivated today belong to *Avena sativa*, but there are also grown some forms of the species *A. strigosa*, the 'bristle-point' oat, so-called from the two long scabrid bristles at the apex of the lemma. Some hybrids between *A. sativa* and *A. sterilis* also have recently come into cultivation.

In almost all the prehistoric material the whole or the apex of the lemma is absent, but Jessen and Helbaek have shewn that even though this is the case, the flowers may often be satisfactorily referred to the *sativa* or *strigosa* group. In the latter the flower base is borne upon a short stalk with a very narrow fracture surface only

some 0.3 mm broad; in the former there is no distinguishable basal stalk and the fracture surface is about 0.6 mm broad. In addition there are certain characters of shape of grain.

With these characters as guide *A. sativa* has been identified from two English Iron Age sites, seven Scottish and one English Roman site, eleven English Anglo-Saxon sites, one Cornish Viking site and two Mediaeval sites, one in England and one in Ireland. Pollen of '*Avena* type' appears also in the pollen analyses of Birks (1965*b*) at the Roman level in Holcroft and Lindow Mosses. It is evident that *A. sativa* must be well represented also in the numerous south of England Iron Age and Roman sites listed under '*Avena* spp.'. Jessen and Helbaek comment that it is remarkable that whilst *A. sativa* is known to have had a considerable range in Europe during the Bronze Age, it has not so far been recognized from the British Isles at this time. *A. sativa* seems to have become established relatively late in British agriculture despite its present widespread use.

Vavilov has suggested that *A. sativa*, which is supposedly derived from the wild oat (*A. fatua*) indigenous to the steppes of eastern Europe, western Asia and north Africa, originally grew as a weed in cultivated fields, especially with emmer, and only afterwards was selected for deliberate cultivation. Jessen and Helbaek point out that the presence of both *A. sativa* and *A. strigosa* in small amounts in larger samples of wheat and barley, in several Iron Age and Roman finds, appears to indicate rather that oats was an accidental contaminant of other crops than a cereal independently cultivated. Two of the Roman sites in Scotland by contrast yield material wholly or largely of oats.

It is hazarded that the concentration of Anglo-Saxon records of oats in East Anglia may reflect the introduction of the crop with the invaders who were familiar with it in their homeland.

Avena strigosa group (including Bristle-pointed oat) (Fig. 138*f, g, h*)

F VIII I.A. *Meare Lake Village, Fifield Bavant, Maiden Castle*; R. *Camp Hill, Forth and Clyde Canal*; M. *Ballingarry Downs*

The records above include three of Iron Age, two of Roman and one of early Mediaeval age. The oat from Meare Lake Village was regarded by Percival as resembling *Avena brevis*. Jessen and Helbaek group together in '*Avena strigosa* group', *A. strigosa, A. brevis* and *A. nudibrevis*, and whilst recording modern views upon the evolution and taxonomy of forms within this complex, write that the existing material does not allow one to go beyond the statement 'that oat-forms of the *strigosa* group grew in England in the Early Iron Age and the following time', a conclusion already expressed by Percival. There is a sparse record from Europe of oats of *strigosa* type from the Bronze Age. The bristle-pointed oat is still, or has recently been, in cultivation on light sandy soils from West Iberia through western Europe to the British Isles; it is still grown on poor soils and in wet climate in the upland and hilly districts of Wales, Scotland and the Orkneys. It is occasionally to be found in Britain outside these areas as a weed of cereal crops.

It is suggested that the bristle-pointed oat arose in south-west Europe from a primitive diploid form of *A. barbata*, in much the same way as did *A. sativa* from *A. fatua* at much the same period. Thence it spread to Britain where both *A. strigosa* and *A. sativa* persisted, at first as weeds and later as crops.

(See also *Avena fatua*, p. 404.)

Hordeum Barley

The taxonomy of the cultivated barleys is a complex matter but a primary division can be made into the forms with six rows of grain (*polystichum* forms) and those with two rows (*distichum* forms). The prehistoric material seems to be concerned entirely with the former group. Since the vast majority of fossil records for the British Isles are due to Jessen and Helbaek (1944) or subsequently to Helbaek, the least confusing presentation is acceptance of their nomenclature.

Pollen identification of the genus has been made by Birks (1965*b*) from the Roman level in two Cheshire mosses.

Hordeum polystichum Doell. Six-rowed barley

F VIIb N. *Maiden Castle*; M.B.A. *Lancing, Pond Cairn*; L.B.A. *Ashford, Totternhoe*; B.A. *Culbin Sands, Rockbarton*; VIII I.A.

Little Solsbury, Fifield Bavant, Maiden Castle, Aldwick Barley; R. *North Leigh, Old Kilpatrick*; M. *Lough Faughan*

'Naked barley'

F VIIb N. *Windmill Hill, West Kennet, Maiden Castle, Whitehawk Hill, Easterton, Eday, Whitepark*; E.B.A. *Culbow Hill, Gorsey Bigbury, Winterbourne Stoke, Thickthorn Down, Somerton, Hitcham, Needham Market, Somersham, Moel Hebog, Goodmanham, Garton Slack, Acklam Wold, Barrow Nook, N. Berwick* and *Archerfield, Chapel of Gairioch, Boyndlie, Gambrie, Buckie* and *Reffley Wood*; M.B.A. *Pedugwinion Point, Chycarne, Treworrick, Tresawsen, Collingbourne Ducis, Handley Hill, Thorney Down, Bere Regis Down, Rockbourne Down, Ibsley, Itford, Beddingham, Oddington, Upper Hare Park, Kilpaison Burrows, Holt, Broughton, Hungry Bentley, Garton Slack, Newbold, Warter Wold, Aldro, Wharram Percy, Grindle Top, Pickering, Blanch, Ford, Papcastle, West Kilbride* and *Wetherhill, Cadder, Westwood* and *Balbirnie, Strathblane, Carmylie* and *Downe Castle, Windyhall, Poopluck, Crehelp, Dunruadeagh, Maze Court, Portstewart, Ballymena, Loughlougan, Mullaghnashmount*; L.B.A. *Knackyboy Cairn, Tan Hill, Rockley Down, Roke, Down, Luxford Lake, Hassocks, Abingdon, Hurdle Drove, Mildenhall Fen, Goodmanham, Amble, Torrs Luce, Largs, Newlands, Longside, Spottiswood, Lintlaw, Craigentinny, Dean Bridge, Musselburgh, Arniston, Magdalen Bridge, Outerston Hill, Drymmiewood, Glenballoch, Chrichie, Ballon, Jamestown, Greenhill, Agfarell, Ballycastle, Amoy Bog, Lyles Hill*; B.A. *Hutton Buscel, Gilling, Sands*; VIII I.A. *Hembury Fort, Worlebury Camp, Fifield Bavant, Maiden Castle, Prae Wood*; R. *Star Villa, Windmill Hill*; E.M. *Hellsevean, West Stow Heath, Girton College Cemetery, Newmark, Kilmoyle*

'Hulled barley'

VIIb N. *?Windmill Hill, Maiden Castle, Unstan*; E.B.A. *Culbour Hill, Winterbourne Stoke, Martinstown, Barton Hill, Chippenham, Longstone Edge, Cowlam, Rudstone, Gilling, Aldro, Painsthorpe Wold, Riggs, Ford, Plumpton*; M.B.A. *Ballowell Cairn, Treworrick, Tresawsen, Beckhampton, Durrington, Amesbury, Winterbourne Stoke, Scrubbity Coppice, Thorney Down, Bloxworth Down, Telscombe, Little Downham, Goodmanham, Huggate, Warter, Landesborough, Aldro, Garton Slack, Pickering, Baskfield, Musselburgh, Strathblane, Doune Castle*; L.B.A. *Elworthy, Battlegore, Collingbourne, Deverel Barrow, Bincombe, Hassocks, Plumpton Plain, Itford Hill, Mildenhall Fen, Snailwell, Outerston Hill, Jamestown*; B.A. *Gilling, Sharpe*; VIII I.A. *Hembury Fort, Worlebury, Glastonbury Lake Village, Meare Lake Village, Little Solsbury, Fifield Bavant, Winkelbury, Maiden Castle, Portland 'Beehives', Prae Wood, Wickbourne, Radley, Chastleton Camp*; R. *Rotherley, Wickbourne, Park Street, St Albans, Verulamium*; E.M. *Hellsevean, St Ives, Abingdon, Sutton Courtenay, Tuddenham, Holywell Row, Moneyport Hill, Western Heath, Lackford, Downham Market, Haslingfield, Girton College Cemetery, St John's College, Little Wilbraham, Stamford*

Hordeum hexastichum Erect six-rowed barley

F VIIb L.B.A. *Itford Hill*; VIII I.A. *Glastonbury Lake Village, Swallowcliffe Down, Fifield Bavant*; E.M. *Barnhapple Loch*

Hordeum vulgare L. (*H. tetrastichum* Kcke) Bere, Lax quadrangular barley

F vIIb L.N./E.B.A. *Stag Field*; L.B.A. *Itford Hill*; vIII I.A. *Glastonbury Lake Village, Meare Lake Village, Lidbury, Fifield Bavant, Maiden Castle*; R. *Stockton, Caerwent, Isca*; E.M. *Larkfield*

Hordeum polystichum Doell.   Six-rowed barley

Helbaek (1966) has made out a strong case for regarding as the progenitor of the cultivated barleys the hulled two-rowed *Hordeum spontaneum*. This species grows wild to-day in the area of Iran, Anatolia and Palestine where it was cultivated as early as 7000 B.C. Shortly thereafter, at Jarmo in Iraqi Kurdistan, the rachis of the two-rowed barley became tough instead of brittle, a sure index of domestication. From about this time also it appears that this stock gave rise in cultivation to naked six-rowed barley, and perhaps rather later, to hulled six-rowed barley. Together with emmer wheat, these naked and hulled six-rowed barleys constituted the main cereal crop of Neolithic people throughout Europe, but there is no evidence that two-rowed barleys were cultivated in this continent until the Middle Ages.

The adherence or non-adherence of the chaff to the grain has been an extremely significant feature to the ethnobotanist. Sarauw had long ago pointed out that both hulled and naked grains were represented in pre-historic Danish pottery impressions and Jessen established that the glumeless condition of the naked grains was a primary character and not due to the conditions of pre-servation. Consequently Jessen and Helbaek have been able to analyse the large bulk of British prehistoric barley records into the hulled and naked categories.

Within the British Isles naked barley has yielded far the greater number of impressions in the early sites, whereas hulled barley has been preponderant after the Late Bronze Age, very strongly so in Anglo-Saxon time (see fig. 140). This sequence parallels that known from Den-mark and is not apparently at variance with the less thoroughly known evidence of prehistoric barley in central Europe.

It is more difficult in fossil material to separate the six-row forms with densely set spikelets and stiff, erect habit (*H. hexastichum* L.) from those with distant spikelets, a lax, nodding habit and four rows of ears. The latter is the nodding, quadrangular barley, *H. vulgare* L., now known as 'bere' and formerly also known as 'byg': this type was still in cultivation in Scotland and northern England in recent historic time. Where fragments of ear have been preserved the length of the rachis internodes serve as a guide to the lax or erect habit of the barley concerned.

It has not proved possible in many instances to recover large enough ear fragments to distinguish between *H. hexastichum* and *H. vulgare*. It is, however, apparent that *H. hexastichum* (both naked and hulled) was present in the Bronze Age and again in the early Christian period.

H. vulgare was present in the Orkneys as early as the Neolithic, and in the Iron Age and Roman time seems to have been more frequent than *H. hexastichum*. According to Schiemann *H. hexastichum* was generally cultivated in the circum-alpine lake-dwelling area in Neolithic time, and remained the dominant cereal in central Europe through the Bronze Age. Gradually there-after, however, it was replaced by *H. vulgare* which had already been present in the Neolithic, and indeed was known from the most ancient Egyptian cultures. It seems possible that the replacement of *H. hexastichum* by *H. vulgare* in the Iron Age was a general European feature.

If we consider the records for barley as a whole and in relation to the records of other cereals it is apparent that it has been much the commonest cereal in the British Isles throughout the period from the Neolithic to the Viking period. It has been recovered from over one hundred and sixty localities, and although the Neolithic records are few they are very widespread, including even the Orkney Islands. The geographical distribution of barley records for successive prehistoric periods has been displayed by Jessen and Helbaek, whose maps indicate several features of interest, particularly the avoidance of the west of Ireland, the central Irish plain and mountainous areas generally, together with concentration in those areas still particularly favourable to cultivation of cereals. In com-paring the Bronze Age and Iron Age maps it is evident that in the latter period there took place a considerable cultivation of barley on the calcareous soils of Somerset, Dorset, the Chilterns and the chalk of East Anglia, where previously it had been slight or absent. Many alternative suggestions might no doubt be offered in explanation of this.

Hordeum sativum   Two-rowed barley

F vIII M. *Larkfield* (Ident. J. R. B. Arthur)

Triticum sp.   Wheat

Macros: F vIIb B.A. *Culbin Sands, Albury*; vIII I.A. *Little Solsbury, Fifield Bavant, Maiden Castle, Aldwick Barley*; R. *Malton, Birrens, York Hill, Old Kilpatrick, Forth and Clyde Canal, Castle Cary*

    Pollen: F vIIb ?*Old Buckenham Mere*; vIII *Old Buckenham Mere, Holcroft Moss, Lindow Moss*

Because of the confusion caused by the failure of previous workers to distinguish the caryopses of naked barley, we have omitted grain identifications prior to those of Jessen and Helbaek (1944). In the given list Culbin Sands and Albury are Bronze Age, these are followed by four Iron Age sites and six Roman.

Pollen identification has been made at the generic level at Old Buckenham Mere from horizons thought to be Iron Age, Roman, Anglo-Saxon and possibly earlier. It is also recorded by H. J. B. Birks from the Roman level in two Lancashire mosses.

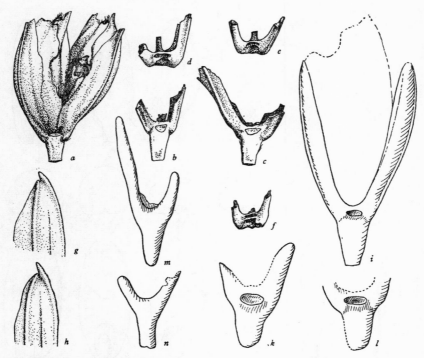

Fig. 142. *a–l*, Spikelets and fragments of emmer (*Triticum dicoccum*) drawn from carbonized remains and plastilina casts; *m, n*, small spelt (*Triticum monococcum*). Material from British prehistoric sites (Jessen & Helbaek, 1944).

Triticum monococcum L.   Small spelt, Eincorn
(Fig. 142*m, n*)

F ᵥɪɪb ɴ. *Hembury Fort, Windmill Hill, Dunloy Cairn, ?Maiden Castle*; ᴇ.ʙ.ᴀ. *Garrowsby Wold*; ᴍ.ʙ.ᴀ. *St Just in Roseland, Long Wittenham*; ʟ.ʙ.ᴀ. *Plumpton Plain, ?Radley, ?Mildenhall Fen, ?Torrs*; ɪ.ᴀ. *Hembury Fort, ?Meare Lake Village*

(The records queried in the list above may, according to the authority of Jessen and Helbaek, be either *T. monococcum* or *T. dicoccum*.)

Jessen and Helbaek point out that three species of *Triticum* have a fragile rachis so that the ear breaks up into segments, each spikelet with a rachis joint attached to it. Whereas in spelt (*T. spelta*) the spikelet is detached with the rachis joint distal to it (which therefore lies alongside the spikelet), in small spelt (*T. monococcum*) and emmer (*T. dicoccum*) the spikelet is detached with the rachis joint distal to itself, so forming a unit with the spikelet sitting upon its subtending rachis joint. Even when the grains have been lost from the spikelets of *T. monococcum* and *T. dicoccum* the empty glumes (fig. 142) remain attached to the rachis joint, giving a structure shaped like the prongs and shaft head of a hay-fork, or tuning-fork. These spikelet-forks have sometimes been preserved and offer a satisfactory basis for distinguishing the two species not only from other *Triticums* but also from one another, since the width of the insertion of spikelet upon the rachis in *T. monococcum* is only 1.8–2.4 mm and that in *T. dicoccum* is about 2.3–3.2 mm.

Naturally, where whole spikelets are preserved, carbonized or as impressions, the distinction between the three fragile rachis species of *Triticum* is much simpler.

T. monococcum has been only sparsely recorded in Britain, nearly always along with larger amounts of *T. dicoccum* as if it were 'a more or less fortuitous component of the Emmer field' and Helbaek (1953*b*) writes 'there is nothing to indicate that the species was ever cultivated intentionally or separately in this country'. The sparse Neolithic occurrences are on the Chalk of Wiltshire and Dorset but there is one also in northern Ireland. In the Bronze Age too it has a wide and sparse range from Scotland to Cornwall and East Anglia; it is even sparser in the Iron Age and has not been found in a subsequent period.

Eincorn is a diploid wheat of the genome *AA* that has been found with its presumptive progenitor, *T. boeoticum*, in the oldest Neolithic settlements of the near East, in deposits at least as old as 6000 to 7000 B.C. It extended widely across Europe in the Neolithic period and although generally less common than emmer, occasionally, as in Sweden, it might have been the more successful crop. Subsequent history of eincorn on the European mainland is comparable with that in Britain.

Triticum dicoccum (Schrank) Schübeler.   Emmer (Figs. 142*a–l*, 143)

F ᵥɪɪb ɴ. *Hembury, Windmill Hill, Maiden Castle, Dunloy Cairn, ?Aston-on-Trent*; ᴇ.ʙ.ᴀ. *Upavon, Somerton, Moel*

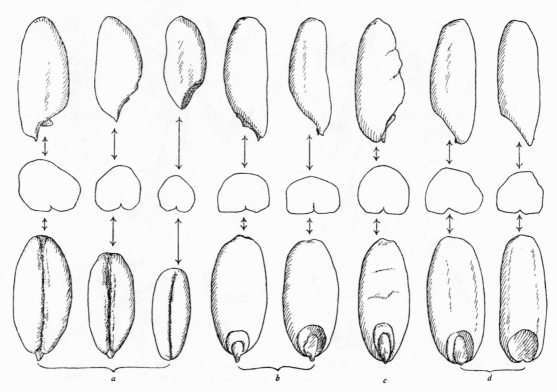

Fig. 143. Carbonized grains of emmer (*Triticum dicoccum*) from prehistoric sites of Bronze Age to Roman Age (Jessen & Helbaek, 1944): *a*, Bronze Age (Culbin Sands); *b*, Early Iron Age (Fifield Bavant); *c*, Early Iron Age (Little Solisbury); *d*, Roman Age (Malton).

Hebog, Hanging Grimstone, Garton Slack; M.B.A. *Boscawen-Un, Tresawsen, Beckhampton, Handley Hill, Thorney Down, Abingdon, Swaffham, Londesborough, Grindle Top*; L.B.A. *Plush, Itford Hill, Albury*; B.A. *Glenluce, Culbin Sands*; I.A. *Trebarvath, Hembury, Small Down Camp, Worlebury, Glastonbury Lake Village, Meare Lake Village, Little Solsbury, Fifield Bavant, Maiden Castle, ?Worth Matravers*; I.A. *or* R. *Abington Pigotts*; R. *Northleigh, Malton, Old Kilpatrick, Castle Cary, Forth and Clyde Canal*; early Christian *Barhapple Loch*

(It is possible that certain records cited under *T. monococcum* with a query are referable to *T. dicoccum*.)

A substantial number of grain and spikelet impressions of *Triticum dicoccum* (emmer) have been recorded by Jessen and Helbaek from the periods Neolithic, Early and Middle Bronze Age and Early Iron Age, and in addition carbonized remains of grain have been identified by these authors and others so as to prove the presence of emmer at each stage of the prehistoric period from the Neolithic onwards into early Christian times. The emmer grains are recognizable by their comparatively slender shape, narrow furrow and somewhat angular cross-section with a flattened ventral face. There is, however, a large category of carbonized grains not referable to *T. compactum* or *T. vulgare* and Jessen and Helbaek think that *T. dicoccum* very likely contributed largely to this category of 'various wheats'.

From these sources there is enough material to warrant an estimate of the relative frequency of emmer in the successive prehistoric periods, and from this it is concluded that emmer and barley appear to have been the most important cereals of Neolithic time, and in the Windmill Hill material emmer much outnumbered both eincorn and the barleys; emmer decreased in frequency in the Bronze Age, only to become dominant again in the Iron Age (at any rate in the south of England whence come all the Iron Age records). It continued in culture in Britain into Roman times and in Scotland into the early Christian period. In Denmark and central Europe generally it seems that emmer was also abundant in the Neolithic and decreased in the Bronze Age but that thereafter its use was discontinued: there are, however, a few limited recent areas of cultivation of emmer.

Triticum dicoccum is a tetraploid wheat with the genome *AA, BB* and is thought to have been derived from the wild *T. dicoccoides* of similar constitution. Emmer is the commonest of all the early domesticated cereals found in the 'fertile crescent' of western Asia. Spreading outwards thence it became, after the barleys, the most important cultivated cereal of central and northern Europe. With the *Bandkeramic* culture it had reached the Netherlands by about 6000 B.P.

Triticum aestivum (*sensu* Helbaek) Naked hexaploid wheats

The hexaploid wheats with a genomic constitution of *AA, BB, DD* include the hulled group of spelt wheats (*T. spelta*), club wheat (*T. compactum*), shot wheat (*T. sphaerococcum*), *T. vavilovii* and common wheat or bread wheat (*T. aestivum* L., *T. vulgare* Host., *T. sativum* Lam.). Of these *T. sphaerococcum* and *T. vavilovii* are probably unrepresented in our historical records, and *T. spelta* has persistently adherent glumes. We are concerned therefore in the naked hexaploid wheats only with two major types, *T. compactum* and *T. vulgare*, recognized by Jessen and Helbaek in their 1944 paper: we retain the terminology they then employed.

Triticum vulgare Host. Bread wheat or **Triticum compactum** Host.   Club wheat

F VIIb N. *Windmill Hill*; E.B.A. *Lambourne Downs, Ackham Wold*, VIII I.A. *Little Solsbury, Fifield Bavant, Ashwell*; R. *Verulamium, North Leigh, Caistor by Norwich, Isca*

There is a considerable percentage of carbonized grains and grain impressions remaining from the records of Jessen and Helbaek after they have separated out from the carbonized wheat grains the slender emmer type, the pointed, ovoid spelt and the very short, thick grains presumed to belong to club wheat, *Triticum compactum*. In this residue, referred to as 'various wheats', the length/breadth ratio of the grains is about 1.9, as in modern material of *T. vulgare* and *T. compactum* which appear inseparable on grain characters alone. There is the separate evidence for the presence of a very round-grained type of *T. compactum*, but it seems likely that *T. vulgare* was also present.

Whereas earlier investigators had suggested that in Great Britain during the greater part of the prehistoric period these two types had been the main cereal crop, it clearly appears from Jessen and Helbaek that impressions of this type amount to less than 1 per cent of all grain impressions collected in British–Irish pottery, whilst *of these* less than one-fiftieth are constituted by the closely recognizable *T. compactum*, chiefly from the Iron Age and Roman periods. In his 1953 paper Helbaek makes it clear that in the past grains of emmer, spelt, hulled barley and especially naked barley have been often identified as 'bread wheat'; this was largely due to the fact that the existence of naked barley had not then been recognized. In view of this we have, upon consideration, omitted all the older records attributed to the 'bread wheats'.

Wheat grains of *T. vulgare–T. compactum* type are infrequent in Denmark as in this country. In both they seem to have been more frequent in Neolithic time and to have diminished later, though the British figures are admittedly small to warrant a firm conclusion. It does, however, seem clear that bread wheat and club wheat only became important bread plants in England during historic time.

Triticum compactum Host.   Club wheat

F VIIb B.A. *?Pond Cairn, Culbin Sands*; VIII I.A. *Fifield Bavant, R. Malton, Castle Cary*

From British sites of Bronze Age, Iron Age and the Roman period Jessen and Helbaek segregated some sixty-five grains with a length/breadth ratio close to 1.5 and very rounded shape which they attributed to an extinct form of *T. compactum* already recognized on the European continent as *T. compactum* var. *antiquorum* Heer, var. *muticum* Heer, or var. *globiforme* Buschan. Percival has pointed out that these grains may be most closely matched among existing wheats with those of *T. sphaerococcum* Perc., an Indian wheat hexaploid like *T. compactum* and *T. vulgare*. Helbaek subsequently took a more tentative view upon the identification of grains from Maiden Castle and Little Solsbury (see preceding entry).

Triticum spelta L.   Spelt

F VIIb I.A. *Hembury Fort, Small Down Camp, Meare Lake Village, Glastonbury Lake Village, Little Solsbury, Casterley Camp, Fifield Bavant, Highfield, Winkelbury, Maiden Castle, Worth Matravers, Portland 'Beehives', Corfe Mullen, Wickbourne, ?Micklemoor*; R. *Rotherley, Iwerne, Wickbourne, Park St. (St Albans), Verulamium, Rivenhall, North Leigh, Isca, Caistor by Norwich*

It has already been mentioned that spelt belongs to the brittle-rachis species of *Triticum* and can be distinguished from small spelt and emmer by the adherent rachis joint lying alongside the separated spikelet. Furthermore the glume of spelt has characteristic strong nervation, blunt truncate apex with short blunt tooth and membranous margin, and the grains are 'slender, somewhat tapering toward both ends, the dorsal side evenly curved, the ventral side flat and the cross-section rounded' (Jessen & Helbaek, 1944). From their own material various spikelet remains and carbonized grains have been referred by Jessen and Helbaek to *Triticum spelta*. In his discussion of the burned Roman grain store at Caerleon, Helbaek (1964) indicates that emmer and spelt, otherwise hard to distinguish, may be separated by the greater width of the glume base in the former species.

The cited records shew that the prehistoric cultivation of spelt, so far as is known, was restricted to the south of England, and to the Iron Age and Roman periods.

Triticum spelta is a hexaploid wheat of the genome *AA, BB, DD* but there is no certain knowledge of its wild ancestry. It did not occur in prehistoric western Asia and its earliest discoveries are from the second millennium B.C. in Switzerland and northern Italy whence it seems to have spread to central and northern Europe, only later reaching north Africa and Asia Minor. It is remarkable that the hexaploid bread wheats (*T. aestivum*) of similar genomic constitution have been found, even in central Europe, in deposits very much older than those with the earliest spelt. It seems unlikely on this evidence that spelt

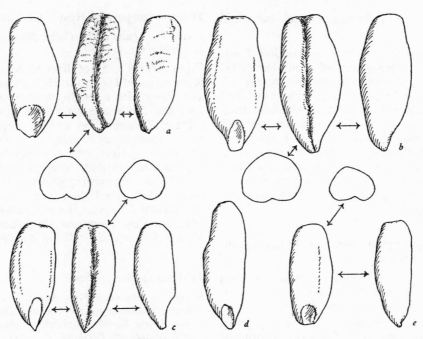

Fig. 144. Carbonized grains of rye (*Secale cereale*) of Roman age, shewing the typical twisted and blunt-ended form of caryopsis (Jessen & Helbaek, 1944): *a*, Forth and Clyde Canal; *b–e*, Castle Cary; all of Roman age.

can have played any part in the origin of the bread wheats and that *per contra* it arose late and in Europe, rather than early and in Asia.

Spelt is not known wild but is cultivated still, especially in mountain areas, in some small areas of the upper Rhine, in the Eifel–Hunsrück–Ardennes region, in Spain, the Siebenbürgen and the Banat. Of these the German area is possibly the recent source of the spelt in the other areas of cultivation.

Secale cereale L. Rye (Fig. 144)

Macros: F VIII I.A. *Fifield Bavant, Winkelbury, Maiden Castle*; R. ?*Wickbourne, Verulamium, Castle Cary, Clyde and Forth Canal, North Leigh,* ?*Isca*

Pollen: F VIII *Hockham Mere*; R. *Lindow Moss, Holcroft Moss*; A.S. and N. *Old Buckenham Mere*

The caryopses of rye are slender and lop-sided, strikingly corrugated and have the upper end broadly truncate, the lower pointed. These characters preserve sufficiently well to allow impressions and carbonized fruits to be recognized, but the whole record for Britain so far is from six Iron Age and Roman sites in southern England, two Roman sites in Scotland, and a single Roman site in south Wales.

In his recent study (1971) of the origin and migration of rye, Helbaek points to the very low percentage frequencies in the European pre-Roman sites, of which the three English are the most westerly. He recalls that the spinous brittle florets are highly adapted to casual dispersal and regards all these early occurrences as due to

accidental importations. By contrast, at the Roman sites of Verulamium and Isca, rye grains were present in large quantity along with spelt, and at Isca in conditions that suggested the accidental burning of a malt-house. He draws attention to the way in which substantial amounts of rye are known only from the Roman garrisons and conjectures a possible importation of rye and spelt grain specifically for brewing, not only in Britain but at Roman sites along the Rhine and Danube. It is agreed that the primary source of *Secale* is in central Asia, and it appears that it was not brought into the near East until the time of the Turkoman invasions of the tenth and eleventh centuries A.D. It seems probable that rye first migrated into Europe in the late Neolithic, advancing north of the Caspian Sea as far as the Polish plain and Bohemia.

It has been suggested that rye originally occurred as a weed and that gradually, in poor soils and bad climates, it ousted the cereals it accompanied, rising in time to the status of crop-plant itself, but it is difficult to identify the times and places where this occurred. In this matter we may ultimately find the pollen evidence decisive. Among the cereals the pollen grains of rye are perhaps the most distinctive, with a protruding, well delimited but narrow poral annulus and a surface covered with scattered minute punctae. In contrast with the wheat, the allogamous wind-pollinated rye has a large production of air-borne pollen so that cultivated crops are strongly represented in the local pollen spectra. Thus at Old Buckenham Mere, Norfolk (fig. 171) *Secale* pollen is present in substantial frequencies throughout the periods conjectured to cover Anglo-Saxon and Norman time and it is accompanied by high frequencies of *Cannabis* pollen and lower, but persistent frequencies of *Linum*, as well as pollen of many weeds of arable land, including *Spergula arvensis* and *Centaurea cyanus* (Godwin, 1967*b*, 1968*b*). In lower frequencies *Secale* pollen was present at Old Buckenham Mere from the Roman period as it was shewn to be by Birks (1965*b*) in two lowland Cheshire mosses. Unpublished analyses by R. Sims at Hockham Mere, also in the East Anglian Breckland, shew continuous if low values for *Secale* pollen (with weeds of arable cultivation) in what is taken to be the Viking or Anglo-Saxon period, although isolated grains occur much earlier.

It seems improbable that rye was a bread crop in the British Isles in prehistoric time, but of course it became of the very greatest importance in mediaeval time, not only in this country but on the Continental mainland, as historic sources and pollen diagrams confirm. It remains uncertain as yet whether the British expansion of rye cultivation stemmed from an older Romano-British tradition or was essentially brought in by the Anglo-Saxon invaders, as we have conjectured for *Cannabis*. Rye is still grown fairly commonly in the dry marginal soils of the East Anglian Breckland where soil poverty and low rainfall exclude most other crops.

VI

PATTERN OF CHANGE IN THE BRITISH FLORA

1. FLORISTIC CHANGES IN THE TERTIARY PERIOD

Evidence of fossil pollen suggests that angiosperms were already evolved in the Jurassic period, and by the early Cretaceous the flowering plants had attained that general predominance in the world's flora that they enjoy today. From that time forward we begin to glimpse the character of the great floristic aggregates, and even the vegetational types which have successively occupied our Islands. And although our analysis of the recorded history of the British flora has been restricted to evidence from the Quaternary period, there is something to be gained by briefly reviewing our knowledge of floristic change in the preceding geological period, the Tertiary. The main divisions of the two periods are set out in table 31 which follows; in considering it, regard must be had to the fact that whereas the Quaternary period lasted approximately one million years, the duration of the Tertiary was of the order of seventy million.

Whereas in the earliest records the Cromer Forest Bed used to be referred to the upper Pliocene, it was subsequently placed, together with some of the marine 'Crag' deposits, in the lower Pleistocene, but this reference in turn has been modified, principally as the result of the work of West and his associates in East Anglia, so that it is now seen to belong to the middle Pleistocene and to have been preceded by a sequence of what are in all probability glacial and interglacial stages. There seems no sufficient case for the retention of the concept 'Holocene' for the period in which we live, and the interglacial stage name 'Flandrian' implies that we exist in the latest of a sequence of such stages within the uncompleted Pleistocene.

The extensive deposits of the London Clay (of lower Eocene age) contain a very large and important fossil flora monographed by E. M. Reid and Miss Chandler who shewed that this consisted of some 250 species, and could be characterized as a 'tropical Indo-Malayan flora' having very few affinities with our present British flora. It formed a tropical evergreen forest vegetation, and included numbers of tropical genera, among them the palm genera *Nypa* and *Sabal*. Analysis of the present-day geographical range of the London Clay flora gave the following striking results:

73 per cent of genera have living relations in the Malayan Islands.

53 per cent of genera have living relations in the Malay Peninsula, India, Ceylon, South China and Burma.

40 per cent of genera have living relations in Australasia.

39 per cent of genera have living relations in tropical Africa.

20 per cent of genera have living relations in America.

Since the days when conclusions were necessarily based only upon macro-fossils, pollen identifications have greatly reinforced, but have not substantially altered, our views on the geographical affinities of the London Clay flora (Chaloner, 1968). The London Clay plant families include some that today are purely tropical, the Nypaceae, Bursaraceae, Icacinaceae, Bixaceae and Sapotaceae, but many more are highly typical of the tropics. From this and much other biological evidence it is apparent that temperate regions of Europe and North America then enjoyed mean annual temperatures of 10 to 12 °C higher than those of today, and that from the Eocene onwards temperatures gradually declined, probably at an increasing rate as the time of the Pleistocene glaciations approached. This progression was illustrated some thirty-five years ago by E. M. Reid, who together with her husband and M. E. J. Chandler had examined and considered all the most important Tertiary and early Quaternary fossil floras then known in western Europe. Mrs Reid (1935) shewed the gradual modification of the British flora from the early Eocene, by setting out the percentage proportions of British genera in Tertiary floras of successive ages, as in table 32.

It is entirely reasonable to include in this comparison the important deposits from Tegelen and Reuver in Holland and the Dutch–German border for these were all part of the deltaic deposits of the lower Rhine; the geological ages, but not the sequence of the three youngest deposits have been modified to present belief.

By the relationships of families, genera and species alike, E. M. Reid thus demonstrated the transformation of an exotic flora into one substantially that of the present day. Here analyses of the geographical ranges of the fossil families and genera demonstrated that the Pliocene flora of western Europe represented an extinct latitudinal link between the living floras of China and Japan on the one hand and of eastern North America on the other. To explain survival in the two regions and extinction in the third, Mrs Reid adopted the old theory of migration of flora from circumpolar regions under the compulsion of a progressively cooling climate which drove the old floras southwards to be extinguished against the great mountain barriers thrown up in the Miocene period. In eastern Asia and in eastern North America the migrating flora could not only move freely from north to south as the climate

TABLE 31

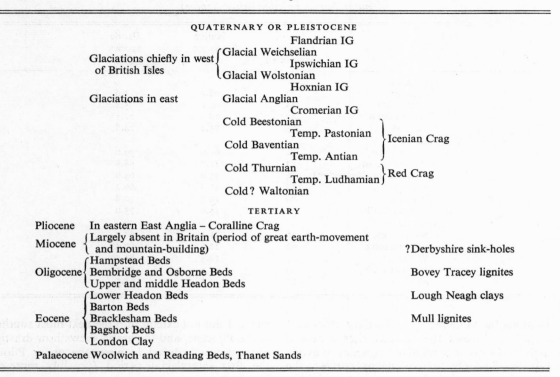

QUATERNARY OR PLEISTOCENE

Flandrian IG
Glaciations chiefly in west of British Isles { Glacial Weichselian
Ipswichian IG
Glacial Wolstonian
Hoxnian IG
Glaciations in east   Glacial Anglian
Cromerian IG
Cold Beestonian
Temp. Pastonian } Icenian Crag
Cold Baventian
Temp. Antian }
Cold Thurnian
Temp. Ludhamian } Red Crag
Cold? Waltonian

TERTIARY

Pliocene	In eastern East Anglia – Coralline Crag	
Miocene	{ Largely absent in Britain (period of great earth-movement and mountain-building)	?Derbyshire sink-holes
Oligocene	{ Hampstead Beds / Bembridge and Osborne Beds / Upper and middle Headon Beds	Bovey Tracey lignites
Eocene	{ Lower Headon Beds / Barton Beds / Bracklesham Beds / Bagshot Beds / London Clay	Lough Neagh clays / Mull lignites
Palaeocene	Woolwich and Reading Beds, Thanet Sands	

TABLE 32

	%	
Middle Pleistocene (Cromer)	97	(98 % of species still living)
Early Pleistocene (Tegelian)	78	(70 % of species still living)
Late Pliocene (Reuverian)	53	(18 % of species still living)
Middle Pliocene	59	(First living species found, 12 % spp. still living)
Middle Oligocene (Bembridge)	—	(12 ex 35 genera are British)
Lower Oligocene	34	
Upper Eocene (Hordle Beds)	23	(Malayan element almost gone: 8 ex 35 genera are British)
Lower Eocene (London Clay)	2	

changed, but was also offered an altitudinal range of accommodation. This type of explanation is now the generally accepted basis for historical phytogeography, more especially in relation to the Glacial period itself. Thus, on the broadest of scales it is understood that whilst in the lowlands of the tropics, the ancient Tertiary flora has persisted relatively unchanged, the effect of periods of general temperature depression has been to cause the temperate Tertiary flora of both the northern and southern hemispheres to penetrate to the equator by way of longitudinal mountain ridges, where the altitude has permitted many of them to remain until the present day. By contrast nothing, or almost nothing, now remains in Europe of the Malayan evergreen forest: it was destroyed

by the gradual climatic changes of the Tertiary even before the drastic changes of the Glacial period.

A more recent general analysis of the evolution of the European Tertiary flora is that given by Szafer as part of his assessment of the extremely rich results of his wartime researches into the Tertiary fossil flora of Krościenko in Poland (Szafer, 1946–7: Harris, 1950). He deals with Europe as composed of three major climatic regions: an 'outer' zone, north of the Alps, Carpathians and Pyrenees, an 'inner' zone south-east of the mountains and including Bulgaria and Rumania, and a 'southern' zone round the Mediterranean. The major part of his comparisons are concerned with Tertiary (and early Pleistocene) fossil floras situated within the outer zone, which he supposes

TABLE 33. *Relative proportions of native and exotic species in fossil floras of the outer European zone (after Szafer)*

	No. of species considered	Native species (%)	Exotic species (%)
Pleistocene			
Cromer	132	98.4	1.6
Vogelheim	55	94.6	5.4
Schwanheim	47	78.9	21.1
Tegelen	85	75.2	24.8
Pliocene			
Castle Eden	41	61.0	39.0
Willershausen	44	41.2	58.8
Rhine delta (Reuverian)	116	35.0	65.0
Krościenko	113	33.8	66.2
Frankfurt	89	29.2	70.8
Pont de Gail	36	25.0	75.0
Miocene			
Sośnice, etc.	46	31.1	68.9
Niederlausitz	62	22.6	77.4
Wieliczka	27	18.5	81.5
Herzogenrath	39	12.8	87.2
Salzhausen	22	4.5	95.5

to have had much the same reality in Tertiary times as it now has. Table 33 shews the various sites arranged approximately in the order of relative frequency of exotic and native species. Szafer by 'native' means 'native to the outer zone', and consequently he groups as 'exotic' the phytogeographical elements: west Asiatic, south European, Balkan–Colchican, east Asiatic, Atlantic-north American, Pacific-north American, Mediterranean, Macaronesian, south-east Asiatic, Neotropic.

This demonstrates for the outer Euorpean zone the same progressive 'Europeanization' of the flora that E. M. Reid has shewn for a more limited area, but over a longer time.

Szafer has further analysed the changes in floristic composition from Miocene time onwards by again arranging the chief known fossil sites in presumed order of age, and setting out graphically the relative proportions in which the four chief geographical groups of the fossil floras are present (see fig. 145). The boundary between the Miocene and Pliocene is then seen to be marked by the big decrease in the North American element with a large rise in the Asiatic component, the sub-tropical element remaining unaffected. There appeared at this time a strong wave of the east Asiatic element, a fact bearing out the supposition of Krishtofovich and Mädler, that the outer zone received its dominant exotic floristic component from the north-east: it may be supposed that this was the old Arctic–Tertiary element.

It was the outer European zone which suffered first and most from the onset of the great Pleistocene glaciations, and it is evident that the longitudinal mountain barrier of Carpathians to Pyrenees was the chief cause of the violent extermination of the east Asiatic forest flora of the outer zone. We know from fossil evidence that this element did not extend into the next most southerly zone in the Pliocene, and the graph shews how drastically this element was reduced between the upper Pliocene and the first interglacial period (possibly represented by Tegelen).

Zagwijn (1959) has later shewn how closely the evidence from the lower Rhine area matches that from western and central Europe in demonstrating the steep decline of the tropical and sub-tropical elements of the flora between the Miocene and late Pliocene, and the big rise of the east Asiatic north American element from the Miocene to a high level in the Pliocene with subsequent decline to low values after the first Pleistocene glaciation.

Fairly detailed summaries of the taxa identified in different deposits of the European Tertiary are given by Walter and Straka (1970) and flora and fauna are jointly considered by Pearson (1964) in a work specifically directed to the Cenozoic era over the whole world. Table 34 indicates merely the type of information extractable from the abundant middle and west European evidence.

Szafer, by comparison of regions where the representatives of the mid-Pliocene flora now grow, estimates that the climate of the European outer zone in the deciduous (and mixed) forest belt differed from that of the present in the following way. It had

a January mean temperature about 11 °C higher (i.e. 6 °C);

a July mean temperature about 9 °C higher (i.e. 25.4 °C);

a yearly mean temperature about 9 °C higher (i.e. 15.7 °C);

a more maritime climate;

about twice the annual rainfall.

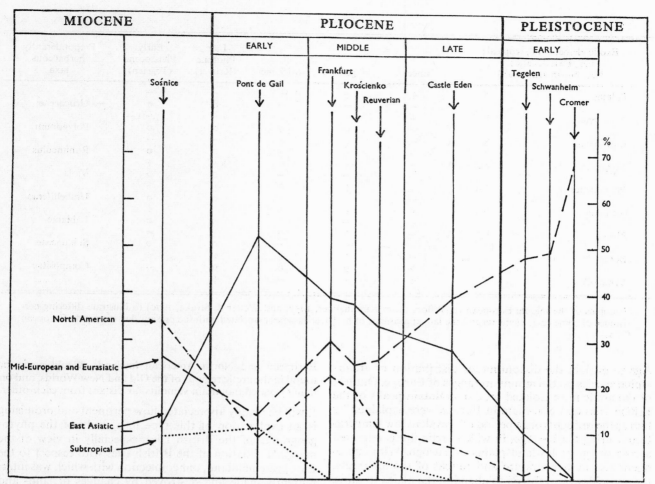

Fig. 145. Analysis of the chief fossil floras of Szafer's 'Outer European' zone, to shew the progressive change in proportion of the main phytogeographic groups throughout the middle and late Tertiary period and up to the middle Pleistocene. The progressive modification of the outer European flora throughout the Pliocene, very long before the first glacial period, is extremely evident, and the diagram gives a good guide to the changes leading up to the composition of the British flora as it was during Cromer Forest Bed time.

A climate of this kind can be matched in very few parts of the world, among them the western Carpathians where, of course, there still exists a remarkable collection of Tertiary species.

The loss of this climate has no doubt been very largely responsible for the loss of the larger part of the great east Asiatic forest flora of Pliocene time.

Although the British Isles lie on the western oceanic margin of the European outer phytogeographic zone, they must have suffered floristic changes broadly in accord with those Szafer has established for that region. We may expect future research to fill in much of the detail of the slow changes of the early Tertiary, and the quickening changes in the Pliocene, leading to the onset of the Glacial period by which time the British flora had a composition not far removed from that of today.

2. THE ICE AGE AND THE INTER-GLACIAL PERIODS: CLIMATE, SOIL AND VEGETATION

In the preceding section of this chapter we have shewn how, by the close of Tertiary time, the bulk of species present in Britain were those which still live here or on the nearby Continent, and in an earlier chapter we have briefly pointed to some of the outstanding features of the Glacial period in the British Isles, and of the sequence of deposits representing that period. There yet remain the most striking effects of all, those which caused the migrations, expansion, retractions and extinctions of our flora so as to yield the complex and yet meaningful distribution patterns in which they exist today.

At a very early stage in the study of phytogeography, Edward Forbes invoked the influence of the Great Ice

TABLE 34

Exotic elements (T, tropical; EA, East Asiatic; NA, North American)		Eocene	Oligocene	Miocene	Late Pliocene (Reuver)	Early Pleistocene (Tegelen)	Preponderantly herbaceous taxa
Palmae	T	+	+	+	−	−	
		−	−	−	o	o	Urticaceae
Nypaceae	T	+	−	−	−	−	
		−	o	o	o	o	Polygonum
Engelhartia	NA+EA	+	+	+	−	−	
		−	−	−	o	o	Ranunculus
Magnolia	NA+EA	+	+	+	+	+	
		−	−	−	o	o	Viola
Phellodendron	EA	+	+	+	+	+	
		−	−	−	o	o	Umbelliferae
Actinidia	EA	+	+	+	+	+	
		−	−	−	o	o	Labiatae
Nyssa	EA+NA	+	+	+	+	−	
		−	−	−	o	o	Solanaceae
Styrax	T+EA	+	+	+	+	−	
		−	o	o	o	o	Compositae
Symplocos	T	+	+	+	+	−	

Examples of the middle European fossil flora (after Kirchheimer, 1957, and Walter & Straka, 1970) to illustrate differing persistence of some early elements and the late appearance of herbaceous types associated with the present-day temperate European flora.

Age to explain the discontinuous distribution of arctic–alpine species on the mountain ranges of Europe. Outliers of the arctic flora isolated on the mountain peaks of the British Isles and elsewhere in Europe were explained by him as the residue from a period of prevalent low temperatures during the Ice Age, in which arctic floras migrated across the great lowland plains of Europe. With subsequent rise in temperature and retreat of the ice-sheets, the arctic flora was forced to abandon the plains to move northwards and to ascend the mountain peaks, leaving behind an indication of former events in the striking discontinuity of the relict populations in disjunct mountain areas. Darwin, in *The Origin of Species*, followed the same explanation:

As the cold came on, and as each more southern zone became fitted for the inhabitants of the north, these would take the places of the former inhabitants of the temperate regions. The latter, at the same time, would travel further and further southwards, unless they were stopped by barriers, in which case they would perish. The mountains would become covered with snow and ice, and their former Alpine inhabitants would descend to the plains. By the time that the cold had reached its maximum, we should have an arctic fauna and flora, covering the central parts of Europe, as far south as the Alps and Pyrenees, and even stretching into Spain...

As the warmth returned, the arctic forms would retreat northward, closely followed up in their retreat by the productions of the more temperate regions. As the snow melted from the bases of the mountains, the arctic forms would seize on the cleared and thawed ground, always ascending, as the warmth increased and the snow still further disappeared, higher and higher, whilst their brethren were pursuing their northern journey. Hence, when the warmth had fully returned, the same species, which had lately lived together on the

European and North American lowlands, would again be found in the arctic regions of the Old and New Worlds, and on many isolated mountain summits far distant from each other.

It will be readily appreciated how pertinent and promising is an explanation of this type in a study of the phytogeography of the British Isles, especially in view of the marginal situation of the British Isles with respect to the European mainland, our connection with which was intermittently made and destroyed by changes of land- and sea-level, and concerning which it might well be held either that the glaciations had destroyed all traces of life therein, or that they had offered refugia for substantial aggregates of species throughout the worst rigours of the cold.

The position is excellently well put by Deevey (1959, p. 1340):

One of the most popular topics in all biogeography is the question whether the whole of the fauna and flora of the British Isles immigrated in postglacial times, or whether some fraction survived from an earlier time. The attractiveness of this problem arises in part from its manageability. For a given taxonomic group the British list is always small in comparison to that of the mainland, British distributions are often exceedingly well-known owing to the inexhaustible supply of amateur and professional naturalists, and the most elementary comparison between British and French lists produces at once some fascinating problems. As a result, the literature is enormous and it is difficult to achieve a balanced presentation. The problem is complicated by the question of the relations of land and sea during the Pleistocene, for these are by no means clear.

We now approach the problem with a much closer appreciation of the detailed stratigraphy of deposits of the

420

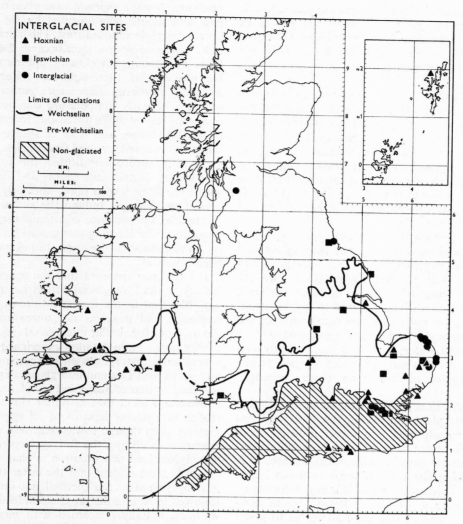

Fig. 146. Map of the British Isles shewing sites of interglacial plant deposits. The Anglian and Wolstonian glaciations have similar southern limits shewn as one line across England: south of this the country was unglaciated, as probably were small mountain areas in southern central Ireland.

Glacial period and much better, though still inadequate information, as to climatic and edaphic conditions than was available before, a fact of the greatest importance since, to quote Deevey again, 'only against a background of this sort of information does an account of existing ranges of species yield any data that could not have been deduced by Darwin and his contemporaries'.

The formation of till or Boulder Clay over substantial areas of the British Isles at each of the main glacial stages indicates the area of actual ice movement. Over such areas the ice directly scraped away the plant cover and the soil which sustained it, generally leaving a blanket of drift material over the affected landscape; such territory must be repopulated entirely by reinvasion when the climate ameliorates. Even during the episode of maximal extension the ice-sheets did not extend south of the Thames in Great Britain (fig. 146), and seem to have left

parts of southern Ireland also unglaciated (Farrington, 1947). The rest of the British Isles, and such parts of the adjacent present sea-bottom as then were exposed, were largely within the region of periglacial climate, where solifluction movements of soil occur (cryoturbatic phenomena) and where there is often a 'permafrost' layer above which the soil movements cause polygon formation, solifluction terraces and related phenomena. Although the range of occupying plant species is no doubt restricted by temperature conditions, they have the advantage that fresh mineral soils are always being made available, leaching is unimportant and competition from taller species within closed communities is insignificant. In periglacial regions there are furthermore enormous volumes of seasonal melt water which cause great deposits of river-borne sand, silt and gravel: from these the finest grades of material (in the grade 0.1–0.01 mm) may be wind-borne for great distances to form the characteristic 'loess' which was apparently bound by the growth of grass-dominated vegetation. Loess is only weakly represented in deposits now remaining in the British Isles, but it is found up to a thickness of 15–20 ft in northern France.

As we have earlier explained, the eustatic fall in absolute level of the oceans is maximal when the ice-sheets are at their largest, and was possibly of the order of 300 ft, which would suffice to have laid bare most of the North Sea. In opposition to this, however, the northern parts of our islands were depressed isostatically by the ice-sheets centred upon them so that the emergent coastlines were chiefly in the south. For us the salient point must be that the changing sea-levels alternatively removed and restored the insularity of our islands.

In interglacial periods, in these latitudes, vegetation covers the hills and flanks of all but the highest and steepest mountains, water supply to the rivers is good and regular, and but little material washes down into the rivers. Under conditions of glaciation on the other hand, the mountains and hills which, because of height, steepness or situation, are not actually covered with ice are rapidly frost-weathered, the debris is moved downhill by solifluxion and melt water, and fresh surfaces are constantly exposed to colonization by such species as can withstand the rigour of life upon these 'nunatak' areas.

Although we can thus picture, with some plausibility, the conditions for plant life during the several glacial stages, it must be admitted that direct evidence from contemporary fossil plant material as to the actual species present outside the range of the ice-sheets themselves is substantially lacking from all of them save the latest, the Weichselian. This we shall consider at a later stage, but already we may say that our knowledge of periglacial fauna and flora of that time allows us to reject the 'tabula rasa' hypothesis according to which the glacial conditions were so severe in the British Isles that the whole of our present fauna and flora must have immigrated since the last glaciation.

In contrast with the sparsity of evidence for the plant life of the glacial stages, we now have very substantial information of such evidence for the intervening temperate periods, in part based upon macroscopic identifications, frequently those of older workers, but increasingly due to the application of pollen analysis to deposits of interglacial age. The late Professor Jessen made the first dramatic generalization when he was able to shew conclusively that of the interglacials of different ages present in Denmark and north-western Germany, each exhibited the effects of a definite climatic cycle, from the sub-arctic conditions at the close of the preceding Glacial period, through a stage of amelioration to a climatic optimum with conditions rather warmer than those of today, and then a gradual recession of temperature ending in subarctic conditions before the onset of the next glaciation. The sparse sub-arctic flora with dwarf willows, arctic

birch and *Dryas octopetala* gave place to forests of birch, pine and aspen, later invaded by the elements of the broad-leaved deciduous forest. The latter, including alder, oak, hazel and linden, then became dominant, whilst in lakes and streams the climatic optimum was marked by the presence of thermophilous water plants such as *Brasenia purpurea*, *Dulichium spathaceum* and *Trapa natans*. The early temperature diminution was accompanied by retreat of the mixed-oak forest, at first in favour of the hornbeam, and later of spruce accompanied by birch and pine in increased frequencies. Next pine, birch, aspen and spruce constituted the forest cover, until replaced by *Betula nana* heaths and sub-arctic swamps, whilst the lakes held a poor, northerly aquatic flora. In one or two sites in Jutland the situation appears complicated by the presence of a temporary return of warm conditions (the Herning oscillation) before the readvance of the Scandinavian ice-sheet, but elsewhere this has not been observed.

It is at once apparent that the course of vegetational history in the Post-glacial period has followed a course very like that in the first two-thirds of this interglacial sequence, and it seems not unreasonable to interpret the evidence for similar interglacial sequences in the light of our much more detailed experience of the Flandrian vegetational history. Jessen's experience of the Danish and north-west German interglacial deposits was already matched by that of Szafer in Poland, and as time went on it became evident that a similar climatic sequence controlled interglacial deposits of all ages in north-western Europe, even though it was relatively rare to find single examples exhibiting the full cycle of change. We have already outlined this situation in chapter III: 1.

It was soon evident that the latter stages of the interglacial cycle after the thermal maximum by no means followed in reverse the progression leading up to it. Thus we were initially concerned with migration, establishment and movement towards climatic climax woodland with its tight competition for light and space, whereas afterwards we witnessed disintegration (recession), incursion of late forest migrants and loss of species intolerant of worsening conditions. To these differences Iversen (1958) added the progressive nature of soil changes in the cycle from the bare shifting mineral soils of the Lateglacial phase, through stabilization as vegetational cover increased, leading to deep brown forest soils under closed deciduous forest, and finally to podsolization with acidic heath and acidic bog development in the ensuing phase of worsening climate. For these successive stages Iversen proposed the terms *Cryocratic*, *Protocratic*, *Mesocratic* and *Telocratic* as indicated in our fig. 147. Subsequently S. T. Andersen (1966) in the light of his own investigations of Danish interglacials, suggested that quite independently of thermal decline, an Oligocratic phase of soil degeneration could be recognized: it precedes or overlaps the Telocratic in Iversen's sense.

The notion of progressive modification through the interglacial cycle is involved also in the concept developed

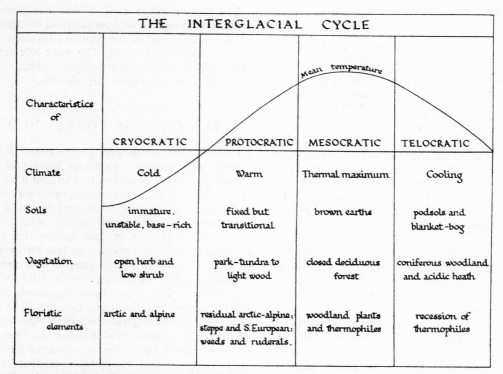

THE INTERGLACIAL CYCLE			
Characteristics of	CRYOCRATIC	PROTOCRATIC	MESOCRATIC	TELOCRATIC
Climate	Cold	Warm	Thermal maximum	Cooling
Soils	immature, unstable, base-rich	fixed but transitional	brown earths	podsols and blanket-bog
Vegetation	open herb and low shrub	park-tundra to light wood	closed deciduous forest	coniferous woodland and acidic heath
Floristic elements	arctic and alpine	residual arctic-alpine; steppe and S. European: weeds and ruderals.	woodland plants and thermophiles	recession of thermophiles

Fig. 147. Schematic representation of the changes of climate, soil, vegetation and floristic elements that accompany the progress of an interglacial period. (Godwin, 1959.)

by Gritchuk (1964), which suggests that the cycles of climatic warmth and wetness are out of phase, so that extreme dryness at the opening of the cycle progresses to extreme wetness at its close.

To facilitate comparison of British interglacials with one another Turner and West (1968), taking account of these principles, proposed that each could be divided into four consecutive sub-stages based on the biostratigraphic zones revealed by pollen analyses. These were I, *Pre-temperate*, characterized by the development and closing in of forest vegetation, usually *Betula* and *Pinus* after a Late-glacial, with abundant herbs and light-demanding shrubs; II, *Early-temperate* (mesocratic), covering establishment and dominance of mixed-oak forest on deep mull soils; III, *Late-temperate* (oligocratic), increasingly dominated by late-appearing trees such as *Carpinus*, *Abies*, sometimes *Picea* (and perhaps *Fagus*) with change to mor soils; and IV, *Post-temperate* with returning dominance of the boreal tree-genera, *Pinus* and *Betula*, the virtual extinction of the temperate deciduous forest elements and often the spread of Ericaceous heaths (which may however persist into the succeeding glacial stage). This quadruple division was effectively employed by West (1970b) in comparison of the five latest British interglacials and it is the basis we have adopted throughout this book.

3. THE EARLY PLEISTOCENE

Knowledge of the Early Pleistocene in Britain is extremely tenuous and restricted to a handful of sites in East Anglia, of which the most important is that at Ludham, east of Wroxham, where a deep borehole was specially made passing through the first boulder clay to a depth of −163 ft (54 m) O.D. (West, 1961b). This encountered a series of sands and silty clays, the latter especially rich in pollen. The disadvantages of working with shallow-water marine sediments (errors of derivation, flotation etc.) were compensated for by the advantages of correlation with lithological changes and of evidence from mollusca and foraminifera, the more important since we are in the area of the classic sequence of Crag deposits. The sparsity of macroscopic plant remains was offset by the fact that the almost continuous pollen-analytic series could be matched with similar series from the Netherlands, the work of Zagwijn and other Dutch Quaternary geologists, where the terrestrial facies of the Rhine–Thames delta was more fully represented and where extensive macroscopic plant identifications had already been made, especially at Tegelen by C. & E. M. Reid. The extensive plant list is copied by Pearson (1964) and from it we note the following species that are of special phytogeographic interest:

Salvinia cf. *natans*	*N. minor*
Azolla tegeliensis	*Stratiotes aloides*
Najas marina	*Dulichium vespiforme*

423

Brasenia purpurea
Abies sp.
Picea sp.
Tsuga sp.
Carya sp.
Pterocarya sp.
Carpinus betulus
Eucommia europaea
Trapa natans
Corema intermedia

Decodon globosus
Laserpitium siler
Physalis alkekengi
Menispermum dahuricum
Magnolia kobus
Actinidia faveolata
Phellodendron elegans
Staphylea sp.
Vitis cf. *sylvestris*

Dutch pollen diagrams presumed to be of this age shew a dominance of temperate forest vegetation with *Pinus*, *Alnus*, *Picea*, *Betula* and the mixed-oak forest trees, but also persisting in low frequency, the Tertiary types that had been abundant in the late Pliocene, namely *Tsuga*, *Pterocarya*, *Carya*, *Eucommia*, *Carpinus*, *Fagus* and the so-called 'Taxodium type'.

The long pollen sequence worked out by West at Ludham disclosed an alternation which, in conjunction with faunistic and lithological evidence, he has attributed to three temperate stages (Ludhamian, Antian and Pastonian) with intervening cold stages (Thurnian and Baventian) – see table 31.

The warm stages in this alternation were marked by high arboreal pollen frequencies and pollen spectra indicative of temperate mixed coniferous–deciduous woodland with high pine (possibly over-represented), substantial *Picea*, *Tsuga*, *Alnus* and *Quercus* and low frequencies of *Betula*, *Ulmus*, *Carpinus* and *Pterocarya*. The *Tsuga* pollen was referable to two types, the commonest to one very close to *T. canadensis*, and the other to *T. caroliniana* (see account for the genus). This genus and the less frequent *Pterocarya* represent all that remains of the Tertiary tree-pollen, an even smaller residue than that in the corresponding Dutch diagrams. The warm stages are also indicated by spores of *Osmunda*, most commonly *O. regalis* but also *O. claytoniana*.

The cold stages in the borehole series were indicated by high non-arboreal to arboreal pollen ratios, by high frequencies of ericoid pollen mainly attributable to *Empetrum nigrum*, high *Sphagnum* and *Lycopodium* spores and emphasis on the less thermophilous trees. The series at Easton Bavents adds useful confirmation to the Ludham sequence, especially as it indicates the near presence of Scandinavian ice: the pollen series with typical *Tsuga* of both types is referable to the preceding temperate Antian stage.

West is certainly correct in emphasizing the extreme oceanicity of these early Pleistocene stages, comparing the high *Empetrum* percentages of the glacial stages with those known from the Late Weichselian of north-western Europe. He believes also that this oceanicity sharply decreased eastwards in Holland and central Europe.

The uppermost temperate stage of the Ludham borehole was subsequently named the Pastonian and was separated from the Cromerian by the cold Beestonian: it has very much the character of the two preceding temperate stages, with a comparable forest composition with *Carpinus* in sub-stages II and III and *Tsuga* in III.

Throughout these early Pleistocene pollen diagrams Chenopodiaceae are well represented but this is probably a reflection mainly of the coastal situations in which the deposits were laid down. The very low representation of *Corylus* in the temperate stages contrasts strongly with its behaviour in the later interglacials.

4. THE CROMER FOREST BED SERIES

Exposed at numerous sites on the coast of Norfolk and Suffolk are deposits collectively known as the Cromer Forest Bed series, described by Clement Reid (1882, 1890) and yielding at his hands a remarkable wealth of evidence including macroscopic plant remains of a great many vascular plants. This series of beds with a recognizably temperate flora sitting just below the earliest East Anglian till, at first regarded as late Pliocene and subsequently as earliest Pleistocene, is now seen as middle Pleistocene. The numerous, often partial and interrupted sequences of beds have been intensively reinvestigated by West and collaborators, notably Miss Duigan and Mrs Wilson. Although an extremely important indication of the trend of these investigations has been briefly given (West & Wilson, 1966), the main body of the stratigraphic, pollen-analytic and macro-fossil evidence is as yet unpublished. What has however emerged is that the Cromer Forest Bed series of Reid, which in his terms included an upper and a lower freshwater bed separated by an estuarine 'Forest Bed', is part of a complex sequence representing two interglacials, the Pastonian and Cromerian, separated by a glacial stage, the Beestonian. The lower interglacial was taken to be identical with the uppermost interglacial stage of the Ludham borehole, and to be equivalent to the Weybourne Crag. The Beestonian cold stage included the Arctic Freshwater Bed of Reid and exhibited ice-wedge casts, involutions and soil polygons at certain horizons: it seemed to shew more pronounced glacial characteristics than preceding glacial stages, and West and Wilson have found in its deposits *Salix polaris*, *S. herbacea*, *Betula nana*, *Saxifraga* spp. and *Oxyria digyna*, together with a number of aquatic plants which, though less obviously northern in range, have actually been consistently found in the Early, Middle and Late Weichselian.

For the preceding temperate stage, the Pastonian, there are as yet no published pollen diagrams, but West indicates that he has evidence to shew that all four stages of the interglacial cycle are recognizable. Sub-stage I dominated by *Pinus*, II by mixed-oak forest with *Carpinus*, III characterized by *Picea* and very low frequencies of *Tsuga*, and IV chiefly dominated by *Pinus* and *Alnus*. The temperate character of the Pastonian is also evidenced by the identification from it of the seeds of *Chamaedaphne caliculata*, a north American plant typical of oligotrophic bogs, the presence of *Fraxinus* among the mixed-oak forest, and of the aquatic *Azolla filiculoides* and the southern strand plant *Frankenia laevis*.

The Upper Freshwater Bed at West Runton was the

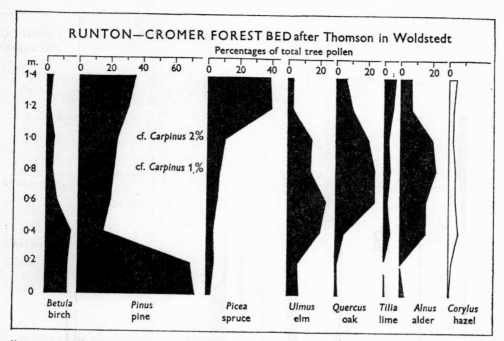

Fig. 148. Pollen diagram compiled by P. Thomson from samples collected by Woldstedt at West Runton, from the Cromer Forest Bed. Although the samples are widely spaced there is an indication of mediocratic mixed-oak forest, and telocratic pine and spruce.

first interglacial to yield a major pollen-analytic sequence: it was produced by P. W. Thomson from samples collected by P. Woldstedt in 1949, and although S. L. Duigan produced many partial sequences, conformable with it, it remains the most comprehensive diagram yet published (fig. 148). The sub-stages shew I, dominant *Pinus*; II, dominant mixed-oak forest with abundant *Alnus* and *Ulmus* with lesser amounts of *Tilia* pollen; III, with returning dominance of conifers, *Pinus* and *Picea*; IV, dominant *Pinus* and *Betula*. *Tilia* was apparently not present in the Pastonian, nor was *Abies* which occurs in low frequency in sub-stage III at some sites; *Corylus*, in low frequency only, was restricted to sub-stages II and III. The onset of the next glacial stage, represented on this coast by the Cromer Till, is indicated by cryoturbation effects and ice-wedge casts, and by the Arctic Freshwater Bed itself. In sub-stage IV the AP/NAP ratios are already lower and Duigan's pollen spectra indicate more open vegetation with genera such as *Armeria*, *Artemisia*, *Centaurea*, *Plantago*, *Polygonum*, *Rumex*, *Scleranthus*, *Limonium*, *Pastinaca* and *Thalictrum*, which have become familiar indicators of the Weichselian periglacial vegetation, and whose presence is conformable with that of much pollen of *Betula nana* type.

Correspondingly at West Runton, the beginning of the interglacial cycle is indicated by a low AP/NAP ratio, *Betula nana* pollen in the prevalent birch-pine trees, and a similar range of herbaceous pollen types to those just indicated for the end of the cycle; they also include *Helianthemum* (Duigan, 1963). Also at this locality the temperate sub-stages yielded abundant megaspores of

Azolla filiculoides, a thermophilic plant whose presence sorts with that of *Fraxinus*, *Viburnum*, *Frangula*, *Juglans*, and, among the water plants, *Najas minor* and *Salvinia natans*. The two last-named aquatics, like *Azolla filiculoides*, are no longer part of the native British flora, a category supplemented by Mrs Wilson's recognition from the Cromerian at Pakefield of seeds of *Corema alba* and fruits of *Trapa natans*. Reid had described both from the Cromer Forest Bed series as a whole, though referring the *Corema* to an extinct species *C. intermedia*, which had also been found in the Dutch Tegelian. *C. alba* at present has a coastal distribution in Spain, Portugal and the Azores. Also in C. Reid's lists are *Najas minor*, *Trapa natans*, *Picea excelsa* and *Hypecoum procumbens*, a Papaveraceous plant of the Mediterranean area.

West and Wilson have a very long list of identified plant taxa from deposits of the Cromer Forest Bed series, and we expect that publication of these, referred to their proper stages and sub-stages, will greatly enlarge our knowledge of middle Pleistocene floristic and vegetational history of this country.

5. THE CORTON INTERSTADIAL AND THE HOXNIAN STAGE: MIDDLE PLEISTOCENE

The first glacial stage of the Pleistocene to be marked by actual glaciations in East Anglia, the Anglian, was represented by two glacial advances, the earlier which brought Scandinavian material from the north and the

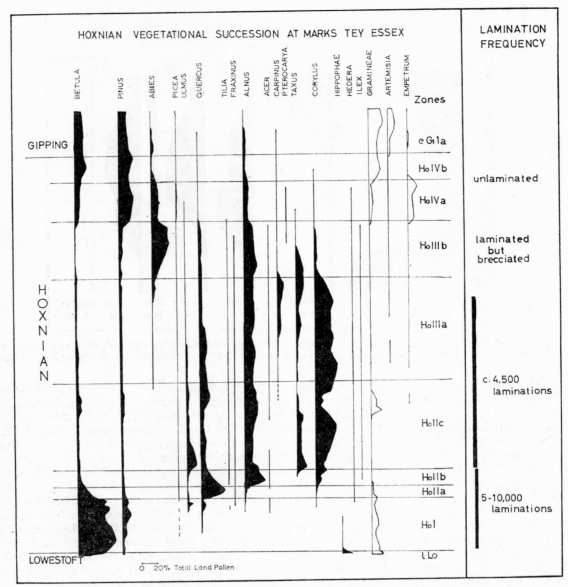

Fig. 149. Pollen diagram from the brick-pit at Marks Tey, Essex displaying vegetational change from the late Anglian to the early Wolstonian through all four sub-stages of the Hoxnian interglacial. In the lower half of the sequence annual laminations strongly indicate that the duration of the interglacial was of the order of 30000 to 50000 years. (After C. Turner, 1970.) The telocratic effects are notably clear.

north-east coast and the later which had a much more western origin, crossing the English Midlands from Wales and the north-west. In the interval between were laid down the shelly sands known as the Corton Beds and these have yielded macro-fossils of *Oxyria digyna, Cerastium alpinum* and *Ranunculus hyperboreus*, plants of a strongly northern aspect suggesting that if a climatic amelioration was indeed present it represented no more than an interstadial event. They appear also to indicate some persistence of arctic–alpine species at no great distance from the ice-front. A number of aquatics were

present, including *Stratiotes aloides* and *Potamogeton crispus* with a mainly southern distribution, and there was also some derived material including megaspores of *Azolla* sp., and a few remains of tree birches, but these do not modify the main vegetational inference (West & Wilson, 1968).

The second interglacial stage of the middle Pleistocene, the Hoxnian, is excellently defined at the type site in Suffolk where it is represented by sediments infilling a basin in the surface of the Lowestoft Boulder Clay that itself overlies the Cromerian interglacial beds of the

Norfolk coast. The site at Hoxne has long been famous for the wonderful series of Late Acheulian flint tools recovered there and for the extensive list of plant identifications by C. Reid. The stratigraphy of the basin has been reconstructed in detail by West who has also made both detailed pollen analyses and macroscopic plant identifications (West, 1956). Hoxnian interglacial deposits have been recovered from a considerable number of East Anglian sites including the Nar Valley, Southelmham and Clacton-on-Sea, where however the cycle was incompletely present. They presented however a very similar picture of vegetational history which can be best appreciated by considering the pollen diagram from Marks Tey, Essex where at the one site, Turner (1970) provides evidence of the whole interglacial cycle (fig. 149). After late-Anglian deposits distinguished by high frequencies of the pollen of *Hippophaë rhamnoides*, the pre-temperate sub-stage I shews prevalence of *Betula* and *Pinus*; in early-temperate sub-stage II dominance has changed to that of mixed-oak forest together with *Alnus* and *Corylus*; *Fraxinus*, *Acer* and *Tilia* are present in small frequency (*Tilia* spp. much under-represented by its pollen) and other thermophiles now present are *Hedera* and *Ilex*. The late-temperate sub-stage III represents a modification of II by the early decrease of *Ulmus* and the expansion first of *Carpinus* and then of *Abies*. This is a displacement perhaps foreshadowing the developments of sub-stage IV with an indication of progressive soil degeneration in high *Empetrum* values and a change to dominance of *Pinus* and *Abies* with some *Betula*, as all the mixed-oak forest genera and thermophiles taper away. Finally the cold conditions of the early Wolstonian glacial stage are reflected in increased pollen frequencies of non-arboreal pollen, notably of grasses and *Artemisia*. In low frequency in sub-stages III and IV pollen of *Pterocarya fraxinifolia* suggests a possible presence locally of this tree now not native nearer than the Caucasus, and both *Buxus sempervirens* and *Vitis* pollen occurred in sub-stage III, indicate persistent high summer temperatures.

The pollen series at Hoxne extended uninterrupted through sub-stages I and II, but III and IV were contracted though recognizable. West was able to shew that the Acheulean industry was stratified into the deposits of late sub-stage II and that it corresponded with a very pronounced phase of forest-clearance comparable with those of the European early Neolithic and indeed so certified by Iversen's own opinion. The temptation to regard this effect as a consequence, direct or indirect, of the presence of Palaeolithic man on the site, has been modified by Turner's discovery of a precisely similar vegetational event in his Marks Tey diagrams, and we are left seeking regional rather than parochial causes for this dramatic and significant effect. The weakness of representation of the late-temperate sub-stage was already compensated by the first-published of British interglacial pollen diagrams, that from Clacton-on-Sea where the *Abies* expansion was very pronounced, and where, from an earlier part of the cycle, S. Hazzledine Warren had recovered palaeo-lithic flints of a culture named after this site (Pike & Godwin, 1953).

The only other extensive Hoxnian pollen diagram from England is from Nechells, Birmingham (Duigan, 1956; Kelly, 1964) where the vegetational development closely resembles that in East Anglia, although early in sub-stage III this site shews very low *Carpinus* values, but a great expansion of *Picea*.

In western Ireland at Gort, Baggotstown and Kildromin and in south-eastern Ireland at Kilbeg, highly important deposits have been investigated, all assumed referable to the Hoxnian stage and exhibiting a vegetational history resembling the English in a general way but with far larger representation of pine in relation to the mixed-oak forest trees. We cite Turner's comment (1970):

The Irish sites, particularly Gort and Kilbeg, give evidence of a highly oceanic climate during the Late- and Post-temperate zones of the interglacial, where exotics such as *Rhododendron ponticum* and *Erica scoparia* flourished, and members of the existing Lusitanian element of the Irish flora (such as *Daboecia cantabrica*, *Erica mackiana* and *E. ciliaris*) occurred well beyond the limits of their present distribution. These records can be set beside the occurrence of *Pterocarya* and *Erica* cf. *terminalis* at Marks Tey as evidence that the climatic conditions over the British Isles in general did indeed become very oceanic towards the end of the Hoxnian interglacial.

In the latter half of sub-stage III the pollen diagrams present us with a picture of a largely evergreen forest with *Pinus*, *Abies*, *Picea*, *Taxus*, *Rhododendron* and *Buxus* accompanied by *Alnus* and sparse mixed-oak forest elements, not indeed unlike the Caucasian mountain forest with which Szafer compared the Pliocene forest from Krościenko. This raises in our mind the conjecture that the Lusitanian plants might conceivably be relics from an old Pliocene vegetational type formerly widespread but now represented by a few species of very disjunct distribution. In this connection it is important to note the description by Birks and Ransom (1969) of an interglacial peat at Fugla Ness, Shetland Is., very plausibly referred to the late Hoxnian and yielding seeds of *Daboecia cantabrica*, *Erica mackiana* and *E. scoparia* var. *macrosperma* with a good deal of *Pinus*, some *Picea* and *Abies* and a wealth of other pollen. As the authors observe, it affords us a view of the northern oceanic facies of the Hoxnian–Gortian vegetational pattern. On the other side of Britain, then scarcely separated from the Continent, there is accepted correlation with the Holsteinian interglacial of western Europe and Poland, and a probable correlation with the Marbellan of the Biscay coast.

Of species no longer growing in the British Isles, besides those already mentioned, we note: from Hoxne, *Vitis vinifera* ssp. *sylvestris* (seed); from Marks Tey, *Acer* cf. *platanoides* (pollen), *Silene* cf. *wahlbergella* (seed); from Hoxne and Marks Tey, *Osmunda* cf. *claytoniana* (spores); from Nechells and Kilbeg, *Eleocharis* cf. *carniolica* (fruits); from Kilbeg, *Eleocharis ovata* (fruits); from Clacton, *Najas minor* (fruit); from Gort, *Astrantia* cf.

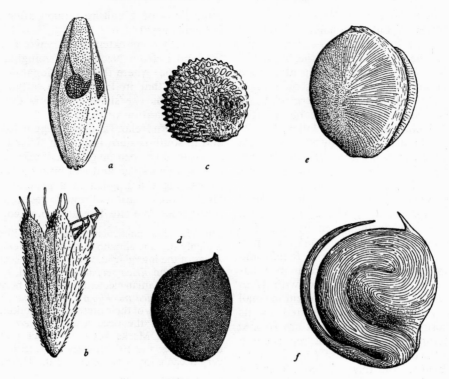

Fig. 150. Fossil material identified and drawn by S. L. Duigan from interglacial deposits in the Summertown Terrace of the Thames Valley at Dorchester near Oxford, samples supplied by Dr K. Sandford: *a*, flower-bud of *Armeria maritima* (×15); *b*, calyx of *Armeria maritima* (×15); *c*, seed of *Cerastium* cf. *arvense* (×22); *d*, achene of *Ranunculus* cf. *hyperboreus* (×22); *e*, fruit of *Potamogeton* cf. *filiformis* (×15); *f*, fruit of cf. *Groenlandia densa* (×15).

minor (pollen), *Brasenia* cf. *purpurea* (pollen), *Lysimachia punctata* (seed), *Nymphoides cordata* (seeds, cf. pollen), see Watts (1971); from most British Hoxnian sites, often in abundance, *Azolla filiculoides* (megaspores, glochidia), see West and account of the species. Several of these species persist in North America, others have a southern European range.

In addition to the taxa now lost from the British flora, the long list of Hoxnian records contains many species of great indicator value, several highly restricted in range today, as for instance *Eriocaulon septangulare*, *Najas flexilis*, *N. marina*, *Euphorbia hyberna*, *E. stricta*, *Viburnum lantana*, *Erica ciliaris*, *Petroselinum segetum*, *Silaum silaus*, *Aphanes microcarpa*, *Linaria vulgaris*, *Valerianella* (three species), *Salix polaris*, *Draba incana*, *Thlaspe alpestre*, *Betula nana*, *Chamaepericlymenium suecicum*. The extensive lists include large numbers of herbs, weeds and ruderals, aquatic and marsh plants, acidicolous plants and a fair number of northern types, especially from sub-stages I and IV.

6. THE IPSWICHIAN INTERGLACIAL

During the course of sewer excavations on the Histon Road, Cambridge, in 1938 and again in 1949 and 1958, botanical evidence was secured pointing to the conclusion that here is an interglacial deposit later than that of Clacton, and very possibly corresponding to the Eemian interglacial of Holland, north Germany and Denmark. Prominent in this evidence is the fact that part of the 1949 and 1958 sequences disclosed a stage of mixed-oak forest trees dominated by the hornbeam (*Carpinus betulus*): the hornbeam reached this order of frequency in western Europe only in Jessen's 'zone f' of the Eemian interglacial.

This early attribution, substantiated by identification also of various macrofossils, was confirmed when West published his very detailed investigation of the interglacial deposits at Bobbitshole, near Ipswich, which then gave its name to the Ipswichian (West, 1958). Subsequently various other partial but important sequences from this interglacial have been described, often in relation to evidence of contemporary sea-level, molluscan and mammalian fauna. These include Wortwell, Stutton, Aveley, Grays and Wretton (East Anglia), Ilford and Trafalgar Square (London), Selsey and Stone (Southern England), and Shortalstown (south-east Ireland). Not even at Bobbitshole was the full interglacial cycle represented, but there is ample evidence to support West's composite pollen diagram reproduced as fig. 151.

The vegetational cycle differs from that of the Hoxnian in several significant features, beginning with the fact that it takes origin in a late-glacial landscape from which

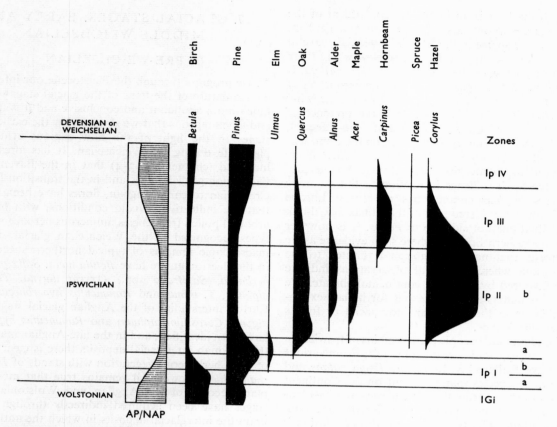

Fig. 151. Composite pollen diagram assembled by West from the evidence of several East Anglian sites, to represent vegetational changes through the sub-stages of the Ipswichian interglacial. High *Carpinus* frequencies in sub-stage III are particularly characteristic of this interglacial.

were absent the abundant growths of the sea-buckthorn, *Hippophaë rhamnoides*, so typical of the preceding interglacial. Sub-stage I is however again dominated by *Betula* and *Pinus*, but in the succeeding sub-stage II *Ulmus* and *Quercus* are prominent first, are later joined by *Corylus* which rises rapidly to high values, and later still by *Alnus* and *Acer*. *Ulmus* is in low frequency only and *Tilia* still scarcer, whilst in contrast with the Hoxnian, *Alnus* (presumably *A. glutinosa*) is abundant only locally in wet sites from which abundant fossil fruits indicate the presence of patches of fen carr. It is also peculiar to this interglacial that pollen of *Acer* occurs in substantial amount in this sub-stage as it does in other west European deposits: it is reasonably associated with the identification of fruits of *Acer monspessulanum*, a species of southern continental range that no longer grows naturally or indeed even fruits in Britain. Sub-stage III is characterized by sharply reduced values of *Alnus*, *Acer*, *Corylus* and continued decrease of *Quercus* as frequencies of *Carpinus* rise to values that sometimes attain 60 per cent of the total tree pollen. This phase of hornbeam dominance also characterizes the equivalent interglacials of western and central Europe, though there it is accompanied and succeeded by *Picea* in fair amount. In sub-stage IV *Betula* and especially *Pinus* constitute virtually all the arboreal pollen in spectra with high NAP/AP ratios.

Whereas the Hoxnian interglacial deposits are very often the infilling of lake basins which present in their pollen catch a generalized picture of the region, those of the Ipswichian are almost all formed in the flood plains of river systems and they reflect the presence of numerous and diverse plant communities, many aquatic and marsh associations, but also varying woodland types responding to local conditions. West (Sparks & West, 1970) calls attention to the likelihood that the high *Corylus* frequencies at Wretton indicate the presence of hazel scrub, and he regards the high frequencies of dry-land herbaceous plants, even in the temperate sub-stages, as shewing the local persistence of open communities, for instance those with abundant grasses, *Artemisia*, and species of *Plantago* (see also Sparks & West, 1968). Both of these phenomena point to instability of the local vegetation and the presence of seral communities equally on the wet and the drier soils. The activities of herds of large mammals, such as hippopotamus and elephant, well attested by the fossil evidence also, are a not unlikely cause making for this environmental diversity.

429

Although the lime is an important constituent of the warm temperate woodlands of the continental Eemian, its pollen is very sparse in the British Ipswichian, save in the extreme south-east, especially at Aveley, where there is enough to suggest the local growth of *Tilia cordata*. It seems likely that the consistent if low representation of *Picea* in this region might likewise be due to trees growing nearby. What is quite incontestable is the presence of extensive forests of *Carpinus betulus* in sub-stage III, representing an importance not attained on the Continental mainland at this time, or indeed in any other interglacial. West (1970*b*) has given much care to the interpretation of the fossil evidence in relation to the extensive present-day *Carpinus* communities of central Europe where it is often considered part of the climax mixed-oak forest in situations not suitable to *Fagus*. It is however a far cry to southern England where such evidence as we have suggests that native hornbeam is limited to the south-east and where ecological observations indicate that it is favoured by the more continental climate. We by no means understand the reasons for its late expansion in this, as in the Flandrian interglacial or for its Ipswichian dominance.

The abundant macroscopic plant remains of Ipswichian fluviatile deposits are of very great interest. From various sites we have fruits of *Najas minor*, *Trapa natans* and *Salvinia natans*, species that now fruit on the Continent but which are absent from the British Isles and its Flandrian records. Not only do these species indicate considerable continentality of climate, but this conclusion is supported by the abundant fruiting of other aquatics still present in the British flora, but restricted in their range and in their dependence on high summer temperatures for effective fruit formation: these include *Lemna minor*, *Hydrocharis morsus-ranae*, *Najas marina*, and *Stratiotes aloides*, which indeed does not now fruit in Britain but of which both pollen and fruit are found fossil. Among the dry-land plants also, the presence of *Acer monspessulanum*, *Buxus sempervirens*, *Thelycrania sanguinea*, *Viburnum lantana*, *Verbena officinalis*, and *Pyracantha coccinea* (native of southern Europe and no longer native here) are indicative of a warmer climate, but one without question far less atlantic than the Hoxnian, a feature very strongly indicated by the far weaker representation of the whole family of Ericaceae and of the genera *Empetrum*, *Lycopodium*, *Selaginella*, *Osmunda* and *Taxus*. It is tempting to attribute the poor representation of both *Hedera and Ilex* in sub-stages III and IV also to this greater continentality of climate. The single Irish interglacial Ipswichian age at Shortalstown contained not only pollen of *Stratiotes* but also a single fruit of *Decodon*, a genus whose one living species is north American, but which has been recorded (as *D. globusus*) from the lower Pleistocene of Holland (Colhoun & Mitchell, 1971).

7. GLACIAL STAGES: EARLY AND MIDDLE WEICHSELIAN

(*a*) PRE-WEICHSELIAN

As we progress through the Pleistocene our information on the nature of the flora of the glacial stages becomes much more abundant and conclusive and it is inevitable and advantageous that we should view the earlier glacial stages in the light of the now copious evidence for plant life during the Weichselian. It has already been indicated (chapter VI: 3–4) that in the Baventian and Beestonian glacial stages and in the transition from late Cromerian to early Anglian, floras have been recorded that are indicative of cold conditions, with high non-arboreal pollen frequencies, numerous herbaceous pollen types recognized in the Weichselian glacial stage and macroscopic remains of typical northern species. Thus in the Beestonian we have *Betula nana*, *Salix polaris*, *S. herbacea*, *Saxifraga* spp., *Oxyria digyna*, *Thalictrum alpinum*, *T. minus* and *Ranunculus hyperboreus*; in the Corton interstadial of the Anglian glacial stage *Oxyria digyna*, *Cerastium alpinum* and *Ranunculus hyperboreus* have been identified; and in the late-Anglian marls below the Hoxnian interglacial deposits there is clear evidence of open herbaceous vegetation with stands of *Hippophaë rhamnoides*. However it remains true that most of our plant records of the Anglian and Wolstonian glacial stages have been acquired indirectly through research upon the interglacial deposits in which the authors have commonly given close attention to the transitional late-glacial beds on which they rest, and the early-glacial beds to which they give place. Very frequently pollen analyses permit these transitions to be reconstructed in vegetational terms, and macroscopic plant identifications usefully link with those of the interglacials themselves: such studies are naturally of especial value where no hiatus in the sequence exists. The situation is quite other in the Weichselian glacial stage for here numerous and prolific plants deposits have been recovered, not only in correlation with the detailed geological history of the glaciation, but co-ordinated by the use of radiocarbon dating which fortunately, at its extreme range, extends backwards almost to the end of the Ipswichian interglacial. The last two thousand years of the glacial stage and the transition to the ensuing Flandrian have been investigated with particular thoroughness. The abundance of evidence for the Weichselian makes it our primary source of information as to vegetational and floristic conditions in this country through glacial stages, and we may only say of the earlier glacial stages, that so far as the sparse evidence shews, they were broadly similar.

We have already outlined in chapter III: 1 & 3 the presumed nature of the soil conditions and the climate encountered by plants during the last glaciation, and in its final stages of retreat. Periglacial conditions prevailed over the southern half of Great Britain for most of this time, and over the larger part of the British Isles during its concluding stages; the nature of such conditions has been again referred to in chapter VI: 2(*a*).

In chapter III: 3 have also been given the bases for sub-division of the Weichselian, the decisive radiocarbon dates and the presumed broad correlation with events of the last glacial stage on the near Continental mainland.

(*b*) EARLY WEICHSELIAN AND CHELFORD

Knowledge of the Early Weichselian in Britain, the time between about 70000 and 50000 years B.P., depends mainly upon three sites, Sidgwick Avenue, Cambridge; Wretton, Norfolk; and Chelford, Cheshire (see p. 16). Some seventy or eighty species of vascular plants were recorded from the Sidgwick Avenue site, representative of a wide range of habitats, although arboreal plants include only *Salix*. They included ruderals and weeds such as *Atriplex hastata*, *Ranunculus repens*, *R. sardous*, *Potentilla anserina* and *Stellaria media*; scree and mountain plants such as *Arenaria ciliata*, *Draba incana* and *Thalictrum alpinum* (cf. Arctic–alpine character); chalk grassland plants of southern range, such as *Helianthemum canum* and *Linum anglicum*; a considerable variety of aquatics including no less than eleven species of *Potamogeton*, and finally a category of halophytes such as *Plantago maritima*, *Potamogeton pectinatus*, *Suaeda maritima*, *Scirpus americanus* and *Armeria maritima*. This variety of habitats is easily associated with the site conditions of braided streams, shifting over the gravels and sands of a terrace solidified by permafrost: locally the gravels were leached and bore acidicoles and shallow pools on the terrace surface became saline through climatic conditions favouring high evaporation.

Similar plant assemblages were recognizable in the organic layers of the terrace of the R. Wissey in Norfolk (Reports of the Sub-department of Quaternary Research for 1961–2 and 1964–5). Abundant skeletal material of large mammalia, particularly bison, reindeer and mammoth, indicate the presence of adequate open pasturage. Detailed publication of this site is especially wanted since it covers the period before, during and after the interstadial episode worked out at Chelford (Simpson & West, 1958) and dated by de Vries at 60800 ± 1500 B.P. This Cheshire site lies within the limits of the Irish Sea ice, whose till overlies it. West's pollen analyses through the laminated organic muds shew strongly preponderant arboreal pollen, mostly of birch, pine and spruce; there were needles, cones and stumps of these conifers so that woodland was undoubtedly present, of a type thought to parallel that growing today in northern Finland where the climate is cool continental with a mean annual temperature of about +2 °C. The evidence of insect remains supports the same conclusion. Though the presence of such genera as *Calluna*, *Empetrum* and *Sphagnum* reflects the acidic quality of the local soils, pollen of some calcicoles such as *Helianthemum* and *Linum catharticum* occurred, and aquatics are well represented.

The more continuous record of this period in western Europe makes it apparent that after this mild interstadial the climate became once more too severe for woodland growth, and thenceforward until the Allerød interstadial tree-growth can only at best have been sparse and localized in the British Isles.

(*c*) MIDDLE WEICHSELIAN

Between about 50000 and 15000 years B.P. fall the plant-bearing deposits of the Middle Weichselian, for which the designation 'Full-glacial' or 'Pleniglacial' seems apt since there is no evidence that it was broken by any climatic phase mild enough to have induced woodland. We may accordingly employ the evidence of the fairly numerous sites jointly to characterize the Full-glacial vegetation and flora, though the range of radio-carbon ages extends from 41900 (Upton Warren) nearly to 13560 years B.P. (Colney Heath). It has to be remembered that all these sites lie in the central and eastern part of England and that all, except that at Dimlington in east Yorkshire, are outside the actual area reached at any time by the Weichselian ice front (fig. 152). It is also important that most of the deposits come from river terrace and outwash deposits with seasonal flooding, cryoturbation and shift of substrate with frequent local changes of the drainage pattern. This involves the likelihood that the identified plant material will include the more durable remains derived from older deposits, and it is virtually certain that records of fruits of *Carpinus betulus* and of *Thelycrania sanguinea* belong to this category, coming indeed from Ipswichian interglacial beds of the region.

The Middle Weichselian plant-bearing beds fall into three major categories. Historically earliest to be investigated were the various deposits of the 'Arctic Plant Bed' in the Lea Valley, north London, where the plant remains were recovered from rafts of silty material embedded deeply in the coarse gravel of the latest of several terraces in the valley; these rafts could only have been embedded in a solidly frozen condition. The bulk of the plant identifications were made by Clement Reid, afterwards revised by Mrs Reid (1949), by J. Allison and, most recently, by F. G. Bell (1969) who also, with C. A. Lambert, revised the identifications made by M. E. J. Chandler on material from the Barnwell Station site in Cambridge (Bell & Dickson, 1971). The geological situation, the presence of large mammalia, such as reindeer and mammoth, of the lemming (*Dicrostonyx henzelii*), the mollusca and mosses all suggested a full glacial context, and this was confirmed subsequently by radiocarbon dates of 28000 for the Lea Valley and 19500 for Barnwell, the latter however from material possibly contaminated by younger intrusive material and therefore of too recent a date.

Next in sequence of investigation were the deposits of the west Midlands investigated from Professor Shotton's department in Birmingham (Upton Warren, Fladbury, Tame Valley and Brandon). Finally there is the group in the area from Leicestershire to Cambridgeshire and Berkshire, investigated most recently by F. G. Bell who applied particularly critical standards to the plant identifications and the evaluation of the evidence they provide (Bell, 1969). They include Syston, Thrapston, Great Billing, Earith and Marlow. Finally at Dimlington in east Yorkshire there is a sparse plant bed, carbon dated at 18 000 B.P., which underlies the tripartite till of the most recent glacial advance in eastern England. At Colney Heath, north London, peat erratics in terrace gravel are so late (13 560 B.P.) that they lie just above the vague boundary between Middle and Late Weichselian (Godwin, 1964).

All these sites display great similarity in floristic and vegetational pattern so that the environment can be characterized in the following terms. Woodland was apparently absent or highly restricted, but there were large communities of dwarf shrubs, particularly the arctic willows, *Salix herbacea*, *S. polaris* and *S. phylicifolia* with the taller *S. viminalis*, *Juniperus* and on acidic soils at Four Ashes, *Empetrum*. There is no evidence of actual presence of *Pinus* for its pollen may have been transported from a distance, but there is some slight evidence for tree birch in small amount. At the same time it is certain that there were extensive areas dominated by terrestrial herbs and one can conjecture the presence of various communities comparable with those of today. Thus at Colney Heath and Earith the 'tall-herb' community (McVean, 1964) has been deduced from the occurrence of such species as *Valeriana officinalis*, *Filipendula ulmaria*, *Mercurialis perennis*, *Urtica dioica* and *Luzula* sp. Wet flushes associated with late snow beds are suggested by species such as *Oxyria digyna*, *Caltha palustris*, *Salix herbacea* and *Saxifraga hypnoides*, and stony flushed habitats by *Potentilla fruticosa*, *Minuartia stricta* and *Primula* cf. *farinosa*. Dry open stony ground was needed for plants like *Cerastium arvense*, *Polygonum aviculare*, *Thalictrum alpinum*, *Diplotaxis tenuifolia*, *Arenaria* sp., *Potentilla anserina*, *Silene furcata*, *Stellaria crassifolia*, *Draba incana*, *Minuartia verna* and poppies of the section *Scapigerum*. Many of these communities were calcicolous with affinities on the one hand with the short open turf of northern limestones, and on the other with the chalk and limestone grasslands of southern Britain.

The openness and instability of soils are indicated especially by the long list of plants that today rank as ruderals and weeds, including such species as *Lapsana communis*, *Polygonum aviculare*, *Urtica dioica*, *Scleranthus*. Bell has emphasized that although this may be taken as evidence for the ancient native status of many ruderals and weeds, some of the species present in the middle Weichselian, such as *Erysimum cheiranthoides* and *Chrysanthemum leucanthemum*, comprise two distinctive taxa of which one might have been the Weichselian plant and the other a subsequent introduction. There is ample evidence again for abundant fen and marsh communities associated with streams and with the shallow pools formed over the permafrost, no doubt transiently in the

changing drainage pattern of the outwash plains and river channels. These extensive herbaceous communities sort well with the prevalence in most of the deposits of the remains of bison, reindeer and mammoth, and of the beetles that lived on their dung. There can be little doubt that grasses and sedges were abundant although specific identifications are still inadequate.

Of particular interest is the recurrent evidence for collections of halophytic plants, both obligate halophytes such as *Glaux maritima*, *Juncus gerardii*, *Suaeda maritima* and *Triglochin maritimum*, and facultative halophytes that include *Armeria maritima*, *Blysmus rufus*, *Atriplex* spp., *Carex arenaria*, *C. maritima*, *Eleocharis uniglumis*, *Najas marina*, *Plantago maritima*, *Zannichellia palustris*. Although it was plausible to invoke local brine springs to explain the presence of such plants in the west Midlands, their widespread abundance is more probably associated with the combination of permafrost in the ground and the big evaporation of a very continental climate, conditions that produce strongly saline habitats and communities today in the Lena valley of Siberia and in northeastern Greenland. Bell, who strongly advances this explanation, also points to the good correlation shewn between numbers of halophytic species and species of southern range. These embrace a component that has affinity with the south-western Russian steppe, such as *Corispermum* (no longer native in Britain), *Linum perenne*, *Helianthemum canum* and *Artemisia* sp. Such species require high summer temperatures such as the continental climate would have provided. This too could be the explanation of the presence, alongside a strong arctic–alpine component, of a long list of 'southern' species that today do not extend in Scandinavia north of the Arctic Circle or indeed are limited to still lower latitudes: such include *Groenlandia densa*, *Mercurialis perennis*, *Onobrychis viciifolia*, *Ranunculus sardous*, *Rumex maritimus*, *Lycopus europaeus*, *Herniaria* and *Damasonium alisma*, the last-named identified originally from the Lea Valley and regarded askance, but now also recorded from Earith and from Wretton. This would also explain the fruiting of species such as *Lemna minor* that responds today by fruiting only in hot British summers. No doubt aspect and other favourable microhabitat conditions encouraged local stands of such species, and also the high solar angle in summer must have heated up shallow bodies of water in the way suggested by Wesenberg-Lund, but so many habitat types are concerned, including especially the saline ones, that only a general climatic explanation suffices, and that one which permitted the co-existence of the species both of northern and of southern range to a degree that is no longer realized. Bell (1969) deduces that at Earith, the most continental of the British sites, whilst the mid-Weichselian summers were only a little cooler than those of today (perhaps a mean July temperature of 12–16 °C), the winters were much cooler (possibly with a mean for January of − 15 °C), whilst the mean annual temperature was just below zero, as would be required to maintain permafrost.

An explanation in these terms varies somewhat from the concepts of Shotton and Coope which take particular account of the coleopteran fossil evidence, and which refer the west Midland sites within the period of 50000 to 25000 years B.P. to an 'Upton Warren Interstadial Complex' (see fig. 12). They suppose that climatic conditions were such as to have been suitable in themselves to the growth of a boreal woodland type and explain the absence of trees by the intense grazing pressure of herbivore mammalia, an explanation without much appeal to ecologists however. With this argument in doubt, it is evident that the interstadial character of this period is less pronounced than that of Chelford or of the Allerød time that follows, although it corresponds with Dutch episodes also regarded as interstadia without having developed boreal woodlands.

It should of course be realized that the climatic conditions we now postulate for the Middle Weichselian do indeed provide *summer* temperatures appropriate in themselves to tree growth, but accompanied by other features strongly adverse such as very cold winters, severe wind exposure, severe spring thaw and floods, and high unstable soils. It seems impossible to match at the present day assemblages of plant communities comparable with all those of the British Middle Weichselian but Bell has drawn attention to the descriptions by Cajander (1903) of vegetation in Siberian river flood-plains, whilst great glacial outwash complexes of braided channels such as the Canterbury Plains of New Zealand illustrate the local variety of edaphic conditions, though not the climate, that many of our sites provided.

To a very large extent the Middle Weichselian flora comprises species within the present flora of Britain. The early identifications from Barnwell and the Lea Valley were subjected to later revision, especially by Bell and Dickson (1971). This has meant the rejection of *Linum praecursor* Reid for seeds now referred to *Linum perenne* agg. or in some instances to *Linum anglicum*, and the rejection of *Silene coelata* Reid for seeds now recognized as those of *S. vulgaris* agg. and a capsule which was in fact a very decayed (and derived) capsule of *Fagus*. There are left no records of plants presumed to be extinct species. Certain non-British species such as *Arenaria biflora*, *A. arctica*, *Gentiana cruciata*, *Potentilla* cf. *nivalis*, *P. nivea* and *Armeria arctica* have been rejected, but there remains the group of non-British taxa from these sites of *Alyssum* type, *Papaver* section *Scapigerum*, *Ranunculus aconitifolius*, *R. hyperboreus*, and cf. *Stellaria crassifolia*. It is not surprising that indeed there should have been losses from the categories of alpine, arctic and arctic–alpine categories with the passing of the cold conditions of the Middle Weichselian, and there have also been lost *Erysimum cheiranthoides*, *Salix polaris*, *Carex capitata*, *Pedicularis lanata*, *P. hirsuta*, *Silene furcata* and *Stellaria crassifolia* all of northern range. In addition however there have been lost *Corispermum* sp., a genus mainly of southern–continental range, and *Dianthus carthusianorum*, a species of central and southern Europe, and it seems possible that these have been affected both by increased competition and by the atlanticity of the succeeding Flandrian. These are issues closely bound up with analysis of the phytogeographic groupings of the Weichselian and Flandrian floras that are dealt with at some length in the following section 9, on 'Phytogeographic considerations'.

British botanists have long sought knowledge of the nature of our vegetation and flora south of the ice-sheets throughout the last glaciation. The sites so far recorded go a great distance to meet this requirement although neither the far west nor extreme south of the British Isles (nor of course the north) is represented. It is scarcely reasonable, in my view, to see each organic deposit as a minor mild interstadial after which the landscape was denuded of vegetation and re-invaded for the next. It could strictly be regarded as unproven that the last glacial advance, post-Dimlington, had not extinguished all or most of the Middle Weichselian flora, but the general similarity of the various sites to one another and to those of Colney Heath, Loch Droma and Roberthill, that post-date the latest glacial episode, as well as the substantial climatic gradient we can assume to the west and south, persuades one that we indeed now have an effective first view of the British flora that persisted through the glacial stage.

8. GLACIAL STAGES: LATE WEICHSELIAN

In order to see the Late Weichselian conditions of the British Isles in relation to those of the nearby Continent we reproduce a short correlation table for three West European areas (table 35).

In the provisional maps given by Firbas (1949) for central Europe it will be seen that during the Allerød period Denmark lay within the birch wood region, that pine predominated south of a line approximately passing through Berlin and Koblenz, whilst between was a region with both birch and pine, the latter more frequent towards the south. In the ensuing zone III, however, the whole north European plain is shown again in the 'waldarm' state which it had during zone I.

This is a picture that has not been substantially altered by later investigation.

It was the work of Jessen in Ireland that firmly established the existence of Late Weichselian deposits in the British Isles comparable in character with, and of the same age as, those known from Denmark. He first made this known in his joint paper with Farrington (Jessen & Farrington, 1938) upon the famous site at Ballybetagh where remains of the giant Irish deer were recovered from what were clearly shown to be deposits of the Allerød period. By the time of publication of Jessen's definitive work upon the Late Quaternary deposits and flora history of Ireland, he and his pupil G. F. Mitchell had established the presence of Late Weichselian deposits in many parts of Ireland.

TABLE 35

Zone	Mid-Jutland (Iversen)	Holstein (Schütrumpf)	Holland (van der Hammen)
	EARLY FLANDRIAN		
IV	Birch forest	Pine forest and birch–pine forest	Birch–pine forest
	LATE WEICHSELIAN		
III	Tundra or park tundra	Park tundra	Park tundra
IIb} IIa}	Park tundra	Birch–pine forest	{ Birch–pine forest { Birch forest
Ic Ib	Tundra } Park tundra }	Park tundra	{ Tundra { Park tundra
Ia	Tundra	Tundra	Tundra (?treeless)

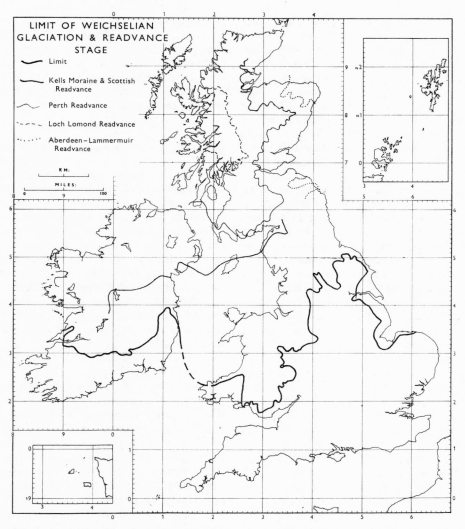

Fig. 152. Map of the British Isles shewing maximum extension of the Weichselian ice and of its major halt or readvance positions. Correlations, especially across the Irish Sea, are often tentative: the Loch Lomond readvance is taken to correspond with pollen zone III of the Late Weichselian.

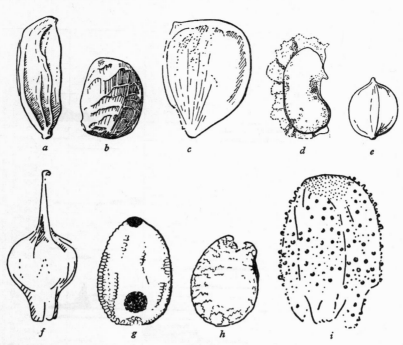

Fig. 153. Fruits and seeds from Late Weichselian layers at Hawks Tor kaolin pit, Bodmin Moor, Cornwall: *a, Thalictrum alpinum* L. (×13); *b, Batrachium* spp. (×13); *c, Ranunculus* sp. (×13); *d, Callitriche hermaphroditica* L. (×13); *e, Rumex acetosella* L. (*non sensu lato*) (×13); *f, Sparganium angustifolium* Mich. (×10); *g, Veronica aquatica* (*sensu lato*) (×50); *h, Sagina subulata* (Sw.) C. Presl. or *S. saginoides* (L.) Karst. (×50); *i, Saxifraga* cf. *granulata* L. (×50).

Investigations of deposits of this age began later in Great Britain, but by now the characteristic tripartite sequence of the Allerød climatic oscillation has been recognized and associated with floristic and vegetational studies at sites well distributed throughout the British Isles (fig. 156). In all of them the mild conditions of the Allerød itself (pollen zone II) had allowed the development of soil stability and sufficiently complete vegetational cover to cause the valley and lake deposits to have a largely organic character. This contrasts strongly with the primarily mineral character of the deposits of zones I and III, the equivalents of the older and younger *Dryas* deposits of the near continent, which reflected the fresh, unstable soils, incomplete vegetational cover and susceptibility to freeze–thaw, cryoturbation, and spring flooding typical of periglacial conditions. Sometimes these mineral deposits exhibited solifluxion characteristics, as at Ballybetagh (Jessen & Farrington, 1938) and Hawks Tor (Conolly, Godwin & Megaw, 1950), and sometimes, especially in deeper lakes, they were represented by clay or silt, perhaps with annual lamination as in Windermere (Pennington, 1947). At some sites, such as some in East Anglia, oscillation was unregistered by changed lithology, perhaps because of gentle topography and highly porous soils or smaller precipitation (Godwin, 1968b).

Whereas all the earlier investigations, and notably those of Jessen in Ireland, depended heavily upon the identification of plant macro-fossils, those carried out later took advantage of the greatly improved quality of pollen identification, more especially after Iversen had shewn, in his classic papers on the late-glacial flora of Denmark (Iversen, 1947a, 1947b, 1954) how effective the recognition of herbaceous pollen types could be. With the registration of a remarkable range of taxa in the pollen spectra, many, such as *Koenigia, Centaurea cyanus* and *Polemonium*, unrepresented at all by macro-fossils, and with large counts from samples at close intervals, it became far easier to define the climatic oscillation in vegetational terms. The triple zonation system that had initially been based to a substantial extent upon the stratigraphic sequence, was increasingly qualified by the palynological evidence and there have been suggestions by various authors (Walker, 1966; Smith, 1961a; Watts, 1963) for redefining the zone boundaries entirely or largely in vegetational terms. For the present there is too little agreement for such proposals to be a basis for a survey such as the present one, which however can safely rest upon the prime fact of the thermal oscillation with maximum in zone II and minimum in I and III, wherever one may ultimately place the separating boundaries.

The application of radiocarbon dating to Late Weichselian deposits quickly confirmed the synchroneity of the phenomena indicative of the main oscillation both within the British Isles and over western Europe as a whole (Godwin & Willis, 1959a). Thus the mild Allerød phase was seen to fall approximately between 12 000 and 10 800 B.P. and the succeeding cold phase between 10 800 and 10 300 B.P. No criteria have been agreed for the

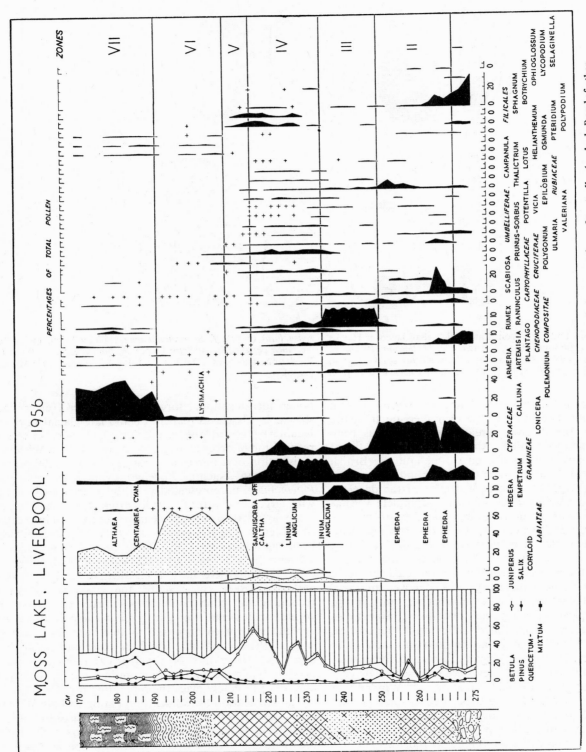

Fig. 154. Pollen diagram through deposits at Moss Lake, Liverpool. They comprise Late Weichselian and early Flandrian organic lake muds (interrupted by a minerogenic layer in zone III), overlaid by ombrogenous peat in Flandrian zones VI and VII. On the left are shewn the relative proportions of tree pollen (unshaded) and of other dry-land plants (shaded); on the right the shrub pollen (dotted) and dry-land herbaceous plants (solid), expressed as percentages of the total pollen of dry-land plants. (After Godwin, 1959.)

opening of zone I; the arbitrary choice of 15000 B.P. places Colney Heath (13560 ± 210), Loch Droma (12420 ± 220) and Roberthill (12940 ± 200) within the Late Weichselian zone I rather than the Middle Weichselian.

At the Danish site of Bølling Sø, Iversen found evidence for a temperate climate phase of less significance than the Allerød, and preceding it. This minor oscillation has been identified at a number of Continental sites and has been radiocarbon dated between 12100 and 12500 B.P. In Britain it has been identified mainly by a temporary increase in the pollen of tree birch in the generally tree-less zone I. It was recognized at Tadcaster by Bartley (1962) who used it to sub-divide Late Weichselian zone I. Evidence for it is sparse and conjectural in the British Isles, though it may be represented at Star Carr (Walker & Godwin, in Clark, 1954) and Hawes Water (Oldfield, 1960).

More recently radiocarbon dating of a close series of samples through lacustrine Late Weichselian deposits has allowed the presentation of pollen-analytic information as absolute rates of pollen incorporation per annum, and this has yielded at Blelham Bog, Westmorland considerably improved characterization of the vegetational history, especially by the extremely low pollen productivity of the *Rumex*-grass communities of zone I in contrast with that of the ensuing *Juniper* and *Betula* dominated communities of zone II (Pennington & Bonny, 1970).

Although the climatic progression and retrogression of the Late Weichselian expresses itself over the whole of the British Isles, it is repeatedly emphasized, as for example by Birks (1973) and Walker (1956) that extreme topographic and edaphic diversity prevailed in each zone and entailed the co-existence of a great range of plant communities. We are still in the earliest stages of attempting the identification of these, although Birks has established that in the Late Weichselian of Skye not only was there a vegetational diversity like that of today, but that many present-day communities of the island were already present. When, for lack of more precision, one employs the words 'tundra' or 'park-tundra' to characterize zonal vegetation it must be understood that a great variety of plant communities is circumscribed by the term. There were also strong vegetational responses to the climatic gradients from north to south and east to west across the British Isles. It is upon this pattern of strong spatial variation that we have to impose the effects of temporal change, especially those caused by the climatic shifts.

(i) WOODLAND

Commonest of trees in the Late Weichselian were the tree birches. In zones I and III they were restricted to more favoured sites but in zone II, over a large central part of the British Isles, there was more or less closed birch woodland with *Betula pubescens* the commoner species, and *B. pendula* less frequent and more typical of the north. Macroscopic and pollen-analytic evidence combine

TABLE 36. *Late Weichselian records of woodland herbs*

Ajuga reptans	cf. *Moehringia trinervia*
Circaea lutetiana	*Oxalis acetosella*
Galium aparine	*Potentilla sterilis*
Linnaea borealis	*Pteridium aquilinum*
Mercurialis perennis	*Stellaria holostea*

to shew that in zone II through Lincoln and Norfolk to Kent and Hampshire and possibly even in Aberdeen, *Pinus sylvestris* grew in substantial amount, the continental eastern side of Britain thus representing the western margin of the mainland pine–birch region that indeed then no doubt extended across the area that is now the North Sea. Further north and at higher altitudes even in this zone the birches were limited to copses in sheltered situations and they were barely present in the exposed Orkneys or the very oceanic west of Ireland. *Populus* (probably *P. tremula*) and *Sorbus* (probably *S. aucuparia*) had a minor role. It is not surprising that the list of woodland ground flora plants (table 36) should be a short one, or that only one of these, *Ajuga reptans*, should have a Middle Weichselian record.

(ii) AQUATIC AND MARSH HABITATS

The list of species falling within the category of marsh or aquatic plants (tables 37 and 38) is strikingly long. This is no doubt due in part to the preference given to such species by the conditions of preservation, but partly also to the wide geographical range of plants within this category and to the abundance of open lakes and fens in the fresh and uneven landscape left behind by retreating ice-sheets and thawing soils. In considering the lists given (tables 36–8) it has to be borne in mind that the Middle Weichselian sites are almost all in the region of sedimentary rocks and till deposits of such fresh and often calcareous quality that the ponds and streams must have been preponderantly eutrophic. Likewise Middle Weichselian conditions allowed little scope for hydroseral development so that fens and marshes did not have extensive development. In the Late Weichselian by contrast, sites have been investigated over the whole country, many are in montane regions of igneous or metamorphic rock yielding shallow, acidic soils susceptible to leaching and often within regions of high rainfall. Accordingly oligotrophic as well as eutrophic lakes abounded. The stable soils and continuous vegetation cover of zone II allowed a long period for development of fen and marsh peats and in many regions there are indications of the beginnings of acidic mire or bog formation. Thus we now have listed *Eriophorum angustifolium*, *E. vaginatum*, *Scirpus cespitosus*, *Rhyncospora alba*, numerous carices such as *C. curta*, *C. dioica*, *C. echinata*, *C. hostiana*, *C. nigra* and *C. pauciflora* which indicate acidic soils, as well as the ericoid plants like *Calluna vulgaris*, *Erica tetralix*, *Vaccinium oxycoccus* and *Andromeda polifolia*.

TABLE 37. *Middle and Late Weichselian records for plants of aquatic habitats*

(M = Middle, L = Late Weichselian)

Alisma plantago-aquatica L	*M. verticillatum* M, L	*P. praelongus* M, L
Apium nodiflorum M	*Najas flexilis* M, L	*P. pusillus* M, L
Butomus umbellatus M	*N. marina* M	*P. trichoides* M, L
Callitriche hermaphroditica L	*Nuphar lutea* M, L	*P. zizii* M, L
C. intermedia L	*Nymphaea alba* M, L	*Ranunculus* subgenus *Batrachium* M, L
C. obtusangula L	*N.* cf. *candida* L	*R. aquatilis* M, L
C. stagnalis L	*Potamogeton acutifolius* M, L	*R. hederaceus* M
Ceratophyllum demersum L	*P. alpinus* M, L	*Rorippa microphylla* L
Damasonium alisma M	*P. berchtoldii* M, L	*Sagittaria sagittifolia* M
Equisetum fluviatile L	*P. coloratus* M	*Scirpus fluitans* L
Groenlandia densa M, L	*P. crispus* M, L	*S. lacustris* L
Hippuris vulgaris M, L	*P. filiformis* M, L	*S. tabernaemontani* L
Isoetes echinospora L	*P. friesii* L	*Sparganium angustifolium* M, L
I. lacustris L	*P. gramineus* M, L	*S. emersum* M
Lemna cf. *minor* L	*P. lucens* L	*S. erectum* M, L
L. cf. *trisulca* L	*P. natans* M, L	*S. minimum* M, L
Littorella uniflora M, L	*P. obtusifolius* M, L	*Subularia aquatica* L
Montia fontana M, L	*P. pectinatus* M	*Typha angustifolia* L
Myriophyllum alterniflorum M, L	*P. perfoliatus* M	*T. latifolia* M, L
M. spicatum M, L	*P. polygonifolius* M, L	*Zannichellia palustris* M, L

TABLE 38. *Middle and Late Weichselian records for herbaceous plants of marshes and fens*

(M = Middle, L = Late Weichselian)

Barbarea vulgaris L	*C. rostrata* M, L	*Parnassia palustris* M, L
Berula erecta L	*C. spicata* L	*Pedicularis palustris* L
Blysmus compressus L	*Cicuta virosa* M, L	*Pilularia globulifera* L
Caltha palustris M, L	*Cladium mariscus* M, L	*Polygonum hydropiper* M
Cardamine amara L	*Elatine hexandra* L	*P. minus* L
C. pratensis L	*Eleocharis multicaulis* L	*Potentilla palustris* M, L
Carex aquatilis ?M, L	*E. palustris* M, L	*Ranunculus flammula* M, L
C. curta L	*E. quinqueflora* M, L	*R. lingua* M, L
C. diandra L	*E. uniglumis* M, L	*R. sceleratus* M, L
C. dioica L	*Epilobium parviflorum* L	*Rhyncospora alba* L
C. distans M	*E. palustre* L	*Rorippa islandica* M, L
C. echinata L	*Eriophorum angustifolium* L	*Rumex aquaticus* L
C. flacca L	*E. vaginatum* L	*Scirpus × carinatus* L
C. flava M, L	*Filipendula ulmaria* M, L	*S. cespitosus* L
C. hostiana L	*Galium uliginosum* L	*Stellaria alsine* M, L
C. lachenalii M	*Geum rivale* L	*S. palustris*
C. laevigata L	*Juncus acutiflorus* L	*Succisa pratensis* M
C. lepidocarpa L	*J. bulbosus* L	*Thalictrum flavum* M, L
C. nigra M, L	*J. effusus* L	*Thelypteris dryopteris* L
C. ?pallescens M	*J. squarrosus* L	*Valeriana dioica* M, L
C. panicea M, L	*Lychnis flos-cuculi* L	*V. officinalis* L
C. paniculata L	*Lycopus europaeus* M, L	*Veronica anagallis-aquatica* L
C. pauciflora L	*Mentha* cf. *aquatica* L	*V. scutellata* L
C. pseudocyperus ?M, L	*Menyanthes trifoliata* M, L	*Viola palustris* M, L
C. pulicaris M, L	*Myosoton aquaticum* M	
C. riparia M, L	*Ophioglossum vulgatum* L	

This increased diversity of ecological range in the Late as compared with the Middle Weichselian may explain why there is so great an increase in the numbers of listed herbs of marsh and fen in the later period (33 to 67). The list of aquatic plants changes far less (39 to 50).

(iii) SHRUB AND HEATH

Shrubs undoubtedly played a great role in all three Late Weichselian zones. In the west and north, *Salix herbacea* was especially abundant in zones I and III, but in southern and eastern England this may have been in part replaced by *Betula nana*. *Salix reticulata*, *S. polaris* and *S. repens*

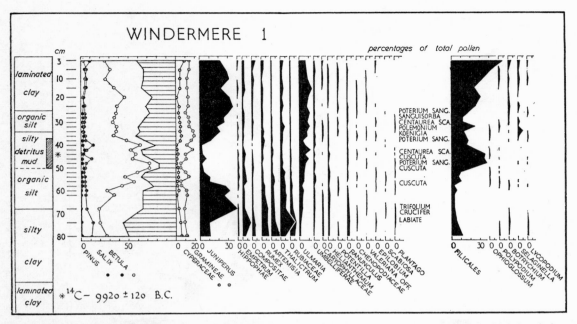

Fig. 155. Pollen diagram through a core in the bed of Windermere. The organic silts between laminated clays were formed in the Allerød (zone II) period, as is confirmed by the radiocarbon date. In the left-hand block tree and shrub pollen is unshaded, herbaceous pollen types shaded. Before and after the high *Betula* frequencies of zone II, *Juniperus* was extremely abundant and there is a great variety of plants of open habitats throughout these Late Weichselian deposits (anal. R. Andrew: Godwin, 1960a).

were also present and taller willows such as *S. phylicifolia*, *S. cinerea* (or *S. atrocinerea*) and, more remarkably, *S. viminalis* occurred often in large amount as in the Middle Weichselian. The large representation of willows in the vegetational cover is conveyed by the large proportion of *Salix* in the total pollen count, particularly in zones I and III, though some sub-arctic willows of the Middle Weichselian have apparently disappeared. *Hippophaë rhamnoides* occurred here as in Continental deposits of the same age, although with low pollen frequency.

The crowberry, *Empetrum nigrum*, was an important and widespread shrub, sensitive both to soil conditions, strongly preferring acid soils, and to climatic conditions. In the phrase of A. P. Brown (1971) it responds to 'a decreasing oceanicity gradient from west to east...operating in association with an increasing density of forest and woodland from north to south'. His map displays far lower *Empetrum* pollen frequencies in southern and central England (as in Denmark and Holland) than those of the north and west, where, especially in zone II, they have been held to indicate extensive development of *Empetrum* heath (Jessen, 1949) in place of the birch woodland in less oceanic regions. Brown has employed the Late Weichselian *Empetrum* pollen frequencies to indicate progressive decrease in continentality in zones I and II and of unsuspected continentality in zones I and III in south-western England, but the interpretation is made difficult by the susceptibility of the crowberry to the tree competition that developed in zone II. Other acidicoles, such as *Andromeda polifolia*, *Vaccinium* spp., *Arctosta-*

phylos uva-ursi, *Calluna vulgaris* and *Erica tetralix* were also present, though in much lower frequencies than *Empetrum*.

Juniperus communis played a role of considerable importance in the Late Weichselian. In zone I it was frequent north of the Wash up to northern Scotland, it more or less doubled in frequency in zone II and suffered much recession in zone III save for special sites in the Lake District and north-eastern Ireland. It was almost absent through the south and east. It seldom behaved as a continuous community dominant, forming a transient stage of scrub between the climatic harshness of zone I and the competition of woodland in zone II. At some sites the sequence is represented by successive maxima of pollen of *Empetrum*, *Juniper* and *Betula*: in Skye the *Juniper* expansion is dated at 12250 and that of birch at 11800 B.P. Birks (1973) comments on the sparsity of any modern analogues of the *Juniperus*, *Betula nana* sub-zone vegetation indicated in the Late Weichselian of Skye.

In a different ecological category from the shrubs already mentioned is *Helianthemum*, pollen of which is very consistently present, though in low frequency, in Late Weichselian deposits. Macroscopic identifications, like those of the Middle Weichselian represent *H. canum*, but both this species and *H. chamaecistus* are calcicolous and typical of open communities. Pollen of *Ephedra* cf. *distachya* has often been found in low frequency but always in a context that does not exclude long-distance wind transport.

439

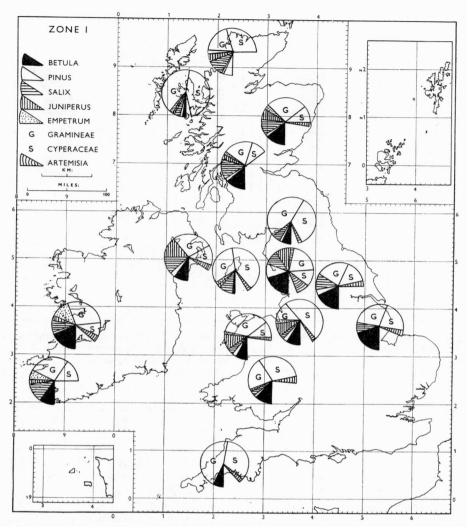

Fig. 156. Vegetational composition in the British Isles through zones I to III of the Late Weichselian, represented at each site by a sectorial diagram displaying the proportions of the major pollen types. The increase in zone II and subsequent recession of the tree component (chiefly *Betula*) is very evident: grass, sedges and *Artemisia* were abundant throughout.

(iv) GRASSLAND

Wherever continuous pollen analyses have been made through the full zone sequence of the Late-glacial, it has always been apparent that a large change in the ratio of tree pollen to non-tree pollen marks the transition from one zone to the next. In zones I and III the non-tree pollen is always relatively high, whilst in zone II it sinks sharply, as also at the transition to zone IV. The high non-tree pollen values are contributed to very largely by the grasses and sedges, but because the organic deposits have necessarily formed in wet situations it is hard to know how far the sedge component may be due to local over-representation. The list of identified sedges is considerable, but the grasses are almost unknown despite their great importance in the more or less open communities of zones I and III, and indeed their apparent dominance in

zone II on the limestone soils in the oceanic climate of Co. Mayo, and in upland and northern situations where trees were sparse. From the Isle of Man however there are recorded *Festuca ovina* or *F. rubra*, *Poa* cf. *pratensis*, *Poa* cf. *trivialis*, as in the Middle Weichselian at Earith there were *Festuca rubra* and *Poa* cf. *trivialis* (Bell, 1970a). Pollen of *Agropyron junceiforme* was recorded at Moss Lake and may indicate the presence of dunes.

(v) OTHER TERRESTRIAL COMMUNITIES

The lists of macroscopic plant fossils alone or together with the pollen evidence have encouraged many investigators to identify numerous further communities of characteristic habitat preferences, some of which we have already ascribed to the Middle Weichselian. The halophytic communities of that time persist but are less

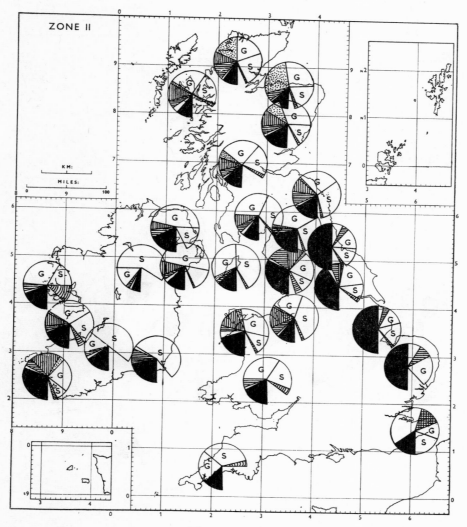

Fig. 156 (cont.).

numerous and extreme: at the Isle of Man however eight species are referred to such vegetation types (C. A. & J. H. Dickson & Mitchell, 1970). The same authors suggest that calcareous sand-dune communities were also present, citing a considerable list of species that occur in this habitat though of them only *Juncus balticus* seems exclusive to it. These authors again suggest the presence of communities of temporarily inundated gravel, sand and clay flats, attributing to them such plants as *Pseudoblitum*, *Rorippa islandica*, *Sagina* sp., *Potentilla anserina*, *Rumex acetosella*, *Juncus bulbosus*, *Ranunculus hyperboreus* and the very characteristic *Koenigia islandica*, so dwarf that its pollen is almost wholly incorporated in flushing. Aggregates of this kind have been envisaged at several Late Weichselian sites including Mapastown (Mitchell, 1953), Moss Lake (Godwin, 1959) and Loch Droma (Kirk & Godwin, 1962–3). 'Tall herb' communities, present already in the Middle Weichselian, continue

to be recognizable with an enlarged species list that includes *Polemonium caeruleum*, represented only by its pollen. Comparable communities remain today on those ledges of deeper soil in the Ben Lawers range where sheep grazing is impossible, and there flourish communities of tall herbs including:

Trollius europaeus	*Filipendula ulmaria*
Geranium sylvaticum	*Hieracium* spp.
Silene dioica	*Rumex acetosa*
Rubus saxatilis	*Oxyria digyna*
Angelica sylvestris	*Alchemilla alpina*
Succisa pratensis	*Luzula sylvatica*
Cirsium heterophyllum	*Carex atrata*
Sedum rosea	*Poa alpina*

Particularly in northern and montane sites communities have been described that are typical of late snow beds: they are represented by such species as *Salix herbacea* and *Rumex acetosa*, *Cryptogramma crispa*, *Polygonum viviparum*, *Athyrium alpestre* and *Lycopodium alpinum*,

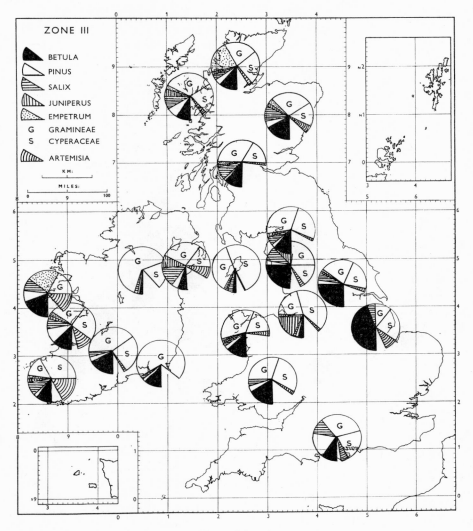

Fig. 156 (*cont.*).

together with a very diagnostic assemblage of bryophytes (Dickson, 1973). Conversely upon occasion the infrequency of long and deep snow cover is indicated by *Juniper* and *Empetrum*, and more specifically by *Artemisia*.

It is also in northern and high altitude sites that one encounters a montane heath or grassland assemblage characterized by high frequencies of spores of the dwarf *Pteridophyta*, *Lycopodium* spp., *Selaginella selaginoides*, *Botrychium lunaria*, and *Ophioglossum* set in a matrix of sedges and grasses. In less exposed situations the grasslands of zones I and III are characterized as assemblages of 'Rumex–grass–willow', or 'Rumex–grass–Artemisia', a very tentative first approach to identification.

Springs and flushes add to the communities that contributed both to the pollen and to macroscopic remains representing the mosaic of terrestrial communities we glimpse as having been present.

(vi) RUDERALS AND WEEDS

The very long list of species that we now regard as ruderals or weeds (table 39) indicates the prevalence of open conditions, bare soil surfaces and freedom from competition in the Middle and Late Weichselian alike.

These identifications make it quite evident that many weed species hitherto thought to be introductions to the British Isles by Neolithic and later agriculturists are of much older standing. Some are especially remarkable: such for instance, is *Linaria vulgaris*, identified by its unmistakable winged seeds; *Centaurea cyanus*, pollen grains of which have been found at several sites in Late Weichselian deposits in western Europe; *Galeopsis tetrahit* and *Pastinaca sativa*. It is indeed true that there is nothing here which disproves the thesis that they may have been subsequently extinguished and later reintroduced, but in several instances repeated discoveries of

TABLE 39. *Middle and Late Weichselian records of weeds and ruderals*

(M = Middle, L = Late Weichselian)

Achillea millefolium M, L	*Crepis capillaris* M	*Polygonum aviculare* M, L
Acinos arvensis L	cf. *Descurania sophia* L	*P. lapathifolium* M
Aphanes arvensis M	*Diplotaxis tenuifolius* M	*P. nodosum* L
Arabis hirsuta L	*Erysimum cheiranthoides* M	*P. persicaria* L
Arenaria leptocladus L	*Galeopsis tetrahit* L	*Potentilla anserina* M, L
A. serpyllifolia L	*Galium aparine* L	*P. reptans* M
Atriplex hastata M, L	*Glechoma hederacea* M	*Ranunculus acris* M, L
A. glabriuscula L	*Heracleum sphondylium* L	*R. bulbosus* M, L
A. patula L	*Lapsana communis* M	*R. repens* M, L
Bunium bulbocastanum L	*Leontodon autumnalis* M, L	*Rumex acetosa* M, L
Capsella bursa-pastoris M	*L. hispidus* L	*R. acetosella* M, L
Carduus acanthoides L	*Linaria vulgaris* L	*R. crispus* M, L
C. nutans M, L	*Linum anglicum* L	*R. obtusifolius* L
Centaurea cyanus L	*L. catharticum* L	*Sagina procumbens* L
C. nigra L	*Lotus corniculatus* L	*Scleranthus annuus* M, L
C. scabiosa L	*Lychnis viscaria* L	*Silene vulgaris* M, L
Cerastium arvense M, L	*Malva sylvestris* L	*Sonchus arvensis* L
C. holosteoides L	*Odontites verna* L	*Spergula arvensis* L
Chaenorrhinum minus L	*Onobrychis viciifolia* M, L	*Stellaria media* M, L
Chamaenerion angustifolium L	*Origanum vulgare* L	*Taraxacum officinale* M, L
Chenopodium album M, L	*Pastinaca sativa* M	*Tripleurospermum maritimum* M
C. rubrum M, L	*Plantago coronopus* L	*Urtica dioica* M, L
Cirsium arvense L	*P. lanceolata* L	*Valerianella dentata* M
C. vulgare L	*P. major* L	*Vicia sepium* L
Clinopodium vulgare L	*P. media* L	*V. sylvatica* M
Corispermum sp. M	*Poa annua* M	*Viola tricolor* M

such species at all stages of the Flandrian render such an hypothesis unnecessary.

These positively dated records bear out in striking manner the opinion expressed by Salisbury as long ago as 1932 when he wrote of 'the considerable area of morainic deposits that must have fringed the European ice front throughout the Pleistocene glaciations', that 'it is not improbable that this was the primary home of species which today are mainly, if not exclusively, associated with the artificial conditions of cultivated and disturbed soil'.

It is apparent that these species, suppressed and restricted by the closed forests which covered the lowlands, and the peat mires which covered much of the uplands, survived the greater part of the Flandrian period on cliffs, beaches, river banks and screes until the destruction of the native forest by prehistoric man gave them opportunity to expand once more. That such species can have a place in the British flora irrespective of human introduction is clearly shown by the numerous ruderals and weeds represented in the river channel deposits of Ipswichian and other interglacials, for although man was present then his numbers and influence were doubtless small.

scree slopes and mountain ledges at the present day. Like the category of present-day ruderals and weeds, these species of the mountains and sub-arctic also demand open habitats and freedom from competition. Whereas at the present day, however, the two categories tend to be altitudinally separate, we must visualize, throughout the lowlands of the British Isles in Middle and Late Weichselian time, a vegetation type in which both were prevalent. It is not to be expected that equivalent communities should necessarily exist today, especially since among the weeds there are many of distinctly southern distribution and limited altitudinal range, as emerges from later consideration of the history of phytogeographic patterns (section 9).

As we have already emphasized, one cannot help but form the picture of an open landscape where changing aspect, slope, drainage and geological formation led to much local variation in vegetation; this remained sparse and open on thin soils, northern slopes, exposed situations and sites affected by solifluxion, but it became rich and closed on colluvial soils with good drainage, and in the fens and marshes round the rather well-stocked pools.

(vii) OPEN MONTANE OR SUB-ARCTIC HABITATS

It is quite apparent from the long and striking list of records (table 40) that in Late Weichselian times the lowlands of the British Isles carried a type of vegetation resembling in some degree that of our sub-alpine meadows,

(viii) CONTRACTION OF RANGE AND EXTINCTION

We have noted in the preceding section that since the Middle Weichselian a number of taxa have become extinct in the British Isles though surviving on the Continental mainland. Since the Late Weichselian the losses

TABLE 40. *Middle and Late Weichselian records for plants of open montane or sub-arctic habitats*

(M = Middle, L = Late Weichselian)

Arabis hirsuta L	*Koenigia islandica* M, L	*S. phylicifolia* M, L
A. cf. stricta L	*Loesleria procumbens* L	*S. polaris* M, L
Arctostaphylos uva-ursi M	*Luzula spicata* M, L	*S. reticulata* M, L
Arenaria ciliata M, L	*Lychnis alpina* M, L	*Saussurea alpina* L
A. gothica M	*Lycopodium alpinum* L	*Saxifraga aizoides* L
Armeria maritima M, L	*L. annotinum* L	*S. cespitosa* L
Artemisia norvegica L	*L. clavatum* M, L	*S. hirculus* L
Astragalus alpinus L	*L. selago* L	*S. hirsuta* M
Betula nana M, L	*Minuartia rubella* M	*S. hypnoides* M, L
Cardaminopsis petraea L	*M. stricta* ?M, L	*S. nivalis* L
Carex atrata M	*M. verna* M	*S. oppositifolia* M, L
C. bigelowii M	*Oxyria digyna* M, L	*S. rosacea* L
C. capillaris M	*Papaver alpinum* (sect. *Scapigerum*) M	*S. stellaris* L
C. capitata M	*Pedicularis lanata* M	*S. subgenus dactyloides* L
C. lachenalii M	*P. hirsuta* M	*S. subgenus nephrophyllum* L
C. norvegica L	*Polygonum viviparum* M, L	*S. tridactyloides* M
Cerastium alpinum M, L	*Potentilla crantzii* M, ?L	*Sedum rosea* L
C. cerastioides L	*Primula farinosa* M	*Selaginella selaginoides* M, L
cf. *Chamaepericlymenum suecicum* M	*P. scotica* M	*Sibbaldia procumbens* L
Cherleria sedoides M	*Ranunculus aconitifolius* M	*Silene acaulis* M, L
Cochlearia officinalis agg. M, L	*R. hyperboreus* M, L	*S. furcata* M
cf. *C. danica* M	*Rubus chamaemorus* L	*S. vulgaris* ssp. *maritima* M
Cryptogramma crispa L	*Salix arbuscula* M	*Stellaria crassifolia* M, ?L
Draba incana M, L	*S. herbacea* M, L	*Thalictrum alpinum* M, L
Dryas octopetala M, L	*S. lanata* M	*Tofieldia pusilla* L
Helianthemum canum M, L	*S. lapponum* M	*Viola lutea* M

have been substantially fewer, including only *Stellaria crassifolia, Ranunculus hyperboreus* (arctic–sub-arctic) and *Salix polaris* (arctic–alpine). At the same time however it is apparent that for a large number of plants the Late Weichselian distribution area was much wider than their present British range. This is easily seen in the distribution maps for *Koenigia islandica* (fig. 78), an arctic–sub-arctic plant, and for *Betula nana* (figs. 87, 88), *Dryas octopetala* (figs. 62, 63), *Oxyria digyna* (fig. 80), *Salix herbacea* (fig. 98), *Thalictrum alpinum* (fig. 45), *Draba incana* (fig. 48), *Saussurea alpina* (fig. 118), and *Saxifraga oppositifolia* (fig. 66), all arctic–alpine plants. It is shewn within the northern montane element by *Potamogeton filiformis* (fig. 123), *Polemonium coeruleum* (fig. 114), *Arctostaphylos uva-ursi* (fig. 104) and *Salix phylicifolia* (fig. 96). The records for many other species, in these phytogeographic categories especially, shew a greater or less restriction and disjunction of range. This constriction must be due to the effect of conditions in the Flandrian period, whether operating directly by climatic change or indirectly. In a number of instances, such as *Rubus chamaemorus*, *Betula nana* (fig. 87), *Selaginella selaginoides* (fig. 34), *Potamogeton praelongus* (fig. 122), *Sibbaldia procumbens* and *Potamogeton filiformis* (fig. 123), the record continues a greater or less distance through the Flandrian, documenting to some extent the progress of constriction.

This process has naturally led to the extinction of many Late Weichselian species from large areas of the British Isles. In particular the arctic, sub-arctic and arctic–alpine elements have largely gone from the English midlands, East Anglia and southern England, we no longer have in Wales *Betula nana, Artemisia norvegica, Loisleuria procumbens, Linnaea borealis* and *Rumex aquatilis*, whilst extinctions from Ireland include *Betula nana, Koenigia islandica, Stellaria crassifolia, Astragalus alpinus, Cerastium cerastioides, Minuartia stricta, Luzula spicata, Ranunculus reptans* and *Juncus balticus*. The list of extinctions from the Isle of Man is considerably longer amounting to some twenty-four Late Weichselian species, of which the great majority are in arctic–alpine or northern range categories. We may conjecture that from a Late Weichselian condition of great floristic similarity, confirmed by the formal studies reported later, the small size, isolation by the early Flandrian rise in sea-level, and the absence of mountain refugia all combined to make the Isle of Man particularly susceptible to the loss of late-glacial species.

(ix) CLIMATIC INFERENCES

Not only do the vegetational and phytogeographic characters of the Middle and Late Weichselian call for explanation in terms of the former climate, but they themselves are in no small degree the basis of conclusions about it. The danger of circular argument is lessened by the necessity to accommodate to the strong evidence provided by stratigraphy and geomorphology, as already indicated (chapter III), and by other groups of organisms represented fossil. G. R. Coope especially (1967, 1970) has drawn attention to the high indicator value of fossil

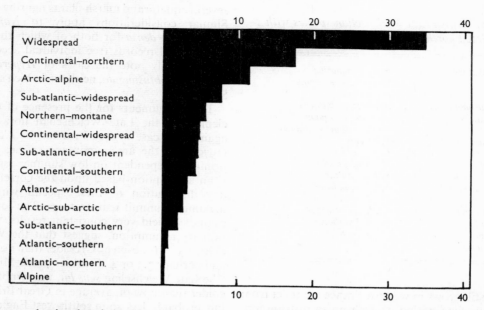

Fig. 157. Diagram shewing the relative percentages of various geographic elements of the present British flora (defined by H. J. B. Birks and J. Deacon), represented by Late Weichselian records in this country.

insect assemblages, especially of the beetles, which can be referred not only to present distribution ranges but also to the present-day altitudinal zones on the Scandinavian mountains. From study of large collections of both Middle and Late Weichselian provenance, conclusions are drawn often strikingly similar to those derived from the plant fossils. There is a parallel occurrence of arctic stenotherms, especially in the colder deposits, with numerous species no longer native in Britain, and this component diminished substantially in the later Weichselian. These are however accompanied by species of southern and oceanic range. Coope also finds that many beetles occur in Weichselian deposits that have exclusively eastern distributions at the present day, avoiding the oceanic fringe of Europe. This is attributed to a former continentality of climate arising from the effect of the Weichselian ice-sheet upon the general climatic pattern. In the summer the intensification of anticyclonic conditions over the ice caused prevalent north-east or easterly winds to cross the British Isles from the ice-cap: their dryness was not ameliorated since the North Sea was then bare. Such cold dry summers, Coope suggests, no longer occur anywhere in Europe. On the other hand cyclonic systems might still have had access to north-western coastal areas. The winter anticyclonic area was of greater extent and the weather clear, calm and extremely cold, whilst instability in autumn and spring allowed ingress of cyclonic depressions to the British Isles with considerable snow- or rainfall. To this glacial climatic regime, not typically oceanic or continental but combining features of both, Coope attributes the curious assemblages of beetles that today seem to be climatically incompatible.

The parallelism with the biogeographic attributes of the fossil plants is striking and it is hard to conclude that similar climatic causes are not involved for all the major groups of organisms. There can be little doubt that both plants and animals found very considerable compensation for a generally unsuitable climate in special habitats where shelter, slope, aspect etc., gave protective local conditions. It is likely also that the climatic gradient across the country altered steeply to the west, a fact suggested by the general lack of cryoturbatic and other frost effects in western Scotland, and by the biological evidence that has persuaded biologists such as Lindroth, Nordhagen and Dahl, of the persistence of ice-free glacial refugia off the western coast of Norway.

We find numerous features of the Weichselian plant record in accord with this reconstruction of former climate. Thus we have already noted how the Middle Weichselian plant records are indicative of steppe conditions and pronounced continentality of climate. Iversen, as early as 1951, had drawn attention to these conditions in the Late Weichselian, citing among the characteristic fauna, records of *Spermophilous rufescens*, *Equus caballus* and *Bison bonasus arbustotundrarum*, and among the plants *Hippophaë rhamnoides*, *Centaurea cyanus*, *Poterium sanguisorba* (*Sanguisorba minor*) that are found also in the British Late Weichselian. The high pollen frequencies of *Artemisia* are also often held to indicate a steppe vegetation. More recently pollen grains of *Ephedra* cf. *distachya* have been found in many Late Weichselian sites in this country. Although it is possible that such grains may have been blown from distant central European Steppe, it remains possible also that it was in fact locally present in small stands as far west as these islands, and that the coastal

TABLE 41. *British Late Weichselian species with restricted northern range in Scandinavia*

Agropyron junceiforme	*Mentha aquatica*
Ajuga reptans	*Mercurialis perennis*
Berula erecta	*Ononis* sp.
Callitriche obtusangula	*Pilularia globulifera*
Carex paniculata	*Plantago coronopus*
C. riparia	*Potamogeton acutiflorus*
Circaea lutetiana	*P. crispus*
Cladium mariscus	*P. trichoides*
Elatine hexandra	*Potentilla sterilis*
Eleocharis multicaulis	*Scirpus fluitans*
Geranium sanguineum	*Stellaria holostea*
Groenlandia densa	*Thalictrum minus*
Helianthemum canum	*Typha angustifolia*
Hypericum elodes	*Valeriana dioica*
Jasione montana	*V. officinalis*
Lycopus europaeus	

sites where it still grows in western France are relict from the Late Weichselian period of widespread distribution. Bell (1969) has given a very thorough analysis of the element of southern distribution plants in the Middle Weichselian alongside the northern ones, and her explanations also apply to the Late Weichselian.

As we shew in the analyses of section 9 and in fig. 157, plants of the southern distribution groups constitute only a small proportion of the total Late Weichselian flora, and of these the 'Continental–southern' contributed much more than the 'Sub-atlantic–southern' and 'Atlantic–southern' together, so that continentality rather than high mean temperatures can explain their occurrence. A larger group however comes from species of wider range which nevertheless shew a southern geographical tendency in the British Isles or on the western mainland of Europe. Thus the species recorded from the Late Weichselian in the British Isles still have a limited northerly range in Scandinavia (table 41).

It is to be noted that among these not only is *Lycopus europaeus* now of southern range in Scandinavia, but that it is clearly shown by fossil finds to have had there a wider northern extent in the Flandrian warmth period. It is a marsh and riverside plant not susceptible to forest competition and apparently therefore directly restricted in its northern range by climatic conditions. That it was present freely in the British Isles in the Late Weichselian period suggests that the climate was (in respect of those attributes affecting the plant) no more severe than the present climate of Stockholm or south Finland.

The example of *L. europaeus* is of particular importance since Scandinavian ecologists (see Dahl, 1951) have urged that many of the species thus restricted in range in Scandinavia and known from their own Late Weichselian deposits, owe this limitation to the geographical barriers and the early spread of forest in the Flandrian period, which effectively prevented northwards migration. This is, however, an argument largely inapplicable to the

several aquatic and marsh plants figuring in the list above. Similar considerations apply to *Najas flexilis* and *Groenlandia densa* for both of which there is a weighty list of fossil records (see individual accounts). We may also specially note the cases of *Hypericum elodes* and *Callitriche obtusangula*, neither of which so much as enters Sweden.

These arguments for the presence of a thermophilous element in the Late Weichselian flora have to be set against the case argued in detail by Conolly and Dahl (1970) that the arctic–montane species of that time are primarily dependent on low summer temperatures. This is an assumption based upon the correlation of present-day distribution of these species with mean annual maximum summit temperatures. The forty-seven species examined yield very consistent results, which, given the primary assumption, suggest that the Weichselian 'was at least 5 °C (or even 6 °C) colder than now in East Anglia, and perhaps 3.5 or 4 °C colder in the Midlands' and that 'the Late Weichselian *sens lat.* was probably at least 3 °C colder than now on average in Great Britain and Ireland but probably less so in south-west England'. The results were not employed to make deductions on the amplitude of temperature through the year, nor on relative continentality or oceanicity.

Whatever the precise climatic deductions may be, it must be accepted as characteristic of this period in Britain that there was admixture of the thermophiles with northern and arctic plants, just as oceanic and steppe elements were also present together. If plants of these diverse elements were then growing side by side, whereas now they are more segregated in our islands, this may be interpreted, in part at least, as due to the fact that the Late Weichselian was a period of rapid floristic change with vigorous persistence and expansion of older elements and vigorous extension of newer ones.

9. PHYTOGEOGRAPHIC SYNOPSIS

The biogeographic classification of the British flora by J. W. Matthews (1937) had the great advantage of being based upon consideration, not merely of the range of British species in this country, but upon the pattern of the whole of their European distribution. This system proved its value in analysis of the history of the British flora (Matthews, 1955; Godwin, 1956) but suffered from the omission of a considerable proportion of the flora of less readily recognizable distribution patterns. This deficiency has now been made good by H. J. B. Birks and Mrs J. Deacon and their extended classification which takes in all British vascular plants is the basis for our consideration of the biogeographic history of the flora from Middle Weichselian time to the present day. The seventeen phytogeographical elements employed will be seen set out in table 42. Since for some regions and in some periods these categories contain rather small numbers of species, it has been convenient for primary

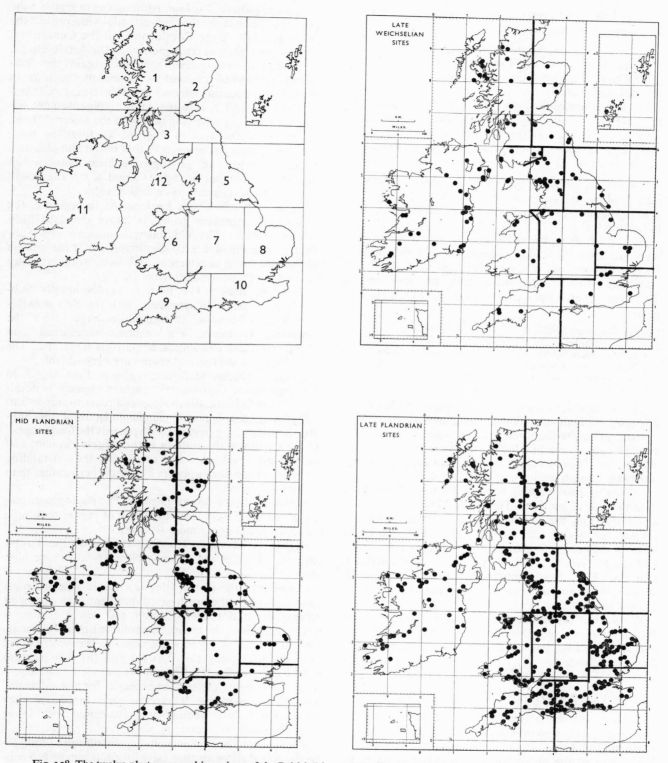

Fig. 158. The twelve phytogeographic regions of the British Isles and the distribution of sites within them at different times within the Quaternary Period: Late Weichselian, middle and late Flandrian.

analysis and visual presentation to aggregate them in five larger groups as follows:

(*a*) Arctic–alpine, Arctic–sub-arctic, Alpine.

(*b*) Northern–montane, Continental–northern.

(*c*) Continental–northern, Continental–southern, Continental–widespread.

(*d*) Sub-atlantic–northern, Sub-atlantic–southern, Sub-atlantic–widespread, Atlantic–northern, Atlantic–southern, Atlantic–widespread, Mediterranean–atlantic.

(*e*) Continental–southern, Sub-atlantic–southern, Atlantic–southern.

A few elements recur in more than one of the five groups, thus Continental–northern in both (*b*) and (*c*), Continental–southern in (*c*) and (*e*) and Sub-atlantic–southern in (*d*) and (*e*).

H. J. B. Birks and Mrs Deacon (1973) have also been responsible for the second exercise, the ordination of the floristic assemblages of the twelve regions of the British Isles that we employ for subsequent historical analysis. These regions are displayed in fig. 158 and can be referred to as:

1 North-western Scotland	7 English Midlands
2 Eastern Scotland	8 East Anglia
3 South-western Scotland	9 South-western England
4 North-western England	10 South-eastern England
5 North-eastern England	11 Ireland
6 Wales	12 Isle of Man

The ordination of the vascular plant taxa at the present day discloses features of considerable interest. In terms of floristic similarity the three Scottish regions are not only close to one another but distinct from the others: likewise regions 6 to 10 are strongly similar, with 4 and 5, the two northern English regions, resembling one another and yet closer to the remaining regions of England and Wales than to those of Scotland. The Isle of Man has strong dissimilarity from the rest and Ireland, though fairly dissimilar, has closest resemblance, not to Scotland, but to western regions of England and Wales plus East Anglia and the Midlands (4, 6, 7, 8, 9). It is of great interest to find that East Anglia and the south-west are highly similar, closer to one another indeed than to the south-eastern region.

We do not reproduce the ordination diagrams published by Birks and Deacon, but subsume their main conclusions in the two series of histograms presented in figs. 159 and 170.

FLORA OF THE PRESENT DAY

Region 1, *North-western Scotland.* The Sub-atlantic–widespread is the dominant element with the Continental–northern and Arctic–alpine well represented. Despite the dominance of the Sub-atlantic–widespread element, the Sub-atlantic–Atlantic aggregate group is smaller here than elsewhere in the British Isles (fig. 159): this follows from the much larger representation of the Sub-atlantic and Atlantic–southern elements further south.

Region 2, *Eastern Scotland.* Although as in region 1 the dominant element is the Sub-atlantic–widespread, the flora is generally more Continental, all the Continental elements having higher frequencies than the Arctic–alpine.

Region 3, *South-western Scotland.* Again the Sub-atlantic–widespread element is dominant, but, as in region 2, Continental elements are well represented. Unlike regions 1 and 2, here the Arctic–alpine element has a low value, partly reflecting no doubt the lower altitude.

Region 4, *North-western England.* In total the Continental aggregate group exceeds that of the Sub-atlantic–Atlantic group. The phytogeographical composition resembles closely that of regions 2 and 3, but the Continental–southern element has a higher value.

Region 5, *North-eastern England.* As in region 4, the Continental aggregate group is larger than the Sub-atlantic–Atlantic. The Continental–southern and Sub-atlantic–widespread are co-dominant: on these criteria the flora is more continental and more southern than that of Eastern Scotland (region 2).

Region 6, *Wales.* The flora is predominantly Sub-atlantic to Atlantic in emphasis, with the Sub-atlantic–widespread dominant, though little larger than the Continental–southern. The Atlantic–widespread and Atlantic-southern elements are relatively large also. The high values for southern elements are very notable.

Region 7, *English Midlands*; *region* 8, *East Anglia.* In both regions the Continental–southern element is dominant but the Sub-atlantic–widespread remains important.

Region 9, *South-western England.* In total the Sub-atlantic–Atlantic aggregate group exceeds the Continental. As the Continental–southern is the largest element, and the Atlantic–southern is also important, the outstanding character of the flora appears as its southern rather than its Atlantic aspect.

Region 10, *South-eastern England.* The phytogeographical groupings closely resemble those of region 9, though the Continental component is larger.

Region 11, *Ireland.* The Sub-atlantic is the largest element, although the Atlantic elements also have high values.

Considering the British Isles as a whole, what emerges most strongly is the north–south gradient. The east–west differentiation is relatively weak and where evident is recognizable more by the increase of Continental elements eastwards than of Atlantic elements to the west. Of the five aggregate categories it is (*a*) and (*b*) that increase northwards, whilst (*c*), (*d*) and (*e*) behave conversely, a result that seems perhaps unexpected for the Atlantic–Sub-atlantic elements: it may be conjectured that they have some habitat requirement associated with lower latitude as well as oceanicity *per se*.

The main features of present-day distribution of phytogeographical elements thus established, it remains to consider whether parallel analysis of the fossil occurrences at known times in the past, can shed light on how the present pattern came into being, and, in a historical sense, what has caused it. Two periods of particular significance

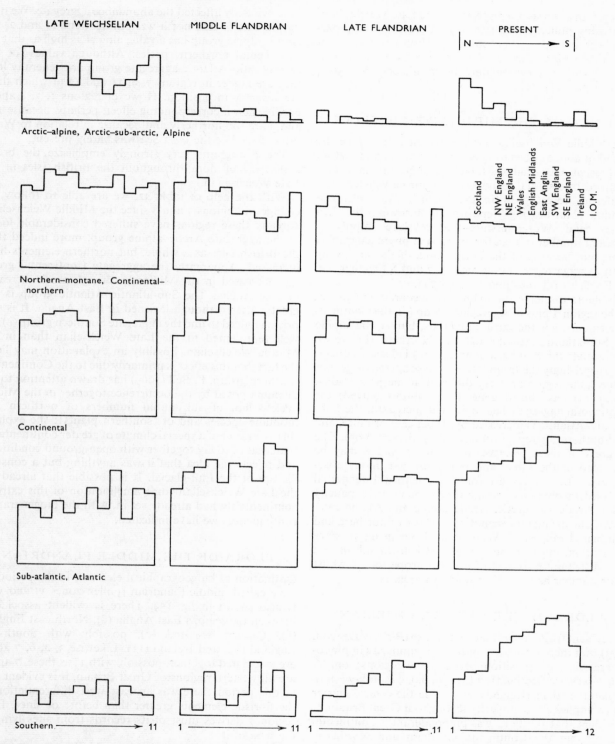

Fig. 159. Regional analysis of fossil plant records for the Late Weichselian, middle Flandrian and late Flandrian, and of the existing flora: all represented as percentages of the total recorded flora in each of five aggregate phytogeographic groups. The analysis is given in terms of 12 regions set out in fig. 158; from the two central columns the last region (Isle of Man) is omitted. The regions for Great Britain are arranged in uniform sequence 1 to 12, left to right, and this places them in roughly north to south order, with Ireland and the Isle of Man (if present) on the far right as shewn in the key on the top right-hand corner.

for such an analysis are respectively the Late Weichselian, embracing pollen zones I to III and extending approximately from 15000 to 11000 radiocarbon years B.P., and the Middle Flandrian, taken to include zones VI and VIIa and extending approximately between 9000 and 5000 radiocarbon years B.P.

FLORA OF THE MIDDLE WEICHSELIAN

The Middle Weichselian may be taken to embrace the period of about 45000 to 15000 radiocarbon years before the present. Our analyses treat separately the important Chelford Interstadial (c. 60000 B.P.) during which boreal coniferous forest grew in the English Midlands. Important plant-bearing deposits at Barnwell, the Lea Valley and Worcestershire–Warwickshire supply a sufficient number of taxa for analysis in phytogeographic terms but, because of the limited area of the sites, only as an average representing regions 7 and 8 together, i.e. the English Midlands and East Anglia.

Table 42 discloses the main phytogeographic features of the region 7 plus 8. Aggregate group (*c*), the Continental elements, are the most numerous, but Arctic–alpine (*a*), Sub-atlantic–Atlantic (*d*) and Northern (*b*) are all substantial: the Southern aggregate (*e*) is least frequent. Not surprisingly the frequencies of these aggregate groups in the same region are very different at the present day. Group (*a*) has almost gone, (*b*) is almost halved; the Continental aggregate has somewhat increased, the Sub-atlantic–Atlantic is increased by 50 per cent or more, and the Southern aggregate (*e*) has enlarged even more. The progress of these changes in regions 7 and 8 may be discerned in the table, and is commented upon subsequently. The species list for the Chelford Interstadial analysed in phytogeographic terms shews great paucity in the Arctic–sub-arctic, Arctic–alpine and Alpine categories, but strong representation in those of northern and continental ranges. As West pointed out in his primary publication, most of the species at Chelford still have a wide distribution in north-western Europe today, whilst a fair number are of rather southern areal type.

FLORA OF THE LATE WEICHSELIAN

It is evident from the ordination exercise (Birks & Deacon, 1973) that there must have been great similarity in phytogeographical composition between the regions; only 1 (north-western Scotland), 7 (Midlands), 9 (south-western England) and 11 (Ireland) lie outside the central cluster which implies close similarity throughout Great Britain. It is evident that the flora was preponderantly Continental (*c*), but that the Continental–northern and Northern–montane aggregate (*b*) (closely related to the Continental and partly overlapping in calculation) has also very large values. The histograms of fig. 159 illustrate certain regional variation: thus the aggregate group (*a*) has very high values in regions 1 and 2, and high ones in 4 and 6. These regions all include high mountains, which even at this

time evidently affected the abundance aggregate. We note that in eastern and north-western Scotland (1 and 2) the Arctic–alpine group has a value almost as high as that for Continental–northern species. Although values for the Sub-atlantic–Atlantic aggregate group are generally low, they are higher in regions 7, 9, 11 and 12 of which three are westerly in position. However regions 1, 3, 4 and 6 hardly shew a corresponding effect, perhaps because 1, 3 and 4 are too northerly and because in relation to Wales (6) at that time the Irish Sea was barely present.

The histograms very strongly emphasize the broad *uniformity* of flora throughout the British Isles in the Late Weichselian.

With the help of table 42, we are able to follow the changes in regions 7 and 8 since the Middle Weichselian. Already these regions have suffered considerable losses in the aggregate Arctic–alpine group, more indeed than the British Isles as a whole, but northern elements have increased. Apparently the aggregate Continental group has decreased in the Midlands, though much increased in East Anglia. The Sub-atlantic–Atlantic group is not greatly altered though lessened in East Anglia. It is surprising indeed to find the aggregate Southern group (*e*) less well represented in the Late Weichselian than in the Middle Weichselian. Possibly an explanation may lie in the fact that this effect is primarily due to the Continental–southern group. F. Bell (1969) has drawn attention to the dilemma posed by the occurrence together in the Middle Weichselian of substantial numbers of northern and montane species and of 'southern plants': she explains this as a result of a special climate of greater continentality than that of today together with open-ground conditions, and rejects the idea that it was anything but a constant feature of the Full-glacial. It is possible that already in the Late Weichselian some amelioration of this extreme continentality had already set in, with the biogeographic consequences we have indicated.

FLORA OF THE MIDDLE FLANDRIAN

Ordination of biogeographical elements in the period we have called middle Flandrian (pollen zones VI and VIIa) is also shewn in fig. 159. There is evident association between the regions East Anglia (8), North-east England (5), Eastern Scotland (2), possibly with South-east England (10) and Ireland (11). Likewise 3, 4, 6, 7 and 9 are associated together, possibly with 1: as these, 7 apart, are all western regions of Great Britain, it is evident that at this period there was an east–west differentiation of the floristic elements greater than before or since. Even region 7 derives most of its records from sites lying to the west of it.

The Continental and the Sub-atlantic–Atlantic aggregate groups are of similar, considerable, importance but in the three eastern regions 2, 5 and 8 the Continental group has the higher values. In the western regions 1, 3 and 4 the two groups are sub-equal, but in regions 6 (Wales), 7 (Midlands), 9 (South-western England) and to

TABLE 42. *Phytogeography of Midlands (7) and East Anglia (8) from Middle Weichselian to present day. Figures are percentages of the total number of species in the recorded flora of the period*

PRIMARY GROUPS	Middle Weichselian			AGGREGATE GROUPS								PRIMARY — Late Weichselian
			Middle Weichselian	Late Weichselian		Middle Flandrian		Late Flandrian		Present Day		
	7+8		7+8	7	8	7	8	7	8	7	8	Br. Is.
Arctic–alpine	14·4											11·4
Arctic–sub-arctic	3·2	(a)	19·5	11·0	7·0	3·4	1·8	0	0	1·2	0·2	1·8
Alpine	1·4											0·6
Northern–montane	5·4											5·7
		(b)	17·6	22	22	16	14·3	13·7	11·9	11·0	9·3	
Continental–northern	12·2											17·6
Continental–southern	7·7	(c)	25·0	21·7	33·4	20·5	28·7	21·0	29·2	34·0	34	4·5
Continental–widespread	4·5											5·0
Sub-atlantic–northern	4·5											4·7
Sub-atlantic–southern	2·7	(e)	11·4	5·4	9·2	5·7	14·3	8·8	15·1	23·4	25·5	1·0
Sub-atlantic–widespread	6·8											7·8
		(d)	19·0	20·7	15	26·9	23·0	23·5	24·7	28·7	31·7	
Atlantic–northern	0·5											0·6
Atlantic–southern	0·9											0·9
Atlantic–widespread	3·2											3·2
Widespread	33											34·9
Number of taxa	335											346

some extent 10 (South-eastern England), the Sub-atlantic–Atlantic group has much the higher values. Thus it seems that the emphasis on 'atlanticity' was pronounced in the south and west with 'continentality' pronounced in the east. It is clear from the histograms that the Sub-atlantic–Atlantic component is far more important than in the Late Weichselian, and that in England the Northern montane plus Continental aggregate is reduced, though it is increased in Scotland. The Arctic–alpine and Arctic–sub-arctic element is even more strikingly reduced, again more strikingly in the south. Even in regions 1, 2 and 4 where members of this group persist to the present, their proportionate representation is low because sampling sites are rarely at the high altitudes they favour. Despite the big decrease in this aggregate group there is only a moderate rise in the proportion of species of the Southern element: only in 4 (north-western England), 7 (Midlands), 9 (south-western England) is it substantially increased and the Atlantic–southern group is very sparse overall.

FLORISTIC CHANGE THROUGH THE FLANDRIAN

The histograms of fig. 159 permit a survey of the main trends of change in the biogeographic elements of the British Isles through from Late Weichselian to present times.

Aggregate group (a) (Arctic–alpine, Arctic–sub-arctic, Alpine) was abundant over the whole country in the Late Weichselian, in the lowlands and the south as well as the mountainous north. The much warmer Middle Flandrian imposed a natural and large decline upon this geographic component, and at the same time gave it a pronounced north–south gradient explicable no doubt by the spread of closed forest over southern and lowland areas as well as by more direct climatic controls. This gradient is persistent in the existing flora, and indeed is emphasized since plant-collectors exploit the montane habitats favoured by this group more effectively than the lake and bog samples of palaeoecologists are likely to have done.

Aggregate group (*b*) (Northern–montane, Continental–northern) was important and equally abundant everywhere in the Late Weichselian, but became subject in the Middle Flandrian to a very pronounced north–south gradient like that of group (*a*) and explicable in similar terms. This gradient is retained in the flora of the present day but is diminished by a loss of this element from the northern regions, only partly a consequence of the expansion of other categories.

Group (*c*), which includes the three Continental elements, is remarkable as retaining the same high and widespread importance from the Late Weichselian through to the present day; a gradient of increasing abundance southwards since the Middle Flandrian is primarily due to the Continental–southern component, and low values of this element characterizing Ireland and the Isle of Man also affect the group histogram strongly.

Group (*d*), which includes seven Atlantic and Sub-atlantic elements, is of great interest as retaining a similar gradient of increasing frequency southwards through the three periods whilst exhibiting a notable overall increase through from the Late Weichselian to the present day. This increase appears broadly to balance the decrease through this period of the Northern aggregate group (*b*), and not any change in group (*c*) the Continental aggregate.

Group (*e*) (Continental–southern, Sub-atlantic–southern, Atlantic–southern) was low in the Late Weichselian, and exhibited in the Middle Flandrian a gradient from north to south associated particularly with big increases in the eastern regions and north-western England. By the present day however this group had come to shew an exceedingly pronounced gradient of increase to the south, and a very great overall increase in importance. One might conjecture that the prevalence of agriculture on the sedimentary rocks and in the warmer conditions of the south had allowed establishment of many weeds and ruderals in this biogeographic group, but this may well not be the whole story. All three elements of the aggregate group share the effect which accordingly must be associated with the quality of 'southernness'.

The Late Weichselian evidently exhibited little regional differentiation in a biogeographic sense, Continental and northern elements preponderated, Arctic–alpine and Sub-atlantic–Atlantic elements were both important and the Southern element was small.

The Middle Flandrian saw decrease in the Arctic–alpine and Northern elements, especially in the south, and increase in the Sub-atlantic–Atlantic and Southern elements, again especially in the south: the Continental element was little affected.

Finally the change to the present day was one broadly extending those tendencies already begun. The Northern element is more reduced but retains, like the Arctic–alpine, its diminution southwards. The Sub-atlantic–Atlantic group has increased its importance, keeping the increase southwards, and the Southern group shews this effect very strongly. For the first time something of a north–south gradient is seen also in the Continental element, a consequence however of the Continental–southern component in it. We have already indicated that east-west gradients have never been nearly so pronounced as those from north to south and have been expressed chiefly through changes in the Continental elements.

Persistence, extinction and immigration have naturally affected the regions differently although the overall drift of change has been similar. In Scotland the high relief has ensured persistence of the Arctic–alpine component and to a less extent this has been true of north-east England, Ireland and Wales. Further south and east species in this category have been lost. Correspondingly it seems that southern elements established themselves from the Middle Flandrian heavily in the south, while they have never reached the north.

Only in regions 7 and 8 do we have information that allows us to commence our survey of phytogeographic change as early as the Middle Weichselian, but this nevertheless gives us the substantial advantage of shewing that some of the major tendencies we have seen to hold for the country as a whole were already operating between the Middle and Late Weichselian, notably losses from the Arctic–alpine aggregate. On the other hand the transitional character of the Late Weichselian appears to be indicated by the rise to a temporary maximum in the Northern and Continental elements and a temporary decrease in the still substantial Southern element.

Our phytogeographical historical analysis seems to conform to the hypothesis that in the Middle Weichselian a strongly Continental climate, induced by prevalent anticyclonic conditions over the ice-sheets, extended over the British Isles, and that this was already less extreme in the Late Weichselian whilst in the Flandrian it was replaced largely by an Atlantic cyclonic climate that persists today. The increased oceanicity was no doubt enhanced by the restoration of ocean level in the early Flandrian and the consequent reappearance of the Irish Sea, North Sea and English Channel. With this climatic change the general climatic uniformity of the British Isles gave place to the regional gradients that are evident today and which express themselves in the north–south and east–west gradients in the biogeographic elements of the Middle Flandrian and the Present. We do not find pronounced gradients of atlanticity and those apparent in the elements associated with atlanticity seem to be as much north–south as east–west. Throughout, the Continental and Widespread elements have been able to maintain themselves substantially.

10. FLORISTIC HISTORY IN THE LIGHT OF WEICHSELIAN RECORDS

The distribution of Weichselian fossil sites of arctic, alpine and northern plants throughout the lowland plains and at the lowest latitudes in the British Isles, together with their relatively high frequency, give clear proof of the assumption made by Forbes, that the history of the European arctic–alpine element was to be understood as including a former phase of widespread occurrence during cold conditions over the whole lowland plains of Europe. Again and again the Weichselian plant records have established the fact that vegetation rich in arctic–alpine and boreal species was characteristic of this time over the whole periglacial region of Europe. This floristic element was at least as rich in the older Middle Weichselian period as in the Late Weichselian, and it may be assumed that species of this category survived the Weichselian Glaciation within the periglacial territory in the south of the British Isles. Some support for this view is apparent in the fact that the Middle Weichselian records include many species now extinct within the British Isles, and unrepresented in the very considerable total of the Late Weichselian records. It appears probable that species such as *Pedicularis lanata*, *P. hirsuta*, *Erysimum cheiranthoides*, *Carex capitata*, *Dianthus carthusianorum* and the genus *Corispermum* represent merely a more ancient stage of this period of widespread lowland distribution of arctic–alpine and boreal plants. It would be distinctly more difficult to explain their history alternatively in terms of a hypothesis of two invasions of such species in the Middle and Late Weichselian respectively, whilst between there was a glacial stage of such severity that it killed out even this element in our flora.

Much dispute prevailed in the past among British botanists over the question of 'perglacial' survival, but it should be understood that a good deal of the discussion turned upon the employment of perglacial survival *in situ* to explain the local and restricted distribution at the present day of certain aggregates of rare species. Wilmott, who was the leading British proponent of this view, pointed to the occurrence in a number of places in Scotland (e.g. Ben Lawers), the north of England (Teesdale), north Wales (Snowdonia) and western Ireland (Ben Bulben), of collections of species whose presence it is difficult to explain in these places in view of their almost complete absence from places of apparently similar character round about. He took the view that these collections of plants survived the Ice Age in these very places, and adduced some evidence (not by any means uncontested however) that these places were unglaciated nunataks during the last glaciation. In support of his views he stressed the fact that the areas in question carry relatively large collections of species and that similar collections recur in the other areas with no graded admixtures of the species in the region between: he recalled the evidence for the present existence of large floras close beside existing glacier fields, and drew attention to the conclusions parallel to his own reached by Fernald in North America and by Nordhagen in Norway. Wilmott sought to clinch his point by concluding that 'any fact of distribution which can be related to topographic fact of one period and one period only of earth history, must be regarded as a product of that period'; this is reasonable enough but does not in fact exclude other explanation than that favoured by its

author, since the 'topographic fact' may plausibly be related to some period other than the glacial one.

Wilmott's views did not lack opposition, for whilst few botanists were prepared to adopt the *tabula rasa* hypothesis of Reid, and were inclined to accept the probability that arctic–alpine species did survive the last glaciation within Britain itself, they were very reluctant to accept Wilmott's extension of the survival *in situ* hypothesis to such aggregates as the Irish Lusitanian flora, which contains species quite evidently intolerant of even moderate coldness. The antipodal viewpoint, set out clearly by Salisbury, is one which reduces the historical factor to relative insignificance: it regards the species of each special locality as present there because it is there that they find suitable conditions of soil and climate for their growth, and it accepts as corollary to this, the assumption that dispersal is naturally so effective that time becomes unimportant as a factor in explaining these facts of geographical distribution. Salisbury was perfectly explicit upon the latter point, and wrote: 'No hypothesis [of the history of the British Flora] can be acceptable which involves directly, or by implication, the assumption of inefficiency of plant distribution.' In its extremest form Salisbury's dictum is clearly untenable, and it is at variance with the judgement of experienced phytogeographers such as Wulff, Cain, Willis, and many more.

It will be found, however, that the arguments based upon Late Weichselian identifications force us to recognize a hypothesis combining in some respects these conflicting views. In the first place it is perfectly clear that there is no point whatever in retaining an explanation of perglacial survival *in situ* for the aggregates of species within a refugium such as Teesdale when we know beyond question that at a distinctly later period the same floristic assemblage was widespread throughout the lowlands of the British Isles. Teesdale may still indeed be a refugium, but it is not a refugium from a past interglacial through a glaciation, so much as a refugium for species of the Weichselian through the warm Flandrian period to the present day. The picture of restriction of range and of extinction among arctic–alpine and boreal species since the Late Weichselian is by now extraordinarily well documented. It is apparent that pronounced fragmentation of area has accompanied withdrawal from the wide ranges of Weichselian time, and this is apparent not only in the disjunct distribution patterns within the British Isles, but in the disjunct European ranges also.

It may very well be that some of the altered range or loss of our Weichselian species has been due to the direct effect of increasing temperatures, and Conolly and Dahl (1970) have indeed suggested that certain northern species are intolerant of high summer temperatures *per se* and survive only in situations where their special needs are still met. Most of the retraction, however, seems more reasonably interpreted in terms of three indirect effects of climatic change acting together in the Flandrian period.

Most important is the effect of dense forest establishment and persistence through the period of the Flandrian Climatic Optimum, an effect foreshadowed as early as the Allerød period by the pronounced diminution of herbaceous vegetation in this first birch woodland period. Secondly there is the development of peat mires, chiefly blanket bog, upon flat and gently sloping uplands at altitudes above the forest limit: such mires were apparently scarce in Boreal and Pre-boreal time, but they had clearly begun to develop extensively on our British mountains in the Atlantic period. Finally we may note that those climatic factors which caused extensive mire formation led, in less extreme conditions of altitude, slope, parent rock and local climate, to more or less pronounced podsolization of the soil, a factor given much importance by Pearsall, and clearly of greater importance for species dependent upon soils of high base status, such as many plants of relict distribution appear to be.

Wherever mountain habitats occur on sites free from trees, where the parent rock is base-rich and local conditions constantly renew the soil surface, where slope and drainage do not allow deep peat formation, and preferably where intensive grazing is absent, there we may expect to find relict populations from the Late Weichselian flora, and this indeed applies with force to all our best-known refugia. Quite recently it has also been shewn that similar circumstances may explain the isolated aggregates of species on calcareous soils at lower altitudes and more southerly localities in the British Isles (Pigott & Walters, 1953).

It must not, of course, be overlooked that, in many mountain refugia, altitude and aspect have also ensured the conditions of coolness and maintained moisture needed by many of the arctic–alpine species, and this is also applicable to the lowland sites on the Atlantic coasts. This factor is emphasized by consideration of the fact that as the Flandrian changes forced upon the Late Weichselian plants changes of range and habitat, it appears to have allowed certain of them to survive separately in both of these present-day habitats. Thus of species present generally in the Late Weichselian landscape, several, such as *Armeria maritima*, *Silene maritima*, *Cochlearia officinalis* and *Plantago maritima*, now grow either in the sub-alpine mountain vegetation or in maritime habitats of sea-cliffs, dunes and salt-marshes. Recent pollen records in Cwm Idwal, North Wales, actually shew two of these very species persisting through the Flandrian warm period. It may be noted that *Hippophaë rhamnoides* has claim to belong to this class, for although now restricted in Great Britain to coastal dunes, on the Continent it has also the status of a sub-alpine shrub.

It will be seen that this view of the nature of relict areas for arctic–alpine and boreal plants does not so much solve the question of perglacial survival as demonstrate that for the phytogeographic questions at issue it is not necessary to invoke perglacial survival. Our conclusions likewise force us to a view of the significance of dispersal rates, which in some degree reconciles the apparently conflicting views of the two schools of thought already referred to.

The unmistakable trend of all our evidence from the Late Weichselian is that it was a period of extremely rapid plant spread. Abundant open habitats were created by melting ice and snow, by solifluxion and by the re-sorting of glacial and periglacial superficial deposits, and into a region of this kind plants must have been able to advance as rapidly as they have done during historic time into areas laid bare by human activity. This rapidity of spread is attested by the remarkable admixture already at this time of elements of the most diverse phytogeographic categories, not all of which are likely to have survived the glaciation in this country. By contrast, in the period of dense afforestation which followed, these species of open habitats were heavily restricted and isolated: their rates of spread must have been very greatly diminished across the sea of deciduous forest in which the refugia became mere islands. Only in recent times of agriculture and forest clearance has spread of open-habitat species become once more rapid and easy. Naturally, within the forest period the species characteristic of deciduous woodlands found *their* opportunity for spread, an opportunity absent alike in the preceding treeless period and in the present period of forest regression and discontinuous woodland.

The concept of succeeding phases of very different dispersal rates gains particular importance from the operation of those substantial natural barriers to plant migration, the North Sea and English Channel, affecting colonization of the British Isles as a whole, and the Irish Sea affecting the colonization of Ireland.

A fall of 300 ft in ocean level at the present day would expose two narrow land bridges between North Wales and the Irish coast, as well as a broader land-belt between north-western Ireland and south-west Scotland. A fall of 400 ft (122 m), however, would mean that apart from a narrow lake off north-east Ireland, the area of the Irish Sea and English Channel would be entirely dry, and 400 ft is indeed within the range of modern estimates of the eustatic fall in ocean level due to the locking up of water in the ice-sheets of the Glacial period, although on the large side. It must be recognized that the same climatic amelioration which rendered the land available for re-colonization by plants and animals also restored by melting of the ice-cover the sea barriers which presently impeded migration. Despite the importance of our doing so, we have not yet been able to establish the precise relationship between these processes. Nevertheless it is apparent from pollen analyses of peat recovered from the bed of the North Sea at depths between 174 and 120 ft (53–56 m) that in Late Weichselian and early Flandrian time a large part of the bed of the North Sea was still dry land. Conditions of open ground thus extended across from north-west Europe in the period of rapid plant distribution: by the time, however, that the North Sea had reached its present extent, which within a few feet of vertical height was the end of the Boreal period (zone VI), forest growth was widespread over the British Isles, and dispersal of all save woodland species had become very restricted.

It is apparent that freedom of movement of plants in the Late Weichselian and early Flandrian was enhanced by the fact that the North Sea was then dry land, whilst difficulty of movement in the ensuing Flandrian period was accentuated by the fact that the Irish Sea and the North Sea had then attained their present size.

It would clearly be of considerable advantage if it were possible to estimate closely the rate and timing of the submergence of the land connection between the British Isles and the Continent. It is not entirely sufficient to rely upon the several radiocarbon-based progress curves for eustatic restoration of ocean level, such as were summarized by Jelgersma (1966), because warping movements have certainly also to be reckoned with. There is evidence of land depression both from the Dutch coast and the English Fenlands during the Flandrian, but its extent is not agreed. Making some allowance for this, as for the direct evidence of submerged freshwater peat beds, it seems that dry land connection with the Continent probably persisted until zone VI of the Flandrian, so recent a time that all thermophilous elements of our flora might be thought to have entered the country overland. This is naturally somewhat later than was the case with the connection between France and southern England, or between Ireland and Great Britain where no corresponding down-warping has to be reckoned with. Indeed the warping effect is the other way in relation to Northern Ireland and Scotland, for the elevation which gave rise to the 25 ft raised beach means that the sea floor between the two countries was deeper than it now is when it was not fully recovered from the burden of the Weichselian ice-sheets.

11. THE BEGINNING OF THE FLANDRIAN PERIOD: THE PRE-BOREAL PERIOD

(10000 to 9000 years B.P.)

An outstanding advantage of the Blytt and Sernander climatic scheme was that apart from its climatic validity (in fact limited) it afforded a time division for the last ten thousand years, when no other was available. It can usefully be retained for this purpose even though radiocarbon dating promises to replace it and has indeed already given to the quasi-climatic periods the fairly precise absolute dates that we now give at the head of discussion on the successive periods of the Flandrian. Nevertheless there is of course a great mass of geological, biological and archaeological evidence that bears upon sub-division and correlation within this, our latest, incomplete, interglacial.

It is appropriate to recall that since we choose to divide the Flandrian chronologically, in any one time division we may expect to find a range of vegetation corresponding to latitudinal, longitudinal and climatic gradients across the British Isles, and, as we have earlier indicated, the pollen zones are not, therefore, identical in different parts of the country in the representation of this vegetation. Thus there is regional parallelism in vegetational history, a conclusion that is being progressively confirmed by radiocarbon dating of widely distributed and detailed pollen diagrams.

The Flandrian period in north-western Europe is taken as beginning when the ice-sheet standing at the Central Swedish Moraine line began its rapid decay under a mild climate in which closed birch forest was able to spread over the ground as quickly as it was set free from ice. In the British Isles the ice of the valley glaciations now decayed. This first stage of amelioration is the Pre-boreal (zone IV) referred to by Firbas as the 'Vorwärmezeit', during which the Yoldia Sea occupied the Baltic Basin, and it is presumed to have begun about 8000 B.C. The north-west German plains were then covered with birch woodlands containing some pine, which tree was found in increasing frequencies southwards and dominated the central European forests.

The boundary between the Late Weichselian and the Flandrian periods is in practice often diagnosed by the sharp recession of the ratio of non-tree pollen to tree pollen in the pollen diagrams, so that it has become a basic assumption that with the onset of closed birch forest the herbaceous and dwarf shrub vegetation of the last part of the Weichselian period must have diminished greatly in frequency. It is in the nature of the evidence that the pollen record should more readily reflect this change than the macro-fossils, for with the coming of a closed vegetation carpet, the cessation of solifluxion and the accelerated colonization of lake margins the chances must have been greatly lessened of incorporating seeds, fruits and other debris of land plants into the accumulating lake-muds. Although it will therefore almost certainly appear that many dry-land plants are recorded less frequently in the early Flandrian than in the Late Weichselian, and although a real diminution in their frequency is undoubtedly to be expected, mere absence from early Flandrian records is not to be interpreted as denoting extinction: when such species have later Flandrian records and persist at the present day it is reasonable to assume that they have in fact persisted through from the Late Weichselian period.

The conjunction of radiocarbon dates and close-interval pollen analyses makes it increasingly apparent that very soon after 10000 years B.P. all parts of the British Isles supported at least open woodland. The most abundant trees were the birches whose pollen is present in high frequency everywhere, though relatively diminished by high pine in the south and east. Macroscopic evidence for the presence of tree birches is plentiful, and indicates the greater prevalence of *Betula pubescens*, a more northerly species than *B. pendula*, which also occurred. There is evidence for the hybrid between *B. pubescens* and the dwarf birch, *B. nana*, which certainly persisted into the Flandrian (see fig. 160). *Pinus sylvestris*, whose pollen

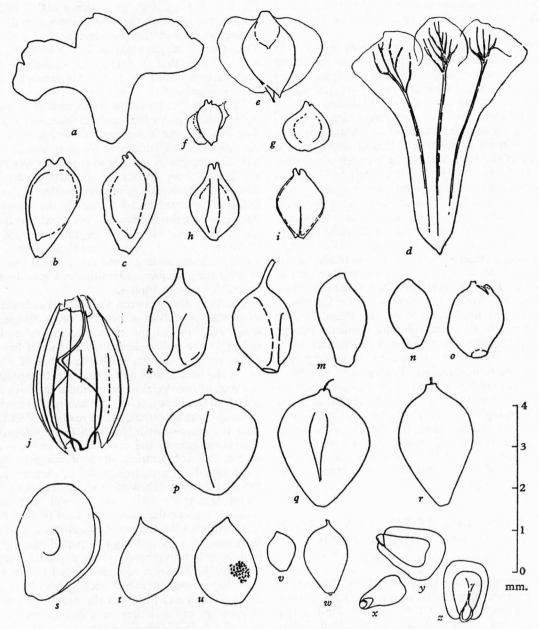

Fig. 160. A collection of macroscopic plant remains recovered from peat at 59 ft (18 m) below sea-level in the English Channel at Weevil Lake, Gosport. They are dated by pollen analysis to the opening of the Post-glacial period (zone IV): *a, Betula pendula* Ehrh., female cone-scale; *b, c, B. pendula* Roth., fruits lacking wings; *d, B. nana × pubescens*, female cone-scale; *e, B.* cf. *nana × pubescens*, fruit; *f, g, B. nana* L., fruit; *h, i, B. nana* L. or *pubescens × nana*, fruit; *j, Carex rostrata* Stokes, fruit; *k, l, C. rostrata* Stokes, nutlets; *m, n, C. disticha* Huds., nutlets; *o, C. nigra* (L.) Reichard, nutlets; *p, q, r, Scirpus lacustris* L., fruits; *s, Potamogeton* cf. *obtusifolius*, stone; *t, u, Ranunculus flammula* L., achenes, *u*, shewing a patch of the surface pitting; *v, Urtica dioica* L., fruit; *w, Scirpus fluitans* L., fruit; *x, y, z, Lycopus europaeus* L., *x* without inflated border, *y* shewing outer face of nutlet, *z* shewing inner face.

is represented at an increasing number of sites from zone I to zone VII, displays high pollen frequencies in the Hampshire basin in zone IV, with macroscopic remains in several sites: high pollen frequencies extend thence north-eastwards as far as Yorkshire and indicate that southern and eastern England were in the European region of birch–pine woodland. Elsewhere in England, Scotland and Wales, scattered sites with variable percentages of pine pollen, often exceeding 10 or 20 per cent of the arboreal pollen, indicate the likely presence of local stands of pine despite the sparseness of macro-scopic evidence. In Ireland pollen frequencies are

PLATE XVII

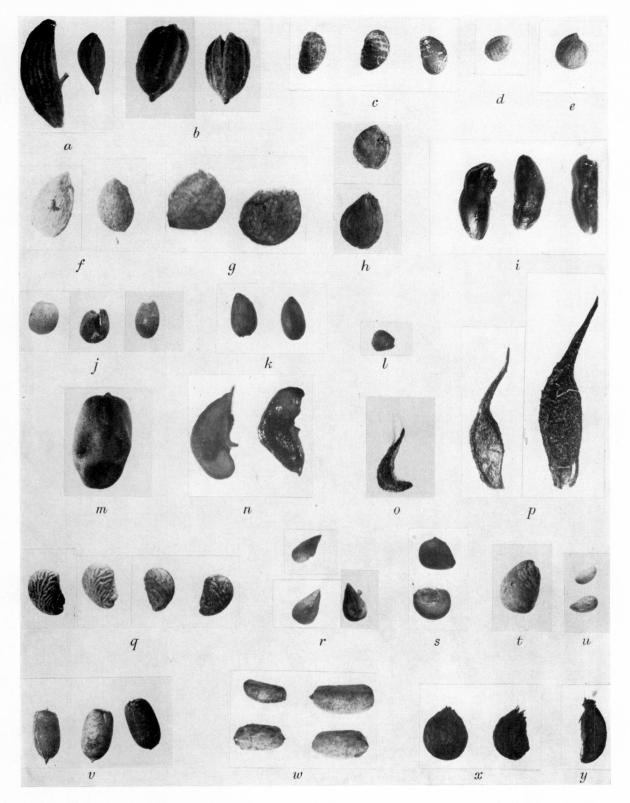

Abbreviations used in legends to Plates XVII–XX: cp., carpel; cs., cone-scale; ct., calyx tube; fr., fruit; gr., pollen grain; l., leaf; m., megaspore; n., nut; p., perianth; r., replum of cruciferous pod; s., seed sh., shoot; st., stone of fleshy fruit; u., utricle.

Macroscopic plant remains from Late Weichselian deposits at Nazeing (Lea Valley), Essex: *a, Thalictrum alpinum* (2 fr.); *b, T. flavum* (2 fr.); *c, Ranunculus Batrachian* sp. (3 fr.); *d, R. sceleratus* (1 fr.); *e, R. flammula* (1 fr.); *f, R. lingua* (2 fr.); *g, Ranunculus* cf. *acris* (2 fr.); *h, Ranunculus* cf. *repens* (2 fr.); *i, Caltha palustris* (3 fr.); *j, Barbarea vulgaris* (3 fr.); *k, Viola* sp. (2 s.); *l, Cerastium holosteoides* (1 s.); *m, Geranium sanguineum* (1 s.); *n, Filipendula ulmaria* (2 fr.); *o,* cf. *Dryas octopetala* (2 fr.); *p, Geum* sp. (2 fr.); *q, Potentilla sterilis* (4 fr.); *r, P. fruticosa* (1 fr., 3 views); *s, Potentilla palustris* (2 fr.); *t, u, Potentilla* spp. (3 fr.); *v, Hippuris vulgaris* (3 fr.); *w, Myriophyllum spicatum* or *M. verticillatum* (4 fr.); *x, Cicuta virosa* (1½ fr.); *y, Petroselinum segetum* (½ fr.). (Magnifications of *a–y,* ×14.)

PLATE XVIII

Macroscopic plant remains from Nazeing (Lea Valley), Essex: *a, Berula erecta* (2 fr.); *b, Daucus carota* ($\frac{1}{2}$ fr.); *c, Galium* cf. *aparine* (2 fr., 2 views); *d, Valeriana dioica* (3 fr.); *e, Carduus* sp. (3 fr.); *f, Cirsium heterophyllum* or *C. palustre* (3 fr.); *g, Taraxacum officinale* (1 fr.); *h, Sonchus arvensis* (2 fr.); *i, Arctostaphylos uva-ursi* (1 st.); *j, Menyanthes trifoliata* (2 s.); *k, Linaria vulgaris* (2 s.); *l, Mentha* sp. (3 fr.); *m, Lycopus europaeus* (2 fr.); *n, Prunella vulgaris* (2 fr.); *o, Galeopsis tetrahit* (2 fr.); *p, Teucrium scordium* (1 fr., 2 views); *q, Chenopodium* cf. *album* (2 s.); *r, Polygonum aviculare* (1 n., 1 n. enclosed in p.); *s, Rumex* sp. (1 n.); *t, Betula*, tree spp. (3 fr.); *u, B. nana* (5 fr.). (Magnifications of *a* to *u*, × 14.) All occur here in Late Weichselian layers except *Daucus carota* (zone V), and *Teucrium scordium* (zone IV).

PLATE XIX

Macroscopic plant remains of Late Weichselian age from Nazeing (Lea Valley), Essex: *a, Empetrum nigrum* (1 st.); *b, Alisma plantago-aquatica* (2 fr.); *c, d, Potamogeton natans* (3 fr., *c.*, with and *d*, without outer skin); *e, Potamogeton* cf. *pusillus* (3 fr.); *f, Potamogeton* cf. *filiformis* (2 fr.); *g, Najas flexilus* (2 fr.); *h, Eleocharis palustris* (1 fr.); *i, Scirpus lacustris* (2 fr.); *j, Scirpus* sp. (1 fr.; *k, Carex paniculata* (4 u.) *l, C. aquatilis* or *bigelowii* (2n.+u., 1 n.); *m, Carex* cf. *laevigata* (3 n.+u.); *n, C. rostrata* (3 n.+u., 1 n.); *o, Betula nana* (3 cs.); *p, Dryas octopetala* (1 fragment); *q, Betula nana* (1) shewing perithecia of ? *Venturia* sp.; *r, Armeria maritima* (5 ct.). (Magnifications of *a–r*, ×14.)

PLATE XX

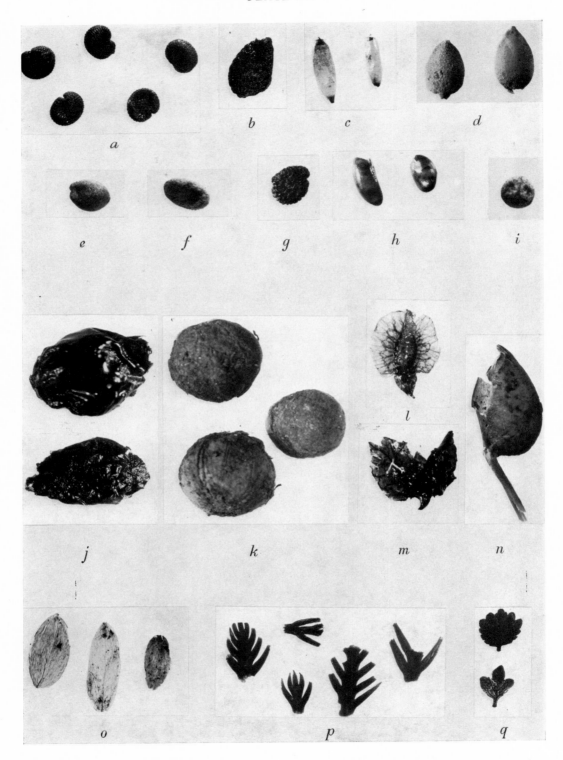

Macroscopic plant remains from Nazeing (Lea Valley), Essex: *a, Arenaria ciliata* (5 s.); *b*, cf. *Stellaria alsine* (1 s.); *c, Typha* sp. (2 s.); *d, Urtica dioica* (2 fr.); *e, Rorippa islandica* (1 s.); *f, Saxifraga* cf. *hypnoides* (1 s.); *g, Cerastium holosteoides* (1 s.); *h, Campanula patula* or *C. rotundifolia* (2 s.); *i, Selaginella selaginoides* (1 m.); *j*, cf. *Sorbus* (2 fr.); *k, Thelycrania sanguinea* (3 st.); *l, Rumex acetosa* (1 p.); *m, Rumex* cf. *crispus* (1 p. + n.); *n, Geranium sanguineum* (1 cp.); *o, Draba incana* (3 r.); *p, Myriophyllum* sp. (5 sh.); *q, Betula nana* (2 l.). (Magnifications of *a–i*, ×26; *j–n*, ×10; *o–q*, ×7.) All occur here in Late Weichselian layers except the *Typha* (zone IV) and *Thelycrania sanguinea* (zone V).

consistently low and actual presence of pine remains uncertain.

The role of the hazel, *Corylus avellana*, is of very great interest for its pollen has now been recorded in surprisingly high frequency at numerous sites in Scotland, north-west England and north-east Ireland, and H. J. B. Birks recognizes a *Corylus–Betula* assemblage zone in Skye that is firmly pinned to the opening Flandrian by radiocarbon dates. For the rest of Ireland, Wales and England far smaller pollen frequencies occur, but in this north-western area it is difficult to avoid the conclusion that hazel shared with birch the early woodland expansion of zone IV (see account for *Corylus* p. 267, and fig. 91). No doubt along with pine, birch and hazel there persisted from the Late Weichselian *Populus tremula, Sorbus aucuparia, S. rupicola* and numerous species of *Salix*.

In contrast with these we note that the zone IV maps for *Quercus* and *Fraxinus* shew such low and such uniform pollen frequencies over the British Isles that they are presumably due solely to long-distance pollen transport. The corresponding evidence for *Alnus* is similar though values of 5–10 per cent A.P. at two sites might reflect the local presence of small stands of the tree. 'Migration' and especially 'immigration' of a species into or across a region is often assumed or asserted to be the outcome of palynological studies, but the evidence of fossil pollen is an insecure basis for such conclusions. This follows from the tremendous transportability of pollen, the magnitude of the distant component in the pollen rain, and the influence on the distant component of the sparsity or abundance of local pollen production. When however macroscopic remains of one tree, for example birch, are present with a high pollen frequency of that tree, local presence is assured and moreover if at the same time pollen of another tree, such as hazel, is present in similar pollen frequency, it is difficult not to attribute local presence also to that species (even though macroscopic evidence is lacking) since long distance transport of hazel pollen could not attain such dimensions against the background of pollen from the local birch woods. This is the situation already discussed for *Corylus* in zone IV in the north-western British Isles.

There is of course evidence of the transition from the open vegetation of zone III to the woodlands of zone IV. Many of the characteristic pollen types of the earlier zone, such as *Helianthemum, Polemonium, Artemisia, Galium* and *Rumex*, are still to be found in zone IV but in diminishing frequency from the base upwards. It is also highly characteristic that near the zone III/IV boundary *Juniper* scrub flourished temporarily after suppression by the exposure to cold in zone III and before shading out by developing woodland. The very large peaks of *Juniperus* pollen that precede birch forest establishment are often associated with smaller peaks of *Filipendula ulmaria* taken by some authors, following Iversen, to be indicative of increased warmth.

Better indication of thermal conditions is given by the aquatic vegetation. Not only are aquatic plants unaffected by forest dominance, but they spread with great facility and not uncommonly set seed within the year of germination. They avoid to some degree the consequences of slope, aspect, soil maturation and parental rock variation that so much affect terrestrial herbs, so that there is better chance that their ranges reflect the thermal climate of a locality. This certainly seems concerned with the striking development of aquatic vegetation exhibited in lakes and pools in the opening stages of the Flandrian, an effect readily seen in the diagram from Moss Lake (fig. 161) and in the limnological studies of Y. and A. Vasari (1968) in the lochs of Scotland. These authors point out that the rapid thermal improvement took place in waters made eutrophic by the raw, often base-rich soils laid bare by erosion in zone III, so that this early limnic phase can be regarded as 'thermoeutrophic'. As, subsequently, vegetation clothed the surroundings and soils matured, more especially in oceanic regions and on acidic parent rock, these conditions passed and the lakes became oligotrophic. It has to be borne in mind that plants such as *Isoetes echinospora* and *I. lacustris* that are often abundant in the zone IV proliferation thrive under eutrophic conditions if not hindered by competition (Seddon, 1965).

In view of the convergence of evidence of rapidly increasing warmth it is not surprising to find that of taxa with few or no records in the Late Weichselian and now appearing in zone IV many have geographic ranges in the British Isles that strongly favour the south and east. These include *Rumex hydrolapathum, Solanum dulcamara, S. nigrum, Eupatorium cannabinum, Ruppia spiralis* and *Carex strigosa*, all of which are well represented in the subsequent Flandrian. Their presence further confirms an early presence of species of southern range as already adduced for the Late Weichselian.

12. THE EARLY WARM PERIOD: THE BOREAL PERIOD

(9000 to 7000 years B.P.)

The relief of the Scandinavian peninsula of its burden of ice was expressed by uplift of the crust which for a while brought the southern regions above sea-level and established the great freshwater *Ancylus* lake over the northern and central Baltic, a lake discharging westwards by the ancient Svea river across central Sweden. This stage is the geological marker for the first stage of the warm period, the 'frühe Wärmezeit' of Firbas, and the Boreal period of Blytt and Sernander. The vegetational history of this period is substantially that conveyed by the consistent and striking results of pollen analysis throughout north-western Europe. It is true that macroscopic plant remains continue to be recorded from deposits of this age, but now that the soil surface has become covered with closed vegetation, and lakes and pools are bordered by extensive fens and not by bare mineral soil, the chance of incorporation of the larger and heavier debris of a land flora has

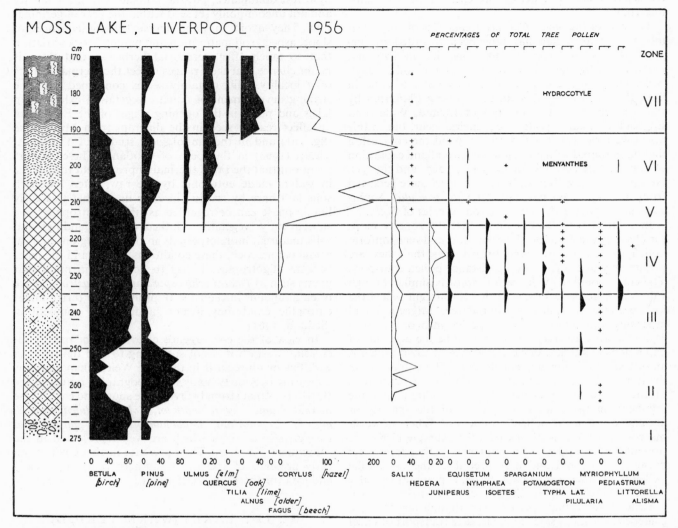

Fig. 161. Pollen diagram through deposits at Moss Lake, Liverpool (cf. fig. 154), shewing trees, shrubs and aquatic plants as percentages of total tree pollen. Abundance of aquatic plants in zone IV may be associated with the eutrophic state of the lake after the minerogenic deposition in zone III, and with rising temperature. Zone IV also exhibits a passing phase of abundant juniper. (After Godwin, 1959.)

become much reduced. On the other hand the wind-borne pollen is relatively unaffected by these circumstances, and indeed once the forest trees have become the dominant element in the vegetation, the pollen rain conveys not only a fairly complete and quantitative estimate of vegetational composition but gains importance from the fact that it records the vegetational *dominants*, those types standing most directly under climatic influence and those exerting maximum influence upon the existence of all other forms of life.

The story of vegetational change throughout the Boreal period is that of forest expansion and migration under the influence of increasing warmth, and Firbas has characterized the middle European forest development as exhibiting outstandingly:

(a) the strong expansion of the hazel, especially in the west and at high altitudes;

(b) the powerful extension of the pine into the birch-dominated Pre-boreal countryside;

(c) accompanying or following extension of the pine, the onset of displacement of the pine by the expanding mixed-oak forest trees, more especially the elm and the oaks;

(d) the first expansion of the spruce in the south-eastern Mittelgebirge.

With the spruce the history of the British flora is no more concerned but the first three categories of change apply broadly to the British Isles as well as to the Continental mainland.

Within England and Wales the standard pollen zonation placed zones v and vi within the Boreal period, the latter sub-divided into sub-zones a, b and c: in Ireland zones v and vi were similarly defined though the termination of zone vi proved hard to identify, and extension of the standard zones to Scotland was considerably more difficult.

In zone v the birches and hazel were the dominant trees of the landscape throughout the British Isles except for the south-eastern corner of England and the Hampshire Basin where pine was strikingly abundant. *Pinus* was already widely and strongly established, especially in England and Wales with fairly high pollen frequencies in northern Scotland and locally in Ireland. *Quercus* pollen is generally present throughout the British Isles, but in low frequencies, so that local presence is uncertain though probable. The case is similar for *Ulmus* though with a few higher values widely spaced in England and Wales only. By contrast *Alnus* pollen, widespread at low values throughout England and Wales, has higher frequencies in west-central Scotland and though present in adjacent north-east Ireland, is unrecorded elsewhere in that country.

In zone vi, as the pollen-frequency maps for the individual genera shew, very large changes in forest composition now occurred. The birches remained very abundant throughout Scotland, northern and western England, Wales and eastern Ireland. This is a pattern one inclines to view as a consequence mainly of the competition elsewhere provided by *Pinus*, *Ulmus* and *Quercus*, forest dominants in the more temperate conditions now prevalent. The widespread abundance of *Corylus* evident in zone v is maintained, and extended even to south-east England in zone vi. This is part of a phase of hazel expansion that in fact characterizes the Boreal period throughout western Europe, though in these islands hazel pollen values exceed any encountered on the western European mainland coasts, often many times exceeding the totals for all other arboreal pollen. In the account for the species we have outlined the view that these very high values must imply the widespread presence of hazel scrub, possibly the consequence of the hazel outstripping the migration of the mixed-oak forest, under highly favourable conditions of soil and climate. It is notable that no such hazel-dominated period occurs in the vegetational development of any preceding interglacial period. It is also worth recalling the possibility that the hazel may have been associated with the aspen (*Populus tremula*), for macroscopic remains of this tree occur rather commonly as early as zone iv: its pollen unfortunately has not yet been systematically recorded. The hazel certainly favours mull soils and tends to be especially associated with fresh calcareous soils, and this agrees with its lower frequencies during the Boreal period in parts of southern and western Ireland where soils are naturally lime-deficient.

As compared with zone v, where in England and Wales the highest hazel pollen values were concentrated in the west, in zone vi in these countries the hazel frequencies tend to be more uniform and one recalls that during this zone the rapid restoration in level of the North Sea was in progress and that this must have extended oceanic conditions eastwards across the British Isles (fig. 162). Both from site to site and from sub-zone to sub-zone the hazel frequencies shew very great variability, and the causes of this are not at all understood. It appears nevertheless that in general the hazel maximum had been passed by the last sub-zone vic.

In zone vi pine had become abundant throughout the British Isles, and in Ireland this may have been a major factor in the much diminished importance of birch. Macroscopic remains also testify to this wide presence. We are not used at present to envisaging the pine as abundant upon the more calcareous soils, but in fact it grows well upon them, and in the absence of the broadleaved forest trees there is no reason to doubt the evidence of the pollen diagrams that it was prevalent upon such soils in Boreal time, and we may indeed have to envisage a passing period of open pine forest with hazel undergrowth. We may also note that at many British sites (as at Flixton, fig. 163) pine pollen frequencies rise particularly high in sub-zone vic, at which time also layers of pine stubs often occur suggesting local response to the drying out of peat mires in the late-Boreal. A feature emerging from the pollen-frequency maps (figs. 39 and 164) is the notable absence of pine from north-western England during both zone v and zone vi: it might be possible that this is due to the dual routes of establishment of pine, one from the south and the other from the north and east. The pine could well not have reached the region before broad-leaved trees had occupied it.

Considerable changes in pollen composition separate zone v from zone vi, for at this time both *Ulmus* and *Quercus* increased rapidly in importance, the elm reaching surprisingly high values on many lowland sites, more particularly in Ireland. In sub-zone a it generally outweighed the oak both in Great Britain and in Ireland, but in sub-zone b in England the oak has become the more frequent pollen type.

Since *Ulmus glabra* alone of the British species is now native in Ireland, whilst most of the rest may be the result of human introduction, it seems evident that this species must have been the major contributor to the elm component of the Boreal forests. In England *Ulmus* shews a steady increase in mean pollen frequency from zone iv to zone viia, where in some western sites it amounts to 30 per cent of the total tree pollen; thereafter it declines. In Ireland it evidently expanded later, shewing no significant trace in zone v, and not attaining maximum values until sub-zones vib and vic: in these sub-zones, however, the elm pollen attained even higher values than those in south-western England; afterwards it declined in frequency. It is by no means clear why the elm shews this pronounced Boreal increase of frequency towards the west, for it is not apparent that the genus favours oceanic conditions at the present day. The genus certainly resembles *Corylus* in susceptibility to late spring frosts and

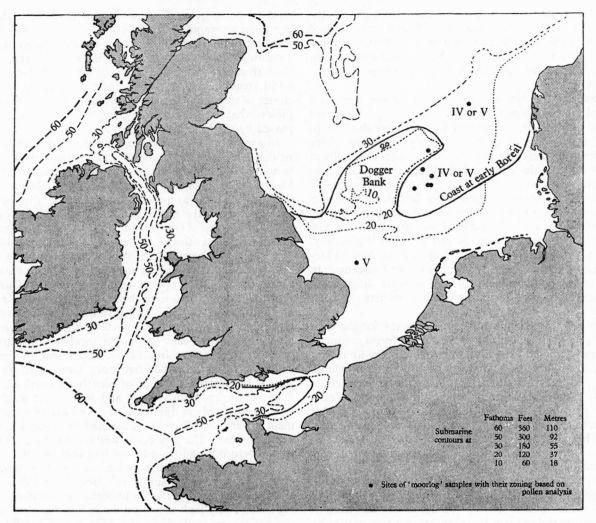

Fig. 162. Map of coastal waters of the British Isles shewing present ocean depths in fathoms, sites of pollen analyses from the floor of the North Sea with their approximate zoning, and an approximate estimate of the coastline between England and the Continental mainland at the opening of the Boreal period.

It will be noted that whereas Ireland must have been cut off from Great Britain and France when the sea-level began to rise from 60 fathoms (110 m), even when the eustatic recovery had progressed by a further 40 fathoms (73 m) a wide belt of land still existed between eastern England and the western mainland of Europe.

The position of the early Boreal coastline has been estimated by supposing that by this time eustatic recovery had brought ocean level to within 20 fathoms (37 m) of its present height, and by taking into account the local subsidence of the southern North Sea since early Boreal time.

Radiocarbon dates confirm the conclusions indicated initially by pollen analysis.

in requiring mull soils for easy regeneration. Both genera shew an apparent lag in extension in Ireland relative to that in England, and it may well be that in the Boreal period they were both expanding freely and to some extent haphazard before the forest cover had become dense, whereas later they came into competition with the oak (and with lime in England) in a phase of dense deciduous woodland.

Quercus pollen, which had been present in low frequency in zone v, increased somewhat in the English sub-zone vIa, and then increased sharply in vIb to equal or exceed the elm frequencies. In vIc it increased slightly and this increase was continued throughout zone vII. The

Irish history is similar, but the highest oak frequencies for zone vI as a whole are concentrated in eastern Ireland, where they are similar to the English and Welsh values. In Scotland and north-east England oak, though present, was in much lower frequency.

Alnus glutinosa was clearly present in low frequency throughout the British Isles in zone vI though certainly its pollen frequency was increasing notably in the closing stages of that zone. The pollen frequencies suggest that it was most frequent in northern England and north-eastern Ireland. *Tilia cordata*, which may already have been present in East Anglia during zone v, established itself throughout England and Wales during zone vI, in

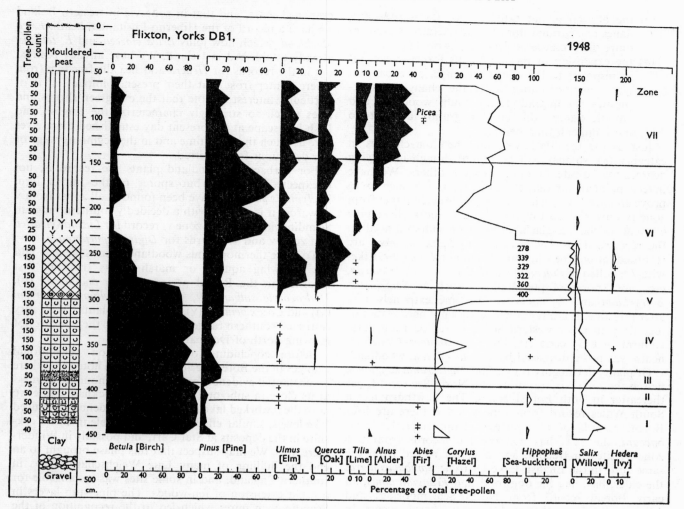

Fig. 163. Pollen diagram through the deposits of Lake Pickering in north-east Yorkshire, shewing the general drift of vegetational alteration from the Late Weichselian to the thermal maximum of the Flandrian. The *Abies* and *Alnus* pollen in zone I must be derived. Close to this site is the prolific Mesolithic settlement site of Seamer, shewn by pollen analysis to be referable to the transition between zones IV and V. Note especially the *Hippophaë* in the Late Weichselian layers, the very high *Corylus* early in zone VI and the high *Pinus* later in zone VI.

considerable frequency south of the line joining the Humber and the Bristol Channel. As with the alder, this English expansion was especially pronounced in subzone c. The low pollen frequencies make its actual presence less certain in Scotland, and it is not represented in Ireland then or subsequently. Records for *Fraxinus excelsior*, the ash, are so infrequent and shew such low pollen frequency that presence remains uncertain, save perhaps in the extreme south-west, a region peculiar also in that it seems not to have had a Boreal pine maximum but to have been essentially dominated by the broad-leaved trees. This prompts the thought, supported by the history of the English Lake District and of Skye, that in regions with mountain topography near our western coasts early expansion might have taken place of elements of deciduous woodland that had long previously found shelter as small stands in particularly favoured valley situations.

As thus disclosed the vegetational history of the Boreal period in the British Isles is one of response to increasing warmth by the migration and expansion of the pine and hazel, closely followed by the broad-leaved forest trees. At first migration rates and local circumstances much affected the woodland development, but by the end of the period closed deciduous forest covered the lowland landscape with vast continuous expanses broken locally only by river valleys, mires, lakes and steep local scarps of rock. The main features of floristic alteration during this period must have been:

(a) the expansion and rise to dominance of the trees themselves;

(b) the suppression and extinction from vast areas of the species requiring open habitats, together with those affected directly and adversely by the increasing warmth;

(c) the accompanying extension of the shrubs and lianes and ground flora which tolerate or even require the presence of deciduous woodland;

(d) the expansion of thermophilous forms in general in response to the climatic change, an expansion now much conditioned by the changed opportunities for migration: no doubt woodland and aquatic plants still found migration at least no more difficult than before.

Just as the records have established the changes in category (a), so they will be found, though in varying degree, to indicate or document the others. We have indeed pointed out that the evidence is not such as to prove absence of a species, but it is quite evident that from zone IV onwards most of the northern and arctic–alpine elements of the late-glacial flora occur seldom if at all in the records. Thus *Betula nana* and *Salix herbacea* are recorded infrequently in zone IV and not thereafter. Likewise the pollen of *Polemonium*, *Helianthemum*, *Artemisia*, etc., and spores of pteridophytes such as *Selaginella*, *Botrychium* and *Ophioglossum* become extremely rare after this time. It is a point which does not need elaborating. Even in north-western Scotland radiocarbon dates confirm that by 9000 B.P. the open *Juniper–Empetrum* heath was being displaced by *Corylus–Betula* woodland.

Of plants in category (c) we find *Thelycrania sanguinea* first at the zone IV/V transition in a North Sea record, but thereafter in both Boreal zones. The hawthorn has a South Wales record from zone VIb and there are four Boreal records of *C. monogyna* from Ireland; *Taxus baccata*, the yew, has macroscopic records from two zone VI sites in Ireland, and a pollen record from the same zone in the English Lake District; *Sambucus nigra*, the elder, has, like the dogwood and *Frangula alnus*, an early Boreal record from North Sea moorlog; and *Viburnum opulus*, the guelder rose, twice occurs in Ireland in zone VI. Pollen of *Ilex aquifolium*, the holly, occurred first rather dubiously in zone V and at the V/VI transition, but there follow three zone VI records before its great expansion in zone VII. Pollen of the ivy, *Hedera helix*, has been more frequent than that of *Ilex*; after an occasional grain in zone IV and very low frequencies in V, it exhibits much increased frequency in zone VI. Although *Lonicera* pollen is present in the south-west in zones IV and V, it is in zone VI that it became more frequent: we assume the records to be of *L. periclymenum*.

Of these plants it is noteworthy that the dogwood, elder, honeysuckle, buckthorn, hawthorn, yew, holly and ivy are species with a very southerly distribution range in Scandinavia, and although the guelder rose has a scattered range up to the Arctic Circle it clearly shews, like the dogwood, much diminished frequencies in the north of the British Isles: the dogwood is indeed possibly not native in Scotland and both species are notable in the British flora as having naked winter buds, a character to which Raunkaier has given decided ecological significance. It will be noted further that the holly, ivy and dogwood are atlantic or sub-atlantic in a phytogeo-graphical sense, and that in zone VIc in western Ireland we have a record of the Hiberno-Lusitanian heath, *Erica mackiana*, which now joins *Erica cinerea* and *E. tetralix* already established in Ireland. We have already pointed out in the records for *Hedera* and *Ilex* the freedom from severe winter frost that their presence implies. It is of particular interest to note that the evergreen shrubs and trees which so strikingly characterize the English and Irish landscape at the present day established themselves here at much the same time and in the beginning stage of our dediceous forest period.

For herbaceous woodland plants the record is, not unexpectedly, sparse, but spores of the marsh fern, *Thelypteris palustris*, have been found in all zones from V onwards: it is a plant with a decidedly southern range in Scandinavia. There is a zone VI record for *Luzula sylvatica* and zone V and VI records for *Digitalis purpurea*.

To these thermophilous woodland species we may add the following aquatic or marsh plants that are first recorded in the Boreal period: *Apium nodiflorum* (V), *Hydrocotyle vulgaris* (V, VI), *Ceratophyllum submersum* (VI) and *Carex pendula* (V). All of these plants have a very restricted southern range in Scandinavia, the two last not growing north of Denmark.

Before concluding our comment on vegetational changes in the Boreal period it should be noted that there is substantial evidence for a phase of pronounced dryness at its close in sub-zone VIc. Thus at Hockham Mere we have the reworked lake muds and the evidence of reduced lake levels, similar effects are recognizable in east Yorkshire in the deposits at Star Carr, and possibly in Windermere also. We may suspect that this phase brought to an end the infilling of many Late Weichselian valleys in southern England, and in some sites was responsible for a rapid extension of fen-woods. The pine-stub layers in Scandinavian mires which led to the recognition of the Boreal as a dry period may, one feels, have been connected with this shorter phase of dryness at its close. In Ireland it is noteworthy that Jessen has pointed to low lake levels in Boreal time and also to the greater prevalence of wood peats then as compared with the succeeding Atlantic period. Likewise in Scotland the Scandinavian interpretation of the data of Giekie and Lewis is that the tree-stump layer beneath the 'Lower Turbarian' must be Boreal in age.

Although this effect has a certain interest as explaining local characteristics of the geological and floristic records and causing an obscure and tiresome hiatus in the sequence of deposits, it may be doubted whether the late-Boreal dry phase made much permanent impression on the present pattern of British vegetation or floristic distribution.

13. THE THERMAL MAXIMUM AND THE ATLANTIC PERIOD

(7000 to 5000 B.P.)

While we may, with some plausibility, use the dates shewn above to give to the Atlantic period convenient chronological limits, it would be very misleading, or at best premature, to equate it precisely with the Thermal Maximum (or 'Climatic Optimum' or 'Hypsithermal Interval') that nevertheless is well expressed within the Atlantic: it remains uncertain (and to some extent a matter of definition) how far one should extend it back into the Boreal or forward into the Sub-boreal period.

The reality of the Flandrian Hypsithermal Interval is borne out by the fact that not only did the European forest belts move northwards to latitudes higher than those they now occupy but at the same time they ascended the great mountain chains to descend later during the climatic degeneration. Evidence of these movements is amply supplied by pollen analysis on the Continental mainland, but it is supplemented by records of macrofossils such as the fruits of hazel investigated by Andersson; by remains of *Trapa natans*, the water chestnut; *Cladium mariscus*, the sword sedge; *Emys orbicularis*, the pond tortoise; and many more species, both plant and animal, which have been shewn to have occupied in Flandrian time regions considerably northwards of their present natural limits. Andersson considered that the growing season during the hazel's widest northern extension must have been longer than it is now, and that it had a mean summer temperature 2.5 °C higher. These conclusions are broadly substantiated by Iversen's analyses of the Danish Flandrian records for *Ilex*, *Hedera* and *Viscum*.

It is quite clear that the same increased warmth had effect in the British Isles, and indeed it was the thermophilous marine mollusca of the Belfast Lough deposits that first led Lloyd Praeger to announce the existence of this former maximum of warmth. As we have already indicated the thermal maximum in the British Isles is characterized by the 'mediocratic' behaviour of certain species which are shown to have been more abundant and more widely spread in the middle Flandrian period than during its early or its later stages. Within this category we have already mentioned *Ceratophyllum demersum*, *Cladium mariscus*, and, in the broadest sense, the deciduous mixed-oak forest trees themselves. It seems likely that the period of highest temperatures included the later part of zone VI (the end of the Boreal period), and zone VIIa (the Atlantic period), for this is the time of greatest extension of the two floristic elements best placed to respond to climatic improvement, the forest trees themselves on the one hand and aquatic plants on the other.

The most evidently thermophilous genus of the Flandrian woodlands, *Tilia*, exhibits a great expansion in zone VIIa: although it had attained its full Flandrian range in zone VI, its frequencies were considerable only south of the Humber–Bristol Channel line, but with the onset of the Atlantic suddenly it is represented over all England and Wales by substantial pollen frequencies that almost certainly under-represent the real frequency of the tree, that now must be reckoned an important component of the deciduous woodland, and perhaps locally dominant with *Quercus*. There seems good evi-

dence, at least for central and eastern England, that the summer lime, *Tilia platyphyllos*, expanded along with *T. cordata* to which the great bulk of pollen records in this country belong. The lindens did not now, or later, establish themselves in Ireland. The ivy, *Hedera helix*, which had already expanded greatly in zone VI, increased further in four of the major geographical areas of the country (fig. 74), only in eastern England exhibiting a small decline. The still more thermophilous holly, *Ilex aquifolium*, expanded substantially in the Atlantic period, and likewise the ash (*Fraxinus excelsior*), a species that is decidedly frost sensitive, increased in both site numbers and pollen frequency from zone VI to VIIa.

Among the aquatic plants the cases of *Najas marina* and *N. flexilis* are outstanding. The former is now very closely restricted to a few Norfolk Broads, but it has been recorded at several sites in Ireland from Boreal deposits and at Skelsmergh Tarn in the English Lake District. Hockham Mere, Norfolk is at no great distance from the Norfolk Broads, but its chief interest in this connection is that Tallantire's numerous borings there have given a clear picture of the former abundance of *N. marina* during the end of zone VI and the beginning of zone VII. Thereafter it seems to have disappeared from the lake or to have become very infrequent.

N. flexilis is a plant now restricted in the British Isles to the extreme Atlantic fringe, with a very sparse and disjunct European range, and it gains especial interest from having a much denser North American frequency, so that it has sometimes been considered that its present range reflects an American origin (see chapter VI: 16). The fossil records shew that *N. flexilis* occurred widely throughout England, Wales and Ireland in the Flandrian period, particularly during the time of the Thermal Maximum. Thus its present range can be considered only as a product of recent Flandrian restriction of range and not as affording evidence of earlier events.

The boundary between the Boreal and Atlantic periods was made sharper by the climatic dryness of the closing stages of the Boreal. It is generally held that at this time the climate of the British Isles became pronouncedly more oceanic, and whether this is indeed the cause it is certain from extending series of pollen analyses that the development of the ombrogenous raised bog and blanket bog was now initiated throughout the land. Thus the Pennine mountain chain, hitherto forest-clad to high altitudes, passed by way of wet *Molinietum* to a continuous cover of acidic oligotrophic bog, and over lakes and coastal flats the oligotrophic *Sphagnum* spp. began to build up their huge domes of brown peat. This widespread process was not reversed until recent historic time, although extension was checked in the later part of the Sub-boreal period.

The spreading of a continuous mantle of peat mire poor in plant nutrients, waterlogged and moderately or highly acidic in reaction, across great stretches of country, especially in the north and west, must have affected profoundly the biogeographic pattern of the land. On the

Fig. 164. Forest composition in the British Isles, as indicated by radiocarbon dated pollen diagrams, at two periods of the Flandrian, respectively 8400 and 5500 years B.P., the former early Boreal, the latter at the end of the Atlantic period and the top of the marine transgression that in Scotland laid down the Carse Clays (shewn black). In both periods the forests were still unaffected, to any serious extent, by human activity, and can be supposed largely determined by climatic and edaphic factors.

one hand it exterminated from vast areas most of the woodland plants, and of all plants, including aquatics and species of open habitats, requiring neutral or alkaline soils of high base status and good drainage. The late-glacial relicts already restricted by the extension of forest were now, for the most part, driven from all mountain habitats save where local effects of steepness, aspect or erosion prevented peat accumulation.

On the other hand the oligotrophic species which so strikingly dominate *Sphagnum* bogs now must have increased dramatically in range and abundance. Our fossil records accord with this; thus *Eriophorum vaginatum* becomes frequent after zone VI, and *Calluna vulgaris* occurs wherever the blanket bog or raised bog peat has been found.

It is of course true that each vegetational type favours, and may act as a refugium for a certain range of plant and animal species, and this remains true of the oligotrophic mires, although the range of phanerogams they support is small. We should not overlook the fact that these mires are devoid of tree-cover, and that provided plants tolerate the extreme soil conditions they can achieve the advantages of a relatively open habitat. Until recently we had no examples to match that of the persistence of *Betula nana* recorded by Overbeck and Schmitz in the Luneberger Heide, where it is found throughout all the layers of the bog from late-glacial time up to the present day, and still remains (truly relict) as a colony of living plants far separated from its next occurrence in the Hartz mountains. In 1966 however Hutchinson reported that

a small colony of the dwarf birch was still growing on Widdybank Fell in upper Teesdale far from all other British sites, and traces of the species were recovered in peat deposits of the Fell shewing that it had persisted there at least from zone V. Many comparable instances have been furnished by the bryophyta (Dickson, 1973).

Whilst pollen analyses have shewn the alder present throughout the British Isles in zone VI, and in such considerable amount that its local presence is certain, nevertheless over a great deal of the British Isles at the opening of zone VIIa it exhibited a very dramatic increase in frequency. In part this was a consequence of the thermal response of the tree, but the sudden massive rise in the alder pollen curves can only be the response to general waterlogging of lowland plains and valleys, so that alder now became part of a forest mosaic, occupying wet glades in the prevalent deciduous mixed-oak forest. The preceding dryness of zone VIc made the contrast more pronounced. In central Ireland the alder expanded later and more slowly, perhaps because waterlogging induced widespread mire formation, and in the Central Highlands of Scotland again it spread later and ineffectively, evidently not generally spread in the pine–birch forests, as it was in the oak woodlands further south. It is difficult to explain the near-synchroneity and widespread expansion of an already established tree such as the alder save by some general climatic shift such as we have postulated.

What is of especial interest is that in the Atlantic period, with migratory movements of the arboreal plants, aside from beech and hornbeam, now completed, we can see

PLATE XXI

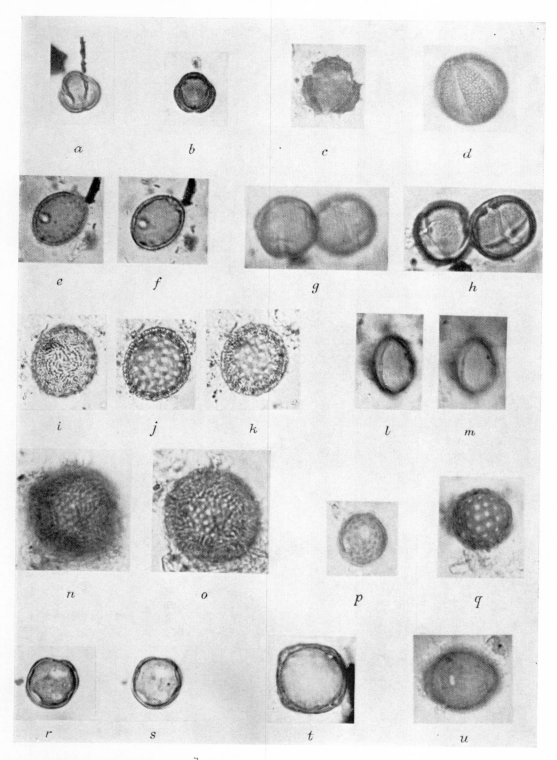

Pollen grains from Late Weichselian deposits at Nazeing (Lea Valley), Essex: *a*, cf. *Filipendula* (1 gr.); *b*, *Artemisia* (1 gr.); *c*, *Matricaria* type (1 gr.); *d*, *Labiate* cf. *Thymus* (1 gr.); *e*, *f*, *Jasione* (same gr., 2 planes of focus); *g*, *h*, cf. *Lysimachia* (2 gr., 2 planes of focus); *i*, *j*, *k*, *Polemonium* (same gr., 3 planes of focus); *l*, *m*, *Helianthemum* (same gr., 2 planes of focus); *n*, *o*, *Polemonium* (same gr., 2 planes of focus); *p*, *Plantago* cf. *media* (1 gr.); *q*, *Chenopodium* (1 gr.); *r*, *s*, *Thalictrum* (same gr., 2 planes of focus); *t*, *u*, *Myriophyllum* cf. *verticillatum* (2 gr., 2 planes of focus). (Magnifications of *i*, *j*, *k*, *l*, *m*, ×520; *a*, *b*, *g*, *h*, *q*, *r*, *s*, ×740; *c*, *d*, *e*, *f*, *p*, *t*, *u*, ×820.)

PLATE XXII

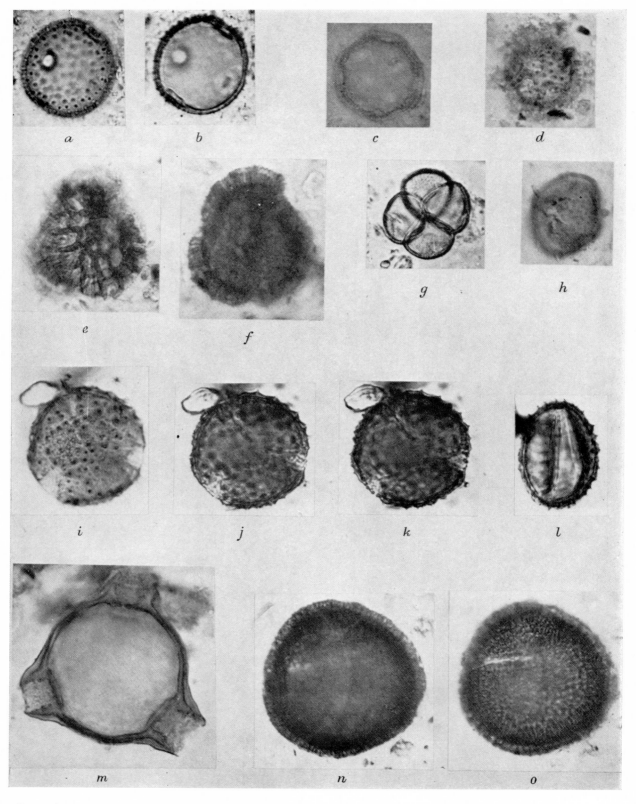

Pollen grains from Late Weichselian deposits at Nazeing (Lea Valley), Essex: *a, b, Campanula* cf. *rapunculoides* (1 gr., 2 planes of focus); *c, Silene* type (1 gr.); *d, Sonchus* type (1 gr.); *e, f, Armeria maritima*, type A (1 gr., 2 planes of focus); *g, Typha latifolia* (tetrad of 4 gr.); *h, Centaurea* cf. *scabiosa* (1 gr.); *i, j, k, l, Valeriana officinalis* (1 gr., 4 planes of focus); *m, Epilobium* (1 gr.); *n, o, Succisa pratensis* (1 gr., 2 planes of focus). (Magnifications of *a–h*, × 740; *i–m*, ×820; *n, o*, ×660.)

PLATE XXIII

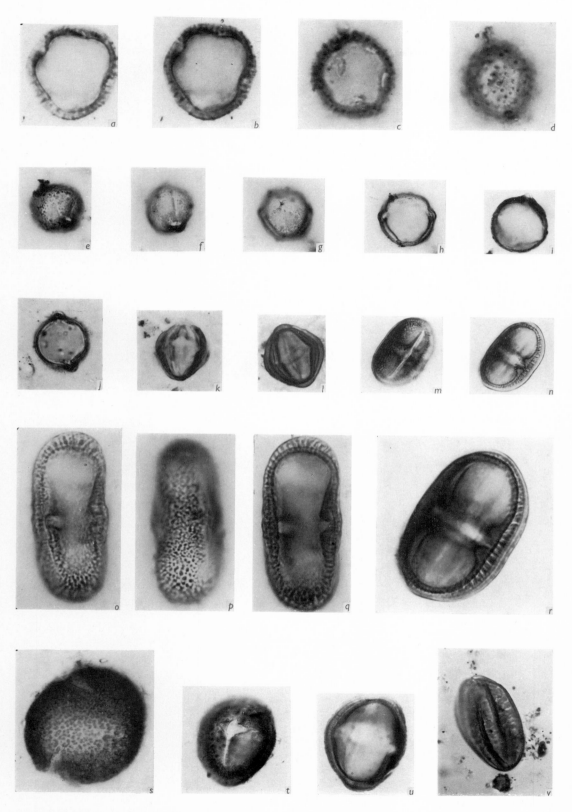

Pollen grains of the Late Weichselian period: *a–d, Koenigia islandica* (Whitrig Bog, zone III); *e–i, Hippophaë rhamnoides* (Hoxne, IG); *j, Poterium sanguisorba* (Drake's Drove, zone VIII); *k, l, Sanguisorba officinalis* (Eastwick, Late-glacial); *m, n, o, Centaurea cyanus* (Drake's Drove, early zone VIII); *p, q, r, Pastinaca sativa* (Cross Fell, zone VII); *s, Succisa pratensis* (Cwm Idwal, zone VI); *t, u, Centaurea scabiosa* (Elstead, zone IV); *v, Ephedra* cf. *distachya* (Whitrig Bog, zone III). (Magnifications of *a–d,* × 1350; *e–n, s–v,* × 600; *o–r,* × 1200.)

PLATE XXIV

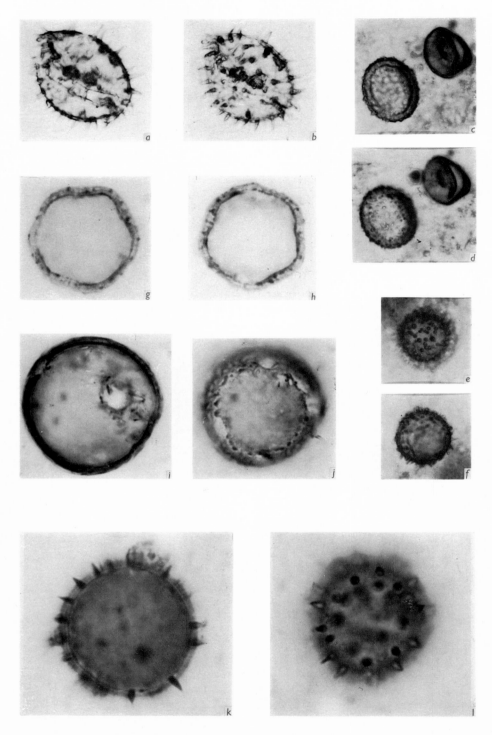

Pollen of various aquatic plants: *a, b, Nymphaea alba* (Flixton, zone v/vi); *c–f, Nuphar lutea* (Old Decoy, Late Weichselian); *g, h, Sagittaria sagittifolia* (Glastonbury Lake Village, zone viii); *i, j, Littorella uniflora* (Cwm Idwal, zone vi); *k, l, Lemna* (Kirkby Thore, zone vi). (Magnifications of *a–h*, ×600; *i, j*, ×1200; *k, l*, ×2300.)

for the first (and possibly last) time the major climax woodlands of the British Isles in stable equilibrium, their distributions determined by natural environmental controls (fig. 164). If we refer back to the zone VIIa pollen frequency maps for the tree genera, a very pronounced regional differentiation is apparent, of which the outstanding feature is the contrast between England and Wales on the one hand, and central and northern Scotland on the other. Deciduous woodland, dominated by *Quercus*, *Tilia* and *Ulmus* with abundant *Corylus* and *Alnus*, and *Fraxinus* in smaller amount, must have extended over the whole of England and Wales attaining higher altitudes than conditions of today allow. In north-western England however the woodlands contained still a great deal of one or other species of birch, a genus by now severely restricted in the closed deciduous woodlands further south.

While this climax vegetation characterized England and Wales, it is now certain that in the central and northern highlands of Scotland, the climax vegetation was pine–birch forest, constituting, as A. G. Tansley had earlier pointed out, the westerly extension of the great northern coniferous forest belt of the Continental mainland. This is now proved to be the ancient natural origin of the present-day native Scottish pine forests centering on the Cairngorm and Spey areas. It is of considerable interest that in the far north of Scotland, pine was less frequent and birch forest was dominant, as if again reflecting the European latitudinal sequence of climax forest. No doubt thin soils over igneous rock and soils developing podsolization favoured the birch and pine forests that persisted in England and Wales only in special pre-climax situations of soil, aspect and drainage. We can be sure that oak, elm, hazel and alder were present locally in Scotland, especially in the west, responding to situations of soil and local climate that favoured post-climax woodland. In some respects, such as the higher frequency of oak, hazel and alder, Scotland south of the Clyde–Forth line was transitional to English conditions and in fact radiocarbon dating of peat above and below the Carse Clays shews beyond doubt that throughout the Atlantic period, Scotland north of this line was indeed almost an island separated by salt water from the rest of Great Britain (fig. 164), a circumstance no doubt enhancing the strong vegetational contrast we have demonstrated. It seems probable that Ireland was, like England and Wales, now mantled with deciduous woodland, from which however lindens, and to a large extent ash, were absent and in which the elm, *U. glabra*, and hazel, *Corylus avellana*, played a larger part. However, pine woods evidently persisted in considerable strength outside the central plain where, as we have said, alder remained surprisingly infrequent. As in England and Wales, birch was not present in high frequency and one could not suggest any equivalence in vegetational terms to the Scottish birch–pine woodlands.

14. PREHISTORIC HUSBANDRY AND THE SUB-BOREAL PERIOD

(5000 to 2800 B.P.)

In the consideration we have thus far developed of the pattern of change in the British flora, and of the vegetation types composed by it, we have had to do essentially with the responses of plants and animals to the 'natural' features of the environment: we have seen the major vegetational types migrating and altering under compulsion from changing climate, and species spreading or retreating according to the circumstances of soil, climate and the presence or absence of effective barriers to dispersal. In this system prehistoric man has played a part no more significant than that of other vertebrates dependent by hunting, fishing, and plant collecting upon the productivity of the environment. This dependent status of man in the ecosystem undoubtedly held for Palaeolithic man and his Mesolithic successors: the forest or the tundra dominated him, dictated the nature of his food, his tools, his clothing, his dwellings, and directly or indirectly controlled almost every aspect of his existence. He, on the other hand, made small local clearances by stone axe and by fire and provided scattered and impermanent habitats for a few species capable of responding quickly to the local openness of vegetation. At Star Carr on the islands in Lake Pickering the colony of Mesolithic red-deer hunters felled the local birches, and seem to have encouraged the growth of nitrophilous plants such as the nettle and members of the Chenopodiaceae, and Sims has detected Mesolithic clearance effects in recent studies of Hockham Mere. Gross has laid stress upon similar activities in Middle Europe as likely to have provided refugia for steppe species but this effect *in toto* cannot have been large.

The role of man within the ecosystem began to change quite decisively as farmer cultures were spread across Europe by the migrations of Neolithic peoples bringing with them the techniques of cultivating cereals and keeping domestic animals. By comparison with these innovations the introduction of polished stone axes as against flaked ones was a matter of small ecological significance. It was now that man first exhibited such power over the environment that he could destroy the natural vegetation, provide himself with sustenance from secondary sources, and eventually dominate the ecosystem instead of remaining a passive and minor reactant within it.

The first documentation of this effect upon the natural vegetation of western Europe was provided by the palynological researches of Iversen, published in 1941 under the title *Land occupation in Denmark's Stone Age*. The changes he describes are first apparent a small distance above an horizon in the pollen diagrams where the pollen of elm (*Ulmus*) and ivy (*Hedera*) shews a consistent and lasting decrease, whilst that of ash (*Fraxinus*) first establishes itself as a continuous curve. This is the generally accepted boundary between the sub-zones VIIa and VIIb, i.e.

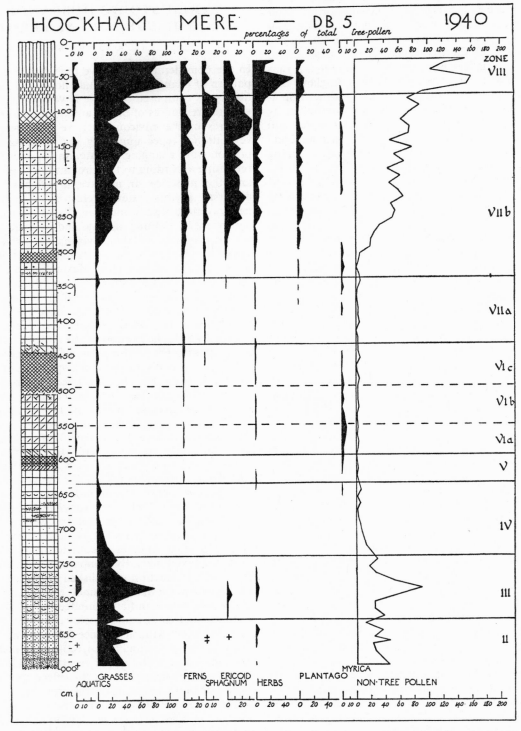

Fig. 165. Non-tree pollen diagram at Hockham Mere, Norfolk. In the Late Weichselian the ratio of non-tree to tree pollen is high, but thereafter to the end of zone VIIA the very low non-tree pollen values indicate a condition of closed forest cover. Shortly above the 340 cm horizon, approximately representing the early Neolithic level, the non-tree pollen curves begin substantial and maintained increases which indicate the presence of the great 'heath' communities of the Breckland. Note that pollen of *Plantago lanceolata* (ribwort plantain) corresponds to the general run of non-tree pollen curves (cf. fig. 26).

between the Atlantic and Sub-boreal climatic periods, and, as we have already indicated, it appears to be roughly contemporary over a large part of north-western Europe. In many Danish pollen diagrams, not far above this boundary the pollen curves demonstrate a sudden and peculiar set of changes in forest composition: this is often most strongly shewn where a layer of charcoal, stratified into the marginal muds of lake or fjord, gives evidence of local burning and thus of local human occupation. In Iversen's words, 'at this level the elements of the high forest, *Quercus*, *Tilia*, *Fraxinus* and *Ulmus* undergo a distinct but temporary decline, while *Betula* reveals a transitory, *Alnus* a more lasting increase in pollen frequency, and at the same time the *Corylus* (hazel) curve reaches a very pronounced maximum'. He takes these shifts to express the local vegetational changes in an area where land-tilling people had occupied territory and had cleared the dense primeval forest by felling and by burning. It is indeed easy to explain the sequence of the pollen curve phenomena in terms of our knowledge of the ecology of the various genera involved. The decline of the curves indicative of high-forest tree pollen is due to local destruction, the rise of birch, alder and hazel afterwards is due to the characteristically rapid regeneration of these species in cleared areas, partly no doubt from stumps, but partly also from seeds, which, especially in birch and alder, are very readily dispersed by wind.

This interpretation is decisively supported by many lines of evidence: there is the local presence of Neolithic remains, the charcoal layers corresponding with the onset of the vegetational disturbance, and the fact that at this level the *absolute* tree-pollen frequencies (i.e. pollen grains per unit area of microscope slide) fall suddenly to very low values and then slowly recover. Finally there is the remarkable evidence of the contemporary pollen of herbaceous land plants from the same sequence of records. As one would expect, *ex hypothesi*, just above the charcoal layer the sum of the non-tree pollen was found to increase suddenly and then to fall away; and, still more significantly, it was possible to identify most of the species constituting this component. Firbas had already pointed out that the pollen grains of cultivated cereals could be distinguished from those of most wild grasses, and Iversen was able to shew that low frequencies of cereal pollen were continuously present in the Danish pollen diagrams above the level of the sudden oak-forest diminution. Even more striking was the fact that from this level there occurred continuous and often substantial curves for the pollen of many plants now recognized as weeds or ruderal plants. Foremost amongst these was the pollen of *Plantago lanceolata* and *P. media*, the latter a species so strongly associated with human settlement as to have earned the name from the North American Indian of the 'white man's footprint'. There was also present in high frequencies the pollen of *Artemisia*, and Iversen suggests that this is probably derived from *A. vulgaris*, the mugwort or common wormwood, a plant which remained a serious weed in Denmark until deep ploughing came

into general practice, and which still persists in the British Isles as a common ruderal plant.

Sometimes the occupation episode seems to have been but brief, and the pollen diagrams suggest that the high forest was soon able to heal the wound in its uniformity. In other places the results suggest that there was a whole series of forest clearances persisting over a large area, but by and large the forests did not shew signs of permanent or widespread modification until the opening of the Iron Age, some two thousand years after the first Neolithic settlement. It seems probable that the Neolithic settlers made use of a clearance fire to establish areas for grain-growing, sowing seed in the wood ash and continuing to crop the same area for a few years, abandoning it as its fertility decreased. A similar usage is known among many aboriginal peoples and indeed continues today in some of the most backward parts of Europe. It is a technique which provides not only the space for cultivating cereals, but gives areas of herbaceous vegetation and tender coppice shoots for fodder of grazing animals. The latter must indeed not be underestimated, for by preventing seedling establishment they provide a most potent and continuing means of arresting regeneration of the forests after their initial destruction by fire and axe.

No sooner had Iversen's work been published than its applicability to others part of Europe was evident, and it clinched the interpretation already formulated for the pollen curves at Hockham Mere near Thetford in East Anglia. The latter is a site of especial importance for English phytogeography because it lies just at the northern margin of the so-called Breckland, an area of open heath, grassland, and broken woodland some few hundred square miles in extent and harbouring a rich variety of plant life, including many species of the Watsonian 'Germanic' type, with their main centre of distribution in central or eastern Europe. Ecologists were very uncertain of the ecological factors responsible for the preponderating treelessness of the region, although Farrow had made a strong case for grazing as the chief cause. The deep muds of Hockham Mere (fig. 165) provided a continuous picture of the Breckland vegetation from the Late Weichselian until post-Tudor time, and it was immediately apparent that throughout zones V, VI and VIIA complete woodland cover must have obtained: throughout this time the total non-arboreal pollen remained a very small fraction of the total tree pollen. Not far, however, above the VIIA/VIIB boundary, the ratio of non-tree pollen to tree pollen rose to high values. From this time forward grass pollen, ericoid pollen (chiefly *Calluna*), fern spores, *Plantago lanceolata*, and miscellaneous herbs including *Rumex* and many types of Compositae contributed largely to the pollen spectra. When we recall that the Breckland was as densely occupied in Neolithic time as any part of Britain, and that the great flint mines of Grimes Graves lie only ten miles or so away, it becomes impossible not to recognize the origin of the present heath communities of Breckland in the forest clearances initiated here, as in Denmark, by the

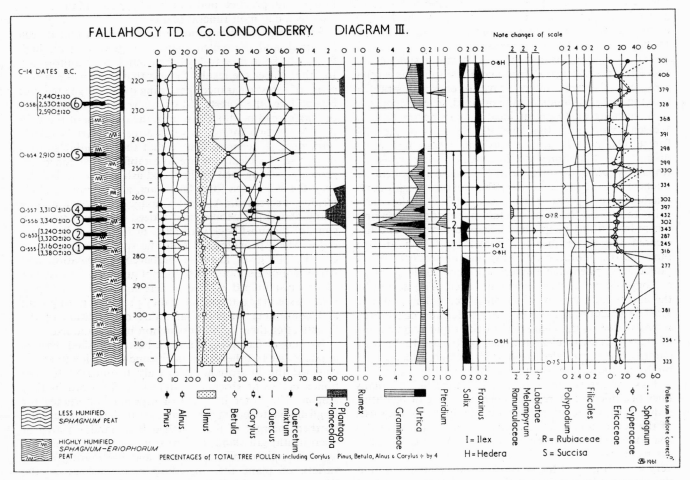

Fig. 166. Pollen diagram from Fallahogy, Co. Londonderry, shewing two forest clearance phases. Values are expressed as percentages of total tree pollen plus *Corylus*, but with values for *Betula, Pinus, Alnus* and *Corylus* divided by four. Six samples for radiocarbon assay date the clearances (in years B.C.). The earliest clearance, Neolithic, is before 3000 B.C., and the radiocarbon dates provide a time scale for the successive clearance, farming and regeneration stages disclosed by the pollen analyses. (After Smith & Willis, 1961–2.)

Neolithic agriculturists. The pollen curves appear to indicate that the heaths once established remained open, thereafter probably extending, as in Denmark again, during the Iron Age. These results have been strongly confirmed by subsequent work in the region.

The feasibility of forest clearance by Neolithic man was illustrated by the enterprise of Iversen and J. Troels-Smith who successfully undertook clearance of an area of Danish deciduous forest using only polished stone axes for felling and ringing trees. The experiment involved burning of the felled timber *in situ*: sowing Neolithic-type wheats in the ash gave successful crops and even shewed the invasion of arable weeds.

There now followed the wide recognition by palyno-logists throughout western Europe of similar effects to those described by Iversen, and intensive investigation by close sampling and large pollen counts, to clarify the vegetational events induced by early agriculture. Troels-Smith (1956) drew attention to the evidence in Danish sites that the steep decline in elm pollen frequency decisively preceded the expansion of cereals, *Plantago lanceolata* and weeds of pasture. This he attributed to a cultural phase in which there was selective gathering of the leafy shoots of *Ulmus* for cattle kept penned in stalls, a practice that persists in many parts of the world up to the present day. Oldfield (1960) subsequently shewed that a 'pre-plantain' clearance stage had been present at Thrang Moss in Lancashire, and the research school in Belfast have used radiocarbon dating to shew that in north-eastern Ireland it had a duration of about 100 to 200 years (fig. 166). Similar exercises shewed that the cycle of clearance, occupation and regeneration indicated by the pollen diagrams had a time span acceptable in ecological terms, viz. about 150 years at Whixall Moss, Shropshire (Turner, 1965) and 300 years at Fallahogy, Co. Derry (Smith, 1958c; Smith & Willis, 1961–2).

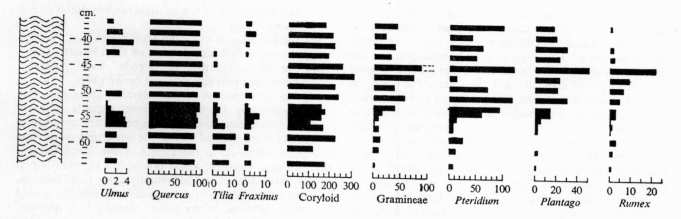

As percentage of (*Ulmus* + *Quercus* + *Tilia* + *Fraxinus*) pollen

Fig. 167. Pollen diagram from the raised bog at Whixall Moss, Shropshire with very closely spaced samples to elucidate changes at the level of the *Tilia* decline. Each sample took about thirty years to form and the episode is dated 3230 ± 115 B.P. Shortly after the *Tilia* decline, *Fraxinus* expanded briefly and then *Fraxinus* and *Ulmus* both declined, as grasses, bracken, plantains and docks greatly increased, reflecting the effect of grazing the cleared woodland area. (After J. Turner in Hutchinson, 1965.)

Miss Turner advanced our capacity to reconstruct post-clearance effects by distinguishing pollen types indicative of pasturage, e.g. Gramineae, *Plantago* and *Rumex*, from those indicative of arable cultivation, such as *Artemisia*, Chenopodiaceae, Compositae (including *Centaurea cyanus*) and Cruciferae. This enabled her to discern that at Tregaron in central Wales temporary small clearances of the kind described by Iversen persisted on and off from the early Neolithic until the Iron Age when extended clearances were accompanied by a pastoral economy that itself gave place to arable farming from just before Norman time. Her results also shewed the intensive agriculture of the seventeenth century and a reversion to pastoralism in the last 100 years, both events corresponding with historical evidence. We shall see further evidence of this kind in discussing the Sub-atlantic period.

The 'elm' decline itself was so conspicuous and so consistent that it had been accepted as a zone boundary in the British Isles and independently on the Continent before there was any appreciation of the association with forest clearance that is now so widely recognized. It was accompanied by other floristic changes such as a decrease in *Pinus* and *Hedera* pollen frequencies and was regarded as presumably a response to general climatic change, difficult though it was to suggest a mechanism acting so selectively upon the elm among forest trees. Some support was lent to this view however by the approximate synchroneity of radiocarbon dates for this event, a very large part of which fell between 5500 and 5000 years B.P. It seemed difficult to envisage such a sudden and extensive a spread of this Neolithic husbandry, and there remained some European areas where the elm decline was evident though apparently Neolithic occupation was absent. The initial resistance of archaeologists to such early dates was overcome by a flood of radiocarbon evidence, and investigations such as those at Shippea Hill (Clark &

Godwin, 1962) and Beaghmore (Pilcher, 1969; Pilcher, Smith, Pearson & Crowder, 1971) tied vegetational and archaeological events firmly together. Ten Hove (1968) has very objectively surveyed the present situation, concluding that the major cause of the European elm decline has been human activity though with some background climatic effect. The strongly selective adverse effect on the elm, explicable by Troels-Smith's hypothesis, implies either extraordinary damage by a short cultural phase or prolonged leaf-fodder collection from the tree after the 'pre-plantain' phase. Morrison (1959) indeed had suggested that in Antrim forest clearance took place preferentially on calcareous areas within the generally basaltic region, and that these areas had supported most of the elm population. A similar effect was proposed by Turner (1964) for the *Tilia* decline in Shropshire (fig. 167) but it cannot be by any means the whole explanation.

There is no doubt that the concept of the elm decline as indicator of the spread of early Neolithic husbandry fits extremely well into the pattern of the extension of that economy from its centres of origin in Northern Iran, Irak, Turkistan, Palestine, Syria and the Caspian from about 8000 B.P. Waterbolk (1968) has convincingly surveyed these events. He draws attention to the way in which the nomadic Mesolithic hunters of the early Boreal sought the abundant large mammalia over the whole countryside, whereas the subsequent closure of the climax deciduous forest diminished the big game and forced man progressively to the shores of seas and lakes where fish and bird life provided a constancy of protein food that allowed at least semi-permanent settlement. It was naturally in these coastal areas then that the new Neolithic economy, probably by cultural diffusion rather than migration, was most readily adopted, a conclusion generally in line with the archaeological evidence. Certainly radiocarbon dates for early Neolithic settlements in western France, Ireland and south-western Sweden indicate an early spread along the Atlantic coastline.

Although human disturbances to the forest communities of the British Isles were extremely widespread in the Sub-boreal period they were, in Iversen's phrase, 'brief and violent' and they did not very greatly modify the general pattern established in the Atlantic period, as may be seen from the frequency maps of the individual tree genera. In the mixed-oak forest the patterns for *Quercus*, *Tilia* and *Corylus* remained unchanged; only *Ulmus* displayed a dramatic decrease in frequency, most especially however in Ireland where it had previously been remarkably abundant (see fig. 83). It was possibly a consequence of this that *Alnus*, which elsewhere in the British Isles retained its former frequency, increased substantially throughout the Central Irish Plain. One wonders whether woodland clearance on the basic soils carrying the elm had induced such a local rise in ground water-level as to admit the alder. *Tilia* held its own in the Sub-boreal, despite the fact that it was susceptible to forest clearance, and Turner (1962) has shewn that it suffered similarly to the elm from local clearances that were demonstrated in the Somerset Levels and in Shropshire to have quite different radiocarbon ages. Throughout England, Wales, Ireland and southern Scotland, but again on basic soils, *Fraxinus* increased decisively, most particularly in southern Scotland and northern England. It appears to have shewn a bigger positive response to clearance than any other tree, and it is possible to see in this phenomenon the origin of the native ash-woods whose ecological status was so long dubious. It is notable how *Betula* remains frequent in Scotland (especially the north-west) but low throughout England and Wales and central Ireland; it possibly increased in Devon and Cornwall but it seemed generally not to be able to take advantage of clearances in the main mixed-oak forest area. *Pinus* retains the frequency pattern of the Atlantic period: the apparent decline in the Scottish Highlands is associated with the fact that zones VIIb and VIII are scarcely separable in pollen diagrams from that area. It is interesting that in north-east Ireland Pilcher (1969) had found it susceptible to clearance along with elm. *Fagus sylvatica*, the beech, exhibits a very clear expansion from zone VIIa and numerous macroscopic identifications confirm the view that it was firmly established south of the Wash–Severn line: its presence is more problematical further north but is not to be excluded. It may well be that it was already taking advantage of the conditions associated with clearing high forest as Iversen (1949) and many others have suggested is commonly the case. The hornbeam, *Carpinus betulus*, was certainly present in England in the Sub-boreal as macroscopic records in the south testify, but it was less frequent than the beech.

The introduction of agriculture into a fully forested land must have affected our flora in a great variety of ways, of which the most immediately obvious are:

(*a*) the destruction of forest trees themselves, perhaps selectively;

(*b*) the diminution of the habitats for woodland undergrowth flora;

(*c*) the introduction of alien crop plants;

(*d*) the accidental introduction of alien weeds;

(*e*) the encouragement of 'weeds' and 'ruderals', plants of open habitats and broken soil;

(*f*) the development of considerable areas of scrub, particularly by those elements capable of rapid dissemination, and resistant to fire and to grazing.

Whilst for the present we confine our attention to these effects as expressed in the events of the Neolithic and Bronze Ages, both of which fall within the Sub-boreal period, it is evident that they equally apply to all periods. The first effect needs no further comment, and the second is scarcely reflected in the pollen records. It is however generally conceded that those domesticated plants of the Neolithic economy were all introductions from southeastern Europe and the near East where they originated (see account of 'Cereals', pp. 405–15).

It is evident that during the Neolithic period wheat and barley were of comparable importance, but that throughout the ensuing Bronze Age barley far exceeded wheat in frequency. Most of the barley in the Neolithic and Bronze Ages was the 'naked' form of *Hordeum hexastichum*, but the hulled form also occurred and *H. vulgare*, the 'bere', was present already in Neolithic time. It is notable that even in the Neolithic the cultivation of barley was practised from Sussex to the Orkneys and County Antrim. Of the wheats the small spelt (*Triticum monococcum*) has been very infrequently recorded in the Neolithic and Bronze Ages, but emmer (*T. dicoccum*) was equal with barley as the dominant cereal of the Neolithic, though diminishing in the Bronze Age. Although bread wheat (*T. vulgare* and *T. compactum*) occurred in both periods it was always in low frequency, and perhaps diminished from the Neolithic period to the Bronze Age. Rye, oats and spelt were later introductions. A number of other economic plants are first recorded in this country during the Sub-boreal, including *Coriandrum sativum*, *Linum usitatissimum*, *Brassica rapa*, *Chenopodium bonushenricus*, *Malus sylvestris*, *Prunus domestica* and *P. cerasus*.

It is naturally difficult, if not impossible, to decide at this late date which species of our present weed flora were introduced along with these prehistoric cereal crops, and which were older inhabitants now re-extending their range. None the less it is of interest to note the list (table 43) of such plants that have not so far been recorded in the British Isles before the Neolithic or the Bronze Ages.

It is of course appreciated that many plants widespread in the open vegetation of the Late Weichselian now re-extended their range, a fact especially evident from the pollen frequency of genera such as *Artemisia* and *Rumex*. It is particularly notable how large a proportion of all plants recorded in zones VIIb and VIII, and having a southerly phytogeographic range (as defined in section 9), fall into the category of weeds and ruderals:

TABLE 43. *Weeds and ruderals first recorded in the Sub-boreal (Neolithic and Bronze Ages)*

Ranunculus bulbosus L.	*Anthriscus caucalis* Bieb.
Papaver rhoeas L.	*Torilis nodosa* (L.) Gaertn
Fumaria officinalis L.	*Polygonum lapathifolium* L.
Brassica rapa L.	*Urtica urens* L.
B. nigra (L.) Koch	*Anagallis minima* (L.)
Sinapis arvensis L.	E. H. L. Krause
Raphanus raphanistrum L.	*Cuscuta epithymum* (L.) L.
Thlaspi arvense L.	*Veronica hederifolia* L.
Cardamine flexuosa With.	*Lamium purpureum* L.
Silene gallica L.	*Sherardia arvensis* L.
Spergula arvensis L.	*Galium spurium* L.
Chenopodium bonus-henricus L.	*Picris echioides* L.
C. botryodes Sm.	*Sonchus asper* (L.) Hill
Medicago lupulina L.	*Bromus sterilis* L.
Trifolium campestre Schreb.	*B. secalinus* L.
Vicia tetrasperma (L.) Schreb.	*Arrhenatherum tuberosum*
Euphorbia amygdaloides L.	(Gilib.) Schultz
Agrimonia eupatoria L.	*Phleum arenarium* L.
Lythrum hyssopifolia L.	

TABLE 44. *Trees and shrubs first recorded in the Sub-boreal (asterisked) or extending then in relation to forest clearance*

*Acer campestre L.	**P. avium* (L.) L.
*Euonymus europaeus L.	**Sorbus aria* (L.) Crantz.
*Rhamnus catharticus L.	**Malus sylvestris* Mill.
*Buxus sempervirens L.	*Crataegus* sp.
*Ulex sp.	**Carpinus betulus* L.
*U. europaeus L.	*Fagus sylvatica* L.
*Rosa sp.	**Salix fragilis* L.
Prunus sp.	*Fraxinus excelsior* L.
P. spinosa L.	*Sambucus nigra* L.

this may reflect how, apart from climate, conditions of arable cultivation favour such species but it will also be due to the tendency for arable cultivation to be concentrated in the south and east of Britain.

Patches of scrub are now such a familiar part of the British landscape that it is difficult to realize that in pre-Neolithic time scrub was unknown, or at most limited to small areas of extreme exposure to wind or to local grazing. Within the constantly moving mosaic of natural high forest the shade-tolerant shrubs play a minor and migratory role, thriving in the local clearances caused by the death of old trees and suppressed by dense competition of subsequent tree regeneration. Likewise it may be supposed that where streams, marshes and fens broke the continuity of the forest canopy, shrubs tolerant of moist soils would occur, locally even abundantly in patches of fen carr. There is good palynological evidence that the temporary forest clearances of the Neolithic at various sites encouraged a passing expansion of such pioneer genera as *Betula*, *Corylus*, *Fraxinus* and *Alnus*, and no doubt shrubs to some extent accompanied them. These occurrences however are not comparable with the scrub communities of today, such as occur throughout the country on sites of clear-felled woodland, and on arable and grassland abandoned from husbandry in times of poor economic returns for agriculture. Such scrub represents the stages of secondary successions leading back towards the forest climax. The abundance of spiny or unpalatable shrubs, especially in the marginal parts, emphasizes the resistance which such scrub offers to grazing, which may however, with the sheep and the rabbit, easily be sufficiently intense to keep the scrub from spreading and to give it a hard unnatural edge.

Our plant records give strong support to the view that such scrub as this was first produced in the Neolithic and Bronze Ages, for the charcoals recovered from settlement sites of these periods not only are the earliest occurrences

for many woody shrubs, but they occur widely, abundantly, and in association with one another.

The list in table 44 gives the woody plants that the records shew to have substantially increased at the onset of prehistoric agriculture, and those marked with an asterisk have not been certainly recorded in the Flandrian period before this time.

The conjunction of dwarf shrub plants was clearly shown by Hyde's identifications (Hyde: in Fox, 1937) at the Middle Bronze Age site at Pond Cairn in South Wales. Here he recorded charcoal of *Ulex* (gorse), *Crataegus* (hawthorn), *Corylus* (hazel) and *Pteridium* (bracken), and it is apparent that they indicate that scrub had already come into existence at this time.

We must of course not expect prehistoric land utilization in the Neolithic and Bronze ages to bear much resemblance to modern intensive agriculture and the earliest leaf-fodder collection and shifting cultivation were no doubt succeeded by many variants in the kind and intensity of exploitation and many transient systems of treating the modified vegetation types that arose successively with the intake of different soils and different landscapes.

Where the woodlands were not fully destroyed they may well have been heavily exploited for grazing, pannage, and leaf fodder as well as for timber. There is clear evidence from the Somerset Levels that extensive hazel coppice already existed on the Liassic limestone hills. More than one Neolithic wooden trackway has been found made of the long straight shoots of *Corylus* that could only have grown as coppice shoots, and various pollen diagrams from peat bogs in the region shew such large and sharply fluctuating pollen curves for *Corylus* that considerable stretches of hazel scrub must be inferred. No doubt this was capable of yielding not only timber but an important source of storable food: hazel nuts have been so used from Mesolithic time.

Pronounced as was the impact of the introduction of plant and animal husbandry upon our native vegetation, its influence reached different regions at very different times. Known occupation sites shew that in Neolithic times settlements were concentrated upon the lighter soils, and in England the densest establishment was to be found upon the ridges of the great calcareous hills of southern England, the Wolds, Salisbury Plain, the

Cotswolds, the North and South Downs, the Mendip, and, as already mentioned, the East Anglian Breckland.

Since we have already had evidence that the Neolithic disforestation of Breckland produced there the open heath communities which have persisted right through to the present day, it is not unreasonable to suggest that the great stretches of grassland upon our chalk hills may have had a similar history. Indeed the radiocarbon dated pollen diagrams from Frogholt and Wingham in Kent, with an abundant weed flora and high non-arboreal/arboreal pollen ratio, suggest that the Chalk of south-eastern England was largely disforested at least as early as 3700 B.P., i.e. by the end of the Neolithic. There must naturally have been large shifts of prehistoric populations and much alternate retreat and readvance of the forests in many areas. Further consequences of this, and other effects of prehistoric agriculture, will be considered in the next section, which concerns the Sub-atlantic period in which they were extended and intensified.

We have finally to consider what is known of the climatic conditions of the Sub-boreal period since a good deal of biogeographic theory has been made to depend upon it. In the schema of Blytt and Sernander the Sub-boreal was described for Scandinavia as warm and dry with Continental characteristics; subsequently Gams and Nordhagen upheld the view that these climatic qualities were retained across north-western Europe during the same period. Phytogeographers were not slow to develop from this the concept of a Flandrian 'xerothermic period' during which, it was supposed, species of 'steppe' affinity from central and south-eastern Europe were able to migrate northwards and westwards. A special impetus was given to this approach by the publications of Gradmann, who, already at the turn of the century, had pointed out that in central Europe there was close correspondence between the sites of prehistoric settlements and the localities where a 'Steppenheide' flora still persisted, mostly sites of open or unwooded character on thin and sterile soils. It was not a far cry from this to the extreme view that the dry conditions of the Sub-boreal period had prevented tree growth on the poorer soils at this time, and had provided a corridor of more or less open landscape across central Europe from the Danube north-westwards, along which Neolithic man had been able to spread with rapidity and ease. It is evident, as has already been pointed out, that clearance of forests upon lighter soils took place in Neolithic time, but it seems unlikely that *general* opening of the forests preceded establishment by farmer peoples. Indeed the evidence by pollen analysis is chiefly of local clearances in a forest cover capable of renewing itself when human disturbance had ceased, and there is now little doubt that the primary cause of forest clearance in Britain was anthropogenic.

We may consider what evidence there remains of the character of the Sub-boreal climate in this country. In Scotland the 'Upper Forestian' of peat bogs, rich in stumps of pine and birch and ascending to high altitudes on the mountains, has been generally held to be Sub-

boreal, and its equivalent is apparently present in the dried-out surfaces of raised bogs below the Sub-atlantic peat. It is nevertheless easy to overestimate both the warmth and dryness of this period, for the tree layers, and layers of *Calluna* and *Eriophorum* in the bogs, are emphasized much more by the sudden climatic shift demonstrated in their overlying layers than by distinctness from layers below them. So far the programme of radiocarbon dating tree stumps in Scottish peat deposits has failed to demonstrate a concentration within the Sub-boreal, nor indeed do the more extensive Scandinavian results do so.

The decline in frequency of elm at the zone VIIa/VIIb boundary, taken with the decrease in eastern Britain of the ivy, is evidence pointing to increased continentality; but the expansion of the ash and beech, both susceptible to spring frosts, points in the contrary direction, and with the strongly thermophilous linden holding its ground it is difficult to accept any suggestion of any considerable fall in mean temperature. We shall note that lake deposits, such as those of Hockham Mere, shew no evidence of any drastic fall in level such as that which marked the close of the earlier Boreal period. We may add to this the fact that in general the peat of Atlantic age in our raised bogs is at least as fully humified as that of the Sub-boreal, and that indeed in some places already within the Sub-boreal period a recurrence surface (less developed than that at its close) was present.

Having in mind the paucity of evidence that the British climate was in fact conspicuously warm or dry, and the fact of our full isolation by sea from the Continental mainland at this time, one is disinclined to invoke a 'xerothermic period' as an important phytogeographic agency, nor so far do the plant records appear to demand it.

15. THE SUB-ATLANTIC PERIOD AND CLIMATIC DETERIORATION

(2800 B.P. to present day)

The Sub-atlantic period which follows the Sub-boreal and continues to the present day is characterized by two effects: in it there became apparent the so-called 'climatic deterioration', and during its progress human influence upon the natural plant cover of our land became ever more extensive and severe. Neither effect acted uniformly, and the operation of the one effect was often obscured or reinforced by that of the other.

The term 'deterioration' is an expression of a turn of climate towards moister, more atlantic conditions with lower summer temperatures and less severe winters, but there were of course climatic fluctuations of some magnitude within the Sub-atlantic period itself. One very pronounced effect of the general climatic change was to cause resumption in growth of the more or less dried surfaces of raised bogs and blanket bogs, which had passed into the standstill phase, and were often *Calluna*-clad, or bore birch woodland. They now rapidly became waterlogged, and in the pools upon their surfaces was laid down a most characteristic layer of aquatic *Sphagnum* spp. (often associated with *Andromeda*), or, in places where calcareous drainage water could overflow them, a layer of *Cladium–Hypnum* peat. This aquatic peat (the Precursor peat) formed the basis for the re-establishment of the active growth of *Sphagnum* peat in a climate highly favourable to it, so that in general this layer separates the lower highly humified dark brown *Sphagnum–Calluna* peat from an upper fresh, unhumified, very pale *Sphagnum* peat. As active peat growth recommenced at this boundary it has become known as a 'recurrence surface'. As we have pointed out earlier, recurrence surfaces have been recognized at several different stages of the Flandrian period, but throughout all the lowlands of north-western Europe the most pronounced was long attributed to the Sub-boreal/Sub-atlantic boundary, and regarded as corresponding closely with the transition from Bronze Age to Iron Age. However, intensive palynological, stratigraphic and radiocarbon investigations have shewn that the most prominent recurrence surface is not always of this age, nor even is such a surface necessarily synchronous in its extent over a single bog (see Godwin, 1966). In the Somerset Levels the boundary between the 'black' and the 'white' peat is marked by numerous wooden trackways built over the bog surface in response to its progressive flooding: not only do they display the mortices and axe-cuts typical of late Bronze Age tools, but many radiocarbon age determinations date them between 2900 and 2400 B.P. The very few other wooden trackways from English peat mires are of similar age, and it seems likely that there was climatic worsening through the Late Bronze Age as well as the succeeding Iron Age: this conclusion agrees well with Jessen's experience that in Irish raised bogs Late Bronze Age artifacts have most commonly been recorded from immediately over the main recurrence surface, his RS:C.

This major change of wetness, directly recorded by the plant remains of the bogs, clearly favoured the extension of some bog species more than others. From this time on, *Sphagnum imbricatum*, a strongly atlantic moss, played a very large part in building up the raised bogs of England, Wales and Ireland, as of northern Germany and Holland. *Scheuchzeria palustris*, a plant which now characteristically forms loose floating mats along with *Carex limosa* in the deep pools of active raised bogs (for example in Komosse, South Sweden), now became abundant in English raised bogs, often forming very conspicuous layers of slender rhizomes and papery leaf-sheaths in a matrix of *Sphagnum cuspidatum* (see plate XI). Similar layers are frequent in the early Sub-atlantic peat layers of Denmark. They have been found at a single site in Ireland, where also the living plant persisted for a short time. Increasing wetness of the bog surfaces may well also explain the fact that *Narthecium ossifragum* (bog asphodel) has been found in the Irish peat bogs at several localities during this zone.

It is evident that the same climatic shift that caused growth to recommence in the raised bogs revitalized also existing blanket bog, and certainly caused mires of both kinds to extend widely over land previously held by woodland. In western Ireland this effect is very notable, and Mitchell (1951) has described Sub-atlantic peat grown over the flank of a large Bronze Age tumulus on Carrowkeel, Co. Sligo. The extended growth of blanket bog on the mountains must have had the effect not only of destroying high-altitude woodland, but of destroying some of the habitats capable until that time of harbouring species of open habitats relict from Late Weichselian time. Many of those relict areas still remaining were now islands in a sea of blanket bog, as formerly they had been in a sea of woodland.

Outside the range of edaphic and climatic circumstances favourable to development of ombrogenous mire, there must have been big stretches of country in which podsolization set in where there had previously been brown earths. This process was possibly most pronounced where *Calluna* had spread into abandoned clearings in woodland on sandy soils, and certainly in Denmark the *Calluna* heaths extended enormously in the Sub-atlantic period. Strong as the probability is that these changes occurred, and important as they must have been in altering the plant distribution of the country, our records are not of a kind to document them save in exceptional instances, and they are also very closely bound up with progressive human destruction of the natural vegetation cover of the landscape.

We may naturally enquire to what extent the climatic worsening of the Sub-atlantic is found to be reflected in the distribution pattern of the forest-dominants, but in practice this is a matter of uncertainty. For a considerable time the decline in *Tilia* pollen frequency, apparent throughout England and Wales, had been taken as a climatic response that indicated the zone VIIb/VIII boundary. But detailed re-examination, by pollen samples at close intervals, at the hand of Miss Turner (1962) shewed each episode of sharp *Tilia* decline to be associated with forest clearance in the very obvious manner illustrated in fig. 165. Moreover radiocarbon dating of these episodes shewed that they were of different ages in different parts of the country, although, as is natural, they tended to fall mostly when clearances of the mixed-oak forest accelerated. Consequently the pollen frequency map for *Tilia* (fig. 53) shews in zone VIII the same range, but with far diminished frequencies, as that of zone VIIb; the Scottish and Irish records are of frequencies low enough to be explained by wind transport from a distance. The other tree to shew a pattern of Sub-atlantic diminution is *Pinus sylvestris*, although severe retraction is specially characteristic of the whole of Ireland, where no doubt it was extensively exploited for fuel and timber especially throughout the long monastic period. Elsewhere in the British Isles both pollen and macroscopic records testify that pine still held in full its former range.

In general the trees of the mixed-oak forest do not shew much change in *relative* pollen frequency, though their absolute abundance, as indicated by low arboreal/non-arboreal pollen ratios, greatly diminished. Oak shews no consistent change in relative amount, nor does the elm, except perhaps for a tendency to increase in the western half of England. The alder is generally unchanged in status save for East Anglia where it appears diminished by widespread drainage on lands converted to arable cultivation. The hazel is not much changed except for a decrease also in East Anglia.

By contrast with these four genera, four others shew relative expansion in the Sub-atlantic, i.e. *Betula, Fraxinus, Fagus* and *Carpinus*. The birches have clearly increased throughout the British Isles save in central and northern Scotland where they were already very frequent: the increases are specially evident in Wales, south-west Ireland, along the length of the Pennines and in most individual pollen diagrams. The ash exhibits a clear increase in north-western and south-western England, south-western Scotland and Ireland, which is certainly a response to extended forest clearance on basic and moist soils. It is too thermophilous to be responding to lowered temperatures, and likewise, although the birches have a generally terminocratic role in the interglacial cycle, it seems likely that they too are mainly responding to woodland destruction by quick establishment of secondary wood on drier and more acidic soils.

Beech and hornbeam present a less straightforward case. *Fagus sylvatica* in the Sub-atlantic apparently increased in frequency in all the areas it had occupied in the Sub-boreal, and macroscopic records again confirm its local presence. It seems to have grown at least as far north as Lancashire, but exhibits its highest frequencies in Kent, East Anglia and Somerset. We have already indicated in the account of this species, that mass expansion of the beech, on the Continent as here, was controlled to a high degree by anthropogenic factors. Accordingly it seems probable that a shift of arable cultivation to heavier and deeper soils with the advent of improved ploughs and heavier plough teams in Roman and Anglo-Saxon time left the thin soils over the Chalk and Oolite (which had previously been cleared) with sparser human populations, and open to swift colonization by the beech. The calcareous hills of southern England and Wales are indeed the areas in which our residual native beech forests still occur.

The species account for *Carpinus betulus* makes it apparent that not only were there more sites of recorded hornbeam in zone VIII than in zone VIIb but that the mean pollen frequency had increased. This increase is most apparent south of a line from the Wash to south Wales, an area within which our residual native hornbeam woodlands are still found. The low hornbeam pollen frequencies at sites further north are less certainly indicative of local presence of the tree. As with the beech, Continental evidence indicates an extremely sensitive reaction of the tree to human forest clearance and exploitation, but its

endurance of lopping and its rapid capacity for spread suggests that it too, like the beech, probably expanded in response to removal of competition by previously undisturbed deciduous forest. The evident responsiveness of both beech and hornbeam to forest clearance should not blind us to the possibility that both may well also be responding in some degree to the Sub-atlantic climatic shift for both have behaved as terminocratic forest elements in earlier interglacials where the vegetation was unaffected by man's activities.

Jessen has made the suggestion that the present treelessness of western Ireland is in part due to strong winds, and that the wind effects increased in the Sub-atlantic period. In this he is following a lead given by Praeger, but it may be doubted whether many British ecologists would give so great weight to it. When woodlands grow naturally the trees lend one another so much cover that they will grow (albeit dwarfed and wind-shaped) to within a short distance of very exposed western shores: it is when the more powerful agencies of forest destruction, felling, burning, grazing and mire formation have reduced the former cover to tatters that wind effects shew themselves most clearly. At high levels on the mountains and on narrow strips of coastal land in the west, wind (aided in the latter case by salt spray) has certainly inhibited tree-growth, but there seems no clear evidence that this effect was widespread inland or was intensified in Sub-atlantic time.

The extension of agriculture and its concomitant forest clearance continued with increasing rapidity in the Sub-atlantic period. The Late Bronze Age and Early Iron Age invaders, entering the country from the south and east chiefly occupied regions already opened by their farming predecessors. Sheep-grazing greatly increased in Iron Age and Romano-British times, as analysis of bones of domestic animals from settlement sites upon the Downs shews. In the earlier periods bones of pig and cattle preponderated, both forest-dwelling animals, but sheep displaced them to a large extent in the Iron Age, a clear indication that the open grasslands of the Chalk were in existence by this time. From now on the pace of disforestation was accelerated by the needs of charcoal for iron smelting, and by the increasing powers conferred by better tools, larger populations and improved techniques. Already by 1944 it was clear that both Late Bronze Age and pre-Roman Iron Age clearances could be identified in pollen diagrams from the Somerset Levels (fig. 168).

Increasingly in recent years palynologists have followed Miss Turner's lead in identifying and characterizing the phases of forest destruction and subsequent land-use at sites throughout the British Isles. Already by 1970, as fig. 169 shews, more than a hundred radiocarbon datings had been specifically applied to palynologically determined clearance episodes, and in sites such as Tregaron (Cardiganshire), Bloak Moss (Ayrshire), Beaghmore (Co. Tyrone) and Leash Fen (Derbyshire) many successive phases were recognized at a single site. In yet other instances, such as Old Buckenham Mere (Norfolk),

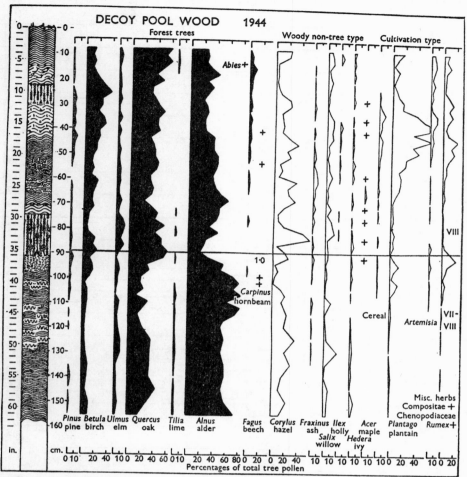

Fig. 168. Composite pollen diagram from the upper layers of the raised bog at Decoy Pool wood, Shapwick Heath, Somerset. Tree pollen curves are blacked in, non-tree pollen (expressed as a percentage of total tree pollen) are open curves. The peat stratigraphy shews a pronounced flooding horizon at *c.* 90 cm, where *Cladium–Hypnum* peat replaces humified *Sphagnum–Calluna* peat, and a less pronounced one at 30 cm. The former is at the Bronze Age/Iron Age transition and the tree pollen curves reflect the climatic deterioration at this time; the latter is of Iron Age or Roman date. Preceding each flooding episode the pollen of plants indicating agricultural activity (cereals, *Artemisia*, *Plantago lanceolata*, and miscellaneous herbs) rises to coincident maxima; these are small in the Late Bronze Age and much larger in the Iron Age. The pollen curves for *Ilex*, *Fraxinus* and *Acer* appear to shew some tendency to follow the weed and cereal pollen curves, and this may indicate that they were favoured by woodland clearance.

although radiocarbon dating was impracticable, it was possible to apply a reasonable time scale to the sequence of lake deposits extending almost up to the present day.

In some respects these exercises are very informative: they give for the vicinity of the pollen diagrams unmistakable evidence of forest clearance at specific times and convey an idea of the vegetation that replaced the woodland, and this can be compared with other evidence for the density and quality of historic or prehistoric occupation. The interpretation in closer vegetational terms is less easy, although one can characterize woodland regeneration phases and distinguish arable cultivation from grazing. It is particularly difficult to know whether a given pollen picture reflects a small but adjacent settlement or wider clearances over a distant, larger territory. These

deficiencies may be removed as research progresses. Meanwhile the dated diagrams provide a generally consistent picture of the impact of husbandry on the vegetational cover of this country (fig. 170). From the opening of the Neolithic with its still enigmatic elm decline, the dated diagrams disclose continuing transient forest clearances. It is in the Bronze Age, the Middle Bronze Age in the Lake District and Shropshire (as also in Huntingdonshire), and the Late Bronze Age in Tregaron and Somerset that clearances became more extensive, in the interest however of economies that were both pastoral and arable, or especially in the mountains of the west, dominantly pastoral. It was in the Late Bronze Age and Early Iron Age that the Urnfield people from Central Europe introduced the ox-drawn scratch plough,

475

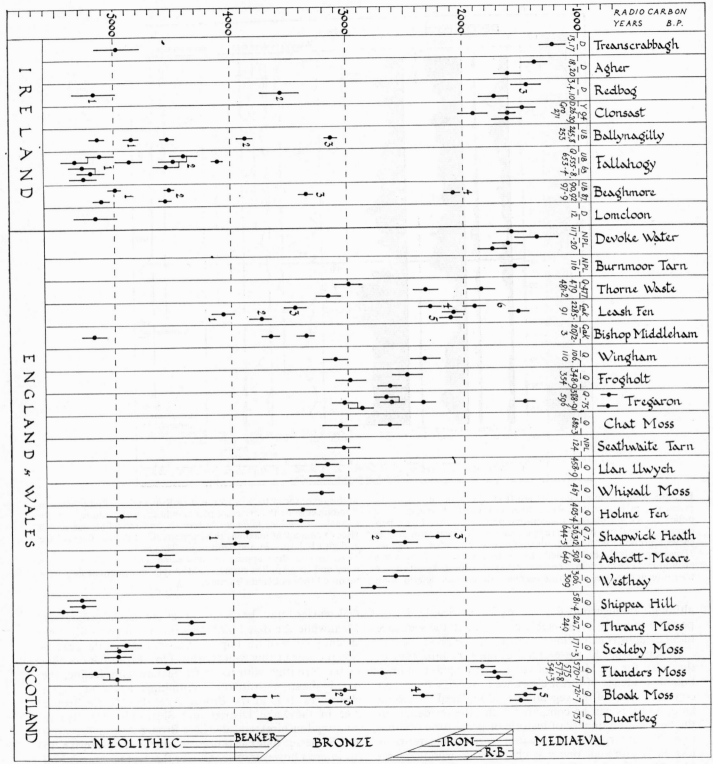

Fig. 169. Radiocarbon dates for forest clearance and agricultural phases identified pollen-analytically. Dates linked by horizontal lines indicate repeated determinations on one sample. At several sites successive indications of agriculture occur over a very long time span and can be referred roughly to the appropriate cultural stage by the radiocarbon dates. Note the consistently early onset in both Great Britain and Ireland. (After Godwin, 1970.)

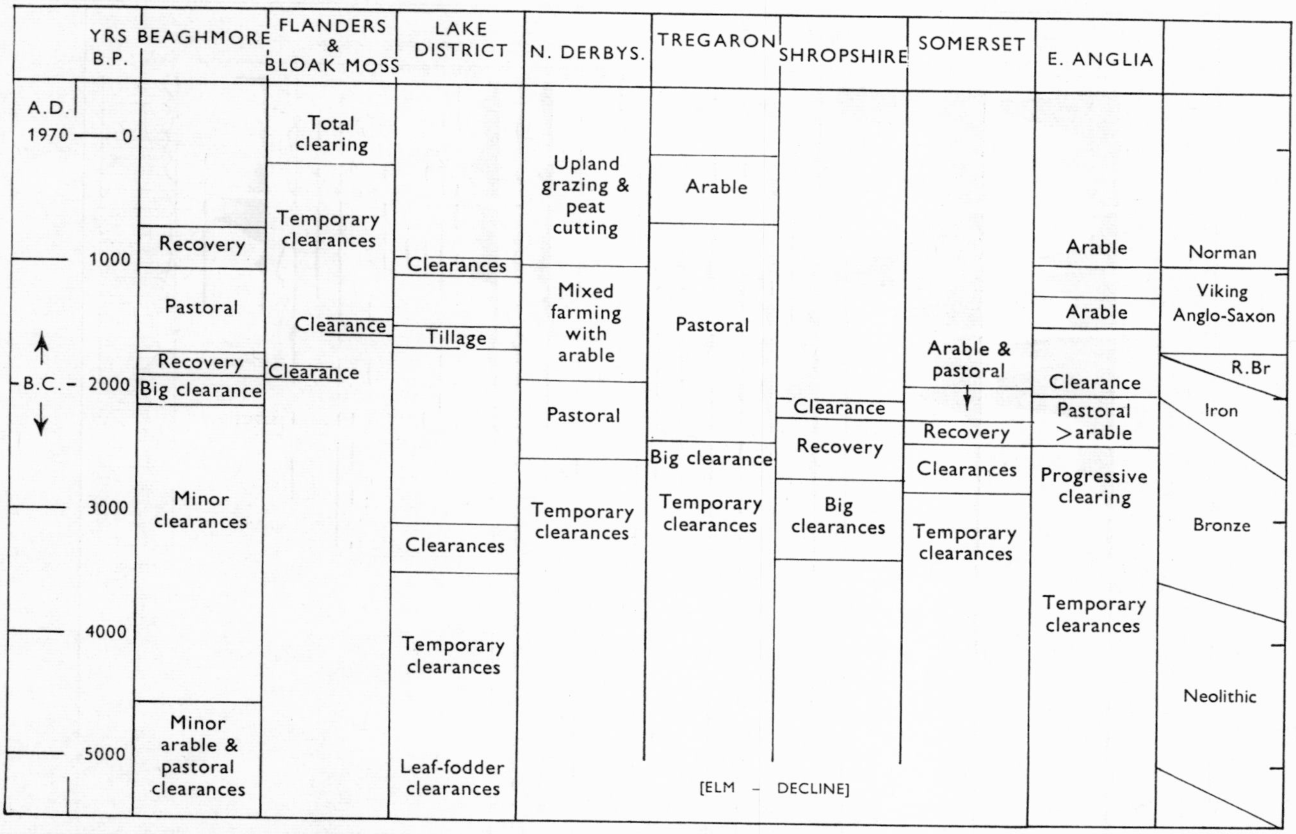

YRS B.P.	BEAGHMORE	FLANDERS & BLOAK MOSS	LAKE DISTRICT	N. DERBYS.	TREGARON	SHROPSHIRE	SOMERSET	E. ANGLIA	
A.D. 1970 — 0		Total clearing		Upland grazing & peat cutting	Arable				Norman
		Temporary clearances						Arable	
1000	Recovery		Clearances	Mixed farming with arable	Pastoral			Arable	Viking Anglo-Saxon
	Pastoral	Clearance	Tillage				Arable & pastoral	Clearance	R.Br
B.C. — 2000	Recovery	Clearance	Clearance			Clearance	Recovery	Pastoral >arable	Iron
	Big clearance			Pastoral	Big clearance	Recovery	Clearances	Progressive clearing	
3000	Minor clearances			Temporary clearances	Temporary clearances	Big clearances	Temporary clearances		Bronze
			Clearances					Temporary clearances	
4000			Temporary clearances						Neolithic
5000	Minor arable & pastoral clearances		Leaf-fodder clearances						

[ELM – DECLINE]

Fig. 170. Composite diagram representing conclusions drawn from sites in the British Isles at which serial pollen analyses allow deduction of the character of prehistoric land usage, and where an effective dating of the major phases has been possible, usually by radiocarbon age determinations at selected levels. Thus indirectly we attain a correlation with successive archaeological cultures on the one hand and realize regional differences on the other.

reconstructions of which have recently been made and tested in Reading University. The length of furrow achieved by the single ox was short and cross ploughing produced the characteristically small and square Celtic fields, which in all probability represent the first permanent arable land. The population increase in Central Europe at this time was considerable and the Urnfield people rapidly moved out to the western fringe of the Continent where their activities are faithfully reflected in the pollen diagrams by sharp increases in the indications of prehistoric agriculture. This type of economy was the basis of the extensive pre-Roman Iron Age cultivation of cereals throughout lowland Britain, south of the Humber–Severn line, and concentrated mainly upon the limestone hills and plateaux. Elsewhere to the north and west, although clearances were accelerated, they were in the interest of pastoral activities; the highland peoples, as the Roman invaders were to record, subsisting mainly on meat and milk and trading in hides. In Belgic time, as Applebaum has indicated, there was the introduction of the 'slip-eye' axe, the mould-board plough and larger cattle and horses, that seemed to promote the beginnings of a movement from the uplands to the deeper loams and stiffer, more productive soils of the lowlands. It was a tendency pursued in Romano-British time, possibly aided by economic and climatic factors. This brings us close to the time when historic records become available. It is also within sight of the introduction of big ox teams drawing the true plough, with the aid of which our Anglo-Saxon ancestors were able to maintain the acre strip system of arable cultivation and with which they were able to exploit the clearances of woodlands from the heavy clays of the English lowlands. From this time onward the forests vanished even more quickly, and our present agricultural landscape began to emerge in recognizable form. The pollen diagram from Old Buckenham Mere (fig. 171) shews clearly the spread of arable cultivation near the East Anglian capital, Thetford, where the heavy boulder clays now yielded crops of wheat, rye, flax and hemp in fields infested with the weeds typical of such cultivation. There is even evidence for selective felling of oak for building. In the English Lake District Mrs Tutin's diagrams reflect the still later influence of Viking intakes in the mountain valleys. Her analyses in this area are based upon coring in a series of tarns at different altitudes and serve to illustrate the difference

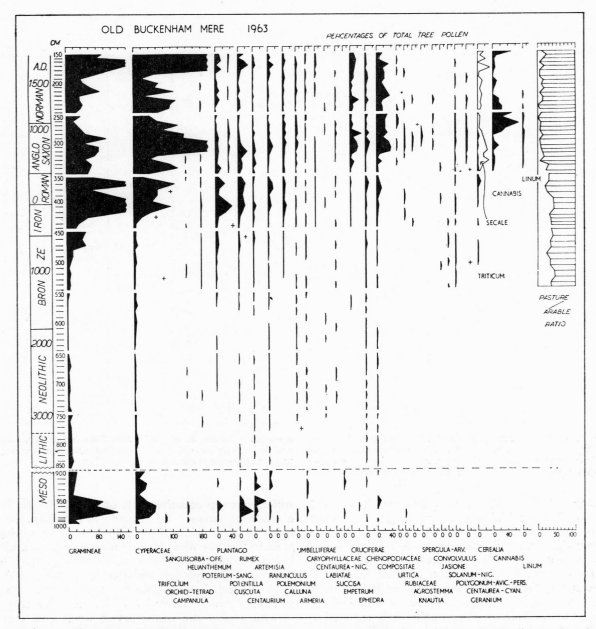

Fig. 171. Non-tree pollen diagram at Old Buckenham Mere, Norfolk: pollen of herbaceous plants associated with forest clearance, pasture indicators on the left and arable on the right. On extreme right the ratio of pasture- to arable-indicator pollen, and on the extreme left the conjectural time scale. The diagram shews increased arable cultivation in Anglo-Saxon and Norman time with crops of *Secale* (rye), *Cannabis* (hemp), *Linum* (flax) and associated weeds. (After Godwin, 1967*b*.)

that clearance had upon vegetation at varying altitudes, forest recovery being less certain at the higher sites which were more likely to be associated with podsolization and with the erosion episodes that have been registered in mineral layers in the tarn deposits (Pennington, 1964).

Several pollen analysts have seen in their diagrams the influence of the foundation of the monasteries, especially those of the Cistercians. Renewed clearance and extensive farming are thus indicated near Strata Florida, Cardigan-

shire, and Mitchell identified a phase of extensive monastic farming in Ireland a good deal earlier than that in Great Britain (Mitchell, 1965). The dissolution of the monasteries has been identified (*c*. A.D. 1550) near Furness Abbey, and we may perhaps hope some time to see diagrams that reflect such disasters as the Black Death, just as Continental diagrams have documented the *Wüstungsperiode* (A.D. 400 to 550) and the Thirty Years War (A.D. 1618 to 1648).

478

TABLE 45. *List of weeds and ruderals first appearing during the Flandrian in a Romano-British context*

Ranunculus sardous Cranz	*Lavatera arborea* L.	*Polygonum rurivagum* Jord. ex Bor.
R. parviflorus L.	*Geranium dissectum* L.	*P. aequale* Lindm.
Papaver argemone L.	*G. molle* L.	*Atropa bella-donna* L.
P. somniferum L.	*Ononis repens* L.	*Verbena officinalis* L.
Chelidonium majus L.	*Vicia hirsuta* (L.) S. F. Gray	*Betonica officinalis* L.
Coronopus squamatus (Forsk.) Aschers.	*V. cracca* L.	*Stachys arvensis* (L.) L.
Isatis tinctoria L.	*V. sativa* L.	*Ballota nigra* L.
Barbarea vulgaris R. Br.	*V. angustifolia* (L.) Reichard	*Nepeta cataria* L.
Erysimum cheiranthoides L.	*Lathyrus aphaca* L.	*Sambucus ebulus* L.
Alliaria petiolata (Bieb.)	*L. nissolia* L.	*Valerianella dentata* (L.) Poll.
Cavara and Grande	*L. cicera* L.	*Senecio sylvaticus* L.
Sisymbrium officinale (L.) Scop.	*Potentilla argentea* L.	*Anthemis cotula* L.
Reseda lutea L.	*Alchemilla vulgaris* L. agg.	*Chrysanthemum segetum* L.
Silene vulgaris (Moench) Garke	*Aphanes microcarpa* (Boiss. and Reut.)	*Arctium lappa* L.
Agrostemma githago L.	Rothm.	*Onopordon acanthium* L.
Chenopodium polyspermum L.	*Chaerophyllum aureum* L.	*Hypochaeris radicata* L.
C. murale L.	*Conium maculatum* L.	*H. glabra* L.
C. hybridum L.	*Bupleurum tenuissimum* L.	*Picris hieracioides* L.
Malva sylvestris L.	*Conopodium majus* (Gouan) Loret	*Tragopogon pratensis* L.
M. pusilla Sm.	*Aegopodium podagraria* L.	

* Recorded also from the Weichselian.

Although these careful pollen sequences are invaluable guides to the general effects of clearance and husbandry for evidence as to the crops themselves we have for the most part to rely upon identifications of macroscopic remains preserved by waterlogging or by charring and recovered in an archaeological context. Only here and there, as with *Secale* (rye) and *Cannabis* (hemp), do we derive substantial support from pollen evidence.

The crops cultivated by the Iron Age peoples shewed a distinct change from those of the preceding Bronze Age, the barleys were no longer dominant, their place as major crop being taken by the wheats, *Triticum dicoccum* (emmer), which had long been grown in this country, *T. compactum* and *T. vulgare* (the bread wheats) also known in the Neolithic and Bronze Ages, together with *T. spelta* (spelt), not previously grown here. Meanwhile apparently the cultivation of *T. monococcum* (small spelt) seems to have ceased. For the first time apparently, oats (*Avena strigosa* and *A. sativa*) now came into cultivation, and in Romano-British time it was followed by rye (*Secale cereale*) which was to become, during Mediaeval time, the staple food of the English peasantry. So far as the somewhat sparse records go, however, it appears that barley and oats were preponderant in the Anglo-Saxon period (see fig. 89). There is some evidence that flax-growing was practised in the British Isles from the Bronze Age, or even the Neolithic, onwards. Now that pollen identification of hemp is practicable it is apparent that it was cultivated in large amount in Anglo-Saxon and Norman time. Later, in the Tudor period its cultivation was enforceable by law. There is some evidence for it as a crop in Roman Britain, but there is a good case for thinking its extensive growth was associated with the immigration of Anglo-Saxon agriculturalists (Godwin, 1967*b*, *c*).

Since both flax (*Linum usitatissimum*) and rye are crops particularly subject to weed infestation and indeed tend to have a closely characteristic weed flora, they may be used to point the very obvious deduction that changes of crops and cropping techniques were most probably associated with the introduction and spread of certain weed species and the retraction or extinction of others. *Agrostemma githago*, the corncockle, which is a weed very closely associated with the rye crop, is first recorded in Romano-British sites; it has very greatly diminished in recent years, partly perhaps because of better seed-cleaning and weed control and partly because of the diminished cultivation of rye. *Thlaspi arvense*, an almost constant associate of the flax crop, has been found with seeds of flax in Roman Silchester: it had previously been recorded also in the Late Bronze Age. *Spergula arvensis*, corn spurrey, which is also a constant weed in flax crops, has seed records in relation to human settlement from Neolithic to early Mediaeval time: its pollen records include that at Old Buckenham Mere at an Anglo-Saxon horizon when rye and flax were both cultivated (fig. 171).

It can hardly be doubted either that chess (*Bromus secalinus*) found in Romano-British, Iron Age and Bronze Age sites, barren brome (*Bromus sterilis*), a Mediterranean, Balkan and near Asiatic plant found in Late Bronze Age, Iron Age and Roman sites, and the wild oat (*Avena fatua*) found in Iron Age, Roman and Anglo-Saxon sites, are weeds which have followed the cultivated cereals northwards and westwards across Europe into the British Isles.

As Jessen and Helbaek have pointed out, it is however difficult to draw any exact line between the cultivated plants and the weeds: thus *Chenopodium album* (white goosefoot), *Polygonum convolvulus* (black bindweed), *Polygonum lapathifolium* (pale persicaria), and *Spergula arvensis* (corn spurrey) appear to have been collected for

TABLE 46. *Economic and crop-plants with Roman records in the British Isles*

Abies sp. (fir)	*Smyrnium olusatrum* L. (alexanders)
Pinus pinea L. (stone pine)	*Apium graveolens* L. (celery)
Raphanus raphanistrum L. (radish)	*Aegopodium podagraria* L. (gout-weed)
Papaver somniferum L. (opium poppy)	*Foeniculum vulgare* Mill. (fennel)
Isatis tinctoria L. (woad)	*Peucedanum graveolens* Benth. and Hook. (dill)
Linum usitatissimum L. (flax) *N, Br, I*	?*Fagopyrum esculentum* Moench (buckwheat)
Vitis vinifera L. (vine)	*Ficus carica* L. (fig)
Vicia faba L. (broad bean) *I*	*Juglans regia* L. (walnut)
Lens esculenta Moench (lentil)	*Morus nigra* L. (mulberry)
Pisum sativum L. (pea)	*Cannabis sativa* L. (hemp)
Prunus domestica L. (cultivated plum) *N, Br, I*	*Castanea sativa* L. (chestnut) ?*N, I*
P. cerasus L. (cherry)	*Atropa belladonna* L. (belladonna)
Mespilus germanica Hook. (medlar)	*Hyoscyamus niger* L. (henbane) *I*
Malus sylvestris (L.) Mill. (apple) *N, Br, I*	*Verbena officinalis* L. (vervain)
Coriandrum sativum L. (coriander) *Br*	

The italicized capitals shew pre-Roman records in Neolithic (*N*), Bronze (*Br*) and Iron (*I*) ages.

food by prehistoric man, if not actually grown as crops, and the occurrence of carbonized tubers of *Arrhenatherum tuberosum* (onion couch) along with barley in a Late Bronze Age site in Wiltshire is very suggestive of collection for food. *Isatis tinctoria*, the woad, which still persists locally as a weed, although traditionally a dye plant of ancient standing, has not been identified before Anglo-Saxon time; it was of course cultivated until recent historic time.

From the stomachs of corpses in the Iron Age bog-burials in Denmark we have long lists of identifications of weed seeds accompanying barley and flax, and it seems highly likely that peasants of this period made much use of weeds collected from fallow and incorporated in a gruel to supplement the cereals.

Although there is a long list of weed species first encountered in Neolithic, Bronze Age or Iron Age deposits, a great number of further species have their earliest Flandrian record in Romano-British sites, as table 45 indicates.

There is no reason to doubt that many of these species will in time be recorded from earlier Flandrian deposits, but the length of the list gives some notion of the extension of our weed population by Romano-British cultivation. Some of the list were undoubtedly survivors in this country from Late Weichselian time now finding opportunity to multiply once more in territories from which they had long been excluded by forest: others were most probably introduced with crops by Roman or earlier agriculturists, and we can hardly doubt that this was the case with *Papaver somniferum* (the opium poppy) or *Chaerophyllum aureum* which still retains no more than the precarious hold of a 'colonist' or 'denizen' in this country. Among the weed species now recorded for the first time many had already found their way to this country in a previous interglacial period, and these include south European species which migrated naturally to Britain during a long and decidedly warm interglacial period, but which perhaps would have hardly reached here in the Flandrian period except for human introduction. The reality of introductions in Roman times can hardly be doubted in view of the list (table 46) of identifications of economically useful plants.

Some of these, particularly the pot-herbs and drug-plants have now the role of established ruderals, others are still found only as cultivated plants or occasionally as garden escapes, whereas the records for *Abies* and *Pinus pinea* may well refer respectively merely to imported timber and cones.

The impetus given to the introduction and extension of weeds in Romano-British time depends upon many coincident factors. Among these are the creation of the Roman road system making dispersal across the country easier, the frequent inland and overseas transfer of soldiery and provisions, a highly developed agricultural system and a capacity for civil engineering which allowed the completion of great drainage works. It is seldom realized that the Romans cut hundreds of miles of wide drains in the East Anglian Fenlands, and that during their occupation farmsteads occupied the silt areas of the Fenlands even more densely than they do today. It is worthwhile to note in passing that these silt-lands, now the most fertile of English farm land, were actually brought into existence by a eustatic rise in sea-level completed in the period of the Roman occupation of Britain, so that the whole plant colonization of this big region has taken place in the last seventeen hundred years and during continued human exploitation. Mrs Hallam has vividly reconstructed its aspect in Romano-British time: 'We may envisage the later R.-B. fenland as cattle-ranching country, not unlike parts of the present-day Camargue, with large farm-buildings housing stock standing in their yards among drained and undrained pasture, and with arable only in the driest parts of the region' (Salway, Hallam & Bromwich, 1970).

The history of post-Roman events upon the plant life of the British Isles is chiefly one of progressive encroachment upon, and destruction of the natural vegetation of

PLATE XXV

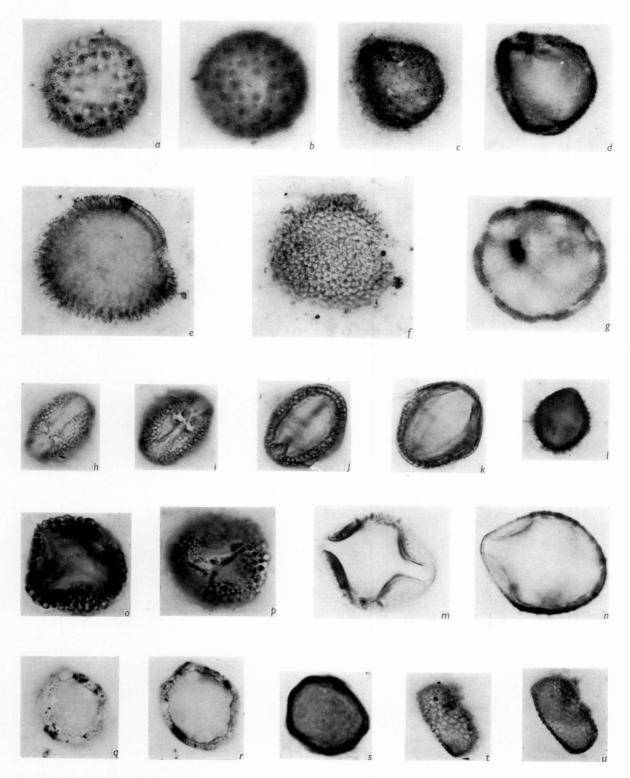

Pollen grains of thermophilous genera: *a, b, Malva* (Drake's Drove, zone VIIb); *c, d, Lonicera* (Glastonbury Lake Village, zone VIIa); *e, f, Geranium* (Eastwick, Late-glacial); *g, Calystegia* (Glastonbury Lake Village, zone VIIa); *h–k, Hedera helix* (Cwm Idwal, zone VI); *l, m, n, Viscum album* (Peacock Farm II, zone V); *o, p, Ilex aquifolium* (Drake's Drove, zone VIIb); *q, r,* cf. *Buxus* (Histon Road, IG); *s, Juglans regia* (Nazeing, derived); *t, u, Impatiens parviflora* (Copse Gate, early zone VIII). (Magnifications of *a–g, l, s, t, u,* × 500; *h–k, m–p,* × 1000; *q, r,* × 1100.)

PLATE XXVI

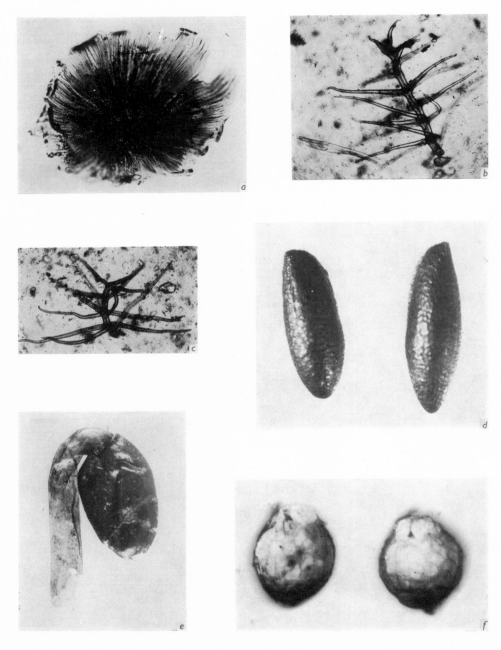

a, Leaf-hair of *Hippophaë rhamnoides* (sub-arctic stage of Hoxne interglacial) ×500; *b, c*, leaf-hairs of *Platanus* (contaminant) ×240; *d*, fruits of *Najas marina* (late Boreal and early Atlantic Hockham Mere) ×8; *e*, tuber of *Potamogeton pectinatus* (Hoxnian interglacial) ×5·5; *f*, megaspores of *Salvinia natans* (Ipswichian Interglacial) ×44.

PLATE XXVII

Outcrop of the resistant peaty deposits of the Cromer Forest Bed exposed on the foreshore at West Runton, Norfolk. Behind are high cliffs cut in the Cromer Till (Anglian glaciation) which overlies it.

The Cromer Forest Bed exposed in cliff section in 1954 at West Runton. It here consists of interbedded clays and sands with peaty layers and above there is a strongly developed cryoturbatic horizon indicative of the onset of the colder conditions of the oncoming Anglian glaciation that laid down the overlying Cromer Till (see fig. 148).

PLATE XXVIII

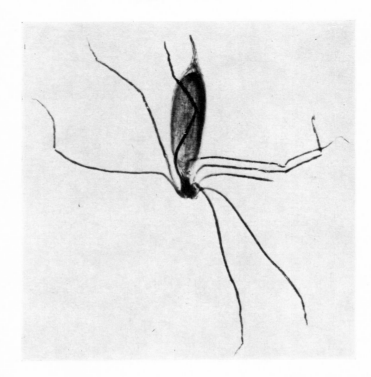

Fruit of *Dulichium arundinaceum* (L.) Britt. recovered from Hoxnian deposits at Quinton, near Birmingham. (Phot. F. G. Bell.)

Macroscopic plant remains from the Middle Weichselian deposits at Earith, Hunts.: *a*, leafy shoot of *Saxifraga oppositifolia*; *b*, fruit of *Onobrychis viciifolia*; *c*, calyx of *Armeria maritima*; *e*, leaf of *Salix polaris*; *f*, capsule of *Helianthemum canum* (Phots. F. G. Bell); *d*, fruit of *Najas marina* from Ipswichian interglacial. (Phot. R. G. West; from *Pleistocene, Geology and Biology*, Longmans, Green & Co.)

the country, and the effects upon plant distribution already considered were increasingly operative through this time. The progress of change was naturally intermittent and broken by changes in economy and technique. Thus it was in Anglo-Saxon times that the lowland forests on the heavy clay soils of areas like the midland plain were first cleared, in Norman times came the reservation of the great royal forests, by no means all woodland, which have served as areas of conservation of flora and fauna from that time. The Norman period saw also the introduction of the rabbit, a very powerful biotic agent comparable in influence with the sheep during the Mediaeval period when wool production was England's staple trade. There was extensive drainage and reclamation of mires coupled with peat-cutting on a scale hardly believable today: records of turbary rights exist in many towns and villages where no trace of peat bogs can now be found, and there is now no doubt that the open-water lakes of the Norfolk Broads have been created by peat-digging. The continuance of drainage has certainly been responsible for the loss of many aquatic species from our regional floras, and peat-digging is no doubt partly responsible for the present extreme sparsity of *Scheuchzeria palustris*, which, as we have seen, was abundant in the early Sub-atlantic (fig. 119). The larger elements of the native fauna, like the forest trees, have not escaped direct destruction by man, especially the large predators and fur-bearers; the effects of this upon the whole ecosystem must have been far-reaching though complex and obscure. It goes without saying that plant introductions, deliberate and accidental, have continued increasingly, and indeed the history of introduction and spread of many of them, such as *Elodea canadensis* (Canadian pondweed) and *Matricaria matricarioides* (rayless mayweed), and *Acorus calamus* (sweet flag), are very well documented (see Salisbury, 1932; Dunn, 1905).

Besides the broad and evident effects of these changes it must be borne in mind that others less immediately apparent may also be exercising great effects upon the flora. Thus the dissection of our woodlands into small separated units creates new difficulties in dispersal of the woodland flora and isolates from interbreeding any populations, such for example as those of *Primula vulgaris* (primrose) and *P. elatior* (oxlip), which may chance to occur alone in a residual piece of woodland. Equally of course the intersection of woodland will permit new opportunities for hybridization; thus the oxlip and cowslip (*P. veris*), or primrose and cowslip, now find far more opportunity for hybridization than before. A similar instance is that of the interbreeding of *Silene alba* (white campion), a cornfield weed of southern origin, with *S. dioica* (red campion) a woodland plant of ancient standing in this country, which has been so clearly demonstrated by H. G. Baker (1947). Likewise A. Bradshaw has acquired much evidence that disforestation facilitated the crossing of *Crataegus oxyacanthoides*, the typical woodland shade hawthorn, by *C. monogyna* which is a plant of open habitats and scrub: this situation

is, as he points out, paralleled by the wave of hybridization among the natural species of *Crataegus* in North America after the opening up of the native forests.

It might be questioned whether *Silene alba* was indeed introduced to Britain with prehistoric agriculture, but there can be no question that *Spartina alterniflora*, one of the parents of *S. townsendii*, was introduced by shipping from South America. The consequence of its hybridization with the native salt-marsh plant *S. maritima* has been the production of the fertile tetraploid, *S. townsendii*, which had the remarkable quality of being fitted for a coastal habitat hitherto unable to be colonized by any species of phanerogam.

It must be remembered that alongside the overwhelming anthropogenic changes wrought upon our vegetation and flora there have also been in progress during the later part of Sub-atlantic time climatic fluctuations of some importance. Thus Jessen has pointed to the existence in many Irish bogs of two recurrence surfaces, RS:B and RS:A, after the most pronounced one which he took to fall at the opening of the Sub-atlantic. In Somerset there is at least one clear flooding horizon between that of the Late Bronze Age and the late Romano-British occupation of the bog surfaces, and in the raised bogs at Tregaron in Cardiganshire there is a 'retardation layer' in the Sub-atlantic peat which appears to indicate a phase of pronounced dryness.

A good deal of evidence of very varying kind points to the occurrence in western Europe of an early mediaeval warm epoch (*c.* A.D. 1150 to 1300) during which, apparently, it was warmer than at any time since the main Flandrian thermal maximum (Lamb, Lewis & Woodroffe, 1966). Between about 1550 and 1700 fell the epoch called 'Little Ice Age' from the prevalent low temperatures shewn by glacial advances and historical records. Finally in the first five decades of this century a general rise in summer temperatures appears to have caused recession of glaciers in all parts of the world, and retraction of the pack ice further into the Polar Basin than has ever been known in historic time. The latter effect has certainly had its ascertained biogeographic consequences, for the cod followed the warmer water northwards in the Atlantic Ocean, and in turn caused a major alteration in the diet and economy of the Greenland Esquimaux. At the same time, and within the period of reliable recording, seabirds have greatly extended their northwards range in this region. It is however a matter of great difficulty, against a background of violent human disturbance, safely to attribute to such climatic shifts recent changes in vegetation or flora.

We may appropriately consider how far the effects we have mentioned are reflected by phytogeographic analyses described in chapter VI: 9, and there carried as far as the middle Flandrian. This has been extended by calculating totals for the late Flandrian, i.e. zones VIIb + VIII. Of the thirteen phytogeographic categories, aside from the Widespread element, the Sub-atlantic–widespread is the most abundant with highest values in the west and south

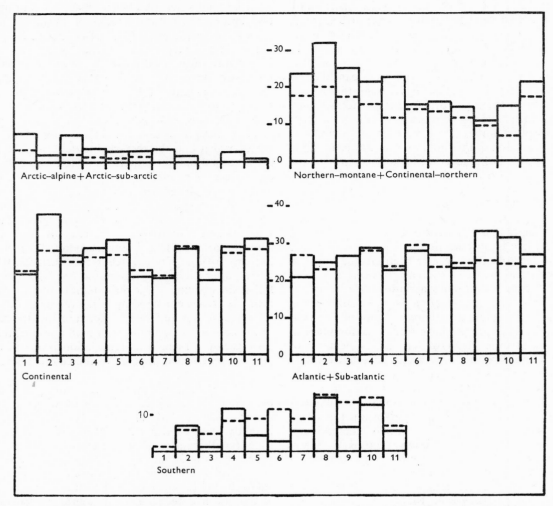

Fig. 172. Percentage representation of five major phytogeographical groups during the middle and late Flandrian (shewn by solid and broken lines respectively) within eleven regions of the British Isles. (Region 12, Isle of Man, has inadequate data for these periods.)

(regions 9, 10 and 11). In the east the Continental elements as a group exceed the Atlantic plus Sub-atlantic. It seems that since the middle Flandrian the flora of the country has become more Sub-atlantic, especially in the west. In the middle Flandrian the Continental–northern element exceeded the Sub-atlantic–widespread but their relations are now transposed: nevertheless the Continental–northern remains a highly important category.

The five aggregate groups shewn in fig. 172 are specially informative. The Continental group has changed little from the middle Flandrian and shews no evident gradients across the country; the Sub-atlantic plus Atlantic, an even larger aggregate, seems to have increased in western Scotland (1 and 3) and to have decreased in Ireland and southern England (9, 10, 11). Two groups both exhibit a decrease from the middle Flandrian and retain the strong north–south gradient they already exhibited: the Arctic–alpine aggregate is now recorded only from regions 1 to 6 (Scotland, northern England and Wales).

The opposite effect is exhibited by the Southern aggregate which has increased everywhere except eastern Scotland and north-western England. Within this aggregate the Continental–southern element has most evidently gained ground, particularly in East Anglia, southern England and Ireland, regions with much agricultural activity: though this demonstrates a strong north–south gradient, in southern England it also exhibits an east–west gradient, Wales, the Midlands and the south-west having lower values than East Anglia and the south-east. It is interesting that the Continental–southern element has increased in this way whilst in the late Flandrian the floristic composition as a whole has become more Sub-atlantic. It seems that the east–west gradient is possibly more marked in the late Flandrian records than at any other stage, including even the present, but it is unlikely, in view of what has been said above, that this is solely a climatic effect.

It is not surprising, in view of these results, to find

(Birks & Deacon, 1973) that the ordination plot for all the Flandrian records is, apart from the placing of the necessary end points, almost a straight line and that it indicates a north–south gradient in the flora. It closely resembles in these features the ordination plot for the flora of the present day, for which however the gradients are distinctly steeper, most especially with regard to the Southern phytogeographic element. It is apparent that by the late Flandrian the pattern of phytogeographic distribution had almost attained that of our existing flora.

Much of our comment upon the events of the Sub-atlantic period, and especially the later part of it, has not, it is true, had the character of evidence from the fossil records of the factual course of floristic events. It is in the nature of the records that they should indeed fall away in frequency and significance as the historic period, with its own more ample information, sets in. What we have sought to underline, however, is that during this time, by the operation of factors utterly different in kind and magnitude from any which had acted previously, the present biogeographic conditions of these islands came into being. It must be apparent that these conditions differ so greatly from those of pre-Neolithic time, that no modern instances of dispersal and establishment of species in this country can be safely adduced as evidence for the nature of these phytogeographical processes in the conditions of undisturbed native vegetation. It will also be evident that the present-day distribution patterns of many species in the British flora are the consequences of the great disturbances of the last two or three millennia, but that other species are of a kind to have escaped these effects and in their range still bear the stamp of more ancient and more natural historic processes.

16. REGIONAL DIFFERENTIATION, MIGRATION AND SURVIVAL

It would be reasonable at this point to enquire how far the fossil plant records set out in chapter V and the generalizations based upon them in the preceding sections of this chapter throw light upon the genesis of the major phytogeographic regions of the British Isles today, and whether indeed they point to unsuspected lines of demarcation. At the same time it would be appropriate to adduce any generalizations this exercise may evoke on the processes of survival, migration and extinction involved in our explanations.

We can achieve some insight into the floristic evolution of this country by resuming the phytogeographic analyses developed in sections 9 and 15 of this chapter, and considering the relative frequency of the phytogeographic elements, in each of seven districts (Scotland, the Lake District, Ireland, Wales, South-western England, East Anglia and South-eastern England) as they change from the Late Weichselian through middle Flandrian to the

late Flandrian, in which period the pattern has become close to that presented by our existing flora. Most of the features commented on are represented in summary in the histogram of fig. 159, that is based on the geographical regions 1 to 12 shown in fig. 158.

SCOTLAND (Regions 1, 2, 3)

(i) *Late Weichselian.* The flora has a strong Continental–northern emphasis, with the Arctic–alpine element well represented, particularly in the north-west and north-east: in the north-west the Sub-atlantic-widespread element has a relatively high value.

(ii) *Middle Flandrian.* In the north-west and south-west the flora is still strongly Continental–northern with the Arctic–alpine element well represented, but the Sub-atlantic–widespread component has greatly increased, so that in the north-west these two elements have equally high representation. In eastern Scotland, although the Sub-atlantic–widespread element has also increased greatly since the Late Weichselian, the Continental–northern element has a high, and the Continental–southern element, a fairly high, value. Thus, in contrast with the Late Weichselian, the Scottish flora has now become less northern and now shews an east–west differentiation.

(iii) *Late Flandrian.* In western Scotland (regions 1 and 3) the Sub-atlantic–widespread is now the largest element, although the Continental–northern is not much less. The Atlantic–widespread element now for the first time shews relatively high values in both these western regions. The Sub-atlantic–widespread element has a particularly high value in the south-west. In eastern Scotland (2) the Sub-atlantic–widespread is the major element, closely approached by the Continental–northern, and in aggregate the Continental group of elements exceeds the Sub-atlantic–Atlantic group; Atlantic elements have only low values, but the Continental–southern element is now substantially present, perhaps a consequence of arable cultivation. These features shew that the east–west gradient recognizable across Scotland in the middle Flandrian has been accentuated, particularly by advancing atlanticity in the west.

ENGLISH LAKE DISTRICT (Region 4)

(i) *Late Weichselian.* The floral analysis is very similar to that of Scotland at this time, that is strongly Continental–northern with Arctic–alpine and Northern-montane elements well represented: the Sub-atlantic–Atlantic group is however better represented than in the other northern regions.

(ii) *Middle Flandrian.* As in western Scotland the Sub-atlantic–widespread element has increased greatly from the Late Weichselian, now only slightly exceeded by the Continental–northern. The Continental–southern element has increased also, with a value like that for eastern Scotland. The Arctic–alpine element is much diminished.

(iii) *Late Flandrian*. As in western Scotland the increased Sub-atlantic–widespread element now exceeds the diminished Continental–northern component: the Continental–southern element retains its middle Flandrian status and is higher than in western Scotland.

IRELAND (Region 11)

(i) *Late Weichselian*. Ireland has the same high values for the Continental–northern element, but higher values for the Sub-atlantic–Atlantic component: the Arctic–alpine element is well represented. The values of individual floristic elements are very similar to those of south-west Scotland and the Lake District, although stronger in the Sub-atlantic–Atlantic elements.

(ii) *Middle Flandrian*. The floristic change from the Late Weichselian is less pronounced than in other regions. As elsewhere the Arctic–alpine element has sharply decreased and the Sub-atlantic–widespread element has increased, though for the latter the rise is less and the value lower than in other regions. The Continental–northern element has not decreased and still exceeds the Sub-atlantic–widespread, but the Continental–widespread has almost doubled. The floristic trend seems towards a less northern and somewhat more atlantic aspect.

(iii) *Late Flandrian*. The Sub-atlantic–widespread and Continental–northern elements are equal, but the Continental aggregate exceeds the Sub-atlantic. The Sub-atlantic–Atlantic elements of the flora are less strong than those of south-west Scotland and the Lake District; perhaps because of the relatively continental area of central and eastern Ireland.

WALES (Region 6)

(i) *Late Weichselian*. The Welsh flora at this time resembles that of Scotland, being strongly Continental–northern with a considerable Arctic–Alpine element, and with the same high values for the Sub-atlantic–widespread element that are found in north-west Scotland. The relatively high value for the Continental–southern element is remarkable, possibly indicative of 'refugia' for southern species.

(ii) *Middle Flandrian*. The flora has now become strongly Sub-atlantic–Atlantic. It does not maintain the northern aspect seen in Scotland: the Continental–northern element has greatly decreased whilst both Northern–montane and Arctic–alpine elements have only low values.

(iii) *Late Flandrian*. Since the middle Flandrian the Welsh flora has become more Sub-atlantic and more southern. The Sub-atlantic–widespread element has increased and the Continental–northern decreased whilst the Continental–southern element has attained a very high value, almost equalling that of the Continental–northern.

SOUTH-WESTERN ENGLAND (Region 9)

(i) *Late Weichselian*. Although the Continental–northern element here, as elsewhere, has a high value, the Sub-atlantic element is also quite important with a value equal to that of the Arctic–alpine. Thus by the floristic indication the general severity of climate seems to have been mitigated by a degree of atlanticity.

(ii) *Middle Flandrian*. The flora is now strongly Sub-atlantic–Atlantic. As in Wales, the Continental–northern, Arctic–alpine and Northern-montane elements have very low values, but here Southern elements are less prominent.

(iii) *Late Flandrian*. In a flora still preponderantly Sub-atlantic–Atlantic there has now been a strong increase in Southern elements, mainly in the Continental–southern group which we tend to regard as favoured by extending cultivation.

EAST ANGLIA (Region 8)

(i) *Late Weichselian*. The most important floristic element is the Continental–northern, but the Continental–southern and Continental–widespread also have high values exceeding those in all other regions of the British Isles. The Arctic–alpine element is poorly represented.

(ii) *Middle Flandrian*. The East Anglian flora remains primarily Continental with the Continental–northern element still predominant, but the values of all these Continental elements have declined whilst the Sub-atlantic–Atlantic components have increased, especially the Sub-atlantic–southern and Atlantic–widespread elements.

(iii) *Late Flandrian*. The Sub-atlantic–widespread is the most numerous element although its value is low in comparison with those in other British regions. The Continental–southern element has a relatively high value and the aggregate Continental group exceeds the Sub-atlantic–Atlantic.

SOUTH-EASTERN ENGLAND (Region 10)

(i) *Late Weichselian*. Although still the most important element of the south-east region, the Continental–northern element has the lowest Late Weichselian values in the British Isles, and the floristic evidence for continentality is less than in East Anglia. The fact that values for Arctic–alpine species are higher than elsewhere in eastern England may well be only a sampling effect.

(ii) *Middle Flandrian*. Although there is a marked increase in the importance of the Sub-atlantic–widespread element, the Sub-atlantic–Atlantic aggregate only slightly exceeds the Continental. The flora has lost its northern character: all northern elements now have very low values. Conversely the Continental–southern and the Sub-atlantic–southern elements have high values in comparison with other regions.

(iii) *Late Flandrian*. The Sub-atlantic–widespread element retains its predominance, but the Continental aggregate group exceeds the Sub-atlantic–Atlantic: very low values of the Continental–northern element are compensated by values for the Continental–southern element that reach a maximum in this region.

NORTH-EASTERN ENGLAND (Region 5)

(i) *Late Weichselian*. The Late Weichselian flora is strongly Continental–northern with the Northern–montane and Arctic–alpine elements equally and substantially present.

(ii) *Middle Flandrian*. Although, as in other regions, the Sub-atlantic–widespread element has now greatly increased in importance, the Continental–northern and Continental–widespread elements have also expanded, so that the flora maintains its strongly continental character. Arctic–alpine and Northern–montane elements now however retain only low frequencies.

(iii) *Late Flandrian*. The flora has kept its overall continental character but there has been a shift of emphasis: the values of the Continental–northern and Continental–widespread elements have decreased, the Sub-atlantic–widespread is substantially unaltered whilst the Continental–southern and Sub-atlantic–southern have both increased.

The pattern through from the Late Weichselian is in fact rather like that for eastern Scotland (region 2).

It has seemed unprofitable to consider individually the other floristic regions. The Midlands have relatively few sites and these badly spaced with a strongly western bias, and the Isle of Man has little but Weichselian records. It seems evident that the strong regional differentiation of our existing flora is no new phenomenon, but is matched by corresponding earlier Flandrian patterns that exhibit progressive differentiation from the general uniformity and prevalent cold continentality of the Weichselian through the increasing warmth and atlanticity of the Flandrian. Even in the Late Weichselian the floras of north-western Scotland and south-western England shew in the values for the Sub-atlantic–widespread element a degree of atlanticity (as they do today), that was not then exhibited in south-western Scotland, the English Lake District (and possibly Wales) to nearly the same extent. We may think this is a reflection of the fact that the Irish Sea bed was still dry and that marine influences affected only our far western shores.

Through the Flandrian, as now, the major gradient affecting the flora has been the north–south climatic control, but considerable east–west differentiation has also been apparent; thus East Anglian continentality was strongly apparent even in the Late Weichselian. The superposition of widespread cultivation upon climatic change in the Late Flandrian has given emphasis to the spread of Continental and Southern elements, especially of course in the south and east.

The results of this purely phytogeographic analysis effectively supplement the conclusions already drawn from the extensive evidence of pollen analysis and the fossil history of individual species. The strong latitudinal gradient has not only given Scotland a strong phytogeographic individuality but it is now apparent that in the middle Flandrian, between about 8000 and 4000 B.P., the Central Highlands were pine dominated, whilst further south a climax of deciduous mixed-oak forest prevailed. It is also increasingly probable that *Corylus*, *Alnus* and *Pinus* became established in Scotland independently of any northward migration from England, the hazel and alder from the west and pine perhaps from the north. The possibility of off-shore refugia for these and other genera corresponds to the ideas of Nordhagen and other biogeographers on survival off the Scandinavian west coast; it is however more readily considered in relation to Ireland (see below).

A feature of unexpected interest is the highly individual history of the northern Irish Sea region. Not only do south-western Scotland, the English Lake District and Wales have much in common in terms of phytogeographic analysis, but also in vegetational history disclosed by pollen analysis, in which regard they relate closely also to north-eastern Ireland. They are all regions with a big altitudinal range and seem to exhibit such early establishment of deciduous forest elements, and locally of pine, that one is tempted to conjecture that the area had thermally favourable localities if not 'refugia'. The Late Weichselian record of the Isle of Man, so far as it goes, fits this regional concept; but the island's small size, limited elevation and subsequent isolation by eustatic ocean rise have greatly depauperated the flora in relation to the surrounding regions of Ireland and Great Britain.

The biogeographic identity of Ireland is particularly well defined and deserves special attention not only because it has been so long recognized and discussed, and because it concerns both animals and plants, but because there now is a substantial volume of information as to its flora in past interglacials, especially the Hoxnian, and considerable knowledge of its Pleistocene geology. The biogeographic features that have occupied the attention of biologists through the last eighty years or so are chiefly (1) the attenuation in numbers of plant and animal species in comparison with Great Britain, (2) the presence in the flora and fauna of the so-called 'North American element', (3) the presence of the 'Lusitanian–Mediterranean element'.

It seems that when compared on the same taxonomic basis, only some 70 per cent of the total British flora is present in Ireland. In considering this we must bear in mind that Britain is a considerably larger island than Ireland, extending much further northwards and southwards, through ten degrees of latitude as against four, and is thus capable of supporting northern and southern species not found in Ireland. It possesses also a climatic range towards continentality outside that of Ireland, and possibly also a wider range of soils and types of habitat. Taking these facts fully into account Praeger (1934, 1939, 1950) nevertheless regards the disproportion between the biotas of the two islands as primarily due to the sea-barriers having prevented immigration of a great many species capable, once introduced, of becoming established in Ireland, and most biologists would accept this contention.

Many of the absentees from Ireland have long attracted notice: thus among the reptiles and amphibia Ireland has none of the British snakes, one lizard, one newt, and the Natterjack toad (*Bufo calamita*); the native status of the frog (*Rana temporaria*) is uncertain. Likewise among the mammals the weasel (*Mustela nivalis*) and the polecat (*M. putorius*) are absent and the Irish stoat (*M. hibernicus*) is held to be different from the English; the Irish hare is also distinguished as *Lepus hibernicus*. The common shrew (*Sorex araneus*) and the mole (*Talpa europaea*) are absent and the only truly native representative of the rats and mice is the field-mouse (*Apodemus sylvaticus*); Ireland has also only seven of the twelve British species of bat. Among the flowering plants the most conspicuous absentees are the lindens, beech and hornbeam, confirmed pollen-analytically as never having grown in Ireland at any time during the Flandrian period.

Among the species common or widespread in Britain but absent from Ireland we have

Genista anglica	*Scabiosa columbaria*
Astragalus glycyphyllos	*Paris quadrifolia*
Ononis spinosa	*Convallaria majalis*
Lathyrus sylvestris	*Helichtotrichon pratense*
Chrysosplenium alternifolium	

Many more are rare in Ireland although widespread in Britain, some with a wide north–south range in the latter country, a few with a northern range, and more with a southern range.

Whilst it is possible to consider that total absentees may have been prevented from colonizing Ireland by failure to cross the sea-barriers, this explanation cannot be offered for species present although uncommon in Ireland: for them the restriction of range must generally be due to factors, past or present, operating within that country itself. It is just possible that restricted range may in a few instances be due to recent introduction, but with others, such as *Sanguisorba officinalis*, *Lycopodium inundatum* and the genus *Helianthemum*, the present sparse and disjunct Irish distribution would seem more likely to be relict from the Late Weichselian period of more widespread occurrence.

Such an explanation has been given much support by the Irish fossil records for species now absent from Ireland. Thus Mitchell's record of *Betula nana* from the Irish Late Weichselian shews that the drastic Flandrian restriction of its range in Britain was represented in Ireland by a total extinction. It seems likely that this happened also to other arctic–alpine or northern species not now growing in Ireland, but which were present in the Late Weichselian, viz. *Astragalus alpinus*, *Minuartia stricta* and *Luzula spicata*.

Where expansion in the Flandrian warm period has been followed by retraction of range it is not surprising to find that this retraction has sometimes also extended to total loss of a species. This is the case, for instance, with *Najas marina*, to which we have already referred, which was present freely in Ireland during the Boreal period, but has since vanished, whilst in Britain it has retreated to a few lakes in East Anglia.

It is apparent from these instances that many of the absences of flowering plants from the Irish as against the British lists must be due to Flandrian extinction, but it is also apparent that other species failed to enter Ireland before it was cut off by rising sea-level.

The North American element consists of species widespread in North America, but in Europe confined either to Ireland alone, or to Ireland and extreme western coasts of Scotland and Scandinavia. These include the two orchids, *Spiranthes gemmipara* and *S. romanzoffiana*, *Sisyrinchium angustifolium* (blue-eyed grass), *Juncus tenuis*, *Najas flexilis*, *Eriocaulon septangulare* (pipewort) and the freshwater sponge *Heteromeyenia ryderi*. The curious distribution range of these organisms has led biogeographers to suppose (since recent natural immigration and incipient spreading seemed unlikely) that they had entered Europe by some transatlantic land bridge now disappeared, and that, if the existence of such a connection had to be put far back into geological time, that these species had survived the subsequent Ice Ages in Ireland.

It seems unlikely that such considerable assumptions are, however, necessary, for it is apparent (as Deevey notes) that the evidence for *Najas flexilis* supplies the key to the whole problem. This water plant, now restricted to lakes in western Ireland, the Lake District, western Scotland and western Scandinavia, is shewn by pollen-dated records of its fruits to have been both abundant and widespread during the Boreal period in all western Europe, Britain and Ireland. In Britain it was already present in the Late Weichselian period. Since it had this great Flandrian extension no hypothesis of perglacial survival *in situ* is called for, and indeed most of the present and fossil range of *N. flexilis* in Europe is within the area of glaciation. It is evidently as Deevey suggests, a widely ranging, perhaps circumpolar species whose present distribution is relict from the early and middle Flandrian. Deevey in this connection points out that the remarkable thing about species of the North American element is not their disjunction of range, for many occur on both sides of the Atlantic, but the *asymmetry* of the distribution, their limited range west of the Atlantic contrasted with their wide occurrence east of it. He comments upon the existence of species with ranges asymmetric in the opposite sense, and is inclined to interpret the asymmetry as most probably due to causes affecting survival rather than means of distribution. Among species found in both Europe and America in Tertiary time the total loss of *Tsuga canadensis* and *Dulichium arundinaceum* from Europe, and of *Trapa natans* and of the horses from North America, lend colour to this argument. The fossil record has in fact shewn *Tsuga* to have persisted in England until at least the Antian and *Dulichium* (along

 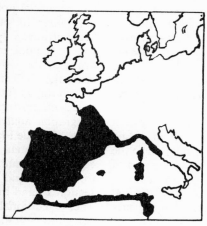 

Erica ciliaris L. Erica scoparia L. Daboecia cantabrica (Huds.) G. Koch

Fig. 173. Present-day European range of three Lusitanian species of Ericaceae, two of which, *Erica ciliaris* and *Daboecia cantabrica*, still grow in western Ireland. All have Hoxnian records in the British Isles.

with *Trapa*) as late as the Hoxnian in England and the Eemian interglacial (=Ipswichian) in Denmark (Bell, 1970*b*).

The discovery of the easily recognizable pollen grains of *Eriocaulon septangulare* from peat of early Atlantic age in western Ireland, at a time before human introduction is acceptable, makes it certain that the species is indigenous in Ireland. Its pollen has twice been more doubtfully identified from a Hoxnian context in Ireland, but all the fossil grains appear to correspond in size more closely with that of the native Irish tetraploid plant than with that from America, which is diploid (see p. 373). In face of this it is difficult to disagree with the conclusion that *E. septangulare* survived the Wolstonian and Weichselian glaciations in Europe if not in Ireland itself. It now seems likely that another Hiberno–American plant of similar distribution pattern, *Sisyrinchium bermudianum*, likewise has cytological differentiation between the Irish and North American populations though they are indistinguishable in gross morphology (Ingram, 1967, 1968). In both species there must have been ancient separation of the populations on the two sides of the Atlantic, and it is plausible to explain their imbalance in terms of long distant events.

A more baffling problem is presented by the Lusitanian–Mediterranean element in the flora of western and south-western Ireland. Among the flowering plants it includes:

Saxifraga spathularis	*Sibthorpia europaea*
S. hirsuta	*Pinguicula grandiflora*
Arbutus unedo	*P. lusitanica*
Erica mediterranea	*Euphorbia hyberna*
E. mackiana	*E. peplis*
E. ciliaris	*Neotinea intacta*
E. vagans	*Simethis planifolia*
Daboecia cantabrica	*Puccinellia foucaudii*

These species have their main areas of distribution in Spain, Portugal, and round the Mediterranean, but they have a northwards extension of range along the Atlantic seaboard as far north as Ireland (fig. 173). There is nothing difficult to comprehend in their distribution apart from their occurrence in Ireland (and to a lesser degree the presence of some of them in south-western England). It seems apparent that these species migrated northwards as far as Ireland at a time when (*a*) the climate was favourable, and (*b*) the coastline was continuous in a way which it now is not.

Although transport by wind or by birds across the English Channel is not altogether out of the question, for some of these species it seems improbable, as also for such animals as the Kerry slug (*Geomalacus maculosus*) and the woodlouse, *Trichoniscus viridis*. This being so the species of the group as a whole may be taken either to have survived the last glaciation (and perhaps previous ones) inside Ireland, or to have entered the country during the phase of eustatic lowering of sea-level. To each alternative the same objection has been urged, to wit, that many of the species are highly intolerant of winter cold and could neither have survived the last glaciation *in situ*, nor have reached Ireland before the rise in sea-level cut off the British Isles from the Continental mainland. It is true that some biologists have accepted the suggestion of perglacial survival of the Lusitanian–Mediterranean species in Ireland, and their case has been somewhat strengthened by recent refutation of the once-accepted view that the whole of Ireland has suffered glaciation. Farrington demonstrates that even were the different Boulder Clays of southern Ireland to be regarded as contemporaneous (which they are not) there would still be a belt of hilly country extending eastwards across southern Ireland from Brandon Mountain, and taking in the Galtees, which had never been ice-covered. And if the ice-sheets were *not* contemporaneous then still larger areas at each glaciation must have been free from ice. It is not hard, with this in mind, to accept the possibility of perglacial survival for some of those plants in the group, such as the two species of saxifrage, which can maintain themselves at high altitudes in the mountains, but it is impossible to consider survival of such frost-sensitive plants as *Arbutus unedo* in a landscape shewing evidence of periglacial climate if not of glaciation.

The problem has gained a new dimension however in the evidence afforded by the considerable and growing amount of fossil evidence that this category was well represented during the Hoxnian interglacial in western Ireland. The situation has been authoritatively discussed by Mitchell and Watts (1970) in particular relation to the Ericaceae. They recall that in the middle and later part of the Hoxnian there grew *Calluna, Daboecia, Erica ciliaris, E. cinerea, E. mackaiana, E. scoparia, E. tetralix* and *Rhododendron ponticum*. They call attention to the presence of a similar flora in a broad belt of Hoxnian age sites extending along the western shores of Europe from the coasts of Portugal to the Shetland Isles, and consider what must have happened to it in the succeeding glacial stage when periglacial conditions extended right to the present western shores of Ireland. The broad areas of continental shelf would have experienced similarly severe conditions, but, as they point out, where the mountain chains of Connemara, Kerry and the Pyrenees extend westwards they must have made promontories still moderated by the Gulf Stream, and might possibly have afforded in their ridges, valleys and well drained soils refugia for these Lusitanian species. Failing this it is necessary to evoke an explanation in terms of reinvasion from Iberia south of the Pyrenees, when climate had ameliorated but the eustatic rise in ocean level had not yet isolated Ireland. Increasingly we get evidence that the thermal rise was earlier than had been supposed (Shackleton, unpublished), and we know that sea-level was at least 180 ft (54 m) below present sea-level off southern Ireland in zone II and that it still lacked some 50 ft (16 m) of its final height in the middle of zone VI. A fall in sea-level of 200 ft would establish a land bridge to France across the eastern end of the English Channel, but would barely halve the present width of St George's Channel between England and Ireland. A fall of 300 ft (which is a figure commonly suggested by geologists) would lay bare, however, most of the English Channel, establish land bridges between Wales and Ireland, Scotland and Ireland, and reduce the intervening channel substantially (fig. 162). Accordingly it does not seem impossible that the greater part of the Lusitanian–Mediterranean element should have reached England and Ireland by coastwise migration during the early Flandrian period, especially as we know how many thermophiles were certainly present in Ireland as early as the Late Weichselian (Mitchell, 1953). Final conclusions are yet to seek, but it appears that in a biogeographic sense Ireland presents no problems different in kind from those of Britain, and that these countries shared a common pattern of floristic history. The western station of Ireland and separation by the sea from Britain have merely given special acuteness to particular problems. The extremely oceanic climate

has permitted juxtapositions and overlaps of range quite unfamiliar in other lands, so that arctic–alpine and Mediterranean species now meet on the western Irish coasts, the one escaping high summer temperatures and drought, the latter escaping winter frosts. It remains to bring these effects into the account of the history of the Irish flora, as future research permits.

It is interesting to compare the results of purely phytogeographic analysis with those derived by pollen analysis for the Middle Flandrian, at which time we have supposed that the climax vegetation had substantially achieved equilibrium and had not yet been modified by disforestation and husbandry. These were outlined in section 13 and the accompanying sketch-maps (fig. 164). The vegetational differentiation of the Atlantic period coincides with the phytogeographic in strongly indicating a north–south climatic gradient, with birch forests in far northern Scotland, and pine–birch in the Central Highlands whilst mixed-oak forest dominated England and Wales. Strong phytogeographic gradients that set off East Anglia and south-eastern England are paralleled by the development there of linden in the mixed oak woodlands and of beech and hornbeam woodlands in the later Flandrian. The special phytogeographic character of south-west England is matched by the fact that it seems to have been the seat of earliest expansion of oak woodland in Britain (see fig. 93). Brown (1971) has drawn attention to other indications of its ecological peculiarity. Ireland, so strikingly distinct in its phytogeography from the rest of the British Isles, has had a vegetational history also distinct, not merely through absence of beech, hornbeam and the lindens, but also in such features as the exceptional prevalence of *Corylus* in the early Flandrian, the strong persistence of pine forest, generally without much birch, into the Atlantic period, and the much delayed and slow rise in the frequency of alder, compared with its behaviour in Great Britain.

The reality of these broad differences has been confirmed by radiocarbon dating, to which we look increasingly, in conjunction with pollen analysis, for resolution of the complexities of our vegetational history, but in making use of it we need the *caveat* that the use of pollen analysis as an index to the former pattern of range of major forest communities and their component tree species is heavily handicapped if we use premature conclusions based on too little considerations of the ecological realities and of the width of uncertainty as to the meaning of the palynological patterns. Invaluable as are the radiocarbon datings of sequential pollen diagrams, conclusions from them still require interpretation in realistic ecological terms. This is especially so in relation to the synchroneity or metachroneity of movement of individual forest genera.

The uncertainty is clearly exposed in current adherence to the terms 'empiric limit' and 'rational limit' for early stages in the pollen curve for a given genus: these terms have no generally understandable English meaning in this context, and also they lack quantitative definition.

The first with intermittent low frequencies, indicates the earliest presence of the pollen type (*advent*), the second (the continuous pollen curve) its continued representation in the diagram (*establishment*); later there often follows the substantial rise in frequency that, in ecological terms, represents *expansion*.

All too often the 'empiric' or 'rational' limit is taken to be the index of *immigration* of the given tree into the area, and accordingly we are presented with radiocarbon data said to represent the *migration* of spruce, beech etc., across a given territory. This overlooks the unacknowledged assumption that we are involved with the advance of the tree (or the forest type) along a narrow well-defined front, as for instance in a sequence of precise advances across the European plain from the Alpine, Carpathian and Pyrenaean glacial refuges. The different genera are assumed to have travelled at different rates *because* they achieved dominance according to a certain sequence in the north and west, but this might equally be the consequence of differential climatic requirements progressively realized with time in the interglacial cycle. Pollen analyses *per se* do not decide between these explanations. More specifically they fail to do so because the concept of narrow zone boundaries is faulty.

Near their limits of tolerance plants tend to be limited to sites with locally favourable conditions provided by local variation of slope, altitude, aspect, drainage, parent-rock and local topography. Thus every margin of distribution is diffuse and irregular with outliers, islands in a sea of different composition. Between the major vegetational belts of north-western Europe there are extremely broad ecotones in which, for example, a southern component extends north beyond its main area in a series of enclaves where it forms 'post-climax' communities in sites of special warmth and shelter such as ravines, cliffs, south-facing slopes, well-drained soils and sea-modulated climate. Such sites are the analogues of the wooded 'kloofs' or 'wadis' that diversify the deserts of South Africa and Egypt. There are of course corresponding inliers of the northern elements as pre-climax stages in vegetation south of the ecotone where 'harsh' conditions favour this element. That this is the ecological reality we may see in the few pieecs of undisturbed natural forest left to us, such as Białowiecza, where even small variations in soil type, slope and drainage allow widely different forest types to co-exist.

Given that this is the case, what are the consequences of a substantial climatic shift, especially within these broad ecotonal belts, and how will the pollen diagrams respond to these changes? In the first place the plant responses can be swift; there are established nuclei from which expansion can occur without waiting for slow advance from a narrow front. If, for example, waterlogging occurs generally, the alder can be expected suddenly to be generally prevalent and the event can be broadly synchronous over big areas. Whether the outliers could be detected in a pollen diagram would depend upon their nearness to the sampling site, the pollen productivity of the taxon, the density of local pollen rain and the total pollen count at the given horizon. It would be extremely hazardous in these circumstances to deduce *absence* of a tree genus from a region. It is much safer to regard the successive empiric and rational 'limits' as merely shewing relative increase in the taxon concerned, accepting that we do not here have the means of proving 'immigration', but only of phases of expansion in frequencies of varying degree.

It is instructive to consider in these terms the case of the late expansion of *Fagus* and *Carpinus* in the British Isles. They were not delayed by the barrier of the English Channel and North Sea for they were already present here in Neolithic time. In the Somerset Levels, though represented still in the Late Bronze Age by a discontinuous pollen curve, *Fagus* had been present a thousand years earlier, as wooden stakes of beech in Neolithic trackways prove. It evidently grew in some favoured sites in the almost unbroken sea of mixed-oak forest, but was able to spread only slowly. This was a situation quite unlike that of *Corylus*, *Quercus* and *Ulmus* spreading across the country in the early warm period against little or no competition from pioneer birch forest. Thus *Fagus*, like *Carpinus*, was able swiftly to expand in response to clearance which removed barriers to extension and competition by established mixed-oak forest. This response was the same on the north-west European lowlands, and affected also the spruce. All three genera have been demonstrated to shew metachroneity in their expansion, but this seems a phenomenon, not of migration rate *per se* (i.e. natural dispersal rates), but of ecesis (establishment) and expansion in changing environmental conditions. One might indeed expect metachroneity to be exhibited specially by species near the limits of their climatic range. That beech, hornbeam and spruce tend to appear in the later stages of other interglacial cycles need mean no more than that only then had conditions of climate and soil become favourable to them. These are all matters to be resolved by increasing volume of accurate historic, genetic and autecological information. For the reasons just given and because of the small volume of research specifically directed to the problem, we remain very much in ignorance of the locality and extent of the glacial refugia of the thermophile mediocratic elements of our interglacials. As Beug (1957) and Lang (1970) have made clear, during the Weichselian glaciation Artemisia- and Chenopod-rich Steppe communities extended over the European plains and south of the Alps and Pyrenees, so that there were no zonal belts of deciduous forest. These mountains however are presumed to have sheltered small and isolated units of such vegetation, but we remain ignorant of the possible presence of similar centres further north and especially, as we have indicated, off the present west European coast. We look forward to the possibility that sediment cores still deeper and older than those already procured in Skye may extend our record there and elsewhere in the western islands of the British Isles and Scandinavia back into the Full Weichselian.

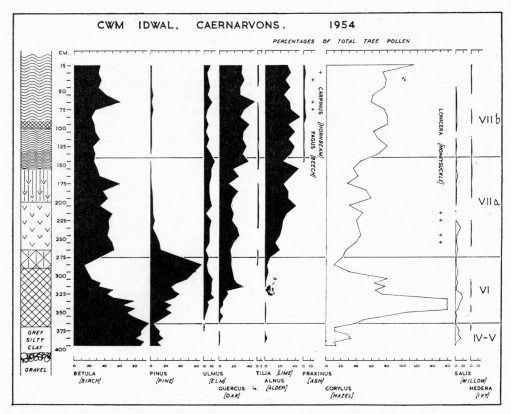

Fig. 174. Tree pollen diagram through deposits at Cwm Idwal, Caernarvonshire, a notable plant refuge in the Snowdonian massif. Altitude 375 m. The diagram shews continuous accumulation from the end of the Late Weichselian through most of the Flandrian. (After Godwin, 1955.)

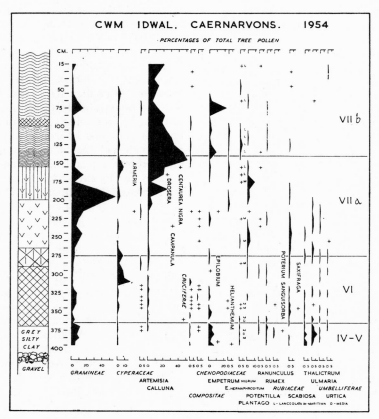

Fig. 175. Non-tree pollen diagram from Cwm Idwal, Caernarvonshire (cf. fig. 174). It shews the persistence locally into the middle Flandrian of several species or genera typical of the Late Weich-selian (*Armeria, Plantago maritima, Artemisia, Saxifraga* etc.). Many of these persist in the adjacent mountain habitats.

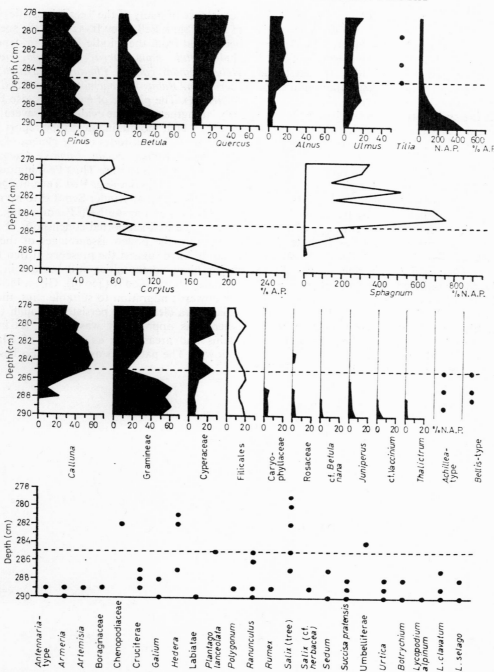

Fig. 176. Pollen analyses through the uppermost mineral soils below the blanket bog on the southern Pennines, altitude 610 m. They disclose a varied assemblage of Late Weichselian and montane taxa, such as *Armeria maritima*, *Thalictrum*, *Juniperus* and possibly *Betula nana* persisting in zone VI, after which time they disappear in conditions of general paludification. The broken line in the diagrams represents the transition from mineral soil to blanket bog peat. (After Tallis, 1964.)

The situation is far more satisfactory however with regard to the refugia which shelter the anathermic species of the Weichselian from the rigours of the Flandrian climate and its vegetational movements. It is apparent that the centres where such species are mainly found at the present day are characterized by open habitats secured by altitude and topography from the competition of closed deciduous forest, by a cool and moist climate, either mountainous or maritime or both, by base-rich soils partly determined by the native rock formation or covering drift, but also by freedom from the growth of acidic peat mire, and finally by freedom from excessive grazing.

Areas such as Ben Lawers, Ben Bulben, the Northern Pennines, Snowdonia, Cader Idris and the Burren exhibit most or all of these features. It was Pigott and Walters (1954) who pointed out how the 'eyebrows' of hard limestone scarps were often sufficiently exposed to be free of arboreal vegetation and to have the qualities of refugia in otherwise wooded regions.

Evidence that the species typical of such refugia have indeed been present there through the thermal maximum is now considerable. In 1955 pollen analyses were published from the deposits of Llyn Idwal, a lake at 1200 ft O.D. (375 m) in Snowdonia lying at the foot of the cliffs of the Devil's Kitchen, a celebrated locality for mountain plants, many of very disjunct range. As will be seen from figs. 174 and 175, pollen of many taxa prevalent in Late Weichselian time persisted into zones VI and VIIa, the time of the thermal maximum. These include *Artemisia*, *Urtica*, *Rumex*, *Thalictrum*, *Scabiosa*, *Saxifraga*, *Campanula*, *Botrychium* and *Lycopodium selago*. More striking still is the continued presence of pollen of *Plantago maritima*, *Armeria maritima* and *Empetrum hermaphroditum*, all of which still grow close by. This proof of the refugial character of Cwm Idwal has subsequently been matched by proof of the persistence of *Polemonium* in the thermal maximum at Malham (375 m) (Pigott & Pigott, 1959), and by the evidence for similar persistence of many of the Teesdale rarities in the vicinity of Widdybank Fell (1600 ft, 525 m): the species that have been listed from the middle Flandrian deposits there include *Betula nana*, *Dryas octopetala*, *Gentiana verna*, *Helianthemum canum*, *Plantago maritima*, *Polemonium coeruleum*, *Rubus chamaemorus*, *Saxifraga aizoides* and *S. stellaris*. The picture of events has also been considerably strengthened by the evidence of several other investigators of high mountain sites; they report the continued presence at high altitudes of the genera of open habitats (*Empetrum*, *Helianthemum*, *Rumex*, *Artemisia* etc.) into zone VI or even into VIIa. Thus Pennington (1970) in the English Lake District (e.g. Red Tarn at 1700 ft, 560 m), and Tallis (1964) at Kinder Scout (610 m) and the Snake Pass (515 m) on the southern Pennines (fig. 176). Finally we note the well-supported conclusions of H. J. B. Birks (1973) that the pollen assemblages of the Late Weichselian in Skye suggest the presence at that time of all the major plant communities still present in the island at altitudes above 1500 ft (490 m). Given local opportunity for upwards migration to suitable sites, the Late Weichselian flora clearly has persisted through the Flandrian: where this opportunity was absent, as in the lowland plains and areas such as the Isle of Man, extinction followed. The pattern is as Edward Forbes conjectured.

VII

CONCLUSION

In drawing this book to a conclusion it is reasonable to ask what progress its evidence and arguments represent, and to what future lines of advance they point. We have already drawn together in chapter VI our main conclusions upon the pattern of the history of our flora, and to summarize this in turn would be to risk over-simplification and premature commitment. Nevertheless it seems worth while to point out various instances in which these conclusions seem to have given a changed emphasis to problems of floristic history, and where they indicate the importance of new angles of approach.

Our records have now become extremely abundant for the Middle and Late Weichselian and for the whole of the Flandrian interglacial which commands our special attention since we live within it and have the full detail of the existing vegetation and flora before us, representing the conclusion to which preceding Pleistocene events have led. In particular they elucidate the effects of the glaciation itself upon our flora, and it may be said at once that the results of our survey of fossil material have shown that many phenomena once thought explicable in terms of local survival during glaciation are now shewn to be the product of post-glacial events. To this extent it has become of less interest to learn of the condition of our flora at the height of the several glaciations, but this problem retains considerable phytogeographic interest nevertheless, and especially for the periglacial areas outside reach of the glacier fields themselves.

From the Weichselian period onwards our records yield a fairly continuous documentation of the history of our flora, and it is instructive to consider how they bear upon the problem on the one hand of immigration and establishment of species in the British Isles, and on the other of the alteration of range within this country after their first establishment. We have examined the first of these problems by recording (see fig. 177) the period of first appearance of each species within our plant lists, and thence constructing an increment graph shewing the progressive building-up of the total recorded flora. In constructing these graphs we have assumed that a species once recorded has subsequently remained in the country, an assumption only approximately true of course. The totals are reached by counting as unity each species identified with or without qualification. Records only at generic level, hybrids, crop-plants, and aliens have been excluded. Fig. 177 gives the result of these estimations, the upper section in the form of absolute numbers, and the lower as percentages of the present existing flora. It includes also the roughly comparable figures available by

1954 and published in the first edition, from which we also present the Irish data of that time. It will be noted that whereas the Irish increment curves begin with the Late Weichselian, those for the British Isles begin with the Middle Weichselian period. The calculations of percentages are based upon a figure of 1500 existing species for the British Isles and 1000 for Ireland, round numbers which exclude microspecies such as are unrepresented in the fossil determinations. The calculations must be taken as approximate only, and they constantly alter as fresh records are made or old ones are revised in the light of new knowledge; nevertheless they convey an all-over picture of considerable interest and importance.

It will be seen that records have now been made of more than 750 species for the British Isles, that is to say, half of the total existing flora. Of these no less than 316 were recorded from the Middle Weichselian and 444 from the Late Weichselian. If it is assumed that the species recorded are representative of the whole flora (a substantial but not unreasonable assumption) then we may conclude that no less than 60 per cent of the present British flora was already in the country by the end of Late Weichselian, and as much as 42 per cent in the Middle Weichselian. We must note moreover that during the process of acquiring records of fresh species, continuing enquiry will also incidentally extend back in time the earliest record of appearance of many species already discovered. Thus, as the number of observations increases it is inevitable that the proportion of species first recorded in the earlier periods will rise, and it is evident that the figures now given will be enlarged. The rapidity of growth of total records is evident by comparison of the present percentage increment curve with that based on the 1954 data, which were in fact based on less exacting standards.

Again assuming that the fossil flora of 50 per cent is representative of the whole existing flora, it would appear that by the close of zone VI (the end of the Boreal period), when the North Sea had finally isolated us from the European mainland, *at least* 67 per cent of our present flora had already reached this country. If the true figure lies only 10 per cent higher we are left to account for no more than 390 new species entering in the last 7000 years. Of these a very large proportion will be attributable to introduction by prehistoric and early historic man with his crops, his animals and his trading. It appears consequently unlikely that any large establishment took place after the expansion of the North Sea of plants crossing from the Continent by natural means.

If we consider in the same way the 1954 figures for

493

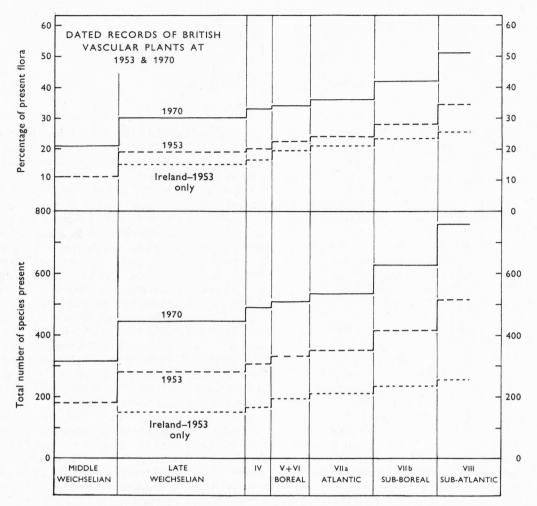

Fig. 177. Increment graphs for all dated records of plant species excluding aliens, hybrids and crop-plants. The 1970 census comprises all identifications at species level, and genera for which no individual species has been recorded up to the time in question. The 1953 figures are those cited in the first edition which rest on a slightly less exacting standard of species identification. The percentages of the existing flora have been based on the assumption for this purpose of a figure of 1500 for the British Isles and 1000 for Ireland. Note the high percentage of all records already in the Middle and Late Weichselian and by the end of the Atlantic period. The subsequent increased rate of rise is, in part at least, a consequence of the extension of disforestation and agriculture.

Although the records have greatly increased between the two censuses, this requires no modification of conclusions already drawn and strengthens the view that they are likely to be applicable to the whole of the present British flora.

Ireland it appears that at least 60 per cent of the present flora was present by the close of the Late Weichselian period, and that by the end of the Boreal period at least 77 per cent of the existing flora had entered the country. For Ireland more especially it is to be noted that there has been a disproportionately heavy analysis of Late Weichselian, as compared with later deposits, but the figures nevertheless conform to those from the British Isles as a whole. For a more modern compilation the percentages would be higher. If the view we have put forward of the general character of establishment of the flora is correct, we should indeed expect that by the time the eustatic rise in sea-level had isolated Ireland from Great Britain a larger proportion of the Irish flora must have already reached that country, than the corresponding proportion for Great Britain. This follows from the greater immunity of Ireland from later plant introductions, a feature partly associated with geographical position, partly with the extensiveness of peat-bog development, and partly with freedom from some of the invasions (e.g. the Roman) which affected Great Britain. The fewer late entries come into the present total of the Irish flora the larger the proportion of that flora we must expect to have reached the country before its isolation from the rest of the British Isles.

It may fairly be taken that the smooth rise of the increment graph for the British Isles up to the end of zone VIIa, the end of the Atlantic period, represents the natural

494

immigration of our flora under conditions hardly affected at all by the presence of man. At this point the curves rise sharply for zone VIIb, and again for zone VIII, the two together embracing the Neolithic, Bronze and Iron Ages followed by Roman, early mediaeval and Norman periods. The character of the individual records has made it evident that the greater part of the increment is due to weeds of cultivation and other plants of open habitat. In part the rise in total numbers must be due to the fact that analysis of archaeological sites has now become a much more important source of records than hitherto, in part to the fact that these sites lie preponderantly in regions not previously so intensively represented by analyses, and in part to the fact that the archaeological material represents types of habitat (such as the Chalk and Oolitic limestones) not well covered by earlier investigations. Granted these qualifications however, there seems every reason still to allow that the practices of disforestation, grazing and agriculture were directly responsible for most of the rise in numbers of identified species: these practices would lead not only to direct introductions deliberately or accidentally along with crops, stock and goods exchanged in trade, but also by affording greatly increased areas of clear soil and open habitats for colonization by such migrules as were able to reach them by natural dispersal processes. If this conjecture is correct, no amount of increased recording will be able to remove the discontinuity of the curves at the transition from the Atlantic to the Sub-boreal section of the record.

To make a brief recapitulation, the oversimplified and overall picture of the history of the arrival of the present flora of the British Isles seems to have followed these lines:

A substantial component (including many arctic–alpine species, but almost no trees) was present whilst ice-fields occupied parts of the country. In the following Late Weichselian period there was extremely rapid invasion across the North Sea (then dry) of a large proportion of our flora including many thermophiles. This was possibly accompanied by invasions from areas of the continental shelf off our western shores. The following Pre-boreal and Boreal periods saw this invasion continued with species still more exacting of warmth until by the time the North Sea had virtually reached its present extent the bulk of our flora (including most woodland plants) had become established. In the Atlantic period natural invasion from the Continent became very slow, not only on account of the sea-barrier, but also because of the almost unbroken forest-cover which migrules would encounter on arrival. The Sub-boreal saw the introduction and expansion of the processes of plant and animal husbandry which encouraged direct and incidental introduction of plants from overseas by man.

Upon this picture of the *entry* of species into these islands must be superposed all the effects of expansion and retraction of range within the country and the associated phenomena of disjunction and extinction. In the Late Weichselian period there was undoubtedly a wide expansion of range of survivors and invaders alike, but as climatic improvement continued, and closed woodlands extended, retraction of the range of many species began, some affected directly by climate, many through the effects of competition. In upland areas further retraction was forced upon many species by the growth, from Atlantic time onward, of a thick mantle of blanket bog suitable for occupation only by a restricted group of specialized oligotrophic plants. Spread of the species characteristic of woodland and peat mire was of course correspondingly improved by the conditions of this time. Certain thermophilous species responded to the conditions of the Flandrian warm period by achieving their widest ranges towards the north and west, and afterwards, when climate deteriorated in the Sub-atlantic period, they suffered corresponding retraction.

The activities of man, pursuing plant and animal husbandry, led not only to the introduction of many species but to large changes in the ranges of others. In particular, those species of bare soil and open habitats which had been prevalent in the Late Weichselian period, and had since then been restricted to localized habitats such as dunes, cliffs, river-banks and mountain-tops, were now presented with increasing opportunities to spread across the country. Whereas they had hitherto existed in localities more or less isolated in the vast stretches of deciduous forest, they now were able to move freely about the country conditioned in their establishment only by the limits of their climatic and soil tolerances. Conversely as the woodlands became split into discrete fragments, the means of natural movement of woodland plants must have been lessened, but this was at least offset if not outbalanced by the effects of human transportation of plants over quite long distances by accident in nursery stock, as well as by deliberate planting.

Our conceptions of the history of the flora have been given perspective by greatly extended knowledge of plant records and geological events in the interglacial and glacial stages that preceded the Weichselian, and in the Hoxnian and Ipswichian interglacials we have instances of the climatic cycle fully completed and effectively unmodified by the vast human disruption that complicates the history of the unfinished, though more familiar, Flandrian. The underlying similarity of pattern is displayed in fig. 178, which sets out recorded frequency of arboreal pollen in successive sub-stages, and, in the case of the Flandrian, successive pollen zones. The two terminocratic tree genera, *Betula* and *Pinus*, recede from high initial values in all three interglacials to low frequencies in the two middle sub-stages: in the Hoxnian and Ipswichian they return to high frequency in the ultimate sub-stage with the return to glacial conditions, but the Flandrian shews little indication of this. However, the more specifically terminocratic genera, *Abies*, *Picea*, *Fagus* and *Carpinus*, shew great increase in the third sub-stage of the Hoxnian and Ipswichian, and a smaller but distinctive rise in the most recent Flandrian.

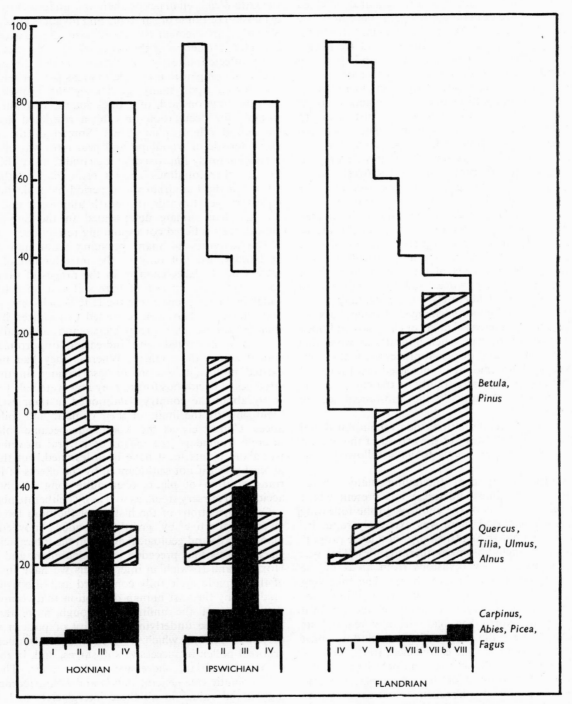

Fig. 178. Relative frequencies of arboreal pollen in three major categories through the three latest interglacials, shewing the essential similarity of pattern in each. The curves for *Betula* and *Pinus*, behaving as terminocratic elements, indicate that the Flandrian cycle is still incomplete, a conclusion supported by the behaviour of the main mediocratic group (*Quercus, Tilia, Ulmus, Alnus*), and the telocratic or late mediocratic group (*Carpinus, Abies, Picea, Fagus*).

CONCLUSION

TABLE 47. *Interglacial plants extinct in present-day British flora*

	Ludhamian	Antian	Pastonian	Cromerian	Hoxnian	Ipswichian
Pterocarya sp.	+	+	.	.	?	?
Tsuga sp.	+	+	?	?	.	.
Picea abies (L.) Kast.	+	+	+	+	+	+
Abies alba Mill.	+	.	+	+	+	+
Rhus sp.	.	+	.	.	.	.
Corispermum sp.	.	.	+	.	.	.
Azolla filiculoides Lam.	.	.	+	+	+	.
Chamaedaphne calyculata Mönch	.	.	?	.	.	.
Najas minor All.	.	.	*	+	+	+
Salvinia natans (L.) All.	.	.	.	+	.	+
Hypecoum procumbens L.	.	.	*	.	.	.
Trapa natans L.	.	.	*	.	+	+
Corema alba D. Don.	.	.	*	+	.	.
Dulichium arundinaceum (L.) Britt.	.	.	.	.	+	.
Osmunda claytoniana L.	.	.	.	.	+	.
Brasenia purpurea (Michx.) Casp.	.	.	.	.	+	.
Rhododendron ponticum L.	.	.	.	.	+	.
Erica scoparia L.	.	.	.	.	+	.
E. terminalis Salisb.	.	.	.	.	?	.
Nymphoides cordata Ell. (Fern.)	.	.	.	.	+	.
Eleocharis ovata (Roth.) R. Br.	.	.	.	.	+	.
E. carniolica Koch	.	.	.	.	?	.
Astrantia minor L.	.	.	.	.	?	.
Vitis vinifera L.	.	.	.	.	+	.
Acer platanoides L.	.	.	.	.	?	.
Pyracantha coccinea M. J. Roem.	.	.	.	?	+	.
Lysimachia punctata L.	.	.	.	.	+	.
Acer monspessulanum L.	.	.	.	.	.	+
Decodon sp.	.	.	.	.	.	+

? = doubtful identification.   * = Cromer Forest Bed series, and possibly the Antian.

The mediocratic *Quercus*, *Ulmus*, *Alnus* and *Tilia*, components of the deciduous mixed-oak forest, were evidently dominant in sub-stage II of the Hoxnian and Ipswichian and co-dominant in III, whereas in the Flandrian their dominance is longer maintained against the beech and hornbeam. The similarity of pattern is unmistakable, making evident the nature of the Flandrian as an incomplete interglacial, and confirming the preponderant control throughout of similar major factors, primarily of course, climate.

This strong parallelism is reinforced by analysis of plant records, on a similar chronological basis, the taxa respectively of southern and northern range (in the terms set out in chapter VI: 9) and of acidophilic plants. Predictably the northern elements display highest values in the early and late Anglian and Wolstonian, and Early, Middle and Late Weichselian: they diminish in the middle of the interglacials, reaching lowest Flandrian values at the present day. The southern elements broadly present the converse of this behaviour although their numbers are surprisingly high in each of the three glacial stages, and their frequency rises only slowly in the interglacials: again the inference is that the present day is only at the middle or late-middle of the Flandrian. Finally the records for taxa of plants favouring or needing acidic soil also display congruity of pattern. In each interglacial their relative frequency rises through the cycle as soils mature and then podsolize, although sub-stages IV of the Hoxnian and Ipswichian shew a decrease perhaps associated with the onset of soil disturbance or the rise of other elements. The Flandrian shews no such fall, but its Sub-boreal/Sub-atlantic frequencies are extremely high and prompt the thought that extreme oceanicity may characterize this particular interglacial, as apparently minimal oceanicity characterized the Ipswichian, at least in eastern England whence come most of our records. It is very striking that acidophilic taxa increased so spectacularly in the Late Weichselian, indicating that the continentality of the Middle Weichselian had gone and that to an unsuspected degree soils were sufficiently stable and locally vegetated to allow podsolization and even peat formation.

These analyses *in toto* confirm that, whatever the influence of man in the last 5000 years, the Flandrian exhibits the broad features of previous interglacials. We may accordingly with some assurance transpose conclusions as to floristic and vegetational evolution backwards or forwards between the Flandrian and preceding interglacials, as we may between the Weichselian and previous glacial stages. For the Flandrian observer the scene is *déjà vu*. This must not however make us neglect the progressive impact of the recurrent cycle of glacial and

TABLE 48. *Interglacial and glacial records of weeds and ruderals*

	Cs	C	An	LAn	H	EWo	LWo	I	EW	MW	LW	F
Ranunculus sardous Cranz	.	.	.	.	.	.	.	+	+	+	.	VIII
R. parviflorus L.	.	.	.	.	+	.	.	+	.	.	.	VIII
Chelidonium majus L.	.	.	.	.	.	.	.	+	.	.	.	VIII
Brassica nigra (L.) Koch	.	.	.	.	.	.	.	+	.	.	.	VIII
Thlaspi arvense L.	.	.	.	.	.	+	.	.	.	.	.	VIIb, VIII
Capsella bursa-pastoris (L.) Medic.	.	.	.	.	.	.	+	.	.	+	.	VIIII
Barbarea vulgaris R. Br.	.	.	.	.	+	.	.	+	.	.	.	VIIb, VII
Hypericum perforatum L.	.	.	.	.	.	.	.	+	.	.	.	VIIb, VIII
H. maculatum Cranz	.	.	.	.	.	.	.	+	.	.	.	VIIb, VIII
H. tetrapterum Fr.	+	.	+	.	+	.	.	+	.	.	.	VIIb, VIII
Silene nutans L.	.	.	.	.	.	.	.	+	.	.	.	
S. dioica (L.) Clairv.	.	.	.	.	+	.	*	+	.	+	+	M+L
Cerastium arvense L.	.	.	.	+	.	.	*	.	+	+	+	
C. holosteoides Fr.	.	.	.	.	+	.	.	+	.	+	+	VIIb, VIII
Stellaria media (L.) Vill.	+	.	+	.	+	.	*	+	.	.	+	VIIb, VIII
S. pallida (Dumont.) Piré	.	.	.	.	.	.	.	+	.	.	.	
S. graminea L.	.	.	.	+	+	.	.	.	+	+	+	VIIb, VIII
S. alsine Murr.	.	.	.	.	.	.	.	+	.	+	+	VIIb, VIII
Arenaria serpyllifolia L.	+	.	.	+	+	.	+	+	.	.	+	VIII
Spergula cf. *arvensis* L.	.	.	.	.	+	.	.	.	.	.	.	
Scleranthus annuus L. s.l.	.	.	.	.	+	+	+	.	.	+	+	IV
Chenopodium bonus-henricus L.	.	.	.	.	.	.	.	+	.	.	.	VIIb, VIII
C. polyspermum L.	.	.	.	.	+	.	.	+	.	.	.	VIII
C. album L.	.	+	.	.	+	.	cf.	+	.	cf.	cf.	M+L
C. murale L. (Beestonian)	.	.	.	.	+	.	.	.	.	.	.	VIII
C. urbicum L.	.	.	.	.	+	.	.	.	.	.	.	VIII
C. hybridum L.	.	.	.	.	+	.	.	+	.	.	.	VIII
C. rubrum L.	+	.	.	.	+	.	.	+	+	.	+	VIIa,b, VIII
C. botryodes Sm.	.	.	.	.	+	.	.	+	.	.	.	VIIb, VIII
Atriplex hastata L. (Beestonian)	.	+	.	.	+	.	+	.	+	+	+	M+L
Linum perenne agg.	+	.	.	.	+	+	.	.	.	+	+	
L. cartharticum L.	.	.	.	+	+	.	+	.	.	+	+	VIIb, VIII
Potentilla sterilis (L.) Garcke	.	.	.	.	+	.	.	+	.	+	+	VIIb, VIII
P. anserina L.	.	.	+	.	+	.	*	+	.	+	.	IV, VIIb, VIII
P. reptans L.	.	.	.	.	+	.	.	+	.	+	.	VIIa, VIII
Agrimonia eupatoria L.	.	.	.	.	+	.	.	+	.	.	.	VIIb
Aphanes arvensis L.	+	.	.	.	+	.	.	+	+	.	.	M+L
A. microcarpa (Boiss. & Reut.) Rothm.	.	.	.	.	+	+	.	.	.	.	.	VIII
Chaerophyllum temulum L.	.	.	.	.	.	.	.	+	.	.	.	
Anthriscus sylvestris (L.) Hoffm.	+	.	.	.	.	.	.	+	.	.	.	
Torilis japonica (Hoult.) DC.	+	.	.	.	+	.	.	.	.	.	.	VIIb
T. nodosa (L.) Gaertn.	.	.	.	.	.	.	.	+	.	.	.	VIII
Conium maculatum L.	+	+	.	.	.	.	.	.	.	.	.	VIII
Apium graveolens L.	.	.	.	.	+	.	.	+	.	.	.	VIIb
Aethusa cynapium L.	+	.	.	.	+	.	.	.	.	.	.	VI, VIIa,b, VIII
Pastinaca sativa L.	+	+	.	.	+	.	.	+	+	+	+	M+L
Heracleum sp.	.	.	.	.	+	.	.	.	.	.	.	
H. sphondylium	+	.	.	.	+	.	.	.	.	.	+	VIII
Polygonum aviculare L. s.l.	+	.	+	.	+	.	*	+	.	+	+	M+L
P. persicaria L.	+	.	.	.	+	.	*	+	.	.	+	M+L
P. lapathifolium L.	.	+	.	.	.	.	+	+	.	+	.	M+L
P. nodosum Pers.	.	+	.	.	+	.	.	.	.	.	+	M+L
P. convolvulus L.	+	.	.	.	.	.	+	+	.	.	.	M+L
Rumex acetosella L.	+	.	+	.	+	+	.	+	.	+	+	M+L
R. acetosa L.	.	.	+	.	+	.	.	+	.	.	+	M+L
R. crispus L.	+	.	.	.	+	+	.	+	.	+	+	VIIb, VIII
R. obtusifolius L.	+	.	.	.	+	.	.	+	.	.	+	VIII
R. conglomeratus Murr.	+	.	.	.	+	.	.	+	.	.	.	VIIa,b, VIII
Urtica urens L.	+	.	.	.	+	.	.	+	.	.	.	VIIb
U. dioica	+	.	.	.	+	+	+	+	.	+	+	M+L
Anagallis arvensis L.	.	.	.	.	+	.	.	+	.	.	+	VIIb, VIII
Verbascum thapsus L.	+	.	.	.	.	.	.	+	.	.	.	
V. cf. *nigrum* L.	.	.	.	.	+	.	.	.	.	.	.	
Linaria vulgaris Mill.	.	.	.	cf.	.	.	.	.	+	+	+	

TABLE 48 (*cont.*)

	Cs	C	An	LAn	H	EWo	LWo	I	EW	MW	LW	F
Lycopus europaeus L.	+	.	.	.	+	+	+	+	.	+	+	M+L
Acinos arvensis (Lam.) Dandy	+	.	.	.	+	.	.	.	.	.	+	VIII
Clinopodium vulgare L.	.	.	.	.	+	.	.	+	.	.	+	
Origanum vulgare L.	.	.	.	.	+	.	.	+	.	.	+	VIIa,b, VIII
Stachys arvensis (L.) L.	+	.	.	.	cf.	.	.	+	.	.	.	VIII
S. palustris L.	+	.	.	.	+	.	.	+	.	.	+	VIIa, VIIb
S. sylvatica L.	+	+	.	+	+	.	.	+	.	.	.	IV, VIIb, VIII
Ballota nigra L.	+	+	.	.	.	.	.	+	.	.	.	VIII
Lamium purpureum L.	.	.	.	.	+	.	.	+	.	.	.	VIIb, VIII
Nepeta cataria L.	.	.	.	.	.	.	.	.	.	cf.	.	VIII
Marrubium vulgare L.	.	.	.	.	.	.	.	+	.	.	.	VIII
Plantago major L.	.	.	.	+	+	.	+	+	.	.	+	M+L
P. media L.	.	+	.	+	+	+	+	+	.	+	+	M+L
P. lanceolata L.	.	+	.	.	+	.	+	+	.	.	+	M+L
Galium aparine L.	+	.	.	.	.	.	.	.	.	.	+	VIIb, VIII
Scabiosa columbaria L.	+	.	.	.	+	.	+	+	.	+	+	V/VI
Tussilago farfara L.	+	.	.	.	.	.	.	+	.	.	.	
Anthemis cotula L.	.	.	.	.	.	+	.	.	.	.	.	VIII
Achillea millefolium L.	.	.	.	+	.	.	.	.	+	+	+	
A. ptarmica L.	.	.	.	.	.	+	.	.	cf.	.	.	
Matricaria recutita L.	.	.	.	.	+	.	.	.	.	.	.	
Chrysanthemum leucanthemum L.	.	.	.	.	.	.	.	.	.	+	.	M+L
Artemisia sp.	.	.	.	.	+	.	+	+	+	+	+	M+L
Arctium lappa L.	.	.	.	.	+	.	.	.	.	.	.	VIII
Carduus pycnocephalus L.	.	.	.	.	.	.	.	.	.	+	.	
C. nutans L.	+	.	.	.	.	.	.	.	.	+	+	M+L
C. acanthoides L.	.	cf.	.	.	.	.	.	.	.	.	cf.	VIII
Cirsium vulgare (Savi) Ten.	+	.	.	+	+	.	.	+	.	.	+	VIIb, VIII
C. arvense (L.) Scop.	.	.	.	+	.	.	+	+	.	.	+	VIIb, VIII
Centaurea nigra L.	.	.	.	.	+	.	.	.	.	.	.	VIII
Lapsana communis L.	+	.	.	.	+	.	.	+	.	+	.	VIII
Leontodon autumnalis L.	+	.	.	+	+	+	.	.	+	+	+	VI, VIIb, VIII
L. taraxacoides (Vill.) Mérat	.	.	.	.	+	.	.	.	.	.	.	V, VIIb, VIII
Picris hieracioides L.	+	.	.	.	+	.	.	+	.	.	.	VIII
Sonchus arvensis L.	.	.	.	+	+	.	.	+	.	.	+	VIII
S. oleraceus L. (Beestonian)	.	+	.	+	+	+	.	.	.	.	.	VI, VIII
S. asper (L.) Hill.	.	.	.	.	+	.	.	.	.	.	.	VIII
Hieracium pilosella L.	.	.	.	.	.	.	.	+	.	.	.	
Crepis capillaris (L.) Wallr.	.	.	.	.	.	.	.	.	.	+	.	VIIb
Taraxacum sp.	.	.	.	.	+	+	.	+	.	+	+	IV, VIII

Cs, Cromer Forest Bed series; C, Cromerian; An, Antian; H, Hoxnian; Wo, Wolstonian; I, Ipswichian; W, Weichselian; F, Flandrian (with pollen zones). * = the Wolstonian unresolved into Early or Late.

interglacial changes, each imposing expansions, contractions and extinctions upon our flora, so that through the Pleistocene as a whole we may expect a weeding out of the less mobile and less tolerant of late Tertiary elements, and impoverishment in some cases of genetic variability. The strong evidence that the Cromerian was preceded by earlier glacial and interglacial stages is accompanied by the evidence, still scanty, that they still retained such genera as *Pterocarya* and *Tsuga* whilst *Abies* and *Picea* grew in subsequent interglacials though not the Flandrian.

These features naturally have particular value in that they allow individual identification of the different interglacials, given adequate palynological sequences. They have differed however in less readily understandable ways, as West has indicated. Thus *Corylus*, the hazel, is of little significance in the Cromerian, low in the Hoxnian, briefly important in sub-stage II of the Ipswichian, but only in the Flandrian exhibits enormous expansion in the early warm period with sustained high frequencies. Was this entirely due to a progressively increasing oceanicity of climate, was the taxon changing, or was its migration route from glacial refugia different in the succeeding interglacials? Likewise we are puzzled to attribute a reason for the strikingly more important role of *Carpinus* in the Ipswichian or of *Abies* in the Hoxnian. Whatever factors affect the trees, no doubt similar effects operate upon the other ecological categories, as can be seen by considering the list of Pleistocene extinctions (table 47).

It must not be forgotten that the physiographic background to migration was probably dissimilar in respective interglacials. Sea-level in relation to southern England probably attained +20 ft (6 m) in the Ipswichian and +80 ft (26 m) in the Hoxnian, as compared with the Flandrian. Likewise the English Channel seems probably to have been broadened to its present width at the Straits of Dover no earlier than the Wolstonian glacial stage when North Sea ice deflected southwards the discharge of the Elbe and Rhine. If this is so, the hindrance to dry-land immigration of flora from the European mainland would have been less in the Cromerian and Hoxnian than in the later interglacials, a reason that might, very speculatively, be adduced for the close similarity of British and West European forest history in the earlier as compared with the two following interglacials. In the early Pleistocene the coastal morphology of the southern North Sea was more substantially different from that of today, Kent and Sussex lying south of the main channel by which the North Sea and the Atlantic Ocean were in communication. Thus the south-eastern Chalk land of England had a long Pleistocene history of continuity with the Continental mainland, a fact to be borne in mind in considering the floristic richness of this region and the affinity of its flora to that of northern France.

One particular instance of the value of cross-reference between interglacials is the case of the category of weeds and ruderal plants. The expectation that in our present flora these were wholly due to introductions with plant and animal husbandry had already been modified by the proof that many were already present in the Late Weichselian, and the strong probability that they had tenuously continued through the forest climax stage in habitats such as cliffs, coastal dunes and river margins. To this we now add not only direct evidence of middle Flandrian presence, but the evidence that they persisted through earlier interglacials, being especially frequent in riparian habitats through the Ipswichian of East Anglia and continuing in the succeeding Weichselian glacial stage.

One of the salient features brought out by table 48 of weeds and ruderals with interglacial and glacial fossil records is how large a number of them (27 species) that were (a) present in the Weichselian, (b) without record in the middle Flandrian, (c) reported once again in the late Flandrian period of expanded husbandry, can now be seen to have a substantial record of occurrence in the middle stages of earlier interglacials. This much strengthens the case for supposing that such species were also able to persist in this country through the middle Flandrian. It can also be seen that no less than nineteen more species that have records in the Weichselian also have records proving their continuous presence through the middle Flandrian.

In comparing the interglacials with one another we have to note not only the reappearance of the same species but also of major vegetational entities. If Lang (1970) is correct, as he must be, in supposing that the major forest communities of temperate Europe were all totally reconstituted after the Weichselian ice age, this makes more remarkable their present wide consistency of composition and morphology, such that phytosociological classification can be effectively used to describe them. This emphasizes the great strength of the factors that determine the structure and constitution of the climax communities, i.e. the common needs of the species concerned and their interlocking requirements and interdependencies. It demonstrates again the quality of A. G. Tansley's concept of the 'quasi-organism' and makes evident the strong persistent individuality and coherence of plant communities that reappear with broad similarity in different interglacials and over great areas. Lang's views are of course strongly reinforced by the North American evidence for the rebuilding of complex forest communities over territories occupied by the Wisconsin ice-sheets, through a period of time measured only in a few thousands of years. The strong integration of the climax forest communities underlies their importance as climatic indicators and as pivots for the palynological exercises envisaged by von Post (1946) in his 'Prospect for pollen analysis in the study of the earth's climatic history'.

In our preoccupation with comparison of the successive interglacial floras we must not overlook the reciprocal effects of the climatic cycle upon the flora and vegetation of the glacial stages themselves. West's work has made it clear that in each of the glacial stages, Beestonian, Anglian, Wolstonian and Weichselian there was present a similar assemblage of full-glacial species, representing the major categories of northern, montane and maritime elements with numerous 'weeds' and a scatter of species of southern geographical range. As he writes, 'the mixed assemblages must derive from the diversity of habitat and microclimate of the periglacial area, with its possibilities for permafrost, waterlogged soil, sunny banks, solifluction slope and so on; and perhaps from minor climatic fluctuations in the periglacial area' (West, 1970b). During the interglacials these species occupied the type of refugia in which they have been found during the Weichselian and indeed where they grow at the present day. We have no clear evidence for any progressive loss of such species through successive glacial stages, nor is this to be expected. The phytogeographic implications of such interstadials as that at Chelford with its boreal forest vegetation must wait upon more extensive information, but Chelford itself has special interest in shewing for the last time native spruce (Picea abies) in this country.

The case we have sought to make in the preceding pages is not that we can, here and now, present a final judgement upon the complex problem of the history of our British flora. It is rather that there is lying unexploited, and in some instances neglected, a very large body of direct evidence of the former presence of the various species making up that flora, at sites and at times which can be adequately dated by reference to the chronological framework of Quaternary history. In collecting together this material and systematizing its records there has been scope for inadequacy and error. There certainly remain fairly numerous records in the literature not yet brought into our survey, and among the records there are, despite careful assessment, undoubtedly many inaccurate identifications of plant material, and some errors of dating. When all due allowance has been made for such deficiencies, there remains nevertheless a very substantial body of factual information which cannot be ignored in phytogeographical study, and which in fact affords a final proof or disproof of specific hypotheses of the historical circumstances responsible for the present ranges of British plants. Many such instances have appeared in our comments upon the records of individual species and in the conclusions and generalizations of chapter VI. Sometimes, as in the proof that many common ruderals and weeds are survivors from the Weichselian period and even from the last interglacial, and the demonstration that the disjunct range of many arctic–alpine plants is due to post-glacial events, the evidence is such that clear conclusions can be established; but for many categories of species and many types of phytogeographical problem, the collected data are yet insufficient. Not only is the total volume of evidence still too small but different regions of the British Isles have at present very different densities of plant records; thus Scotland as a whole, Wales and the Midland Plain of England are deficient, the Downs and the Cotswolds are rich in archaeologically dated material from Neolithic times onwards, and south-eastern England is rich in records of interglacial age. Much remains to be done both by pollen analysis and identification of macro-fossils to enlarge our knowledge in the regions where information is sparse. Some regions deserve much more extensive investigation because of their critical importance for resolution of some particular phytogeographic problem, and here one thinks at once of the south-western peninsula still rich in Mediterranean–Lusitanian species. The possibility is strong that Late Weichselian or early Flandrian deposits from Devon or Cornwall could yield decisive evidence of the past history of this intriguing element in our flora. The case for such regional studies in depth could not be more convincingly made than by reference to J. H. B. Birks' recently published work (1973) on the Island of Skye.

Throughout the progress of these studies our view has been strengthened that it is essential to consider the facts of phytogeographic history in the light of all possible information about the circumstances of the Pleistocene period throughout its eventful progression. That we need to be able to evaluate former climates is shewn by the sensitiveness of the range and frequency of species such as *Najas marina* to the Flandrian Thermal Maximum, by the past changes in abundance of plants like the ivy and the linden, and by the vast co-ordinated movements of the forest belts disclosed by pollen analysis for the Flandrian period throughout western Europe. The rapid migration and establishment of so many species in the Weichselian period is inexplicable without knowledge of the soil conditions in periglacial regions; and the general prevalence of dense deciduous forest in Boreal and Atlantic times must be recognized as having had the profoundest influence in encouraging the spread of some plants and animals and inhibiting that of others. We may evidently point also to the necessity for knowing the extent of glacier fields of different ages, and for knowing the nature and timing of those changes of relative land- and sea-level which alternately isolated the British Isles from the Continental mainland and re-established connection with it. Likewise we need to know the history of successive archaeological cultures, and of successive prehistoric inhabitants of these islands, whence and at what time they came and how their activities affected their environment.

At every turn the need for such information is apparent, and as the analysis of floristic history advances the need becomes apparent for increasingly more detailed and more accurate knowledge of the conditions of the Pleistocene period, in the first instance of the Flandrian period for which the information is more readily attainable, and subsequently of the more remote and obscure glacial and interglacial periods. For the Flandrian and Weichselian periods radiocarbon dating will be of inestimable value, as for the Pleistocene as a whole will be the vastly increased interest in the geology and climate of this era.

Of equal importance is the supplementation of our meagre knowledge of the autecology of the plants themselves, so that direct experimental work can assure us, for instance, whether or not high summer temperatures can directly inhibit the growth of arctic–alpine species (as Dahl's field correlations appear to indicate), so that we can have reliable estimates of mechanisms and rates of dispersal, and so that critical modern taxonomy of particular groups of plants can be considered in the light of phytogeographic history. Despite the encouraging progress of the *Biological Flora of the British Isles* under the British Ecological Society and of publications such as those of C. D. Pigott on the factors controlling species near to the limits of their range, our lack of suitable ecological knowledge is extreme. In these circumstances it is worth noting how our historical studies in many instances point the way to experimental attack upon crucial points of autecology. Thus it would evidently be of interest to determine the temperature control of those species apparently responding strongly to former changes of climate; among them many species of water plants suggest themselves as particularly suitable for experimental treatment, among them *Trapa natans*, and the various species of *Najas*, *Myriophyllum*, *Stratiotes* and *Lemna*.

The evident applicability of the arguments adduced from the biogeography of flowering plants to other groups of organisms whose remains are preserved in suitable frequency, is illustrated by J. H. Dickson's book on *Bryophytes of the Pleistocene* (1973), and by the impressive stream of publications, chiefly from the Birmingham University Department of Geology, on Pleistocene

Coleoptera. For many more groups there are promising initial studies.

Since the first publication of the hypotheses that arise when fossil plant records are systematically examined, and especially since publication of the first edition of this book, the validity and promise of this approach have been very widely accepted. Vastly more plant records have been accumulated, the standards of identification and interpretation have been much improved, especially in palynology, radiocarbon dating has underpinned all the late Pleistocene chronology, and knowledge of Pleistocene geology and archaeology have expanded remarkably. With these advantages and increased numbers of active investigators we evidently shall make accelerated progress in enlarging the 'factual basis for phytogeography'.

BIBLIOGRAPHY

AALTO, M., 1970. Potamogetonaceae fruits. I. Recent and subfossil endocarps of the Fennoscandian species. *Acta bot. fenn.*, **88**, 3.

AARIO, R., 1965. Die Fichtenverhäufigung in Lichte von C14-Bestimmungen und die Altersverhältnisse der finnischen Pollenzonen. *C. r. Soc. géol. Finl.*, **37**, 215.

AARTOLAHTI, T., 1966. Über die Einwanderung und die Verhäufigung der Fichte in Finnland. *Acta bot. fenn.*, **3**, 368.

ALLISON, J., 1947. *Buxus sempervirens* in a late Roman burial in Berkshire. *New Phytol.*, **46**:1, 122.

ALLISON, J. & GODWIN, H., 1949. Bronze Age plant remains from Wiltshire. Data for the study of post-glacial history, XII. *New Phytol.*, **48**:2, 253.

ALLISON, J., GODWIN, H. & WARREN, S. H., 1952. Late-Glacial deposits at Nazeing in the Lea Valley, North London. *Phil. Trans.*, B **236**, 169.

ANDERSEN, S. T., 1961. Vegetation and its environment in Denmark in the Early Weichselian Glacial. *Danm. Geol. Unders.*, R II, **75**, 1.

ANDERSEN, S. T., 1966. Interglacial vegetational succession and lake development in Denmark. *Palaeobotanist*, **15**, 117.

ANDERSEN, S. T., 1969. Interglacial vegetation and soil development. *Medd. Dansk G. För.*, **19**:1, 91.

ANDERSEN, S. T., 1970. The relative pollen productivity and pollen representation of North European trees, and correction factors for tree pollen spectra. *Danm. geol. Unders.*, R II, **96**, 1.

ANDREW, R., 1970. The Cambridge pollen reference collection. In Walker & West, 1970.

ANDREW, R., 1971. Exine pattern on the pollen of British species of *Tilia. New Phytol.*, **70**, 683.

ANNUAL REPORTS OF THE SUBDEPARTMENT OF QUATERNARY RESEARCH. *Cambridge Reporter.*

ASHBEE, P., 1958. The excavation of Trequllard Barrow, Treneglos Parish, Cornwall. *Antiqu. J.*, **38**, 174.

ASHBEE, P., 1966. The Fussel's Lodge Long Barrow excavations, *Archaeologia*, **100**, 1.

ASHBY, T., 1906. Report on excavations at Caerwent, 1904–5. *Rep. Brit. Ass.*, p. 401.

ATKINSON, R. J. C., 1946–7. A Middle Bronze Age barrow at Cassington, Oxon. *Oxoniensia*, **11**, 5.

BACKMAN, A. L., 1941. *Najas marina* in Finnland während der Postglazialzeit. *Acta bot. fenn.*, **30**, 1.

BACKMAN, A. L., 1948. *Najas flexilis* in Europa während der Quartärzeit. *Acta bot. fenn.*, **43**, 3.

BACKMAN, A. L., 1950. *Najas tenuissima* (A. Br.) Magnus einst und jetzt. *Comm. Biol. Soc. Sci. Fennica*, **10**, 19.

BACKMAN, A. L., 1951. *Najas minor* All. in Europa einst und jetzt. *Acta bot. fenn.*, **48**, 3.

BAILLIE-REYNOLDS, P. K., 1936. Excavations on the site of the Roman fort at Caerhun. *Arch. Cambrensis*, **91**, 240.

BAKER, H. G., 1947. Biological Flora of the British Isles: *Melandrium. J. Ecol.*, **35**, 271.

BAKER, H. G., 1948a. Dimorphism and monomorphism in the Plumbaginaceae. I. A Survey of the Family. *Ann. Bot., Lond.*, n.s., **12**, 207.

BAKER, H. G., 1948b. Significance of pollen dimorphism in Late Glacial *Armeria. Nature, Lond.*, **161**, 770.

BARKER, P. A., 1961. Excavation on the Town Wall, Roushill, Shrewsbury. *Med. Arch.*, **5**, 181.

BARTLEY, D. D., 1960. Rosgoch Common, Radnorshire: stratigraphy and pollen analysis. *New Phytol.*, **59**, 238.

BARTLEY, D. D., 1962. The stratigraphy and pollen analysis of lake deposits near Tadcaster, Yorkshire. *New Phytol.*, **61**, 277.

BARTLEY, D. D., 1964. Pollen analysis of organic deposits in the Halifax region. *Naturalist, Lond.*, **890**, 77.

BARTLEY, D. D., 1966. Pollen analysis of some lake deposits near Bamburgh in Northumberland. *New Phytol.*, **65**, 141.

BAYNES, E. N., 1908. The excavations at Din Lligwy. *Arch. Cambrensis*, **8**, 197.

BAYNES, E. N., 1920. A smelting floor at Penrhos Lligwy. *Arch. Cambrensis*, **20**, 91.

BEATSON, M. E., 1955. Sub-fossil pollen of *Lemna* in Quaternary deposits. *New Phytol.*, **54**, 208.

BEAUMONT, P., TURNER, J. & WARD, P. F., 1969. An Ipswichian peat raft in Glacial till at Hutton Henry, Co. Durham. *New Phytol.*, **68**, 747.

BEHRE, K. E. & MENCKE, B., 1969. Pollenanalytische Untersuchungen an einen Bohrkern der südlichen Doggerbank. *Deutsche Akad. Wissenschaften zum Berlin Inst. für Meers kunde*, p. 122.

BELL, A. M., 1904. Implementiferous sections at Wolvercote (Oxfordshire). *Quart. J. geol. Soc. Lond.*, **60**, 120.

BELL, F. G., 1968. Weichselian glacial floras in Britain. Ph.D. thesis, Cambridge.

BELL, F. G., 1969. The occurrence of southern steppe and halophyte elements in Weichselian (full glacial) floras from southern England. *New Phytol.*, **68**, 913.

BELL, F. G., 1970a. Late Pleistocene floras from Earith, Hunts. *Phil. Trans.*, B **258**, 347.

BELL, F. G., 1970b. Fossil of an American sedge, *Dulichium arundinaceum* (L.) Britt, in Britain. *Nature, Lond.*, **227**, 629.

BELL, F. G. & DICKSON, C. A., 1971. The Barnwell Station Arctic flora: a reappraisal of some plant identifications. *New Phytol.*, **70**, 627.

BELLAMY, D. J., BRADSHAW, M. E., MILLINGTON, G. R. & SIMMONS, I. G., 1966. Two Quaternary deposits in the lower Tees Basin. *New Phytol.*, **65**, 429.

BENNIE, J., 1895. On the occurrence of peat with Arctic plants in boulder clay at Faskine, near Airdrie, Lanarkshire. *Trans. geol. Soc. Glas.*, **10**:1, 148.

BENNIE, J., 1896. Arctic plant beds in Scotland. *Ann. Scot. nat. Hist.*, No. 17, 53.

BENNIE, J. & SCOTT, A., 1893. The ancient lake of Elie. *Proc. R. phys. Soc. Edinb.*, **12**, 148.

BERGLUND, B. E. & MALMER, N., 1971. Soil conditions and Late-Glacial stratigraphy. *G.F.F.*, **93**, 575.

BERTSCH, A., 1961. Untersuchungen an rezenten und fossilen Pollen von *Juniperus. Flora Jena*, **150**, 503.

BERTSCH, K., 1951. *Geschichte des deutschen Waldes*, 3rd edition. Fischer, Jena.

BEUG, H. J., 1957. Untersuchungen zur spätglazialen und früh-postglacialen Floren und Vegetationsgeschichte einige Mittelgebirge. *Flora, Jena*, **145**, 167.

BEUG, H. J., 1961. *Leitfaden der Pollenbestimmung*. Gustav-Fischer, Stuttgart.

BEUG, H. J., 1971. *Leitfaden der Pollenbestimmung*, parts II and III. Gustav-Fischer, Stuttgart.

BIBBY, H. C., 1940. The submerged forests at Rhyl and Abergele, North Wales. *New Phytol.*, **39**:2, 220.

BIRKS, H. H., 1969. Studies in the vegetational history of Scotland. Ph.D. Thesis, Cambridge.

BIRKS, H. H., 1970. Studies in the vegetational history of Scotland. I. A pollen diagram from Abernethy Forest, Inverness-shire. *J. Ecol.*, **58**, 827.

BIRKS, H. H., 1972. Studies in the vegetational history of Scotland. III. A radiocarbon-dated pollen diagram from Loch Maree, Ross and Cromarty. *New Phytol.*, **71**, 731.

BIRKS, H. J. B., 1963–4. Chat Moss, Lancashire. *Mem. Proc. Manchr. lit. phil. Soc.*, **106**, 1.

BIRKS, H. J. B., 1965a. Late-glacial deposits at Bagmere, Cheshire and Chat Moss, Lancashire. *New Phytol.*, **64**, 270.

BIRKS, H. J. B., 1965b. Pollen analytical investigations at Holcroft Moss, Lancashire and Lindow Moss, Cheshire. *J. Ecol.*, **53**, 299.

BIRKS, H. J. B., 1968. The identification of *Betula nana* pollen. *New Phytol.*, **67**, 309.

BIRKS, H. J. B., 1969. The Late Weichselian and present vegetation of the Isle of Skye. Ph.D. thesis, Cambridge.

BIRKS, H. J. B., 1973. *The Present and Past Vegetation of the Isle of Skye – a Palaeoecological Study*. Cambridge University Press, London.

BIRKS, H. J. B. & DEACON, J., 1973. A numerical analysis of the past and present flora of the British Isles. *New Phytol.*, **72**, 877.

BIRKS, H. J. B. & RANSOM, M. E., 1969. An Interglacial peat at Fugla Ness, Shetland. *New Phytol.*, **68**, 777.

BISHOP, W. W., 1963. Late-glacial deposits at Lockerbie, Dumfriesshire. *Trans. J. Proc. Dumfries. Galloway nat. Hist. Antiq. Soc.*, **40**, 117.

BISHOP, W. W. & DICKSON, J. H., 1970. Radiocarbon dates related to the Scottish Late-glacial sea in the Firth of Clyde. *Nature, Lond.*, **227**, 480.

BLACKBURN, K. B., 1946. On a peat from the Island of Barra, Outer Hebrides. Data for the study of post-glacial history, X. *New Phytol.*, **45**:1, 44.

BLACKBURN, K. B., 1947. *Pollen Analysis, 1944, of Wellaugh Flow, N. Tyne Border District*. Forestry Commission Pamphlet.

BLACKBURN, K. B., 1952. The dating of a deposit containing an Elk skeleton found at Neasham, near Darlington, County Durham. *New Phytol.*, **51**:3, 364.

BLACKBURN, K. B., 1953. A long pollen diagram from Northumberland. *Trans. nth. Nat. Un.*, **2**:1, 40.

BÖCHER, T., HOLMEN, K. & JAKOBSEN, K., 1968. *The Flora of Greenland*. Haare & Son, Copenhagen.

BRADSHAW, A. D., 1953. Human influence on hybridization in *Crataegus*. In Lousley, 1953.

BRAMWELL, D., 1964. The excavations at Elder Bush Cave, Wetton, Staffs. *N. Staffs. J. Fld. Studies*, **4**, 46.

BRAUN-BLANQUET, J. & RÜBEL, E., 1932. Flora von Graubünden I u. II. *Veröff. geobot. Inst. Rübel*, H. 7. Hüber, Bern and Berlin.

BROWN, A. P., 1971. The *Empetrum* pollen record as a climatic indicator in the Late Weichselian and early Flandrian of the British Isles. *New Phytol.*, **70**, 841.

BRUCE, J. R. & CUBBON, W., 1930. Cronk yn How: An Early Christian and Viking site at Lezayre, I.O.M. *Arch. Cambrensis*, **85**, 267.

BULLEID, A. & GRAY, H. St G., 1911. *The Glastonbury' Lake Village*, vol. I. Taunton Castle: private publication.

BURCHELL, J. P. T., 1927. A final account of the investigations carried out at Lower Halstow, Kent. *Proc. prehist. Soc. E. Angl.*, **5**, 289.

BURCHELL, J. P. T., 1935. Some Pleistocene deposits at *Kirmington and Crayford. Geol. Mag.*, **72**, 327.

BURCHELL, J. P. T., & ERDTMAN O. G. E., 1950. Indigenous *Tilia platyphyllos* in Britain. *Nature, Lond.*, **165**, 411.

BURCHELL, J. P. T. & PIGGOTT, S., 1939. Decorated prehistoric pottery from the bed of the Ebbsfleet, Northfleet, Kent. *Antiqu. J.*, **19**, 407.

BURNETT, J. H., 1964. *The Vegetation of Scotland*. Oliver & Boyd, Edinburgh and London.

BURRELL, W. H., 1924. Pennine Peat. *Naturalist, London*, **808**, 145.

CAIN, S. A., 1944. *Foundations of Plant Geography*. Harper, New York.

CAJANDER, A. K., 1903. Beiträge zur Kenntnis der Vegetation der Alluvionem des nördlichen Eurasiens. I. *Acta Soc. Sci. fenn.*, **32**, 1.

CALKIN, J. B., 1947. Neolithic pit at Southbourne. *Proc. Dorset nat. Hist. Fld Cl.*, **69**, 32.

CALLOW, W. J. & HASSALL, G. I., 1968. National Physical Laboratory radiocarbon measurements. V. *Radiocarbon*, **10**, 115.

CAMPBELL, J. et al., 1970. The Upper Palaeolithic period. In *The Mendip Hills in Prehistoric and Roman Times*. Bristol Archaeological Research Group, Bristol.

CAMPO, M. VAN & ELHAI, H., 1956. Etude comparative des pollens de quelques chênes. Application à une tourbière Normande. *Bull. Soc. Bot. Fr.*, **103**, 254.

CASE, H. J., DIMBLEBY, G. W., MITCHELL, G. F., MORRISON, M. E. S. & PROUDFOOT, V. B., 1969. Land use in Goodland Td., Co. Antrim, from Neolithic times until today. *J. R. Soc. Antiqu. Ireland*, **99**, 39.

CATON, L. L. F., 1914–15. Spade work in north west Suffolk. *Proc. prehist. Soc. E. Angl.*, **2**:1, 35.

CHALONER, W. G., 1968. The paleoecology of fossil spores. In *Evolution and Environment*, ed. E. T. Drake. Yale.

CHANDLER, M. E. J., 1889. Observations on some undescribed lacustrine deposits at Saint Cross, South Elmham, in Suffolk. *Quart. J. geol. Soc. Lond.*, **45**, 504.

CHANDLER, M. E. J., 1921. The Arctic flora of the Cam valley at Barnwell, Cambridge. *Quart. J. geol. Soc. Lond.*, **77**, 4.

CHANDLER, M. E. J., 1923. Geological history of the genus *Stratiotes. Quart. J. geol. Soc.*, **79**:2, 117.

CHANDLER, M. E. J., 1946. Note on some abnormally large spores formerly attributed to *Isoetes. Ann. Mag. nat. Hist.*, 11th Series, **13**, 684.

CHAPMAN, C., 1969. The Post-glacial lacustrine deposits at Sinfin Moor, S. Derbyshire. *Mercian Geologist*, **3**, 151.

CHAPMAN, S. B., 1964. The ecology of Coom Rigg Moss, Northumberland. *J. Ecol.*, **52**, 299.

CHENEY, H. J., 1935. An Aeneolithic occupation site at Playden, near Rye. *Antiqu. J.*, **15**:2, 160.

CHILDE, V. G., 1934. Final report on the excavation of the stone circle at Old Keig, Aberdeenshire. *Proc. Soc. Antiqu. Scot.*, **68**, 372.

CHILDE, V. G., 1936. A promontory fort on the Antrim coast. *Antiqu. J.*, **16**, 179.

CHILDE, V. G., 1954-6. Maes Howe. *Proc. Soc. Antiqu. Scot.*, **88**, 157.

CHRIST, H., 1910. *Die Geographie der Farne*. Fischer, Jena.

CHRISTY, Miller, 1924. The hornbeam (*Carpinus betulus* L.) in Britain. *J. Ecol.*, **12**, 39.

CHURCHILL, D. M., 1962. The stratigraphy of the Mesolithic sites III and V at Thatcham, Berkshire, England. *Proc. prehist. Soc.*, **28**, 362.

CHURCHILL, D. M., 1963. A report on the pollen analysis of the muds from the medulla tissue of two fossil human skeletons: Tilbury Man and Thatcham Man. *Proc. prehist. Soc.*, **29**, 427.

CHURCHILL, D. M., 1965. The displacement of deposits formed at sea level 6500 years ago in southern Britain. *Quaternaria*, **7**, 239.

CHURCHILL, D. M., 1970. Post Neolithic to Romano British sedimentation in the southern Fenlands of Cambridgeshire and Norfolk. In Salway, Hallam & Bromwich, 1970.

CHURCHILL, D. M. & WYMER, J. J., 1965. The Kitchen Midden site at Westward Ho!, Devon, England: ecology, age and relation to changes in land and sea level. *Proc. prehist. Soc.*, **31**, 74.

CLAPHAM, A. R., 1953. Human factors contributing to a change in our flora. In Lousley, 1953, p. 26.

CLAPHAM, A. R. & B. N., 1939. The Valley Fen at Cothill, Berkshire. *New Phytol.*, **38**:2, 167.

CLAPHAM, A. R. & GODWIN, H., 1948. Studies of the post-glacial history of British vegetation. VIII. Swamping surfaces in peats of the Somerset levels. IX. Prehistoric trackways in the Somerset levels. *Phil. Trans.*, B **233**, 233; 275.

CLAPHAM, A. R., TUTIN, T. G. & WARBURG, E. F., 1962. *Flora of the British Isles*, 2nd edition. Cambridge University Press, London.

CLARK, J. G. D., 1933. Report on an Early Bronze Age site in the south-eastern Fens. *Antiqu. J.*, **13**, 266.

CLARK, J. G. D., 1934. A Late Mesolithic settlement site at Selmeston, Sussex. *Antiqu. J.*, **14**, 140.

CLARK, J. G. D., 1936a. Report on a Late Bronze Age site in Mildenhall Fen, West Suffolk. *Antiqu. J.*, **16**:1, 29.

CLARK, J. G. D., 1936b. The timber monument at Arminghall. *Proc. prehist. Soc.*, **2**, 1.

CLARK, J. G. D., 1938a. A Neolithic house at Haldon, Devon. *Proc. prehist. Soc.*, **4**, 222.

CLARK, J. G. D., 1938b. Microlithic industries from Tufa deposits at Prestatyn, Flintshire and Blashenwell, Dorset. *Proc. prehist. Soc.*, **4**, 330.

CLARK, J. G. D., 1952. *Prehistoric Europe: the economic basis*. Methuen, London.

CLARK, J. G. D., 1954. *Excavations at Star Carr*. Cambridge University Press, London.

CLARK, J. G. D., 1960. Excavations at the Neolithic site at Hurst Fen, Mildenhall, Suffolk. *Proc. prehist. Soc.*, **26**, 202.

CLARK, J. G. D., 1965. Radiocarbon dating and the spread of farming economy. *Antiquity*, **39**, 45.

CLARK, J. G. D., 1970. Star Carr; a case study in bioarchaeology. *Addison Wesley Modular Publication*, **10**, 1.

CLARK, J. G. D. et al., 1933. Report on an Early Bronze Age site in the south eastern Fens. *Antiqu. J.*, **13**:3, 266.

CLARK, J. G. D. & FELL, C. I., 1953. The Early Iron Age site at Micklemoor Hill, West Harling. *Proc. prehist. Soc.*, **19**, 1.

CLARK, J. G. D. & GODWIN, H., 1940. A Late Bronze Age find near Stuntney, Isle of Ely. *Antiqu. J.*, **20**, 52.

CLARK, J. G. D. & GODWIN, H., 1956. A Maglemosian site at Brandesburton, Holderness, Yorkshire. *Proc. prehis. Soc.*, **22**.

CLARK, J. G. D. & GODWIN, H., 1962. The Neolithic in the Cambridgeshire Fens. *Antiquity*, **36**, 10.

CLARK, J. G. D., GODWIN, H. & M. E. & CLIFFORD, M. H., 1935. Report on Recent Excavations at Peacock's Farm, Shippea Hill, Cambridgeshire. *Antiqu. J.*, **15**:3, 284.

CLARK, J. G. D. & RANKINE, W. F., 1939. Excavations at Farnham, Surrey, 1937-8. *Proc. prehist. Soc.*, **5**:1, 111.

CLARKE, R., 1938. An Iron Age hut at Postwick and an earth-work on East Wretham Heath, Norfolk. *Norfolk Norw. Arch. Soc. (for 1937)*, **26**:3, 271.

CLARKE, R. & APLING, H., 1935. An Iron Age tumulus on Warborough Hill, Stiffkey, Norfolk. *Norfolk Norw. Arch. Soc. (for 1934)*, **25**:3, 408.

CLARKE, R. H., 1970. Quaternary sediments off south-east Devon. *Quart. J. geol. Soc. Lond.*, **125**, 278.

CLARKE, W. G., 1914-15. Two north-west Suffolk floors. *Proc. prehist. Soc. E. Angl.*, **2**:1, 39.

CLAY, R. C. C., 1924. An Early Iron Age site on Fifield Bavant Down. *Wiltsh. archaeol. nat. Hist. Mag.*, **42**, 457.

CLAY, R. C. C., 1925. An inhabited site of La Tène I date on Swallowcliffe Down. *Wilts. Arch. Mag.*, **43**, 59.

CLIFFORD, E. M., 1937. Notgrove Long Barrow, Gloucestershire. *Archaeologia (for 1936)*, **86**, 151.

CLIFFORD, E. M., 1938. The excavation of Nympsfield Long Barrow, Gloucestershire. *Proc. prehist. Soc.*, **4**, 208.

CLIFFORD, E. M., 1944. Graves found at Hailes, Gloucestershire. *Trans. Bristol. Glos. Arch. Soc.*, **65**, 187.

CLIFFORD, E. M. & DANIEL, C. E., 1940. The Rodmarton and Avening Portholes. *Proc. prehist. Soc.*, **6**, 139.

CLIFFORD, M. H., 1936. A Mesolithic flora in the Isle of Wight. *Proc. Is. Wight nat. Hist. Soc.*, **2**:7, 582.

COLES, J. M., COULTS, H. & RYDER, M. L., 1964. Late Bronze Age find from Pyotdykes, Angus, Scotland, with associated gold, cloth, leather and wood remains. *Proc. prehist. Soc.*, **30**, 186.

COLES, J. M. & HIBBERT, F. A., 1968. Prehistoric roads and tracks in Somerset, England. 1. Neolithic. *Proc. prehist. Soc.*, **34**, 238.

COLES, J. M., HIBBERT, F. A. & CLEMENTS, C. F., 1970. Prehistoric roads and tracks in Somerset, England. 2. Neolithic. *Proc. Prehist. Soc.*, **36**, 125.

COLHOUN, E. A., DICKSON, J. H., McCABE, A. M. & SHOTTON, F. W., 1972. A Middle Midlandian freshwater series at Derrymee, Maguiresbridge, County Fermanagh, Northern Ireland. *Proc. R. Soc.*, B **180**, 273.

COLHOUN, E. A. & MITCHELL, G. F., 1971. Interglacial marine deposits and Late Weichselian freshwater deposits at Shortalstown, Co. Wexford. *Proc. R. Ir. Acad.*, **71**, 211.

COLLINGWOOD, E. F. & COWEN, J. D., 1948. A prehistoric grave at Haugh Head, Wooler. *Arch. Aeliana*, **26**, 54.

CONOLLY, A. P., 1941. A report of plant remains at Minnis Bay, Kent. *New Phytol.*, **40**:4, 299.

CONOLLY, A. P., 1961. Some climatic and edaphic indications from the Late-glacial flora. *Proc. Linn. Soc. Lond.*, **172**, 56.

CONOLLY, A. P. & DAHL, E., 1970. Maximum summer temperature in relation to the modern and Quaternary distributions of certain arctic–montane species in the British Isles. In Walker & West, 1970.

CONOLLY, A. P. & DICKSON, J. H., 1969. A note on a Late Weichselian *Splachnum* capsule from Scotland. *New Phytol.*, **68**, 197.

CONOLLY, A. P., GODWIN, H. & MEGAW, E. M., 1950. Studies in Post-glacial history of British vegetation. XI. Late-glacial deposits in Cornwall. *Phil. Trans.*, B 234, 397.

CONWAY, E., 1953. Spore and Sporeling survival in bracken (*Pteridium aquilinum* (L.) Kuhn). *J. Ecol.*, **41**:2, 289.

CONWAY, V. M., 1936. Studies in the autecology of *Cladium mariscus* R.Br. I. Structure and development. *New Phytol.*, **35**, 177.

CONWAY, V. M., 1938. Studies in the autecology of *Cladium mariscus* R.Br. V. The Distribution of the Species. *New Phytol.*, **37**:4, 312.

CONWAY, V. M., 1942. Biological flora of the British Isles: *Cladium mariscus* (L). R.Br. *J. Ecol.*, **30**:1, 211.

CONWAY, V. M., 1947. Ringinglow Bog, near Sheffield. *J. Ecol.*, **34**:1, 149.

CONWAY, V. M., 1954. Stratigraphy and pollen analysis of Southern Pennine blanket peats. *J. Ecol.*, **42**:1, 117.

COOK, C. D. K., 1961. *Sparganium* in Britain. *Watsonia*, 5, 1.

COOPE, G. R., 1959. A late Pleistocene insect fauna from Chelford, Cheshire. *Proc. R. Soc.*, B 151, 70.

COOPE, G. R., 1962. Coleoptera from a peat interbedded between two boulder clays at Burnhead near Airdrie. *Trans. geol. Soc. Glasg.*, **24**:3, 279.

COOPE, G. R., 1967. The value of Quaternary insect faunas in the interpretation of ancient ecology and climate. *Proc. VII Congr. I.N.Q.U.A.*, 7, 359.

COOPE, G. R., 1968a. An insect fauna from Mid-Weichselian deposits at Brandon, Warwickshire. *Phil. Trans. R. Soc.*, B 254, 425.

COOPE, G. R., 1968b. Fossil beetles collected by James Bennie from Late-glacial silts at Corstorphine, Edinburgh. *Scot. J. Geol.*, 4, 339.

COOPE, G. R., 1969. The response of Coleoptera to gross thermal changes during the Mid-Weichselian interstadial. *Mitl. Internat. Verein. Limnol.*, 17, 173.

COOPE, G. R., 1970. Interpretations of Quaternary insect fossils. *A. Rev. Ent.*, 15, 97.

COOPE, G. R., MORGAN, A. & OSBORNE, P. J., 1971. Fossil Coleoptera as indicators of climatic fluctuations during the last glaciation in Britain. *Palaeogeography. Palaeoclimatol., Palaeoecol.*, 10, 87.

COOPE, G. R., SHOTTON, F. W. & STRACHAN, I., 1961. A late Pleistocene fauna and flora from Upton Warren, Worcs. *Phil. Trans. R. Soc.*, B 165, 389.

COUCHMAN, J. E., 1914. A Roman well at Hassocks. *Sussex Arch. Coll.*, 56, 197.

CRABTREE, K. & ROUND, F. E., 1967. Analysis of a core from Slapton Ley. *New Phytol.*, **66**, 255.

CRA'STER, M. D., 1965. Aldwick barley: recent work at the Stone Age site. *Proc. Camb. Antiqu. Soc.*, **38**, 1.

CREE, J. E., 1908. Prehistoric kitchen-midden and superimposed medieval stone floor found at Tusculum, North Berwick. *Proc. Soc. Antiqu. Scot.*, **42**, 253.

CRUDEN, S. H., 1939–40. The ramparts of Traprain Law; excavations in 1939. *Proc. Soc. Antiqu. Scot.*, **74**, 48.

CUNNINGTON, M. E., 1917. Lidbury Camp. *Wilts. archaeol. nat. Hist. Mag.*, **40**, 12.

CUNNINGTON, M. E., 1929. *Woodhenge*. Devizes.

CURWEN, E. C., 1927. Prehistoric agriculture in Britain. *Antiquity*, 1:3, 261.

CURWEN, E. C., 1929. Excavations in the Trundle, Goodwood, 1928. *Sussex Arch. Coll.*, **70**, 33.

CURWEN, E. C., 1931. Excavations in the Trundle, Goodwood, 1930. *Sussex Arch. Coll.*, **72**, 100.

CURWEN, E. C., 1932. Excavations at Hollingbury Camp, Sussex. *Antiqu. J.*, **12**, 12.

CURWEN, E. C., 1934a. Excavations in Whitehawk Neolithic Camp, Brighton, 1932–3. *Antiqu. J.*, **14**:2, 130.

CURWEN, E. C., 1934b. A Late Bronze Age farm and Neolithic pit-dwelling. *Sussex Arch. Coll.*, **75**, 168.

CURWEN, E. C., 1946. Excavations in Whitehawk Camp, Brighton, third season 1935. *Sussex Arch. Coll.*, **77**, 60.

CURWEN, E. C. & E., 1926. Harrow Hill flint-mine excavation 1924–5. *Sussex Arch. Coll.*, **67**, 103.

CURWEN, E. C. & ROSS-WILLIAMSON, R. P., 1931. The date of Cissbury Camp. *Antiqu. J.*, **11**, 14.

DĄBROWSKI, M. J., 1959. Late-glacial and Holocene history of Białowieza primeforest. *Acta Soc. Bot. Pol.*, **28**, 197.

DAHL, E., 1951. On the relation between summer temperature and the distribution of alpine vascular plants in the lowlands of Fennoscandia. *Oikos*, 3:1, 22.

DANDY, J. E., 1958. *List of British Vascular Plants*. British Museum (Natural History) and Botanical Society of the British Isles, London.

DANIEL, J. E., EVANS, E. E. & LEWIS, T., 1927. Excavations on the Kerry Hills, Montgomeryshire. *Arch. Cambrensis*, 7, 159.

DANIELSEN, A., 1969. Pollen-analytical Late Quaternary studies in the Ra district of Østfold, Southeast Norway. *Arb. Univ. Bergen Nat-Naturv.*, **14**, 7.

DAVEY, N., 1935. The Romano-British cemetery at St Stephens, near Verulamium. *Trans. St Albans Herts. Archit. Arch. Soc. (for 1935)*, p. 243.

DAVIES, J., 1905. The find of British urns near Capel Cynon in Cardiganshire. *Arch. Cambrensis*, 5, 62.

DAVIES, O., 1950. Excavations at Island MacHugh. *Supplement to Proc. Belfast nat. Hist. Soc.*

DAVIS, A. J., 1967. Notes on peat deposits in the West Hyde area, lower Colne valley. *Ruislip Dist. Nat. His. Soc.*, p. 54.

DEACON, J., 1972. A data bank of Quaternary plant fossil records. *New Phytol.*, **71**, 1227.

DEEVEY, E. S., Jnr. 1949. Biogeography of the Pleistocene. Part I. Europe and North America. *Bull. geol. Soc. Amer.*, 60, 1315.

DEGERBØL, M. & KROG, H., 1951. Den europaeiske Sumpskildpadde (*Emys orbicularis* L.) i Danmark. *Danm. geol. Unders.*, R II, no. 78.

DENNELL, R. W., 1970. Seeds from a medieval sewer in Woolster Street, Plymouth. *Econ. Bot.*, 24, 151.

DEWAR, H. S. L. & GODWIN, H., 1963. Archaeological discoveries in the raised bogs of the Somerset Levels, England. *Proc. prehist. Soc.*, 29, 17.

DICKSON, C. A., 1970. The study of plant macrofossils British Quaternary deposits. In Walker & West, 1970. in

DICKSON, C. A., DICKSON, J. H. & MITCHELL, G. F., 1970. The Late Weichselian flora of the Isle of Man. *Phil. Trans.*, B 258, 31.

DICKSON, J. H., 1973. Bryophytes of the Pleistocene. The British record and its chronological and ecological implications. Cambridge University Press, London.

DIMBLEBY, G. W., 1952. The historical status of Moorland in north-east Yorkshire. *New Phytol.*, **51**, 349.

DIMBLEBY, G. W., 1967. *Plants and Archaeology*. John Baker, London.

DODD, P. W. & WOODWARD, A. M., 1922 Excavations at Slack, 1913–15. *Yorks. Arch. J.*, **26**, 1.

DODDS, J. G., 1953. Biological flora of the British Isles: *Plantago coronopus* L. *J. Ecol.*, **41**, 467.

DONNER, J. J., 1956–7. The geology and vegetation of the Late-glacial retreat stages in Scotland. *Trans. R. Soc. Edinb.*, **63**, 221.

DONNER, J. J., 1958. Loch Mahaick, a Late-glacial site in Perthshire. *New Phytol.*, **57**, 183.

DONNER, J. J., 1960. Pollen analysis of the Burn of Benholme peat bed, Kincardineshire, Scotland. *Commentat. biol.*, **12:1**, 22.

DONNER, J. J., 1962. On the Post-glacial history of the Grampian Highlands of Scotland. *Commentat. biol.*, **24**, 5.

DONNER, J. J., 1972. Pollen frequencies in the Flandrian sediments of Lake Vakojarvi, south Finland. *Commentat. biol.*, **53**, 1.

DRAKE, H. C. & SHEPPARD, T., 1909. Classified list of organic remains from the rocks of the East Riding of Yorkshire. *Proc. Yorks. geol. Soc.*, **17:1**, 4.

DREW, C. D., 1929. Early Iron Age site at West Parley. *Proc. Dorset nat. Hist. Fld Cl.*, **51**, 232.

DREW, C. D. & PIGGOTT, S., 1936. The excavation of long barrow 163a on Thickthorn Down, Dorset. *Proc. prehist. Soc.*, **2**, 94.

DUIGAN, S. L., 1953. Plant remains from gravels of the Sommerton–Radley Terrace near Dorchester. *Quart. J. geol. Soc. Lond.*, **111**, 225.

DUIGAN, S. L., 1956. Interglacial plant remains from the Wolvercote Channel, Oxford. *Quart. J. geol. Soc. Lond.*, **112**, 373.

DUIGAN, S. L., 1963. Pollen analysis of the Cromer Forest Bed series in East Anglia. *Phil. Trans.*, B **246**, 149.

DUNLOP, E., 1888. Note on a section of boulder-clay containing a bed of peat. *Trans. geol. Soc. Glasg.*, **8:2**, 312, 314.

DUNN, S. T., 1905. *Alien Flora of Britain*. West, Newman & Co., London.

DUNNING, G. C., 1932. A Saxon hut at Bourton-on-the-Water, Gloucestershire. *Antiqu. J.*, **12**, 292.

DUNNING, G. C., 1943. A stone circle and cairn on Mynydd Epynt, Brecknockshire. *Arch. Cambrensis*, **97**, 194.

DURNO, S. E., 1956. Pollen analysis of peat deposits in Scotland. *Scot. Geog. Mag.*, **72**, 177.

DURNO, S. E., 1957. Certain aspects of vegetational history in north-east Scotland. *Scot. Geog. Mag.*, **73**, 176.

DURNO, S. E., 1958. Pollen analysis of peat deposits in eastern Sutherland and Caithness. *Scot. Geog. Mag.*, **74**, 127.

DURNO, S. E., 1959. Pollen analysis of peat deposits in the eastern Grampians. *Scot. Geog. Mag.*, **75**, 102.

DURNO, S. E., 1961. Evidence regarding the rate of peat growth. *J. Ecol.*, **49**, 347.

DURNO, S. E., 1965. Pollen analytical evidence of 'Landnam' from two Scottish sites. *Trans. Bot. Soc. Edinb.*, **40**, 13.

DURNO, S. E. & MCVEAN, D. N., 1959. Forest history of the Beinn Eighe Nature Reserve. *New Phytol.*, **58**, 228.

ERDTMAN, O. G. E., 1924. Studies in the micropalaeontology of Post-glacial deposits in Northern Scotland and the Scotch Isles. *J. Linn. Soc. (Bot.)*, **46**, 449.

ERDTMAN, O. G. E., 1925. Pollen statistics from the Curragh and Ballaugh, Isle of Man. *Proc. Lpool geol. Soc.*, **14:2**, 158.

ERDTMAN, G., 1928. Studies in the Postarctic history of the forests of north-western Europe. I. Investigations in the British Isles. *Geol. Fören. Stockh. Förh.*, **50**, 123.

ERDTMAN, G., 1935. *In*: Some Pleistocene deposits at Kirmington and Crayford; by J. P. T. Burchell. *Geol. Mag.*, **72**, 327.

ERDTMAN, G., 1940. Flower dimorphism in *Statice Armeria* L. *Svensk. Bot. Tidskr.*, **30**, 377.

ERDTMAN, G., 1943. *An Introduction to Pollen Analysis*. Chronica Botanica, Waltham, U.S.A.

EVANS, A. C. & FENNICK, V. H., 1971. The Graveney Boat. *Antiquity*, **45**, 89.

EVANS, E., 1936. Note on Doey's Cairn, Dunloy. *Antiqu. J.*, **16**, 208.

Evans, G. H., 1970. Pollen and diatom analysis of Late Quaternary deposits in the Blelham Basin, N. Lancashire. *New Phytol.*, **69**, 821.

FAEGRI, K. & IVERSEN, J., 1964. *Textbook of pollen analysis*. 2nd edition. Blackwell, Oxford.

FARRINGTON, A., 1947. Unglaciated areas in southern Ireland. *Irish Geog.*, **1**, 89.

FELL, C. I., 1954. Further notes on the Great Langdale axe factory. *Proc. prehist. Soc.*, **20**, 238.

FIELDS, N. H., MATTHEWS, C. L. & SMITH, F. I., 1964. New Neolithic sites in Dorset and Bedfordshire with a note on the distribution of Neolithic storage pits in Britain. *Proc. prehist. Soc.*, **30**, 352.

FIRBAS, F., 1934. Zur spät- und nacheiszeitlichen Vegetationsgeschichte der Rheinpfalz. *Beih. bot. Zbl.*, **52** B, 119.

FIRBAS, F., 1937. Der pollenanalytische Nachweis des Gretreide-baus. *Z. bot.*, **31**, 447.

FIRBAS, F., 1949. *Spät- und nacheiszeitliche Waldgeschichte Mitteleuropas nördlich der Alpen*, vol. 1. Fischer, Jena.

FISHER, M. J., FUNNELL, B. M. & WEST, R. G., 1969. Foraminifera and pollen from a marine interglacial deposit in the western North Sea. *Proc. Yorks. Geol. Soc.*, **37**, 311.

FLENLEY, J. R. & PEARSON, M. C., 1967. Pollen analysis of peat from the Island of Canna. *New Phytol.*, **66**, 299.

Flora Europaea. 1964, 1968, 1972, ed. T. G. Tutin *et al.*, vols. 1, 2 and 3. Cambridge University Press, London.

FLORIN, 1963. The distribution of conifer and taxad genera in time and space. *Acta Horti Bergiani*, **20**, 312.

Fox, A., 1940. The legionary fortress at Caerleon, Monmouthshire. Excavations in Myrtle Cottage orchard, 1939. *Arch. Cambrensis*, **95**, 152.

Fox, A., 1951. Eighteenth report on the archaeology of Devon. *Trans. Devons. Ass.*, **83**, 38.

Fox, A., 1954. Celtic fields and farms on Dartmoor, in the light of recent excavations at Kester. *Proc. prehist. Soc.*, **20**, 87.

Fox, C., 1922–3. A settlement of the Early Iron Age at Abington Pigotts, Cambridgeshire. *Proc. prehist. Soc. E. Angl.*, **4:1**, 211.

Fox, C., 1926. The Ysceifiog Circle and Barrow, Flintshire. *Arch. Cambrensis*, **6**, 77.

Fox, C., 1937. Two Bronze Age cairns in South Wales. *Archaeologia*, **87**, 172.

Fox, C., 1941. Stake-circles in turf barrows: a record of excavation in Glamorgan, 1939–40. *Antiqu. J.*, 21, 97.

Fox, C., 1943a. A beaker barrow, enlarged in the Middle Bronze Age, at South Hill, Talbenny, Pembrokeshire. *Arch. J.*, 99, 32.

Fox, C., 1943b. A Bronze Age barrow (Sutton 268) in Llandow Parish, Glamorganshire. *Archaeologia*, 89, 117.

Fox, C. & Hyde, H. A., 1939. A second cauldron and an iron sword from the Llyn Fawr hoard, Rhigos, Glamorganshire. *Antiqu. J.*, 19:4, 369.

Fox, C. & Wolseley, G. R., 1928. The Early Iron Age site at Findon Park, Findon, Sussex, *Antiqu. J.*, 8, 449.

Fox, C. F., 1928. A Bronze Age refuse pit at Swanwick, Hants. *Antiqu. J.*, 8, 331.

Fox, C. F., 1930. The Bronze Age pit at Swanwick, Hampshire. *Antiqu. J.*, 10, 30.

Francis, A. G., 1931. A west Alpine and Halstatt site at Southchurch, Essex. *Antiqu. J.*, 11, 410.

Franks, J. W., 1960. Interglacial deposits at Trafalgar Square, London. *New Phytol.*, 59, 145.

Franks, J. W. & Johnson, R. H., 1964. Pollen analytical dating of a Derbyshire landslip: Cown Edge landslides, Charlesworth. *New Phytol.*, 63, 209.

Franks, J. W. & Pennington, W., 1961. The Late-glacial and Post-glacial deposits of the Esthwaite Basin, N. Lancashire. *New Phytol.*, 60, 27.

Fraser, G. K. & Godwin, H., 1955. Two Scottish pollen diagrams: Carnwarth Moss, Lanarkshire and Strichen Moss, Aberdeenshire. *New Phytol.*, 54, 216.

Frenzel, B., 1966. Climatic change in the Atlantic/Sub-Boreal transition on the northern hemisphere: botanical evidence. Proc. Int. Sym. on World Climate 8000 to 0 B.C. *Proc. Roy. Met. Soc.*, 99.

Frenzel, B., 1968. *Grundzüge der Pleistozänen Vegetationsgeschichte Nord-Eurasiens*. F. Steiner, Wiesbaden.

Frere, S. S., 1944. An Iron Age site at West Clandon, Surrey and some aspects of Iron Age and Romano-British culture in the Wealden area. *Arch. J.*, 101, 55.

Funnell, B. M. & West, R. G., 1962. The early Pleistocene of Easton Bavents, Suffolk. *Quart. J. geol. Soc. Lond.*, 118, 125.

Gams, H. & Nordhagen, R., 1923. Postglaziale Klimaänderungen und Erdkrustenbewegungen in Mitteleuropa. *München, Geogr. Gesellsch. Landesk. Forschungen.*, 25.

Gardner, W., 1913. Excavation of tumuli, Eglwys Bach, Denbighshire. *Arch. Cambrensis*, 13, 337.

Gaunt, G. D., Coope, G. R. & Franks, J. W., 1970. Quaternary deposits at Oxbow opencast coal site in the Aire valley, Yorks. *Proc. Yorks. geol. Soc.*, 38, 175.

Gaunt, G. D., Coope, G. R., Osborne, P. J. & Franks, J. W., unpublished. An Interglacial site near Austerfield, South Yorkshire.

Gay, P. A., 1957. Distribution and variation of *Erica mackiana*. In *Progress in the Study of the British Flora*, ed. J. F. Lousley. Botanical Soc. Br. Is., Arbroath.

Geikie, J., 1881. *Prehistoric Europe*. Stanford, London.

Gepp, A., 1895. Fossil plant-remains in peat. *J. Bot.*, 33, 180.

Glenn, T. A., 1915. Prehistoric and historic remains at Dyserth Castle. *Arch. Cambrensis*, 15, 47.

Godwin, H., 1934. Pollen analysis. An outline of the problems and potentialities of the method. *New Phytol.*, 33, 278.

Godwin, H., 1940a. Pollen analysis and forest history of England and Wales. *New Phytol.*, 39:4, 370.

Godwin, H., 1940b. A Boreal transgression of the sea in Swansea Bay. *New Phytol.*, 39:3, 308.

Godwin, H., 1940c. Studies of the Post-glacial history of British vegetation. III. Fenland pollen diagrams. IV. Post-glacial changes of relative land- and sea-level in the English fenland. *Phil. Trans.*, B 230, 239.

Godwin, H., 1941. Studies of the Post-glacial history of British vegetation. VI. Correlations in the Somerset Levels. *New Phytol.*, 40:2, 108.

Godwin, H., 1943. Coastal peat beds of the British Isles and North Sea. *J. Ecol.*, 31:2, 199.

Godwin, H., 1944. Age and origin of the 'Breckland' heaths of East Anglia. *Nature, Lond.*, 154, 6.

Godwin, H., 1945a. Coastal peat-beds of the North Sea region, as indices of land- and sea-level changes. *New Phytol.*, 44, 29.

Godwin, H., 1945b. A submerged peat bed in Portsmouth Harbour. *New Phytol.*, 44:2, 152.

Godwin, H., 1946. The relationship of bog stratigraphy to climatic change and archaeology. *Proc. prehist. Soc.*, 12, 1.

Godwin, H., 1948. Studies of the Post-glacial history of British vegetation. X. Correlations between climate, forest composition, prehistoric agriculture and peat stratigraphy in Sub-Boreal and Sub-Atlantic peats of the Somerset levels. *Phil. Trans.*, B 233, 275.

Godwin, H., 1954. Recurrence-surfaces. *Danm. geol. Unders.*, R II, 80, no. 19.

Godwin, H., 1955. Vegetational history at Cwm Idwal: a Welsh plant refuge. *Svensk bot. Tidskr.*, 49, 35.

Godwin, H., 1956. Studies of the Post-glacial history of British vegetation. The Meare Pool region of the Somerset Levels. *Phil. Trans.*, B 239, 161.

Godwin, H., 1958. Pollen analysis in mineral soil, an interpretation of a podzol pollen-analysis by Dr G. W. Dimbleby. *Flora, Jena*, 146, 321.

Godwin, H., 1959. Plant remains from Hartford, Hunts. *New Phytol.*, 58, 85.

Godwin, H., 1960a. Radiocarbon dating and Quaternary history in Britain. *Proc. R. Soc.*, B 153, 287.

Godwin, H., 1960b. Studies of the Post-glacial history of British vegetation. XIV. Late-glacial deposits at Moss Lake. *Phil. Trans.*, B 242, 127.

Godwin, H., 1960c. Prehistoric wooden trackways in the Somerset Levels: their construction, age and relation to climatic change. *Proc. prehist. Soc.*, 26, 1.

Godwin, H., 1962. Vegetational history of the Kentish Chalk Downs as seen at Wingham and Frogholt. *Veröff. geobot. Inst. Zürich*, 37, 83.

Godwin, H., 1964. Organic deposits at Colney Heath, Herts. *Proc. R. Soc.*, B 160, 258.

Godwin, H., 1966. Introductory address. Proc. Int. Sym. on World climate 8000 to 0 B.C. *Proc. Roy. Met. Soc.* 99.

Godwin, G., 1967a. Discoveries in the peat near Shapwick Station, Somerset. *Proc. Somerset Arch. Nat. His. Soc.*, 3, 20.

Godwin, H., 1967b. Pollen-analytic evidence for the cultivation of *Cannabis* in England. *Rev. Palaeobotan. Palynol.*, 4, 71.

Godwin, H., 1967c. The ancient cultivation of hemp. *Antiquity*, 41, 42.

Godwin, H., 1968a. The development of Quaternary palynology in the British Isles. *Rev. Palaeobotan. Palynol.*, 6, 9.

GODWIN, H., 1968b. Organic deposits at Old Buckenham Mere, Norfolk. *New Phytol.*, 67, 95.

GODWIN, H., 1968c. Terneuzen and buried forests of the East Anglian fenland. *New Phytol.*, 67, 733.

GODWIN, H., 1969. The value of plant materials for radio-carbon dating. *Amer. J. Bot.*, 56, 723.

GODWIN, H., 1970. The contribution of radiocarbon dating to archaeology in Britain. *Phil. Trans. Roy. Soc. Lond.*, A 269, 57.

GODWIN, H. & BACHEM, K., 1962. In 'Excavations in Hungate, York', ed. K. M. Richardson. *Arch. Journ.*, 116.

GODWIN, H. & CHAMBERS, T. C., 1961. The fine structure of the pollen wall of *Tilia platyphyllos*. *New Phytol.*, 60, 393.

GODWIN, H. & CHAMBERS, T. C., 1971. Scanning electron microscopy of *Tilia* pollen. *New Phytol.*, 70, 687.

GODWIN, H. & CLAPHAM, A. R., 1951. Peat deposits on Cross Fell, Cumberland. *New Phytol.*, 50:2, 167.

GODWIN, H. & CLIFFORD, M. H., 1938. Studies of the Post-glacial history of British vegetation. I. Origin and strati-graphy of fenland deposits near Woodwalton, Hunts. II. Origin and stratigraphy of deposits in southern fen-land. *Phil. Trans.*, B 229, 323.

GODWIN, H. & M. E., 1933. British Maglemose harpoon sites. *Antiquity*, 7, 36.

GODWIN, H. & M. E., 1934. Pollen analysis of peats at Scolt Head Island, Norfolk. In *Scolt Head Island*, ed. J. A. Steers. Heffer, Cambridge.

GODWIN, H. & M. E., 1936. *In*: Archaeology of the submerged land-surface of the Essex coast; by S. H. Warren *et al. Proc. prehist. Soc.*, 2:2, 178.

GODWIN, H. & M. E., 1940. Submerged peat at Southampton. *New Phytol.*, 39:3, 303.

GODWIN, H. & M. E., CLARK, J. G. D. & CLIFFORD, M. H., 1934. A Bronze Age spear-head found in Methwold Fen, Norfolk. *Proc. prehist. Soc. E. Angl.*, 7:3, 395.

GODWIN, H. & M. E. & CLIFFORD, M. H., 1935. Controlling factors in the formation of Fen deposits, as shown by peat investigations at Wood Fen, near Ely. *J. Ecol.*, 23:2, 509.

GODWIN, H. & MITCHELL, G. F., 1938. Stratigraphy and de-velopment of two raised bogs near Tregaron, Cardigan-shire. *New Phytol.*, 37:5, 425.

GODWIN, H. & NEWTON, L., 1938. The submerged forest at Borth and Ynyslas, Cardiganshire. Data for the study of Post-glacial history. *New Phytol.*, 37:4, 333.

GODWIN, H., SUGGATE, R. P. & WILLIS, E. H., 1958. Radio-carbon dating of the eustatic rise in ocean level. *Nature, Lond.*, 181, 1518.

GODWIN, H. & SWITSUR, V. R., 1966. Cambridge University natural radiocarbon measurements VIII. *Radiocarbon*, 8, 390.

GODWIN, H. & TALLANTIRE, P. A., 1951. Studies of the Post-glacial history of British vegetation. XII. Hockham Mere, Norfolk. *J. Ecol.*, 39:2, 285.

GODWIN, H. & TANSLEY, A. G., 1941. Prehistoric charcoals as evidence of former vegetation, soil and climate. *J. Ecol.*, 29:1, 117.

GODWIN, H., WALKER, D. & WILLIS, E. H., 1957. Radio-carbon dating and Post-glacial vegetational history: Scaleby Moss. *Proc. R. Soc.*, B 147, 352.

GODWIN, H. & WILLIS, E. H., 1959a. Radiocarbon dating of the Late-glacial period in Britain. *Proc. R. Soc.*, B 150, 199.

GODWIN, H. & WILLIS, E. H., 1959b. Cambridge University natural radiocarbon measurements I. *Radiocarbon*, 1, 63.

GODWIN, H. & WILLIS, E. H., 1960. Cambridge University natural radiocarbon measurements II. *Radiocarbon*, 2, 62.

GODWIN, H. & WILLIS, E. H., 1961. Cambridge University natural radiocarbon measurements III. *Radiocarbon*, 3, 60.

GODWIN, H. & WILLIS, E. H., 1962. Cambridge University natural radiocarbon measurements V. *Radiocarbon*, 4, 57.

GODWIN, H. & WILLIS, E. H., 1964. Cambridge University natural radiocarbon measurements VI. *Radiocarbon*, 6, 116.

GODWIN, H., WILLIS, E. H. & SWITSUR, V. R., 1965. Cam-bridge University natural radiocarbon measurements VII. *Radiocarbon*, 7, 205.

GOULD, J., 1964–5. Excavations in advance of construction at Shenstone and Wall. *Lichfield & S. Staffs. Arch. & Hist. Soc.*, 6, 17.

GRANLUND, E., 1932. De svenska högmossernas geologi, (German summary.) *Sverig. geol. Unders. Afh.*, 26:1, Ser. C., 373, S. 5.

GRAY, H. ST G., 1913. Excavations at Maumbury Rings. *Proc. Dorset nat. Hist. Fld Cl.*, 34, 106.

GRAY, H. ST G., 1926. Excavations at Ham Hill, South Somerset. *Proc. Somerset Arch. nat. Hist. Soc.*, 72, 55.

GRAY, H. ST G., 1934. The Avebury excavations 1908–22. *Archaeologia*, 84, 159.

GRAY, H. ST G., 1935. Excavations at Combe Beacon, Combe St Nicholas, 1935. *Proc. Somerset Arch. nat. Hist. Soc.*, 81, 83.

GRAY, H. ST G., 1936. Discovery of Neolithic pottery on Meare Heath, Somerset. *Proc. Somerset Arch. nat. Hist. Soc.*, 82, 160.

GREENHILL, B., 1971. An ancient vessel brought to light: the Graveney Boat. *Country Life*, 149, 948.

GRESWELL, R. K., 1958. The post glacial raised beach in Furness and Lyth, North Morecambe Bay. *Trans. Inst. Br. Geogr.*, 25, 79.

GRIMES, W. F., 1938. A barrow on Breach Farm, Llanbleddian, Glamorgan. *Proc. prehist. Soc.*, 4, 120.

GRIMES, W. F., 1948. Pentre-ifan burial chamber, Pembroke-shire. *Arch. Cambrensis*, 100, 23.

GRIMES, W. F. & HYDE, H. A., 1935. A prehistoric hearth at Radyr, Glamorgan, and its bearing on the nativity of beech (*Fagus sylvatica* L.) in Britain. *Trans. Cardiff Nat. Soc.*, 68, 46.

GRITCHUK, 1964. Comparative Study of the Interglacial and Interstadial flora of the Russian Plain. *Proc. VI. Congr. I.N.Q.U.A.*, 2, 395.

GROHNE, U., 1957. Die Bedeutung des Phasenkontrast-verfahrens für die Pollenanalyse, dargelegt am Beispiel der Gramineenpollen vom Getreidetyp. *Photographie Forsch.*, 7, 237.

GROOT, J. J. & C. R., EWING, M., BURKLE, L. & CONOLLY, J. R., 1967. Spores, pollen, diatoms and provenance of the Argentine Basin sediments. *Progress in Oceanography*, 4.

GROSE, J. D. & SANDELL, R. E., 1964. A catalogue of pre-historic plant remains in Wiltshire. *Wilts. Arch. Nat. Hist. Mag.*, 59, 58.

GROVE, R. L. A., 1963. Researches and discoveries in Kent – Larkfield. *Arch. Cantiana*, 78, 192.

HAFSTEN, U., 1970. A Sub-division of the late Pleistocene period on a synchronous basis, intended for global and universal usage. *Palaeogr. Palaeoclon., Palaeoecol.*, **7**, 279.

HAMMEN, T. VAN DER, 1949. De Allerød-oscillatie in Nederland. *Proc. Acad. Sci. Amst.*, **52:1**, 69; **52:2**, 169.

HAMMEN, T. VAN DER, 1951. Late-glacial flora and periglacial phenomena in the Netherlands. *Leid. geol. Meded.*, **17**, 71.

HAMMEN, T. VAN DER, MAARLEVELD, G. C., VOGEL, J. C. and ZAGWIJN, W. H., 1967. Stratigraphy, climatic succession and radiocarbon dating of the last glacial in the Netherlands. *Geologie Mijnb.*, **46**, 79.

HANSEN, I., 1950. Die europäischen Arten der Gattung *Erica*. *Bot. Jb.*, **75**, 1.

HARDY, E. M., 1939. Studies in the Post-glacial history of British vegetation. V. The Shropshire and Flint Maelor mosses. *New Phytol.*, **38:4**, 364.

HARRIS, T. M., 1950. A great Pliocene flora (Review of W. Szafer, 1946–7). *New Phytol.*, **49**, 421.

HARRISON, J. W. H. & BLACKBURN, K. B., 1946. The occurrence of a nut of *Trapa natans* L. in the Outer Hebrides, with some account of the peat bogs adjoining the loch in which the discovery was made. *New Phytol.*, **45**, 124.

HARTZ, N. & MILTHERS, V., 1901. Det senglaciale ler i Allerød Teglvaerksgrav. *Medd. dansk. geol. Foren.*, **8**, 31.

HAWKES, C. F. C., 1936. The excavations at Buckland Rings, Lymington, 1935. *Proc. Hunts. Fld Cl. Arch. Soc.*, **13:2**, 124.

HAWKES, C. F. C., 1939. The excavations at Quarley Hill, 1938. *Proc. Hants. Fld Cl. Arch. Soc.*, **14:2**, 193.

HAWKES, C. F. C., 1940. A site of the Late Bronze–Early Iron Age transition at Totternhoe, Bedfordshire. *Antiqu.*, **20**, 487.

HAWKES, C. F. C., MYERS, J. N. L. & STEVENS, C. G., 1929. St Catherine's Hill, Winchester. *Proc. Hants. Fld Cl. Arch. Soc.*, **11**, 137.

HAYTER, A. G. K., 1921. Excavations at Segontium. *Arch. Cambrensis*, **1**, 19.

HEDBERG, O., 1946. Pollen morphology in the genus *Polygonum* L. *s.lat.* and its taxonomical significance. *Svensk bot. Tidskr.*, **40:4**, 371.

HELBAEK, H., 1952a. Preserved apples and panicum in the prehistoric site at Nørre Sandegaard in Bornholm. *Acta. Archaeol.*, **23**, 107.

HELBAEK, H., 1952b. Spelt (*Triticum spelta* L.) in Bronze Age Denmark. *Acta Archaeol.*, **23**, 7.

HELBAEK, H., 1953a. Archaeology and agricultural botany. *Annu. Rep. Lond. Univ. Inst. Archaeol.*, 44.

HELBAEK, H., 1953b. Early crops in southern England. *Proc. prehist. Soc.*, **18**, 194.

HELBAEK, H., 1957. Note on carbonised cereals. In The Late Bronze Age settlement on Itford Hill, Sussex; by G. F. Burston & G. A. Holleyman. *Proc. prehist. Soc.*, **23**, 167.

HELBAEK, H., 1959. Notes on the evolution of *Linum. Kuml.*; *Årbok f. Jysk Ark. Selsk.*, 103.

HELBAEK, H., 1964. The Isca grain: a Roman plant introduction in Britain. *New Phytol.*, **63**, 158.

HELBAEK, H., 1966. Commentary on the Phylogenesis of *Triticum* and *Hordeum*. *Econ. Bot.*, **20**, 350.

HELBAEK, H., 1971. The origin and migration of rye, *Secale cereale* L.; a palaeo–ethnobotanical study. In *Plant Life in South-West Asia*, ed. P. H. Davis *et al.* Bot. Soc. Edin., Edinburgh.

HEMP, W. J., 1931. The chambered cairn of Bryn Celli Ddu. *Arch. Cambrensis*, **86**, 249.

HEMP, W. J., 1935. The chambered cairn known as Bryn yr Hen Bobl near Plas Newydd, Anglesey. *Archaeologia*, **85**, 281.

HENCKEN, H. O'N., 1942. Ballinderry crannog. No. 2. *Proc. R. Irish Acad.*, **47 C 1**, 1.

HENCKEN, T. C., 1939. The excavation of the Iron Age camp on Bredon Hill, Gloucestershire, 1935–6. *Arch. J.*, **94**, 40.

HENDERSON, E., 1914–15. Opening of a barrow at Salthouse, Norfolk. *Proc. prehist. Soc. E. Angl.*, **2:1**, 155.

HENSHALL, H., 1963–4. A dagger-grave and other cist burials at Ashgrove, Methilhill, Fife. *Proc. Soc. Antiqu. Scot.*, **97**, 166.

HESLOP-HARRISON, Y., 1955. *Nymphaea* L. em. Sm. *J. Ecol.*, **43**, 719.

HIBBERT, F. A., SWITSUR, V. R. & WEST, R. G., 1971. Radiocarbon dating of Flandrian pollen zones at Red Moss, Lancashire. *Proc. R. Soc.*, B **177**, 161.

HICKS, H., 1892. On the discovery of mammoth and other remains in Endsleigh Street. *Quart. J. geol. Soc. Lond.*, **48**, 453.

HIRST, J. M. & HURST, G. W., 1967. Long-distance spore transport. *Symp. Soc. gen. Microbiol.*, **17**, 309.

HOGG, A. H. A., 1940. A long barrow at West Rudham, Norfolk. *Norfolk Norwich Arch. Soc.*, **27:2** (*for* 1939), 315.

HOGG, A. H. A., O'NEIL, B. H. ST J. & STEVENS, C. E., 1941. Earthworks on Hayes and West Wickham Commons. *Arch. Cantiana*, **54**, 28.

HOGG, A. H. A. & STEVENS, C. E., 1937. The defences of Roman Dorchester. *Oxoniensia*, **2**, 71.

HOLLEYMAN, G. A., 1936. An early British agricultural village site on Highdole Hill, near Telscombe. *Sussex Arch. Coll.*, **77**, 202.

HOLLEYMAN, G. A., 1937. Harrow Hill excavations 1936. *Sussex Arch. Coll.*, **78**, 230.

HOLLEYMAN, G. A. & CURWEN, E. C., 1935. Late Bronze Age Lynchet-settlements on Plumpton Plain, Sussex. *Proc. prehist. Soc.*, **1**, 16.

HOLLINGWORTH, S. E., 1962. The climatic factor in the geological record. *Quart. J. geol. Soc. Lond.*, **118**, 1.

HOLLINGWORTH, S. E., ALLISON, J. & GODWIN, H., 1950. Interglacial deposits from the Histon Road, Cambridge. *Quart. J. geol. Soc. Lond.*, **105** (*for* 1949), 495.

HOOD, S. & WALTON, H., 1948. A Romano-British cremating place and burial ground on Roden Downs, Compton, Berkshire. *Trans. Newbury Dist. Fld Cl.*, **9:1**, 47.

HOPF, M., 1967. Einige Bemerkungen zu römerzeitlichen Fässern. *Jn. des Römisch-Germanischen Zentralmuseums, Mainz*, **14**.

HOULDER, C. H., 1961. The excavation of a Neolithic stone implement factory on Mynydd Rhiw in Caernarvonshire. *Proc. prehist. Soc.*, **28**, 108.

HOVE, H. A. TEN, 1968. The *Ulmus* fall at the Transition Atlanticum–Subboreal in pollen diagrams. *Palaeogeography, Palaeoclimatol., Palaeoecol.*, **5**, 359.

HOWARD, H. W. & LYON, A. G., 1952. Biological flora of the British Isles: *Nasturtium officinale* R. Br. *J. Ecol.*, **40:1**, 228.

HULTÉN, E., 1950. *Atlas of the Distribution of Vascular Plants in North-west Europe*. Generalstubens Litografiska Anstalts, Stockholm.

HULTÉN, E., 1958. *The Amphi-Atlantic Plants and their Phyto-geographical Connections.* Almqvist & Wicksell, Stockholm.

HUTCHINSON, J., 1965. *Essays on Crop Plant Evolution.* Cambridge University Press, London.

HUTCHINSON, T., 1966. The occurrence of living and sub-fossil remains of *Betula nana* in Upper Teesdale. *New Phytol.*, **65**, 351.

HYDE, H. A., 1930. Appendix to: The chambered cairn of Bryn Celli Ddu. *Archaeologia*, **80**, 214.

HYDE, H. A., 1936. On a peat bed at the East Moors, Cardiff. *Trans. Cardiff Nat. Soc.*, **69**, 39.

HYDE, H. A., 1939. On the date of an axe-hammer from Llangeitho, Cardiganshire. *Proc. prehist. Soc.*, **5**, 166.

HYDE, H. A., 1940. On a peat bog at Craig-y-llyn, Glam. *New Phytol.*, **39**:2, 226.

HYDE, H. A., 1950-2. *46th Annual Report of the National Museum of Wales.* Mynydd Llysworney.

HYDE, H. A., 1952. Studies in atmospheric pollen. V. A daily census of pollens at Cardiff for the six years 1943-8. *New Phytol.*, **51**:3, 281.

HYDE, H. A., 1953. The excavation of a Neolithic dwelling and a Bronze Age cairn. *Rep. Trans. Cardiff Nat. Soc.*, **81**, 91.

HYDE, H. A., 1969. Aeropalynology in Britain – an outline. *New Phytol.*, **68**, 579.

INGRAM, R., 1967. On the identity of the Irish populations of *Sisyrinchium. Watsonia*, **6**, 283.

IVERSEN, J., 1929. Studien über die pH-Verhältnisse dänischer Gewässer und ihren Einfluss auf die Hydrophyten-Vegetation. *Bot. Tidsskr.*, **40**:4.

IVERSEN, J., 1936. Secundäres Pollen als Fehlerquelle. Eine Korrektionsmethode zur Pollenanalyse minerogener Sedimente. *Danm. geol. Unders.*, R IV, **2**, no. 15.

IVERSEN, J., 1940. Blütenbiologische Studien, I. Dimorphie und Monomorphie bei *Armeria. K. danske vidensk. Selsk., Biol., Meddel.*, **15**:8, 1.

IVERSEN, J., 1941. Landnam i Danmarks Stenalder. En pollenanalytisk Undersøgelse over det første Landbrugs Indvirkning paa Vegetationsudviklingen. *Danm. geol. Unders.*, R II, no. 66.

IVERSEN, J., 1944a. *Viscum, Hedera* and *Ilex* as climate indicators. *Geol. Fören. Stockh. Förh.*, **66**, 463.

IVERSEN, J., 1944b. *Helianthemum* som fossil Glacialplante i Danmark. *Geol. Fören. Stockh. Förh.*, **66**, 774.

IVERSEN, J., 1946. Geologisk datering af en senglacial Boplads ved Bromme. *Aarbøger Nord. Oldkynd. Hist.*, **198**.

IVERSEN, J., 1947a. *Centaurea cyanus* pollen in Danish Late-glacial deposits. *Medd. dansk geol. Foren.*, **11**:2, 197.

IVERSEN, J., 1947b. Plantevaekst, Dyreliv og Klima i det senglaciale Danmark. *Geol. Fören. Stockh. Förh.*, **69**, 67.

IVERSEN, J., 1949. The influence of prehistoric man on vegetation. *Danm. geol. Unders.*, R IV, **3**, no. 6.

IVERSEN, J., 1951. Steppeelementer i den senglaciale Flora og Fauna. *Medd. dansk geol. Foren.*, **12**, 174.

IVERSEN, J., 1954. The late-glacial flora of Denmark and its relation to climate and soil. *Danm. geol. Unders.*, **96**, 87.

IVERSEN, J., 1958. The bearing of Glacial and Interglacial epochs on the formation and extinction of plant taxa. *Uppsala Univ. Årsskr.*, **6**, 210.

IVERSEN, J., 1960. Problems of the early Post-glacial forest development in Denmark. *Danm. geol. Unders.*, R IV, **4**, no. 3.

IVERSEN, J., 1964. Retrogressive vegetational succession in the post-glacial. *J. Ecol.*, **52**, 59.

IVERSEN, J., 1969. Retrogressive development of a forest ecosystem demonstrated by pollen diagrams from fossil mor. *Oikos*, **12**, 35.

JACOB-FRIESEN, K. H., 1949. Grosswildjäger des Eiszeitalters in Niedersachsen. *Kosmos, Stuttgart,*. **11**, 408.

JANSSEN, C. R., 1960. On the Late-glacial and Post-glacial vegetation of South Limburg (Netherlands). *Wentia*, **4**, 112.

JANSSEN, C. R. & HOVE, H. A. TEN, 1971. Some late Holocene pollen diagrams from the Peel raised bogs. *Rev. Palaeobotan. Palynol.*, **11**, 7.

JEFFERIES, R. L., WILLIS, A. J. & YEMM, E. W., 1968. The Late-and Post-glacial history of the Gordano Valley, North Somerset. *New Phytol.*, **67**, 335.

JELGERSMA, S., 1966. Sea-level changes during the last 10 000 years. Proc. Int. Symp on World Climate 8000 to 0 B.C. *Proc. Roy. Met. Soc.*, **99**.

JENNINGS, J. N., 1952. The origin of the Broads. *R.G.S. Res. Ser.*: no. 2.

JENNINGS, J. N., 1955. Further pollen data from the Norfolk Broads. *New Phytol.*, **54**, 199.

JERMAN, H. N., 1935. Oak piles from bog at Trallwm Farm, Abergwesyn, Breconshire. *Antiqu. J.*, **15**, 68.

JESSEN, K., 1948. *Rhododendron ponticum* L., in the Irish Inter-glacial Flora. *Irish Nat. J.*, **9**, 174.

JESSEN, K., 1949. Studies in Late Quaternary deposits and flora-history of Ireland. *Proc. R. Ir. Acad.*, **52** B 6, 85.

JESSEN, K., 1955. Key to subfossil *Potamogeton. Bot. Tidsskr.* **52**, 1.

JESSEN, K., ANDERSEN, S. & FARRINGTON, A., 1959. The Inter-glacial deposit near Gort, Co. Galway, Ireland. *Proc. R. Ir. Acad.*, B **60**, 3.

JESSEN, K. & FARRINGTON, A., 1938. The bogs at Ballybetagh, near Dublin, with remarks on Late-Glacial conditions in Ireland. *Proc. R. Ir. Acad.*, **44** B 10, 205.

JESSEN, K. & HELBAEK, H., 1944. Cereals in Great Britain and Ireland in prehistoric and early historic time. *K. danske vidensk. Selsk.*, **3**:2, 1.

JESSEN, K. & MILTHERS, V., 1928. Stratigraphical and Paleontological studies of Inter-glacial fresh-water deposits in Jutland and northwest Germany. *Danm. geol. Unders.*, R II, no. 48.

JESSEN, K. & MITCHELL, G. F., 1942. A Pollen diagram from Ballinderry No. 2. *Proc. R. Ir. Acad.*, **47**, 1.

JESSUP, R. F. & COOK, N. C., 1936. Excavations at Bigberry Camp, Harbledown. *Arch. Cantiana*, **48**, 151.

JOHNSEN, J., DANSGAARD, W. & CLAUSEN, H. B., 1970. Climatic oscillations A.D. 1200–2000. *Nature, Lond.*, **227**, 482.

JONES, E. W., 1944. Biological flora of the British Isles: *Acer* L. *J. Ecol.*, **32**:2, 215.

JØRGENSEN, S., 1963. Early Postglacial on Aamosen *Danm. geol. Unders.*, R II, **87**, 105.

KASSAS, M., 1951. Studies in the ecology of Chippenham Fen. I and II. *J. Ecol.*, **39**, 1.

KEEF, P. A. M., 1940. Flint-chipping sites and hearths on Bedham Hill, near Pulborough. *Sussex Arch. Col.*, **81**, 215.

KELLY, M. R., 1964. The middle Pleistocene of north Birmingham. *Phil. Trans.*, B **247**, 533.

KELLY, M. R., 1968. Floras of middle and upper Pleistocene age from Brandon, Warwickshire. *Phil. Trans.*, B **254**, 401.

KELLY, M. R. & OSBORNE, P. J., 1965. Two faunas and floras from the alluvium at Shustoke, Warwickshire. *Proc. Linn. Soc. Lond.*, **176**, 37.

KENNARD, A. S., 1944. The Crawford brickearths. *Proc. Geol. Ass., Lond.*, **55**:3, 121.

KENNARD, A. S. & WARREN, S. H., 1903. On a section of the Thames alluvium in Bermondsey. *Geol. Mag.* IV, **10**, 456.

KERNEY, M. P., 1963. Late-glacial deposits on the chalk of south-east England. *Phil. Trans.*, B **246**, 203.

KERNEY, M. P., 1964. Late-glacial and Post-glacial history of the chalk escarpment near Brook, Kent. *Phil. Trans.*, B **248**, 135.

KILBRIDE-JONES, H. E., 1935. An account of the excavation of the stone circle at Loanhead of Daviot, Aberdeenshire. *Proc. Soc. Antiqu. Scot.*, **69**, 168.

KING, D. G., 1966. The Lanhill Long Barrow, Wiltshire, England: an essay in reconstruction. *Proc. prehist. Soc.*, **32**, 73.

KIRCHEIMER, F., 1957. *Die Laubgewächse der Braunkohlenzeit.* Knapp, Halle/Saale.

KIRK, W. & GODWIN, H., 1962–3. The Late-glacial site at Loch Droma, Ross and Cromarty. *Proc. R. Soc. Edin.*, **65**, 225.

KNÖRZER, K.-H., 1966. Über Funde römischer Importsfruchte in Noraesium. *Bonner Jb.*, **166**, 433.

KNÖRZER, K.-H., 1971. Genützte Wildpflanzen in vorgeschichtlicher Zeit. *Bonner Jb.*, **171**, 1.

KNOX, E. M., 1954. Pollen analysis of a peat at Kingsteps Quarry, Nairn. *Trans. Proc. Bot. Soc. Edin.*, **36**, 224.

KOPEROWA, W., 1964. *Koenigia islandica* L. in the late Pleistocene deposits of Poland. *Proc. VI Congr. I.N.Q.U.A.*, **2**, 447.

KÖRBER-GROHNE, U., 1964. *Probleme der Küstenforschung im südlichen Nordseegebiet. 7. Bestimmungschlüssel für subfossile* Juncus-*Samen und Graminen-Früchte.* August Lax, Hildersheim.

KÖRBER-GROHNE, U., 1967. Geobotanische Untersuchungen auf der Feddersen Wierde. Wiesbaden, F. Steiner.

KUBITZKI, K., 1961. Zur Synchronisierung der nordwesteuropäischen Pollendiagramme. *Flora, Jena*, **150**, 43.

KUBITZKI, K. & MUNNICH, K. O., 1960. Neue C. 14 Datierung zur nacheiszeitlichen Waldgeschichte Nordwest deutschlands. *Ber. dt. bot. Ges.*, **4**, 137.

LACAILLE, A. D., 1951. A stone industry from Morar, Inverness-shire: its Obanian (Mesolithic) and later affinities. *Archaeologia*, **94**, 103.

LAMB, H. H., 1968. Volcanic dust, melting of ice caps, and sea levels. *Palaeogeography, Palaeoclim., Palaeoecol.*, **4**, 219.

LAMB, H. H., LEWIS, R. P. W. & WOODROFFE, 1966. Atmospheric circulation and the main climatic variables between 8000 and 0 B.C.: meteorological evidence. Proc. Int. Sym. on World Climate 8000 to 0 B.C. *Proc. Roy. Met. Soc.*, **99**.

LAMBERT, C. A., PEARSON, R. G. & SPARKS, B. W., 1963. A flora and fauna from late Pleistocene deposits at Sidgwick Avenue, Cambridge. *Proc. Linn. Soc. Lond.*, **174**, 13.

LAMBERT, J. M., JENNINGS, J. N., SMITH, C. T., GREEN, C. & HUTCHINSON, J. N., 1960. The making of the Broads. A reconsideration of their origin in the light of new evidence. *R.G.S. Res. Ser.*, no. 3.

LAMPLUGH, G. W., GIBSON, W., SHERLOCK, R. L. & WRIGHT, W. B., 1908–9. The geology of the country between Newark and Nottingham. *Mem. geol. Surv., Sheet Memoirs*, section 2, 87.

LANG, G., 1951. Nachweis von Ephedra im südwestdeutschen Spätglazial. *Naturwissenschaften*, **38**:14, 334.

LANG, G., 1970. Florengeschichte und Mediterran–Mitteleuropäisch Florenbeziehungen. *Reprium nov. Spec. Regni veg.*, **81**, 315.

LAWRENCE, G. H. M., 1940. Armerias, native and cultivated. *Gentes Herbarum*, **4**, 391

LEEDS, E. T., 1929. Bronze Age urns from Long Wittenham, Berkshire. *Antiqu. J.*, **9**, 153.

LEEDS, E. T., 1947. A Saxon village at Sutton Courtenay, Berkshire. *Archaeologia*, **92**, 79.

LEESON, J. R. & LAFFAN, G. B., 1894. On the geology of the Pleistocene deposits in the valley of the Thames at Twickenham. *Quart. J. geol. Soc. Lond.*, **50**, 453.

LEOPOLD, E. B., 1967. Late Cenozoic patterns of plant extinction. In: Pleistocene Extinctions. *Proc. VII Cong. I.N.Q.U.A.*, **6**.

LETHBRIDGE, T. C., FELL, C. I. & BACHEM, K. E., 1951. Report on a recent dug-out canoe from Peterborough. *Proc. prehist. Soc.*, **17**, 232.

LEWIS, F. J., 1905–7, 1911. The plant remains in the Scottish peat mosses. *Trans. roy. Soc. Edinb.*, **41, 45, 46, 47.**

LID, J., 1952. *Norsk Flora.* Norske Samlaget, Oslo.

LIDDELL, D. M., 1930. Report on the excavations at Hembury Fort, Devon, 1930. *Proc. Devon Arch. Explor. Soc.*, **1**:2, 40.

LIDDELL, D. M., 1933. Excavations at Meon Hill. *Proc. Hants. Fld. Cl. Arch. Soc.*, **12**:2, 127.

LIDDELL, D. M., 1935. Excavations at Meon Hill. *Proc. Hants. Fld Cl. Arch. Soc.*, **13**:1, 37.

LOUSLEY, J. E., 1939. *Rumex aquaticus* L. as a British Plant. *J. Bot.*, **77**, 149.

LOUSLEY, J. E., 1953. The changing flora of Britain. *Report 1952 Conference.* Botanical Soc. Br. Is., Arbroath.

LÖVE, A. & D., 1958. The origin of the North Atlantic flora. *Aquilo Ser. Botanica*, **6**, 52.

LÖVE, A. & SARKAR, P., 1957. Heat tolerances of *Koenigia islandica. Bot. Notiser*, **110**.

LÜDI, W., 1944. Die Waldgeschichte des südlichen Tessin seit dem Rückzug der Gletscher. *Ber. geobot. ForschInst. Rübel, Zürich (for 1943)*, **20**, 12.

LUNDQUIST, G., 1962. Geological radiocarbon datings from the Stockholm station. *Sveriges Geol. Undersok. Arsbok.*, **56**, 1.

LUTHER, H., 1945. Über die rezenten Funde von *Najas flexilis* (Willd.) Rostks & Schmidt in Ostfennoskandien. *Mem. Soc. F. Fl. Fenn.*, **21**, 60.

LYELL, A. H., 1907a. Note in: Excavations on the site of the Roman City at Silchester, Hants., by W. H. St J. Hope. *Archaeologia*, **60**:2, 449.

LYELL, A. H., 1907b. Note in: Excavations at Caerwent, Monmouthshire, by T. Ashby. *Archaeologia*, **60**:2, 463.

LYELL, A. H., 1909. Note in: Excavations on the site of the Roman city at Silchester, Hants, in 1908, by W. H. St J. Hope. *Archaeologia*, **61**:2, 485.

LYELL, A. H., 1911. Note in: Excavations at Caerwent, Monmouthshire, in the years 1909 and 1910, by T. Ashby, A. E. Hudd & F. Hing. *Archaeologia*, **62**:2, 448.

LYELL, A. H., 1912. Seeds and woods found in various London excavations. *Archaeologia*, **63**, 334.

LYELL, A. H. & REID, C., 1906. Note in: Excavations at Caerwent, Monmouthshire, by T. Ashby. *Archaeologia*, **60**:1, 123.

LYELL, A. H. & REID, C., 1909. Note in: Excavations at Caerwent, Monmouthshire, by T. Ashby, A. E. Hudd & F. King. *Archaeologia*, 61:2, 568.

LYELL, A. H. & REID, C., 1910. Note in: Excavations at Caerwent, Monmouthshire, in the year 1908, by T. Ashby, A. E. Hudd & F. King. *Archaeologia*, 62:1, 19.

LYELL, A. H., REID, C. & NEWTON, E. T., 1906. Note in: Recent discoveries in connexion with Roman London, by P. Norman & F. W. Reader. *Archaeologia*, 60:1, 216.

MABY, J. C., 1930. Appendix to: Excavations at Kingsdown Camp. Mells, Somerset, 1927–9, by H. St G. Gray. *Archaeologia*, 80, 97.

MCAULAY, I. R. & WATTS, W. A., 1961. Dublin radiocarbon dates I. *Radiocarbon*, 3, 26.

MACFAYDEN, W. A., 1955. The microfauna of four samples of a postglacial silt and clay from New Cut, Spalding, Lincs. *Fenland Foraminifera*, note 24.

MCVEAN, D. N., 1953. Biological flora of the British Isles: *Alnus glutinosa* (L.) Gaertn. *J. Ecol.*, 41:2, 447.

MCVEAN, D. N., 1961. Flora and vegetation of the islands of St Kilda and North Rona in 1958. *J. Ecol.*, 49, 39.

MCVEAN, D. N., 1964. Herb and fern meadows. In *Vegetation of Scotland*, ed. J. H. Burnett. Oliver & Boyd, Edinburgh.

MCVEAN, D. N. & RATCLIFFE, D. A., 1962. *Plant Communities of the Scottish Highlands*. HMSO, London.

MAHR, A., 1934. A wooden cauldron from Altartate, County Monaghan. *Proc. R. Ir. Acad.*, B 42.

MALMSTRÖM, C., 1920. *Trapa natans* L. i Sverige. *Svensk bot. Tidskr.*, 14, 39.

MAMAKOVA, K., 1968. Lille Bukken and Lerøy – two pollen diagrams from Western Norway. *Arbok Univ. Bergen*, 4, 42.

MARKGRAF, V., 1969. Moorkundliche und vegetationsgeschichtliche Untersuchungen an einen Moorsee an der Waldgrenze in Wallis. *Bot. Jb.*, 89, 1.

MARTIN, P. S., 1967. Prehistoric overkill: the search for a cause. In: Pleistocene Extinctions. *Proc. VIIth Cong. I.N.Q.U.A.*, 6, 75.

MATTHEWS, J. R., 1937. Geographical relationships of the British flora. *J. Ecol.*, 25:1, 1.

MATTHEWS, J. R., 1955. *Origin and Distribution of the British Flora*. Hutchinson, London.

MEDLAND, M. H., 1894–5. An account of Roman and Medieval remains found in the site of the Tolsey at Gloucester in 1893–4. *Trans. Bristol Glos. Arch. Soc.*, 19, 55.

MERCER, J. H., 1969. The Allerød oscillation: a European climatic anomaly? *Arctic and Alpine Res.*, 1, 227.

MITCHELL, G. F., 1940. Studies in Irish quaternary deposits: 1. Some lacustrine deposits near Dunshaughlin, Co. Meath. *Proc. R. Ir. Acad.*, 46 B 2, 13.

MITCHELL, G. F., 1941a. Studies in Irish quaternary deposits: 2. Some lacustrine deposits near Ratoath, Co. Meath. *Proc. R. Ir. Acad.*, 46 B 13, 173.

MITCHELL, G. F., 1941b. Studies in Irish quaternary deposits: 3. The reindeer in Ireland. *Proc. R. Ir. Acad.*, 46 B 14, 183.

MITCHELL, G. F., 1942. A Late-glacial Flora in Co. Monaghan, Ireland. *Nature, Lond.*, 149, 502.

MITCHELL, G. F., 1948a. Late-glacial deposits in Berwickshire. *New Phytol.*, 47:2, 262.

MITCHELL, G. F., 1948b. Two interglacial deposits in southeast Ireland. *Proc. R. Ir. Acad.*, 52 B 1, 1.

MITCHELL, G. F., 1951. Studies in Irish Quaternary deposits: no. 7. *Proc. R. Ir. Acad.*, 53 B 11, 111.

MITCHELL, G. F., 1952. Late-glacial deposits at Garscadden Mains, near Glasgow. *New Phytol.*, 50:3, 277.

MITCHELL, G. F., 1953. Further identifications of macroscopic plant fossils from Irish Quaternary deposits, especially from a Late-glacial deposit at Mapastown, Co. Louth. (Studies in Irish Quaternary Deposits: no. 8.) *Proc. R. Ir. Acad.*, 55 B 12, 225.

MITCHELL, G. F., 1954. A pollen diagram from Lough Gur, Co. Limerick. (Studies in Irish Quaternary Deposits: no. 9.) *Proc. R. Ir. Acad.*, 56, 481.

MITCHELL, G. F., 1956. Post Boreal pollen diagrams from Irish raised bogs. *Proc. R. Ir. Acad.*, 57, 185.

MITCHELL, G. F., 1958. A Late-glacial deposit near Ballaugh, Isle of Man. *New Phytol.*, 57, 256.

MITCHELL, G. F., 1961. The palynology of Ringneill Quay, a new Mesolithic site in Co. Down, Northern Ireland. *Proc. R. Ir. Acad.*, 61, 171.

MITCHELL, G. F., 1965. Littleton Bog, Tipperary: An Irish vegetational record. *Proc. VII Congr. I.N.Q.U.A.*, 7.

MITCHELL, G. F., 1970. The Quaternary deposits between Fenit and Spa on the north shore of Tralee Bay, Co. Kerry. *Proc. R. Ir. Acad.*, B 70, 141.

MITCHELL, G. F., LODER, A. M. & ANDREW, R., 1970. Interglacial deposits at Marros: Appendix to 'Southeast and central south Wales' by D. Q. Bowen. In *The Glaciation of Wales and Adjoining Regions*, ed. C. A. Lewis. Longmans, London.

MITCHELL, G. F., O'LEARY, M. & RAFTERY, J., 1941. On a Bronze Age halberd from Co. Mayo and a bronze spear from Co. Westmeath. *Proc. R. Ir. Acad.*, 46 C 6, 287.

MITCHELL, G. F. & O'RIORDAIN, S., 1942. Early Bronze Age pottery from Rockbarton Bog, Co. Limerick. *Proc. R. Ir. Acad.*, 48 C 6, 255.

MITCHELL, G. F. & PARKES, H. M., 1949. The giant deer in Ireland. *Proc. R. Ir. Acad.*, 52 B 7, 291.

MITCHELL, G. F. & WATTS, W. A., 1970. The history of the Ericaceae in Ireland during the Quaternary epoch. In Walker & West, 1970.

MITTRE, V., 1959. Post-glacial history of the Whittlesey Mere region of East Anglian fenland. Ph.D. thesis, Cambridge.

MITTRE, V., 1971. Fossil pollen of *Tilia* from the East Anglian fenland. *New Phytol.*, 70, 693.

MOAR, N. T., 1964. The history of the Late-Weichselian and Flandrian vegetation in Scotland. Ph.D. thesis, Cambridge.

MOAR, N. T., 1969a. Late Weichselian and Flandrian pollen diagrams from south-west Scotland. *New Phytol.*, 68, 433.

MOAR, N. T., 1969b. A radiocarbon dated pollen diagram from north-west Scotland. *New Phytol.*, 68, 209.

MOAR, N. T., 1969c. Two pollen diagrams from the mainland Orkney Islands. *New Phytol.*, 68, 201.

MOE, D., 1970. The Post-glacial immigration of *Picea abies* into Fennoscandia. *Bot. Notiser*, 123, 61.

MOORE, P. D., 1968. Human influence upon vegetational history in north Cardiganshire. *Nature, Lond.*, 217, 1006.

MOORE, P. D., 1970. Studies in the vegetational history of mid-Wales. II. The Late-glacial period in Cardiganshire. *New Phytol.*, 69, 363.

MOORE, P. D. & CHATER, E. H., 1969a. Studies in the vegetational history of mid-Wales. I. The Post-glacial period in Cardiganshire. *New Phytol.*, 68, 183.

MOORE, P. D. & CHATER, E. H., 1969b. The changing vegetation of west-central Wales in the light of human history. *J. Ecol.*, 57, 361.

MORGAN, A., 1969. A Pleistocene fauna and flora from Great Billing, Northants, England. *Opusc. Ent.*, **34**, 111.

MORGAN, W. L., 1907. Report on the excavations at Coelbren. *Arch. Cambrensis*, **7**, 173.

MORRISON, M. E. S., 1955a. Carbonised grain from the Roman villa of North Leigh, Oxon. *Oxoniensia*, **24**.

MORRISON, M. E. S., 1955b. Carbonised grain from Lough Faughan Crannog. *Ulster J. Arch.*, **18**.

MORRISON, M. E. S., 1959. Evidence and interpretation of 'Landnam' in the north-east of Ireland. *Bot. Notiser*, **112**, 185.

MORRISON, M. E. S., 1961. The palynology of Ringniell Quay, a new Mesolithic site in Northern Ireland. *Proc. R. Ir. Acad.*, **61**, 171.

MORRISON, M. E. S. & STEPHENS, N., 1960. Stratigraphy and pollen analysis of the raised beach deposits at Ballyhalbert, Co. Down, Northern Ireland. *New Phytol.*, **59**, 153.

MORRISON, M. E. S. & STEPHENS, N., 1965. A submerged Late Quaternary deposit at Roddans Port on the north-east coast of Ireland. *Phil. Trans.*, B **249**, 221.

MOSELEY, F. & WALKER, D., 1952. Some aspects of the Quaternary period in North Lancashire. *Naturalist, Hull*, 41.

MOVIUS, H. L., 1936. A Neolithic site on the River Bann. *Proc. R. Ir. Acad.*, **43** C 2, 17.

MOVIUS, H. L., 1940. An early post-glacial archaeological site at Cushendun, Co. Antrim. *Proc. R. Ir. Acad.*, **46** C 1, 1.

MOVIUS, H. L., 1950. A wooden spear of Third Interglacial age from Lower Saxony. *Sthwest. J. Anthrop. (Univ. New Mexico)*, **6**, 139.

MÜLLER, I., 1947. Der Pollenanalytische Nachweis der Menschlichen Besiedlung im Federsee und Bodenseegebiet. *Arch. Wiss. Bot.*, **35**, 70.

MUNAUT, A. V., 1967. Recherches paleo-écologiques en Basse et Moyenne Belgique. *Acta geogr. Iovaniensia*, **6**, 191.

NASH-WILLIAMS, V. E., 1929. The Roman legionary fortress at Caerleon in Monmouthshire: report on the 1926 excavations. *Arch. Cambrensis*, **84**, 306.

NASH-WILLIAMS, V. E., 1932. The Roman legionary fortress at Caerleon in Monmouthshire. Part III. *Arch. Cambrensis*, **87**, 265.

NASH-WILLIAMS, V. E., 1939. An Early Iron Age coastal camp at Sudbrook, near the Severn Tunnel, Monmouthshire. *Arch. Cambrensis*, **94**, 77.

NEAVERSON, E., 1936. Recent observations on the Postglacial peat beds around Rhyl and Prestatyn (Flintshire). *Proc. Lpool geol. Soc.*, **17**:1, 45.

NEWEY, W. W., 1966. Pollen analysis of subcarse peats of the Forth Valley. *Trans. Inst. Br. Geogr.*, **53**.

NEWEY, W. W., 1968. Pollen analyses from south-east Scotland. *Trans. Proc. bot. Soc. Edinb.*, **40**, 424.

NEWEY, W. W., 1970. Pollen analysis of Late Weichselian deposits at Corstorphine, Edinburgh. *New Phytol.*, **69**, 1167.

NEWSTEAD, R., 1909. On a Roman concrete foundation in Bridge Street, Chester. *J. Chester Archit. Arch. Hist. Soc.*, **16**:1, 25.

NEWSTEAD, R. & DROOP, J. P., 1932. The S.E. corner of the Roman fortress, Chester. *Chester Archit. Arch. Hist. Soc.*, **29**, 41.

NEWSTEAD, R. & NEAVERSON, E., 1938. The Post-glacial deposits of the Roman site at Prestatyn, Flints. *Proc. Lpool geol. Soc.*, **17**:3, 243.

NICHOLS, H., 1967. Vegetational change, shoreline displacement and the human factor in the late Quaternary history of south-west Scotland. *Trans. R. Soc. Edinb.*, **67**:6, 145.

NILSSON, T., 1964. Standard pollendiagramme und C^{14} Datierungen aus dem Ageröds Mosse im mittleren Schonen. *Lunds Univ. Arsskrift*, **59**:7, 1.

NORDBORG, G., 1966. *Sanguisorba* L., *Sarcopoterium* Spach, and *Bencornia* Webb et Berth. *Op. Bot. (Lund.)*, **11**:2, 1.

OAKLEY, V. P., RANKINE, W. F. & LOWTHER, A. W. G. 1939. A survey of the prehistory of the Farnham district (Surrey). Part I. Geology and prehistoric studies. *Surrey archaeol. Soc.*, Guildford.

OBERDORFER, E., 1957. Süddeutsche Pflanzengesellschaften. *Pflanzensoziologie*, **10**, 564.

ØDUM, S., 1965. Germination of ancient seeds. Floristic observations and experiments with archaeologically dated soil samples. *Dansk bot. Ark.*, **24**, 1.

OLDFIELD, F., 1959. The pollen morphology of some of the west European Ericales. *Pollen Spores*, **1**, 19.

OLDFIELD, F., 1960. Studies in the Post-glacial history of British vegetation: lowland Lonsdale. *New Phytol.*, **59**, 192.

OLDFIELD, F., 1964a. Late Quaternary deposits at Le Moura, Biarritz, south-west France. *New Phytol.*, **63**, 374.

OLDFIELD, F., 1964b. Late Quaternary vegetational history in south-west France. *Pollen Spores*, **6**, 157.

OLDFIELD, F., 1966. The Palaeoecology of an early Neolithic waterlogged site in north-western England. *Rev. Palaeobotan. Palynol.*, **4**, 67.

OLDFIELD, F. & STATHAM, D. C., 1963. Pollen analytical data from Urswick Tarn and Ellerside Moss, north Lancashire. *New Phytol.*, **62**, 53.

OLDFIELD, F. & STATHAM, D. C., 1964-5. Stratigraphy and pollen analysis on Cockerham and Pilling Moss. North Lancashire. *Mem. Proc. Manchr. lit. phil. Soc.*, **107**, 1.

O'NEIL, B. H. ST J., 1934. Excavations at Titterstone Clee Hill Camp, Shropshire, 1932. *Arch. Cambrensis*, **89**, 110.

O'NEIL, B. H. ST J., 1936. Excavations at Caerau Ancient Village, Clynnog, Caernarvonshire, 1933 and 1934. *Antiqu. J.*, **16**:3, 303-4.

O'NEIL, B. H. ST J., 1937. Excavations at Breiddin Hill Camp, Montgomeryshire, 1933-5. *Arch. Cambrensis*, **92**, 111.

O'NEIL, B. H. ST J., 1943. Excavations at Ffridd Faldwyn Camp, Montgomery, 1937-9. *Arch. Cambrensis*, **97**, 54.

O'NEIL, B. H. ST J. & FOSTER-SMITH, A. H., 1936. Excavations at Twyn y Cregen, Llanarth, Monmouthshire. *Arch. Cambrensis*, **91**, 258.

O'NEIL, H. E., 1945. The Roman villa at Park Street near St Albans, Hertfordshire. *Arch. J.*, **102**, 102.

O'NEIL, H. E., 1966. Sale's Lot Long Barrow, Withington, Gloucestershire, 1962-5. *Trans. Bristol Glos. Arch. Soc.*, **85**, 5.

O'RIORDAIN, S. P., 1938. A Bronze Age find from Oldtown Kilcashel, Co. Roscommon. *J. Galway Arch. Hist. Soc.*, **18**, 40.

O'RIORDAIN, S. P. & HARTNETT, P. J., 1943. The excavation of Ballycatteen Fort, Co. Cork. *Proc. R. Ir. Acad.*, **49** C, 1.

O'RIORDAIN, S. P. & LUCAS, A. T., 1946-7. Excavation of a small crannog at Rathjordan, Co. Limerick. *N. Munster Antiqu. Journ.*, **5**, 68.

OVERBECK, F., MUNNICH, K. O., ALETSEE, L. & AVERDIECK, Fr., 1957. Das Alter des 'Grenzhorizonts' Norddeutscher Hochmoore nach Radiocarbon Datierungen. *Flora, Jena*, **145**, 37.

OVERBECK, F. & SCHMITZ, H., 1931. Zur Geschichte der Moore, Marschen und Wälder Nordwestdeutschlands. I. Das Gebiet von der Niederweser bis zur unteren Ems. *Mitt. d. Provinzialst. f. Naturdenkmalpfl. Hannover*, **3**, 1.

OVERBECK, F. & SCHNEIDER, S., 1938–9. Mooruntersuchungen bei Lüneburg und bei Bremen und die Reliktnature von *Betula nana* L. in Nordwestdeutschland. *Z. bot.*, **33**, 1.

PAEPE, R. & VANHOORNE, R., 1967. The stratigraphy and palaeobotany of the late Pleistocene in Belgium. *Mem. Expl. Cartes. Géol. et Min. de la Belgique* no. 8.

PARKER, H. H., 1934. *The Hop Industry*. King & Son, London.

PEAKE, A. E., 1913. An account of a flint factory, with some new types of flints: Excavations at Peppard Common, Oxon. *Arch. J.*, **70**, 33.

PEAKE, A. E., 1915–16. Recent excavations at Grimes Graves. *Proc. prehist. Soc. E. Angl.*, **2**:2, 268.

PEARS, N. V., 1968. Post-glacial tree lines of the Cairngorm Mountains, Scotland. *Trans. Bot. Soc. Edinb.*, **40**, 361.

PEARSON, M. C., 1960. Muckle Moss, Northumberland: historical. *J. Ecol.*, **48**, 647.

PEARSON, R., 1964. *Animals and Plants of the Cenozoic Era*. Butterworths, London.

PENNINGTON, W., 1947. Lake sediments: pollen diagrams from the bottom deposits of the north basin of Windermere. *Phil. Trans.*, B **233**, 137.

PENNINGTON, W., 1962. Late-glacial moss records from the English Lake District: data for the study of Post-glacial history. *New Phytol.*, **61**, 28.

PENNINGTON, W., 1964. Pollen analysis from the deposits of six upland tarns in the Lake District. *Phil. Trans.*, B **248**, 205.

PENNINGTON, W., 1965. The interpretation of some Post-glacial vegetational diversities at different Lake District sites. *Proc. Roy. Soc.*, B **161**, 310.

PENNINGTON, W., 1969. *The History of British Vegetation*. English Universities Press Ltd, London.

PENNINGTON, W. & BONNY, A. P., 1970. Absolute pollen diagram from the British Late-glacial. *Nature, Lond.*, **226**, 871.

PENNINGTON, W., HAWORTH, E. Y., BONNY, A. P. & LISHMAN, J. P., 1972. Lake sediments in northern Scotland. *Proc. R. Soc.*, B **264**, 191.

PERCIVAL, J., 1934. *Wheat in Great Britain*. Privately printed (Reading and London, 1943).

PERRING, F. H. & SELL, P. D., 1968. *Critical Supplement to the Atlas of the British Flora*. Thomas Nelson, London.

PERRING, F. H. & WALTERS, S. M., 1962. *Atlas of the British Flora*. Thomas Nelson, London.

PHILLIPS, C. W., 1935. The excavation of Giants' Hills long barrow, Skendleby, Lincolnshire. *Archaeologia*, **85**, 103.

PHILLIPS, C. W., 1936. An examination of the Ty Newydd chambered tomb, Llanfaelog, Anglesey. *Arch. Cambrensis*, **91**, 98.

PICTON, H., 1912. Observations on the Bone Bed at Clacton. *Proc. prehist. Soc. E. Angl.*, **1**:2, 158.

PIGGOTT, C. M., 1938. A middle Bronze Age barrow and Deverel-Rimbury urnfield at Latch Farm, Christchurch, Hants. *Proc. prehist. Soc.*, **4**, 183.

PIGGOTT, C. M., 1943a. Excavation of fifteen barrows in the New Forest 1941–2. *Proc. prehist. Soc.*, **9**, 25.

PIGGOTT, C. M., 1943b. Three turf barrows at Hurn, near Christchurch. *Proc. Hants. Fld Cl. Arch. Soc.*, **15**:3, 261.

PIGGOTT, C. M., 1947–8. The excavations at Hownam Rings, Roxburghshire, 1948. *Proc. Soc. Antiqu. Scot.*, **82**, 193.

PIGGOTT, S., 1937. The excavation of a long barrow in Holdenhurst Parish, near Christchurch, Hants. *Proc. prehist. Soc.*, **3**, 13.

PIGGOTT, S., 1947–8. The excavations at Cairnpapple Hill, West Lothian. *Proc. Soc. Antiqu. Scot.*, **82**, 68.

PIGGOTT, S. & DIMBLEBY, G. W., 1955. A Bronze Age barrow on Turner's Puddle Heath. *Proc. Dorset Nat. His. Arch. Soc.*, **75**, 34.

PIGOTT, C. D., 1958. Biological flora of the British Isles. (*Polemonium caeruleum* L.). *J. Ecol.*, **46**, 507.

PIGOTT, C. D., 1969. The status of *Tilia cordata* and *Tilia platyphyllos* on the Derbyshire limestone. *J. Ecol.*, **57**, 491.

PIGOTT, C. D. & WALTERS, S. M., 1953. Is the box-tree a native of England? In Lousley, 1953.

PIGOTT, C. D. & WALTERS, S. M., 1954. On the interpretation of the discontinuous distributions shown by certain British species of open habitats. *J. Ecol.*, **42**:1, 95.

PIGOTT, M. E. & C. D., 1959. Stratigraphy and pollen analysis of Malham Tarn and Tarn Moss. *Field Studies* **1**:1, 1.

PIKE, K. & GODWIN, H., 1953. The Interglacial at Clacton-on-Sea, Essex. *Quart. J. geol. Soc. Lond.*, **108**:3 (*for* 1952), 261.

PILCHER, J. R., 1969. Archaeology, palaeoecology and radio-carbon dating of the Beaghmore Stone Circle site. *Ulster J. Arch.*, **32**, 73.

PILCHER, J. R., 1970. Palaeoecology and radiocarbon dating of sites in Co. Tyrone, Northern Ireland. Ph.D. thesis, Queen's University, Belfast.

PILCHER, J. R., SMITH, A. G., PEARSON, G. W. & CROWDER, A., 1971. Land clearance in the Irish Neolithic: new evidence and interpretation. *Science, N.Y.*, **172**, 560.

PITT-RIVERS, A., 1881–5. *Excavations in Cranbourne Chase*. Privately printed, London.

PLANCHAIS, N., 1967. Analyse pollinique de la tourbière de Gizeux (Indre-et-Loire) et étude duchêne vert à l'optimum climatique. *Pollen spores*, **9**, 505.

POLACH, H. A. & GOLSON, J., 1966. *Collection of Specimens for Radiocarbon Dating and Interpretation of Results*. Aus. Inst. Aboriginal Studies, Canberra.

POLAK, B., 1959. Palynology of the 'Uddeler Mere'. *Acta bot. neerl.*, **9**, 547.

POLAK, B., 1963. A buried Allerød pine-forest. *Acta bot. neerl.*, **12**, 533.

POLUNIN, N., 1939. Arctic plants in the British Isles. *Nature, Lond.*, **144**, 352.

POLUNIN, N., 1940. *Botany of the Canadian eastern arctic*, vol. I. Dept. of Mines and Resources, Canada.

POORE, M. E. D. & ROBERTSON, V. C., 1949. The vegetation of St Kilda in 1948. *J. Ecol.*, **37**:1, 82.

POST, L. VON, 1916. Om skogsträdspollen i sydsvenska torf-mosselagerföljder (föredragsreferat). *Geol. Fören. Förh.*, **38**, 384.

POST, L. VON, 1924. Some features of the regional history of forests of southern Sweden in post-arctic time. *Geol. Fören. Stockh. Förh.*, **46**.

POST, L. VON, 1925. Gotlands-agen (*Cladium mariscus* R.Br.) i Sveriges postarktikum. *Ymer* **45**, 295.

POST, L. VON, 1930. Problems and working-lines in the Post-arctic forest history of Europe. *Report Int. Bot. Congr.*, Cambridge.

POST, L. VON, 1946. The prospect for pollen analysis in the study of the earth's climatic history. *New Phytol.*, **45**:2, 193.

PRAEGER, R. L., 1934. *The Botanist in Ireland*. Hodges, Figgis & Co., Dublin.

PRAEGER, R. L., 1939. The relations of the flora and fauna of Ireland to those of other countries. *Proc. Linn. Soc. Lond.*, **151**, 192.

PRAEGER, R. L., 1950. *Natural History of Ireland*. Collins, London.

PRAGLOWSKI, J. R., 1962. Notes on the pollen morphology of Swedish trees and shrubs. *Grana Palynol.*, 3, 45.

PRAGLOWSKI, J. R. & WENNER, C. G., 1968. The two *Alnus* species – in varve chronology, pollen analysis and radiocarbon dating. *Stockh. Contr. Geol.*, **18**, 75.

PRECHT, J., 1953. On the occurrence of the 'Upper Forest Layer' around Cold Fell, N. Pennines. *Trans. nth. Nat. Un.*, **2**:1, 40.

PRENDERGAST, E. & MITCHELL, G. F., 1960. Amber necklace from Co. Galway. *Proc. R. Soc. Antiqu. Ireland*, 90, 61.

PRING, M. E., 1961. Biological flora of the British Isles: *Arabis stricta*. Huds. *J. Ecol.*, 49, 431.

PROCTOR, M. C. F. & LAMBERT, C. A., 1961. Pollen spectra from recent *Helianthemum* communities. *New Phytol.*, **60**, 21.

PROUDFOOT, E., 1965. Bishops Canning, Roughridge Hill. *Wilts. Arch Mag.*, 60, 133.

PULL, J. H., 1953. Further discoveries at Church Hill, Findon. *Sussex Mag.*, 27, 1.

RAISTRICK, A. & BLACKBURN, K. B., 1931. Pollen analysis of the peat on Heathery Burn Moor, Northumberland. *Proc. Univ. Durham phil. Soc.*, 8, 351.

RAISTRICK, A. & BLACKBURN, K. B., 1932. The Late-glacial and Post-glacial periods in the North Pennines. Part III. The Post-glacial peats. *Trans. nth Nat. Un.*, 1:2, 79.

RAISTRICK, A. & BLACKBURN, K. B., 1938. Linton Mires, Wharfedale. Glacial and Post-glacial History. *Proc. Univ. Durham phil. Soc.*, 10, 24.

RAISTRICK, A. & WOODHEAD, T. W., 1930. Plant remains in post-glacial gravels near Leeds. *Naturalist, Lond.*, No. 877, 39.

RALSKA-JASIEWIEZ, M., 1964. Correlation between the Holocene history of the *Carpinus betulus* and prehistoric settlement in North Poland. *Acta Soc. Bot. Pol.*, 33, 461.

RANKINE, W. F. & W. M. & DIMBLEBY, G. W., 1960. Further excavations at a Mesolithic site at Oakhanger, Selbourne, Hampshire. *Proc. prehist. Soc.*, 26, 246.

RATCLIFFE, D., 1959. The habitat of *Koenigia islandica* in Scotland. *Trans. bot. Soc. Edinb.*, 37, 272.

READER, F. W., 1907-9. Report of the Red Hills Exploration Committee 1906-7. *Proc. Soc. Antiqu.*, 22, 165.

READER, F. W., 1909-11. Report of the Red Hills Exploration Committee 1908-9. *Proc. Soc. Antiqu.*, 23, 84.

READER, F. W., 1910-11. A Neolithic floor in the bed of the Crouch River, and other discoveries near Rayleigh, Essex. *Essex Nat.*, 16, 249.

REANEY, D., 1966. A beaker burial at Aston on Trent. *Derbys. Arch. J.*, **186**, 103.

REID, C., 1882. The geology of the country around Cromer. *Mem geol. Surv., Sheet memoirs*, **68** E.

REID, C., 1890. The Pliocene deposits of Britain. *Mem. geol. Surv.*

REID, C., 1893. A fossiliferous Pleistocene deposit at Stone, on the Hampshire Coast. *Quart. J. geol. Soc. Lond.*, **49**, 325.

REID, C., 1897. The Palaeolithic deposits at Hitchin and their relation to the Glacial epoch. *Proc. roy. Soc.*, **61**, no. **369**, 40.

REID, C., 1899. *The Origin of the British Flora*. Dulau & Co., London.

REID, C., 1901. Further note on the Palaeolithic deposits at Hitchin. *Trans. Herts. nat. Hist. Soc.*, **11**:2, 63.

REID, C., 1901-9. Notes on the Plant-remains of Roman Silchester, in articles: Excavations on the site of the Roman City at Silchester, Hants., by W. H. St J. Hope.

    1901*a*. *Archaeologia*, 57:2, 252.
    1902. *Archaeologia*, 58:1, 34.
    1903. *Archaeologia*, 58:2, 425.
    1905. *Archaeologia*, 59:2, 367.
    1906. *Archaeologia*, 60:1, 164.
    1907*a*. *Archaeologia*, 60:2, 449.
    1908. *Archaeologia*, 61:1, 210.
    1909. *Archaeologia*, 61:2, 485.

REID, C., 1907. Note in: Excavations at Caerwent. Monmouthshire, in the year 1906, by T. Ashby. *Archaeologia*, **60**:2, 463.

REID, C., 1911. Note in: A late Celtic and Romano-British cave-dwelling at Wookey Hole, Somerset, by H. E. Balch & R. D. R. Troup. *Archaeologia*, **62**:2, 590.

REID, C., 1916. The Plants of the Late glacial deposits of the Lea Valley. *Quart. J. geol. Soc. Lond.*, **71** (*for* 1915), 155.

REID, C. & FLETT, J. S., 1907. The geology of the Land's End district. *Mem. geol. Surv., Sheet Memoirs*, Section 1, 82.

REID, C. & E. M., 1908. On the Pre-glacial flora of Britain. *Proc. Linn. Soc. Lond.*, **38**, 206.

REID, C. & E. M., 1914. A new fossil, *Corema. J. bot. Lond.*, **52**, 113.

REID, E. M., 1920. On two Preglacial floras from Castle Eden, County Durham. *Quart. J. geol. Soc. Lond.*, **76**:2, 104.

REID, E. M., 1921. Appendix to: Some recent excavations in London, by F. Lambert. *Archaeologia*, **71**, 111.

REID, E. M., 1935. British floras antecedent in the Great Ice Age. In: Discussion on the origin and relationship of the British flora. *Proc. Roy. Soc. Lond.*, B 118. 197.

REID, E. M., 1949. The Late-glacial flora of the Lea Valley. *New Phytol.*, **48**:2, 245.

REID, E. M. & CHANDLER, M. E. J., 1923*a*. The fossil flora of Clacton-on-Sea. *Quart. J. geol. Soc. Lond.*, **79**:4, 619.

REID, E. M. & CHANDLER, M. E. J., 1923*b*. The Barrowell Green (Lea Valley) arctic flora. *Quart. J. geol. Soc. Lond.*, **79**:4, 604.

REID, E. M. & CHANDLER, M. E. J., 1925. On the occurrence of *Ranunculus hyperboreus* Rottb. in Pleistocene beds at Bembridge, Isle of Wight. *Proc. Is. Wight nat. Hist. Soc.* (*for* 1924), **1**:5, 292.

REID MOIR, J., 1917. On some human and animal bones, flint implements, etc., discovered in two ancient occupation-levels in a small valley near Ipswich. *J. R. Anthrop. Inst.*, **47**, 367.

REID MOIR, J., 1920. An early Neolithic 'floor' discovered at Ipswich. *Man*, **20**, 84.

REID MOIR, J., 1930. Ancient man in the Gipping-Orwell Valley, Suffolk. *Proc. prehist. Soc. E. Angl.*, **6**:3, 182.

RENFREW, C., 1970. The tree-ring calibration of radiocarbon: an archaeological evaluation. *Proc. Prehist. Soc.*, 36, 280.

RICHARDSON, K. M., 1961. Excavations in Hungate, York. *Arch. J.*, **116**, 51.

RIDGWAY, M. H. & LEACH, G. B., 1946–7. Prehistoric flint workshops site near Abersoch, Caernarvonshire. *Arch. Cambrensis*, **99**, 78.

RIETZ, G. E. DU, 1935. Discussion on the origin and relationship of the British flora. *Proc. R. Soc.*, B **118**, 197.

ROBARTS, N. F., 1905. Notes on a recently discovered British camp near Wallington. *J. R. Anthrop. Inst.*, **35**, 387.

ROBERTSON, D., 1881. On the Post-tertiary beds of Garvel Park, Greenock. *Trans. geol. Soc. Glasg.*, **7**, 1.

ROEDER, C., 1899. Recent Roman discoveries in Deansgate and on Hunt's Bank, and Roman Manchester re-studied. *Trans. Lancs. Ches. Antiqu. Soc.*, **17**, 87.

ROGERS, I., 1908. On the submerged forest at Westward Ho!, Bideford Bay. *Rep. Devon. Ass. Adv. Sci.*, **40**, 249.

ROISIN, P., 1969. *La Domaine phytogéographique Atlantique d'Europe*. J. Dunculot, S. A., Gembloux.

RONA, E. & EMILIANI, C., 1969. Absolute dating of Caribbean cores P 6304·8 and P 6304·9. *Science, N.Y.*, **163**, 66.

ROSS-WILLIAMSON, R. P., 1930. Excavations in Whitehawk Neolithic Camp, near Brighton. *Sussex Arch. Coll.*, **71**, 82.

ROYAL METEOROLOGICAL SOCIETY, 1966. Proceedings of the International Symposium on World Climate 8000 to 0 B.C. *Proc. Roy. Met. Soc.*, **99**.

RUDOLPH, K., 1936. Mikrofloristische Untersuchung tertiärer Ablargerungen in nördlichen Böhmen. *Beih. bot. Zbl.*, **54**, 244.

SALISBURY, E. J., 1932. The East Anglian Flora. *Trans. Norfolk Norw. Nat. Soc.*, **13:3**, 191.

SALISBURY, E. J. & JANE, F. W., 1940. Charcoals from Maiden Castle, etc. *J. Ecol.*, **28:2**, 310.

SALWAY, P., HALLAM, S. J. & BROMWICH, J., 1970. The Fenland in Roman Times. *R.G.S. Res. Ser.*, no. 5.

SALZMANN, L. F., 1908. Excavations on the site of the Roman Fortress at Pevensey, 1907–8. *Arch. J.*, **65**, 125.

SALZMANN, L. F., 1909. Excavations at Pevensey, 1907–8. *Sussex Arch. Coll.*, **52**, 83.

SAMUELSSON, G., 1910. Scottish peat mosses. A contribution to the knowledge of the late-quaternary vegetation and climate of north-western Europe. *Bull. geol. Instn. Univ. Uppsala*, **10**, 197.

SAMUELSSON, G., 1934. Die Verbreitung der höheren Wasserpflanzen in Nordeuropa (Fennoskandien und Dänemark). *Acta Phytogeogr. suec.*, **6**.

SANDEGREN, R., 1943. *Hippophaë rhamnoides* L. i Sverige under senkvartär tid. *Svensk bot. Tidskr.*, **37**, 1.

SAVORY, H. N., 1940. A middle Bronze Age barrow at Crick, Monmouthshire. *Arch. Cambrensis*, **95**, 186.

SAVORY, H. N., 1948. Two Middle Bronze Age palisade barrows at Letterston Pembrokeshire. *Arch. Cambrensis*, **100**, 81.

SCHOENICHEN, W., 1933. *Deutsche Waldbäume und Waldtypen*. Fischer, Jena.

SCHOFIELD, J. C., 1960. Sea-level fluctuations during the past four thousand years. *Nature, Lond.*, **185**, 836.

SCHOFIELD, J. C., 1964. Post-glacial sea-levels and isostatic uplift. *N.Z. Jl. Geol. Geophys.*, **7:2**, 359.

SCOTT, W. L., 1934. Excavation of Rudh'an Dunain Cave, Skye. *Proc. Soc. Antiqu. Scot.*, **68**, 200.

SCOTT, W. L., 1935. The chambered cairn of Clettraval, North Uist. *Proc. Soc. Antiqu. Scot.*, **69**, 480.

SEAGRIEF, S. C., 1959. Pollen diagrams from southern England: Wareham, Dorset and Nursling, Hampshire. *New Phytol.*, **58**, 316.

SEAGRIEF, S. C., 1960. Pollen diagrams from southern England. Cranes Moor, Hampshire. *New Phytol.*, **59**, 73.

SEAGRIEF, S. C. & GODWIN, H., 1960. Pollen diagrams from southern England: Elstead, Surrey. *New Phytol.*, **59**, 84.

SEALY, J. R., 1949. *Arbutus unedo*. *J. Ecol.*, **37**, 365.

SEDDON, B., 1961. Report on the organic deposits in the pool at Dinas Emrys. *Arch. Cambrensis*, **109**, 72.

SEDDON, B., 1962. Late-glacial deposits at Llyn Dwythwch and Nant Ffrancon, Caernarvonshire. *Phil. Trans.*, B **244**, 459.

SEDDON, B., 1965. Submerged peat layers in the Severn channel near Avonmouth. *Proc. Bristol. Nat. Soc.*, **31**, 101.

SEROCOLD, O. P. & MAYNARD, G., 1949. A Dark Ages settlement at Trebarveth, St Keverne, Cornwall. *Antiqu. J.*, **29**, 169.

SHACKLETON, N. J., 1967. Oxygen isotope analysis on Pleistocene temperatures reassessed. *Nature, Lond.*, **215**, 15.

SHACKLETON, N. J. & TURNER, C., 1967. Correlation between marine and terrestrial Pleistocene successions. *Nature, Lond.*, **216**, 1079.

SHAW, C. T., 1933–6. Bronze Age urns from Honiton. *Proc. Devon Arch. Explor. Soc. (for 1935)*, **2:3**, 191.

SHAWCROSS, F. W. & HIGGS, E. S., 1961. The excavation of a *Bos primigenius* at Lowe's Farm, Littleport. *Proc. Cam. Antiqu. Soc.*, **54**, 3.

SHEPHARD, F. P., 1963. Thirty-five thousand years of sea-level. In *Essays in Marine Geology*. University of South California Press, Los Angeles.

SHEPPARD, T., 1906. On a section of the post-glacial deposit at Hornsea. *Naturalist, Lond.*, 420.

SHEPPARD, T., 1910. The prehistoric boat from Brigg. *Trans. E. Riding Antiqu. Soc.*, **17**, 33.

SHORE, T. W. & ELWES, J. W., 1889. The New Dock excavation at Southampton. *Proc. Hants. Fld Cl.*, **113**, 43.

SHOTTON, F. W., 1967. Investigation of an old peat Moor at Moreton Morrell, Warwickshire. *Proc. Coventry & dist. Nat. His. Sc. Soc.*, **4**, 13.

SHOTTON, F. W. & STRACHAN, I., 1956. The investigation of a peat moor at Rodbaston, Penkridge, Staffs. *Quart. J. geol. Soc. Lond.*, **115**, 1.

SHOTTON, F. W., SUTCLIFFE, A. J. & WEST, R. G., 1962. The flora and fauna from the brickpit at Lexden, Essex. *Essex Naturalist*, **31:1**, 15.

SHOTTON, F. W. & WEST, R. G., 1969. Stratigraphical table of the British Quaternary. In: 'Recommendations on stratigraphical usage'. *Proc. geol. Soc. Lond.*, no. 1656, 139.

SIMMONDS, N. W., 1945. Biological flora of the British Isles: *Polygonum*. *J. Ecol.*, **33:1**, 117.

SIMMONS, I. G., 1964. Pollen diagrams from Dartmoor. *New Phytol.*, **63**, 165.

SIMMONS, I. G., 1969. Pollen diagrams from the north York moors. *New Phytol.*, **68**, 807.

SIMPSON, I. M. & WEST, R. G., 1958. The stratigraphical palaeobotany of a late Pleistocene deposit at Chelford, Cheshire. *New Phytol.*, **57**, 239.

SIMS, R. E., 1973. The anthropogenic factor in East Anglian vegetational history: an approach using A.P.F. techniques. In *Quaternary Plant Ecology*, p. 223, ed. H. J. B. Birks & R. G. West. Blackwell, Oxford.

SINGH, G., 1963. Pollen analysis of a deposit at Roddans Port, Co. Down. N. Ireland, bearing reindeer antler fragments. *Grana Palynologica*, **4**, 466.

SINGH, G., 1970. Late-glacial vegetational history of Lecale, Co. Down. *Proc. R. Ir. Acad.*, **69**, 189.

SMEDLEY, N. & OWLES, E. J., 1966. A Romano-British bath house at Stonham Aspel. *Proc. Suff. Inst. Arch.*, 30, 221.

SMITH, A. G., 1956. The pollen spectra from the cooking place sites at Ballycroghan, Co. Down. *Ulster J. Arch.*, 18, 26.

SMITH, A. G., 1957. Note of the pollen analysis of some deposits at Knockiveagh (Co. Down) Neolithic cairn. *Ulster J. Arch.*, 20, 27.

SMITH, A. G., 1958a. Post-glacial deposits in south Yorkshire and north Lincolnshire. *New Phytol.*, 57, 19.

SMITH, A. G., 1958b. Two lacustrine deposits in the south of the English Lake District. *New Phytol.*, 57, 363.

SMITH, A. G., 1958c. Pollen analytical investigations of the mire at Fallahoghy Td., Co. Derry. *Proc. R. Ir. Acad.*, 59, 329.

SMITH, A. G., 1959a. The mires of south-western Westmorland: stratigraphy and pollen analysis. *New Phytol.*, 58, 105.

SMITH, A. G., 1959b. Clea Lake Crannog. Appendix. *Ulster J. Arch.*, 22, 101.

SMITH, A. G., 1961a. Cannons Lough, Kilrea, Co. Derry: stratigraphy and pollen analysis. *Proc. R. Ir. Acad.*, 61, 369.

SMITH, A. G., 1961b. Problems in the study of the earliest agriculture in Ireland. *Proc. VI Congr. I.N.Q.U.A.*, 6.

SMITH, A. G. & WILLIS, E. H., 1961–2. Radiocarbon dating of the Fallahogy Landnam Phase. *Ulster J. Arch.*, 24–25, 16.

SMITH, C. N. S., 1946. A prehistoric and Roman site at Broadway. *Trans. Worcs. Arch. Soc.*, 23, 57.

SMITH, R. A., 1907. Recent and former discoveries at Hawkshill. *Surrey Arch. Coll.*, 20, 119.

SPARKS, B. W., 1958. The non-marine Mollusca of the Interglacial deposits at Bobbitshole, Ipswich. *Phil. Trans.*, B 241, 33.

SPARKS, B. W. & LAMBERT, C. A., 1961. The Post-glacial deposits at Apethorpe, Northants. *Proc. malac. Soc. Lond.*, 34, 302.

SPARKS, B. W. & WEST, R. G., 1959. The palaeoecology of the Interglacial deposits at Histon Road, Cambridge. *Eiszeitalter Gegenw.*, 10, 123.

SPARKS, B. W. & WEST, R. G., 1963. The Interglacial deposits at Stutton, Suffolk. *Proc. Geol. Assoc. Lond.*, 74, 419.

SPARKS, B. W. & WEST, R. G., 1968. Interglacial deposits at Wortwell, Norfolk. *Geol. Mag.*, 105, 471.

SPARKS, B. W. & WEST, R. G., 1970. Late Pleistocene deposits at Wretton, Norfolk. I. Ipswichian interglacial deposits. *Phil. Trans.*, B 258, 1.

SPARKS, B. W. & WEST, R. G., 1972. *The Ice Age in Britain*. Methuen, London.

SPARKS, B. W., WEST, R. G., WILLIAMS, R. G. B. & RANSOM, M., 1969. Hoxnian Interglacial deposits near Hatfield, Herts. *Proc. Geol. Ass. Lond.*, 80, 243.

SPURRELL, F. C. J., 1889. On the estuary of the Thames and its alluvium. *Proc. Geol. Ass. Lond.*, 11, 210.

SRODON, A., 1968. A survey of botanical and palaeobotanical research of the Polish Spitzbergen Expeditions. 1957–60. *Q. Rev. Publs Pol. Acad. Sci.*, 3.

STAPLETON, P., 1908. The Bryngwyn tumuli. *Arch. Cambrensis*, 8, 369.

STATHER, J. W., 1904. The drift deposits of Kirmington and Great Limber, Lincolnshire. *Trans. Hull geol. Soc.*, 6, 28.

STATHER, J. W., 1906–9. The Bielsbeck fossiliferous beds. *Trans. Hull. geol. Soc.*, 6:2, 103.

STATHER, J. W., 1912. Shelly clay dredged from the Dogger Bank. *Quart. J. geol. Soc. Lond.*, 68, 324.

STEEVES, M. W. & BARGHOORN, E. S., 1959. The pollen of *Ephedra*. *J. Arnold Arbor.*, 40, 221.

STEFFEN, H., 1931. Vegetationskunde von Ostpreussen. *Pflanzensoziologie*, 1. Jena.

STELFOX, A. W., KUIPER, J. G. J., McMILLAN, N. F. & MITCHELL, G. W., 1972. The Late-glacial and Post-glacial Mollusca of the White Bog, Co. Down. *Proc. R. Ir. Acad.*, B 72, 185.

STEVENS, F., 1934. 'The Highfield pit dwellings', Fisherton, Salisbury. *Wilts archaeol. nat. Hist. Mag.*, 46, 579.

STEVENS, L. A., 1960. The Interglacial of the Nar valley, Norfolk. *Quart. J. geol. Soc. Lond.*, 115, 291.

STONE, J. F. S., 1935. Some discoveries at Ratfyn, Amesbury and their bearing on the date of Woodhenge. *Wiltsh. archaeol. nat. Hist. Mag.*, 47, 67.

STONE, J. F. S., 1938. An Early Bronze Age grave in Fargo Plantation near Stonehenge. *Wiltsh. archaeol. nat. Hist. Mag.*, 48, 357.

STRAHAN, A., 1896. On submerged land-surfaces at Barry, Glamorganshire, with notes on the fauna and flora by Clement Reid. *Quar. J. geol. Soc. Lond.*, 52, 474.

STRAHAN, A., 1907. The geology of the South Wales Coalfield. The county around Swansea. *Mem. geol. Surv.*, *Sheet Memoirs*, p. 142.

STRAKA, H., 1969. Die Sporen- und Pollenmorphologie als Grundlage angewandt-palynologischer Forschungen. *Ber. dt. bot. Ges.*, 81, 471.

STRAKA, H., 1970. Le *Cornus suecica* L., relicte glaciaire dans le N.-W. de l'Europe. *Monde des Plantes*, 65, no. 366, 1.

STRAKER, E. & LUCAS, B. H., 1938. A Romano-British bloomery in East Sussex. *Sussex Arch. Coll.*, 79, 224.

STUART, J. D. M. & BIRKBECK, J. M., 1936. A Celtic village on Twyford Down. *Proc. Hants. Fld Cl. Arch. Soc.*, 13:2, 207.

SUB-DEPARTMENT OF QUATERNARY RESEARCH: Annual reports from 1948/9. *Cambridge University Reporter*, 80–.

SUESS, H., 1967. Bristle-cone pine calibration of the radiocarbon time scale from 4100 B.C. to 1500 B.C. In *Radiocarbon Dating and Methods of low-Level Counting*. Int. Atomic Energy Agency, Vienna.

SUGGATE, R. P. & WEST, R. G., 1959. The extent of the last glaciation in eastern England. *Proc. R. Soc.*, B 150, 263.

SWINNERTON, H. H., 1931. The Post-glacial deposits of the Lincolnshire coast. *Quart. J. geol. Soc. Lond.*, 87, 360.

SWITSUR, V. R., HALL, M. A. & WEST, R. G., 1970. University of Cambridge: natural radiocarbon measurements IX. *Radiocarbon*, 12(2), 590.

SZAFER, W., 1945. Plejstocen w Łękach Dolnych Koło Tarnowa. [The Pleistocene in Łęki Dolne near Tarnow.] *Pol. Akad. Umiejet. Starunia*, No. 19.

SZAFER, W., 1946–7. The Pliocene flora of Krościenko in Poland. *Rozpr. Wydz. mat. przyr. Akad. Um.*, 72.

SZAFER, W., 1952. Schyłek plejstocenu w Polsce. [Decline of the Pleistocene in Poland.] *Panstwowy instytut geologiczny*, 65.

TAGG, H. F., 1911. Appendix: vegetable remains. In: *A Roman frontier post and its people – the Fort of Newstead in the Parish of Melrose*, by J. Curle. Maclehose & Sons, Glasgow.

TALLANTIRE, P. A., 1953. Studies of the Post-glacial history of British vegetation. XIII. Lopham Little Fen, a Late-glacial site in central East Anglia. *J. Ecol.*, 41:2, 361.

TALLANTIRE, P. A., 1954. Old Buckenham Mere. Data for the study of Post-glacial History XIII. *New Phytol.*, **53**, 131.

TALLIS, J. H., 1964. Studies on the southern Pennine peats. I. The general pollen record. *J. Ecol.*, **52**, 323.

TALLIS, J. H., 1965. Studies on southern Pennine peats. IV. Evidence of recent erosion. *J. Ecol.*, **53**, 509.

TALLIS, J. H. & BIRKS, H. J. B., 1965. The past and present distribution of *Scheuchzeria palustris* in Europe. *J. Ecol.*, **53**, 287.

TANSLEY, A. G., 1939. *The British Islands and their Vegetation.* (Reprinted, 1949) Cambridge University Press, London.

TAUBER, H., 1965. Differential pollen dispersion and the interpretation of pollen diagrams. *Danm. geol. Unders.*, R II, **89**, no. 1.

TERASMÄE, J., 1951. On the pollenmorphology of *Betula nana*. *Svensk bot. Tidskr.*, **45**, 358.

THOMAS, K. W., 1965. The stratigraphy and pollen analysis of a raised peat bog at Llanllwch near Carmarthen. *New Phytol.*, **64**, 101.

THOMPSON, P. A., 1970. Characterisation of the germination response to temperature of species and ecotypes. *Nature, Lond.*, **225**, 827.

TOHALL, P., DE VRIES, H. L. & VAN ZEIST, W., 1955. A trackway in Corlona Bog, Co. Leitrim. *J. R. Soc. Antiqu. Ireland*, **86**, 77.

TRALAU, H., 1961. De europeiska arktiskt-montana växternes areal utveckling under Kvartärperioden. *Bot. Notiser*, **114**(2), 213.

TRALAU, H., 1962. *Najas tenuissima* (A.BR.) Magnus during the Late Cainozoic period in Europe. *Bot. Notiser*, **115**, 421.

TRALAU, H., 1963. The recent and fossil distribution of some boreal and arctic-montane plants in Europe. *Ark. Bot.*, ser. 2, **5**, 533.

TRALAU, H. & ZAGWIJN, W. H., 1962. Fossil *Salix polaris* Wahlb. in the Netherlands. *Acta bot. neerl.*, **11**, 425.

TRAVIS, C. B., 1913. Geological notes on recent dock excavations at Liverpool and Birkenhead. *Proc. Lpool geol. Soc.*, **11**, 267.

TRAVIS, C. B., 1926. The peat and forest bed of the south-west Lancashire coast. *Proc. Lpool geol. Soc.*, **14**:3, 263.

TRAVIS, C. B., 1929. The peat and forest beds of Leasowe, Cheshire. *Proc. Lpool geol. Soc.*, **15**:2, 157.

TRECHMANN, C. T., 1947–9. The submerged forest beds of the Durham coast. *Proc. Yorks. geol. Soc.*, **27**:1, 23.

TROELS-SMITH, J., 1954. Pollenanalytische Untersuchungen zu einigen Schweizerischen Pfahlbau-problemen. *Das Pfahlbau–problem.* Meier, Schaffhausen.

TROELS-SMITH, J., 1956. Neolithic period in Switzerland and Denmark. *Science, N.Y.*, **124**, 876.

TROELS-SMITH, J., 1964. The influence of prehistoric man on vegetation in central and north-western Europe. *Proc. VI Congr. I.N.Q.U.A.*, **2**, 487.

TURNER, C., 1966. Middle Pleistocene in East Anglia. Ph.D. thesis, Cambridge.

TURNER, C., 1968a. A Lowestoftian Late Glacial flora from the Pleistocene deposits at Hoxne. *New Phytol.*, **67**, 327.

TURNER, C., 1968b. A note on the occurrence of *Vitis* and other new plant records from the Pleistocene deposits at Hoxne, Suffolk. *New Phytol.*, **67**, 333.

TURNER, C., 1970. The middle Pleistocene deposits at Marks Tey, Essex. *Phil. Trans.*, B **257**, 373.

TURNER, C. & KERNEY, M. P., 1971. The age of the freshwater beds of the Clacton channel. *J. geol. Soc. Lond.*, **127**, 87.

TURNER, C. & WEST, R. G., 1968. The subdivision and zonation of interglacial periods. *Eiszeitalter Gegenw.*, **19**, 93.

TURNER, J., 1962. The *Tilia* decline: an anthropogenic interpretation. *New Phytol.*, **61**, 328.

TURNER, J., 1964. The anthropogenic factor in vegetational history. I. Tregaron and Whixall Mosses. *New Phytol.*, **63**, 73.

TURNER, J., 1965. A contribution to the history of forest clearance. *Proc. R. Soc.*, **161**, 343.

UCKO, P. J. & DUMBLEBY, G. W., 1969. *The Domestication and Exploitation of Plants and Animals.* Duckworth, London.

VARLEY, W. J., 1938. The Bleasdale Circle. *Antiqu. J.*, **18**:2, 155.

VASARI, Y., 1962. A study of the vegetational history of the Kuusarno district (north east Finland) during the Late-quaternary period. *Ann. Bot. Soc. Vanamo*, **33**(1), 140.

VASARI, Y. & A., 1968. Late and Post-glacial macrophytic vegetation in the lochs of northern Scotland. *Acta bot. fenn.*, **80**, 5.

VUORELA, I., 1970. The indication of farming in pollen diagrams from southern Finland. *Acta bot. fen.*, **87**, 1.

WAINWRIGHT, G. J., 1963. A reinterpretation of the microlithic industries of Wales. *Proc. prehis. Soc.*, **29**, 99.

WALKER, D., 1953. The Interglacial deposits at Histon Road, Cambridge. *Quart. J. geol. Soc. Lond.*, **108**:3 (*for* 1952), 273.

WALKER, D., 1955a. Late-glacial deposits at St Bees. *Quart. J. geol. Soc. Lond.*, **112**, 93.

WALKER, D., 1955b. Studies in the Post-glacial history of British vegetation. Skelsmergh Tarn and Kentmere, Westmorland. *New Phytol.*, **54**, 222.

WALKER, D., 1955c. Late-glacial deposits at Lunds, Yorkshire. *New Phytol.*, **54**, 343.

WALKER, D., 1956. A site at Stump Cross, near Grassington, Yorkshire and the age of the Pennine microlith industry. *Proc. prehist. Soc.*, **22**, 23.

WALKER, D., 1964. Post-glacial deposits at Tarn Wadling, Cumberland. *New Phytol.*, **63**, 232.

WALKER, D., 1965. The Post-glacial period in the Langdale Fells, English Lake District. *New Phytol.*, **64**, 488.

WALKER, D., 1966. The Late Quaternary history of the Cumberland lowland. *Phil. Trans.*, B **251**, 1.

WALKER, D. & LAMBERT, C. A., 1955. Boreal deposits at Kirkby Thore, Westmorland. *New Phytol.*, **54**, 209.

WALKER, D. & WEST, R. G., 1953. Some Late-glacial plants. In Lousley, 1953.

WALKER, D. & WEST, R. G., 1970. *Studies in the vegetational history of the British Isles.* Cambridge University Press, London.

WALTER, H. & STRAKA, H., 1970. *Arealkunde: floristisch-historische Geobotanik.* Eugen Ulmer, Stuttgart.

WARD, J., 1902. Prehistoric interments near Cardiff. *Arch. Cambrensis*, **2**, 25.

WARDLE, P., 1961. Biological flora of the British Isles: *Fraxinus excelsior* L. *J. Ecol.*, **49**, 739.

WARREN, S. H., 1912. On a Late Glacial stage in the valley of the River Lea. *Quart. J. geol. Soc. Lond.*, **68**, 213.

WARREN, S. H., 1914. On certain botanical and geological observations made during the opening of the Romano-British barrow on Mersea Island. *Essex Nat.*, **17**, 261.

WARREN, S. H., 1916. Further observations on the Late-glacial stage of the Lea Valley. *Quart. J. geol. Soc. Lond.*, **71**:2 (*for* 1915), 164.

WARREN, S. H., 1921. Excavations at the stone-axe factory of Graig-Lwyd, Penmaenmawr. *J. R. anthrop. Inst.*, **51**, 165.

WARREN, S. H., 1923a. The *Elephas antiquus* bed at Clacton-on-Sea and its flora and fauna. *Quart. J. geol. Soc. Lond.*, **79**, 606.

WARREN, S. H., 1923b. The Late-glacial stage of the Lea Valley. (Third report.) *Quart. J. geol. Soc. Lond.*, **79**, 603.

WARREN, S. H., CLARK, J. G. D., GODWIN, H. & M. E. & MACFAYDEN, W. A., 1934. An early Mesolithic site at Broxbourne sealed under boreal peat. *J. R. anthrop. Inst.*, **64**, 101.

WATERBOLK, H. T., 1954. De praehistorische mens en zijn milieu. Ein palynologisch onderzoek naar de menslijke insloed op de plantengroi van de diluviale gronden in Nederland. *Diss. Groningen* (Engl. Zuz.), **153**, 16.

WATERBOLK, H. T., 1968. Food production in prehistoric Europe. *Science, N.Y.*, **162**, 1093.

WATERBOLK, H. T. & VAN ZEIST, W., 1966. Preliminary report on the Neolithic bog settlement at Niederwil. *Palaeohistoria*, **12**, 559.

WATTS, W. A., 1959a. Pollen spectra from the Interglacial deposits at Kirmington, Lincs. *Proc. Yorks. Geol. Soc.*, **32**, 145.

WATTS, W. A., 1959b. Interglacial deposits at Kilbeg and Newtown. *Proc. R. Ir. Acad.*, **60**, 79.

WATTS, W. A., 1963. Late Glacial pollen zones in western Ireland. *Ir. Geogr.*, **4**, 367.

WATTS, W. A., 1964. Interglacial deposits at Baggotstown, near Bruff, Co. Limerick. *Proc. R. Ir. Acad.*, **63**, 167.

WATTS, W. A., 1967. Interglacial deposits in Kildromin Td, near Herbertstown, Co. Limerick. *Proc. R. Ir. Acad.*, **65**, 339.

WATTS, W. A., 1971. The identity of *Menyanthes microsperma* n.sp. foss. from the Gort Interglacial, Ireland. *New Phytol.*, **70**, 435.

WEBB, D. A., 1950. A revision of the dactyloid saxifrages of north-western Europe. *Proc. R. Ir. Acad.*, **53** B 12, 207.

WELTEN, M., 1944. Pollenanalytische, stratigraphische und geochronologische Untersuchungen aus dem Faulenseemoos bei Spiez. *Veröff. geobot. Inst. Rübel*, **21**, 201.

WELTEN, M., 1952. Pollenanalytische Stichproben über die subrezente Vegetationsentwicklung in Bergland von Kerry. *Veröff. geobot. Inst. Rübel*, **25**.

WELTEN, M., 1957. Über das glaziale und spätglaziale Vorkommen von *Ephedra* am nordwestlichen Alpenrand. *Ber. schweiz. bot. Ges.*, **67**, 33.

WELTEN, M., 1967. Ein Brachsenkraut, *Isoetes setacea* Lam., fossil im Schweizerischen Molasseland. *Bot. Jb.*, **86**, 527.

WENNER, C. G., 1947. Pollen diagrams from Labrador: A contribution to the geology of Newfoundland–Labrador with comparisons between North America and Europe. *Geogr. Ann., Stockh.*, **29**, 137.

WEST, R. G., 1953. The occurrence of *Azolla* in British Interglacial deposits. *New Phytol.*, **52**:3, 267.

WEST, R. G., 1956. Interglacial site at Hoxne. *Phil. Trans.*, B **239**, 265.

WEST, R. G., 1958. Interglacial site at Bobbitshole. *Phil. Trans.*, B **241**, 1.

WEST, R. G., 1961a. Symposium on Quaternary ecology II – Interglacial and Interstadial vegetation in England. *Proc. Linn. Soc. Lond.*, **172**, 81.

WEST, R. G., 1961b. Vegetational history of the early Pleistocene of the Royal Society borehole at Ludham, Norfolk. *Proc. R. Soc.*, **155**, 437.

WEST, R. G., 1964a. The Interglacial vegetation of Britain and continental Europe compared. *Proc. VII. Congr. I.N.Q.U.A.*, **2**, 503.

WEST, R. G., 1964b. Inter-relations of ecology and Quaternary palaeobotany. *J. Ecol.*, **52**, 47.

WEST, R. G., 1967. The Quaternary of the British Isles. In *Geologic Systems: The Quaternary*, ed. K. Rankama, vol. 2, p. 1. Interscience, New York.

WEST, R. G., 1968. *Pleistocene geology and biology*. Longmans, Green and Co. Ltd, London.

WEST, R. G., 1969a. Pollen analyses from Interglacial deposits at Aveley and Grays, Essex. *Proc. Geol. Ass. Lond.*, **80**, 271.

WEST, R. G., 1969b. A note on pollen analyses from the Speeton Shell Bed. *Proc. Geol. Ass. Lond.*, **80**, 217.

WEST, R. G., 1970a. Pollen zones in the Pleistocene of Great Britain and their correlation. *New Phytol.*, **69**, 1179.

WEST, R. G., 1970b. Pleistocene history of the British flora. In Walker & West, 1970.

WEST, R. G., LAMBERT, C. A. & SPARKS, B. W., 1964. Interglacial deposits at Ilford, Essex. *Phil. Trans.*, B 247, 185.

WEST, R. G. & SPARKS, B. W., 1960. Coastal Interglacial deposits of the English Channel. *Phil. Trans.*, B 243, 95.

WEST, R. G. & WILSON, D. G., 1966. Cromer Forest Bed series. *Nature, Lond.*, **209**, 497.

WEST, R. G. & WILSON, D. G., 1968. Plant remains from the Corton Beds, Lowestoft, Suffolk. *Geol. Mag.*, **105**, 116.

WESTELL, W. P., 1937. Excavation of an uncharted Romano-British occupation-site at Great Wymondley, Herts. *East Herts. Arch. Soc.*, **10**:1, 11.

WHEELER, R. E. M., 1943. Maiden Castle, Dorset. *Rep. 12 Res. Comm. Soc. Antiqu.*, p. 374.

WHITAKER, E. D., 1921. Peat problems. *Trans. Leeds geol. Ass.*, Part 18, 23.

WHITE, G. M., 1934. A settlement of the South Saxons. *Antiqu. J.*, **14**:4, 393.

WHITEHEAD, H., 1920. More about 'Moorlog': A peaty deposit from the Dogger Bank in the North Sea. *Essex Nat.*, **19**, 242.

WHITLEY, M., 1943. Excavations at Chalbury Camp, Dorset, 1939. *Antiqu. J.*, **23**, 98.

WIERMANN, R., 1962. Botanisch-moorkundliche Untersuchungen in Nordfriesland. *Meyniana*, **12**, 97.

WIERMANN, R., 1965. Moorkundliche und vegetationsgeschichliche Betrachtungen zum Aussendeichsmoor bei Sehestedt. *Ber. dt. bot. Ges.*, **78**, 269.

WILLERDING, U., 1969a. Pflanzenreste aus frühgeschichtlichen Siedlungen des Göttinger Gebietes. *Neue Ausgrabungen und Forschungen in Niedersachsen*. 4, 391.

WILLERDING, U., 1969b. Ursprung und Entwicklung der Kulturpflanzen in Ur- und frühgeschichtlicher Zeit. *Dt. Agrargeschichte*, **1**, 188.

WILLERDING, U., 1970. Vor- und frühgeschichtliche Kulturpflanzenfunde in Mitteleuropa. *Neue Ausgrabungen und Forschungen in Niedersachsen*, 5, 287.

WILLIAMS, A., 1940a. A Megalithic tomb at Nicholaston, Gower, Glamorgan. *Proc. prehist. Soc.*, **6**, 178.

WILLIAMS, A., 1940b. The Excavation of Bishopston Valley, Promontory Fort, Glamorgan. *Arch. Cambrensis*, **95**, 17.

WILLIAMS, A., 1941. The excavation of High Penard Promontory Fort, Glamorgan. *Arch. Cambrensis*, **96**, 29.

WILLIAMS, A., 1945. Two Bronze Age barrows on Fairwood Common, Gower, Glamorgan. *Arch. Cambrensis*, **98**, 62.

WILLIAMS, A., 1946. A homestead moat at Nuthampstead, Hertfordshire. *Antiqu. J.*, **26**, 138.

WILLIAMS, A., 1947. Bronze Age barrows near Chewton Mendip, Somerset. *Proc. Somerset Arch. nat. Hist. Soc.*, **93**, 39.

WILLIAMS, A., 1950. Excavations at Allard's Quarry, Marnhull. *Proc. Dorset nat. Hist. Fld. Cl.*, **72**, 20.

WILLIAMS, A. & FRERE, S., 1948. Canterbury excavations, Christmas 1945 and Easter 1946. *Arch. Cantiana*, **61**, 1.

WILLIAMS, H., 1921. Excavation of Bronze-Age tumulus near Gorsedd, Holywell, Flintshire. *Arch. Cambrensis* Series 7, **1**, 275.

WILLIS, E. H., 1961. Marine transgression sequences in English fenlands. *Ann. N.Y. Acad. Sci.*, **951**, 368.

WILLOCK, E. H., 1933–6. A Neolithic site on Haldon. *Proc. Devon Arch. Explor. Soc. (for 1936)*, **2**:4, 244.

WILSON, D. G., 1968a. Roman Britain in 1967. I. Sites explored. *J. Rom. Studies*, **58**, 176.

WILSON, D. G., 1968b. Plant remains from the Roman well at Bunny, Notts. *Trans. Thoroton Soc. Notts.*, **72**, 42.

WILSON, J. W., 1952. Vegetation patterns associated with soil movement on Jan Mayen Island. *J. Ecol.*, **40**:2, 249.

WISEMAN, J. D. H., 1965. The changing rate of calcium carbonate sedimentation on the equatorial Atlantic floor and its relation to continental Late Quaternary stratigraphy. *Rep. Swed. deep Sea Exped.*, **2**.

WODEHOUSE, R. P., 1935. *Pollen Grains*. Hafner, New York and London.

WOLDSTEDT, P. W., 1950. Das Vereisungsgebiet der Britischen Inseln und seine Beziehungen zum festländischen Pleistozän. *Geol. Jahrb.*, **65**, 621.

WOODHEAD, T. W., 1929. History of the vegetation of the Southern Pennines. *J. Ecol.*, **17**, 1.

WOODHEAD, T. W. & ERDTMAN, O. G. E., 1926. Remains in the peat of the Southern Pennines. *Naturalist, Lond.*, **835**, 245.

WOODS, G. & R. M., 1934. Excavations in a dry valley in Beer, south-east Devon. *Proc. prehist. Soc. E. Angl.*, **7**:3, 355.

WRIGHT, E. V. & CHURCHILL, D. M., 1965. The boats from Ferriby, Yorkshire, England, with a review of the origins of the sewn boats of the Bronze Age. *Proc. prehist. Soc.*, **31**, 1.

WRIGHT, E. V. & C. W., 1947. Prehistoric boats from North Ferriby, East Yorkshire. *Proc. prehist. Soc.*, **13**, 114.

WULFF, E. V., 1943. *An Introduction to Historical Plant Geography*. Chronica Botanica, Waltham, Mass.

WYMER, J., 1962. Excavations at the Maglemosian sites at Thatcham, Berkshire, England. *Proc. prehist. Soc.*, **28**, 329.

YOUNG, A. & MITCHELL, M. C., 1939. Report on excavation at Monzie. *Proc. Soc. Antiqu. Scot.*, **73**, 62.

YOUNG, W. E. V., 1950. A beaker interment at Beckhampton. *Wilts. Mag.*, **53**, 311.

ZAGWIJN, J. & PAEPE, R., 1968. Die Stratigraphie der Weichselzeitlichen Ablagerungen der Niederlande und Belgiens. *Eiszeitalter Gegenw.*, **19**, 129.

ZAGWIJN, W. H., 1959. Zur stratigraphischen und pollenanalytischen Gliederung der pliozänen Ablagerungen in Roental-Graben und Venloer Graben der Niederlande. *Fortschr. Geol. Rheinld. Westf.*, **4**, 5.

ZAGWIJN, W. H., 1960. Aspects of the Pliocene and early Pleistocene vegetation in the Netherlands. *Med. Geol. Sticht* (Ser. C-III), **1**, no. 5.

ZANDSTRA, K. J., 1966. The occurrence of *Salvinia natans* in Holocene deposits of the Rhine delta. *Acta bot. neerl.*, **15**, 389.

ZEIST, W. VAN, 1959. Studies on the Post-boreal vegetational history of south-eastern Dienthe (Netherlands). *Acta bot. neerl.*, **8**, 156.

ZEIST, W. VAN, 1964. A palaeobotanical study of some bogs in western Brittany (Finistère) France. *Palaeohistoria*, **10**, 157.

INDEX

The historical plant records of chapter V are picked out from other references to plants by numbers in **bold type**.

Reference to text-figures is indicated by a page number and asterisk.

Topics and authors are indexed throughout only for the most important references.

Chapter IV (recorded sites) has not been indexed at all.

Chapter V (the plant record) has not been indexed for authors, places, climatic periods or archaeological periods, but a few important general topics have been included.

black poplar, see *Populus nigra*
black saltwort, see *Glaux maritima*
black sedge, see *Carex atrata*
blackberry, see *Rubus fruticosus* agg.
BLACKBURN, K. B., 19, 46
blackthorn, see *Prunus spinosa*
bladder campion, see *Silene vulgaris*
bladder sedge, see *Carex vesicaria*
bladder wort, see *Utricularia* sp.
blaeberry, see *Vaccinium myrtillus*
blanket bog, 7, 30, 32, 35, 252, 260f., 263, 295, 390, 402, 454, 472f., 491*, pl. v
Blashenwell, Dorset, 56
Blechnum spicant, 92f.
blinks, see *Montia fontana*
bloody cranesbill, see *Geranium sanguineum*
bluebell, see *Campanula rotundifolia* and *Endymion non-scripta*
bluebottle, see *Centaurea cyanus*
blunt-flowered rush, see *Juncus subnodulosus*
Blysmus compressus, **390**, 438
 rufus, **390**, 432
BLYTT, A. and SERNANDER, R., climatic system, 27, 33, 53, 55, 106, 455, pl. XXVI
bog asphodel, see *Narthecium ossifragum*
bogbean, see *Menyanthes trifoliata*
bog heather, see *Erica tetralix*
bog myrtle, see *Myrica gale*
bog pimpernel, see *Anagallis tenella*
bog pondweed, see *Potamogeton polygonifolius*
bog rush, see *Schoenus nigricans*
bog sandwort, see *Minuartia stricta*
bog stitchwort, see *Stellaria alsine*
bog whortleberry, see *Vaccinium uliginosum*
Bølling interstadial, 18, 437
Boraginaceae, **315**
Boreal period, 27, 34f., 37, 51ff., 55
Borth Bog, Cardigans., 31*, 50*, 103f., pl. II
Botrychium lunaria, **98**f., 442, 462, 492, pl. XV
bottle sedge, see *Larex rostrata*
boundary surface (boundary horizon), *see* recurrence surface *and* 'Grenzhorizont'
bows, 116
box, see *Buxus sempervirens*
bracken, see *Pteridium aquilinum*
brandy-bottle, see *Nuphar luteum*
Brasenia purpurea, **128**, 424, 428, 497
Brassica sp., **130**
 napus, 130
 nigra, 130, **131**, 471, 498
 oleracea, 130
 rapa, 40, **131**, 470f.
bread wheat, see *Triticum vulgare*
Breckland, 92, 467
bristle scirpus, see *Scirpus setaceus*
bristle-pointed oat, see *Avena strigosa*
bristly ox-tongue, see *Picris echioides*
Bristol rock-cress, see *Arabis stricta*
broad bean, see *Vicia faba*
broad blysmus, see *Blysmus compressus*
broad-leaved cotton-grass, see *Eriophorum latifolium*
broad-leaved dock, see *Rumex obtusifolius*
broad-leaved marsh dandelion, see *Taraxacum spectabile*
broad-leaved pondweed, see *Potamogeton natans*
broad-leaved willow-herb, see *Epilobium montanum*
Bromus sp., **403**
 secalinus (or *mollis*), **403***, 471, 479
 sterilis, **403***, 471, 479
Bronze Age, 1, 7, 8, 24, 33f., 39–42, 52f., 82, 406–13 *passim*, 470–81 *passim*, pls. VIII, IX, XIII
brook saxifrage, see *Saxifraga rivularis*
broom, see *Sarothamnus scoparius*

Brørup interstadial, 15, 28*
brown beak-sedge, see *Rhynchospora fusca*
brown sedge, see *Carex disticha*
Bryonia dioica (bryony), **229**
Bryophyta, 10
buckbean, see *Menyanthes trifoliata*
buck's-horn plantain, see *Plantago coronopus*
buckthorn, see *Rhamnus catharticus*
buckwheat, see *Fagopyrum esculentum*
bud-scale identifications, 7
Bufo calamita (natterjack toad), 486
bugle, see *Ajuga reptans*
bulbous buttercup, see *Ranunculus bulbosus*
bulbous rush, see *Juncus bulbosus*
bullace, see *Prunus insititia*
bulrush, see *Typha* sp. and *Scirpus lacustris*
Bunium bulbocastanum, **224**, 443
Bupleurum sp., **223**
 tenuissimum, **223**, 479
bur chervil, see *Anthriscus caucalis*
bur marigold, see *Bidens* sp.
burdock, see *Arctium* sp.
burials, 40, 175f., 220, 233, 480
buried forests, 23*, 35f., 50**, 55, pls. II, V
burnet rose, see *Rosa pimpenellifolia*
burnet saxifrage, see *Pimpinella saxifraga*
bur-reed, see *Sparganium* spp.
bush vetch, see *Vicia sepium*
Butomus umbellatus, **355**, 438
buttercup, see *Ranunculus* sp.
Buxus sempervirens, 40, **175**f., 427, 430, 471, pl. XXV
byg, see *Hordeum vulgare*

CAIN, S., 292
Calamagrostis canescens, **404**
Calamintha ascendens, **322**f.
calcareous muds, 32*
calcareous tufa, 15, 25
Callitriche sp., **215**
 hermaphroditica, **216**f., 435*, 438
 intermedia, **216**, 438
 obtusangula, **216**, 438, 446
 platycarpa, **215**
 stagnalis, **215**f., 438
Calluna heath, 473
Calluna vulgaris, 53, 252, 292f., **294-6***, 385, 431, 437, 439, 464, 467, 472, 488, pl. VI
Caltha palustris, **119**, 432, 438, pl. XVII
 radicans, 119
Calystegia sp., **316**, pl. XXV
Camelina linicola, 40
Campanula sp., **322**, 492, pl. XXII
 patula, 333, pl. XX
 rotundifolia, **338**, pl. XX
campion, see *Silene* sp., *Lychnis* sp., *Melandrium* sp.
Cannabis sativa, **242**f., 415, 478*ff.
Capsella bursa-pastoris, **131**, 443, 498
Cardamine sp., **134**
 amara, **134**, 438
 flexuosa, **135**, 471
 impatiens, **134**
 pratensis, **134**, 438
Cardaminopsis petraea, **135**, 444
Carduus sp., **346**, pl. XVIII
 acanthoides, **347**, 443, 499
 nutans, **347**, 443, 499
 pycnocephalus, **347**, 499
Carex sp., 6
 acuta, **396**, 444
 acutiformis, **395**, pl. IV
 appropinquata, **397**